"This ambitious book provides a comprehensive the remarkable Bronze Age Civilization of the Ox tion of the settlement systems, the archaeological unique culture, as well as discussions of relations in Iran, the Indus valley, and Mesopotamia."

Holly Pittman, *Un*

"*The World of the Oxus Civilization* brings to life one of the great 'lost civilizations' of the ancient world, lost only because, while extraordinarily rich and extensive in its reach, its story has long been hidden from the broader world of scholarship by extreme political and linguistic barriers. This volume, telling the full story of the Oxus Civilization for the first time, is a magnificent gift for anyone interested in the rich complexity of the ancient world and the early rise of internationalism across Asia. Its comprehensive coverage, presented as a series of specialist essays by a range of international scholars, will be warmly welcomed as a much needed authoritative work of reference filling a critical gap in the early history of Old World development, linking Eurasia with the greater Iranian world, the Indus valley with Mesopotamia, and placing Central Asia in its rightful position as a wellspring of early cultural development."

Alison Betts, *University of Sydney, Australia*

THE WORLD OF THE OXUS CIVILIZATION

This collection of essays presents a synthesis of current research on the Oxus Civilization, which rose and developed at the turn of the 3rd to 2nd millennia BC in Central Asia.

First discovered in the 1970s, the Oxus Civilization, or the Bactria-Margiana Archaeological Complex (BMAC), has engendered many different interpretations, which are explored in this volume by an international group of archaeologists and researchers. Contributors cover all aspects of this fascinating Bronze Age culture: architecture; material culture; grave goods; religion; migrations; and trade and interactions with neighboring civilizations, from Mesopotamia to the Indus, and the Gulf to the northern steppes. Chapters also examine the Oxus Civilization's roots in previous local cultures, explore its environmental and chronological context, as well as the available sources, and look into the reasons for its decline.

The World of the Oxus Civilization offers a broad and fascinating examination of this society, and provides an invaluable updated resource for anyone working on the culture, history, and archaeology of this region and on the multiple interactions at work at that time in the ancient Near East.

Bertille Lyonnet is Directrice de Recherches Emeritus at the French National Center for Scientific Research (CNRS), Paris, as an archaeologist. She has worked in Central Asia (Afghanistan, Tadjikistan, and Uzbekistan), northeastern Syria, the northern Caucasus, and Azerbaijan. A specialist in ceramics, she has always shown a particular interest in the interrelations between the different areas of the world where she has worked. She is the author of several books and over 150 articles.

Nadezhda A. Dubova is a main researcher and head of the Center of Human Ecology at the Institute of Ethnology and Anthropology (IEA) of the Russian Academy of Sciences, Moscow, Russia. She was the head and/or a member of more than 60 physical anthropological and archaeological expeditions in Russia, Central Asia, Iran, and the Moldova Republic. She now heads the Russian-Turkmen Margiana archaeological expedition. She gives lectures at different universities (Lomonosov Moscow, Ufa (Bashkortostan), Voronezh, South Kazakhstan (Shimkent), and Bern (Switzerland)). She is the author and editor of more than 400 publications on physical anthropology, human ecology, Eurasian archaeology, and theoretic problems of anthropology.

THE ROUTLEDGE WORLDS

THE ETRUSCAN WORLD
Edited by Jean MacIntosh Turfa

THE SWAHILI WORLD
Edited by Stephanie Wynne-Jones and Adria LaViolette

THE MEDIEVAL WORLD, SECOND EDITION
Edited by Peter Linehan, Janet L. Nelson, and Marios Costambeys

THE ELAMITE WORLD
Edited by Javier Álvarez-Mon, Gian Pietro Basello, and Yasmina Wicks

THE FIN-DE-SIÈCLE WORLD
Edited by Michael Saler

THE GNOSTIC WORLD
Edited by Garry W. Trompf, Gunner B. Mikkelsen, and Jay Johnston

THE ANDEAN WORLD
Edited by Linda J. Seligmann and Kathleen Fine-Dare

THE SYRIAC WORLD
Edited by Daniel King

THE FAIRY TALE WORLD
Edited by Andrew Teverson

THE MELANESIAN WORLD
Edited by Eric Hirsch and Will Rollason

THE MING WORLD
Edited by Kenneth M. Swope

THE GOTHIC WORLD
Edited by Glennis Byron and Dale Townshend

THE IBERIAN WORLD
Edited by Fernando Bouza, Pedro Cardim, and Antonio Feros

THE MAYA WORLD
Edited by Scott Hutson and Traci Ardren

THE WORLD OF THE OXUS CIVILIZATION
Edited by Bertille Lyonnet and Nadezhda A. Dubova

THE GRAECO-BACTRIAN AND INDO-GREEK WORLD
Edited by Rachel Mairs

THE UMAYYAD WORLD
Edited by Andrew Marsham

THE BUDDHIST WORLD
Edited by John Powers

THE CRUSADER WORLD
Edited by Adrian Boas

THE POSTCOLONIAL WORLD
Edited by Jyotsna G. Singh and David D. Kim

THE SUMERIAN WORLD
Edited by Harriet Crawford

THE OCCULT WORLD
Edited by Christopher Partridge

THE WORLD OF INDIGENOUS NORTH AMERICA
Edited by Robert Warrior

THE WORLD OF THE REVOLUTIONARY AMERICAN REPUBLIC
Edited by Andrew Shankman

THE SHAKESPEAREAN WORLD
Edited by Jill L. Levenson and Robert Ormsby

THE WORLD OF COLONIAL AMERICA
Edited by Ignacio Gallup-Diaz

THE ORTHODOX CHRISTIAN WORLD
Edited by Augustine Casiday

THE MODERNIST WORLD
Edited by Allana Lindgren and Stephen Ross

THE EARLY CHRISTIAN WORLD, SECOND EDITION
Edited by Philip F. Esler

www.routledge.com/Routledge-Worlds/book-series/WORLDS

THE WORLD OF THE OXUS CIVILIZATION

Edited by

Bertille Lyonnet and Nadezhda A. Dubova

Routledge
Taylor & Francis Group
LONDON AND NEW YORK

First published 2021
by Routledge
4 Park Square, Milton Park, Abingdon, Oxon OX14 4RN
605 Third Avenue, New York, NY 10017

First issued in paperback 2023

Routledge is an imprint of the Taylor & Francis Group, an informa business

British Library Cataloguing-in-Publication Data
A catalogue record for this book is available from the British Library

Library of Congress Cataloging-in-Publication Data
Names: Lyonnet, Bertille, editor. | Dubova, Nadezhda, editor.
Title: The world of the Oxus Civilization / edited by Bertille Lyonnet and Nadezhda Dubova.
Identifiers: LCCN 2019044889 (print) | LCCN 2019044890 (ebook)
Subjects: LCSH: Asia, Central – History. | Asia, Central – Civilization. | Asia, Central – Antiquities. | Civilization, Ancient.
Classification: LCC DS329.4 .W67 2020 (print) | LCC DS329.4 (ebook) | DDC 939/.6 – dc23
LC record available at https://lccn.loc.gov/2019044889
LC ebook record available at https://lccn.loc.gov/2019044890

ISBN: 978-1-032-57003-7 (pbk)
ISBN: 978-1-138-72287-3 (hbk)
ISBN: 978-1-315-19335-9 (ebk)

DOI: 10.4324/9781315193359

Typeset in Sabon
by Newgen Publishing UK

Publisher's Note
The publisher has gone to great lengths to ensure the quality of this reprint but points out that some imperfections in the original copies may be apparent.

CONTENTS

PART II: THE "CORE AREA" 143

FIGURES

TABLES

CONTRIBUTORS

Valérie Andrieu-Ponel has been Maître de Conférences at the Aix-Marseille University (AMU), France, since 2000. She teaches plant biology, ecology, and paleoecology (AMU, Tunis, Sfax), and her research focuses on the reconstruction of ancient ecosystems and environments in Europe, the Mediterranean, the Middle East, and Central Asia over the last five years. She has published 113 articles in peer-reviewed journals, 35 book chapters, and participated as either leader or participant in 71 research programs (Industry, Europe, USA).

Elena V. Antonova is a main researcher in the Department of History and Culture of the Ancient East, at the Oriental Institute of the Russian Academy of Sciences (Moscow), Russia. She is the author of more than 100 publications on antiquity historical and cultural data, Bronze Age cultures of Central Asia and neighboring regions, theoretical problems, the village-to-city social structure evolution, rituals and mythological representations, culture formation, and the relations between Central Asia and the Near East.

Nona A. Avanesova is a specialist of the Chalcolithic and Bronze Age in the Eurasian steppes, the Zeravshan, and southern Uzbekistan. She is a professor in the Department of Archaeology at Samarkand University, Uzbekistan, and an associated professor in the Department of Theory and History of Culture at Samara State University, Russia. She was an invited professor at the École des Hautes Études en Sciences Sociales in Paris, France. She has discovered and investigated more than 20 archaeological sites in the Zeravshan valley, eight of which have been excavated. She is a member of international archaeological expeditions in southern West Siberia and Altai regions in Russia, Kazakhstan, and Tajikistan. She has published more than 150 scientific publications, including four monographs and four tutorials.

Julio Bendezu-Sarmiento is a senior archaeologist and was director of the French Archaeological Delegation in Afghanistan (DAFA) from 2014 to 2018. Trained in funeral archaeology and biological anthropology, he first worked on Bronze and Iron Age populations in Kazakhstan. He was a researcher at the French Institute for Iranian studies (IFRI) in 2005–2007, deputy director and director of the French Institute for

Central Asian Studies in Uzbekistan (IFEAC) in 2007–2010, and has been involved for more than two decades in archaeological fieldwork in Afghanistan, Iran, Uzbekistan, Karakalpakistan, Kazakhstan, Tajikistan, Turkmenistan, and the United Arab Emirates (UAE). A member of the French National Center for Scientific Research (CNRS/MNHN), Paris, he is currently mainly working on Bronze and Iron Age periods in Central Asia. He has contributed to numerous publications on archaeology (methods, techniques, fieldwork) and on biological anthropology and archaeothanatology.

Raffaele Biscione is an archaeologist and senior researcher emeritus at Consiglio Nazionale delle Ricerche (CNR), Rome, Italy. He has worked in Iran at Shahr-i Sokhta, and is currently excavating at Tepe Chalow, a Bronze Age site in the north-eastern part of Iran. He also worked on Urartian fortresses in Armenia. He is the author of a number of publications.

Gian Luca Bonora has taught at the University of Bologna, Italy, and for four years at the L.N. Gumilev Eurasian National University in Astana, Kazakhstan. He participated in numerous archaeological expeditions in Central and Middle Asia led by the International Association of Mediterranean and Oriental Studies (ISMEO), the Italian Institute for Africa and the Orient (ISIAO), and the University of Bologna. He has been the director of the Italian Archaeological Mission in Kazakhstan, on behalf of the Italian Ministry of Foreign Affairs, since 2007. In 2012, he became a member and scientific researcher for ISMEO in Rome, Italy, and he is currently associate professor of cultural anthropology at the Antonianum University of Vatican City. He is the author of more than 80 publications in Italian, English, Russian, and Kazakh, mainly dedicated to the study and analysis of the prehistoric and protohistoric communities of Middle and Central Asia, among which a two-volume *Multilingual Dictionary of Archaeological Terms* in four languages (Kazakh, Russian, English, and Italian) (2014).

Natalja D. Burova is a scientific researcher, now responsible for radiocarbon dating and is head of the Laboratory of Archaeological Technology, at the Institute for the History of Material Culture, Russian Academy of Sciences (RAS), St. Petersburg, Russia.

Annie Caubet is a French archaeologist and museum curator. She is currently a professor at Ecole du Louvre and honorary curator at the Musée du Louvre, Paris, France, where she was the former head of the Department of the Ancient Near East (1988–2006). She was also a Mellon fellow (2013–2014) at the Metropolitan Museum of Art, New York, USA. She has done extensive fieldwork in Syria, Cyprus, and various archaeological sites in the Near East, and has worked at Ulug Depe (Turkmenistan). Her main areas of interest are luxury goods in the ancient Near East and Central Asia, especially faience, ivory, precious stones, the archaeology of music, and coroplastic studies.

Barbara Cerasetti has been the project director of the Italian Archaeological Mission in Turkmenistan since 2006. Her main specialization is the reconstruction of ancient landscape in prehistoric Central Asia, in correlation with climate change through advanced technology, especially remote sensing. She is particularly interested in the interaction and integration between nomad and sedentary cultures in Turkmenistan and in Central Asia during the Bronze Age. She has taken part in numerous field research projects worldwide, often in collaboration with European and American institutions. She is the author of many contributions in peer-reviewed journals and has a long experience as peer reviewer for specialized journals as well as funding sponsor.

Lucie Cez is a PhD student in geoarchaeology at the University of Pantheon-Sorbonne (Paris 1), France, and a member of the UMR 7041 ArScAn (Nanterre), France. Her doctoral research is devoted to the study of ancient irrigation canals discovered in the vicinity of the Late Chalcolithic site of Sarazm occupied in the 4th and 3rd millennium BC (Zeravshan valley, Tajikistan). The sedimentary archives of several fills in these canals allow the documentation of relations between paleoenvironmental conditions and human occupation in this Central Asia semiarid valley highly dependent on tectonic constraints. She has developed geoarchaeological approaches based on geomorphology and micromorphology of soils in semiarid to arid regions in East Africa (Egypt and Sudan).

Nadezhda A. Dubova is a main researcher and head of the Center of Human Ecology at the Institute of Ethnology and Anthropology (IEA) of the Russian Academy of Sciences, Moscow, Russia. She was the head and/or a member of more than 60 physical anthropological and archaeological expeditions in Russia, Central Asia, Iran, and the Moldova Republic. She now heads the Russian-Turkmen Margiana archaeological expedition. She gives lectures at different universities (Lomonosov Moscow, Ufa (Bashkortostan), Voronezh, South Kazakhstan (Shimkent), and Bern (Switzerland)). She is the author and editor of more than 400 publications on physical anthropology, human ecology, Eurasian archaeology, and theoretic problems of anthropology.

Michel Fontugne is a geochemist affiliated to the Commissariat à l'énergie atomique et aux énergies alternatives (CEA) and to CNRS (Paris and Aix-en-Provence), France. His research mainly concerns the dynamics of paleoenvironmental changes and their relationship with the cultural development of human societies in the last 50,000 years. He is the author or co-author of over 350 publications.

Eric Fouache is a professor of physical geography and geoarchaeology at the University of Paris-Sorbonne (Paris 4), France, and vice-chancellor of Paris-Sorbonne University in Abu Dhabi, UAE, since 2012. His research concerns geoarchaeological, paleoenvironmental, and paleogeographical issues in the Mediterranean, the Black Sea, the Near and Middle East, and Central Asia (UMR ENeC 8185, France). He has published two books as a single author and more than 100 scientific papers and book chapters, including more than 50 in international peer-reviewed journals. In 2000 and 2007, he received an award from the French Société de Géographie in Paris for his books. In 2012, he was awarded "Professeur Honoris Causa" from Bucharest University in Romania and in 2015, "Chevalier dans l'ordre des Palmes académiques."

Jennifer Garner is a research staff member of the Deutsches Bergbau-Museum Bochum, Germany, in the division of mining archaeology. Her main interests are Bronze Age mining and metallurgy, La Tène iron production, and lead-silver exploitation during the Middle Ages.

Michaël Guichard is a professor at the École Pratique des Hautes Études, Paris, France, since 2014. He teaches Mesopotamian history as well as Sumerian and Akkadian literature. Among others, he participates in the edition of the Royal Archives of the Mari Palace. He is the author of many publications, including *La Vaisselle de luxe des rois de Mari* (vol. 31 of the Mari Royal Archives, 2005), and the *Epopée de Zimri-Lim* (2014).

Kai Kaniuth is a senior lecturer at Ludwig Maximilian University, Munich, Germany. He has taught previously at the universities of Jena and Moscow (RSUH), Russia. He has conducted fieldwork in Uzbekistan, Turkey, Iraq, Iran, and, most recently, Azerbaijan, with an emphasis on the 2nd and 1st millennia BC.

Ljubov' B. Kircho is a professor in history and a leading researcher in the Department of Archaeology of Central Asia and the Caucasus, at the Institute for the History of Material Culture of the Russian Academy of Sciences (IHMC) in St. Petersburg, Russia. She was the scientific secretary of the IHMC and is currently the scientific secretary of the yearbook *Zapiski IIMK RAN* (Transactions of the IHMC, RAS). She has participated in and led more than 50 archaeological expeditions in Central Asia, the European plain, southern Siberia, and the Crimea. She is the author of more than 170 publications on the archaeology of the Chalcolithic and Bronze Age of Central Asia, the Middle East, and South Asia, and on chronology, cultures, and cultural interactions.

Steffen Kraus studied archaeometry at the University of Freiberg, Germany, and did his PhD in natural science archaeology at the University of Tübingen, Germany. He works as a scientist at the Curt-Engelhorn-Zentrum Archäometrie GmbH in Mannheim, Germany, with his main focus on prehistoric mining and metallurgy in Central Asia.

Vladimir V. Kufterin has a PhD in biological sciences and was a researcher at the Anuchin Institute and Museum of Anthropology, Lomonosov Moscow State University, Russia. Since 2019 he has been a senior researcher at the Institute of Ethnology and Anthropology (Russian Academy of Sciences), Moscow. He has taught at the Akmullah Bashkir State Pedagogical University (Ufa), Russia. His scientific interests include bioarchaeology, paleopathology, archaeology, and physical anthropology of Central Asia and adjacent regions. He is a member of the European Anthropological Association and has published around 120 scientific publications.

C.C. Lamberg-Karlovsky is a Stephen Philips research professor of archaeology, the director of the American School of Prehistoric Studies, and the past director of the Peabody Museum of Archaeology and Ethnology, Harvard University, USA. He has directed archaeological excavations and surveys in Syria, Iran, Saudi Arabia, Pakistan, and Turkmenistan. His interests focus upon the cultural interactions that brought different archaic states of the ancient Near East and Central and South Asia in the Bronze Age into contact.

Pierre Lombard is a senior researcher at CNRS, Paris, France, mostly specializing in the archaeology of the eastern Arabian Peninsula. His fieldwork started in the UAE (Jebel Hafit, Hili, Rumeilah), and, since 1989, he has been leading the French Archaeological Mission to Bahrain. He is notably in charge of the excavations at

Qal'at al-Bahrain, the former capital and main urban center of the Dilmun kingdom. He has also been in charge of several international exhibitions devoted to the archaeology of Bahrain (Dilmun and Tylos) in Europe, Russia, and the Gulf countries.

Elise Luneau is an archaeologist and presently a researcher at the Deutsches Archäologisches Institut, Eurasia Department, Berlin, Germany, and associate researcher at CNRS, UMR 8215, Paris, France. She has been a researcher at CNRS, UMR 9993, Nanterre, a fellow researcher at the French Institute of Central Asia Studies (IFEAC) and the Berlin Topoi Atlas of Innovations, and a visiting scholar at the Cotsen Institute, UCLA, USA. Her primary research interests are the Bronze Age societies of Central Asia and northern Iran (3rd–2nd millennium BC). She leads a research project on Bronze Age mobile populations and interactions in two valleys of Uzbekistan and Tajikistan funded by the Deutsche Forschunggemeinschaft. She is also currently involved in several international research programs in Uzbekistan, Turkmenistan, and Kyrghyzstan.

Bertille Lyonnet is Directrice de Recherches Emeritus at the French National Center for Scientific Research (CNRS), Paris, as an archaeologist. She has worked in Central Asia (Afghanistan, Tadjikistan, and Uzbekistan), northeastern Syria, the northern Caucasus, and Azerbaijan. A specialist in ceramics, she has always shown a particular interest in the interrelations between the different areas of the world where she worked. She is the author of several books and over 150 articles.

Ruslan G. Muradov is a professor at the International Academy of Architecture, Moscow branch (IAAM), Russia, and vice-president of the Turkmen Union of Architects, Turkmenistan. He specializes in the history of architecture of Central Asia. He has a diploma in architectural design from the Turkmen Polytechnic Institute and a certificate of postgraduate study termination in fine art history from the Institute of History of the Academy of Sciences of Turkmenistan. He is the author of more than 100 articles. His writings have been published in international journals on archaeology, architecture, and design, and are presented at various conferences.

Benjamin Mutin is a research associate at CNRS-UMR7041, ArScAn (Archaeologies and Sciences of Antiquity), France, Archaeology of Central Asia, and at the Department of Anthropology at Harvard University, USA. He has worked in Iran, Pakistan, Afghanistan, Tajikistan, Oman, and Jordan, and he currently is co-directing a new field research program in southeastern Iran.

Vincent C. Pigott is a consulting scholar in the Asian Section of the University of Pennsylvania Museum, USA, and a research associate at the Institute for the Study of the Ancient World (ISAW) at New York University, USA. As an archaeologist he is focused on the prehistoric origins, transmission, and societal impact of metallurgy across Eurasia. His areas of particular focus are Southwest and Southeast Asia and he currently co-directs the Thailand Archaeometallurgy Project for the University of Pennsylvania Museum. He spent his career at the museum as a senior research scientist in the Museum Applied Science Center for Archaeology (MASCA), as a specialist in archaeometallurgy, and also served as the museum's associate director. Upon retiring from the University of Pennsylvania Museum he became a visiting professor at University College London's Institute for Archaeology, UK. He is a fellow of the

Society of Antiquaries in London, UK, and has edited or co-edited several volumes and published numerous articles.

Rocco Rante is an archaeologist at the Louvre Museum, France, and associated member of the UMR 7041 (ArScAn). His interests focus on the archaeology of the Middle East and the western part of Central Asia, from the end of the 1st millennium BC to the Early Islamic period. He is co-director of the Franco-Uzbek Archaeological Mission in the Bukhara Oasis (MAFOUB) and is currently working on *The Bukhara Oasis 1: Population, Depopulation and Settlement Evolution* (forthcoming), the first of the three volumes concerning the latest research on this territory.

Shereen Ratnagar was a professor at the Jawaharlal Nehru University, Delhi, India. Her interests lie in the Harappa Civilization and its state organization, and in trade, ethnoarchaeology, nomads, and the political appropriations of archaeology. She is the author of numerous articles and books.

Robert M. Sataev has a PhD in biology and is author of more than 100 publications on Eurasian archaeology, archaeozoology, and historical human ecology. He is the founder and director of the Museum of Natural History at Ufa in Bashkortostan, Russia. He has taught at the Bashkir Academy of Public Administration and Bashkir State Pedagogical University (Department of Ecology and Nature Management). He has participated in more than 50 scientific expeditions in Russia, Turkmenistan, Tajikistan, and Iran.

Mike Teufer is a research associate at the Eurasia Department of the German Archaeological Institute in Berlin, Germany, specializing in the Bronze and Early Iron Age in Middle Asia. He has conducted field research in Uzbekistan and Iran. Since 2008, his research has focused around Tajikistan, where he has conducted several projects financed by the German Research Foundation. He currently heads excavations at the Bronze Age settlement of Saridjar in southern Tajikistan.

Ali A. Vahdati is deputy director in the provincial office of Iranian Cultural Heritage and Tourism Organization, northern Khorasan branch, Iran. He received his MA in 2005 from the University of Tehran and is currently a PhD candidate in protohistoric archaeology at the University Paris 1, France. Since 2007 he has conducted several field projects in Khorasan and has been co-director of a number of international expeditions, among which the most important results came from the joint Iranian–Italian excavations at the BMAC necropolis of Tepe Chalow in the plain of Jajarm.

Massimo Vidale is a professor of Near Eastern archaeology and the archaeology of craft production in the Department of Cultural Heritage at the University of Padua, Italy. He is involved in several archaeological and ethnoarchaeological projects in Italy and in Middle and South Asia dealing with the formation of early states and the organization of ancient craft production in the development of social complexity. He has published several articles in specialized journals (including *Nature*, *Antiquity*, *Paléorient*, *Cambridge Archaeological Journal*, *World Archaeology*, and *Radiocarbon*), and also authored several books, which include *The Archaeology of Indus Crafts: Indus Craftpeople, and Why We Study Them* (2000, Rome, IsIAO Reports and Memoirs), *Lapis Lazuli Bead Making at Shahr-i Sokhta* (edited with A. Lazzari, 2017), and *Treasures from the Oxus* (2017).

Natal'ja M. Vinogradova has a PhD in history and is a lead researcher at the Institute of Oriental Studies of the Russian Academy of Sciences in Moscow, Russia, and corresponding member of the German Archaeological Institute, Berlin, Germany. She is a specialist in the history and culture of Central Asia during the 3rd–2nd millennium BC. She is co-director of the joint project of the Institute of Oriental Studies and Institute of History, Archaeology and Ethnography of the Tajik Academy of Sciences. She has directed many excavations in southwestern Tadjikistan and is the author of several books and numerous articles.

Sylvia Winkelmann is a research associate at the University of Bern, Switzerland, a member of the Margiana archaeological expedition, and head of the archaeological expedition of the Institute for Archaeological Sciences of the University of Bern in Gonur Depe (southern Turkmenistan). She has worked and taught at the Universities of Halle, Leipzig, Mainz, and Bochum. She specializes in the Bronze Age and Parthian art of Iran and Central Asia, with a focus on glyptic art.

Ganna I. Zajtseva has a PhD in chemical sciences and is a leading specialist in radiocarbon dating. She works at the Institute for the History of Material Culture of the Russian Academy of Sciences where she was head of the Laboratory of Radiocarbon Dating (now Laboratory of Archaeological Technology). She is the author of over 200 publications.

FOREWORD

The transcription system from the Russian to the English alphabet used in the book is that adopted by the United Nations Educational, Scientific and Cultural Organization (UNESCO). It also concerns names from original Persian or Turkish languages that have been transformed due to the history of ex-USSR. Only a few exceptions were admitted, either because the name is usually written otherwise, like Tajikistan, or because publications concerning some sites have already been done in English, like Djeitun or Adji Kui. The Persian names from Iran have their own transcription (Chapter 19).

Russian	*Translit.*	*Russian*	*Translit.*	*Russian*	*Translit.*
А	a	К	k	Х	kh
Б	b	Л	l	Ц	ts
В	v	М	m	Ч	ch
Г	g	Н	n	Ш	sh
Д	d	О	o	Щ	shch
Е	e	П	p	Ы	y
Ё	ë, jo	Р	r	Ъ	‘
Ж	zh	С	s	Э	e
З	z	Т	t	Ю	ju
И	i	У	u	Я	ja
Й	j	Ф	f		

To avoid too many diacritic signs, we also have preferred to write sh instead of š (for Marhashi for instance) except for the text of M. Guichard (Chapter 2).

INTRODUCTION

Bertille Lyonnet and Nadezhda A. Dubova

The Oxus Civilization, also named the Bactria-Margiana Archaeological Complex (or Culture) (BMAC), developed in southern Central Asia during the Middle and Late Bronze Age[1] and lasted for about half a millennium (ca. 2250–1700 BC) until its decline (ca. 1700–1500 BC). It is one of the recent highlights of archaeology and history of art in the ancient East but it is still mysterious as it bloomed and disappeared rather suddenly.

Its "core area" – where the bulk of its remains comes from – is situated in southern Central Asia (nowadays Turkmenistan, southern Uzbekistan, and Afghanistan), but similar cultural features are known on its outskirts, in Tadjikistan, central Uzbekistan (Zeravshan valley), and also far beyond, in a large part of eastern Iran and Baluchistan. In these surrounding "outer" zones most of the finds come from graveyards – now more and more numerous as research goes on – and only very few settlements have been recognized up to now.

Though fully developed, with a well-designed and planned architecture, monumental buildings, and a wealth of splendid artifacts made of the most precious stones or metals depicting a fantastic pantheon, it did not have a writing system to shed light on its overall culture, language, society, or economy, or on the reasons for its development.

A large amount of evidence shows that it was included in a huge network of relationships, most probably of commercial nature, which had developed in the last centuries of the 3rd and the first centuries of the 2nd millennium between Anatolia, Mesopotamia, the Gulf, Elam, and the Indus valley. Unfortunately, its name in the Mesopotamian sources is unknown, although several proposals have been advanced.

Its "core area" was later known as Bactria and Margiana, two provinces that played a major role during the Achaemenid and Hellenistic periods when these names are mentioned in written sources for the first time, and that form the origin of both denominations that archaeologists have given it.

Many articles and books have already been published, but a great number are in Russian and/or are difficult to find. Among those in foreign languages, several deal with exceptional masterpieces of art coming from illegal excavations and now in museums or private collections. Though, of course, we all condemn vigorously the pillage of archaeological heritage and the definitive loss of the context from where these artifacts come and that modern analyses could now have exploited so incisively, we must admit that these publications give access to a material that would otherwise mostly be hidden away. Furthermore, some authors have made a real scientific work out of these market pieces and we constantly use their studies for our research. We will not, therefore, forbid the use of some of them in this volume.

This book intends to provide the last data available on the Oxus Civilization. However, it would have been impossible to develop all the different aspects of this culture even in several hundreds of pages, and we thus admit that, necessarily, this book is limited. The balance of research is unequal within the "core area" itself, and we also regret not to present here in a more developed way the BMAC-related finds coming from areas further away like Elam, Mesopotamia, North Syria, or Anatolia with which this culture was related, though many references will be made to these regions.

We made the choice to let a number of specialists illustrate parts of this civilization. Some deal with their most recent results coming either from excavations, or from studies on the material or the environment. Others reconsider its myths and beliefs. Syntheses have been made on the surrounding areas related to the BMAC (Zeravshan valley, steppes, Indus, eastern Iran), and on its final phase. Finally, several chapters deal with the ongoing research about metallurgy and mining areas, which may have been one of the reasons for its wealth.

The book is divided in four parts.

Part I gives the general setting and historical background of this civilization:

- Bertille Lyonnet and Nadezhda Dubova (Chapter 1) present an updated overview of the BMAC where they raise questions on several aspects of this culture and on the different interpretations that have been given to explain its composite character: they underline the importance of trade.
- Michaël Guichard (Chapter 2) gives the epigraphist's point of view as for its possible name in the Mesopotamian texts.
- Eric Fouache, Lucie Cez, Valérie Andrieu-Ponel, and Rocco Rante (Chapter 3) deal with the environment on the basis of recent studies in the plain of Balkh and in the lower Zeravshan.
- Ljubov' Kircho (Chapter 4) outlines the local development of the Namazga/Anau culture in southern Central Asia from the Late Chalcolithic to the rise of the Oxus Civilization.

Part II constitutes the main bulk of the book and presents different studies on the "core area":

- Ruslan Muradov (Chapter 5) presents the monumental and casual architecture of Bactria and Margiana.
- Elena Antonova (Chapter 6) analyzes the local beliefs known in Central Asia before the rise of the BMAC but that are still present during its bloom as shown by part of the artifacts.
- Annie Caubet (Chapter 7) considers the new pantheon at work in the Oxus Civilization.

- Sylvia Winkelmann (Chapter 8) gives a detailed analysis of the seals coming mostly from the Margiana excavations.
- Massimo Vidale (Chapter 9) studies chlorite vessels from different proveniences.
- Nadezhda Dubova (Chapter 10) presents an analysis of a group of elite graves from the "Royal Necropolis" at Gonur Depe.
- Nadezhda Dubova (Chapter 11) makes a technical and comparative study of the mosaics found in the elite graves.
- Robert Sataev (Chapter 12) presents the animal burials found at Gonur Depe and the rituals around them.
- Julio Bendezu-Sarmiento (Chapter 13) reports on his archaeothanatological studies based upon the data from Dzharkutan and Ulug Depe.
- Vladimir Kufterin (Chapter 14) presents bioanthropological data from Bactria and Margiana.
- Robert Sataev (Chapter 15) gives the results of his archaeozoological studies on the animal bones found at Gonur Depe.
- Kai Kaniuth (Chapter 16) proposes an insight into the small rural settlement he excavated in northern Bactria.
- Barbara Cerasetti (Chapter 17) gives her preliminary results about a mobile part of the Margiana population.
- Elise Luneau (Chapter 18) presents the end of the Oxus Civilization – often viewed as a collapse – and the new orientations this culture took at that time.

Part III is dedicated to the "surrounding areas." The border line is, however, blurred as some of them are considered to be part of the Oxus Civilization/BMAC by those who deal with them, as in eastern Iran, the Zeravshan valley, and Tadjikistan. These areas are not as well known and still have a lack of urban features up to now since no settlement has been excavated there, at least on a sufficient scale. The frequent presence of another clearly identified local culture beside these graves with obvious BMAC material distinguishes these regions from the "core area." They challenge our comprehension of the relations between the "core" and the "periphery" and raise questions on the way of life of these populations and/or on the reasons for their movement or relations. We have arbitrarily divided Part III into two main geographical zones, one to the west and south, the other to the east and north.

- Raffaele Biscione and Ali Vahdati (Chapter 19) deal with northeastern and eastern Iran up to Seistan and advance the idea of the Great Khorasan Civilization.
- Benjamin Mutin and C.C. Lamberg-Karlovsky (Chapter 20) give an overview of the BMAC presence in eastern Iran and Baluchistan.
- Shereen Ratnagar (Chapter 21) analyzes the BMAC relations with the Indus Civilization.
- Pierre Lombard (Chapter 22) presents the BMAC interactions with the Gulf.
- Natal'ja Vinogradova (Chapter 23) reports on her recent results from excavations at cemeteries in southern Tajikistan and raises questions on the BMAC formation.
- Nona Avanesova (Chapter 24) studies the interrelations with the Zeravshan valley mainly occupied by a "steppic" population.
- Mike Teufer (Chapter 25) reconsiders the local Vakhsh culture in Tadjikistan and its connections with the BMAC in the light of new discoveries.

- Gian Luca Bonora (Chapter 26) analyzes the relationships with the northern steppes.

Finally, Part IV concerns the research about metals and metal deposits in relation to the BMAC:

- Steffen Kraus (Chapter 27) gives the last results of archaeometallurgical studies.
- Jennifer Garner (Chapter 28) presents the known metal sources of tin and copper in Central Asia.
- Vincent Pigott (Chapter 29) reconsiders the origin of the tin used in the Near East in the light of recent discoveries in Iran.

The Appendix at the end of the book gives most of the known radiocarbon dates from southern Central Asia and the surrounding areas related to the BMAC. They have been gathered by Nadezhda Dubova for Gonur Depe, and many new ones, analyzed by Ganna I. Zajtseva and Natalja Burova, are presented here. The dates for the other sites have been collected by Bertille Lyonnet. Unfortunately, those published without their BP dates could not be included. All have been recalibrated by Michel Fontugne according to the last OxCal program.

Most of the illustrations were provided by the authors. Some needed revisions and we would like to acknowledge Ardeth Anderson, Emmanuel Baudouin, Vérène Chalendar, Carolin Frank, and Jean-Michel Roynard for their help in this work.

NOTE

1 There is no harmonization in the period denominations over different areas. In Central Asia, the Middle Bronze Age corresponds to Near East Early Bronze IV/Akkadian and Ur III periods, and the Late Bronze Age to Near East Middle Bronze Age/Isin-Larsa and Paleo-Babylonian up to the Mitanni periods.

PART I

THE OXUS CIVILIZATION BACKGROUND

CHAPTER ONE

QUESTIONING THE OXUS CIVILIZATION OR BACTRIA-MARGIANA ARCHAEOLOGICAL CULTURE (BMAC)

An overview

Bertille Lyonnet and Nadezhda A. Dubova[1]

INTRODUCTION

The Oxus Civilization, also named the Bactria-Margiana Archaeological Complex (or Culture) (BMAC), developed in southern Central Asia during the Middle and Late Bronze Age[2] and lasted for about half a millennium (ca. 2250–1700 BC) until its decline (ca. 1700–1500 BC) (see Table 1.1). It bloomed and disappeared rather suddenly but did not have a writing system to shed light on its overall culture, language, society, or economy, and/or on the reasons for its development. It is still a mysterious phenomenon, and the presence of exceptional artifacts in its material culture, combined with the lack of knowledge on its background, has led to various hypotheses about its origins. Though there is numerous evidence that it was included in a vast and intensive network of exchanges going on in the ancient Near East from the Mediterranean to the Indus at the turn of the 3rd to 2nd millennium BC (see Figure 1.1), we have no hint as yet as to what its name was in the Bronze Age (see Chapter 2).

It was first recognized in northern Afghanistan and southern Uzbekistan and soon after in eastern Turkmenistan, areas that played a major role during the Achaemenid and Hellenistic periods when they are mentioned for the first time in written sources as Bactria and Margiana. The discoveries made in these regions are plentiful and concern both graveyards and settlements, including the largest known up to now, Gonur Depe, they are therefore considered to be the "core area." A number of long known "stray finds" (mainly from hoards or graves) scattered in a much wider geographical zone all around, up to the frontiers of the Indus valley, in Baluchistan, or in northeastern Iran were seen as part of a "periphery area." However, recent finds in Tajikistan (see Chapter 23), in the Zeravshan valley (Avanesova 2010; see also Chapter 24), in eastern Iran (see Chapter 19), and in Baluchistan (see Chapter 20) now question the validity of a distinction between a core and a periphery, or, at least, the position of the dividing line. This question will be tackled again later on.

Table 1.1 Chronological table for sites in Central Asia, the Indus area, Iran, the Gulf, and Mesopotamia

Dates	*Anau/ NMG*	*Gonur*	*Steppes*	*Hissar*	*Sarazm*	*Mundigak*	*Mehrgarh*	*Nausharo*
1500		III						
1600			Tazabag'jab Fedorovo					
1700		II						
1800	VI							
1900		I	Petrovka/ Sintashta				VIII/Sibri	IV
2000	V			III C			Quetta hoard	
2100							Hiatus	III
2200			Poltavka					
2300								II
2400	IV			III B	IV			
2500						IV		
2600				III A	III		VII C	I D
2700							VII, A, B	I A–C
2800	III				II			
2900				II B				
3000					I	III	VI	
3100								
3200	II						V	
3300						II		
3400				II A			IV	
3500						I, 3–4		
3600								
3700								
3800								
3900	I					I, 1–2		
4000							III	

Based upon: for Anau/NMG – Kircho, this volume; Gonur – Zajtseva *et al.* 2008, see Appendix, this volume; Steppes – Bonora, this volume; Hissar – Dyson and Howard (eds.) 1989; Sarazm – Lyonnet 1996 and ^{14}C dates; Mundigak – Jarrige *et al.* 2011, Salvatori and Tosi 2005; Mehrgarh – Jarrige *et al.* 2011; Nausharo – Jarrige *et al.* 2011; Harappa – Meadow and

Harappa	*Makran*	*Sh.-i Sokhta*	*Tepe Yahya*	*Fars*	*Susa*	*Dilmun*	*Kültepe Kanesh*	*Mesopotamia (middle chrono)*
5/ cemetery H			IV A					
							Ib 1833–1720 II 1935?–1835	1760 end of Mari
4				Mid. Kaftari	V B			Samsi Addu
		IV, 0	IV B1			II b		Isin-Larsa
3 C		Hiatus		Early Kaftari	V A	II a		Ur III
		IV, 1						Akkad
3 B	IV	III	IV B2–6					
					IVA2			ED III
3 A	III c	II						
		I			III B, C			ED I–II
2/ Kot Diji	III b		IV C					
	III a							Djemdet Nasr
					III A			
1 A, B Ravi/ Hakra								LC 5/Uruk
	Hiatus		V		II			LC 4
	II							
								LC 3
			VI					LC 2
								LC 1

Kenoyer 2005, Kenoyer 2005; Makran – Besenval 2005; Shahr-i Sokhta – Salvatori and Tosi 2005; Tepe Yahya – Hiebert and Lamberg-Karlovsky 1992, Mutin and Lamberg-Karlovsky, this volume; Fars – Petrie *et al.* 2005; Susa – Salvatori and Tosi 2005; Dilmun – Laursen 2010; Kültepe/Kanesh – Dercksen 2008; Mesopotamia – Middle Chronology.

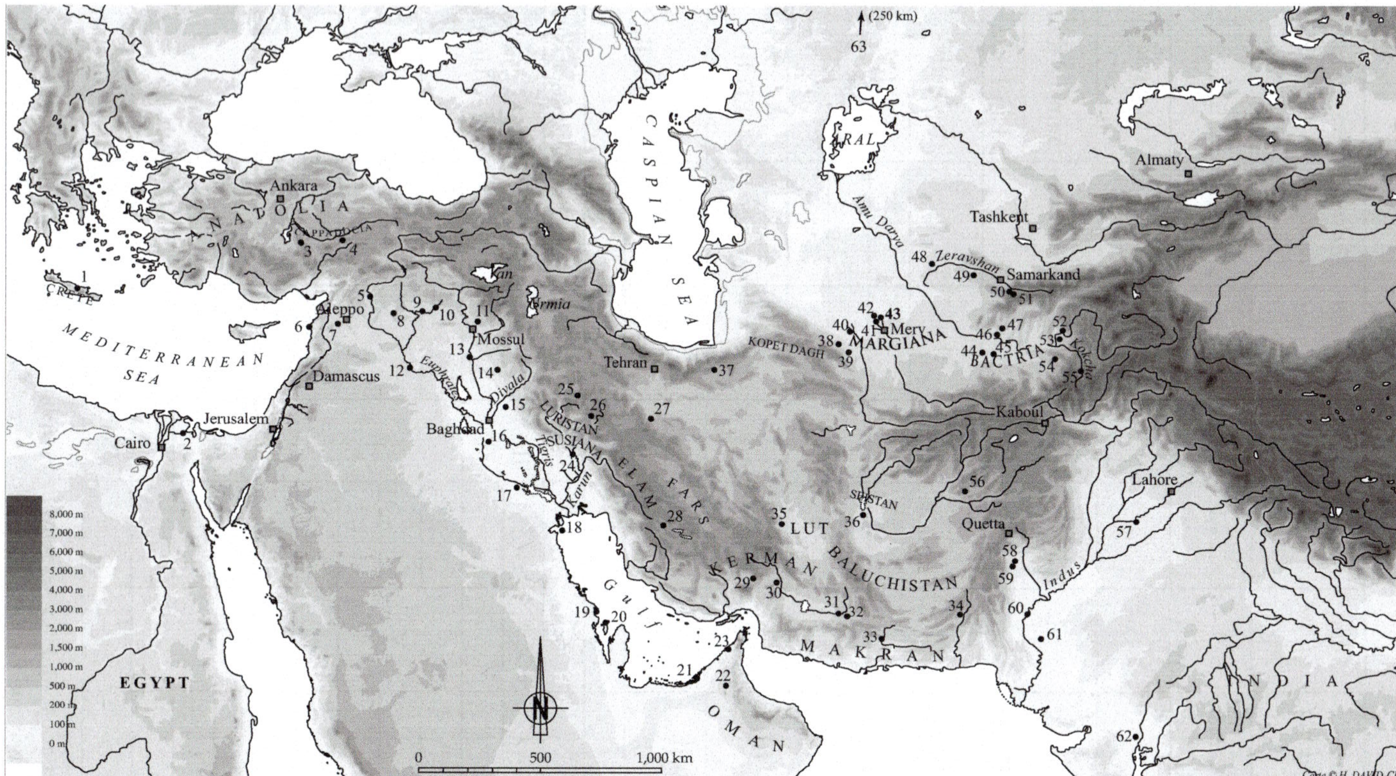

Figure 1.1 The world of the Oxus Civilization: main sites mentioned in the book (adapted from a map by H. David). 1 Knossos; 2 Tell el-Dab'a; 3 Acemhöyük; 4 Kültepe/Kanesh; 5 Karkemish; 6 Ugarit; 7 Ebla; 8 Tell Chuera; 9 Tell Brak; 10 Tell Leilan; 11 Tepe Gawra; 12 Mari; 13 Assur; 14 Nuzi; 15 T. Asmar/Eshnunna; 16 Babylon; 17 Ur; 18 Failaka; 19 Tarut; 20 Bahrain/Dilmun; 21 Umm an-Nar; 22 Hili; 23 Tell Abraq; 24 Susa; 25 Godin-Tepe; 26 Deh Hosein; 27 Tepe Sialk; 28 Anshan; 29 Tepe Yahya; 30 Jiroft; 31 Bampur; 32 Khurab; 33 Miri Qalat; 34 Mehi; 35 Shahdad; 36 Shahr-i Sokhta; 37 Tepe Hissar; 38 Namazga-Depe; 39 Altyn-Depe; 40 Geoksjur; 41 Togolok; 42 Kelleli; 43 Gonur-Depe; 44 Dashly-Tepe; 45 Bactra; 46 Sapallitepa; 47 Dzharkutan; 48 Zamanbaba; 49 Karnab (tin mines); 50 Sarazm; 51 Mushiston (tin mines); 52 Farkhor; 53 Shortughai; 54 Taluqan; 55 Sar-i Sang (lapis-lazuli mines); 56 Mundigak; 57 Harappa; 58 Mehrgarh and Sibri; 59 Nausharo; 60 Mohenjo Daro; 61 Chanhu Daro; 62 Lothal; 63 Sintashta.

This large geographical distribution explains the different names this civilization was given. V.I. Sarianidi (1974) made most of the discoveries in Bactria and Margiana and is the author of the appellation "Bactria-Margiana Archaeological Complex" (BMAC), a way for him to distinguish it from the Namazga culture that developed further west in Turkmenistan, and to show where it was implanted. This appellation has been slightly modified later as "Bactria-Margiana Archaeological Culture" (see Dubova *et al.* 2018: 8). After a series of brilliant discoveries, Sarianidi (2006, 2008a) maintained this name and considered it as the fifth civilization after Sumer, Egypt, the Indus, and China. H.-P. Francfort (1984, 2016), who excavated the Harappan site of Shortughai in eastern Bactria, equaled it to the Indus Civilization and, because of its position in the Oxus River catchment area, gave it the more glorious name of the "Oxus Civilization." The new discoveries in northeastern Iran have recently led the authors to consider that, with its extension from there up to Tajikistan, it more or less corresponds to the area of the Sasanian and Early Islamic Greater Khorasan province: they therefore propose to call it the "Greater Khorasan Civilization" (see Chapter 19). However, as already mentioned by S. Salvatori (2016) for the first two names, to which we can add this third one, neither is satisfactory because they all are too geographical and exclude the areas lying outside. G.L. Possehl (2002), from his Indo-Baluchistan point of view, had proposed to call it the "Middle Asia Interaction Sphere" and it might be the best name – but rarely used – since it covers all the areas where it is known and also implies a circulation of people, goods, and ideas. Here, we will use these names interchangeably, though more frequently BMAC or Oxus Civilization (or Oxus culture).

Publications and books on the Oxus Civilization are plentiful, but a great number are in Russian and/or are difficult to find. Among those in foreign languages, several deal with exceptional masterpieces of art coming from illegal excavations and now in museums or private collections (e.g., Pittman 1984; Ligabue and Salvatori 1989; Aruz and Wallenfels 2003; Vidale 2017b). As mentioned in the Introduction to this volume, we of course condemn in the most vigorous manner the pillage of archaeological heritage and the terrible losses it leads to. This is especially true as far as this civilization is concerned. Nevertheless, these publications give access to a material that would otherwise mostly be hidden away, and most of these authors made a real scientific work out of these market pieces that we constantly use for our research (e.g., Amiet 1977, 1978, 1986, 1988, 1989a, 1989b, 2007; Pottier 1984; Sarianidi 1998b; Winkelmann 2004). We will not, therefore, forbid the use of some of them in this volume.[3]

This book intends to provide the last data available on this civilization, but, as briefly mentioned in the Introduction to this volume, we are aware of its limits. At Gonur Depe, a major site where excavations are still going on today, a lot of new research and analysis is going on that we cannot report in detail here.[4] On the other hand, some may consider that we give this site too much weight with several articles and numerous references. Certainly, the fact that it was the major center of Margiana, had a special function (administrative, commercial, and/or religious) – as highlighted in Chapters 10, 12, and 15 – and also that it is the only site excavated on a very large scale, could be considered to bias the data, but the same could be said for many other civilizations where the emphasis has mostly been done on the major cities. We also regret not to present here in a more developed way areas like Mesopotamia, North

Syria, or Anatolia with which the BMAC was related, albeit in an indirect manner. From these extreme regions, only written sources (see Chapter 2) and tin-mining areas (see Chapter 29) are considered here, but we will refer to them several times in this overview. However, a chapter is dedicated to the Gulf (see Chapter 22).

As mentioned in the Introduction, we made the choice to let a number of specialists illustrate parts of this civilization.

In the following pages, we wish to give the readers a short updated general presentation (or overview) of the BMAC, but, in so doing, our aim is to raise questions on some of its most intriguing cultural features and on the different interpretations that have been made about them.

After a brief review of: (1) the first discoveries, and (2) their successive chronological attributions, we will present (3) the BMAC and the "Middle Asia Interaction Sphere," (4) the BMAC main cultural features (successively environment, cultural background, architecture, material culture, economy, way of life, political and religious systems, funerary rituals, population, and actual chronology), and, finally, (5) we will try to explain the origins of the "exotic" finds (imports or local products, wider "Elamite *koinè*" and migrations, trade). An appendix, at the end of the book, gives all the known ^{14}C dates[5] of the BMAC sites as well as those from other related sites, and a few on the preceding period. All these dates have been recalibrated according to the last OxCal program to make them comparable.

THE FIRST DISCOVERIES

The first discoveries about the BMAC were only made about half a century ago, at the very end of the 1960s and beginnings of the 1970s on both sides of the Amu Darya, the Oxus River of the Greeks. A.A. Askarov (1973) excavated the settlement of Sapallitepa in southern Uzbekistan, and Sarianidi (1977) excavated several others in the Dashly oasis in northern Afghanistan, an area that was later to become the rich province of Bactria under the Achaemenids and the Greeks. In Afghanistan, most of these sites were in an almost abandoned area not reached by modern irrigation. This greatly facilitated the discoveries of numerous remains, especially those of cemeteries, the pit contours of which could easily be figured out on the ground. The first graves excavated were rich of material and this, unfortunately, gave rise to their disastrous pillage by the very same people who were working with Sarianidi. Back to Kabul after the end of his field season, Sarianidi saw the same metal artifacts, beads, and pots that he was excavating exhibited by the hundreds on the sidewalks of Shahr-i Nao, while the best pieces were in the hands of antique dealers waiting to be sold out to foreigners. Indeed, a few years later, the political situation of the country further dramatically increased the looting. This explains why the focus of this Oxus Civilization has now mainly turned to another part of its core area, Margiana.

It was also during the 1970s that other surveys and excavations started under the direction of V.I. Sarianidi and I.S. Masimov at the fringe of the Kara Kum Desert in Turkmenistan, in the abandoned area of the Murghab delta. This region was named Mouru in Young Avestan, Marghu in Old Persian, and Margiana by the Greeks (Struve 1949: 15; Sarianidi 1981b, 1986a, 1993; Hiebert 1994a: 12), but its name during the Bronze Age is unknown. In fact, earlier expeditions had been made in this area during the 1950–1960s, and some sites, like Auchin or Takhirbaj 3, had then been discovered, but their importance and integration into a culture of their own was not yet understood and, at that time, they were considered as the Murghab variant of the Kopet Dagh Late Bronze Age (period Namazga (NMG) VI and possibly also the earlier NMG V) (see below and pp. 30–32) (Masson 1959). The new expeditions brought to light, half hidden under the sand dunes, several dozen settlements and graveyards, most of which had similar material to that of Bactria. This area, together with southern Uzbekistan (Askarov and Shirinov 1993), became the main field of excavations after the war started in Afghanistan. It is in Margiana that Sarianidi discovered and began to excavate the largest known settlement of this culture, Gonur Depe, which will be often referred to in this volume. It is there, too, that cooperation with several Western archaeologists started after the collapse of USSR (e.g., Salvatori 1993; Hiebert 1994a; Tosi's team for the Murghab's survey in Gubaev *et al.* 1998; Rossi Osmida 2007, 2011; Salvatori and Tosi 2008).

Since then, BMAC sites and, above all, graves or cemeteries, have been discovered in many other areas, from northeastern Iran to Seistan and Baluchistan,[6] and to Tajikistan, pointing at a very large territorial expansion of this civilization (see Part III). As mentioned above, the growing number of these discoveries outside of the "core area" questions the basis of such a distinction between a core and a periphery. Did this civilization cover this huge territory all the time, or are these outer sites just temporary or later occupations? The reasons and the chronology of the events that would have led part of the BMAC population to move are still to be investigated (nomadism, trade, migration) (see later in this chapter and also Chapters 19, 20, and 21), but we need to keep this in mind in our analysis.

THE SUCCESSIVE CHRONOLOGIES

The artifacts and the overall material culture brought to light in Bactria and Margiana comes from sites generally founded on virgin soil and apparently of rather short duration, never making huge *tepes* in size or height. From the very beginning, this material had been related to that discovered earlier in Turkmenistan at the foothills of the Kopet Dagh on large sites like Namazga Depe and Altyn Depe, in layers attributed to the NMG V – or Middle Bronze Age – and NMG VI periods – or Late Bronze Age (Masson 1959, 1981). In the Kopet Dagh, these layers topped a long series of occupation going back to the beginning of the Chalcolithic period (NMG I), if not to the Neolithic (Djeitun). The Bactria-Margiana assemblage was considered to have started a little after that of NMG V and was thus chronologically dated to the late NMG V and NMG VI periods, and most researchers explained its apparent sudden formation by a migration from the Kopet Dagh. An "urban crisis" during the NMG V period, possibly linked with environmental problems, would have forced the population to move in successive waves first to Margiana at the end of the NMG V period, then to Bactria (NMG VI period), and later (end of NMG VI) to eastern Bactria (see, e.g., Kohl 1984).[7] At that time, radiocarbon dates were very rare and, in coherence with their local stratigraphic scenario, the absolute dates proposed by our colleagues were very late, between 1700/1600 and 1100/1000 BC (Sarianidi 1990: 74) or 2000 and 750 BC (Sarianidi 1998a: 76–78). These late dates long impeded Sarianidi to make exact correlations with the distant areas, which he was the first, however, to have detected (see pp. 39–40): he finally accepted the results from calibrated radiocarbon dates only after the discovery of an Akkadian seal in 2001 (Sarianidi 2001: 67, 197; 2007: 188).

Though these areas (the Kopet Dagh foothills, Margiana, and Bactria) do share similar cultural features, especially in pottery and artifacts, or in many of the burial rituals, they also present clear regional, chronological, or genetic differences, and it soon became clear that this scenario and the dates associated with it had to be revised (see Francfort 1984, 1989: 339–376; Kohl 1984: 209–236). One of the reasons for the questioning of the BMAC chronology was also that, at the same time when Sarianidi was excavating the first Bactrian Bronze Age sites in northwestern Afghanistan, a survey made further to the east by a French team directed by J.-C. Gardin was making a just as astonishing find with the discovery of several sites related to the Indus Civilization (Lyonnet 1977) and to the BMAC (Lyonnet 1997: 57–82), among which the Harappan settlement of Shortughai, later excavated by Francfort (1989). Recent finds in southern Tajikistan prove, if it were needed, that the scenario of successive waves of migrants from west to east is wrong (see Chapters 23 and 25). We will return to these questions later on.

THE BMAC AND THE MIDDLE ASIA INTERACTION SPHERE

From the very beginning, the Oxus Civilization created a turmoil among archaeologists and art historians because of the astonishing discoveries related to a variety of geographical and cultural horizons that were popping up in such a remote place. Its fame comes above all from these "exotic" artifacts – imported or, more probably, made locally under strong foreign influences by itinerant craftsmen (see p. 39) – which point at an "interaction sphere" of unprecedented level. The BMAC specific character comes from these artifacts and the extraordinary melange of relations and influences behind them. They are at the core of its rise and development and the basis of all the questions around it. It is therefore important to underline them right away before we give an overview and question all the other aspects of the BMAC. We will, at the end, make attempts to explain these interactions.

The material brought to light in Bactria and Margiana, or in the surrounding areas, is extraordinarily rich in objects and materials of all kinds, and several chapters in this volume will deal with some of them (e.g., Chapters 7, 8, 9, 11, 13, 16, 17). We will not present these artifacts here, but will rather investigate the areas they are linked with. Suffice it to say that these objects range from chlorite decorated containers and mysterious composite seated women statuettes wearing *kaunakes*, to alabaster vessels, metal ceremonial axes with fantastic creatures, metal compartmented seals, stone amulets with a pantheon yet totally unknown in Central Asia, painted and inlayed mosaics with griffins and other mythological creatures, cosmetic metal flacons and their pins with beautiful tops of various animal shapes, elegant pottery, strange polished stone objects in the shape of small columns and discs with (or without) a handle (weights?), long rods (scepters?), ivory game boards and toilet items, lapis beads and artifacts, and exquisite gold and silver vessels in great number, including some with beautiful low-relief or engraved human and/or animal compositions.

It should not be forgotten, however, that most of this exceptional material comes from offerings in graves (see p. 27), extremely numerous in Central Asia at that time and exceptionally well furnished even though most (over 80%) had already been looted not long after the burials (Pottier 1984; Salvatori 1995; Sarianidi 2001, 2007; see also Chapters 10 and 13). There is no doubt that this funerary provenience biases our understanding of this civilization, especially because few comparisons are possible, be it with the preceding and following periods in Central Asia when graves and cemeteries are rare or totally absent, or with other contemporaneous cultures where they are far less frequent.

Beside its wealth, the most intriguing character of this material is the extraordinary extent of relationships it underlines with regions far away (on these matters, see also Part III).

Connections with the world of the *Indus Civilization* were among the first to have been pointed out, thanks to Indus items found at Altyn Depe (Masson 1977, 1981) and, conversely, to a few artifacts from Central Asia excavated long ago in the Indus valley (Vats 1940; During-Caspers 1994; Possehl 2002: 215–236). The presence of an Indus outpost at Shortughai in eastern Bactria confirmed tight relations between the two areas, though the first two periods at this site have not given any evidence of BMAC material and the last two are linked to its late and final phases (Francfort 1989). Other Indus-related discoveries gave more evidence of the interrelations between the two areas. They come from Dashly (Sarianidi 1981a, 1982), Dzharkutan (Kaniuth 2010), and especially Gonur, and include a number of ivory items, etched carnelian beads or faience bangles, stone seated small sculptures (Sarianidi 2008a: figs. 108–109; Salvatori 2010; Frenez 2017, 2018), an inscribed Indus seal (Sarianidi 2005: 258, fig. 114) (see Chapter 23), and a glazed faience figurine of a monkey (Sarianidi *et al.* 2012, 2014). Salvatori (2000) has also mentioned an obvious inspiration from the Harappan world on several pottery shapes found in the Gonur Large Necropolis. The high jars discovered in hypogeum 3210 (and others) of the "Royal Necropolis," though more elongated, can be compared with a well-known type in the Indus (e.g., Kenoyer 1998: 72, fig. 4.8a). The links between the two areas were further confirmed by the discoveries of a BMAC cemetery, settlement, and hoard close to the Indus valley in the Kachi plain and at Quetta (Santoni 1984; Jarrige and Hassan 1989), while other isolated cemeteries are known in different areas of Baluchistan (see Chapters 20 and 21). As we mentioned above, the correlations established between both areas are at the basis of the name "Oxus Civilization," as a pendant of the "Indus Civilization," given by Francfort (1984, 2016).

But the Oxus Civilization was, undoubtedly, also related to many other cultural worlds.

From the very beginning, Sarianidi (1972, 1974, 1976, 1977, 1983, 1985, 1988, 1996, 1998a: 138) has pointed out a number of similarities among the BMAC artifacts with those found in *northeastern Iran*, *Kerman*, and *Elam*. This was largely confirmed by later discoveries, for example the small stone columns and handled disks, metal trumpets, precious metal vessels, etc., known from Tepe Hissar or the Astrabad Treasure in northeastern Iran. The existence of direct relations between the two areas has been confirmed by the discovery of several items, like that of a most probably imported gray ware vessel with a burnished decoration typical of Hissar IIIC at Gonur North grave 4150 (Sarianidi *et al.* 2012, 2014), of another at the citadel of Dzharkutan (Bendezu-Sarmiento and Mustafakulov 2013), and also in a looted grave from the Dashly oasis (Sarianidi 1977, fig. 29), to mention just a few. The recent discoveries in eastern Iran tighten even more the two areas (see Chapter 19).

At that same time when Sarianidi was outlying a vast area of interactions, P. Amiet underlined in detail connections with the Elamite world, among which Susa, especially at the time of the *sukkalmahs*. He drew parallels between the two areas for a number of artifacts, from beads to stone vessels, metal ceremonial axes and *harpé*, seals and themes on them, female figurines, etc. (see e.g., Amiet 1977; 1986: 184–207; 1988; 2007). He considered that all these eastern areas on the Iranian Plateau were tied into a large "Elamite *koinè*," or evoke the notion of "Greater Iran," though it is now clear that within this large area, all these cultures are different and only share some specific attributes due to intensive relationships. A silver goblet found in the vicinity of Persepolis in Fars and decorated with a "princess in *kaunakes*," close to those found in Central Asia, further confirmed the tightness of the relations between Elam and the BMAC (Potts 2008b). Excavations at Shahdad in Iran, where material associated to Elam, Oman, the Indus, and the BMAC had been found together, have highlighted one of the possible roads between the two areas (Hakemi 1997; Salvatori and Tosi 1997; Salvatori 2008).

Recent excavations and surveys in eastern Iran and Seistan (Biscione and Vahdati 2011; see also Chapter 19), or the excavations at Tepe Yahya and their revisions (Hiebert and Lamberg-Karlovsky 1992; Potts 2001; see also Chapter 20), and, indirectly, the recent discoveries made at Jiroft (Majidzadeh 2003; Perrot and Majidzadeh 2005, 2006; Pittman 2013, 2014a, 2014b, 2018, 2019)[8] have since enlarged even more these relations. The BMAC seal styles testify of evident connections with the Jiroft Civilization of southeastern Iran as is well demonstrated by S. Winkelmann (see Chapter 8), and the same is true for some of the chlorite products (Salvatori 2010; see also Chapter 9). A "hut motif" graffito on a local jar sherd, or other chlorite flacons with the same "hut"/temple shape found recently at Gonur Depe bring further evidence (Fribus and Grushin 2016).

BMAC artifacts have also been found in the *Gulf*, especially at T. Abraq and Bahrain (ancient *Dilmun*), or, conversely though more rarely, typical Gulf artifacts – especially chlorite ones – have been discovered in the BMAC (Salvatori 2010; see Chapters 9 and 22). The analysis of metal artifacts from T. Abraq (ca. 2200–1900 BC) has shown that tin bronze was then making its first appearance in the Gulf, leading us to question the provenience of tin since it is not present in Arabia, and pointing at a probable source in Central Asia (Weeks 2007). Undeniable connections between the two regions at a time when intensive international trade was reported in the Mesopotamian cuneiform sources have been underlined by a number of authors (During-Caspers 1992, 1994, 1998; Potts 1993, 1994, 2008a, 2008b; Crawford 1998; Kohl and Lyonnet 2008).

Further away, Sarianidi was one of the first to raise several comparisons with *Mesopotamia* due to similarities either in the architecture (the "blind windows," i.e., recessed niches or pilastered walls) or in some of the themes on seals, like that of the master of animals, winged spirits, or horned deities. But he also underlined their local interpretation (see, e.g., Sarianidi 1982, 1986c, 2007: 100–108). He was followed by many authors and other comparisons have been made, for instance in jewelry (Francfort and Tremblay 2010: 117–118).

Numerous parallels have been underlined, also for the architecture but especially for the seal iconography, with *northern Syria/Anatolia*. This area has been pointed out by Amiet (e.g., 1978, 1989b), Collon (1987), Salvatori (2000), and Lyonnet (2001, 2005), but it has mainly been Sarianidi's focus of interest. He especially noticed similarities in the bird-headed anthropomorphic kneeling figure, winged or not, or the appearance of monkeys, which he mainly compared, then, with Hittite or Mittani seals due to his late dating of the BMAC (Sarianidi 1976; 1989b; 1994; 1998a: 142–148; 1999). However, he sometimes did refer to the period of the *kārum* of Kültepe/Kanesh, and the more so after he admitted an older date (Sarianidi 1992, 2007). He also was the first to make interesting comparisons between BMAC and northern Syria or Anatolia concerning cult vessels with applied animals (see p. 26). Other very close comparisons in ceramic shapes can be added (Lyonnet 2001: 63; 2019).

Sarianidi also made comparisons with *Cyprus* or *Crete* for, among others, the theme of the bull leaping depicted on seals, or specific architectural features (Sarianidi 1996; 1999; 2007: 156–159). Many of his books are filled with comparative illustrations between the BMAC artifacts and those from these areas (see especially Sarianidi 1998a, 2007). Relations with *Egypt* can also be underlined, probably passing through the Gulf, Mari, or Ebla (Lyonnet 2001).

Finally, the "quiet" presence of mobile people from the *northern steppes* has long been noticed. The relationships between the BMAC and some "Andronovo" groups (mainly the Tazabag'jab culture) have often been underlined due to the presence of their typical incised pottery on small areas close to, or within, the BMAC sites (Sarianidi 1975; Hiebert 1994a: 69–70; Cerasetti 1998; Hiebert and Moore 2004; Kohl 2007: 203–211; Cattani 2008; see also Chapter 17). The late date of this cultural group (usually not before ca. 1700 BC), makes it only contemporary with the final part of the BMAC, when its decline started (see Chapter 18). However, further research in Margiana itself tends to show that the coexistence between the BMAC sedentary and mobile groups, whose origins is not yet clearly identified, has started earlier (see Chapter 17). Furthermore, a long series of discoveries in the Zeravshan valley shows that various steppe groups have constantly lived at the edge of the BMAC since the middle of the 4th millennium (see Chapter 26). The oldest evidence (Afanasievo) is at Sarazm (see Isakov 1992; and, for the date of the grave, Lyonnet 1996: 47, type III, 11). It is followed by various steppe groups gathering at Zhukov (Avanesova 2013b), a metallurgical site at Tugai (Avanesova 1996, 2015), a grave at old Samarkand/Afrasiab (Avanesova 1991), and by many chance finds around Samarkand (Avanesova 2004; see also Chapter 24). In southern Tajikistan, the presence of kurgans in the Vakhsh culture, now dated on firm ground and clearly contemporary with the BMAC, points at possible links with the Altai and Xinjiang (see Chapter 25). These pastoralist groups certainly used the rich pasturelands of the Zeravshan valley and the piedmont of the Pamirs, but they were mainly skilled metallurgists and their interference with the BMAC on that matter is undeniable (see pp. 23–24 and Chapters 24, 26 and 28).

This short review intended to underline the evidence of relations and interactions between the BMAC and a wide sphere around it, including regions very far away.

This raises questions as to what, in this civilization, could have been the basis for such relationships. We will now give an overview of its main cultural features.

THE BMAC CULTURAL FEATURES

Clearly, aside from the "foreign," "influenced," or "related" rich artifacts that strike us from this civilization, it is also evident that there was also a specific local culture in a particular environment.

Environment

The sudden appearance of a regular settlement in the Balkh and Murghab delta plains certainly questions the environment at the time of the BMAC. Today, most of the Murghab delta is heavily covered with sand dunes, and the BMAC bloomed at the time of the worldwide 4.2 kyr mega-drought event, which, according to H. Weiss (Weiss *et al.* 1993; Weiss 2016) led to the collapse of the Akkadian empire and to many changes in the population over Mesopotamia. There is no doubt, however, that people lived for several centuries in these delta plains so that the environmental conditions must have been more favorable than today, when life is made possible only through colossal works with the construction of the Kara Kum canal, with a huge feeder taking its water from the very distant Amu Darya.

From the excavated field weeds, different environments have been recognized, represented by plants from moist zones, salty marshes, or desert steppe. Similarly, the presence of both wild boar and gazelle or onager bones points at two different kinds of environment, one swampy and the other arid. Recent anthracological studies show that the wood or plants used for fuel also come from both riverside (saksaul, *Salix, Tamarix*) or from semidesert areas like camel-thorn plants (*Alhago*), and that poplar and reeds were used for roofs. The wheels of the carts found in the "Royal Necropolis" (see Chapter 10) were made of elm wood (*Ulmus*), which is not attested in the area and probably came from the Kopet Dagh (Sataeva 2016, 2018).

Different research has been made on the surroundings of the BMAC settlements, but their results lead to as many interpretations.

From his first surveys in Margiana, Sarianidi concluded that the occupation occurred in oases separated by sand dunes and that it retracted successively towards the south due to an increasing shortage of water. This was based upon paleohydrological investigations that started in the early 1960s in the Murghab delta but were published much later (Ljapin 1983, 1990a, 1990b, 1991), including in a postmortem summary (Ljapin 2014). In these works, many questions were addressed concerning the Holocene climate fluctuations and the related formation of paleosoils during humid phases and use of irrigation canals during dry ones. Ancient canals – still visible both on aero-photos and in the field in these early years of investigations – are mentioned and were tentatively dated to the Bronze Age though the basis for such dates is dubious. No general irrigation system for the whole Murghab was pointed out, the author considering that each oasis had its own small system.

Further research made by an Italian team led our colleagues to consider that, during Gonur period I (the earliest), sand was not yet present and a great number of natural streams were active so that the whole delta was available for cultivation and occupation (Gubaev *et al.* 1998; Salvatori and Tosi 2008). However, at the Large Necropolis, a number of graves are dug in the sand (Sarianidi 2007: 26–27, 212), and the same is true for the "graveyard of the temenos" (i.e., Gonur South) (Dubova and Muradova 2008).[9] At the Gonur North palace, two rooms were filled with sand especially brought in, maybe as a basis for an upper floor (Sarianidi 2002: 216, 221; 2005: 68–69, 81–92; Artemiev and Urmanova 2010). Sand, therefore, was close by from the very beginning of the BMAC, but the dunes were probably anchored by shrub vegetation and not as active as they became at the end of the Late Bronze Age (Rouse and Cerasetti 2017).

Artificial irrigation is constantly mentioned in the work of our Italian colleagues, but no built-up canals like those known in eastern Bactria (Gentelle 1989) are clearly visible in the different sections that were studied, or were followed to a certain length. This has been recently confirmed by Rouse and Cerasetti (2017). The often quoted work of Cremaschi (1998: 17, 19) specifically states: "The canals do not seem to have an irrigation function, but rather to regulate the watercourses … artificial canals are concentrated in the area in which the main Bronze Age sites are located and slightly north of it." The irrigation function of the only artificial canal that has been partly excavated at Gonur itself (Sataev 2008) is also questionable since it is said to bring water to the Large Pond. Moreover, it dates to a late period since it destroyed part of the rooms of Area 16, which cannot be dated earlier than period II[10] since the pond also partly destroyed an earlier palace close to the "Royal Necropolis" (Sarianidi 2007: 58; see also Chapter 10). No specific investigations have yet been made around the site to judge for the presence or not of artificial irrigation. The only evidence comes from the first period of Gonur in the eastern part of the site, where it is said to have consisted in simple derivations of the water from the natural channels into the fields (Sataeva and Sataev 2016; Dubova *et al.* 2017).

The main sites are all situated along the old natural channels (Salvatori 2008) and numerous scatters of sherds on very low elevation found in between the channels could attest to many ephemeral occupations (Cattani *et al.* 2008) rather than an important population (Salvatori and Tosi 2008: x). In the Balkh plain, recent work shows that the sites also seem to be all situated at the end of the natural channels, and irrigation is considered to have been limited to a strip of land along them (see Chapter 3).

The intrusion of aeolian sand was, however, starting at that time (end of the 3rd millennium) and became more widespread by the beginning of the 2nd millennium since Tazabag'jab sites are positioned over the sand dunes (Cremaschi 1998; Cattani 2008).

The deterioration of the environment is often considered as an explanation for the abandonment of the area and, more or less, for the end of the BMAC. It is usually combined with another natural phenomenon related to tectonics where a general uplift of the whole area from the northeast to the southwest, still continuing nowadays, irremediably pushed westwards the course of the Murghab (Marcolongo and Mozzi 1998). This would have first led to the construction of irrigation canals, and in the end to the abandonment of the settlements (like Gonur) for new ones (like Togolok) located closer to the new course of the river. In principle, this settlement displacement could have been endless. Thus, factors other than just the environment must have played a major role both in the foundation and the end of the BMAC.

Recent research by a French team in the Balkh and lower Zeravshan area points out that, if the climatic 4.2 ky drought and cooling event is undeniable at a global level, a great number of other factors may have led to eventual contrary effects on the environment at the time of the BMAC. This is especially the case with the important role played by the Central Asia mountains (see Chapter 3). Further work is obviously needed on all these questions.

The Namazaga/Anau and southern Hindukush background

Though the BMAC is certainly distinct from the Namazga/Anau culture, there is no doubt that a large part of its cultural features (e.g., pottery, figurines, part of the geometric designs on seals, etc.) comes from the Namazga/Anau culture that developed in the foothills of the Kopet Dagh since the Neolithic period (see Chapters 4 and 6). This Namazga culture had certainly greatly benefitted from the Proto-Elamite/Baluchistan expansion. It had reached a milestone during the NMG III period (ca. 3000 BC) with the rapid development of the Geoksjur oasis in the Tedjen delta (west of the Murghab delta) and that of the site of Sarazm in the upper Zeravshan valley (see, e.g., Bonora and Vidale 2013). At that time too, there is strong evidence of interrelations between Central Asia and central Iran (Tal-i Malyan), northeastern Iran (Tepe Hissar area), Seistan (Shahr-i Sokhta), Mundigak, and the Quetta area of Baluchistan (see pp. 41–42 and Chapter 20).

There is no doubt that the Proto-Elamite expansion that developed immediately after the sudden collapse of the Uruk world system (Algaze 1993) initiated these interactions and the transformations that followed. However, if the presence of Proto-Elamite tablets and beveled-rim bowls is attested all over Iran (except at Tepe Hissar), and up to Seistan and Baluchistan (Besenval 2005; Mutin 2013: 61–64, 169–172; Petrie 2013), none has been discovered in Central Asia. We will return later in the chapter to this period of the early interaction sphere.

A few painted sherds in the Geoksjur style of this NMG III period have been discovered in Margiana to the north of the Kelleli oasis (Masimov 1979) and at Adji Kui 9 (Rossi Osmida 2007: 125), and others are mentioned much further to the south but at great depth (from 4.5 to 6 m) under the actual surface of the plain (Ljapin 2014). Two sherds have also been found at Gonur North itself under the first period of construction (Sataev 2018). Three radiocarbon dates from Gonur Depe also fall within the Geoksjur/Sarazm period (see the Appendix at the end of the volume), but only narrow walls and a short-term occupation related to this period have been brought to light (Sataev 2018). As yet, these isolated finds cannot be considered as an occupation of the Murghab delta that would be buried under alluvium, but just as occasional temporary settlements, possibly from groups on their way to the Zeravshan valley. Though some authors consider that the Oxus Civilization could be an ultimate development of the Namazga culture (see Chapter 4), there is in fact a gap in our knowledge of a few hundred years in Margiana between the end of the NMG III period and the beginning of the BMAC, and the great majority of the sites in Bactria and Margiana are founded upon virgin soil.

More importantly, even if the BMAC shares with the Namazga/Anau culture a number of features, it is clearly not its pure offshoot (Sarianidi 1984: 6–9; 1990: 75–77; 2007: 187; Askarov 1985; Jarrige 1985, 1994). It differs in several points that present closer ties with southern Hindukush and the Helmand Civilization (Askarov 1973: 123–124; Lyonnet 1988, 1997: 74–75; Jarrige 1994). This will be discussed further later in the chapter.

Architecture

The architecture brought to light is often monumental and fortified (see Chapter 5). The buildings were undoubtedly laid out from well-mastered preconceived plans. They consist mostly of rectangular fortified areas with an exterior pilastered thick wall doubled by another thinner one separated from it by a "corridor." In the early phase, these "fortresses" had square towers (Gonur North, Kelleli 3, Dzharkutan "Palatial complex"), transformed into round towers in the late phase (Gonur South, Togolok 21, Dzharkutan "Temple"). Other strange gamma- and T-shaped pilastered corridor-like constructions are known but rare (Sapalli, Dashly 3 "palace"), while there is one case of a round building also doubled with a corridor, with square towers (Dashly 3 "temple").[11] The outside walls do not have any evidence of slit windows (but see p. 148) and the "corridors" inside are usually partitioned into "rooms," making them seemingly useless for archery protection against exterior enemies.[12] Even small settlements often present the same "fortified" layout. Nothing similar is known in the Kopet Dagh, though a small part of the city wall excavated at Altyn has been said to possibly point at similar architecture (Salvatori 2008), but this construction is rather considered as a retaining wall than a fortification (see Chapter 4).

According to Sarianidi (1998a: 87–89), the plan of the Gonur palace reflects the influence of Syro-Anatolian traditions transformed in Central Asia. However, it is rather to the south of the Hindukush, at Mundigak, that the origins of this BMAC architecture have to be searched for, especially in the large pilastered building of period IV.1 (ca. 2800 BC) and the rampart with its square bastions and thick exterior wall doubled by a thinner one enclosing a "corridor" partitioned into rooms (Casal 1961: 56 ff., and fig. 21). Some buildings at Sarazm can also be considered as a source model, as mentioned by R. Muradov (see Chapter 5). At both sites these features are very probably strongly pervaded by the earlier Proto-Elamite foreign influence and ultimately by a Mesopotamian one.

A debated question is the exact function of these "fortresses," and the identification of "palaces" rather than "temples" or vice versa. The ramparts certainly aimed at protection. In this sense, it has to be underlined that most, if not all, these fortresses had important storage rooms (so-called cell rooms) at their very heart (see p. 45).

Material culture

Some of the artifacts are presented in different chapters of the book (for the figurines, see Chapters 6 and 7; for the stamped seals, see Chapters 6 and 8; for the chlorite vases, see Chapter 9).

There is a marked difference between the casual artifacts that have their origin in the two local cultural backgrounds mentioned above and the "exotic" ones that have no roots in the previous imagery (see Chapter 6). The shapes and decoration of the casual artifacts add more arguments to the input of the southern background in the formation of the Oxus culture. It is, for instance, the case for ceramic shapes (the so-called fruit stands and footed goblets; see Udeumuradov 2002) that are clearly related to earlier ones in southeastern Iran (on chlorite vessels, see Perrot and Majidzdeh 2005: pl. 1) and in southern Hindukush (Mundigak IV.1 "verres à pied," Casal 1961: fig. 64). These shapes developed there around the middle of the 3rd millennium, while nothing similar is known in the contemporary NMG IV period in Turkmenistan.

Compartmented seals are a typical feature of Central Asia glyptic (see Chapter 8). At the beginning of the 3rd millennium they are present in small number at Kara Depe and in the Geoksjur oasis in Turkmenistan and, apparently, they were only made of stone (Kohl 1984: 100, 102; Bonora *et al.* 2014). They are more numerous at Mundigak in the slightly earlier period III (Casal 1961: 256–257) and at the contemporaneous Mehrgarh V (Jarrige 1988: 98). All of them at that time had only geometric designs. It is important to underline here that a stronger differentiation should be made within the BMAC glyptic between the seals with a typical local geometric style – probably often used as apotropaic amulets, or having a shamanistic origin as proposed by Francfort (1994) – and the less frequent "exotic" ones with anthropomorphic deities, hybrid/mixed beings, or developed scenes of master/mistress of animals – which relate to a previously totally unknown pantheon that can only be foreign even if it has been "adapted" to part of the BMAC society (see Chapter 6).

Economy

We still know little about the economy or the way of life of its population. Most of the excavations have been done on the largest sites, and small ones that could have given a glimpse at the everyday life have not been studied as intensively (but see Chapters 5 and 16). On the large sites we lack stratigraphic precisions and a mapping of the finds that could help understand the overall function of the buildings or that of some of the rooms. Moreover, the fact that Gonur was certainly an administrative center biases the data.

As yet, only a few production workshops have been recognized that would show how and where the casual artifacts were made. Pottery workshops are the most frequently mentioned due to the presence of numerous two-chamber kilns and pottery slags. A few other discoveries are linked with the production of luxury finds (see p. 38).

Isolated remains of metalworking have been identified at a number of sites, among which Togolok 1 (Sarianidi 1990: 192), Sapalli, Dashly, and Dzharkutan, but only recently has a whole metal workshop been excavated at Gonur in Area 9 (Dubova 2008; Papakhristu 2008, 2010, 2014, 2016) with evidence of the production of bronze cruciform compartmented seals and other utilitarian tools. It reuses previous buildings and dates to the 19th century according to radiocarbon dates (see the Appendix). It seems, however, that small sites could also have been engaged in metalworking, as shown by the Italian survey (Cattani 2008). Early metallurgical analyses from the BMAC artifacts had pointed at a difference between those from Bactria (using more tin alloys) and from Margiana (using less) (Chernykh 1992: 179–182). But recent ones made on the Gonur metal objects show that about half are made of a copper-arsenic or copper-lead alloy, and that pure copper and tin bronze are also present (Pigott 2018; see also Chapter 27).

If Margiana and Bactria themselves are large open plains devoid of mineral resources, all the mountains around are extremely rich in ores of different kinds, from metal (copper, tin, lead) (see Chapters 24, 26, and 28) to semiprecious stones (lapis, turquoise, chlorite, alabaster) (see Chapter 9).

We lack direct evidence for relations between the BMAC and the Afghan mineral sources (Thomalsky *et al.* 2013), but we have those with the northern Zeravshan sources mainly occupied by an Andronovo-related population involved in ore exploitation and metal production. It is in this area that the major tin resources of Central Asia have been identified (see Chapter 28). In the lower Zeravshan valley, pottery, terracotta models of houses, stepped crosses, and other artifacts discovered at the site of Zamanbaba testify of tight relations with local "steppe" or Kel'teminar groups in the immediately pre- or very early period of the BMAC (Guljamov *et al.* 1966). In the middle Zeravshan valley, the site of Tugai, also related to a steppe "Andronovo" group and dated to the beginning of the BMAC period (see the Appendix), gave the unique evidence we have of a furnace (Avanesova 2015; see Chapters 24 and 26).[13] Another proof of metallurgical relation is indirect and dated slightly later, and is due to the presence of a nozzle in a grave at Sazagan in the same area (Avanesova 2010; see also Chapter 24). Along the Zeravshan valley, the tin mine of Mushiston radiocarbon dates confirm its exploitation since the end of the 3rd millennium and those from the Karnab area are dated to ca. 17th century (Parzinger and Boroffka 2003; Garner 2013; see also Chapter 28).

We will return later in the chapter to production activity as far as foreign-influenced artifacts are concerned.

Way of life: agriculture, husbandry/sedentariness, mobility

There is little doubt that the people living in these monumental buildings and "fortresses" had a sedentary life.

A great deal of what we know about the flora and the fauna in the BMAC area comes from Gonur Depe, and might be biased by the special function of the site (see Chapters 12 and 15). Other studies from Shortughai (e.g., Desse 1989; Willcox 1989), Ulug Depe and Dzharkutan (Miller 1993; Mashkour 2013; Tengberg 2013), or from Tilla Bulak (see Chapter 16) complete them. This research received great impetus at Gonur in the 1990s with the excavations of a midden deposit or deep sounding (Miller 1993; Moore 1993; Hiebert 1994a: 132–134). New and regular studies are now being made (Sataev and Sataeva 2014; Sataev 2016; Sataeva 2016, 2018; see also Chapter 15).

All point at a variety of cereals, among which wheat and barley are the most frequent, but millet is also present, with pulses. Grapes (*Vitis vinifera*) – appear for the first time in Central Asia, and many other fruit remains are attested. The presence of particularly dense morphotypes of wheat and barley on all these sites is considered by some authors as proof of irrigation while, on the contrary, others think it is due to their growth in arid conditions (Tengberg 2013).

Sheep and goats are the most common domestic animals consumed throughout the occupation, while bovids are less represented in number, but not in the amount of meat. Pigs are also an unexpected staple resource (see Chapter 15). Other animals were used as pack or draft animals, like donkeys or two-humped camels, and (very few) horses are also present, the first ever attested in southern Central Asia (see Chapters 15 and 26). The presence of humped bovids is less clear (except in the very last levels of Gonur South) but they are represented sometimes on seals and frequently on a number of figurines, alone or drawing carts (see Chapter 15).

Beside horse bones, other proof of the presence of domesticated and mounted equids in BMAC contexts is indirect. They have been discovered in northern Bactria (at Dzharkutan, see Khuff and Shajdullaev 1999) and in the Zeravshan valley (at Sazagan, see Chapter 24; and at Zardcha Khalifa, see Bostongukhar 1998 and Chapter 26), and consist of bone cheek-pieces similar to those known in the steppes, at Sintashta for instance.

Beside this sedentary population, evidence of mobility is constantly growing. B. Cerasetti (Cerasetti *et al.* 2019; Chapter 17) raises very interesting questions on the relationships between sedentary and mobile groups within the Margiana heartland. It is not yet clear, however, whether the mobile groups were related to steppe groups or to local BMAC herders in a dimorphic-like society (Rowton 1973). Nevertheless, the steppe mobile population was intensively interacting with the BMAC, the evidence of which grows especially in the 2nd millennium. Developments on these connections with the steppe peoples are given in Chapters 18, 24, 25, and 26. Often these relations are considered as being linked with trading activities (Frachetti and Rouse 2012; Lyonnet 2016).

S. Ratnagar (2004; see also Chapter 21) underlines the importance of mobility in the southern Hindukush world and the impact it may have had in the dispersion of some material like lapis lazuli or turquoise over a huge area. This accords well with the recent new interpretation of the Shahr-i Sokhta lapis lazuli and turquoise workshops where their production is now considered to be done by part-time specialists working on a seasonal basis for an internal market, rather than for a long-distance one (Vidale 2017a: 303–321). The recent discovery in eastern Iran of a number of BMAC cemeteries apparently devoid of settlement nearby (see Chapter 19) could also be part of the same phenomenon. The earlier discoveries of such implantations outside of the "core area" had at first been explained by migrations from Central Asia to south-eastern Iran and Baluchistan corresponding to a BMAC expansion in its later phase (Hiebert and Lamberg-Karlovsky 1992; Lamberg-Karlovsky 1993, 1994). The rapid BMAC development within a Khanate system would have led to the search for new land or natural resources in a model similar to that well known at the end of the Uruk period in Mesopotamia (see Chapter 20).

The answer to this question will be left for the future, when precise chronological and bio data (way of life, ancient-DNA, etc.) is obtained from large enough excavations in these new zones. For now, we cannot decide whether they constitute a contact zone between different cultures within the Middle Asia Interaction Sphere, or an area of seasonal movements, or a migration/extension zone.

Political and religious systems

The wealth and variety in the remains and the imposing architecture have led researchers to consider that this newly discovered civilization was at a proto-urban stage, exhibiting all the required elements for such a qualification, except for the lack of a writing system.[14]

The size estimation of Gonur, the major site of the BMAC, has tremendously varied and is now estimated at roughly 50 ha,[15] including its vast necropolis, but some consider that the overall settlement is much larger and reaches 130 ha (see Chapter 5). Many consider it as the capital of a proto-state (Sarianidi 1998a, 2002; Salvatori and Tosi 2008; Salvatori 2016). The discovery of very rich tombs within the settlement with specific constructions (hypogea) unknown anywhere else has further led Sarianidi and Dubova to see there the "kings" of the BMAC (see later in the chapter and Chapter 10).

Others, however, consider that the fortified settlements and their regular distribution over a large area rather evoke the *qala* constructions of a Khanate system with independent local/tribal powers (Lamberg-Karlovsky 1994, 2003, 2012; see also Chapter 20).

Furthermore, this culture underwent important changes in the course of its existence and the political system may have changed meanwhile. This is especially underlined by Salvatori (2010, 2016), who considers that only during its first period (well represented at Gonur North), from ca. 2300/2250 up to 1900 BC, there was a centralized power controlling an array of smaller settlements and the international web of relations. Yet, for unknown reasons, this system suddenly disappeared and Gonur North was abandoned as were many other settlements in the Murghab delta. Though there seems to be no time gap with the following period, or population decrease, Salvatori considers that the new political system was much looser, with no site ever attaining the size of Gonur, and that regionalism increased. He also points at a considerable drop in commercial interactions underlined by the disappearance of compartmented seals and other exotic artifacts. The final period, between ca. 1700 and 1500 BC, often considered as a "collapse" (but see Chapter 18), would just be the dramatic continuation of this previous decline.

Certainly, the brilliant achievements presented above are all assignable to a rather complex society. However, they are of little help in determining the political and social system of the BMAC in the absence of written sources, together with the lack of, or as yet insufficient, studies on some aspects of the economy. Many questions are left unsolved and one may also question Salvatori's period II decline in political control and trade, when the architecture with its round towers is so homogeneous all over Margiana and Bactria, and when exterior relations are still attested.[16]

Thus, the political system and religious beliefs in the BMAC area are still at the level of personal interpretations based upon convictions. Though many of the buildings excavated were given names like "palaces," "temples," "king's or priest's graves," we have no real evidence as to their exact function, or how and by whom this country was ruled. Nothing allows for qualifying the BMAC more as a proto-state than a complex chiefdom. This vagueness about the institutions is not specific to the BMAC. The same questions remain at many other places lacking written sources, and even the Indus Civilization with its huge and perfectly organized towns still defies us.

From the differences visible in the grave materials, Sarianidi (2007: 125) considered that three main social groups could be identified, the elite (ca. 4%), the middle class (85.5%), and the lower class (10.5%).

The main buildings certainly had an ostentatious function and were related to an elite, be it political, economic, or religious (possibly one and the same, if one is to judge from Mesopotamia).

During the early period of the BMAC, only clearly identified on a large scale at Gonur North, several complexes of buildings, courtyards, and rooms are called temples ("fire," "water," "sacrificial" temples). The fire temple contains a series of four rectangular ovens highly fired, or rooms filled with pure white ashes that are at the basis of its name. Within the complex, a single-room construction (no. 100) with indented niches on four sides does share architectural features with Mesopotamian temples and was rebuilt several times (Sarianidi 1998a: 120–132; 2007: 172–174), but there is no trace of any cult inside.

Most of these buildings have a two-chamber hearth/chimney built in one of the walls considered as sacred.[17] Altogether, however, more than 300 such hearths (a few have only one chamber) are attested at Gonur North, so they could not have had only a sacred function and other interpretations can certainly be given including a heating function in winter times. Hearths/chimneys built within the walls are present at all the sites in Margiana and Bactria and in all the main rooms. At Adji Kui or Tilla Bulak, they are not considered as sacred (Taganowich 2007; see also Chapter 16).

In the second major period, some Margiana fortresses (Togolok 1, 21, or Gonur South) are considered as temples due to the presence of "white rooms" and podiums with plastered jars (see Sarianidi 2007: 174–178) that have been compared with libation-plastered containers at the Ninni-Zaza Mari temple. Fire is also said to have played a major role in these buildings. Many have raised doubts on these statements, but the presence of seeds of hemp and pollen of ephedra in jars cannot be denied and testifies that these installations were dedicated to the confection of hallucinogenic drinks (Mejer-Melikjan 1990, 1998; Meyer-Melikyan and Avetov 1998). At Dzharkutan, a similar fortress with round towers is also assigned to a temple, but there the plastered installations are linked to "wine" production on the basis of similarities with much later periods, while a central altar on a platform has been related to later known fire altars (Askarov and Shirinov 1993: 104–113). This last "construction" has, however, been denied by further excavations (see Teufer 2015: 18).

Nevertheless, ritual vessels with a decoration made of various animals and of a few anthropomorphic figurines applied along the rim and the sides, as well as incised other motifs (Sarianidi 1998a: 36, fig. 10) have many parallels in northern Syria and Iraq, as noticed by Sarianidi (1998a: 43) and confirmed by Lyonnet (2001, 2019). Recent publications show that, at Tell Brak, some of these vessels may have been used for liquids (Oates *et al.* 2001: 167–168, fig. 407). At Mari, they are considered as dedicated to some kind of cult due to their position near temples and to some of their incised designs with horned deities (Margueron 2004: 391–392). At both these sites they are dated to the post-Akkadian and *shakkanakku* (Ur III) periods. At Tell Chuera, where they are also frequent, they are dated slightly earlier (i.e., to the early EJ IV period, between ca. 2300 and 2100) (Orthmann and Pruss 1995). They seem to continue into the Old Babylonian period, at least as shown by a vessel from T. Rijim in northern Iraq (Kolinski 2000, fig. 28). At the BMAC sites, decorated bone tubes with an eye motif are supposed to be made for drinking the *haoma*, the sacred drink of the Avesta, but also for a far more prosaic use (see Chapter 15). Similar tubes, also decorated (with dotted circles) are known at the Kültepe *kārum* II where they are considered as handles for unknown artifacts (Özgüç 1986a: 70, and pl. 121, 1–8). They are also frequent in Syria, Palestine, and the Aegean, but date earlier (EB II) and are not linked there with any sacred ritual (Zarzecki-Peleg 1993; Saliari and

Draganits 2013). More important, in the Near East hallucinogenic drinks are never mentioned and it would then be necessary to understand the origin of this practice in Margiana, equally unknown up to now in the previous Namazga culture and never mentioned south of the Hindukush. It is known later among the Scythians, and ephedra has been found with the Tarim basin mummies who had been considered as being related to the BMAC population, but this hypothesis has not been confirmed by craniometric studies (Hemphill and Mallory 2004).

The presence of fire constructions and hallucinogenic drinks have led Sarianidi to constantly claim that the population was at an early or pre-stage of Zoroastrianism and they gave arguments for practices or rituals that do have a touch of it (Sarianidi 1986b; 1989a; 1998b: 168–175; 2008b; 2010). But specialists contest any link, at least with the later known forms of Zoroastrianism (see, e.g., the comments at the end of Sarianidi 1989a). As already mentioned, if one accepts these proto-steps towards Zoroastrianism, it would be important to trace where it stems from. In Anatolia and northern Iraq where at least part of the BMAC population would be coming from according to Sarianidi, nothing similar is mentioned. No allusions to such practices or beliefs are made either in the Namazga/Anau or Mundigak/Shahr-i Sokhta cultures.

Was this land Shimashki mentioned in Mesopotamian texts, as proposed by Potts (2008b), or rather Marhashi, as claimed by Francfort and Tremblay (2010)? There is no agreement with either of these attributions and many other authors consider Marhashi to be in the present-day Iranian Kerman province (Steinkeller 1982, 2014). M. Guichard (see Chapter 2) reexamines the Mesopotamian written sources and considers that neither Marhashi or Tukrish or Shimashki fit with what we know of the Oxus Civilization. He considers that such a remote region may not have been really known to the Mesopotamians. This, however, has no effect on the reality of the exchanges, and they were certainly not direct, as mentioned earlier.

Graves and funerary rituals

Graves are in outstanding numbers.[18] The majority had already been grabbed not long after the burial (Salvatori 1995; Sarianidi 2001, 2007; Rossi Osmida 2002; Bendezu-Sarmiento 2013), possibly by the family of the deceased (see Chapter 13); looking at the wealth of what is preserved, one can only wonder about the discoveries that would have been made had they not been looted.

Large cemeteries outside the settlements (and contemporaneous with them) have been excavated at Gonur and Dzharkutan, dozens others have been grabbed by looters in northern Afghanistan (especially the Dashly oasis), and smaller ones are known in Tajikistan (see Chapters 23 and 25). Individual graves within the buildings have also been uncovered and are sometimes said to be contemporaneous with the settlement (at Sapalli[19]) but usually were made after its abandonment, at least partially, since they are intrusive in the walls of the buildings (see Chapter 13). Finally, as mentioned above, a small group of specific imposing tombs or "hypogea," mainly built underground and only partly visible on the surface, have been excavated. They are exclusively known at Gonur, within the settlement, but close to its borderline. Their house-like construction and outstanding material even though they have been grabbed several times made them attributed to the "kings" of the city ("Royal Necropolis") (Sarianidi 2005: 204–259; see also Chapters 10 and 11).

Graves have various shapes (Askarov 1973; Askarov and Abdullaev 1983; Salvatori 1993, 1995; Sarianidi 2001, 2007; Rossi Osmida 2002; Vinogradova 2004; Dubova and Rykushina 2007; Avanesova 2013a; Bendezu-Sarmiento 2013; Sarianidi and Dubova 2016; Dubova *et al.* 2018). Most of them consist of simple pits or of pits with a chamber on the side (*podboj*) called shaft tombs by some authors and pseudo-catacombs by others. Cist tombs (i.e., graves surrounded by a brick wall and covered by a vault, or house-like hypogea),[20] are rare and only known at Gonur Depe up to now. Considering the weight of traditions attached to the rituals surrounding death, such a variety is unusual and would point at a mixed population (Sarianidi 1998a: 71; Babakov *et al.* 2001; Dubova and Rykushina 2004, 2007), but this is difficult to prove and it could also be linked to various social positions (Sarianidi 2007: 30). The discovery of obvious foreign items in them, like one Old Akkadian and three southeastern Iranian cylinder seals, have led some to consider that these graves are those of foreigners (Pittman 2014a, 2019; see also Chapter 8), but there is no certainty since these items could also have been transmitted, as suggested by Guichard (see Chapter 2). Moreover, the reuse of seals is known in Middle Asia, for instance at Kültepe/Kanesh (Teissier 1994: 46). Nevertheless, considering that Gonur was more an administrative and cult center than a living place, a fact confirmed by the meager daily life living remains like kitchen fauna (see Chapter 15), one may wonder who were the people buried there. The possibility of "exterior people" – whatever the distance – cannot be excluded until more analyses are done: In the Indus valley, where cemeteries are few, isotopic analyses have demonstrated that the deceased were not local people but came from the Indus hinterland (Valentine *et al.* 2015).[21]

Furthermore, other various rituals unknown before have been identified. This is the case for pits fired before the inhumation (Sarianidi 2007: 36–41), the presence of ashes and charcoals in the tombs,[22] followed later by cremation (Avanesova 2013a), the presence of a few fractional burials (Sarianidi 1998a: 72–73), or for that of a supposed *dakhma*[23] for exposition of the bodies at Gonur North, room 92 (Sarianidi 2007: 163–167). The origin of these new rituals is as yet unknown. Moreover, in the "Royal Necropolis" "accompanying persons" (i.e., sacrificed) have been buried with an elite personage and/or in a ditch nearby (see Chapter 10). In the Near East, this practice is known only in a few cases; namely, in the Royal Cemetery of Ur, or during the Ur III period at least with Shulgi and perhaps already with Ur-Namma (Michalowski 1977), as well as, most probably, in the Dilmun Royal Mounds of Amorite rulers (Laursen 2017: 291), and it is still a puzzling question. It is better known in the Caucasus area (Maikop Kurgan, "Royal Tomb" of Arslantepe, or Martkopi Kurgans; see Lyonnet 2016).

The burial of complete animals – Bactrian camels, bovids, donkeys, dogs, rams or lambs, and even pigs – is also a typical phenomenon of the BMAC, especially at Gonur (Sarianidi 1990: 141; 1998a: 73–75; Sarianidi and Dubova 2008, 2015; Dubova 2012, 2015). It is attested in a lesser degree at other sites in Bactria (Sapallitepa, Dashly, Buston) where it mainly concerns sheep (or lambs) and dogs (see Chapter 12). This ritual, totally unknown before in Central Asia, certainly comes from the Near East as suggested by Sataev in Chapter 12. In a few cases, rather than a burial it seems that a sacrifice had taken place (see Chapter 12).

Most burials (except in the "Royal Necropolis") are individual, though a few contain two or three skeletons. However, a detailed analysis of the teeth has shown that some graves may have contained more people, probably buried successively. Finally, some tombs are cenotaphs (Sarianidi 2007: 32; Dubova and Rykushina 2007), which become more numerous at the very end of the BMAC (see Chapter 18).

It should also be mentioned that while the great majority of the tombs has nothing to signal them on the surface, some brick constructions have been recently noticed (see Chapter 13), and a few cemeteries, all situated in southern Tajikistan and related to the contemporary Vakhsh culture, present a low kurgan-like construction (see Chapter 25).

Population

Anthropological studies from Gonur (Babakov *et al.* 2001; Babakov 2004; Dubova and Rykushina 2004, 2007) have identified a rather homogeneous, gracile population of Mediterranean type though many variations are pointed out within it, with some individuals being of a bigger size, with increased height, and large bones. This population is similar to the local one known since the Neolithic period all over southern Central Asia, as shown by many previous studies. No evidence of any Andronovo feature has been noticed. Only a few brachicephal skulls are considered as possible markers of intrusions from Iran, while a certain number of "Veddoid" features on others links them with the Indus areas.

A recent a-DNA analysis made on 39 samples from Gonur Depe points to different results. Two individuals are possibly related to the Kel'teminar group – a northern "steppe-like" culture, that developed around the Aral sea and probably constitutes the ancestors of Tazabag'jab – three others are related to the steppe group, and one with a profile similar to that from samples of the Swat valley about a millennium later (Narasimhan *et al.* 2019). However, in this study there is no data on the origin and precise date of the individuals submitted to the analyses, so these results do not help much: As we saw earlier, connections with Zamanbaba/Kel'teminar are probable in a very early phase and genes are transmitted from one generation to the other; similarly, relations between the BMAC and Andronovo/Tazabag'jab are attested by archaeological artifacts in a rather late phase of the BMAC. If there is little doubt that all this new ancient-DNA research will give excellent results in the future, it still needs to enter a lot more data, from many different areas, and to work on more precise chronological grounds.

In her recent study, Dubova (2017) does not see craniometric differences in the overall population between the different cemeteries of the site (Large Necropolis, "Royal Necropolis," and graves in the ruins of the site). Though this could be viewed as a fact pointing at the absence of any massive intrusion of migrants during the BMAC, she cautiously considers that it only shows that the same general population was buried in these different cemeteries. The "rulers" apparently also belonged to the same local population.

Nevertheless, Dubova's last anthropological studies point out that there is an evident connection between the construction type of the grave and its wealth (Dubova and Kufterin 2017). It is especially clear with cist graves and chamber tombs (including hypogea), which are the richest. In the hypogea, the main individual for whom the grave was built cannot be clearly identified because of the looting that disturbed the original arrangement and because of the number of "accompanying" persons (see Chapter 10). In the cists, however, which contain only one person (a few cases with two or three), the skull measurements testify of differences with those of the other graves and point at a specific sociocultural group. More research is obviously needed on this.

Paleobiological research on a few skeletons tends to show familial connections between the individuals (Sperduti *et al.* 2002). Studies on the pathologies are still going on (see Chapter 14) and other research just starting hopefully will give more data on the diet, way of life, or the genetic relations between these people.

Other anthropological studies have been done, especially in northern Bactria, on the material from Sapallitepa and Dzharkutan (Khodzhajov 1977; Khodzhajov and Khalilov 1978, 1984; Dubova and Kufterin 2015; see also Chapter 13). Specific craniometrics comparative research has been done by Hemphill (1999) to trace the origins of the northern Bactrian population: it points at a local origin with a possible gene flow from the previous Geoksjur component that the proximity with Sarazm could explain.

A very interesting observation made long ago is that women's graves are wealthier than men's graves, especially in seals (Askarov 1977: 42; Sarianidi 2001: 184, 2007: 52; Luneau 2008). This phenomenon recalls in many ways the situation of the Assur society at the time of the merchants when men were abroad leaving their wives to deal with trade at home (see Teissier 1994: 49).

Chronology

As mentioned earlier, the question on the exact chronology of the Oxus Civilization has been debated since the 1980s, and it is still not yet totally solved, especially as far as the different sites and their levels within the BMAC are concerned. It relies both upon the relative chronology, the absolute radiocarbon dates, and the obvious relations it had with a very large surrounding area, as shown above. Unfortunately, in the Near East at least, the absolute chronology is not better ascertained and the debate is still unsolved between the early, middle, or late chronologies (Gasche *et al.* 1998a, 1998b; Pruzsinszky 2004). The middle chronology, however, seems to fit better with the absolute dates gathered for the BMAC and other related areas (see the Appendix).

Relative chronology

As said earlier, for a long time and because of the whole Central Asia chronological sequence built up by our colleagues from the Neolithic down to the Greek period and the fact that the BMAC was considered as issued from the NMG V period of the Kopet Dagh area, our Russian and Uzbek colleagues dated it exclusively within the 2nd millennium, with an end around 1000 BC. The presence of a few iron artifacts – apparently from non-meteoric origin (cf. Sarianidi 1998a: 77) – confirmed these late dates.

Most of the small sites seem to be of short duration, no more than one period, and the scatter of sherds on the surface of the desert may indicate constant settlement shifts

(Hiebert 1994a: 28–29; Cattani 2008). For the largest settlements, the first architectural plan was maintained for a certain length of time with several phases of repairs and some modifications. At Gonur North, there is evidence of at least three main periods of occupation (Sarianidi 1998a: 83–85; Sarianidi and Dubova 2014a),[24] followed by one where the site is used as a graveyard. The first is related to the construction of the *kremlin* and the buildings associated to it on its four sides, including the "Royal Necropolis"; it ended in a huge fire dated to the 20th century BC in median radiocarbon dates (see the Appendix). A second phase shows the construction of the outside wall over the hypogea of the "Royal Necropolis" and a reuse of some of the rooms for a metallurgical workshop. A third one shows the decline and reuse of the palace for ordinary houses as well as the construction of many water channels that destroyed part of the site (Sarianidi and Dubova 2012: 33–39; 2014b: 93, fig. 1). Finally, after the abandonment of this part of the site, a fourth phase shows that intrusive graves were dug within the walls. It is not yet clear, however, where Gonur South fits in this scheme.

On the same site, buildings are often scattered, and not rebuilt over several times, so that the stratigraphy is more horizontal than vertical, and their exact chronological correlation is unclear. At Gonur, for instance, the South mound dates to a later period than the North one and has been subdivided into five phases, the last one being related to the Takhirbaj 3 period or end of the Late Bronze Age (Hiebert 1994a: 113–114). Nonetheless, soundings show that below the wall of the fortress with round towers there are earlier levels almost untouched.

Many studies have attempted to establish the precise internal chronology at the major sites. A great deal of material has been published but the publications do not give all the necessary details to understand the stratigraphy. Further works should shed light on these questions. Specific chrono-typological studies have been attempted on the ceramics but they mostly concern that from graves (P'jankova 1993; Udeumuradov 1993, 2002; Luneau 2014; Teufer 2015).

All these problems have led to the constitution of separate chronological charts for Margiana and northern Bactria. In both areas, successive phases have been distinguished (in Margiana, the Kelleli, Gonur, Togolok, and Takhirbaj phases; in northern Bactria, those of Sapalli, Dzharkutan, Kuzali, Molali, and Bustan). To make the question more complex, in the last region, the succession has been contested with the Kuzali phase placed at the end, and other denominations are used (Late Bronze Age I and II) (see Kaniuth and Teufer 2001; *contra* Vinogradova in Vinogradova *et al.* 2008: 220–228) (see Chapters 16 and 13).

In general, the BMAC is divided into three main large periods, with conventional dates. The oldest period (pre-2000 BC) is mainly identified in Margiana and probably also at Sapalli and Dashly 3. The middle period (after 2000 BC) covers all the areas of the BMAC until ca. 1800–1700 BC. The last period shows signs of disaffection, abandonment, and isolation, and runs until ca. 1500 BC (see Luneau 2014; see also Chapter 18). Steppe intrusions with campsites around the BMAC settlements may have started already within the early period (see Chapter 17) but they mainly grow in importance during the middle and last periods (Cattani 2008; Salvatori and Tosi 2008: xi; Rouse and Cerasetti 2017).

This chronological and geographical distribution, however, is still debated and recent finds in southern Tajikistan show that this area was also related to the BMAC from the very beginning (see Chapters 23 and 25).

Absolute chronology

Up until 20 years ago, ^{14}C dates were still rare, isolated, and rarely calibrated. As already mentioned, our Russian colleagues considered that the BMAC extended mainly within the 2nd millennium up to around 1000 BC. Salvatori has often dated its beginning very early (ca. 2400 BC), to make it match with Shahdad where a large amount of material similar to that of the BMAC has been discovered. With the start of international cooperation and the multiplication of analyses, the dates now admitted by all place the Oxus Civilization between 2250 and 1700 BC, while its final phase extends until ca. 1500 BC.

The constant improvement in the calibration of the dates have led us to collect once more[25] most of the known dates of the major sites linked with the BMAC and to recalibrate them all according to the last standards (see the Appendix).

On the basis of the 129 radiocarbon dates made at Gonur itself, and from their median date, we can see that five of them point at an early date (first third of the 3rd millennium), seven are situated around 2500–2400 BC, five in the 2300s, 12 between 1700–1500 BC, one is an anomaly, and all the rest (99) are situated between the 2200–1700s BC. Only further excavations will eventually bring to light architecture and material related to the earlier 3rd millennium, which is for the moment lacking.

In a very broad scheme, using the available information on the provenience of the ^{14}C samples, these radiocarbon dates point out that Gonur North (including the Large Necropolis), or period I, existed mainly between the 23rd and the 19th centuries BC. It also shows that it was in part contemporaneous with Sapallitepa, the Citadel and the pre-temple levels of Dzharkutan, Tillja Bulak, Tugai, Gelot and Darnajchi, Shahr-i Sokhta IV, Hissar IIIC, and Tal-i Malyan (Kaftari period). It is contemporaneous with the end of the Indus urban phase (IIIC) and the beginning of the Late Harappan phase (IV).[26] The second BMAC major period, with round towers fortresses is placed during the 18th century BC at Gonur South and the Dzharkutan Temple (VI). Consequently, a major change seems to have happened in the second half or end of the 19th century BC (see Tables 1.1, 1.2, and 1.3).

Thus the changes between periods I and II are contemporaneous to those observed in the "international" sphere (decline of the Harappan Civilization, endless wars within Mesopotamia, change in the Gulf traffic from Magan to Dilmun) and we cannot rule out that they may have had an impact on the BMAC (Lyonnet 2019).

EXPLAINING THE "EXOTIC" FINDS AND THE WEALTH OF THE BMAC

Undoubtedly, there was a local basis for the extraordinary *floruit* visible in the BMAC, but the novelties that make it so specific imply a very strong external impulse, mainly visible through the exotic finds found in the wealthiest graves or the largest buildings and therefore probably related to an elite. More puzzling is the multidirectional directions these exotic artifacts underline.

Different explanations have been advanced and contested. Each author has usually presented his answer as unique, but we will see that they are not mutually exclusive.

Imports or local production?

As we have seen, the intriguing character of these luxury artifacts raises questions with regard to their provenience.

We have no hint at what could have been the local "Bactrian" art or religious/ mythological pantheon before the appearance of the BMAC. The few pieces of art of the preceding periods in the Namazga culture are the terracotta figurines (see Chapter 6), which are far from the sophisticated low-relief animal and human representations on silver or gold vessels, or the ceremonial axes, composite figurines, mosaics, or some of the compartmented seals with anthropomorphic designs that highlight the Oxus Civilization. At Mundigak, however, sculpture seems to have been more developed (Casal 1961, vol. II, pl. XLIII–XLIV).

Furthermore, these exotic artifacts are already in far fewer numbers or even disappear after ca. 1900, as already pointed out by Salvatori (2016); that is, well before the "end" of this civilization (see Chapter 18).

Only the geometric designs like the stepped-cross compartmented seals are undoubtedly issued from decorations made on pottery (and possibly also on other perishable items) of the previous NMG III and IV periods, and they were also used at Mundigak before the BMAC developed (see Chapter 6). Some motifs based upon animals like snakes and eagles may reveal regional mythological beliefs since they are also attested in southeastern Iran on the "*série ancienne*" of the chlorite vessels. These seals, which probably are rather amulets with an apotropaic function, are indeed much more numerous than those with anthropomorphic deities.

On the other hand, if many of the luxury items point at "foreign" elements in their style, material, fabric, etc., only rarely does it seem that we are dealing with real imports (Hiebert 1994b), though it is not always easy to distinguish those with certainty.

Uncontested imports at Gonur consist only in a few handfuls of artifacts. This is the case for a few seals like the Indus stamp seal mentioned above, an Akkadian cylinder seal with an inscription from grave 2550 at the Large Necropolis (Sarianidi 2007: 107), or the three southeastern Iranian cylinder seals with a vegetation goddess: two from graves 23 and 1393 at the Large Necropolis (Sarianidi 2007: 105–106); and one from Area 6, room 11 (Sarianidi 2005: fig. 137; Pittman 2014a, 2019; see also Chapter 8). To these seals we can add a few Harappan artifacts (etched or long carnelian beads, faience figurine, and bangles), some gray streak-polished vessels from northeastern Iran (see p. 16), a few chlorite artifacts from the Gulf or southeastern Iran (Potts 2008a; Salvatori 2010; (see also Chapter 9 and 22), and possibly a necklace from grave 2900, the preserved thread of which with a S-spun has been identified as a typically Egyptian-type twist (Tsareva 2007). It is also said, from xylological studies, that the wood (*Exochorda sp.*) used as a sheath for an unknown object from grave 4014 (Area 18) comes from the eastern mountains (Pamir, Alai, Tian Shan) (Sataeva 2016).

Sometimes, doubts have been raised on the manufacture place of these luxury artifacts, as for the "Persepolis silver vase" decorated with two female figures in *kaunakes* and an inscription in Linear Elamite. It was first considered to have been made in Elam but the BMAC area has finally be retained due to several different items

Table 1.2 Approximative chronological table for Gonur Large Necropolis and Gonur North (different areas and graves), based upon the median ^{14}C dates (see the Appendix at the end of this volume)

Dates	Gonur Large Necropolis	Gonur Nth "Royal Necropolis"	Gonur Nth Kremlin	Gonur Nth Palace	Gonur Nth Area 5 rooms	Gonur Nth Area 5 graves	Gonur Nth Area 6	Gonur Nth Area 8 rooms	Gonur Nth Area 8 graves	Gonur Nth Area 9 rooms	Gonur Nth Area 9 grave	Gonur Nth Area 10 fortif. wall
1450												
1500												
1550												
1600												
1650												
1700												
1750												
1800												
1850												
1900												
1950												
2000												
2050												
2100												
2150												
2200												
2250												
2300												
2350												
2400												
2450												
2500												
2550												
2600												
2650												
2700												
2750												
2800												
2850												
2900												
2950												

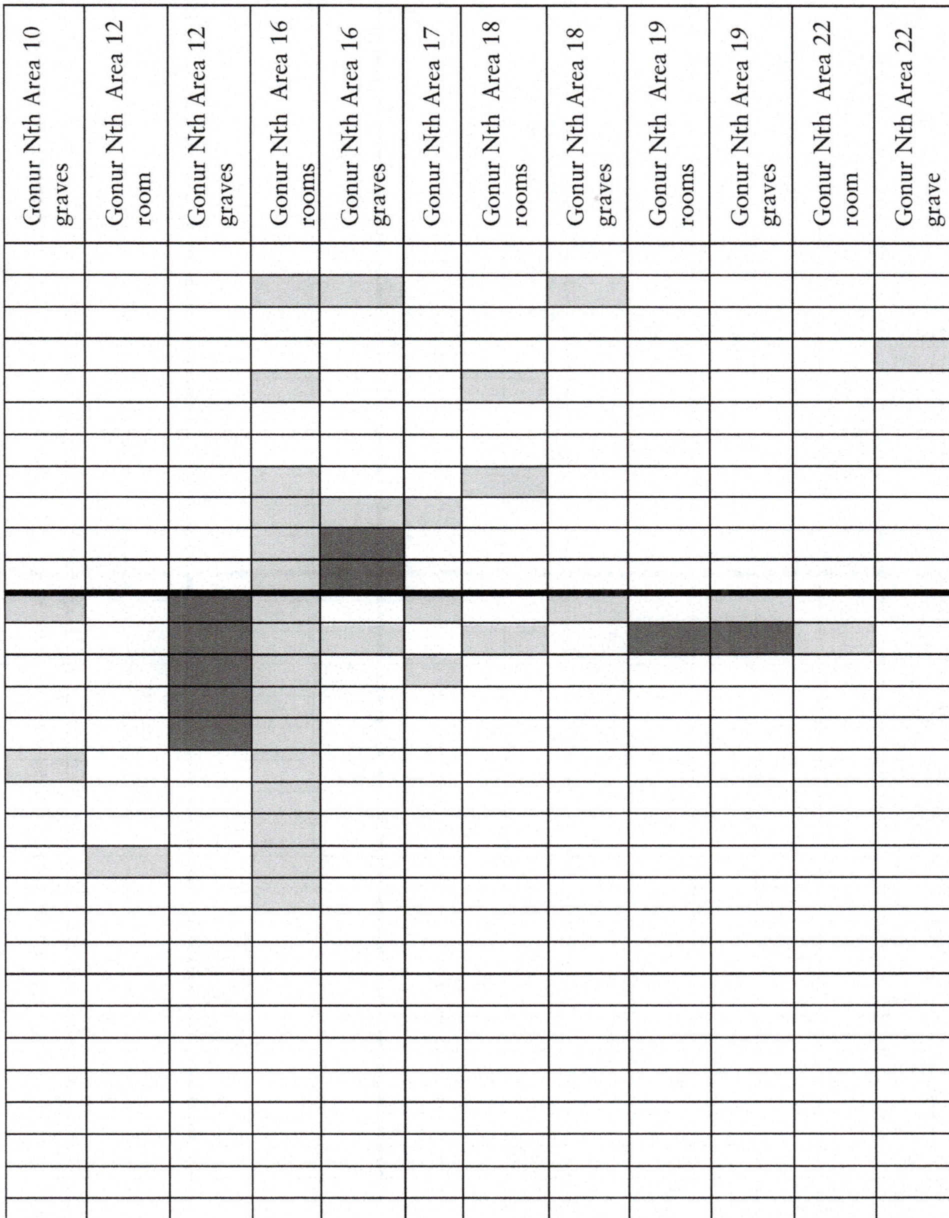
Gonur Nth Area 10 graves
Gonur Nth Area 12 room
Gonur Nth Area 12 graves
Gonur Nth Area 16 rooms
Gonur Nth Area 16 graves
Gonur Nth Area 17
Gonur Nth Area 18 rooms
Gonur Nth Area 18 graves
Gonur Nth Area 19 rooms
Gonur Nth Area 19 graves
Gonur Nth Area 22 room
Gonur Nth Area 22 grave

Table 1.3 Approximative chronological table for sites in Margiana, Turkmenistan, and Bactria, based upon the median ^{14}C dates (see the Appendix at the end of this volume)

Dates	Gonur South	Gonur 20 room	Gonur 20 graves	Gonur 21 graves	Adji Kui 9	Togolok 21	Takhirbaj 1	Ojakly	Namazga-Depe V	Namazga-Depe VI	Altyn-D. NMG V	Shortughai Mnd A	Shortughai Mnd B	Dara-i Kur Goat Cult
1450														
1500														
1550														
1600														
1650														
1700														
1750														
1800														
1850														
1900														
1950														
2000														
2050														
2100														
2150														
2200														
2250														
2300														
2350														
2400														
2450														
2500														
2550														
2600														
2650														
2700														
2750														
2800														
2850														
2900														
2950														

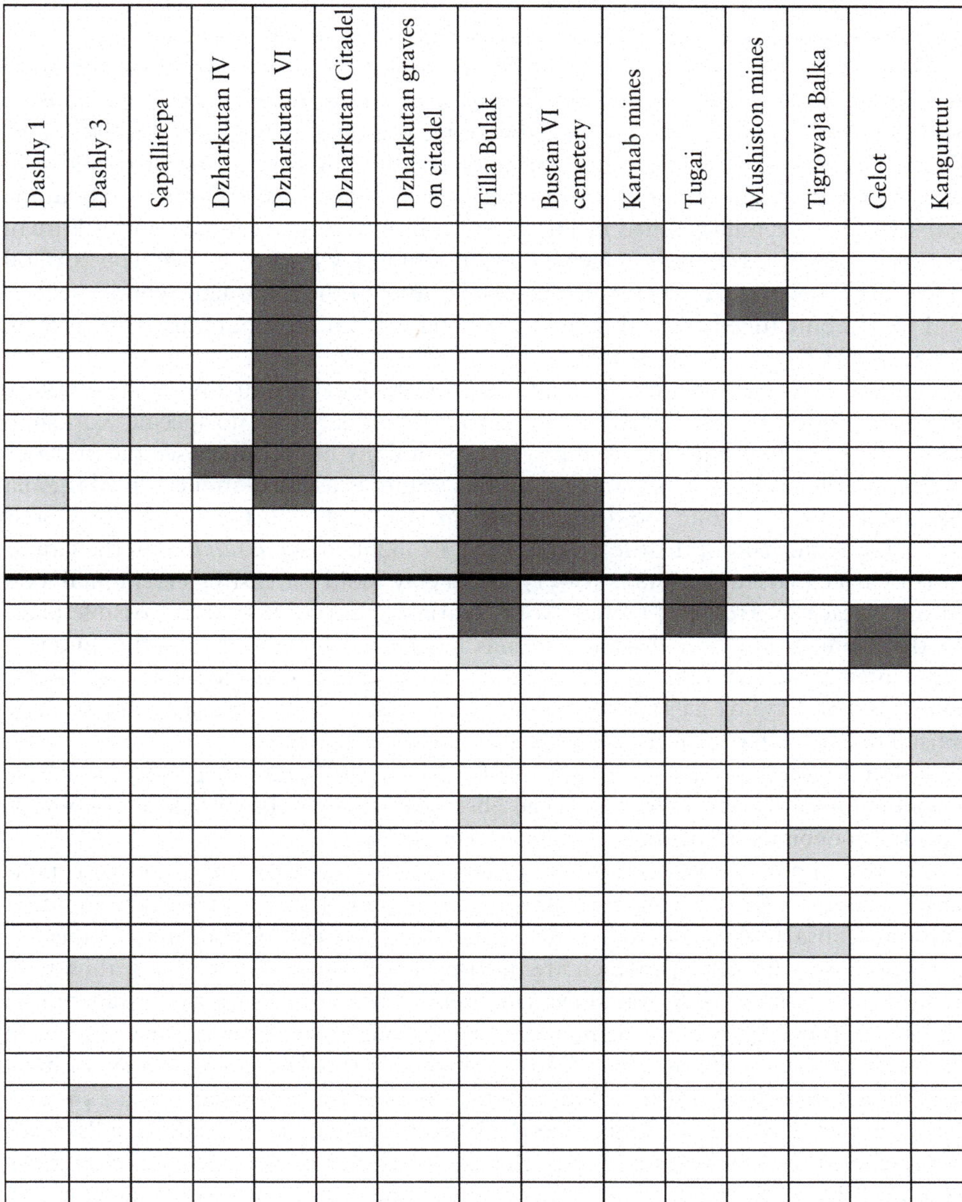
Dashly 1
Dashly 3
Sapallitepa
Dzharkutan IV
Dzharkutan VI
Dzharkutan Citadel
Dzharkutan graves on citadel
Tilla Bulak
Bustan VI cemetery
Karnab mines
Tugai
Mushiston mines
Tigrovaja Balka
Gelot
Kangurttut

made in the same style (Potts 2008b), like the vase with the camels in the image on the cover of this volume.

These vessels possibly remind us of the famous "GAL" vessels described in the Mari archives, usually in gold or silver, with various decoration (often animals), but of which none has been discovered in Mesopotamia (see Durand 1998: 249–253; Guichard 2005).

Most of the time, the style and/or the themes of the objects are a combination of Elamite and/or Gulf, Near East, Kerman, or Indus elements with an unusual treatment or a reinterpretation and, therefore, presumably, they are locally made. This is especially true for the seals (Sarianidi 1998b), but also for the silver and gold vessels (Francfort 2005b). It is worth noticing here that a similar large variety in the seal styles has been underlined in southeastern Iran at Konar Sandal South (Pittman 2014a, 2018), as well as in Anatolia at the time of *kārum* II of Kültepe (Collon 1987: 41; Teissier 1994: 55) where the iconography of the Anatolian cylinder seals is said to combine together local themes with others from Mesopotamia, Old Assyria, Syria, and/or Egypt (Özgüç 1986b).

Up until now, however, the available data on the local production of such luxury or exotic products in the BMAC area is extremely reduced and no specific workshop for them has been found (for local production of daily life products see the previous section on the economy). Sarianidi mentions semifinished seal-amulets in Margiana (Sarianidi 1994), a stone mold for casting bronze eagles on the surface of Gonur North (Sarianidi 1996), steatite fragments of a roughly made *kaunakes* in the Gonur North Palace (Sarianidi 1998a: 50–51), and a clay mold for casting ceremonial axes on the surface of Area 5 at Gonur North (Sarianidi 2007: 78–79). A possible place for the production of small stone columns has been pointed out recently (Juminov and Tjutev 2016). Possible traces of stone working are also mentioned at Gonur South, and at Togolok 21 (Hiebert 1994a: 127–128, 151), but the last ones are now considered as a later recovery for reuse (see Salvatori 2008).

Still at Togolok 21, a large fragment of a cowry shell and other Indian seashells probably for inlays or jewelry have been observed, while on the surface of Togolok 3, lapis lazuli debitage is mentioned (Hiebert 1994a: 153).

It is also important to mention two tombs, one from the Gonur Large Necropolis (grave 1200, see Rossi Osmida 2002: 98–101; Sarianidi 2007: 118–119), the other from Sazagansaj in the Zeravshan valley at the edge of the BMAC core area (Avanesova 2010; see also Chapter 24), which are most probably those of jewelers dealing with semiprecious stones. They contained semifinished beads and raw stone minerals as well as the tools to produce them, and, in the Gonur grave, several finished arms in alabaster for composite figurines. Worth noticing in this last grave, is a small duck weight undoubtedly of southern Mesopotamian origin (see Ratnagar 2004: 246–248, where they are dated before Shu-Sin or the Ur III Dynasty; and Ascalone and Peyronel 2001, where they are mentioned up to the Paleo-Babylonian period).[27] Another grave from the "Royal Necropolis" (no. 3245) yielded one piece of uncarved elephant tusk together with several ivory artifacts (Sarianidi 2008b: 305, fig. 46; 2010: 110, 156; Salvatori 2010) (for one of the artifacts of this grave, see Chapter 7, and for the others see Chapter 22).

An interesting recent study on the Gonur ivory artifacts (mainly linked with game boards or "divination" tools, and with toiletry items), previously considered as imports from the Indus valley, shows that nothing similar is actually known in the subcontinent. These objects must then have been cut and carved especially to the taste of the Bactrian elite, though almost certainly by Indian specialized craftsmen considering the technical know-how that this material needs (Frenez 2017, 2018). This author is inclined to consider that Indian specialists had moved to Bactria with their raw tusks, a fact that might be confirmed by the discovery mentioned above of a piece of uncarved elephant tusk in grave 3245, and by a cache of raw ivory in a wall at Altyn Depe (see Chapter 21). But no workshop especially dedicated to this material has been found up to now.

This ivory example could be extended to almost all the BMAC luxury artifacts, including the mosaics discovered in one of the "Royal" tombs that find close iconographic comparisons in the wall paintings of the Mari Palace (see Chapter 11): who made them, and where were they made? The presence of mosaic pieces in a storeroom at Dashly 3 palace (see Chapter 5) suggests they were possibly an imported production, at least at a regional scale (from Margiana?). Vidale (see Chapter 9) also evokes skilled craftsmen coming from abroad to carve the chlorite vessels. The possibility of itinerant craftsmen proposed long ago by Amiet (1986: 138, 212; 1988; 1989a) therefore seems highly relevant. This question, however, only concerns the highly sophisticated production. There is no doubt that the rest of the production was made *in situ* and most probably by local craftsmen.

The Elamite *koinè* and migrations: questions on the Indo-Aryans and the BMAC population

The foreign influence on some artifacts could also be explained by different movements of populations directed to or from the BMAC, as advanced by some authors.

Thus, the Oxus Civilization is often considered as belonging to a large "Elamite *koinè*," associated with movements of people, which would explain the likeness of many artifacts within all this area. This is mainly defended by Amiet (1986: 171–214; 2007), but many others have supported it, or transformed it into a politico-economical east-Iranian entity (Marhashi?) sharing the same ideology (Salvatori and Tosi 1997). If, in this case, the movement is rather from Elam to Central Asia, some consider that a group of Bactria-Margians may have migrated to Anshan and Susa and founded a dynasty there (Potts 2008b).

Sarianidi was convinced of an Indo-Aryan population movement from Iran and Anatolia to Central Asia and he based most of his work trying to confirm an Aryan presence in the BMAC. This movement would explain part of the pantheon on the seals and other artifacts related to the Syro-Anatolian world. In so doing, however, he rejected the well-established conception of the Indo-Aryans as mobile pastoralists living in the steppe invading at some moment the southern areas up to India (Mallory 1989; Kuz'mina 2007). Most of the comparisons he made were with western areas (at first northeastern Iran, then Anatolia and Kurdistan), using the myths represented on the seals, the burial rituals in house-like hypogea, the presence of horses, chariots, trumpets, and gray ware, etc. According to him, all these elements were linked in one way or another with the Avesta, the Rig Veda, the Hittites, or the Mittani worlds (Sarianidi 1998a: 149–166; 1999; 2007: 172–183). The west to east direction was dictated at first by his very late dating of the BMAC, but he maintained his position even after he finally accepted a much earlier date for it. His interpretation received Parpola's support (1993, 2004). A recent study also indirectly supports this view. It concerns the "Balafré," one of the enigmatic hybrid man–dragon representations of the BMAC, unfortunately never found in regular excavations: This one-eyed scarred figure with a fish-scale skin holding a vase is considered to be the expression of an old Indo-European myth when cosmic waters are released after the dragon has been laid down by a one-eyed hero (Vidale 2017b: 196–200; see also Chapter 7).

As we mentioned earlier, migrations have also been considered, this time from Central Asia to Iran and Baluchistan, to explain the BMAC expansion in its later phase (Hiebert and Lamberg-Karlovsky 1992; Lamberg-Karlovsky 1993, 1994; see also Chapter 20). The rapid BMAC development within a Khanate system would have led to the search for new land or natural resources in a model similar to that well known at the end of the Uruk period in Mesopotamia. The recent discovery of a wide distribution of BMAC populations in eastern Iran, though mainly attested by cemeteries without associated built sites (see Chapter 19), could also point to a similar trend, but we still lack chronological data to anchor it with a BMAC late phase. The Quetta hoard mentioned earlier, dates to the full bloom of the BMAC. In any case, these "migrations" would explain the presence of BMAC artifacts outside of its core area.

Migrations of parts or of whole populations are difficult to prove. A series of research studies based upon anthropometrics studies, mainly skulls, do not testify large movements (Hemphill 1998, 1999; Hemphill and Mallory 2004; Dubova and Rykushina 2007; Dubova 2014). But, on the contrary, ancient-DNA analyses point at a sufficiently important movement from the steppes to have modified the genes of part of the Indian population (Narasimhan *et al.* 2019), though its date is still to be confirmed.

It cannot be excluded, however, that small groups of people moved at that time because of the intense international relations between the Mediterranean and the Indus (see, e.g., Salvatori 2010), and perhaps also because of the general environmental crisis. We have evident proof of the movement of some pastoralist/nomadic groups like the Amorites who became extremely powerful and within two or three centuries took control over all of Mesopotamia (Charpin 2004) up to Dilmun (Crawford 1998; Glassner 2002; Laursen 2017). It was at the same time, which also corresponds to the BMAC, that an extraordinary network of interactions and trade took shape with the Assyrian merchants, implying caravans going from place to place. Most probably, itinerant craftsmen did so too. We cannot therefore exclude that among them some were Indo-Iranian speakers and that they became powerful enough to influence many aspects of the society and beliefs, or impose their language in the BMAC (Lyonnet 2001). Such a situation is documented at Kanesh where Hittites (Indo-European speakers), said to have arrived in Central Anatolia during the last centuries of the 3rd millennium, had already imposed their language by the 19th century (Kanesh Karum II) (Dercksen 2007, quoted in Michel 2014).

Trade

Trade is of course a good option to explain the multidirectional flow of "exotic" artifacts, though we know from Mesopotamian sources that many exceptional and foreign items can also be explained by booties, tribute, gifts, marriages, or diplomatic exchanges. These other possibilities, however, do not seem likely for the BMAC prestigious artifacts: Texts do not mention incursions at Anshan or Susa from "easterners," nor do they say that Elam were under the control of eastern rulers. On the contrary, the period of the BMAC *floruit* corresponds to the period after Marhashi is said to have been conquered by Sargon, and mainly when Elam is at its zenith and was able, first, to put an end to the Ur III kingdom and, second, to control all the Amorite kingdoms of Mesopotamia (see Potts 1999/2016). As for gifts, marriages, and diplomatic exchanges, they were not without ulterior motive, and the donators expected gifts or peace in return (Lerouxel 2002), so that, again, one would expect the BMAC – if more or less united in a kind of chiefdom – to be mentioned one way or another in the Elamite written sources.[28] Furthermore, most of the exceptional foreign-influenced artifacts come from graves and not from hoards in palaces or temples, and very similar items have been found in other areas like Shahdad, Hissar IIIC, or Quetta, so it is difficult to consider them as due only to diplomatic activity.

Trading activities seem therefore a more plausible explanation, but we have no idea of how it actually functioned: Was it centralized or not? Who were the merchants (foreigners, local people)? What were these artifacts exchanged for? We will discuss these questions further later in the chapter, but looking first into the preceding period will give more weight to this trade hypothesis.

A previous model

The BMAC is not a novelty in Central Asia. A similar phenomenon of an intensive and wide network of relations had occurred ca. 1,000 years earlier, during the Proto-Elamite period and the first half of the 3rd millennium, leading to the rapid development of the Geoksjur oasis in the Tedjen delta and to that of the site of Sarazm in the upper Zeravshan valley. Similar motifs or pottery are shared from central (Tal-i Malyan) and northeastern Iran to Sarazm in the upper Zeravshan valley, and from Geoksjur to Shahr-i Sokhta, Mundigak, and the Quetta area of Baluchistan. Some authors consider that the Oxus Civilization could be its ultimate development (see Chapter 4). Another contemporary and large interregional network, also visible through pottery diffusion – Late Shahi-Tump gray ware – existed between Baluchistan/Makran, Seistan, and Kerman (Mutin *et al.* 2017). Both networks intermingled at a few sites, such as Shahr-i Sokhta.

There is no doubt that these transformations were linked with the Proto-Elamite expansion that developed immediately after the sudden collapse of the Uruk world system (Algaze 1993). The last one had mainly been oriented towards the north (eastern Anatolia and Caucasus),[29] most probably for the procurement of raw material though evidence of what exactly was looked for is lacking, and it possibly broke down because of the intrusion of Kura-Araxes and other groups coming from the north of the Caucasus. The new Proto-Elamite network was definitely oriented eastwards. The presence of Proto-Elamite tablets and beveled-rim bowls is attested all over Iran (except at Tepe Hissar), up to Seistan and Baluchistan (Besenval 2005; Mutin 2013: 61–64, 169–172; Petrie 2013). Though there is no direct evidence of Proto-Elamite influence over the Kopet Dagh Mountains, a few hints show that Central Asia was also involved in this vast network.

The settlement of Sarazm (Isakov 1991), in the Zeravshan valley, probably best expresses the international connections of that period. If no tablet has been found there, a cylinder seal with a standing bovid (Isakov 1996: fig. 10) and some of the pottery shapes evoke Mesopotamia (Lyonnet 1996: 47 fn31). An Afanasievo-like burial in the earlier level, ca. 3500 BC, contained together pottery dated to the end of the NMG II period (Lyonnet 1996: 35, 47, type III.11), shell bracelets from the Indian Ocean, hundreds of lapis lazuli beads, gold ones, and silver artifacts (Isakov 1992). The overall pottery found at the site (Isakov and Lyonnet 1988; Lyonnet 1996) presents evident relations with the Geoksjur area of the end of the NMG II and during the NMG III periods, with only one NMG IV-like unpainted sherd. It is also clearly related to Mundigak (periods III.6 to IV.2) south of the Hindukush, to other Baluchistan sites at the edge of the Indus valley, and to the Kot-Diji (pre-Harappan) period, while a smaller group of pots is comparable to Hissar IIIB, and a few other sherds are Kel'teminar or steppe-like. Recent excavations have confirmed these comparisons (Mutin and Razzokov 2014). Interestingly, if relations with all these areas are attested from the beginning, their relative importance changes in the course of time, the Geoksjur pole being far more important in the first two periods than the southern Hindukush one, and vice versa for the last period. The contemporary sites found at Taluqan in eastern Bactria, close to the lapis mines of Sar-i Sang, could be intermediate outposts to this southern Hindukush intrusion northwards (Lyonnet 1981). These changes in orientation point at the appearance of a new commercial road, probably the first step towards the Gulf maritime route that became so important in the second half of the 3rd millennium between Mesopotamia, Magan, and Meluhha (Lyonnet 1988, 1996: 68–69). This new southern maritime road could explain the sudden abandonment at Ilgynly Depe at the foot of the Kopet Dagh during the NMG III period (Salvatori *et al.* 2009) and the replacement of Sarazm by Shortughai. It is at the same time, or slightly later ca. 2400 BC, that the so-called "Intercultural Style" develops in southeastern Iran (Kohl 1975; Lamberg-Karlovsky 1988; see also Chapter 20), now better identified in the Halil River basin around Jiroft, or the "Middle Asia Interaction Sphere" (Possehl 2002).

What all these people were looking for at Sarazm is yet unknown, but the position of the site, close to abundant metal ores (Razzokov 2008: 12–14 and 66–67 map 2), together with the presence of numerous stone tools for crushing ore or for metalwork (Razzokov 2008: 34–37), clay molds (Isakov 1991: fig. 24), and many metal finds (Isakov *et al.* 1987) suggests that these may have been the main coveted products. The presence of a handful of sherds similar to some of Sarazm ca. 20 km westward at Tugai, where a metallurgical furnace has been found, supports this hypothesis (Avanesova 2015), as does the importance metallurgy seems to have played at Ilgynly Depe before its abandonment (Salvatori *et al.* 2009). But which specific metal was looked for in this remote area is unknown. The Proto-Elamite tablets found in Iran on contemporary settlements only point at local affairs (see Mutin 2013: 170).

A trade impetus to explain the BMAC?

Trade could have been the main motor behind the wealth and exotic luxury products that are one of the characteristics of the BMAC. Though no one has ever denied the reality of exchanges between the BMAC and other areas (e.g., Salvatori and

Tosi 1997; Sarianidi in all his publications; Kaniuth 2010), rare were those (Amiet 1977, 1986; Pottier 1984; Lamberg-Karlovsky 1993; Lyonnet 2001, 2005; Kohl and Lyonnet 2008) who considered it as the main motor of its wealth. The ongoing discoveries made in the course of excavations have, however, strengthened this proposal and it is now put forward by many more authors (Francfort 2005a; Sarianidi 2005: 259; Rossi Osmida 2007; Potts 2008a, 2008b; Salvatori 2008, 2010; Francfort and Tremblay 2010; see also Chapter 20).

But what was exchanged is as yet unsolved and, moreover, may have varied in the course of time. An extraordinary number of seals has been found, but, comparatively little bullae or seal impressions (see Sarianidi 1998b: 316–321; see also Chapter 8), so many consider that these seals may have had other functions than purely administrative (see Chapter 16). The products that were the matter of this trade have, of course, disappeared, and, in the absence of writing, we are left with pure hypothetical solutions.

This situation is not different from that of all the surrounding regions. If it were not for the Mesopotamian or the Assyrian merchants' written documents, we would have little understood the importance of trade in the Indus Civilization, the Gulf, or Anatolia, and never would have guessed which products were concerned. The extremely reduced number of foreign artifacts found in the Indus (During-Caspers 1972; Kenoyer 1998: 97–98; Ratnagar 2004: 106–107), in the Gulf (Crawford 1998: 152–153; see Chapter 22), or at Kültepe/Kanesh (Özgüç 1986a: 39–83) compared to the quantities of imports/exports mentioned in the texts has been constantly underlined, and a similar pattern is visible in Central Asia (e.g., Kaniuth 2010). Concerning the Old Assyrian trade, all the products involved were convertible (metals) or perishable (textiles).

Undoubtedly, a strong impeding factor for the BMAC is that it is not yet clearly identified in the Mesopotamian sources as to what was exported from it. As seen above, different propositions have been advanced for its ancient name (Steinkeller 1982, 2014; Potts 2008b; Francfort and Tremblay 2010) but none is totally convincing (see Chapter 2).

Among the possible resources that all these areas could have searched for in Central Asia, is, of course, *lapis lazuli*, a stone highly prized at that time and coming from Badakhshan in northeastern Afghanistan. But its trade had started much before the BMAC (Herrmann 1968; Sarianidi 1968, 1970; Casanova 2001, 2013) and, even if it was still very much in demand in the Near East at the beginning of the 2nd millennium (Guichard 1996; Michel 2001), and, at Mari at least, reserved for the king (Durand 2018), it cannot alone explain the importance of the exchanges we observe. Both archaeological finds and Near Eastern texts show that the stone was exported in raw blocks and manufactured in the western cities.

Metals are also one among the major possible traded goods, since Central Asia is extremely rich in different ores, especially copper and tin (Lyonnet 2001, 2005; see also Chapter 28). *Copper* ore from these "local" sources was certainly known and used, though it also seems that importations were made from central Iran to Gonur (see Chapter 27).[30]

There is no agreement either on the use of *tin* in Central Asia, with some claiming that it was little used during the first two main periods of the Oxus Civilization (Terekhova 1990; Kaniuth 2006), while others see its importance growing in the Namazga sequence already from the NMG IV period (Ruzanov 1999; Pigott 2018;

see also Chapter 27). All agree that it was more widespread in Bactria, especially at Sapallitepa (Askarov and Ruzanov 1977; Hiebert and Killick 1993; Kaniuth 2007; Ruzanov 2013). Recent research on one of the closest tin-producing areas along the Zeravshan valley (Parzinger and Boroffka 2003; Garner 2013) has not produced proof of an exploitation of the mines at Karnab before the 17th century BC – though we cannot exclude that earlier traces of work would have been erased. But there is evidence further east at Mushiston in Tajikistan (22th–20th century BC) (see Chapter 28 and the Appendix for the dates). At Tugai, however, situated not very far from these mines and the unique place contemporary with the early phase of the BMAC where a metallurgical furnace has been found, only meagre traces of tin have been noticed (Avanesova 2015) (see note 11). At Sazagan, burial 1, "bronze" ingots are nevertheless mentioned (see Chapter 24). In addition, new research in Iran puts forward the possibility that the tin used in Mesopotamia could have come from much closer areas in northern Zagros (see Chapter 29). This, however, contradicts the written sources of the 19th–18th centuries BC (from Kanesh Karum II-Ib and Mari) that clearly mention or leave little doubt for its southern provenience: For Mari, the merchants had to go to Eshnunna/T. Asmar or Susa to buy it together with lapis (Joannès 1991; Guichard 1996; Reiter 1997: 222–251; Durand 2018); for the Assyrian merchants, the tin they gathered at Assur in exchange of silver and gold at Kanesh and other Anatolian trade centers came by caravans organized by non-Assyrian merchants from the southeast ("the Lower Country") (Dercksen 2005, 2014). In a recent overview on tin bronze in Central Asia, Pigott (2018) considers it highly probable that the BMAC played a major role in the transmission of tin both to the northern steppes and to the Iranian Plateau.

Gold is also abundant in Central Asia and could have been searched for there before the gold from Anatolia was introduced through the Old Assyrian trade: recent analyses on the gold from the Royal Ur graves seem to point towards it originating in eastern Bactria (Jansen *et al.* 2016).

Silver and *lead* are equally present in Central Asia and it is interesting to notice that five "plaques" of the latter, which could well be ingots considering the shape described for the tin ingots in the Assyrian texts (Dercksen 2005), have been discovered "hidden" under a wall of the Togolok 21 temple (Sarianidi 1998a: 57).

Of course, we will never know of all the *perishable goods*. Fragments of *silk* clothes had been discovered at Sapallitepa by Askarov (1973: 133–134) together with millet grains but it long stayed an isolated find. A more recent discovery of this precious textile comes from the Indus Civilization (Good *et al.* 2009), and other finds in Europe are now emerging dated to the 2nd millennium BC (Good 2010), opening possibilities for its trade at the time of the BMAC. It should be noticed here that destroyed graves related to the BMAC found a decade ago at Shagym in eastern Ferghana (Amanbaeva *et al.* 2006), close to Soch where a chlorite handled "weight" of the Intercultural "*série ancienne*" artifacts had been discovered, also point at a "Chinese" direction.

Finally, regarding *chlorite* containers, beside the probable import of some of them, Vidale (see Chapter 13) also mentions a possible trade of their content (i.e., cosmetic products).

This brief list of possible traded items should also include, though they are almost never mentioned for the BMAC, the eventual search for *human labor forces* (see Algaze 2018).

If this important issue (i.e., which products were looked for at the BMAC sites), still escapes us, we do have several indirect but clear testimonies that merchandise was transported and that it represented the main wealth of this area. The following proof is evident:

(1) The multidirectional iconographic parallels shown in the themes and shapes of part of the seals that almost all the authors have underlined and that correspond to the areas where some of the BMAC artifacts have been found and that were important trading centers. As said earlier, the same variety has been noticed at Konar Sandal South and at Kültepe/Kanesh.

(2) The numerous cell rooms or storage places of the BMAC that must have kept very rich goods since the largest ones are situated at the heart of almost all the main buildings in each settlement. These grill-plan structures are not interpreted as storage places by Sarianidi though he made parallels with Hattusa and Mari. He first considered that they were underfloor heating structures (Sarianidi 1977: 43), then sacred places and related to temples (Sarianidi 1998a, 2005: 83–86). A recent study shows that they are numerous and at different places, with several small ones beside much larger ones (Urmanova 2014); most are empty, except at Adji Kui 1 where storage jars and sealings have been found, and at Dashly 3 Palace where they still contained fragments of mosaic inlays (see Chapter 5). At least some of them were probably supporting a platform (considered as a terrace with a roof cover by Urmanova (2014)). At Dzharkutan, similar constructions in the temple are seen as storage areas (Askarov and Shirinov 1993: 106–107), while near the palace, storage was made in large pits and in jars among which were found a probably imported Hissar IIIC vessel and seal-impressed jars (Bendezu-Sarmiento and Mustafakulov 2013).

(3) The wheeled chariots buried in the "ditches" exclusively associated with the "Royal Necropolis" where much of the luxury artifacts found at Gonur Depe comes from.[31] The vehicles, their drivers, and other caravan members were most probably buried in these ditches, as well as the pack/draw animals (i.e., camels and donkeys; see Chapter 10). Interestingly, the wheels of these chariots have the same copper "tires" as those known at two of the main commercial "entries" to the Near East, Susa (Pottier 1984: 49 and 178 no. 326;[32] Dubova 2004: 278), and Assur (Kaniuth 2010).

CONCLUSION

The BMAC or Oxus Civilization testifies of intensive relationships with a wide array of countries from the last third of the 3rd millennium to the end of the 19th century BC. Their apex is manifest in the "Royal Necropolis" that was of somewhat short duration – within the Ur III period – as shown by the goods it contained. These connections seem to diminish afterwards but are not totally interrupted until ca. 1700 BC. Thereafter, a real break is clearly visible and all the previous prestigious artifacts totally disappear – and with them the representations of myths and gods that figured on seals and luxury items (see Chapter 18) – though we clearly would need larger excavations on such late sites as Takhirbaj or Dzharkutan to absolutely confirm this. Undoubtedly, the tight relations between Bactria, Margiana, and all the other areas where the BMAC had been identified in eastern Iran or Baluchistan do not exist anymore. After a period of international relations, with common material culture, this is a time when local regional cultures develop (each with its own material and spiritual culture).

With the data at hand nowadays, we consider that trade best explains many of the peculiarities of the BMAC, including the presence of small groups of newcomers bringing with them their specific cultural features (burial rituals, religion, pantheon, language, etc.).

These commercial exchanges most probably played a major role in its *floruit*, though we do not know what was looked for so intensively. The extraordinary economic growth and the hybrid cultural testimonies visible in part of the material culture of this Oxus Civilization can be compared with what has been discovered at Kültepe/Kanesh during the *kārum* period (Dercksen 2008).

Different roads were used at the same time and/or successively, depending on the political situation along each of them. The road through the Indus and the Gulf was certainly the earliest one, following a previous route that had developed between Sarazm and Baluchistan around the middle of the 3rd millennium, shortly before the Indus Civilization's major development. But a road through the Iranian Plateau (Shahdad/Kerman and then the Magan territories) also probably started at the same time and led to the southern part of the Gulf. At the time of Ur III, Shimashki, and the *sukkalmahs*, the road to Susa and Elam became more obvious, although the Gulf maritime road was still functioning, as shown by the importance of T. Abraq, Tol-e Peytul/Liyan, and Bahrain/Dilmun at that time. We still lack data on northern Iranian sites at the turn of the 3rd to 2nd millennium to clearly identify another route through northern Iran that may have existed and led to Assur at the time of the Old Assyrian trade to Anatolia. The presence of a site like Dinkha Tepe near Hasanlu to the southwest of the Urmia Lake points to its existence, although texts are mute on this subject (Lyonnet 2019). Finally, a road through the steppes should not be ignored, although we also have only few hints at it, mainly with the spread of light, spoke-wheeled chariots and specific horse cheek-pieces.

In the absence of texts, it is not possible to establish direct relations with the events in the demanding countries in the West and those seen in Central Asia. There is no doubt, however, that the end of the Old Assyrian trade together with the endless wars between the kings in Mesopotamia at the beginning of the 2nd millennium had a serious impact on the countries at the source of the trade. This, together with environmental problems and possible other internal unknown difficulties, rang the tolling bell for the BMAC. Interestingly, it did not take very long for trade to turn to another direction; that is, to the Levant, with Qatna, Ugarit, Egypt, and Crete becoming then the major centers of the Near East, leaving Central Asia isolated and without "income" any more. Tin in the Near East comes then obviously from another source, and lapis also disappears.

The importance of trade that we propose here to explain the wide range of provenience of exotic finds, ideas, possibly religion and myths, and moreover language in the Oxus Civilization, should not hide the reality of the development of the local population and its own skills. The local society and its casual production, its beliefs and political system, has unfortunately been little studied because of the focus instead on its wealth and prestigious goods.

NOTES

1 This chapter relies in part on: (1) an unpublished work (but that has largely circulated) presented by Bertille Lyonnet as an habilitation thesis in 2001, a reduced abstract of which was published in 2005; and (2) the excavations made in Central Asia during the last 17 years, a great part of which was done at Gonur Depe where Nadezhda Dubova played a very important role at Sarianidi's side. She is now the head of excavations at this site.

2 There is no harmonization in the period denominations over different areas. In Central Asia, the Middle Bronze Age corresponds to the Near East Early Bronze IV/Akkadian and Ur III periods, and the Late Bronze Age to the Near East Middle Bronze Age/Isin-Larsa and Paleo-Babylonian up to the Mittani periods.

3 We are all aware that some of the art market pieces published in these books might be fakes, sometimes difficult to recognize.

4 These investigations are published in a collection called (in English) *Transactions of the Margiana Archaeological Expedition* of which seven volumes have been published since 2004 (the first volume was called *U Istokov Tsivilizatsii*, translated as *Near the Sources of Civilizations*).

5 We could not enter those for which we lacked the BP dates, unfortunately.

6 Some of the discoveries were made long before, but were not understood at that time (Stein 1931, 1937). A good number of BMAC artifacts, presumably coming from the Hari Rud area in Western Afghanistan, are also stored at the Herat Museum (Franke 2016).

7 For an excellent presentation in English of all the different Bronze Age sites of Central Asia, the work published by Phil Kohl in 1984 is still one of the best documented (Kohl 1984). For a shorter updated synthesis, see Kohl (2007: 182–243). For Margiana itself, see also Hiebert 1994a.

8 The Jiroft Civilization, related to the "*série ancienne*" of the "Intercultural Style," is slightly earlier than the BMAC.

9 These graves had almost no offerings and are thus difficult to date precisely, and the Temenos graveyard tombs contained typical BMAC material. Certainly this is not enough to give a precise date either, but it raises doubts as to an area totally void of sand during period I.

10 Another channel has destroyed part of the metal workshop in Area 9. Its date is unknown.

11 Other possible cases of circular buildings are mentioned in Chapter 5.

12 Much later however, during the Kushano-Sasanian period (end of the 3rd century, beginning of the 4th century CE) at Pajkend near Bukhara, a similar structure of the rampart with a succession of inner rooms is considered to have been the place for the garrison in charge of the defense of the area (Omel'chenko 2012).

13 In this article, Avanesova mentions chemical analyses made on 14 different metal items from the site and the furnace or around. They show that ca. 76% are in "pure" copper, up to 10% are in tin bronze (tin content from 1.6 to 3.8%), and that copper with a low content of arsenic (<0.2%) and or lead (>1%) is constantly found. She adds that these results are close to those from the analyses of the nine ore samples found at the site (chalcopyrite, malachite, and azurite), and she proposes an origin for them in the Kyzyl Kum where polymetallic ores are well known.

14 Only a handful of tokens bearing incised signs on them (counting items?) are known (Sarianidi 1998a, fig. 41). Two other signs incised on a sherd of a local vessel are considered

as possibly Linear Elamite (Klochkov 1995, 1998). Research is actually in progress on all the signs incised on pottery.

15 The site is considered only 22 ha by Hiebert (1994a: 24). Gonur North itself, within its outer more or less circular wall, was considered to be only ca. 10 ha (Sarianidi 2005: 37), but this has now been corrected to 25 ha after other areas have been discovered.

16 Salvatori had a tendency to date the BMAC foreign relations too early (see, e.g., Salvatori 2010). There are still obvious external relations during the 18th century BC (see Lyonnet 2019).

17 The separation in two chambers, one for the fire and the other for cooking, is viewed as being linked to religious beliefs.

18 Graves had been estimated at about 5,000 in the Gonur Large Necropolis (Rossi Osmida 2002), but only 2,860 have been discovered and excavated. Nevertheless, altogether over 5,000 graves are known from different graveyards at Gonur Depe (Sarianidi and Dubova 2016). They are also estimated at over 5,000 in the four cemeteries of Dzharkutan (Bendezu-Sarmiento 2013).

19 This is contested by Salvatori (2008) who considers that they are all intrusive.

20 Lamberg-Karlovsky (2012) had wondered whether these could not be previous houses reused as graves. There is no doubt, however, as to their unique use as tombs.

21 As for close distance, a possibility that they partly come from the sites around Gonur has been proposed in by Sarianidi and Dubova (2016).

22 In the case of the hypogea, chimneys in the walls (as is the case in the palace itself) contained ashes and charcoals.

23 Dakhma are known in Zoroastrianism as places where the corpse is placed after death to be defleshed by birds of prey. Up to now, very few are identified (Grenet 2015).

24 A re-analysis of the stratigraphy at Gonur North palace points at a possible earlier pre-palace occupation (see Sataev 2018).

25 Similar attempts have been done by Kohl (1984), Possehl (1993), Hiebert (1994a: 75–87), Kircho and Popov (1999, 2005), and Kaniuth (2006). The regular changes and precision in the calibration make this update necessary.

26 Unfortunately, we have only two dates for this last phase, coming probably from its beginning (see the Appendix).

27 We are grateful to G. Chambon who mentioned to us that these duck weights are usually related to the Mesopotamian shekel of 8.4g. This artifact is mentioned by Kaniuth (2010), without information on its context, and by Francfort and Tremblay (2010: 135–136).

28 Which may be the case if Marhashi is not only focused on southeastern Iran but also includes the BMAC.

29 Only two Uruk sites are known in Eastern Iran, Tal-i Iblis, and Mahtoutabad (Mutin 2013: 185).

30 Copper ingots have been mentioned for the "Togolok phase" at Adam Basan (Terekhova 1990) and Gonur South (Hiebert 1994b).

31 Rich graves with chariots have often been related to "kings" (Royal Cemetery of Ur, Martkopi Kurgans, "Royal Necropolis" here at Gonur), but it cannot be excluded that they were those of rich merchants, see Lyonnet (2016). Nevertheless, the chariots can also be related to a more mystical consideration, that of transporting the soul of the deceased (Laursen 2017: 285).

32 Pottier was referring to chance finds coming from looted graves in Bactria.

REFERENCES

Algaze, G. 1993. *The Uruk World System: The Dynamics of Expansion of Early Mesopotamian Civilization*. Chicago: University of Chicago Press.

Algaze, G. 2018. "Entropic Cities: The Paradox of Urbanism in Ancient Mesopotamia." *Current Anthropology* 59(1): 23–54.

Amanbaeva, B.E., A.E. Rogozhinskij, and D. Merfi 2006. "Mogil'nik Shagym: novyj pamjatnik epokhi bronzy vostochnoj Fergany (Kyrgyzstan)" [The cemetery of Shagym: a new Bronze Age site in eastern Ferghana (Kirghizstan)]. *Arkheologicheskie Issledovanija v Uzbekistane 2004–2005 godu* 5: 256–265.

Amiet, P. 1977. "Bactriane proto-historique." *Syria* LIV(1–2): 89–121.

Amiet, P. 1978. "Antiquités de Bactriane." *La Revue du Louvre, Etudes* 3: 153–164.

Amiet, P. 1986. *L'âge des échanges inter-iraniens, 3500–1700 avant J.-C.* Paris: Réunion des musées nationaux.

Amiet, P. 1988. "Elam et Bactriane." In *L'Asie centrale et ses rapports avec les civilisations orientales des origines à l'Age du Fer. Actes du colloque franco-soviétique, Paris 1985*, ed. J.-C. Gardin. Paris: de Boccard, pp. 27–30.

Amiet, P. 1989a. "Elam e Battriana." In *Battriana. Una antica civiltà delle oasi dalle sabie dell'Afghanistan*, ed. G. Ligabue and S. Salvatori. Venice: Erizzo, pp. 127–140.

Amiet, P. 1989b. "Le antichità della Battriana e dell'Iran esterno al Museo del Louvre." In *Battriana. Una antica civiltà delle oasi dalle sabie dell'Afghanistan*, ed. G. Ligabue and S. Salvatori. Venice: Erizzo, pp. 159–180.

Amiet, P. 2007. "Les échanges inter-iraniens 3500–1700 avant J.-C." In *Sulla Via delle Oasi. Tesori dell'Oriente Antico*, ed. G. Ligabue and G. Rossi Osmida. Padova: Il Punto, pp. 64–87.

Artemiev, V. and A. Urmanova 2010. "Predvaritel'nye itogi arkhitekturnogo issledovanija Severnogo Gonura v 2007 g." [Preliminary results of architectural research at North Gonur in 2007]. In *Na puti otkrytija tsivilizatsii* [On the track of uncovering a civilization]. *Trudy Margianskoj Arkheologicheskoj Ekspeditsii*, vol. 3, ed. P.M. Kozhin, M.F. Kosarev, and N.A. Dubova. St. Petersburg: Aletejja, pp. 172–203.

Aruz, J. and R. Wallenfels 2003. *Art of the First Cities: The Third Millennium BC from the Mediterranean to the Indus.* New Haven, CT and London: Metropolitan Museum of Art.

Ascalone, E. and L. Peyronel 2001. "Two Weights from Temple N at Tell Mardikh-Ebla, Syria: A Link Between Metrology and Cultic Activities in The Second Millennium BC?" *Journal of Cuneiform Studies* 53: 1–12.

Askarov, A.A. 1973. *Sapallitepa.* Tashkent: Fan.

Askarov, A.A. 1977. *Drevnezemledel'cheskaja kultura juga Uzbekistana* [Bronze Age ancient farming culture of southern Uzbekistan]. Tashkent: Fan.

Askarov, A.A. 1985. "Traditions et innovations dans la culture du nord de la Bactriane à l'Age du Bronze." In *L'archéologie de la Bactriane ancienne: actes du Colloque de Dushanbe 1982*, ed. J.-C. Gardin. Paris: Editions du CNRS, pp. 119–124.

Askarov, A.A. and B.N. Abdullaev 1983. *Dzharkutan (k probleme protogorodskoj tsivilizatsii na juge Uzbekistana)* [Dzharkutan (on the question of a proto-urban civilization in southern Uzbekistan)]. Tashkent: Fan.

Askarov, A.A. and V.D. Ruzanov 1977. "Nekotorye dannye o metallicheskikh splavakh s poselenija Sapallitepa" [Data on the metal alloys from Sapallitepa]. *Obshchestvennye Nauki Uzbekistana* 6: 56–59.

Askarov, A.A. and T.Sh. Shirinov 1993. *Rannjaja gorodskaja kul'tura epokhi bronzy juga Srednej Azii* [The Bronze Age early urban culture in southern Central Asia]. Samarkand: Institute of Archaeology of Uzbekistan.

Avanesova, N.A. 1991. *Kul'tura pastusheskikh plemen epokhi bronzy aziatskoj chasti SSSR.* [The Bronze Age pastoral groups culture in the Asian part of USSR]. Tashkent: Fan.

Avanesova, N.A. 1996. "Pasteurs et agriculteurs de la vallée du Zeravshan (Ouzbékistan) au début de l'âge du Bronze: relations et influences mutuelles." In *Sarazm (Tajikistan) céramiques (Chalcolithique et Bronze Ancien)*, ed. B. Lyonnet. Paris: de Boccard, pp. 117–124.

Avanesova, N.A. 2004. "Sluchajnye nakhodki epokhi bronzy iz fondov Gosmuzeja-zapovednika Samarkanda." In *U Istokov Tsivilizatsii* [Near the sources of civilizations], ed. M.F. Kosarev, P.M. Kozhin, and N.A. Dubova. Moscow: Staryj Sad, pp. 405–414.

Avanesova, N.A. 2010. "Zeravshanskaja kul'turnaja provintsija Baktrijsko-Margianskoj Tsivilizatsii" [The Zeravshan cultural province of the Bactria-Margiana Civilization]. In *Na puti otkrytija tsivilizatsii* [On the track of uncovering a civilization]. *Trudy Margianskoj Arkheologicheskoj Ekspeditsii*, vol. 3, ed. P.M. Kozhin, M.F. Kosarev, and N.A. Dubova. St. Petersburg: Aletejja, pp. 334–364.

Avanesova, N. 2013a. *Buston VI: nekropol' ognepoklonnikov dourbanisticheskoj Baktrii.* [Buston VI: a necropolis of fire worshippers in proto-urban Bactria]. Samarkand: MITSAI.

Avanesova, N.A. 2013b. "Zhukov, un 'sanctuaire' énéolithique d'anciens nomades dans la vallée du Zeravshan (Ouzbékistan)." *Paléorient* 39(2): 85–108.

Avanesova, N. 2015. "Keramika poselenija gornjakov-metallurgov Zarafshana" [Pottery from a miners-metallurgists' settlement of the Zeravshan valley]. *Istorija Material'noj Kul'tury Uzbekistana* 39: 47–62.

Babakov, O. 2004. "Naselenie Gonur-Depe v epokhu bronzy (v svete antropologicheskikh dannykh). In *U Istokov Tsivilizatsii* [Near the sources of civilizations], ed. M.F. Kosarev, P.M. Kozhin, and N.A. Dubova. Moscow: Staryj Sad, pp. 337–348.

Babakov, O., G.V. Rykushina, N.A. Dubova, S.V. Vasiliev, A.P. Pestryakov, and T.K. Khodzhayov 2001. "Human Skeletal Remains of the Necropolis of Gonur-Depe. Appendix 1." In *Necropolis of Gonur and Iranian Paganism*, ed. V. Sarianidi. Moscow: Mir-Media, pp. 219–240.

Bendezu-Sarmiento, J. 2013. "Archéologie funéraire et bio-anthropologie à Ulug dépé et Dzharkutan: Age du Bronze au Turkménistan et en Ouzbékistan." *Cahiers d'Asie centrale* 21–22: 501–532.

Bendezu-Sarmiento, J. and S. Mustafakulov 2013. "Le site proto-urbain de Dzharkutan durant les âges du bronze et du fer. Recherches de la Mission archéologique franco-ouzbèque-Protohistoire." *Cahiers d'Asie centrale* 21–22: 207–236.

Besenval, R. 2005. "Chronology of Protohistoric Kech-Makran." In *South Asian Archaeology 2001*, vol. I, ed. C. Jarrige and V. Lefèvre. Paris: Editions Recherche sur les Civilisations, pp. 1–9.

Biscione, R. and A.A. Vahdati 2011 "Excavation at Tepe Chalow, Northern Khorasan, Iran." *Studi Micenei ed Egeo-Anatolici* 53: 236–241.

Bonora, G.L. and M. Vidale 2013. "The Middle Chalcolithic in Southern Turkmenistan and the Archaeological Record of Ilgynly-Depe." In *Ancient Iran and Its Neighbours: Local Development and Long-Range Interaction in the Fourth Millennium BC*, ed. C.A. Petrie. Oxford: Oxbow Books, pp. 140–165.

Bonora, G.L., M. Vidale, M. Mariottini, and G. Guida 2014. "On the Use of Tokens and Seals Along the Kopet Dagh Piedmont, Turkmenistan (ca. 6000–3000 BCE)." *Paléorient* 40(1): 55–71.

Bostongukhar, S. 1998. *Verkhov'ja Zarafshana vo II tysjacheletii do n.e.* [The upper Zeravshan valley during the 2nd millennium BC]. Dushanbe: Donish.

Casal, J.-M., 1961. *Fouilles de Mundigak*, 2 vols. Mémoires de la Délégation Archéologique Française en Afghanistan, t. XVII. Paris: Klincksieck.

Casanova, M. 2001. "Le lapis-lazuli, la pierre précieuse de l'Orient ancient." *Dialogues d'histoire ancienne* 27(2): 149–170.

Casanova, M. 2013. *Le lapis-lazuli dans l'Orient ancient. Production et circulation du Néolithique au IIe millénaire av. J.-C.* Paris: CTHS.

Cattani, M. 2008. "The Final Phase of the Bronze Age and the 'Andronovo Question' in Margiana." In *The Bronze Age and Early Iron Age in the Margiana Lowlands: Facts and Methodological Proposals for a Redefinition of the Research Strategies*, ed. S. Salvatori and

M. Tosi. BAR International Series 1806: Archaeological Map of the Murghab Delta Studies and Reports, vol. II. Oxford: Archaeopress, pp. 133–151.

Cattani, M., B. Cerasetti, S. Salvatori, and M. Tosi 2008. "The Murghab Delta in Central Asia 1990–2001: The GIS from Research Resource to a Reasoning Tool for the Study of Settlement Change in Long-Term Fluctuations." In *The Bronze Age and Early Iron Age in the Margiana Lowlands: Facts and Methodological Proposals for a Redefinition of the Research Strategies*, ed. S. Salvatori and M. Tosi. BAR International Series 1806: Archaeological Map of the Murghab Delta Studies and Reports, vol. II. Oxford: Archaeopress, pp. 39–45.

Cerasetti, B. 1998. "Preliminary Report on Ornamental Elements of 'Incised Coarse Ware.'" In *The Archaeological Map of the Murghab Delta: Preliminary Reports 1990–95*, ed. A. Gubaev, G. Koshelenko, and M. Tosi. Rome: IsIAO, pp. 67–74.

Cerasetti, B., R. Arciero, M. Carra, A. Curci, J. De Grossi Mazzorin, L. Forni, E. Luneau, L.M. Rouse, and R.N. Spengler III 2019. "Bronze and Iron Age Urbanisation in Turkmenistan: Preliminary Results from the Excavation of Togolok 1 on the Murghab Alluvial Fan." In *Urban Cultures of Central Asia from the Bronze Age to the Karakhanids. Learnings and Conclusions from New Archaeological Investigations and Discoveries*, ed. C. Baumer and M. Novák. Wiesbaden: Harrassowitz, pp. 63–72.

Charpin, D. 2004. "Histoire politique du Proche-Orient amorrite (2002–1595)." In *Mesopotamien: Die altbabylonische Zeit*, ed. P. Attinger, W. Sallaberger, and M. Wäfler. Orbis Biblicus et Orientalis 160/4. Fribourg-Göttingen: Academic Press/Vandenhoeck and Ruprecht, pp. 25–480.

Chernykh, E.N. 1992. Ancient Metallurgy in the USSR: The Early Metal Age. Cambridge: Cambridge University Press.

Collon, D. 1987. *First Impressions: Cylinder Seals in the Near East*. London: British Museum Press.

Crawford, H. 1998. *Dilmun and Its Gulf Neighbours*. Cambridge: Cambridge University Press.

Cremaschi, M. 1998. "Palaeohydrography and Middle Holocene Desertification in the Northern Fringe of the Murghab Delta." In *The Archaeological Map of the Murghab Delta: Preliminary Reports 1990–95*, ed. A. Gubaev, G. Koshelenko, and M. Tosi. Rome: IsIAO, pp. 15–25.

Dercksen, J.G. 2005. "Metals According to Documents from Kültepe-Kanish Dating to the Old Assyrian Colony Period." In *Anatolian Metal III*, ed. Ü. Yalçin. *Der Anschnitt* 18: 17–34.

Dercksen, J.G. 2008. "The Assyrian Colony at Kanesh." In *Ḫattuša-Boğazköy. Das Hethiterreich im Spannungsfeld des Alten Orients: 6th International Colloquium der Deutschen Orient-Gesellschaft*, ed. G. Wilhelm. Wiesbaden: Harrassowitz Verlag, pp. 109–124.

Dercksen, J.G. 2014. "The Old Assyrian Trade and its Participants." In *Documentary Sources in Ancient Near Eastern and Greco-Roman Economic History. Methodology and Practice*, ed. H.D. Baker and M. Jursa. Oxford: Oxbow Books, pp. 59–112.

Desse, J. 1989. "Etude archéozoologique." In *Fouilles de Shortughaï. Recherches sur l'Asie centrale protohistorique*, ed. H.-P. Francfort. Paris: de Boccard, pp. 187–206.

Dubova, N.A. 2004. "Mogil'nik i tsarskij nekropol' na beregakh Bol'shogo Bassejna Severnogo Gonura" [Graveyard and Royal Necropolis on the edge of the Large Pool at Gonur North]. In *U istokov tsivilizatsii* [Near the sources of civilizations], ed. M. Kosarev, P. Kozhin, and N. Dubova. Moscow: Starij Sad, pp. 254–281.

Dubova, N.A. 2008. "Masterskaja po proizvodstvu splavov na osnove medi Severnogo Gonura (zapadnaja chast' Raskopa 9)" [Workshop for the production of copper alloys at Gonur North (western part of Area 9)]. In *Trudy Margianskoj Arkheologicheskoj Ekspeditsii*, vol. 2, ed. V.I. Sarianidi. Moscow: Staryj Sad, pp. 94–104.

Dubova, N.A. 2012. "Pogrebenija zhivotnykh v strane Margush" [Animal burials in Margush]. In *Trudy Margianskoj Arkheologicheskoj Ekspeditsii*, vol. 4, ed. V.I. Sarianidi. Moscow: Staryj Sad, pp. 101–139.

Dubova, N.A. 2014. "Anthropological Essay on the Bronze Age Migrations in Central Asia (case of Gonur Depe, Turkmenistan)." In *"My Life Is Like the Summer Rose": Maurizio Tosi e l'Archeologia come modo de vivere – Papers in Honour of Maurizio Tosi for His 70th Birthday*, ed. C.C. Lamberg-Karlovsky, B. Genito, and B. Cerasetti. BAR International Series 2690. Oxford: Archaeopress, pp. 193–202.

Dubova, N.A. 2015. "Animal Burials and Their Cults in Margiana." In *Defining the Sacred*, ed. N. Laneri. Oxford: Oxbow books, pp. 13–23.

Dubova, N.A. 2017. "Antropologicheskie osobennosti naselenija Gonur Depe i ego okrugi" [Anthropological peculiarities of Gonur Depe population and its neighborhood]. In *XII Kongress antropologov i etnologov Rossii* [Congress of anthropologists and ethnologists of Russia], ed. A.E. Zagrebin and M.Ju. Martynova. Izhevsk: IEA RAN, UIIJaL UrO RAS, p. 136.

Dubova, N.A., E.V. Antonova, R.G. Muradov, R.M. Sataev, and A.A. Tishkin 2018. "Foreword." In Trudy Margianskoj Ekspeditsii, vol. 7, ed. N.A. Dubova. Moscow: Starij Sad, p. 8.

Dubova, N.A. and V.V. Kufterin 2015. *Antropologija naselenija Juzhnogo Uzbekistana epokhi pozdnej bronzy (po materialam necropolja Buston VI)* [Anthropology of the Late Bronze Age southern Uzbekistan population (materials from the Buston VI necropolis)]. Moscow: Staryj Sad.

Dubova, N.A. and V.V. Kufterin 2017. "Dannye arkheologii i antropoligii po sotsial'noj stratifikatsii drevnezemledel'cheskogo obshchestva Margiany (Turkmenistan)" [Archaeological and anthropological data on the social stratification of the Margiana society]. In *Trudy V (XXI) vserossijskogo arkheologicheskogo c'ezda v Barnaule-Belokurikhe* [Transactions of the fifth (XXI) All Russian archaeological congress], ed. A.P. Derevjanko and A.A. Tishkin. Barnaul: Izd. Altajskogo gosudarstvennogo universiteta III, pp. 31–37.

Dubova, N.A., V.V. Kufterin, R.M. Sataev, and L.V. Sataeva 2017. "Izmenenija estesvennoj sredy obitanija i adaptivnye processy u naselanija epokhi bronzy Juzhnogo Turkmenistana" [Dynamics of the environment and adaptive processes of the southern Turkmenistan Bronze Age population]. In *Paleopochvy, paleoecologija, paleoekonomika. Materialy vserossijskoj mezhdistsiplinarnoj nauchnoj konferentsii s mezhdunarodnym uchastiem* [Paleosoils, paleoecology and paleoeconomy: international interdisciplinary scientific conference], ed. A.V. Borisov, L.N. Plekhanova, and S.N. Udal'tsov. Pushchino: Tovarishestvo nauchnykh izdanij KMK, pp. 53–57.

Dubova, N.A. and E.A. Muradova 2008. "'Mogil'nik temenosa' Gonura" [The Gonur temenos graveyard]. In *Trudy Margianskoj Ekspeditsii*, vol. 2, ed. V.I. Sarianidi. Moscow: Staryj Sad, pp. 105–111.

Dubova, N.A. and G.V. Rykushina 2004. "Necropolis and Area 5 of Gonur-Depe: some anthropological data." In *U Istokov Tsivilizatsii* [Near the sources of civilizations], ed. M.F. Kosarev, P.M. Kozhin, and N.A. Dubova. Moscow: Staryj Sad, pp. 317–336.

Dubova, N.A. and G.V. Rykushina 2007. "New Data on Anthropology of the Necropolis of Gonur-Depe." In *Necropolis of Gonur*, by V. Sarianidi. Athens: Kapon Editions, pp. 296–329.

Durand, J.-M. 1998. *Documents épistolaires du palais de Mari*, vol. I. Littératures Anciennes du Proche-Orient. Paris: Editions du Cerf.

Durand, J.-M. 2018. "Le commerce à longue distance de Mari d'après les textes du règne de Zimri-Lim." *Pasiphae* XII: 69–87.

During-Caspers, E.C.L. 1972. "Harappan Trade in the Arabian Gulf in the 3rd millennium BC." *Mesopotamia* VII: 167–191.

During-Caspers, E.C.L. 1992. "Intercultural/Mercantile Contacts Between the Arabian Gulf and South Asia at the Close of the Third Millennium BC." *Proceedings of the Seminar for Arabian Studies* 22: 3–28.

During-Caspers, E.C.L. 1994. "Further Evidence for 'Central Asian' materials from the Arabian Gulf." *Journal of the Economic and Social History of the Orient* XXXVII: 33–53.
During-Caspers, E.C.L. 1998. "The BMAC and the Harappan Script." *Ancient Civilizations from Scythia to Siberia* 5(1): 40–58.
Dyson, R.H. Jr. and S.M. Howard (eds.) 1989. *Tappeh Hesār: Reports of the Restudy Project 1976*. Monografie di Mesopotamia. Firenze: Casa Editrice Le Lettere.
Frachetti, M.D. and L.M. Rouse 2012. "Central Asia, the Steppe, and the Near East, 2500–1500 BC." In *A Companion to the Archaeology of the Ancient Near East*, ed. D.T. Potts. Oxford: Wiley-Blackwell, pp. 687–705.
Francfort, H.-P. 1984 "The Early Periods of Shortughaï (Harappan) and the Western Bactrian Culture of Dashly." In *South Asian Archaeology 1981*, ed. B. Allchin. Cambridge: Cambridge University Press, pp. 170–175.
Francfort, H.-P. 1989. *Fouilles de Shortughaï. Recherches sur l'Asie centrale protohistorique*. Paris: de Boccard.
Francfort, H.-P. 1994. "The Central Asian Dimension of the Symbolic System in Bactria and Margiana." *Antiquity* 68(259): 406–418.
Francfort, H.-P. 2005a. "La civilisation de l'Oxus et les Indo-Iraniens et Indo-Aryens." In Aryas, Aryens et Iraniens en Asie centrale, ed. G. Fussman, J. Kellens, H.-P. Francfort, and X. Tremblay. Publications de l'Institut de Civilisation Indienne LXXII, Collège de France. Paris: de Boccard, pp. 253–328.
Francfort, H.-P. 2005b. "Observations sur la toreutique de la Civilisation de l'Oxus." In *Afghanistan, ancien carrefour entre l'est et l'ouest*, ed. O. Bopearachchi and M.-F. Boussac. Turnhout: Brepols, pp. 21–63.
Francfort, H.-P. 2016. "How the Twins Met: Indus and Oxus Bronze Age Civilizations in Eastern Bactria. Shortughaï Revisited Forty Years Later." In *Trudy Margianskoj Arkheologicheskoj Ekspeditsii*, vol. 6, ed. N.A. Dubova. Moscow: Staryj Sad, pp. 461–475.
Francfort, H.-P. and X. Tremblay 2010. "Marhaši et la Civilisation de l'Oxus." *Iranica Antiqua* 45: 51–224.
Frenez, D. 2017. "Manufacturing and Trade of Asian Elephant Ivory in Bronze Age Middle Asia: Evidence from Gonur Depe (Margiana, Turkmenistan)." *Archaeological Research in Asia* 15: 13–33. https://doi.org/10.1016/j.ara.2017.08.002.
Frenez, D. 2018. "Torgovlja slonovoj kost'ju iz bivnej Asiatskogo slona i proizvodstvo predmetov iz nee v epokhu bronzy v Srednej Azii. Svidetel'stva Gonur-depe (Margiana, Turkmanistan)" [Manufacturing and trade of Asian elephant ivory in Bronze Age Middle Asia: Evidence from Gonur Depe (Margiana, Turkmenistan)]. In *Trudy Margianskoj Arkheologicheskoj Ekspeditsii*, vol. 7, ed. N.A. Dubova. Moscow: Staryj Sad, pp. 106–133.
Franke U. 2016. "Prehistory." In *Herat Through Time: The Collections of the Herat Museum and Archive*, ed. U. Franke and M. Müller-Wiener. Ancient Herat, vol. 3. Berlin: Staatliche Museen zu Berlin, pp. 20–57.
Fribus, A.V. and S.P. Grushin 2016. "Keramicheskij sosud s izobrazheniem 'fasada khrama' s jugo-zapadnoj territorii (raskop 12) Gonur-depe" [A pottery vessel with a representation of a temple from the southwestern zone (Area 12) at Gonur Depe]. In *Trudy Margianskoj Arkheologicheskoj Ekspeditsii*, vol. 6, ed. N.A. Dubova. Moscow: Staryj Sad, pp. 327–334.
Garner, J. 2013. *Das Zinn der Bronzezeit in Mittelasien*, vol. II. Die Montanarchäologischen Forschungen der Zinnlagerstätten. Archäologie in Iran und Turan 12, Veröffentlichungen aus dem Deutschen Bergbau-Museum 194. Bochum: von Zabern.
Gasche, H., J.A. Armstrong, S.W. Cole, and V.G. Gurzadyan 1998a. *Dating the Fall of Babylon: A Reappraisal of Second-Millennium Chronology (a Joint Ghent–Chicago–Harvard Project)*. Mesopotamian History and Environment, Series II, Memoirs IV. Ghent and Chicago: University of Ghent and the Oriental Institute of the University of Chicago.

Gasche, H., J.A. Armstrong, S.W. Cole, and V.G. Gurzadyan 1998b. "A Correction to Dating the Fall of Babylon: A Reappraisal of Second-Millennium Chronology Ghent and Chicago 1998." *Akkadica* 108: 1–4.

Gentelle, P. 1989. *Prospections archéologiques en Bactriane orientale (1974–1978)*. Vol. I, *Données paléogéographiques et fondements de l'irrigation*. Paris: de Boccard.

Glassner, J.-J. 2002. "Dilmun et Magan: Le Peuplement, l'Organisation Politique, la Question des Amorrites et la Place de l'Ecriture. Point de Vue de l'Assyriologue." In *Essays on the Late Prehistory of the Arabian Peninsula*, ed. S. Cleuziou, M. Tosi, and J. Zarins. Rome: IsIAO, pp. 337–381.

Good, I. 2010. "When East Met West: Interpretative Problems in Assessing Eurasian Contact and Exchange in Antiquity." *Archäologische Mitteilungen aus Iran und Turan* 42: 23–45.

Good, I., J. M. Kenoyer, and R. H. Meadow 2009. "New Evidence for Early Silk in the Indus Civilization." *Archaeometry* 51(3): 457–466.

Grenet, F. 2015. "Zoroastrianism in Central Asia." In *The Wiley Blackwell Companion to Zoroastrianism*, ed. M. Stausberg and Y. Sohrab-Dinshaw Vevaina: John Wiley & Sons, pp. 129–146.

Gubaev, A., G. Koshelenko, and M. Tosi (eds.) 1998. *The Archaeological Map of the Murghab Delta: Preliminary Reports 1990–95*. Rome: IsIAO.

Guichard, M. 1996. "A la recherche de la pierre bleue." *NABU* 1: 30–32.

Guichard, M. 2005. *La vaisselle de luxe des rois de Mari. Matériaux pour le Dictionnaire de Babylonien de Paris*, vol. II. Archives Royales de Mari XXXI. Paris: Editions Recherche sur les Civilisations.

Guljamov, Ja.G., U. Islamov, and A. Askarov 1966. *Pervobytnaja kul'tura i vozniknovenie oroshaemogo zemledelija v nizov'jakh Zaravshana* [Indigenous culture and origin of farming irrigation in the Low Zeravshan valley]. Tashkent: Fan.

Hakemi, A. 1997. *Shahdad: Archaeological Excavations of a Bronze Age Center in Iran*. Rome: IsMEO.

Hemphill, B.E. 1998. "Biological Affinities and Adaptations of Bronze Age Bactrians: III. An Initial Craniometric Assessment." *American Journal of Physical Anthropology* 106: 329–348.

Hemphill, B.E. 1999. "Biological Affinities and Adaptation of Bronze Age Bactrians: IV. A Craniometric Investigation of Bactrian Origins." *American Journal of Anthropology* 108: 173–192.

Hemphill, B.E. and J.-P. Mallory 2004. "Horse-Mounted Invaders from the Russo-Kazakh Steppe or Agricultural Colonists From Western Central Asia? A Craniometric Investigation of the Bronze Age Settlement of Xinjiang." *American Journal of Physical Anthropology* 124: 199–222.

Herrmann, G. 1968. "Lapis Lazuli: The Early Phases of Its Trade." *Iraq* 30: 21–57.

Hiebert, F.T. 1994a. *Origins of the Bronze Age Oasis Civilization in Central Asia*. American School of Prehistoric Research, Bulletin 42. Cambridge, MA: Peabody Museum of Archaeology and Ethnology, Harvard University.

Hiebert, F.T. 1994b. "Production Evidence for the Origins of the Oxus Civilization." *Antiquity* 68(259): 372–387.

Hiebert, F.T. and D. Killick 1993. "Metallurgy of Bronze Age Margiana." *Unesco IASCCA Information Bulletin* 19: 186–204.

Hiebert, F.T. and C.C. Lamberg-Karlovsky 1992. "Central Asia and the Indo-Iranian Borderlands." *Iran* XXX: 1–15.

Hiebert, F.T. and K.M. Moore 2004. "A small Steppe Site Near Gonur." In *U Istokov Tsivilizatsii* [Near the sources of civilizations], ed. M.F. Kosarev, P.M. Kozhin, and N.A. Dubova. Moscow: Staryj Sad, pp. 294–302.

Isakov, A.I. 1991. *Sarazm*. Dushanbe: Donish.

Isakov, A.I. 1992. "Bogatoe zhenskoe pogrebenie iz Sarazma (Tadzhikistan)" [A rich woman burial at Sarazm (Tajikistan)]. *Arkheologicheskie Vesti* 1: 64–75.

Isakov, A.I. 1996. "Sarazm, an Agricultural Center of Ancient Sogdiana." *Bulletin of the Asia Institute* 8: 1–25.

Isakov, A.I., P.L. Kohl, C.C. Lamberg-Karlovsky, and R. Maddin 1987. "Metallurgical Analysis from Sarazm, Tajikistan SSR." *Archaeometry* 29(1): 90–102.

Isakov, A.I. and B. Lyonnet 1988. "Céramiques de Sarazm (Tajikistan, URSS): problems d'échanges et de peuplement à la fin du Chalcolithique et au début de l'âge du Bronze." *Paléorient* 14(1): 31–47.

Jansen, M., S. Aulbach, A. Hauptmann, H.E. Höfer, S. Klein, M. Krüger, and R.L. Zettler 2016. "Platinum Group Placer Minerals in Ancient Gold Artifacts: Geochemistry and Osmium Isotopes of Inclusions in Early Bronze Age Gold from Ur/Mesopotamia." *Journal of Archaeological Science* 68: 12–23.

Jarrige, J.-F. 1985. "Les relations entre l'Asie centrale méridionale, le Baluchistan et la vallée de l'Indus à la fin du 3e et au début du 2e millénaire." In *L'archéologie de la Bactriane ancienne: actes du Colloque de Dushanbe 1982*. Paris: Editions du CNRS, pp. 105–118.

Jarrige, J.-F. éd. 1988. *Cités oubliées de l'Indus. Archéologie du Pakistan*. Paris: Musée Guimet.

Jarrige, J.-F. 1994. "The Final Phase of the Indus Occupation at Nausharo and Its Connection with the Following Cultural Complex of Mehrgarh VIII." In *South Asian Archaeology 1993*, ed. A. Parpola and P. Koskikallio. Helsinki: Suomalainen Tiedeakatemia, pp. 295–313.

Jarrige, J.-F. and M.U. Hassan 1989. "Funerary Complexes in Baluchistan at the End of the Third Millennium in the Light of Recent Discoveries at Mehrgarh and Quetta." In *South Asian Archaeology 1985*, ed. K. Frifelt and P. Sorensen. Copenhagen: Curzon Press, pp. 150–166.

Joannès, F. 1991. "L'étain, de l'Elam à Mari." In *Mésopotamie et Elam: Actes de la 36e Rencontre Assyriologique*, ed. L. de Meyer and H. Gasche. Ghent: University of Ghent, pp. 67–76.

Juminov, A.M. and Ja.M. Tjutev 2016. "Kamennye orudija iz 'masterskoj' na Raskope 8 administrativno-kul'tovogo kompleksa Gonur-depe" [Stone tools from a workshop in Area 8 of the administrative and cult complex of Gonur Depe]. In *Trudy Margianskoj Arkheologicheskoj Ekspeditsii*, vol. 6, ed. N.A. Dubova. Moscow: Staryj Sad, pp. 264–271.

Kaniuth, K. 2006. *Metallobjekte der Bronzezeit aus Nordbaktrien*. Archäologie in Iran und Turan, 6. Mainz: Ph. von Zabern.

Kaniuth, K. 2007. "The Metallurgy of the Late Bronze Age Sapalli Culture (Southern Uzbekistan) and Its Implications for the 'Tin Question.'" *Iranica Antiqua* XLII: 23–40.

Kaniuth, K. 2010. "Long Distance Imports in the Bronze Age of Southern Central Asia: Recent Finds and Their Implications for Chronology and Trade." *Archäologische Mitteilungen aus Iran und Turan* 42: 3–22.

Kaniuth, K. and M. Teufer 2001. "Zur Sequenz des Gräberfeldes von Rannij Tulchar und seiner Bedeutung für die Chronologie des spätbronzezeitlichen Baktrien." *Archäologische Mitteilungen aus Iran und Turan* 33: 89–113.

Kenoyer, J.M. 1998. *Ancient Cities of the Indus Valley Civilization*. Oxford and New York: Oxford University Press.

Khodzhajov, T.K. 1977. *Antropologicheskij sostav naselenija epokhi bronzy Sapallitepa* [Anthropological structure of the Bronze Age population of Sapallitepa]. Tashkent: Fan.

Khodzhajov, T.K. and Kh.Kh. Khalilov 1978. "Cherepa epokhi bronzy iz mogil'nikov Dzharkutan i Buston (raskopki 1975 g.)" [Bronze Age skulls from the Dzharkutan and Bustan cemeteries (1975 excavations)]. *Istorija Material'noj Kul'tury Uzbekistana* 14: 43–50.

Khodzhajov, T.K. and Kh.Kh. Khalilov 1984. *Naselenie verkhov'ev Amu Dar'i po dannym paleoantropologii.* [The population of the upper Amu Darya according to paleoanthropological data]. Tashkent: Fan.

Khuff, D. and Sh. Shajdullaev 1999. "Nekotorye rezul'taty rabot uzbeksko-germanskoj ekspeditsii na gorodishche Dzharkutan" [Some results from the Uzbek-German expedition work at the site of Dzharkutan]. *Istorija Material'noj Kul'tury Uzbekistana* 30: 19–26.

Kircho, L.B. and S.G. Popov 1999. "K voprosu o radiouglerodnoj khronologii drevnejshikh tsivilizatsij Srednej Azii" [Questions on the radiocarbon chronology of the ancient civilizations of Central Asia]. *Stratum Plus* 2: 350–361.

Kircho, L.B. and S.G. Popov 2005. "K voprosu o radiouglerodnoj khronologii arkheologicheskikh pamjatnikov Srednej Azii V-II tys. do n.e." [Questions on the radiocarbon chronology of archaeological sites of Central Asia, V–II millennium BC]. *Khronologija epokhi pozdnego eneolita-srednej bronzy Srednej Azii (pogrebenija Altyn-depe).* St. Petersburg: Nestor-Istorija.

Klochkov, I.S. 1995. "Znaki na cherepke s Gonura. K voprosu o margianskoj pis'mennosti" [Signs on a sherd from Gonur: Questions on a Margiana writing]. *Vestnik Drevnej Istorii* 2: 54–60.

Klochkov, I.S. 1998. "Signs on a Potsherd from Gonur (on the Question of the Script Used in Margiana)." *Ancient Civilizations from Scythia to Siberia* 5(2): 165–175.

Kohl, P.L. 1975. "Carved Chlorite Vessels: A Trade in Finished Commodities in the Mid-Third Millennium." *Expedition* 18(pt. 1): 18–31.

Kohl, P.L. 1984. *Central Asia: Palaeolithic Beginnings to the Iron Age.* Paris: Recherche sur les Civilisations.

Kohl, P.L. 2007. *The Making of Bronze Age Eurasia.* Cambridge World Archaeology. New York: Cambridge University Press.

Kohl, P.L. and B. Lyonnet 2008. "By Land and By Sea: The Circulation of Materials and Peoples, ca. 3500–1800 BC." In *Intercultural Relations Between South and Southwest Asia: Studies in Commemoration of E.C.L. During-Caspers (1934–1996)*, ed. E. Olijdam and R.H. Spoor. BAR International Series 1826. Oxford: Archaeopress, pp. 29–42.

Kolinski, R. 2000. *Tell Rijim, Iraq: The Middle Bronze Layers.* BAR International Series 837. Oxford: Archaeopress.

Kuz'mina, E.E. 2007. *The Origin of the Indo-Iranians.* Leiden and Boston: Brill.

Lamberg-Karlovsky, C.C. 1988. "The 'Intercultural-Style' Carved Vessels." *Iranica Antiqua* XXIII: 45–95.

Lamberg-Karlovsky, C.C. 1993. "Reflections on the Central Asia Bronze Age." *Unesco IASCCA Information Bulletin* 19: 29–40.

Lamberg-Karlovsky, C.C. 1994. "The Bronze Age Khanates of Central Asia." *Antiquity* 68(259): 398–405.

Lamberg-Karlovsky, C.C. 2003. "Civilization, State, or Tribes? Bactria and Margiana in the Bronze Age." *Review of Archaeology* 24(1): 11–19.

Lamberg-Karlovsky, C.C. 2012. "The Oxus Civilization (aka: the Bactria-Margiana Archaeological Complex)." *Review of Archaeology* 30: 59–75.

Laursen, S.T. 2010. "The Westward Transmission of Indus Valley Sealing Technology: Origin and Development of the 'Gulf Type' Seal and Other Administrative Technologies in Early Dilmun, *c.*2100–2000 BC." *Arabian Archaeology and Epigraphy* 21: 96–134.

Laursen, S.T. 2017. *The Royal Mounds of A'ali in Bahrain: The Emergence of Kingship in Early Dilmun.* Aarhus: Narayana Press.

Lerouxel, F. 2002. "Les échanges de présents entre souverains amorites au XVIIIe siècle av. n. è. d'après les archives de Mari." In *Florilegium marianum VI,* Mémoires de N.A.B.U., vol. 7, ed. D. Charpin and J.-M. Durand. Paris: Sepoa, pp. 413–463.

Ligabue, G. and S. Salvatori 1989. *Battriana, una antica civiltà delle oasi dalle sabbie dell'Afghanistan.* Venice: Erizzo.

Ljapin, A.A. 1983 "Irrigatsija drevnemurgabskoj delty" [Irrigation of the ancient Murghab delta]. *Pamjatniki Turkmenistana* 2: 20–23.

Ljapin, A.A. 1990a. "K paleogeografii delty Murgaba (epokha bronzy, zheleznyj vek)" [On the paleogeography of Murghab delta (Bronze and Iron Age)]. *Problemy osvoenija pustyn'* 3: 57–67.

Ljapin, A.A. 1990b. "Pochvovedenie i arkheologija Murgaba" [Soil science and archaeology of the Murghab]. *Izvestija Akademii nauk Turkmenskoj SSR* 6: 24–33.

Ljapin, A.A. 1991. "Cherty paleogeografii del't Murgaba i Tedzhena" [Paleogeographical features of the Murghab and Tedjen deltas]. *Problemy osvoenija pustyn'* 2: 63–70.

Ljapin, A.A. 2014. "K istorii oroshenija v del'te Murgaba" [On the history of irrigation in the delta of the Murghab River]. In *Trudy Margianskoj Arkheologicheskoj Ekspeditsii*, vol. 5, ed. V.I. Sarianidi. Moscow: Staryj Sad, pp. 60–91.

Luneau, E. 2008. "Tombes féminines et pratiques funéraires en Asie centrale protohistorique. Quelques réflexions sur le 'statut social' des femmes dans la civilisation de l'Oxus." *Paléorient* 34(1): 131–157.

Luneau, E. 2014. *La fin de la Civilisation de l'Oxus: Transformations et recompositions des sociétés de l'Age du Bronze final en Asie centrale méridionale (1800–1500/1400 avant n.è.).* Paris: de Boccard.

Lyonnet, B. 1977. "Découverte de sites de l'Age du Bronze dans le N.E. de l'Afghanistan: leurs rapports avec la Civilisation de l'Indus." *Annali* 37: 19–35.

Lyonnet, B. 1981. "Etablissements chalcolithiques dans le Nord-Est de l'Afghanistan: leurs rapports avec les civilisations du bassin de l'Indus." *Paléorient* 7(2): 57–74.

Lyonnet, B. 1988. "Les relations de la Bactriane orientale avec le monde indo-baluche à partir des données céramologiques, du 3e au 2e millénaire avant notre ère. In *L'Asie centrale et ses rapports avec les civilisations orientales des origines à l'Age du Fer. Actes du colloque franco-soviétique, Paris 1985*, ed. J.-C. Gardin. Paris: de Boccard, pp. 143–151.

Lyonnet, B. 1996. *Sarazm (Tajikistan) céramiques (Chalcolithique et Bronze Ancien).* Paris: de Boccard.

Lyonnet, B. 1997. *Prospections archéologiques en Bactriane orientale (1974–1978).* Vol. 2, *Céramique et peuplement, du Chalcolithique à la conquête arabe.* Paris: Editions Recherche sur les Civilisations.

Lyonnet, B. 2001. *Mari et la Margiane ou 'la circulation des biens, des personnes et des idées' dans l'Orient ancient à la fin du 3e et au début du 2e millénaire avant notre ère.* Mémoire d'Habilitation à diriger des recherches. Paris I Sorbonne (unpublished).

Lyonnet, B. 2005. "Another Possible Interpretation of the Bactro-Margiana Culture (BMAC) of Central Asia: The Tin Trade." In *South Asian Archaeology 2001*, vol. I, ed. C. Jarrige and V. Lefèvre. Paris: Editions Recherche sur les Civilisations, pp. 191–200.

Lyonnet, B. 2016. "A Grave with a Wooden Wagon in Transcaucasia (Azerbaijan): Its Relations with Central Asia." In *Trudy Margianskoj Arkheologicheskoj Ekspeditsii*, vol. 6, ed. N.A. Dubova. Moscow: Staryj Sad, pp. 378–385 and separated color plates.

Lyonnet, B. 2019. "La 'céramique du Khabur' et ses relations orientales: les routes de l'étain." In *De l'argile au numérique. Mélanges en honneur de Dominique Charpin*, ed. G. Chambon, M. Guichard, and A.-I. Langlois, Th. Römer and N. Ziegler. Leuven: Peeters, pp. 613–637.

Majidzadeh, Y. 2003. *Jiroft: The Earliest Oriental Civilization.* Tehran: Ministry of Culture and Islamic Guidance, Iran Cultural Heritage Organization Research Center Printing and Publishing.

Mallory, J.P. 1989. *In Search of the Indo-Europeans. Language, Archaeology and Myth.* London: Thames & Hudson.

Marcolongo, B. and P. Mozzi 1998. "Outline of Recent Geological History of the Kopet-Dagh Mountains and the southern Kara-Kum." In *The Archaeological Map of the Murghab*

Delta: Preliminary Reports 1990–95, ed. A. Gubaev, G. Koshelenko, and M. Tosi. Rome: IsIAO, pp. 1–13.
Margueron, J.-C. 2004. *Mari, metropole de l'Euphrate au IIIe et au début du IIe millénaire av. J.-C.* Paris: Picard/Erc.
Mashkour, M. 2013. "Sociétés pastorales et économies de subsistence au nord-est de l'Iran et au sud du Turkménistan." *Cahiers d'Asie centrale* 21–22: 533–544.
Masimov, I.S. 1979. "K voprosu ob osvoenii nizovij Murgaba drevnezemledel'cheskimi plemenami" [About the lower Murghab development by ancient groups of farmers]. *Izvestija Akademii Nauk Turkmenskoj SSR* 2: 72–77.
Masson, V.M. 1959. *Drevnezemledel'cheskaja kul'tura Margiany* [The ancient farmers' culture of Margiana]. Materialy i Issledovanija po Arkheologii 73. Moscow: Nauka.
Masson, V.M. 1977. "Pechaty protoindijskogo tipa iz Altyn-Depe" [Seals of a proto-Indus type at Altyn Depe]. *Vestnik Drevnej Istorii* 4: 147–155.
Masson, V.M. 1981. *Altyn-Depe*. Trudy JuTAKE, vol. XVIII. Leningrad: Nauka.
Mejer-Melikjan, N.R. 1990. "Opredelenie rastitel'nykh ostatkov iz Togolok 21" [Identification of vegetal remains from Togolok 21]. In *Drevnosti Strany Margush*, by V.I. Sarianidi. Ashkhabad: Ylym, pp. 203–205.
Meyer-Melikyan, N.R. 1998. "Analysis of Floral Remains from Togolok 21." In *Margiana and Protozoroastrism*, by V.I. Sarianidi. Athens: Kapon editions, pp. 178–179.
Meyer-Melikyan N.R. and N.A. Avetov, 1998. "Analysis of Floral Remains in the Ceramic Vessel from the Gonur Temenos." In *Margiana and Protozoroastrism*, by V.I. Sarianidi. Athens: Kapon editions, pp. 176–177.
Michalowski, P. 1977. "The Death of Shulgi." *Orientalia* 46: 220–225.
Michel, C. 2001. "Le lapis-lazuli des Assyriens au début du IIe millénaire av. J.-C." In *Veenhof Anniversary Volume: Studies Presented to K.R. Veenhof on the Occasion of his Sixty-Fifth Birthday*, ed. W.H. van Soldt, J.G. Dercksen, N.J. Kouwenberg, and Th.J. Krispijn. PIHANS, vol. 89. Leiden: Nederlands Instituut voor het Nabije Oosten, pp. 341–359.
Michel, C. 2014. "Central Anatolia in the Nineteenth and Eighteenth Centuries BC." In *Constituent, Confederate and Conquered Space. The Emergence of the Mittani State*, ed. E. Cancik-Kirschbaum, N. Brisch, and J. Eidem. Berlin: de Gruyter, pp. 111–136.
Miller, N. 1993. "Preliminary Archaeobotanical Results from the 1989 Excavation at the Central Asian Site of Gonur Depe, Turkmenistan." *Unesco IASCCA Information Bulletin* 19: 149–163.
Moore, K. 1993. "Animal Use at Bronze Age Gonur Depe." *Unesco IASCCA Information Bulletin* 19: 164–176.
Mutin, B. 2013. *The Proto-Elamite Settlement and Its Neighbors: Tepe Yahya Period IVC*. American School of Prehistoric Research Publications. Cambridge MA: Peabody Museum, Harvard University.
Mutin, B., L.D. Minc, C.C. Lamberg-Karlovsky, and M. Tosi 2017. "Regional and Long-Distance Exchange of an Emblematic 'Prestige' Ceramic in the Indo-Iranian Borderlands. Results of Neutron Activation Analysis." *Paléorient* 43(1): 141–162.
Mutin, B. and A. Razzokov 2014. "Contacts Across the Hindu Kush in the Bronze Age: Additional Insights from Sarazm – Soundings 11–11A (Tajikistan). *Archäologische Mitteilungen aus Iran und Turan* 46: 123–147.
Narasimhan V.M., N.J. Patterson, P. Moorjani, N. Rohland, R. Bernardos, S. Mallick, I. Lazaridis *et al.* 2019. "The Formation of Human Populations in South and Central Asia." *Science* 365(6457), eaat7487. doi: 10.1126/science.aat7487.
Oates, D., J. Oates, and H. McDonald 2001. *Excavations at Tell Brak*. Vol. 2, *Nagar in the Third Millennium BC*. London: British School of Archaeology in Iraq-McDonald Institute.
Omel'chenko, A.V. 2012. "On the Question of the Sasanian Presence in Sogdiana. Recent Results of Excavations at Paykand." *Journal of Inner Asian Art and Archaeology* 7: 79–107.

Orthmann, W. and A. Pruss 1995. "Der Palast F." In *Ausgrabungen in Tell Chuera in Nordost-Syrien*. Vol. I, Vorbericht über die Grabungskampagnen 1986 bis 1992. Saarbrück: Saarbrücker Druckerei und Verlag, pp. 121–150 and figs. 65–86.

Özgüç, T. 1986a. *Kültepe-Kaniš II: New Researches at the Trading Center of the Assyrian Trade Colonies*. Türk Tarih Kurumu Yayinlarindan, vol. 19. Ankara: Türk Tarih Kurumu Basimevi.

Özgüç, T. 1986b. "New Observations on the Relationship of Kültepe with Southeast Anatolia and North Syria during the Third Millennium BC." In *Ancient Anatolia. Aspects of Change and Cultural Development. Essays in Honor of M.J. Mellink*, ed. J.V. Canby, E. Porada, B.S. Ridgway, and T. Stech. Madison: University of Wisconsin Press, pp. 31–47.

Papakhristu, O.A. 2008. "K reconstruktsii tiglej dlja plavki splavov na osnove medi iz masterskoj Gonur Depe" [Reconstruction of the crucible for melting copper alloys from the Gonur Depe workshop]. In *Trudy Margianskoj Arkheologicheskoj Ekspeditsii*, vol. 2, ed. V.I. Sarianidi. Moscow: Staryj Sad: pp. 196–224.

Papakhristu, O.A. 2010. "Litejnye formy iz masterskoj stolichnogo Gonura (predvaritel'noe issledovanie)" [Casting molds from a workshop at the capital city of Gonur (preliminary report)]. *Na puti otkrytija tsivilizatsii* [On the track of uncovering a civilization], ed. P.M. Kozhin, M.F. Kosarev, and N.A. Dubova. St. Petersburg: Aletejja, pp. 265–284.

Papakhristu, O.A. 2014. "K voprosu o masterskoj po proizvodstvu predmetov iz splavov na osnove medi na Gonur Depe (Turkeministan)" [About a workshop for the production of copper-based alloys at Gonur Depe (Turkmenistan)]. In *Drevnejshie tsivilizatsii Tsentral'noj Azii* [The most ancient civilizations of Central Asia], ed. T. Mkrtychev and S. Bolelov. Moscow: State Oriental Museum, pp. 105–106.

Papakhristu, O.A. 2016. "Opyt sopostavlenija: masterskaja metallistov na gorodishe Gonur-depe (Turkmenistan) i kvartal metallistov na gorodishe Shakhdad (Iran)" [A comparative essay: a metalworkers' workshop at Gonur Depe (Turkmenistan) and the metalworkers' quarter at Shahdad (Iran)]. In *Trudy Margianskoj Arkheologicheskoj Ekspeditsii*, vol. 6, ed. N.A. Dubova. Moscow: Staryj Sad, pp. 232–256.

Parpola, A. 1993. "Margiana and the Aryan Problem." *Unesco IASCCA Information Bulletin* 19: 41–77.

Parpola, A. 2004. "Gandhara Graves and the Gharma Pot, the Nasatyas and the Nose: in Pursuit of the Chariot Twins." In *U Istokov Tsivilizatsii* [Near the sources of civilizations], ed. M.F. Kosarev, P.M. Kozhin, and N.A. Dubova. Moscow: Staryj Sad, pp. 102–128.

Parzinger, H. and N. Boroffka 2003. *Das Zinn der Bronzezeit in Mittelasien I: Die siedlungsarchäologischen Forschungen im Umfeld der Zinnlagerstätten*. Archäologie in Iran und Turan 5. Mainz: von Zabern.

Perrot, J. and Y. Madjidzadeh 2005. "L'iconographie des vases et objets en chlorite de Jiroft (Iran)." *Paléorient* 31(2): 123–152.

Perrot, J. and Y. Madjidzadeh 2006. "À travers l'ornementation des vases et objets en chlorite de Jiroft." *Paléorient* 32(1): 99–112.

Petrie, C.A. 2013. "Ancient Iran and Its Neighbours: Emerging Paradigms and Future Directions." In *Ancient Iran and Its Neighbours: Local Developments and Long-Range Interactions in the Fourth Millennium BC*, ed. C.A. Petrie. British Institute of Persian Studies. Oxford: Oxbow Books, pp. 385–411.

Pigott, V. 2018. "The Bactria-Margiana Archaeological Complex (BMAC), the Seima-Turbino Horizon and a Possible Eastward Transmission of Tin-Bronze Technology in Later Third and Early Second Millennium BCE Inner Asia." In *How Objects Tell Stories: Essays in Honor of E.C. Bunker*, ed. K.M. Linduff and K.S. Rubinson. Inner and Central Asian Art and Archaeology I. Turnhout: Brepols, pp. 191–221.

Pittman, H. 1984. *Art of the Bronze Age, Southeastern Iran, Western Central Asia and the Indus Valley*. New York: Metropolitan Museum of Art.

Pittman H. 2013. "Eastern Iran in the Early Bronze Age." In *The Oxford Handbook of Ancient Iran*, ed. D.T. Potts. Oxford: Oxford University Press, pp. 304–324.

Pittman H. 2014a. "Anchoring Intuition in Evidence: A Continuing Discussion of Cylinder Seals from Southeastern Iran." In *Edith Porada zum 100. Geburstag. A Centenary Volume*, ed. E. Bleibtreu and H.U. Steymans. Orbis Biblicus et Orientalis 268. Göttingen: Vandenhoeck & Ruprecht, 375–393.

Pittman H. 2014b. "Hybrid Imagery and Cultural Identity in the Age of Exchange: Halil River Basin and Sumer Meet in Margiana." In *"My Life Is Like the Summer Rose": Maurizio Tosi e l'Archeologia come modo de vivere – Papers in Honour of Maurizio Tosi for His 70th Birthday*, ed. C.C. Lamberg-Karlovsky, B. Genito, and B. Cerasetti. BAR International Series 2690. Oxford: Archaeopress, pp. 625–636.

Pittman, H. 2018. "Administrative Role of Seal Imagery in the Early Bronze Age: Mesopotamian and Iranian Traders on the Plateau." In *Seals and Sealings in the Ancient World: Case Studies from the Near East, Egypt, the Aegean, and South Asia*, ed. M. Ameri, S.K. Costello, G. Jamison, and S.J. Scott. Cambridge: Cambridge University Press, pp. 13–35.

Pittman H. 2019. "Bronze Age Interaction on the Iranian Plateau: From Kerman to the Oxus Through Seals." In *The Iranian Plateau During the Bronze Age. Development of Urbanisation, Production and Trade*, ed. J.-W. Meyer, E. Vila, M. Mashkour, M. Casanova, and R. Vallet. Lyon: Mom éditions, pp. 267–288.

Possehl, G.L. 1993. *Radiometric Dates for South Asian Archaeology.*

Possehl, G.L. 2002. *The Indus Civilization: A Contemporary Perspective.* Lanham, MD: Rowman & Littlefield.

Pottier, M.-H. 1984. *Matériel funéraire de la Bactriane méridionale de l'Age du Bronze.* Paris: Editions Recherche sur les Civilisations.

Potts, D.T. 1993. "A New Bactrian Find from Southeastern Arabia." *Antiquity* 67: 591–596.

Potts, D.T. 1994. "South and Central Asian Elements at Tell Abraq (Emirate of Umm al-Qaiwain, United Arab Emirates), c.2200 BC–AD 400." In *South Asian Archaeology 1993*, ed. A. Parpola and P. Koskikallio. Helsinki: Suomalainen Tiedeakatemia, pp. 615–628.

Potts, D.T. [1999] 2016. *The Archaeology of Elam: Formation and Transformation of an Ancient Iranian State.* Cambridge World Archaeology. New York: Cambridge University Press.

Potts, D.T. 2001. *Excavations at Tepe Yahya, Iran 1967–1975: The Third Millennium.* American School of Prehistoric Research Bulletin 45. Cambridge, MA: Peabody Museum of Archaeology and Ethnology, Harvard University.

Potts, D.T. 2008a. "An Umm an-Nar-Type Compartmented Soft-Stone Vessel from Gonur Depe, Turkmenistan." *Arabian Archaeology and Epigraphy* 19(2): 168–181.

Potts, D.T. 2008b. "Puzur-Inšušinak and the Oxus Civilization (BMAC): Reflections on Šimaški and the Geo-Political Landscape of Iran and Central Asia in the Ur III Period." *Zeitschrift für Assyriologie und Vorderasiatische Archäologie* 98: 165–194.

Pruzsinszky, R. 2004. "Evidence for the Short Chronology in Mesopotamia? The Chronological Relationship Between the Texts from Emar and Ekalte." In *Mesopotamian Dark Age Revisited*, ed. H. Hunger and R. Pruzsinszky. Wien: Österreichische Akademie der Wissenschaften, pp. 43–50.

P'jankova, L.T. 1993. "Pottery of Margiana and Bactria in the Bronze Age." *Unesco IASCCA Information Bulletin* 19: 109–127.

Ratnagar, Sh. 2004. *Trading Encounters, from the Euphrates to the Indus in the Bronze Age.* New Delhi: Oxford University Press.

Razzokov, A. 2008. *Sarazm.* Dushanbe: Echod.

Reiter, K. 1997. *Die Metalle im Alten Orient unter besonderer Berücksichtigung altbabylonischer Quellen.* Münster: Ugarit Verlag.

Rossi Osmida, G. 2007. *Adji Kui Oasis: La cittadella delle statuette*, vol. I. Padova: Il Punto.

Rossi Osmida, G. 2011. *Adji Kui Oasis: La cittadella delle statuette*, vol. II. Padova: Il Punto.

Rossi Osmida, G. (ed.) 2002. *Margiana. Gonur-Depe Necropolis: 10 Years of Excavations by Ligabue Study and Research Centre*. Trebaseleghe: Il Punto.

Rouse, L.M. and B. Cerasetti 2017. "Micro-Dynamics and Macro-Patterns: Exploring New Archaeological Data for the Late Holocene Human–Water Relationship in the Murghab Alluvial Fan, Turkmenistan." *Quaternary International* 437: 20–34.

Rowton, M.B. 1973. "Autonomy and Nomadism in Western Asia." *Orientalia* 42: 247–258.

Ruzanov, V.D. 1999. "Zum frühen Auftreten der Zinnbronze in Mittelasien." In *The Beginnings of Metallurgy*, vol. 9, ed. A. Hauptmann, E. Pernicka, T. Rehren, and U. Yalçin. Bochum: Deutsches Bergbau-Museum, pp. 103–106.

Ruzanov, V.D. 2013. *Metalloobrabotka na juge Srednej Azii v epokhu bronzy* [Bronze Age metalworking in southern Uzbekistan]. Samarkand: Institute of Archaeology Ja.G. Guljamov.

Saliari, K. and E. Draganits 2013. "Early Bronze Age Bone Tubes from the Aegean: Archaeological Context, Use and Distribution." Archeometriai Mühely X(3): 179–192.

Salvatori, S. 1993. "The Discovery of the Graveyard of Gonur-depe I (Murghab Delta, Turkmenistan): 1992 Campaign Preliminary Report." *Rivista di Archeologia* XVII: 5–13.

Salvatori, S. 1995. "Gonur-Depe 1 (Margiana, Turkmenistan): The Middle Bronze Age Graveyard – Preliminary Report on the 1994 Excavation Campaign." *Rivista di Archeologia* XIX: 5–37.

Salvatori, S. 2000. "Bactria and Margiana Seals: A New Assessment of Their Chronological Position and a Typological Survey." *East and West* 50: 97–145.

Salvatori, S. 2008. "Cultural Variability in the Bronze Age Oxus Civilization and its Relations with the Surrounding Regions of Central Asia and Iran." In *The Bronze Age and Early Iron Age in the Margiana Lowlands: Facts and Methodological Proposals for a Redefinition of the Research Strategies*, ed. S. Salvatori and M. Tosi. BAR International Series 1806. Oxford: Archaeopress, pp. 75–98.

Salvatori, S. 2010. "Thinking Around Grave 3245 in the 'Royal Graveyard' of Gonur (Murghab Delta, Turkmenistan). In *Na puti otkrytija tsivilizatsii* [On the track of uncovering a civilization]. Trudy Margianskoj Arkheologicheskoj Ekspeditsii, vol. 3, ed. P.M. Kozhin, M.F. Kosarev, and N.A. Dubova. St. Petersburg: Aletejja, pp. 244–257.

Salvatori, S. 2016. "Bactria-Margiana Archaeological Complex: How Terminology Hides Historical Processes." In *Trudy Margianskoj Arkheologicheskoj Ekspeditsii*, vol. 6, ed. N.A. Dubova. Moscow: Staryj Sad, pp. 449–460.

Salvatori, S. and M. Tosi 1997. "Postscriptum. Some Reflections on Shahdad and Its Place in the Bronze Age of Middle Asia." In *Shahdad. Archaeological Excavations of a Bronze Age Center in Iran*, ed. A. Hakemi. Rome: IsMEO, pp. 121–138.

Salvatori, S. and M. Tosi (eds.) 2008. *The Bronze Age and Early Iron Age in the Margiana Lowlands: Facts and Methodological Proposals for a Redefinition of the Research Strategies*. Archaeological Map of the Murghab Delta Studies and Reports, vol. II. BAR International Series 1806. Oxford: Archaeopress.

Salvatori, S., M. Vidale, G. Guida, and E. Masioli 2009. "Ilgynly-Depe (Turkmenistan) and the 4th Millennium BC Metallurgy of Central Asia." *Paléorient* 35(1): 47–67.

Santoni, M. 1984 "Sibri and the South Cemetery of Mehrgarh: Third Millennium Connections Between the Northern Kachi Plain (Pakistan) and Central Asia." In *South Asian Archaeology 1981*, ed. B. Allchin. Cambridge: Cambridge University Press, pp. 52–60.

Sarianidi, V.I. 1968. "O velikom lazurotovom puti na Drevnem Vostoke." *Kratkie Soobshchenija Instituta Arkheologii* 114: 3–9.

Sarianidi, V. 1970. "The Lapis Lazuli Route in the Ancient East." Archaeology 24(1): 12–15.

Sarianidi, V.I. 1972. "Izuchenie pamjatnikov epokhi bronzy i rannego zheleza v severnom Afganistane" [Studies of Bronze and Iron Age sites in northern Afghanistan]. *Kratkie Soobshchenija Instituta Arkheologii* 132: 16–22.

Sarianidi, V.I. 1974. "Baktrija v epokhu bronzy" [Bactria in the Bronze Age]. *Sovetskaja Arkheologija* 4: 49–71.
Sarianidi, V.I. 1975. "Stepnye plemena epokhi bronzy v Margiane" [Bronze Age steppe groups in Margiana]. *Sovetskaja Arkheologija* 2: 20–29.
Sarianidi, V.I. 1976. "Pechaty-amulety Murghaboskogo stilja" [Seal-amulets of the Murghab style]. *Sovetskaja Arkheologija* 1: 42–68.
Sarianidi, V.I. 1977. *Drevnie zemledel'tsy Afganistana* [Ancient farmers of Afghanistan]. Materialy Sovetsko-Afganskoj ekspeditsii 1969–1974 gg. Moscow: Nauka.
Sarianidi, V.I. 1981a. "Drevnjaja Baktrija: novye aspekty staroj problem" [Ancient Bactria: new aspects of an old problem]. In *Etnicheskie problem istorii Tsentral'noj Azii v drevnosti (II tysjacheletie do n. e.)*. Moscow: Nauka, pp. 180–191.
Sarianidi, V.I. 1981b. "Margiana in the Bronze Age." In *The Bronze Age Civilization of Central Asia, Recent Soviet Discoveries*, ed. P.L. Kohl. New York: Me Sharpe, pp. 165–193.
Sarianidi, V.I. 1982. "Ob odnoj gruppe drevnebaktrijskoj gliptiki" [About one group of ancient Bactrian glyptic]. *Drevnjaja Indija i nauka*. Moscow: Nauka, pp. 297–306.
Sarianidi, V.I. 1983. "Traditions and Innovations in the Bronze Age Culture of Southern Bactria." *Unesco IASCCA Information Bulletin* 4: 50–51.
Sarianidi, V.I. 1984. "Southern Turkmenia and Margiana in the Bronze Age." *Unesco IASCCA Information Bulletin* 7: 5–17.
Sarianidi, V.I. 1985. "Les origines de la culture agricole primitive de la Bactriane." In *L'archéologie de la Bactriane ancienne. Actes du colloque franco-soviétique de Dushanbe 1982*, ed. J.C. Gardin. Paris: Editions du CNRS, pp. 125–135.
Sarianidi, V.I. 1986a. "Bronzovij vek Margiany" [Margiana Bronze Age]. *Kratkie Soobshchenija* 188: 83–91.
Sarianidi, V.I. 1986b. "Le complexe cultuel de Togolok 21 en Margiane." *Arts Asiatiques* XLI: 5–21.
Sarianidi, V.I. 1986c. "Mesopotamija i Baktria vo vtorom tys. do n.e." [Mesopotamia and Bactria in the 2nd millennium BC]. *Sovetskaja Arkheologija* 2: 34–46.
Sarianidi, V.I. 1988. "Cult Symbolism of Bactrian and Margiana Amulets." In *Orientalia Iosephi Tucci Memoriae Dicata*, ed. G. Gnoli and L. Lanciotti. Serie Orientale Roma LVI.3. Rome: IsMEO, pp. 1281–1294.
Sarianidi, V.I. 1989a. "Protozoroastrijskij khram v Margiane i problema vozniknovenija Zoroastrizma" [A protozoroastrian temple in Margiana and the question of the birth of Zoroastrism]. *Vestnik Drevnej Istorii* 1: 152–169 (and discussion by F. Grenet, V.A. Gaibov, G.A. Koshelenko, V.A. Livshits, I.M. Steblin-Kamenskij, and B.A. Litvinskij pp. 170–179).
Sarianidi, V.I. 1989b. "Siro-Khettskie bozhestva v Baktrijsko-Margianskom panteone" [Syro-Hittite deities in the Bactria-Margiana pantheon]. *Sovetskaja Arkheologija* 4: 17–24.
Sarianidi, V.I. 1990. *Drevnosti strany Margush* [Ancient Margush]. Ashkhabad: Ylym.
Sarianidi, V.I. 1992. "Ottisk tsilindricheskoj pechati 'mitannijskogo stilja' iz Margiany" [Sealings of cylinder seals in Mitanni style from Margiana]. *Vestnik Drevnej Istorii* 3: 90–93.
Sarianidi, V.I. 1993. "Margiana in the Ancient Orient." *Unesco IASCCA Information Bulletin* 19: 5–28.
Sarianidi, V.I. 1994. "Margiana and the Indo-Iranian World." In *South Asian Archaeology 1993*, vol. II, ed. A. Parpola and P. Koskikallio. Helsinki: Suomalainen Tiedeakatemia, pp. 667–680.
Sarianidi, V.I. 1996. "Aegean-Anatolian Motifs in the Glyptic Art of Bactria and Margiana." *Bulletin of the Asia Institute* 8: 27–36.
Sarianidi, V. 1998a. *Margiana and Protozoroastrism*. Athens: Kapon Editions.
Sarianidi, V. 1998b. *Myths of Ancient Bactria and Margiana on Its Seals and Amulets*. Moscow: Pentagraphic.

Sarianidi, V.I. 1999. "Siro-Khettskoe proiskhozhdenie Baktrijsko-Margianskoj gliptiki" [The Syro-Hittite origin of Bactria-Margiana glyptic]. *Vestnik Drevnej Istorii* 1: 53–73.
Sarianidi, V.I. 2001. *Necropolis of Gonur and Iranian Paganism*. Moscow: World Media.
Sarianidi, V.I. 2002. *Margush: Ancient Oriental Kingdom in the Old Delta of the Murghab River*. Aşgabat: Türkmendöwlethabarlary.
Sarianidi, V.I. 2005. *Gonurdepe, Türkmenistan: City of Kings and Gods*. Aşgabat: Miras.
Sarianidi V.I. 2006. "Turkmenistan as a Central Asian Center of Ancient Oriental Civilization." In *Ancient Margiana Is the New Centre of the World Civilization*. International Scientific Conference. Aşgabat: Miras, pp. 71–87.
Sarianidi, V.I. 2007. *Necropolis of Gonur*. Athens: Kapon Editions.
Sarianidi, V.I. 2008a. *Margush: Mystery and True of the Great Culture*. Aşgabat: Türkmendö wlethabarlary.
Sarianidi V.I. 2008b. *Zoroastrianism: A New Motherland for an Old Religion*. Athens: Kyriakidis Brothers S.A.
Sarianidi, V.I. 2010. *Long Before Zaratushtra*. Moscow: Staryj Sad.
Sarianidi, V.I., N.G.O. Boroffka, and N.A. Dubova 2012. "Cultural Contacts of Margiana, Turkmenistan, in the 3rd Millennium BC: New Evidence from Gonur Depe, Burial 4150." *Gandhāran Studies* 6: 1–17.
Sarianidi, V.I, N.G.O. Boroffka, and N.A. Dubova 2014. "Kul'turnye kontakty Margiany (Turkmenistan) v III tys. do n.e. Novye dannye po Gonur-Depe (pogrebenie no. 4150)" [Margiana (Turkmenistan) cultural contacts in the 3rd millennium BC: new data from Gonur Depe (tomb no. 4150)]. In *Trudy Margianskoj Arkheologicheskoj Ekspeditsii*, vol. 4, ed. V.I. Sarianidi. Moscow: Staryj Sad, pp. 127–137.
Sarianidi, V.I. and N.A. Dubova 2008. "Arkheologicheskie raboty na jugo-zapadnom kholme Gonur-depe" [Archaeological work on the southwestern hill of Gonur Depe (Area 16)]. In *Trudy Margianskoj Arkheologicheskoj Ekspeditsii*, vol. 2. ed. V. Sarianidi. Moscow: Staryj Sad, pp. 28–49.
Sarianidi, V.I. and N.A. Dubova 2012. "Arkheologicheskie raboty Margianskoj arkheologicheskoj ekspeditsii v 2008–2011 gg" [Works of the Margiana archaeological expedition in 2008–2011]. In *Trudy Margianskoj Arkheologicheskoj Ekspeditsii*, vol. 4, ed. V.I. Sarianidi. Moscow: Staryj Sad, pp. 29–55.
Sarianidi, V.I. and N.A. Dubova 2014a. "Gonur Excavations in 2000–2010." In *Proceedings of the 8th International Congress on the Archaeology of the Ancient Near East 2012*, vol. 2, ed. P. Bieliński, M. Gawlikowski, R. Koliński, D. Ławecka, A. Sołtysiak, and Z. Wygnańska. Wiesbaden: Harrassowitz, pp. 503–518.
Sarianidi, V.I. and N.A. Dubova 2014b. "Raboty Margianskoj arkheologicheskoj ekspeditsii v 2011–2013 gg." [Works of the Margiana archaeological expedition in 2011–2013]. In *Trudy Margianskoj Arkheologicheskoj Ekspeditsii*, vol. 5, ed. V.I. Sarianidi. Moscow: Staryj Sad, pp. 92–111.
Sarianidi, V.I. and N.A. Dubova. 2015. "Kul't zhivotnykh v Margiane" [Cult of animals in Margiana]. In *Akademicheskoe Vostokovedenie v Rossii i stranakh blizhnego zarubezhja (2007–2015). Arkheologija. Istorija. Kul'tura*, ed. V.P. Nikonorov and V.A. Alekshin. Academic Oriental Studies in Russia and the Neighbor Countries (2007–2015). Archaeology. History. Culture. St. Petersburg: Contrast, pp. 521–558.
Sarianidi, V.I. and N.A. Dubova 2016. "Types of Graves at Gonur Depe Bronze Age Site in Turkmenistan." In *Proceedings of the 9th International Congress on the Archaeology of the Ancient Near East 2014*, vol. 3, ed. A.R. Stucky, O. Kaelin, and H.-P. Mathys. Wiesbaden: Harrassowitz, pp. 631–642.
Sataev, R.M. 2008. "Raskopki drevnego irrigatsionnogo kanala." [Excavations of the ancient irrigation canal] In *Trudy Margianskoj Arkheologicheskoj Ekspeditsii*, vol. 2, ed. V.I. Sarianidi. Moscow: Staryj Sad, pp. 65–66.

Sataev, R.M. 2016. *Zhivotnye v kulture Drevnej Margiany* [Animals in the Ancient Margiana culture]. Moscow: Staryj Sad.

Sataev, R.M. 2018. "Predvaritel'nye itogi izuchenija stratigrafii dvortsa Severnogo Gonura" [Preliminary results from the stratigraphic analysis of the Gonur North Palace]. In *Trudy Margianskoj Arkheologicheskoj Ekspeditsii*, vol. 7, ed. N.A. Dubova. Moscow: Staryj Sad, pp. 80–86.

Sataev, R. and L. Sataeva 2014. "Results of Archaeozoological and Archaeobotanical Research at the Bronze Age Gonur Depe Site (Turkmenistan)." In *Proceedings of the 8th International Congress on the Archaeology of the Ancient Near East*, ed. P. Bieliński, M. Gawlikowski, R. Koliński, D. Ławecka, A. Sołtysiak, and Z. Wygnańska. Wiesbaden: Harrassowitz, pp. 369–372.

Sataeva, L.V. 2016. "Rezul'taty issledovanij ostatkov drevesiny iz raskopok Gonur-depe" [Results from research on wood remains from Gonur Depe excavations]. In *Trudy Margianskoj Arkheologicheskoj Ekspeditsii*, vol. 6, ed. N.A. Dubova. Moscow: Staryj Sad, pp. 307–314.

Sataeva, L.V. 2018. "Arkheobotanicheskie issledovanija na Gonur-Depe v 2015 g." [Archaeobotanical studies at Gonur Depe in 2015]. In *Trudy Margianskoj Arkheologicheskoj Ekspeditsii*, vol. 7, ed. N.A. Dubova. Moscow: Staryj Sad, pp. 156–158.

Sataeva, L.V. and Sataev R.M. 2016. "Zemledelie v Drevnej Margiane" [Agriculture in ancient Margiana]. *Vestnik Bashkirskogo gosudarstvennogo agrarnogo universiteta* 3: 30–36.

Sperduti, A., L. Bondioli, and M. Macchiarelli 2002. "Paleobiology of the Third Millennium BC: Human Skeletal Remains from the Graveyard of Gonur-Depe. A Preliminary Report (Oct. 1995)." In *Margiana, Gonur Necropolis: 10 Years of Excavation by Ligabue Study and Research Centre*, ed. G. Rossi Osmida. Padova: Il Punto, pp. 165–179.

Stein, M.A. 1931. *An Archaeological Tour in Gedrosia*. Memoirs of the Archaeological Survey of India 43. New Delhi: Cosmo Publications.

Stein, M.A. 1937. *Archaeological Reconnaissances in the North-Western India and South-Eastern Iran*. London: Macmillan & Co.

Steinkeller, P. 1982. "The Question of Marhasi: A Contribution to the Historical Geography of Iran in the Third Millennium BC." *Zeitschrift für Assyriologie*: 237–265.

Steinkeller, P. 2014. "Marhaši and Beyond: The Jiroft Civilization in a Historical Perspective." In *"My Life Is Like the Summer Rose": Maurizio Tosi e l'Archeologia come Modo di Vivere – Papers in Honour of Maurizio Tosi for His 70th Birthday*, ed. C.C. Lamberg-Karlovsky and B. Genito. BAR International Series 2690. Oxford: Archeopress, pp. 691–707.

Struve, V.V. 1949. "Vosstanie v Margiane pri Darii I" [Insurrection against Darius I in Margiana]. In *Materialy Juzhno-Turkmenistanskoj kompleksnoj arkheologicheskoj expeditsii*, vol. 1. Ashkhabad: Turkmeniskij filial akademii nauk SSSR, pp. 9–34.

Taganowich, O.A. 2007. "The Architecture of the Adji Kui Oven-Fireplaces." In *Adji Kui Oasis. La cittadella delle statuette*, vol. I, ed. G. Rossi Osmida. Padova: Il Punto, pp. 203–212.

Teissier, B. 1994. *Sealing and Seals on Texts from Kültepe Kārum Level 2*. Istanbul: Nederlands Historisch-Archaeologisch Instituut Te Istanbul.

Tengberg, M. 2013. "Economies végétales et environnements en Asie centrale, du Néolithique à l'époque sassanide." *Cahiers d'Asie centrale* 21–22: 545–558.

Terekhova, N.N. 1990. "Obrabotka metallov v drevnej Margiane" [Metalworking in ancient Margiana]. In *Drevnosti strany Margush*, by V.I. Sarianidi. Ashkhabad: Ylym, pp. 177–202.

Teufer, M. 2015. *Spätbronzezeitliche Grabfunde aus Nordbaktrien und benachbarten Regionen. Studien zur Chronologie zwischen Aralsee und Persischem Golf*. Berlin: Dietrich Reimer Verlag GmbH.

Thomalsky, J., B. Bräutigam, M. Karaucak, and S. Kraus 2013. "Early Mining and Metal Production in Afghanistan: The First Year of Investigations." *Archäologische Mitteilungen aus Iran und Turan* 45: 199–230.

Tsareva, E. 2007. "Gonur City Linens of the Second Millennium BC." In *Necropolis of Gonur*, by V.I. Sarianidi. Athens: Kapon Editions, pp. 330–334.

Udeumuradov, B.N. 1993. *Altyn-depe i Margiana: svjazi, khronologija, proiskhozhdenie* [Altyn Depe and Margiana: connections, chronology, origin], ed. V.M. Masson. Ashkhabad: Ylym.

Udeumuradov, B.N. 2002. "Ceramic Material from Gonur Depe Necropolis." In *Margiana: Gonur Depe Necropolis: 10 Years of Excavations by Ligabue Study and Research Centre*, ed. G. Rossi Osmida. Venice: Il Punto, pp. 133–143.

Urmanova, A.M. 2014. "K problem sooruzhenij tipa "kel'i" [About the cell-type constructions]. In *Trudy Margianskoj Arkheologicheskoj Ekspeditsii*, vol. 5, ed. V.I. Sarianidi. Moscow: Staryj Sad, pp. 138–150.

Valentine, B., G.D. Kamenov, J.M. Kenoyer, V. Shinde, V. Mushrif-Tripathy, E. Otarola-Castillo, and J. Krigbaum 2015. "Evidence for Patterns of Selective Urban Migration in the Greater Indus Valley (2600–1900 BC): A Lead and Strontium Isotope Mortuary Analysis." *PLoS ONE* 10(4): e0123103. doi: 10.1371/journal.pone.0123103.

Vats, M.S. 1940. *Excavations at Harappa*. 2 vols. Delhi: Government of India.

Vidale, M. 2017a. "New Interpretations." In *Lapis-Lazuli Bead Making at Shahr-i Sokhta*, ed. M. Vidale and A. Lazzari. Rome: ISMEO, pp. 303–321.

Vidale, M. 2017b. *Treasures from the Oxus: The Art and Civilization of Central Asia*. London: I.B. Tauris.

Vinogradova, N.M. 2004. *Jugo-Zapadnyj Tadzhikistan v epokhu pozdnej Bronzy* [Southwestern Tajikistan in the Bronze Age]. Moscow: Institut Vostokovedenija.

Vinogradova, N.M., V.A. Ranov, and T.G. Filimonova 2008. *Pamjatniki Kangurttuta v Jugo-Zapadnom Tadzhikistane (epokha neolita i bronzovogo veka)* [The site of Kangurttut in southwestern Tajikistan (Neolithic and Bronze Age)]. Moscow: Institut Vostokovedenija.

Weiss, H., M.-A. Courty, W. Wetterstrom, F. Guichard, R. Meadow, L. Senior, and A. Curnow 1993. "The Genesis and Collapse of Third Millennium North Mesopotamian Civilization." *Science* 261: 999–1004.

Weiss, H. 2016. "Global Megadrought, Societal Collapse and Resilience at 4.2–3.9 ka BP Across the Mediterranean and West Asia." *Pages Magazine* 24(2): 62–63.

Weeks, L.R. 2007. "Prehistoric Metallurgy at Tell Abraq, U.A.E. 1." *Arabian Archaeology and Epigraphy* 8(1): 11–85.

Willcox, G.H. 1989. "Etude archéobotanique." In *Fouilles de Shortughaï. Recherches sur l'Asie centrale protohistorique*, ed. H.-P. Francfort. Paris: de Boccard, pp. 175–185

Winkelmann, S. 2004. *Seals of the Oasis: From the Ligabue Collection*. Venice: Il Punto.

Zarzecki-Peleg, A. 1993. "Decorated Bones of the Third Millennium BCE from Palestine and Syria: Stylistic Analysis." *Israel Exploration Journal* 43: 1–22.

CHAPTER TWO

THE OXUS CIVILIZATION AND MESOPOTAMIA

A philologist's point of view

Michaël Guichard

INTRODUCTION

Do the Mesopotamian cuneiform texts mention the famous "Oxus Civilization" situated in Margiana and Bactria? Some Assyriologists thought they could identify this cultural area with place names mentioned in the Akkadian or Sumerian texts. Since contradictory opinions have been formulated on that subject and since the proposed identifications – often fluctuating ones – are not compelling, the answer could be negative. Nevertheless, the importance gained by this civilization in the last decades and the anticipation of its relations – direct or not – with Mesopotamia explain why many scholars try to find its mention in cuneiform texts. The mineral resources of the area show that it was the origin of much appreciated materials, such as lapis lazuli and tin (Lyonnet 2005; Steinkeller 2016: 129). But the question is: Were the Mesopotamians aware of its existence and was their knowledge of it sufficient for a factual notion of its territory and its people?

At first, an a priori positive answer could be given due to the discovery of a seal with an Akkadian inscription in a grave at Gonur Depe, one of the major Margiana sites (Potts 2008). The object seems to be of the end of the 3rd millennium, a period of intense contacts between Mesopotamia and its oriental neighbors. This small object is, until now, the most oriental testimony of cuneiform ever found. But the way seals could circulate is unknown, not considering the fact that this made them then devoid of their original administrative and legal function (see Lyonnet 2005: 196). Last, but not least, the clumsiness of this seal makes some think of a rather Elamite artifact (see Potts 2008: fn40), though others consider it a real Old Akkadian seal of Naram Sin's time after his administrative reforms (Pittman 2018). Anyhow, even if this discovery is certainly important by itself, it does not bring proof of a direct relation between Mesopotamia and Margiana.

The great distance between Margiana and Mesopotamia is not in itself an obstacle for movements of goods and men. The greatest difficulties were neither technical nor stemming from topographic impediments such as mountains or deserts. Camels were fundamental for the transport of goods through Margiana and Mesopotamia and they were already domesticated by the end of the 3rd millennium. By 2050 the kingdom of Ur possessed a small herd (Steinkeller 2009). Real barriers were largely human and political. Different peoples lived in the countries stretching between

Margiana and Mesopotamia, among which rose political states such as that of the Gutis by 2200 BC, Elam by 2100 and 2000 BC, or Marhaši during the whole second half of the 3rd millennium. The relations between Central Asia and the Tigris or the Euphrates plains depended on what happened on the Iranian Plateau. The formation of the Proto-Elamite kingdom in the second part of the 3rd millennium played a determining role for the development of the Mesopotamian cities and their relations with the eastern countries. Iran was essentially a real screen preventing or reducing the possibilities of direct contact between the Mesopotamians and the eastern, more remote countries. But Iran itself certainly was in tight relations with them, as shown by the number of Oxus Civilization archaeological finds discovered there.

Some Iranian states were the unavoidable commercial "partners" of Sumer, a country devoid of mineral deposits and the related places of production. Moreover, it seems that during the 3rd millennium "oriental" goods mainly come by sea, along the Indian Ocean Coast and the Persian Gulf, for the greatest profit of Sumer and its main harbors, such as Gu-abba in the South of Lagaš and Ur (Laursen and Steinkeller 2017). It is probably thanks to the opportunities offered by this maritime traffic that Meluhha (the Indus Civilization), though even more remote than Margiana and Bactria, became so prosperous, and most certainly constituted another intermediary between Central Asia and the Mesopotamians. The commercial and diplomatic networks as much as the regularity and intensity of the "business" between the main cultural areas varied a lot over time according to the historical hazards.

To these constraints other cultural and anthropological factors could be added, such as the occasional curiosity towards other peoples and unknown foreign cultures.

The origin of long-distance trade, however, is not to be found in the desire of knowing one's neighbors, but rather in the willingness to become wealthier, as only the concentration of world goods at home meant that a ruler could become powerful. This search for raw materials and luxury objects created a long-distance relational network, which led to the development of numerous settlements, the formation of elites, and even contributed to the apparition of proto-states. But the rise of exchanges also exacerbated conflicts.

Thanks to these exchanges, countries separated by huge distances were related, and may even have been partially interdependent, according to the importance this international trade had upon their own economy. It is now, however, minimized as far as Mesopotamia is concerned. Nevertheless, we cannot consider in a global way the whole Middle East and Central Asia because, clearly, communication networks were de facto segmented.

Neither the Mesopotamian nor the Oxus Civilization chronology are yet established in absolute dates with certainty. The Oxus Civilization *floruit* would be between 2300/2200 and 1800/1700 BC. At Gonur Depe, a historical break around 2000 BC has been advanced (Salvatori 2008; but see Chapter 1 for its dating rather in the 19th century BC). During the second phase it seems that the presence of nomadic populations (the so-called Andronovo culture) becomes more visible. This scenario would place the beginning of the second phase of the urban Oxus Civilization more or less at the same moment as the fall of the Third Dynasty of Ur in Mesopotamia. This event was certainly a turning point in international relations, with the end of direct trade with Oman (= Magan) and Meluhha. In Mesopotamia it marks the end of a long prosperous period when silver was abundant and semiprecious stones were exchanged in large quantities, sometimes as blocks (lagab) weighing several minas, or even more. However, international trade did not disappear, it radically changed. The relative concomitance of the observable "crises" or "transformations" in various regions of the Near East and Central Asia and Pakistan was probably no accident, but this complex question would be beyond the scope of the current chapter (on the interactions between the Arabian Peninsula and Gujarat/Indus, see Uesugi 2019: 349).

This study aims at presenting the main written information at hand coming from Mesopotamia between the end of the 3rd and the beginning of the 2nd millennium, related, directly or not, with Bactria and Margiana; this far and remote Orient that cuneiform sources possibly accounted for or simply mentioned more or less accurately.

Before entering the difficult question of the possible designation of this area in the documentation, it seems useful to examine the way the former inhabitants of Mesopotamia perceived the world to their east. The best way, it seems, is to read once more some passages of the classical Sumerian literature. This documentation to a large extent originated in the beginning of the 2nd millennium, but the context it refers to mostly dates to the previous millennium, regarded as the age of prosperity.

CLASSICAL SUMERIAN LITERATURE AND THE EAST

In these sources, the point of view is always ethnocentric. A center of gravity (be it Sumer, Akkad, Dilmun, or city states such as Lagaš, Nippur, or Babylon) prevails over countries more or less obedient, exotic, strange, rich, and even magnificent, which make its periphery in a broad sense. These surrounding countries have in common the fact that they were submitted to the world order claimed by the Mesopotamian gods, and some texts assert that the purpose of their existence is to serve them. This is the way Sargon of Akkad summarizes his grip over the traffic in the Persian Gulf (from Frayne 1993: 28):

> Sargon king of Kiš (understand: "supreme king") gained a victory on 34 cities(?), and he destroyed their bulwarks until the borders of the Sea (the Gulf). He moored to the quays of Akkad the ships of Meluhha, Magan and Tilmun.

His control of the sea trade was the consequence of his victorious war. He expressed his power by the capacity of having wealth converging towards his capital city. This idea was again mentioned in the *Curse of Akkad*, which alludes to the joyful harmony of Akkad universal dominion (Cooper 1983: 52–53):

> That foreigners would cruise about like unusual birds in the sky,
> That (even) Marhaši would be reentered on the (tribute) rolls,
> That monkeys, mighty elephants, *water buffalo*, exotic animals,
> would jostle each other in the public squares –
> (as well as) thoroughbred dogs, lions, mountain ibexes, and *alum* sheep with long
> wool, (so that all this might happen),
> Holy Inana did not sleep.
> She then filled Akkad's stores for emmer wheat with gold,
> She filled its stores for white emmer wheat with silver;

> She delivered copper, tin, and blocks of lapis lazuli to its granaries and sealed its silos from outside ...
> Inside the city were tigi drums; outside it, flutes and *zamzam* instruments.
> Its harbour where ships moored was full of joy.
> All foreign lands rested contentedly, and their people experienced happiness ...
> It was like the Tigris flowing into the sea as holy Inana opened the portals of its city-gates and made Sumer bring its own possessions upstream by boats.
> The highland Martu, people ignorant of agriculture, entered (into the temple) before her with perfect oxen and perfect kids.
> The Meluhhans, the people of the black land, brought exotic wares (var. wares of foreign countries) up to her. Elam and Subir loaded themselves with goods for her as if they were pack asses.

Similar words to those of Sargon's royal inscription mentioned above are taken up by Gudea, Lagaš's ruler, around 2100 (Edzard 1997: 42): "Thanks to the strength of Nanše and thanks to the strength of Ningirsu, Magan, Meluhha, Gubi(n) and the land of Dilmun submitted themselves to Gudea who obtained the scepter from Ningirsu. All the ships came to him to Lagaš." His B statue ("the architect") evokes this in a more general and peaceful way (as for the extract of the *Curse of Akkad* the opening of the trade routes according to god Ningirsu's will). This inscription resumes the very image of a spontaneous convergence of raw materials from the whole world to his town, Girsu, for the construction of its temple Eninnu. This long and precise description is followed however by the evocation of a victorious military campaign against Anšan and the kingdom of Elam. This version of the Cylinder A (col. xv) confirms the relation between the opening of the Gulf route and the military victory (Edzard 1997: 78):

> The Elamites came to him from Elam. The Susians came to him from Susa. Magan and Meluhha submitted themselves to him from their country. They gathered in Girsu the city of Gudea to build the temple of Ningirsu. He gave instructions to Ninzaga (= Enzag): he did have to provide Gudea, the builder of the temple, with copper as if it were simply a question of getting enormous quantities of wheat. He gave instructions to Ninsikila: she did have to provide the leader who has built the Eninnu with very large oaks, ebony and "wood of the sea."

Ninzaga and Ninsikila, the Dilmun gods, represented in this text as the Persian Gulf masters (Heimpel 1987: 45; Marchesi 2017: 432), submitted themselves to Ningirsu, perhaps as a result of the victory over Elam. The access to the oriental exotic products was made possible by a decisive military operation, the success of which was ensured by Nanše and Ningirsu according to Gudea's words. The similarity between these words and those of Ur-Namma, the founder of the Ur III "empire," who asserted that he himself overcame Elam and made the Magan and Meluhha's merchant ships come back to Sumer (in his discourse there is no mention of Dilmun Island), may be explained by the fact that both of them use the same rhetorical stock. This wording led modern scholars to the supposition that Elam's defeats, as recalled by Gudea and Ur-Namma, were in reality one and the same event (see Steinkeller 2013: 294). This ideology goes back, as mentioned above, at least to the time of Sargon of Akkad.

Thus, the appropriation of oriental, exotic products depended upon the control over Elam (Susa and Anšan) and the Persian Gulf. Moreover, within the voluminous corpus of the Sumerian literature, three works contain a list of countries that were considered as typical suppliers of wealth (UET 6.1, G hymn to Ninurta, and ISET 1.211/212).

One of the oldest is certainly the G hymn to Ninurta. The dating of this work – of which we possess only a Nippur version (around 1700 BC) – is tricky, but it was probably a third millennium composition:

> That day, thanks to the gold from Harali (Ninurta), you fully merited to become the Lord. Thanks to the carnelian and lapis-lazuli from Meluhha, you became the Lord par excellence. Thanks to the chlorite(?) from Marhaši, you became the Lord par excellence. Thanks to the silver from the fifteen cities, you became the Lord par excellence. Thanks to the copper and to the tin [from Magan] you became the Lord par excellence. Thanks to the bronze from ... you became the Lord par excellence. Thanks to the silver from Dilmun, you fully merited to become the Lord. Thanks to the limestone [extracted] from the *mouth* of the Mountain, you became the Lord par excellence. Thanks to the gypsum from the white mountain you became the Lord par excellence.
>
> (SLTNi 61, text from Nippur N2478, see Kramer 1944: pl. 30 and ETCSL, c.4.27.07)

The products concerned in this poetic inventory, on the anaphoric mode, are stones and metals. Maybe the composition is not unusual, as the well-known Marhašian origin of *duhšia* stone (chlorite, steatite, or serpentine?), a topic that appears at the time of the Akkad kings' conquests. But in many ways this text is unique. The group including Harali, Meluhha, Marhaši, the 15 cities (an oriental city confederacy or Elam itself in a broad aspect before its unification?), Magan (the mention of which is only a conjecture), and Dilmun – meaning Oman and Bahrain – certainly forms a unity, the common denominator of which could be the Persian Gulf. The joint mention of Magan(?) and Marhaši pleads for a description of the commercial routes at the time of Akkad or that of the Ur III Dynasty. Magan, indeed, does not play a central role anymore after 2000 BC. From a historical point of view, this country is eclipsed by Dilmun, but the memory of its golden age is still vivid in the school traditions. The poem makes an artificial distinction between lapis lazuli and tin, which may nevertheless have followed the same circuit from Afghanistan/Central Asia. Magan and Meluhha (another possible tin-exporting country) (Reiter 1997: 212) had no tin mines. Producing countries and intermediate ones were confused or hardly distinguished in Mesopotamia proper. This part of the hymn to Ninurta makes a probable reference to an age when Magan was a hub for copper and tin traffic. Though a very ancient emporium (Foster 2016: 113), mention of Dilmun seems to be ancillary or symbolic. Lastly, the hymn praises the origin of the materials by giving the *raison d'être* of the international trade: The ownership of this wealth allows Ninurta to exercise its lordly authority (*nam-en*) and to assert it. He concretely enjoys his prestige thanks to his ornaments (in particular the emblem of its divinity known as *suhgir*, which was supposed to be fastened around his neck), or his monuments and cultic objects.

The accumulation of goods allows him to redistribute them as he wants, a fact that reinforces his symbolic superiority.

Another list of exotic countries, still more often mentioned, appears in the famous narrative *Enki and Ninhursag*, which celebrates the holy island of Dilmun and Ninsikila, its goddess (Attinger 2015). After evoking the creation of Nin-sikila's sanctuary, its pure water, and its quay that became the "the storeroom on the quay of the Land (= Sumer)," the Ur version adds a description of the prosperous trade that came out of it (UET 6.1.1) (Gadd and Kramer 1963: pl. 2, emphasis added):

> May the land of Tukriš hand over to you (Ninsikila, goddess of Dilmun) gold *from* Harali and *beautiful* lapis lazuli!
> May the land of Meluhha send you big ships (loaded) with carnelian, object of greed so precious, the rosewood (mes šagan° for mes-wood of Magan!) and the "wood of the sea" of first quality!
> May the land of Marhaši come before you with precious stones (including) duhšia-chlorite!
> May the land of Magan send you copper, hard materials, very solid such as the diorite, the emery(?) and the šu.min-rock!
> May the land of the sea hand over to you the ebony that fits so well to the king!
> May the land (where people lived in) tents send you fat-tailed sheep of high-quality!
> May the land of Elam hand over to you yellow wool as tribute.
> May the sanctuary of Ur, the seat of the kingship (and) *the cities of Sumer*, send you big ships (loaded) with barley, oil, magnificent fabrics and high-quality clothes!
> May the abundance of the vast sea stream towards you!

This text – with Ur as the kingship seat, and the mention of Marhaši, Magan, and Elam, this last country being a mere vassal state – reflects the historical situation of the Ur Third Dynasty. Nevertheless, this passage appears only in the Ur version that was found in the house of a priest and was written during the reign of Rim-Sin, a king of Larsa (19th–18th centuries BC). It could be – according to some proposals (e.g., Komoróczy 1977) – a mere addition: either a peculiar review from the Ur tradition (because unattested in Nippur texts, see Delnero 2012: 64–73), or an individual and isolated interpolation that could be from the middle of Rim-Sin's reign. This interpolation could show not only the *chauvinism* of the scribe, a member of Nanna's temple, but also a conservative point of view, even a backward-looking one, because mention of Meluhha and Magan hark back to a bygone era. One could think that positing Dilmun as the center of this vast traffic is better suited to the economic situation at the beginning of the 2nd millennium, that of the kingdom of Larsa. The hymn illustrates well the Dilmun prosperity and that of the Larsa kingdom, which controlled Ur in the second part of the 19th century. However, Dilmun had been integrated into the Ur empire, a situation that the poem could echo here.

In this version of *Enki and Ninhursag*, the origin of lapis lazuli is not from Meluhha but from Tukriš, where the Harali gold also came from. Tukriš is also mentioned first in the Ninurta hymn (G) but without any reference to Tukriš. An earlier mention of Tukriš is found in a Sumerian poem from Nippur – generally attributed at the time of Ur III (ISET 1.211) (Çiğ *et al.* 1969: pl. 153 [no. 2126 and 4178]) – and so represents an eventual inspiration for UET 6.1: "Tukriš ..., the land that ... has brought to you gold, silver, metal kug-NE-a, lapis lazuli and a lot of great resources" (see pp.75–76). One may explain the lack of mentions of tin and silver in the Ur list quoted above by the fact that these products no longer participated in the Old-Babylonian Dilmun prosperity, nor therefore in that of Ur.

POSSIBLE CANDIDATES FOR THE OXUS CIVILIZATION

Countries such as Marhaši (*duhšia* = chlorite?), Harali (gold), and Tukriš (lapis lazuli, gold) or even the land of Arata, are still beyond our knowledge and are possible candidates for the Oxus Civilization. As mentioned above, tin and lapis lazuli should be good indicators to identify this country among the candidates. According to the narrative *Enmerkar and the Lord of Arata*, tin and lapis lazuli were celebrated as coming from Arata. But Arata was a somewhat legendary country, believed to be situated somewhere to the east of Sumer (Mittermayer 2009: 36–39). Actually, the origin of the materials is never mentioned: they could come via Meluhha (Gudea Cylinder B) or via Magan (G Ninurta hymn), that is to say through the Persian Gulf, but also via Niqqum in the Zagros Mountains, a possible stopover for tin according to a scribal letter that, doubtless, preserved part of the real situation at the beginning of the 2nd millennium.[1] During his campaign to Šimaški, Šū-Suen took away tin as booty, a fact that shows a continental route at that time (Frayne 1997: 305, RIME 3/2.1.4.3). As for lapis lazuli, which some texts associate directly with tin, it could be imported from Meluhha (G Hymn to Ninurta) or from Tukriš via Dilmun (*Enki* and *Ninhursag*). Indications about networks are present in administrative texts of the 2nd millennium, which prove the existence of a market at Dilmun. At their entrance to Ur, lapis lazuli blocks and other supplies such as *šuba* stones coming from Magan were submitted to an import tax (UET 5.292). Other centers such as Ešnunna and Susa played an even more important role. In other words, the two routes, the maritime one (Persian Gulf) and the continental one (Iran) through Susa or Ešnunna, the gateways to central Mesopotamia, still existed, even if the second one was gradually replacing the first. The fact that tin and lapis lazuli are apart in literary texts and attributed to countries where they were only goods in transit, such as Meluhha and Magan, is proof of the variety and volatility of the routes and shows the difficulty of the scribes in identifying the specific origin of such exotic supplies that were brought home. The scribes to whom we owe this information were priests or members of the administration and who are often somewhat indifferent to the way these supplies were obtained, as long as they were available. Most of the celebrated countries were those with which the palace and the temples had actual relations and the name or the reputation of which had acquired some prestige. We can presume that the travelers and merchants themselves knew more about the real situation, but this is not certain.

Between the Akkadian empire and the Old-Babylonian period, the alluvial-plain states formed complex and at times turbulent relationships with the political entities of the east. The two main states of these eastern countries were initially the remote Marhaši and Meluhha, both most certainly in direct contact with Margiana.

Meluhha

Relations with Meluhha – the Indus Civilization that flourished from around 2500 to 1900 BC – were almost primarily commercial in nature and its geographical remoteness represented an important obstacle to possibly more intense and diversified exchanges. Due to his very aggressive policy for the control of the Persian Gulf routes, Sargon and his successors provoked reactions all the way to Meluhha, the latter forming an alliance with Marhaši. According to Rimuš, this led to a confrontation between the Mesopotamian and Indus armies, although the latter had only come, together with other foreign troops, to join the Marhaši army (RIME 2.1.2.8).

During the Akkadian period, Meluhha products could have arrived by sea all the way to the port of Akkad itself. However, Magan was a compulsory transit point, which explains why, for example, in the ISET 1.212 (// ISET 1.211) Sumerian hymn they form a pair: "Magan and Meluhha have submitted to you!" As for the land of Marhaši, it also received part of the products coming from Meluhha, as concretely shown by the case of a Meluhhan "multicolored dog" (a felid, maybe a cheetah; see Potts 2002), which was part of the Marhaši tribute received by Ibbi-Sîn, king of Ur (RIME 3/2.1.5.4; UET 8.37). It may also be the case for the "Marhashian carnelian," the very name of which was introduced and preserved in a standard lexical list (Schuster-Brandis 2008: 407–409). The Meluhhan merchants and their ships could arrive directly to Sumer (at least until around 2200 BC) and occasionally go up the Tigris. While the Meluhhan ships are attested (the Sumerian expression has not yet made it possible to know whether it was a type of ship chartered by the Indus, Sumer, or some other maritime country, its mission being simply to travel to this area) (see Heimpel 1987), there is no proof that Sumerian ships and merchants had themselves reached the mouth of the Indus or the famous port of Lothal. In fact, this trade would have first and foremost been the affair of the Meluhhans themselves (Parpola 2015: 214–217). On the other hand, not only are Indus artifacts archaeologically attested in several Mesopotamian sites, but the presence of the Meluhhans is documented by texts, albeit this has to be revised as doubts have recently appeared as to the existence of a Harappan "colony" in southern Lagaš (Laursen and Steinkeller 2017). As we have seen, the products associated with those from Meluhha were likely to come from even further afield. And since close links between the Indus and Bactria have been archaeologically attested, it is hardly surprising that raw materials and artifacts originating from Central Asia should reach Sumer via the Indus valley, which, as we have seen, could explain a possible confusion with actual Meluhhan exports.

Marhaši

As for Marhaši, the country was initially one of Akkad's main eastern adversaries before finally becoming partners under the reign of Šar-kali-šarri. After the Gutian interlude, relations with the kings of Ur were restored, being both diplomatic and economic in kind. Despite their numerous military expeditions in the east, the kings of Akkad and those of Ur never ventured very far into the Iranian Plateau and never, it seems, as far as Marhaši, despite their numerous contacts (but see Potts 1999: 103). The most important of all the operations we know about so far was led by Šū-Suen and affected only the land of Šimaški bordering on the Caspian Sea (RIME 3/2.1.4.3).

Troop movements between Marhaši and Sumer have left traces in the Ur kings' archives. The network that linked Sumer with Marhaši is not clearly apparent from the written records, but it must have been mainly terrestrial and the major trade route was to pass through Susa and Anšan. An economic document from Girsu has been found that (fortuitously?) associates individuals from Marhaši, Magan, and Kimaš, giving a possible clue as to the links between Iran and the Gulf via Sumer (Sigrist and Ozaki 2013).[2] There were regular and diverse contacts between the periods of Šulgi and Ibbi-Sîn: a dynastic marriage, gift exchanges, possible military cooperation as perhaps indicated by the presence of men from Marhaši in the Ur army's ranks. The case of a Marhashian singer who went to Ur for training bears witness to the cultural exchanges and transfer of knowledge between the two entities (Koslova 2009).[3]

We do not know whether any Mesopotamian king ever planned or even tried to take over a territory with extraordinary and appealing wealth. While this did not materialize into an actual historical event, an undoubtedly Old Babylonian scribe imagined it when composing (drawing on an old legend?) a fictitious royal inscription narrating the victorious expedition of an ancient king of Adab named Lugal-ane-mundu – who is attested in the Sumerian king list that places him before Sargon (RIME 1.1.8.1). This king is divinely chosen to get back Marhaši – which has become a city in this story – to its vassal status – a theme clearly inspired by *The Curse of Akkad* cited above. Marhaši is a regional superpower counting no less than 13 vassals. No doubt, its fall allows the seizure of incredible booty that reflects its real or imaginary prosperity, but the text is incomplete.

A relative consensus has been reached around the identification suggested by Steinkeller between Marhaši and the Jiroft Civilization in the southeastern Kerman province of today's Iran, which was only discovered in the early 2000s (Steinkeller 2006; but see, e.g., Potts 1999: 103; and Potts 2002: 344). However, H.-P. Francfort and X. Tremblay (2010) have subsequently tried, in a very in-depth study, to prove that the name Marhaši is at the source of that of Margiana (Margush). While they were right to emphasize the continental character of Marhaši's power, rightly noting that communication with this country occurred primarily by land, an argument against their thesis stems from the fact that Meluhha never exported the *duhšia* stone characteristic of Marhaši, which should have been the case if it were Margiana, along with the circulation of lapis lazuli. As for Steinkeller's hypothesis, it relies mainly on the identification of the *duhšia* stone with chlorite, but this has not yet been definitively proven.

Whatever the case may be, the arrival of merchandise from this territory in Sumer was seen as a veritable sign of the economic and political power of the center. The actual economic activity tends to be ignored if not taboo in literature, which prefers to describe supplying countries as tributaries. The term for "tribute," gun_2, is actually quite vague, meaning simple merchandise that may have been meant for export. While, at the end of the 3rd millennium, Marhaši was a partner that played a decisive role in the circulation of wealth coming from the east, it could also be depicted as one of the most emblematic tributary countries. However, its name is not in any way associated in our sources with tin or lapis lazuli, therefore, it was known well enough and not considered as the starting point of these materials.

Regular contacts between Sumer and Marhaši stopped with the beginning of Ibbi-Sîn's reign, following the rise of Elam and the arrival of a new dynasty called Šimaški.

Thus, the formation of this new state that caused Ur to collapse cut off Marhaši from Mesopotamia. This must have gone along with its political decline which coincided with that of Meluhha. To this phenomenon we should add the concomitant sudden decline of Magan, which led diorite imports into Mesopotamia to dry up (Ramez 2019: 868). According to the *Lament for Sumer and Ur* (166–173), the Elamites were responsible for the cessation of trade in the Gulf (Michalowski 1989: 46–47):

> Enlil brought down the Elamites, the enemy, from the highlands. Nanše, the noble daughter, was settled outside the city. Fire approached Ninmarki in the shrine Gu-aba. Large boats were carrying off its silver and lapis lazuli. The lady, sacred Ninmarki, was despondent because of her perished goods. On that day he [Enlil] decreed a storm blazing like the mouth of a fire. The province of Lagaš was handed over to Elam. And then the queen also reached the end of her time.

Nevertheless, trading in the Gulf resumed quickly, albeit then under the domination of Dilmun (van de Mieroop 1992: 194–197). Raw materials such as tin seem to have been mainly controlled by Elam. In 1750 BC, the king of Mari was convinced that the agent he was sending to Larsa would be able to get lapis lazuli there for him (Guichard 1996). He learnt to his own cost that the only existing market was that of Susa, which controlled the tin. Thus, Elam replaced Marhaši to its own advantage and benefitted from Magan's decline.

Šimaški and Tukriš

Marhaši's identification with southeastern Iran has made possible other hypotheses on Margiana's identification.

- First, Potts (2008) has suggested Šimaški (mentioned above) as a candidate – an undoubtedly eastern territory that was subject to a powerful raid by Šū-Suen, who provides the most detailed description of it. However, the absence of this name from the list of countries known for their luxury products immediately poses a problem. It was in fact located south of the Caspian Sea. Obviously, this position would have been favorable for direct links with Margiana and Bactria, as possibly shown by the existence of the so-called gu_2-$guru_5$ animals that its elite may have owned and offered to the kings of Ur. These may have been the very first camels of Central Asia documented by cuneiform texts (Steinkeller 2009). Their Sumerian name, which is attested only during a very short period between Šulgi and Amar-Suen, may describe them as having "a nape (gu_2) that splits ($guru_5$)."
- Second, Tukriš, through a process of elimination, is one of the last plausible candidates (Fuchs 2014; Archi 2016). Steinkeller (2016: 129) has recently attempted to establish a link between it and Margiana since, as we have seen, this country was known in southern Mesopotamia to have lapis lazuli and gold, deposits of which could have been close to each other. No source has so far established a link between Tukriš and tin. Its reputation was old enough to be mentioned in the ISET 1.211 hymn, which is generally dated to Ur III (a concept that can be understood in the wider meaning established by O. Edzard, which

includes the First Dynasty of Isin). We should note that the mention of Tukriš in ISET 1.211 is a rehashing of ISET 212 (incomplete and older in style). The former can only date from the First Dynasty of Isin (around 2000–1900 BC) since the name of Tukriš is completely absent from the administrative records of Ur III. This tends to show that it could only have appeared around or after 2000 BC. The issue of its possible connection with the place name of Dugurasu found in the pre-Sargonic texts from Ebla remains at present uncertain given the gap of several centuries that separates them (Archi 2016). Beside gold and lapis lazuli, the land of Tukriš also supplied kug-NE-a: a kind of silver the identity of which is disputed and that was documented only in the 3rd millennium (Reiter 1997: 85). This list of materials is not so different from the one attributed to Meluhha. Indeed, with the exception of tin and carnelian, lapis lazuli blocks and NE-silver are, according to Gudea Cyl. B (col. 14), Meluhhan imports. It could thus be that the name of Tukriš replaced that of Meluhha at a time when the fame of the latter began to wane.

There have been several attempts to locate Tukriš, including Moorey's Luristan hypothesis (Moorey 1995). This uncertainty underlines the difficulty of the research and the fragility of the suggestions put forward. Until Steinkeller's hypothesis, the proposals at least all agreed on the region where it is supposed to have been located, namely Iran. The intrinsic fragility of these speculations comes from the fact that the supposed connection between Tukriš and lapis lazuli is based on a single text (UET 6.1) to which ISET 1.211 must now be added. However, we are not in any event dealing with primary sources. ISET 1.211 counts Tukriš as a distant land (kur-ša$_3$) comprising Magan, Meluhha, Arata, and even the heavens (utaḫ) that supply translucent lapis lazuli. As for UET 6.1, it provides two more precise clues. On the one hand, its products arrived at Dilmun via the Gulf before reaching Ur and, on the other hand, Tukriš exported the so-called "Harali" gold. According to the hymn to *Ninurta G*, Harali was a mountain famous for its metal and was placed on a par with Meluhha and Magan. In Ur III records there appears a place called Arali, which supplies Sumer with "sea wood," identification of which is uncertain (AAS 128, Grégoire 1970: 158). According to literary sources, this type of wood is supposed to come from the Indus valley. It is tempting to make a connection between the land of Harali and this rare mention of Arali, a territory no doubt located along the Indian Ocean coast more or less near the Indus valley. However, this would exclude all links with Central Asia (but see Chapters 20 and 21). It should be added that the term "(H)arali" is known in religious literature as a mythical mountain where gold ore was extracted and that, above all, corresponded to one of the names for hell. It could be that, in the poem of *Enki and Ninhursaĝ*, the Harali gold is a poetic way of referring to a type of gold famous for its quality and is ultimately a reinterpretation of the term, the geographic meaning of which was lost in the 19th century BC.

The main problem is the very mention of Tukriš in relation to Dilmun or Sumer exclusively in religious texts: in these contexts, the country should have a primarily mythical dimension. Historiography has played a decisive role in the interpretation of the data, leading to an overestimation of a classic literary source that had the advantage of being "Babylonian."

However, it is only in the information from the "periphery" that clues about the historical Tukriš can be found. Indeed, it is only in Assyrian sources from the 19th century BC that Tukriš enters history as we currently know it. The most important piece of information comes from Samsi-Addu, the king of Ekallatum, who had taken control over upper Mesopotamia and obtained tributes from the kings of Tukriš and a mysterious "king of the upper land" (= Zalmaqqum?) (see Grayson 1987: 50).[4] This statement, which he inscribed in foundation documents belonging to the temple of Assur, also mentions, without making an explicit connection with the previous information, the abundance of riches he introduced in the temple, such as gold, silver, lapis lazuli, and carnelian. This discourse merely recycles and adapts well-rehearsed themes of the 3rd millennium royal ideology. It is clear that, at the time of the Old-Assyrian trade, Tukriš (= Anatolia?) was seen as a land of plenty, bursting with wealth, which spontaneously submitted to the conquering Mesopotamian king just as Magan and Meluhha had done in their own time in relation to Sargon, Gudea, or Ur-Namma. It is obviously a similar theme that the story of *Enki and Ninhursaĝ* echoes. However, in the geopolitical context of the kingdom of upper Mesopotamia, it seems improbable that the Tukriš of Samsi-Addu would refer to East Asia (Charpin 2004: 190–191).

It is only in the Mari palace texts, most of which date to the early 18th century BC, that more precise details can be gleaned about Tukriš. The Mari kingdom had not established direct links with this territory no doubt because of the distance that separated them and, above all, because Tukriš was not part of its network of relations. However remote it may have been, a visit made by individuals coming from Tukriš is nevertheless attested, not to mention a letter fragment showing that a Mari agent was able to personally go there (Guichard 2012: 25–29). However, it was first and foremost the artifacts of this country or its natural products such as timber that made their way into the coffers of the Mari king (Guichard 2005: 320–324). Most of them were obtained by means of diplomatic gifts from intermediary countries. In this way, Tukriš supplied wood, vases made of the so-called Tukriš stone, silver objects such as beakers in the shape of a bull's head, ivory, or faience artifacts. These Tukriš goods generally came from the west, either Aleppo or Carchemish. There is no concrete relationship between Tukriš and gold, and even less so with lapis lazuli, since the latter explicitly came from Susa all the way to Mari.

Thus, the reason why the author of *Enki and Ninhursag* associated the name of Tukriš with this stone remains a mystery. It is not an invention as ISET 1.211 tends to show. In any event, the mention of Tukriš betrays the fact that, at the time these works were written (i.e., between 2000 and 1900 BC) – and UET 6.1 too dates from the same period as that of Samsi-Addu – Tukriš was a country internationally famous for its products such as its stones and artistic works. For the inhabitants of Sumer, the increasingly popular land of Tukriš was such a distant region that it is possible they believed it to be vaguely neighboring Meluhha and Marhaši. We are forced to conclude that the Tukriš of Sumer does not correspond to that of upper Mesopotamia. But the records at our disposal do not allow us to know what its precise historical substance was.

CONCLUSION

This leads us to the inevitable conclusion that the Mesopotamian elites were not aware of the society that developed in the Oxus area. This ignorance is partly connected to the reality of trade networks and a lack of interest in countries overseas, especially on the part of the literate classes and administrations. In the 18th century BC, a Bedouin leader claimed that merchants were able to travel everywhere, even in wartime (Sasson 2015: 7; Durand 2018: 72–75). Reality, though, was very different or, at least, more complex. Trade networks were very well organized and sometimes subject to monopolies. Mesopotamian traders seem to have rarely ventured away from Mesopotamia. In the Gulf, we do not know whether they ever went any further than Magan. At the beginning of the 2nd millennium, the horizon even tended to narrow down, with Dilmun representing at that time a limit beyond which the ships of Ur did not dare travel. Mesopotamian merchants used to buy goods in Ešnunna and went as far as Niqqum, which was further east, or even all the way to Susa. But there must have been very few who went further than that.[5] In Anatolia, on the other hand, at least from the beginning of the 2nd millennium, Assur merchants ventured all the way to the heart of Anatolia in search of silver and thus established a complex network of trading posts that gave them a trading monopoly well beyond their home. The Assyrians themselves, however, were prevented from exploiting other areas such as the Konya plain region (Larsen 2015: 148). Nevertheless, the treaties they made with the Anatolian cities show that they could occasionally pursue an aggressive policy towards their Akkadian (= Babylonian) competitors (Veenhof 2008: 211). In the 18th century BC, the Mari king who wanted to obtain luxury goods from the king of Kaneš had no choice but to go through Assyrian merchants who actually duped him (Guichard 2008: 50–52).

To sum up, knowledge about Margiana and its brilliant civilization could only have been indirect. The Mesopotamians could only have known about it through people from Meluhha and Marhaši in the 3rd millennium, and then through Elamites in the 2nd millennium. They may have indirectly learnt about it through the arrival of exotic animals such as Bactrian camels or the importation of artifacts representative of the art nowadays recognized as being "Intercultural" in nature. But texts alone do not allow us to identify it.

ABBREVIATIONS

AAS	Archives Administratives Sumériennes
CDLI	Cuneiform Digital Library Initiative (https://cdli.ucla.edu)
ETCSL	Electronic Text Corpus of Sumerian Literature (http://etcsl.orinst.ox.ac.uk)
ISET	Istanbul arkeoloji müzelerinde bulunan Sumer edebi tablet ve parçalari [Sumerian literary tablets and fragments in the Archaeological Museum of Istanbul]
PPAC	Periodic Publications of Ancient Civilizations
RIMA	Royal Inscriptions of Mesopotamia: Assyrian Period, Volume 1
RIME	Royal Inscriptions of Mesopotamia: Early Periods

SLTNi Sumerian Literary Texts from Nippur
UET (Serie) Ur Excavations Texts

NOTES

1 Letter from Ilak-nu'id; cf. Kleinerman and Gadotti (2014) and Guichard (2016: 31).
2 PPAC 5 59 (= BM 12878).
3 See CDLI: P212186.
4 RIMA 1, A.0.39.1.
5 At Konar Sandal South in Kerman, one impression of a city seal from Ur on a door sealing (ca. 2900 BCE) points at a southern Mesopotamian presence at that time on the site, but there is no such evidence dated to the Akkadian or Ur III periods (Pittman 2018).

REFERENCES

Archi, A. 2016. "Egypt or Iran in the Ebla Texts?" *Orientalia* 85: 1–49.

Attinger, P. 2015. "Enki und Ninchursaĝa." In *Erzählungen aus dem Land Sumer*, ed. K. Volk. Wiesbaden: Harrassowitz, pp. 5–20.

Charpin, D. 2004. "Histoire Politique du Proche-Orient Amorrite (2002–1595)." In *Mesopotamien. Die altbabylonische Zeit*, ed. P. Attinger, W. Sallaberger, and M. Wäfler. Orbis Biblicus et Orientalis 160/4. Friburg/Göttingen: Academic Press/Vandenhoeck & Ruprecht, pp. 25–480.

Çiğ, M., H. Kizilyay, and S.N. Kramer 1969. *Istanbul Arkeoloji müzelerinde bulunan. Sumer edebî tablet ve parçalari*, vol. I. Türk Tarih kurumu yayinlarindan, VI. seri, Sa. 13. Ankara: Türk Tarih Kurumu Basimevi.

Cooper, J.S. 1983. *The Curse of Agade*. Baltimore, MD and London: Johns Hopkins University Press.

Delnero, P. 2012. *The Textual Criticism of Sumerian Literature*. Journal of Cuneiform Studies Supplemental Series, No. 3. Atlanta, GA: American School of Oriental Research.

Durand, J.-M. 2018. " Le commerce à longue distance de Mari d'après les textes de Mari du règne de Zimrî-Lîm." *Pasiphae* XII: 69–87.

Edzard, D.O. 1997. *Gudea and His Dynasty: The Royal Inscriptions of Mesopotamia, Early Periods*, vol. 2. Toronto, ON: University of Toronto Press.

Foster, B.R. 2016. *The Age of Agade: Inventing Empire in Ancient Mesopotamia*. London and New York: Routledge.

Francfort, H.-P. and X. Tremblay 2010. "Marhaši et la civilisation de l'Oxus." *Iranica Antiqua* 45: 51–224.

Frayne, D. 1993. *Sargonic and Gutian Periods* (2334–2113 BC): *The Royal Inscriptions of Mesopotamia, Early Periods*, vol. 2. Toronto, ON: University of Toronto Press.

Frayne, D. 1997. *Ur III Period (2112–2004 BC)*: *The Royal Inscriptions of Mesopotamia Early Periods*, vol. 3.2, Toronto, ON: University of Toronto Press.

Fuchs, A. 2014. "Tukriš." In *Reallexikon der Assyriologie und Vorderasiatischen Archäologie*, ed. Bruno Meissner. Berlin: De Gruyter, pp. 174–175.

Gadd, C.J. and S.N. Kramer 1963. *Ur Excavations Texts VI, Literary and Religious Texts: Part I*. London: Trustees of the Two Museums.

Grayson, A.K. 1987. Assyrian Rulers of the Third and Second Millennia BC *(to 1115 BC)*. Royal Inscription of Mesopotamia, Assyrian Periods, vol. 1. Toronto, ON: University of Toronto Press.

Grégoire, J.-P. 1970. *Archives Administratives Sumériennes*. Paris: Geuthner.

Guichard, M. 1996. "À la recherche de la pierre bleue." *Nouvelles Assyriologiques Brèves et Utilitaires* 36: 30–32.

Guichard, M. 2005. La Vaisselle de luxe des rois de Mari, Matériaux pour le Dictionnaire de Babylonien de Paris 2. Archives Royales de Mari, t. XXXI. Paris: Editions Recherche sur les Civilisations.

Guichard, M. 2008. "Naḫur et la route des marchands assyriens à l'époque de Zimrî-Lîm." In *Anatolia and the Jazira During the Old Assyrian Period (*OAAS *3)*, ed. J.G. Dercksen. Publications de l'Iinstitut Historique-Archéologique Néerlandais de Stamboul 111. Leiden: Nederlands Instituut voor het Nabije Oosten, pp. 43–53.

Guichard, M. 2012. "Relations entre Carkémish et Mari: nouveaux fragments." *Semitica* 54: 25–29.

Guichard, M. 2016. "Guerre et diplomatie: lettres d'Iluni roi d'Ešnunna d'une collection privée." *Semitica* 58: 29–31.

Heimpel, H. 1987. "Das Untere Meer." *Zeitschrift für Assyriologie* 77: 22–91

Kleinerman, A. and A. Gadotti 2014. "Two Sumerian Literary Letters About Merchants." *Nouvelles Assyriologiques Brèves et Utilitaires* 65: 73–78.

Komoróczy, G. 1977. "Tilmun als 'Speicher des Landes' im Epos 'Enki und Ninhursag.' " *Iraq* 39: 67–70.

Koslova, N. 2009. *Ur III- Texte der St Petersburger Ermitage III*. Unpublished manuscript, State Hermitage Museum.

Kramer, N. 1944. "Sumerian Literary Texts from Nippur in the Museum of the Ancient Orient at Istanbul." In *The Annual of the American Schools of Oriental Research*, vol. 23. New Haven, CT: American School of Oriental Research.

Larsen, M. T. 2015. *Ancient Kanesh: A Merchant Colony in Bronze Age Anatolia*. New York: Cambridge University Press.

Laursen, S. and P. Steinkeller 2017. *Babylonia, the Gulf Region and the Indus*. Winona Lake, IN: Eisenbrauns.

Lyonnet, B. 2005. "Another Possible Interpretation of the Bactro-Margiana Culture (BMAC) of Central Asia: The Tin Trade." In *South Asian Archaeology 2001*. Vol. 1, Prehistory, ed. C. Jarrige and V. Lefèvre. Paris: Editions Recherche sur les Civilisations, pp. 191–200.

Marchesi, G. 2017. "Appendix 5: Inscriptions from the Royal Mounds of A'ali (Bahrain) and related texts." In *The Royal Mounds of A'ali in Bahrain. The Emergence of Kingship in Early Dilmun*, ed. S.T. Laursen. Aarhus: Narayana Press, pp. 425–437.

Michalowski, P. 1989. *The Lamentation Over the Destruction of Sumer and Ur. Mesopotamian Civilizations*. Winona Lake, IN: Eisenbrauns.

Mittermayer, C. 2009. *Enmerkara und der Herr von Arata: Ein ungleicher Wettstreit*. Orbis Biblicus et Orientalis 239. Friburg/Göttingen: Academic Press/Vandenhoeck & Ruprecht, pp. 26–39.

Moorey, P.R.S. 1995. "The Eastern Land of Tukrish." In *Beiträge zur Kulturgeschichte Vorderasiens: Festschrift für Rainer Michael Boehmer*, ed. U. Finkbeiner, R. Dittman, and H. Hauptmann. Mainz: von Zabern, pp. 439–449.

Parpola, A. 2015. *The Roots of Hinduism: The Early Aryans and the Indus Civilization*. Oxford: Oxford University Press.

Peterson, J. 2015. "The Fat-Tailed Sheep (Gukkal) in *Enki and Ninhursaĝa* B II 10." *Notes Assyriologiques Brèves et Utilitaires* 4: 5.

Pittman, H. 2018. "Administrative Role of Seal Imagery in the Early Bronze Age: Mesopotamian and Iranian Traders on the Plateau." In *Seals and Sealings in the Ancient World: Case Studies from the Near East, Egypt, the Aegean, and South Asia*, ed. M. Ameri, S.K. Costello, G. Jamison, and S.J. Scott. Cambridge: Cambridge University Press, pp. 13–35.

Potts, D.T. 1999. *The Archaeology of Elam: Formation and Transformation of an Ancient Iranian State*. Cambridge World Archaeology. Cambridge: Cambridge University Press.

Potts, D.T. 2002. "Total Prestation in Marhashi-Ur Relations." In *Iranica Antiqua*, vol. 37, ed. D.T. Potts. Leiden: Brill, pp. 343–357.

Potts, D.T. 2008. "Puzur-Inšušinak and the Oxus Civilisation (BMAC): Reflections on Šimaški and the Geo-Political Landscape of Iran and Central Asia in the Ur III period." *Zeitschrift für Assyriologie* 98: 184.

Ramez, M. 2019. " 'Tu représenteras parfaitement la force de mon héroïsme!' La diorite comme vecteur de l'idéologie royale au pays de Sumer et d'Akkad (IIIe-IIe millénaires av. J. C." In *De l'argile au numérique. Mélanges assyriologiques en l'honneur de Dominique Charpin*, vol. 2, ed. G. Chambon, M. Guichard, and I. Langlois. Publications de l'Institut du Proche-Orient Ancien du Collège de France 3. Leuven-Paris-Bristol: Peeters, pp. 841–877.

Reiter, K. 1997. *Die Metalle im Alten Orient unter besonderer Berücksichtung altbabylonischer Quellen*. Alter Orient und Altes Testament, Band 249. Münster: Ugarit Verlag.

Salvatori, S. 2008. "Cultural Variability in the Bronze Age Oxus Civilization and Its Relations with the Surrounding Regions of Central Asia and Iran." In *The Bronze Age and Early Iron Age in Margiana Lowlands*, vol. II, ed. S. Salvatori and M. Tosi. BAR International Series 1806. Oxford: Archaeopress, pp. 75–98.

Sasson, J. 2015. *From the Mari Archives: An Anthology of Old Babylonian Letters*, Winona Lake, IN: Eisenbrauns.

Schuster-Brandis, A. 2008. *Steine als Schutz- und Heilmittel*. Alter Orient und Altes Testament, Band 46. Munster: Ugarit-Verlag.

Sigrist, M. and T. Ozaki 2013. *Administrative Ur III texts in the British Museum*. Periodic Publications of Ancient Civilizations 5. Shangchun: Institute for the History of Ancient Civilizations, Northeast Normal University.

Steinkeller, P. 2006. "New Light on Marhaši and Its Contacts with Makkan and Babylonia." *Journal of Magan Studies* 1: 1–17.

Steinkeller, P. 2009. "Camels in Ur III Babylonia?" *Exploring the Longue Durée. Essays in Honor of Lawrence E. Stager*, ed. J. David Schloen. Winona Lake, IN: Eisenbrauns, pp. 415–418.

Steinkeller, P. 2013. "Puzur-Inšušinak at Susa: A Pivotal Episode of Early Elamite History Reconsidered." In *Susa and Elam: Archaeological, Philological, Historical and Geographical Perspectives*, ed. K. de Graef and J. Tavernier. Mémoires de la Délégation en Perse, vol. 58. Leiden and Boston: Brill, pp. 293–317.

Steinkeller, P. 2016. "The Role of Iran in the Inter-Regional Exchange of Metals: Tin, Copper, Silver and Gold in the Second Half of the Millennium BC." In *Ancient Iran: New Perspectives from Archaeology and Cuneiform Studies*, ed. K. Maekawa. Ancient Text Studies in the National Museum, vol. 2. Iran and Japan: National Museum of Iran and Kyoto University, pp. 127–150.

Uesugi, A. 2019: "A Note on the Interregional Interactions Between the Indus Civilization and the Arabian Peninsula During the Third Millennium BCE." In *Decades in Deserts: Essays on the Near Eastern Archaeology in Honour of Sumio Fujii*, ed. S. Nakamura, T. Adachi, and M. Abe. Japan: Rokuichi Syobou, pp. 337–355.

van de Mieroop, M. 1992. *Society and Enterprise in Old Babylonian Ur*. Berliner Beiträge zum Vorderen Orient, Band 12. Berlin: Dietrich Reimer Verlag.

Veenhof, K.R. 2008. "The Old Assyrian Period." In *Mesopotamia. The Old Assyrian Period*, ed. M. Wäfler. Orbis Biblicus et Orientalis 160/5. Friburg/Göttingen: Academic Press/ Vandenhoeck & Ruprecht, pp. 13–264.

CHAPTER THREE

ENVIRONMENTAL CHANGES IN BACTRIA AND SOGDIANA (CENTRAL ASIA, AFGHANISTAN, AND UZBEKISTAN) FROM THE NEOLITHIC TO THE LATE BRONZE AGE

Interaction with human occupation

Eric Fouache, Lucie Cez, Valérie Andrieu-Ponel, and Rocco Rante

INTRODUCTION

The Holocene climate change involving a general aridification and the timeline of its fluctuations are well established in Central Asia (Staubwasser and Weiss 2006; Chen *et al.* 2008). The Early Holocene climate was dry, when it changed to wetter or less dry from the Early to Mid-Holocene and finally moderately wet in the Late Holocene period (Mayewski *et al.* 2004; Chen *et al.* 2008). These rapid climatic changes or "RCC events" have been identified in the northern hemisphere in many fossil sequences (Bond *et al.* 1997; Mayewski *et al.* 2004). They led to significant environmental modifications and had a socioeconomic impact on ancient societies (Weiss and Bradley 2001; Weiss 2016), especially during the Neolithic and Bronze Age. However, while the correlation between abrupt climatic changes and collapse of civilizations is a fact for some authors (e.g., Weiss *et al.* 1993; Possehl 1997; Cullen *et al.* 2000; DeMenocal 2001; Madella and Fuller 2006; Kaniewski *et al.* 2013; Welc and Marks 2014), it is less obvious for others (e.g., Coombes and Barber 2005; Knapp and Manning 2016). During the last few years, the relation between climate change and societal evolutions and the causal links between paleoclimates, environments, and culture changes have gained a new interest (Butzer 2012; Tang *et al.* 2013; Kuzucuoğlu and Tsirtsoni 2016).

In southwestern arid Central Asia, the strong dependance of societies on water resources is well illustrated by the intertwined spatial distribution of ancient settlements and paleochannels as well as by the importance of runoffs for their subsistance, mainly provided by irrigated agriculture (Lewis 1966; Andrianov 1985, 2016; Francfort and Lecomte 2002). In this context, two phenomena should be considered that had a possible impact on the modifications in the settlement pattern: a decrease in the river flows and a change in the river courses. In order to render the spatio-temporal river course modifications, our research has focused on two deltaic floodplains, the Balkh River in northern Afghanistan and the Zeravshan River in Uzbekistan (Figure 3.1). What is the impact of aridification on the annual discharge of these rivers? What are the other factors controlling the changes of river courses?

The avulsion process is one of them, diverting flows from previous channels and leading to the abandonment of all or part of a meander belt (Allen 1965, 1978). It is a natural process in response to a variety of factors, such as topography, floodplain slopes, water table level, vegetation cover, and runoff energy. Some of these factors, exacerbated in the frame of a climatic change, result in an increase of the sedimentation rate, a change in the water discharge peak, and the degradation of the vegetation cover to varying extents according to the specific features of each river catchment (Mohrig *et al.* 2000). Avulsion is also an answer to exterior disruptions like local tectonics promoting a rise of the flow base level and obstructions of the drainage valley (Törnqvist 1994; Jones and Schumm 1999; Korup 2004; Stouthamer and Berendsen 2007; Strom and Abdrakhmatov 2009; Wang *et al.* 2015). In Central Asia, tectonics have been demonstrated as a major agent of fluvial instability during the Quaternary (Létolle and Mainguet 2004), and more light can be shed on other phenomena induced by seismicity such as dammed lakes linked to rockslides in the upper part of mountainous fluvial valleys or outburst floods (Strom 2013).

Our research[1] has focused on the lower deltaic part of the Balkh River near the Aqcha oasis (Afghanistan) and on that of the Zeravshan River, near the adjacent Bukhara and Qaraqöl oases (Uzbekistan). Both are situated on the desertic Turan plain. Considering the sparse water resources and the sensitive environments of arid southwestern Central Asia, fluvial metamorphose is a step in the hydro-climatic and environmental degradation induced by aridification and tectonic activity. We argue that the geographical allocation of the successive paleochannels as well as their dating and the characterization of their flow dynamic through time make it possible to settle back the successive human occupations in their climatic context since they depended, for a significant part, on water availability. This, therefore, should enable us to understand their co-evolution. In order to explore and reproduce the river changes during the Holocene, we applied a geoarchaeological approach based upon the diagnostic role of fluvial geomorphology, optically stimulated luminescence (OSL) dating, sedimentology studies, and their correlation with archaeological data. These two case studies show that channel avulsions are paleoenvironmental clues that reflect the evolution of human occupation in regions later known as Bactria and Sogdiana.

MORPHO-TECTONIC AND ENVIRONMENTAL SETTINGS IN SOUTHWEST CENTRAL ASIA

The Balkh region in northern Afghanistan

The Balkh plain is bordered to the north by the Amu Darya River and to the south by the limestone ridges of the Albruz Mountains (1572 m asl), which are the western end of the Hindukush mountain range (Figure 3.2). The Balkh plain is a subsided basin, edged to the south by a southeast/northwest overlap scarp and infilled by coarse debris and Quaternary alluvium. The Balkh River has shaped a vast cone between Mazar-i-Sharif

Figure 3.1 The studied areas (Balkh plain and lower Zeravshan).

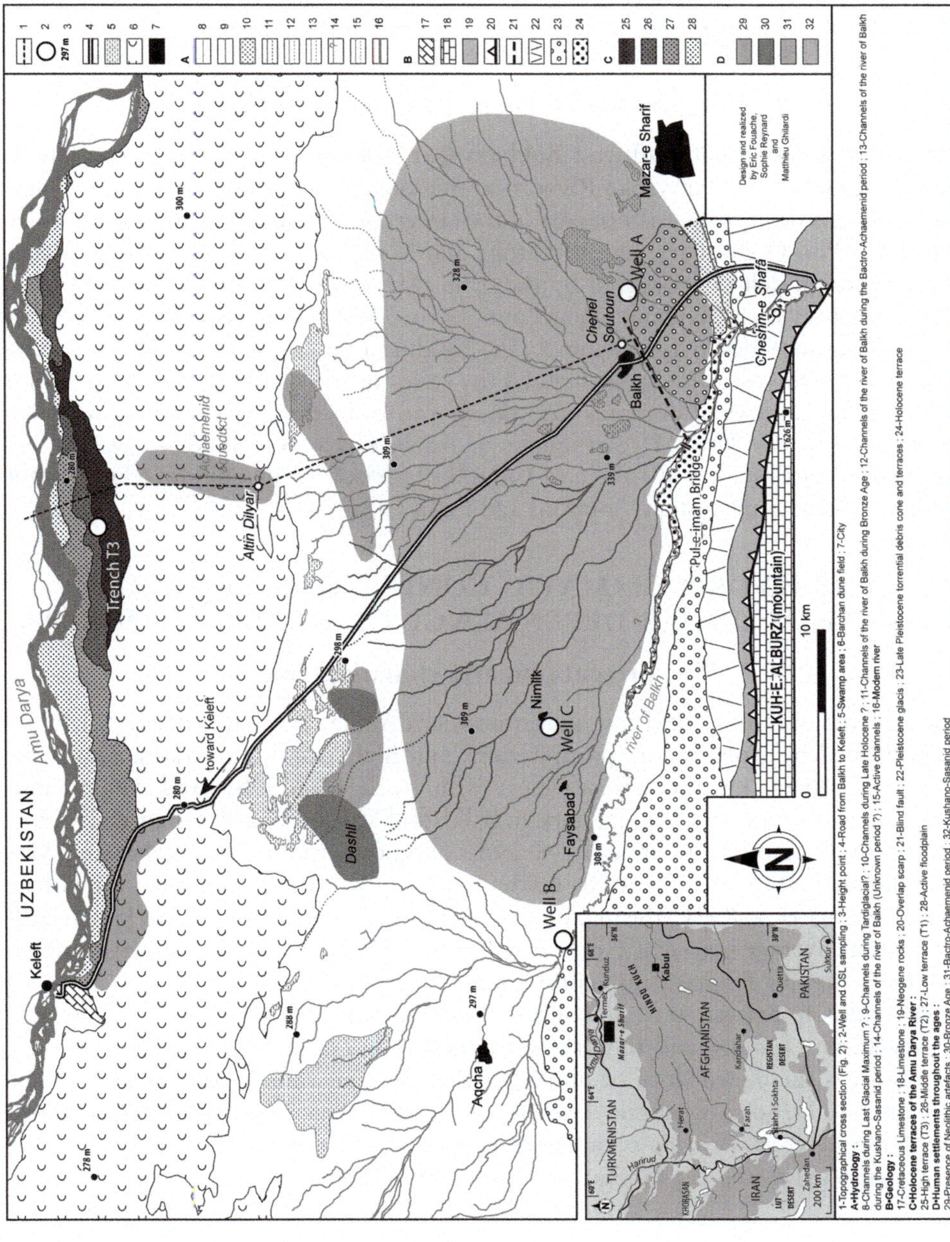

Figure 3.2 Geomorphological map of the Balkh plain (northern Afghanistan).

and Aqcha. Northward, the alluvial cone is bordered by a 10 km wide desert strip up to the Amu Darya and, beyond the river, is connected to the arid Turan plain. The Balkh region is a tectonically active area located in front of a fault propagation and folds controlled by the Albruz Marmul fault system (Ruleman *et al.* 2007) and related to the ongoing Indian/Eurasian plates collision. This is illustrated by the 918 AD Balkh earthquake (Ambraseys 1961) and the recent 1948 Mazar-i-Sharif 6.3 Mw earthquake.

The Zeravshan delta in southeastern Uzbekistan

The adjacent northeast–southwest Bukhara-Qaraqöl oases extend on 5,100 km² in the deltaic end of the Zeravshan River in southeastern Uzbekistan (Figure 3.3). These alluvial plains occupy the eastern part of the larger Turan plain, and actively subsided during the Plio-Quaternary period (Létolle and Mainguet 2004). The southern edge of the Qaraqöl delta is bordered by the Amu Darya River in which the Zeravshan River was previously flowing (Bensidoun 1979; Kohl 1984). To the west lies the Kyzyl Kum sandy desert. To the north, the Kulduktu range corresponds to an active southeast–northwest slip fault flanked by two north–south overlapping fronts in a tectonic context of compression (Bossu 1996). In this region, active tectonics are attested by the 818 AD earthquake at Gazli, a city near Bukhara, which still experienced earthquakes in 1976 and 1984, the latter of 7.0 magnitude related to the Bukhara-Hissar fault system (Simpson and Leith 1985).

THE HYDRO-CLIMATIC CONDITIONS: SEMIARID LOWLANDS UNDER DEPENDENCE OF WATER RESOURCES

The present semiarid climatic frame

Our research area in Uzbekistan and in northern Afghanistan encompasses semiarid lands defined by globally low precipitations and high evaporation rates. The Balkh plain experiences hot and dry summers with +31.4°C in July,[2] and cold winters between −1.5°C and +9.1°C in January. In Balkh and Mazar-i-Sharif, the annual rainfall is respectively about 180 mm and 190 mm, concentrated between October and May with maximal falls in March. In the Bukhara oasis, July is the warmest month with an annual average of +28.8°C, the maximal temperature reaching +36.5°C. In January, the temperature average is +0.8°C. At the fringe of the Kyzyl Kum Desert, annual rainfall is about 156 mm for Bukhara, also concentrated between October and May.

The rainfall pattern (i.e., low levels and inter-annual variability), determines the vegetation scarcity and the dry ecosystems. The consequence of these climatic conditions is the dependence of human settlements on water resources and annual river discharges. Socioeconomic development is partly based on irrigated agriculture. The dense network of canals implemented nowadays in the Balkh plain (Fouache *et al.* 2012: 3417) and the Bukhara oasis illustrates this point.

Rivers and hydrological data

The Balkh River (460 km long) flows from the north side of the Hindukush, which ranges up to 4,700 m asl and drains a catchment area of 18.700 km². It is a perennial

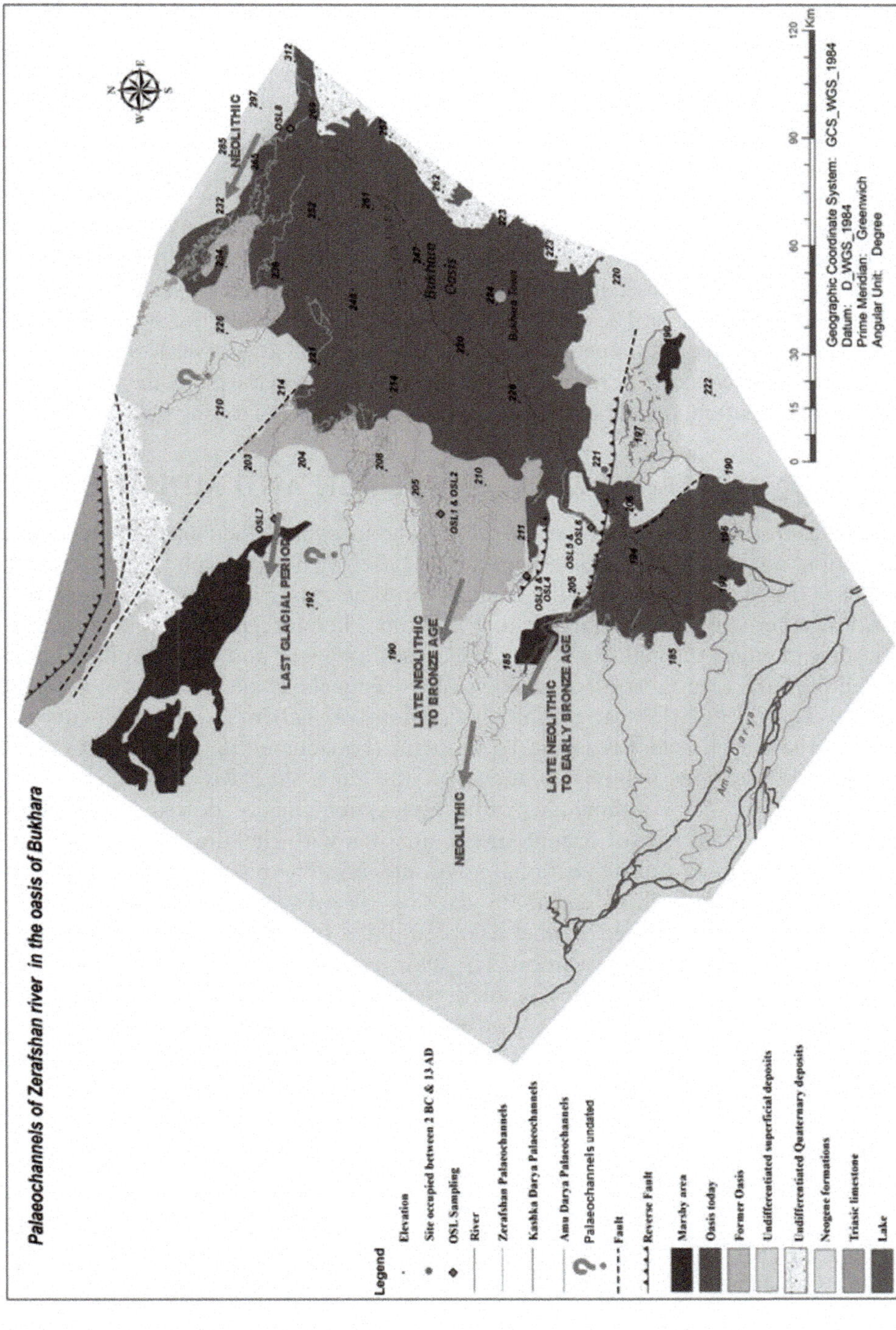

Figure 3.3 Paleochannels of the Zeravshan River in the Bukhara oasis (Uzbekistan).

river with 49.2 m^3/sec of annual discharge.[3] Coming from the high-altitude Hindukush range, it is fed both by snow and by a significant rainfall amount of 300 to 400 mm/year (Humlum 1959). It has a high water level from the beginning of spring until the end of summer, between March and July. As it opens in the Turan plain, the main channel of the Balkh River flows westward and branches into four arms ending in an alluvial fan around the Aqcha oasis. It disappears in the sandy desert, which borders the left bank of the Amu Darya.

The Zeravshan River (741 km long), the *Polytimetos* of the Greeks, constitutes the backbone of ancient Sogdiana. It flows in an east–west direction in Tajikistan and the Uzbekistan territories. Its sources are in the Pamir-Alai Mountains and it covers a 143,000 km^2 drainage basin. It is a perennial snow- and ice-fed river originating in the Zeravshan glacier, with an annual discharge of 170 m^3/sec[4] and a high water level between May and the end of September. The snow melting occurs from the outset of spring followed by the glacier melting in summer (Suslov 1961), with a discharge never less than 30 m^3/sec, despite the irregularity of rainfall amounts. When it enters the plain, the river flow is diverted for irrigation, mainly in the Samarkand and Bukhara farming oasis.

GEOARCHAEOLOGICAL APPROACH AND METHODS

In order to document the changes of the river courses and their impact on human occupation, our geoarchaeological approach is based on a geomorphological study. It comprises the identification of paleochannels and the dating of their implementation period together with the alluvial fans they built up. Therefore, it is possible to identify the course changes through the Holocene. Spatial imagery and topographical maps have allowed the identification of paleochannels (Fouache *et al.* 2012: 3418; Fouache *et al.* 2016: 154, fig. 4). Stratigraphic description was performed on sediments visible in modern well trenches in the Balkh plain (Fouache *et al.* 2012: 3423–3424) and from different fossil fluvial branches of the Zeravshan River (Fouache *et al.* 2016: 157–159). Each sedimentary unit was systematically described and bulk sediments were collected for granulometric analysis with the aim of characterizing the alluvial dynamics and depositional environments. The results of sedimentology analyses were linked with paleoclimatic data for the western centrasiatic region. In parallel, each identified paleochannel has been dated by OSL in order to place it in the frame of the floodplain evolution. The OSL method is based on the measure of time elapsed since the last exposure of the sediments to daylight (Aitken 1985).[5] The sampling strategy and laboratory protocol have been detailed in previous papers (Fouache *et al.* 2016: 155–156; Zink *et al.* 2017 for the Zeravshan paleochannels;[6] Fouache *et al.* 2012: 3419–3420 for the Balkh paleochannels).

On the whole, the geomorphological map presents the chronological steps of the floodplain evolution with the main paleochannel diversions and settlement displacements through the Holocene.

CONTINUITY OF HUMAN OCCUPATION IN THE TWO ALLUVIAL PLAINS OF BACTRIA AND SOGDIANA

In semiarid Central Asia, the development of past societies was closely related to water resource availability and to hydraulic works for irrigation (Gardin and Gentelle

1976, 1979; Gentelle 1989; Francfort and Lecomte 2002). Archaeological data shows a continuous human occupation within the deltaic plains as shown by a number of research studies in southwestern Central Asia (e.g., Lisitsyna 1965; Gubaev *et al.* 1998; Rouse and Cerasetti 2017 for the Tedjen and Murghab deltas; and Brunet and Debaine-Francfort 2012; Brunet *et al.* 2013 for the lower Zeravshan areas).

Chronology of settlements in the alluvial plain of the Balkh River

Along the Amu Darya riverbanks, only prehistoric occupation is documented (Vinogradov 1979). Except for this sandy zone, in the Balkh plain no remains earlier than the Bronze Age have been discovered up to now. Those of the Dashly oasis (Sarianidi 1976, 1977; Kohl 1981) are now dated between the 3rd and the first half of the 2nd millennium BC, but occupation continued there until Antiquity. These sites are located to the north of the contemporary oases of Aqcha, Fayzabad, and Balkh, about 10 km from the southern current end of the Balkh River channels near the village of Mardiana. They form two groups of oases: the western group (Dashly 1 to 6 and 9 to 14) and the eastern one (Dashly 7, 8, and 14 to 17) all related to the Oxus Civilization. The Dashly sites belong to a settlement network extending from west to east in northern Afghanistan (e.g., Daulatabad, Nichlin, and Farukhabad), which altogether represent a significant Bronze Age occupation (Kohl 1984), and that continued later on. Bactria was to become a rich and famous country during the successive Bactrian (10th–6th century BC), Achaemenid (6th–4th century BC), Hellenistic (3rd–2nd century BC), and Kushan periods (1st–3rd century AD). It regained its importance during the Early Islamic period (9th–12th century AD) and under the Timurids and their successors (15th–17th century AD).

The Zeravshan River delta occupation

In the lower Zeravshan area, a Neolithic occupation is especially related to the Kel'teminar culture well known around the Aral Sea and extending up to the Ajakagytma Lake (northwest of Bukhara), the Kyzyl Kum Desert, and in the Kum Sultan region in the eastern part of the Bukhara-Qaraqöl oasis (Gul'jamov *et al.* 1966: 19, fig. 3; Brunet 2005, 2011; Brunet *et al.* 2013). Gul'jamov's studies in the southwest of the Bukhara oasis and at Zamanbaba have documented a Chalcolithic and Early Bronze Age culture based upon farming and cattle breeding (Askarov 1965, 1981; Gul'jamov *et al.* 1966: 17–20, figs. 3 and 4). Zamanbaba settlements and cemeteries have also brought to light remains of the Tazabag'jab steppe Bronze Age culture, related to a southern branch of the Andronovo complex. However, the Zamanbaba culture chronology is unclear, and its dates have widely fluctuated. As for Bactria, later occupations are attested and Bukhara became one of the highlights of Sogdiana during the Early Middle Ages and during the Early Islamic period up to the Mongol conquest.

These brief overviews on both the Balkh and Zeravshan deltaic floodplains highlight a continuity of human occupation since the Neolithic period and attest to undeniable relationships between the sedentary settlements and the fluvial network. Indeed, deltas floodplains constitute valuable and attractive environments due to their higher rainfall regimes and humidity rates than in the surrounding deserts. *Tugaï* thickets and marsh patches areas also provide vegetal and animal resources.

A complete understanding of the archaeological and historical evolution of these endorheic deltas requires the reconstruction of the appearance of rivers and floodplains throughout the centuries.

HOLOCENE CLIMATE FRAME FOR CENTRAL ASIA

The Holocene, dated from 11500 cal. year BP to today, experienced several "rapid" climatic changes (RCC) (Denton and Karlén 1973; O'Brien *et al.* 1995; Bond *et al.* 1997; Mayewski *et al.* 1997, 2004). The "RCC events" refer to six cold intervals of low amplitude with a 1,450-year periodicity related to global cooling anomalies during the last Glacial period and extending up to historical time (Weninger *et al.* 2009). Mayewski *et al.* (2004) identified RCC events from more than 50 fossil sequences over the planet. These events took place at 9–8 ky, 6–5 ky, 4.2–3.8 ky, 3.5–2.5 ky, 1.2–1 ky, and 600–150 cal. year BP. Most of them are associated to high-latitude cooling and low-latitude tropical aridity. In Central Asia, paleoclimatic studies demonstrate a variability in rainfall regime at different scales through the Holocene with a clearly dry climate during the Early Holocene (11–8 ky), a humid climate during the Mid-Holocene (8–5 ky), and a moderately wet climate for the Late Holocene (Table 3.1).

Holocene climatic phases in Central Asia

The Early Holocene, between 11 ky and 8 ky, was a dry period in mid-latitude Central Asia (which includes the arid Xinjiang region of China) due to a widespread and severe desertification (Feng *et al.* 2006; Chen *et al.* 2008; Jin *et al.* 2012). Before 8 ky, this dry climate corresponds to a change in the westerlies regime due to a diminution of ice sheets in low latitudes and orbital forcing decrease in the North Atlantic Ocean (Jin *et al.* 2012).

The climate of the Mid-Holocene period extending between 8 ky and 5 ky is characterized by a greater humidity culminating around 6 ky: it corresponds to the Holocene Climatic Optimum (HCO). Wet conditions are related to an increase of precipitation volume, which led to the almost total disappearance of desert landscapes in the temperate zone. Due to regional geography, significant time gaps exist at the start and at the highest level of these wettest conditions. In the western lowlands of Turkmenistan and Uzbekistan, this humid period corresponds to the "Ljavljakan Pluvial phase," which occurred between 8 and 4 ky, with a maximum around 6.5 ky (Vinogradov and Mamedov 1975; Lioubimtseva *et al.* 1998). Varushchenko *et al.* (1987) argued that the annual rainfall average was up to 300 mm, thus allowing a forest development along river valleys. In the northern hemisphere, the Mid-Holocene climate was also globally warmer (Ilyashuk *et al.* 2005; Braconnot *et al.* 2012; Liu *et al.* 2014). Paleoclimatic reconstructions indicate that the air temperature during the summertime in the interior of the western part of Central Asia was 2°C to 6°C warmer than at present (Joussaume *et al.* 1999). These warmer and wetter conditions favored a subhumid Artemisia and Graminea steppes in the Kyzyl Kum Desert (Tarasov *et al.* 1998). In the Aral Sea basin, steppe conditions developed during the "Drevnjaja Aralskaja phase" and superseded former desert landscapes. In the Caspian Sea basin, forest replaced shrublands from 8.8 to 4.4 ky (Leroy *et al.* 2014).

Table 3.1 Synthetic table of climatic evolution during the Holocene in arid Central Asia

Period	*Ages cal. year BP*	*"RCC events" in northern hemisphere*	*Paleo-moisture levels in Central Asia*	*Climatic phases in Turkmenistan and Uzbekistan*	*Landscapes and vegetation*	*Lakes chronologies (AS: Aral Sea; CS: Caspian Sea)*
Younger Dryas	12800 – 11500		Cold and dry climate	Djanak arid phase	Steppe dry vegetation	(AS) Paskevich phase (12–9 ky) regression of the lake (CS) low level before 8 ky
Early Holocene	11000 – 8000	9–8 ky	Dry to extremely dry climate	Warmer conditions	Desert vegetation with regional timing lags	(AS) Drevnye Aral'skaja phase (9–5 ky): warm and wet climate for the basin with increased water level of the lake ("Great Aral Sea stage"); (CS) low level before 8 ky
Middle Holocene	8000 – 5000	6–5 ky	Wet or less dry climate (HCO)	Ljavljakan pluvial phase (8–4 ky and culminating at 6.5 ky)	Subhumid steppe in deserts (ergs) and forest vegetation along the river valleys	(AS) high level of the lake (CS) high level culminating at 6 ky and forest vegetation
Late Holocene	5000 – 2000	4.2–3.8 and 3.5–2.5 ky	Moderately wet climate	Short-term arid periods known: 4.8 to 4.4; 3.5 to 2.9; 2.5 to 2.2		(AS) Pozdnye Aral'skaja phase (3–2 ky) regression of the lake level (CS) low level
Last two Millennia	From 2000	1.2–1 ky and 600–150 y	Slightly wet climate	Short humid phases in western Turkmenistan from 9th–14th century AD		(AS) "Kerderi Regression": the lowest level during 12th–15th centuries AD Numerous cold and arid climatic phases

Adapted from Lioubimtseva *et al.* 2005; data are from Vinogradov and Mamedov 1975; Varushchenko *et al.* 1987; Lioubimtseva *et al.* 1998; Tarasov *et al.* 1998; Boomer *et al.* 2000, 2009; Krivonogov *et al.* 2010; Mayewski *et al.* 2004; Chen *et al.* 2008.

After the prevalence of humid climatic conditions in Central Asia during the Mid-Holocene period, especially between 6 and 5 ky, the subsequent drying trend is a response to reducing summer insolation and the gradual reduction of the mid-latitude westerlies intensity (Feng *et al.* 2006; Chen *et al.* 2008; Lauterbach *et al.* 2014). Several lakes of the arid and semiarid China have their highest levels until 5 ky before a clear decrease after 4 ky (An *et al.* 2006; Wünnemann *et al.* 2006; Hartmann and Wünnemann 2008). This moderately humid climate of the Late Holocene (it remained higher than during the Early Holocene) was interrupted by a pronounced dry interval between 4.9 and 3.9 ky in southeast Kyrgyzstan and northwest China (Beer *et al.* 2007; An *et al.* 2012; Lauterbach *et al.* 2014). After 3.9 ky, a general cooling trend associated with more humid conditions is observed in overall western Central Asia (e.g., Mischke *et al.* 2010; Aichner *et al.* 2014). The rest of the Late Holocene period is characterized by a moderate and gradual drying of climate until the present and again interrupted by short-term warmer and drier spells, mainly from 4.8 to 4.4 ky, from 3.5 to 2.9 ky, and from 2.5 to 2.2 ky (Tolstov and Kes 1960). Aichner *et al.* (2014) give 3 to 2.7 ky, 2.5–1.9 ky, and 1.4–0.6 ky warm/dry intervals.

To sum up, the Turanian lowlands and deserts experienced multiple shifts from hyper-arid conditions to subhumid shrublands correlated to moisture-level fluctuations throughout the Holocene (Varushchenko *et al.* 1987; Kes *et al.* 1993; Tarasov *et al.* 1998).

THE BALKH RIVER FLOODPLAIN RECONSTRUCTION SINCE THE PLEISTOCENE

Reconstruction of the timeline for the hydrographic changes in the Balkh River delta is based on the identification and the dating of fossil paleochannels. In the Balkh plain, a series of coalescent fans corresponds to an alluvium accumulation from the foot of the Albruz Mountains in the Hindukush to the Amu Darya River since the end of the last Ice Age (see Figure 3.2). These fans, created by the channels activity, are located at the top of an older Pleistocene cone.

Alluvial cone generations and fluvial networks

The Pleistocene cone is made of coarse alluviums, which have been incised by several paleochannels probably at the beginning of the post-glacial warming period, when fluvial beds were stabilized. The cone apex underwent a movement about 30 km westwards and progressed further to the west until it reached its current position, the upstream Aqcha. As suggested by Fouache *et al.* (2012: 3418), this shift is probably due to a thrust fault northeast–southwest activity joining the Balkh city and the epicenter of the 1948 Mazar-i-Sharif earthquake, located 15 km northeast of the city (Ruleman *et al.* 2007: 26, fig. 11a). Four alluvial fans were identified from west to east and are characterized by different fluvial networks. For each network, alluvial deposit sequences were identified in wells (wells A, B, and C), described, and sampled for luminescence dating. The aim is to reconstruct the different timelines of the operating alluvial network generations.

The first network flows to the west and opens onto the Fayzabad village. At the Aqcha fan apex, 100 m away from the right bank of the Balkh River, well B presents a 5.5 m high alluvial sequence composed of fine and coarse deposit layers recording recurrent flood events from the Balkh River.

The second fluvial network, flowing northwest, was occupied at its terminal end by the Bronze Age Dashly sites, dated from the late 3rd to the middle of the 2nd millennium BC, suggesting an operating hydrographic network during the Bronze Age. Later Iron Age up to antique period sites are also present, which could have been supplied either by natural streams or by artificial canals. Well C, located in the village of Nimlik, shows a 9.5 m high sequence made only of fine deposits.

The northward termination of the third network corresponds to a densely occupied area during the Iron Age (Bactrian and Achaemenid periods). Small archaeological sites testify of agriculture activity along these paleochannels and the remains of an artificial aqueduct in the vicinity of the Altyn Dilyar fortress further north, in the current *barkhan* desert area, confirms the availability of water at that time.

Two main channels encircling the ancient city of Balkh make the fourth northeast hydrographic network. Excavations at Chehel Soutoun revealed an eastern paleochannel dated to the Kushan period (1st–3rd centuries AD).[7] Well A, located 4.5 km east of Chehel Soutoun, shows a 6 m high sequence of alternating silty and clayish deposits with a few sand beds.

Paleochannels dynamic and human occupation

The Bronze Age sites, in the Dashly area to the south of Keleft, occupy the second fluvial network flowing northwest. It seems that, at that time, agriculture was limited to an irrigated strip along the active fluvial channels of the Balkh River. The Balkh River moved eastward during the Iron Age. As mentioned above, archaeological evidence shows that Bronze Age sites were reoccupied during the Iron Age within this second fluvial network. The 1st millennium BC settlements extended further east, over the third fluvial network, but also occupied the upper part of the second fluvial system. During the Iron Age, the hydraulic engineering mastery allowed irrigation to develop over the whole Balkh plain, exploiting both active and abandoned channels. The construction of a 4 m wide aqueduct has been identified near the Altyn Dilyar fortress only along 3 m (its total length is unknown), and its extension north towards the Amu Darya, testify of the development of irrigation at that time (Fouache *et al.* 2012: 3423, fig. 7). This suggests that the continuous human settlements attested in this area were based upon increasing agricultural production. This could be related to the impressive defensive system extending from the Cheshme Shafa fortress, on the southern foothills, to the Altyn Dilyar citadel, at the edge of the northern dune zone. This defensive system was probably created to protect the extensive irrigation network.

Later (i.e., during the Kushan period – first half of the 1st millennium AD), the active fluvial system shifted eastward, forming a new fluvial network (the fourth described above) on both sides of the city of Balkh. It was maintained until the Early Islamic period, as shown by the excavations at Chehel Soutoun. At that time, archaeological settlements are located along the active channels of the Balkh River but also in the vicinity of abandoned ones. Hellenistic blocks reused along a channel edge, bridge pillars found *in situ*, as well as several meters of sediments deposited at this location, prove that the channel was in use at least from the 4th century BC until the ancient Islamic period. Within this area, we note an unequal erosion dynamic as shown in the upper part of the well A stratigraphic sequence (i.e., 2 m high of alternating silty and clayish deposits taking place in 6.1 +/– 0.6 ky (pA OSL-2)), while several meters of sediments were deposited in the channel at the location of the bridge pillar in just a few centuries. This point suggests both an extreme spatial mobility of the fluvial channels network during the Holocene and a significant variability in the runoff regime of the streams.

Finally, a significant change of the Balkh River is observed, from its previous eastern position towards the west, flowing until Fayzabad and in the Aqcha oasis (the first network described). Luminescence dates obtained in well B for the most western network range from 6 to 0.4 ky and is younger than those from wells C and A, respectively dated from 13 to 10 ky and from 21 to 6 ky. The silty deposits at the base of well B were buried around 5.8 ky (pB OSL-1), whereas the top coarse deposits, characteristics of flood dynamics, were buried at around 0.7–0.4 ky intervals (pB OSL-2, pB OSL-3, pB OSL-5). This channel was then active at around 6 ky and dried up until it became reactivated during the last millennium. The last change in flow direction probably took place during the Early Islamic period. Archaeological artifacts from the Early Islamic period and later are only found in and around Aqcha (Ball 1982). To sum up, this irrigation network flourished even more during the Kushan period and this state persisted up to the Islamic era.

To conclude, our research proves that the active channel network was spatially mobile and shifted from east to west. Tectonics could have initiated such drastic changes in the drainage direction. Unequal heights of sedimentary sequences also suggest avulsion phenomena. Finally, fan-head entrenchment is a common process linked to avulsion and resulting from the internal adjustment of the river to a new slope gradient (Bridge 2003).

EVOLUTION OF THE ZERAVSHAN RIVER IN THE BUKHARA-QARAQÖL OASIS

The Zeravshan built up successive coalescent alluvial fans since the last Ice Age. The Neogene plateau that separates the Bukhara plain from the Qaraqöl one is occupied by a 5 km wide and mostly north-south oriented paleovalley. A 80 km progressive westward shift of the active fan's apex is observed throughout the Holocene. The treatment[8] of the Landsat ETM+ image enabled the identification of eight paleochannels, which were digitalized. In the field, six paleochannels have been studied thanks to trenches where stratigraphic studies and sampling for sedimentological and OSL dating could be realized. We argue that the location of the human settlements indicates a period when the channels were operating and can be used for the relative dating of some generations of paleochannels. Thus, our geochronological approach is based on data from the archaeological map and supplemented by OSL absolute dating on paleochannels sediments (Zink *et al.* 2017).

Mapping and dating paleochannels

As shown in other studies that have investigated the past human–environmental relationship in arid Central Asia (e.g., Cattani *et al.* 2008; Tang *et al.* 2013; Rouse and Cerasetti 2017), we consider that the geographical position of human settlements in relation to the ancient watercourses can help in addressing the socio-environmental interactions relative to water resources through time. The correlation between the successive human occupation and the paleochannels' operating periods has shed light both on the geomorphological evolution of the Bukhara-Qaraqöl plains since the end of the upper Pleistocene and on the evolution of the Holocene occupation.

In the upper part of the alluvial fan of the Zeravshan River, to the north of Bukhara, an east–west paleochannel has been identified near Vabkent and Zandana Tepe. The OSL sample (OSL 7) performed in the bottom pebble layer corresponds to 29.2 ky (i.e., the last Ice Age). There is no corresponding archaeological settlement in this area. The active channel presented high-energy flows that conveyed coarse deposits (pebbles and sands) identified at the bottom, while fine deposits could be related to low-energy dynamics like overflows in the riverbank or floodplain. The diminishing grain size of the deposits observed in this sequence can be interpreted as a decrease in the intensity of the runoff, or can also suggest a spatial mobility of this fluvial branch.

Still north of the Bukhara oasis, another paleochannel oriented southwest–northeast has been identified downstream a bifurcation of the Zeravshan River. The 6.58 ky age (OSL 8) obtained for this paleochannel shows that it was active during the Neolithic period. At the southern edge of the Bukhara oasis, another paleochannel was active during the Neolithic as shown by the 9.15 ky (OSL 3) and 7.89 ky dates (OSL 4), and characterized by a low-energy fluvial dynamic. The fine and laminated deposits of this section (Tarab trench) indicate a weak energy flow. In this western part of the oasis, a series of Neolithic sites have been discovered in the terminal part of the Zeravshan to the north,[9] as well as remains of the Kel'teminar culture along this latter southern paleochannel.[10]

In the southwestern part of the oasis, an area explored during the Sovietic period (Gul'jamov *et al.* 1966) and more recently by the Italian Archaeological Mission directed by C.S. Antonini, surveys have highlighted human occupation related to the Neolithic and Bronze Age (Cerasuolo 2009: fig. D48). The paleochannel close to the Romitan and Bashtepa sites group was active between 6.2 (OSL 1) and 4.97 ky (OSL 2). The stratigraphy corresponds to a low-energy alluvial sequence covered by massive wind-deposit layers. At the same time, the paleochannel located a few kilometers to the north of the east–west reverse fault that marks the northern edge of the Qaraqöl oasis was active. OSL dating performed in this trench provided consistent ages of 5.99 ky (OSL 5) and 5.82 ky (OSL 6), corresponding to the Late Neolithic and the Bronze Age. It should be noted that remains of both cultures were found along the channel crossing the so-called "Zamanbaba area" and further to the west (Gul'jamov *et al.* 1966: figs. 3–4). The Zamanbaba trench, located upstream from the paleochannel branch, presents the same stratigraphy as the previous one, made of fine sediments featuring a very weak and decreasing fluvial dynamic. Mineral grain examination of the top massive layer (which is similar in color and grain size to that of the top Tarab branch sequence) revealed mat to blunt quartz grains, suggesting an aeolian origin for these sands. It should also be noted that the stratigraphic contact between the silty-clay and sandy layer with the top sandy aeolian layer presents an irregular and distorted surface. A liquefaction process on nonsaturated to semi-lithified deposits is involved in this type of soft sediment deformation (SSD). Nevertheless, more investigations should be carried out in order to confirm and characterize the deformation process at a microscopic scale, but also to establish the origin of this phenomenon. Indeed, the closeness of the water table to the surface and the seismic origin of this soft deformation cannot be asserted at this stage,[11] even if earthquakes are a possible triggering cause, especially given the faulted area in proximity.

Paleocourse generations and human settlement chronology

Except for the northern paleochannel dated to 29.2 ky (OSL 7) and related to the last Ice Age, all the paleochannels of the Zeravshan River were dated within the Holocene period, between 9.15 and 4.89 ky. The active fluvial channels and human settlements correspond to two major periods: the Neolithic and the transition from Neolithic to the Bronze Age. The Neolithic period is represented by two active fluvial channels on the northern part of the oasis (OSL 8) and on the southwestern part of the Bukhara oasis with the Tarab branch (OSL 3–4). In both areas, the Kel'teminar Neolithic culture was identified. The period ranging from the Late Neolithic to the Bronze Age is documented by two active paleochannels within the inhabited western areas of Zamanbaba and Bashtepa. These two Zeravshan branches were active between 6.66–4.89 ky (OSL 1–2) and 5.99–5.82 ky (OSL 5–6). Given the uncertain margins around 10% for OSL results (Zink *et al.* 2017), it remains difficult to estimate the operating time for these channels. Nevertheless, they do not seem to be active more than one to two millennia. All the studied stratigraphic sequences – except the oldest and the far-upstream OSL 7 one – are produced by weak-energy alluvial dynamics with a silty-deposit enrichment characteristic of a downstream end for the channels' course. The aeolian sands that cover all these sequences – up to 60 cm thick for the Zamanbaba trench – prove the inactive state of these different fluvial branches.

Our reconstruction at the oasis scale shows a shift of the main active channel of the Zeravshan River, from an active alluvial fan centered on the Bukhara oasis to a more recent one centered on the Qaraqöl oasis. The last major shift of the Zeravshan delta and its final orientation towards the southwest is established to the 4th century BC according to the archaeological data obtained on the sites of Pajkend and Iskijat, both located on a natural branch of the main course of the Zeravshan River.[12] The oldest archaeological sites within the Bukhara and Qaraqöl oasis are dated to the 3rd and 2nd centuries BC, and therefore suggest that the Qaraqöl oasis has been active since the 4th century BC. The reason why the inner oasis was occupied so late might rest in the marshy nature of this territory, driving the population to settle mainly at the end of channels. From this period onwards, urbanization and demographical growth depended on several factors, among which migration waves coming from the northeast, but probably also from the southeast. The urbanization process may have led to the opening of the northeastern branch of the "Silk Road," relating Merv directly to Bukhara and avoiding Bactra.

TOWARDS A BETTER UNDERSTANDING OF THE HOLOCENE SOCIO-ENVIRONMENTAL DYNAMICS IN SOUTHWEST ARID CENTRAL ASIA

The research we led in southwest arid Central Asia demonstrates the extreme mobility of the Balkh River and the Zeravshan River[13] and emphasizes the avulsion process. Understanding the triggering mechanisms of the changes in the river course remains a challenge. We can, however, discuss the probable factors initiating the avulsion process on the basis of the scientific data obtained in both study cases (i.e., climatic change involving drought and tectonic activity).

Holocene climate change and the avulsion process

Even if climate change in arid Central Asia is still poorly understood, and if all the earth surface processes involved are insufficiently documented (Lioubimtseva *et al.* 2005; Chen *et al.* 2008), these climatic fluctuations certainly influenced ecosystems and their resources and generated changes in the river dynamics. A significant part of the drainage network extended in currently desert environment, as shown for the Zeravshan River on the outskirts of the Bukhara oasis, and this demonstrates the existence of more humid conditions during the HCO, until ca. 5 ky (Joussaume *et al.* 1999; Chen *et al.* 2008). After this warmer and wetter climate phase in Central Asia, the climate evolution in western Central Asia for the Late Holocene period is more complex. While western Central Asia was subjected to a global drought associated with a cooler climate, many lakes of high altitude still had high water levels until 4 ky. Such is the case, for instance, of high-altitude lakes in eastern Pamir and Tian Shan (Xinjiang, China) under the westerlies circulation influence. Ricketts *et al.* (2001) have shown that the Issyk Kul' lake (Kyrgyzstan) had a high water level from 4.9 ky due to a more humid period, reflecting an enhanced influence of westerly winds bringing winter storms. Mischke *et al.* (2010) have determined a high water level for the Karakul' lake (Tajikistan) as a result of warm climatic conditions from 4.2 to 3.5 ky generating ice melting. In this region, the wet phase seems to be characterized by late-winter and spring precipitation (Aichner *et al.* 2014). At the southern rim of the Ferghana valley (Kyrgyzstan), the environmental conditions of the Late Holocene, inferred from a cave stalagmite record at ca. 1,500 m asl, highlight more complex events with a wet climate around 4.2 ky followed by an abrupt shift to drier climate between 4 and 3.5 ky (Wolff *et al.* 2017). In general, climatic changes could have an impact on river dynamics, especially on allogenic rivers originating in remote regions with more favorable climatic conditions. The case is illustrated by the Aral Sea endorheic lake situated in a more continental part of Central Asia but supplied by the ice- and snow-fed Syr Darya, which experienced a high water level before 3.5 ky (Boomer *et al.* 2000). At ca. 3.5 ky, the onset of significant drier conditions in the mountainous regions are inferred on the basis of the reduced meltwater discharge into Lake Karakul' and into the Aral Sea region (Mischke *et al.* 2010: 22).

The impact of these climatic changes are likely to be seen in hydrographic network modifications (bifurcations and abandoned meanders), or changes in the hydrological regime implying local runoff decrease or flood increase.

In the Balkh plain, the geomorphological dynamics of the vast torrential debris cone reflect the climatic evolution since the Last Glacial Maximum, 20,000 to 15,000 years ago, when morphoclimatic conditions had increased the river flow, leading to a braided system implementation at the scale of the cone. These cold and dry conditions – known as the "Djanak arid phase" during the Younger Dryas period in Uzbekistan and Turkmenistan – stimulated mechanical erosion on the rock outcrops at the southern mountainous edge of the Balkh plain. It provided the coarse materials of the vast debris cone. The stratigraphic sequences of wells A and C reflect the deposition conditions during these cold and dry periods, when active channels were presumably characterized by weak to middle energy and when abundant fine-sized grain sediment was transported. During the Tardiglacial period, from 14,000 to 11,000 years ago, the incremental rise of temperatures induced an extension of the vegetation cover, which contributed to stabilize the slopes and therefore reduced the sediment amount transferred downstream by the river. As a consequence, the active channels incised their fluvial beds in the surface of the debris cone remains from the Tardiglacial period, while further new alluvial fans were built up downstream.

During the Holocene, warmer conditions probably had an impact on the hydrological regime of the river. The southern mountainous range, less prone to erosion, provided a limited volume of clastic materials downstream. Indeed, the luminescence dating from the sequences of wells A and C show the Balkh River alluvium only reached 6 m high during the Holocene, and is dominated by fine sediments. Our stratigraphic and sedimentological study of the Balkh River paleochannels revealed significant variations in the flow regime. Deposition conditions[14] in wells A, B, and C correspond to more or less regular flood episodes, which could correspond to shifts in active channels of the river. The decrease of runoff intensity observed in well A, as well as the top of the sequence composed by a 1 m layer of homogenous silts, and the fluvial dynamic reactivation of the upper part of the well B sequence, all support the hypothesis of shifts in the Balkh's channels. This is also visible in the high variation between the sediment height in well A (6 m) and at Chehel Soutoun (12 m). Nowadays, the Balkh River has sometimes torrential runoffs and disastrous floods[15] that occur when snowmelt is well advanced, especially after heavy rainfall in March/April (Shobair and Alim 2004). Particularly important floods may have caused successive shifts of the channels on the cone's surface and thus created several hydrographic network generations.

For the Zeravshan delta, the fluvial branches dated from the Late Neolithic and Bronze Age show low-energy and decreasing runoffs, gradually becoming inactive. The climatic aridification process could have resulted in shifts of these river courses, but also in modifications of the hydrological regime of the rivers depending on the cryosphere (snow and glaciers).

Due to the climatic warming occurring now in Central Asia and the rapid recession of glaciers,[16] a significant increase of the annual discharge of the Zeravshan flow regime has been observed during the last decades (Olsson *et al.* 2010: 195). The Zeravshan streamflow is closely influenced by rainfall, snowmelt, and glacier melt in the high mountain regions, as shown recently by discharge peaks in 2002 and 2003. Exceptional floods caused by outbursts from damned lakes should also be considered in tectonically active high mountain regions, where mass movements caused by earthquakes are frequent (Havenith *et al.* 2003). Climatic and geomorphological conditions, among which the snow and glacier meltwater in the high mountains basin drainage with steep slopes weakened by seismic shocks, played a primary role in favor of a mass debris flows outbreak. In the Zeravshan valley, a rock debris flow blocked the entire valley width near the town of Aini in 1964, creating a damned lake with a cumulative discharge of both the Zeravshan River and its Fandarya tributary (Strom 2013: 23–25). This author states that much larger prehistoric debris flows had preceded the 1964 one. The breaking of temporary dams and the release of enormous quantities of water and debris generate significant modifications for streamflow (aggradation, avulsions) and over long distances downstream (Strom 2010).

The role of tectonic activity on fluvial network organization

Except for exceptional floods, tectonic activity probably played a significant role in the long-term network organization of channels. This fact is particularly visible in the Zeravshan delta, where a global shift of channels due to successive avulsions was conducted by neotectonic deformation (active land subsidence) and an uplift movement during the Quaternary (Bossu 1996). Such a geotectonic process is also observed in Turkmenistan on the southern edge of the Turan plain with a regional Holocene tilt (Thomas *et al.* 2009). It caused hydrological shifts of both the Murghab and Tedjen Rivers, but unevenly at the Murghab fan scale (Rouse and Cerasetti 2017: 3–4). The active northeast–southwest thrust fault is suggested as a factor that initiated the Balkh River shifts and, generally, the fan-head entrenchment corresponds to a classic process of avulsion caused by a new slope gradient at the cone surface. Such a regularization process of the runoff base level may have been emphasized by the thrust fault activity and the slope's instability on the edge of the cone.

The role of tectonics as a major agent of fluvial instability during the Quaternary has already been demonstrated (e.g., Nikonov and Penkov 1973; Dodonov 1980; Létolle and Mainguet 2004). Our research brings back an increased interest to regional tectonic paroxysms, climatic fluctuations, and their coupled impact on the geography of active channels that trigger the variability of the water resources.

River course changes and human settlements

To address the question of the impact of river course changes on past human communities, the relation between archaeological settlements and water resources (paleochannels) is considered. The characterization of the spatial position and the operating period of paleochannels, as well as their hydrological regime, contribute to document this relation at any given time.

In the Zeravshan delta, Neolithic human concentrations have been identified mainly along and at the end of active channels, where diversified fauna and flora *tugaï* corridors are attested. The Neolithic human occupation matches with the more humid period documented in the Russian literature as the "Ljavljakan pluvial phase" between 8 and 4 ky. Maman *et al.* (2011) confirm the stabilization of the sandy cover of the Kara Kum and Kyzyl Kum ergs from 7 to 5 ky. During this Mid-Holocene wet phase, a shrubby vegetation growth is observed in Margiana, to the north of the Murghab alluvial fan (Maman *et al.* 2011). In the southwestern and western part of the Zeravshan delta, Bronze Age occupations have been identified at the end of the fluvial system and along the active channels. In the inner oasis, there are no remains dated before the 4th century BC. This questions the hydrological conditions in the uninhabited inner oasis. Before the 4th century BC, it was probably a marshy land due to the proximity of the Zeravshan and Kashka Darya Rivers and therefore not attractive for settlement. The transformation during the 4th century BC fixes the last change of the Zeravshan delta and its final current orientation towards the south-west. More suitable conditions from the 3rd and 2nd centuries BC are suggested by archaeological data.

For the Balkh River, the same occupation is observed for the Bronze Age periods with the concentration of settlements at the end of paleochannels. The dense sedentary occupations in the Dashly oasis during the Bronze Age settled along the paleochannels banks that extended further north but also their floodwater terminal sectors, where marshy lands are nowadays attested.[17] Generally in southern Bactria, settlements fall into dense clusters in the terminal parts of the deltaic floodplains with the Daulatabad,[18] Dashly, and Farukhabad[19] sites and, further to the west, with the Iron Age Tillja Tepe[20] site, where Bronze Age artifacts have also been identified. Cultivated fields should be situated in the terminal areas of channels or immediately on the riverbank of these channels and should probably be irrigated by canals. Artificial canals – riverbeds artificially modified and man-made linear canals – have been observed in Margiana in the Murghab delta and related to Bronze Age sites, but their function for irrigation is not ascertained (Cremaschi 1998: 17). Moreover, numerous *takyrs* (ancient small flat areas with stagnant water) located in the humid periphery of deltas and at the desert fringe, also testify of seasonal settlements as shown by ceramic scatters on their clayish surfaces (e.g., *takyrs* to the west of Dal'verzin[21] in northern Afghanistan; see Sarianidi (1977: 28)). A similar occupation scheme has been established for the Tedjen oasis, where human occupation in the so-called "Geoksjur" oasis was concentrated on the edges of the main branches of channels within the delta and in the foothill zone. But there, numerous irrigation canals are dated to the 3rd millennium BC (Lisitsyna 1965).

The transition towards the drier climatic conditions of the Late Holocene is featured by an aridification that started 5,000 years ago. This period is a complex one with a moderate wet climate and more arid oscillations. In this context, the "rapid" changes depend on the type and time of the hydrological response in regard to the characteristics of the drainage basin. At first, the aridification generates more water resources upstream due to the river dependence on snow amounts and the extension of glaciers. The greater water inflow also generates rapid avulsions, as our research documents. However, the sharp character of these modifications remains difficult to apprehend. In the Zeravshan delta, the sedimentary sequence of the Bashtepa paleochannel (OSL 1–2), contemporaneous with the Late Holocene period, reveals an incremental decrease of runoff energy over two millennia, although the Zamanbaba branch (OSL 5–6) shows a more rapid decline. These observations demonstrate that the delta shrinkage due to the climatic drought is an uneven process through time and space due to avulsions.

The location of the settlements, for the Neolithic and Bronze Age, illustrates their dependence to water streams for domestic and agricultural uses. General reconstructions for both the Balkh and Zeravshan Rivers show that the mobility of the active channels matches with the population displacements. This is particularly clear for the Balkh River through the Holocene, with the actual foundation of the Aqcha oasis dated only from the 9th century AD after a shift of the Balkh River. For the Qaraqöl delta, the opening of a northward new Silk Road itinerary corresponds to the deviation of the Zeravshan River in an east–west direction (Rante *et al.* 2016).

CONCLUSION

In the lowlands and deserts of southwestern Central Asia, the Holocene climatic evolution is characterized by alternating periods of cold-arid, warmer and wetter semiarid, and finally dryer climate from 5 ky onwards. The environmental conditions of the human societies that inhabited and exploited this dryer environment are, however, complex to document, as they require a thorough review of the geo-hydrological processes implied in the variability of water resources over time and space. Our research in northern Afghanistan and Uzbekistan between 2009 and 2016 highlights the significant mobility of the Balkh and Zeravshan river courses in their two deltaic floodplains, related to successive avulsions under both climatic and tectonic control. These paleodrainage systems, located in a transition area between the mountains and the arid Turan plain, constitute sensitive sectors to understand both the modalities of sedentary occupation and the paleoenvironmental constraints. On a regional scale, these deltaic floodplains concentrate a dense and continuous occupation of human settlements throughout the Holocene period. At a local scale, the people favored settlements along active channels and their terminations, although some variability in ecological conditions must have existed within these deltas. Paradoxically, the drought led to greater volumes of surface water being carried by perennial snow- and ice-fed rivers. Exceptional floods have probably generated multiple avulsions in turn. In this active alluvial context, the adaptation capacity of Neolithic and Bronze Age populations is demonstrated by their successive displacements in different parts of the deltas.

NOTES

1 Our research was carried out within the MAFouz-Protohistory (Uzbek-French Archaeological Mission, Protohistory), DAFA (French Archaeological Delegation in Afghanistan), MAFAC (French Archaeological Mission in Central Asia), and MAFOUB (Uzbek-French Archaeological Mission in the Bukhara Oasis) archaeological missions from 2009 to 2016.
2 Climate data for the Balkh and Bukhara cities are taken from https://fr.climate-data.org.
3 At Rabat-e-Bala station (Olson and Williams-Sether 2010: 345–346) for the 1965–1978 period.
4 At Ravathodja station, in Uzbekistan for the 1923–2007 period.
5 Equivalent Dose (De) corresponds to energy accumulation related to natural radioactivity (uranium, thorium, and potassium decay products) and trapped in the crystal lattice of quartz minerals since the last light exposition. The age equation is equal to the equivalent dose (De) divided by the dose rate (Da), which represents the amount of energy deposited in the crystal lattice per year.
6 OSL dating of the Zeravshan paleochannels has been performed by the C2RMF (Centre of Research and Restoration of Museum of France) and of the Balkh River samples by the Montreal University of Quebec.

7 At Chehel Soutoun, Hellenistic blocks were reused and disposed along the edges of a fluvial channel. The channel remained in use until the Early Islamic period, as shown by the recent discovery of a bridge pillar from that era, found *in situ* (Besenval and Marquis 2008).

8 The identification of paleochannels has been made easier by the use of middle infrared band 7 (MIR) treatment of the Landsat ETM+ (water surface and humid areas) associated to Laplacian filter (edge enhancement) (Fouache *et al.* 2016: 154, fig. 4).

9 Surveys led by the French-Uzbek Archaeological Mission (MAFANAC-Ajakagytma 2005–2015, since 2015 MAFAC) under the direction of F. Brunet.

10 The French Archaeological Mission of the Bukhara Oasis (MAFOUB) directed by R. Rante led recent surveys where Kel'teminar culture sites have been identified thanks to comparisons with F. Brunet's previously published material (Brunet 2005, 2011).

11 In 2016, further examination was conducted on the Zaman Baba trench and two blocks of sediments were sampled by L. Cez, who leads the micromorphological study (in progress).

12 Through stratigraphic data coming from the Pajkend excavations. The site is located on a naturally oriented branch of the main Zeravshan course, from which the Qaraqöl micro oasis also originates further down. The stratigraphic data also comes from Iskijat, located along an ancient meander of the same branch in the upper part of the oasis. Both sites have been ^{14}C and thermoluminescence dated.

13 In the case of the Zeravshan River, in 2012 Brunet orally advanced the hypothesis of river course changes in the northern part of the delta (Brunet and Debaine-Francfort 2012).

14 Sedimentological samples from wells A, B, and C were transferred into Passega's diagram (1977). They are mostly situated in the RS segment, which characterizes transportation by particle suspension and a floodplain-type deposition environment. Some very fine samples belong to zone T in which decantation is the principal mode of deposition, other coarser ones belong to segment QR in which transportation by saltation dominates.

15 At the peak flow period, the Balkh River flows into the Kelif Uzboj ancient riverbed of the Amu Darya River and even into Turkmenistan (Ahmad and Wasiq 2004: 12–13).

16 In the global climate change, Central Asia is experiencing a global warming above average: +0.7°C (Christensen *et al.* 2007), while the temperature rise is at +0.5°C on average (Giese *et al.* 2007). This more severe situation is exemplified by an average temperature gain of 1.2–2.1°C since the 1950s, and a 1–2°C increase observed during the last century in Central Asia (Christensen *et al.* 2007; Giese *et al.* 2007).

17 The top of the water table is close to the surface and varies about 1 m through the year.

18 Daulatabad is located on the left bank of the Shirin-Tagab River, in the Faryab province. Three Late Bronze Age sites have been identified in a nearby river valley: Tikar 1, Tikar 2, and Tikar 3 (Sarianidi 1977).

19 The Farukhabad sites are near the contemporary localities of Farukhabad and Kultag, to east of the Balkh plain.

20 Tillja Tepe is located 3 km to the east of modern Shibergan town, on the Band-i-Turkestan foothills and in the alluvial plain of the Darya-i-Safid and Darya-i-Siakh Rivers, left bank tributaries of the Amu Darya River.

21 The name of the site, Dal'verzin (Tepe), was later transformed into Dil'berdzhin in order not to confuse it with another site of the same name and same period in northern Bactria (Uzbekistan).

REFERENCES

Ahmad, M. and M. Wasiq 2004. *Water Resource Development in Northern Afghanistan and Its Implications for Amu-Darya Basin*. World Bank Working Paper. Washington, DC: World Bank.

Aichner, B., S.J. Feakins, J.E. Lee, U. Herzschuh, and X. Liu 2014. "High Resolution Leaf Wax Carbon and Hydrogen Isotopic Record of Late Holocene Palaeoclimate in Arid Central Asia." *Climate Past Discussion* 10: 4385–4424.

Aitken, M.J. 1985. *Thermoluminescence Dating*. London: Academic Press.

Allen, J.R.L. 1965. "A Review of the Origin and Characteristics of Recent Alluvial Sediments." *Sedimentology* 5: 89–191.

Allen, J.R.L. 1978. "Studies in Fluviatile Sedimentation: An Exploratory Quantitative Model for the Architecture of Avulsion-Controlled Alluvial Suites." *Sedimentary Geology* 21: 129–147.

Ambraseys, N.N. 1961. "On the Seismicity of South-West Asia: Data from the XV Century Arabic Manuscript." *Revue pour l'étude des calamités (*Genève*)* 37: 18–33.

An, C.-B., Z.D. Feng, and L. Barton 2006. "Dry or Humid? Mid-Holocene Humidity Changes in Arid and Semi-Arid China." *Quaternary Science Reviews* 25: 351–361.

An, C.-B., Y. Lu, J. Zhao, S. Tao, W. Dong, H. Li, M. Jing, and Z. Wang 2012. "A High-Resolution Record of Holocene Environmental and Climatic Changes from Lake Balikun (Xinjiang, China): Implications for Central Asia." *Holocene* 22(1): 43–52.

Andrianov, B.V. 1985. *History of Irrigation and Drainage in the USSR*. Vol. I, *History of Irrigation, Drainage, Flood Control and River Engineering*, ed. USSR National Committee on Irrigation and Drainage and International Commission on Irrigation and Drainage (ICID). New Delhi: UNESCO.

Andrianov, B.V. 2016. *Ancient Irrigation Systems of the Aral Sea Area: The History, Origin and Development of Irrigated Agriculture*, ed, S. Mantellini, C.C. Lamberg-Karlovsky, and M. Tosi. American School of Prehistoric Research Monograph Series. Oxford: Oxbow Books.

Askarov, A.A. 1965. "Novye nakhodki andronovskoj kul'tury v nizov'jakh Zarafshana" [New Andronovo culture discoveries in the lower Zeravshan]. *Istorija Material'noj Kul'tury Uzbekistana* 6: 53–60.

Askarov, A.A. 1981. "K peredatirovke kul'tury Zamanbaba" [Re-dating the Zamanbaba culture]. In *Kul'tura i iskusstvo drevnego Khorezma* [Culture and art in ancient Khorezm], ed. M.A. Itina and Ju. A. Rapoport. Moscow: Nauka, pp. 99–110.

Ball, W. (with J.-Cl. Gardin) 1982. *Archaeological Gazetteer of Afghanistan. Catalogue des sites archéologiques d'Afghanistan*. Paris: Éditions Recherche sur les civilisations.

Beer, R., O. Heiri, and W. Tinner 2007. "Vegetation History, Fire History and Lake Development Recorded for 6,300 Years by Pollen, Charcoal, Loss on Ignition and Chironomids at a Small Lake in Kyrgyzstan (Alay Range, Central Asia)." *The Holocene* 17: 977–985.

Bensidoun, S. 1979. *Samarcande et la vallée du Zerafshan. Une civilisation de l'oasis en Ouzbekistan-URSS*. Paris: Éditions Anthropos.

Besenval, R. and P. Marquis 2008. "Les travaux de la Délégation Archéologique Française en Afghanistan (DAFA): résultats des campagnes de l'automne 2007 au Printemps 2008 en Bactriane et à Kaboul." *Comptes rendus de l'Académie des Inscriptions et Belles Lettres* 152–153: 973–995.

Bond, G., W. Showers, M. Cheseby, R. Lotti, P. Almasi, P. deMenocal, P. Priore, H. Cullen, I. Hajdas, and G. Bonani 1997. "A Pervasive Millennial-Scale Cycle in the North Atlantic Holocene and Glacial Climates." *Science* 278: 1257–1266.

Boomer, I., N. Aladin, I. Plotnikov, and R. Whatley. 2000. "The Palaeolimnology of the Aral Sea: A Review." *Quaternary Science Reviews* 19, 1259–1278.

Boomer, I., B. Wünnemann, A.W. Mackay, P. Austen, P. Sorrel, C. Reinhardt, D. Keyser, F. Guichard, and M. Fontugne. 2009. "Advances in Understanding the Late Holocene History of the Aral Sea region." *Quaternary International* 194: 79–90.

Bossu, R. 1996. "Étude de la sismicité intraplaque de la région de Gazli (Ouzbekistan) et localisation de la déformation sismique." Unpublished PhD dissertation. University Joseph Fourier, Grenoble.

Braconnot, P., S.P. Harrison, M. Kageyama, P.J. Bartlein, V. Masson-Delmotte, A. Abe-Ouchi, B. Otto-Bliesner, and Y. Zhao 2012. "Evaluation of Climate Models Using Palaeoclimatic Data." *Nature Climate Change* 2: 417–424.

Bridge, J.S. 2003. *Rivers and Floodplains: Forms, Processes, and Sedimentary Records.* Oxford: Blackwell.

Brunet, F. 2005. "Pour une nouvelle étude de la culture néolithique de Kel'teminar, Ouzbékistan." *Paléorient* 31: 87–105.

Brunet, F. 2011. "Comment penser la néolithisation en Asie Centrale (X^e^–IVe millénaires)? L'émergence de nouveaux modèles de sociétés entre sédentaires et nomades." *Paléorient* 37(1): 187–204.

Brunet, F. and C. Debaine-Francfort 2012. "L'espace temporairement apprivoisé, Étude de cas (Ouzbékistan, Chine)." In *L'archéologie à découvert, Quatrième partie: espaces et Environnements, Questions d'échelles*, ed. S.A. de Beaune and H.-P. Francfort. Paris: Éditions CNRS, pp. 160–168, 310–311.

Brunet, F., M. Hudžanazarov, and K. Szymczak 2013. "Le site d'Ajakagytma et le complexe culturel de Kel'teminar au sein des processus de néolithisation en Asie centrale." *Cahiers d'Asie centrale* 21–22: 191–205.

Butzer, K.W. 2012. "Collapse, Environment and Society." *Proceedings of the National Academy of Sciences of United States of America* 109: 3632–3639.

Cattani, M., Cerasetti B., Salvatori S., and Tosi M. 2008. "The Murghab Delta in Central Asia 1990–2001: The GIS from Research Resource to a Reasoning Tool for the Study of Settlement Change in Long-Term Fluctuations." In *The Bronze Age and Early Iron Age in the Margiana Lowlands: Facts and Methodological Proposals for a Redefinition of the Research Strategies*, ed. S. Salvatori, M. Tosi, and B. Cerasetti. Oxford: Archaeopress, pp. 39–45.

Cerasuolo, O. 2009. "Indagine topografica nel territorio di Varakhsha." In *Gli scavi di Uch Kulakh (oasis di Bukhara)*, ed. C.H. Antonini and D. Mirzaakhmedov. Rome: Fabrizio Serra Editore, pp. 189–216.

Chen, F., Z. Yu, M. Yang, E. Ito, S. Wang, D.B. Madsen, X. Huang *et al.* 2008. "Holocene Moisture Evolution in Arid Central Asia and Its Out-of-Phase Relationship with Asian Monsoon History." *Quaternary Science Reviews* 27: 351–364.

Christensen, J.H., B. Hewitson, A. Busuioc, A. Chen, X. Gao, I. Held, R. Jones *et al.* 2007. "Regional Climate Projections." In *Climate Change 2007: The Physical Science Basis Contribution of Working Group I to the Fourth Assessment Report of the Intergovernmental Panel on Climate Change*, ed. S.D. Solomon, M. Qin, Z. Manning, M. Chen, K.B. Marquis, M. Averyt, and H.L. Miller. Cambridge and New York: Cambridge University Press, pp. 879–886.

Coombes, P. and K. Barber. 2005. "Environmental Determinism in Holocene Research: Causality or Coincidence?" *Area* 37: 303–311.

Cremaschi, M. 1998. "Palaeohydrography and Middle Holocene Desertification in the northern fringe of the Murghab Delta." In *The Archaeological Map of the Murghab Delta, Preliminary Reports 1990–95*, ed. A. Gubaev, G.A. Koshelenko, and M. Tosi. Rome: IsIAO, pp. 15–25.

Cullen, H.M., P.B. deMenocal, S. Hemming, G. Hemming, F.H. Brown, T. Guilderson, and F. Sirocko 2000. "Climate Change and the Collapse of the Akkadian Empire: Evidence from the Deep Sea." *Geology* 28: 379–382.

DeMenocal, P.B. 2001. "Cultural Responses to Climate Change During the Late Holocene." *Science* 292: 667–673.

Denton, G.H. and W. Karlén. 1973. "Holocene Climatic Variations, Their Pattern and Possible Causes." *Quaternary Research* 3(2): 155–174.

Dodonov, A.E. 1980. "Late Pliocene-Quaternary Stage of Tectogenesis in the Tadjik Depression." *International Geology Review* 22(1): 11–21.

Feng, Z.D., C.B. An, and H.B. Wang 2006. "Holocene Climatic and Environmental Changes in the Arid and Semi-Arid Areas of China: A Review." *The Holocene* 16(1): 119–130.

Fouache, E., R. Besenval, C. Cosandey, C. Coussot, M. Ghilardi, S. Huot, and M. Lamothe 2012. "Palaeo-Channels of the Balkh River (Northern Afghanistan) and Human Occupation Since the Bronze Age Period." *Journal of Archaeological Science* 39: 3415–3427.

Fouache, E., R. Rante, D. Mirzaakhmedov, R. Ragala, M. Dupays, C. Vella, J. Fleury *et al.* 2016. "The Role of Catastrophic Floods Generated by Collapse of Natural Dams Since the Neolithic in the Oases of Bukhara and Qaraqöl: Preliminary Results." *International Journal of Geohazards and Environment* 2(3): 150–165.

Francfort, H.-P. and O. Lecomte. 2002. "Irrigation et société en Asie centrale des origines à l'époque achéménide." *Annales. Histoire, Sciences Sociales* 57(3): 625–663.

Gardin, J-C. and P. Gentelle. 1976. "Irrigation et peuplement dans la plaine d'Aï-Khanoum, de l'époque achéménide à l'époque musulmane." *Bulletin de l'Ecole Française d'Extrême Orient* 63: 59–99.

Gardin, J.-C. and P. Gentelle. 1979. "L'exploitation du sol en Bactriane antique." *Bulletin de l'École française d'Extrême-Orient* 66: 1–29.

Gentelle, P. 1989. *Données paléogéographiques et fondements de l'irrigation, Prospections archéologiques en Bactriane orientale (1974–1978)*. Vol. I, *Prospections archéologiques en Bactriane orientale (1974–1978)*, ed. J.C. Gardin. Paris: De Boccard.

Giese, E., I. Mossig, D. Rybski, and A. Bunde 2007. "Long-Term Analysis of Air Temperature Trends in Central Asia." *Erdkunde* 61(2): 186–202.

Gubaev, A., G. Koshelenko, and M. Tosi 1998. *The Archaeological Map of the Murghab Delta: Preliminary Reports 1990–95 – Reports and Memoirs*. Rome: IsIAO.

Gul'jamov, J.G., U.I. Islamov, and A.A. Askarov. 1966. *Pervobytnaja kul'tura i vozniknovenie oroshaemogo zemledelija v nizov'jakh Zarafshana* [Indigenous culture and origins of irrigated agriculture in the lower Zeravshan valley]. Tashkent: Fan.

Hartmann, K. and B. Wünnemann 2008. "Hydrological Changes and Holocene Climate Variations in NW China, Inferred from Lake Sediments from Juyenze Palaeolake by Factor Analysis." *Quaternary International* 194: 137–147.

Havenith, H.-B., A. Strom, D. Jongmans, A. Abdrakhmatov, D. Delvaux, and P. Tréfois 2003. "Seismic Triggering of Landslides, Part A: Field Evidence from the Northern Tien Shan." *Natural Hazards and Earth System Science* 3(1/2): 135–49.

Humlum, J. 1959. *La géographie de l'Afghanistan. Étude d'un pays aride*. Copenhagen: Scandinavian University Books, Gyldendal.

Ilyashuk, E.A., B.P. Ilyashuk, D. Hammarlund, and I. Larocque 2005. "Holocene Climatic and Environmental Changes Inferred from Midge Records (Diptera: Chironomidae, Chaoboridae, Ceratopogonidae) at Lake Berkut, Southern Kola Peninsula, Russia." *The Holocene* 15(6): 897–914.

Jin, L., F. Chen, and C. Morrill 2012. "Causes of Early Holocene Desertification in Arid Central Asia." *Climate Dynamics* 38(7–8): 1577–1591.

Jones, L.S. and S.A. Schumm 1999. "Causes of Avulsion: An Overview." *Fluvial Sedimentology* VI: 169–178.

Joussaume, S., K.E. Taylor, P. Braconnot, J.F.B. Mitchell, J.E. Kutzbach, S.P. Harrison, I.C. Prentice *et al.* 1999. "Monsoon Changes for 6000 Years Ago: Results of 18 Simulations from the Paleoclimate Modeling Intercomparison Project (PMIP)." *Geophysical Research Letters* 26(7): 859–862.

Kaniewski, D., E.V. Campo, J. Guiot, S.L. Burel, T. Otto, and C. Baeteman 2013. "Environmental Roots of the Late Bronze Age Crisis." *PLoS ONE* 8(8): e71004. https://doi.org/10.1371/journal.pone.0071004.

Kes, A.S., E.D. Mamedov, S.O. Khondkarjan, G.N. Trofimov, and K.V. Kremenetskij 1993. "Stratigrafija i paleogeografija ravninnikh oblastej Srednej Azii v pozdnem pleistocene i golocene" [Stratigraphy and paleogeography in the plains of Central Asia at the end of Pleistocene and in the Holocene]. In *Razvitie landshaftov i klimata severnoje Evrazii: Podznij Pleistotsen, Golotsen, Elementy prognoza* [Evolution of landscapes and climates of northern Eurasia: Late Pleistocene, Holocene, predicting elements], ed. A.A. Velichko. Moscow: Nauka, pp. 82–87.

Knapp, A.B. and S.W. Manning 2016. "Crisis in Context: The End of the Late Bronze Age in the Eastern Mediterranean." *American Journal of Archaeology* 120: 99–149.

Kohl, P.L.A. 1981. *The Bronze Age Civilization of Central Asia: Recent Soviet Discoveries*. Armonk, NY: M.E. Sharpe.

Kohl, P.L.A. 1984. *Central Asia Palaeolithic Beginnings to the Iron Age*. Paris: Editions Recherche sur les civilisations.

Korup, O. 2004. "Landslide-Induced River Channel Avulsions in Mountain Catchments of Southwest New Zealand." *Geomorphology* 63: 57–80.

Krivonogov, S.K., Y.V. Kuzmin, G. Burr, S.A. Gusskov, L.B. Khazin, E.Y. Zhakov, A.N. Nurgizarinov, R.K. Kurmanbaev, and T.I. Kenshinbay 2010. "Environmental Changes of the Aral Sea (Central Asia) in the Holocene: Major trends." *Radiocarbon* 52(2–3): 555–568.

Kuzucuoğlu, C. and Z. Tsitsoni 2016. "Changements climatiques et comportements sociaux dans le passé: quelles corrélations?" *Les nouvelles de l'archéologie* 142: 49–55.

Lauterbach, S., R. Witt, B. Plessen, P. Dulski, S. Prasad, J. Mingram, G. Gleixner *et al.* 2014. "Climatic Imprint of the Mid-Latitude Westerlies in the Central Tian Shan of Kyrgyzstan and Teleconnections to North Atlantic Climate Variability During the Last 6000 Years." *The Holocene* 24(8): 970–984.

Leroy, S.A.G., L. Lopez-Merino, A. Tudryn, F. Chalié, and F. Gasse 2014. "Late Pleistocene and Holocene Palaeoenvironments in and Around the Middle Caspian Basin as Reconstructed from a Deep-Sea Core." *Quaternary Science Review* 101: 91–110.

Létolle, R. and M. Mainguet 2004. "Histoire des rivières d'Asie centrale depuis deux millions d'années: certitudes et spéculations." *Cahiers d'Asie centrale* 11–12: 317–341.

Lewis, R.A. 1966. "Early Irrigation in West Turkestan." *Annals of the Association of American Geographers* 56: 467–491.

Lioubimtseva, E., R. Cole, J.M. Adams, and G. Kapustin 2005. "Impacts of Climate and Land-Cover Changes in Arid Lands of Central Asia." *Journal of Arid Environments* 62: 285–308.

Lioubimtseva, E., B. Simon, H. Faure, L. Faure-Denard, and J.M. Adams 1998. "Impacts of Climatic Change on Carbon Storage in the Sahara-Gobi Desert Belt Since the Late Glacial Maximum." *Global and Planetary Change* 16–17: 95–105.

Lisitsyna, G.N. 1965. *Oroshaemoe zemledelie epokhi eneolita na juge Turkmenii* [Irrigated agriculture during the Chalcolithic period in southern Turkmenia]. Moscou: Nauka.

Liu, Z., J. Zhu, Y. Rosenthal, X. Zhang, B. Otto-Bliesner, A. Timmermann, R.S. Smith, G. Lohmann, W. Zheng, and O.E. Timm 2014. "The Holocene Temperature Conundrum." *Proceedings of the National Academy of Sciences* 111(34): 3501–3505.

Madella, M. and D.Q. Fuller 2006. "Palaeoecology and the Harappan Civilisation of South Asia: A Reconsideration." *Quaternary Science Reviews* 25: 1283–1301.

Maman, S., H. Tsoar, D.G. Blumberg, B. Mamedov, and N. Porat 2011. "Central Asian Ergs: A Study by Remote Sensing and Geographic Information Systems." *Aeolian Research* 3(3): 353–366.

Mayewski, P.A., L.D. Meeker, M.S. Twickler, S. Whitlow, Q. Yang, W.B. Lyons, and M. Prentice 1997. "Major Features and Forcing of High-Latitude Northern Hemisphere Atmospheric Circulation Using a 110,000-Year-Long Glaciochemical Series." *Journal of Geophysical Research* 102: 26345–26366.

Mayewski, P.A., E.E. Rohling, J. Curt Stager, W. Karlén, K.A. Maasch, L.D. Meeker, E.A. Meyerson *et al.* 2004. "Holocene Climate Variability." *Quaternary Research* 62: 243–255.

Mischke, S., I. Rajabov, N. Mustaeva, C. Zhang, U. Herzschuh, I. Boomer, E.T. Brown *et al.* 2010. "Modern Hydrology and Late Holocene History of Lake Karakul, Eastern Pamirs (Tajikistan): A Reconnaissance Study." *Paleogeography, Palaeoclimatology, Palaeoecology* 289: 10–24.

Mohrig, D., P.L. Heller, C. Paola, J., and W.J. Lyons 2000. "Interpreting Avulsion Process from Ancient Alluvial Sequences: Guadalope-Matarranya System (Northern Spain) and Wasatch Formation (Western Colorado)." *Geological Society of America Bulletin* 112(12): 1787–1803.

Nikonov, A.A. and A.V. Penkov 1973. "Pliocene and Early Pleistocene Geochronology of Central Asia and Kazakhstan." *International Geology Review* 16(10): 1187–1100.

O'Brien, S.R., P.A. Mayewski, L.D. Meeker, D.A. Meese, M.S. Twickler, and S.I. Whitlow 1995. "Complexity of Holocene Climate as Reconstructed from a Greenland Ice Core." *Science* 270(5244): 1962–1964.

Olson, S.A. and T. Williams-Sether 2010. *Streamflow Characteristics at Streamgages in Northern Afghanistan and Selected Locations: US Geological Survey Project Product Number 179*. US Department of the Interior Data Series 529. Reston, VA: US Department of the Interior.

Olsson, O., M. Gassmann, K. Wegerich, and M. Bauer 2010. "Identification of the Effective Water Availability from Streamflows in the Zerafshan River Basin, Central Asia." *Journal of Hydrology* 390: 190–197.

Passega, R. 1977. "Significance of CM Diagrams of Sediments Deposited by Suspension." *Sedimentology* 24: 723–733.

Possehl, G.L. 1997. "The Transformation of the Indus Civilization." *Journal of World Prehistory* 11: 425–472.

Rante, R., E. Fouache, and D. Mirzaakhmedov 2016. "Dynamics of Human Settlements Ensuing from River Transformation and Changes in Commercial Behavior: The Birth of the 'North-Eastern Silk Road.'" *Journal of Archaeological Science: Reports* 9: 437–447.

Ricketts, R.D., T.C. Johnson, E.T. Brown, K.A. Rasmussen, and V.V. Romanvsky 2001. "The Holocene Paleolimnology of Lake Issyk-Kul, Kyrgyzstan: Trace Element and Stable Isotope Composition of Ostracodes." *Paleogeography, Palaeoclimatology, Palaeoecology* 176: 207–227.

Rouse, L.M. and B. Cerasetti 2017. "Micro-Dynamics and Macro-Patterns: Exploring New Archaeological Data for the Late Holocene Human–Water Relationship in the Murghab Alluvial Fan, Turkmenistan." *Quaternary International* 437(Part B): 20–34.

Ruleman, C.A., A.J. Crone, M.N. Machette, K.M. Haller, and K.S. Rukstales 2007. *Map and Database of Probable and Possible Quaternary Faults in Afghanistan*. US Geological Survey Open-File Report 2007-1103. Washington, DC: US Department of the Interior.

Sarianidi, V.I. 1976. "Recherches des sites de l'oasis de Dashly" (abstract in French). In *Drevnjaja Baktrija* [Ancient Bactria]. Vol. I, *Materialy Sovetsko-Afganskoj Arkheologicheskoj Ekspeditsii 1969–1973* [Materials of the Soviet–Afghan expedition 1969–1973], ed. I.T. Krugilova. Moscow: Nauka, pp. 21–86.

Sarianidi, V.I. 1977. *Drevnie zemledel'tsy Afganistana* [Ancient farmers of Afghanistan]. Moscow: Nauka.

Shobair, S.S. and A.K. Alim 2004. "The Effects of Calamities on Water Resources And Consumption in Afghanistan." Rome: Food and Agriculture Organization of the United Nations (FAO). www.nourin.tsukuba.ac.jp/~tasae/2004/Afghanistan.pdf.

Simpson, D.W. and W. Leith 1985. "The 1976 and 1984 Gazli, USSR, Earthquakes. Were They Induced? " *Bulletin of the Seismological Society of America* 75: 1465–1468.

Staubwasser, M. and H. Weiss 2006. "Holocene Climate and Cultural Evolution in Late Prehistoric–Early Historic West Asia." *Quaternary Research, Holocene Climate and Cultural Evolution in Late Prehistoric-Early Historic West Asia* 66: 372–387.

Stouthamer, E. and H.J.A. Berendsen 2007. "Avulsion: The Relative Roles of Autogenic and Allogenic Processes." *Sedimentary Geology* 198(3–4): 309–325.
Strom, A. 2010. "Landslide Dams in Central Asia region." *Landslides: Journal of the Japan Landslide Society* 47(6): 309–324.
Strom, A. 2013. "Geological Prerequisitses for Landslides Dam's Disaster Assessment and Mitigation in Central Asia." In *Progress of Geo-Disaster Mitigation Technology in Asia, 17 Environmental Science and Engineering*, ed. F. Wang, M. Miyajima, T. Li, W. Shan, and T.F. Fathani. Berlin: Springer-Verlag, pp. 17–53.
Strom, A.L. and K.E. Abdrakhmatov 2009. "International Summer School on Rockslides and Related Phenomena in the Kokomeren River Valley, Tien Shan, Kyrgyzstan." In *Landslides-Disaster Risk Reduction*, ed. K. Sassa and P. Canuti. Berlin: Springer, pp. 223–227.
Suslov, S.P. 1961. *Physical Geography of Asiatic* Russia, ed. J.E. Williams. San Francisco, CA and London: W.H. Freeman & Co.
Tang, Z., D. Chen, X. Wu, and G. Mu 2013. "Redistribution of Prehistoric Tarim People in Response to Climate Change." *Quaternary International* 308–309: 36–41.
Tarasov, P.E., T. Webb III, A.A. Andreev, N.B. Afanas'eva, N.A. Berezina, L.G. Bezusko, T.A. Blyakharchuk *et al.* 1998. "Present-Day and Mid-Holocene Biomes Reconstructed from Pollen and Plant Macrofossil Data from the Former Soviet Union and Mongolia." *Journal of Biogeography* 25: 1029–1053.
Thomas, J.-C., J.-R. Grasso, R. Bossu, J. Martinod, and B. Nurtaev 2009. "Recent Deformation in the Turan and South Kazakh Platforms, Western Central Asia, and Its Relation to Arabia–Asia and India–Asia Collisions." *Tectonics* 18: 201–214.
Tolstov, S.P. and A.S. Kes 1960. *Nizov'ja Amudari, Sarykamysh, Uzboj: istorija formirovanija i zaselenija. Materialy Khorezmskoj Ekspeditsii* [Lower Amu Darya, Sarykamysh, Uzboi: history of their formation and of settlement], vol. III. Moscow: Nauka.
Törnqvist, T. 1994. "Middle and Late Holocene Avulsion History of the River Rhine (Rhine-Meuse delta, Netherlands)." *Geology* 22: 711–714.
Varushchenko, S.I., A.N. Varushchenko, and R.K. Klige 1987. *Izmenenija rezhima Kaspijskogo morja i besstochnikh vodoemov v paleovremeni* [Changes in the Caspian Sea level and reservoirs without discharge in paleotimes]. Moscow: Nauka.
Vinogradov A.V. 1979. "Issledovanija pamjatnikov kammenogo veka v severnom Afganistane" [Researches on Stone Age sites in northern Afghanistan]. In *Drevnjaja Baktrija* [Ancient Bactria], vol. 2, ed. I.T. Kruglikova. Moscow: Nauka, pp. 7–62.
Vinogradov, A.V. and E.D. Mamedov 1975. *Pervobytnyj Ljavljakan, etapy drevnejshego zaselenija i osvoenija vnutrennikh Kyzylkumov* [Indigenous Ljavljakan: ancient occupation and development stages of the inner Kyzylkum], ed. M.A. Itina. Vol. 10, *Materialy Khorezmskoj Ekspeditsii*. Moscow: Nauka.
Wang, J., Z. Jin, R.G. Hilton, F. Zhang, A.L. Densmore, G. Li, and A.J. West 2015. "Controls on Fluvial Evacuation of Sediment from Earthquake-Triggered Landslides." *Geology* 43: 115–118.
Weiss, H. 2016. "Global Megadrought, Societal Collapse and Resilience at 4.2–3.9 ka BP Across the Mediterranean and West Asia." *Climate Change and Cultural Evolution* 24: 62–63.
Weiss, H. and R.S. Bradley 2001. "What Drives Societal Collapse?" *Science* 291: 609–610.
Weiss, H., M.A. Courty, W. Wetterstrom, F. Guichard, L. Senior, R. Meadow, and A. Curnow 1993. "The Genesis and Collapse of Third Millennium North Mesopotamian Civilization." *Science* 261: 995–1004.
Welc, F. and L. Marks. 2014. "Climate Change at the End of the Old Kingdom in Egypt Around 4200 BP: New Geoarchaeological Evidence." *Quaternary International* 324: 124–133.
Weninger, B., L. Clare, E.J. Rohling, O. Bar-Yosef, U. Böhner, M. Budja, M. Bundschuh *et al.* 2009. "The Impact of Rapid Climate Change on Prehistoric Societies During the Holocene in the Eastern Mediterranean." *Documenta Praehistorica* XXXVI 902(4–5): 631–637.

Wolff, C., B. Plessen, A.S. Dudashvilli, S.F.M. Breitenbach, H. Cheng, L.R. Edwards, and M.R. Strecker 2017. "Precipitation Evolution of Central Asia During the Last 5000 years." *The Holocene* 27(1): 142–154.

Wünnemann, B., S. Mischke, and F. Chen 2006. "A Holocene Sedimentary Record from Bosten Lake, China." *Paleogeography, Palaeoclimatology, Palaeoecology* 234: 223–238.

Zink, A.J.C., E. Porto, E. Fouache, and R. Rante 2017. "Paléocours du delta du Zerafshan (oasis de Boukhara, Ouzbékistan): premières datations par luminescence." *L'Anthropologie* 121: 46–54.

CHAPTER FOUR

THE RISE OF THE EARLY URBAN CIVILIZATION IN SOUTHWESTERN CENTRAL ASIA

From the Middle Chalcolithic to the Middle Bronze Age in southern Turkmenistan

Ljubov' B. Kircho

INTRODUCTION

Archaeological studies of Neolithic and Chalcolithic sites in southwestern Central Asia have shown that South Turkmenistan, and possibly northeastern Khorasan, was one of the ancient farming cradles of this area, an improvised northern outpost of the Ancient East (Figure 4.1). Its development has led to the formation of large early urban centers of the Bronze Age in the last third of the 3rd millennium.

The birth and development of settled farming communities in southwestern Central Asia relied significantly on the favorable natural environment of the Kopet Dagh foothill plain (Lisitsina 1978: 23–35), where the earliest farmers and cattle breeders of the Djeitun Neolithic culture settled at the end of the 7th–6th millennium (Masson 1971: 76; Harris 2010).

The 6th–4th millennia correspond to the first half of the Holocene Climatic Optimum (HCO) (i.e., the warmest and most humid one) (Vinogradov and Mamedov 1975; Mamedov 1980; see also Chapter 3). The majority of researchers agree that a dry period began towards the end of the Atlantic period (Khotinsky 1980; Dolukhanov 1984: 44–46) and continued up until now (Stepanov 1980: 31). A steady paleoecological deterioration during the 3rd millennium is underlined by the Altyn Depe palynological data (Levkovskaja and Kircho 2004: 90). Modern studies of paleoclimate have revealed the presence during the Holocene of three global cooling episodes for the northern regions, and dry ones during the end of the flourishing time of the Harappan Civilization for southern areas across most of Europe. These are the "8200 cal. BP," "5300 cal. BP," and "4200 cal. BP" events (Ivanova *et al.* 2011; see also Chapter 3). In the Near East, this last event had considerable consequences visible within the archaeological data, like the fall of the Akkadian kingdom, the abandonment of large settlements in ancient Iran, the end of the flourishing time of the Harappan Civilization, etc.

In southwestern Central Asia, these climatic "events" match with the beginning of three major cultural and production stages: (1) the rise and development

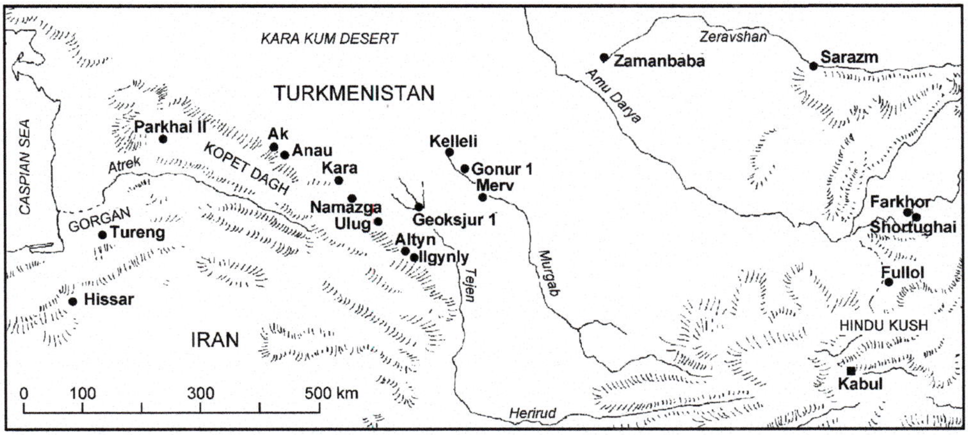

Figure 4.1 Schematic map of southern Central Asia with the sites mentioned in the chapter.

of ancient farming cultures during the Neolithic and the Early and Middle Chalcolithic; (2) the apparition of the Geoksjur archaeological complex at the late stage of the Middle Chalcolithic, which led to the development of the Early Bronze Age, both on a cultural and on a technological point of view, and, later, to the earliest urban civilization at Altyn Depe during the Middle Bronze Age; (3) the decline of the Kopet Dagh piedmont settlements in the last centuries of the 3rd/early 2nd millennia. This was combined with a great and prosperous development in the ancient delta of the Murghab River and the formation of a Near Eastern type of proto-state with Gonur Depe as its capital city (Sarianidi 1990, 2002, 2005, 2007, 2008).

The archaeological periodization of the Chalcolithic and Bronze Age[1] sites of Central Asia is based on the stratigraphy of multilayer settlements (*depe*), which serves as reference (i.e., the northern and southern mounds at Anau[2] and Namazga Depe). On these sites, in the first half of the 20th century AD, successive phases of development were determined for the Chalcolithic (Anau/Namazga) culture, which followed the Djeitun Neolithic culture. These periods are Anau IA (beginning of the Early Chalcolithic), Namazga (NMG) I or Anau IB (Early Chalcolithic), NMG II or Anau II (Middle Chalcolithic), and NMG III (Late Chalcolithic). The cultures of the Bronze Age (Anau III) are divided into three periods on the basis of the material from Namazga Depe: NMG IV (Early Bronze), NMG V (Middle Bronze), and NMG VI (Late Bronze Age) (Kuftin 1956; Kohl 1984: 65, 73–76; Masson 2006: 43–44, 61–71; Kircho 2014b: 359). This periodization is primarily based on the continuous but successive changes of the most visible features in the pottery complexes, like technological changes or modification in the decorative system. Because the foundations of this periodization rest directly on stratigraphy and on the fundamental patterns of production development, it remains the basis for all the chronological and cultural determinations of the Central Asian Chalcolithic and Bronze Age finds, despite many attempts at revision. However, the correlation of the different complexes within each main period, their longevity, and their absolute dates are constantly revised (Kircho 2005b).

THE MIDDLE CHALCOLITHIC PERIOD (CA. 3650–3200/3100 BC)

The economic and cultural developments and the expansion of the Early Chalcolithic communities during the second third of the 4th millennium led to the formation of two local variants of the Anau culture during the Middle Chalcolithic. Two stages of development have been stratigraphically identified at several sites: Kara Depe, the settlements of the Geoksjur oasis, and Ilgynly Depe.

The early phase (ca. 3650–3300 BC)

In the central part of the Kopet Dagh piedmont, this phase corresponds to early NMG II (Kara 7–4 levels). In the southeastern part, and in the ancient delta of the Tedjen River, it is represented by the Jalangach assemblage (respectively, levels VI–III at Ilgynly Depe and levels 7–4 at Geoksjur 1) (Masson 1962; 1982: 26–34, tabs. XIV–XVIII; 1989: 15–20; Khlopin 1969). In the southwestern part of the Kopet Dagh (i.e., in the Sumbar valley), a specific cultural complex "Southwestern Turkmenistan VI" (or SWT VI) has been identified, characterized by its gray-polished ware (Khlopin 1997: 67) sharing many common features with the Gorgan valley sites in the southeastern Caspian region.

Pottery

This phase is generally characterized by handmade painted and polished ware with organic (chopped straw or grass) temper and by non-painted red-slip ware with mineral temper (sand and ground limestone). It also includes jar-like large containers heavily tempered with organic material and sometimes with a monochrome decoration. The diagnostic pottery during the early NMG II period in the central part of the piedmont (Namazga Depe, Kara Depe, Elen Depe, Anau North, Ak Depe) is a polished light-slipped ware with a geometric bichrome dark-brown and red-painted decoration (Figure 4.2, *18–21*). In the eastern area where the Jalangach variant developed (i.e., in the Geoksjur oasis), at Ulug Depe, Ilgynly Depe, and Altyn Depe, the typical pottery is a red-slip ware with black painted horizontal stripes (sometimes connected by triangles and dashes; see Figure 4.2, *22–24*), together with light-slip cylinder-shaped vessels decorated with vertical stripes of hatched rhombs or triangles (Figure 4.2, *25*, *26*). The tableware is fired in double-chamber kilns. The earliest kiln of this type, dated to the end of the Early Chalcolithic period, has recently been discovered in a specialized workshop at Ilgynly Depe and was used to fire small items (Solovjova 2015, 2019). Both the tableware and the large containers with organic temper usually have a dark core at the section due to low firing.

The late phase (ca. 3300–3200/3100 BC)

Pottery

In the central area of the foothills zone (late NMG II or Kara Depe levels 3–2), the painted pottery is mostly a polished light-slip ware with painted dark-brown (almost black) geometric and zoomorphic (usually depicting quadrupeds, probably goats) decoration (Figure 4.2, *27–30*) (Masson 1962). In the southeastern zone, the

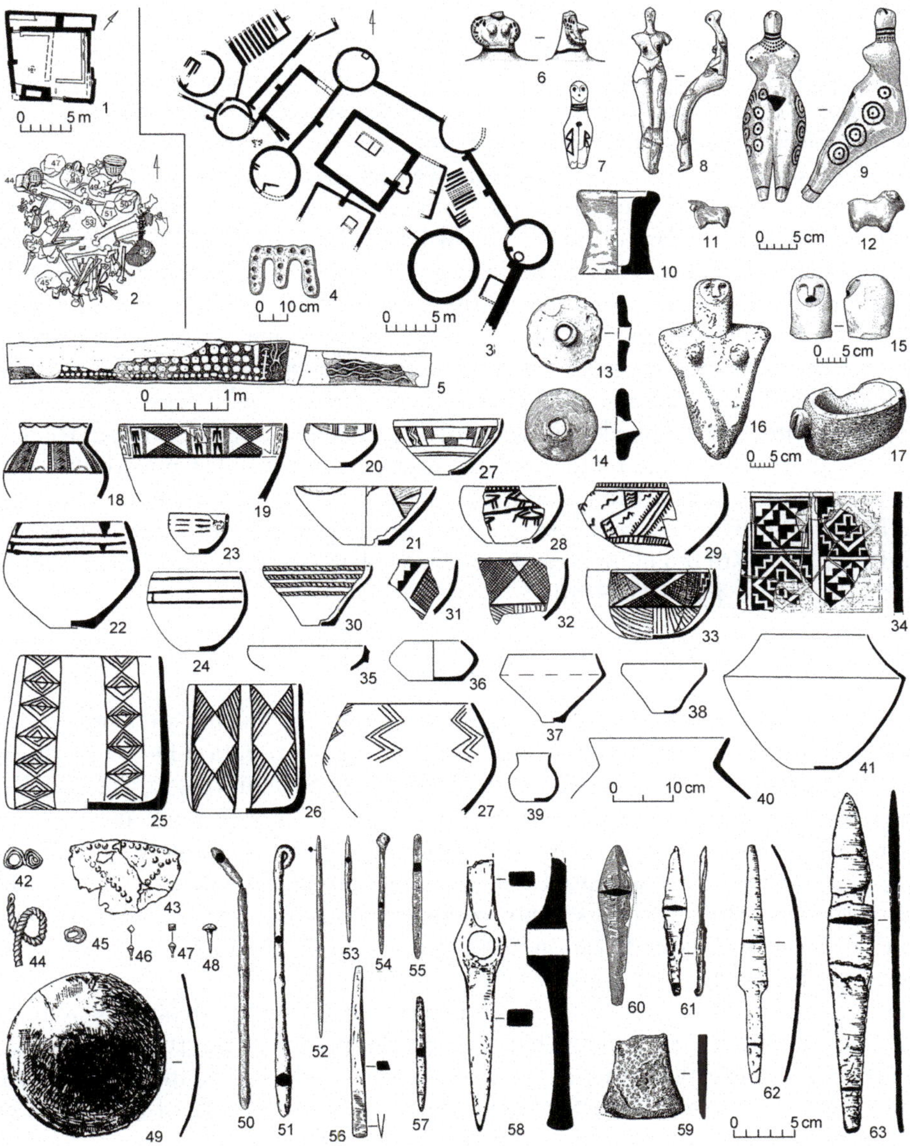

Figure 4.2 Southern Turkmenistan, Middle Chalcolithic period (NMG II, ca. 3650–3200/3100 BC). Ilgynly–D.: *1*, plan of a house; *2*, collective successive burials; *5*, painted murals (sgraffito technique). Jalangach–D.: *3*, *4*, plan of the settlement and clay motif stick on the wall; *6–14*, terracotta artifacts; *15–17*, stone artifacts; *18–41*, pottery; *42–63*, metal artifacts.

Early Geoksjur period starts, with a typical "Geoksjur-style" pottery without organic tempers and with double-sided stepped geometric painted motifs. These motifs appear suddenly and probably copy the basketry decoration of double-sided mats or containers. Sherds of one large cylinder-shaped vessel with organic temper and decorated with "the largest (monochrome – L.K.) Geoksjur ornamental composition discovered so far" have been found at Ilgynly Depe (Figure 4.2, *34*) (Berezkin and Solovjova 1998: 113, fig. 16, *1*, *2*).

The Geoksjur-type painted tableware is made of a dense clay with fine mineral temper. The pots were probably shaped using a conical or hemispherical mold and finished with the "anvil and spatula" method. The bichrome Geoksjur-style decoration was a complex process applied along different stages. After shaping, the pot was first covered with a thin layer of light slip, then its interior was slipped with red paint, and, on the exterior, special sectors were drawn in red on its lower part and bottom (Figure 4.2, *32*, *33*; Figure 4.4, *29*, *39*, *47*, *48*, *82*, *88*; Figure 4.8, *88–90*). On the upper part (in case of bowls – at two-thirds of their height) the main motifs were red-painted, that is, a horizontal band of plain (Figure 4.8, *87*) or crosshatched stepped rhombs, triangles, or other various geometrical motifs (vertical, inclined, zigzag), limited on top and bottom by horizontal red lines. After it had dried, it was polished, and over the polished surface, the red-painted geometric motifs were outlined with thin black lines, and between them were painted black crosshatched or stepped rhombs, cross-shaped figures, and dividers made of opposed stepped triangles. The typical Geoksjur cross motif is usually inscribed into a red steppedrhomb figure and consists of a central rhomb and four black crosshatched triangles. Black lines also frame the ornamental frieze on top and bottom. It ends up in alternating stepped motifs outlined with red filling, crosshatched, or plain (see Figure 4.4, *54*, *57*, *80–83*, *85*, *87*, *88*; Figure 4.8, *84*, *88–90*).

Economy

The main economic basis were: (1) agriculture, already "exclusively irrigated and based both on high-water flooding periods, and regular irrigation fed by canals and ditches" (Lisitsina 1978: 209; see also Masson and Sarianidi 1972: fig. 24; Kohl 1984: map 10); (2) cattle breeding (with a complex strategy for keeping herds of small ruminants and cattle) and specialized hunting on ungulates (Kasparov 2006: 129–130); (3) technologically advanced domestic craft production including the production of various artifacts made of clay, copper (Figure 4.2, *42–63*) (Solovjova *et al.* 1994), or stone and bone (Masson 1982: 54, tab. XV, *14–19*; XVI, *38–40*; XVIII, *11–17*, *32*, *34*, *41*, *43*, *44*).

This was related to a high social development and a complex ideological system. It is reflected in the architectural and painted decoration of the main rooms (Figure 4.2, *4*, *5*) (Berezkin and Solovjova 1998; Solovjova 1998: 125, fig. 2), in the cult of bulls (Kasparov 2006: 60–62, 130; Solovjova 1998), and in the cult of a female deity with numerous variants of terracotta painted female figurines (Figure 4.2, *6–9*) and her first stone anthropomorphic representation (Figure 4.2, *15*, *16*) (Masson and Korobkova 1989; Solovjova 2005).

Architecture

The cultural and social-level development during the Middle Chalcolithic period is attested in different ways, like a settlement hierarchy, visible in the Geoksjur oasis with the formation of large centers between 8–14 ha (Geoksjur 1, Ilgynly Depe, Kara Depe and Namazga Depe, and, probably, Altyn Depe and Ulug Depe). It is also visible through the construction of defensive walls with round structures (towers?) within their perimeter at Jalangach Depe (Figure 4.2, *3*) and Mullali Depe, and that of uninterrupted constructions with individual households, separated by narrow alleys or courtyards. At Ilgynly Depe, such households consist of rather large single-room houses with a uniform interior plan, opening on an adjacent courtyard (both for economical and formal use). Behind their back wall (opposite to the entrance), were usually two or three narrow chambers (utility compartments?). To the left of the entrance, a structure against the wall, or a compartment, was separated by a ledge (Figure 4.2, *1*).

In the Geoksjur oasis settlements, double hearths were usually located at the corner on the left of the entrance, raised up to 0.4 m above the floor level (Khlopin 1964: 24, 74). At Ilgynly Depe, inside the main room, there were often two to four clay (or wooden in one case) red-painted benches, and, on the left of the room, usually a hearth pit or a disk-shaped hearth altar layered with clay with a burnt pit in the center (Figure 4.2, *1*). The Geoksjur settlements of the Jalangach period (Akcha Depe, Jalangach Depe, Mullaly Depe, Ajna Depe) also have on the floor a mudbrick rectangular raised hearth surrounded by a clay border and divided into two parts by a clay lump or a small step (Sarianidi 1962: 45–46; Khlopin 1964: 77–78) (Figure 4.2, *3*). The earliest circular hearth altars surrounded by a clay border appear in the early NMG II period (Sarianidi 1962: 49, fig. 3; 4, *3*; Khlopin 1964: 87–88; 1969, tab. XXI, 4; XXIV, 2; Masson 1982: 30, 44, fig. 8, 2). At Ilgynly Depe such hearths (without border) are present in most of the main rooms (Berezkin and Solovjova 1998: 90, fig. 2, *1*, *2*, *5*, *8*, *11*, *13*, *20*, *22*; 3–5, *12*, *19*, *20*).

Burials

It is also during the early Geoksjur period that the earliest collective burial has been found, with successive inhumations accompanied with red-polished vessels (Figure 4.2, 2). At Kara Depe (levels 4–2), a series of individual burials, some of them lined with bricks (rectangular cist-like shape) have been excavated.

Artifacts

At that time, several modifications and novelties appear. This is the case with the iconography and the technique of underlining the anthropomorphic figurine details (Solovjova 2005: 24) or with the first copper cosmetic sticks in the inventories of individual burials at Ilgynly Depe and Kara Depe (Figure 4.2, *50*, *51*) (Masson 1962: tab. X, *23*; Solovjova *et al.* 1994, fig. 2, *1*, *3*). It is also then that the first artifacts probably made of artificial (heated) minerals appear (Masson 1962: tab. XVI, 2), called gypsum or paste. They consist of white large or small beads and of schematic anthropomorphic figurines and amulets(?), found in burials of the Late NMG II–Early NMG III at Kara Depe (Kara 3–Kara 1B complexes) (Masson 1960: 409–420, tab. XVI, *1–14*; 1962: 9, 42, tab. X, *1*, *2*) (see Figure 4.7, *28*). Their microscopic study made by T.A. Sharovskaja has shown that they are made of a heated ground stone including greenish and bluish crystals with a pearl-color polish on the surface, possibly burnt (and glazed?) steatite.

Thus, during the Middle Chalcolithic period, the population of South Turkmenistan had reached a cultural level comparable with that of the developed centers in the ancient Near East, possibly because of regular influences and contacts with foreign cultures as shown in the artifacts of the middle/second half of the 4th millennium.

Contacts and influences

Long ago, general and very wide parallels had been noticed between the crosshatched geometric motifs on subcylindrical vessels (Figure 4.2, *25*, *26*) from settlements of the Geoksjur oasis with Mesopotamian prototypes, as well as between the linear drawings of goats and Z-shaped painted motifs with those from southern Iran (ceramics of the 5th millennium Djowi culture in Susiana) (Masson 1960: 376–377). Closer analogies have been underlined between the bi-color ornamentation of the NMG II tableware with the Tal-i Iblis IV complex (ca. 4400–3300 BC) (Masson 1982: 34).

The main innovation in the ceramic complexes of the Middle Chalcolithic Anau culture is that of unpainted red-polished ware with carbonate temper (Kircho and Kovnurko 2001: 139). Such vessels (Figure 4.2, *35–41*) are especially frequent in the settlements of the central part of the Kopet Dagh foothills (Masson 1962: 14; 1982: 29, tab. 2, 3; Kurbansakhatov 1987: 53–55, 95, 99, tab. 1; Kircho 1999: 61–64, fig. 3). They are technologically close to the Lapui phase pottery of southeastern Iran (first half of the 4th millennium), which precedes the proto-Elamite culture (Banesh phase) (Sumner 1988). There is also some gray ware within the NMG II settlements (Masson 1962: 14; 1982: 29, tab. 3; Khlopin 1969: 29, tab. XXIV, *3*, *29*), attesting contacts with the SWT VI culture.

The major geometric decorative motifs of the Geoksjur tableware, the round hearths surrounded by a clay border, and the successive inhumations in collective burials have their distant prototypes in ancient Susiana and Mesopotamia (Sarianidi 1965b: 29–30, 48). Close analogies to the Geoksjur motifs are found in the wall paintings of ancient proto-Elamite Anshan; namely, Tal-i Malyan III (middle stage of the Banesh phase, 3300–2900 BC) (Amiet 1986: 107, fig. 52; Sumner 2003: figs. 15, 18). Unexpected matches exist between the motifs on cylindrical seals of Susa Cc–Da (Amiet 1972: vol. I, 144; vol. II, pl. 26, 119, 120 nos. 1184–1190 and 1194–1197) and one of the most common types of geometrical painted pottery decoration of Kara 1B (see Figure 4.5, *69*) (Masson 1960: tab. VIII, *16*; XXXVI, *11*).

Round-shaped terracotta and unfired clay items supposed to be wheel models with a one-side hub already appear in the early stage of the NMG II period (Figure 4.2 *13*) (Khlopin 1969: tab. VII, *45–47*; XXIII, *8*, *9*; XXVI, *48–50*; Kircho 2009a: 26, fig. 1). It is also the case for lapis lazuli beads and pendants (Kurbansakhatov 1987: 91, fig. 44, *1*; Masson 1962: 8, 22, tab. X, *3*), for copper round mirror and socketed adze-axe (Figure 4.2, *49*, *58*) (Solovjova *et al.* 1994: fig. 1, 4 and 6), or a terracotta model of a mold for casting such axes (Berezkin and Solovjova 1998: 93). Similar casting molds for axes are known in Iran at Tepe Ghabristan in the first half of the 4th millennium (Amiet 1986: 41–42, fig. 7). In the late stage of the NMG II period, 28 lapis lazuli beads accompanied the burial of a child at Kara Depe (Kara 2 period; Masson 1960: 415, tab. IX, *3*). Many lapis lazuli beads and pendants were found on the surface of Geoksjur 1 and of other settlements in the Geoksjur oasis of the late NMG II–early NMG III period (Sarianidi 1965b: tab. III; 1968, fig. 1, *1–6*, *10*, *13*, *15–17*, *21*).

Stone "weights" dated to the Middle Chalcolithic period have been found at Anau North and at Ilgynly Depe (Masson 1962: tab. XI, *10–13*; Korobkova and Sharovskaja 1994: 29; Kircho 2007: fig. 3, *3–6*). Similar items are known at sites in Iran and Afghanistan from the 4th millennium, like Sialk III.4 (Ghirshman 1938: 142, pl. LXXXV, S223), Hissar IC (Schmidt 1937: 404, pl. XVIII, A, H 2095), and Mundigak I.5

and II.1–3 (Casal 1961: vol. I, 235; vol. II, fig. 135, 4; see also Chapter 26). A "weight" from Sialk is made of gypsum, while the "weights" from Hissar, Mundigak, and the sites southwest of Central Asia are made of various kinds of limestone.

Cosmetic sticks have even more distant analogies. The same objects were used to apply make-up in pre-Dynastic Egypt (Freed 1982: 221, nos. 269–271). At sites dating from the 4th millennium in Iran, cosmetic sticks with droplet-shaped bulges on both sides have been found at Sialk III.1–2, 4, 7 (Ghirshman 1938: pl. LXXXIV, S402, S311, S225, S1804, S1745, S1643) and Tepe Gabristan (Tallon 1987: vol. 1: 250). From Mundigak II.3, comes a bi-spiral headed cosmetic stick with the spirals wound inwards. From the same site periods III.1 and III.3, are rods with drop-like bulges on both ends (Casal 1961: vol. 1: 248; vol. 2: fig. 139, 4, 7).

Another line of cultural interactions can be traced in the materials of late NMG II–early NMG III when the westernmost sites of the Anau culture are abandoned temporarily or permanently (Tilkin Depe, Ak Depe, Anau North), while Ak Depe near Ashkhabad is destroyed by fire (Kircho 1999: 53).

At the end of the 4th millennium, in the central area of the Kopet Dagh foothills, a new system of decoration appears for painted ceramics, the Kara Depe style of the NMG III period (Kara 1B–1A). The goat representations on the Kara 1B pottery (see Figure 4.5, *61*) are extremely close to those of the north Iranian site of Ismailabad (Maleki 1968: 53, fig. 60), and spotted animals figures ("leopards") on the Kara 1A pottery find parallels in the zoomorphic images on the Sialk III.7–Hissar IC–IIA pottery (Masson 1960: 376). At the same time, finds from Hissar IIA levels, like terracotta anthropomorphic tokens and female figurines, are also typical Anau artifacts of the Late NMG II–Early NMG III periods (Schmidt 1937: pl. XXVIIB, H 3735, H 3634). Furthermore, from the very beginning of the Late Chalcolithic, both at Kara Depe and Geoksjur 1, gray pottery, characteristic of the southeastern Caspian region, is found not only in the cultural layer, but also among the burial inventories (Masson 1960: 331, tab. XXX, *1–3*; Sarianidi 1965b: fig. 11, 4). Altogether, these analogies point at an increasing influence from central and northern Iran in the late 4th millennium and of a migration of populations from the southeastern Caspian region towards the northeast, up to the Ashkhabad oasis (Masson 1982: 51; Khlopin 1997: 119–120).

In turn, part of the South Turkmenistan population makes its way eastwards and to the northeast. In the ancient delta of the Murghab River the first settlements with ceramics and figurines of the late NMG II make their appearance (Masimov 1979: 72–77, fig. 2). Contemporaneous gray-polished pottery was recently discovered at the Zhukov sacred complex near Samarkand (Avanesova and Dzhurakulova 2008: 19, fig. 2, *14*, *15*), and the early pottery from Sarazm (Sarazm I–II) in the upper Zeravshan valley demonstrates that the foundation of this settlement directly involved communities from South Turkmenistan (Isakov 1991: 137; Lyonnet 1996: 55–57; Kircho 2007).

In general, the innovations seen in the Anau culture of the middle/second half of the 4th millennium point at three probably interrelated processes resulting from the Late Uruk Mesopotamian colonization (Lamberg-Karlovsky 1985: 56–58). The first is an indirect influence from the most developed areas of western Asia; the second is related to the beginning of long-distance trade, primarily based upon lapis lazuli (Sarianidi 1968, 1971); and the third is due to the activation of contacts and relocations of population groups from central, northern, and especially southwestern Iran, the proto-Elamite heartland. The latter must have been related to the rise of the Geoksjur complex.

Thus, the Middle Chalcolithic Anau culture lies at the basis of the Late Chalcolithic in South Turkmenistan, but an influx and, probably, an infiltration of population from Iranian centers led to the development of wide interregional cultural interactions.

THE LATE CHALCOLITHIC PERIOD (3200/3100–2800/2700 BC)

As shown above, this period, which corresponds to NMG III in southeastern Turkmenistan, is very tightly related to the previous phase.

Environment

The change in the course of the riverbeds (Marcolongo and Mozzi 2000), together with, probably, a gradual climatic aridity, led the population to concentrate in a more limited number of medium-sized and large settlements (Geoksjur 1, Geoksjur 5, Altyn Depe, Khapuz Depe, Ulug Depe, Namazga Depe, Kara Depe, Gara Depe). At that time, Namazga Depe, Altyn Depe, and possibly Ulug Depe gained their maximum sizes, while Anau North, Elen Depe, and Ilgynly Depe are abandoned. By the middle of the Late Chalcolithic, the Geoksjur oasis was completely abandoned and its population moved upwards along the Tedjen where Khapuz Depe was founded (Sarianidi 1964, 1969). Almost at the same time, the center of the Artyk oasis, Kara Depe, was also abandoned, as well as the nearby Gara Depe at the beginning of the Early Bronze Age (Shchetenko 1968: 24–26). Their population had probably gradually transferred to the mountain valley of Darreh Gaz (Yarym Tepe) (Kohl and Heskel 1980). In sum, the end of the Late Chalcolithic period is represented almost exclusively at the largest settlements of the Kopet Dagh piedmont area, and of the ancient Tedjen River valley (i.e., at Namazga Depe, Gara Depe, Ulug Depe, Khapuz Depe, and Altyn Depe).[3]

Economy

At that time, which corresponds to NMG III, the economy passes a new step. More complex irrigation canals are built to bring water closer to Geoksjur 1 after the Tedjen started to change direction (Lisitsina 1966), and bulls are probably now used as draft animals for tilling the soil. The finds at Altyn Depe levels 12–10 of single-axis cart terracotta models with one draft beam (see Figure 4.3 and Figure 4.8, *32*, *40*), and those at Kara Depe level Kara 1A of bull figurines with a painted probable harness or with a through hole in the shoulder (see Figure 4.7, *18*, *19*), point at a farming system with a two-animal plowing team (Kircho 2009a: 27).

Pottery

As during the previous stage, two local cultural variants are identified. One, the NMG III cultural complex, develops in the central area of the foothills zone and has been best studied at Kara Depe and in stratigraphic soundings at Namazga Depe and

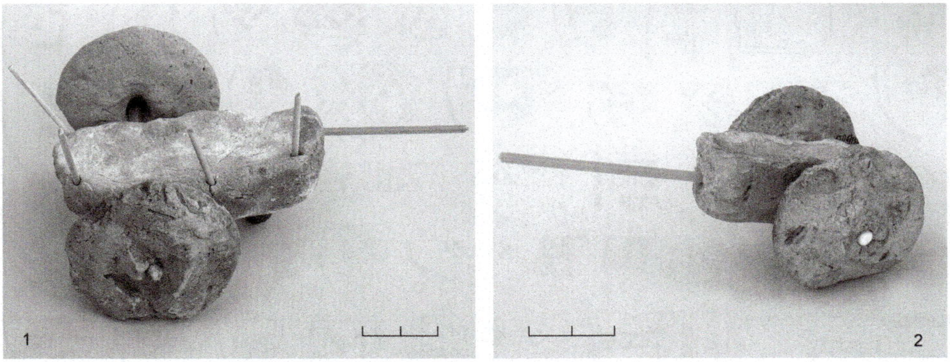

Figure 4.3 Altyn Depe, cart models of the Late Chalcolithic period.
1, unbaked clay; 2, terracotta.

Ak Depe. The other, in southeastern Turkmenistan, corresponds to the Geoksjur and post-Geoksjur cultural complexes known from the sites of Geoksjur 1, Geoksjur 5 (Chong Depe) and Altyn Depe, as well as from Khapuz Depe and Ulug Depe.[4]

The typical NMG III pottery is mostly made of a fine vegetal temper decorated with painted motifs in the Kara Depe style. During the early and short phase, Kara 1B pottery types continue and follow the technological tradition of the end of the Middle Chalcolithic period. They consist of relatively thick vessels rather heavily vegetal tempered with a grayish unevenly fired core, covered with a thick layer of light (cream-colored) or, much rarer, red slip. Their surface is entirely polished and painted over with thin dark-brown (almost black) geometric motifs consisting of large stepped motives, crosshatched or plain triangles, vertical and diagonal bands, concentric or crosshatched rhombs, and circles with central dots (Figure 4.4, *19*, *56*, *62*, *68*; Figure 4.5, *61*, *65–76*, *81*, *82*). Most of the Kara 1B decoration motifs have direct analogies and prototypes in the Geoksjur style. This ware was quickly replaced by the Kara 1A painted ceramics (i.e., a thin-walled tableware heavily tempered with thin vegetal elements, covered with a light, often greenish slip and well fired). Vessels are decorated with a monochrome thick brown-blackish paint. This most common NMG III tableware, associated with various degenerated geometric and zoomorphic ornaments (Figure 4.4, *9–11*, *18*, *20*; Figure 4.5, *1–60*, *62–64*), determines the very complex of NMG III itself. Popular zoomorphic motifs include goats in profile, spotted animals ("leopards"), birds, and also "eagles" with spread wings (Figure 4.5, *8*, *9*, *13*, *15*, *21–24*, *30*, *36–38*, *45*, *50*, *52*, *56*, *58*, *64*, *77*). From the analysis of its technological features and the development of the decoration system, an intermediary stage from the late Kara 1A to the late NMG III, represented mostly in the materials of Ak Depe, Namazga Depe, and Altyn Depe has been pointed out (Kircho 1999: 43–45).

In southeastern Turkmenistan, pottery at that time is typically fine mineral tempered, decorated with monochrome or bichrome geometric motifs mainly of the late Geoksjur and post-Geoksjur styles, including numerous geometric representations of goats (see Figure 4.4, *3*, *4*; Figure 4.8, *64–82*, *87*) (Kircho 2000a: 192; Kircho *et al.* 2008: 119–123, figs. 19–5). In addition, a foreign polychrome painted tableware has been discovered at Kara Depe, Ak Depe, and Altyn Depe (Kircho 2000a: 187, 190,

Figure 4.4 Geoksjur 1 and Geoksjur 5 (Chong Depe), Geoksjur period: pottery of the end of the Middle Chalcolithic, and the early and middle phases of the Late Chalcolithic (end NMG II, early/middle NMG III, ca. 3200–3000 BC). *1–37*, Geoksjur 5 (*9*: Kara 1A type; 10, *11, 18, 20*: early Kara 1A type; *17, 19*, Kara 1B type); *38–88*, Geoksjur 1 (*56, 63, 68, 69*: Kara 1B type; *62, 86*, Kara 2 type).

193, fig. 2; 5, *18, 23*) (Figure 4.5, *33, 78*; Figure 4.8, *59*), and gray pottery from the southeastern Caspian region has been found at all the South Turkmenistan sites of the NMG III period (Figure 4.5, *74, 79, 80*) (Sarianidi 1965b: 23, tab. V, group 2, *12–18*; Masson 1982: 43, tab. XIX, *26, 29, 32*; XXI, *37, 41*, 42, *44–50*; Kircho *et al.* 2008: 112, group 13, tabs. V–VIII).

Architecture

During the Late Chalcolithic period, Altyn Depe expanded almost 1.5 times, reaching 25 ha. Almost at the same time a peripheral (retaining) wall was built around the newly built territory, forming a straight line in the western part of the settlement (see Figure 4.8, *1–3*). The main street, still visible on the actual relief, connected the

Figure 4.5 Evolution of NMG III–type painted pottery complex (3200/3100–2800/2700 BC). *1–5*, Gara D.; *6–17*, Ak D.; *18–82*, Kara D. (*1–9*, *13*, *18–32*: late Kara 1A; *33–53*, Kara 1A; *54–60*, *62–64*: early Kara 1A; *61*, *65–82*: Kara 1B (*74*, *79*, *80*: grey ware)).

southwestern and northeastern entrances to the settlement, and was made probably at the end of the Middle and during the Late Chalcolithic period. In all, the layers of the late NMG II–NMG III period reach a total thickness of 4.6 m and represent six to seven construction levels.

The NMG III period settlements, like Kara Depe and Geoksjur 1 and 5, which have been excavated over a significant area, show tightly packed adjacent households,

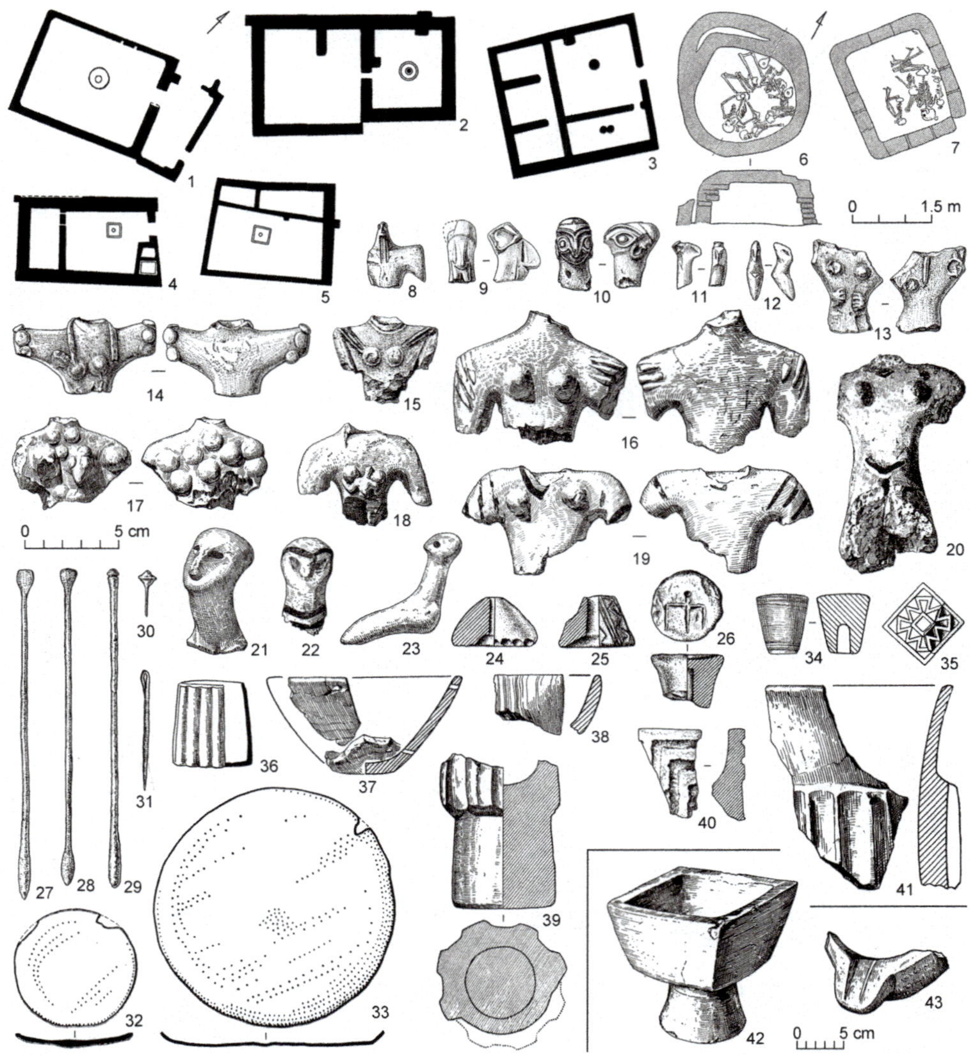

Figure 4.6 Geoksjur 1 and Geoksjur 5 (Chong Depe), Geoksjur period: architecture, terracotta, metal, and stone artifacts of the end of the Middle Chalcolithic, and the early and middle phases of the Late Chalcolithic (end NMG II–early/middle NMG III, ca. 3200–3000 BC). *1–5*, plans of houses; *6, 7*, burial chambers; *8–26*, terracotta artifacts; *27–33*, metal artifacts; *34–43*, stone artifacts.

forming quarter-blocks divided by narrow streets (Masson 1982: 56). Within these blocks, each residential complex consisted mostly of two rooms, sometimes with additional utility compartments along one of the largest, main-room walls. As was the case previously, the house entrances from the courtyard or the streets led directly into the main (largest) room. In the Geoksjur oasis settlements, fireplaces with a round (rarely square) clay base were placed on the floors of the largest rooms (Figure 4.6, *1, 2, 4, 5*) (Sarianidi 1965b: figs. 1, 6, 7, 9; 1966: figs. 32, 33; Masson 1982: 36, 38,

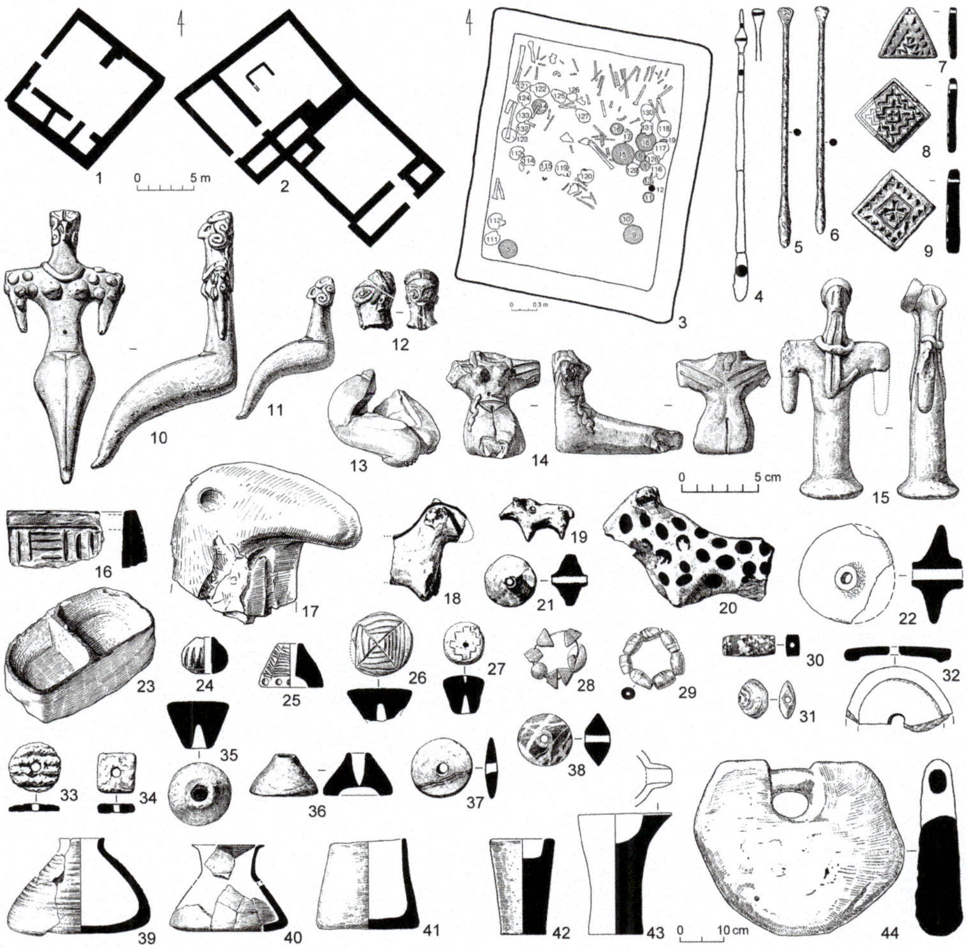

Figure 4.7 Kara Depe, main components of the cultural complex of the Late Chalcolithic period (early and middle NMG III, ca. 3200/3100–2900 BC). *1, 2*, plans of houses; *3*, burial chamber; *4–6*, metal artifacts; *7, 8, 28–44*, stone artifacts; *9–27*, terracotta artifacts.

44; Kircho 2007: fig. 1, *3*, *4*, *6*, *7*, *9*). V.I. Sarianidi considered such rooms as domestic chapels, although most probably, these hearths are just the continuation of the Middle Chalcolithic traditional ones. At Kara Depe, hearths in the center of large rooms were often made with bottomless kitchen vessels dug into the floor (Masson 1960: fig. 15; 1982: 36). However, at this site, excavation 4, among the two cases of double-rooms houses (i.e., with a narrow entrance and a spacious room), one had a square hearth with vertical borders placed in its center (Figure 4.7, *2*) (Khlopin 1971: 222–224, figs. 1, 2; Masson 1982: 39, tab. XX). A similar layout was found in double-room houses of the late NMG III period at Altyn Depe (Figure 4.8, *4–6*), two of which had central hearths on a round clay base (Kircho *et al.* 2008: 53, 59–70, figs. 5a, 11–13, 16).

The creation of a special small entrance may have been due to the deterioration of the climatic conditions in the 3rd millennium. In addition, special raw-brick structures made of several low parallel walls that served as bases for granary platforms were found in the courtyards at Altyn Depe and especially at Kara Depe (Masson 1982: 37–38; Kircho *et al.* 2008: 67–68, tab. 166). This granary interpretation is indirectly supported by the absence of large container vessels in the pottery complexes of the NMG III period at these two sites.

Burials

At Geoksjur 1 in the Late Chalcolithic period, the collective circular (rectangular in one case) burial chambers built above ground with successive inhumations (Figure 4.6, *6, 7*) (Sarianidi 1965a; 1966: 91, fig. 32; Masson and Sarianidi 1972: fig. 15) were to one side of the settlement. Similarly, at Kara Depe in the early NMG III period, the individual burials dug within the abandoned cultural layers created a small separate cemetery (Masson 1982: 39). This separation points at the beginning of a settlement division into specialized areas. Successive inhumations of at least 23 individuals in a rectangular burial chamber are also attested at Kara Depe, in the late Kara 1A period (Figure 4.7, *3*) (Masson 1982: 39).

Artifacts

Among the characteristic components of the Late Chalcolithic cultural complex, anthropomorphic (mostly terracotta) figurines with details in relief, like elaborate hairstyles, headdresses, jewelry, and signs on the shoulders or thighs, played a significant role (Figure 4.6, *9–18*; Figure 4.7, *10–15*; Figure 4.8, *8–18*). The female figurines were depicted in sitting position, and standing men always had double-braided beards (Figure 4.6, *8*; Figure 4.7, *15*; Figure 4.8, *16, 17*) (Masson and Sarianidi 1972: fig. 28–31; Masson 1982: 60–61, fig. 9, tab. V, type III and 4; XXI, *12, 14–18*; XXII, *15–18, 28–30*; Kircho 2009b: 380, fig. 3, *13–36*; Brunet 2011: fig. on pp. 250 and 254).

It is also during the Late Chalcolithic period that bone, stone, and terracotta seal-amulets and button seals, as well as their imprints on clay bullae (Figure 4.8, *33–35*), appear in South Turkmenistan (Figure 4.6, *35*; Figure 4.7, *7–9*; Figure 4.8, *36–38*). They point out the beginning of property and counting (Masson and Sarianidi 1972: fig. 25; Kircho *et al.* 2008: 129–131, pl. 119, *1, 2*; 145, *13–21*; 168).

At Altyn Depe and Ulug Depe, typical terracotta boxes/"reliquaries" decorated with relief geometrical ornaments (the earliest square vessels are sometimes painted) are considered as containers for the anthropomorphic figurines (Figure 4.8, *42–44*) (Kircho *et al.* 2008: 127–128, pl. 12, *15*; 23, *22*; 30, *18–20*; 46, *5*; 73, *10, 11*; 96, *14–17*; 109, *32*; 120; 121, *1–12*; 147, *25–17*; 148).

Stone (mostly alabaster) cosmetic vessels of various shapes are widespread, often with an engraved decoration (Figure 4.6, *36, 38, 39–41*; Figure 4.7, *39–41*; Figure 4.8, *48–55*) (Masson 1960: tab. XVII, *1–6, 8, 9*; XVIII, *14, 15*; Sarianidi 1965b: fig. 11, *1*, pl. XXVI, *2, 13, 15*; Kircho *et al.* 2008: 171–172, fig. 28; Kircho 2009b: fig. 6). Production areas where such vessels were made have been discovered at Altyn Depe, and the different technological steps for their production have been identified. Gypsum or alabaster was also used for massive bull figurines (Masson and Sarianidi 1972: fig. 22) and miniature anthropomorphic ones (Figure 4.8, *46, 47*) (Kircho *et al.* 2008: pl. 141, *5–7*).

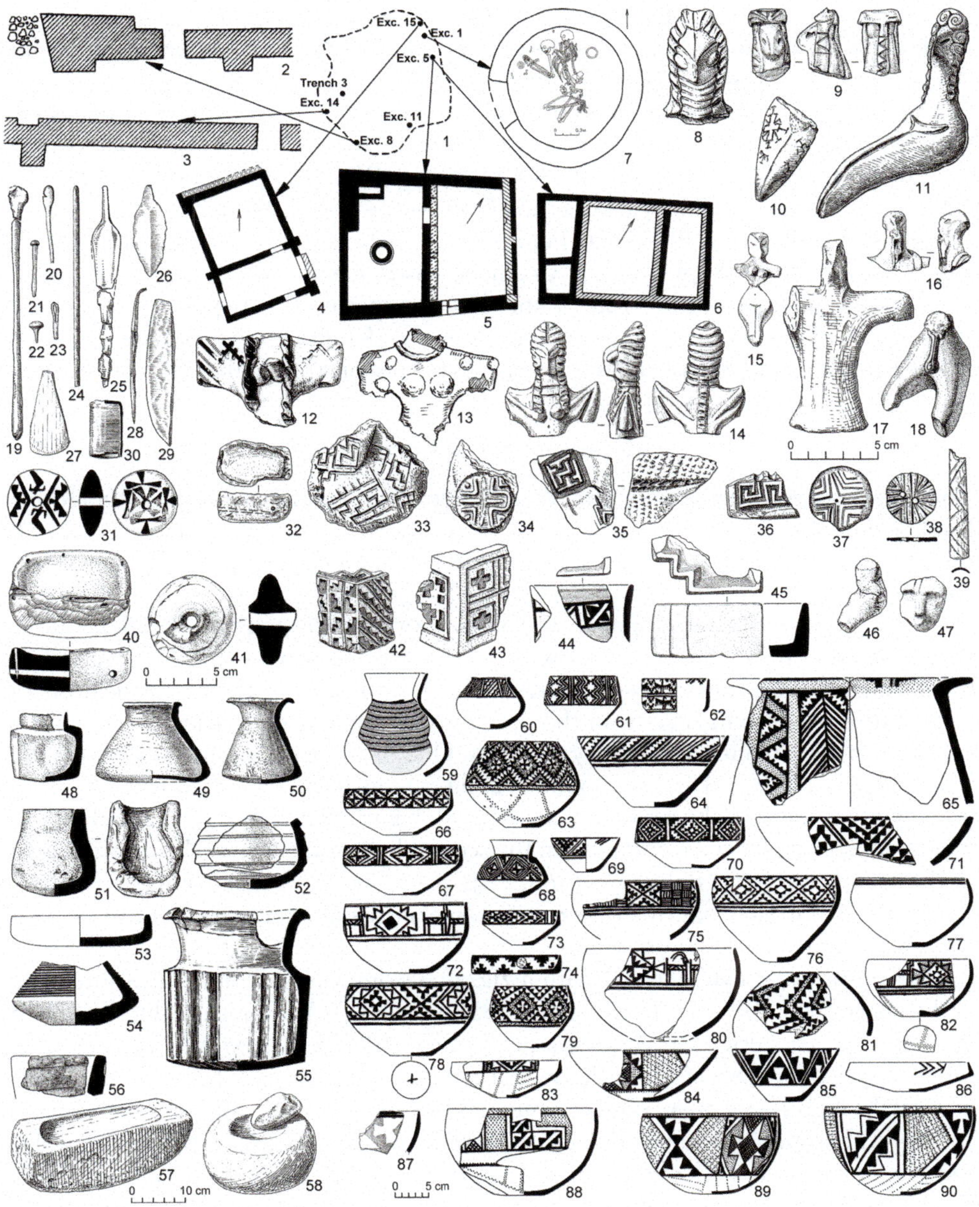

Figure 4.8 Altyn Depe (Altyn 14–9 levels, ca. 3300–2800/2700 BC), main components of the cultural complex of the last phase of the Middle Chalcolithic period (83–90) and Late Chalcolithic period (1–82). *1*, schematic plan of the location of the excavations; *2*, *3*, the retaining walls; *4–6*, plans of houses; *7*, burial chamber; *8–18*, *31*, *32*, *36*, *42–45*, terracotta artifacts; *19–30*, metal artifacts; *33–35*, slightly baked clay; *37*, *40*, *41*, unbaked clay; *38*, *39*, bone artifacts; *46–58*, stone artifacts; *59–90*, pottery.

Since the end of the Middle Chalcolithic period, relatively large mortar vessels made of solid limestone are also known in South Turkmenistan, either zoomorphic (Figure 4.2, *17*) (Masson *et al.* 1994: fig. 6, 2), with spouts (Figure 4.6, *43*; Figure 4.7, *43*), or with relief geometric decoration (Figure 4.8, *56*). Unique footed stone vessels with a subsquare upper part have been found in the "sanctuary" at Geoksjur 1 (Figure 4.6, *42*) (Masson 1960: tab. XXXI, *3, 6, 9*; XXXII, *3*; Sarianidi 1965b: tab. XXVI, *7–9*; Kircho and Sharovskaja 2001: 120–124, fig. 1). Stone "weights" with a distinctive handle are still present (Figure 4.7, *44*) (Masson 1960: 373, tab. XXXII, *1, 4*; Kircho *et al.* 2008: pl. 100, *1*).[5] Cosmetic sets in female burials include alabaster vessels for make-up, copper cosmetic sticks with a spatula, a flat spiral, or cone head (Figure 4.6, *27–29*; Figure 4.7, *4–6*; Figure 4.8, *19*) (Kircho 2007: 201, fig. 4, *16–18, 20–22*), and a round mirror (Figure 4.6, *32, 33*) (Masson 1982: tab. VIII, *1–6, 18–20*).

Relations and influences

Contacts between the populations of southwestern Turkmenistan and the Kopet Dagh piedmont area are visible thanks to the presence of small alabaster vessels found among the burials items from the SWT V period (Khlopin 1997: 110, tab. 68, 2).

The general cultural development and innovations seen during the Late Chalcolithic period are due to an economic growth with a development of production and of the social organization (Masson 1982: 55–57), and also to the intensification of interrelations. The influence from the developed centers of the ancient Near East is visible primarily in cult and luxury artifacts. One of the female figurines from Kara Depe (Figure 4.7, *17*) (Kircho 2000b: 112, fig. 1) has been compared to those from pre-Dynastic Egypt (Ucko 1968: no. 48, fig. 38). Figurines and alabaster small vessels with complex shapes were common in the second half of the 4th–early 3rd millennia at Susa (Le Brun 1971: fig. 54, *1–4*; 68, *1, 3, 4–12, 13, 14*; Amiet 1986: 57, fig. 19). A miniature alabaster figurine of a ram is also known at Mehrgarh VII in northern Baluchistan (Jarrige 1988: 95, fig. 61).

The imprint of a cylinder seal with a zigzag decoration was found at Altyn Depe in the latest NMG III level (Kircho *et al.* 2008: pl. 168, *10*) and recalls the seals from the Jemdet Nasr period in Mesopotamia (Amiet 1983: 206, 207, fig. 1g, *15*). One cylinder seal with a bovid is also known at Sarazm (Isakov 1996: fig. 10).

Hearths on a rectangular base surrounded by a clay border, double-room houses, and conical stone vessels are already known in the early phases Mundigak I.4–I.5 and II.1–II.3 (Casal 1961: vol. I, 30–36, 233; vol. II, fig. 6–12, 134, *1, 1a*; pl. IVB, VA, VIA, VII). In southeastern Turkmenistan and the Zeravshan valley (Sarazm II), these cultural components appear almost simultaneously at the end of the 4th millennium (Kircho 2007: 196, fig. 1, *7, 8, 10*), and probably reflect the influence of the Mundigak cultural tradition. A similar hearth is also known at Hissar IIB (Schmidt 1937: 107, figs. 62, 63).

A cosmetic stick with a spatula head, similar to the South Turkmenistan ones, was found at Tepe Yahya IVC (around 3300–2900 BC) in southeastern Iran (Lamberg-Karlovsky and Tosi 1973: fig. 124, bottom).

The decoration on the handmade and "wheel-turned" Mundigak III pottery (end of the 4th to the first quarter of the 3rd millennium) is close to the geometric motifs of the late NMG II–NMG III periods. The closest similarities with the Geoksjjur-style or Kara 1B pottery (Masson 1960: pl. VII, *13*, *21*, *23*; XXX, *4*, *7*, *9*) concern Mundigak III.6 (Casal 1961: vol. II, fig. 52, *44*, *45*; 53, *50*, *57*, *59*, *60*; 54, *60*; 54, *65*; 55, *72*, *73*, *81*; 57, *95*; 59, *117*, *124–128*; pl. XXXA). Anthropomorphic figurines with South Turkmenistan features were also found in Mundigak II.2–III.1–III.2, and round and square stone button seals in Mundigak II.2–III.4–III.6 (Casal 1961: vol. I, 253, 256; vol. II, pl. XLI, *1–3*, pl. XLVA, *2–6*). Stone, terracotta, bone, or ivory button seals are also present at Mehrgarh V–VII (ca. 3300–2800 BC) (Jarrige 1988: 98–100, figs. 71–73, 75–82). Mundigak III.4 and IV.1 have provided footed rectangular containers (Casal 1961: vol. 1, 227–228; vol. II, fig. 130, *4*, *6*), resembling the Altyn Depe level 8 "reliquary" from the beginning of the Early Bronze (see Figure 4.9, *37*).

These analogies, together with the unique round hearth altar with a surrounding clay border and the successive inhumations in collective rectangular brick chambers at Mundigak III.6 (Casal 1961: vol. I, 41, 44–46; vol. II, fig. 18, room CXXI; pl. XB–D), testify to direct contacts between the two populations in the late 4th/early 3rd millennium.

The geometric motifs on the golden and silver vessels of the Fullol Hoard (northeastern Afghanistan) (Tosi and Wardak 1972: figs. 3, 4a, 4b, 8) have their prototypes in the Geoksjur and Kara 1B pottery decoration (Figure 4.2, *34*; Figure 4.5, *82*). Finally, the seal-amulet recently discovered at Sarazm (Razzokov *et al.* 2006: fig. 134) is identical to the Kara Depe ones, both by the technique of its manufacturing, and by its red-brown stone (Kircho 2014a: 63).

Striking analogies with Shahr-i Sokhta, the largest center of Iranian Seistan, have long been pointed out (Biscione 1973; Sarianidi 1983). The monochrome painted decoration of Shahr-i Sokhta I, phases 10–9, is similar to the Geoksjur and Kara 1B ones. The question on the role played by the Geoksjur cultural tradition in the formation of this early Shahr-i Sokhta complex has been raised in many publications. Direct movements of population groups from South Turkmenistan have often been proposed (see, e.g., Masson 2006: 58–59). A Geoksjur monochrome decorated bowl and a conical stone vessels have been recently discovered in graves of Shahr-i Sokhta I (Sajjadi 2003: 66–74, tab. 4, figs. 22c, 33i, 35). Period I levels also contained numerous bullae with seal impressions, including those of Jemdet Nasr/Early Dynastic I cylindrical seals (Amiet and Tosi 1978; Amiet 1983; Tosi 1969: 283–285, fig. 285). In the late Shahr-i Sokhta I period, a unique semi-dug round-shaped brick burial chamber contained the successive burials of 13 people (Mariani and Tosi 1987: tab. XXXII). At the turn of periods I–II (around 2750 BC), cosmetic sticks and stone containers[6] appear (Piperno and Tosi 1975: 186–197, fig. p. 195 bottom; Sajjadi 2003: fig. 36c, type 2; 40, type 1). In addition, from the very start of period II, the Shahr-i Sokhta houses usually contained hearths on a square clay base and stone button seals (Lamberg-Karlovsky and Tosi 1973: figs. 19, 32–45).

The extensive evidence of interaction between South Turkmenistan and the neighboring or more distant regions in the late 4th/early 3rd millennium probably reflects two main processes. On the one hand, active contacts existed, related to the proto-Elamite expansion and explained by the mining and trade of valuable mineral raw materials – lapis lazuli, alabaster, metals, etc., that were delivered through the mediation of the Central Asian, Afghan, and Iranian populations to the centers of Western Asia (see also Chapters 1 and 21). During the 3rd millennium, Shahr-i

Sokhta was a large processing center for lapis lazuli and the manufacture of stone vessels, while Tepe Yahya near Kerman produced steatite items (Lamberg-Karlovsky and Tosi 1973). On the other hand, the diffusion of shared similar traditions – be it in house construction (size of mudbricks, low manhole entrances, fireplaces on a clay base), in the burial types, and in the set of the funerary artifacts (cosmetic stone vessels and copper sticks, seals) – probably does not only reflect contacts between populations, but also the formation of a kind of unified subculture in which the Geoksjur and Mundigak cultural traditions were also taking part. Anyhow, this Late Chalcolithic cultural background developing in southern Turkmenistan served as a basis for the early urban civilization of the Bronze Age in Central Asia.

THE EARLY BRONZE AGE (CA. 2800/2700–2400/2300 BC)

This period corresponds to NMG IV. In southern Turkmenistan, the large settlements of Namazga Depe, Ulug Depe, Khapuz Depe, and Altyn Depe concentrate most of the population. In the central part of the Kopet Dagh foothills, only one large center, Namazga Depe, has been identified. At Ak Depe near Ashkhabad, the 3rd millennium material synchronous with late NMG III and NMG IV is dominated by gray ware with applied decoration (but light-slip painted pots are also present), terracotta anthropomorphic ladles, double-sided seal-like artifacts, and a specific type of female figurine (Sarianidi 1976: 95, fig. 15, *1*, *9*, *14*; Kircho 1999: 113, fig. 21, *9*, *10*, *12*; 2014b: fig. 4, *1*), unknown at sites located to the east of Ashkhabad. The Ak Depe complex is original in that, during the first half of the 3rd millennium BC, it testifies to contacts both with the Namazga–Altyn cultural area and with the southeastern Caspian one.

Architecture

At Altyn Depe, within the settlement, an organized system of communications appears with large streets almost intersecting in the center of the mound and, accordingly, four entrances (Kircho 2011: 83). Monumental perimeter walls bolstered with counterforts are built (Figure 4.9, *2*). During the whole period, the density of constructions constantly rises. In the early stage, residential quarters consist of several double-room houses (i.e., a main room and an entrance hall) with a common utility courtyard. The center of the main room is occupied by a hearth on a round or square clay base (Figure 4.9, *3*). The burial chambers and production structures like kilns are built in special remote areas on the outskirts or inside the courtyards (Kircho 2001b: 31–33, 35). Important social changes are visible in building works like the erection of powerful retaining walls needing a cooperation between different groups.

Burials

In addition to the round or rectangular burial chambers with collective inhumations (Figures 4.9, *4*; 4.10, *5*), individual burials dug into the cultural layers are known, as well as a few cases of three simultaneous inhumations. "Standard" sets of burial artifacts for rich female graves are now well established (Kircho 2005a: 418–421, figs. 3, 4, 6, 18–21).

Material culture

An intense development of pottery (Figures 4.9, *66–88*; 4.10, *50–79*) and the production of valuable stone items (Figures 4.9, *39–65*; 4.10, *28–49*) are probably primarily

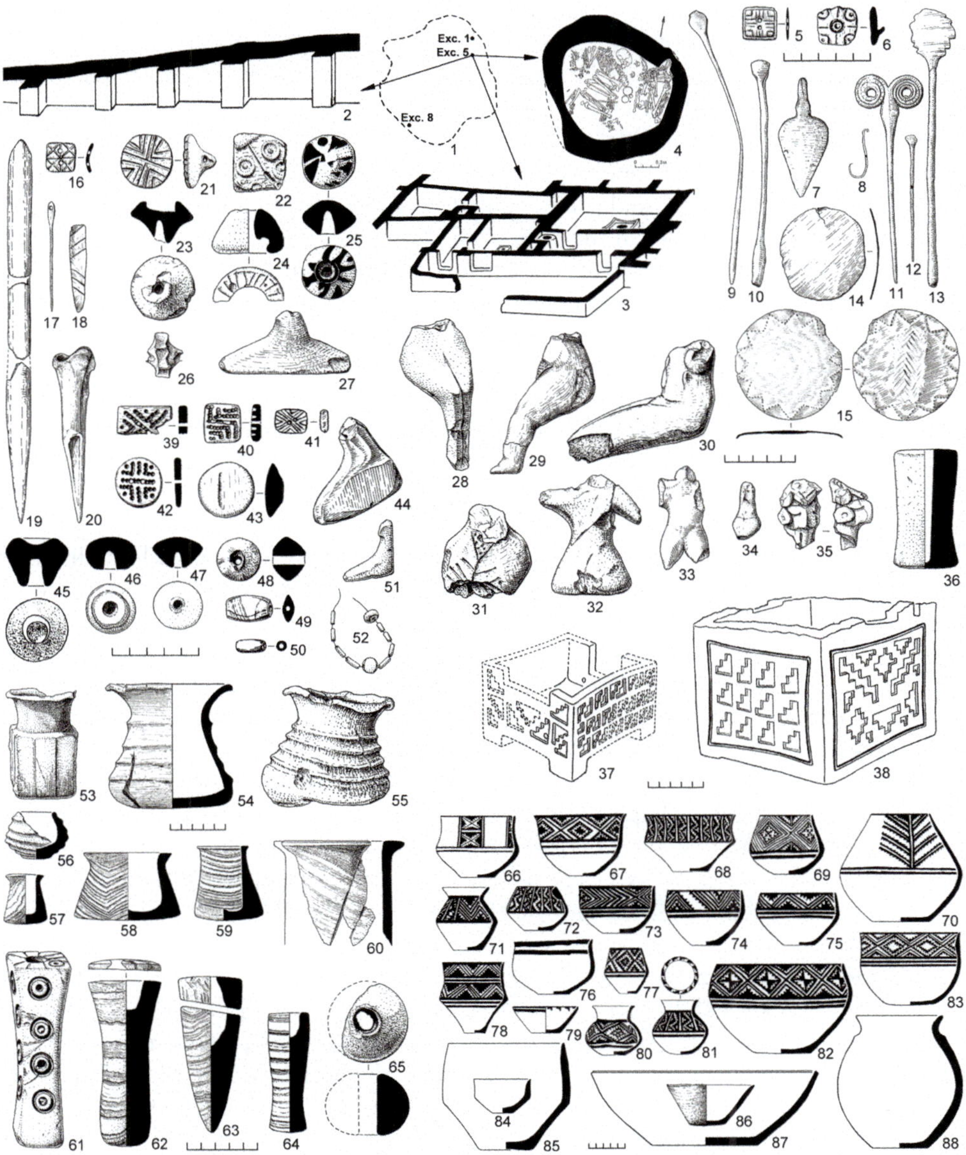

Figure 4.9 Altyn Depe (8–6 levels), main components of the cultural complex of the early phase of the Early Bronze Age (ca. 2800/2700–2500 BC). *1*, schematic plan of the location of the excavations; *2*, retaining wall; *3*, isometric view of a house; *4*, burial chamber; *5–15*, metal artifacts; *16–20*, bone artifacts; *21, 23–27, 29–31, 33–38*, terracotta artifacts; *22*, slightly baked clay; *28, 32*, unbaked clay; *39–50, 53–65*, stone artifacts; *51, 52*, heated-stone artifacts; *66–88*, pottery.

due to the introduction of rotational tools (potter's wheel and mounted drills), brought in from the more developed Iranian and south Afghan centers. Metalworking testifies to a significant evolution, as shown in the stamped or engraved decoration that appears on mirror and seal, and on the steppedrhomb shaped heads of the cosmetic sticks (Figure 4.9, *6*, *13*, *15*). Use-wear analyses of tools and artifacts have shown that metal instruments are increasingly used and that almost all the areas of production are now specialized (Korobkova 2001: 190).

The Early Bronze Age material culture continues and develops the Late Chalcolithic traditions (Masson and Kircho 1999: 65–68, figs. 1, 2). Almost all the major components of the early NMG IV (levels 8–6 at Altyn Depe), like dwellings and graves, pottery shapes and their geometrical decoration, anthropomorphic figurines, button seals, "reliquaries," alabaster cosmetic vessels, copper mirrors, and cosmetic sticks, have their prototypes in the Late Chalcolithic materials (Kircho 2004: 387).

However, there are also innovations. At the very beginning of the Early Bronze Age, the first copper and terracotta stamp seals appear (Figure 4.9, *6*, *21*) (Kircho 2004: 388, fig. 4, 2, *12*) and marble-like stone cosmetic vessels spread out (Figure 4.9, *57–60*) (Kircho 2014a: 67). The main innovation of the NMG IV period is the presence of stone make-up containers (previously considered as lamps) (Figure 4.9, *61–64*) in the inventories of rich female burials, clearly related to a specific burial rite or to a social group.

Relations and influences

The relations with the surrounding areas also follow the previous ones. Hearths on a square clay base, storage vessels and cosmetic containers, metal items, and seals are all known at Mundigak IV.1–IV.2 (Casal 1961: vol. I, 229, 233, 234, 236, 249, 257; vol. II, fig. 131, *16*; 134, 4, 9, *10*, *15*; 135, *11*, *11a–d*, *12*; 139, *17*; 140, *19*, *20*; pl. XLVA, 9). The applied decoration on boxes – "reliquaries" of southeastern Turkmenistan (Figure 4.9, *37*, *38*; Figure 4.10, *23*) has direct analogies with that on Jiroft containers (Perrot and Majidzadeh 2005, fig. 10) and with the sculptured frieze of the monumental "palace" at Mundigak IV.1 (Dumarçay 1984: 56–61, figs. 7–9). At Shahr-i Sokhta II.5 a wooden "reliquary" is identical in shape and decoration with those of South Turkmenistan, but painted in blue (Tosi 1969: fig. 125). Similarities can also be underlined in the shapes of the stepped niched walls at Altyn Depe 6 "sanctuary" and at Hissar IIIB "burnt building" (Schmidt 1937: 165, fig. 93; Masson 1988: pl. I, *1*; Dyson and Remsen 1989: 93–95, figs. 19, 22). Close analogies to a painted wheel-turned goblet sherd (Kircho *et al.* 2008: pl. 44G, 4) are visible in the Mundigak IV.1 pottery (Casal 1961: vol. II, fig. 74, 237, 237c) and at Shahr-i Sokhta II–III (Tosi 1969: figs. 40, 41, 42c; Sajjadi 2003: figs. 33c, e–h; 35l; 36a, burials 1505 and 1900; 36c, type 4; 40d, type 1, 2b, burial 1605, 3). Flat trapeze-shaped lapis lazuli beads are present among the burials of Shahr-i Sokhta II and Altyn Depe in the early NMG IV period (cf. Sajjadi 2003: fig. 40a, type 8; and Kircho 2005a: 393, fig. 23, type IIB2). Among the direct imports from SWT IV period found within the inventory of female burials at Altyn Depe are a double spiral pin (Figure 4.9, *11*) and a gray-ware footed bowl (Figure 4.10, *71*) (Kircho 2005a: 352, 401, fig. 4, 4; 5, *34*; cf. Khlopin 2002: tab. 7, *5*, *11*; 13, 4; 14, 7; 51, 9; 56, 2, 3).

In the middle of the Early Bronze Age (around 2600/2500 BC), the appearance of soft rock heating to increase the hardness of the items testifies of an influence (possibly indirect) from northern Baluchistan (Mehrgarh) and from the Kot Dijian pre-Indus

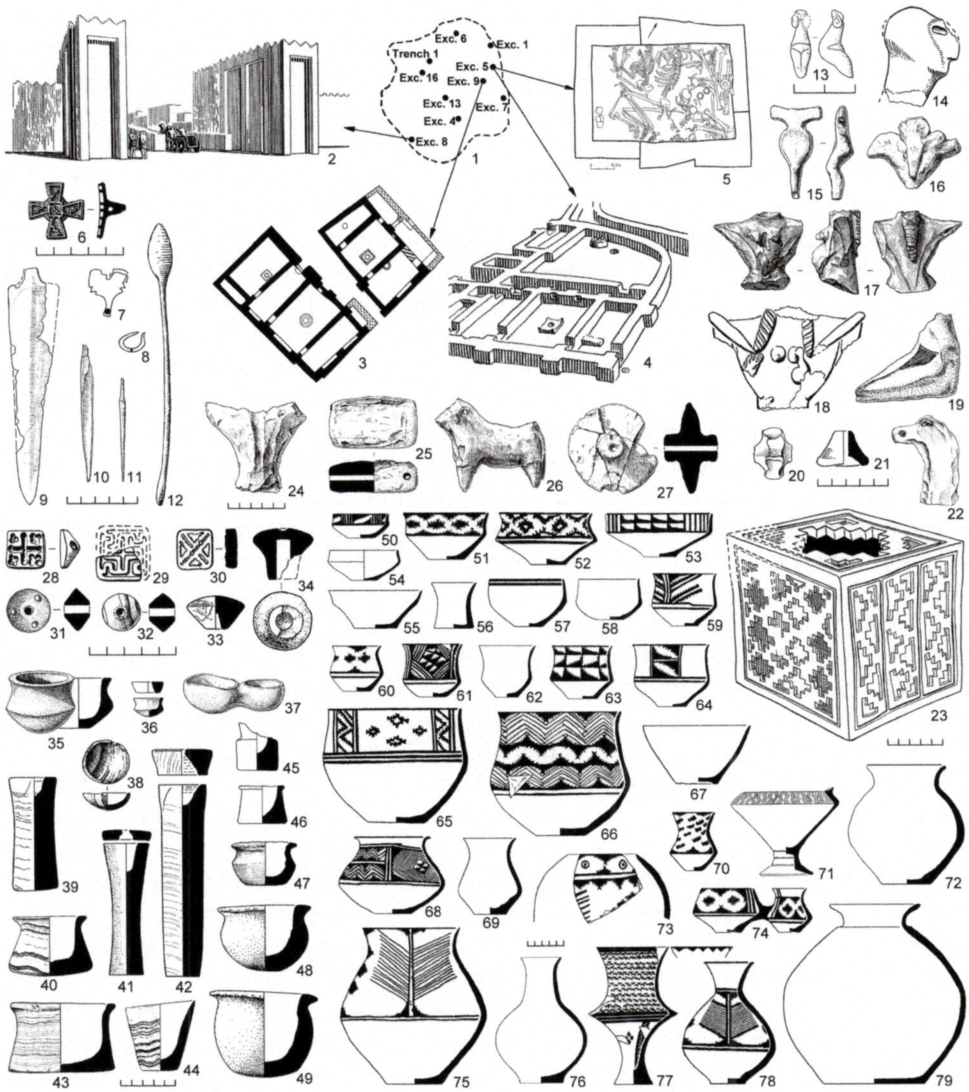

Figure 4.10 Altyn Depe (5–4 levels), main components of the cultural complex of the late phase of the Early Bronze Age (late NMG IV, ca. 2500–2350 BC). *1*, schematic plan of the location of the excavations; *2*, south–western town entrance (reconstruction, after Masson 1988, Fig. 11); *3*, *4*, plan and isometric view of houses; *5*, burial chamber; *6–12*, metal artifacts; *13–23*, terracotta artifacts; *24–27*, unbaked clay artifacts; *28–49*, stone artifacts; *50–79*, pottery.

Civilization through the borrowing of a specific technology for steatite artifacts (see Vidale 2000: 58–66). However, besides the flat-ring and small cylindrical beads typical of the Indian production, one anthropomorphic figurine (Figure 4.9, *51*), a seal and heated Turkmenistan-type beads made of local minerals were found at Altyn Depe, making clear that it was technology that was imported rather than the artifacts themselves (Kircho and Kovnurko 2003: 109–110, fig. 1, *1*, *4*, *5*, *10*).

At the end of the Early/beginning of the Middle Bronze Age, the evolution process of specialized crafts production, which is the economic basis of a proto-urban civilization, had reached its highest level. There was an outbreak in the field of firing equipment like the high-temperature heating of stone items, the double-chamber pottery kilns, and the lost-wax casting. At Altyn Depe, the main southwestern entrance is now framed by two pylons/towers and has a Near Eastern appearance (Figure 4.10, *2*); it is divided into a pedestrian and a wheel-cart part (Masson 1988: 31–33, fig. 11).

In the late phase of the Early Bronze in addition to the double-room houses, new types of houses appear, made of three or four rooms, each with its own courtyard (Figure 4.10, *3*, *4*) (Kircho 2001b: 31). Streets paved with large broken pottery sherds separate quarters made of several houses (Kircho 2001b: 31). At the end of the Early Bronze period, several new burial rituals are introduced, like a double (successive) burial in an underground chamber, burial constructions with several rooms, the earliest jar child burials, and a ritual where skulls are placed in a special niche within a room wall (Kircho 2005b: 513, 517). Anthropomorphic terracotta figurines become flatter and more schematic (Figure 4.10, *15*) (Kircho 2009b: fig. 3, *1*, *3*). The visible differentiation in the households and the increasing complexity in the burial rites probably reflect the beginning of change processes within the social structure.

The domestication of camels, around the middle of the 3rd millennium, triggers an entirely new stage in the development of ancient commercial routes in Central Asia (see also Chapter 26). With the developing climatic aridity, this "desert ship" becomes the main means of transport, greatly increasing the mobility of the population and playing an essential role in the continuous development of long-distance trade. A new type of transport vehicle appears; namely, four-wheeled wagons drawn by camels, or, less commonly, by bulls (Figure 4.11) (Kircho 2009a: 28–30, figs. 5, 8, 9). At the very end of the Early Bronze Age, colonization starts in the ancient delta of the Murghab River (Udeumuradov 1993: 68, fig. 27, *11*).

The finds of late NMG IV painted pottery pots at Shahr-i Sokhta III.4 prove direct migrations from South Turkmenistan (Tosi 1983a: 138, pl. LXI, fig. 21; 1983b: pl. XLIV, fig. 90). The same can be said for NMG IV artifacts found in burials at Zamanbaba in the lower reaches of the Zeravshan River (Guljamov *et al.* 1966: tab. V, *1*), like cosmetic sticks with a spatula head, mirrors, and a flattened terracotta figurine (Guljamov *et al.* 1966: tab. V, *4*, *5*; XVI, *1–5*, *7*, *9*, *10*, *14*, *18*). An unpainted biconical vessel of the early NMG V type has also been discovered at Sarazm period IV (Lyonnet 1996: 49, fig. 38, *3*). Steatite artifacts begin to be imported probably from the Kerman area in southern Iran (Kircho 2014a: 78–79), as shown by the presence at Altyn Depe of the earliest known steatite artifacts in Turkmenistan – a seal, vessels, bi-conical beads/spindle whorls, and a "lamp"/make-up container (Figure 4.10, *28*, *31*, *36*) (Kircho 2005a: 418; 2014a: 74, fig. 1, *1*; 2, *37*; 3, *4*; 5, *4*, *6*, *8*; 7, *17*, *18*). A honey-color onyx marble is said to come from the Karljuk deposit on the right bank of the Amu Darya to produce cosmetic vessels (Kircho 2014a: 80, fig. 8, *1–8*).

THE MIDDLE BRONZE AGE (CA. 2350–2000/1900 BC) AND THE OUTBURST OF INTERNATIONAL RELATIONS

With the NMG V period, South Turkmenistan enters a new level of development. From the very beginning, Altyn Depe (and, probably, Namazga and Ulug Depe)

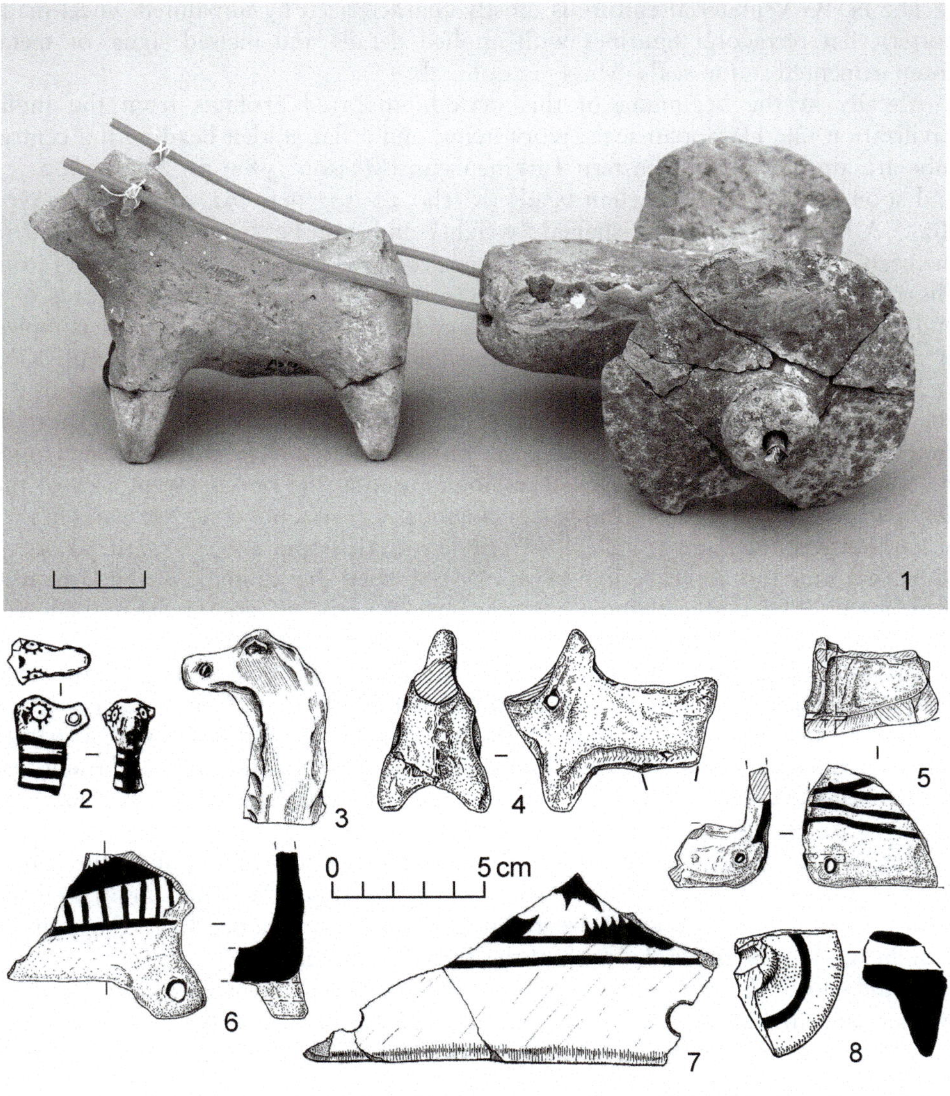

Figure 4.11 Altyn Depe and Taichanak Depe (2), cart clay models of the late phase of the Early Bronze Age (late NMG IV, ca. 2500–2350 BC). *1*, unbaked clay model of a bull-drawn cart (partial reconstruction); *5–7*, fragments of cart models; *8*, wheel model (*2–8*, terracotta).

becomes a large early urban center with developed communication networks, differentiated areas, a specialized potters' district, and a monumental stepped cult center for the whole city (Masson 1988: 97–110; 2006: 68–69). Such monumental stepped buildings are also known at Mundigak IV.1 (Dumarçay 1984: 58–59, figs. 9, 10) and Tureng Tepe in the southeastern Caspian Region (Deshayes 1977: 95–98, figs. 1, 2). A little later, in the last centuries of the 3rd millennium, a new type of architectural structure appears in the ancient delta of the Murghab River; namely, large rectangular or square "fortresses" with double perimeter walls protecting residential buildings (Mamedov 2003: 16–46; see also Chapter 5).

The NMG V material culture is mostly characterized by unpainted wheel-made pottery, flat terracotta figurines with applied details and incised signs, or metal compartmented stamp seals (Masson 2006: tab. 12).

Already at the beginning of this period, imported artifacts from the Indus Civilization like Harappan seals, ivory items, and a flat golden bead with a central tube are attested in southeastern Turkmenistan (Masson 1988: pl. XXII, *1b*, 2, *5*), and soon after, etched carnelian beads (Kircho 2005a: 391–392, 395, fig. 22, type IB2).[7] A handled stone disc-shaped "weight" and a stone "scepter" discovered in the early NMG V "sanctuary" at Altyn Depe can well be compared to objects from Hissar IIIB (Schmidt 1937: pl. LXI, H 2895; LXIV, H 2894). The gold heads of a bull and a wolf (or, more probably, a dog) and bone "sticks" from the same complex are probably related to ancient eastern traditions (Masson 1988: 75–77; pl. XIX, 2; XXIII). A South Turkmenistan metal stamp seal has also been discovered in the Hissar IIIB levels (Schmidt 1937: fig. 118, H 2697). Beads made of artificial (heated) minerals discovered in Middle Bronze Age burials at Altyn Depe (Kircho 2005a: 392, fig. 22, type IA'2) find their closest analogies among the burial inventories of the Shahdad cemetery in central Iran (e.g., Hakemi 1997: 303, no. 2757, etc.) and in that of Parkhai II in the Sumbar valley (SWT II period) (Khlopin 2002: 63, tab. 94, 4).

At that time too, typical Altyn Depe NMG V bead shapes made of the same raw stone material have been found among the jewelry items at Zamanbaba in the lower Zeravshan valley (Kircho 2005a: 400). Finally, recently, other striking analogies with Altyn Depe (and Zamanbaba) Early and Middle Bronze Age material have been discovered at the Farkhor cemetery in southern Tajikistan (see Chapter 23); namely, stone cosmetic containers decorated with incised circles, copper sticks with a spatula and a steppedrhomb head, round mirrors, cross-shaped and steppedrhomboidal beads, stone "scepters," and clay "house models" (Bobomulloev *et al.* 2017: fig. 8, *1*; 10, *5*; 14, *1*; 15, *1*; 16, 2, *3a*, 4, *5*; 18, *5*, *6*, *6c*; 19, *1*; 10, *1*; 24, 4, *5*; 24B, *10*; 25, *1*, 2, *2a*; 26, *1*; 27, *1*, 4). The location of this cemetery, on the right bank of the Pjandj River (considered nowadays as the major source stream of the Amu Darya), and the fact that it was used for a long time, point at the presence of a nearby ford opening a straight road to the mineral wealth of the Tajik Mountains.

At the end of the 3rd millennium (late NMG V period), new categories of metal artifacts appear like vessels, "small shovels," mirrors with handles, cosmetic sticks with elaborate figured heads (Kircho 2001a: 71), and also tin bronze artifacts (Terekhova 2001: 113; see also Chapter 27).

Historically, the fast development of the Murghab delta in the last centuries of the 3rd millennium matches with the dramatic intensification of interregional contacts (see Chapter 1). It is basically conditioned both by the high technical, technological, and cultural development of South Turkmenistan, and by the involvement of new groups of population. Extensive evidence of direct contacts at that time with western Asia, Elam, and ancient India has been demonstrated by the finds from the main center of ancient Margiana, Gonur Depe (Sarianidi 2005, 2007, 2008). The basis for the economic bloom of ancient Margush was probably the key role it played on the ancient route to natural resources of northern Central Asia (Lyonnet 2005). Gonur Depe in the late 3rd/early 2nd millennium was not only an administrative and religious center, but also a production and trade center, just like Ancient Merv later on. At that time, southern Turkmenistan played an original role of transmission between the civilizations of southern and western Asia and the Eurasian steppe.

NOTES

1 It has to be underlined here that the Central Asian periodization does not correspond to that used in the Near East where the Chalcolithic period ends before 3000, the Early Bronze Age around 2000, and the Middle Bronze Age around 1500 BC (see Chapter 1).

2 Excavations at Anau resumed in 1977–1982, under the direction of K. Kurbansakhatov (Kurbansakhatov 1987), and in 1997 under the direction of F. Hiebert and K. Kurbansakhatov (Hiebert and Kurbansakhatov 2003).

3 It cannot be ruled out that part of the small settlements dated to the Chalcolithic and Bronze Age in the foothills of the Kopet Dagh have been covered by later alluvial depositions. Material of the Late Chalcolithic period has been found at Serakhs along the Tedjen River (Adykov and Masson 1960), and at Ak Depe (end of the LC period) (Kircho 1999: 77).

4 Unfortunately, very little from Ulug Depe and Khapuz Depe has been published.

5 They are numerous at Sarazm (Isakov 1991: 134 and fig. 31).

6 Their content analyses have shown that what we used to consider as lamps were actually cosmetic containers (Vidale *et al.* 2016).

7 For further information on the relations between Altyn Depe and the Indus Civilization, see Masson (1977) and Shchetenko (1970).

REFERENCES

Adykov, K.A. and V.M. Masson 1960. "Drevnosti Tedjen-Murgabskogo mezhdurech'ja" [Protohistory between the Tedjen and the Murghab River]. *Izvestija Akademii nauk Turkmenskoj SSR. Serija obshchestvennykh nauk* 2: 58–66.

Amiet, P. 1972. *Glyptique susienne des origines à l'époque des Perses Achéménides. Cachets, sceau-cylindres et empreintes antiques découverts à Suse de 1913 à 1967*. Vol. I–II, *Mémoires de la Délégation Archéologique en Iran XLIII*. Paris: Librairie Orientaliste P. Geuthner.

Amiet, P. 1983. "The Archaic Glyptic at Shahr-i Sokhta (Period I)." In *Prehistoric Sistan* 1, ed. M. Tosi. Rome: IsMEO, pp. 199–210, pl. LXXXIV–LXXXVII.

Amiet, P. 1986. *L'âge des échanges inter-iraniens 3500–1700 avant J.-C.* Paris: Éditions de la Réunion des musées nationaux.

Amiet, P. and M. Tosi 1978. "Phase 10 at Shahr-i Sokhta: Excavations in Square XDV and the Late 4th Millennium B.C. Assemblage of Sistan." *East and West* 28(1–4): 9–31.

Avanesova, N. and D.M. Dzhurakulova 2008. "Drevnejshie nomady Zeravshana" [The oldest nomads in Zeravshan]. In *Kul'tura nomadov Tsentral'noj Azii* [Central Asia nomads' culture], ed. Sh.R. Pidaev. Samarkand: Mezhdunarodnyj Institut Tsentralnoaziatskikh issledovanij, pp. 13–33.

Berezkin, Y.E. and N.F. Solovjova 1998. "Paradnye pomeshchenija Ilgynly-depe (predvaritel'naja tipologija)" [Ceremonial rooms at Ilgynly Depe (preliminary typology)]. *Arkheologicheskie Vesti* 5: 86–123.

Biscione, R. 1973. "Dynamics of an Early South Asian Urbanization: the First Period at Shahr-i Sokhta and its Connections with Southern Turkmenia." In *South Asian Archaeology 1971*, ed. N. Hammond. London: Duckworth, pp. 105–118.

Bobomulloev, S., N.M. Vinogradova, and B. Bobomulloev (with contributions by M. Teufer and G. Lombardo) 2017. "Preliminary Results of Research of the Burial Ground of Farkhor, the Middle Bronze Age Monument in the South of Tajikistan." In *Materialy Farkhorskoj arkheologicheskoj ekspeditsii* [Materials of the Farkhor expedition], vol. I, ed. S. Bobomulloev. Dushanbe: Institute of History, Archaeology and Ethnography of Academy of Sciences of Republic of Tajikistan, pp. 68–114.

Brunet, F. 2011. "A Clay Figurine Workshop 6000 Years Ago." In *Monuments of History and Culture of Turkmenistan*, ed. M. Mamedov. Ashgabat: Turkmen State Publishing Service, pp. 250–255.

Casal, J.-M. 1961. *Fouilles de Mundigak*, 2 vols. Paris: Librairie C. Klincksieck.

Deshayes J. 1977. "A propos des terrasses hautes de la fin du IIIe millénaire en Iran et en Asie Centrale." In *Le plateau iranien et l'Asie Centrale des origines à la conquête islamique. Leurs relations à la lumière des documents archéologiques*, ed. J. Deshayes. Paris: Éditions du Centre National de la Recherche Scientifique, pp. 95–111.

Dolukhanov, P.M. 1984. *Razvitie prirodnoj sredy i khozjajstvo pervobytnogo naselenija Vostochnoj Evropy i Perednej Azii v pozdnem plejstocene i golocene* [Environmental and economical evolution in Eastern Europe and Middle Asia at the end of the Pleistocene and during the Holocene]. Moscow: Institut arkheologii Akademii nauk SSSR.

Dumarçay, J. 1984. "L'architecture de Mundigak." *Bulletin de l'Ecole française d'Extrême-Orient* 73: 47–66.

Dyson, R.H. and W.C.S. Remsen 1989. "Observations on Architecture and Stratigraphy at Tappeh-Hesār." In *Tappeh-Hesār: Reports of the Restudy Project, 1976*, ed. R.H. Dyson and S.M. Howard. Monografie di Mesopotamia II. Firenze: Casa Editrice le Lettere, pp. 69–109.

Freed, R.E. 1982. *Egypt's Golden Age: The Art of Living in the New Kingdom 1558–1085 BC: Catalogue of the Exhibition*. Boston: Museum of Fine Arts.

Ghirshman, R. 1938. *Fouilles de Sialk près de Kashan 1933, 1934, 1937*. Paris: P. Geuthner.

Guljamov Ja.G., U. Islamov, and A. Askarov 1966. *Pervobytnaja kul'tura i vozniknovenie oroshaemogo zemledelija v nizovjakh Zarafshana* [Indigenous culture and origin of irrigated agriculture in lower Zeravshan]. Tashkent: Fan.

Hakemi, A. 1997. *Shahdad: Archaeological Excavations of a Bronze Age Center in Iran*. IsMEO Reports and Memoirs 27. Rome: IsIAO.

Harris, D.R. 2010. *Origins of Agriculture in Western Central Asia: An Environmental–Archaeological Study*. Philadelphia: University of Pennsylvania.

Hiebert, F.T. and K. Kurbansakhatov 2003. *A Central Asian Village at the Dawn of Civilization: Excavations at Anau, Turkmenistan*. University Museum Monograph, vol. 116. Philadelphia: University of Pennsylvania Museum of Archaeology and Anthropology.

Isakov, A.I. 1991. *Sarazm (k voprosu stanovlenija rannezemledel'cheskoj kul'tury Zeravshanskoj doliny (raskopki 1977–1983 gg.))* [Sarazm (questions on the formation of farming culture in the Zeravshan valley (1977–1983 excavations)]. Dushanbe: Donish.

Isakov, A.I. 1996. "Sarazm, an Agricultural Center of Ancient Sogdiana." *Bulletin of the Asia Institute* 8: 1–25.

Ivanova, S.V., D.V. Kiosak, and E.I. Vinogradova 2011. "Modeli zhiznedejatel'nosti Severo-Zapadnogo Prichernomor'ja i klimaticheskie anomalii (6200–2000 let do n. e.)" [Occupation models and climatic anomalies in the northwest Black Sea]. *Stratum Plus* 2: 101–140.

Jarrige, J.-F. (ed.) 1988. *Les Cités oubliées de l'Indus: archéologie du Pakistan*. Paris: Musée National des Arts Asiatiques Guimet.

Kasparov, A.K. 2006. *Skotovodstvo i okhota epokhi neolita – paleometalla v Juzhnom Turkmenistane* [Cattle breeding and hunting during the Neolithic and Chalcolithic in southern Turkmenistan]. St. Petersburg: Evropejskij Dom.

Khlopin, I.N. 1964. *Geoksjurskaja gruppa poselenij epokhi eneolita* [The Chalcolithic period Geoksjur group of settlements]. Moscow: Nauka.

Khlopin, I.N. 1969. *Pamjatniki razvitogo eneolita Jugo-Vostochnoj Turkmenii* [Mature Chalcolithic settlements in southeast Turkmenia]. Svod arkheologicheskikh istochnikov B3-8: Eneolit juzhnykh oblastej Srednej Azii III [B3-8 Archaeological data codex: the Chalcolithic period in the southern regions of Central Asia III]. Moscow: Nauka.

Khlopin, I.N. 1971. "Temenos na Kara-Tepe" [The Kara Tepe temenos]. *Sovetskaja Arkheologija* 3: 221–228.

Khlopin, I.N. 1997. *Eneolit Jugo-Zapadnogo Turkmenistana* [The southwest Turkmenistan Chalcolithic period]. Trudy Juzhno-Turkmenistanskoj arkheologicheskoj kompleksnoj ekspeditsii 20. St. Petersburg: Evropejskij Dom.

Khlopin, I.N. 2002. *Epokha bronzy Jugo-Zapadnogo Turkmenistana* [The southwest Turkmenistan Bronze Age]. Trudy Juzhno-Turkmenistanskoj arkheologicheskoj kompleksnoj ekspeditsii 21. St. Petersburg: Peterburgskoe Vostokovedenie.

Khotinsky, N.A. 1980. "Tri tipa izmenenija klimata Severnoj Evrazii v golocene" [Three kinds of climate changes in northern Eurasia during the Holocene]. In *Kolebanija uvlazhnennosti Aralo-Kaspijskogo regiona v golocene* [Humidity oscillations in the Aral-Caspian region during the Holocene], ed. B.V. Andrianov. Moscow: Nauka, pp. 5–12.

Kircho, L.B. 1999. *K izucheniju pozdnego eneolita Juzhnogo Turkmenistana* [Southern Turkmenistan Late Chalcolithic studies]. St. Petersburg: Institut istorii material'noj kul'tury Rossijskoj akademii nauk.

Kircho, L.B. 2000a. "Mikrostratigrafija pozdnego eneolita Juzhnogo Turkmenistana" [Chalcolithic southern Turkmenistan micro-stratigraphy]. In *Sud'ba uchenogo. K 100-letiju so dnja rozhdenija B.A. Latynina* [A scholar's fate: tribute to B.A. Latynin on the 100 years of his birthday], ed. N.G. Gorbunova, N.K. Kachalova, and Ju.Ju. Piotrovskij. St. Petersburg: Gosudarstvennyj Ermitazh, pp. 177–194.

Kircho, L.B. 2000b. "Novyj tip statuetki iz Kara-depe" [A new statuette type from Kara-depe]. In *Vzaimodejstvie kul'tur i tsivilizatsij: V chest jubileja V.M. Massona* [Culture and civilization interrelations: in honor of V.M. Masson anniversary], ed. Ju.E. Berezkin. St. Petersburg: Institut istorii material'noj kul'tury Rossijskoj akademii nauk, pp. 110–115.

Kircho, L.B. 2001a. "Metallicheskie izdelija Altyn-depe" [Metal artifacts from Altyn Depe]. In *Osobennosti proizvodstva poselenija Altyn-depe v epokhu paleometalla* [Production specificities at Altyn Depe during the Paleometal Age], ed. V.M. Masson and L.B. Kircho. Materialy Juzhno-Turkmenistanskoj arkheologicheskoj kompleksnoj ekspeditsii 5. St. Petersburg: Institut istorii material'noj kul'tury Rossijskoj akademii nauk, pp. 60–84.

Kircho, L.B. 2001b. "Osnovnye tipy sooruzhenij i tekhnologija stroitel'stva Altyn-depe v epokhu pozdnego eneolita i rannej bronzy" [Main types of buildings and construction technology at Altyn Depe during the Late Chalcolithic and Early Bronze Age]. In *Osobennosti proizvodstva poselenija Altyn-depe v epokhu paleometalla* [Production specificities at Altyn Depe during the Paleometal Age], ed. V.M. Masson and L.B. Kircho. Materialy Juzhno-Turkmenistanskoj arkheologicheskoj kompleksnoj ekspeditsii 5. St. Petersburg: Institut istorii material'noj kul'tury Rossijskoj akademii nauk, pp. 5–39.

Kircho, L.B. 2004. "Formirovanie drevnejshej protogorodskoj tsivilizatsii bronzovogo veka Srednej Azii (po materialam Altyn-depe)" [Formation of the ancient proto-urban Bronze Age civilization in Central Asia (from Altyn Depe data)]. In *U istokov tsivilizatsii* [Near the sources of civilizations], ed. M.F. Kosarev, P.M. Kojin, and N.A. Dubova. Moscow: Staryj Sad, pp. 142–160.

Kircho, L.B. 2005a. "Pogrebal'nyj inventar Altyn-depe" [Altyn Depe funerary inventory]. In *Khronologija epokhi pozdnego eneolita -srednej bronzy Srednej Azii (pogrebenija Altyn-depe)* [Chronology of Central Asia Late Chalcolithic-Middle Bronze Age (Altyn Depe burials)], ed. V.M. Masson and Ju.E Berezkin. Trudy Instituta istorii material'noj kul'tury Rossijskoj akademii nauk 16. St. Petersburg: Nestor-Istorija, pp. 347–421.

Kircho, L.B. 2005b. "Zakljuchenie" [Conclusion]. In *Khronologija epokhi pozdnego eneolita-srednej bronzy Srednej Azii (pogrebenija Altyn-depe)* [Chronology of the Central Asia Late Chalcolithic–Middle Bronze Age (Altyn Depe burials)], ed. V.M. Masson and Ju.E Berezkin. Trudy Instituta istorii material'noj kul'tury Rossijskoj akademii nauk 16. St. Petersburg: Nestor-Istorija, pp. 512–519.

Kircho, L.B. 2007. "Drevnie svjazi naselenija Juzhnogo Turkmenistana i doliny Zeravshana (nachalo formirovanija torgovykh putej v Srednej Azii)" [Ancient population relations between southern Turkmenistan and the Zeravshan valley]. *Zapiski Instituta istorii material'noj kul'tury Rossijskoj akademii nauk* 2: 193–208.

Kircho, L.B. 2009a. "The Earliest Wheeled Transport in Southwestern Central Asia: New Finds from Altyn-Depe." *Archaeology, Ethnology & Anthropology of Eurasia* 37(1): 25–33.

Kircho, L.B. 2009b. "Osnovnye napravlenija i kharakter kul'turnykh vzaimodejstvij naselenija Juzhnogo Turkmenistana v V–III tys. do n.e." [Main southern Turkmenistan cultural interrelations during the 5th–3rd millennium: direction and character]. *Stratum plus 2005–2009* 2: 374–392.

Kircho, L.B. 2011. "Nekotorye problemy arkheologii Juzhnogo Turkmenistana i izuchenie Altyn-depe v 1991–2001 gg." [Archaeological questions about southern Turkmenistan and Altyn Depe studies (1991–2001)]. In *Pamjatniki istorii i kul'tury Turkmenistana* [Turkmenistan historical sites and cultures], ed. M.A. Mamedov. Aşgabat: Türkmen döwlet neşirýat gullugy, pp. 81–99.

Kircho, L.B. 2014a. "Izdelija iz kamnja i istochniki syr'ja v period rannej bronzy (po materialam Altyn-depe, Juzhnyj Turkmenistan)" [Stone artifacts and their sources during the Early Bronze Age (from the Altyn Depe material, southern Turkmenistan]. In *Arkheologija drevnikh obshchestv Evrazii: khronologija, kul'turogenez, religioznye vozzrenija. Pamjati V.M. Massona* [The archaeology of Eurasia ancient societies: chronology, formation, religious conceptions. In memory of V.M. Masson], ed. V.A. Alekshin, L.B. Kircho, M.T. Kashuba, and V.Ja. Stegantseva. Trudy Instituta istorii material'noj kul'tury Rossijskoj akademii nauk 42. St. Petersburg: Art-Xpress, pp. 62–85.

Kircho, L.B. 2014b. "Problems in the Periodic Division and Chronology of Sites of the Palaeometallic Age of Southern Turkmenistan and the Stratigraphic Sequence at Altyn-Depe." In *"My Life Is Like the Summer Rose": Maurizio Tosi e l'archeologia come modo di vivere – Papers in Honour of Maurizio Tosi for His 70th Birthday*, ed. C.C. Lamberg-Karlovsky, B. Genito, and B. Cerasetti. BAR Series 2690. Oxford: Archaeopress, pp. 359–364.

Kircho, L.B., G.F. Korobkova, and V.M. Masson 2008. *The Technical and Technological Potential of the Eneolithic Population of Altyn-Depe as the Basis of the Rise of an Early Urban Civilization*. St. Petersburg: Evropejskij Dom.

Kircho, L.B. and G.M. Kovnurko 2001. "Osobennosti proizvodstva keramiki epokhi eneolita i bronzy Altyn-depe po dannym petrografii" [Pottery production during the Chalcolithic and Bronze Age at Altyn Depe: its specificities from petrographic studies]. In *Osobennosti proizvodstva poselenija Altyn-depe v epokhu paleometalla* [Production specificities at Chalcolithic and Bronze Age Altyn Depe], ed. V.M. Masson and L.B. Kircho. Materialy Juzhno-Turkmenistanskoj arkheologicheskoj kompleksnoj ekspeditsii 5. St. Petersburg: Institut istorii material'noj kul'tury Rossijskoj akademii nauk, pp. 118–141.

Kircho, L.B. and G.M. Kovnurko 2003. "Prestizhno-kul'tovye predmety i ukrashenija Altyn-depe iz drevnikh iskusstvennykh mineralov (po dannym rentgenometricheskogo analiza)" [Prestigious-cultic artifacts and jewelry made of artificial minerals at Altyn Depe (from X-ray analysis)]. *Arkheologicheskie Vesti* 10: 108–113.

Kircho, L.B. and T.A. Sharovskaja 2001. "Ornamentirovannyj kamennyj sosud iz Altyn-depe (k voprosu o postgeoksjurskom komplekse Juzhnogo Turkmenistana)" [A decorated stone vessel from Altyn Depe (question on the post-Geoksjur complex in southern Turkmenistan)]. *Arkheologicheskie Vesti* 8: 120–124.

Kohl, P.L. 1984. *Central Asia: Palaeolithic Beginnings to the Iron Age*. Paris: Éditions Recherche sur les Civilisations.

Kohl, P.L. and D.L. Heskel 1980. "Archaeological Reconnaissance in the Darreh Gaz Plain. A Short Report." *Iran* 18: 160–172.

Korobkova, G.F. 2001. "Funktsional'naja tipologija orudij truda i drugikh nemetallicheskikh izdelij Altyn-depe" [A functional typology of non-metallic tools and artifacts from Altyn

Depe]. In *Osobennosti proizvodstva poseleniya Altyn-depe v epokhu paleometalla [Production Specificities at Chalcolithic and Bronze Age Altyn-depe]*, ed. V.M. Masson and L.B. Kircho. Materialy Juzhno-Turkmenistanskoj arkheologicheskoj kompleksnoj ekspeditsii 5. St. Petersburg: Institut istorii material'noj kul'tury Rossijskoj akademii nauk, pp. 146–212.

Korobkova, G.F. and T.A. Sharovskaja 1994. "Stone Tools from Ilgynly-depe (Turkmenistan): The Evidence from Use-Wear Analysis." *Archaeological Studies* 16: 27–30.

Kuftin, B.A. 1956. "Polevoj otchet o rabote XIV otrjada JuTAKE po izucheniju kul'tury pervobytno-obshchinnykh osedlozemledel'cheskikh poselenij epokhi medi i bronzy v 1952 g." [Field report of JuTAKE XIV's team: studies on the first farming cultures of the Chalcolithic and Bronze Age in 1952]. In *Trudy Juzhno-Turkmenistanskoj arkheologicheskoj kompleksnoj ekspeditsii VII*, ed. M.E. Masson. Ashhabad: Akademija nauk Turkmenskoj SSR, pp. 260–290.

Kurbansakhatov, K. 1987. *Eneolit Anau* [Anau during the Chalcolithic period]. Ashhabad: Ylym.

Lamberg-Karlovsky, C.C. 1985. "The Longue Durée of the Ancient Near East." In *De L'Indus aux Balkans: Recueil à la mémoire de Jean Deshayes*, ed. J.L. Huot, M. Yon, and Y. Calvet. Paris: Recherche sur les civilisations, pp. 55–72.

Lamberg-Karlovsky, C.C. and M. Tosi 1973. "Shahr-i Sokhta and Tepe-Yahya: Tracks on the Earliest History of the Iranian Plateau." *East and West* 23(1–2): 15–23.

Le Brun, A. 1971. "Recherches stratigraphiques à l'Acropole de Suse (1969–1971)." *Cahiers de la Délégation Archéologique Française en Iran* 1: 163–216.

Levkovskaja, G.M. and L.B. Kircho 2004. "K voprosu o paleoekologii Juzhnogo Turkmenistana v III tys. do n. e. (predvaritel'naja palinologicheskaja kharakteristika otlozhenij Altyn-depe)" [Paleoecology in southern Turkmenistan during the 3rd millennium BC (preliminary characteristics of the Altyn Depe palinological data)]. In *Orudija truda i sistemy zhizneobespechenija naselenija Evrazii* [Eurasian populations' tools and their supply system], ed. G.F. Korobkova. St. Petersburg: Evropejskij Dom, pp. 87–91.

Lisitsina, G.N. 1978. *Stanovlenie i razvitie oroshaemogo zemledel'ija v Juzhnoj Turkmenii* [Formation and evolution of irrigated agriculture in southern Turkmenia]. Moscow: Nauka.

Lisitsina, G.N. 1966. "Izuchenie geoksjurskoj orositel'noj seti v Juzhnoj Turkmenii v 1964 g." [Studies of the South Turkmenia Geoksjur irrigation system in 1964]. *Kratkie Soobschenija Instituta Arkheologii Akademii Nauk SSSR* 108: 96–100.

Lyonnet, B. (with A. Isakov) 1996. *Sarazm (Tadjikistan) céramiques (Chalcolithique et Bronze Ancien)*. Paris: de Boccard.

Lyonnet, B. 2005. "Another Possible Interpretation of the Bactro-Margiana Culture (BMAC) of Central Asia: The Tin Trade." In *South Asian Archaeology 2001*, ed. C. Jarrige and V. Lefèvre. Paris: ERC, pp. 191–200.

Maleki, Y. 1968. "Art abstrait et décor animalier chez les céramistes de la région de Téhéran." *Archéologie Vivante* 1(1): 42–55.

Mamedov, E. 1980. "Izmenenija klimata sredneaziatskikh pustyn' v golocene" [Climatic change in the Central Asia deserts during the Holocene]. In *Kolebanija uvlazhnennosti Aralo-Kaspijskogo regiona v golocene [Humidity oscillations in the Aral-Caspian region during the Holocene]*, ed. B.V. Andrianov. Moscow: Nauka, pp. 170–175.

Mamedov, M. 2003. *Drevnjaja arkhitektura Baktrii i Margiany* [Ancient architecture in Bactria and Margiana]. Ashkhabad: Cultural Center of the Embassy of Iran.

Marcolongo, B. and P. Mocci 2000. "Geomorfologicheskaja evoljutsija predgornoj ravniny vostochnogo Kopetdaga v epokhu golocena: Predvaritel'nyj geoarkheologicheskij obzor" [Geomorphological evolution of the valleys in the eastern Kopet Dagh foothills during the Holocene: a preliminary geo-archaeological review]. *Arkheologicheskie Vesti* 7: 33–40.

Mariani, L. and M. Tosi 1987. "L'universo familiare a Shahr-i Sokhta attraverso le attività domestisce e le strutture residentali." In *Orientalia Josephi Tucci memoriae dicata*, ed. G. Gnoli and L. Lanciotti. Rome: IsMEO, pp. 853–879.

Masimov, I.S. 1979. "K voprosu ob osvoenii nizovij Murgaba drevnezemledel'cheskimi plemenami" [Questions on the lower Murghab ancient farmers' development]. *Izvestija Akademii nauk Turkmenskoj SSR. Serija obshchestvennykh nauk* 2: 72–77.

Masson, V.M. 1960. "Kara-depe u Artyka" [The Artyk Kara Depe]. In *Trudy Juzhno-Turkmenistanskoj Arkheologicheskoj Kompleksnoj Ekspeditsii X*, ed. M.E. Masson. Ashkhabad: Akademija nauk Turkmenskoj SSR, pp. 319–463.

Masson, V.M. 1962. *Pamjatniki razvitogo eneolita Jugo-Zapadnoj Turkmenii* [Mature Chalcolithic settlements in southwest Turkmenia]. Svod arkheologicheskikh istochnikov B3-8: Eneolit juzhnykh oblastej Srednej Azii II [B3-8 Archaeological data codex: the Chalcolithic period in the southern regions of Central Asia II]. Moscow: Nauka.

Masson, V.M. 1971. *Poselenie Dzhejtun (problema stanovlenija proizvodjaschej ekonomiki)* [The Djeitun settlement (problem of the formation of production economy)]. Materialy i Issledovanija po Arkheologii SSSR 180. Leningrad: Nauka.

Masson, V.M. 1977. "Pechati protoindijskogo tipa iz Altyn-depe" [Proto-indus seals from Altyn Depe]. *Vestnik Drevnej Istorii* 4: 147–155.

Masson, V.M. 1982. "Eneolit Srednej Azii" [The Central Asia Chalcolithic]. In *Eneolit SSSR*, ed. V.M. Masson and N.Ja. Merpert. Arkheologija SSSR. Moscow: Nauka, pp. 9–92.

Masson, V.M. 1988. *Altyn-depe*. University Museum Monograph 55. Philadelphia: University Museum University of Pennsylvania.

Masson, V.M. 1989. "Ilgynly-depe – novyj tsentr eneoliticheskoj kul'tury Juzhnogo Turkmenistana" [Ilgynly Depe – a new Chalcolithic center in southern Turkmenistan]. *Izvestija Akademii nauk Turkmenskoj SSR. Serija obshchestvennykh nauk* 6: 15–24.

Masson, V.M. 2006. *Kul'turogenez Drevnej Tsentralnoj Azii* [The culture formation in ancient Central Asia]. St. Petersburg: Filologishcheskij fakul'tet Sankt-Peterburgskogo gosudarstvennogo universiteta.

Masson, V.M., Yu. E. Berezkin, and N.F. Solovyeva 1994. "Excavations of Houses and Sanctuaries at Ilgynly-Depe Chalcolithic Site, Turkmenistan." In New Archaeological Discoveries in Asiatic Russia and Central Asia, ed. A. Kozintsev and V.M. Masson. St. Petersburg: Institute for the History of Material Culture Russian Academy of Sciences, pp. 18–26.

Masson, V.M. and L.B. Kircho 1999. "Izuchenie kul'turnoj transformatsii rannezemledel'cheskikh obshchestv (po materialam novykh raskopok na Altyn-depe i Ilgynly-depe)" [The cultural transformations of the early farming societies: studies from Altyn Depe and Ilgynly Depe on newly excavated material]. *Rossijskaja Arkheologija* 2: 61–76.

Masson, V.M. and G.F. Korobkova 1989. "Eneolithic Stone Sculpture in South Turkmenia." *Antiquity* 63(238): 62–70.

Masson, V.M. and V.I. Sarianidi 1972. *Central Asia: Turkmenia Before the Achaemenids*. Ancient Peoples and Places 79. London: Thames & Hudson.

Perrot, J. and Y. Madjidzadeh 2005. "L'iconographie des vases et objets en chlorite de Jiroft (Iran)." *Paléorient*, 31(2): 123–152.

Piperno, M. and M. Tosi 1975. "The Graveyard of Shahr-i Sokhta, Iran." *Archaeology* 28(3): 186–197.

Razzokov, A.R., G.R. Karimova, Sh. Kurbanov, and S.G. Bobomulloev (eds.) 2006. *Sarazm*. Dushanbe: Echod.

Sajjadi, S.M.S. 2003. "Excavations at Shahr-i Sokhta: First Preliminary Report of the Excavations of the Graveyard 1997–2000." *Iran* XLI: 21–97.

Sarianidi, V.I. 1962. "Kul'tovye zdanija poselenija anauskoj kul'tury" [Cult buildings in Anau culture settlements]. *Sovetskaja Arkheologija* 1: 44–56.

Sarianidi, V.I. 1964."Khapuz-depe kak pamjatnik epokhi bronzy" [Khapuz Depe, a Bronze Age settlement]. *Kratkie Soobshchenija Instituta Arkheologii Akademii Nauk SSSR* 98: 60–65.

Sarianidi, V.I. 1965a. "Geoksjurskij nekropol'" [The Geoksjur necropolis]. In *Novoe v sovetskoj arkheologii*, ed. E.I. Krupnov. Materialy i Issledovanija po Arkheologii SSSR 130. Moscow: Nauka, pp. 102–105.

Sarianidi, V.I. 1965b. *Pamjatniki pozdnego eneolita Jugo-Vostochnoj Turkmenii* [Late Chalcolithic settlements in southeastern Turkmenia]. Svod arkheologicheskikh istochnikov B3-8: Eneolit juzhnykh oblastej Srednej Azii IV [B3-8 Archaeological data codex: the Chalcolithic period in the southern regions of Central Asia IV]. Moscow: Nauka.
Sarianidi, V.I. 1966. "Raskopki v jugo-vostochnykh Karakumakh v 1964 g." [The 1964 excavations in southeastern Karakum]. *Kratkie Soobshchenija Instituta Arkheologii Akademii Nauk SSSR* 108: 89–95.
Sarianidi, V.I. 1968. "O velikom lazuritovom puti na Drevnem Vostoke" [On the great lapis lazuli route in the ancient East]. *Kratkie Soobshchenija Instituta Arkheologii Akademii Nauk SSSR* 114: 3–9.
Sarianidi, V.I. 1969. "Novye raskopki na Khapuz-depe" [New excavations at Khapuz Depe]. *Kratkie Soobshchenija Instituta Arkheologii* 115: 116–123.
Sarianidi, V.I. 1971. "The Lapis Lazuli Route in the Ancient East." *Archaeology* 24: 12–15.
Sarianidi, V.I. 1976. "Material'naja kul'tura Juzhnogo Turkmenistana v period rannej bronzy" [The southern Turkmenistan material culture during the Early Bronze Age]. In *Pervobytnyj Turkmenistan* [*Indigenous Turkmenistan*], ed. V.M. Masson and E. Atagaryev. Ashkhabad: Ylym, pp. 82–122.
Sarianidi, V.I. 1983. "The Pottery of Shahr-i Sokhta I and Its Southern Turkmenian Connections." In *Prehistoric Sistan*, vol. 1, ed. M. Tosi. Rome: IsMEO: 183–198.
Sarianidi, V.I. 1990. *Drevnosti strany Margush* [Ancient Margush]. Ashkhabad: Ylym.
Sarianidi, V.I. 2002. *Margush: Ancient Oriental Kingdom in the Old Delta of the Murghab River*. Aşgabat: Türkmendöwlethabarlary.
Sarianidi, V.I. 2005. *Gonur-Depe: City of Kings and Gods*. Aşgabat: Türkmenistanyň milli medeniýet "Miras" merkesi.
Sarianidi, V.I. 2007. *Necropolis of Gonur*, 2nd ed. Athens: Kapon Editions.
Sarianidi, V.I. 2008. *Margush: Mystery and Truth of the Great Culture*. Aşgabat: Türkmendöwlethabarlary.
Schmidt, E.F. 1937. *Excavations at Tepe-Hissar Damghan*. Philadelphia: University of Pennsylvania Press.
Shchetenko, A.Ja. 1968. "Raskopki pamjatnikov epokhi eneolita i bronzovogo veka v Kaakhkinskom rajone" [Chalcolithic and Bronze Age settlements excavations in the Kaakhka region]. In *Karakumskie drevnosti*, vol. 1, ed. V.M. Masson. Ashkhabad: Ylym, pp. 18–29.
Shchetenko, A.Ja. 1970. "O torgovykh putjakh epokhi bronzy po materialam turkmenistano-harappskikh parallelej" [On the Bronze Age trade routes from comparisons between Turkmenistan and Harappa artifacts]. *Kratkie Soobshchenija Instituta Arkheologii Akademii Nauk SSSR* 122: 59–63.
Solovjova, N.F. 1998. "Stenopis' Ilgynly-depe" [Ilgynly Depe wall paintings]. *Arkheologicheskie Vesti* 5: 124–130.
Solovjova, N.F. 2005. *Anthropomorphic Figurines from Ilgynly-Depe, South Turkmenistan: Classification, Analysis and Catalogue*. BAR International Series 1336. Basingstoke, UK: Basingstoke Press.
Solovjova, N.F. 2015. "Spetsializirovannaja masterskaja epokhi rannego eneolita v Karakumakh" [A specialized workshop in Karakum during the Early Chalcolithic period]. *Zapiski Instituta istorii material'noj kultury Rossijskoj akademii nauk* 12: 76–89.
Solovjova, N.F. 2019. "Objigatelnie pechi Ilgynly-depe perioda rannego eneolita" [Early Chalcolithic kilns at Ilgynly-depe]. *Zapiski Instituta istorii material'noj kultury Rossijskoj akademii nauk* 20: 147–162.
Solovjova, N.F., A.N. Jegorkov, V.A. Galibin, and Ju.E. Berezkin 1994. "Metal Artifacts from Ilgynly-Depe, Turkmenistan." *Archaeological Studies* 16: 31–35.
Stepanov, I.N. 1980. "Periodicheskaja povtorjaemost' pochvoobrazovanija v plejstocene-golocene Aralo-Kaspijskogo regiona" [Periodical repetition of the soil formation during the

Pleistocene and Holocene in the Aral-Caspian region]. In *Kolebanija uvlazhnennosti Aralo-Kaspijskogo regiona v golocene* [Humidity oscillations in the Aral-Caspian region during the Holocene], ed. B.V. Andrianov. Moscow: Nauka, pp. 22–32.

Sumner, W.M. 1988. "Prelude to Proto-Elamite Anshan: The Lapui Phase." *Iranica Antiqua* XXXIII: 23–43.

Sumner, W.M. 2003. *Early Urban Life in the Land of Anshan: Excavations at the Tal-e Malyan in the Highlands of Iran*. Philadelphia: University of Pennsylvania Museum of Archaeology and Anthropology.

Tallon, F. 1987. *Métallurgie susienne I. De la fondation de Suse au XVIIIe siècle avant J.-C.* 2 vols. Paris: Editions de la Réunion des Musées Nationaux.

Terekhova, N.N. 2001. "Tekhnika metalloobrabotki na poselenii Altyn-depe" [The metal-working techniques at Altyn Depe]. In *Osobennosti proizvodstva poselenija Altyn-depe v epokhu paleometalla* [Production specificities at Chalcolithic and Bronze Age Altyn Depe], ed. V.M. Masson and L.B. Kircho. Materialy Juzhno-Turkmenistanskoj arkheologicheskoj kompleksnoj ekspeditsii 5. St. Petersburg: Institut istorii material'noj kul'tury Rossijskoj akademii nauk, pp. 104–117.

Tosi, M. 1969. "Excavations at Shahr-i Sokhta: Preliminary Report on the Second Campaign, September–December 1968." *East and West* 19(3–4): 283–386.

Tosi, M. 1983a. "Development Continuity and Cultural Change in the Stratigraphical Sequence of Shahr-i Sokhta." In *Prehistoric Sistan*, vol. 1, ed. M. Tosi. Rome: IsMEO, pp. 127–180.

Tosi, M. 1983b. "Excavations at Shahr-i Sokhta 1969–1970." In *Prehistoric Sistan*, vol. 1, ed. M. Tosi. Rome: IsMEO, pp. 73–125.

Tosi, M. and R. Wardak 1972. "The Fullol Hoard: A New Find from Bronze Age Afghanistan." *East and West* 22(1–2): 9–17.

Ucko, P.J. 1968. *Anthropomorphic Figurines of Predynastic Egypt and Neolithic Crete with Comparative Material from the Prehistoric Near East and Mainland Greece*. Occasional paper of the Royal Anthropological Institute of Great Britain and Ireland 24. London: A. Szmidla.

Udeumuradov, B.N. 1993. *Altyn-depe i Margiana. Svjazi, khronologija, proiskhozhdenie.* [Altyn Depe and Margiana: relations, chronology and origins]. Ashkhabad: Ylym.

Vidale, M. 2000. *The Archaeology of Indus Crafts: Indus Craftspeople and Why We Study Them*. Rome: IsIAO.

Vidale, M., G. Salviulo, F. Zorzi, and I. Mocchiutti 2016. "Cosmetics and Cosmetology at Shahr-i Sokhta." *Iran* LIV(II): 1–24.

Vinogradov, A.V. and E.D. Mamedov 1975. *Pervobytny Ljavljakan. Etapy drevnejshego zaselenija i osvoenija Vnutrennikh Kyzylkumov* [Indigenous Ljavljakan: ancient occupation and development stages of the inner Kyzylkum]. Moscow: Nauka.

PART II

THE "CORE AREA"

CHAPTER FIVE

THE ARCHITECTURE OF THE BACTRIA-MARGIANA ARCHAEOLOGICAL CULTURE

Ruslan G. Muradov

INTRODUCTION

The history of architecture in southern Central Asia has been entirely renewed by the discovery, at the end of last century, of ca. 150 settlements of various sizes in the ancient delta of the Murghab River, together with several more or less synchronous sites on both sides of the Amu Darya in northern Afghanistan and southern Uzbekistan. They are dated to the Middle and Late Bronze Age: the Namazga (NMG) V and VI periods.[1] They all belong to the Bactria-Margiana Archaeological Culture (BMAC), and their characteristic architectural features serve as a marker for such an attribution. Some of them demonstrate clear signs of administrative or religious character, and sharply contrast with the surrounding households.

These remains of architecture, especially the monumental one, provide a firm ground for the analysis of the plans and construction methods that highlight their specific Central Asian features, thus providing impetus to search for their origins and to underline their continuity into the Iron Age. These issues have already been investigated by various scholars (Sarianidi 1977a, 1981, 1986, 1990a, 1990b, 1990c, 1991, 1994, 2002a, 2002b; Masimov 1981; Hiebert 1994; P'jankova 1994; Mamedov 1993, 1996, 2002, 2003; Zadneprovsky 1995; Huff 2001, 2005). They each have outlined the influences, described the evolution process of the local construction practices, and given their point of view on this BMAC Bronze Age architecture. M. Mamedov (1994, 2003) has also attempted to determine the scale of its borrowings to the Mesopotamian and Syro-Anatolian architectures, and to determine the extent of its influence into the antique and Middle Ages architecture of Central Asia. Here it must be underlined that the excavations at the main settlements of Altyn Depe or Namazga Depe (at the foothills of the Kopet Dagh) never reached the large scale of those in Margiana, which may explain why no similar monumental buildings have been brought to light there and prevents comparisons with them. The discoveries made at Sarazm in the upper Zeravshan valley, however, show that already during the NMG III and IV periods (ca. 3500–2500 BC) the architecture tended towards geometrical plans and monumental constructions and to the frequent use of pilasters (Razzokov 2016).

All the structures mentioned in this chapter only survive in plans. They fall into the category of archaeological artifacts and this makes proper architectural analysis extremely difficult due to the lack of knowledge of the volumetric-spatial parameters. Nevertheless, even at a superficial glance, one can see preconceptions behind the plan schemes (i.e., a sharp trend towards geometric shapes and a desire to stick to the rules of symmetry).

The exact function of most of the monuments is a matter of debate. In the absence, up to now, of the full data concerning the finds discovered in them and their exact position in each room, any proposal remains hypothetical and we will only touch on that question. However, we will still use the names given by V.I. Sarianidi and the other diggers (i.e., "temple," "palace") to differentiate them.

This chapter will present successively the monumental and domestic architecture. The buildings are built mostly with sun-dried mudbricks and had flat-roofs resting on wooden beams, though a few vaulted constructions are possible. Some rooms were entirely plastered with a white mortar, but usually floors and walls were covered with a mud plaster. Details on the building materials and the construction methods are given at the end of the chapter. Funerary architecture is not analyzed here, except for some hints concerning the construction methods, since it is dealt with separately later in the volume (see Chapter 10).

MONUMENTAL ARCHITECTURE

In Margiana, the settlements and their pattern of distribution have been thoroughly studied (Gubaev *et al.* 1998), but this is not the case in Bactria.[2] In Afghanistan, only a small area has so far been studied, covering only four micro oases on the left bank of the Amu Darya River, while in Uzbekistan the damages caused by intensive modern agriculture have preserved only a few isolated sites. Therefore, for these last two areas, it is not possible to make a reconstruction of their ancient architecture on the same level as that of Margiana.

The most distinctive character of the BMAC architecture is the presence of a totally new type of structure unknown before; namely, square or rectangular "fortresses" with rectangular or round towers at the corners and along the walls, which remained the dominant architectural type in southern Central Asia until the last century (the so-called *qala*) (Lamberg-Karlovsky 1994).

For V.M. Masson (1979: 32–33) "the square fortress type is an entirely new phenomenon," unknown in the settlements of the foothills of the Kopet Dag (Altyn Depe, etc.), and this led him to suppose an influence from the Harappan world. Since he wrote these lines, much more evidence of contacts with the Indus valley has been found, but one may nevertheless question this Harappan origin since, as formulated by G. Pugachenkova (1971: 238), architecture, unlike other historical artifacts, "knows neither export nor import and thus is most tightly connected with its native land, both in a literal and figurative sense." Another new layout is that of a round fortress, though it has been discovered only in Bactria up to now, at the Dashly 3 "temple." The importance of this type of round settlement mainly rests on its further development in Central Asia, during the pre-Achaemenid and Achaemenid period (in Margiana at Erk Kala; in Afghanistan at Kutlug Tepe in the Farukhabad oasis, or At Chapar 1 and 2 in the Dashly oasis; in Uzbekistan at Talashkantepa), and later during the Kushan period. Some have tried to look for the origin of such round settlements in prehistoric architecture, but this is doubtful. Similar shapes, contemporary to the BMAC Dashly

3, are known further to the north in the Urals – for example, at Arkaim (Zdanovich and Zdanovich 2002). In Mesopotamia, after the isolated late 5th millennium (level XIA/B) Gawra Round Building, or the Oval Building of the end of the 4th millennium at Godin Tepe, a number of circular structures are known during the Early Dynastic period in the Hamrin area of Iraq and in the Syrian steppe of the Djezirah (the so-called Kranzhügel). Later, at the time of the BMAC, they are still present in the more arid inner Syrian steppe at T. al-Rawda, and in Cappadocia at Kültepe/Kanesh. There too, the origins of this circular plan remain unknown. Sarianidi pointed out long ago a possible schematic reproduction of the plan of the Dashly 3 "temple" on some of the round bronze compartmented seals in the shape of concentric circles (Sarianidi 1977a: 29). This proposal can also be applied to some of the cross-shaped seals that usually have a central square associated with two "courtyards" and with two- or three-stepped sides on the exterior, an outline reminiscent of the building layouts of the nearby Dashly 3 "palace" (Pugachenkova 1982: 24).

In Margiana, numerous settlements excavated since the early 1980s have provided remains of ancient architecture, fragmentary or complete.[3] In Bactria, as mentioned above, much fewer Bronze Age settlements have been excavated and the main center of this area has not yet been discovered.[4] A short summary of this BMAC monumental architecture is provided next.

Margiana

Kelleli is the northernmost oasis in Margiana. It is considered as the oldest occupied zone, in relation with a colonization from the Kopet Dagh during the Middle Bronze Age. It consists of a central mound Kelleli 1, clearly the residence of a local nobleman, surrounded by ten smaller settlements. Only three of them have been partially excavated, among which Kelleli 3, a monumental square-shaped fortress (128 x 128 m), the exterior wall of which is made of two rows of mudbrick separated by a corridor (Figure 5.1). Entrances are positioned in the middle of each side and are framed by nearly square towers. Similar towers are placed at equal distance on each side of the entrances, so that altogether there were six towers on each facade. These features provide a good picture of the fortified architecture at the end of the 3rd millennium BC, but the interior of the settlement is as yet not studied. Its digger considered that its origins were to be find in the rampart remains of Altyn Depe (Masimov 1986: 179).

Gonur Depe must have been the main central settlement of Margiana, judging by its scale and the grandeur of its structures (Figure 5.2a). It presents both a defensive, religious, administrative, living place and a funerary character. Altogether, including a very Large Necropolis of about 10 ha, its excavated surface is ca. 40 ha but the overall built up area is estimated at ca. 130 ha. Conventionally, it has been divided into two areas, Gonur North and South.

Gonur North is a fortress with an inner fortified "palace" (Sarianidi 2005: 58–97) (Figure 5.2b). The whole northern settlement (ca. 15 ha) is enclosed within an oval not fortified (without towers) and less than 1 m wide wall with inner buttresses. It should be mentioned, however, that, outside of it, vernacular buildings have also been discovered, though there is as yet no information on their precise date. The area inside the oval wall contains a large rectangular fortification (named *kremlin*) made of a simple wall with

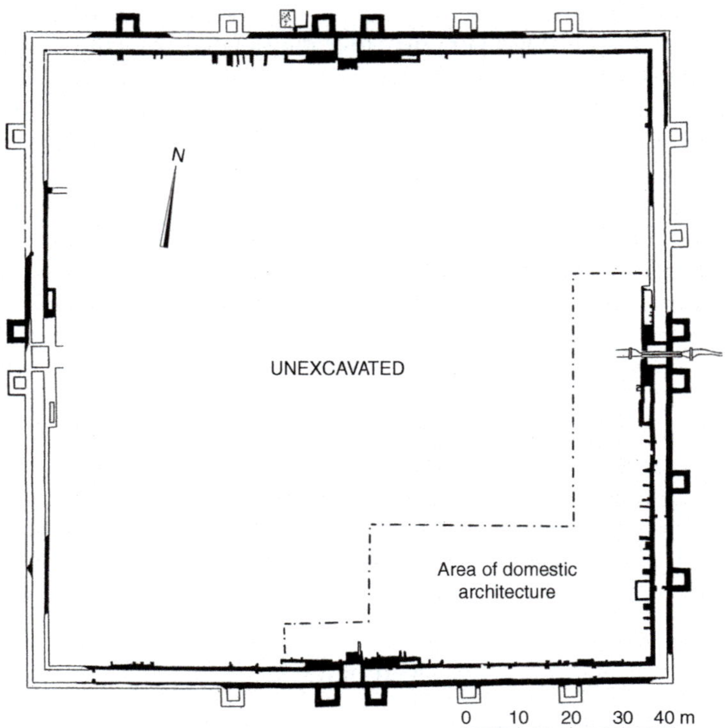

Figure 5.1 Kelleli 3. Plan (from Masimov 1986: 174).

rectangular towers at the corners and along the sides. Three entrances in the middle of the northern, western, and southern walls were protected by towers on each side. Inside this rectangular fortification, another monumental and almost square fortified construction is the "palace." Its defensive walls (128, 115, 117, and 118 m long), like at Kelleli 3, consist of a double wall with an opened corridor inside, associated with rectangular corner towers. Small triangular openings were used either for archers or for light (Artemiev and Urmanova 2010). Three entrances in the middle of three of the walls were protected by a rectangular tower on each side. On the fourth (eastern) side, between the towers, was a building considered to be a fire temple. The "palace" itself consisted of a series of large rooms interconnected by corridors. Judging from their size, they may have been audience halls for official ceremonies.

The "fire temple" outside the eastern wall was rebuilt four times at the same place, always with the same layout (i.e., an interior courtyard with a storage part on one side, and cultic rooms with sanctuaries and fire altars on the other). The best preserved is the third temple, with its interior courtyard inserted into a rectangle of peripheral chambers (12 on each side). This layout is known in the Hittite temple architecture. The Gonur sanctuaries are only provided with "blind windows" i.e. stepped-niched walls creating false windows and underlining their special purpose compared to other rooms. Similar "blind windows" have also been found in northern Mesopotamia and western Iran temples. The space between the rectangular and the square fortress was tightly packed with "cult" buildings (according to Sarianidi).

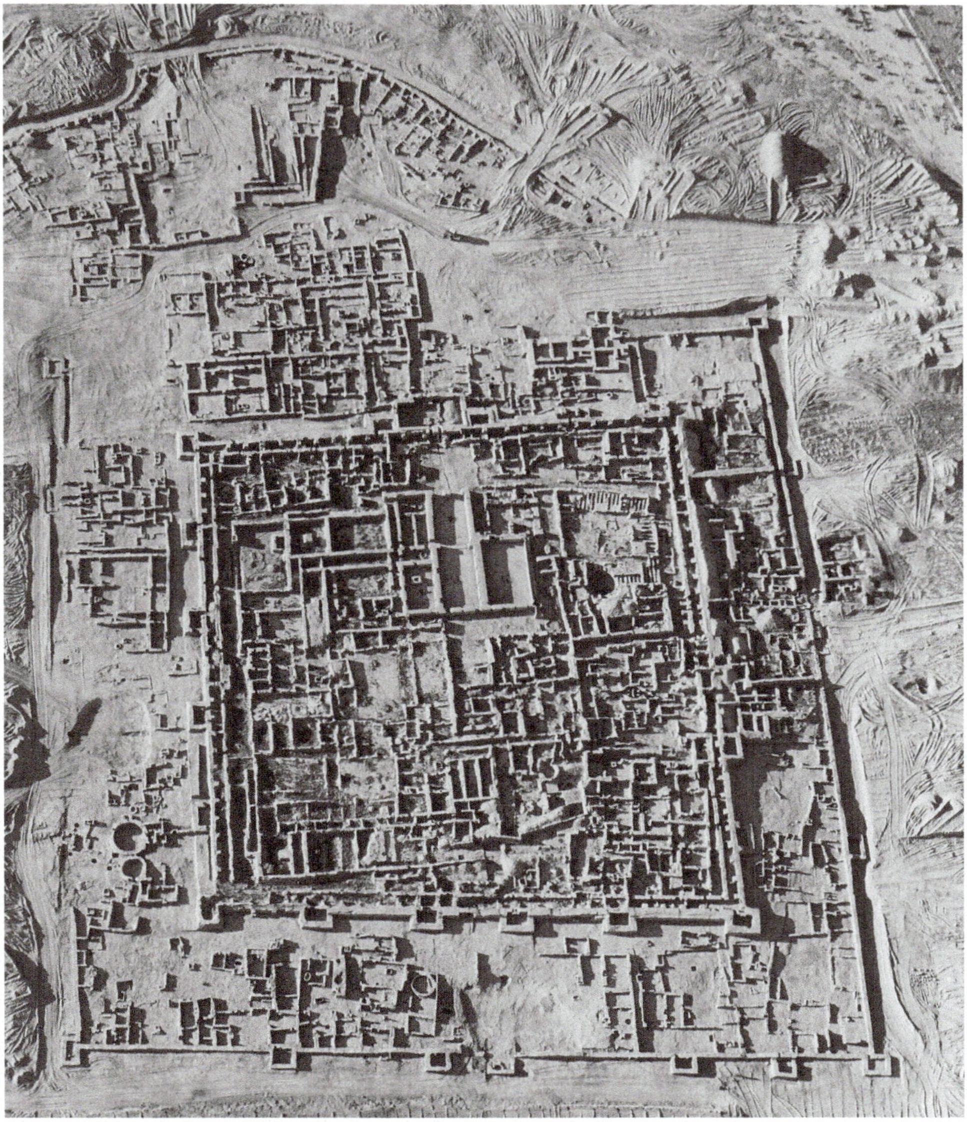

Figure 5.2a Gonur Depe North. Palace-temple complex. Air photo (taken in 2004, from Sarianidi 2005: 88–89).

Outside the square "palace" and *kremlin*, and within the encircling wall, the north, east, and west parts of the settlement were entirely built with domestic and craft production quarters, while in the south, there were two water pools, large and small, surrounded with buildings considered as sanctuaries dedicated to the cult of water. On the northern side of the large pool was a small necropolis consisting exclusively of several multichambered hypogea that had been looted several times but still contain extremely rich burial artifacts, including unique mosaic panels for their inner decoration (see Chapter 11). No other traces of interior decoration have been found

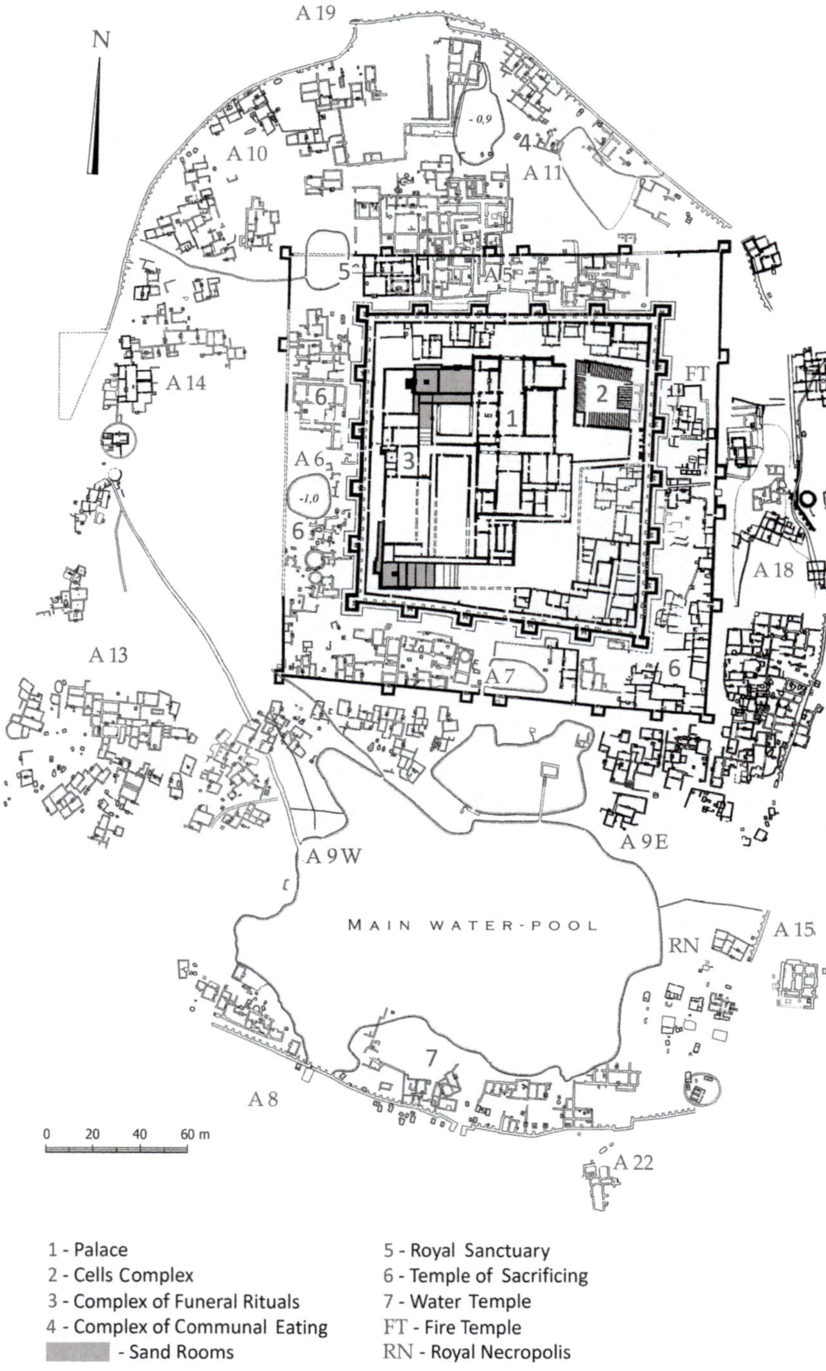

Figure 5.2b Gonur Depe North. Palace-temple complex. Plan (2018).

anywhere else in the Margiana buildings.[5] The exceptional discoveries made in these tombs have led their diggers to consider that it was a "Royal Necropolis" (Dubova 2004; Sarianidi and Dubova 2010; see also Chapter 10).

The most spectacular architecture is undoubtedly that of the double fortification with a "palace" and a "temple." Its layout consists of rectangular rooms and halls, narrow corridors, and a series of interior courtyards. Columns were sometimes located right at the center and inside the wide doorways of some rooms. The presence of a column to support the doorways or placed inside a courtyard testifies of problems of exceptional range for the beams supporting the roof (usually over 4 m). No room is the same at the Gonur "palace." One is decorated with regularly spaced deep rectangular niches along its walls, while another has the niches arranged at the very corners of the room, and the walls of one of the central halls are dissected with deep stepped niches: With the same basic element, different decorative results were produced. Many doorways have stepped ramps. Two groups of rooms inside the palace were completely filled with sand: We cannot exclude that sand was used as a construction method to create solid and high platforms for some unknown rituals because these parts dominated in height the surrounding structures, as shown by the presence of a wide staircase that probably led to the top of the platforms. The most mysterious construction in the "palace" is the comb-like structure in its northeastern part, which Sarianidi (2005: 83–87) designated as a "cell complex." Its purpose is not yet totally clear. Similar groups of very narrow and low empty chambers with vaulted ceilings and with carefully brick-sealed entrances are known in almost every monumental BMAC building: at Gonur South, Adji Kui 1, Togolok 21, Dashly 3, Dzharkutan, and Sapallitepa. For some authors, it is even considered as the architectural marker of the BMAC (Urmanova 2014). These structures, however, are well known in Mesopotamia, Syria, and Anatolia, starting with the Neolithic period, and are considered to have been "granaries" or storage rooms.

The "palace" is not without a certain resemblance to the palace of Mari on the banks of the Euphrates in northern Mesopotamia, which was famous for its size, number of rooms, and rich decoration (Parrot 1959; Margueron 2004: 459ff.). Rumors of its glory spread far beyond the borders of its realm and people from Ugarit, Yamhad, and Babylon came to see it as revealed in a survived document where the Ugarit ruler asks the king of Alalakh to let him visit the Mari royal palace, of which he had heard (Woolley 1968). Could the same rumors have reached Margiana and led to the building of a comparable construction? However, as Leonard Woolley once wrote, architectural ideas cannot travel with trade caravans, and transmission requires direct professional contacts. Sarianidi has long promoted the idea of a migration of population from northern Mesopotamia and Asia Minor into Margiana, bringing with it its Western architectural traditions (Sarianidi 1996: 326–329).[6] However, despite the presence of real novelties in the BMAC architecture, like corner rectangular towers in the fortification walls or grilled-plan storage rooms, the rest of the constructions mostly follow the traditional layout scheme. The compositional core consisted of a square courtyard framed by peripheral corridor, a symmetric decoration of the four facades of the defensive walls, and a precise orientation to the cardinal directions.

Gonur South was built significantly later than the northern complex, in the first centuries of the 2nd millennium BC. It covers an area of 1.5 ha and has been totally excavated (Sarianidi 1993, 1997). It has been considered by its digger as a

"Temenos," framed along its outer perimeter by a thick defensive wall with round towers at the corners and semicircular ones along the walls. Inside the sacred area, a small "temple" has been found in one corner, said to have dealt with the cult of hallucinogenic beverages and fire worship. Tightly packed residential houses and storage premises were built around the temple, where the officiants who devoted their lives to the rituals practiced in this building lived. When the "temple" was abandoned, another monumental (but smaller) building was constructed on top of its ruins; namely, a cross-like fortress with round towers at the corners and two other such towers in the middle of each wall guarding the entrances, but its construction was never finished. Its layout presents a slight deviation in orientation compared to the outer "Temenos" wall (Figure 5.3). This "fort" is the prototype of Bactrian Late Bronze/Early Iron Age temples such as those at Tillja Tepe in Afghanistan (Sarianidi 1989a; 1989b: 113, fig. 4).

In the Adji Kui oasis next to the Gonur one, nine settlements have been identified during a survey, including Adji Kui 8, where the walls were visible immediately under

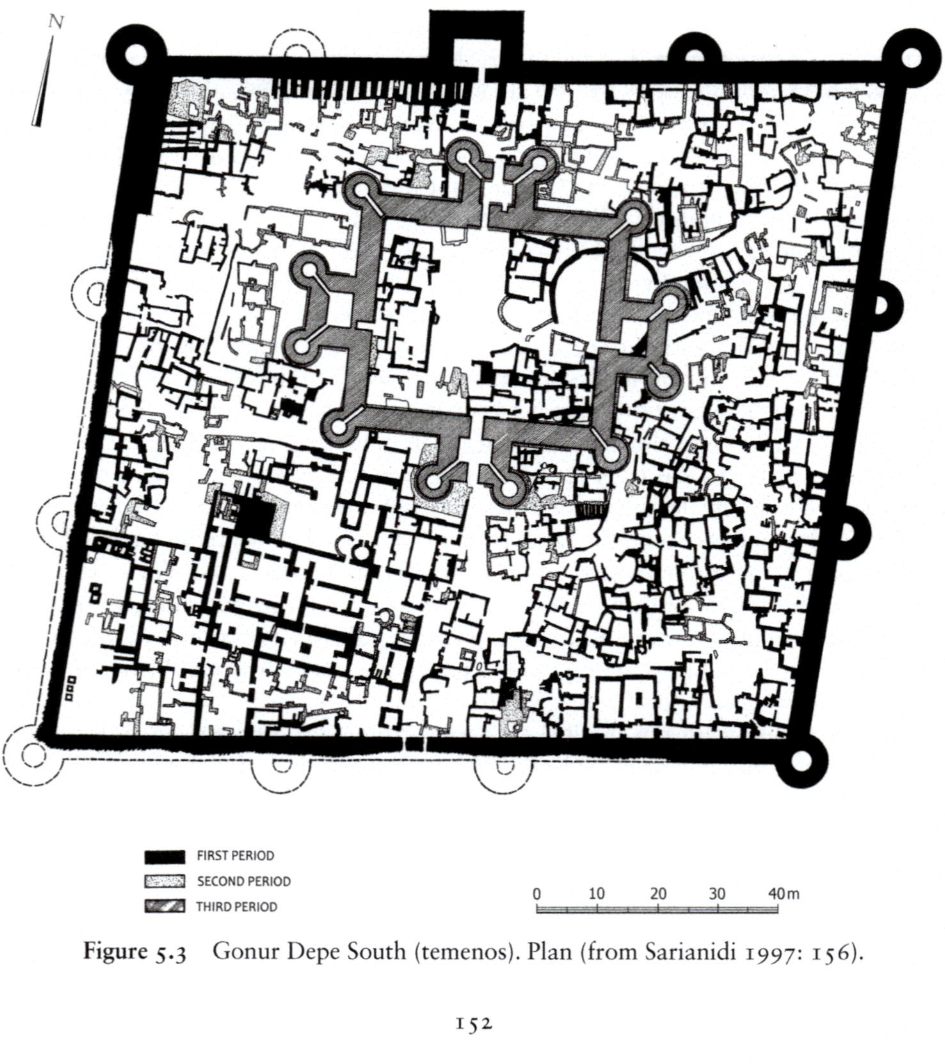

Figure 5.3 Gonur Depe South (temenos). Plan (from Sarianidi 1997: 156).

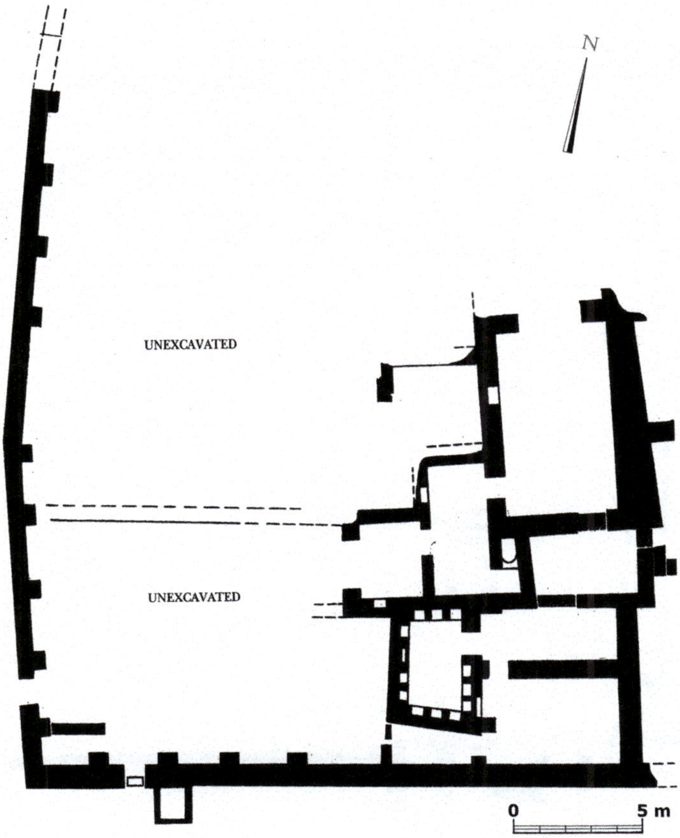

Figure 5.4 Adji Kui 8. Plan (from Sarianidi 1990a: 8).

the surface. Two peripheral walls ca. 1 m thick were outlined, but not excavated. The walls present a regular pattern of buttresses on the interior, and form the southwestern corner of a building (Figure 5.4). In the excavated area, a small room has three in-wall niches on each side at 60 cm from the floor. A rectangular structure, probably a tower, was found at the southern entrance into the fortress (Sarianidi 1990a: 7–11). From the layout of these remains, Adji Kui 8 seems a typical example of a BMAC fortified "square house."

Some 23 years after these survey soundings, full-scale excavations were made at the neighboring settlements of Adji Kui 1 and 9 and lasted for ten seasons. It then became clear that Adji Kui 1 was indeed the largest site of this oasis. Its area is estimated at almost 5 ha. The core of the settlement has the shape of a fortress ("citadel") with thick peripheral walls (from 1 to 3.2 m) (Figure 5.5). The central entrance in the southern wall is curved and flanked by two rectangular towers. Inside the "citadel," groups of small residential and storage rooms were excavated along the walls. The storage rooms contained a great number of large pottery vessels, possibly for the whole community, as indicated by the discovery of clay sealings. Nothing in this "citadel" presents any military function and the

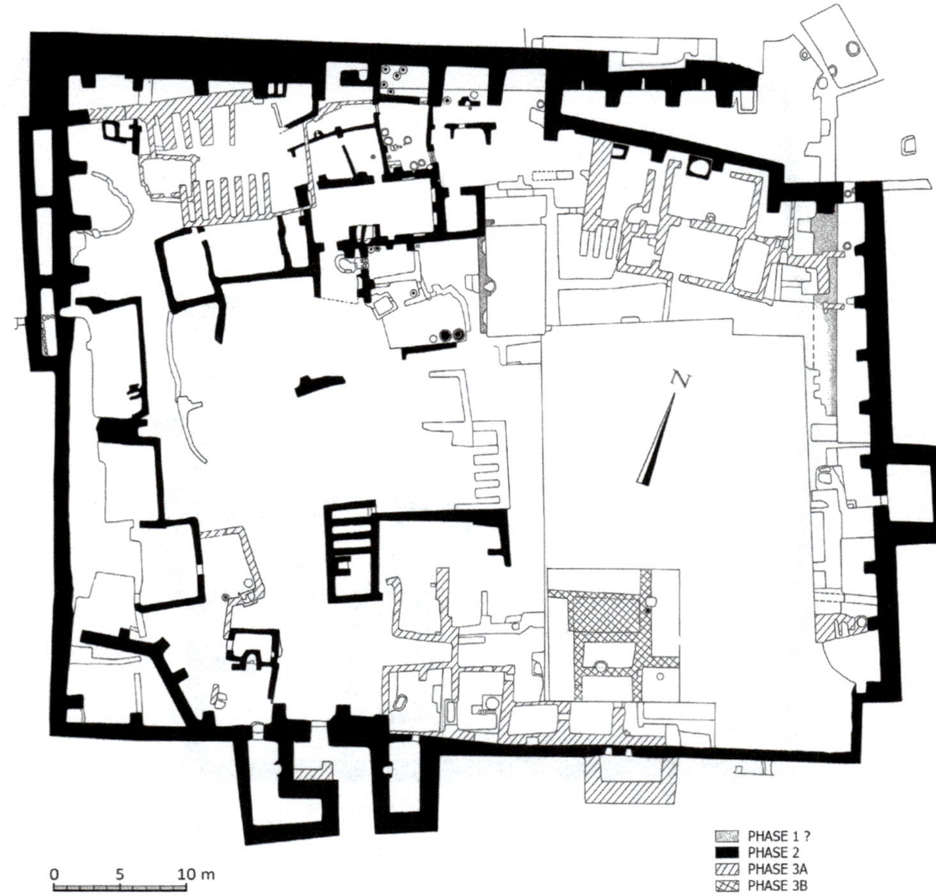

Figure 5.5 Adji Kui 1. Plan (from Udeumuradov 2011: 213).

settlement was more likely a center for trading agricultural and other manufactured products with close and distant neighbors (Rossi Osmida and Udeumuradov 2003; Udeumuradov 2011).

The settlement of Adji Kui 9 is only 500 m to the east. It is a rhomb-shaped fortress of ca. 0.5 ha (Rossi Osmida 2007, 2011). It is surrounded by a wall (67 × 76 × 65 × 77 m) 1 m thick with rectangular towers. A second wall, separated from the first by a 2 m wide corridor divided into rooms, runs inside along the first one (Figure 5.6). Considering the twin towers found at the southeastern corner and shaped as a "dovetail," it is possible that similar constructions were located at all four corners of the fortress. In such cases, the total number of towers could reach 20, with five on each side of the fortification. The only narrow entrance into the fortress, flanked by two towers, has a high doorstep (20–27 cm) and therefore was primarily meant for pedestrians and not for carts or carriages. A large interior courtyard was later occupied by buildings with the typical BMAC layout (i.e., separate households located along peripheral corridors and grouped around interior courtyards). Craft activities dealing with metalwork and pottery have been identified outside the fortress, just as at Adji Kui 1 and Gonur Depe. Despite S. Salvatori's

Figure 5.6 Adji Kui 9. Subphase 3B. Plan (from Rossi Osmida 2007: 115).

(2007) severe criticism on the excavation methods and the publication (Rossi Osmida 2007), the data acquired from this site gives many specific details on the BMAC monumental and household architecture.

The excavations made in 1972 by Sarianidi at Auchin 1 also revealed the presence of a two construction periods fortress with round corner bastions, and additional round towers along the walls (Sarianidi 1990a: 10–11). A similar layout has been excavated at the "temple" complexes of Togolok 21 (Sarianidi 1990b, 1990d) and Togolok 1, though, on this last site, it was not as well built up (Sarianidi 1990c) (Figure 5.7).

The central core of Togolok 1 is a fortified structure with a square plan (ca. 29 x 30 m) and round towers at the corners on the northern side. The exterior perimeter wall is ca. 1.5 m thick. The interior consists of rectangular and square rooms and small courtyards. Except on its northern side, the fortress is surrounded first by a

Figure 5.7 Togolok 1. Temple. Plan (from Sarianidi 1991: fig.37).

spacious courtyard then by a more or less oval wall with a deep angular recess on the southern part. Remains of constructions are visible in the northeastern part of the courtyard. During the second period, the southeastern part of the exterior wall was strengthened by thickening and the addition of five semicircular solid or hollow towers. The fortress itself was bolstered with additional rows of bricks, and a hollow round tower was added to its southwestern corner. Due to the bad conditions of preservation, entrances have not been always recognized.

The Togolok 21 architectural complex is represented by three rectangles of different sizes built one within the other (Sarianidi 1990b, 1990d) (Figure 5.8).

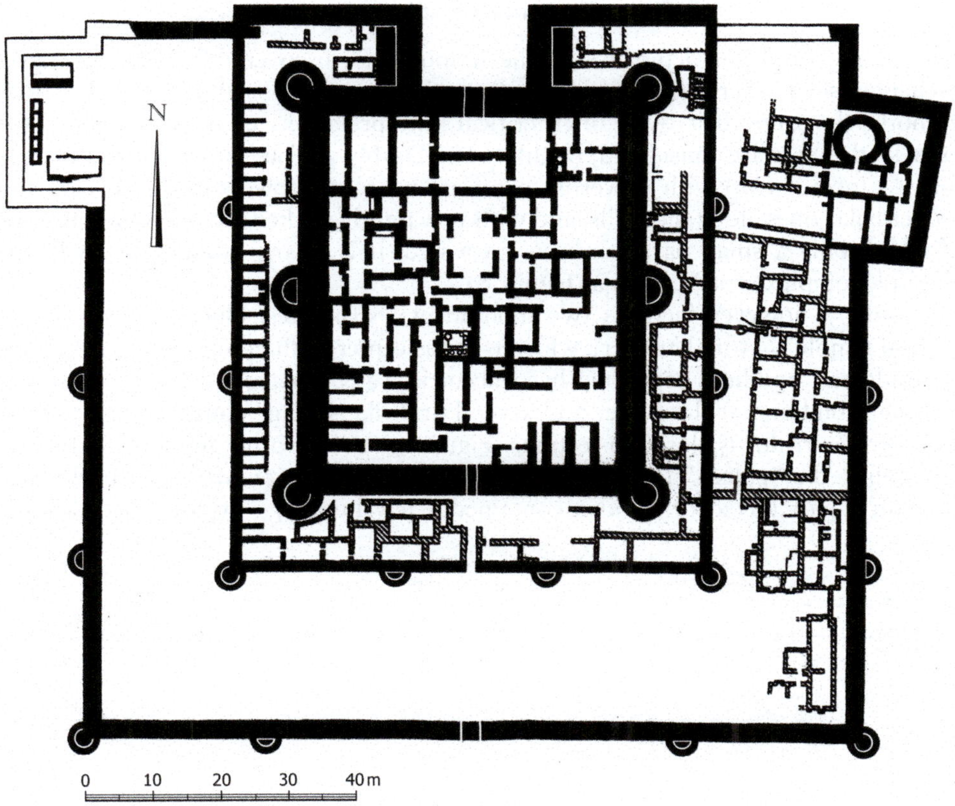

Figure 5.8 Togolok 21. Plan (from Sarianidi 1990a: 126).

The external fortified wall is 140 × 100 m, with round towers in its southeastern and southwestern corners and with two semicircular ones along the sides, except for the northern one. Two special platforms step out of the perimeter lines and violate the strict geometry of the construction in its northeastern and northwestern parts. The middle construction presents the same layout as that of the fortified wall and the towers. A series of small cells run inside along the western wall. The third, inner construction is bolstered with the largest round towers of this complex that are positioned at the corners and in the middle of its western and eastern walls. The building contained rectangular rooms around a central courtyard, as well as cell rooms.

The walls of the inner fortress are much thicker than those of the two others, and they probably were also higher so that this central structure dominated the rest of the complex. The same layout is known at Gonur South and testifies of the existence of planning rules during the first half of the 2nd millennium. A similar perspective is attested later, during the 8th century BC on Assyrian reliefs at Dur-Sharrukin, though the towers there are quadrangular (Albenda 1986: pl. 112, 124–126, 128, 136, 138). According to Sarianidi, this specific plan was mainly dedicated to religious architecture.

Southern Bactria

The architectural characteristics of the monuments discovered by Sarianidi have been published several times, by him (Sarianidi 1977a, 1977b) and also by other authors who gave their own, often critical, interpretations of these remains. Two main settlements are considered, Dashly 1 and Dashly 3. The former is a rectangular fortress (99 x 85 m) with powerful fortifications, round towers in the corners and semicircular ones along the walls, and with very tightly packed internal constructions of which only a small part has been excavated. It is clearly comparable with the Togolok complexes of Margiana (Figure 5.9).

Dashly 3, however, includes two monumental structures. One is known as the "round temple" (35 m diameter), which is at the center of three surrounding sections of residential buildings separated by concentric walls. The round fortress contains one main building at the center and several grill-like storage constructions, and its peripheral corridor is closed in sections (Figure 5.10). The other monumental structure called the "palace" is a complex cross-like construction with pilastred-walls. Its central part (40 x 38 m) also contained a few buildings together with a large

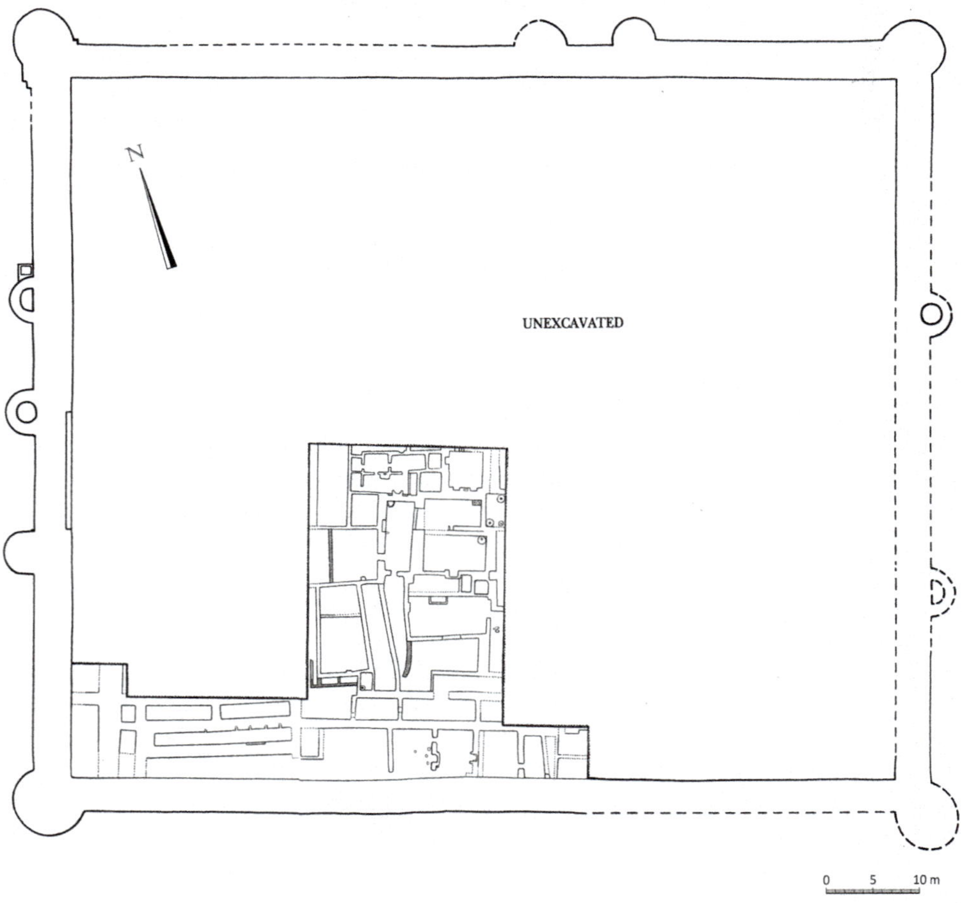

Figure 5.9 Dashly 1. Plan (from Sarianidi 1974: 52).

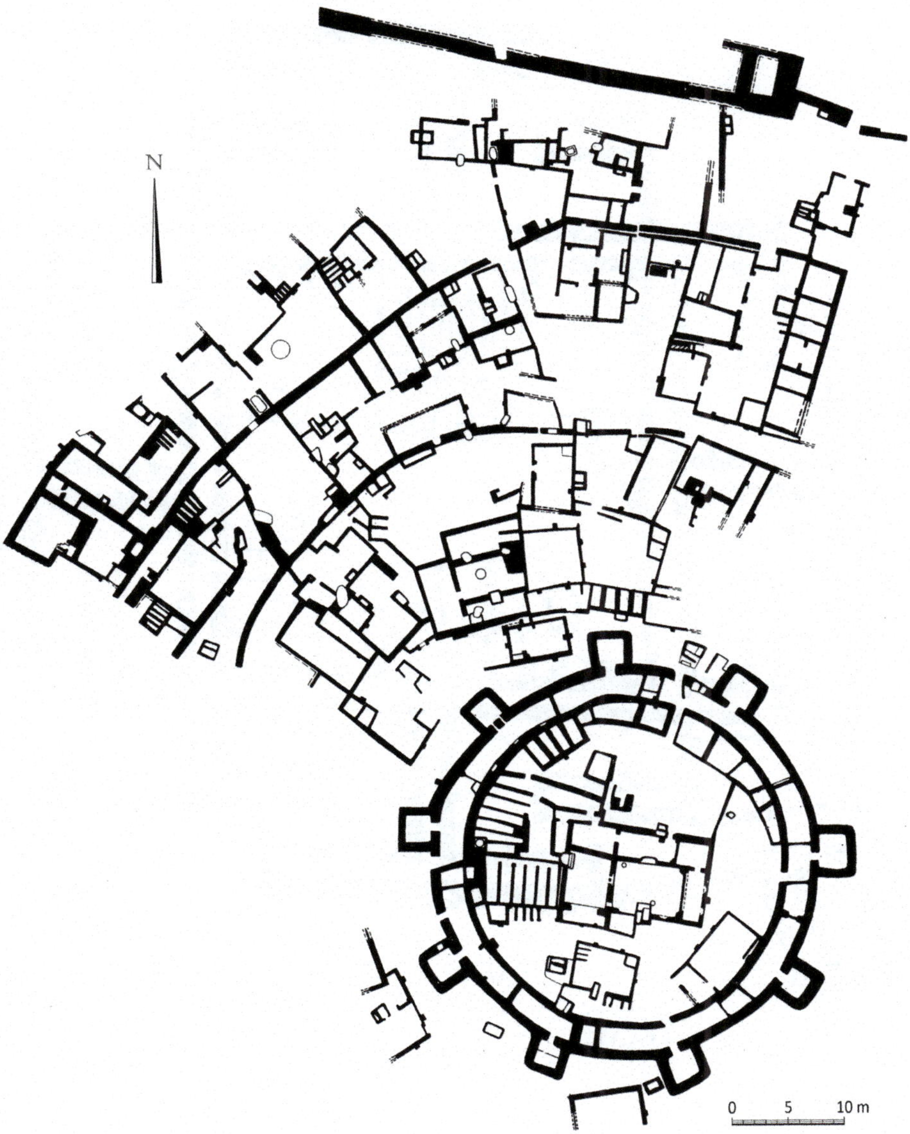

Figure 5.10 Dashly 3. "Round temple." Plan of the excavated part (from Sarianidi 1977a: fig.11).

grill-plan storage structure. T- and G-shaped corridors with pilasters on their outside and a complex entrance give it a fortified appearance (Figure 5.11). These two buildings have no parallels in Margiana.[7] They have led to various hypotheses as to their function (see, e.g., Brentjes 1981; Pugachenkova 1982; Antonova 1984: 78–82; Bulatov 1988: 28–35, 44–46; Chmelnizkiy 1993: 304–307). The round building evokes northern Mesopotamian structures, as mentioned above.

Both are the result of the skillful realization of a pre-planned program and demonstrate the importance of geometric principles based on the square, the circle, or the

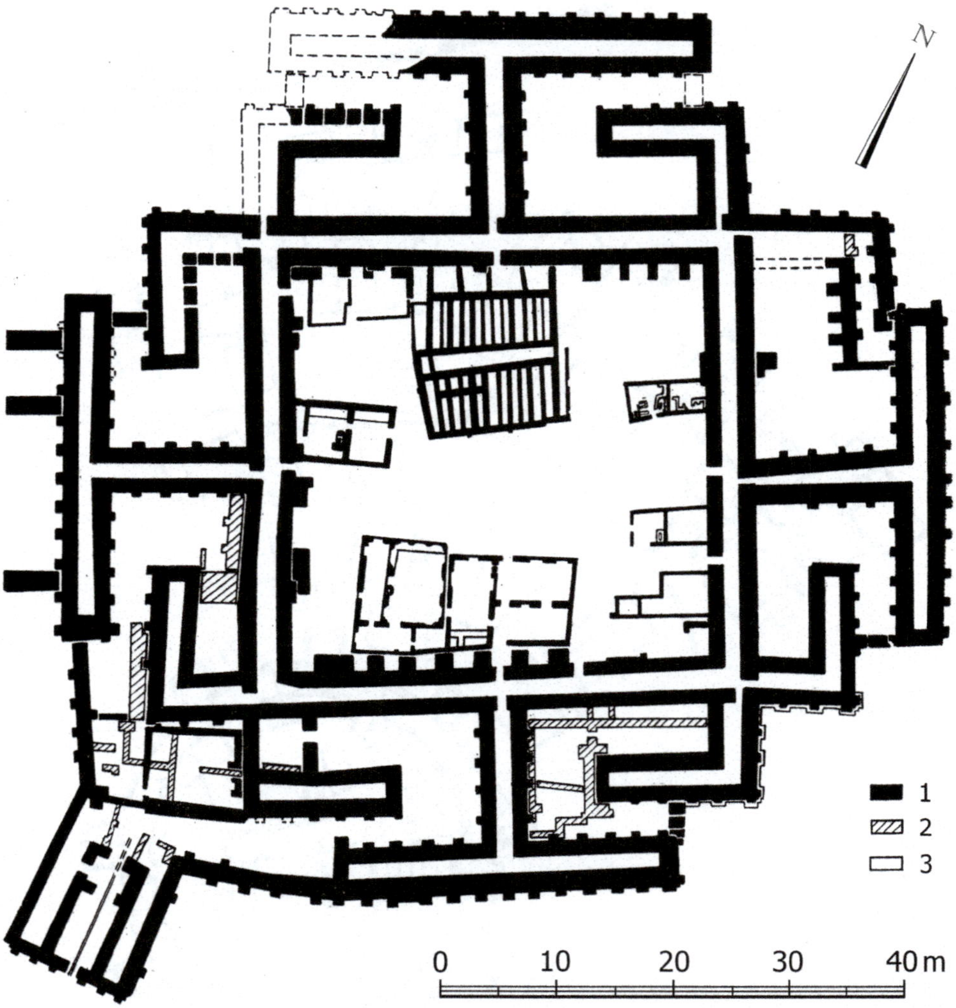

Figure 5.11 Dashly 3. Palace. Plan (from Sarianidi 1977b: fig.6).

rectangle. As in Margiana, the monuments have a strict orientation to the cardinal directions, a central courtyard, a peripheral corridor (or corridor system), cell rooms, and defensive walls with rectangular towers (for the round "temple"). The "palace" has a complex T-shaped defensive system.

Sarianidi (1977a: 50) considered that the combination of round and square buildings cannot be accidental. These building structures could reflect cosmological ideas, based on the fight against the realm of "evil and lies" representing chaos, and on the creation of a "blessed celestial dwelling place" on earth. Some authors clearly state that, for example, the round building at Dashly 3 "undoubtedly has a religious-cosmological origin, but that its full meaning has not been deciphered yet" (Kartzev 1986: 98–99).

Northern Bactria

The settlement of Sapallitepa is situated on the right bank of the Amu Darya in the Surkhan Darya valley (Uzbekistan) and also testifies to a high level of fortification skill. The settlement consists of a fortress, shaped as a regular square (82 × 82 m) (Figure 5.12). The core of the building is a square courtyard (58 × 58 m) framed by a mudbrick wall, with two rectangular corridor-like and more or less T-shaped rooms on each of its four sides regularly disposed, one 26 m, the other 36 m long, connected through small square anterooms. The position of the eight exterior corridor-like rooms is perfectly planned. These rooms create similar but reversed T-shaped (or L-shaped in the corner of the fortress) dead-end corridors accessible from the exterior, which, according to their digger A. Askarov, were false "trap" entrances. The true entrance

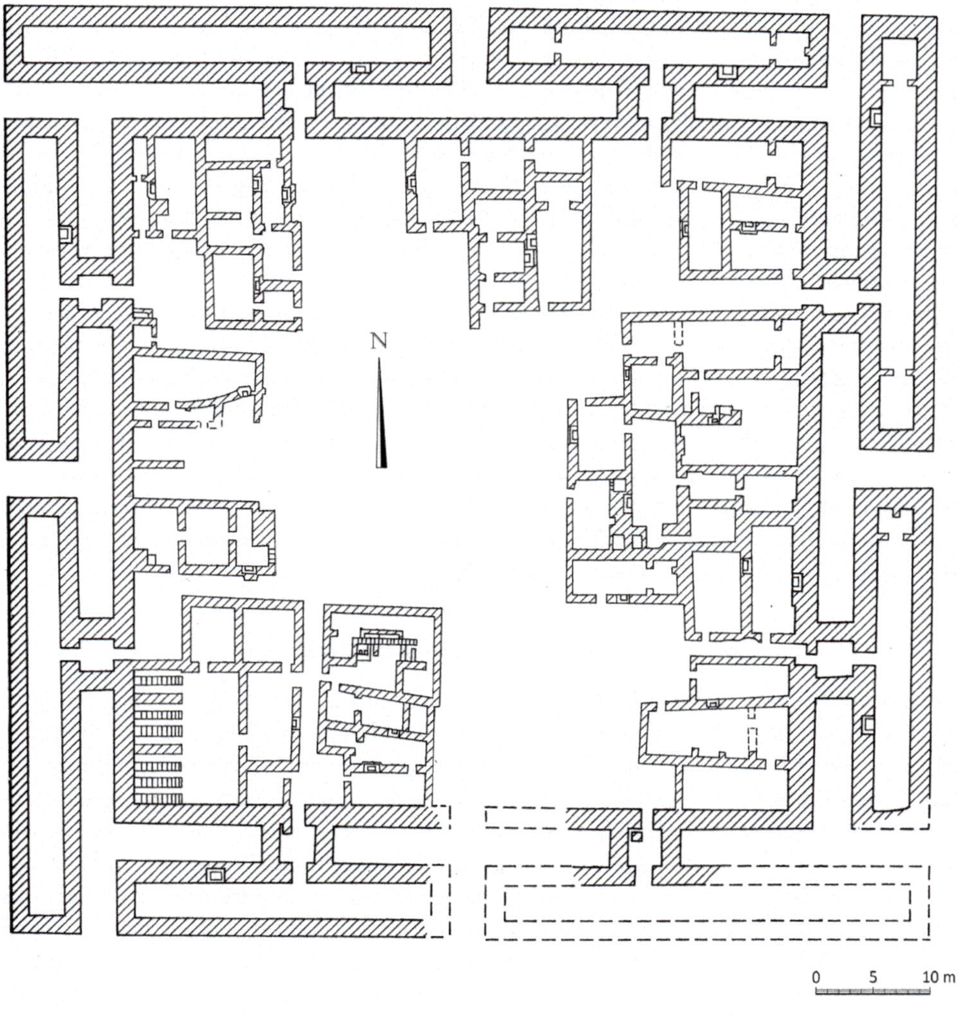

Figure 5.12 Sapallitepa. Plan (from Askarov 1977: fig.4).

to the complex is supposed to have been in the center of the southern side as reflected by the layout of the constructions inside. These were apparently individual multiroom houses each with their own courtyards, and only one with cell rooms. All are located along the perimeter wall and most of the center of the building remained free of constructions (Askarov 1973).

By its architectural-planning composition, the fortress of Sapallitepa belongs to the "courtyard framed by a single row of rooms" layout. Such a compositional scheme was probably created at the end of the 3rd/early 2nd millennium BC in Margiana as shown at Kelleli 4 (see p. 165). However, at Sapallitepa, this composition developed further and was reinterpreted leading to the creation of an original system of fortifications. Here, as at the Dashly 3 "palace," corridor-like rooms protruding beyond the outline of the main square and creating a system of labyrinths also play the role of peculiar elongated defensive towers.

The main administrative and religious center of northern Bactria was probably the large settlement of Dzharkutan (about 100 ha altogether),[8] located 20 km to the north of Sapalli (Askarov and Shirinov 1993, 1994). There, a "citadel" with a "palace," living quarters, craft activities areas, a huge necropolis, and a fortified rectangular "temple" with nine round and semicircular towers (44.5 × 60m) have been excavated.[9] The western and eastern facades of the "temple," each with two semicircular bastions between the corner towers, mirror each other (Huff 1997) (Figure 5.13). The central part of the building is rectangular (13 × 31 m) and paved with bricks. As in most of these monumental structures, it contained a grill-storage structure. Thus, the general shape of the construction is typical of the BMAC religious buildings.

Finally, though preceding the BMAC and without demonstrated direct genetic links with it, a few words need to be said on the monumental buildings at Sarazm in the upper Zeravshan valley (Tajikistan) (Isakov 1991). The site is dated between ca. 3500 and 2500 BC and attests to strong influences from cultures very far away, especially Geoksjur in the Murghab delta and Baluchistan sites. Some of the constructions are decorated with pilasters/buttresses, some of which supported the roof beams. One of them is of the "square courtyard framed with corridors" layout that became typical of the BMAC (Figure 5.14). Their high-quality building techniques and geometric plans testify of pre-planned constructions (Razzokov 2016).

Conclusion on the monumental architecture

To sum up, the representative BMAC architecture of the end of the 3rd/beginning of the 2nd millennium BC was innovative and original. The predominance of "temple" complexes with "granaries," if this religious function is demonstrated, would indicate that during the BMAC, just like at Altyn Depe before that, the main role in the management of the production activities was played by a religious elite. This elite could also have headed the Gonur "palace" complex with its surrounding "temples" and necropolis (Masson 2011). At one point during the early 2nd millennium BC, this religious elite disappeared, leaving the "temples" in abandonment, and they were then replaced by fortified "palace-like" structures. Even after a long period of decline and changes, many of the features of this monumental architecture will reappear during the Iron Age and later.

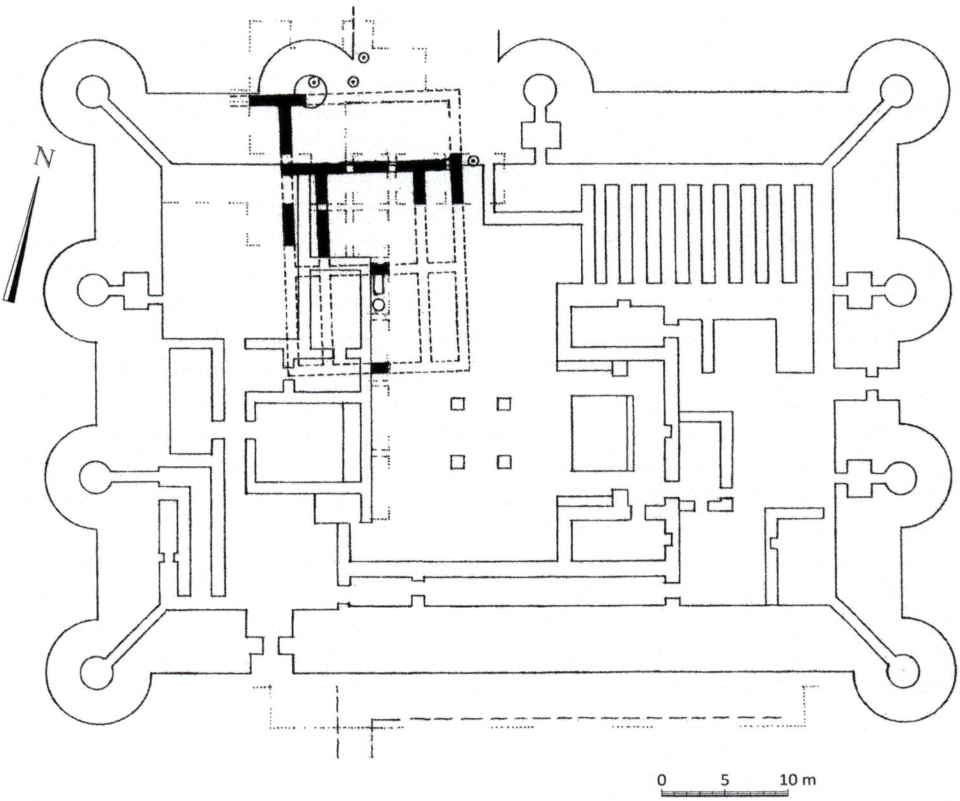

Figure 5.13 Dzharkutan. Tepe VI. Plan of the later building covering an earlier building underneath; in black: the excavated wall of the latter with hypothetical reconstructions in dashes (from Huff 2000: 67).

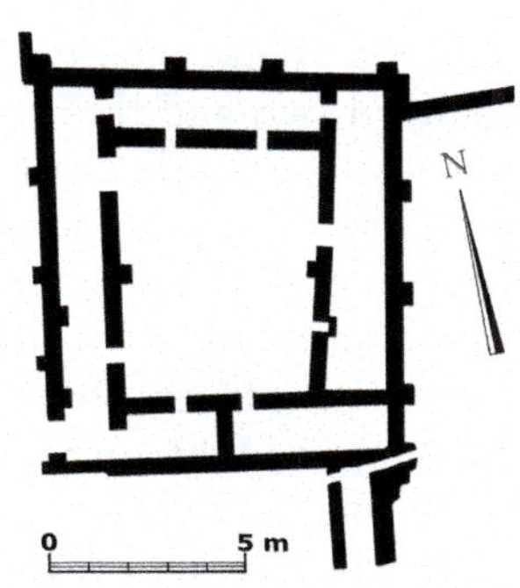

Figure 5.14 Sarazm. Plan (from Razzokov 2016: fig. 66).

DOMESTIC ARCHITECTURE

The BMAC period is considered as that of a proto-early urban civilization. Besides the monumental buildings, other types of casual domestic architecture are known.

Ordinary living complexes are represented by groups of houses (blocks) either outside the monumental buildings mentioned above or isolated but within a square fortified courtyard. In a number of cases, domestic structures correspond to a reoccupation phase of earlier monumental buildings, as the descriptions that follow clearly show.

The earliest known examples of BMAC domestic architecture were found among the Kelleli settlements mentioned earlier. Thus, the partially excavated settlement of Kelleli 6 is constituted of a series of 12 rectangular rooms built within in a semicircle (Masimov 1986: 176–178). Their distinctive feature is the presence of projections of unknown purpose at the end of the walls. These are made of beaten earth (*pakhsa*), coated with a clay plaster. In the western part of the settlement, at some distance from the rooms, remains of a straight wall have been discovered, slightly deviating from the north–south axis. By its planning composition, Kelleli 6 resembles the general layout of the South Bactrian settlement around the Dashly 3 "temple" (see p. 159), where the whole circular development around the central core is inscribed into a huge square formed by an exterior perimeter wall. It also presents certain similitudes with sites of the Neolithic Djeitun culture, in particular with Chagylly Depe, despite the several thousand years between them.

On the basis of the distinctive features of Kelleli 6 that set it aside from the other sites of the Kelleli oasis, and on the basis of the "archaic" character of the domestic structures and their rudimentary construction material (beaten earth), Masimov (1986: 177) considered that the site either reflected the lower social status of its population, or a different chronological period. He considered it as one of the first settlements in the Murghab delta founded by migrants from the Kopet Dagh.

Remains of a residential complex with a particular layout have been discovered at Kelleli 4 (Masimov 1984). The settlement is square (29.5 × 29.5 m) strictly oriented along the cardinal directions and built of mudbricks usually plastered. The exterior wall is not very thick (1 m) and is associated with single rectangular towers placed in the middle of each facade, except for the southern one where two such towers frame the entrance to the site.[10] Along the exterior wall, runs a second thinner wall (0.3m) separated from it by a corridor divided into several rooms by transversal partitions (Figure 5.15). The interior of the square is divided into 40 rectangular rooms within which can be identified two and three interconnected rooms making residential houses, with a fireplace built into the wall in at least one of the rooms. All the towers at Kelleli 4 have an interior room and are connected to the "corridor rooms" by doorways. The whole complex seems to have been built at one time and was never significantly rebuilt in the course of its existence (Masimov 1984: 17). This clearly prefigures the architectural plans of later (Iron Age and Hellenistic) periods, when rooms are built along the perimeter of the courtyard (Pugachenkova 1982: 35). Here the central courtyard is very small, and the constructions around it are very dense. It is unknown whether there was an exterior wall further away, protecting the whole settlement.

The Togolok 21 monumental "temple" fortress described above has a second known period of existence marked by the construction of domestic structures on the eastern side between the first exterior wall and the second one. These are adjacent multiroom houses, clearly separated one from the other by walls perpendicular to those of the previous fortress (see Figure 5.8). Similarly, in the second period, a series of rooms were built between the southern wall of the two inner fortresses, the walls of which stand on

Figure 5.15 Kelleli 4. Plan (from Masimov 1984: 16).

top of refuse layers of the previous occupation. The walls were built carelessly, often with one row of bricks, but still with fireplaces in them. Almost all the rooms are rectangular in shape. These constructions seem to continue on the eastern side.

Unlike the earlier sites of the Kelleli oasis, the domestic constructions at Togolok 21 do not show any distinct plan. Who was living in these residential buildings has been a matter of debate because of the differences in the building quality with the "temple," which was considered as the most important one in all Margiana (Sarianidi 1990a: 151). The fact that the houses and the "temple" are attributed to two different periods partially solves the question.

The site of the round "temple" of Dashly 3 in Bactria also presents domestic buildings outside the main round construction, between it and the perimeter walls around (see Figure 5.10). According to Sarianidi, the buildings may have housed priests and officiants (Sarianidi 1977a: 37). In this case, the domestic buildings and the "temple" existed and functioned at the same time.

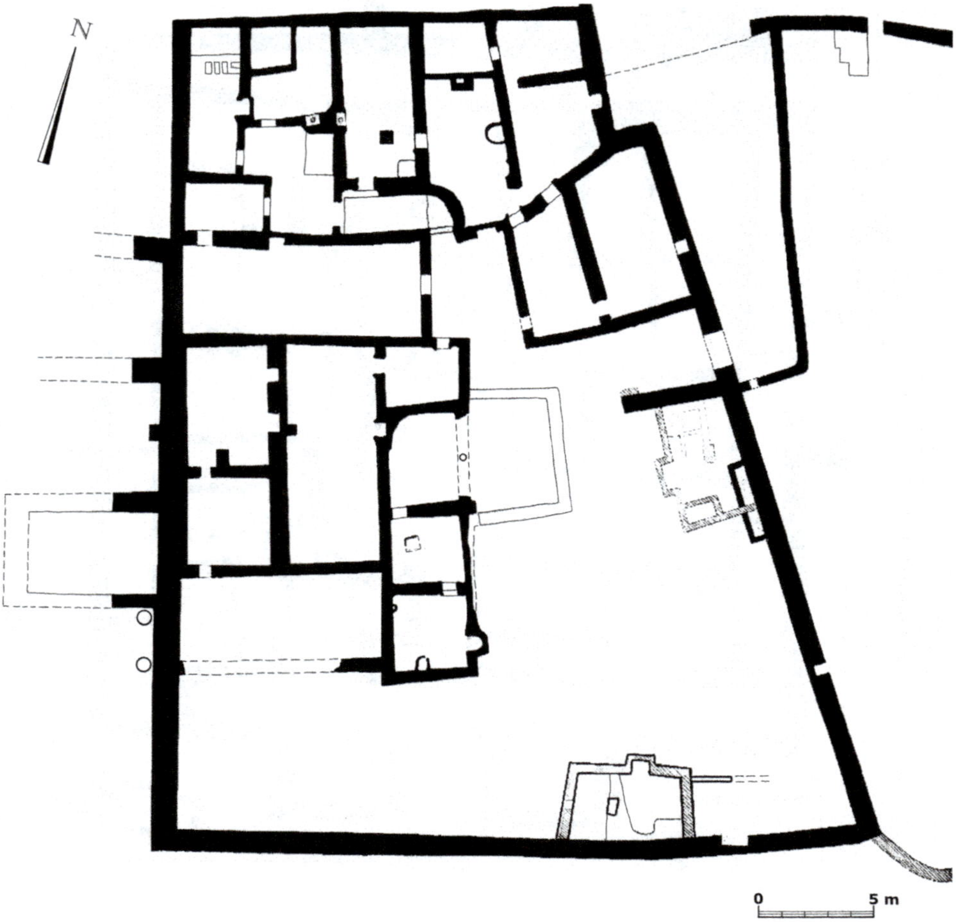

Figure 5.16 Adji Kui 1. Residential house (from Rossi Osmida and Udeumuradov 2003: 60).

Modest constructions are known to have been built in a later phase at Togolok 1. Dwellings also appeared inside the fortress wall of Gonur South in the second period of its existence, and remains of buildings associated with a rectangular courtyard with a round kiln very close to modern Turkmenistan tandoors have been excavated at the late BMAC site of Takhirbaj 3 (Masson 1959: 15–16).

At Adji Kui 1, a large house (472 m^2) has been excavated in the western part of the site and is in many ways typical of the BMAC domestic architecture. It is a multiroom building consisting of five "apartments" varying in size from two to seven rooms, or 30 to 105 m^2 (Figure 5.16). Each of them has one or two doorways opening on the common courtyard, and each of them has one to six in-wall fireplaces for heating and cooking represented by a niche with a vertical chimney, as well as one to two storage rooms. Curiously, these last rooms were the largest and covered over 28% of the total area of the estate, apart from the courtyard, and there is no doubt as to their function as shown by the great number of large pottery containers found in them. In most cases, the rooms are rectangular and the whole building testifies to a well pre-established plan. The doorways between the rooms are 50–70 cm wide and were completed with doorsteps 10 to 30 cm high. Some rooms had a rectangular or semicircular podium made of sun-dried mudbricks carefully plastered with a clay mortar. No evidence of windows has been discovered. Several trash pits filled with fragments of kitchenware, animal bones, and layers of humus were found in a large household courtyard. This kind of large building may have housed a large family group where each smaller entity had its own section. Nothing in the architecture or the finds testifies of any difference in the social status of the inhabitants. No evidence of religious purpose was found either, probably because of the presence of another specific building for religious practices (Rossi Osmida and Udeumeradov 2003).

Other types of domestic architecture include the craft complexes identified at the satellite settlements around Gonur Depe. At Gonur 20, several single- and double-room houses have been excavated, with separate ovens and double-chambered chimney/hearths built into the walls, as well as large kilns and traces of a canopy. Gonur 21 is almost square (25 × 26 m) with a thick surrounding wall and a single entrance on the northern side (Figure 5.17). It contained four sectors of rooms opening on a central courtyard, bent at right angles. The set and configuration of the rooms, as well as the type of oven or hearth, is different in each sector. In one of them, and in a rather small room, a rare case of a three-chambered hearth/oven has been found. Graves have also been discovered under the floors of some rooms (Sarianidi

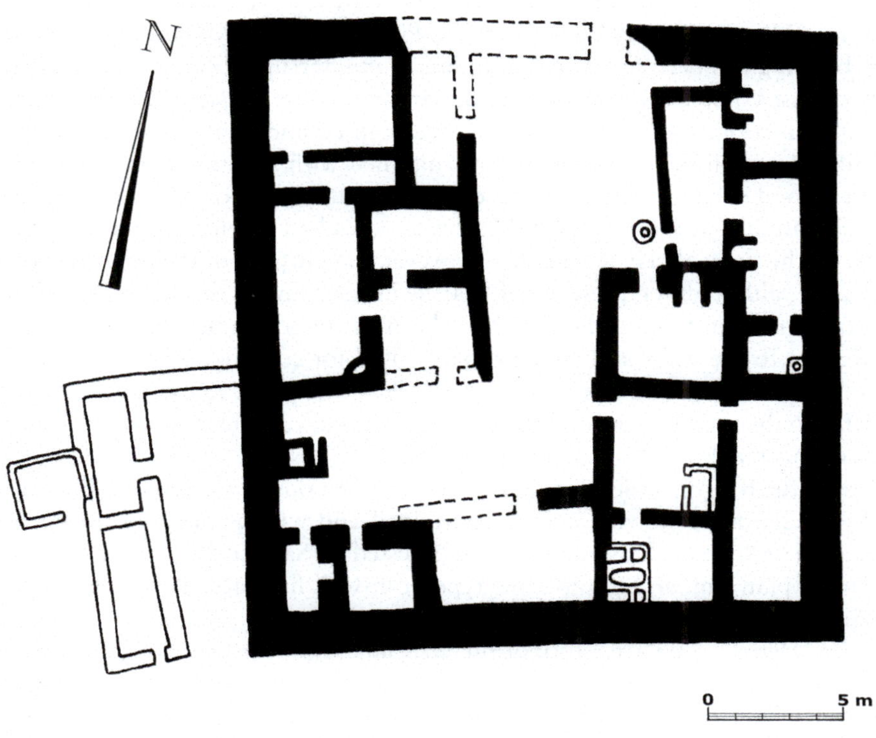

Figure 5.17 Gonur 21. Residential house. Plan (from Sarianidi and Dubova 2012: 43).

and Dubova 2012: 40–45). These features are not those of an ordinary large-family community, but rather those of a "multifunctional manor" of one privileged family, where separate economical units were associated with the living quarters.

To sum up, although the excavated domestic structures of the BMAC in the Murghab region are still scarce, some of their characteristic features can be outlined. The multiroom houses making dwelling units can be traced back to the Late Chalcolithic and Early Bronze Age in southern Turkmenistan (see Chapter 4). The genetic links between the BMAC and the Kopet Dagh foothills are clearly visible in the layout of Kelleli 1, very similar to the houses of the "noble quarter" at Altyn Depe (Masson 1989: 173). However, quoting F. Hiebert (1994: 101–102):

> Gonur's domestic architecture is different from that of the contemporary late Namazga V urban sites of the Kopet Dag foothills plain. At Altyn Depe, the rooms in the domestic architecture of excavation 10 are smaller than the rooms at Gonur and more complex than at Gonur, Kelleli 4 and Togolok 21, and more complexly interconnected, with many doors and corridors. In contrast, the household architecture of Gonur is similar to that of other Period 1 sites of Margiana. Room size, number of hearths per complex, and the number of rooms connected by doorways at Kelleli 3 and 4 are very similar to Gonur domestic architecture. Unlike Gonur, however, these domestic units are isolated from other architecture by an enclosing set of walls and towers. The domestic architecture at Gonur is similar to patterns of household units at Kelleli 4.

In Bactria, domestic architecture has been mainly studied inside the Dashly 1 fortress (Sarianidi 1976: 21–26). The plan (see Figure 5.9) shows that only a small part of the building has been excavated. It points at uninterrupted constructions, without distinct house complexes, enclosed in a square space. It is possible that they each had a specific function. The narrow rooms directly in contact with the fortress exterior wall differ from all the others as they are all filled with a burnt layer and were possibly destroyed by fire. From their unusual configuration and location near a defensive tower, Sarianidi (1976) considered that these rooms had a defensive function further proven by the finds of clay sling balls. However, the purpose of the other rooms is as yet obscure, although the presence of hearths inside some of them can provide some answers. Two main types of hearths have been excavated: some (hearths/oven-like), are adjacent to the walls and were probably used for cooking; others are fireplaces, sometimes within the walls, probably used to heat the main rooms. Excavations have unearthed only one-fifth of the fortress, which impedes reconstruction of the general layout of this block.

Most of the BMAC settlements in Bactria actually consist of a fortress and an adjacent living quarter, following a well-determined and widespread settlement pattern (Sarianidi 1976: 26). The same model is attested at Sapallitepa, where not only is the general plan similar, but the same type of in-wall fireplaces have been identified (Askarov 1973: 41).

At the Dashly 3 "palace," too, ordinary residential buildings have been brought to light (see Figure 5.11). After its abandonment, a small settlement was built on its ruins, which later grew in size reaching in the latest, third period, an area of 40 × 40 m inside the former courtyard (Figure 5.18). This way of rebuilding domestic buildings

Figure 5.18 Dashly 3. Domestic architecture. Plan (from Sarianidi 1977b: fig. 4-3).

over deserted "palaces" and "sanctuaries" is another characteristic feature of the BMAC: As shown above, it happened at Togolok 1 and 21, and also at Gonur South.

The third construction level at the Dashly 3 "palace" (see Figure 5.11) consists of uninterrupted construction of residential and utility rooms, with, notably, a narrow north–south corridor alley conventionally dividing a unit into two parts. On the west of the alley, almost all the rooms contain a heating structure. Sarianidi (1977a) considered that those in the shape of a constructed fireplace were probably used for heating, and therefore that the rooms were residential, and those with in-wall fireplaces were probably used for cooking. Storage rooms still contained well-preserved jars dug into the floors. On the other side of the alley, a similar system of interconnected rooms has been identified. The total area is limited by the walls of the former courtyard, but the dual division of this last period settlement might give a clue to understand the social life at that time (Sarianidi 1977a: 31–34).

According to Askarov, the living quarters at Sapallitepa and Dzharkutan also consisted mainly of two- and three-room "apartments." They included a kitchen with a fireplace and a bedroom with sofas (benches) and brick pavement. Some houses consisted of a single large room containing an oven-fireplace and a sofa. Storage rooms were often located next to the living ones. They probably contained grain since their floors were carefully finished with a clay or limestone plaster, or, in other cases paved with pottery fragments. It is considered that common storage places existed elsewhere since only a few such places are attested in each residential complex. At Sapallitepa, the walls are considered to be not over 3 m high, without windows but with small holes for light and air in the upper part of the peripheral rooms (Askarov 1973: 150–151).

To sum up, studies concerning the BMAC domestic architecture underline the main principles of architectural plans in southern Central Asia, most of which will still be in use later, in more developed forms, in cultic and mundane buildings of greater Central Asia.

BUILDING MATERIALS AND CONSTRUCTION METHODS

Clay (*pakhsa* and bricks)

Early BMAC buildings were erected either with *pakhsa* (beaten earth), for example, at Kelleli 6, or, more often, with molded sun-dried mudbricks of rectangular shape (proportions 1:2:4 or 9–12 × 18–28 × 42–58 cm) weighing up to 25 kg. The bricks were used for the construction of both small household structures, and public and monumental complexes, including long perimeter walls. They are nearly the same everywhere. The earliest use of square bricks (24 × 24 × 10 cm) has been identified at Namazga Depe in NMG VI layers (Late Bronze), as well as at Kelleli 4 (22 × 22 × 10–12 cm) and Altyn Depe.

Baked bricks are extremely rare and used only as isolated elements: for floor pavement, for lining underfloor waste water(?) channels, or for double rows in perimeter walls. Such baked bricks have been found at Kelleli 3, Adji Kui 1, and Gonur 1, and many of them are of very high quality (Orazov 2016). Some of them, with a straw temper, have a specific format (46 × 21 × 12 cm), others (34 × 17 × 6.5 cm) are made of pure clay with a glossy surface.

Sun-dried mudbricks as a construction material are known in Central Asia since the Neolithic Djeitun culture. At that time, however, they looked more like "clay loaves" than distinctly shaped bricks, except at Chagylly Depe where rectangular examples are attested. Due to their technical properties, and most importantly, to their easy production with local loess, mudbricks are extremely popular all over the arid zone and are still used today in rural areas. Their production process has remained almost unchanged for millennia. In most cases, the clay/loess is first washed and then soaked into a pit where chopped straw or *yandaq* (camel thorn) is added to improve the quality of the bricks. The mixed clay is then poured into special wooden bottomless molds lying on the ground and the bricks are ready the next day. To build the walls a clay mortar is used to bind the bricks.

The floors and walls of the living places were coated with a mixture of clay and straw, often renewed as shown by the numerous layers found in the excavations. Often doorways have brick steps and, in some cases, are associated with stone door-sockets. The walls of adjacent rooms were usually bonded rather than built one against the other, a fact that points at the existence of a general plan before the construction started. The roofs of the buildings were flat, made of horizontal beams resting on the walls. The beams were covered with a thick layer of reeds, coated with clay. This system is still used today.

Gypsum ("ganch")/alabaster

Gypsum was used only for the finishing of specific places at some buildings in Bactria and Margiana. At Togolok 21, gypsum totally covered the walls and floors of several rooms that are supposed to be those where the *haoma* beverage was made. This has led researchers to consider this complex as a "temple."

In the Dashly 3 palace, fragments of alabaster mosaic were found in the grilled-plan area for storage, and were probably intended for the decoration of the walls of special rooms or places (Sarianidi 1977a: 43; see also Chapter 11).

Vaults

Due to the presence of square rooms, or of the round altars at Togolok 21, some authors (Mamedov 2003: 111) have considered the possibility of domed roofs, but no proof for them has yet been found. There are, however, examples of "false vaults" (i.e., when parallel rows of bricks are laid successively with a gradual progression until they make a vault). This is often the case for pottery kilns at Namazga Depe (Pugachenkova 1958: 219–220; Baimatowa 2008: 144–148), Khapuz Depe, Ulug Depe, etc.

It is also the case for the exterior "corridor rooms" of the "palace" at Dashly 3. As these corridors are T-shaped, one has to assume that the builders knew how to deal with the binding of the two domes. Similar false vaults are said to have covered the corridor around the interior courtyard of Togolok 21 (Mamedov 2003: 114).

It has also been argued that domes made of brick rings covered the round towers at Togolok 21, Gonur South, Dashly 1, or the Dzharkutan temple (Huff 2001: 188–190). It is not excluded that such domes, after collapse, were partially replaced with flat wooden roofs.

Finally, some of the burial cist-like brick structures can be reconstructed with a vaulted roof at Gonur Depe and Adji Kui. The vault was made of three to five bricks with a special trapezoidal one as the key at the top. Large fragments of pottery were also added between the bricks, to increase the structural strength of the whole construction (Orazov 2016: 225).

The Gonur Depe chamber tombs are smaller copies of real houses, except for the presence of an entrance hall (Sarianidi 2002b: 22). The two chambers usually followed a 1:2 proportion and were constructed of sun-dried mudbricks (10–12 × 22–24/20–24 × 42–44/42–45–48 cm). None of them is totally preserved, but remains of fallen brick vaults have been found in all of them. The roof-cover of the hypogea is unknown.

Wood

During the excavations at the Gonur North palace, fragments of wooden beams that had fallen from the ceiling were discovered on the floor, especially along the whole length of the fortress corridor and at the entrance of the towers where they were probably used as lintels (Artemiev and Urmanova 2010).

Doors, attested by the presence of stone door-sockets, were probably made of wood. Due to the position of the door-socket, the door was usually placed inside the rooms so that the wooden panels had to be much larger than the doorways themselves. In order to close the gap between the wall, the door panel, and the doorstep, semicircular clay moldings were applied on the walls, as shown at Adji Kui 1. The average height for the doorways was ca.1.50 m, but reached 1.70 m at the Gonur South Temenos. Generally, compared to the low reduced height of other areas doorways during the Early Bronze Age, like at Altyn Depe (only 65–70 cm, see Berezkin 2001: 50–51), those of the BMAC buildings are higher (Mamedov 2003: 112).

Finally, there is some evidence for the use of wooden columns supporting roofs, as shown on the floors in some of the Gonur North palace rooms and multicolumn halls may have existed.

CONCLUSION

Nothing but architecture can reflect so well the economic level, the labor relationship, the social and political organization, and the ideological system of cultures without writing. The presence or not of specific and large buildings like fortresses, palaces, or temples immediately enlightens the general function of the settlements, even before the whole accompanying set of material culture is brought to light, although this is of course needed to decide between a religious or administrative one. Unlike movable artifacts (figurines, seals and amulets, jewelry, etc.), which can be imported and there can be a matter of debate as to their origins, architectural constructions are undoubtedly autochthonous, even if the plan conceptions and building methods can be influenced from abroad. If this was possibly the case for the BMAC architecture, these borrowings, nevertheless, have been transformed, adapted, and merged with traditional features.

NOTES

1 For the absolute dates of the sites mentioned in this chapter, see Chapter 1, Table 1.3 and the Appendix at the end of this volume.

2 Bactria is considered here to include northern Afghanistan, southern Uzbekistan, and Tajikistan.

3 Auchin 1 (Sarianidi 1990a: 10–13); Takhirbaj 3 (Sarianidi 1990a: 53–56); Kelleli 4 (Masimov 1984; Mamedov 1994: 263–264); Kelleli 3 (Masimov 1986: 174–176); Adji Kui 8 (Sarianidi 1990a: 7–10); Adji Kui 1 (Rossi Osmida and Udeumuradov 2003); Adji Kui 9 (Rossi Osmida 2007, 2011); Uch Depe (Sarianidi 1990a: 58, pl. LXIII); Togolok-21 (Sarianidi 1986; 1990a: 102–153); Togolok-1 (Sarianidi 1990b); Taip 1 (Muradova and Orazov 2013); and, finally, Gonur 1 with two truly unique ensembles, designated as Gonur South ("temenos") (Sarianidi 1997) and Gonur North ("palace-temple" complex) (Sarianidi 1993, 2002a, 2002b, 2005; Sarianidi and Dubova 2010, 2012, 2014). The latter is still being studied, and is not yet totally published.

4 In northern Afghanistan these are Dashly 1 and Dashly 3, the last with two buildings designated as the "palace" and the "round temple" (Sarianidi 1977a, 1977b, 1989b). Three settlements are known in southern Uzbekistan: Sapallitepa (Askarov 1973),

Dzharkutan (Askarov and Shirinov 1994; Huff 2000; Širinov 2002), and Tilla-Bulak (see Chapter 16).

5 A few mosaic fragments have been found at the Dashly 3 palace, see p. 171.

6 Instead of a movement of population from northern Mesopotamia/Anatolia to Central Asia, others advocate the influence due to the international trade going on at that time, which included the displacement of merchants and craftsmen as well as the transfer of ideas (see Chapter 1). Population movements at the time of the BMAC appear more and more clearly, but the lack of precise chronology often prevents knowledge of which direction the arrow has to be drawn in and when (see Chapters 19 and 20).

7 A possible round building may exist at Kelleli 6 but only its outskirts have been partially excavated, see p. 164.

8 Only ca. 50 ha are preserved (see Bendezu-Sarmiento and Mustafakulov 2013).

9 The towers were not discovered during the first excavations, except for one on the northern side that was considered then not as a fortification tower, but as an "atashgah" (sacred place for fire), built right into the body of the perimeter wall (Askarov and Shirinov 1994: 63). All the other towers were found during Huff and Shajdullaev's later work.

10 The same type of entrance had been described for Kelleli 3, see p. 148.

REFERENCES

Albenda, P. 1986. *The Palace of Sargon, King of Assyria: Monumental Wall Reliefs at Dur-Sharrukin, from Original Drawings Made at the Time of Their Discovery in 1843–1844 by Botta and Flandin*. Paris: Editions Recherche sur les civilisations.

Antonova, E.V. 1984. *Ocherki kul'tury drevnikh zemledel'tsev Perednej i Srednej Azii* [Essays on the Middle and Central Asia ancient farmers' cultures]. Moscow: Nauka.

Artemiev, V. and A. Urmanova 2010. "Predvaritel'nye itogi arkhitekturnogo issledovanija Severnogo Gonura v 2007 godu" [Preliminary results of architectural research at Gonur North in 2007]. In *On the Track of Uncovering a Civilization*, ed. P.M. Kozhin, M.F. Kosarev, and N.A. Dubova. Trudy Margianskoj Arkheologicheskoj Ekspeditsii, 3. St. Petersburg: Aletejja, pp. 172–203.

Askarov, A.A. 1973. *Sapallitepa*. Tashkent: Fan.

Askarov, A.A. 1977. *Drevnezemledel'cheskaja kul'tura epokhi bronzy juga Uzbekistana* [Bronze Age ancient farming culture of southern Uzbekistan]. Tashkent: Fan.

Askarov, A.A. and T. Shirinov 1993. *Rannjaja gorodskaja kul'tura epokhi bronzy juga Srednej Azii* [The Early Bronze Age urban culture of southern Central Asia]. Samarkand: Institute of Archaeology.

Askarov, A.A. and T. Shirinov. 1994. "The 'Palace,' Temple, and Necropolis of Jarkutan." *Bulletin of the Asia Institute* 8: 13–25.

Baimatowa, N. 2008. *5000 Jahre Architektur in Mittelasien. Lehmziegelgewölbe vom 4./3. Jt. v. Chr. bis zum Ende des 8. Jh. n. Chr.* Archäologie in Iran und Turan, 7. Mainz: Philipp von Zabern.

Bendezu-Sarmiento J. and S. Mustafakulov 2013. "Le site proto-urbain de Dzharkutan durant les âges du bronze et du fer. Recherches de la Mission archéologique franco-ouzbèque-Protohistoire." *Cahiers d'Asie centrale* 21–22: 207–236.

Berezkin, Ju.E. 2001. "Kvartal'nij tsentr epokhi bronzy na Altyndepe" [The Bronze Age central district at Altyn Depe]. In *Osobennosti proizvodstva poselenija Altyndepe v epokhu paleometalla* [Production specificities at Altyn Depe in the Paleometal Age], ed. V.M. Masson. St. Petersburg: Institute of the History of Material Culture, pp. 40–59.

Brentjes, B. 1981. *Die Stadt des Yima, Weltbilder in der Architektur*. Leipzig: E.A. Seemann Verlag.

Bulatov, M.S. 1988. *Geometricheskaja garmonizatsija v arkhitektura Srednej Azii IX-XV vv* [Geometric harmonization in the architecture of Central Asia 9th–15th century]. Leningrad: Nauka.

Chmelnizkiy, S. 1993. "Das problem der kultischen merkmale einiger altertümlichen bauten in Mittelasien." In *Actes du XIIe Congres International des Sciences Préhistoriques et Protohistoriques*, vol. 4. Bratislava: Institut archéologique de l'Académie slovaque des sciences, pp. 304–307.

Dubova, N.A. 2004. "Mogil'nik i tsarskij nekropol' na beregakh Bol'shogo bassejna Severnogo Gonura" [Graveyard and "Royal Necropolis" on the edge of the large pool at Gonur North]. In *U istokov tsivilizatsii* [Near the sources of civilizations], ed. M. Kosarev, P. Kozhin, and N. Dubova. Moscow: Staryj Sad, pp. 254–281.

Gubaev, A., G. Koshelenko, and M. Tosi (eds.) 1998. *The Archaeological Map of the Murghab Delta: Preliminary Reports 1990–1995*. Reports and Memoirs: Series Minor, III. Rome: IsIAO.

Hiebert, F.T. 1994. *Origins of the Bronze Age Oasis Civilization in Central Asia*. American School of Prehistoric Research Bulletin, 42. Cambridge, MA: Peabody Museum Press.

Huff, D. 1997. "Deutsch-Uzbekische Ausgrabungen auf dem Džandaulattepe und in Džarkutan, Süd-Uzbekistan, 1993–1995." *Mitteilungen der Berliner Gesellschaft für Anthropologie, Ethnologie und Urgeschichte* 18: 83–95.

Huff, D. 2000. "Djarkutan: Archaeological Research on Tepe VI." *History of Material Culture of Uzbekistan* 31: 58–69.

Huff, D. 2001. "Bronzezeitliche Monumentalarchitektur in Zentralasien." In *Migration und Kulturtransfer. Der Wandel Vorder- und Zentralasiatischer Kulturen im Umbruch vom 2. zum 1. vorchristlichen Jahrtausend*, ed. R. Eichmann and H. Parzinger. Kolloquien zur Vor- und Frühgeschichte 6. Bonn: Habelt, pp. 181–197.

Huff, D. 2005. "Urbanisierungsansätze in Zentralasien." In *Wege zur Stadt. Entwicklung und Formen urbanen Lebens in der alten Welt, (Vergleichende Studien zu Antike und Orient)*, ed. H. Falk. Bremen: Hempen, pp. 82–120.

Isakov, A.I. 1991. *Sarazm (K voprosu stanovlenija drevnezemledel'cheskoj kul'tury Zeravshanskoj doliny)* [Sarazm (the formation of ancient farming culture in the Zeravshan valley)]. Dushanbe: Donish.

Kartzev, V.N. 1986. *The Architecture of Afghanistan*. Moscow: Stirojizdat.

Lamberg-Karlovsky, C.C. 1994. "The Bronze Age 'khanates' of Central Asia." *Antiquity* 68(259): 398–405.

Mamedov, M. 1993. "The Architectural Layout of the Togolok-21 Complex: An Analysis." *IASCCA Information Bulletin* 19: 228–231.

Mamedov, M. 1994. "Architectural Planning in Margiana." *Mesopotamia* XXIX: 263–277.

Mamedov, M. 1996. "Monumental Architecture in the Delta of the Murghab River." In *Turkmenistan: Environment. History. Monuments. Ethnography*, ed. G. Rossi Osmida. Venice: Centro Studi Ricerche Ligabue, pp. 94–106.

Mamedov, M. 2002. "Architecture of the 'Palace' on Gonur-1." In *Margiana. Gonur Depe Necropolis*, ed. G. Rossi Osmida. Venice: Il Punto Edizioni, pp. 51–54.

Mamedov, M. 2003. *Drevnjaja arkhitektura Baktrii i Margiany* [Ancient architecture of Bactria and Margiana]. Ashgabat: Cultural Center of the Embassy of Iran.

Margueron, J.-Cl. 2004. *Mari, métropole de l'Euphrate aux IIIe et au début du IIe millénaire av. J.-C.* Paris: Picard/Editions Recherche sur les Civilisations.

Masimov, I.S. 1981. "The Study of Bronze Age Sites in the Lower Murghab." In *The Bronze Age Civilization of Central Asia: Recent Soviet Discoveries*, ed. P.L. Kohl. New York: M.E. Sharpe, pp. 194–220.

Masimov, I.S. 1984. "Zhiloj dom na Kelleli" [A casual house at Kelleli]. *Pamjatniki Turkmenistana* [Monuments of Turkmenistan] 2(38): 16–18.

Masimov, I.S. 1986. "Novye issledovanija pamjatnikov epokhi bronzy na Murgabe" [New research on Bronze Age sites in the Murghab)]. In *Drevnie tsivilizatsii Vostoka* [Ancient civilizations of the East], ed. V.M. Masson. Tashkent: Fan, pp. 171–181.
Masson, V.M. 1959. *Drevnezemledel'cheskaja kul'tura Margiany* [The ancient farming culture of Margiana]. Materialy i Issledovanija po Arkheologii SSSR 73. Moscow: Nauka.
Masson, V.M. 1979. "*Fortifikatsija Srednej Azii v bronzovom veke*" [Bronze Age Central Asian fortifications]. In *Etnografija i arkheologija Srednej Azii* [Ethnography and archaeology of Central Asia], ed. A.V. Vinogradov, M. Vorob'eva, M. Itina, L. Levina, E. Nerazik, and Ju. Rapport. Moscow: Nauka, pp. 28–34.
Masson, V.M. 1989. Pervye tsivilizatsii [The first civilizations]. Leningrad: Nauka.
Masson, V.M. 2011. "Formation and Blossom of a Civilization of Ancient Oriental Type in the South of Turkmenistan." In *Historical and Cultural Sites of Turkmenistan*, ed. M. Mamedov. Ashgabat: Turkmen State Publishing Service, pp. 74–75.
Muradova, E. and A. Orazov. 2013. "Studying of Fortification of Taip Depe." In *Materials of the International Scientific Conference "Thousand-Year Sources of Building Culture of Turkmenistan."* Ashgabat: Turkmen State Publishing Service, pp. 140–141.
Orazov, A.T. 2016. "Epokha bronzy v Margiane: stroitel'nye materialy i konstruktsii" [Bronze Age in Margiana: building materials and constructions]. In *Trudy Margianskoj Arkheologicheskoj Ekspeditsii*, vol. 6, ed. N.A. Dubova. Moscow: Staryj Sad, pp. 224–231.
Parrot, A. 1959. *Mission Archéologique de Mari. Le Palais*. Paris: CNRS Editions Alpha.
P'jankova, L. 1994. "Central Asia in the Bronze Age: Sedentary and Nomadic Cultures." *Antiquity* 68(259): 355–372.
Pugachenkova, G.A. 1958. Puti razvitija arkhitektury Juzhnogo Turkmenistana pory rabovladenija i feodalizma [The ways of southern Turkmenistan architecture development at the time of slavery and feudalism]. Moscow: Nauka.
Pugachenkova, G.A. 1971. "Zodchestvo Srednej Azii i Irana v istoricheskikh svjazjakh" [The architecture of Central Asia and Iran in their historical relations]. In *Istorija Iranskogo gosudarstva i kultury* [History of the Iranian state and culture], ed. B.G. Gafurov. Moscow: Nauka, pp. 237–245.
Pugachenkova, G.A. 1982. "K tipologii monumental'nogo zodchestva drevnikh stran sredneaziatskogo regiona" [Typology of Central Asia ancient monumental architecture]. *Iranica Antiqua* XVII: 21–42.
Razzokov, F.A. 2016. *Stroitel'nye kompleksy drevnezemledel'cheskogo poselenija Sarazm v IV-III tys. do n.e.* [Building complexes at Sarazm, an ancient farming settlement in the 4th–3rd millennia BC]. St. Petersburg: OOO HKT.
Rossi Osmida, G. 2007. *Adji Kui Oasis: The Citadel of the Figurines*, vol. I. Venice: Il Punto.
Rossi Osmida, G. 2011. *Adji Kui Oasis: The Citadel of the Figurines*, vol. II. Venice: Il Punto.
Rossi Osmida, G. and B. Udeumuradov 2003. "Preliminary Report on Results of Study into the Settlements and Necropolis Adji Kui 1 and 9." In *Central Asia Cultural Values*, vols. 1–2, ed. R.G. Muradov. Padova: Il Punto, pp. 53–94.
Salvatori, S. 2007. "About Recent Excavations at a Bronze Age Site in Margiana (Turkmenistan)." *Rivista di Archeologia* XXXI: 11–28.
Sarianidi, V.I. 1974. "Baktria v epokhu Bronzy" [Bactria in Bronze Age]. *Sovetskaja Arkheologija* 4: 49–71.
Sarianidi, V.I. 1976. "Issledovanija pamjatnikov Dashlinskogo oazisa" [Research of settlements in the Dashly oasis]. In *Drevnjaja Baktrija*, ed. I.T. Kruglikova. Moscow: Nauka, pp. 21–86.
Sarianidi, V.I. 1977a. *Drevnie zemledel'tsy Afganistana. Materialy Sovetsko-Afganskoj ekspeditsii 1969 – 1974* [Ancient farmers of Afghanistan: materials of the Soviet-Afghan expedition 1969–1974]. Moscow: Nauka.

Sarianidi, V.I. 1977b. "Pamjatniki monumental'noj arkhitektury Baktrii" [Sites with monumental architecture of Bactria]. *Sovetskaja Arkheologija* 1: 203–224.
Sarianidi, V.I. 1981. "Margiana in the Bronze Age: The Study of Bronze Age sites in the lower Murghab." In *The Bronze Age Civilization of Central Asia: Recent Soviet Discoveries*, ed. P.L. Kohl. New York: M.E. Sharpe, pp. 165–220.
Sarianidi, V.I. 1986. "Le complexe culturel de Togolok 21 en Margiane." *Arts Asiatiques* XLI: 5–21.
Sarianidi V.I. 1989a. *Khram i nekropol' Tillja-tepe* [The Tillja Tepe temple and necropolis]. Moscow: Nauka.
Sarianidi V.I. 1989b. "Soviet Excavations in Bactria: The Bronze Age." In *Bactria, an Ancient Oasis Civilization from Sands of Afghanistan*, ed. G. Ligabue and S. Salvatori. Venice: Erizzo, pp. 107–123.
Sarianidi, V.I. 1990a. *Drevnosti strany Margush* [Ancient Margush]. Ashkhabad: Ylym.
Sarianidi, V.I. 1990b. "Ein aussergewöhnlicher Tempelkomplex von Togolok in Turkmenien." *Das Altertum* 36: 167–176.
Sarianidi, V.I. 1990c. "Sel'skij khram Togolok-1 v Margiane" [The Togolok 1 rural temple in Margiana]. *Vestnik Drevnej Istorii* 2(193): 161–167.
Sarianidi, V.I. 1990d. "Togolok-21: An Indo-Iranian Temple in the Karakum." *Bulletin of the Asia Institute* 4: 159–165.
Sarianidi, V.I. 1991. "Temples of Ancient Margiana." *Pakistan Archaeology* 26: 175–187.
Sarianidi, V.I. 1993. "Excavations at Southern Gonur." *Iran* XXXI: 25–37.
Sarianidi, V.I. 1994. "Temples of Bronze Age Margiana: Traditions of Ritual Architecture." *Antiquity* 68(259): 388–397.
Sarianidi, V.I. 1996. "Proiskhozhdenie iranskikh khramov ognja" [Origin of the Iranian fire temples]. *La Persia e l'Asia Centrale da Alessandro al X secolo*. Rome: Accademia Nazionale dei Lincei, pp. 319–329.
Sarianidi, V.I. 1997. "Temenos Gonura" [The temenos of Gonur]. *Vestnik Drevnej Istorii* 1(220): 148–168.
Sarianidi, V.I. 2002a. "The Fortification and Palace of Northern Gonur." *Iran* XL: 75–87.
Sarianidi, V.I. 2002b. *Margush: Ancient Oriental Kingdom in the Old Delta of the Murghab River*. Ashgabat: Türkmendöwlethabarlary.
Sarianidi, V.I. 2005. *Gonur Depe: City of Kings and Gods*. Ashgabat: Türkmendöwlethabarlary.
Sarianidi, V.I. and N.A. Dubova. 2010. "Novye grobnitsy na territorii 'tsarskogo nekropol'ja' Gonura" [New graves at the area of the Gonur "Royal Necropolis"]. In *On the Track of Uncovering a Civilization. Trudy Margianskoj Arkheologicheskoj Ekspeditsii*, vol. 3, ed. P.M. Kozhin, M.F. Kosarev, and N.A. Dubova. Moscow: Staryj Sad, pp. 144–171.
Sarianidi, V.I. and N.A. Dubova. 2012. "Arkheologicheskie raboty Margianskoj arkheologicheskoj ekspeditsii v 2008–2011 gg." [Archaeological excavation of the Margiana archaeological expedition during 2008–2011 years]. In *Trudy Margianskoj Arkheologicheskoj Ekspeditsii*, vol. 4, ed. V.I. Sarianidi. Moscow: Staryj Sad, pp. 29–55.
Sarianidi, V.I. and N.A. Dubova. 2014. "Raboty Margianskoj arkheologicheskoj ekspeditsii v 2011–2013 gg." [Works of the Margiana archaeological expedition]. In *Trudy Margianskoj Arkheologicheskoj Ekspeditsii*, vol. 5, ed. V.I. Sarianidi. Moscow: Staryj Sad, pp. 92–111.
Širinov, T. 2002. "Die frühurbane Kultur der Bronzezeit im südlichen Mittelasien. Die vorgeschichtliche Siedlung Džarkutan." *Archäologische Mitteilungen aus Iran und Turan* 34: 1–170.
Udeumuradov, B.N. 2011. "Drevnezemledel'cheskie poselenija Adzhikuinskogo oazisa" [Ancient farming settlements of Aji-Kui oasis]. In *Historical and Cultural Sites of Turkmenistan*, ed. M. Mamedov. Ashgabat: Turkmen State Publishing Service, pp. 205–213.

Urmanova, A. 2014. "K probleme sooruzhenij tipa "kel'i" [About the question of the cell-type constructions]. In *Trudy Margianskoj arkheologicheskoj ekspeditsii*, vol. 5, ed. V.I. Sarianidi. Moscow: Staryj Sad, pp. 138–150.
Woolley, C.L. 1968. *A Forgotten Kingdom.* New York: Penguin.
Zadneprovsky, Y.A. 1995. "Early Urban Developments in Central Asia." *Iran* XXXIII: 155–159.
Zdanovich, G.B. and D.G. Zdanovich 2002. "The 'Country of Towns' of the Southern Trans-Urals and Some Aspects of Steppe Assimilation in the Bronze Age." In *Interaction: East and West in Eurasia*, ed. K. Boyle, C. Renfrew, and M. Levine. Cambridge: McDonald Institute, pp. 249–264.

CHAPTER SIX

SOME THOUGHTS ON THE IMAGINARY REPRESENTATIONS IN THE BACTRIA-MARGIANA ARCHAEOLOGICAL CULTURE

Elena V. Antonova

INTRODUCTION

Despite the relatively short time of excavations on the remains of the Bactria-Margiana Archaeological Culture (BMAC), many of its finds are already well known due to numerous publications. Its material culture appears as extremely diverse and heterogeneous. This is not a surprise since its bloom, at the end of the 3rd and in the early 2nd millennium BC, corresponds to a period of intensive contacts with the Near East, Iran, and the Indus (see Chapter 1). The seemingly sudden appearance of this culture has given rise to different hypotheses about its origins. V.I. Sarianidi has long defended its Western components. However, a number of its features, like anthropomorphic figurines, decoration on pottery, or a number of motifs on seal-amulets, provide evidence for its deep roots within the previous Anau-Namazga culture that developed at the foothills of the Kopet Dagh (Antonova 2011; see also Chapter 4).

This chapter proposes to go beyond the images represented by these artifacts and to track the local mythologies at work behind them. As we will see, this study also underlines a difference between the elite imagery and that of the rest of the population, which is important to better understand the BMAC society.

Attempts towards an understanding of the world as an organized system with a specific order have been known since prehistoric times, especially during the Neolithic period, as mentioned long ago by I.M. D'jakonov (1977: 16):

> Human thought, even lacking the required knowledge, intuitively registers the regularity and apparent consistency of natural phenomena ... Isn't it possible to perceive human interaction with the world as in cadence with the first universal generalization in the history of human culture? ... In the images of plants, animals and people we can see a universal rhythmic that for the first time appears as a repeating structuring pattern.

Numerous iconographic data as well as ancient written sources confirm this assumption (Antonova 1984: 59ff).

A specific feature of the early eastern farming societies is their non-hierarchical, kinship- or ancestral-/tribal-based relationships. In such societies, information was orally transmitted or anchored on specific and conventional visual symbols, the meaning of which was known to all their members. According to Bajburin (1989: 70), the limited number of semantic links within the original "archaic cultures, together with the lack of specific means ensuring the transmission of information within the society, resulted in the fact that each cultural element had a more intense and precise meaning than in modern societies." Thus, it seems possible to trace back the imaginary world of early anepigraphic farming cultures on the basis of these universal signs. Anthropomorphic figurines and decorative patterns on containers or on seals made of materials of long duration (terracotta, metal, or stone) have long been our best source of information for this type of research.

ANTHROPOMORPHIC FIGURINES

Clay anthropomorphic figurines of the Anau-Namazga culture in southern Turkmenistan are especially worthy of interest because of their long duration, with successive chronological stages (Solov'eva 2005), and because the BMAC has a great part of its roots anchored in this culture (see Chapter 4). For many centuries, the inhabitants of the region preserved and transformed the traditional symbols conveying their representation of the world.

At the beginning of the Chalcolithic period (from the 5th to the beginning of the 3rd millennium), when there are yet no signs of social hierarchy, generic figurines of sitting women are attested in the domestic houses, some of them being found walled up at the base of fireplaces, pointing at their close relationship with their inhabitants (Antonova 1990: 150).[1]

At that time, they had rather primitive and heavy features (Figure 6.1, *1*). Without proper justification, they are considered as "mother-goddesses," a very general, a-temporal, and not very explanatory terminology. More probably, they depict mythical female progenitors. In fact, they sometimes have additional features, like multiple breasts, which also point at their role as foster-mothers.

The Namazga (NMG) III period corresponds in Mesopotamia to the time of the urban revolution and the "Uruk expansion." There, the first city-states appear, as well as writing. At that time, in all the Near East, the ties between Mesopotamia and the surrounding world strengthen and many changes occur. In Central Asia, the figurines became somewhat more graceful. Sitting or standing figures are represented with broad shoulders and narrow waists, and some are depicted with headwear or a hairdo, and/or with round pellets applied to their shoulders (Figure 6.1, *3*). It is also at this time when, for the first time, a few new male figurines appear – at Ilgynly Depe representing only 3–5% – beside the much larger number of female ones (Berezkin 1994: 33).

The applied round pellets may have had the same mythic function as that of the breasts of the foster-mothers. Interestingly, however, male figures sometimes also have such additional features. Like the female ones, they probably convey the images of mythologized ancestors and we cannot exclude that the ancestors were perceived as androgynous (Antonova 1990: 153–154). These ancestors could be the protectors of the clan and its members. The applied ornaments they exhibit may represent symbols

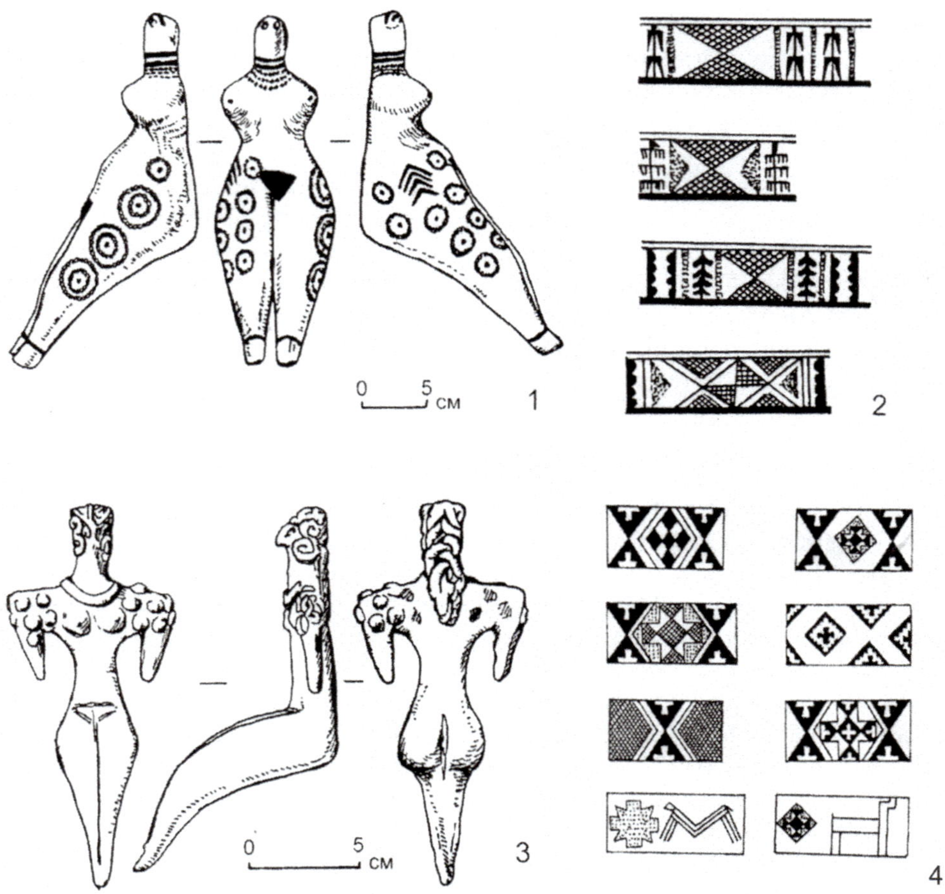

Figure 6.1 (1) Namazga II figurines; (2) Namazga II pottery decoration; (3) Namazga III figurines; (4) Namazga III pottery decoration.

of fertility and prosperity. Such a system has been identified in the Early Chalcolithic period (4th millennium) but may even have started earlier.

With the beginning of the Bronze Age (NMG IV–V periods), numerous innovations are introduced. It is the time when metalworking technologies increasingly develop and when the potter's wheel appears, changing the pottery styles and putting an end to the painting tradition. At the same time, the anthropomorphic figurines significantly change (Figures 6.2, *1–3*; 6.3; 6.5, *2*): they become totally flat, with the lower part turning into a triangle bent forward, and their arms are raised to the sides, making a cross-like silhouette; a "tree"-like decoration covers the front side while other symbolic signs (triangular etched motifs, rhombs) cover the shoulders and the back; their face is now flat with a prominent nose; details of the hairdo, female breasts, and male genitals are applied as additional elements. These motifs are a variation of those that, possibly, conveyed a schematic representation of the world on the Late Chalcolithic pottery (see pp. 183–184). Their location is also the same: in the center, a symbolic "tree" replaces the cross-shaped motif.

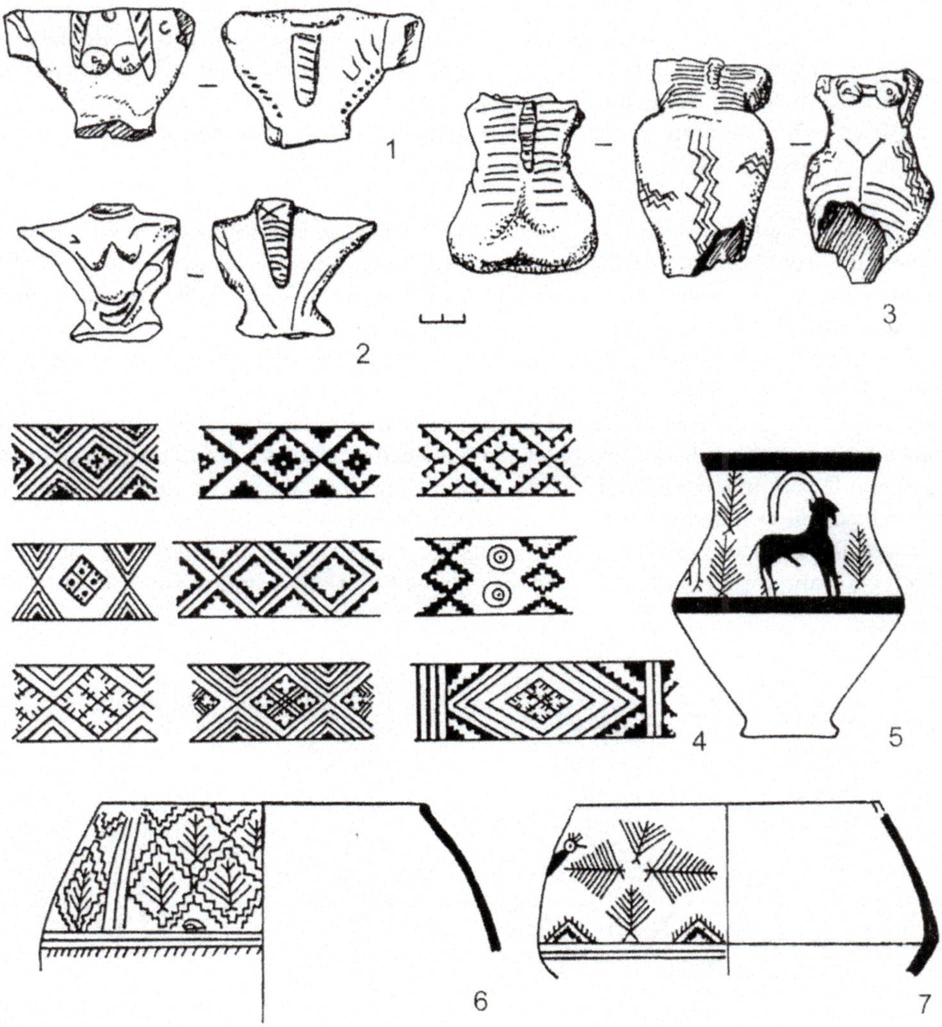

Figure 6.2 (1–3) Namazga IV figurines; (4–7) Namazga III–IV pottery decoration.

As during the NMG III period, the number of male figurines stays in the same low proportion compared to that of the female ones. They share with them the same features: Not only their bodies are geometrized, but they are also covered with symbols akin to those visible on a few clay vessels, indicating a similar representation of the world.

The male figurines' sex is depicted in an erotic raised position, while that of the females underlines fertility. The shapes of both male and female figurines suggest that there was no difference between the macro- and the microcosm.

From the Early Chalcolithic period (4th millennium), and possibly earlier, a system of belief where importance is given to the female side is thus taking shape and must be due to the social organization based upon family/clan relationships. According to Y.E. Berezkin (1994: 33):

> The system of beliefs in the Namazga V period and earlier, with worship focused on little individualized ancestors, was common and widespread in all the pre-state societies and partially in the farming communities ... Such a "pantheon" is often reduced to elementary personification of basic opposed notions like those of male–female, "mine–other," or it splits into countless persons almost completely dissolved into real, recently deceased relatives.

The NMG V figurines with their symbols are images of the universal "fertile space," where they represent a "cosmogram" in which the world is identified as a mythic female creature. A similar representation is visible in the Vedic Aditi, identified with the earth and sky and the guardian of the cosmic order.

It has been said sometimes that the female figurines, with their various symbols, depicted different goddesses. The symbols would point out their functions: goddesses of plants, of water (see Figure 6.3), or of earth (Figure 6.2, *3*), etc. This understanding of the conventional symbols, however, comes from alleged analogies with the Mesopotamian pantheon (Masson and Sarianidi 1973). Actually, the signs drawn on the figurines stem rather from the ornamental patterns on local Chalcolithic vessels (Antonova 1972). In archaeology, as in folk art, universal motifs are regularly encountered, "the thread of which you cannot pull out of the skein, or only half of it" (Nekljudov 1977: 193)

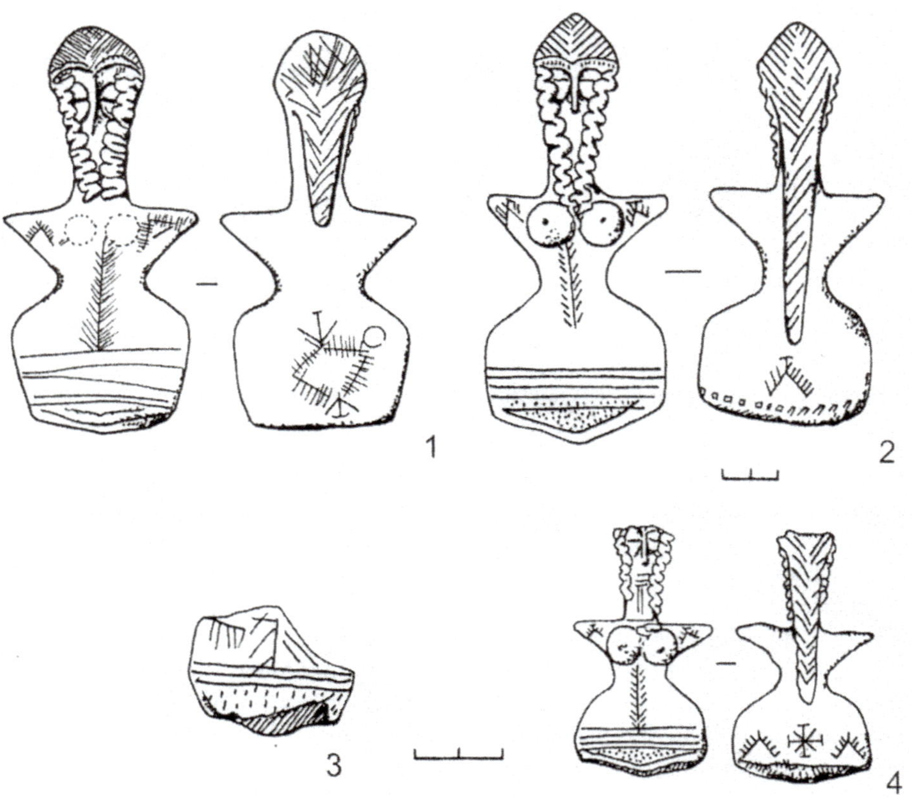

Figure 6.3 Namazga V figurines.

At the time of the BMAC, these traditional anthropomorphic figurines continued to be produced and are found in settlements, but simultaneously, new and totally different ones – more realistic – also made their appearance, such as the composite female statuettes made of dark and white stones (Figure 6.5, *1*), first discovered in the looted material from Afghanistan (see Chapter 7). P. Amiet traced their genesis in the images of Elamite queens, at the turn of the 3rd–2nd millennia and shortly after, with whom they present close similarities. If, in Elam, statues were intended for temples, in Bactria they are mostly associated with burials (Amiet 1986: 200; 1997; Vidale 2017: 98). This is confirmed by the discoveries made in Margiana: At Gonur, for instance, one was associated with a "cultic set" placed inside a hand-made unfired clay container in the rich cist burial 2900 from Area 5 (Sarianidi 2001: 187–188; 2007: 73, 147–153). There is evidence that these stone statuettes were made at the site itself since some of their elements, like hands, have been found in other burials in the Gonur Necropolis (like in the grave of a stone carver no. 1200) (Sarianidi 2007: 90).

Another unusual statuette, in a very bad state of preservation, has been discovered in tomb 3200 of the "Royal Necropolis" (see Chapter 10, Figures 10.21a and 10.21b). It represents a probable female figure standing and dressed in a gold foil with ears of wheat and barley made of gypsum/faience protruding from her body. Two big elements made of same material could be fruits (Sarianidi 2005: 217, fig. 79; 2006: 158ff). On the one hand, this statuette is reminiscent of depictions on seals found at Gonur and in Kerman (see p. 186), but on the other hand, it could also be a reminiscence of the Anau-Namazga female figurines, mother and protector of plants and a cosmic representation.

POTTERY DECORATION

The origins of the symbols made on the NMG V figurines possibly come from the painted decoration on the Geoksjur-style pottery (period NMG III) consisting in a repetition of triangular motifs around a rhomb- and/or a cross-shaped central one (Figure 6.1, 4). These triangles often create a step motif. The same motif is also present in the contemporary shapes of the most ancient seal-amulets found at Altyn Depe (Kircho 1990: 176) (see p. 184).

Later, during the NMG IV period, the stepped rhombs are sometimes replaced by "trees" organized in crosses (Sarianidi 1965: tab. IV) (Figure 6.2, 4–7). Next to the central figure in some cases stands a goat or an unidentified animal with a saggy belly.

The geometric cross-shaped motifs might just be a pure decoration, but the fact that they are sometimes found together with goat-like animals (Figure 6.2, *5*) and later replaced by "trees" raises some doubts, as these animals are frequently depicted next to a tree in the art of the ancient Near East. This change, from a symbolic geometric (cross-shaped) motif to a figurative (tree) one, underlines the importance of the central figure (Figure 6.2, 6–7).

These painted motifs on the pottery have, however, disappeared at the time of the BMAC, but they are maintained on the figurines who still wear comparable signs/symbols. They are also still present in some of the stamp-seal shapes (see pp. 184–186).

The origins of the symbols used as ornamental motifs on the figurines have been discussed several times (Masson and Sarianidi 1973: 115; Antonova 1981). Cross-shaped

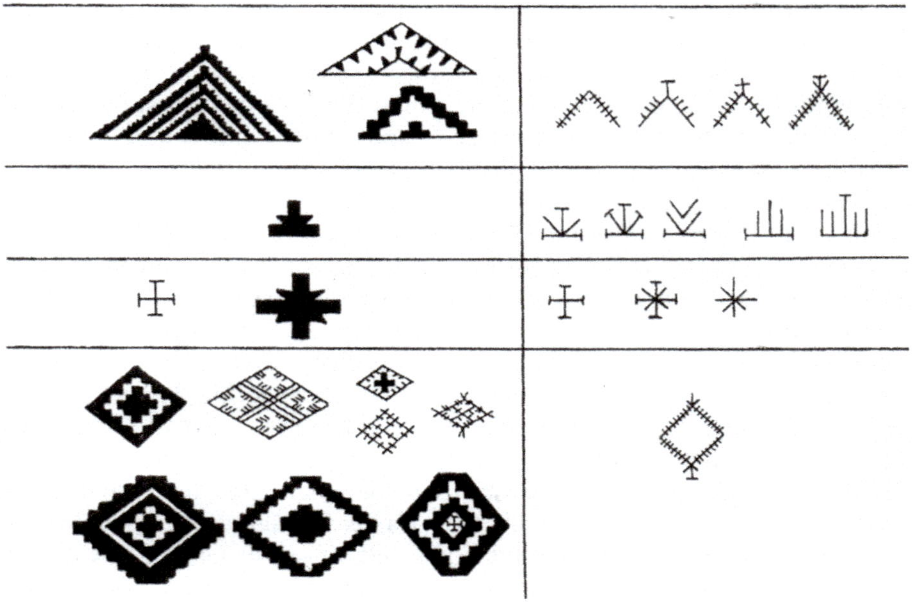

Figure 6.4 Namazga IV pottery decoration (left). Signs on Namazga V figurines (right).

anthropomorphic figurines and their signs, just like these motifs on the Chalcolithic pottery (Figure 6.4), are considered to represent the world as a whole. Pottery vessels themselves, as food storage containers, obviously symbolize the concept of the world.

SEAL-AMULETS AND CYLINDER SEALS[2]

There is no doubt that most of the *realia* of the Chalcolithic/Early Bronze Age Anau-Namazga culture and those of the BMAC are very different. Nevertheless, one can hardly deny a continuation of some traditions behind the novelties, especially in the world representations, since all these items and symbols belong to a culture that developed without interruptions for millennia.

In the graves excavated at Middle Bronze Age Altyn Depe, about a quarter of the metal seal-amulets are cross-shaped like in the previous Chalcolithic period (Masson and Berezkin 2005: 410).

Thousands of seals and amulets have been discovered in the BMAC area and Sarianidi has already written widely about them, especially in a large catalogue (Sarianidi 1998). This catalogue concerns about 1,800 seals, 1,500 of which come from illegal excavations most probably in Afghanistan and are kept in various collections, while the rest come from regular excavations in Turkmenistan and Uzbekistan. Many of the seals Sarianidi published, especially those said to come from Bactria (Afghanistan) are actually of unknown provenance since they were bought on the art market making uneasy their use for further research. But their motifs are well in accordance with those from secure sites.

Beside a very low number of cylinder seals (see p. 224), the vast majority of the others are stamp seals, made locally of copper/bronze or stone. They had both a

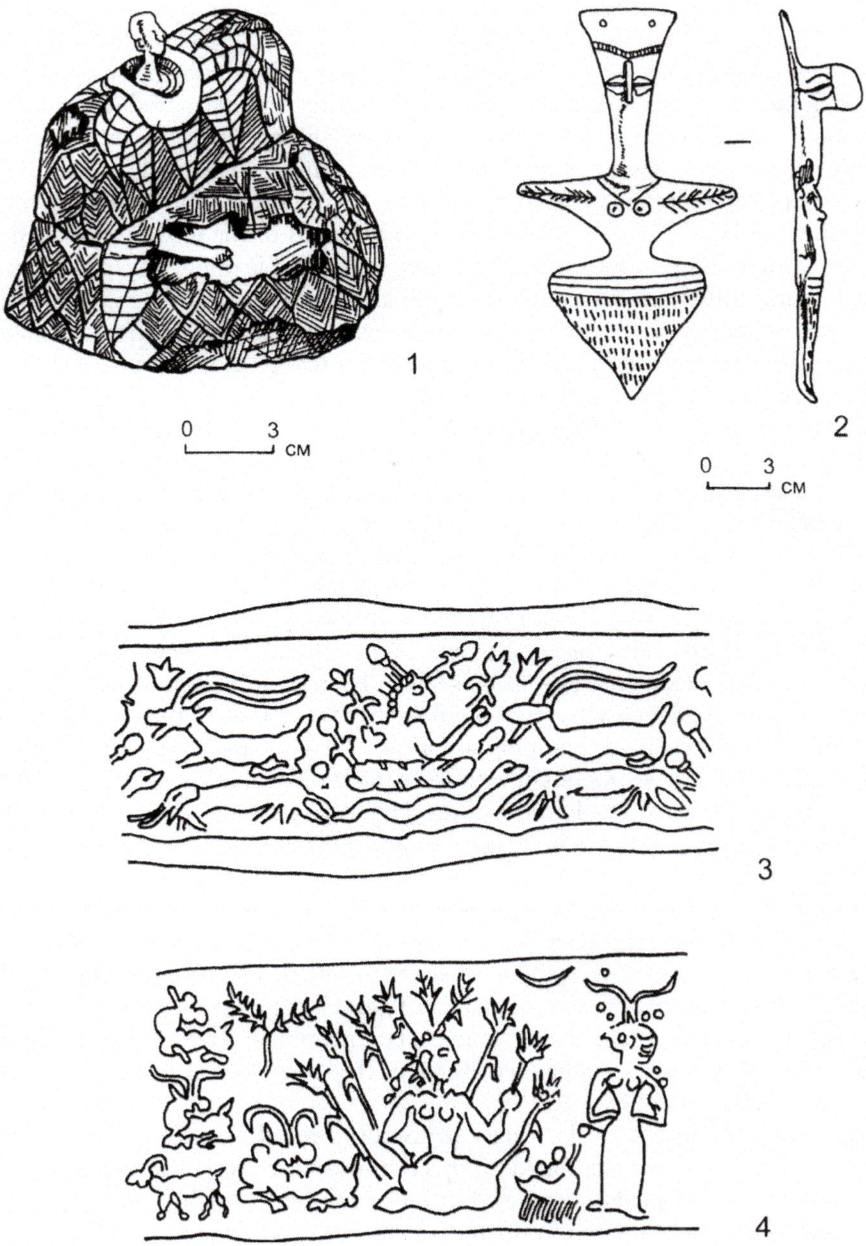

Figure 6.5 (1) Stone composite figurine, Gonur Depe; (2) terracotta figurine, Gonur Depe; (3–4) trans-Elamite cylinder seals, from Gonur-Depe (3), from Tepe Yahya (4).

function of ownership and of apotropaic amulet. It is symptomatic that they continue and develop the ornamental forms known previously, following the canonical rules of the former society with their conventions and symbolism. As Sarianidi (1998: 23–47) noticed, the images of animals, of supposed deities and their different compositions, are much less frequent than the geometric motifs. Is it then possible to extract from these a representation of the BMAC imaginary world?

Most of these stamp seals are circular, while stepped rhombs or stepped crosses are less numerous. On some of them geometric figures are associated with schematic floral motifs. The central cross-shaped design may "bloom" in different ways, sometimes into a tree-like design. This transformation can be explained by the multiplication of spokes radiating from the center or covering the whole surface, making spoke-like branches. On a very frequent type, the arms of the cross are divided with segments radiating up to the edge. The center of the seal is not always the focus, but a division into four parts is nevertheless present (Figure 6.6).

From Sarianidi's (1998) catalogue it can be evaluated that seals with cross-shaped rosette motifs are particularly numerous in Bactria (over 60, nos. 301–363) but rare in Margiana (only 9, nos. 1568, 1597, 1602–1605).[3]

As was the case for the figurines with signs and symbols, the seal maker had a tendency to fill in the surface with additional elements, sometimes very densely. In extreme cases, round and rectangular seals are entirely filled with a grill-like motif without any other additional element (Sarianidi 1998: nos. 733, 815–819). A number of them, however, have a cross motif in the middle (nos. 810–814).

The motif is created at minimum by two elements (i.e., a central circle and radial spokes), but there are designs with three or four elements – a central circle, a cross, spokes or other shapes, and the circular edge. A very frequent type has eight elements, either figurative or geometric (about 80 seals in Bactria, and 20 in Margiana).

Some of the rosettes clearly resemble flowers of different shapes: Their center is represented by a circle if the stamp is round and by a square/lozenge if the stamp is quadrangular. It is sometimes possible to trace the transformation from a stepped lozenge to an eight-petal rosette. In the slightly geometrized designs, Sarianidi recognized the shapes of local flowers – poppy, tulip, and ephedra (Sarianidi 1998: 45; nos. 861–863).

The edges of the metallic seals often present a typical braid motif (see Figure 6.7, 2) that can be related to "movement" as do other representations on the seals themselves, like swastikas (nos. 799–805); spiral-shaped rosettes made of snakes grouped by three, four, or six (nos. 255–262, 795–796, 798); or figures of goats (nos. 100–106). Some of the animal preferences (goat, snake) seem to be intentional: They are probably to be explained by the positions of these creatures in the mythological imaginary world of the BMAC.

Considering the BMAC seals, it may be useful to recall here the veneration many societies have towards protecting gods. This is, for instance, a typical Indian behavior, known since the most ancient times and still present today (Volchok 1972: 293–298). The BMAC seals can be related to the Indian custom of wearing metal plates/amulets ("*chakra*" or "*jantra*"), at the neck for women and at the belt or on the arm for men. These plate-amulets always bear on them the symbols of the eight sides of the world combined in various ways on both sides. Those the Tamil children wear have 16 squares on one side and on the other a circle divided into eight segments. These charms protect them from misfortune coming from the eight directions (Volchok 1972: 271). We consider it very likely that the BMAC population had a similar belief in these eight directions since a remarkably high number of seals present figures where the number of segments is a multiple of four. As shown by the position of the

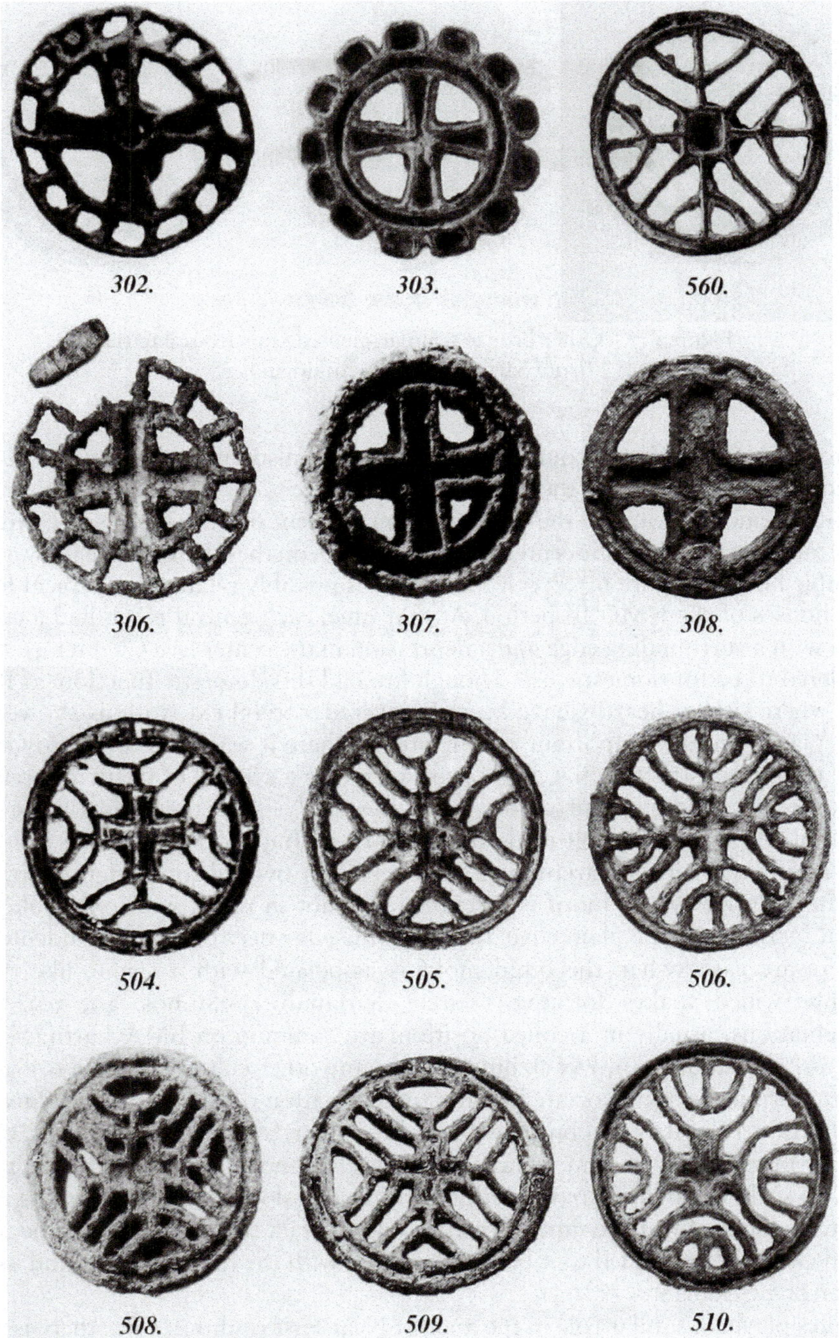

Figure 6.6 Bronze compartmented seals from Bactria, from Sarianidi's 1998 catalogue.

733. 775. 246.

Figure 6.7 Other bronze compartmented seals from Bactria (from Sarianidi's 1998 catalogue).

seals found in graves in the course of excavation, we also know that most were worn around the men's necks and the women's waist.

As mentioned previously, the most frequent element on the stamp seals with geometric motifs is a circle in the center. It is surrounded, on the edge of the seal, by a single or double border (Figure 6.7, 733). This shape is possibly related to a typical feature in the houses of the NMG III period. At that time, each central room had a circular hearth with a surrounding edge and a depression in the center (see Chapter 4). All are considered to be for domestic use. Though fire had this domestic function, at Ilgynly Depe, where similar hearths have been discovered, careful excavations showed that it also played another important role in rituals where it was used to destroy houses before their reconstruction (i.e., it was considered as a symbol of death and rebirth).

Another interesting point comes from the analysis of the edge of the seals. Usually, it consists of a simple or double plain ridge, but a twisted braid is also a frequent motif with several variations reminiscent of a twisted rope (Figure 6.7, 775). Sometimes, however, the motif is simplified and not in relief, with only oblique or straight incisions on the plain edge. In others, the edge may be wavy or indented.

There are cases when the braid motif is associated with a snake, like that of two intertwined snakes forming a circle (Sarianidi 1998: nos. 49, 50). Snake representations, usually in a coiled position, are common on BMAC artifacts (nos. 232–249, 265–266, 1649, 1655), but also in many other cultures (Figure 6.7, 246).

This motif is clearly associated with a representation of the world and thus plays an important role. It is combined with anthropo- or zoomorphic creatures, plants, water, or images of springtime (Francfort 1992; Antonova 2000). The snake phallic shape, its scaly outlook, its rejuvenation through the sloughing of its skin, its ability to entwine especially at the time of mating, and the fact that it belongs to the underground/chthonic world, led to it being associated with the rebirth of life, and used as an apotropaic element.

Snakes played a similar role in the ancient Near East tradition since there is a specific series of images and mythical motifs where they are represented together with the same array of images and, also, as a phallic symbol (Antonova 1984: 146–161).

Thus, anything that encircles could be a sign of protection. Elements of costume and jewelry like necklaces were probably also endowed with such a function. These artifacts, of course, are universal but they have been found in impressive quantity in the Gonur graves. They are also polysemantic, but their specific purpose as a protecting element is highlighted by the fact that the terracotta figurines wear them.

In connection with the braid-like motifs, it is also interesting to look at some Mesopotamian rituals known from literary texts dealing with weaving, cutting the umbilical cord, or fortifying the veins (Afanas'eva 1997: 74, 76, 331, 338). Such texts include images of intertwining, constraint, but also of tearing off the bonds (Antonova 2005: 201). The braided rope could refer to the veins or the umbilical cord. The sphere of meaning of these elements of the human body includes concepts of constraint and delivery in the rituals involving male power and childbirth (Afanas'eva and D'jakonov 1981: 200; Afanas'eva *et al.* 2000: 218, 226).

In their rituals, the Mesopotamians used straps of wool, placed on different parts of the body; they were told to treat them as knots to be tied and untied (i.e., fastening the presence of evil and getting rid of adversity). According to one of the Assyrian rituals, Ishtar gave orders to the hierodules to spin a thread of white and black wool: manipulations with this thread were supposed to discard the spell (Fosse 2001: 133–134, 217, 234). Noteworthy are the indications of various threads used during medical rituals and circular weaving (i.e., the creation of a magical border).

The role of threads in rituals has been much emphasized by M. Eliade. He particularly underlined the constant use of the image of the thread in the Indian tradition, in the rituals of the transition and ecstatic rites: Cosmic strings support the universe and their cutting means the end of the world. The activity of cosmic forces and of many gods is related to the fact that they act as "masters of the threads." The most adequate representation of life is a thread, a spider, a weft, or the act of weaving (Eliade 1998: 287, 291–303). Having considered the data from different traditions, Eliade came to the conclusion that the rope is the key image in people's ideas about the cosmos, life, human existence, or secret knowledge. In archaic cultures it always implies a rise in the sky and the ability to fly. He concluded by saying:

> Images of a thread or a rope constantly excite one's imagination and thought. Thus, these images correspond to an extremely deep experience and, ultimately, reveal a person's situation in the world that cannot be expressed by other symbols and images.
>
> (Eliade 1998: 325)

The position of the rope, circular or vertical, marks the different facets of its value in rituals, either "to close" or "to connect." B.A. Litvinskij, exploring the "rope" design at the rims of Iron Age copper pots from Hindustan and Pamir, interpreted it as having the value of the magic circle in shamanic rituals: "It looks like they believed that, during the ritual process, it opens and, changing direction, rushes upwards, connecting the pole of sacrifice with the heavenly pole" (Litvinskij, 2000: 86–87). Thus, cord-like motifs on the BMAC stamp seals suggest the existence of mythological concepts that give rise to images of snakes surrounding the earth. In this culture, the themes of a limit, of a beginning and infinity, of the vertical connections between the various zones, were relevant. In its rituals, the universal mark of the circle was combined with the ideas of mobility, of spinning, and thus, very likely, of time.

Sarianidi (1986; 1988: 118) pointed out that the rituals of the BMAC inhabitants may have had an ecstatic nature with the use of special drinks, and J. Aruz (1998: 82, after Francfort 1989: 413–421), on the basis of more general considerations, also underlined shamanistic pre-farming notions in this culture. According to ethnographers, such rituals were widely practiced by various ethnic groups, among which those of Central Asia and Kazakhstan (Basilov 1992).

Among the seals, as we already mentioned, only a few are cylinder seals (see Figure 6.5, 3–4 and also Chapter 8, Figure 8.10). These are not local and clearly convey foreign cultural features with their multifigures in different attitudes, sometimes accompanied by inscriptions (Sarianidi 1998: no. 1786 = Figure 6.5, 3; Sarianidi 2004: 241, fig. 17). They are considered as trans-Elamite (Amiet 1986: 165–167) and could come from a BMAC-related environment under a strong influence of Elam or Mesopotamia like Shahdad or Tepe Yahya. They differ from Mesopotamian seals because of the presence of a motif on their base (Francfort 1997). Amiet (1997: 100) considered that:

> The pantheon of Iran depicted on trans-Elamite seals is less extensive and less varied, and accordingly less developed, than that of Mesopotamia … This simpler pantheon most likely comes from a weak political system, which includes various ethnic formations or tribes, each of which must have had its own divine protector, at that time usually of female gender.

This statement could well be attributed also to the similar representations found in BMAC.

Interestingly, though foreign, it is possible to see on the carved scenes a probably local interpretation of these deities: At Gonur, for instance, a seated horned goddess with two standing female fertility deities on both sides is enclosed in an "aura" reflecting the "encircling" local belief (Sarianidi 2004: 239 and fig. 17); on another cylinder seal, a tulip above a goat, probably before its sacrifice, is also viewed as a local feature by V.J. Krjukova (2012). On the basis of the different images of tulips represented on BMAC artifacts, this author considers that they probably point at the existence of a spring fertility festival related to a female deity where these flowers could be a symbol of the representation of the world.

CONCLUSION

The representations hidden behind the clay figurines during the NMG V/BMAC period has a universal, undifferentiated character. The world as a whole is behind these images. On the figurines present in Margiana, the number of conventional signs they bear became limited to that of a vegetation symbol, and they were probably used only at the level of family-kin groups. The elite, apparently, had no relations with these clay figurines, but rather were involved with the newly introduced stone composite statuettes.

It is likewise noteworthy that it is at the time of the BMAC and through the communities with whom its bearers were in contact that images of deities, earlier unknown, appeared on cylinder seals. A few of these images spread onto the stamp seals while the great majority of these artifacts, probably more apotropaic amulets than seals, go on transmitting local symbolic representations that clearly point at an oral-based culture.

At this point it should not be forgotten that we have at our disposal only the visual, or more specifically the material, image that had a deep meaning to all members of the society. As E.S. Novik (1984: 126–127) wrote in her monograph on the rituals and folklore in Siberian shamanism and wondered how the messages, conveyed by special means, were perceived in this oral culture:

> Is it right to speak about a specific perception of thought, speech and representation? Isn't this a generalization of our own that stems not so much from the ethnographic data, but rather from our writing-oriented culture which does not draw a distinct line between visual, acoustic or mental codes of information, but reacts very acutely to the pragmatic side of communication processes ... Behind the thought, the image, the words and their material equivalent, there is some kind of illusory reality.
>
> (Novik 1984: 126–127)

In oral cultures, the visual image is but one of the options that people had for communication.

The nature of the sources available to us who study this oral culture forbids making solid statements upon the "forces" acting within it and the images behind them, since we lack the acoustic or mental codes operating with them. Comparisons with more or less contemporaneous societies with written data can provide only a partial understanding.

NOTES

1 Unfortunately, except for this mention concerning figurines at the foot of hearths, no specific study has ever been done as to the exact position of the others, in houses or in other buildings.
2 Seals and their classification (from another point of view) are dealt with in Chapter 8.
3 The larger number of seals said to come from Bactria is due to illegal excavations. Certainly more graves have been looted than regularly excavated and this may explain this difference in numbers.

REFERENCES

Afanas'eva, V.K. 1997. *Ot nachala nachal. Antologija shumerskoj poezii/Vstupitel'naja stat'ja, kommentarii, slovar' V.K. Afanas'evoj* [From the beginning of time: anthology of Sumerian epics/introduction, commentaries, dictionary by V.K. Afanas'eva]. St. Petersburg: Peterburgskoe vostokovedenie.

Afanas'eva, V.K. and I.M. D'jakonov 1981. *Ja otkroju tebe sokrovennoje slovo. Literatura Vavilonii i Assirii* [I shall unveil the sacred word: literature of Babylon and Assyria], ed. V.K. Afanas'eva and I.M. D'jakonov. Moscow: Khudozhestvennaja literatura.

Afanas'eva, V.K., I.M. D'jakonov, and I.S. Klochkov 2000. *Kogda Anu sotvoril nebo. Literatura drevnej Mesopotamii* [When Anu created the sky: literature of ancient Mesopotamia], ed. V.K. Afanas'eva, I.M. D'jakonov, and I.S. Klochkov. Moscow: Aletejja.

Amiet, P. 1986. *L'âge des échanges inter-iraniens: 3500–1700 avant J.-C.* Paris: Réunion des Musées Nationaux.

Amiet, P. 1997. "Bogini i tsaritsy Elama" [Elamite goddesses and queens]. *Vestnik Drevnej Istorii* 1: 97–100.

Antonova, E.V. 1972. "K voprosu o proiskhozhdenii i smyslovoj nagruzke znakov na statuetkakh anauskoj kul'tury" [Questions on the origin and on the semantics of the signs on statuettes of the Anau culture]. *Sovetskaja Arkheologija* 4: 7–18.

Antonova, E.V. 1981. "Ornamenty na sosudakh i znaki na statuetkakh anauskoj kul'tury (k probleme znachenija)" [Ornaments on pottery and signs on statuettes of the Anau culture: problem of their meaning]. In *Srednjaja Azija i ee sosedi v drevnosti i srednevekov'e* [Central Asia and its neighbors in ancient and medieval times], ed. B. Gafurov and B. Litvinskij. Moscow: Nauka, pp. 5–21.

Antonova, E.V. 1984. *Ocherki kul'tury drevnikh zemledel'tsev Perednej i Srednej Azii. Opyt rekonstruktsii mirovosprijatija* [Overview on the cultures of ancient farmers of Western and Central Asia: an attempt to reconstruct their world perception]. Moscow: Nauka.

Antonova, E.V. 1990. *Obrjady i verovanija pervobytnykh zemledel'tsev Vostoka* [Rituals and beliefs of the ancient farmers of the East]. Moscow: Nauka.

Antonova, E.V. 2000. "'Zmeja' i 'Orel' v gliptike 'tsivilizatsii Oksa' ["Snake" and "Eagle" in the glyptics of the "Oxus Civilization"]. *Vestnik Drevnej Istorii* 2: 46–52.

Antonova, E.V. 2005. "K interpretatsii zhgutovidnykh motivov na veshchakh Baktrijsko-Margianskogo arheologicheskogo kompleksa" [On the interpretation of braid-like motives on the items of the Bactria-Margiana Archeological Complex]. *Vestnik Drevnej Istorii* 1: 194–208.

Antonova, E.V. 2011. "K interpretatsii veshchestvennyh istochnikov: sinkhronija i diakhronija" [On the interpretation of material sources: synchrony and diachrony]. *Vestnik Drevnej Istorii* 4: 8–29.

Aruz, J. 1998. "Obrazy sverkhchuvstvennogo mira: baktriisko-margiansie pechati v ikh svjazi s Blizhnim Vostokom i dolinoj Inda" [Images of the supersensible world: Bactria-Margiana seals in their connections with the Middle East and the Indus valley]. *Vestnik Drevnej Istorii* 2: 81–94.

Bajburin, A.K. 1989. "Semioticheskie aspekty funktsionirovanija veshchej" [Semiotic aspects of item functions]. In *Etnograficheskoe izuchenie znakovykh sredstv kul'tury* [Ethnographic study of the sign/symbols in cultures], ed. A.S. Myl'nikov. Leningrad: Nauka, pp. 63–101.

Basilov, V.N. 1992. *Shamanstvo u narodov Crednej Azii i Kazakhstana* [Shamanism of the peoples of Central Asia and Kazakhstan]. Moscow: Nauka.

Berezkin, Y.E. 1994. "'Gorod masterov' na drevne vostochnoj periferii. Planirovka poselenija i sotsial'naja struktura Altyn-depe v III tys. do n.e." [A "city of craftsmen" at the periphery of the ancient East: settlement layout and social structure at Altyn Depe in the 3rd millennium BC]. *Vestnik Drevnej Istorii* 3: 14–36.

D'jakonov, I.M. 1977. "Vvedenie" [Introduction]. In *Mifologii drevnego mira* [Mythologies of the ancient world], ed. V.A. Jakobson. Moscow: Voctochnaja Literatura: 5–54.

Eliade, M. 1998. *Mefistofel' i androgin* [Mephistopheles and Androgyne]. St. Petersburg: Aletejja.

Fosse, Sh. 2001. *Assirijskaja magija* [Assyrian magic]. St. Petersburg: Evrazija.

Francfort, H.-P. 1989. *Fouilles de Shortughaï: Recherches sur l'Asie centrale protohistorique.* Paris: de Boccard.

Francfort, H.-P. 1992. "Dungeons and Dragons: Reflections on the System of the Iconography in Prehistoric Bactria and Margiana." In *South Asian Archaeology Studies*, ed. G. Possehl. New Delhi: Oxford and IBH Publishing Co., pp. 179–208.

Francfort, H.-P. 1997. "Pechati: raznoobrazie form i izmenchivost' funktsij" [Seals: variety of forms and variability of functions]. *Vestnik Drevnej Istorii* 4: 60–68.

Kircho, L.B. 1990. "Drevnejshie pechati i ikh ottiski iz Altyn-Depe" [The most ancient seals and their prints from Altyn Depe]. *Sovetskaja Arkheologia* 3: 176–184.

Krjukova, V.J. 2012. "O gonurskikh tjul'panakh" [On the tulips of Gonur]. In *Trudy Margianskoj Arkheologicheskoj Ekspeditsii*, vol. 4, ed. V.I. Sarianidi. Moscow: Staryij Sad, pp. 222–236.

Litvinskij, B.A. 2000. "Mednye kotelki iz Indostana i Pamira (drevnie svjazi dvukh regionov)" [Copper pots from Hindustan and Pamir (ancient relations of the two regions)]. In *Arkheologija, paleoekologija i paleodemografija Evrazii* [Archaeology, paleoecology and paleodemography of Eurasia]. Moscow: GEOS, pp. 277–294.

Masson, V.M. and Y.E. Berezkin (eds.) 2005. *Khronologija epokhi pozdnego eneolita i srednej bronzy Srednej Azii (pogrebenija Altyn-depe)* [Chronology of the Chalcolithic and Middle Bronze Age (Altyn Depe burials)]. Vol. 16, *Trudy Institut a Istorii Material'noj Kul'tury RAN*. St. Petersburg: Nestor-Istorija.

Masson, V.M. and V.I. Sarianidi 1973. *Sredneaziatskaja terrakota epokhi bronzy. Opyt klassifikatsii i interpretatsii* [Central Asian terracotta of the Bronze Age: an attempt of classification and interpretation]. Moscow: Nauka.

Nekljudov, S.Ju. 1977. "O funktsional'no-semanticheskoj prirode znaka v povestvovatel'nom fol'klore" [On the functional-semantic nature of the sign in narrative folklore]. In *Semiotiika i khudozhestvennoe tvorchestvo* [Semiotics and artistic creativity], ed. Ju.Ja. Barabash Moscow: Nauka, pp. 193–228.

Novik, E.S. 1984. *Obrjad i fol'klor v sibirskom shamanizme* [Ritual and folklore in Siberian shamanism]. Moscow: Vostochnaja Literatura.

Sarianidi, V.I. 1965. *Pamjatniki pozdnego eneolita Jugo-Vostochnoj Turkmenii* [Late Chalcolithic sites of southeastern Turkmenia]. Svod arkheologicheskih istochnikov B3-8. Moscow: Nauka.

Sarianidi, V.I. 1986. "Monumentalnyj kompleks Togolok 21 v Margiane" [The Togolok 21 monumental complex in Margiana]. In *Arkheologicheskie otkrytija 1984*. Moscow: Nauka, pp. 473–474.

Sarianidi, V.I. 1988. "Protozoroastrijskij khram v Karakumakh" [A protozoroastrian temple in Margiana]. *Priroda* 11 : 117–118.

Sarianidi, V.I. 1998. *Myths of Ancient Bactria and Margiana on Its Seals and Amulets*. Moscow: Pentagraphic Ltd.

Sarianidi, V.I. 2001. *Necropolis of Gonur and Iranian paganism*. Moscow: World Media.

Sarianidi, V.I. 2004. "Strana Margush otkryvaet svoi tajny. Dvortsovo-kul'tovyj ansambl' severnogo Gonura" [The country of Margush opens its secret: the palace-cultic ensemble of Gonur North]. In *U istokov tsivilizatsii* [Near the sources of civilizations], ed. M.F. Kosarev, P.M. Kozhin, and N.A. Dubova. Moscow: Staryj Sad, pp. 229–253.

Sarianidi, V.I. 2005. *Gonurdepe, Turkmenistan: City of Kings and Gods*. Aşgabat: Miras.

Sarianidi, V.I. 2006. "Tsarskij nekropol' na Severnom Gonure" [A "royal necropolis" at Gonur North]. *Vestnik Drevnej Istorii* 2: 155–192.

Sarianidi, V.I. 2007. *Necropolis of Gonur*. Athens: Kapon Editions.

Solov'eva, N.F. 2005. *Antropomorfnye izobrazhenija i kul'tovye kompleksy Juzhnogo Turkmenistana pory srednego eneolita (po materialam raskopok Ilgynly-Depe). Avtoreferat kand. dissertatsii* [Anthropomorphic images and religious complexes of Middle Chalcolithic southern Turkmenistan (from excavated materials of Ilgynly Depe)]. Abstract of PhD thesis. St. Petersburg: History of Material Culture Institute.

Vidale, M. 2017. *Treasures from the Oxus: The Art and Civilization of Central Asia.* London and New York: Tauris.

Volchok, B.Ja. 1972. "Protoindijskie bozhestva" [Proto-Indian deities]. In *Soobshchenija ob issledovanijakh protoindiiskikh tekstov* [Review on the studies of proto-Indian texts/Proto Indica]. Moscow: Nauka, pp. 246–305.

CHAPTER SEVEN

MYTHS AND GODS IN THE OXUS CIVILIZATION

Annie Caubet

INTRODUCTION

Writing about myths and gods in the Bactria-Margiana Archaeological Complex (BMAC) is a hazardous undertaking since this civilization is anepigraphic and we have no hint at what the mythological thoughts of its bearers might have been. Elements of myths preserved in later literary sources, like the Indo-Iranian heritage and Early Islamic narratives, were transformed over the course of time. The present chapter will therefore approach the issue over the *longue durée*, and on a supra-regional scale. It postulates, as a working hypothesis, that the wide area including the Iranian Plateau (with elements known from Susa and the Kerman region), Margiana, and Bactria constitute a multifaceted *koinè*, about which Holly Pittman (1984) and Pierre Amiet (1986) have theorized. The main source to reconstruct this mythology resides in the iconography carried by artifacts related to the Oxus Civilization, but it is difficult to delimitate their geographical space since most of the available artifacts come from illegal excavations and have neither context nor provenance (see, e.g., Pottier 1984). When found *in situ*, like the seals excavated at Gonur Depe, they present traits of style and motifs shared with southeastern Iran, Elam,[1] Syria-Mesopotamia, the Indus, and even the Mediterranean, thus broadening the horizon of the BMAC *stricto sensu*: Long-distance circulation of goods and men engendered the supra-regional metaphysical concepts expressed in that iconography.

Chronology is another difficulty caused by the absence of context and the discrepancy in the dates of the known production centers: Chlorite vessels from Jiroft are dated "in the course of the 3rd millennium" (Perrot and Madjidzadeh 2005, 2006), while excavations at Gonur point to the last centuries of the 3rd and early 2nd millennium (see the Appendix at the end of the volume). The end of the Oxus Civilization, around 1700/1500 BC (Luneau 2014; see also Chapter 18), leaves about 500 years for mythological constructions to evolve and vary across time and space, hence our *longue durée* approach.

The few known epigraphic sources are equally tantalizing and unsatisfactory since the writing systems used in southwestern Iran during the late 3rd–early 2nd millennium remain generally undeciphered (Desset 2018a, 2018b). Oxus artifacts exported to the southwest were inscribed in the local Elamite script, like the so-called Persepolis vessel (Desset 2018a). Those exported to Mesopotamia were appreciated by Early Dynastic rulers: Inscriptions in Sumerian and Akkadian engraved on "Intercultural" chlorite vessels served as the basis for the chronology of this production (Amiet 1986: fig. 63; Potts 2010: 178).

Last challenge, the frequent lack of reliable context makes it difficult to determine the final destination of the artifacts we are concerned with: cultic place, public building, palace or center of power, or tombs.

THE MATERIALS OF THE IMAGERY

The Oxus imagery is associated with a range of characteristic materials, local or imported, which made their appearance (or became routinely used) at the beginning of the Oxus Civilization. The most numerous ones are the *green soft stones* (steatite, chlorite, enstatite, etc.), from various geological deposits (David 1996; Kohl 2001; see also Chapter 9). They were used to carve composite statuettes and figurative vessels, in a style known as "Intercultural" and widely distributed throughout Central Asia, Iran, the Gulf, and Syria-Mesopotamia (Lamberg-Karlovsky 1988). Individual production centers and regional styles have been identified, notably at Tepe Yahya and Jiroft (Perrot and Madjidzadeh 2005).

Calcite alabaster, used to make plain vessels (Casanova 1991) was also employed to carve elements of the composite statuettes.

Lapis lazuli, a stone extracted from Badakhshan and that widely circulated throughout Syria-Mesopotamia, Egypt, and the Mediterranean during the 3rd millennium (Herrmann 1968; Casanova 2014), was worked in the Oxus for the production of stamp seals, personal ornaments, and, sometimes, elements of composite statuettes.

Gold and *silver* were lavishly used to produce seals and vessels engraved or embossed with figurative decoration (Sarianidi 2005: 235, fig. 94).

Working with *faience*, an artificial vitreous compound based on the plastic properties of silica or quartz, was developed in Egypt, Syria-Mesopotamia (Bouquillon *et al.* 2007), Baluchistan, and the Indus. It was adopted by the Oxus Civilization to produce small vessels, some composite statuettes, and personal ornaments (Benoit 2007; Sarianidi 2007: 152, figs. 34–35); faience small plaques were assembled into figurative mosaic panels (Sarianidi 2007: 122, fig. 240; 184–185; see also Chapter 11).

Raw *elephant tusks* were imported from India (Caubet 2016; Frenez 2017), alongside finished products like combs and fortune-telling dices of clearly Harappan origin (Sarianidi 2005: 244, fig. 104; Frenez 2017: figs. 4–5, 8), and other artifacts carved with mythological scenes in a local style (Figure 7.1).

Only *clay* was used long before the appearance of the BMAC period: terracotta figurines were produced since the Neolithic period, displaying during the Bronze Age a notable unity of types and styles throughout most of Central Asia (Luneau and Shirazi 2019). But clay vessels ceased to receive a painted decoration during the BMAC period, in contrast with the preceding times, adopting instead sometimes three-dimensional motifs probably imitated from metal vessels decorated with riveted figures.

TYPES OF ARTIFACTS AND THEIR ICONOGRAPHY

All the categories of artifacts that display a symbolic imagery made their appearance at the time of the Oxus Civilization, except the terracotta figurines, present since the Neolithic period (see Chapter 6).

Figure 7.1 Handle of a spoon: winged dragon swallowing a calf, elephant ivory, Gonur Depe © Margiana Archaeological Expedition, courtesy N. Dubova.

Composite statuettes

The composite statuettes appeared on the art market in the early 1960s. They are all characterized by their manufacture technique combining several materials, mainly green soft stone contrasting with white limestone or calcite (Figure 7.2). Lapis lazuli (Caubet 2018a: cat. 87–88) or copper is sometimes also used (Caubet 2018a: cat. 76), while some were molded in faience (Sarianidi 2007: fig. 56 from burial 1022;[2] Caubet 2018a: cat. 90).The different elements of the body and costume were carved separately and joined, as in a puzzle, by tenon and mortice or with a kind of glue. This composite technique is also known at Ebla in Syria, palace G (Matthiae 2009: figs. 9–14), and at Harappa in the Indus Valley (Meadow 2002). For lack of conclusive laboratory analysis to eliminate forgeries, one has to rely on the examination of materials and techniques to attempt the reconstruction of a coherent iconographic ensemble.

The majority of the composite statuettes depict the "Oxus Lady" (a term we prefer to that of "Bactrian princess"), with many variants, probably made in different workshops (Benoit 2010; see also Chapter 6). Only a few examples, complete or fragmentary, were found in the course of regular excavations at Gonur, Adji Kui, and Togolok, all in a funerary context (Vidale 2017: 98); one fragment comes from Baluchistan and was found in the so-called "Quetta treasure," also a funerary deposit (Jarrige 1988: no. 162). More than 100 such statuettes, not counting numerous probable fakes, are known but undocumented. The *Oxus Lady* is characterized by her dress, which hides most of her body except the head and hands, made of a woolly material closely derived from the Sumerian fleece-like *kaunakes*. She may be standing, seated on a stool, or squatting sideways on her hips. One type reduces the body to a flat triangle.[3] The woolly material of the dress is arranged in an infinity of designs that seem significant: Thin wavy stripes flow like a stream of water on the flat triangular figures, making her perhaps a water spirit(?); long tufts radiate from her shoulder like the rays of a star (an astral spirit?), or like feathers, which make her a spirit of the air, comparable to the winged Lady on stamp seals (see p. 206). A unique figure has the body of a bird with extended wings and a human head, but has no provenance (Caubet 2018a: cat. 75).

Figure 7.2 Seated Lady, chlorite and calcite. Courtesy Ligabue collection, Venice.

The head is usually carved from white calcite or shell, and rarely occurs in a copper alloy (Caubet 2018a: cat. 76); its colors contrast with the dark dress. Nose and mouth are finely detailed; the eyes may have been originally painted, but their place is sometimes left blank, perhaps aiming at the depiction of a blind "seer," endowed with mysterious powers. On top of the head, cut from a dark stone contrasting again with the white face, is an elaborate coiffure, braided hair, turban, or high comb-like crown (Vidale 2017: no. 91; Caubet 2018a: cat. 83).

The "*Scarface*," so named on account of the deep gash that cuts across one side of his face (Ghirshman 1963; Amiet 1986: 200–201, fig. 206; Benoit 2010: fig. 30; Vidale 2017: chap. X; Caubet 2018a: cat. 93–94) is a statuette also carved of several elements in contrasting colors, one for the legs joined at the thighs, one for the kilt, and one for the torso, arms, and head. All the elements are assembled by the tenon and mortice technique, comparable to that of the Oxus Lady, but the color contrast is in reverse (dark skin features, white dress elements) (Figure 7.3). Less than a dozen Scarface statuettes are known and none has a provenance. The first figure to appear on the market was said to be from Fars (Ghirshman 1961: no. 12). Though its association with the Oxus Lady is not attested in context so far, the Scarface is taken into account in this survey of mythological figures.

The body, stocky and muscular, standing on his slightly apart legs, the bulging muscles of the torso, all convey the impression that the Scarface figure is gathering up his strength. The head is very large, about a sixth of the total height of the figure, sunk in the torso, with almost no neck. The face is stern, the nose long and triangular, the eye sockets hollowed out for incrustation (one or the other is often – intentionally? – missing). A collar beard reaches down to the torso, leaving visible the tightly closed, unsmiling lips. Piercings in the upper and lower lips are often incrusted in colored

Figure 7.3 Scarface, chlorite and calcite. Musée du Louvre AO 21104 © author, courtesy Département des Antiquités Orientales, Musée du Louvre.

stones. The low forehead is circled with a ring diadem incrusted in metal, iron, or gold. The long hair falls down the back almost to the waist, ending in a point. Only one Scarface figure, the first that appeared on the art market, wears a "top hat" in stone (Ghirshman 1963). One permanent feature on the Scarface figures is the small globular pot that they hold horizontally under the arm, as if ready to flow, suggesting that the figure is part of a narrative implying the use of liquids. Some of the statuettes wear hoof-shaped shoes, but most are cut at the ankles and are without feet. The kilt, a short tubular skirt belted at the waist, is decorated with regularly spaced strips hanging from the belt and ending each in a cupula or disk. The strips are deeply engraved and often retain traces of an adhesive or incrustations in colored stone or gold. A striking characteristic of all the Scarface statuettes is their skin covered with a thin network of engraved scales, quadrangular or curvilinear, sinuously underlining the muscles. This disquieting figure is generally interpreted as an anthropomorphized form of a dragon-snake (Francfort 1982), possibly a cosmic demon, forerunner of Indra who released the subterranean waters (Vidale 2017: chap. X).

Another iconographic type of composite statuettes, the *kneeling youth*, in stone or faience (Caubet 2018a: cat. 95–98) is preserved in only a few examples, none with a provenance. Like the Scarface, it is added here to the Oxus corpus on the ground of specific technique and iconographic details. These clean-shaven figures are shown in repose, their legs folded under their skirt, their feet shod with pointed boots carved in low relief under the base of the figure. The arms are resting on the thighs or hold a vessel. Their hair is very long, with dreadlocks that frame the face down to the chest, and fall down the back to the waist, with a pointed end. The kneeling youths are also made of the composite technique, using stones in contrasting colors including lapis lazuli, as the Oxus Lady and Scarface. Like the Scarface, the kneeling youths have the muscular torso and arms, the kilt incrusted with hanging stripes and disks, the lip ornaments, the thin metallic diadem, and in some cases a dark body contrasting with a light kilt. Unlike the Scarface, however, they are not bearded, their face is serious but engaging, their gaze calm and confident instead of fierce. One kneeling youth is offering a cylindrical pot, reminiscent of the pot tucked under the arm of the Scarface.

Seals

Seals are among the best carrier of symbolic imagery in miniature (see Chapter 6), and the most reliable as a large number was found in the course of regular excavations. Whether the appearance of sealing practices in the urban civilization of the Oxus was dependent or not on Mesopotamia remains an open question, but the compartmented seal cast in metal is a development entirely original to Central Asia (Amiet 1986: figs. 174–188; Francfort 1998; see also Chapter 8). Across time and space, the glyptic of the Oxus presents a distinct character, comparable with the level of the iconography displayed on the "Intercultural" vessels in chlorite.

Motifs on stamp and compartmented seals, the most numerous in the Oxus Civilization, tend to be isolated, on account of the small format, while rare cylinder seals allow for complex scenes. The iconographic repertoire includes: a goddess, often winged and associated with animals; "real" animals (bull, goat, camel, feline, bird, scorpion, snake); hybrids mixing human and animal (bull-man, snake-man) (Figure 7.4) or mixing two or more animals (snake-dragon, lion-dragon); mountain landscape; plants (tulip, poppy); astral motifs; geometric motifs (notably the stepped cross). Complex narrative scenes with multiple figures offer close parallels with Syrian-Mesopotamian compositions. For instance, mythological combats involve animals, hybrids, and heroes or masters of animals (Figure 7.5); banquet scenes include one exchange of drinking vessel between a kneeling guest and a goddess, also illustrated in the composite statuettes. One bull-leaping scene highlights the wide diffusion of this ritual, from the Mediterranean to the Indus (Francfort 2005; Sarianidi 2007: 156–159)

Vessels

Mythological themes are found on precious metal and soft-stone vessels. Among the metal vessels in gold and especially silver, the high conical cups have a shape known in the local pottery repertoire, but their rim supports small figures in the round (Amiet 1986: fig. 197). Other silver metal vessels, in the shape of cylindrical bowls, are embossed and/or engraved with multiple figures related to various narratives (Figures 7.6a and 7.6b): a banquet scene, a music concert, hunting, war, or agricultural activities. Quadrangular-shaped boxes are engraved with similar figurative scenes or covered with geometric motifs, especially the stepped cross. Rare examples

Figure 7.4 Stamp seal: master of the snakes, copper alloy. Courtesy Ligabue collection, Venice.

Figure 7.5 Stamp seal: Lady seated on a unicorn with caprids, silver. Musée du Louvre AO 30226 © author, courtesy Département des Antiquités Orientales, Musée du Louvre.

of plastic vases are hammered into figures in the round, like a silver kneeling bull-man holding a spouted vessel (Harper *et al.* 1992: 5, fig. 5).

Soft-stone vessels in the "Intercultural" style were produced in various different centers up to now all located in the Kerman region, each with their distinctive styles, as is the case for Tepe Yahya and Jiroft (Lamberg-Karlovsky 1988; Kohl 2001; Perrot and Majidzadeh 2006). They were widely exported to Syria-Mesopotamia (Aruz and Wallenfels 2003: 325–334), as seen especially at Mari (Butterlin 2014), and to

Figure 7.6a Silver pyxis with a banquet feast (photo), Musée du Louvre AO 31881. © author.

Arabia (Zarins 1978; Cotty 2010). The most typical shapes are the low cylindrical bowl, similar to the silver ones mentioned above; the ovoid jar; the globular footed cup; and the high goblet with concave sides recalling the BMAC repertoire of calcite vessels. As a whole, the style of the relief decoration is highly distinctive with clear local specificities (Perrot and Majidzadeh 2006). The rich mythological repertoire of anthropomorphic genies interacting with animal and hybrid monsters, though probably dating slightly earlier, is closely matched by the iconography of the Oxus glyptic (see Chapter 8). The symbolic significance of the relief decoration suggests that these vessels were part of cultic rituals involving liquids, a function alluded to in some of the composite statuettes (the Scarface and kneeling youth) and in the glyptic.

Handled stones

Handled stones (also called weights or handbags) in chlorite, are closely related to the soft-stone vessels by their material, distribution, and rich iconography. The images are carved in bas relief on the quadrangular side, or occupy part of the space between the handle and the base, in open work sculpture (Aruz and Wallenfels 2003: no. 236; Perrot and Madjidzadeh 2005: 128, fig. 11j; 143, fig. 12b). Many of these artifacts were excavated at Jiroft (Perrot and Madjidzadeh 2005: pl. IIf–k.). Only one such item comes from the margins of the BMAC area (Soch in the Ferghana valley, Uzbekistan) and is decorated with two snakes (Brentjes 1971). The iconographic repertoire on those from southeastern Iran includes hybrid creatures (scorpion-man), eagles, architectural motifs; the date-palm tree, also typical of the Jiroft style, is found

Figure 7.6b Silver pyxis with a banquet feast (drawing), Musée du Louvre AO 31881. © Caroline Florimont, courtesy Département des Antiquités Orientales Musée du Louvre.

on many undocumented pieces (Caubet 2018b), as is the master of the snake motif. The handled stones may have been used in the course of cultic celebrations, perhaps alongside other non-utilitarian blanked artifacts such as the stone "scepters" and the miniature columns (Amiet 1986: 165, 195).

Game boards

Game boards are also closely related to the "Intercultural" vessels. Known as the "game of 20 cases," an educational pastime designed to sharpen the wits of the players, it was developed in ancient Sumer, and adopted all over the Near East and Central Asia, retaining its symbolism of fate and destiny (Dunn-Vaturi and Schädler 2016). Boards from Jiroft are decorated with animal or hybrid figures, eagle, scorpion-man, and snake (Perrot and Madjidzadeh 2005: pl. III) (for game boards at Gonur Depe, see also Chapter 11).

Weapons

The weapons from Central Asia belong to a large morphological repertoire common to the ancient Near East (Gernez 2017). Exceptional examples may carry a figurative decoration engraved or added in high relief (Figure 7.7), sometimes gilded and depicting mythological creatures comparable to those found on seals and soft-stone vessels (Amiet 1986: nos. 166–173; Aruz and Wallenfels 2003: no. 264).

Musical instruments

Complex musical instruments, like the harp and the lyre, were invented in Mesopotamia during the 3rd millennium. They spread to Egypt, the Mediterranean, and Iran. No such instrument has survived in Central Asia, but they appear in the iconography of the silver and soft-stone vessels (Lawergren 2018), together with images of drums and trumpets. A type of small metal trumpet specific to the Oxus is a figurative tube in the shape of a human or animal head (Lawergren 2003; Sarianidi 2007: 81, figs. 73–74).

Personal ornaments

Among the jewelry and personal ornaments in precious metal, fine stone, or composite materials, it is mostly the head of cosmetic pins that presents a figurative motif, either anthropomorphic or animals, notably the Bactrian camel (Amiet 1986: fig. 127; Sarianidi 2005: fig. 56; Benoit 2007).

Terracotta figurines

Terracotta figurines, the mainstream expression of popular faith in antiquity, precede the emergence of the Oxus Civilization (see Chapter 6). Those from the Bronze Age in Iran and Central Asia present a remarkable unity of types, over the *longue durée*, with minor local variants (Masson 1981: pl. VIII–XIII; Sarianidi 2007: 70, figs. 47–49; Luneau and Shirazi 2019) (Figure 7.8). Their iconographic repertoire is markedly different from that of the chlorite vessels and the seals. The majority are naked anthropomorphic female

Figure 7.7 Ceremonial axe: seated wise elder, copper alloy and silver. Courtesy Ligabue collection, Venice.

figures and animals, mostly bovines. Unambiguously sexed male anthropomorphic figures are much rarer (Rossi Osmida 2007; Sarianidi 2007: 68, fig. 43). This significant ratio is also observed in the coroplastic productions of ancient Mesopotamia and the Levant. Next to the mass production of small figurines, which belong to a tradition rooted in earlier periods, a few larger figures in clay are a novelty of the BMAC period. They are male and female, come exclusively from funerary contexts and present facial features similar to those of the stone composite figurines (Amiet 1986: 296–297, figs. 129–130; Sarianidi 2007: 71–72, figs. 50–53). Whether they represent substitute images of the dead or mythological beings is a matter of debate.

MYTHOLOGY RECONSTRUCTED FROM THE ICONOGRAPHY

The available corpus of images depicted on these artifacts shows that the iconography is dominated by a female figure interacting with hybrids and animals in a

Figure 7.8 Female figurine, terracotta. Private collection. © author.

nature consisting of mountains, plants, astral representations, and symbols of the natural elements. It is unclear if the figures of the Oxus Lady represent one or several entities, and are of divine character or not, though we are inclined to think that she is a goddess. She is characterized by a flounced skirt inspired from the Mesopotamian *kaunakes* and appears on composite statuettes (Figure 7.2), seals (Figure 7.5), mosaics, and metal vessels (Figures 7.6a, 7.6b) (Amiet 1986: fig. 184; Aruz and Wallenfels 2003: 370, no. 261; Benoit 2010). She is often endowed with wings (Amiet 1986: figs. 135, 181–182; Caubet 2018a: cat 75), a frequent feature of Mesopotamian hybrids, expressing the swift lightness of mythological spirits, or her domination over the air. When seated on a dragon-snake or a lion-unicorn (Figure 7.5), she displays her power over the forces of earth and the waters from the underworld. As mistress of the animals, she finds long-distance parallels with the Mediterranean *Potnia Theron* (Sarianidi 2007: 156–157). Other traits are strangely reminiscent of the later Minoan/Mycenaean elegant gracefulness, up to the ritual curls of hair – especially apparent in the "bath scenes" framed by coils of snakes depicted on chlorite vessels (Perrot and Madjidzadeh 2005: fig. 11d; Musée du Louvre, inventory no. AO 31883) where female heads emerge from the rim of a pithos of similar gigantic size, shape, and wavy decoration as those stored in the storerooms of Mycenaean and Hittite palaces.

Power over the underworld may take a masculine shape. The master of the snakes, an anthropomorphic genie holding a snake in each hand (Figure 7.4) (Aruz and Wallenfels 2003: nos. 224, 227; Perrot and Madjidzadeh 2005: 143, fig. 12b) is sometimes crowned with the bovine horns that distinguish gods from humans in Mesopotamia (Aruz and Wallenfels 2003: no. 235; Sarianidi 2007: 177, fig. 10). The master of the snake motif is an ancient mythological concept common to most of the ancient Near East, reaching as far as the Mediterranean and especially important in Iran (Miroschedji 1981). The Scarface, covered with reptilian scales (Figure 7.3), seems to combine in one entity two antithetic forces, the snake and the anthropomorphic hero (Francfort 1982). The origin of the gash across his face is the object of a seductive hypothesis, unfortunately impossible to prove, where the scar would have been inflicted in the course of combat, perhaps with the Oxus Lady, and/or be a penalty to obtain his freedom and power (Vidale 2017: chap. X), a parallel to the one-eyed Wotan, "paid" to gain his divine kingdom in the German mythology. Whatever its signification, the scar is the material evidence for a lost epic narrative that implied the action of several participants, possibly the Oxus Lady, a male warrior, or a kneeling youth. One common element between these figures is the small vessel, indicating an offering of liquids during the action.

Many of the male figures popular in the Oxus imagery seem to be borrowed from the Syrian-Mesopotamian depictions of the Gilgamesh saga, created in the early 3rd millennium, where the main character interacts with a strange creature. The Mesopotamian hero is typically depicted naked, or wearing a simple girdle, so as to display his strength that resides in his muscles and his flowing and curly hair. Identified as Gilgamesh, the legendary king of the city of Uruk in Sumer, the hero engages in combats with wild lions and bulls, competes with the hybrid figure of the bull-man (Ornan 2010; see also Durand *et al.* 2010). He conquers the grimacing, hirsute Huwawa/Humbaba, incarnation of wilderness and faraway mountains opposed to the civilized and urbanized plains (Graff 2012). In the Oxus iconography, the hero and strong man is a frequent occurrence on metal vessels, chlorite vases, and handled stones. He likewise is shown naked or clad in a short kilt, with a muscular body, long hair hanging loose down his back and on his chest. A Humbaba-like figure, reduced to a fierce head haloed by bushy hair and beard, could be a forerunner of the Avestic Kersaspa (Sarianidi 2007: 177; Winkelmann 2016). The bull-man, frequently represented on the chlorite vessels, is seen holding giant snakes, like the master of the snakes. He appears in complex scenes where the background is densely occupied by lions, bovines, and other animals, surrounded by stylized plants. Hybrids combining human and feline features, unknown in Mesopotamia, are specific to the "Intercultural" style.

Warfare, hunting, and jousts involving warriors are represented in narrative scenes engraved on silver vessels. Mature men and young warriors are distinguished by the presence or absence of a beard, the shape of the kilt, or the hair worn long with dreadlocks for the younger men. It is unclear whether these scenes take place in the real world or not. Bloodshed and wounds are in evidence (Francfort 2003), as is the gash of the Scarface; they may allude to brutal rites of passage. Bull-leaping, confronting man and beast, is another violent practice shared by the Near East, Iran, and the Mediterranean (Francfort 2005).

These dynamic and brutal scenes contrast with peaceful narratives where the protagonists are seated at a banquet, holding a drinking cup, and engaged in conversation, often with the accompaniment of musical instruments (see Figures 7.6a and 7.6b) (Aruz and Wallenfels 2003: no. 228; Francfort 2003: no. 12). They are seated on stools or squatting on the ground, their legs and feet hidden under the skirt, the bag-like appearance of which – a typical feature of Oxus iconography – has parallels in Harappan statuettes (Winkelmann 1994). Age distinction between the participants is carefully notified, as in the combat scenes. Women appear unfrequently on the vessels and are in groups (Vidale 2017: fig. 17) or in a more intimate duo, facing a male (Amiet 1986: figs. 126, 128). Banquet scenes may be associated with hunting, jousting, or agricultural activities (Amiet 1986: figs. 201–202; Benoit 2010: fig. 29), making it unclear whether they take place in the real world, among the elite, or on a supra-human level. Taken together, the "action" scenes and the "quiet" scenes illustrate a dual vision of the world.

Images of animals, plants, and abstract symbols in the Oxus Civilization may derive from the iconography of Syria-Mesopotamia, but subtle mutations take place on the way from west to east: some species tend to change and disappear or give way to others. The snake, a universal symbol of the underground and eternity on account of the yearly renewal of its skin, is depicted in the Oxus with a horn and a leonine jaw (Azarpay 1991). A giant leonine snake is also mentioned in Sumerian literature, but this visual aspect is specific to the Oxus; the belief in this forerunner of Azhi Dhaka in the Avesta (Francfort 1982; Ligabue and Rossi Osmida 2008) survived the arrival of Islam, and endured into the *Shahnameh*. In some cases, the coils of the snake are undistinguishable from the coils of the spiral, a common visual device for the depiction of water in Mesopotamia (see p. 209). A snake coiled so as to provide a seat for a deity, is an iconographic motif of *longue durée* and large distribution, known in Elam for the epiphany of the god Inshushinak (Miroschedji 1981), and adopted for the meditation of Buddha. In the Oxus, the throne supports the Oxus Lady, heralding her domination of the underworld and underground water. The scorpion and the scorpion-man are other frequent rampant creatures on seals and chlorite vessels. Snake, scorpion, and scorpion-man give their shape to boards of the "20 cases game," symbolizing fate and death (Dunn-Vaturi and Schädler 2016: pl. 3; Frenez 2017: figs. 7, 5).

The dragon-lion or unicorn (see Figure 7.5) (Aruz and Wallenfels 2003: nos. 261, 264; Goodnick Westenholz 2004; Sarianidi 2007: 184–185, fig. 11), combining the body of a feline with the head of the horned viper is specific to the Oxus Civilization and the Indus (Possehl 1982; Parpola 2011). It endured into the Persian period and in Chinese iconography. Like the snake, it is frequently associated on stamp seals with the Oxus Lady, mistress of the underworld.

As opposed to these symbols of the underground, birds belong to air, the other domain of the winged goddess in her dual nature. The opposition may be visually materialized by a combat between bird and snake (Perrot and Madjidzadeh 2005: fig. 9a), a combat famously narrated in the Sumerian myth of Etana (Winkelmann 2008). Depicted alone – on jewels or stamp seals (Sarianidi 2007: 104, figs. 171–174; 118, fig. 222; 123, fig. 241), or in complex scenes – seen *en face* in heraldic compositions, or in profile – birds belong mostly to predatory species (eagles, vultures, owls) and may therefore symbolize death, associated with air instead of earth, as the vultures in charge of disposing of the impure corpses in Zoroastrian practices. Like the snake and the scorpion, eagles are made into boards for the "20 cases game" (Dunn-Vaturi and Schädler 2016: pl. 1a–b).

Caprids of various species are depicted on each side of a vegetal element, in antithetic compositions directly inspired from the iconography of Syria-Mesopotamia, but the fantastic and prolific plants have no longer the carefully balanced aspect of the Near Eastern sacred tree. The case of the lions and the felines is exemplary of the transition between the species popular in Mesopotamia and those originating in India or Central Asia. The lion is ubiquitous in the imagery of the Near East as symbol of kingship, fire, and the attribute of Ishtar, goddess of sexuality. It is present in the Oxus imagery (Sarianidi 2005: 255–256, fig. 111; Hiebert and Cambon 2011: 74–75) but Asiatic felines with spots (panther), or stripes (tiger), take over (Amiet 1986: fig. 172). The phenomenon of replacement, observed for the bull-man and the feline-man hybrids, also takes place with the bull and zebu: The bull (*bos taurus*), like the lion, is a dominant animal figure in Mesopotamia where its majestic appearance is enhanced by the beard and the long curls that cover his head and shoulders. It is present in the Oxus iconography, as on the gold vessels from the Fullol treasure (Amiet 1986: no. 196), but it is superseded by the humped zebu, as in the Shahdad standard (Amiet 1986: fig. 126; Hakemi 1997; Helwing 2017: 79), and on chlorite vessels, where it is often associated with water (Perrot and Madjidzadeh 2005: n21). The Bactrian camel, as might be expected, is depicted only in the Oxus; its preeminence on precious vessels and jewelry (see the image on the cover of this book) (Sarianidi 2005: 236–238, figs. 95–97; 2007: 151, fig. 30) might be an indication of its mythological importance (but see Chapter 1 for a more prosaic view). Equids are present in the bestiary to express kingship and status (Dolce 2014), adding arguments to the debated question of the domestication of the horse and the symbolic force of the animal (on the domestication of these animals and their presence in the BMAC, see Chapter 26).

Plants are depicted with a luxuriant abundance, a *horror vacui*, which is significant and meant to convey the energy of nature. The date-palm tree (Caubet 2018b) occupies a place of choice, especially on "Intercultural" vessels and handled stones. It grows alone or in association with the master of the snake, the bull-man, and other genies. This imagery is shared with Elam, where the date-palm, a female symbol in Mesopotamia (Michel-Dansac and Caubet 2013), appeared in the course of the Akkad period, and serves as a symbol of the river valley in opposition to the mountainous plateau.

Images of wheat are specific to southeastern Iran and the Oxus, in association with a "wheat goddess" (Amiet 1986: fig. 132, 8; Sarianidi 2005: 217, fig. 79) depicted as a seated Lady with wheat ears sprouting from her shoulders. The tulip, an indigenous plant to Central Asia, is also frequently associated with the goddess (Sarianidi 2007: 104, fig. 175; Francfort and Tremblay 2010: 46).

Abstract and geometric symbols abound. The spiral is a frequent motif, finely striated to convey an impression of fluidity. In Mesopotamia, it is a common visual convention to represent running or still water, the primary ocean surrounding the earth, or the rain falling down from heavens. The spiral motif blends easily with that of the coiled snake, in a vivid illustration of the association of the snake with the humid underground. The spiral is often combined with the snake on artifacts of the transcultural style like the game boards. Purely geometric designs particular to the Oxus, like the stepped cross or merlon, may perhaps reflect abstract concepts of identity, such as belonging to a clan or a territory.

CONCLUSION

Many of the figurative motives found in the Oxus culture were present in Syria-Mesopotamia and Elam long before they appear in Central Asia. How they were introduced remains an open question, probably alongside trade for desirable commodities that developed at that time (see, e.g., Pittman 1984; Pottier 1984; Lyonnet 2005; see also Chapter 1). The process of adoption/adaptation is difficult to follow, as there are multiple variants in time, space, and various individual hands or regional styles. Each individual production source and cultural center may have brought its own touch, as demonstrated by the stamp seals, which are both indefinitely varied but immediately distinguishable as coming from the Oxus Civilization. The chlorite vessels and handled stones in "Intercultural" style borrow many of their figures from the Syrian-Mesopotamian imagery but develop these motifs with a new vividness, by setting the scene into a dense environment of plants and beasts. The camel, the *bos indicus*, or the tulip allude to a local environment. Hybrids like the bull-man are inspired by the West, but mixing anthropomorphic creatures with spotted felines is a local avatar. The Scarface and the composite statuettes of the Oxus Lady borrow their technique from Syria-Mesopotamia,[4] but they are given a prevalence and a force entirely of their own, and the Scarface has no equivalent outside the Oxus-southern Iranian sphere.

Parallels with mythological creatures of the Indus may easily be explained by trade. Affinities with Anatolia and the Mediterranean world, such as the mistress of animals theme, or the bull-leaping practice, are intriguing and debated.

In fine, the originality of the Oxus iconography resides in the coherent and dynamic construction of its own universe. The complex scenes and the isolated figures are parts of a homogeneous ensemble, where each different motif plays a role in the broader narrative. Duality of mankind and the gods is expressed in symmetrically opposite and complementary figures: male/female, youth/elderly, civilized/wild, beautiful/ugly, human/animal. Opposite and complementary elements of the universe are materialized and combined: earth (snake, dragon-snake), air (birds), water (snake, zebu, spiral), fertile valley (palm tree), mountain (caprids and bull-man), fire (lion).

In the broader perspective of the cosmic speculations elaborated by the Mesopotamians and the Indo-Iranians, Central Asia acted both as an intermediary and as a creative center in the construction of a dualistic universe in action, a universe tamed and pacified by the everlasting opposition between man and animal, nature and culture, life and death. These concepts endured to inform the beliefs of the Zoroastrians (Sarianidi 1998). The arrival of Islam did not cause the gods and genies of the old world to immediately disappear; they were transmuted from powerful and beneficent entities into malevolent spirits and tyrants.

NOTES

1 Although the terms "Elam" and "Elamite" are now considered anachronisms for the 3rd millennium (Desset 2017), we still use them for convenience sake to designate the land and culture of southwestern Iran.

2 This is a personal observation made at the Mary Regional History Museum on the artifact itself. It is made of faience (i.e., a siliceous paste) and clearly shows traces of the molding seams.

3 This is reminiscent of the clay figurines of this period, with their bottom part made into a triangle bent forward (see Chapter 6).

4 This statement depends on the long-disputed date of the destruction of Palace G at Ebla, under Sargon or Naram-Sin. If the latter is responsible for it, then the Ebla and Oxus statuettes are more or less contemporary.

REFERENCES

Amiet, P. 1986. *L'âge des échanges inter-iraniens, 3500–1700 avant J.-C.* Paris: Réunion des Musées Nationaux.

Aruz, J. and R. Wallenfels (eds.) 2003. *Art of the First Cities: The Third Millennium BC from the Mediterranean to the Indus.* New York: Metropolitan Museum of Art.

Azarpay, G. 1991. "The Snake-Man in the Art of Bronze Age Bactria." *Bulletin of the Asia Institute* 5: 1–10.

Benoit, A. 2007. "Les bijoux et parures de la Civilisation de l'Oxus." In *Sulla Via delle Oasi: Tesori dell'Oriente Antico*, ed. G. Ligabue and G. Rossi Osmida. Padova: Il Punto, pp. 128–153.

Benoit, A. 2010. *Princesse de Bactriane.* Collection Solo, Département des Antiquités Orientales. Paris: Musée du Louvre éditions/Somogy éditions d'Art.

Bouquillon, A., A. Caubet, A. Kaczmarczyk, and V. Matoian 2007. *Faïences et matières vitreuses de l'Orient ancien. Étude physico-chimique et catalogue des œuvres du département des Antiquités orientales.* Paris: Musée du Louvre/Editions Snoek.

Brentjes, B. 1971. "Ein Elamitischer Streufund aus Soch, Fergana (Usbekistan)." *Iran* 9: 55.

Butterlin, P. 2014. "Les vases en chlorite du temple d'Ishar et le 'système monde' sumérien." In *Voués à Ishtar. Syrie, janvier 1934, André Parrot découvre Mari*, ed. S. Cuzan and P. Butterlin. Guides archéologiques de l'Institut Français du Proche-Orient no. 11. Paris: IFPO-IMA, pp. 175–188.

Casanova, M. 1991. *La vaisselle d'albâtre de Mésopotamie, d'Iran et d'Asie centrale aux III^è et II^è millénaires avant J.C.* Mémoires de la Mission Archéologique Française en Asie. Centrale, V. Paris: ADPF-ERC.

Casanova, M. 2014. "Luxuries of Precious Materials: The Royal Cemetery of Ur (Iraq) and the Lapis Lazuli, Witnesses of Intercultural Relations in the Near East." In *Les produits de luxe au Proche-Orient ancien aux âges du bronze et du fer*, ed. M. Casanova and M. Feldman. Paris: Travaux de la maison de l'archéologie et de l'ethnologie – René Ginouvès no. 19, pp. 31–44.

Caubet, A. 2016. "Gonur and the Ivory Road." In *Transactions of Margiana Archaeological Expedition*, vol. 6, ed. N. Dubova. Moscow: Staryj Sad, pp. 356–360.

Caubet, A. 2018a. *Idoli: Il Potere dell Imagine.* Milan: Skira, Venise: Fondazione Ligabue.

Caubet, A., 2018b. "Le palmier dattier dans l'Antiquité, de la Mésopotamie à l'Indus." In *L'arbre en Asie*, ed. J.-S. Filiozat. Paris: Académie des Inscriptions et Belles Lettres, pp. 39–60.

Cotty, M. 2010. "Les vases en chlorite de Tarut." In *Routes d'Arabie. Archéologie et histoire du royaume d'Arabie Saoudite*, ed. A. al-Ghabban, B. André-Salvini, F. Demange, M. Cotty, and C. Juvin. Paris: Somogy éditions d'Art/Louvre éditions, pp. 184–185, nos. 40–67.

David, H. 1996. "Styles and Evolution: Soft Stone Vessels During the Bronze Age in the Oman Peninsula." *Proceedings of the Seminar for Arabian Studies* 26: 32.

Desset, F. 2017. "Here Ends the History of 'Elam': Toponomy, Linguistics and Cultural Identity in Susa and SouthWestern Iran, ca. 2400–2800 BC." *Studia Mesopotamica* 4: 55–86.

Desset, F. 2018a. "Linear Elamite Writing." In *The Elamite World*, ed. J. Alvarez-Mon, G.-P. Basello, and Y. Wicks. Routledge World Series. London and New York: Taylor & Francis, pp. 397–415

Desset, F. 2018b. "Nine Linear Elamite Texts Inscribed on Silver 'Gunagi' Vessels (X, Y, Z, F', H', I', J', K' and L'): New Data on Linear Elamite Writing and the History of the Sukkalmah Dynasty." *Iran* 56(2): 105–143.

Dolce, R. 2014. "Equids as Luxury Gifts at the Centre of Interregional Economic Dynamics in the Archaic Urban Cultures in the Ancient Near East." *Syria* 91: 55–75.

Dunn-Vaturi, A.-E. and U. Schädler 2016. "Nouvelles perspectives sur les jeux à la lumière des plateaux du Kerman." *Iranica Antiqua* XLI: 1–29.

Durand, J.-M., T. Römer, and M. Langlois (eds.) 2010. *Le jeune héros: Recherches sur la formation et la diffusion d'un thème littéraire au Proche-Orient ancien.* Fribourg: Vandenhoeck & Ruprecht/Academic Press.

Francfort, H.-P. 1982. "Dungeons and Dragons: Reflections on the System of Iconography in Prehistoric Bactria and Margiana." In *South Asian Archaeology Studies*, ed. G. Possehl. New Dehli: Oxford/IBH Publishing Co., pp. 179–208.

Francfort, H.-P. 1998. "Les sceaux de l'Oxus: diversité de formes et variabilité de fonctions." *Ancient Civilizations from Scythia to Siberia* 5(1): 59–71.

Francfort, H.-P. 2003. "La civilisation de l'Asie Centrale à l'âge du Bronze et à l'âge du fer." In *De l'Indus à l'Oxus. Archéologie de l'Asie Centrale*, ed. O. Bopearachchi, C. Landes, and C. Sachs. Lattes: IMAGO, pp. 29–34.

Francfort, H.-P. 2005. "Note on the 'Acrobat and Bull' Motive in Central Asia." In *Tsentral'naja Azija. Istochniki, Istorija, Kul'tura* [Central Asia: sources, history, culture], ed. E.V. Antonova and K. Mkrtychev. Moscow: Vostochnaja Literatura, pp. 711–716.

Francfort, H.-P. and X. Tremblay 2010. "Marhaši et la civilization de l'Oxus." *Iranica Antiqua* 45: 51–224.

Frenez, D. 2017. "Manufacturing and Trade of Asian Elephant Ivory in Bronze Age Middle Asia: Evidence from Gonur Depe (Margiana, Turkmenistan)." *Archaeological Research in Asia* 15: 13–33. http://dx.doi.org/10.1016/j.ara.2017.08.002.

Gernez, G. 2017. *Les armes du Proche-Orient ancien: Des origines à 2000 avant J.-C.* Collection des Hespérides. Paris: Editions Errance.

Ghirshman, R. 1961. *Sept mille ans d'art en Iran.* Paris: Petit Palais.

Ghirshman, R. 1963. "Statuettes archaïques du Fars (Iran)." *Artibus Asiae* 26: 151–160.

Goodnick-Westenholz, J. (ed.) 2004. *Dragons, Monsters and Fabulous Beasts.* Jerusalem: Bible Land Museum.

Graff, S.B. 2012. "Humbaba/Huwawa." PhD thesis, Institute of Fine Arts, New York University, New York.

Hakemi, A. 1997. *Shahdad: Archaeological Excavations of a Bronze Age Center in Iran.* Reports and Memoirs vol. XXVII. Rome: IsMEO.

Harper, P., J. Aruz, and F. Tallon (eds.) 1992. *The Royal City of Susa: Ancient Near Eastern Treasures in the Louvre.* New York: Metropolitan Museum of Art.

Helwing, B. (ed.) 2017. *Iran: Frühe Kulturen zwischen Wasser und Wüste.* Bonn: Hirmer.

Herrmann, G. 1968. "Lapis Lazuli: The Early Phases of Its Trade." *Iraq* XXX: 21–57.

Hiebert, F. and P. Cambon (eds.) 2011. *Afghanistan: Crossroads from Ancient World.* Washington, DC: National Geographic Magazine.

Jarrige, J.-F. éd. 1988. *Cités oubliées de l'Indus: Archéologie du Pakistan.* Paris: Musée Guimet.

Kohl, P. 2001. "Reflections on the Production of Chlorite at Tepe Yahya: 25 Years Later." In *Excavations at Tepe Yahya, Iran. 1967–1975: The Third Millennium*, ed. C.C. Lamberg-Karlovsky and D. Potts. Cambridge. MA: Peabody Museum of Archaeology and Ethnology, Harvard University, pp. 209–230.

Lamberg-Karlovsky, C.C. 1988. "The 'Intercultural Style' Carved Vessels." *Iranica Antiqua* 23: 45–95.

Lawergren, B. 2003. "Oxus Trumpets, ca. 2200–1800 BCE: Material Overview, Usage, Societal Role and Catalog." *Iranica Antiqua* 38: 41–78.

Lawergren, B. 2018. "I suoni dell'Antichità: l'arpa attraverso il tempo e lo spazio." In *Idoli. Il Potere dell Imagine*, ed. A. Caubet. Milan: Skira/Venise: Fondazione Ligabue, pp. 53–57.

Ligabue, G. and G. Rossi Osmida (eds.) 2008. *Animali e Mito nel Vicino Oriente.* Padova: Il Punto.

Luneau, E. 2014. *La fin de la civilisation de l'Oxus: Transformations et recompositions des sociétés de l'âge du Bronze final en Asie centrale méridionale.* Paris: de Boccard.

Luneau, E. and R. Shirazi 2019. "Figurines féminines nues protohistoriques d'Iran oriental et d'Asie centrale: typologie, usages et symbolique." In *Figurines féminines nues: Proche-Orient, Egypte, Nubie, Méditerranée orientale, Asie centrale (VIIIe millénaire av. J.-C–IVe s. ap. J.-C.)*, ed. S. Donat, R. Hunzker-Rodewald, and I. Weygand. Paris: de Boccard, pp. 151–173.

Lyonnet, B. 2005. "Another Possible Interpretation of the Bactro-Margiana Culture (BMAC) of Central Asia: The Tin Trade." In *South Asian Archaeology 2001*, ed. C. Jarrige and V. Lefèvre. Paris: ERC, pp. 191–200.

Masson, V.M 1981. *Altyn-Depe*. Trudy JuTAKE, t. XVIII. Leningrad: Nauka.

Matthiae, P. 2009. "The Standard of the *Maliktum* of Ebla in the Royal Archives Period." *Zeitschrift für Assyriologie* 99 III: 270–311.

Meadow, R. 2002. "The Chronological and Cultural Significance of a Steatite Wig from Harappa." *Iranica Antiqua* 36: 191–202.

Michel-Dansac, F. and A. Caubet 2013. "L'iconographie et le symbolisme du palmier dattier dans l'Antiquité (Proche-Orient, Égypte, Méditerranée orientale)." *Revue d'ethnoécologie* 4: n.p. https://doi.org/10.4000/ethnoecologie.1275.

Miroschedji, P. 1981. "Le dieu élamite au serpent et aux eaux jaillissantes." *Iranica Antiqua* 16: 1–23.

Ornan, T. 2010. "Humbaba, the Bull of Heaven and the Contribution of Images to the Reconstruction of the Gilgameš Epic." In *Gilgamesch Bilder eines Helden: Ikonographie und Überlieferung von Motiven im Gilgameš-Epos*, ed. H.-U. Steymans. Orbis Biblicus et Orientalis 245. Fribourg/Göttingen: Academic Press/Vandenhoek & Ruprecht, pp. 229–260 and 411–424.

Parpola, A. 2011. "The Harappan Unicorn in Eurasian and South Asian perspectives." In *Linguistics, Archaeology and the Human Past*, ed. T. Osada and H. Endo. Occasional Paper 12, Indus Project Research. Kyoto: Institute for Humanity and Nature, pp. 125–188.

Perrot, J. and Y. Madjidzadeh 2005. "L'iconographie des vases et objets en chlorite de Jiroft (Iran)." *Paléorient* 31(2): 123–152.

Perrot, J. and Y. Madjidzadeh 2006. "À travers l'ornementation des vases et objets en chlorite de Jiroft." *Paléorient* 32(1): 99–112.

Pittman, H. 1984. *Art of the Bronze Age: Southeastern Iran, Western Central Asia and the Indus Valley*. New York: Metropolitan Museum of Art.

Possehl, G. (ed.) 1982. *Harappan Civilization: A Recent Perspective*. New Delhi: Oxford/IBH Publishing Co.

Pottier, M.-H.1984. *Matériel funéraire de la Bactriane Méridionale de l'Age du Bronze*. Mémoire no. 36. Paris: Editions Recherche sur les Civilisations.

Potts, D. 2010. "L'Arabie du nord-est." In *Routes d'Arabie. Archéologie et histoire du royaume d'Arabie Saoudite*, ed. A. al-Ghabban, B. André-Salvini, F. Demange, M. Cotty, and C. Juvin. Paris: Somogy éditions d'Art/Louvre éditions, pp. 173–183.

Rossi Osmida, G. 2007. *Adji Kui Oasis*. Vol. I, *La Citadella delle Statuette*. Padova: Edizioni Il Punto.

Sarianidi, V.I. 1998. *Margiana and Protozoroastrism*. Athens: Kapon Editions.

Sarianidi, V.I. 2005. *Gonurdepe Türkmenistan, City of Kings and Gods*. Ashgabat: Miras.

Sarianidi, V.I. 2007. *Necropolis of Gonur*. Athens: Kapon Editions.

Vidale, M. 2017. *Treasures from the Oxus: The Art and Civilization of Central Asia*. London and New York: Tauris.

Winkelmann, S. 1994. "Intercultural Relations Between Iran, Central Asia and Northwestern India in the Light of Squatting Stone Sculptures from Mohenjo-Daro." In *South Asian Archaeology*, ed. A. Parpola and P. Koskikallio. Helsinki: Suomalainen Tiedeakatemia, pp. 815–831.

Winkelmann, S. 2008. "Animali e miti nel Vicino Oriente Antico: un'analisi attraverso i sigilli." In *Animali e Mito nel Vicino Oriente Antico*, ed. G. Ligabue and G. Rossi Osmida. Venice: Il Punto, pp. 39–61.
Winkelmann, S. 2016. "Some Thoughts About the Wild Haired Snake-Man on BMAC Seals." In *Trudy Margianskoj Arkheologicheskoj Ekspeditsii*, vol. 6, ed. N.A. Dubova. Moscow: Staryj Sad, pp. 482–491.
Zarins, J. 1978. "Steatite Vessels in the Riyadh Museum." *Atlal* 2: 65–93.

CHAPTER EIGHT

BMAC GLYPTICS
Typology, context, function, and background

Sylvia Winkelmann *

INTRODUCTION

The Bactria-Margiana Archaeological Complex (BMAC) was a highly developed urban culture with large cities, huge cemeteries, and a fascinating material culture stretching over present-day southern Turkmenistan, southern Uzbekistan, and Tajikistan, as well as over northern Afghanistan. This culture was based on developed irrigation agriculture (but see Chapter 1, pp. 18–20), specialized handicrafts, mining, and treatment of ores and stones including semiprecious and precious ones. Long-distance trade that connected it with surrounding contemporaneous cultures, like the Harappan culture in the Indus valley up to Baluchistan, the partially older Jiroft culture in Kerman province of southeastern Iran (Majidzadeh 2003, 2008; Piran 2012), and the Elamite culture in Susiana and Fars (Anshan) (Amiet 1986).

These cultures did not only coexist, they also formed a separate and indigenous cultural *koinè* that existed simultaneously with, but independently of, the Mesopotamian one in the late Early Dynastic, Akkadian, Neo-Sumerian, and Old Babylonian periods in the ancient Near East. This *koinè* is characterized by typical artifacts made of stone and metal that appear in similar typological forms and with a specific iconography underlined by a female pantheon and characteristic representations of hybrid/mixed creatures associating anthropomorphic and animal features.

Among all the artifacts from this *koinè*, the glyptic illustrates most strikingly an independent way of development over a long period of time, with a characteristic iconography culminating at the time of the Oxus Civilization.

This chapter is an attempt to present this glyptic (types, styles, iconography) on the basis of the 545 seals excavated in Margiana up to 2014. We will also refer to seals from other BMAC sites or coming from illegal excavations when they have been published and if they give information not yet available from regular excavations.

GENERAL CONSIDERATIONS ON SEALS

Seals are unique objects made to create an impression on a soft surface (mainly clay). They were used in four different ways: (1) in administrative processes, to seal containers, boxes, pottery vessels, or doors, the surface of pottery, or bullae to represent ownership; (2) in legal processes, to seal the envelopes of cuneiform letters or for signing contracts, representing the owner; (3) as a kind of ID card, allowing the owner to be identified: normally, he/she was a natural person, but from Mesopotamia we also know seals representing a god, temple, community, or city; (4) finally, they served to protect the owner, warding off negative influences like illness, disaster, or malediction: Mesopotamian seal owners believed that their protection emanated from the depicted divinities, the carved inscriptions, or from the stone itself.

Ancient Near Eastern seals consist of two main types, the stamp seal and the cylinder seal. The stamp seal produces a restricted picture (a stamp) bounded by the seal surface. Most of the stamp seals have only one face and a handle on the back. They represent the oldest form of seals, known since the Neolithic period.

The cylinder seal is engraved on the round surface of the cylinder, producing an unlimited sequence of identical images when it is rolled over a surface. Normally, cylinder seals are longitudinally perforated. Cylinder seals appear around the middle of the 4th millennium in Mesopotamia. The cylinder seal can be combined with a stamp seal: The lower flat side is engraved like a stamp seal while the upper side has a handle. This specific form is called a stamp cylinder and was used, though rarely, in Mesopotamia[1] and Susa at the end of the 4th and beginning of the 3rd millennium. It is also known in the Hittite world.

Both types, the stamp and cylinder seals, have an engraved design carved in intaglio, that means cut as a negative into the surface producing a positive relief when the seals are pressed into or rolled over a surface.

Seals can be produced from nearly all kinds of material: stone, precious and semi-precious stone, clay, wood, faience, shell, and metal. While the earliest seals consisted of soft material like unbaked clay, terracotta, wood, or soft stones like alabaster, chlorite, or steatite, Near Eastern seals of the 3rd and 2nd millennium were mostly made of harder stones.

In the ancient Near East, cylinder seals largely replaced stamp seals during the 3rd millennium.

A NEW TYPE OF SEAL ON THE IRANIAN PLATEAU: THE COMPARTMENTED SEALS

While stamp seals were an international phenomenon, cylinder seals are a younger variety that developed in Mesopotamia and spread from there over the Iranian Plateau up to Central Asia and the Indus valley. On the Iranian Plateau, cylinder seals were found mainly in the Elamite area, a contact zone with Mesopotamia. They are also attested, in smaller numbers, into the Jiroft culture and at Shahdad in Kerman (Porada 1962: fig. 12; 1993: fig. 31; Hakemi 1972: no. 323; 1997b: Ib.1, 2, 3, fig. on p. 355, Xg; Lamberg-Karlovsky 1973: pl. XXVIc; Potts 1981: fig. 1; Amiet 1986: fig. 132, 1–12; 1994: fig. 1 and 3; Sarianidi 1998a: fig. 27.3; Winkelmann 1997a: fig. 1; Pittman 2008: figs. 29–32). Finally, they are known up to the Harappan culture in the Indus valley and to the BMAC culture of Central Asia.

However, on the Iranian Plateau, the most typical seals were of a totally different type, the compartmented seals unknown elsewhere. In the 3rd and 2nd millennium, compartmented seals are mainly made from metal, which is ideally suited to create this special type, but it is assumed that they were earlier made of stone (Shah and Parpola 1991: Mr-5, 7, 9).

Strictly speaking, they are affiliated with stamp seals since they also create a limited impression. But the manner of producing the motif and creating the impression is completely different from that of classical stamp seals. As mentioned above, stamp seals and cylinder seals both have their motifs engraved as negatives into the surface – creating a positive and three-dimensional image. On compartmented seals, however, the motif is created by small metal strips of the same height soldered onto a massive or open-work base. A handle is usually found on the back. The strips form the outlines of the motifs, dividing the surface of the seal into many small compartments. In contrast to stamp and cylinder seals, the impression created by compartmented seals presents only the outlines of the motif. It is this division of the seal surface into small compartments that led Stuart Piggott to name them compartmented seals (Piggott 1943: 179). As mentioned above, in the BMAC they are mostly made of metal, but some are also made of stone and faience.

The tradition of metal compartmented seals goes back to Chalcolithic traditions on the Iranian Plateau and Baluchistan. The oldest known copper compartmented seals have been found in Mehrgarh III levels in northern Baluchistan (Samzun 1987: 76, 77), suggesting that this type of metal seal was already in use by the end of the 5th or the beginning of the 4th millennium. Other evidence for early compartmented seals is known in the Kerman area: single seals at Tal-i Iblis IV (second half of the 4th millennium) (Caldwell 1967: 24, 186, fig. 38, 7) and at Tepe Yahya IVC (Proto-Elamite period) (Lamberg-Karlovsky and Tosi 1973: fig. 124). All these seals belong to a simple open-work type with a geometric motif.

At the beginning of the 3rd millennium, metal compartmented seals appear all over the Iranian Plateau: in northern and southern Baluchistan (Mehrgarh, Nal, Bampur), in the Kandahar and Seistan areas (Deh Morasi Ghundai, Mundigak, Gardan Reg 6, Shahr-i Sokhta), in Kerman (Shahdad, Jiroft), etc. At that time, the shapes and motifs become more diversified. Beside the old open-work type, new compartmented seals with a massive base and a soldered handle start to appear. The new seals often have a frame surrounding the motif, usually simple and geometric, but figurative seals without frame also make their first apparition. The typical motifs are animals like goats, bulls, and ducks (in Baluchistan, at Nal, see Hargreaves 1929: pl. 15d, f; and at Nausharo IC, see Shah and Parpola 1991: no. Ns-1; in southeast Iran at Bampur, see De Cardi 1970: fig. 47, *15*; Jarrige 1986: pl. 39b), but also many varieties of crosses and whirl motifs (Shah and Parpola 1991: Mr-10–13), as, for instance, the typical Proto-Elamite animal-head whirl or stepped pyramids.

During the first half of the 3rd millennium, the main production area of compartmented seals was located in Seistan and Kandahar regions where flourished the Helmand culture. From 2800 BC onwards (Shahr-i Sokhta II), figurative copper compartmented seals with zoomorphic, plant, and geometric designs appear. It is also at Shahr-i Sokhta (level II–7) that the first figurative compartmented seal with an anthropomorphic design has been found. This seal is particularly important since it is also the oldest representation of a winged horned female on a seal from the Iranian Plateau (i.e., of a female hybrid creature/"mixed being" or goddess) (Winkelmann 1996). It is also the direct predecessor of the later depictions of goddesses well known on BMAC seals. Copper compartmented seals also increase in number at Mundigak in the Kandahar area, in levels IV.1 to IV.3 around the middle of the 3rd millennium (Casal 1961: pl. XLVb).[2] Compartmented seals are also found at Susa, Konar Sandal – the main site of the Kerman culture – Tepe Yahya, and Shahdad where cemetery A (dated to the Early Dynastic III period) produced a great number of compartmented seals and seal impressions showing a whole range of motifs.[3]

It is possible that this development led to the appearance of similar seals in northeastern Iran – at Tepe Hissar IIIB and IIIC – in Anatolia (Salvatori 1981: 123–126, 159–168 pl. I–X),[4] as well as in southern Turkmenistan. There, a rising number of compartmented seals is attested in cities like Anau, Namazga Depe (Masson 1959: pl. XII), and especially Altyn Depe during the Namazga (NMG) V period. They show a broad range of motifs, from the traditional cross and stepped pyramid motifs (issued from pottery painted ornaments, see Chapter 6) to animals, plants, and humans (Masson 1981: pl. XVI, XVII), as well as the first appearance of "mixed creatures" (Masson 1981: pl. XXII, 1a, l b, three-headed feline). Both the seals on the Iranian Plateau and those in the piedmont area of the Kopet Dagh in southern Turkmenistan, seem to have culminated in the BMAC, which starts to develop from 2300 BC onwards. During the BMAC, the production of compartmented seals reached its peak and they constitute the main type of seals for several hundreds of years in Bactria and Margiana.

THE BMAC GLYPTIC: GENERAL REMARKS

BMAC seals started to be known only after the discovery of the first Bronze Age remains in northern Afghanistan in the late 1960s and 1970s. From the later Russian excavations in southern Uzbekistan and southern Turkmenistan three facts emerge: (1) the seals clearly belong to a newly discovered urban culture; (2) the seals belong to the large "stamp seal-province" of the Iranian Plateau; (3) their very large number makes them the best conveyors of myths, rituals, gods, environment, flora, fauna, and a good provider of the social status of their owners in the absence of writing in the BMAC culture.

Seals are found on all the sites and in all the cemeteries of this culture, albeit in different quantities (Table 8.1). Important sites are Dashly Tepe 1 and 3 in Afghanistan, Sapallitepa, and Dzharkutan in southern Uzbekistan, but the greatest number of excavated seals has been found in ancient Margiana in southern Turkmenistan. It was V.M. Masson and V.I. Sarianidi who discovered the first examples of seals during their surveys in the Kara Kum Desert in the 1950s, without being able to yet understand their character, however (Masson 1959; Sarianidi 1965, 1966; compilation of seals: Masimov and Salvatori 2008). But with the new surveys made in the 1970s by Sarianidi (1976, 1981), I.S. Masimov (1979, 1981), and B. Udeumuradov (1993), and with the beginning of excavations at the newly discovered sites in the Taip, Togolok, Adji Kui, and Kelleli oases of Margiana, Sarianidi was able to verify the affiliation of these seals to the BMAC.

Most of the Margiana seals come from the 40 years of excavations made at Gonur Depe, the major site of the BMAC in modern Turkmenistan. The second area of the Murghab delta that provided many seals is the Togolok oasis, mainly settled during the younger phase of the BMAC (i.e., Gonur periods II and III, see Table 8.1).[5] In this oasis, the seals come from excavations at Togolok 1 (16 ex.), Togolok 2 (1 ex.),

Table 8.1 Published BMAC seals from different sites, except Gonur

Site	*Publication*	*Figure or seal number*
Dashly 1	Sarianidi 1976	Terracotta seal: fig. 21, *5*; stone stamp seals: fig. 28, *1–2*; compartmented seals: figs. 17.4, 21.6
	Sarianidi 1984	Compartmented seal: fig. 27.2
Dashly 3	Sarianidi 1977a	Fig. 54
	Sarianidi 1976	Figs. 51.1, 51.3, 51.4, 51.5
Sapalli-T.	Askarov 1977	Pl. XLIV, XLV, XLVI
Dzharkutan	Askarov and Abdullaev 1983	Pl. XXXII, *1*, *9*; XXXIV, XXXV, XLV, *1*
Dzharkutan	Askarov and Shirinov 1993	Fig. 24.II, *11–12*; 50, *1–2*; 63, *5*, *13*; 64, *7*, *11*
Togolok 1	Sarianidi 1998b	Surface: 1627, 1636, 1676, 1684, 1688, 1691, 1708, 1716, 1727, 1767. Rooms: 1666: room 17; 1667: room 26; 1677: room 99; 1682: room 42; 1738: room 68. Graves: 1620: burial 9; 1629: burial 10; 1635: burial 10; 1765: burial 10. Upper level: 1683, 173
Togolok 21	Sarianidi 1998b	Surface: 1554, 1621, 1630, 1632, 1641, 1664, 1687, 1695, 1699, 1706, 1719. Rooms: 1570: room 218; 1571: room 218; 1622: room 17; 1651: room 186; 1686: room 168; 1690: room 175; 1697: room 61; 1705: room 132; 1729: room 175; 1747: room 116; 1755: room 133; 1769: room 534. Temple: 1634, 1644, 1645, 1653, 1675, 1731
Togolok 24	Sarianidi 1998b	Graves: 1649: burial 25; 1696: burial 10; 1707: burial 7
Togolok 2	Sarianidi 1998b	1711: room 139
Togolok 3	Sarianidi 1998b	1642: surface
Togolok 7	Sarianidi 1998b	1652: surface
Togolok 9	Sarianidi 1998b	1689: surface
Togolok 11	Sarianidi 1998b	1780: surface
Togolok 13	Sarianidi 1998b	1640: surface
Togolok 15	Sarianidi 1998b	1757: surface
Togolok 29	Sarianidi 1998b	1647: surface
Togolok 30	Sarianidi 1998b	1567: surface
Auchin 1	Sarianidi 1998b	1658: no information; 1670: burial 4
Kelleli 4	Sarianidi 1998b	1557: room 7

(continued)

Table 8.1 Cont.

Site	*Publication*	*Figure or seal number*
Taip-D. 1	Sarianidi 1998b	1770–1774, 1776
Kara Kum survey, without precise location	Sarianidi 1998b	1555, 1558–1559, 1563, 1569, 1575, 1582, 1602, 1628, 1655, 1673, 1678, 1681, 1694, 1714–1715, 1722–1724, 1728, 1781
Adji Kui 1	Rossi Osmida 2011: 227–231	Nos: S 02: room S1; 05: room 129; 10: room 228; 11: room 238; 14: room 280; 15: room 45; 16: room 72; 17: room 165; 18: room 190; 19: room 5; 20: room 6; 21: room 19; 22: room 47; 24: room 152 NM: 1808/62–68, 1809/127/138–139, 1811/110–112/153/155, 1812/45/51/53/57, 1869/7, 1870/2/4
Adji Kui 9	Rossi Osmida 2011	S 13: room 2; S 04: room 66; S 23: room 152; S06: room 159; S 25: room 160; S 26: room 166; S 08 room 198; S 09: room 203; S 03: room 248; S12: room 277
Sibri	Santoni 1984 Jarrige 1985	Fig. 8.4 Fig. 6. = Shah and Parpola 1991: Sb1, 2
Harappa	Joshi and Parpola 1987 Vats 1940: vol. II Shah and Parpola 1991	Nos. H-154, H-166, H-162, H-349 Pl. XCI H-630, H-631, H-1018
Mohenjo-daro	Joshi and Parpola 1987 Shah and Parpola 1991	M-435 Nos. M-1238, M-1239, M-1243–1249, M-1251–1254, M-1255, M-1390
Lothal	Joshi and Parpola 1987	L-173–175
Chanhu-daro	Joshi and Parpola 1987	C-42, C-49, C-50
Mehi	Joshi and Parpola 1987	Mehi-1
Shahi-Tump	Joshi and Parpola 1987	Sht-1
Luristan	Amiet 1988	Fig. 18d = Sarianidi 1998b: no. 995
Susa	Amiet 1986	Fig. 105
Persian Gulf	Crawford and al Sind 1995	Figs. 1–2

Note: NM = National Museum.

Togolok 21 (33 ex.), and Togolok 24 (5 ex.). Other sites of the oasis only produced one seal found on the surface, like Togolok 3, 7, 9, 11, 13, 15, 29, and 30. These occurrences reflect the different sizes of the sites but also that only the larger sites have been excavated.

A total of 52 seals and many impressions come from rooms and graves in a third area of the Murghab delta, the Adji Kui oasis. Some 40 seals have been found in Adji Kui 1, the largest site in the oasis, excavated by G. Rossi Osmida (2011: 226–237 and personal communication).[6] The Adji Kui 1 seals consist of 27 compartmented seals made of "bronze", faience, and stone, and 14 classical stamp seals made of metal, stone, and unbaked clay. To them must be added seal impressions on pottery, and 26 cretulae, among them one cretula with the impression of a compartmented seal covering the mouth of a large vessel.[7] Adji Kui 9 produced 11 seals, among which ten are compartmented seals, one is a simple round stamp seal in the "drilled style" with a long handle. From Adji Kui 9 also come 18 cretulae closing the opening of containers and bags.

Other sites in Margiana that provided seals in far smaller number are Auchin 1 (two seals), Kelleli 4 (one seal), and Taip 1 (six seals from the surface, see Table 8.1). Taip Depe is especially interesting in so far as all the six seals are cylinder seals, which thus make this site the place where the largest amount of cylinder seals has been found up to now. Unfortunately, it has not been further excavated. Finally, 21 seals of different types are surface finds from surveys in the Kara Kum (see Table 8.1).[8]

All the seals from Margiana are today stored at the National Museum of Turkmenistan (NM) and the State Museum of Fine Art (MFA) in Ashgabat, as well as in the Mary Regional History Museum (MM). These seals are so far not scientifically published, but the author was able to study them thanks to the kind allowance of the Ministry of Culture of Turkmenistan and of Victor Sarianidi and Nadezhda Dubova, directors of the Margiana archaeological expedition.

Furthermore, some BMAC seals come from excavated sites outside Margiana and Bactria proper (see Table 8.1). Two stamp cylinders and one "Murghab-style" (see p. 245) stamp seal were found in a cenotaph at Sibri (Pakistan) together with BMAC material in Mehrgarh VIII levels dated between 2000 and 1800 BC. Other seals concern compartmented seals and their impressions, and three "Murghab-style" stamp seals. We will return to them later in the chapter, but we can say here that they were found in the following places: Harappan cities (Harappa, Mohenjo-daro, Chanhu-daro, and Lothal); Mehi, a site of the Kulli culture in eastern Baluchistan, for one "drilled-style" stamp seal; Shahi Tump in western Baluchistan for one compartmented seal; Luristan for one "Murghab-style" stamp seal; Susa for two compartmented seals; and finally Bahrain/Dilmun in the Persian Gulf for one "Murghab-style" stamp seal.

At Gonur Depe, the majority of the seals comes from graves. Most were found in the Large Necropolis, excavated since 2002 by the Margiana archaeological expedition in cooperation with the Ligabue Research Center (see Salvatori 1993, 1994a, 1994b, 1995, 2000; Rossi Osmida 2002). This necropolis, where 2,944 graves have been excavated, was mainly used during Gonur periods I and II. Some 123 graves contained seals (126 pieces in total) (Table 8.2). Of these graves 13 belong to males (13 graves with 16 seals), 77 to females, while for 33 graves the sex could not be determined (cenotaphs, double graves male/female, or unclear sex). The 126 seals consist of 121 compartmented or simple stamp seals and three cylinder seals made

Table 8.2 Gonur Depe Large Necropolis: distribution of the 126 seals discovered between male, female, or unknown sex graves (from Sarianidi 2007)

Male grave number	*Female grave number*	*Unknown sex, grave number*
308, 351, 397 (two seals), 555 (two seals), 945, 1446, 2029 (two seals), 2038, 2043, 2184, 2300, 2328, 2550	75 (two seals), 114, 165, 178, 237, 333, 358, 393, 406, 425, 560, 715, 727, 735, 740, 791, 805, 812, 821, 823, 837, 855, 856, 869, 921, 939, 1033, 1063, 1091, 1156, 1157, 1227, 1270, 1296, 1299, 1343, 1393, 1409, 1437, 1442, 1465, 1495, 1496, 1532, 1581, 1615, 1643, 1645, 1653, 1679, 1697, 1699, 1710, 1770, 1783, 1770, 1799, 1824, 1889, 1965, 2013, 2050, 2219, 2228, 2280, 2311, 2357, 2373, 2412, 2450, 2490, 2705, 2790, 2804, 2827	23, 55, 94, 164, 216, 238, 262, 318, 324, 375, 527, 570, 717, 788, 971, 1080, 1106, 1207, 1550, 1597, 1725, 1774, 2202, 2337, 2344, 2353, 2359, 2410, 2474, 2528, 2598, 2655, 2809

of different materials. Some 74 seals are in "bronze", 14 in silver, 4 in steatite, 12 in faience, 2 in lead, 19 in gypsum, and 2 in stone (Table 8.3, compiled after Sarianidi 2007: 202–289).

Two bronze compartmented seals have been found in the later graves dug into the abandoned palace of Gonur North (graves 7 and 15; Sarianidi 2007: 290). Smaller areas with graves at Gonur North as well as at Gonur South also produced seals. Up to now, the oldest datable seals have been found in the small "Royal Necropolis" related to the earliest occupation period at Gonur North (Sarianidi 2009: 204, 206, fig. 105, grave 3220: 1 seal; grave 3230: 2 seals on the shoulder of women) (for the "Royal Necropolis," see Chapter 10).

Beside graves, seals have been discovered also in rooms, temples, and in cult areas (Sarianidi 2009: 142 and 129 fig. 33).

Sarianidi (2007: 99) first noticed the interesting fact that seals from the Large Necropolis come mostly from women's graves (60.6% against 12.8% in men's, the rest being undetermined). This percentage is even higher in favor of women when one considers only the graves with a sure sex identification. Seals were mainly found at the waist of the deceased, but also at the neck, wrist, head, and shoulder. During the author's excavations in Area 12,[9] it was found out that seals were sometimes stored under the floor of newly built houses, maybe as a founding ritual. A further revision of the earlier excavated seal contexts within the settled areas is necessary to confirm if this was a general rule or not.

Beside seals coming from scientific excavations, an increasing number of seals come from illicit diggings. It was again Sarianidi who, over many decades, compiled more than 1,500 seals and sealings from the art trade and published them together with

Table 8.3 Gonur Depe Large Necropolis: type and material of the seals (from Sarianidi 2007)

Type of seal	*Grave number*	*Material*
Cylinder seal	2550? (Akkadian)	Stone (dolomite)
Cylinder seal	0023? (southeast Iranian with vegetation goddess)	White stone
Cylinder seal	1393 f (southeast Iranian with complex composition)	Light-colored stone
Compartmented or stamp seals	55?; 75.1 and 75.2 f; 94?; 94?; 165 f; 216?; 237 f; 308 m; 318 –; 324?; 351 m; 358 f; 375?; 393 f; 406 f; 425 f; 560 f; 727 f; 735 f; 740 f; 788 –; 791 f; 805 f; 823 f; 837 f; 855 f; 856 f; 869 f; 921 f; 939 f; 971?; 1063 f; 1080?; 1091 f; 1106? 1156 f; 1157 f; 1207?; 1227 f; 1296 f; 1299 f; 1409 f; 1437 f; 1442 f; 1446 m; 1465 f; 1495 f; 1496 f; 1532 f; 1550 double?; 1581 f; 1615 f; 1679 f; 1697 f; 1699 f; 1725 –; 1770 f; 1965 f; 2013 f; 2029 m; 2038 m; 2228 f; 2280 f; 2300 m; 2328 m; 2337?; 2344?; 2373 f; 2410?; 2474; 2528?; 2598?; 2705 f	Copper alloy
	164?; 262?; 555 m; 570?; 1033 f; 1710 f; 1799 f; 1824 f; 2029 m; 2311 f; 2412 f; 2655?; 2790 f.	Silver
	178 f; 2043 m; 1270 f; 1393 f.	Steatite
	1597?; 2184 m; 2202?; 2219 f; 2353 –; 2357 f; 2359?; 2450 f; 2490 f; 2804 f; 2809?; 2827 f	Faience
	238?; 1889 f	Lead
	114 f; 333 f; 397–392; 527?; 555 m; 715 f; 717 –; 812 f; 821 f; 945 m; 1343 f; 1643 f; 1645 f; 1653 f; 1774?; 1783 f; 2050 f	Gypsum

Note: ? = sex uncertain; f = female; m = male; – = cenotaph.

the 253 seals known then from excavations in his seminal work *Myths of Ancient Bactria and Margiana on Its Seals and Amulets* (Sarianidi 1998a). This book contains 1,802 seals and sealings.[10] At the same time, Susanne Baghestani (1997) presented in another book all the known compartmented seals from Iran, India, and China. Since then, I have been able to examine hundreds more seals in private collections. Thus, the overall number of seals related to the BMAC largely exceeds 2,000, making glyptic its most important type of artifact after pottery. My studies in the museums in Turkmenistan and elsewhere now allow a detailed classification of BMAC glyptic. As said in the introduction, this chapter is an attempt to give an overview of the 535 seals found in Margiana up to 2013.[11] Some additional seals were excavated in 2013 and 2014, making a total of 545 pieces. But references will also be made to seals from other sites, or from illegal excavations when published. The reader will find in the last table (Table 8.14) the correlation between the numbers of the seals presented in the figures, the museum's, and those given by Sarianidi (1998b).

THE BMAC TYPES OF SEALS: PRELIMINARY REMARKS

There are three major BMAC types of seals: stamp seals, compartmented seals, and cylinder seals. While the first two follow the traditional seal production of the Iranian Plateau and of the Namazga culture in the southern Turkmenistan Piedmont area, cylinder seals were introduced from Mesopotamia, through the Elamite and the southeastern Iranian Kerman cultures. However, cylinder seals developed into a new variety that is more frequent, the stamp cylinder.[12] Although the classical variety of longitudinally perforated cylindrical seals is attested, the most typical BMAC cylinder seals have an additional stamp seal on one of the flat sides and a loop handle on the opposite side. Nevertheless, compared with the other two types, cylinder seals and stamp cylinders are relatively rare.

All these different types of seals had various functions: as amulets, to denote rank, family, or household affiliation, as well as ownership. They have been found as impressions on bullae, vessels, and cretulae. As noted above, they may also have served as founding deposits. Furthermore, seals seem to have had a high value. This is not only suggested by the precious metals or stones some are made of, or by their elaborate designs, but also by the traces of repair and reuse some of them present, and the new perforations for suspension done after the loss of their original handle (Figures 8.2b, j; 8.4e, m) (perforated after loss of handle: MM 111, 1025, 1775, 3727, 4352; NM 224, 240, 1090a, e, 1274, 1808/78, 1866/261, 2287/15; face reworked: MM 2009).

Seals were locally produced as shown by the 2005 discovery at Gonur North Area 9, room 107, in a series of rooms reused for metallurgical activities, of a mold for casting a compartmented seal with a four-arms cross framed by a guilloche (Figure 8.18a). A second mold for a seal unfortunately has no provenience indications (Figure 8.18b).

We will consider next, successively, the stamp seals, the compartmented seals, the handles of these two types, the cylinder seals and stamp cylinders, and finally the seal impressions.

THE BMAC STAMP SEALS

There is a great variety of stamp seals, with different numbers of faces and in various styles, distinguishing them from the classical one-side stamp seals of the Mesopotamian tradition. Although there are numerous one-side stamp seals in the BMAC, the most characteristic stamp seal is two-sided and has multifigured motifs.

Stamp seals were made of different materials (Table 8.4). Craftsmen used various soft stones like alabaster, chlorite, or steatite; marble; white stone; or gypsum. Sometimes they were made of a violet or a dark red stone, or lapis lazuli. Some stamp seals were made of faience and bear remains of a greenish glaze. Craftsmen also used metals (mainly

Table 8.4 Margiana stamp seals: materials and store museums

Stamp seal material	*Museum number*
Chlorite or steatite	NM: 49, 225, 227, 229, 236–237, 241, 244, 247, 357, 1262/48, 1139, 1264–1265, 1815/238, 1866/134/142, 1868/39, 2286/49/69/122; MM: 575, 578, 580–581, 937, 999, 1022–1023, 1026–1032, 1035, 1037–1039, 1041–1044, 1202, 1425, 1625, 1775–1777, 1887, 1994, 1996–1999, 2002, 2004–2005, 2008, 2011–2013, 3725, 3727, 3730, 4074, 4168, 4325, 4352, 4362, 4378, 4410; MFA: 16/33, 19/42, 20/45, KEK 176, no. 43, KEK 153, no. 30, KEK-187, KEK-174, KEK-196
Marble	NM: 239, 243, 243/79, 1269; MM: 3635
White stone	NM: 1261, 1266, 1269, 1271, 2287/70; MM: 4379; MFA: 21/84.
Alabaster	NM: 245; MM: 2006
Gypsum	NM: 2286/45; MM: 4231, 5328
Faience (often with greenish glaze)[a]	NM: 1188–1889, 1292, 1815/ 2, 1866/116; MM: 1033, 1442, 1769
Violet stone	NM: 223, 228, 1282
Dark red stone	NM: 231, 2287/88, 2305/26; MM: 1036
Red limestone	MM: 1024, 1040, 2000–2001
White limestone	MM: 579, 938, 2009, 4411
Brownish-gray limestone	MM: 571, 2003; MFA: 18/38
"Bronze" (copper alloy)	MM: 585, 1002, 1664, 1766, 1993, 4239, 4368, 5346, 5282; MFA: 13/74
Terracotta or unbaked clay	Disc: MM: 2013, 2022; oval stamp seal: MM: 1440; round stamp seal: MM: 1009, 1645; NM: 1125/A3; cone seals: MM 615–616, 935, 1009/1–6, 1436–1437, 1439–1440, 1449, 1453, 1885, 2643, 4159, 5206, 5399; NM: 1125/a2/a3, 1270/1, 1275, 1813/250, 1815/43.1/43.2/69/110, 1866/122/123, 1867/5/68, 1886/150, 2286/34/60
Kaolin	MM: 1774
Jade	MM: 2010
Lapis lazuli	MM: 1995
Beige-reddish stone	MFA: 14/25

Note: MFA = Museum of Fine Art; MM = Mary Regional History Museum; NM = National Museum.

[a] Most faience seals are compartmented seals. Only a small number belongs to the type of stamp seals.

"bronze," and exceptionally gold), and, finally, they produced many stamp seals in terracotta or unbaked clay. This clay material mainly applies to conical seals. These clay seals, fired or not, mostly originate from later sites or later periods, like Gonur Depe South, Adam Basan, Taip Depe, Togolok, Adji Kui, and others. Finally, there are also some rare seals made of lapis lazuli, kaolin, or jade (found at Togolok 1 and Gonur).

Most stamp seals are relatively small, between 1 and 2 cm in length or diameter. Flat two-sided stamp seals are often only 2 mm high. Stamp seals with a handle can reach 1 cm or more in height.

The stamp seal motifs were produced in two main different styles (see pp. 243–247 for these styles and Table 8.11): the simple "drilled style" and the "Murghab-style." Only one stone stamp seal in the "evolved style" has been found during excavations. Other stamp seals are attested with roughly engraved, drilled, or scratched motifs, or combinations of the three techniques and constitute the "simple style."

Stamp seals appear in seven different types. The three main ones are: (1) one-sided stamp seals with a handle on the back (type 1, Figures 8.1a, 8.2a–c, 8.2f–h, 8.2j); (2) two-sided stamp seals either with a longitudinal perforation or, when cross-shaped, with perforations on one or two arms (type 2, Figures 8.1c–j; 8.2d–e; 8.2i, k; 8.5a–b, e–f; 8.6a, f); and (3) conical stamp seals with a solid perforated back (type 7, Figure 8.3). They all can be divided into subtypes according to the shape of the face. Conical ones can be divided into subtypes according to the shape of the back. For the handles, see p. 235.

Type 1: one-sided stamp seals with handle on the back (with nine subtypes)

Most of the type 1 stamp seals (Table 8.5) are circular (type 1.1), followed by seals in form of a rosette (type 1.7) or star (type 1.8). But other shapes are possible: oval (type 1.2), square (type.1.3), rectangular (1.4), stepped pyramidal (1.5), lozenge-like (1.6), or figured stamp seals (type 1.9) Figured ones can have a human shape, or represent only a human foot or hand, or an animal shape. Two seals are unusual but important: The first is a "Murghab-style" seal in the form of a woman giving birth (NM 1867/27), which later developed in the "drilled style" into the very often depicted "anchor" motif; the second is a "Murghab-style" stamp seal in the form of two crossing camels, divided in two registers on its stamp side: The upper one represents a cultic scene with a man jumping over a bull attacked by three eagles (Figure 8.7b), the lower row shows a squatting and a walking man. On its base, the seal has an engraved hunting scene where a man with a bow hunts goats in a mountain. This seal is very important as it points to relations to both west and east. While the hunting scene is stylistically very close to Akkadian seal depictions, bull leaping is known from the pre-Harappan cultures of Baluchistan (stamp seal from Mehrgarh VII: Urban 1987: A68), from the Harappan Civilization (Joshi and Parpola 1987: M-312, M-492; Shah and Parpola 1991: H-858, M-1430), as well as from the Kerman culture (Majidzadeh 2003: 65). Of course, this very specific rite is also known from the Minoan culture, so the overall data suggests common rituals in all these cultures.

A very small number of stamp seals have additional ornaments in the shape of inlayed drillings or incised motifs on the reverse side (MM 999). Sometimes, their

Table 8.5 Stamp seals type 1: one-sided with a handle on the back

Stamp-seal type	*Shape*	*Museum or publication number*[a]
Type 1.1	Circular	MM: 51, 233–234, 236–240, 244, 358, 571, 573, 585, 937, 999, 1002, 1009.5/6, 1021–1023, 1025, 1027, 1029, 1033/2, 1034, 1125a3, 1126a2, 1202, 1270, 1273, 1276–1280, 1424, 1441, 1645, 1769, 1774–1777, 1813/250, 1815/2/43.1, 2/69/110, 1866/122/123, 1867/5/68, 1868/150, 1990, 2001, 2009–2010, 2011/5, 2286/34/60, 2305/26, 3028/6, 3635, 3643, 3727, 3744, 4231, 4325, 4352, 4379, 4410–44111, 5228, 5282, 5399, 5752, 5769; NM: 234/70, 237, 238/74, 239–240, 243–245, 263, 1090a, 1189, 1266/82, 1269, 1277–1280, 1438, 1808/62/63/64, 1808/65/76/78, 1809/137, 1811/110/153, 1866/115/117/261/283, 1866/286, 1867/65, 1868/157, 2286/69, 2287/15/70/88, 2305/26; Sarianidi 1998b: 1238, 1241, 1243, 1245, 1247–1253, 1256–1264, 1266–1267, 1273, 1276–1283, 1302–1310, 1312–1319, 1321, 1344, 1815/2
Type 1.2	Oval	MM: 1027, 1440, 1625, 3643, 4410, 4074; NM: 1815/238, 1866/143; Sarianidi 1998b: 1246
Type 1.3	Square	MM: 575, 1028, 1031–1032, 2012; NM: 231–232, 241, 1815/238, 1866/142, 2286/49; Sarianidi 1998b: 1300, 1345–1350
Type 1.4	Rectangular	MM: 572, 1625, 2012, 2026, 3725; NM: 224; Sarianidi 1998b: 1299
Type 1.5	Stepped pyramidal	NM: 2286/17; Sarianidi 1998b: 914
Type 1.6	Lozenge	MM: 938
Type 1.7	Rosette	NM: 190/96, 233, 235/71, 1024, 1026, 1287, 1815/134, 1868/39, 1870/42/184, 2286/45/122; MM: 574, 1030, 4239; MFA: 21/84; Sarianidi 1998b: 1257, 1323, 1341, 1695, 1698, 1702, 1706, 1721
Type 1.8	Star	MM 45, 584, 1028; NM: 45, 1257/13, 1280, 2286; Sarianidi 1998b: 1234, 1237, 1265, 1268–1272, 1274–1275, 1284–1285, 1311, 1320, 1694, 1701
Type 1.9	Figured seals with figured base	Figurative stamp seal in form of a snake whirl from Gonur North, MM: 4239; in form of woman giving birth, NM: 1867/27; in form of the head of lion-man, NM: 357/1437; hand, Sarianidi 1998b: 1614; foot, MM 2018; in form of a camel, NM: 79 = Sarianidi 1998b: 1634; in form of a boar, Sarianidi 1998b: 1636; cross-shaped, Sarianidi 1998b: 1244

Note: MFA = Museum of Fine Art; MM = Mary Regional History Museum; NM = National Museum.

[a] The publication number (from Sarianidi 1998b) is only used for seals from outside Margiana.

rim is additionally embellished with small indentations, which produces a serrated or toothed rim (Figure 8.2a–b, f–h, j).

Type 2: two-sided stamp seals (with 13 subtypes)

Two-sided stamp seals (Table 8.6) are the most typical form of the "Murghab-style" and "drilled-style" seals (for these styles, see pp. 243–247). According to their shape

Table 8.6 Stamp seals type 2: two-sided from various areas of the BMAC

Type	*Section*	*Shape*	*Museum or publication number*[a]
Type 2.1	Flat	Circular	MM: 579–580, 1036–1038, 1999–2001, 2006, 2008–2009, 2112, 3730, 4362, 4378; NM: 49, 1262/48, 1264/50; Sarianidi 1998b: 895, 900, 902–903, 905, 907, 922, 924, 931–932, 942–943, 954–956, 962, 966? 971, 973, 977, 984–985, 991, 994, 1001, 1008, 1024–1025, 1035, 1037–1038, 1040–1043, 1045–1046, 1049, 1060–1067, 1075–1085, 1086–1088, 1096–1097, 1100, 1119–1122, 1125, 1128–1142, 1147–1163, 1168–1176, 1179–1181, 1183, 1191–1201, 1209–1214, 1218–23, 1226, 1232–1233, 1254, 1286, 1289–1298
Type 2.2	Flat	Square	MM: 1035, 2004; NM: 221, 222; Sarianidi 1998b: 889, 906, 909, 912, 916, 925, 929–930, 935, 949–950, 1005, 1019, 1032, 1345–1350, 1352
Type 2.3	Flat	Rectangular	MM: 577, 1044, 2002–2005, 2007, 2026; NM: 230, 1152, 1261/16; Sarianidi 1998b: nos. 886, 901, 917–919, 936, 956, 999, 1009, 1023, 1026, 1034, 1044, 1068, 1089, 1091–1092, 1105, 1113, 1127, 1143–1146, 1202, 1204–1208
Type 2.4	Flat	Stepped cross	MM: 570, 578, 1994, 1996, 2003; NM: KP 225, 228, 232; MFA: 18/38; Sarianidi 1998b: 888, 890–891, 941, 951, 953–955, 963, 968–970, 979, 995, 1014, 1033, 1039, 1094–1095, 1098, 1101, 1103, 1126, 1177–1178, 1190, 1647–1648
Type 2.5	Flat	Triangular	MM: 1039
Type 2.6	Flat	Trapezoid	MM: 568 = Masimov 1981: fig. 5.5
Type 2.7	Flat	Stepped-pyramidal[b]	Sarianidi 1998b: 914
Type 2.8	Flat	Lozenge	MFA: KEK 196; Masimov 1981: fig. 5.2 (Karakum); Sarianidi 1998b: 998–999, 1017, 1020, 1050, 1111–1112, 1123, 1203, 1224, 1665, 1667, 1673, 1692, 1714

Table 8.6 Cont.

Type	*Section*	*Shape*	*Museum or publication number*[a]
Type 2.9	Flat	Lanceolate	NM: 226; Sarianidi 1998b: 945, 972
Type 2.10	Convex	Cushion shape	MM: 223–224, 569–570, 581, 937, 1041–1042, 1886–1887, 1993–1995, 1997–1998, 2007, 4168, 4368; MFA: 14/25, 16/33, 19/42, 20/45, KEK 153/30; KEK 176/43, KEK 187; NM: 223–224, 227, 1265, 1271, 1282; Sarianidi 1998b: 892–894, 897, 903, 910, 920–921, 923, 926–927, 933–934, 937–940, 944, 946, 948, 959–960, 964–965, 967, 974, 981–82, 986, 988, 990, 996, 1002–1004, 1007, 1011–1012, 1015–1016, 1022, 1057, 1059, 1069, 1104, 1106–1009, 1114–1118, 1165–1167, 1185, 1187–1189, 1235, 1357
Type 2.11	Convex	Lentil shape	Sarianidi 1998b: 952, 958, 993, 1010? 1047?; Joshi and Parpola 1987: C-49, C-50
Type 2.12	Concave	Reel	Sarianidi 1977: fig. 50, 1 (Dashly 1); Sarianidi 1998b: 1029, 1182–1183
Type 2.13	Concave	Double stamp	MM: 1453; clay seal found at Gonur Depe North, Area 1, room 11

Note: MFA = Museum of Fine Art; MM = Mary Regional History Museum; NM = National Museum.

[a] The publication number (from Sarianidi 1998b) is only used for seals from outside Margiana.

[b] Form only presented in Sarianidi (1998b), so far not attested by excavations in Margiana.

(face and profile) they can be divided into 13 subtypes. Subtypes 1 to 9 are flat seals with a small rectangular cross-section (Figures 8.1c–f; 8.2d–e, i, k; 8.5a–b, e–f; 8.6d, e), types 10 and 11 have convex cross-sections (Figure 8.6a–c, f), types 12 and 13, concave ones. Thus, two-sided stamp seals have the following shapes: circular (type 2.1, Figures 8.1c–e, h; 8.2d; 8.5e–f), square (type 2.2, Figures 8.1i; 8.5a–b, with toothed rim), rectangular (type 2.3, Figure 8.1g), stepped cross (type 2.4, Figure 8.6d), triangular (type 2.5), trapezoid (type 2.6), stepped pyramid (type 2.7), lozenge (type 2.8, Figure 8.2i), or lanceolate (type 2.9). Some rectangular seals have highly convex "cushion-shaped" sides (type 2.10, Figures 8.1f; 8.2e; 8.6a–c, f) and the same feature concerns the rare lentil-shaped seals (type 2.11). Reel-shaped seals (type 2.12) and double stamp seals (type 2.13) are especially rare forms. The latter consists of two joint cone-shaped stamp seals. Among these shapes, the cushion-shaped seals are the most typical and, together with the round discs, the most numerous seals produced in "Murghab-style." As for the one-sided stamp seals, the two-sided ones are also mostly made of stones, but some pieces are in metal (MM 1993).

Two-sided BMAC stamp seals have also been found outside the BMAC area. The oldest "exported" such seal has the shape of a stepped cross with an eagle attacked by snakes on one face and a four-armed cross on the other, and it was found during excavations at Harappa (Vats 1940: vol. II, 255, 324, pl. XCI). Others were discovered in Luristan (Amiet 1988: fig. 18 = Sarianidi 1998b: no. 995) and in the Persian Gulf, in the Hamad Bronze Age cemetery at Bahrain/Dilmun (Brunswig *et al.* 1983: 103, fig. 5; Crawford and al Sindi 1995: 1–4, figs. 1–2 = Sarianidi 1998b: no. 941; see also Chapter 22).

Other stamp seals: types 3–7

Beside these two main types, a smaller number of seals with several sides also bears motifs made in the "drilled" and "Murghab-style" (Table 8.7). These are: three-sided prisms (type 3, Figure 8.7a); four-sided prisms (type 4); and rectangular brick-like seals with four carved faces, two small and two large ones (type 5). The remaining types are seals in the shape of cones. These include cones with a round or oval base (type 6, Figure 8.3), the back of which presents many variations: with a classical tip (6.1), a rounded tip (6.2), and a cut-off tip (6.3), with a conical back and additional high rectangular handle (6.4), as well as hemispherical ones (type 7). Type 6 is relatively abundant while types 5 and 7 are rare.

Table 8.7 Stamp seals types 3 to 7 from Margiana and Bactria

Type	*Shape*	*Museum or publication number*[a]
Type 3	Three-sided prism	"Murghab-style": MM: 1040, 1043, 5246; Sarianidi 1998b: 1499, 1511–1513, 1516, 1518–1520, 1524, 1536–1538, 1540–1548; "drilled style": NM: 1261; Sarianidi 1998b: 1500–1508, 1510, 1514–1515, 1517, 1522–1523, 1525–1535, 1537, 1539; "evolved style": MFA: 13/74
Type 4	Four-sided prism	NM: 247/119
Type 5	Four-sided rectangular brick-like with four carved faces, two small and two large ones	MM: 1044
Type 6	One-sided conical with a round or oval base	MM: 614, 616, 935, 1009.1–4, 1442, 1436–1437, 1439, 1885, 3643, 4159, 4325; NM: 117, 1270, 1276–1280, 1813/250, 1815/43.1–2/69/110, 1866/123, 1867/5/68, 1868/150, 2286/34/60
Type 7	One-sided hemispherical	MM: 1034

Note: MFA = Museum of Fine Art; MM = Mary Regional History Museum; NM = National Museum.

[a] The publication number (from Sarianidi 1998b) is only used for seals from outside Margiana.

The prism-shaped seals form an interesting group as they are a very specific form without antecedents in the BMAC area. It is possible that they developed independently as they were found at Gonur Depe, Sapallitepa, and Dashly 1 (Sarianidi 1977b: figs. 50, 64; 1986a: figs. 5, 10; 1986b: fig. 92; 1990: pl. LXXXI, 4, *9–11*; 1992: 24, fig. 22; 1993: pl. IXc; 1994: fig. 57, *1c*). The westernmost example of such a prism-shaped seal was found at Anau IIIC (Pumpelly 1908: pl. 45, 8, SK 73). What is most interesting to note is that they seem to have spread as far as the neighboring Harappan Civilization (Joshi and Parpola 1987: M-486–495, H-250–251, H-350–360, H-363–367; Shah and Parpola 1991: M-1428–1431, H-843–858, H-974–987). These Harappan prisms are not perforated like the BMAC prisms, but some show motifs related to the BMAC glyptic (Joshi and Parpola 1987: M-489, M-492; Franke Vogt 1991: nos. 262, 252). A further four-sided prism was excavated at Tepe Yahya IVA (Lamberg-Karlovsky and Tosi 1973: fig. 121). This may be linked to a three-sided prism found in the Persian Gulf at Maysar-1 in Oman (with zebu und goats) (Weisgerber 1980: fig. 15; During-Caspers 1994: 97) and with another one from al-Hajjar in Bahrain/Dilmun with zebu, goat, and Harappan inscriptions (Weisgerber 1981: fig. 54). E. During-Caspers, as early as 1983, assumed that these prisms reflected a BMAC influence over the Indus valley and beyond (During-Caspers 1983, 1997). This assumption has been confirmed by the increasing number of BMAC finds in the Gulf region (Kjaerum 1983: no. 335; Potts 1993a, 1993b, 1994; During-Caspers 1994: 104; but see Chapter 22).

THE BMAC COMPARTMENTED SEALS

By far, the compartmented seals are the most numerous, with various shapes and types. They are mostly casted in metal with the lost-wax method, a technique producing elaborate seals of very high quality and testifying of high skills in metallurgical production. They were produced locally as shown by broken molds found *in situ* at Gonur Depe. They are mostly made of arsenical copper, but some are made of gold, silver, and lead.

Nine silver seals come from Gonur Depe and seem to have been produced in the same workshop (MM 1767, 3474, 3508, 4329; NM 1246, 348, 360, 361, 1250). Other examples from precious metal come from the art trade, among them rare gold seals. The only excavated lead seal comes from Gonur (MM 3029).

Compartmented seals were also made of stones, especially chlorite or steatite (MM 111, 114, 572, 576, 1111, 1023, 1031, 1425, 3725, 4010, 4325; no. 1031 is a curious seal that combines a compartmented seal with "drilled style"); of faience (MM 1033, 1423, 1427, 1430, 1435, 1438, 1446, 1770–1773, 1881–1883, 1886, 3023, 3027, 3061, 3062, 3612; NM 1274, 1866/113/114/117/118/261/282–284/322/250, 1868/93/157/191, 2287/15); and, particularly in the latest levels, of terracotta or unbaked clay (from Togolok 7: MM 241).

Compartmented seals were found in numerous graves attributed to all the BMAC periods, in houses, under the floors of houses, in temples, and cult areas. It should be noticed that in the Gonur Large Necropolis, only ca. 20% of the deceased owned a compartmented seal, suggesting that these objects denoted a high social rank.

Compartmented seals can be classified according to (1) their appearance, (2) the technique used to produce them, and (3) their back.

From their appearance, two main groups can be distinguished: whether they are unframed (1.1) or framed (1.2) (Table 8.8). Group 1.1 concerns seals with

Table 8.8 Types of compartmented seals from Margiana and Bactria

Type	*Description*	*Museum or publication number*[a]
1.1	Unframed, figurative	NM: 74, 109, 169 (1147), 271/52, 272, 558, 1090, 1247, 1250–1251, 1255, 1257–1258, 1267/83, 1426, 1635/3/50, 1815/5/134/275/278, 1866/ 275/278/324, 1868/123, 2286/86; MFA: 11/11, 28/2, 29/3; MM: 353a, 353c, 584, 586, 1203, 1426, 1432, 1635, 1664, 1763, 1876, 1890, 1986, 1988–1989, 1991–1992, 3028, 3485, 3496, 3500, 3503, 3508, 4110, 4240, 4363
1.2	Framed	–
1.2.1	Round frame	NM: 179, 181, 270/1391, 275, 297, 348, 353s, 360, 1000, 1090, 1104, 1123–1124, 1139, 1148, 1151, 1187–1188, 1195, 1246, 1248–1250, 1252, 1254/10, 1256, 1259, 1270, 1273–1274, 1290–1291, 1294, 1405, 1431, 1434, 1760, 1813/50/53/196/212/384/394/434/437, 1815/108/114/ 139/ 278/291, 1866/27/88/113–115/117/ 250/261/269/270/ 272/275–277/ 279–281/283/284/302/323, 1867/ 26/49/ 52/77/106/107/134, 1868/8/9/65/123/124/131/157/192, 1989, 2286, 2287/42/43/47, 2305/20, 2305/20; MM: 353b, 353d, 558, 576, 582–583, 936, 999–1001, 1201–1202, 1423, 1427–1431, 1433–1435, 1443–1446, 1448–1450, 1758–1760, 1762, 1764–1767, 1770–1773, 1777, 1877, 1879, 1881–1883, 1886, 1888, 1893, 1921, 1986–1987, 1990, 2015, 3019–3021, 3023–3027, 3029, 3062, 3442, 3461, 3486–3488, 3491–3493, 3495, 3497, 3499, 3501–3510, 3608, 3644–3645, 3774, 3778, 3786, 4037, 4073, 4107–4108, 4161, 4228–29, 4267, 4329, 5219, 5223, 5282, 5428
1.2.2	Rectangular or square frame	NM: 12, 1809/146, 1867/2; MM: 1447, 3489, 3494, 4031, 4069; Sarianidi 1998b: 364–369, 371, 374–377, 396–397, 399, 428, 496, 536–538, 652, 744–745, 800–801, 803, 805, 812–814, 817–820, 823
1.2.3	Triangular frame[b]	Sarianidi 1998b: 144
1.2.4	Guilloche frame	NM: 297, 1866/273; 1866/281; MM: 1767, 3511, 3490, Gonur 2012.63[c]; Sarianidi 1998b: 1, 47, 137, 156–158, 160, 165, 191, 199, 206, 613
1.3	Pins ending in a compartmented seal	NM: 1167/70 (copper-alloy pin with four-leaf rosette), 168 = Sarianidi 1998b: 1549 (silver pin with enthroned goddess)

Note: MFA = Museum of Fine Art; MM = Mary Regional History Museum; NM = National Museum.

[a] The publication number (from Sarianidi 1998b) is only used for seals from outside Margiana.

[b] Only known from art trade.

[c] Found in Gonur in 2012, find no. 63, today in the Mary Regional History Museum, but inventory number unknown.

one figurative motif without a frame (Figure 8.12a, c, f). Hundreds of shapes are possible, representing all the typical motifs of the BMAC. According to the motif, they can be divided into: (1.1.1) seals with one single motif (human, divinity, hero, animal, mixed creatures made of combined animals or of animal and human, plant, symbols, landscape); (1.1.2) seals with one scenic depiction (animal fights, animal–human fights, hunting, acting humans, humans and divinities); (1.1.3) whirl motifs; (1.1.4) geometric motifs.

Group 1.2 represents the same motifs set inside a single or double frame (Figures 8.12b, d–e; 8.13–8.16). The latter is sometimes divided into compartments as well. These framed seals also feature a wide range of shapes and frames. According to the shape of the rim, we can divide them into seals: with round frames (1.2.1.); rectangular or square frames (1.2.2.); triangular frames (1.2.3.); and with frames in the form of guilloche (1.2.4) (Figures 8.13g and 8.15).

Most compartmented seals are round with single or double frames. In the last case, the outer rim is sometimes wavy, serrated, or toothed (Figure 8.12e). Frames that are made of a guilloche motif (1.2.4) form a special group, this motif being linked to entwined snakes that most frequently surround a central eagle motif. Those seals are, strictly speaking, figurative seals as they represent a specific stylization of the motif of an eagle attacked by snakes. However, it seems that this guilloche motif was also adopted to surround scenic depictions or symbols of goddesses, perhaps representing the snakes attributed to the goddess (Figures 8.13g and 8.15). Seals with simple square, rectangular, or triangular frames are rare.

A very special group within the BMAC seals consists of long pins the head of which has the shape of a compartmented seal (type 1.3). Most are made of "bronze," exceptional pieces of silver. They have a long and heavy stem and the compartmented seal is soldered directly at a right angle on its tip. The seals themselves can have many shapes, but most are simple round-framed compartmented ones with geometric, flower, or star motifs. Many come from illegal excavations (Sarianidi 1977a: fig. 62, *66*; Pottier 1984: fig. 24, *186*, *187*; Ligabue and Salvatori 1988: pl. 79), but one was found at Gonur Depe (NM 1867/70). Among this group, an exceptional pin is topped with a figurative compartmented seal made of silver, discovered in grave 18, unfortunately not dated, at the entrance of the Gonur North palace (Sarianidi 1998b: no. 1549).[13] The seal represents an enthroned goddess dressed in *kaunakes*, sitting on a chair with slanting back and hoof-shaped legs (Figure 8.19i). The stem of the pin is in the prolongation of the slanting chairback. This sitting goddess belongs to the most beautiful and most sophisticated products of BMAC art. The same type of goddess also appears on silver vessels, compartmented seals, and in composite figurines. A very similar and high-quality representation is known on the famous silver vase from Persepolis (Hinz 1969; Calmeyer 1989; Winkelmann 1998), long considered as an Elamite production, though it must be classified as a BMAC piece of art.

Compartmented seals can be classified according the technique used to produce them: (2.1) open-work seals (Figures 8.12–8.16), or (2.2) closed-back seals, with the motif soldered directly on a thin plate (Figure 8.14f). These last seals are relatively rare in Margiana. A third division can be made according to the treatment of the back that can be plain (3.1) or ornamented (3.2).

Seals with a plain back are mainly those made of faience, or figurative seals with animal motif, as well as most of the seals with geometric motifs (Figures 8.4; 8.12a, c, f; 8.14c). The seals with an ornamented reverse (Figures 8.13g–h; 8.14a–b; 8.15; 8.16a, c) form a small but significant group of compartmented seals providing important information about the stylistic features of BMAC art and testifying to their connection with earlier glyptic art from southeastern Iran. The stylized motif of the stamp side is reproduced on their reverse side, but in a highly elaborate manner, either in high or low relief and often additionally incised (Figures 8.13g; 8.15; 8.16c), or with fine carved incisions showing a multitude of details, be it for the human representations (hairstyles, facial features, and details of clothing), or for the animal ones (details of their heads, legs, horns, skin, mane, etc.) (Figures 8.13h; 8.14a–b; 8.16a). These seals with an elaborate reverse side are especially sophisticated ones, have an unusually large size, or special multifigured motifs, and are often made of gold and silver. They must be related to high-ranked persons.

With the decline of BMAC cities, the production of compartmented seals decreased. While wax-cast bronze seals became fewer in number and less elaborated, conversely stone and terracotta seals increased and images became less elaborate, featuring mostly geometric motifs. But the tradition of compartmented seals did not suddenly disappear. In Baluchistan, at Pirak, over 100 years after the BMAC presence had been noticed at Sibri/Mehrgarh VIII, compartmented seals are still produced (Jarrige and Santoni 1979; Shah and Parpola 1991: Pk 16, 17, 25, 26, 49).[14] These seals were found even as far as Gilund, in Gujarat India (Shinde *et al.* 2005).[15] Compartmented seals with cross motifs (the so-called "Nestorian seals") are also known further east in China and have been directly connected to the BMAC ones by some authors, although this has been contested (Hambis 1954, 1956; Biscione 1985).

Compartmented and stamp seals with elaborate animal handles that appear during the second half of the 2nd millennium BC at Marlik and other sites of western Iran, may also be related to former BMAC glyptic art. It is possible that simple compartmented seals with mainly geometric motifs dated to the second millennium BC in Anatolia developed from the BMAC ones (Salvatori 1981: 123–125, pl. III–X), but their high stems are unique. We cannot exclude interrelations between the BMAC and Anatolia at the beginning of the second millennium, since the Karum Kanesh seals and old Syro-Hittite seals share some common motifs with BMAC, like the double-headed eagle with guilloche, the bird-man, and others (Contenau 1877: pl. XI, XII). Finally, in Central Asia itself, some types of compartmented seal with motifs made from embossed stripes survived until the Parthian period at least. Such seals or their impressions have been found at Iron Age Ulug Depe in Turkmenistan (Xin and Lecomte 2012), and it seems plausible that they continued to be produced in the following Achaemenid period as suggested by seal impressions with tamga-like motifs that were found at Dahan-e Ghulaman in eastern Iran (Sajjadi 2007: fig. 13). Sarianidi included in his book some seals with conical backs and pierced handles with such tamga-like motifs, soldered on a round seal, which he considered as dating to the Bronze Age (Sarianidi 1998b: nos. 842–844, 848). New excavations in Parthian settlements in Turkmenistan (Gobekly Depe) as well as the analysis of older Parthian documents from Mesopotamia clearly show the use of such seals alongside Greek-style ring stones and other seals in the Parthian empire (Gaibov 1996, 2007). How this seal type survived is not entirely clear. What does appear certain is that this Bronze Age technique continued to exist for over 2,000 years. Nevertheless, except at Pirak, the total number of these post-BMAC seals found in excavations is extremely reduced.

HANDLES OF STAMP SEALS AND COMPARTMENTED SEALS

All compartmented seals and metal stamp seals with one face have ribbon or string handles. These handles often show clear signs of wear, as they are very thin in the middle.

The handles can have a triangular (type A, Figure 8.4k), semicircular (type B, Figure 8.16a, c), rectangular (type C, Figures 8.4c–f; 8.13d–e), or trapezoidal (type D, Figures 8.14e; 8.15b) profile. Triangular and rectangular handles sometime have a rounded top (types E and F). A very small number of seals has a long handle with a pierced knob or a high stem with a round or a hammer-like top (type G) (Sarianidi 1998b: no. 43, 111; the latter is possibly not from the BMAC period but from a later one). Those handles are also known at Tepe Hissar IIIB, a level in which other BMAC material has been found (Schmidt 1937: 414, pl. CXXIX, H 720, H 458; 258, fig. 118, H 4886).

Stone and clay seals also feature many varieties of solid perforated handles, with the same A–D profiles (Figure 8.3). High handles are often identical to the metal ones. This especially concerns faience seals that most frequently have high triangular handles (MM 1423, 1427, 1435, 1883, 3023, 3061, 3612, 5328; NM 1189, 1292, 1866/118/282, 1868/191; Figure 8.4c–f). However, most stone stamp seals have low handles with semicircular (MM 571 1029, 1202, 2011, 3635; Figure 8.2c, h) or rectangular (Figure 8.2g), but also trapezoidal cross-sections. Finally, there are also some rare solid perforated handles with semispherical, rectangular, conical, or gutter profiles. Some unbaked clay seals with conical backs attest to an added cubical or rectangular handle (MM1009, 1886).

CYLINDER SEALS AND STAMP CYLINDERS

Cylinder seals and stamp cylinders form a specific group that seems to be linked to the Mesopotamian tradition. Most are local products, but a handful are from foreign origin (see pp. 238–239).

Among the local ones, two groups are attested. The first group consists of classical cylinder seals with an engraved surface and a longitudinal perforation for suspension. They can be divided into three types (Figure 8.8; Table 8.9 = NM 1281, 245, 2029): seals with cylindrical body and straight wall (type 1), cylinders with concave sides (type 2), and the very rare cone-shaped (type 3, only from Togolok 21).

The second group consists of stamp cylinders that, in contrast to the standard cylinder seals, have an additional stamp on the lower face and a handle on the upper face. The typical stamp cylinder is small and has a high loop handle. Some of them also have a conical shape (Figure 8.9).

The majority of known BMAC cylinder seals and stamp cylinders come from illicit diggings in the cemeteries of Bronze Age northern Afghanistan in ancient Bactria. They are mostly of the "drilled style." By contrast, in Margiana, the total number of cylinder and stamp cylinder seals is relatively small and carved in the "Murghab-style." In Margiana, the local cylinder and stamp cylinder seals are in black chlorite and in copper "bronze" (see Table 8.10).

In Margiana, eight cylinder seals and stamp cylinders have been found at Gonur Depe (Figure 8.9a–b), four at Taip Depe 1, four in the Togolok oasis (Figure 8.9c), and one on site 1220 near the Togolok oasis (Figure 8.11a). Of these 17 seals, 5 belong to local cylinder seals, that from site 1220 is a cylinder seal in a mixed style and will be discussed later, 4 are imported cylinder seals from Iran and Mesopotamia, and 7 are local stamp cylinders.

Outside of Margiana, two BMAC stamp cylinders were found at Sibri in northern Baluchistan in a cenotaph with BMAC material (Santoni 1984: 57, fig. 8, *4a*; Jarrige 1985: fig. 6; 1988, no. 139). Another stampcylinder seal in clear BMAC style was found at Akra in Pakistan (Lambert 1986: 12, *1a–d*; Collon 1987: no. 598).

The Gonur Depe finds

Eight cylinder and stampcylinders seals were found at Gonur Depe, four of which are of the local style, and the four others are undoubtedly foreign (Table 8.9).

Table 8.9 Cylinder seals and stampcylinder seals from Margiana and Bactria

Shape	*Type*	*Description*	*Museum or publication number*[a]
Cylinder seal	Type 1	Straight wall	1. MM: 2029 = Sarianidi 1998b: 1777[b] (Gonur South, room 592) Seals with unknown storage place: 2. Sarianidi 1998b: 1767[b] (Togolok 1 surface) 3. Sarianidi 1998b: 1770[b] (Taip Depe 1 surface) See also from Bactria: Sarianidi 1998b: 1427, 1429–1431, 1451, 1453–1458, 1466–1469, 1472–1475, 1478, 1485, 1495
	Type 2	Concave sides	1. MM: 2028 = Sarianidi 1998b: 1778[b] (Gonur North, surface)
	Type 3	Conical	1. NM: 1272 (Togolok 21, temple) = Sarianidi 1998b: 1730
	Foreign	Concave sides Straight walled	1. Akkadian: NM: 1196 = Sarianidi 2007: 182–183 2. Kerman: NM: 174, vegetation goddess = Sarianidi 1998b: 1786 (Gonur North, grave 23) 3. Kerman: MFA: 25/21 = Sarianidi 2007: fig. 181, p. 106, multifigured scene with bird-headed goddess, snakes, animal–human mixed beings, and humans adoring a temple (Gonur North, grave 1393) 4. Kerman: MFA: 26/20, two goddesses before a bull-headed goddess in a circle (Gonur North Area 6, room 11)

Table 8.9 Cont.

Shape	*Type*	*Description*	*Museum or publication number*[a]
	Mixed style		1. Salvatori 2008: from site 1220 (near Togolok oasis) 2. Porada 1993: pl. 29 (art trade)
Local stamp-cylinder seals			1. MM: 2027 = Sarianidi 1998b: 1769[b], broken (Togolok 21) 2. NM: 246 = Sarianidi 1998b: 1768, bronze (Gonur South, room 617) 3. NM: 1281 = Sarianidi 1998b: 1766, bronze (Gonur South, room 655) Seals with unknown storage place: 4. Sarianidi 1998b: 1765[b] (Togolok 1, grave 10) 5. Sarianidi 1998b: 1771[b] (Taip Depe 1 surface) Seals with unknown storage place, but impressions in MM: 6. MM: 567 = Sarianidi 1998b: 1772[b], man before tigers (Taip Depe 1 surface) 7. MM: 566 = Sarianidi 1998b: 1773[b], kneeling persons (Taip Depe 1 surface) 8. See also from Bactria: Sarianidi 1998b: 1424, 1428, 1437, 1448, 1459, 1461–1464, 1470

Note: MFA = Museum of Fine Art; MM = Mary Regional History Museum; NM = National Museum.

[a] The publication number (from Sarianidi 1998b) is only used for seals from outside Margiana.

[b] These seals are in black chlorite.

Among those in local style, two are classical longitudinally perforated cylinder seals and two are stamp cylinders. All of them were found at Gonur South and are made in three different styles. The two cylinder seals are completely different in shape and style. The very small slightly concave cylinder seal with rounded rims made of black steatite depicts only one standing slim animal (lion or horse) in an obscure style (Figure 8.8b). The second cylinder seal, also in black steatite, is squat and roughly cut, depicting a winged lion, a snake, a scorpion, a tulip, and a dog arranged at right angles to the surface (Figure 8.8a).[16] The two stampcylinder seals from Gonur South were cast in "bronze" (Figure 8.9a–b), a very unusual material for this kind of artifact, and have very deeply incised motifs. The roughly worked motif of the first piece (from Gonur South, room 617) is not easy to interpret but seems to consist of coiled snake-dragons with open mouths (Figure 8.9b). The second seal (from Gonur South, room 655) bears a mythological motif of three snakes, with the middle snake devouring a man (Figure 8.9a). The stamp on the lower face depicts a whirl of animal heads.

Foreign cylinder seals

Four cylinder seals discovered at Gonur Depe during excavations are of foreign origin (Table 8.9). All these seals are highly significant as they are vivid proof of contacts between BMAC, southeastern Iran, and Mesopotamia in the second half of the third millennium BC (see also Pittman 2014a, 2019). These contacts are probably related to trade connections that existed between the urban Bronze Age cultures of the 3rd millennium BC, but they could also point to marriage and family alliances between BMAC and Kerman, or diplomatic gifts.[17]

The Akkadian cylinder seal from the grave 2550 (Gonur Large Necropolis) traveled the furthest because it came from Mesopotamia. The seal is made from dolomite and bears a classical Akkadian motif: a male hero fighting with a bull and a lion (Figure 8.8c). The cuneiform Akkadian inscription was arranged in three rows and is dated from the time of Naram Sin around 2200 BC (Sarianidi 2007: 108–109, figs. 182, 183).

The other three cylinder seals are made in the genuine southeast Iranian style of the Kerman culture with motifs and representations undoubtedly similar to those on seals from Shahdad, Tepe Yahya, and Konar Sandal (Figures 8.10 and 8.20a–g).

This applies, for example, to the white steatite cylinder seal from grave 23 of the Gonur Large Necropolis, which depicts the typical vegetation goddess of the Iranian Plateau in a squatting position with her lower body made into a stylized block.[18] Wheat ears sprout from her body as she sits on top of a crawling snake. A goat with long curved horns rests on a pedestal or platform in front of the goddess and a tulip flower is depicted above its head (Figure 8.10a).[19] Identical representations are known on six seals from the Kerman culture (Figure 8.20a–i) (Hakemi 1972: no. 323 = Hakemi 1997b: Ib.2; Lamberg-Karlovsky 1973: pl. XXVIc; Potts 1981: fig. 1a; Amiet 1986: fig. 132, 7; 1994: fig. 3; Winkelmann 1997a: fig. 1a–c). The second cylinder seal, made from white calcite stone, was found in grave 1393 of the Gonur Large Necropolis. The owner of this seal was a woman. The seal is adorned with numerous figures in two rows that depict so far unknown mixed creatures, blending several motifs (Figure 8.10b). The upper row represents a woman with two eagle heads. She is flanked by two snakes. The snakes turn their heads outward, attacking neighboring figures. The right snake tackles an eagle, which is attacked from behind by another snake. The snake on her left is assaulting a mythological group of two mixed creatures flanking an eagle. The first one is an eagle with a human head and bull horns, while the second has the lower body and talons of an eagle, the upper body of a zebu, and a bearded human head with goat's horns. The lower row shows two motifs: an eagle beside a resting zebu or bull, both turning their head towards two kneeling women adoring a temple (Sarianidi 2007: 106, fig. 181). The third cylinder seal in Kerman style was found at Gonur North in 2003 between the two round altars on the palace southwest corner (Area 6, room 11). It features twin goddesses (a vegetation and a winged horned one) before another with a bull head enthroned inside a circle surrounded by leaves (Figure 8.10c).

These imported cylinder seals mostly come from the Gonur Depe Large Necropolis where they date to period II. The owners of these seals may have been foreigners buried far from their homeland. These foreign elements are evidence for strong relations, probably related to trade, between Margiana, Iran, and Mesopotamia. A chronological problem needs, however, to be underlined here. Unless the seals had been thesaurized for a long time, these contacts could go back to the Akkadian/Naram Sin period. Furthermore, the oldest southeastern Iranian seals belong to a pre-Akkadian style that already existed around the middle of the third millennium BC (at Konar Sandal: see Pittman in Majidzadeh 2008: 100, fig. 29a, b, fig. 30a, b, g, fig, 31a, b, c, d, all dated around 2400 BC;[20] similar seals at Shahdad, cemetery Takab III A: Hakemi 1997b; Ib.1, 2, 3, 4, fig. on p. 355); at Tepe Yahya IVB6: TY 33: Potts 1981: fig. 2, TY 38: Lamberg-Karlovsky 1973: pl. XXVIc; Tepe Yahya IVB5: TY 32: Lamberg-Karlovsky 1973: pl. XXXIc; Tepe Yahya IVB4–2: TY 34, Lamberg-Karlovsky 1973: pl. XXVIc). These seals, however, lasted until the younger level of Tepe Yahya IVA, contemporary with the Akkadian and post-Akkadian period: Amiet 1986: fig. 132, 3). We will see below that southeastern Iran had a great influence on BMAC glyptic art, and we may assume here that the development and use of cylinder seals in the BMAC is due to these direct contacts with the Kerman culture (see also Chapter 20).

A further cylinder seal found on the surface near the Togolok oasis (Salvatori 2008), and discussed later in the chapter, is made of bone in a mixed southeastern Iranian, Early Dynastic, and BMAC style (Figure 8.11a). The engraved motif is that of the "master of animals and crossed figures," suggesting that a number of other styles could have been developed in the BMAC.

Beside these imported cylinder seals, and though it is not a cylinder, we must add here the discovery at Gonur Depe of a genuine Harappan steatite stamp seal representing an elephant with several Indus signs above it. Together with other Harappan artifacts, like the statuette of a ram, ivory sticks and plates, this seal underlines the long-distance connections of the BMAC culture (Sarianidi 2005: 231, fig. 92; 233, fig. 93; 256, fig. 114; see also Chapters 1 and 21).

SEAL IMPRESSIONS

Seal impressions are not numerous but were found at several sites (Table 8.10). They were used for the sealing of vessels, doors, bullae, and cretulae (Figure 8.17a–g). The latter ones were mainly found at Adji Kui 1 and 9 (Rossi Osmida 2011: 226). Sealings are made from all types of seals: cylinder seals, stamp cylinders, compartmented seals, and stamp seals, and in all the styles: "drilled style," "Murghab-style," or rough seals of unbaked clay. Evidence for this practice come from Gonur Depe South, Taip Depe 1, and the Adji Kui oasis, as well as from surface finds of surveys in the Kara Kum.

At Gonur Depe, sealings from stamp seals, stamp cylinders, cylinder seals, and compartmented seals on bullae were found nearly almost exclusively on the southern (later) mound. Impressions on sherds were found both in Gonur North and South. Compartmented seals were mainly used to seal bullae and cretulae, but also to mark stoppers of vessels and vessels themselves. Cylinder-seal impressions were found also directly on vessels, as shown by the fragment from room 18 at Gonur South depicting

Table 8.10 Impressions of stamp, compartmented, and cylinder seals from Margiana

Site	*Sealed object*	*Number of impressions*	*Impression of*	*Motif*	*Museum no.*	*Sarianidi 1998b no.*
Gonur North exc.1, room 16/6?	Pottery	1	CoS round framed	Eight-rays star with inner circle	MM: 1494	1749
Gonur North surface	Unbaked clay/ bullae	1	CoS round	Heraldic eagle	–	1760
Gonur North surface	Pottery	1	CoS figurative	Cross motif	–	1748
Gonur 9, room 146[a] or Gonur South, room 591?	Bullae	4	2 SS 2 Cyl.	Four eagles, man holding a camel on a line, bull	MFA: without inv. no.	1761
Gonur South, room 591	Bullae	1	CoS figurative	foot	MM: 2018	1754
Gonur South, room 591	Bullae	5	1 Cyl. 4 SS	Snakes attacking eagles/ twisted snakes	MM: 2025	1762
Gonur South, room 453	Cretula	1	CoS round	Eight-petals flower	MM: 1796	–
Gonur South, room 436	Pottery	1	CoS figurative	Nine-rays star	–	1745
Gonur South, room 438?	Pottery Sherd	1	CoS figurative	Eight-petals flower	MM: 2014	1746
Gonur South, room 293	Bullae conical	5	1 SS rectangular 4 SS round	Rectangular: bird-man with surrounding figurines. Round: stylized fish?	MFA 24/111	1763

Gonur South, room 207	Vessel stopper	1	CoS figurative	Eight-petals flower	MM: 2017	1753
Gonur South, room 207	Bullae	6	SS bifacial rectangular	Snake-dragon, attacking eagle	MM: 2026	1764
Gonur South, room 207	Vessel stopper	1	CoS	Heraldic eagle?	–	1752
Gonur South, room 117	Bullae	4	CoS trapezoid Cyl.	Whirl with five arms, network	MM: 2020	1756
Gonur South, room 114/214?[a]	Bullae	1	Cyl.	Two goats flanking a tulip and scorpion	MM: 2023	1759
Gonur South, room 116	Bullae	2 incomplete	CoS round and rectangular	Not clear	MM: 2015	1751.1
Gonur South, room 18	Vessel fragment	1	Cyl.	Bird-man as master of animals holding goats, eagle	–	1775
Gonur South surface	Bullae	1	CoS round	Flower?	MM: 2022	–
Gonur South surface	Pottery	1	CoS figurative	Seven-rays star	–	1750
Gonur South surface	Bullae	1	CoS	Snake	–	1758
Taip Depe 1 surface	Vessel fragment	1	Cyl.	Row 1: lion or lion-dragon and bull flanking a tree; Row 2: animal protomes and eagle	MM: 613	1776
Taip Depe 1 surface	Vessel fragment	1	Cyl.	Attacking bulls in the mountain, attacking eagles, tulips	MM: 613	1774

(*continued*)

Table 8.10 Cont.

Site	*Sealed object*	*Number of impressions*	*Impression of*	*Motif*	*Museum no.*	*Sarianidi 1998b no.*
Togolok 21, room 133	Bullae	2	SS "drilled style"	Stylized mountains and cross	MM: 2019	1755
Togolok 21, room 116	Vessel	1	CoS round	Malteser cross	–	1747
Togolok 15 surface	Bullae	1	SS stepped cross "drilled style"	Cross completely filled with drillings	MM: 2021	1757
Adji Kui[b]	Pottery	1	CoS figurative	Four-armed cross	Rossi Osmida 2011: no. S101	–
Adji Kui	Pottery	1	CoS figurative	Flower	Rossi Osmida 2011: no. S102	–
Adji Kui	Pottery	1	CoS figurative	Flower	Rossi Osmida 2011: no. S103	–
Adji Kui	Cretula	1	CoS figurative	Guilloche/twisted snakes	Not published	–

Note: CoS = compartmented seal; Cyl. = cylinder seal; MFA = Museum of Fine Art; MM = Mary Regional History Museum; NM = National Museum; SS = stamp seal.

[a] The information from the museum and from Sarianidi (1998b) differs.

[b] Not all published, not seen in the museums.

a multifigured mythological scene with a bird-man as animal holder together with an eagle (Sarianidi 1998b: no. 1775).[21] They also appear on bullae: One small cylinder "Murghab-style" seal impression on a piece of unbaked clay represents two goats flanking a tulip and a scorpion (Gonur South, room 214); a cylindrical flat bulla from Gonur South was impressed with an cylinder seal with a snake and eagle motif together with at least four impressions of a round stamp seal (Gonur South, room 591). Another bulla found in the same room, with two impressions of a cylinder seal and two of a round stamp seal both in "Murghab-style," depicts a man holding a camel on a line and a dog (from the cylinder seal), and four eagles (from the stamp seal) (Sarianidi 1998b: no. 1761).[22]

Other sites also produced sealings made from various types of seals. One impression of a cylinder seal on a vessel fragment was found on the surface of Taip Depe 1 (MM 613 = Sarianidi 1998b: no. 1776). The motif was executed in two registers, divided by a guilloche. The above register depicts a lion or lion-dragon and bull flanking a tree, and the register below represents animal *protomés* and an eagle. A second surface find from Taip Depe 1 presents eagles and tulips over two attacking bulls in the mountain (MM 613 = Sarianidi 1998b: no. 1774).

Fragments of broken bullae come from Togolok 21, showing impressions of two circular stamp seals in "drilled style"; from Togolok 15 (MM 2021 = Sarianidi 1998b: no. 1757), impressions of one stepped cross stamp seal and on the back of the bullae visible traces of the strings to fix the sealing to a container); and from Gonur South (stamp seal impressions on bullae). Cretulae with sealings were found at Adji Kui 9 (Rossi Osmida 2011: 226, 235–237), as well as impressions on pottery (Rossi Osmida 2011: 231, nos. SI01–103).

BMAC GLYPTIC STYLES

Two main styles were used for engraving the stamp and cylinder seals, the "drilled style" and "Murghab-style" (Sarianidi 1976; Winkelmann 1999, 2004). A smaller group of stamp seals was produced in other styles, not all well defined. Cylinder seals also show some mixed styles that are so far represented only on single specimens. We will consider them successively (Table 8.11).

The "drilled style"

The "drilled style" consists of small but deep drilled holes set one beside the other, or sometimes overlapping one another to form the design (Figure 8.2). It is attested on stamp seals in Margiana and Bactria and on cylinder or cylinder-stamp seals said to come from Afghanistan (Sarianidi 1998b: nos. 1430, 1431, 1437, 1454–1458, 1461, 1462, 1464, 1467, 1469, 1470, 1472, 1474). Various motifs are represented, humans, animals, mixed creatures, plants, and mountains. Bodies, joints, heads, details like tails, feet, horns, mouths, etc., are featured through drill holes of different sizes, producing astonishingly precise images where there are clearly visible, for instance, men or mixed creatures with characteristic broad shoulders, arm muscles, or strong thighs with small lower legs.

A great number of seals in this style represent crosses, rosettes, plants (Figure 8.2d, i), or stylized mountains (Figure 8.2a–c, f–h, j) made with drillings arranged into

Table 8.11 Margiana stamp seals: styles and store museums

Stamp-seal style	*Museum number*
"Drilled style"	NM: 232, 234/70, 235/71, 236, 237/73, 238/742, 239–240, 243, 245, 1261, 1266/82, 1269, 1280, 1815/2, 1868/39, 2184, 2286/45/122, 2287/70 MM: 575, 579, 938, 1009.5, 1024, 1027, 1031–1032, 1035, 1037, 1202, 1775, 2000, 2007, 2009–2010, 2019, 2021, 4352, 4379, 4441
"Murghab-style"	NM: 13/7419/42, 20/40, 30, 43/176, 49, 153/ 30, 163/33, 174, 187, 196, 221–222, 224–227, 229–231, 241, 247, 1152, 1262, 1264/50, 1265, 1271, 1282, 1866/115, 1867/65, 2098, 2286/69; MM: 571–573, 576–578, 580–581, 585, 937, 999, 1021–1024, 1028–1030, 1033–1034, 1036, 1038, 1041–1044, 1423, 1625, 1776, 1887, 1993–1999, 2001–2006, 2008, 2011–2012, 2026, 3635, 3730, 3744, 4074, 4168, 4239, 4325, 4362, 4368, 4378, 4410, 5246, 5399; MFA: 14/25, 16/33, 18/38, 19/42, 20/45, 21/84, 29/3, KEK 153, KEK 174, KEK-176, KEK-187, KEK-196
"Simple style": roughly cut without clear style (clay seals)	NM: 1125/a2, 1275, 1279, 1276/92, 1277, 1278/94, 1279/95, 1280/96, 1813/250, 1815/43.1/43.2/69/110, 1866/123, 1867/5/68, 2286/34/60, 2287/88 MM: 616, 1009.2–4, 1439, 1885, 3643, 4159, 5206
"Evolved style"	MFA: 13/74
Seal with two styles	MM: 2007 ("drilled style" and "Murghab-style")

Note: MFA = Museum of Fine Art; MM = Mary Regional History Museum; NM = National Museum.

a triangle. A second group of motifs consists of animals represented alone, like scorpions, or attacking snakes (Figure 8.2i), eagles, or lions. Scenes of animal fights are more rarely found (Figure 8.2d, k). Humans or human–animal mixed creatures like the bird-man (Figure 8.2k), bull-man, or goat-man form the third great group of motifs in this style, while depictions of women or goddesses are so far lacking. Very frequent and characteristic are males or male mixed creatures sitting on their calves (Figure 8.2k) (Sarianidi 1998b: nos. 1040, 1041–1055, 1427–1501–1508, 1429), sometimes with one raised knee. Walking men or mixed creatures in the attitude of "master of animals" fighting with snakes, holding snakes, or being attacked by snakes are also represented.

As described above, probably due to the limited surface of most "drilledstyle" seals, and to the technique that was used, most of the seals bear only one motif. This is especially valid for one-sided stamp seals. However, cylinder seals and seals with two or more sides may have multifigured scenic compositions. The same motifs known in other styles and on other artifacts are represented but with more simple features. While cylinder seals have a greater surface to arrange full scenic depictions, stamp seals with two or more sides use a simple trick where the motif is divided into the single actors of a scene arranged on the different faces. For instance, the motif of a fight between an eagle and two snakes on a bi-sided stamp seal is depicted with the eagle with its head turned on one side and the attacking snake on the other. Bearing in mind the ubiquity of these kinds of battle motifs, it is possible to recognize the same ones even on this comparatively less sophisticated style.

Technically, this style has clear antecedents on the Iranian Plateau and southern Central Asia, especially in the Helmand and the Namazga cultures. It is known in Mundigak III–IV and at Shahr-i Sokhta II–III where it is the most characteristic style of stone stamp seals of the Helmand culture (Salvatori and Vidale 1997: 77–78, figs. 251–262: seals and impressions). This style was also known in Baluchistan, but was replaced by Harappan seals after 2500 BC. Similar seals with geometric motifs made from smoothed drillings set against each other are also known at Altyn Depe and Ulug Depe in the Piedmont area of Turkmenistan in NMG III and IV levels (Masson 1988: pl. XVII, 2, 7; XXXVI, *3*; XXXVII, *6*; XXXVIII, *1*; Sarianidi 1986b: fig. 93). While most of these earlier seals had two perforations for their suspension, a handle was added on their back during the NMG V period directly leading to the typical BMAC seals. At that time too, an increase in figurative representations is noticeable (Masson 1988: pl. XVI, *11*, *17*; XVII, *1*, 2, *3*, *4*). These figurative depictions with their many heroes and mixed creatures form the characteristic new feature of the BMAC seals that developed from an older Iranian style. Seals in the "drilled style" seem to have a long duration, from the urban phase to the late Dzharkutan phase (Askarov 1977: pl. XLV; Sarianidi 1977a: fig. 59; Askarov and Abdullaev 1983: fig. 49; Sarianidi 1986a: fig. 5, *11*; 8, *23*; 1986b: figs. 102, 103; 1988: pl. 1).

The "Murghab-style"

The second main BMAC style, also found on stamp, cylinder, and stamp cylinder seals is particularly interesting as it is an indigenous development of the BMAC. Sarianidi (1976) was the first to identify it in the Murghab area and to give it this name.

The motifs are created by flat cuts and well-smoothed flat drillings. Drillings were used for the eyes, mouths, hind legs of animals, or arms and bodies of humans, while all important details like hair, skin, mane, feathers, talons and claws, tongues, teeth, wings, inner ornamentation, tails, and foreleg/forearms were executed by flat cuts (Figures 8.1, 8.5, 8.6, 8.7).

Two subtypes can be distinguished according to the use of drillings and cuts, and to the plastic (Figures 8.1i–j; 8.6.a–e; 8.7a, 2; 8.8a) and graphical finish (Figures 8.1f; 8.5a–b, e–f; 8.6f).

The best plastic finish variety of this style uses smoothed drillings and deeper curved cuts, creating a soft, in relief, impression of the motif. This "softer" variety underlines better the muscular bodies and the details in general, and is stylistically reminiscent of Iranian Plateau seals, especially in southeastern Iran, and of the so-called "Intercultural Style" representations. It concerns specific stylistic features such as the muscular shaping of the arms and legs, the broad shoulders of men (Figure 8.5e), the detailed paws, talons, noses, or heads of specific animals, like felines, snakes or snake-dragons. The last two, especially, repeatedly feature the protruding eyes, the pronounced nose holes and "ears" that are so typical of snakes in southeastern Iranian art (Figures 8.1f and 8.6a–b). Eagles, snakes, and lions with this more plastic finish style also often have a specific band between their head and neck, typical of Kerman culture depictions on stone vessels (Figures 8.6a and 8.7a). This plastic finish subtype mainly comes from Bactria and seems to be the older variety of the "Murghab-style," but it also occurs in Margiana.

In the second subtype (Figures 8.1f; 8.5a–h; 8.6f), the motifs are executed in a more graphical manner, avoiding drillings or working with very flat drillings, but using thin cuts instead. The impression is thus very flat. The human arms and legs, for instance, turn into simple lines, the ornamentation of the human or animal bodies is either strongly stylized or is completely missing, and the motifs generally become more stylized and reduced. This graphical variety is more typical of the seals from Margiana.

Whatever the technique used, the motifs are often animated, vivid, and strikingly detailed. They mainly consist of scenic depictions with many actors: humans, animals, human–animal mixed creatures, landscapes, and plants. Mythological scenes are represented: fights among animals (eagle and snake, Figure 8.6a; lion and snake, Figures 8.6.a and 8.8a; bull, camel, or goat and snake, Figure 8.5a–b), but also between a man or male mixed creature and snakes, holding snakes, lions, or hoofed animals (Figure 8.5e–f), or men attacked or eaten by a snake or lion. Women or goddesses are not represented in this style, but men, male, or animal mixed creatures and mythological heroes appear very often. The most frequent one is the bird-man. He, as well as the male hero, is mostly shown in the very specific attitude of kneeling or sitting on his calves, which is also known in the "drilled style" (Figures 8.2k and 8.6d). Sometimes these male figures are also walking or standing with raised arms. Their upper body is shown in front and their lower body and heads in profile (Figures 8.5e and 8.6d), as in the "drilled style" and on compartmented seals, which will be discussed later. They also repeat the depiction of men with broad shoulders. Animals are represented strictly in profile, having thin lower legs but pronounced thighs and additional ornamentation on their hip or breast (Figures 8.5 and 8.8a). Animals of prey are always depicted in the act of attacking, jumping, running, or diving, and hoofed animals and felines often have their head turned back (Figures 8.1f, j; 8.6f). Lions have a pronounced mane with the tail held over their back (Figures 8.6a, e; 8.8a). The characteristic way of depicting the eagle is to show the body with outstretched wings from the front while the head is turned to the side looking at the attacking snakes (Figure 8.6a). Often filler motifs, such as single plants, leaves, tulips, or mountains (Figures 8.5f and 8.8a) are set beside the main actors. But they also appear alone (Figure 8.1d–e, g). Additionally, many animals and mixed creatures are equipped with wings (Figures 8.1f; 8.6c, e–f; 8.7a; 8.8a). One striking feature is the appearance of the whirl motifs; that is, compositions of snakes, animal heads, various animals, or plants rotating around a center (Figures 8.1a–c, e, h; 8.9d). This very specific composition is shared both by the "Murghab-style" seals and by compartmented seals (Figure 8.12d–e).

"Drilled style" and "Murghab-style" seals were not used concurrently, and there is a small number of two-sided seals that bear both styles (MM 1031, 2007).

Other styles on stamp and cylinder seals

Beside these two clearly defined styles, a few other stylistic groups have been identified though their classification as independent styles is somewhat doubtful.

Some clay and stone seals from Margiana feature only very roughly cut geometric motifs or drillings without clear pictures and cannot be clearly assigned to one of either styles. They could represent a simple group of "Murghab-style" seals or could be defined as "simple style" (Figure 8.3).

Some stamp seals coming from the art market are in a style that I earlier named "developed style" (Winkelmann 2004: 134). Seals in this style have very plastic figures with elaborated muscles, created by the use of deep drillings and deep curved cuts well-smoothed out. Acting figures are vivid and depicted in motion. Bodies have anatomically correct dimensions. All the details are well visible. The depictions are reminiscent of the finest Akkadian or southeastern Iranian seals. Our BMAC attribution is due to their motifs and to the seal shapes. However, the date and diffusion of this style are not yet clear. The very small number of such seals may imply that they were the product of a specific workshop and that they can be related to the "Murghab-style."

In the same publication I had defined another style: the "soft style," where seals have very flat figures cut in outlines, with the body created by flat and well-smoothed drillings, while arms or legs are produced by finely cut lines. One of their typical features is the bagginess of the bodies, which lacks details. However, their attribution to the BMAC is not certain. I have now come to believe that these seals belong to the 1st millennium BC.

Cylinder seals and stamp cylinders in mixed styles

Beside "drilled style" and "Murghab-style" seals, there are cylinder seals and stamp cylinders with styles unknown so far. These styles combine elements, motifs, and stylistic features from Mesopotamian Early Dynastic III and Akkadian seals, from southeastern Iran Kerman Culture seals, and from the Harappan Civilization, together with BMAC elements. These seals are especially significant for they underline intercultural relations between the four most important urban cultures of the Bronze Age in the ancient Near East and the high creativity of BMAC artist to seize and merge different options into a new style.

One bone cylinder seal found on site 1220 near Togolok during a survey is engraved in a mixed southeastern Iran, Early Dynastic, and Bactrian style (Figure 8.11a) (Salvatori 2008). This seal represents a male hero holding two bull-men, whose heads are turned back to two chiastically arranged lions attacking them. As noted by Salvatori (2008: 111), the cylinder seal represents a well-known Early Dynastic III and Early Akkadian seal motif. However, it has been transformed in many ways (see also Pittman 2014b). The originally nude Mesopotamian hero now wears a long flounced skirt, but the triangular upper body with pronounced shoulders and a wasp waist, the outlines of the skirt, as well as the typical alternation of figures in profile and in front view, follow the stylistic features of southeastern Iran seals and BMAC art. Finally, the short beard of the bull-men as well as the long side curl is typical of the BMAC male hairstyle.

Another cylinder seal, published by Edith Porada (1993: pl. 29), clearly features influences from southeastern Iran cylinder seals together with motifs and stylistic features appearing on BMAC "Murghab-style" and compartmented seals (Figure 8.11b). The seal represents two different groups of figures. A kneeling male person holding a snake is represented in the center. A lion tail arises from his neck and two snake heads grow out of his head. The man is depicted inside a circle with eleven prongs that hovers over a mountain with growing-out trees. The mountain and person in the circle are flanked by two kneeling men with raised arms. Both men wear caps adorned with animal heads of zebu and lion. The left man seems to be a mythological creature because of two snakes arising from his armpits, a type known from southeastern Iran as well as from BMAC (Winkelmann 2013: fig. 5; 2014: 210, 211, fig. 4; 2016b). The right man wears a bow and a quiver. Beside this group, a horned goddess in "*kaunakes* skirt" sits on the tail of a typical horned, bearded, and winged snake-dragon of BMAC art. She holds a wheat ear in her hands.

Two other BMAC stamp cylinders in a mixed style were found at Sibri in Pakistan (Santoni 1984: 57, fig. 8, *4A*; Jarrige 1985: fig. 6). While the first one represents the battle between bull and lion, the complicated battle scene on the second one depicts a lion-headed mixed creature standing on a crocodile fighting a lion attacking a zebu. While the shape of seals as well as the scenes and the mixed creature are reminiscent of BMAC seals, the crocodile demonstrates another environment. The seals from Sibri are especially important as they point to a merging of BMAC seal features with Harappan influences. They can be linked to six other cylinder seals from the Harappan culture, which are either genuine BMAC seals or mixed BMAC–Harappan seals rather than original Harappan ones.[23] A cylinder seal with eagles and snakes found at Akra in Pakistan also belongs to this group of exported BMAC seals (Lambert 1986: 12, *1a–d*; Collon 1987: no. 598).

I would also add to this group of seals with BMAC influence one from Ur (Isin-Larsa levels) with a bull jumping carried out by a bird-man (Gadd 1932: pl. I, *6*).

Finally, the presence of a mixed style is also suggested by a stamp cylinder with a bull-jumping scene coming from the art market, which combines Akkadian and BMAC elements (Winkelmann 2001; 2004, III, *1-01*).

The compartmented seals style

The identification of styles on compartmented seals is difficult since their seal side consists only of the rough outlines of the motif. For most compartmented seals, we can therefore only say that the motif is executed in a more or less blunt or fine manner, depending on the size of the seal or the width and depth of the ribbons. This applies both to geometric and figurative representations.

However, there is a small group of elaborated compartmented seals where the motifs in the face are repeated on the back. In these cases, the back motif is very elaborate and exhibits details that are lacking on the face. Its implementation was carried out in three ways: (1) made in higher relief and with very fine incised details (Sarianidi 1998b: nos. 16, 18, 20, 24, 26, 36, 46, 47, 48, 49, 55, 56, 57); (2) worked on a flat relief, which mainly applies to seals with closed-back and flat multifigured depictions on the face (Sarianidi 1998b: nos. 99, 184, 229–231, 238, 239, 248, 249,

250–256, 264, 265, 290, 295, 296); (3) finely engraved and with finely incised details on a flat back.

Most of these seals are made of precious metal, like silver or gold. They were often found in Margiana and seem to come from one workshop (Sarianidi 1998b: no. 1). For the three groups, we have been able to establish a common way of depicting humans, goddesses, and animals that is stylistically reminiscent of the "Murghab-style" seals. The main difference with it is the range and execution of the motifs. First, many of these seals depict goddesses, representing the BMAC female pantheon. Second, beside multifigured representations, there are many symbolic motifs, reducing a multifigured scene to one typical acting figure. This concerns many animals, originally acting in multifigured scenes, or attributes of goddesses like scorpion, feline, goat, star, tulip, or rosette (Figures 8.4; 8.12a, f; 8.13g–h). Third, here are ligatures that combine together the actors of a multifigurative scene into fantastic creatures where, for instance, the bodies of bird-men or goddesses are merged with the bodies of lions. Similarly, the original motif of a man holding snakes is transformed into a man with snake arms (Sarianidi 1998b: nos. 22, 24, 52, 53–55). We also observe a rising number of whirl motifs (Figure 8.12b, d–e).

Generally, figures are comparatively slim and show finely incised or plastically elaborated details, especially of their head and attributes like horns, wings, etc. The most characteristic way of human representation is the front view for the upper body and the profile for the lower part.[24] Waists and hips are narrow, legs, feet, and hands are slim, the latter two often reduced to fine lines. Heads appear in profile (Figures 8.14b, f; 8.15; 8.16). This applies to nearly all depictions of kneeling, sitting, walking, or enthroned persons and mixed creatures.

Males and male mixed creatures appear in exactly the same manner, in the same attitudes and in the same pictorial context. They often have broad shoulders with a pronounced shoulder joint, arm, and breast muscles, and the same narrow hips, slim legs and arms, which also appear in "drilled" and "Murghab-" styles. Heads are often very detailed, featuring large almond-shaped eyes and finely executed short hair and beards. Male mixed creatures have detailed horns, ears, tails, paws, or wings with incised details. Men and animal-men are mostly represented in the nude, wearing short skirts or even only a belt around the waist; however, in ceremonial depictions, they sometimes wear a long skirt or a dress with one bare shoulder.[25]

Male animal creatures like the bird-man, goat-man, bull-man, ram-man, or lion-man generally follow the same manner. Only one of them, the lion-headed man with wild hair, has the face depicted from the front almost in all the cases. Animal–man mixed creatures generally have an animal head and can exhibit additional animal elements like hoofs, wings, or snakes. Only the bird-man, the most frequently depicted and most important creature in BMAC art, appears in many varieties: with and without wings, with and without a bird head, with a human body or a bird body, with wings and arms, or only with arms or only wings, etc. The most frequent and most characteristic position of male or of male mixed creatures is sitting on their knees or calves (Sarianidi 1998b: nos. 1040, 1041–1055, 1427–1430, 1499–1508), sometimes with one knee raised. Men also appear walking (Sarianidi 1998b: nos. 916, 995, 1424, 1431, 1500, 1616, 1618), enthroned (only on compartmented seals, Sarianidi 1998b: nos. 4, 9, 12.26), or standing (as "master of animals," Sarianidi 1998b: nos. 25, 28–32, 36, 42, 65, 904). Walking or standing men or male mixed beings often appear as "master of animals" holding two-hoofed animals, felines, or snakes (Figure 8.16c).

"Ordinary" females are rarely depicted on seals. They only appear in ritual banquet scenes together with a man, mostly sitting on a chair and holding a drinking vessel.[26] They wear long skirts and long hair. As generally for humans, their upper body is shown in front view, with the lower body and head depicted in profile. Their most remarkable traits are their large eyes, often together with a pronounced nose and a detailed hairstyle.

Females mainly appear as goddesses,[27] always slim with a wasp waist and a long skirt covering their feet (Figure 8.19a–d). The skirt is usually plain, but sometimes is "*kaunakes*-like" or decorated with scales (Figures 8.16a and 8.19e–i). The upper and lower body is most frequently shown from the front, the head appears in profile. Heads can be remarkably elaborated, including skillfully executed hairstyles. The most frequently recurring hairstyle is long hair coiled and pinned up in the nape, forming a characteristic roll running from one ear to the other (Figures 8.15a; 8.16a–b; 8.19g, i). Sometimes goddesses have a characteristic sidelock beside the ears (Figure 8.14b). They have slender arms, while their hands are often not visible. The most frequent attitude is the standing goddess with hanging arms and her hands clasped in front of the waist (Figure 8.19a–d) (also Sarianidi 1998b: nos. 23, 24, 34, 35, 37, 38, 39, 40). There are also numerous enthroned goddesses, sitting on a characteristic chair with a slanting back (Sarianidi 1998b: nos. 1, 5–7, 27, (Figures 8.15c and 8.19i), or enthroned on their attribute animals, such as lions, panthers, snakes, snake-dragons, or other creatures (Figures 8.14b; 8.19e, g) (see also Sarianidi 1998b: nos. 14, 15, 17–20, 22, 24). In rare cases, the goddess appears kneeling, with the lower body stylized into a block or with visible feet (Figure 8.16a) (see also Sarianidi 1998b: nos. 64, 66, 67). These three attitudes are also known in the characteristic composite figurines as well as on BMAC silver vessels (Ghirshman 1968; Pottier 1984: nos. 295–303; Amiet 2004: fig. 7; Benoit 2010; Winkelmann in press).

Animals are represented in strict profile (Sarianidi 1998b: nos. 63, 76–85, 89–98, 107–122, 215–231), with detailed heads and bodies. Ears, eyes, noses, mouths, tongues, beards, tails, horns, wings, manes, scales, and paws as well as all the muscles are carefully executed with fine incisions. Especially remarkable are the goggled eyes, noses, and "ears" of the snakes and snake-dragons, which seem to have been directly drawn from the "Intercultural-Style" stone vessels from southeastern Iran. This also applies to the lines separating the head from the neck or for the characteristic manner to depict shoulders and thighs in the form of an irregular oval (Figure 8.14a–b). The animals are usually standing or lying with their head turned back (Sarianidi 1998b: nos. 76, 92, 112, 292, 899, 917, 922, 924, 925, 929, 933, 945, 985, 1073, 1553, 1617). Beasts of prey are mainly depicted in an attacking position. Snakes especially always attack other animals or humans with their open mouth (Sarianidi 1998b: nos. 49, 50, 52–56, 905, 912, 925, 931, 932, 933, 938, 939, 953, 962, 967, 990). Their most typical representation is them fighting with an eagle: It is the most numerous and most varied motif on compartmented and stamp seals (Figures 8.6a and 8.14a) (Sarianidi 1998b: fully developed motif: nos. 127, 172–175, 295, 889, 927, 931, 933, 938, 951, 957, 958, 982, 984, 986, 988, 990, 991, 993, 1073, 1097, 1629, 1646; transformed into eagle with guilloche: nos. 137, 156–158, 160, 165, 170, 191 199, 206; as ligature: no. 125).[28] Eagles appear with their body and wings depicted in front view but with their head turned. Their wingtips are upwards in the battle between eagle and snakes, but also downwards, an attitude coming from the eagle diving on a hoofed animal. The only animals that are always represented in top view are the scorpion (Figure 8.13b) and the human-headed snake (Sarianidi 1998b: nos. 268–289 (scorpions) and nos. 265, 266 (human-headed snake)). Sometimes, we also find entwined snakes in top view (Sarianidi 1998b: nos. 238–243, 248).[29]

THE CHARACTERISTIC BMAC MOTIFS

The art of Bactria and Margiana is overwhelmingly a miniature art where glyptic constitutes the most elaborated group in which the art and iconography is best represented. Glyptic is the best carrier of the BMAC myths, rituals, gods, and mixed creatures, environment, flora, and fauna. It also gives important information on the social rank of the owners.

In all the styles mentioned earlier, the same motifs are present. They reflect the characteristic BMAC representations.

Single figures (Table 8.12)

1.1. *Animals*: hoofed animals like bull, zebu, goat, ram, and camel; felines like lion, leopard, and tiger; eagle; snake; scorpion; monkey; hare; wolf; fish (Figures 8.1f, j; 8.2i; 8.4g; 8.6b, f; 8.7a; 8.9a, d; 8.12a, f; 8.13b).

1.2. *Humans*: man and woman, or parts of them like hand or foot (Figure 8.14f).

1.3. *Animal–human mixed creatures*: bird-man (Figures 8.1j, 8.2k, 8.17c), bull-man, goat-man and ram-man, lion-man, snake with human head, monkey-man, man with snakes arising from his body or with snake arms.

1.4. *Animal mixed creatures*: snake-dragon, lion-dragon, winged lions, lion-eagle-goat-dragons, and winged hoofed animals (Figures 8.1f; 8.6c, f; 8.7a; 8.8a–b; 8.9b; 8.11b; 8.14b).

1.5. *Goddesses*: often with different attributes and attribute animals, enthroned, standing, sitting, and kneeling (Figures 8.14b and 8.16b).

1.6. *Plants*: tulips, leaves, tree, poppies (Figures 8.1d–e, g; 8.2d, i; 8.3g; 8.13a).

1.7. *Landscape elements*: mainly mountains (Figure 8.2a–c, f–h, j).

1.8. *Luminaries*: sun, star, rosette, and crescent (Figures 8.3f; 8.4a–c, e–f, i; 8.17d–e, g).

1.9. *Various geometric motifs*: especially four-armed and stepped crosses, single or in combination with bows or crescents, stepped pyramids, rhombs, and others (Figures 8.4d and 8.17a–b).

Scenic depictions (Table 8.13)

2.1. *Animal fights*: battle between eagle and snake (the most favorite motif, Figures 8.6a3 and 8.14a); between snake and feline (Figure 8.6a1); between snake and other animals (Figures 8.2d, k; 8.5a–b); between feline and hoofed animal (Figure 8.7a); eagle diving on a hoofed animal (Figures 8.7.b and 8.12c); eagle and snake in a tree; eagle with its fledglings; eagle with snake in its beak.

Table 8.12 Characteristic BMAC single motifs

Motif	*Species or detail*	*Museum or publication number*[a]
Single animals	Hoofed animals bull, zebu, goat, ram, and camel	Camel: MM: 588; NM: 79, 225; MFA: 18/38, Sarianidi 1998b: 21, 108–111, 888, 919; bull: MM: 1635, 2011; MFA: 14/25, 13/74; Sarianidi 1998b: 82–85, 88, 962, 1040; goat: MM: 1759; NM: 271/52; Sarianidi 1998b: 89–93, 98, 100–106, 119–121, 292, 922, 963, 1052, 1058, 1067, 1070, 1072, 1080, 1084, 1685; zebu: MFA: 18/38; Sarianidi 1998b: 940, 1056, 1451, 1459, 1639–1640; ram: Gonur North: Sarianidi 1998b: 1616
	Felines	Single felines: MM: 937a, b, 1994, 4168, 4368; NM: 169 (1147), 221; MFA: 13/74, 19/42
	Eagle	MM: 571, 1997, 2008, 3497, 4110, 4378; NM: 229–230, 297, 1250, 1867/26, 1867/65, 1815/114, 1250; MFA: 20/45, 16/33, KEK 153
	Snake	MM: 578–579, 1035, 2010, 2026, 2112; MFA: 19/42; NM: 241, 1261/16
	Scorpion	MM: 1444, 3774; MFA: 43/176; NM: 74, 179, 1148, 1187, 1252–1253, 1270, 1866/277, 1867/106, 1868/93
	Hare	NM: 1249/5; Sarianidi 1998b: 117–118.
	Monkey	Sarianidi 1998b: nos. 68–73
	Wolf	NM: 226; MM: 2028; Sarianidi 1998b: 207
	Fish	MM: 353a/1
Humans or human elements	Woman giving birth	NM: 239, 579, 938, 1867/27, 2287/70; MM: 1035, 1037, 2000
	Hand	NM: 1992, 1262/48; MM: 1887, 2018
	Human face	NM: 357, 1262/48; MM 3730; Sarianidi 1998b: 873–879
Human mixed creatures	Bird-man	MM: 1995; Sarianidi 1998b: 8, 10–11, 13, 22, 28, 33, 41, 57–58, 60–61, 63, 1038, 1040–1047, 1049, 1051–1054
	Bull-man	Sarianidi 1998b: 26, 32, 42, 50, 64, 88, 148, 159, 912
	Goat- or ram-man	Sarianidi 1998b: 54–56, 1037, 1502; ram-man: 1039
	Lion-man	Sarianidi 1998b: 51–53, 905–907, 909, 920, 1235–1236; discussion on figure, see Winkelmann 2016a

Table 8.12 Cont.

Motif	*Species or detail*	*Museum or publication number*[a]
	Snake with human head	Sarianidi 1998b: 265, 266; this is maybe a ligature of the snake devouring a man
	Monkey-man	MM: 1994; Sarianidi 1998b: 68–73, 1050
	Man with snakes arising from his body	Sarianidi 1998b: 32, 52–53, 55, 1235
Animal mixed creatures	Snake-dragon	MM: 4168, 4378, 4386; NM: 224; MFA: KEK 153; Sarianidi 1998b: 19
	Lion-dragon/ winged lions	NM: 222–223, 227; MFA: 33/163, 13/74; MM: 937, 1997, 2029; Sarianidi 1998b: 40, 55, 63, 77–78, 80, 224–225, 857, 898, 904, 907, 917, 927, 929–930, 933, 937, 938–941, 944–951, 953–955, 957, 985, 1237–1238, 1518, 1540, 1625–1632, 1642
	Lion-eagle-goat-dragon	Sarianidi 1998b: 16, 18, 224, 226–227
Goddesses	Enthroned goddess	NM: 174, 360; Sarianidi 1998b: 2, 3, 5–7, 14–20, 23–24, 27, 34, 35, 37–40, 43, 45
	Kneeling goddess	NM: 1815/1405
Plants	Tulip	MM: 573, 577, 586, 587/2, 1767, 1887, 1998, 2001, 3023, 3032, 3484, 5223; NM: 2306/47, 1264/50; Sarianidi 1998b: 859, 1022, 1050, 1659, 1662
	Leaf/tree	MM: 578, 616; 1036, 1038, 1998, 2003, 3493; NM: 224, 226, 230, 232, 1265, 1272; MFA: KEK 196; Sarianidi 1998b: 861–863, 866, 1101; MM: 2000, 2003; NM: 231, 1265
	Poppy	Sarianidi 1998b: 939, 963, 966, 985, 1017–1021, 1023, 1042, 1098, 1126, 1663, 1664, 1666; NM: 1123, 1189, 1257–1258, 1267/83, 1441, 1759/11, 1815/134, 1866/250/269/278/283/284
	Rosette	MM: 574, 1026, 1030, 1423, 1425, 1435, 1645, 1672, 1760, 1771, 1796, 1876, 1986, 1988–1989, 2017, 3500, 3778, 4073–4074, 4325, 4363, 5219; MFA: KEK 30

(*continued*)

Table 8.12 Cont.

Motif	*Species or detail*	*Museum or publication number*[a]
Landscape	Mountains as triangle or heap of drillings	MM: 1027, 1031, 1445, 1759, 1815/2, 2007, 2009, 3635, 4352, 4379, 4410; NM: 237, 240, 243, 275, 297, 1815/2, 1868/39, 2286/122
Luminaries (star/lunate)		MM: 353c, 574–575, 584, 999, 1024, 1026, 1028, 1030, 1201, 1423, 1425, 1435, 1438–1439, 1441, 1446, 1494, 1645, 1664, 1672, 1760, 1763, 1771, 1876, 1986, 1988–1989, 2013–2014, 2017, 3028–3029, 3485–3486, 3488, 3495–3496, 3500, 3502, 3643, 3778, 4073–4074, 4231, 4325, 5211, 5328, 5399, 5428; NM: 244, 1090a,1189, 1294, 1815/134, 1866/118/250/283/284, 1868/191, 2286/45
Heart		NM: 1188, 1866/143
Geometric motifs	Four-armed cross and stepped crosses, single or in combination with bows or crescents	MM: 353d, 576, 936, 999–1000, 1023, 1032, 1042, 1426, 1432–1434, 1442, 1444, 1447, 1449–1450, 1762, 1766, 1772–1774, 1777, 1813/394, 1867/3, 1890, 1921, 2003, 2006, 2009, 2021, 2999, 3019–3020, 3027, 3061–3062, 3495, 3499, 3509, 3644, 3727, 4031, 4037, 4109, 4161, 4227–4228, 4329, 4362, 4410, 5206, 5282; NM: 1254, 1256, 1259, 1273, 1290, 1813/384/394/434, 1815/5, 1866/256, 1868/124/130/150, 2004, 2305/20
	Swastika	NM: 12, 1868/131; MM: 580, 1022
	Stepped pyramids	MM: 353c/1
	Rhombs	MM: 1029/2.17, 4159/10

Note: MFA = Museum of Fine Art; MM = Mary Regional History Museum; NM = National Museum.

[a] The publication number (from Sarianidi 1998b) is only used for seals from outside Margiana.

2.2. *Hero and animals acting together/fighting with each other*: man/mixed creature as "master of animals" (as snake-holder, feline-holder, or holder of hoofed animals; Figures 8.2k, 8.5e, 8.16c); man and eagle (man kneeling before an eagle, man sitting on an eagle); man/mixed creature attacked/killed or eaten by snakes or lions (Figures 8.1i, 8.2k, 8.6e–f, 8.9a); man jumping over a bull (bull-leaping, Figure 8.7b).

2.3. *Goddesses acting together with animals or mixed creatures*: goddesses surrounded by animals, sitting on animals or animal mixed creatures, holding

Table 8.13 BMAC characteristic scene depictions

Scene	*Description*	*Museum or publication number*[a]
Animal fights	Eagle and snake	MM: 581, 1997, 3490; NM: 222, 297, 1250, 2286/69/86; MFA: 43/176; Sarianidi 1998b: 127, 137, 156–158, 160, 165, 170, 172–175, 191, 199, 206, 931, 938, 957–958, 986, 988–993, 1073
	Snake and feline	MM: 4368; Sarianidi 1998b: 112, 924, 927, 930–933, 935, 939–940, 951, 953, 962, 967, 1454
	Snake and other animals	Animal fights: MM: 2029/5.51; lion-dragon and snake: MM: 4386/10.81; NM: 223; lion and snake: Sarianidi 1998b: 924, 927, 931–933, 938–940, 949–951, 953, 957, 964, 967; snake and camel: MFA: 18/38; Sarianidi 1998b: 919, 936, 958; snake and goat/bull: MFA: 18/38; Sarianidi 1998b: 921–922, 925, 936, 940, 990, 1461
	Feline and hoofed animals	Sarianidi 1998b: 176–183, 185, 189, 925, 936
	Eagle and snake in a tree	NM: 230; Sarianidi 1998b: 223, 888, 989
	Eagle with its fledglings	Sarianidi 1998b: 128, 184, 190–191, 214, 1010
	Eagle with snake in its beak	NM: 2286/86, Sarianidi 1998b: 984, 995
	Hoofed animal with eagle	MFA: 4977
Hero and animal	Man/mixed creature "master of animals"	Man and snakes: MM: 2027/75.50; NM: 1152, 1815/139; MFA: 14/25; Sarianidi 1998b: 25–26, 28, 30–31, 46, 50; bull-man and snake: MM 2026/5.50; Sarianidi 1998b: 32, 42, 912/4, 1039; bird-man as master of animals: Sarianidi 1998b: 28–29, 904, 910, 912/2, 916–919; goat-man and snake: Sarianidi 1998b: 54–56, 1037; ram-man and snake: Sarianidi 1998b: 1039
	Man and eagle	Man kneeling before an eagle, man sitting on an eagle: MM. 4227; Sarianidi 1998b: 914, 995, 1234

(*continued*)

Table 8.13 Cont.

Scene	*Description*	*Museum or publication number*[a]
	Man/mixed creature attacked by animals	Battle between man/mixed being and feline: MM: 1994; man/mixed creature attacked/eaten by snakes: MFA: 20/45; MM: 1996; NM: 1262/48; Sarianidi 1998b: 56, 924–925, 1010, 1040, 1044, 1046, 1052–1053, 1427–1428, 1430
	Man bull-leaping	NM: 79
Goddess	Standing or sitting and surrounded or flanked by animals	Goddess and snakes: MM: 579/2.16; NM: 270, 1195; Sarianidi 1998b: 1, 23, 27, 38; goddess, lions, and snakes: Sarianidi 1998b: 40; goddess and lions/birds: Sarianidi 1998b: 22, 24, 34–35, 37, 39, 41, 45
	Sitting on animals or mixed creatures	NM: 1760; Sarianidi 1998b: 14–20
	Enthroned on a chair	Sarianidi 1998b: 1–7
Human and goddess	Young man, goddess and drinking vessel	NM: 1866/281
	Kneeling man or bird-man as shortened depiction of this motif	Sarianidi 1998b: 58–62, 917–919, 822–8925, 1043–1044, 1049, 1051
Banquet scene		NM: 1246; Sarianidi 1998b: 47–49
Animals flanking a tree		MM: 1996, 2023; Sarianidi 1998b: 93, 96
Whirl and knot motifs	Snake whirls or knots	MM: 1025, 1036, 1038, 1770, 3730, 3744, 4239, 4362, 5246/12; MFA: 29/3; NM: 221, 228, 348, 1250, 1277, 1291, 1755/7, 1866/115/270/276/342, 1867/52, 2286/86; Sarianidi 1998b: 250–262, 891, 966, 1001, 1081, 1086
	Eagle whirl, Eagle and snake whirl, Eagle, goat, and snake whirl	MM: 582/2, 1443, 1758, 3491, 3493, 4378; NM: 225, 1151, 1866/273, 1867/70; Sarianidi 1998b: 190, 192–195, 197–198, 200–204, 221, 222, 891, 927, 995, 1098; NM: 1815/1405; MM: 4240; MM: 3510

Table 8.13 Cont.

Scene	Description	Museum or publication number[a]
	Animal–humans, humans,	NM: 1866/323
	Human whirl	Sarianidi 1998b: 47
	Tulip whirl	NM: 1866/323; MM: 586, 3488, 5223
	Poppy/leaf whirl	Sarianidi 1998b: 968, 1101
	Mountain whirl	NM: 234–237, 1866/275, 1868/39, 2286/34/122; MM: 1027, 1445, 2009, 2012, 2014, 4352; Sarianidi 1998b: 1160, 1174, 1255–1256, 1259, 1263–1264, 1267, 1276–1278, 1293–1296, 1304–1205, 1307–1213, 1316–
	Rosette and mountain whirl	1319, 1321, 1867/5/134, 1868/8, 2286/45; MM: 3502
	Geometric whirl	Crosses: MM: 1203, 1888; NM: 1139; 1866/275, 1868/123, 2305/26; bows: MM: 1022, 1025, 1038, 1450, 1764, 1770, 3025, 3036, 3498, 3730; NM: 1868/123, 1866/115; MFA: 28/2; cross and bows: NM: 1294, 1813/434/437, 1866/279, 1867/107, 2305/20; MM: 3019, 3442, 3507, 3509; stepped pyramid: NM: 1866/280

Note: MFA = Museum of Fine Art; MM = Mary Regional History Museum; NM = National Museum.

[a] The publication number (from Sarianidi 1998b) is only used for seals from outside Margiana.

animals, kneeling goddess (goddess in a skirt made of mountain scales with snakes growing out from her body) (Figures 8.14b and 8.16a).

2.4. *Humans and goddesses acting together*: an important scene here is that of a standing or kneeling man being offered(?) a drinking vase from a goddess, a famous motif of ancient Near Eastern art (Figure 8.15).[30] This motif occurs both as fully developed and reduced to the kneeling man or bird-man.

2.5. *Banquet scene* (Figure 8.14f).

2.6. *Animals flanking a tree* (Figure 8.1i).

3. *Whirl motifs and knot motifs* (Figures 8.1a–c, e, h; 8.2a–c; 8.12d–e).

THE BACKGROUND OF THE BMAC ICONOGRAPHY

The most striking point is the strong stylistic similarity between the animal depictions on BMAC compartmented seals or other types of seals in "Murghab-style" and southeastern Iranian "Intercultural-Style" vessels. These vessels are embellished with

Table 8.14 Correlation between the numbers of the figures, those of the museums, and those in Sarianidi (1998b)

Figure no.	*Museum number*	*Sarianidi (1998b) no.*
8.1a	MM 1022	–
8.1b	MM 3744	–
8.1c	MM 3730	–
8.1d	MM 1038	–
8.1e	MM 1036	1666
8.1f	MM 1997	1624
8.1g	MM 2002	–
8.1h	MM 580	1681
8.1i	MM 1996	1621
8.1j	MM 1994	1617
8.2a	MM 4379	–
8.2b	MM 1775	1710
8.2c	MM 3635	1719
8.2d	MM 2000	–
8.2e, see also 8.2k	MM 1995	1620
8.2f	MM 1032	–
8.2g	MM 4411	–
8.2h	MM 1202	–
8.2i	MM 1035	–
8.2j	MM 4352	–
8.2k = 8.2e enlarged	MM 1995	–
8.3a	MM 1009.4	–
8.3b	MM 1009.3	–
8.3c	MM 1009.2	–
8.3d	MM 1009.5	–
8.3e	MM 1439	–
8.3f	MM 3643	–
8.3g	MM 616	–
8.4a	NM 1868/191	1802
8.4b	NM 1866/32	–
8.4c	NM 1189	–
8.4d	NM 1866/282	–

Table 8.14 Cont.

Figure no.	*Museum number*	*Sarianidi (1998b) no.*
8.4e	NM 1292	–
8.4f	NM 1866/118	–
8.4g	NM 1868/93	–
8.4h	NM 1866/114	–
8.4i	NM 1866/284	–
8.4j	NM 1188	–
8.4k	MM 3612	–
8.4l	MM 1883	–
8.4m	MM 1430	–
8.5a, b, c, d	MFA 18/38	1635
8.5e, f, g, h	MFA 14/25	1618
8.6a (obverse and reverse)	MFA 19/42	1643.1
8.6a3	MFA 176/43	1648.1
8.6b	MFA 19/42	1643.3
8.6c	MFA 163/33	1625.1
8.6d	MM 1994	1617.1
8.6e	MM 1996	1621.1
8.6f1–3	MFA 20/45	11654
8.7a	MFA 13/74	1784
8.7b	NM 79	1634
8.8a, d1	MM 2029	1777
8.8b, d2	MM 2028	1778
8.8c, d3	NM 1196	–
8.9a	NM 1281/115	1766
8.9b	NM 246	1768
8.9c	MM 2029	1769
8.10a	NM 174/1141	1786
8.10b	MFA 25/21	–
8.10c	MFA 26/20	–
8.11a	NM 174	–
8.12a	NM 169/1147	–
8.12b	MM 1758	–

(continued)

Table 8.14 Cont.

Figure no.	*Museum number*	*Sarianidi (1998b) no.*
8.12c	MFA 4977	–
8.12d	NM 271/52	–
8.12e	NM 121	1787
8.12f	MM 3510	–
8.13a, d, g	MM 1767	–
8.13b, e, h	MM 3774	–
8.13c, f	MM 4329	–
8.14a, c, e	NM 361	–
8.14b, d	NM 360	–
8.14f	NM 1246	1550
8.15	NM 1866/281	–
8.16a	NM 1815/1405	–
8.16b	NM 270/3191	–
8.16c	NM 1815/139	–
8.17a	MFA 23/109	1761.1
8.17b	MM 2025	1763.1
8.17c	MFA 24/111	1762.2
8.17d	MM 2014	–
8.17e	MM 1796	1746
8.17f, g	MM 2017	1753
8.18a	NM 1868/130	–
8.18b	NM 1090e	–
8.19f	NM 1815/1405	–
8.19g	NM 1866/281	–
8.19h	NM 360	–
8.19i	NM 168	1549

Note: MFA = Museum of Fine Art; MM = Mary Regional History Museum; NM = National Museum.

multifigured motifs of animal and human actors in a very specific manner, and the same variations appear on the BMAC seals.

In both of them the animal imagery is featured with nearly identical specific stylistic details. It is the case for the depiction of eagles, snakes, or lions, especially of their nostrils, the lions' paws and protruding eyes, or for the partition between the neck and the head with small lines or a band. The same can be said for the attitudes of the animals, with their head turned back (for felines and hoofed animals), the frontally depicted eagle with outstretched wings and turned head, or the diving eagle with its wingtips turned downwards. This has already been discussed in many articles (Winkelmann 1997b, 2003, 2005, 2006, 2008b, 2009, 2010, 2013, 2014).

The parallels do not only concern stylistic features, but also the motifs, suggesting a supra-regional concept. However, in BMAC art, the arrangement of the motifs or of single-picture elements underwent specific transformations and variations already discussed (see esp. Winkelmann 2008a, 2009, 2013, 2014).

Thus, the fully developed motifs known on southeastern Iranian stone vessels occur on BMAC seals as new arrangements. They are, for instance, often reduced to the central figure of a multifigured scene,[31] or they appear as a whirl motif or as ligature of the acting figures, as well as in various other forms.

The reduction of a multifigured motif to that of a single animal in a specific posture is especially characteristic of the compartmented seals, in which the images are extremely reduced but have a high symbolic meaning. This also applies to inanimate motifs like stars, rosettes, tulips (Figures 8.4a, b, f, i; 8.13a, c), mountains, or hundreds of varieties of the stepped cross, which were increasingly represented on BMAC seals. These symbols can be considered as aniconic attributes or appearances of the goddesses or of temples (ziggurat).

The "reduction" to single-picture elements has also been used to create a typical and specific arrangement in BMAC art: whirls of one element or of two or three figures acting together (Figure 8.12b, d, e). Sometimes these whirls only consist of heads or upper bodies of persons or animals.

Another variation concerns mixed creatures. Unlike central and southern Mesopotamia, where these creatures are human headed with an animal body, but together with Syria and Anatolia (for the bird-man, see Amiet 1978: 162 or Collon 1987: 142), BMAC art and Iranian art share the creation of animal-headed mixed creatures (von der Osten-Sacken 1992) combining animals and men as well as different animals (Sarianidi 1998b: nos. 196, 224–227). This fusion results in new compositions in BMAC art. This applies especially to the many variations of the BMAC snake-dragon where snakes, lions, eagles, and goats merge into a completely new, specific creature (Figures 8.14b; 8.19e, h) (Winkelmann 2013: figs. 9, 10). Another specific feature in the art of BMAC is the winged animal. Lions, snakes, or hoofed animals are transformed into lion-dragons, snake-dragons, or winged bulls, goats, rams, or camels (Figures 8.6c, f; 8.7a; 8.8a; 8.14b).

Both Iranian and BMAC art have created animal-headed human mixed creatures like the bird-man, ram-man, bull-man, or goat-man, following a tradition that goes back to 4th millennium Iranian art. All these mixed beings act in the same pictorial context, fighting as "master of animals" with snakes or lions, or being attacked or devoured by snakes or lions. This would suggest a similar mythological background. However, in BMAC art, the bird-man has a dominant position as he is the most frequently depicted creature on seals and has numerous appearances, and as one type of the (rare) male representation in terracotta art can also been compared to a bird-man with wings and a pointed beak.[32] Next to him, a new mixed creature, the monkey-man, is found on seals representing a further appearance of a mythological person, maybe influenced by contact with the Harappan Civilization.

The same common features are visible in the depictions of humans, human–animal mixed creatures, and goddesses on BMAC compartmented seals and southeastern Iranian cylinder seals (see also Pittman 2014a, 2019) (Figure 8.20). On the Iranian seals, humans and mixed creatures have the upper body, arms, attributes like wings, plants, or horns depicted in a front view, while the lower body and the heads are in profile. The slim figures have wasp waists and thin arms, legs, and feet. Southeastern Iranian women or goddesses have triangular upper bodies with small breasts. The representation generally is flat, but many details like hairstyle, hands, arms, ornaments, or wings are additionally rendered through finely cut lines. Humans have faces with big almond-shaped eyes without eyelids, a hooked nose, and only rarely a small mouth. They have long necks and long hair falling to the shoulders at the back. Often, a well-smoothed drill hole is used to depict the shoulder joints. Goddesses are presented standing with their hands clasped, sitting on a throne with slanting back, kneeling and sitting on their knees, sometimes with swinging skirts, sometimes with a stylized lower body in the shape of a box. Goddesses can be completely anthropomorphic or have animal heads (for definition of stylistic features, see Winkelmann 1997a; for parallels with southeastern Iranian seals of the Kerman culture, see Porada 1962: fig. 12; 1993: fig. 31; Hakemi 1972: no. 323 = Hakemi 1997: 355, no. 2263, Ib 2 and no. 3, Ib 3; Lamberg-Karlovsky 1973: pl. XXVIc, XXXIc; Potts 1981: fig. 1; Amiet 1986: fig. 132, *1*, *2*, *3*, *4*, *7*, *9*, *11*, *12*; 1994: figs. 1, 3; Collon 1997: fig. 1; Winkelmann 1997a: fig. 1; Sarianidi 1998b: fig. 27, *3*). The same features have been underlined for the BMAC goddess(es).

Another similarity is the pantheon itself: it consists overwhelmingly of female deities or one goddess under various appearances. Although many male heroes or mixed beings appear, we have so far not a single clearly recognizable male god in southeastern Iranian and BMAC art. The goddesses of southeastern Iran and BMAC widely share the same attributes (Figures 8.19 and 8.20), be they animals like scorpion, snake, hoofed animals, felines, and eagle; plants like tulip or wheat; symbols like star and rosette; and aniconic attributes like vessels or mountains that surround the goddesses (Winkelmann 2010: figs. 5, 6, 8–13; 2012: fig. 3). In both regions, the goddesses merge with their animal attributes, wearing horns or wings, or having snakes and lions' tails growing from their bodies. They also merge with plant attributes, for example having wheat ears protruding from body and head, or with a mountain, represented as a skirt with mountain scales (Figures 8.16a and 8.19f).[33] They also seem to share a common mythological subject: Both southeastern Iran seals and miniature work as well as BMAC seals depict a scene where a young man receives a drinking vase from a sitting goddess (Figures 8.15; 8.19g; and 8.20k) (Kerman seal, see Porada 1962: fig. 12; see also Winkelmann 2010: fig. 4i). As this enthroned goddess is surrounded by other divinities, it seems that both regions shared the same central female deity.

It also seems that nearly all BMAC goddesses have clear antecedents in southeastern Iranian art, although not all southeastern Iran deities reappear in identical form in BMAC. Thus, goddesses enthroned on dragons or snake-dragons are a new phenomenon found only in BMAC, and the Iranian goddess holding water streams is not found in the BMAC art. Nevertheless, the general similarity between the pantheons of both regions is clear enough.

We can therefore conclude that the main impetus for the glyptic BMAC art comes from older southeastern Iranian mythological motifs and artifacts, as well as from many of its stylistic features, but that they were transformed into this fascinating Central Asian iconography of Bronze Age art.

Figure 8.1 "Murghab-style" seals, different shapes. a, b: round stamp seals with one face and handle; c–e, h: two-sided disc-shaped stamp seals; f: cushion-shaped two-sided stamp seal; g: rectangular two-sided stamp seal; i: square two-sided stamp seal; j: two-sided seal in shape of a stepped cross. (a: MM 1022; b: MM 3744; c: MM 3730; d: MM 1038; e: MM 1036; f: MM 1997; g: MM 2002; h: MM 580; i: MM 1996; j: MM 1994.)
All photos in this chapter are made by the author on behalf of the Margiana Archaeological Expedition and published with the kind permission of Prof. Victor Sarianidi and Prof. Nadezhda Dubova (© Margiana Archaeological Expedition). MM: Mari Museum; NM: National Museum, Ashgabat; MFA: Museum of Fine Arts, Ashgabat.

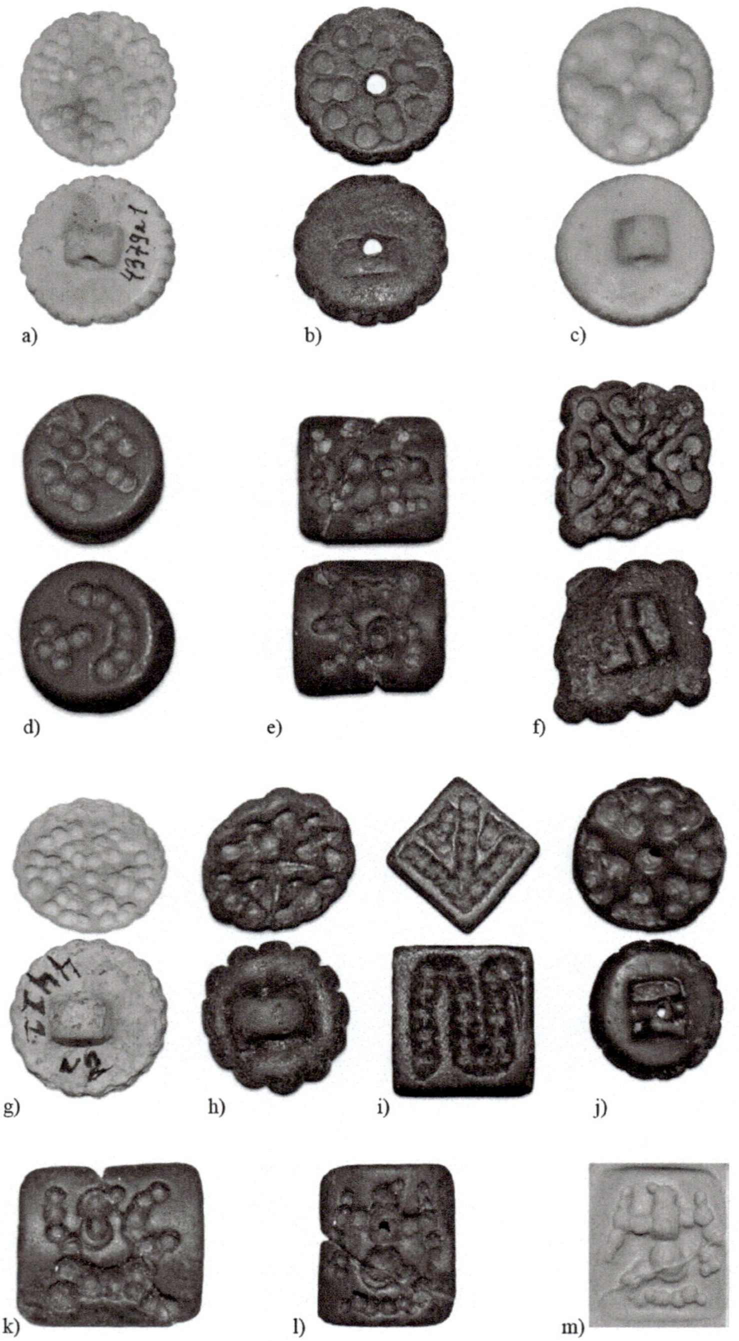

Figure 8.2 "Drilled-style" seals, different shapes. a–c, g, h, j: round stamp seals with one face and handle; f: square stamp seal with one face and handle; e, k: cushion-shaped stamp seal with impression of the picture of kneeling bird-man holding snakes; i: two-sided lozenge-shaped stamp seal. (a: MM 4379; b: MM 1775; c: MM 3635; d: MM 2000; e: MM 1995; f: MM 4352; g: MM 4411; h: MM 1202; i: MM 1035; j: MM 4352; k = e.)

Figure 8.3 Clay stamp seals from Gonur Depe: face and side views. (a: MM 1009,4; b: MM 1009,3; c: MM 1009,2; d: MM 1009,5; e: MM 1439; f: MM 3643; g: MM 616.)

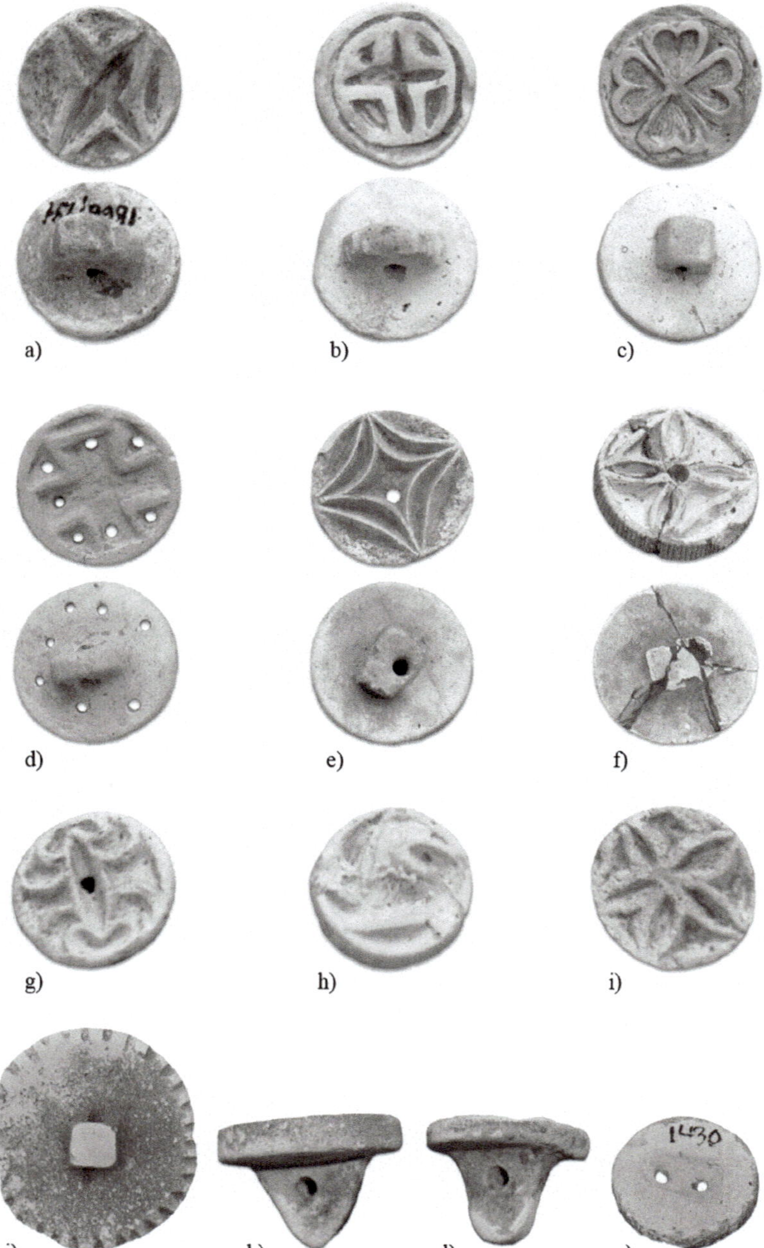

Figure 8.4 Faience seals with remains of greenish glaze, partly reworked with secondary perforation: face, back and handle. a, b, e: star-motif; c: flower; d: four-armed cross; f and i: rosette; g: scorpion; h: whirl. a, b, k: with triangular handle; c–f, j: with rectangular handle; l: triangular handle with rounded top; e, g, m: secondary perforation. (a: NM 1868/191; b: NM 1866/32; c: NM 1189; d: NM 1866/282; e: NM 1292; f: NM 1866/118; g: NM 1868/93; h: NM 1866/114; i: NM 1866/284; j: NM 1188; k: MM 3612; l: MM 1883; m: MM 1430.)

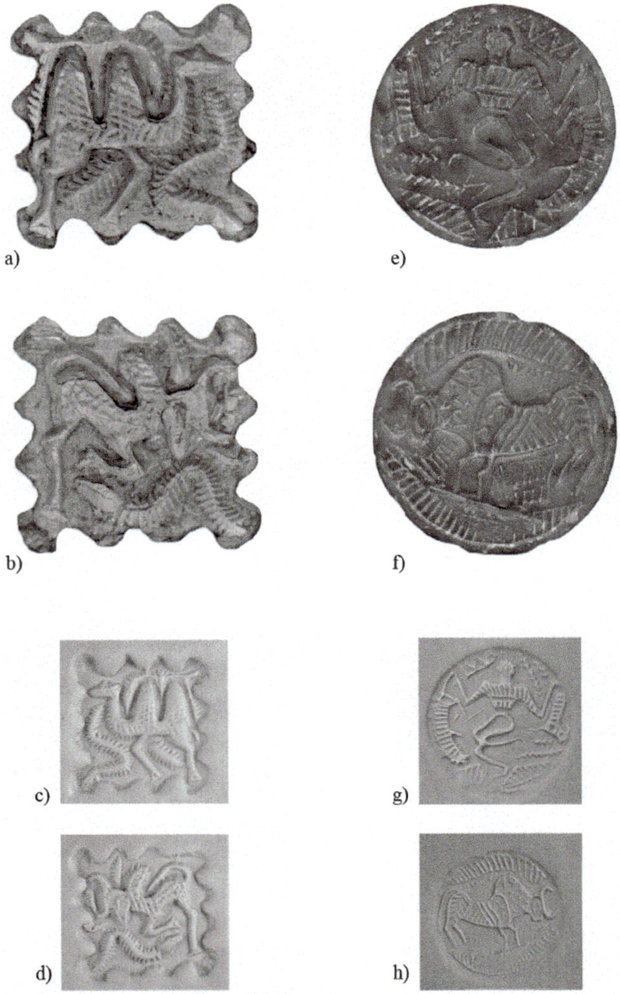

Figure 8.5 Two-sided "Murghab-style" seals in graphic substyle. a–d: two-sided square stamp seal with toothed rim and impressions with motifs of snake attacking a zebu-bull; b, d: and snake attacking a camel (a, c); e–h: two-sided disc-shaped stamp seal and its impressions with the motifs: walking man as "Master of animals" between mountains and plants, holding a goat and a quadruped (e, g) and attacking zebu-bull between mountains (f, h). (a–d: MFA 18/38; e–h: MFA 14/25.)

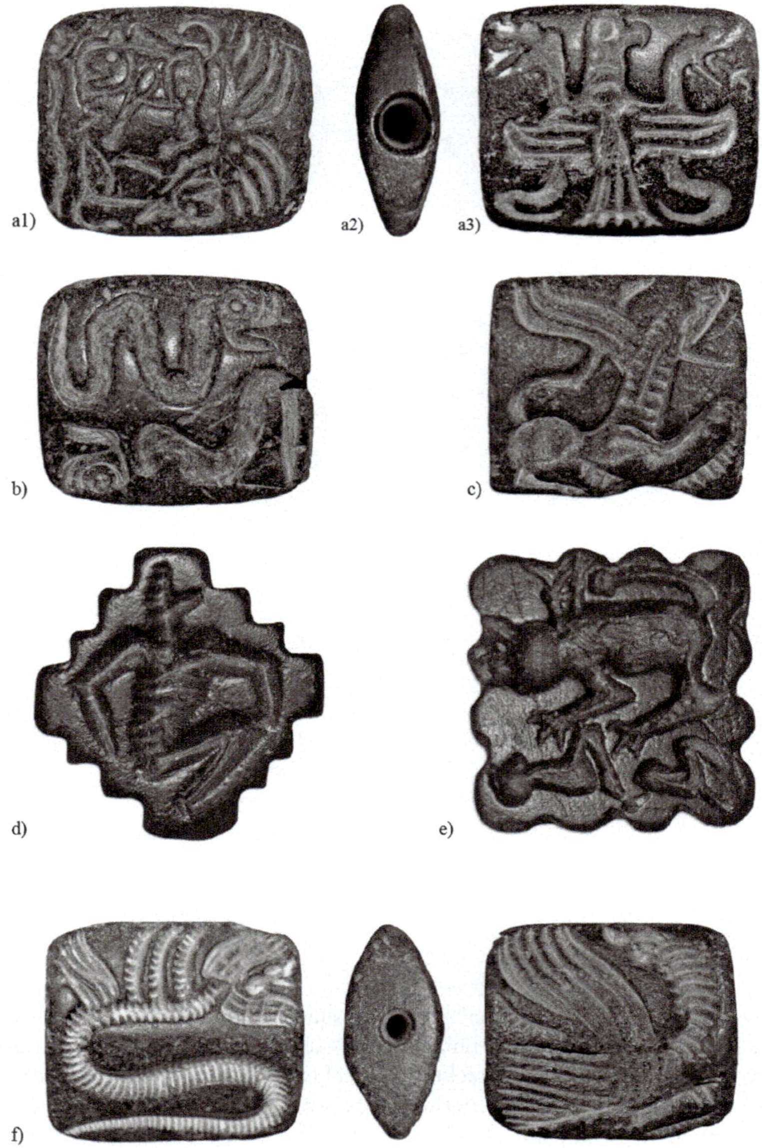

Figure 8.6 Different "Murghab-style" seals. a–e: plastic substyle; f: graphic substyle. a: obverse, side view and reverse of a cushion-shaped seal with battle between two lions and a snake and battle between eagle and snakes; b: cushion-shaped seal with two snakes; c: cushion-shaped seal with attacking lion-dragon and snake under the belly; d: two-sided seal in shape of a stepped cross with kneeling monkey-man; e: square two-sided seal with wavy rim and winged lion devouring a kneeling man; f: obverse, side view and reverse of a cushion-shaped seal with snake-dragon devouring a man and eagle with turned head. (a1–a2 and b: MFA 19/42; a3: MFA 176/43; c: MFA 163/33; d, e: MM 1994; f: MM 1996.)

a)

b)

Figure 8.7 Unusual "Murghab-style" seals. a: three-sided longitudinally perforated prism with side view and its three faces, showing a winged sheep, two crossing bulls and a running lion; b: figurative seal in form of a camel with turned head with stamp face and its impression, representing bull-leaping in the upper part (man jumping over the back of a running bull and three eagles) and one walking and one sitting man in the lower part, separated by guilloche. (a: MFA 13/74; b: NM 79.)

a)
b)
c)

Figure 8.8 Local cylinder seals and Akkadian cylinder seal from Gonur Depe. a: cylinder seal in local Murghab-style and its impression, representing a dog, winged lion, tulip and snake; b: concave cylinder seal in undefined style and its impression with running animal; c: Akkadian cylinder seal from Gonur Depe Large Necropolis: seal and impression. (a: MM 2029; b: MM 2028; c: NM 1196.)

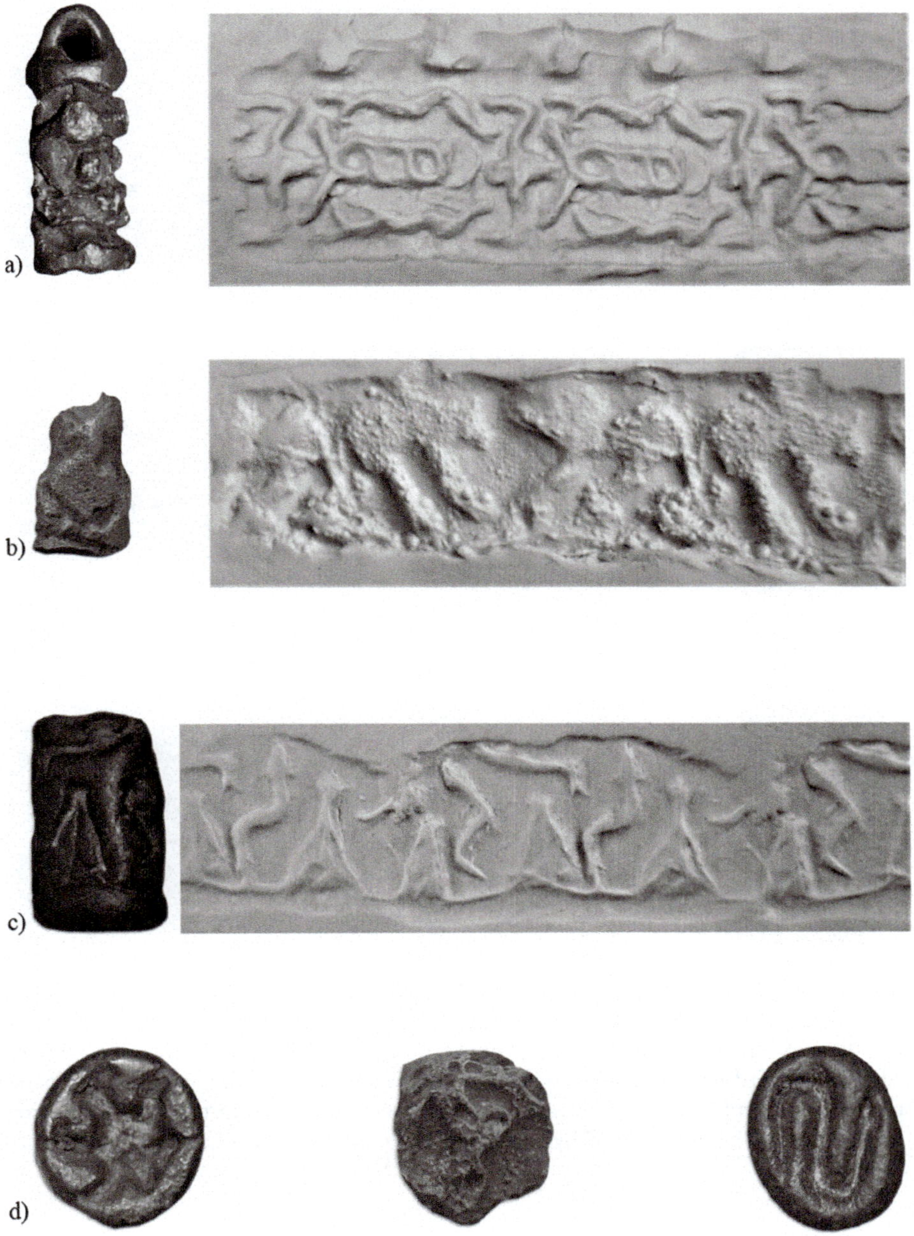

Figure 8.9 Local stamp cylinders from Gonur Depe and Togolok 1. a: stamp cylinder from Gonur Depe, in bronze, deeply engraved, and its impression, with three snakes, the central snake devouring a man; b: fragment of a stamp cylinder from Gonur Depe, in bronze, deeply engraved, and its impression, with three attacking snakes, one with open mouth; c: fragment of a stamp cylinder from Togolok 1, in chlorite, deeply engraved, and its impression with three snakes and one scorpion, flanking mountains; d: stamp faces of the stamp cylinders a–c. (a/d1: NM 1281/115; b/d2: NM 246; c/d3: MM 2029.)

Figure 8.10 Three imported southeastern Iranian cylinder seals found at Gonur Depe. a: cylinder seal and impression with vegetation goddess with wheat ears and tulip over a snake and goat on a platform; b: seal and impression with multifigured scenes with different mixed-beings, a bird-headed goddess with two heads flanked by snakes, two kneeling persons adoring a temple and lying bull and eagle; c: chlorite cylinder and impression with enthroned horned deity in a circle flanked by other goddesses. (a: NM 174/1141; b: MFA 25/21; c: MFA 26/20.)

Figure 8.11 Two cylinder seals of mixed styles. a: cylinder seal, impression and drawing, in a mixed style of Early Dynastic III motif in southeastern Iranian style and technique together with some BMAC features, found near Togolok (Salvatori 2008 = NM 174); b: seal in a mixed southeastern Iranian and BMAC style, found in Iran (Porada 1993: p. 29).

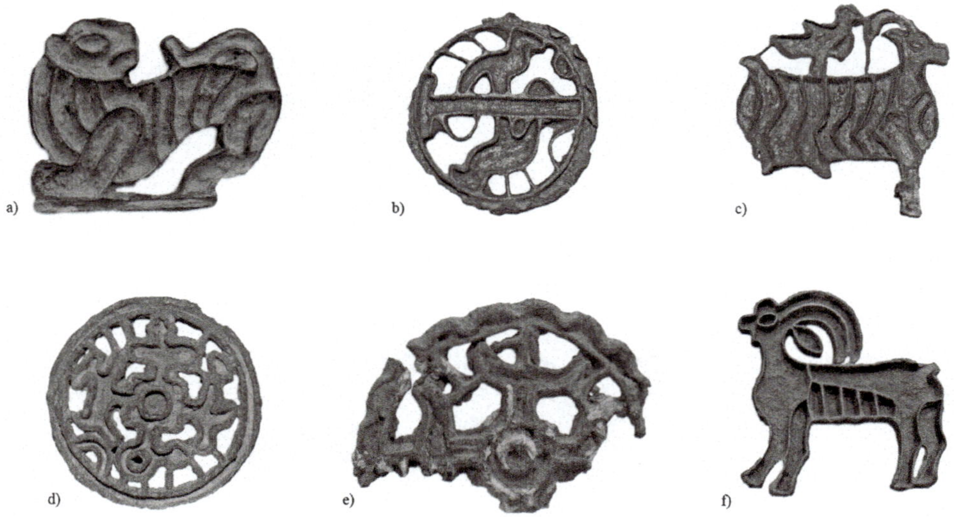

Figure 8.12 Compartmented seals with animal depictions. a: figurative seal in the shape of a tiger with turned back head; b: open-work seal with double frame with two sitting eagles in the mountains; c: figurative seal in the shape of an eagle sitting on the back of a goat; d: open-work compartmented seal with double frame with whirl of eagles, snake, and goat; e: open-work with double frame with eagle whirl; f: figurative seal in the shape of a standing goat. (a: NM 169 /1147; b: MM 1758; c: MFA 4977; d: NM 271/52; e: NM 121; f: MM 3510.)

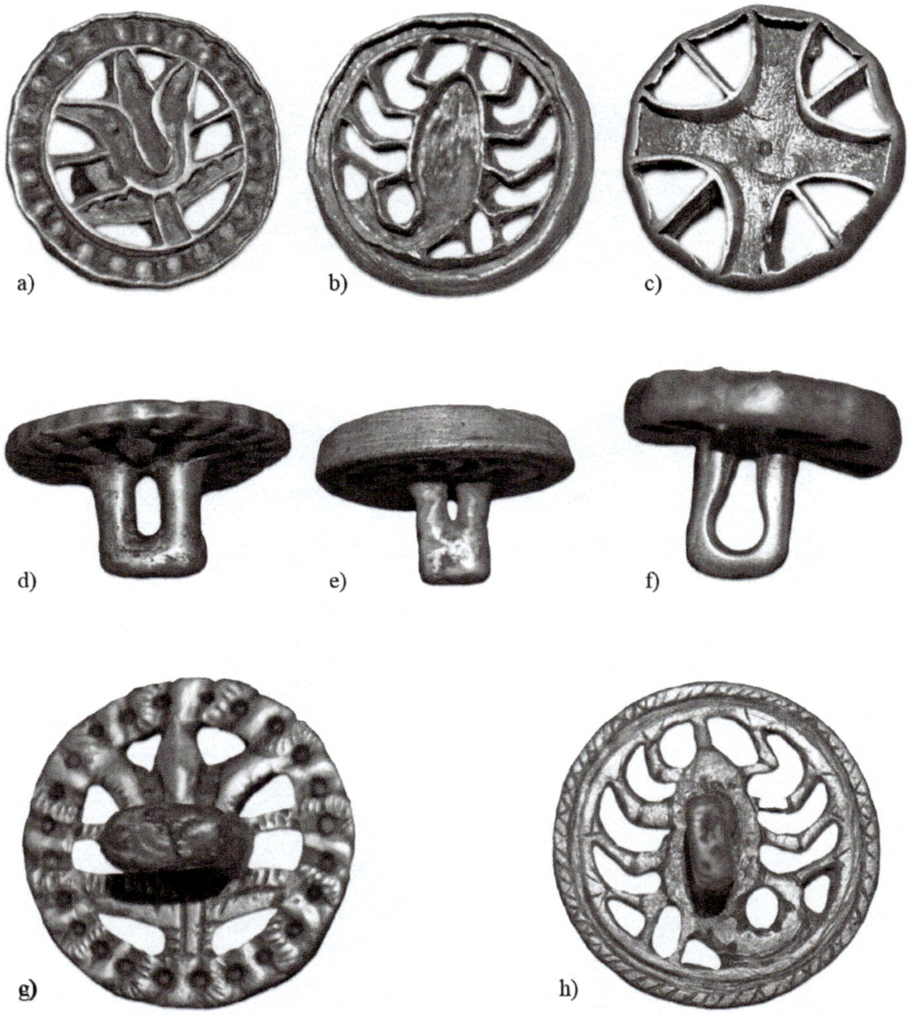

Figure 8.13 Open-work compartmented silver seals with rectangular handles and plastically rendered back or/and incisions coming from the same workshop. a, d, g: face, handle, and plastically rendered back of seal with double rim with tulip, surrounded by stylized snakes (guilloche); b, e, h: face, handle, and plain but incised back of seal with double rim with scorpion; c, f: face and handle of seal with simple rim and crossmotif. (a, d, g: MM 1767; b, e, h: MM 3774; c, f : MM 4329.)

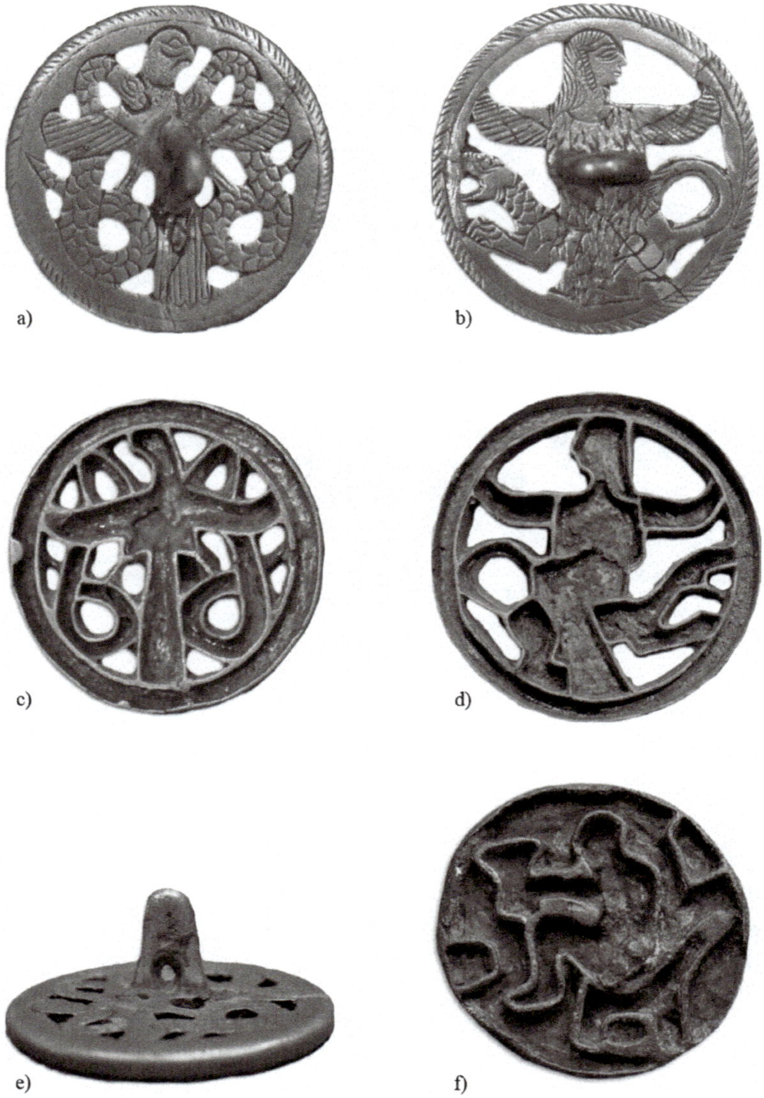

Figure 8.14 Compartmented seals with flat back and mythological depictions. a and b: open-work seals with flat but incised back from the same workshop; f: solid seal; a, c, e: reverse, face, and handle of a seal with battle between eagle and snake; b, d: reverse and face of a seal with goddess on a lion-dragon; f: solid seal with abbreviated banquet scene: man with a beaker. (a, c, e: NM 361; b, d: NM 360; f: NM 1246.)

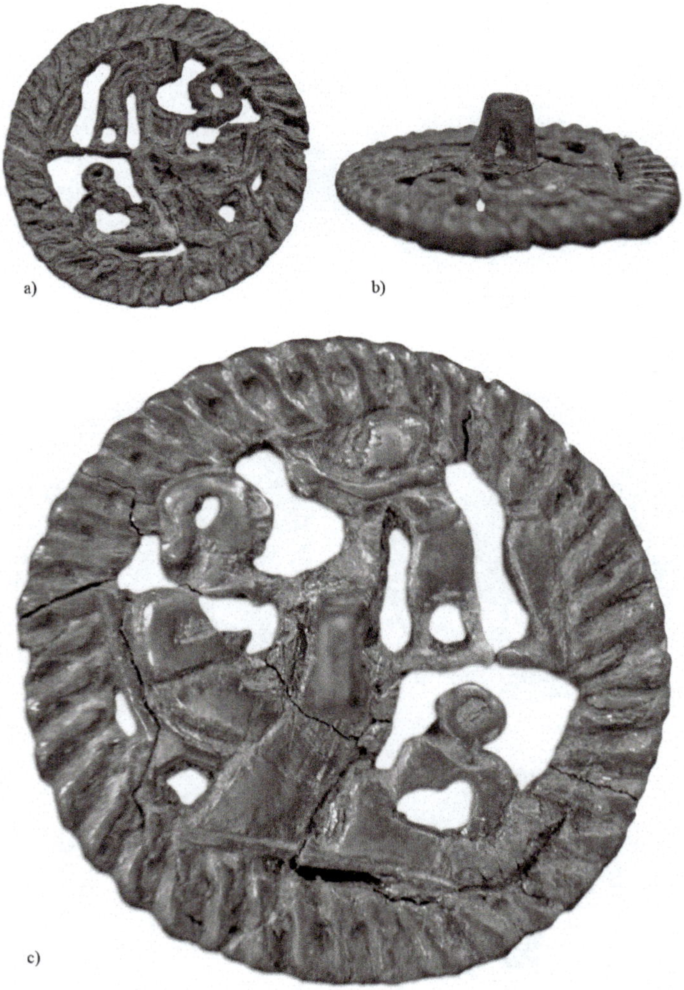

Figure 8.15 Compartmented open-work silver seal with scenic depiction: young man being offered a drinking vase from a goddess. a: obverse; b: handle; c: reverse. (NM 1866/281.)

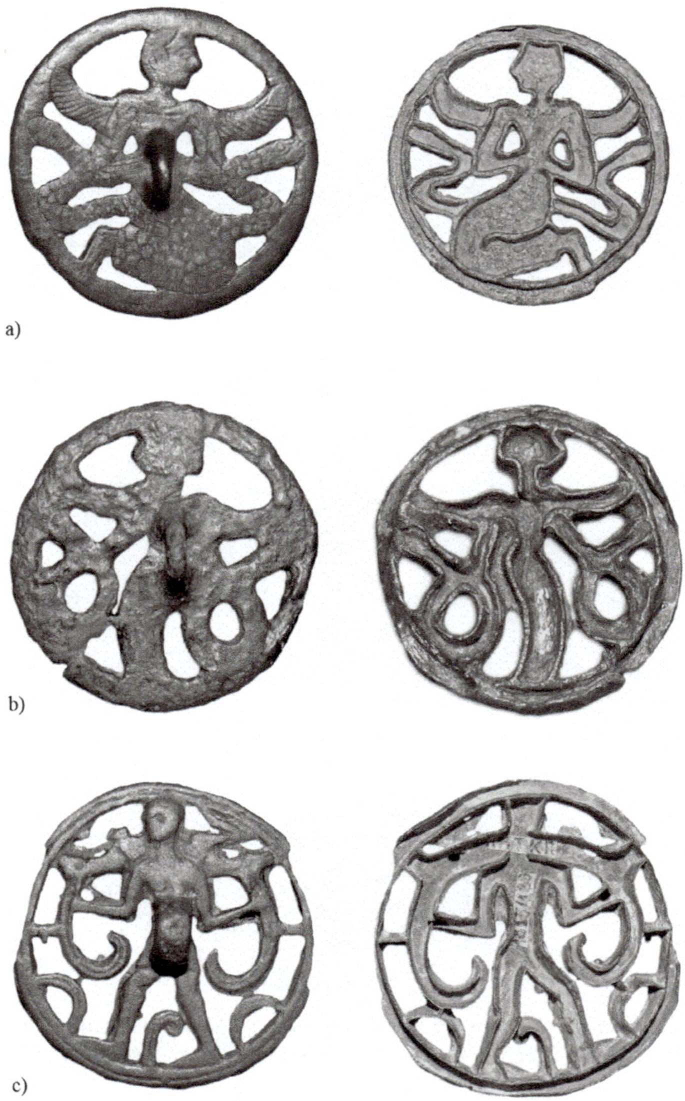

Figure 8.16 Obverse and reverse of open-work compartmented seals with mythological depictions. a: winged goddess with snakes and skirt made of mountain scales. Reverse flat and incised; b: standing winged goddess with snakes; c: walking man as snake-holder. (a: NM 1815/1405; b: NM 270/3191; c: NM 1815/132.)

Figure 8.17 Bullae, cretula, sherd, and stoppers with seal impressions. a: bulla with impression of cylinder seal with man, dog, and two-humped camel; b: bulla with many impressions of round stamp seal with snakes; c: bulla with impressions of different seals. Bird-man as master of animals, plants; d: sherd with impression of a figurative compartmented seal; e: cretula with impression of a compartmented seal; f, g: side view and face of stopper with impression of a figurative stamp seal. (a: MFA 23/109; b: MM 2025; c: MFA 24/111; d: MM 2014; e: MM 1796; f, g: MM 2017.)

Figure 8.18 Two casting molds for seals from Gonur Depe. (a: NM 1868/130; b: NM 1090e.)

Figure 8.19 Standing, kneeling, and enthroned goddesses on compartmented seals of the BMAC. a–g: round framed open-work compartmented seals; i: figurative seal. a: standing winged goddess flanked by snakes; b: standing winged goddess flanked by lions; c: standing goddess flanked by eagles and mountains; d: standing goddess flanked by felines and eagles; e: goddess in *kaunakes*-skirt with goats enthroned on a lion-dragon; f: kneeling, winged goddess with skirt made of mountains scales and with protruding snakes; g: enthroned goddess in *kaunakes*-skirt with second kneeling goddess and young man holding a beaker, surrounded by guilloche; h: winged goddess in *kaunakes*-skirt sitting on a lion-dragon; i: head of silver pin form of a compartmented seal, depicting an enthroned goddess in *kaunakes*-skirt. a–d: Sarianidi 1986b: fig. 1, or Sarianidi 1998a: nos. 38, 36, 39, 37; e: drawing after Sarianidi 1998: no. 16.2; f: drawing after Sarianidi 2002: fig. 3 on p. 261 (NM 1815/1405); g: Sarianidi 2007: fig. 161 (NM 1866/281); h: drawing after Sarianidi 2002: fig. 3 on p. 261 (NM 360); i: Sarianidi 2007: fig. 159 (NM 168).

Figure 8.20 Minor art from southeastern Iran with depictions of goddesses on seals (a-j) and on lapis lazuli discs (j, k). a: standing vegetation goddess with wheat ears; b: standing winged and horned goddess; c: standing winged goddess flanked by snake and quadruped; d: standing goddess with sheep horns; e: enthroned, winged, horned goddess; f: enthroned divinity with snakes, bull head and mountain symbol; g: enthroned goddess with goat head and stars; h: kneeling vegetation goddess with wheat ears; i: kneeling, winged, and horned goddess; j: kneeling goddess over zebu-bulls, associated with snake and eagles; k: two sides of disc: obverse with man kneeling before an enthroned goddess giving him a drinking vessel; reverse: banquet scene with enthroned man and kneeling woman drinking. (a: Amiet 1994: fig. 3 a; b: Amiet 1994: fig. 3 b; c: Amiet 1986a: fig. 132.7; d: Hakemi 1972: no. 323; e: Lamberg-Karlovsky 1973: pl. XXVI c; f: Porada 1962: fig. 12; g: Winkelmann 1997: fig. 1.; h: Hakemi 1972: no. 323; i: Potts 1981: fig. 1a.; j: Winkelmann 2010: fig.5; k: Porada 1993: fig. 19.)

NOTES

* I wish to warmly thank Bertille Lyonnet for her constructive, critical review and comments on this chapter as well as her careful and patient reviewing that greatly improved its presentation.

1 Stamp cylinders and cylinder seals with a handle already existed in the Uruk and Jemdet Nasr period in Mesopotamia and Iran, but disappeared later: See Amiet 1961: pl. 19, no. 315; pl. 20, no. 323 (Susa); pl. 26. nos. 426–427; pl. 27, no. 439; pl. 41, no. 619; pl. 43. nos. 636–637; pl. 44, no. 439, 640; pl. 51, no. 706; pl. 56, no. 770; 1972: pl. 90, no. 727; pl. 92. no. 770; pl. 99, nos. 917, 923; pl. 112, no. 1042. Not all their lower face is drawn or with a picture, but some are clearly stamped. So far, we have no evidence of their presence in the Jiroft culture, but we cannot exclude that this type was still used during the 3rd millennium in Iran, leading to the BMAC ones.

2 A compilation of all the compartmented seals from Shahr-i Sokhta and Mundigak is published in Baghestani (1997: 384–385).

3 Konar Sandal, Kerman culture: Pittman in Majidzadeh (2008: figs. 28, 33); Shahdad: Hakemi (1997b: cemetery A: figs. Ia, *1–5*, *7–9*, *11*, *12*, *14*, *16*; surface: Xi, m, n, o, p, q, l; impressions: figs. on pp. 184–185, 190–193, 238, 244, 261, 295, 303, 310, 315, 370, 446, 674–677 (nos. 92, 115, 190–192, 253), 686 (no. 223), 684 (no. 173), 674 (no. 102), 681 (no. 69), 675 (no. 136), 686 (no. 218), 675 (no. 141), 686 (no. 219), 675 (nos. 126, 131), 676 (no. 154), 675 (nos. 117, 127); Baghestani (1987: fig. 64, no. 306); Tepe Yahya: Lamberg-Karlovsky (1973: pl. XXXb, fig. 1); Susa: Baghestani (1997: figs. 87, 117); Ligabue and Salvatori (1988: pl. 51).

4 Salvatori (1981: 126) gives only a rough estimation for the dating of the seals (pl. I.1–3 in the 3rd millennium, and the seals on pl. II–X in the beginning and first half of the 2nd millennium BC).

5 For the chronology, see Chapter 1, Tables 1.2 and 1.3, and for radiocarbon dates of some of the Gonur graves mentioned in this chapter, see the Appendix at the end of the book.

6 I am very grateful to Gabriele Rossi Osmida who sent me the list and photos of the seals from his excavations.

7 Rossi Osmida (2011: 226) mentions 26 impressions from dated contexts but does not give further information. The very interesting cretulae are on pp. 236–237.

8 Despite intensive investigations in the archives it was not possible to find out the exact location of the places where the seals came from.

9 Thanks to the kind invitation of Professor Dubova, the University of Bern took part in the Margiana archaeological expedition excavations at Gonur Depe in 2014 and 2015. The results were published in Winkelmann (2015, 2016a).

10 The book also contains some interesting seals that belong to later periods, especially to the Achaemenid and post-Achaemenid until the Parthian period, for instance nos. 976, 978 1362–1365, 1367–1423. As shown by the recent excavations at Dahan-e Ghulaman, an Achaemenid site in Iran (Sajjadi 2007: fig. 13), stamp seals with a high knob-handle and tamga-like motifs (which Sarianidi named the "Hittite type") belong to later periods. This would concern seals like nos. 830–834 and 842–848.

11 First attempts: Winkelmann (1999; Salvatori 2000). In this main article Salvatori tried to classify the seals according to their chronological position and their type.

12 At the beginning of the 3rd millennium, in Mesopotamia and Elam, cylinder seals, stamp cylinders, and stamp seals were used together. But, during the first half of 3rd millennium cylinder seals had totally replaced stamp cylinders, and, by the end of the millennium, they also almost completely took over stamp seals.

13 The grave is considered as late by Sarianidi (Gonur period III) but the style of the pin is undoubtedly much earlier (Gonur period I). It may imply a very long use of such items.

14 Pirak was occupied between ca. 1700 and 800 BC and seems to correspond to new settlers who do not belong to the post-Harappan culture. The finds of horses and the pottery hint at some connections with the Central Asia Yaz I culture.

15 The numerous seal impressions from Gilund Rajasthan (northwest India) belong to the end of the middle Ahar Banas culture dated to 2100–1700 BC and thus contemporary with the late and post-Indus Civilization. These impressions suggest links between the two cultures, and perhaps also with the BMAC or BMAC tradition.

16 The circumstances of the find are uncertain. While the inventory book of the Mary Regional History Museum mentions Gonur North as its location, Sarianidi writes that it comes from Gonur South, room 592. The seal is stored under MM no. 2029.

17 B. Brentjes (1971) speculated about inter-marriages between Central Asia and southeastern Iran after the discovery of a typical "intercultural-style" weight in the Soch valley (Ferghana) in Central Asia.

18 This kind of representation is a stylized way to figure a person with a long skirt covering legs and feet and sitting on her knees. In my earlier publications, I defined this type of stylized squatting persons as "fußlose Hocker" (Winkelmann 1993, 1994). It appears in Iranian art already in the Late Uruk time at Susa and elsewhere and continued to exist over the whole 3rd millennium in the glyptic and minor art of Elam and Kerman. Later, it spread to BMAC too, becoming one of the most typical appearances of the composite figurines of goddesses (see Winkelmann, in press).

19 Teufer (2007) and Ascalone (2008) both wrote about tulips as the symbol of a goddess in BMAC and Iran. For tulips on southeastern Iranian seals from the Jiroft culture, see Amiet (1986: fig. 132, 2, 4, *11*, *12*; 1994: fig. 3), and Hakemi (1997b: 676–677 seal impression nos. 190–192). Tulips as the symbol of a goddess appear already on 4th-millennium seals from Luristan and continue until the BMAC.

20 An even earlier date, ED I–II is advanced by Pittman (2013) for a fragmentary seal impression found at Konar Sandal South from a trash deposit (Trench V).

21 Unfortunately not found in any of the museums.

22 This impression is now in the Museum of Fine Art, but it has no number.

23 No. 1: seal found at Tell Asmar: Collon (1987: no. 610, animal row of rhinoceros, elephant, and crocodile). No. 2: seal from Tell Suleimeh, level VII: Collon (1987: no. 609, bird-man, wild ox (gaur), unicorn, and fish). No 3: seal from Ur: Collon (1987: no. 612, unicorn and tree). No. 4: Mohenjo-Daro: Joshi and Parpola (1987: M-418, snake-dragon, goat on a tree, attacking eagle). No. 5: Mohenjo-daro: Franke Vogt (1991: no. 229, SV 20–129, tiger with head turned back and scorpion). No. 6: "Corbiau-seal": Collon (1987: no. 614, seal with two registers presenting a god with a crown flanked by snake-dragons, master of animals with tiger, and wild ox (gaur)).

24 Some rare seals show males completely in profile.

25 This dress appears especially often on BMAC toreutic, for instance on silver vessels: Amiet (1986: fig. 202); on seals: Sarianidi (1998b: nos. 48, 49).

26 Reduced representation on a silver seal from Gonur Depe where a man holds a beaker in his hand: National Museum: KP 1246. Fully developed scene: Sarianidi (1998b: nos. 47–49). Clear antecedents are to be found in the art of Kerman: cf. a lapis lazuli disc with banquet scene in Porada (1993: fig. 19); stamp seal with banquet scene: Collon (1997: fig. 1e); standard from Shahdad: Hakemi (1997b: 270, 271, 649, Gt, object no. 1049); pinhead with banquet scene: Amiet (1986: fig. 127); Ligabue and Salvatori (1988: pl. 101).

27 I consider a goddess a woman who is either recognizable as a supernatural being or a numinous force through her attributes: animals, plants, stellar symbols, water, mountains, vessels, etc., or through her elevated position in a scene – enthroned or adored by humans or other beings.

28 As nearly all the motifs in BMAC art, the battle between an eagle and a snake comes directly from the "intercultural style." The different manners of transfer and transformation are discussed in Winkelmann (1997b, 2000, 2003, 2005, 2008b, 2013, 2014; see especially: Winkelmann 2013: fig. 7).

29 The motif of two entwined snakes comes directly from the "intercultural style" (Brentjes 1971: pl. Ib; Amiet 1986: fig. 74.1; Hakemi 1997a: figs. 20, 49, 51, 58).

30 The connection of this depiction and others with a myth known as the "Etana myth" in Mesopotamia has been demonstrated in a few articles (Winkelmann 2003, 2004, 2008b).

31 Many compartmented as well as all stamp seals showing a single scorpion are a reduced version of a woman holding scorpions, a well-known motif in the Early Dynastic II and III glyptic of Mesopotamia and Elam as well as in the "intercultural style" of Iran. The same is valid for the eagle, which appears with its pinions upwards as part of the "eagle with snakes" motif, or with its pinions downwards, as part of the "eagle diving on a hoofed animal" motif.

32 Rossi Osmida (2007: 173–175, 185, type 3/E, F; and figs. on pp. 144, 174, 175, 189, 190). The same type of figurine has also been found at Gonur Depe.

33 Discussed in Winkelmann (2003, 2005, and 2010). In the latter article the Iranian antecedents of the very important seal of the kneeling goddess in a skirt made of mountain scales and snakes protruding from her body are discussed. The seal is today in the National Museum, inv. no. 1915/1405.

REFERENCES

Amiet, P. 1961. *La glyptique mésopotamienne archaïque*. Paris: CNRS.

Amiet, P. 1972. *Glyptique susienne des origines à l'époque des Perses achéménides. Cachets, sceaux-cylindres et empreintes antiques découverts à Suse de 1913 à 1967*. Paris: Librairie orientaliste Paul Geuthner.

Amiet, P. 1978. "Antiquités de Bactriane." *La Revue du Louvre et des Musées de France* 3, 153–164.

Amiet, P. 1986. *L'âge des échanges inter-iraniens 3500–1700 avant J.-C.* Paris: Édition de la Réunion des musées nationaux.

Amiet, P. 1988. "Antiquities of Bactria and Outer Iran in the Louvre Collection." In *Bactria: An Ancient Oasis Civilization from the Sands of Afghanistan*, ed. G. Ligabue and S. Salvatori. Venice: Erizzo, pp. 159–180.

Amiet, P. 1994. "Un sceaux trans-élamite à Suse." *Revue d'Assyriologie et d'archéologie orientale* LXXXVIII: 1–4.

Amiet, P. 2004. "De l'Elam à la Margiane: Le renouvellement des problèmes archéologiques." In *Seals of the Oasis*, ed. S. Winkelmann. Venice: Il Punto, pp. 9–21.

Ascalone, E. 2008. "Cultural Interactions Among Mesopotamia, Elam, Transelam and the Indus Civilization: The Evidence of a Cylinder Stamp Seal from Jalalabad (Fars) and Its Significance in the Historical Dynamics of Southeastern Iran." In *Proceedings of the 4th ICAANE*. Vol. 1, *The Reconstruction of the Environment: Natural Resources and Human Interrelations Through Time – Art History, Visual communication*, ed. H Kühne. Wiesbaden: Harrassowitz, pp. 254–275.

Askarov, A.A. 1977. *Drevnezemledel'cheskaja kul'tura epokhi bronzy juga uzbekistana* [The Bronze Age ancient farmers' culture of southern Uzbekistan].Tashkent: Fan.

Askarov, A.A. and B.N. Abdullaev 1983. *Dzharkutan (k probleme protogorodskoj tsivilizatsii na juge uzbekistana)* [Dzharkutan (on the question of a proto-urban civilization in southern Uzbekistan)]. Tashkent: Fan.

Askarov, A.A. and T.Sh. Shirinov. 1993. *Rannaja gorodskaja kul'tura epokhi bronzy juga Srednej Azii* [The Bronze Age ancient urban culture of southern Central Asia], Samarkand: Fan.

Baghestani, S. 1997. *Metallene Compartimentsiegel aus Ostiran, Zentralasien und Nordchina.* Rahden: VML.
Benoit, A. 2010. *Princesse de Bactriane.* Paris: Musée du Louvre, Département des Antiquités orientales.
Biscione, R. 1985. "The So-Called 'Nestorian Seals': Connections Between Ordos and Middle Asia in Middle–Late Bronze Age." In *Orientalia Iosephi Tucci Memoriae Dicata I*, ed. G. Gnoli and L. Lanciotti, Serie Orientale Roma 56, 3. Rome: IsMEO, pp. 1281–1294.
Brentjes, B. 1971. "Ein elamitischer Streufund aus Soch, Ferghana (Usbekistan)." *Iran* IX: 155.
Brunswig, R.M., A. Parpola, and A. Potts 1983. "New Indus Type and Related Seals from the Near East." In *Dilmun*, ed. D. Potts. Berlin: Reimer-Verlag, pp. 101–115.
Caldwell, J.R. 1967. *Investigations at Tal-i-Iblis.* Springfield: Illinois State Museum.
Calmeyer, P. 1989. "Beobachtungen an der Silbervase von Persepolis." *Iranica Antiqua* XXIV: 79–85.
Casal, J.-M. 1961. *Fouilles de Mundigak.* Paris: Klincksieck.
Collon, D. 1987. *First Impressions: Cylinder Seals in the Ancient Near East.* London: British Museum.
Collon, D. 1997. "Lazurit iz vostoka, pechat'-shtamp iz Britanskogo Muzeja" [Eastern lapis lazuli, stamp seal in the British Museum]. *Vestnik Drevnej Istorii* 1997(2): 123–129.
Contenau, G. 1877. *La glyptique syro-hittite.* Paris: Librairie Orientaliste.
Crawford, H. and K. al Sind 1995. "A Seal in the Collection of the National Museum of Bahrain." *Arabian Archaeology and Epigraphy* 6: 1–4.
De Cardi, B. 1970. *Excavations at Bampur, a Third Millennium Settlement in Persian Baluchistan, 1966.* Anthropological Papers of the American Museum of Natural History APAMNH vol. 51(3). New York: American Museum of Natural History.
During-Caspers, E.C.L. 1983. "Triangular Stamp Seals from the Arabian Gulf and Their Indus Valley connections." *Annali* 43: 661–670.
During-Caspers, E.C.L. 1994. "Triangular Stamp Seals from the Arabian Gulf, Once Again." *Proceedings of the Seminar for Arabian Studies* 24: 97–114.
During-Caspers, E.C.L. 1997 "Bactrian Elements in the Harappan Glyptic Repertoire." In *South Asian Archaeology 1995*, ed. B. and R. Allchin. Oxford: Science Publishers, pp. 253–263.
Franke Vogt, U. 1991. *Die Glyptik aus Mohenjo Daro.* Mainz: Zabern.
Gadd, C.J. 1932. *Seals of Ancient Indian Style Found at Ur.* Proceedings of the British Academy 18. London: British Academy.
Gaibov, V. 1996. "Bullae from Göbekly-depe (Margiana): Bronze Age Traditions in Parthian sphragistics." In *Archives et sceaux du monde hellénistique: Bulletin de correspondance hellénique, Supplément 29*, ed. M.-F. Boussac and A. Invernizzi. Paris: de Boccard, pp. 385–393.
Gaibov, V. 2007. "The Bullae of Gobekly-depe." In *After Alexander: Central Asia Before Islam*, ed. J. Cribb and G. Herrmann. Oxford: Oxford University Press, pp. 285–294.
Ghirshman, R. 1968. "Notes Iraniennes XVI. Deux statuettes élamites du Plateau Iranien." *Artibus Asiae* XXX (2–3): 237–248.
Hakemi, A. 1972. *Catalogue de l'exposition: Lut-Xabis (Shahdad) – Rāhnamā-i namāyišgāh-i Dašt-i Lūt: Ḫabīṣ "Šahdād."* Symposium annuel de la recherche archéologique en Iran 1. Teheran: Čapḫāna-i Bīst wa panğum-i Šahrīwar.
Hakemi, A. 1997a. "Kerman: The Original Place of Production of Chlorite Stone Objects in the Third Millennium BC." *East and West* 47: 11–40.
Hakemi, A. 1997b. *Shahdad: Archaeological Excavations of a Bronze Age Center in Iran.* IsMEO Reports and Memoirs XXVII. Rome: IsMEO.
Hambis, L. 1954. " Notes sur quelques sceaux-amulettes nestoriens en bronze." *Bulletin de l'École Française d'Extrême Orient* 44: 483–525.

Hambis, L. 1956. "A propos de sceaux-amulettes 'Nestoriens.'" *Revue des Arts Asiatiques* 3: 279–286.

Hargreaves, H. 1929. *Excavations in Baluchistan 1925, Sampur Mound, Mastung and Sohr Damb, Nāl.* Calcutta: Government of India, Central Publication Branch.

Hinz, W. 1969. "Eine neu gefundene altelamische Silbervase." In *Altiranische Funde und Forschungen*, ed. W. Hinz. Berlin: De Gruyter, pp. 11–44.

Jarrige, J.-F. 1985. "Les relations entre l'Asie Centrale méridionale, le Baluchistan et la vallée de l'Indus à la fin du 3ème et au début du 2ème millénaire." In *L'archéologie de la Bactriane ancienne*, ed. J.-F. Jarrige. Paris: Éditions du CNRS, pp. 105–120.

Jarrige, J.-F. 1986. "Excavations at Mehrgarh-Nausharo." *Pakistan Archaeology* 10–22: 63–131.

Jarrige, J.-F. (ed.) 1988. *Les cités oubliées de l'Indus: archéologie du Pakistan.* Paris: Musée national des arts asiatiques-Guimet.

Jarrige, J.-F. and M. Santoni 1979. *Fouilles de Pirak.* Paris: de Boccard.

Joshi, J.P. and A. Parpola 1987. *Corpus of Indus Seals and Inscriptions. 1. Collections in India.* Annales Academiae Scientarum Fennicae, Sarja-Ser. B, Nide-Tom. 239. Memoirs of the Archaeological Survey of India, No. 86. Helsinki: Suomalainen Tiedeakatemia.

Kjaerum, P. 1983. *Danish Archaeological Investigations on Failaka, Kuwait., Failaka/ Dilmun: The Second Millennium Settlement.* Vol. 1, *The Stamp and Cylinder Seals.* Jutland Archaeological Society Publications XVI, 1. Aarhus: Aarhus University Press.

Lamberg-Karlovsky, C.C. 1973. "Urban Interaction on the Iranian Plateau: Excavations at Tepe Yahya 1967–1973." *Proceedings of the British Academy* 59: 283–319.

Lamberg-Karlovky, C.C. and M. Tosi 1973 "Shahr-i Sokhta and Tepe Yahya: tracks on the Earliest History of the Iranian Plateau." *East and West* 23: 21–53.

Lambert, W.G. 1986 "Seals from West Central Asia and Adjacent Regions." *Archäologische Mitteilungen aus dem Iran* 19: 33–40.

Ligabue, G. and S. Salvatori (eds.) 1988. *Bactria: An Ancient Oasis Civilization from the Sands of Afghanistan.* Venice: Erizzo.

Majidzadeh, Y. 2003. *Jiroft: The Earliest Oriental Civilization.* Tehran: Ministry of Culture and Islamic Guidance Printing and Publishing Organization.

Majidzadeh, Y. 2008. "Excavations at Konar Sandal in the Region of Jiroft in the Halil Basin: First Preliminary Report (2002–2008). With a Contribution on Glyptic Art by Holly Pittman." *Iran* 46(1): 69–104.

Masimov, I.S. 1979. "Izuchenie pamjatnikov epokhi bronzy nizovij murgaba" [Study of Bronze Age sites in the lower Murghab]. *Sovetskaja Arkheologija* 1: 111–132.

Masimov, I.S. 1981. "Novye nakhodki pechatej epokhi bronzy iz nizovij murgaba" [New discoveries of Bronze Age seals from the lower Murghab]. *Sovetskaja Arkheologija* 2: 132–150.

Masimov I.S. and S. Salvatori 2008. "Unpublished Stamp Seals from the North-Western Murghab Delta." In *The Bronze Age and Early Iron Age in the Margiana Lowlands*, ed. S. Salvatori and M. Tosi. BAR International Series 1806. Oxford: Archaeopress, pp. 99–110.

Masson, V.M. 1959. *Drevnezemledel'cheskaja kul'tura Margiany* [Margiana Ancient Farmers' culture]. Materialy i Issledovanija po Arkheologii SSSR 73. Moscow: Nauka.

Masson, V.M. 1981. *Altyn Depe.* Trudy JuTAKE XVIII. Leningrad: Nauka.

Masson, V.M. 1988. *Altyn-Depe* (English translation). Philadelphia: University Museum, University of Pennsylvania.

Piggott, S. 1943. "Shah-i Tump Stamp Seals." *Antiquity* 17: 179–181.

Piran, S. 2012. *Objects from the Jiroft Treasury: Soft Stone and Alabaster Objects (Recovered Collection) from the Halil River Basin.* Tehran: Pazineh.

Pittman, H. 2008. "IX. Glyptic Art." *Iran* 46: 95–101.

Pittman, H. 2013. "New Evidence for Interaction Between the Iranian Plateau and the Indus Valley: Seals and Sealings from Konar Sandal South." In *Connections and Complexity: New*

Approaches to the Archaeology of South Asia, ed. S.A. Abraham, P. Gullapalli, T.P. Raczek, and U.Z. Rizvi. London and New York: Left Coast Press, 63–89.

Pittman, H. 2014a. "Anchoring Intuition in Evidence: A Continuing Discussion of Cylinder Seals from Southeastern Iran." In *Edith Porada zum 100. Geburstag. A Centenary Volume*, ed. E. Bleibtreu and H.U. Steymans. Orbis Biblicus et Orientalis 268. Fribourg: Academic Press/Göttingen: Vandenhoeck & Ruprecht, pp. 375–393.

Pittman, H. 2014b. "Hybrid Imagery and Cultural Identity in the Age of Exchange: Halil River Basin and Sumer Meet in Margiana." In *"My Life Is Like the Summer Rose": Maurizio Tosi e l'Archeologia come modo di vivere – Papers in Honour of Maurizio Tosi for His 70th Birthday*, ed. C.C. Lamberg-Karlovsky, B. Genito, and B. Cerasetti. BAR International Series 2690. Oxford: BAR, pp. 625–636.

Pittman, H. 2019. "Bronze Age Interaction on the Iranian Plateau: From Kerman to the Oxus Through Seals." In *The Iranian Plateau During the Bronze Age: Development of Urbanisation, Production and Trade*, ed. J.-W. Meyer, E. Vila, M. Mashkour, M. Casanova, and R. Vallet. Lyon: Mom éditions, pp. 267–288.

Porada, E. 1962. *Alt-Iran*. Baden-Baden: Holle.

Porada, E. 1993. "Seals and Related Objects from Early Mesopotamia and Iran." In *Early Mesopotamia and Iran: Contact and Conflict ca. 3500–1600 BC*, ed. J. Curtis. London: British Museum Press, pp. 44–53.

Pottier, M.-H. 1984. *Matériel funéraire de la Bactriane méridionale de l'âge du bronze.* Paris: Edition Recherche sur les Civilisations.

Potts, D. 1981. "Echoes of Mesopotamian Divinity on a Cylinder Seal from South-Eastern Iran." *Revue d'assyriologie et d'archéologie orientale* LXXV: 135–142.

Potts, D.T. 1993a. "A New Bactrian Find from Southeastern Arabia." *Antiquity* 67: 591–596.

Potts, D.T. 1993b. "Rethinking Some Aspects of Trade in the Persian Gulf." *World Archaeology* 24(3): 423–440.

Potts, D.T. 1994. "South and Central Asian elements at Tell Abraq (Emirate of Umm al-Qaiwain, United Arab Emirates), c.2200 BC–AD 400." In *South Asian Archaeology 1993*, ed. A. Parpola and P. Koskikallio. Helsinki: Suomalainen Tiedeakatemia, pp. 615–628.

Pumpelly, R. 1908. *Explorations in Turkestan: With an Account of the Basin of Eastern Persia and Seistan. Expedition*. Washington, DC: Carnegie Institution of Washington.

Rossi Osmida, G. 2002. *Margiana: Gonur-Depe Necropolis – 10 Years of Excavations by Ligabue Study and Research Centre*. Venice: Il Punto.

Rossi Osmida, G. 2007. *Adji Kui oasis*. Vol. I, *La cittadella delle statuette*. Trebaseleghe: Il Punto.

Rossi Osmida, G. 2011. *Adji Kui oasis*. Vol. II, *La cittadella delle statuette*. Trebaseleghe: Il Punto.

Sajjadi, M. 2007. "Wall Painting from Dahaneh-ye Gholaman (Seistan)." *Ancient Civilizations from Scythia to Siberia* 13: 129–154.

Salvatori, S. 1981. "Sigilli anepigrafi." In *La collezione Anatolica di E. Borowski*, ed. M. Poetto and S. Salvatori. Pavia: GJES edizioni.

Salvatori, S. 1993. "The Discovery of the Graveyard of Gonur Depe 1 (Murghab Delta, Turkmenistan): 1992 Campaign Preliminary Report." *Rivista di Archeologia* XVII: 5–13.

Salvatori, S. 1994a. "Excavations at the Namazga V Late Graveyard of Gonur 1 (Murghab Delta, Turkmenistan). Preliminary Report on the 1993 Field Season." *Rivista di Archeologia* XVIII: 15–39.

Salvatori, S. 1994b. "A Late Third Millennium Graveyard at Gonur Depe 1 (Murghab Delta, Turkmenistan)." In *South Asian Archaeology 1993*, vol. II, ed. A. Parpola and P. Koskikallio. Helsinki: Suomalainen Tiedeakatemia, pp. 657–666

Salvatori, S. 1995. "Gonur-depe 1 (Margiana, Turkmenistan): The Middle Bronze Age Graveyard: Preliminary Report on the 1994 Excavation Campaign." *Rivista di Archeologia* XIX: 5–37.

Salvatori S. 2000. "Bactria and Margiana Seals, a New Assessment on Their Chronological Position and a Typological Survey." *East and West* 50: 97–145.

Salvatori, S. 2008. "A New Cylinder Seal from Ancient Margiana: Cultural Exchange and Syncretism in a 'Worldwide Trade System' at the End of the 3rd Millennium BC." In *The Bronze Age and Early Iron Age in the Margiana lowlands*, ed. S. Salvatori and M. Tosi. Oxford: Archaeopress, pp. 11–118.

Salvatori, S. and M. Vidale 1997. *Shahr-i Sokhta 1975–1978: Central Quarters Excavations*. Rome: IsIAO.

Samzun, A., 1987. "Die frühe Kupferzeit: Mehrgarh Periode III." In *Vergessene Städte am Indus*, ed. G. Urban. Mainz: Zabern, pp. 74–80.

Santoni, M. 1984. "Sibri and the South Cemetery of Mehrgarh: 3rd Millennium Connections Between the Northern Kachi Plain (Pakistan) and Central Asia." In *South Asian Archaeology 1981*, ed. B. and R. Allchin. Cambridge: Cambridge University Press, pp. 52–60.

Sarianidi, V.I. 1965. *Pamjatniki pozdnego eneolita jugo-vostochnoj Turkmenii* [Late Chalcolithic sites of southeastern Turkmenia]. Svod Arkheologicheskikh istochnikov B3–8, IV. Moscow: Nauka.

Sarianidi, V.I. 1966. "Raskopki v jugo-vostochnykh karakumakh v 1964 g" [The 1964 excavations in southeastern Karakum]. *Kratkie Soobshchenija Instituta Arkheologii* 108: 89–95.

Sarianidi, V.I. 1976. "Pechaty amulety murgabskogo stilja" [Seal-amulets of the Murghab style]. *Sovetskaja Arkheologija* 1: 42–68.

Sarianidi, V.I. 1977a. "Bactrian Center of Ancient Art." *Mesopotamia* I: 97–142.

Sarianidi, V.I.1977b. *Drevnie zemledel'tsy Afganistana* [Ancient farmers of Afghanistan]. Moscow: Nauka.

Sarianidi, V.I. 1981. "Seal-Amulets of the Murghab Style." In *The Bronze Age Civilization of Central Asia: Recent Soviet Discoveries*, ed. P. Kohl. New York: Routledge, pp. 221–255.

Sarianidi, V.I. 1984. "Raskopki monumental'nykh zdanij na Dashly 3" [Excavations of monumental buildings at Dashly 3]. In *Drevnjaja Baktrija: materialy Sovetsko-Afganskoj arkheologicheskoj ekspeditsii*, vol. 3, ed. I.T. Kruglikova. Moscow: Nauka, pp. 5–32.

Sarianidi, V.I. 1986a. "The Bactrian Pantheon." *Unesco IASCCA* Information Bulletin 10: 5–20.

Sarianidi, V.I. 1986b. *Die Kunst des alten Afghanistan: Architektur, Keramik, Siegel, Kunstwerke aus Stein und Metall*. Leipzig: Seemann Verlag.

Sarianidi, V.I. 1988. "Cult Symbolism of Bactria and Margiana amulets." In *Orientalia Iosephi Tucci Memoriae Dicata III*, ed. G. Gnoli and L. Lanciotti. Serie Orientale Roma, LVI, 3. Rome: ISMEO, pp. 1281–1294.

Sarianidi, V.I. 1990. *Drevnosti strany Margush* [Ancient Margush]. Ashgabad: Ylym.

Sarianidi, V.I. 1992. *Drevnejshij Merv* [The most ancient Merv]. Moscow: Nauka.

Sarianidi, V.I. 1993. "Excavations at Southern Gonur." *Iran* XXXI(1): 25–37.

Sarianidi, V.I. 1994. "Margiana and the Indo-Iranian World: Objects from Margiana with Parallels in Bactria, Baluchistan and the Indus Valley." In *South Asian Archaeology*, ed. A. Parpola and P. Koskikallio. Helsinki: Suomalainen Tiedeakatemia, pp. 667–680.

Sarianidi, V.I. 1998a. *Margiana and Protozoroastrism*. Athens: Kapon Editions.

Sarianidi, V.I. 1998b. *Myths of Ancient Bactria and Margiana on Its Seals and Amulets*. Moscow: Pentagraphic Ltd.

Sarianidi, V.I. 2005. *Gonur-Depe, City of Kings and Gods*. Ashgabat: Turkmendöwlethabarlary.

Sarianidi, V.I. 2007. *Necropolis of Gonur*. Athens: Kapon editions.

Sarianidi, V.I. 2009. *Margush. Secret and Truth of the Great Culture*. Ashgabad: Turkmendöwlethabarlary.

Schmidt, E.F. 1937. *Excavations at Tepe Hissar, Damghan*. Philadelphia: University of Pennsylvania Press.

Shah, S.G. and A. Parpola 1991. *Corpus of Indus Seals and Inscriptions*. Vol. 2, *Collections in Pakistan*. Annales Academiae Scientarum Fennicae, Ser. B, Tom. 240. Memoirs of the Department of Archaeology and Museums, Government of Pakistan, vol. 5. Helsinki: Suomalainen Tiedeakatemia.

Shinde, V., G.L. Possehl, and M. Ameri 2005. "Excavations at Gilund 2001–2003: The Seal Impressions and Other Finds." In *South Asian Archaeology 2003*, ed. U. Franke-Vogt and H.-J. Weisshaar. Aachen: Linden Soft Verlag, pp. 159–169.

Teufer, M. 2007. "Zu bronzezeitlichen Tulpendarstellungen in Zentralasien und im Iran." In *Scripta praehistorica in honorem Biba Teržan*. Situla 44, ed. M. Blečić, M. Črešnar, B. Hänsel, A. Hellmuth, and E. Kaiser. Metzner-Nebelsick. Ljubljana: Narodni Muzei Slovenije, pp. 183–207.

Udeumuradov, B.N. 1993. *Altyn Depe i Margiana: svjazi, chronologija, proiskhozhdenie*. Ashkhabad: Ylym.

Urban, G. ed. 1987. *Vergessene Städte am Indus*. Mainz: Zabern.

Vats, M.S. 1940. *Excavations at Harappa, Being an Account of Archaeological Excavations at Harappa Carried Out Between the Years 1920–1921 and 1933–34*. Calcutta and Delhi: Manager of Publications.

von der Osten-Sacken, E. 1992. *Der Ziegen-'Dämon': 'Obed- und Urukzeitliche Götterdarstellungen*. Münster: Ugarit-Verlag.

Weisgerber, G. 1980. "Und Kupfer im Oman: Das Oman Projekt des Deutschen Bergbau-Museums." *Der Anschnitt* 2–3: 62–110.

Weisgerber, G. 1981. "Mehr als Kupfer im Oman. Ergebnisse der Expedition 1981." *Der Anschnitt* 5–6: 174–263.

Winkelmann, S. 1993. "Elam, Baktrien, Belutschistan: wo liegen die Vorläufer der Hockerplastiken der Induskultur? Erste Gedanken." *Iranica Antiqua* 28: 57–96.

Winkelmann, S. 1994. "Intercultural Relations Between Iran, Central Asia and Northwestern India in the Light of Squatting Stone Sculptures from Mohenjo-Daro." In *South Asian Archaeology 1993*, ed. A. Parpola and P. Koskikallio. Helsinki: Suomalainen Tiedeakatemia, pp. 815–831.

Winkelmann, S. 1996. "Ein ungewöhnliches figürliches Kompartmentsiegel aus Shahr-i Sokhta, Iran." In *Spurensuche in der Vorderasienarchäologie*. Herausgegeben von W. Beltz und M. Mode. Hallesche Beiträge zur Orientwissenschaft 21. Halle: Universität, pp. 153–172.

Winkelmann, S. 1997a. "Ein neues transelamisches Siegel." *Archäologische Mitteilungen aus Iran und Turan* 28: 135–146.

Winkelmann, S. 1997b. "Southeast-Iranian and Indo-Iranian Elements on Bactrian and Murghab Style Seals." In *South Asian Archaeology*, ed. R. and B. Allchin. Enfield, NH: Science Publishers, pp. 512–523.

Winkelmann, S. 1998. "Bemerkungen zum Grab 18 und den Silbernadeln von Gonur depe." *Archäologische Mitteilungen aus Iran und Turan* 30: 1–16.

Winkelmann, S. 1999. "Von Formen und Stilen – ein Versuch der Gliederung baktrischer Stempel- und Rollsiegel." *Hallesche Beiträge zur Orientwissenschaft* 28: 112–211.

Winkelmann, S. 2000. "Intercultural Relations Between Iran, the Murghabo-Bactrian Archaeological Complex (BMAC), Northwest India and Failaka in the Field of Seals." *East and West* 50: 43–95.

Winkelmann, S. 2001. "Ein neuer baktrischer Stempelzylinder." In *Beiträge zur Vorderasiatischen Archäologie. Winfried Orthmann gewidmet*, ed. J.-W. Meyer, M. Novak, and A. Pruß. Frankfurt/Main: Johann Wolfgang-Goethe-Universität, Archäologisches Institut, pp. 484–493.

Winkelmann, S. 2003. "Berliner Schlangenbecken, Trichterbecher und Cincinnati-Mann: verkannte Schlüssel Objekte der altorientalischen Archäologie?" In *Altertumswissenschaften im Dialog. Festschrift für Wolfram Nagel zur Vollendung seines 80. Lebensjahres*. Alter Orient und Altes Testament (AOAT) 306, ed. R. Dittmann, R.C. Eder, and B. Jacobs. Münster: Ugarit-Verlag, pp. 567–678.

Winkelmann, S. 2004. *Seals of the Oasis: From the Ligabue Collection, with an Introduction by Pierre Amiet*. Venice: Il Punto.

Winkelmann, S. 2005. "Deciphering the Intercultural Style? Remarks on the Southeast Iranian Bronze Age Art." In *South Asian Archaeology 2003*, ed. U. Franke-Vogt and H.-J. Weisshaar. Aachen: Linden Soft Verlag, pp. 185–197.

Winkelmann, S. 2006. "Le dee dell'altopiano iranico e della Battriana." In *Dea Madre*, ed. G. Rossi Osmida. Milano: Electa, pp. 193–207.

Winkelmann, S. 2008a. "Animali e miti nel Vicino Oriente Antico: un'analisi attraverso i sigilli." In *Animali e Mito nel Vicino Oriente Antico*, ed. G. Ligabue and G. Rossi Osmida. Venice: Il Punto, pp. 39–61.

Winkelmann, S. 2008b. "Ein baktrischer Edelmetall-Hortfund und noch einmal zur Frage der Quellen baktrischer Compartimentsiegel." In *Intercultural Relations Between South and Southwest Asia: Studies in Commemoration of E. C. L. During-Caspers 1934–1999*, ed. E. Olijdam and R.H. Spoor. BAR International Series 1826. Oxford: Archeaopress, pp. 184–199.

Winkelmann, S. 2009. "Animals and Myths in Ancient Near East. A General Overview from the Seals." *Central Asia Cultural Values* 4: 27–54.

Winkelmann, S. 2010. "Heimkehr in den Berg: Paradiesvorstellungen im 3. Jt. v. Chr. in Iran und Baktrien." In *Sehnsucht nach dem Paradies. Paradiesvorstellungen in Judentum, Christentum, Manichäismus und Islam. Beiträge des Leucorea-Kolloquiums zu Ehren von Walter Beltz*, ed. J. Tubach, A. Drost-Abgarjan, and S. Vashalomidze. Wiesbaden: Harassowitz, pp. 15–37.

Winkelmann, S. 2012. "Wovor flüchtet Gilgamesch? Betrachtungen zum frühen mesopotamischen Unterweltverständnis." In *Höllen- und Unterweltsvorstellungen in Orient und Okzident. Beiträge des Leucorea-Kolloquiums 2010*, ed. J. Tubach, A. Drost-Abgarjan, and S. Vashalomidze. Studies in Oriental Religions No. 63. Wiesbaden: Harassowitz, pp. 249–259.

Winkelmann, S. 2013. "Transformation of Near Eastern Animal Motifs in Murghabo-Bactrian Art." In *Animals, Gods and Men from East to West: Papers on Archaeology and History in Honour of Roberta Venco Ricciardi*, ed. A. Perruzetto, F.D. Metzger, and L. Dirven. BAR International Series 2516. Oxford: Archaeopress, pp. 47–64.

Winkelmann, S. 2014. "Trading Religions from Bronze Age Iran to Bactria." In *Trading Religion: Religious Formation, Transformation, and Cross-Cultural Exchange between East and West, Dynamics in the History of Religions*, ed. P. Wick and V. Rabens. Leiden: Brill, pp. 199–229.

Winkelmann, S. 2015. "Bericht über die Ausgrabungskampagne 2014 der Russisch-Schweizer Expedition in Gonur Depe, Turkmenistan." In *Jahresbericht der SLSA (Schweizer-Lichtensteiner Stiftung für Archäologie 2014*, pp. 135–166.

Winkelmann, S. 2016a. "Bericht über die Berner Ausgrabungen 2015 in Gonur Depe, Turkmenistan." *In Jahresbericht der SLSA (Schweizer-Lichtensteiner Stiftung für Archäologie 2015)*, pp. 179–210.

Winkelmann, S. 2016b. "Some Thoughts About the Wild Haired Snake-Man on BMAC Seals." In *Trudy Margianskoj Arkheologicheskoj Ekspeditsii*, vol. 6, ed. N. A. Dubova. Moscow: Staryj Sad, pp. 482–491.

Winkelmann, S. in press. "Baktrisch-margianische Göttinnen aus Stein: Betrachtungen zu den bronzezeitlichen Kompositfiguren aus Mittelasien." In *Festschrift für Burchard Brentjes*, ed. K. Rührdanz and H. Neumann.

Xin, W. and O. Lecomte 2012. "Clay Sealings from the Iron Age citadel at Ulug Depe." In *Archäologische Mitteilungen aus Iran und Turan* 44: 313–328.

CHAPTER NINE

CHLORITE CONTAINERS FROM THE OXUS CIVILIZATION

Between technical choices and iconographic codes

*Massimo Vidale**

INTRODUCTION

Were "soft-stone" vessels actually manufactured in Margiana and Bactria, the two cultural poles that we currently ascribe to the Oxus Civilization or Bactria-Margiana Archaeological Complex (BMAC)? And if so, which types precisely? In addition, the precise mineralogical nature of these "soft stones" remains problematic. Chlorite, schist, mica-schist, steatite, talc, soapstone, soft stone: These names, in the past literature, have been freely applied to a wide range of important artifacts of the Bronze age of Middle Asia by archaeologists and writers of archaeological narratives without any consistency, or any proper petrographic or other mineralogical investigation. Chlorites, properly speaking, are a group of approximately ten minerals containing mostly silicon, magnesium, aluminum, and iron belonging to the wider family of phyllosilicates. It is an unsatisfactory term, because *stricto sensu* it does not identify a precise mineral, nor a well-defined rock. In the common archaeological lexicon, however, the term has been since long used for defining a variety of base materials of prehistoric artifacts that include both southeastern Iranian and Central Asian specimens – that is, green-to-gray and blackish soft stones that form in a wide variety of geological conditions (mainly medium temperature metamorphic, igneous, and hydrothermal contexts). Thus, the term "chlorite" in this chapter is to be understood as purely conventional.

Besides the huge geographical extent of the involved area and the expected variability of the stone outcrops, the absence of similar studies explains why the earliest attempts at sourcing the procurement areas of Iranian "chlorites" mainly in chemical terms did not succeed (Kohl 1975, 1976, 1978, 1979). Far-ranging theories on "Intercultural Styles" and globalized "world economy" models (besides Kohl's articles, and among others, see Aubet 2013: 182–192) were actually based more on the absence of archaeological data than on material evidence (see Vidale and Lazzari 2017, on lapis lazuli bead making at Shahr-i Sokhta).

A growing awareness of the complexity of the problem, together with the discovery of the first ancient mining and working sites in Oman (David 1996, 2001;

Harrower *et al.* 2016), and with recent geological and archaeometric work on the "soft stones" of the Jiroft area near Kerman (Emami *et al.* 2017) confirming an old hypothesis of multiple loci of manufacture (since Hakemi 1997b) and local production/distribution, is now changing the picture. In this chapter, the base material of the vessels under scrutiny, in the absence of more relevant information, is simply defined as "chlorite," a conventional term for relatively soft rocks rich in ferrous and magnesiac talc silicates,[1] rather common in the extensive metamorphic formations of various parts of the great subduction zone of Middle Asia (see the appendix at the end of this chapter).

This is only one of the several inherited conditions of fuzziness that affects the subject of this contribution. In the case of the Oxus Civilization, uncertainty is amplified by the lootings, as generalized as they are notorious, which have taken place since the 1970s in Afghanistan, and later (so, at least, it seems) in other regions, among which southern Turkmenistan, and probably Uzbekistan and Tajikistan.

In 2001, the dramatically unfortunate discovery of the graveyards of the Halil Rud Civilization (ancient Marhashi?) near Jiroft in Kerman (Iran), followed by a short season of mass illegal diggings, led to the recovery and the immediate loss and dispersion of thousands of precious artifacts carved out of these soft rocks (Madjidzadeh 2003; Piran 2013; see also Hakemi 1997a for early identifications and interest).

Thus, in short, in the last 50 years, an unbelievable number of "chlorite" artifacts have spilled onto the market, most ending up in private collections and in only a few rare cases going to official museums. To make things worse, some excavation reports and parallel publications are not systematic and scarcely detailed. The drawings of finds from Central Asia, in particular, are usually much less satisfactory than those from the other regions, and this makes the task of comparing them difficult. Contexts are absent or loosely defined, there is a general lack of inventory lists, and the soapstone objects that are the focus of this chapter have been illustrated (often more than once) without drawings and scale in groups from such an angle that the shapes cannot be perceived in a reliable fashion. When the various sides of the same objects have different patterns, often only one or two are illustrated. No wonder, therefore, that "soft-stone" vessels were not considered in Kaniuth's (2010) review of the long-distance contacts of Bronze Age southern Central Asia.

Finally, another obvious complication for the provenience studies of chlorite vessels is the intensive, albeit discontinuous, protohistoric trade. It is no surprise that vessels made in eastern Arabia were found in Sindh, Central Asia, eastern Iran, and Khuzistan (see, e.g., Potts 2003); and that in the Oxus area, as well as at Susa (Miroschedji 1973), many vials circulated that were most probably manufactured in one or more centers of southeastern Iran (see Figure 9.1).

LOCAL TRADITIONS AND INTERACTION SPHERES

The fact remains that: "A quantification of the totality of remains from any site in Iran – if such a thing were possible – would show the imports to represent but a very small percentage of any assemblage studied" (Potts 2003: 77), and this should also be true of the other main cultural areas of the Middle Asian macro region. In short, according to D.T. Potts, every region of Middle Asia where chlorite objects have been found should testify that the bulk was manufactured locally, and imports represent a distinguished minority of the cases.

The Oxus inventory includes square and round compartmented boxes, different types of vials used as cosmetic holders, bowls and pots, cylinder and stamp seals, incised spindle whorls, beads, rods with decorated finials, wigs and bodies of the so-called "Bactrian Ladies," and even shoes of composite figures with limestone or marble legs. Possibly, even objects made with "fired steatite" – a family of pyrotechnological treatments of soapstone that seems to have originated in the Neolithic of Baluchistan and spread westwards during the 6th and 5th millennia – are on record: for example, figurative and geometric inlays, and (apparently) rare stone vessels of the previously called "*série récente*".

Against the mentioned odds, the bright side of the story is that highly competent scholars, judging from papers published in the last three decades (e.g., David 1996; Potts 2003; Salvatori 2003, 2008), agree on a series of important points, allowing us to fix at least part of these Middle Bronze age productions in space and probably in time. These issues can be summarized as follows.

1. The term "Intercultural Style" formerly used for part the Susa collection of "steatite" vessels ("*série ancienne*") as well as for stray finds of the Kerman region and the Tepe Yahya workshops, has to be recognized as a melange of peculiar carving modes or styles of the Halil Rud valley. The recovery of fragments of carved chlorite objects bearing almost every known pattern at the plundered graveyard of Mahtoutabad, near Konar Sandal, demonstrates that these stone containers were primarily meant for local consumption. This style is distinguished by a careful low-relief carving that often covers the entire surface of the objects.

   Moreover, in the narrative scenes, this carving style gets close to the use of perspective for expressing relationships of time between actors in the depicted events (Vidale 2015). It is possible that, at least from this point of view, it was influenced by the best examples of late ED III and early Akkadian seal carving. According to Ascalone (2012: 22), these products (Halil Rud = "*série ancienne*") should be placed between 2600 and 2400–2300 BC, their spread in Mesopotamia taking place near the lowermost chronological limit, and possibly later. H. David (1996: 32) proposed the term "figurative style" for this craft.
2. Pierre de Miroschedji's "*série récente*" of Susa, as demonstrated by Potts (2003), was a simpler production based in Oman and the United Arab Emirates (UAE), and to a great extent limited to small globular bowls and rectangular compartmented boxes decorated with rows of dotted circles. Ascalone (2012) places these eastern Arabian objects between 2150–2100 BC and 1900 BC, in a well-known extended "Interaction Sphere" in which circulation axes seem to link the Oxus to the Persian Gulf, and Khuzistan to Sindh. However, the find at Ra's al Jinz of small hemispherical chlorite pots of the same description, together with Indus material culture more similar to Harappa IIIB than to IIIC (Cleuziou and Tosi 2000) suggests that these objects may have been made and circulated also one or two centuries earlier, at least since ca. 2300–2250 BC. H. David (1996: 34) has labeled this production "Umm an-Nar style."

   The distinctive concentric and dotted circles were carved with flat drill bits or rotating gouges with multiple points,[2] still used by Omani wood carvers.

> This tool, known locally as *zahrah* in Arabic, consists of a three-pointed steel end affixed to a wooden handle with the fragment of a brass bullet shell. The "end" of the tool is flat and contains three sharp points, one longer than the other two. The points of the tool are placed atop a piece of wood. The person then rotates the tool between the palms of the hands, thus "drilling" or "rotating" the design into the surface, similar to the action of a compass.
>
> (Ziolkowski and Al-Sharqi 2006: 154–155)

Interestingly, the same authors also report the use of a traditional tool of a similar description called *ma'z* or *alzer* made of a hard, shiny black rock available locally (Ziolkowski and Al-Sharqi 2006: 156). Although this class of decorative techniques was evidently spread from Mesopotamia to the Indus, no actual tool of this description seems to have ever been found in a dig.

3. The square-based soft-stone cosmetic vials with a trapeze-like or rectangular contour – with both geometric and floral decoration, and, exceptionally, silhouettes of human-like actors and animals, together with low semicircular and kidney-shaped vessels – decorated with pendant, incised, and alternating triangles and sometimes circles, were probably the most common products of the Bactria and Margiana workshops (Potts 2003: 83). In an important summary, S. Salvatori (2008: 84) also accepts that his phiales types C1–C2 – with "quadrangular-section body, cylindrical neck and a narrow, slightly projecting rim," were manufactured in Oxus workshops, dating them from the second half of the 3rd millennium BC to (possibly) the Late Bronze age. Less common were high-stemmed cups and other complex forms.

Freehand, light incision sometimes seconded by the occasional use of simple tubular copper drills are the most evident and diagnostic techniques. In fact,

> It is clear from the published photographs that the vast majority of soft-stone vessels from Margiana – kidney-shaped dishes, round dishes, deep-sided bowls with floral patterns, and square-bodied "cosmetic" flasks – were incised only lightly. Whether one looks at the floral patterns on finds from Gonur or the geometric patterns on the material from Gonur and the Togolok sites, most of the decoration is incised.
>
> (Potts 2003: 83)

4. The zigzag chlorite bowls, a southeastern Iranian production, are distinguished by the use of a negative excision process for making bands of zigzag lines that run below the rim or occur in redundant compositions all over the body of the vessels. That excision was part of the technical and cognitive background of the craft communities of southeastern Iran in the 3rd millennium is confirmed by sherds of ceramic vessels found in the Jazmurian basin, unusually treated in the same way (Heydari *et al.* 2015: fig. 5g). Although excision of vertical bands occurs on the vessels of architectural inspiration or "hut-motive" chlorite pots (see, e.g., Madjidzadeh 2003: 70–74), Potts (2003) considers that, at the workshop dumps of Tepe Yahya, zigzag chlorite items postdate the "*série ancienne*" (i.e., the Halil Rud style). They overlap in time and probably even postdate the

"*série récente*," a production of the late 3rd and early 2nd millennium BC of the ancient land of Magan (Oman and the UAE).

"Shark's-teeth" patterns – rows of tiny angular signs made by incision or light hammering, usually attached to lines, filling segments and elements of vegetal inspiration, are also considered as coming from southeastern Iranian origin (Potts 2003: 9).

The enormous metamorphic area accompanying the subduction zone of southern Eurasia, where talc deposits are originated from dolomitic and ultramafic rocks, must have provided, at the margins of the Oxus Civilization, multiple accesses to outcrops of chloritic and talcose rocks (see the appendix at the end of this chapter) that are so far not identified.

5. More information comes from Salvatori's analysis of the associations, in the graveyard of Shahdad, of variants of his types A and B. Type A has parallel lines on the neck, sometimes excised zigzag lines, incised designs, or dotted circles; type B is a small massive container with cubic body, short concave neck, plain walls, or exterior bearing groups of dotted circles. His study supports that these goods were made in southeastern Iran, perhaps also with the specific purpose of being traded to the Oxus region (Salvatori 2008: fig. 6.3). In the second half of the 3rd millennium, variants of type B coming from the Kerman region were traded westward to Khuzistan and northeastwards to Bactria.

To sum up, the available information suggests that these different production areas (although outlined in a preliminary way) also correspond to the different degrees of popularity for such techniques as surface finishing and decoration (Figure 9.1): In Bactria and possibly in Margiana, shallow geometric incisions freely traced with some liberty on the surface of the chlorite pieces; in the "Marhashi(?)" workshops, low-relief narrative or naturalistic carvings;[3] in Oman and along the eastern coasts of the Persian Gulf, decorating polished surfaces with sequences of dotted circles; in the chlorite-producing centers of southeastern Iran, which seem to have borrowed from the other side of the Gulf, the use of the specialized tools needed for tracing the same dotted circles, the excision of zigzag bands.

POSSIBLE IMPORTS FROM SOUTHEASTERN IRAN AND THE PERSIAN GULF

Figure 9.2 shows some examples of decorative patterns visible on chlorite vessels recovered in Margiana and Bactria that, on the basis of such technical information, and waiting for more objective petrographic and chemical information on the nature and provenience of the stones, are good candidates for imports. Unfortunately, only containers bearing patterns nos. 1, 8, 9, and 13 in Figure 9.2 were found in regular excavations at Gonur.

Hemispherical vessels bearing rows of dotted circles running along the rim, like pattern no. 1 (Sarianidi and Dubova 2013: 190) and no. 2 (Ligabue and Salvatori 1988: 182, fig. 1), following Potts, should be considered as imports from the Persian Gulf area. Entirely covered with these bands is the square compartmented box that

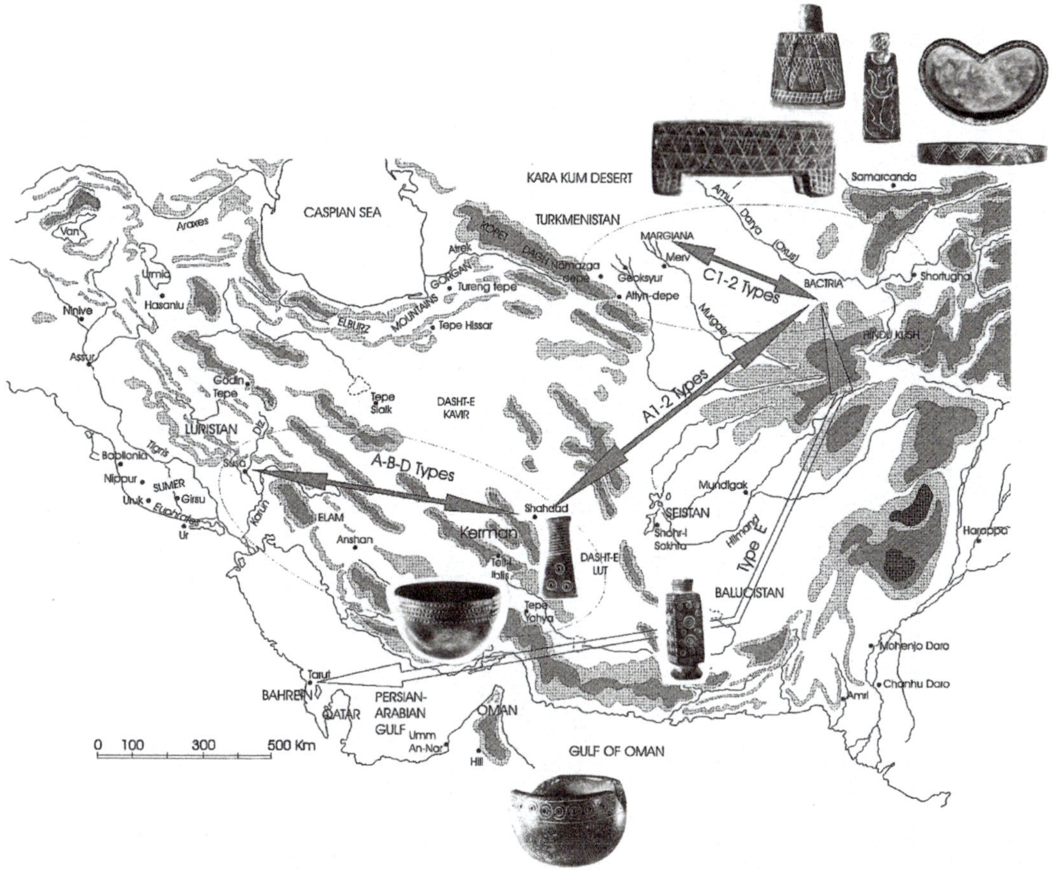

Figure 9.1 A map of Middle Asia modified after an original drafted by S. Salvatori (1998), to describe the production areas of small-sized chlorite containers in the second half of the 3rd millennium BC. At top right, it visualizes the main types of such stone vessels produced in the Oxus region: kidney-shaped low trays (together with crescent and semi-circular variants), square-section rectangular to trapeze-like vials, and footed rectangular trays. Chlorite vessels decorated with dotted circles were likely made in Oman and along the Persian Gulf coasts; the same drilled dotted circles, together with excised lines, distinguish the workshops of southeastern Iran trading their products to the Oxus, to western Iran and to Tarut.

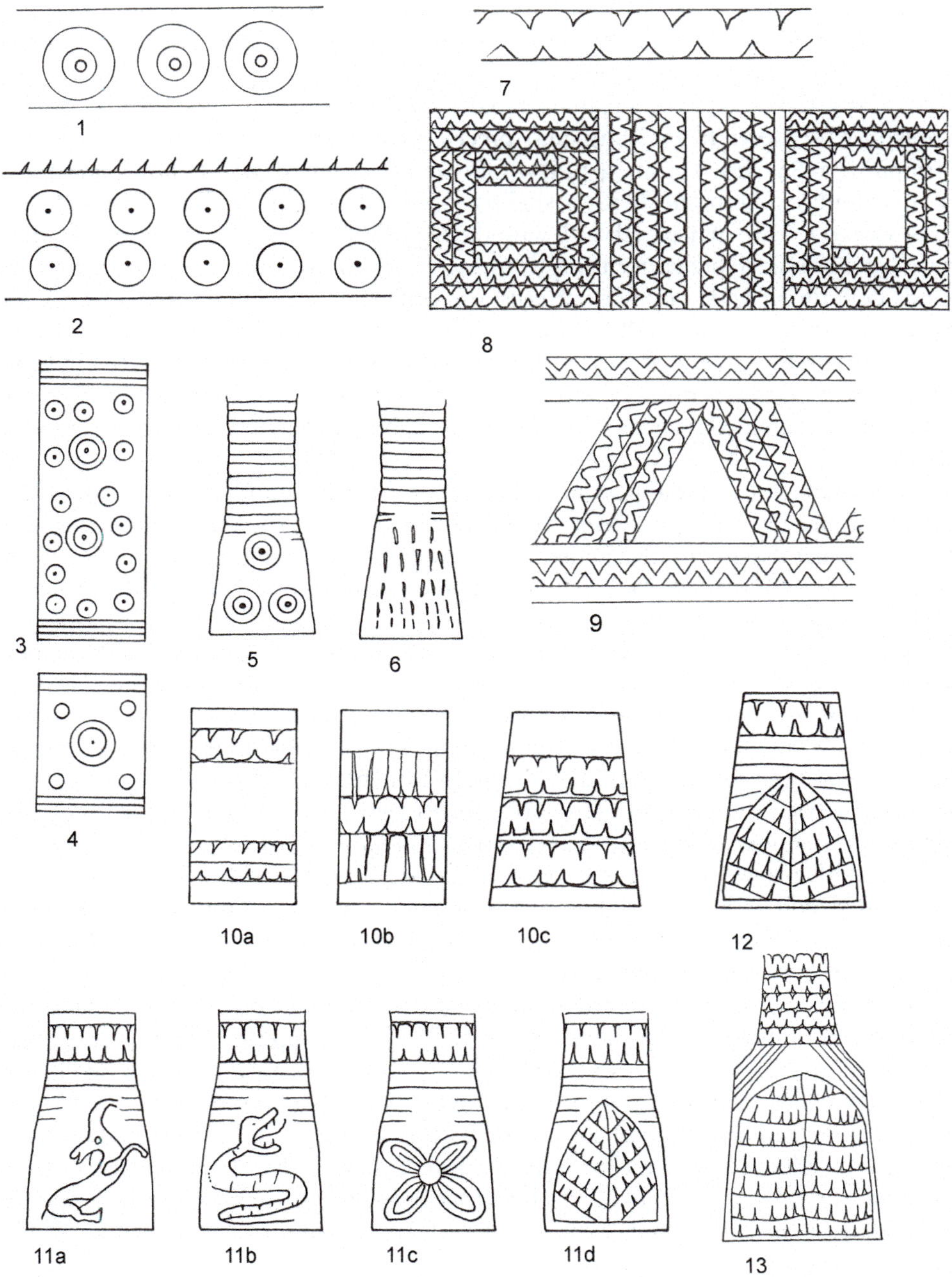

Figure 9.2 Decorative patterns on chlorite containers (drilled and excised, see text) presumably imported in Bactria and/or Margiana from southeastern Iran or further south from the workshop areas of the Gulf coasts (this and the following drawings by M. Vidale).

bears pattern no. 9 (Ligabue and Salvatori 1988: fig. 85), presumably inspired by an architectural pattern with windows, looted from a grave in Bactria, and possibly imported from the Kerman region.

Similarly, the excised zigzag bands in pattern no. 7 visible on some vessels (another hemispherical bowl from Margiana, precise provenience unknown, published in Sarianidi 1998a: figs. 17, 20, 47), no. 8 (a square compartmented box), and no. 9 (a subcylindrical container with a slightly tapered base from Gonur, Sarianidi and Dubova 2013: 191) may match well with the hypothesis of imports from southeastern Iran – if Potts is right. Pattern no. 9, in fact, is decorated with two superimposed bands containing wide angular patterns, each made of three lines of zigzag excised patterns, framed within parallel bands with the same motif. These patterns match the "*Bols à décor de lignes ondulées*" in Miroschedji (1973: 28, see also pl. III) and those ascribed to a "zigzag genres" in Potts (2003: 78). Probably it is the same vase that was found in the "Dog Mausoleum" at Gonur (Ligabue and Rossi Osmida 2007: 164, fig. 13).

The dotted circles that appear in Figure 9.2 on vial nos. 3 and 4 (Pottier 1984: nos. 152, 153, and fig. 21, upper left) are Salvatori's type E, probably manufactured in the same general area and distributed both towards Bactria and to the west to Tarut. These flacons might point to a technical sphere of interaction between the coasts of the Persian Gulf and southeastern Iran.

Vial nos. 5 and 6 in Figure 9.2 (Pottier 1984: nos. 147 and 148), respectively bearing dotted circles and "shark's-teeth" designs, could also be southeastern Iranian products made for long-distance export. The same can be said for nos. 11a–11d (Pottier 1984: 149), a vial with a complex figuration (a rampant goat, snake, flower) that is might be southeastern Iranian in inspiration, plus a highly stylized "tree" with branches(?) bearing thorns like the above-mentioned "shark's teeth" (this "tree" also appears on vial nos. 12 and 13, respectively from Ligabue and Salvatori 1988: fig. 10; and Sarianidi 2007: 93, fig. 117, at Gonur; and at Susa, Miroschedji 1973: pl. VIIa). A chlorite flacon with a very similar and varied figuration was found at Susa (Ligabue and Salvatori 1988: fig. 10), where the tree is substituted by a kind of pipal leaf, and the snake by a second climbing goat.

Vial nos. 10a, 10b, 10c, 12, and 13 in Figure 9.2 share the co-occurrence of excised bands and containers with a shape related to local Oxus models. Vial nos. 10a, 10b, and 10c (Pottier 1984: 153, respectively nos. 140, 141, and 142) show excised bands on shapes marketable to the north. In these cases, Oxus workshops might have adopted a southeastern Iranian pattern, or, on the contrary, southeastern Iranian ones may have produced vials in a form highly appreciated in the Oxus regions. In both cases, we should think of north-south technical interactions along the eastern edge of the Plateau, but only specific petrographic and chemical provenience studies might solve the question.

A REPERTORY OF OXUS PATTERNS INCISED ON CHLORITE CONTAINERS

Here, I consider a selection of chlorite containers bearing the mentioned incised decoration that: certainly, or most likely, were excavated or found in Margiana (Gubaev *et al.* 1988; Sarianidi and Dubova 2010, 2013); or that were recovered in the 1970s and 1980s on the Kabul market as plundered from unknown graveyards (Pottier 1984; Ligabue and Salvatori 1988; Ligabue and Rossi Osmida 2007; see also, Vidale 2017); or the Oxus objects recently found in a regular excavation in northern Khorassan by the Iranian-Italian project at Tepe Chalow near Jajarm in Iran (Vahdati and Biscione 2014). Pottier (1984) remains a highly valuable source, because it was written while the lootings were still occurring and it documents with a great deal of probability "Bactrian" assemblages before their dispersal and contamination with artifacts coming from other illegally excavated sources.

The focus of this work is on decoration rather than on shapes. The frames of each design on Figures 9.2–9.11 represent the field of decoration rather than the shape of the container. The goal is to observe in details within each main group of the Oxus stone containers, if and which types of motifs occur on the surfaces, and to which extent this supports or denies the validity of Salvatori's and Potts's conclusions. Another aspect taken into consideration is the relation or not between a pattern, or a family of patterns, observed on containers with those appearing on other classes of Oxus artifacts, especially the chlorite components of the much more famous "Bactrian Ladies."

For this purpose, I drafted in comparable sketches the decorative patterns visible on the chlorite containers found in Bactria or Margiana in the literature mentioned above. All these sketches (presented by types in Figures 9.3–9.12) have the limits of a similar approach: They come from pictures and former drawings, and even though I tried to count and reproduce as faithfully as possible each trait, they cannot claim exactness, and are not shown in the original scales. Moreover, when only one view of the objects is available while more designs occur on the other sides, the information is presumably partial. On the other hand, it is only by looking at designs in meaningful groups presented in a standardized way that the variability I am seeking can be appreciated.

The objects that will be discussed are:

- low-walled trays with a crescent-like or semicircular shape;
- low-walled trays with a kidney-like shape;
- various types of open, subcylindrical containers of variable height;
- small or miniature restricted subglobular pots;
- square-based, rectangular containers;
- square-based, trapeze-like containers;
- an exceptional composite stemmed cup.

In addition, I discuss a less common class of presumably Bactrian or Oxus chlorite products (i.e., inlaid vessels). They are known since Pottier's (1984) and Ligabue and Salvatori's (1988) publications. Chlorite containers inlaid with white, red, and even light-blue stones are a distinctive feature of the production of the Halil Rud-Jiroft carved artifacts, in which inlays usually envelop the bodies of animals and other mythic actors interlaced in convulsive fights. In contrast, the evident taste for tiny, very regular geometric inlays made of a white material (limestone or some kind of faience?) and other stones carefully arranged in complex and sometimes very elaborated geometric patterns, might be distinctive of the most refined Oxus workshops.

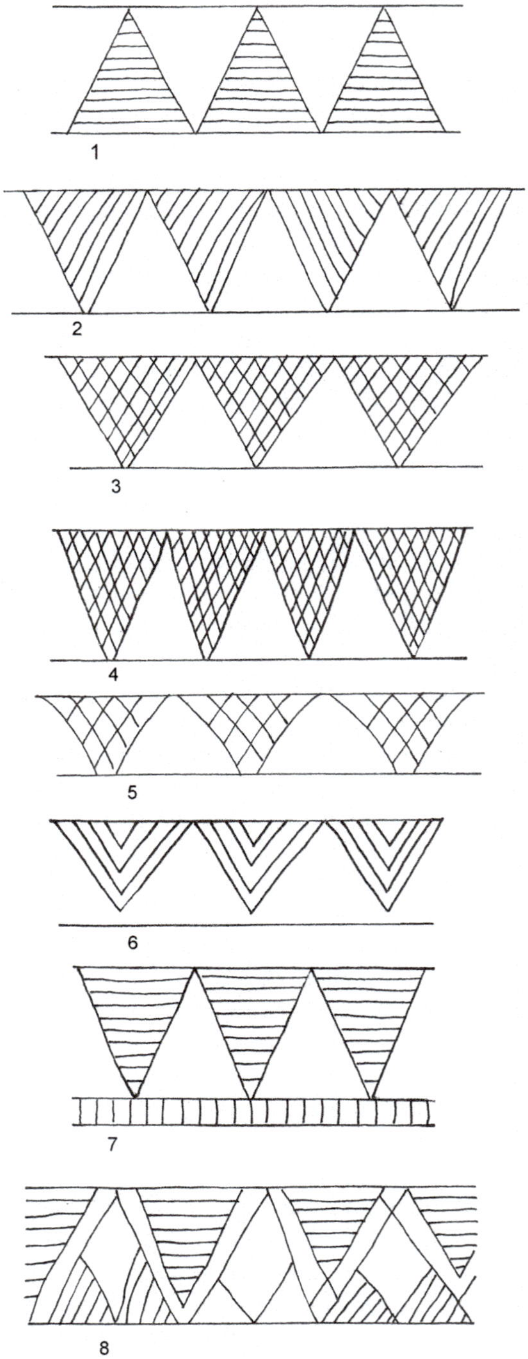

Figure 9.3 Incised patterns, made of a series of triangles on the outer surface of low trays (crescent-like, semicircular, kidney-shaped and, in the only case of number 5, circular).

Table 9.1 Kidney-shaped or semicircular low trays: rows of triangles on the exterior (see Figure 9.3)

No.	*Shape*	*Decoration*	*Provenience*	*Publication*
1	Semicircular low container	Row of horizontally hatched triangles	Grave 247, Adji Kui, Margiana	Ligabue and Rossi Osmida 2007: 206, fig. 5
2	Semicircular low container	Band of inverted triangles filled with oblique hatchings in different directions	Perhaps from Bactria, unknown provenience	Ligabue and Salvatori 1988: fig. 84, upper left and middle right
3	Kidney-shaped low container	Band of hanging crosshatched triangles	Margiana, unknown provenience	Sarianidi 1998a: 50, fig. 18, *1*
3	Kidney-shaped low container	Band of hanging crosshatched triangles	Perhaps from Bactria (unknown provenience)	Ligabue and Salvatori 1988: fig. 84, lower row, left and (coarser fabric) fig. 85, lower row, left
4	Crescent-like low container	Band of hanging elongated triangles filled with a very dense crosshatching	Most probably from Bactria	Pottier 1984: 224, no. 309
5	Circular low tray	Row of hanging triangles, with concave sides and a simple crosshatched filling	Margiana, Gonur North, palace	Sarianidi 1998a: 49, fig. 17, *14*
6	Kidney-shaped low container	Row of hanging nested triangles. It contained a red powder (ochre?)	Most probably from Bactria	Pottier 1984: 99 and 224, no. 307
7	Semicircular low tray	Row of horizontally hatched triangles hanging from the rim; on the base, a clumsy band of short vertical lines (such bands sometimes appear along the base or on the rim of Oxus chlorite vessels)	Most probably from Bactria	Pottier 1984: 284, no. 310
8	Kidney-shaped low container	Complex and irregular frieze of hanging, horizontally hatched triangles; on the lower triangular field are incised or scratched couples of triangles left blank or filled with oblique lines	Probably from Adji Kui(?)	Ligabue and Rossi Osmida 2007: 269, fig. 10

Comments on Table 9.1 and Figure 9.3

Rows of triangles, most of which hang from the rim of the trays, are the most common and rather unimpressive incised decoration of the outer wall of this type of container. The degree of care invested in these incised patterns is variable, but in general is not very high. Although the robes of some of the chlorite "Bactrian Ladies" so far published may be decorated with lozenges and triangles that, at least in terms of geometric templates, are not very different from those appearing on these containers, the latter seem to belong to a different, perhaps lower-rank production. The closest match, admittedly indirect, is with the "Bactrian Ladies" of the style defined as "Graffito" by Vidale (2017), tentatively ascribed, in the light of a preliminary hypothesis of stylistic evolution, between the 21st and the 19th centuries BC. Only in a few instances do the triangle-based designs get more complex (see Figure 9.4, *1–3*).

Comments on Table 9.2 and Figure 9.4

Comparing the incised designs of Figures 9.3 and 9.4, the impression is that while monotonous rows of (hanging) triangles frequently appear on crescent and semicircular containers, kidney-like trays were decorated, besides the same triangles, with a wider range of incised patterns, like the various geometric bands, scales, scallops, leaves, "plumes" (the label is purely conventional), and others (Figures 9.4 and 9.5, see also p. 308). The scale-like designs of pattern nos. 4 and 5 in Figure 9.4 recall (always indirectly) the rendering of the robes of the "Bactrian Ladies" that in Vidale (2017) is defined as "Languettes style," hypothetically considered two or three centuries earlier than the "Graffito style." Similar scales and intertwined sinuous lines occur also in other classes of Oxus artifacts, for example on the rim of silver vessels uncovered at Gonur (Sarianidi and Dubova 2013: 74, 76, 77, 105; see also this book cover). Pattern no. 3 in Figure 9.4 combines a zigzag field partition with the use of two different tubular drills, without central dot. This pattern may be compared with the unusual decoration of a "Bactrian Lady" with stretched body (Vidale 2017: 113, fig. 90). The hatched pipal leaves obviously recall one of the most popular designs on the pottery and other artifacts of Baluchistan and the Indus valley in the 3rd millennium BC.

Comments on Table 9.3 and Figure 9.5

Various transformations of rows of triangles still represent the most common decorative choice on Pottier's (1984) partial selection of kidney-shaped low containers. Pattern no. 3 in Figure 9.5 is identical to that incised on a rectangular footed container (see Figure 11. 3) and might link the two chlorite types in a homogeneous functional couple.

Pattern no. 7 in Figure 9.5 looks like an Oxus version of the elaborated excised pattern visible in a cylindrical vase found at Gonur (compare Figure 9.2, *9* with Sarianidi and Dubova 2013: 191) probably imported from southeastern Iran. While pattern no. 10 in Figure 9.5 shows the freedom with which the Oxus carvers re-elaborated designs inspired by the repertories of the Indus, the "plumes" hanging from the edge of pattern no. 11 have a resemblance with those that appear on the

Figure 9.4 Incised patterns, made of a series of more complex triangles and squares (1–3), scale-like designs (4–5), oblique fringed designs (6), intertwined spirals (7), and pipal leaves (8) on the outer surface of low trays (semicircular, kidney-shaped).

Table 9.2 Kidney-shaped or semicircular low trays: more triangles, curvilinear designs, and others (see Figure 9.4)

No.	*Shape*	*Decoration*	*Provenience*	*Publication*
1	Kidney-shaped low container	Double frieze of superimposed hatched triangles	Margiana, provenience unknown	Sarianidi 1998a: 50, fig. 18, *6*
2	Kidney-shaped low container	Series of crosshatched lozenges	Perhaps from Bactria, unknown provenience	Ligabue and Salvatori 1988a: fig. 84, lower right
3	Kidney-shaped low container	Band containing a zigzag line; circles of two different diameters were drilled in the resulting triangles	Margiana, unknown provenience	Sarianidi 1998a: 50, fig. 18, *3*
4	Kidney-shaped low container	Band including three rows of opposed, in part overlapping, scale-like designs	Margiana, unknown provenience	Sarianidi 1998a: 50, fig. 18, 2
5	Kidney-shaped low container	Two rows of opposed, overlapping scales filled with a crosshatched pattern	Most probably from Bactria	Pottier 1984: 224, no. 306
6	Kidney-shaped low container	A frieze of oblique lines bearing at left series of "shark's-teeth" traits	Most probably from Bactria	Sarianidi 1998a: 50, fig. 18, *4*
7	Kidney-shaped low container	Intertwined spirals with dotted circles in between	Most probably from Bactria	Pottier 1984: 224, no. 308
8	Kidney-shaped low container	Row of hatched pipal leaves	Perhaps from Bactria	Ligabue and Salvatori 1988: fig. 84, middle left

copper hammer-axe with Shulgi's inscription at the Louvre, whose style recalls many Bactrian axes (Vidale 2017: 30, fig. 25). Another iconographic link might be established with the beard hair of bulls in Mesopotamian iconography (perhaps the most famous cases at Ur, see Aruz and Wallenfels 2003: 105, no. 58), well known in Bactrian iconography (e.g., Ligabue and Salvatori 1988: fig. 76) and in the embossed images of the Fullol hoard (Tosi and Wardak 1972; Jarrige 2006).

Figure 9.5 Additional patterns incised on the outer wall of kidney-shaped low trays. 1–3, various kinds of triangles; 4, hatched hourglass design; 5–7, bands filled with oblique lines or angles; 8, oblique lines and shark's teeth; 9, vertical "ladders"; 10, composition of pipal leaves, arranged in a cross-like scheme with various fills; 11, row of "plumes" (for the possible meanings see text) hanging from the rim. Presumably found in Bactria. All re-drafted from Pottier 1984: 176, Fig. 42.

Table 9.3 Additional motives on kidney-shaped trays (re-drafted from Pottier 1984: 176, fig. 42, as a *Tableau Récapitulatif* of such designs) (see Figure 9.5)

No.	*Shape*	*Decoration*	*Provenience*	*Publication*
1	Kidney-shaped low container	Isolated concentric triangles or angles, coarsely traced	Most probably from Bactria	Pottier 1984: 176, fig. 42, *Tableau Récapitulatif*
2	Kidney-shaped low container	Row of horizontally hatched triangles, capped by a horizontal band filled with short vertical lines. Compare with Figure 9.3, 7 (inverted)	Most probably from Bactria	Pottier 1984: 176, fig. 42, *Tableau Récapitulatif*
3	Kidney-shaped low container	Three superimposed rows of hanging triangles, filled with oblique lines	Most probably from Bactria	Pottier 1984: 176, fig. 42, *Tableau Récapitulatif*
4	Kidney-shaped low container	Two rows of triangles opposed in an hourglass form, filled with oblique lines	Most probably from Bactria	Pottier 1984: 176, fig. 42, *Tableau Récapitulatif*
5	Kidney-shaped low container	A horizontal band filled with angular motifs	Most probably from Bactria	Pottier 1984: 176, fig. 42, *Tableau Récapitulatif*
6	Kidney-shaped low container	Two parallel horizontal bands filled with oblique lines with opposed direction	Most probably from Bactria	Pottier 1984: 176, fig. 42, *Tableau Récapitulatif*
7	Kidney-shaped low container	A more complex zigzag band, whose segments are filled with angular motifs bisected by a median line	Most probably from Bactria	Pottier 1984: 176, fig. 42, *Tableau Récapitulatif*
8	Kidney-shaped low container	A wide horizontal band filled with oblique "feathers" (oblique lines bearing short sloping ones). Compare with Figure 9.4, 6 (inverted)	Most probably from Bactria	Pottier 1984: 176, fig. 42, *Tableau Récapitulatif*
9	Kidney-shaped low container	A horizontal band filled with "ladder" elements	Most probably from Bactria	Pottier 1984: 176, fig. 42, *Tableau Récapitulatif*

Table 9.3 Cont.

No.	*Shape*	*Decoration*	*Provenience*	*Publication*
10	Kidney-shaped low container	Couples of pipal leaves with various fillings opposed horizontally by the points; in the resulting fields, two smaller vertical same leaves are opposed by the base	Most probably from Bactria	Pottier 1984: 176, fig. 42, *Tableau Récapitulatif*
11	Kidney-shaped low container	A continuous row of spiral-like or "plume"-like elements hanging from the container's rim	Most probably from Bactria	Pottier 1984: 176, fig. 42, *Tableau Récapitulatif*

Comments on Table 9.4 and Figure 9.6

If the white pot 2[4] was actually made of fired steatite, this would be intriguing given the fact that this technique is from Indus origin. The fact that the pot seems to have been reworked, with a lowering of its edge, and that the geometric incised pattern is an Oxus one, add more problems. The spread of triangle-based designs to these various restricted small forms probably associate them to the low trays and possibly to the preparation of cosmetics(?). In pattern no. 5 in Figure 9.6, pipal leaves with radiating "shark's teeth" perhaps point to an import from the south. Branches of pipal leaves probably filled with a crosshatched pattern also appear on twin cylindrical bowls found in a looted grave in southern Bactria, poorly preserved and not completely illustrated in Ligabue and Salvatori (1988: figs. 8, 9, not illustrated here). In no. 6, similar pipal leaves alternating with indented branches at the top and bottom of the frieze reflect the common Oxus representation of tulip flowers; the design looks like a local adoption of the original Indus motive. The two snake designs (nos. 7 and 8) appear on the same kind of vessel. In the iconographic codes of the ancient Near East and Middle Asia, from Mesopotamia to the Halil Rud valley, superimposed arches like those that appear in no. 8 are generally interpreted as landscape notations for "mountains." The same meaning is fully confirmed by other artworks of the Oxus (e.g., Sarianidi and Dubova 2013: 95). The organization of the incised design resembles that of the late Akkadian Ibni-Sharrum's cylinder seal at the Louvre (among others, see Aruz and Wallenfels 2003: 208–209, no. 135), where it is a river that winds among similar "mountains."

Comments on Table 9.5 and Figure 9.7

Judging from the complete patterns like nos. 2a–2d in Figure 9.7 and others in the following illustrations, some vials may have freely combined different geometric

Figure 9.6 Patterns recorded on different types of unrestricted, hemispherical to subcylindrical containers of variable height, and small restricted pots.

choices, while in others the repetition of similar motives was interrupted by windows (incised decoration or real opening in the wall). On the whole, Figure 9.7 suggests a rather free elaboration and combination of three different motifs: groups of superimposed triangles, hourglasses, and windows framed with hatched bands. While hatched triangles link these vials stylistically to the designs recorded for kidney-like low containers and the like, the inversion visible in nos. 6b, 6c, and 7b, where triangles are left blank and the background is crosshatched, is specific to these receptacles.

Table 9.4 Different types of unrestricted, hemispherical to subcylindrical containers of variable height, and small restricted pots (see Figure 9.6)

No.	*Shape*	*Decoration*	*Provenience*	*Publication*
1	Small subglobular pot	Double frieze of opposed, crosshatched triangles	Most probably from Bactria	Pottier 1984: 165, no. 228
2	Cylindrical vessel in whitish fired steatite, probably reworked	Rows of isosceles triangles of variable height, bisected and filled with alternating opposed hatchings. The motifs are cut on the edge, with a darker core visible along the abraded rim	Margiana (Gonur?)	Sarianidi and Dubova 2013: 65
3	Small restricted truncated-cone-like container, with a rounded rim	Zigzag partition, the resulting triangles filled with drilled dotted circles	Most probably from Bactria	Pottier 1984: 165, no. 232
4	Small subglobular pot with a rounded rim and projecting lugs	Variable frieze of alternating triangles filled with horizontal hatchings	Most probably from Bactria	Pottier 1984: 165, no. 229
5	Open subcylindrical vase, with a rounded rim	Double row of hatched pipal leaves surrounded by "shark's-teeth" designs	From grave 2858 at Gonur North, Margiana	Sarianidi 2007: 112, fig. 199
6	Hemispherical bowl	Row of alternating pipal leaves, filled with oblique lines and flanked by symmetric branches that bear continuous "shark's-teeth" traits. The design is framed between a horizontal band with short vertical segments below the rim and another filled with longer oblique hatchings along the base	From Gonur South, temenos	Sarianidi 1998a: fig. 17, 6

(*continued*)

Table 9.4 Cont.

No.	*Shape*	*Decoration*	*Provenience*	*Publication*
7	Low subcylindrical vase, with a concave profile	Horned(?) snake with a forked tongue slithering all around the pot	Presumably from Bactria	Pottier 1984: 165, no. 231; Ligabue and Salvatori 1988: fig. 84, center
8	Subcylindrical vase with an open profile and a strongly projecting rim	Powerful horned(?) snake with a forked tongue running among superimposed arches. The body of the reptile is covered with star-like elements, the tip of the tail with dots	Presumably from Bactria	Pottier 1984: 164, no. 225

Nos. 3a, 4a, and 6a had a window sealed by a central rectangular lid, cut after the incisions were made (Pottier 1984: 28), for a better extraction of the content, which in such cases should be imagined as a solidified product that could not be extracted from the neck.

Comments on Table 9.6 and Figure 9.8

Figure 9.8 suggests some association between hourglass-like designs and window-like frames. When a window had to be cut from one of the sides, these incised designs might have guided the secondary cutting. The well-carved design of no. 1b closely resembles the decoration of some white beads or seals, rectangular in shape and lentoid in section, which arrived on the antique market as looted in southern Bactria. An example is also visible in Ligabue and Rossi Osmida (2007: 274).

Comments on Table 9.7 and Figure 9.9

The pointed tulip flowers that appear in Figure 9.9, nos. 2c, 3d, 4a, 4c, 5a, 5b, 6a, and 6b, are considered as "a typically Central Asian floral decoration" (Teufer 2007; Potts 2008: 171) representing wild ones. Unmistakable tulip designs also appear on bronze stamp seals (e.g., Sarianidi 1998b: nos. 859, 860). The designs on nos. 3a and 3b in Figure 9.9 have an obvious similarity with the patterns on the robes of the "Bactrian Ladies" of the "Lozenges style" (Vidale 2017). Vial no. 6[5] also had a small hemispherical inlay (lost) on the upper part of each side (Pottier 1984: 28fn25) so that it should be considered as an inlaid container (see p. 324).

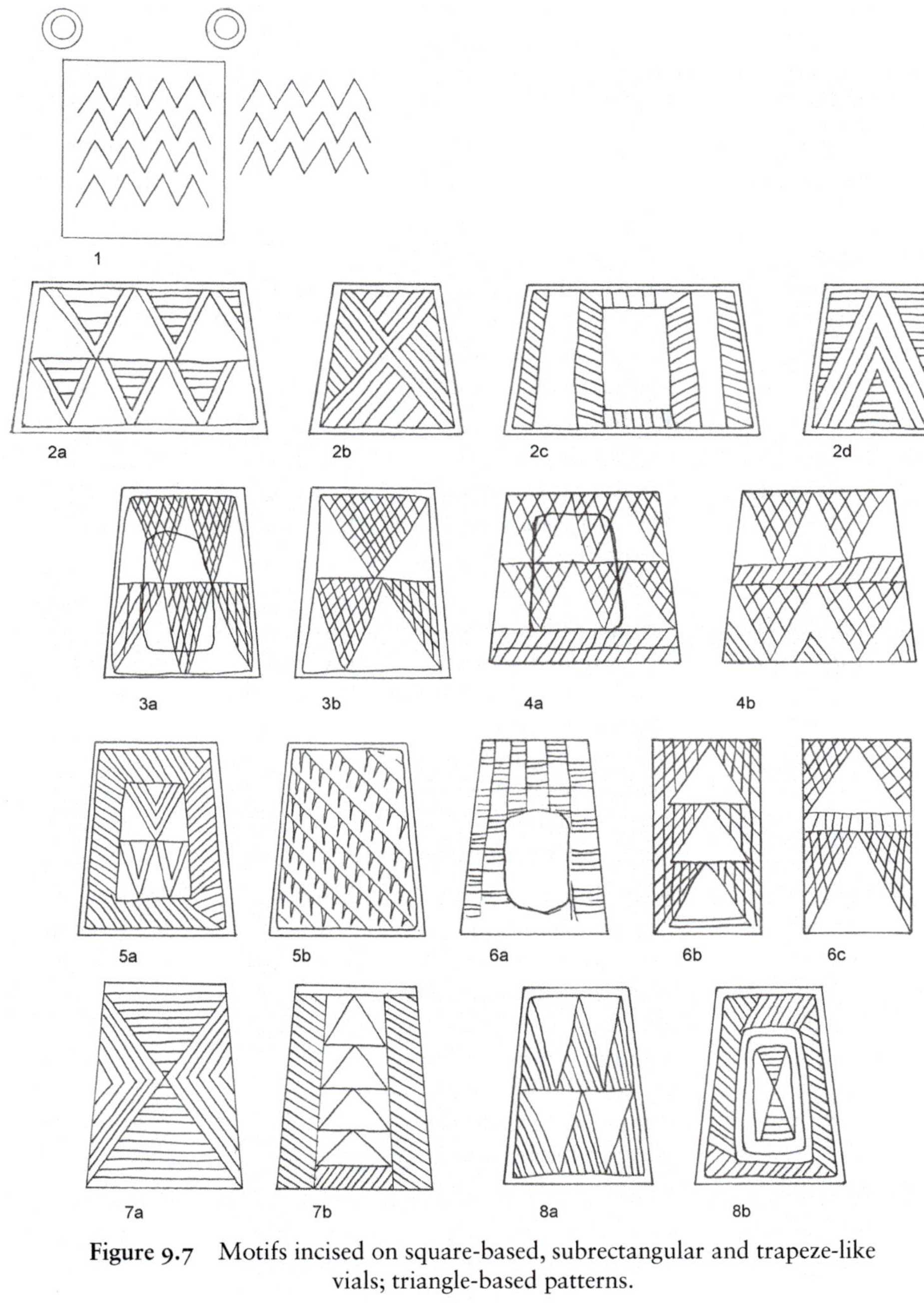

Figure 9.7 Motifs incised on square-based, subrectangular and trapeze-like vials; triangle-based patterns.

Comments on Table 9.8 and Figure 9.10

Vials with such figurative motifs seem to be rather uncommon. Curiously, they combine three iconographic elements ("Bactrian Lady," winged deity, decorated stemmed cup) that unmistakably belong to the Oxus tradition, with southeastern

Table 9.5 Square-based, subrectangular, or trapeze-like vials; triangle-based patterns (see Figure 9.7)

No.	*Shape*	*Decoration*	*Provenience*	*Publication*
1	Square-based, rectangular vial	Rows of zigzag lines and concentric circles on the upper corners	Margiana (Gonur?)	Sarianidi and Dubova 2013: 60
2a–2d	Square-based, trapeze-like vial	Angular or triangular motives in various arrangements on the four sides, including an hourglass-like design and a window	Most probably from Bactria	Pottier 1984: 152, no. 131
3a,3b	Square-based, subrectangular vial	Groups of hanging crosshatched triangles. The two motifs are repeated twice. A window is cut on one side	Most probably from Bactria	Pottier 1984: 153, no. 136
4a, 4b	Square-based, trapeze-like vial	Groups of hanging crosshatched triangles combined with horizontal bands filled with oblique lines and minor modifications. A window is cut on one side; it contained a gray powder	Most probably from Bactria	Pottier 1984: 153, no. 137
5a, 5b	Square-based, subrectangular vial	A window framed by bands with oblique traits, contains three hanging triangles (on the opposite side, the window is cut); the two other sides are entirely filled with oblique lines with series of "shark's-teeth" traits	Most probably from Bactria	Pottier 1984: 152, no. 133
6a–6c	Square-based, subrectangular vial	A chessboard-like design, with boxes filled with horizontal traits, with an open window; on the other three sides, two panels with three and two superimposed triangles, surrounded by a crosshatched background	Most probably from Bactria	Pottier 1984: 153, no. 138

Table 9.5 Cont.

No.	*Shape*	*Decoration*	*Provenience*	*Publication*
7a, 7b	Square-based, subrectangular vial	The two visible sides have an hourglass motif filled with horizontal lines; the lateral fields are filled with concentric angles	Most probably from Bactria	Ligabue and Salvatori 1988: 116, fig. 8
8a, 8b	Square-based, subrectangular vial	Double line of three triangles filled with oblique traits (on three sides); on the last, a window framed with oblique traits encloses a hatched hourglass motif	Most probably from Bactria	Pottier 1984: 153, no. 139

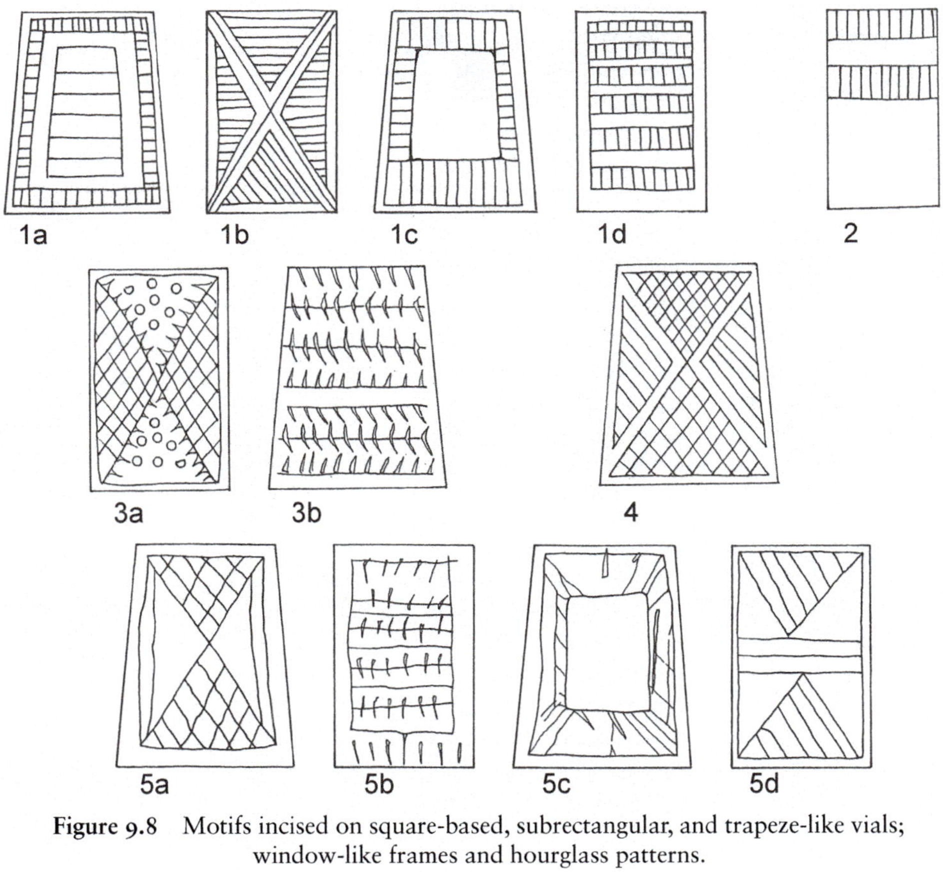

Figure 9.8 Motifs incised on square-based, subrectangular, and trapeze-like vials; window-like frames and hourglass patterns.

Table 9.6 Square-based, subrectangular, or trapeze-like vials; hourglass-based patterns (see Figure 9.8)

No.	*Shape*	*Decoration*	*Provenience*	*Publication*
1	Square-based, subrectangular to trapeze-like vial	Two window-frame designs (1a, 1c), an hourglass design filled with horizontal and oblique lines (1b), and a last side (1d) framed with parallel bands filled with vertical lines; 1c hosts a cut window	Margiana	Sarianidi 1998a: 50, fig. 18, *10*
2	Square-based, subrectangular vial	Parallel bands filled with vertical lines (one side available)	Margiana	Sarianidi 1998a: 50, fig. 18, *13*
3	Square-based, subrectangular vial	On two opposed sides, hourglass motif filled with round dots, crosshatched lateral triangles. Within the hourglass, series of "shark's teeth." On the other two sides, horizontal lines with opposed series of "shark's teeth"	Most probably from Bactria	Pottier 1984: 152, no. 134
4	Square-based, subrectangular to trapeze-like vial	Crosshatched hourglass with lateral triangles filled with oblique lines. Same motif on the four sides. A window is cut on one of them	Most probably from Bactria	Pottier 1984: 152, no. 132
5a–5d	Square-based, subrectangular to trapeze-like vial	Crosshatched hourglass on one side; motif similar to 3b on another; coarse window frame that originally enclosed a removable rectangular lid on the third side; two opposed hatched triangles on the fourth	Most probably from Bactria	Pottier 1984: 152, no. 135

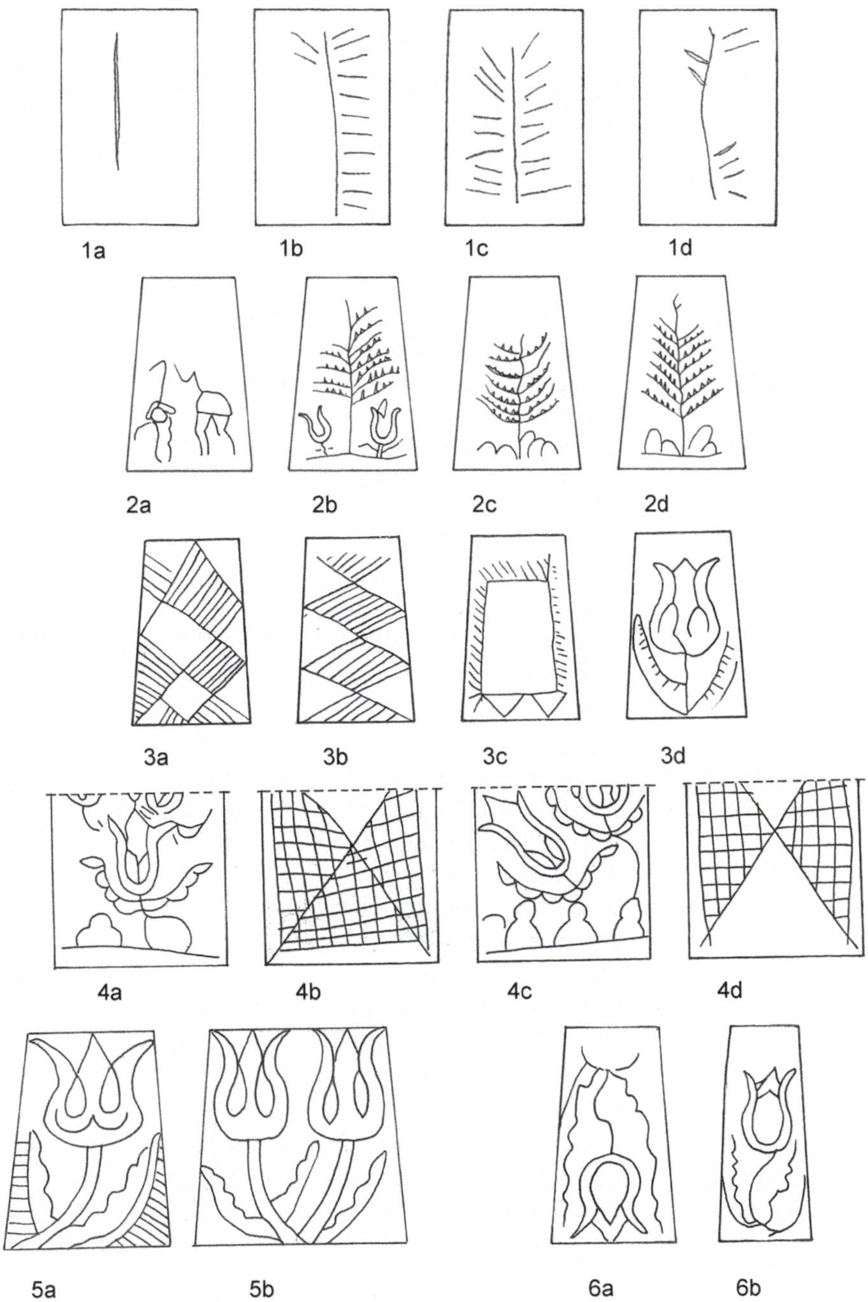

Figure 9.9 Vegetal designs (presumably trees and tulip flowers) on square-section chlorite vials. Flacon 6a, 6b has on top small hemispherical inlays (not enhanced in the design), so it should also be considered as an inlaid container (Pottier 1984: 28fn25).

Table 9.7 Square-based, subrectangular, or trapeze-like vials; trees and tulips (see Figure 9.9)

No.	*Shape*	*Decoration*	*Provenience*	*Publication*
1a–1d	Square-based, subrectangular vial, very worn	Seems to have included three or four sketchy trees	Adji Kui, Margiana	Ligabue and Rossi Osmida 2007: 142, fig. 22
2a–2d	Square-based, subrectangular or trapeze-like vial	2a and 2b with trees among arch-like features possibly representing "mountains"; in 2c, a similar tree is flanked by two symmetric tulip-like flowers; the last side (2d) has a more complex figuration, with a personage standing in front of an unidentified object. This latter image is poorly preserved	Most probably from Bactria	Pottier 1984: 154, no. 145
3a–3d	Square-based, subrectangular or trapeze-like vial	3a and 3b with chains of poorly rendered lozenges filled with oblique lines; 3c with a central rectangular window, originally closed by a lid, surrounded by a coarse decorative window-frame; 3d with a three-pointed stylized tulip flower flanked by two arched branches or leaves	Most probably from Bactria	Pottier 1984: 153, no. 144
4a–4d	Square-based, subrectangular vial, probably heavily reworked (shoulder and neck were re-carved and 1–2 cm lowered down, so that the decoration of the four sides is truncated on top	Sides 4a and 4c show one main tulip with its typical wavy-edged leaves, and other smaller tulips above it, all rising from "mountains" among curved wavy-edged leaves; 4b and 4c are coarsely incised with modified crosshatched hourglass designs	Gonur North, palace	Sarianidi 1998a: fig. 19, *1*

Table 9.7 Cont.

No.	*Shape*	*Decoration*	*Provenience*	*Publication*
5a, 5b	Square-based, trapeze-like vial	The two sides available show respectively one and two tulips with wavy-edged leaves	Margiana	Sarianidi and Dubova 2013: 185
6a, 6b	Square-based, subrectangular, or trapeze-like vial. On the bottom, an oval hole (without lid). On the upper part of both sides, a small hemispherical inlay	The two sides show three-pointed tulip flowers amid its wavy-edged leaves, hanging from the top or rising from the base	Most probably from Bactria	Pottier 1984: 28 and 153, no. 143

Figure 9.10 Three examples of square-section chlorite vials with figurative motifs.

Iranian elements: for example, the dotted circles and "shark's teeth" in vial no. 1 in Figure 9.10, and the hints at the "hut" architectural model and excised lines in nos. 2 and 3.

For the debate on the possible divine identity of the "Bactrian Ladies" see Vidale (2017: 99–101; but see also Chapter 7). Arguments for interpreting the winged figure no. 2 in Figure 9.10 as a winged female deity are presented in Winkelmann (2014). The hypothesis is supported by the architectural feature on which she stands, close to the "hut" or "temple" iconography redundantly appearing on chlorite items

Table 9.8 Square-based holders: three vials with figurative motifs (see Figure 9.10)

No.	*Shape*	*Decoration*	*Provenience*	*Publication*
1	Part of a damaged square-based, subrectangular or trapeze-like vial	Around the neck, rows of dotted circles, and on the sides a "Bactrian Lady" in profile	Site 392 in Margiana, found on surface	Gubaev *et al.* 1998: 44, fig. 4, *1*
2	Subrectangular or trapeze-like vial	The four sides show a winged personage with the arms joined to the chest, standing on the curved top of a building. Aside this figure rise two symmetric tulip flowers with curved indented leaves	Most probably from Bactria	Pottier 1984: 155, lower right; 202, fig. 150
3	Necked cosmetic container carved above a base fashioned as a "hut" or "temple" architectural model	On the neck runs a row of triangles filled with oblique hatchings framed between two excised zigzag lines. On the side of the base a unique depiction of a stemmed subcylindrical cup	Presumably from a looted context in Bactria	Ligabue and Salvatori 1988: 183, fig. 2

of the Kerman area. The curved body and the wings are comparable with the features of a similar image on a bronze stamp seal from Gonur (Sarianidi 2007: 103, fig. 64), and others coming from unknown Bactrian contexts (Sarianidi 1998b: nos. 24.2, 34–40).

Containers similar to no. 3 (discussed as type D in Salvatori 2008: fig. 6.2) were also found at Shahdad (Hakemi 1997b: 621–624), Susa (Miroschedji 1973: fig. 12, pl. V, 4), and among the looted artifacts recovered near Jiroft (Madjidzadeh 2003: 140).

As similar stemmed cups are known at Gonur both in pottery (Sarianidi 2007: 64, fig. 23) and in chlorite, made in two separate pieces (Sarianidi 2007: 112, fig. 198, from grave 1750), as well as, in Bactria, in bronze (Pottier 1984: 211, no. 244), their incised outline may be considered as an Oxus creation.

Comments on Table 9.9 and Figure 9.11

Footed trays no. 1–4 in Figure 9.11 are decorated with motifs quite similar to those encountered on the low kidney-shaped trays and their relative variants, as well as on the square-section rectangular-to-trapeze like vials. Thus the idea that these three kinds

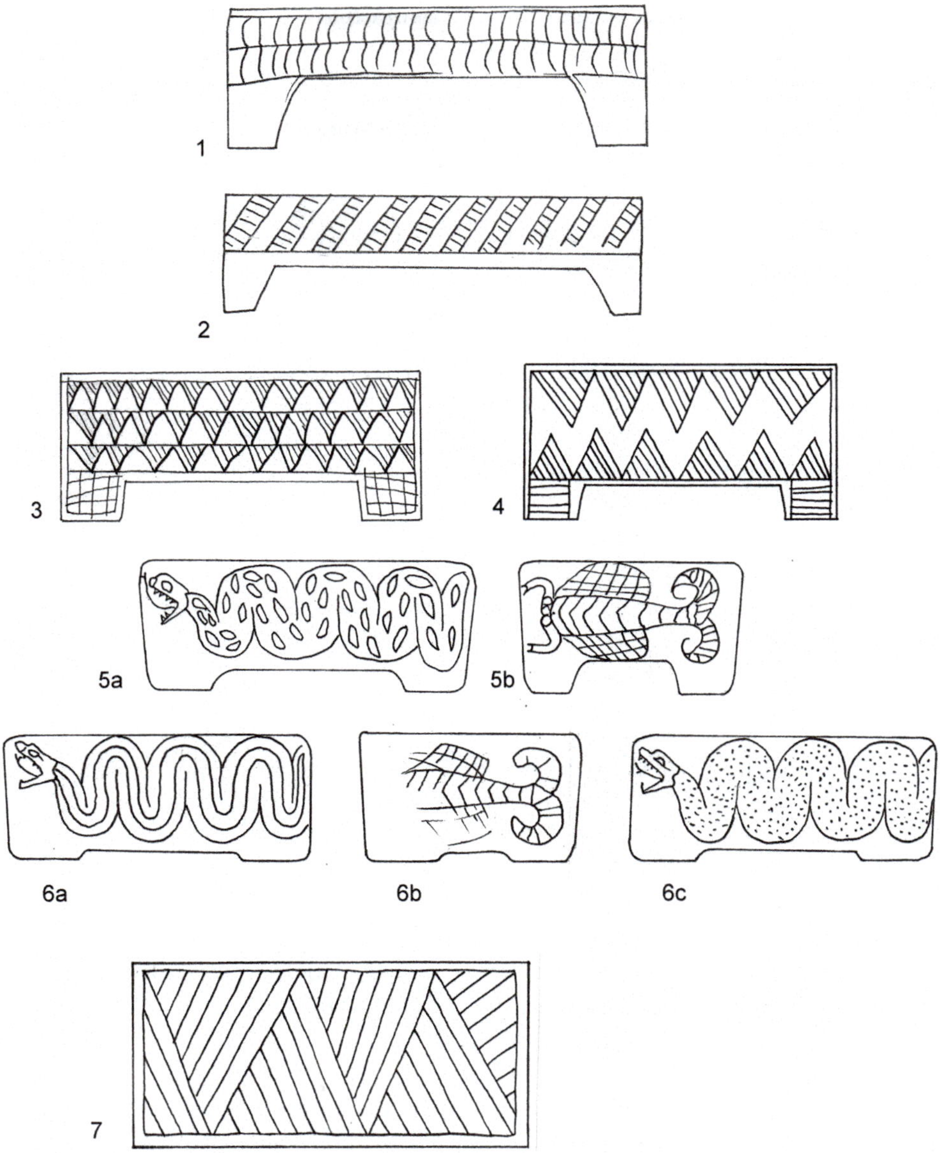

Figure 9.11 Motifs recorded on rectangular footed trays (1–6) and on a square compartmented box (7).

of chlorite containers (and those similar to the compartmented box in no. 7) formed "sets" for storing cosmetics or medicine preparation is supported. Pottier's (1984) tray (nos. 6a–6c), found in a looted context in Bactria(?), and the one excavated at Tepe Chalow in northeastern Iran (nos. 5a and 5b), are so similar that they might have been crafted by the same person in the same workshops, probably (judging from the associated pottery) in the late Namazga (NMG) V period (i.e., in late 3rd millennium BC) (Vahdati and Biscione 2014; G. Bonora, personal communication).

Table 9.9 Footed rectangular trays and a compartmented box (see Figure 9.11)

No.	*Shape*	*Decoration*	*Provenience*	*Publication*
1	Rectangular footed tray	Double band of arches	Probably from Bactria	Ligabue and Salvatori 1988: fig. 85, right
2	Rectangular footed tray	Row of oblique "ladders"	Margiana	Sarianidi 1998a: 50, fig. 18, 14
3	Rectangular footed tray, divided into two inner compartments	Three lines of hatched hanging triangles. The feet are hatched (small orthogonal grids). Remains of red powder in both compartments	Most probably from Bactria	Pottier 1984: 224, no. 311
4	Rectangular footed tray	On the long wall that is visible, a zigzag band defines two opposed series of triangles, filled with oblique traits. On the feet, sequences of horizontal parallel segments	Margiana	Sarianidi and Dubova 2013: 156
5a, 5b	Thick, rectangular footed tray	Snakes with open fangs and forked tongue on the long sides, and stylized two-tailed scorpions on the short ones	From a grave at Tepe Chalow, Khorasan, Iran	Vahdati and Biscione 2014
6a, 6b, 6c	Thick, rectangular footed tray	Snakes and stylized two-tailed scorpions, arranged as described for 5a and 5b	Most probably from Bactria	Pottier 1984: 176, nos. 312a–312b; 224, nos. 312a–312b
7	Square box divided into four compartments, originally covered with a lid	Outer wall divided by a zigzag line; the resulting triangles and half-triangles are filled with oblique lines	Most probably from Bactria	Pottier 1984: 165, no. 236

Figure 9.12 A sketch summarizing the friezes on a composite stemmed cup in black steatite or chlorite, found in the Large Necropolis of Gonur (grave 1750).

Table 9.10 A composite stemmed cup (see Figure 9.12)

No.	Shape	Decoration	Provenience	Publication
1	A stemmed cup made of two separate pieces, cup and stem, black steatite	Series of incised designs: (a) and (d) multiple spirals; (b) triangles or zigzag lines; (c) row of double arches; (e) on the foot, a "landscape" with tulips aside an arborescence and near "mountains" (only partially visible)	Margiana	Sarianidi 2007: 112, fig. 198

Comments on Table 9.10 and Figure 9.12

The series of designs that appear on this cup (Figure 9.12) belong coherently to the Oxus repertory.

INLAID CHLORITE CONTAINERS (NOT ILLUSTRATED)

Most probably, this poorly known class of elaborated stone vessel should be considered a distinctive Oxus production. A preliminary list may include:

- A small globular container, with a single row of 13 more or less ogival inlays made successively of lapis and bone under the tapering rim. Presumably from Bactria (Pottier 1984: 35 and 165, no. 227).
- A square-based, subrectangular vial inlaid with three rows of alternating triangles (the inlays are lost), with a pierced base like the more common chlorite items (Pottier 1984: 28 and 154, no. 146).
- Other finely inlaid chlorite items from Pottier's publication are a low plate and/or a lid (Pottier 1984: 177, no. 324 and 325). As mentioned earlier, four small inlays are also present on the upper part of the square-based vial incised with tulips (Pottier 1984: 28 and 153, no. 143; see also Figure 9.9, *6a–6b*).
- We may also add a square-based, subrectangular, or trapeze-like vial bearing rows of white ogives (on top, hanging down; on the base in the opposite setting) framing two rows of double ogives, published in Vahdati and Biscione (2014).
- A small globular container with a short cylindrical neck, bearing, from top to bottom, rows of inlays of different shapes: hearths, drops, semicircles; again, hearths and drops, flowers(?), presumably looted in Bactria (Ligabue and Salvatori 1988: 117, figs. 9, 1; 219, fig. 86);
- Another series of possible examples from a private collection was recently published in Vidale (2017: fig. 32), but the provenience and date are unknown.

Comments on inlaid containers

Inlaid, sometimes composite, chlorite containers are not common, and are the fruit of an outstanding level of technological elaboration. Their rarity and technical complexity suggests the need to do a first-hand examination before more could be said on this matter. Stone containers with geometric inlays belong to a technological family that also includes mosaics with tiny pieces made of stones of different forms and contrasting colors applied on wooden boxes and furniture, as well as on built surfaces (Sarianidi 2007: 114, figs. 204–206, from burial no. 194; Sarianidi and Dubova 2010; 2013: 145, 116–117; Veresotskaja 2016). Notoriously, mosaics made of larger and more elaborated pieces of contrasting colors, figurative or geometric, were also fitted to chests, wall panels, and even to plaques used as breastplates for camels (Sarianidi and Dubova 2013: 26–35, 125–133; see also Chapter 11).

The taste for chlorite containers inlaid with tiny geometric pieces was also shared in the Halil Rud valley, both on "hut"-like chlorite building models (Madjidzadeh 2003: 68, 119; Piran 2013: 69, 70) and on walls of graves (unpublished excavation from the graveyard of Qaleh Kutcheh near Jiroft). It was also the case on clay "hut" models deposited in the graves of Shahdad (Hakemi 1997b: 662, fig. Ja6). The variety in the shapes of the inlay pieces, and the patterns that they compose, reveal a keen interest in geometry, which also transpires from the templates on which bronze stamp seals were cast.

Apparently, in ancient Margiana chlorite inlaid vessels were even imitated in cheaper wheel-thrown pots (Sarianidi 2007: 57, fig. 13), a detail that reveals how these items were involved in the processes of social confrontation.

DISCUSSION

In spite of the intrinsic problems mentioned in the introduction to this chapter, this preliminary review of the decorative patterns on the chlorite containers points to their local production in Bactria and Margiana or in the nearby areas, even if there is not yet archaeological evidence for it. It also does not contradict and even strengthen Salvatori's and Potts's models. Incision by the means of pointed copper tools was the distinctive technique used in the BMAC, and the decorative patterns are cross-linked by their redundancy, with the same designs appearing with variations in different formal groups of containers. In particular, as stated above, square-section vials, kidney-like trays and their variants, some miniature pots, and part of the footed trays share the same set of simple motives and suggest that they were used on the same tables, for keeping and mixing the same chemicals. The style was essentially geometric, relatively simple and angular; with the only and rare exception of spirals, the curvilinear motifs such as scales, leaves, "plumes," "mountains," flowers, snakes, and scorpions maintained their figurative reference rather than being generated by combining minor abstract components.

These objects were mainly produced for local distribution and use, and when we have more and better analytical evidence on their contents (in two cases, lead-based mixtures, see p. 327; in many cases, some kinds of thick pastes, judging upon the lateral or lower openings) we will learn much on the customs of the elites in the main Oxus centers.

Oxus vials and trays, in fact, did not travel much. Hakemi (1997b) makes clear that they are absent at Shahdad, and the artifacts confiscated near Jiroft in Iran fall short of them, while at Susa only a couple of fragments *might* share the most evident features of the chlorite items reviewed here (cf. Miroschedji 1973: fig. 9.1; perhaps also figs. 11, 15, and 16). The elites of the Oxus did not seem to have much appreciated the richest figurative iconography of the Halil Rud (possibly Marhashi) chlorite vessels: The few stone vessels carved in the figurative Halil Rud style that *might* have been found in southern Central Asia are still doubtful, coming from the antiquity market (see note 4). It is clear that the two decorative repertories (geometric and figurative) were totally independent and impermeable to each other. Possibly, the snake (e.g., in Figure 9.11, *5a*), apparently associated to a late NMG V period, witnesses some distant Halil Rud influence in the Oxus graphics.

In contrast, Susa abounds in Halil Rud and many other southeastern Iranian artifacts; the latter are also widely spread from the Persian Gulf to Margiana and Bactria (decorative patterns are summarized in Figure 9.2; for a full discussion see Salvatori 2003).

Besides the import of finished chlorite pots and vials, the stonecutters of the Oxus adopted specific techniques from the southeastern craft communities, and the formal models that these latter involved (e.g., the dotted circles, "shark's teeth," and excised zigzag lines decoration). As we have seen, even icons that are purely Bactrian or Margian were sometimes created through southeastern Iranian contaminations. From east and southeast (Baluchistan and the Indus valley) came the motif of the pipal leaf, constantly re-elaborated with variable solutions.

In this light, Bactria and Margiana seem to have attracted not only products, but – probably – also people, with their tools and craft skills. If the political and cultural boundaries with Marhashi were a barrier, the possible routes would have been along the northern edge of the Dasht-i Lut towards southeast and southwest, with specialized manufacturing centers like Tepe Yahya (Lamberg-Karlovsky 1970; 1988, 2001; Lamberg-Karlovsky and Potts 2001) acting as pivots. Chlorite and talc deposits were probably available around the lower valleys of the Amu Darya and the Zeravshan (see the appendix at the end of this chapter), but there is no evidence so far that these stones were extensively worked or traded before the middle of the 3rd millennium. Unless foreign craftsmen came from the south with lumps of unworked base materials (as Frenez (2017) implies for the ivory trade and craft), newcomers had to learn to master the performance of local stones through a difficult process of adaptation.

Ultimately, such attractivity and pervasive sphere of sociotechnical interaction with southeastern Iran, in spite of the present lack of meaningful contexts and a reliable chronology, stresses the dimension of political centrality of the early states of Bactria and Margiana in one or more critical moments of the late 3rd millennium (Hiebert 1994: 117).

CONCLUSION

The frequency with which the Oxus cosmetic flacons were reopened by cutting windows on the sides and in the bottom, probably when the content could not be extracted from the neck any longer, shows that, although the holders maintained their prestigious formal identity, the contents were more valuable than the containers. So, what were the contents?

A recent archaeometric study of the cosmetic powders sampled from a series of alabaster cone-like phiales of the 3rd millennium found at Shahr-i Sokhta (Vidale *et al.* 2016) showed that mineral contents were complicated mixtures of white and green-to-blue pigments lightened with quartz and bone powder, made with artificial substances probably obtained by the means of wet chemical processing, and/or possibly gathered in fumaroles along the partially active volcanic environments of inner Baluchistan. The list of these rare minerals include laurionite (lead halide), phosgenite (lead chlorocarbonate), palmierite (lead sulfate), pseudoboleite and cumengeite (lead-copper chlorures hydroxides), plus the commoner galena (lead sulfide) and cerussite (lead carbonate).

As the blue components of the cosmetic powders of Shahr-i Sokhta do not match the composition of lapis lazuli, it was argued that this color was obtained from alternative materials like pseudoboleite and cumengeite, possibly obtained in southeastern Iran, where the alabaster needed for making the phiales is, still nowadays, quite abundant. The idea is that, in the Bronze Age trade of cosmetics, containers were as important as they are today, and that therefore the spatial co-occurrence of the stone for the containers, and of the base materials and technical know-how for pigments and perfumes, was a critical issue in the establishment and development of important commercial routes.

If containers and contents were actually linked, as I suppose, in the form of standardized branded productions (Wengrow 2008), the very partial scientific information on the contents of the Oxus flacons is another very serious handicap. In the report published in Ligabue and Rossi Osmida (2007: 262–269), for instance, lumps of colored substances were found in grave 247 of the Adji Kui graveyard (Ligabue and Rossi Osmida 2007: 266, fig. 5), but holders and cosmetic contents have been manipulated and we do not know which color was found in which container. In the same graveyard lumps of jarosite, a yellow hydrous sulfate of potassium and iron, were found, but the grave number is missing. Brushite, a yellowish calcium phosphate, found in room R66 of site AK9, is also reported as a base material. Galena, cerussite and laurionite (all lead-based substances, see p. 325) are also mentioned, but without context information. More valuable, in contrast, is the information provided by Pottier (1984: 28fn27), where she states that,

> The use of these cosmetic flacons is, obviously, similar to that of the copper small bottles; moreover, some of them still have a pin stuck in their neck. Two samples from flacons nos. 145 and 137 have revealed the existence of a white powder in the first one, and gray for the second one. In both cases, their chemical basis is a lead carbonate. (Translated from the French original by B. Lyonnet).

Presumably, the cosmetic mixtures were quite complex, and might have also included scents obtained from herbs and spices. Chloritic rocks are porous and, even in the absence of the content residues, they might be chemically tested. For example, archaeometric studies have demonstrated that cooking chlorite pots of the Sasanian to medieval periods (5th–13th centuries AD) can absorb and preserve organic residue and reflect the original processing of foodstuffs (Namdar *et al.* 2009).

Sarianidi (2007: 92) reports that a Moscow laboratory found in one of the containers from Margiana residues of artemisia, a plant with nutrients and antioxidants that might explain its use as a medical or herbal tonic (Ebiamadon *et al.* 2009); unfortunately, he does not provide further details. A chlorite flacon from the late 4th–early 3rd millennium found on the surface at Shahdad (Kerman, Iran) was still full of cerussite and hydrocerussite (i.e., of a lead white base), with traces of the aromatic substance 8-metilcoumarine (Vidale *et al.* 2012).

We may reasonably expect that, in the future, when systematic attention is paid to the contents of these flacons and analogous containers of the second half of the 3rd millennium, we will also unlock important data concerning a sophisticated, international wet chemistry know-how, which might have originally developed in Middle Asia, before it spread to Egypt in the early 2nd millennium BC (Martinetto *et al.* 2000a, 2000b).

APPENDIX: CHLORITE AND TALC SOURCES IN CENTRAL ASIAN COUNTRIES

The following was compiled following www.mindat.org, and, for Afghanistan, following Orris and Bliss (2002). It is important to note that here the terms "chlorites" and "talc" refer not to rocks but to minerals, which might be secondary components of ore outcrops or other mineralization, and therefore do not necessarily refer to substantial veins of rocks exploitable for the production of vessels or other large artifacts. Also, obviously enough, the lists were compiled according to the potential modern interests that do not reflect those of Bronze Age economic enterprises. However, it shows that there are several potential locations around the Amu Darya basin, in the contemporary countries of Afghanistan, Tajikistan, Kyrgyzstan, and probably also Uzbekistan (for which information is not available on mindat.org).

"Chlorites" in Afghanistan are mentioned at the Hajigak barite deposits, Shibar district (Bamiyan province), and at the Jagdalek mines, Surobi district, where rubies are found in marble and phlogopite; the two locations are respectively west and east of Kabul. More distant from the Bactrian plains is a fourth possible source in the Kanneshin complex, Dushan district (Helmand Province). Coming to talc occurrence, the closest source to Mazar-i Sharif reported nowadays is Dahan-i Ghori in Baghlan province, near the town of Pol-e Khomri. Other talc deposits of modern interest are found near Kabul, like the Lalandar deposits, Chahar Asyab district (Kabul province), and possibly in the Ferenjal asbestos outcrops, Ghorband district (Parwan province). Still others are located along the present Afghan-Pakistani borderline at Sar-i Sang, Khash, and Kuran Wa, Munjan district (Badakhshan province); further south, in the Nari district deposits (Kunar province, Afghan Nuristan); and again in the Achin, Sherzad, and Shinwar districts (Nangarhar province), where Renaud (2013) mentions a still active small-scale talc mining industry.

In Tajikistan, talc and chlorites are mentioned in the Pamir Mountains, at the border with Afghanistan, east of Darmadar, like the deposits of Kuh-i-Lal in the Pjandzh River valley (Viloyati Mukhtori Gorno-Badakhshan). Reports concerning chlorites for this region include Snezhnoe and Rangkul, in the same mountains; the Maykhura Mine, Kuljabskij Raion, Viloyati Khatlon; the Trezubets deposit, Bazardara ore cluster, Pamir Silver district, still in the Viloyati Mukhtori, Gorno-Badakhshan; and again, the mines of Adrasman, Kokbulak, and the gold- and silver-bearing deposits of Shkol'noe, Kandjol ore field, Kurama Mountains (all in the Viloyati of Sogd).

In Kyrgyzstan, too, chlorites of modern interest are reported in several copper-gold mining locations, in the oblasts of Talas (Aktyuz and Kutessay, Andash in the Karacol district, Taldy-Bulag mines), Issyk-Kul'skaya (Kumtor), Jalal-Abad (Pervalnoe), Naryn (Makmal mine).

Bushmakin (2007) places outcrops of "talc" in the lower valley of the Zeravshan, Uzbekistan; massive deposits of talc and magnesite at the Zinelbulak mines, in the Karaozaq district of Karakalpakstan, Uzbekistan, north of the actual border between Turkmenistan and Uzbekistan, are mentioned in specialized literature (e.g., Hojamberdiev 2010).[6]

NOTES

* A previous version of this chapter was reviewed by B. Lyonnet, who pointed out some errors and several cases of inconsistency, noticeably improving the text. I also warmly thank, as usual, S. Salvatori and G. Bonora for their continuous help. I am also very grateful to my

colleague G. Artioli (Department of Geosciences, University of Padua) for his help on the geological information on the Central Asian sources of chloritic rocks and talc.

1 Here, I use "steatite" only for very rare objects that seem to have been whitened through pyrotechnological treatments, see p. 309.

2 See also David (2002: fn2).

3 After this definition, I would not consider a Bactrian product the beautiful square chlorite box reproduced in Francfort (2003: 58, no. 17). Carved in a naturalistic low relief, it is decorated with what looks like a chest made of a frame of reeds (straight and bent) holding in place a mat. This motif seems to be quite rare in the Bactrian repertories. The possibility of a provenience from the south is strengthened by the presence of very similar boxes in the artifacts seized at Jiroft when the pilfering of the Halil Rud graveyards had come to an end (Madjidzadeh 2003: 42, 43). Thus, if this piece actually came from looted Bactria, it would witness a contact with Kerman. See also a picture published in Ligabue and Rossi Osmida (2007: 269, fig. 10), where three Halil Rud carved chlorite containers are labeled as found in Bactria; but they come from the antique market and the information is meaningless.

4 A fired steatite pot with a more common carved design in the Halil Rud style is presently visible at the Jiroft Museum, Kerman, Iran.

5 Sarianidi (2007: 92) reports the find, in burial no. 421 of Gonur, of a bone receptacle with a similar lower opening closed by a bone plate.

6 More information and maps are available at www.mindat.org/photo-721340.html, www.mindat.org/minlocsearch.php, and https://minerals.usgs.gov/minerals/pubs/country.

REFERENCES

Aruz, J. and R. Wallenfels 2003. *Art of the First Cities: The Third Millennium BC from the Mediterranean to the Indus*. New York: Metropolitan Museum of Art.

Ascalone, E. 2012. "A New South-Eastern Iranian Glyptic Evidence." In *Susa and Elam: Archaeological, Philological, Historical and Geographical Perspectives*, ed. K. De Graef and J. Tavernier. Leiden and Boston: Brill, pp. 4–25.

Aubet, M.E. 2013. *Commerce and Colonization in the Ancient Near East*. Cambridge: Cambridge University Press.

Bushmakin, A.G. 2007. "Minerals and Metals of Bactria and Margiana." In *Sulla Via delle Oasi: Tesori Dell'Oriente Antico*, ed. G. Ligabue and G. Rossi Osmida. Trebaseleghe: Il Punto, pp. 178–189.

Cleuziou, S. and M. Tosi 2000. "Ra's al-Jinz and the Prehistoric Coastal Cultures of the Ja'alan." *Journal of Oman Studies* 11: 19–73.

David, H. 1996. "Styles and Evolution: Soft Stone Vessels During the Bronze age in the Oman Peninsula." *Proceedings of the Seminar for Arabian Studies* 26: 31–46.

David, H. 2001. "Soft Stone Mining Evidence in the Oman Peninsula and Its Relation to Mesopotamia." In *Essays on the Late Prehistory of the Arabian Peninsula*, ed. S. Cleuziou, M. Tosi, and J. Zarins. *Serie Orientale Roma* XCI. Rome: Istituto italiano per l'Africa e l'Oriente, pp. 317–335.

David, H. 2002. "Soft Stone Vessels from Umm an-Nar Tombs at Hili (UAE): A Comparison." *Proceedings of the Seminar for Arabian Studies* 32: 175–185.

Ebiamadon, A.B., E. Umoren, F. Brisibe, P.M. Magalhães, J.F.S. Ferreira, D. Luthriag, X. Wu, and R.L. Priorh 2009. "Nutritional Characterisation and Antioxidant Capacity of Different Tissues of Artemisia Annua L." *Food Chemistry* 115(4): 1240–1246.

Emami, M., M. Razani, N. Alidadi Soleimani, and Y. Madjidzadeh 2017. "New Insights into the Characterization and Provenance of Chlorite Objects from the Jiroft Civilization in Iran." *Journal of Archaeological Science* 16: 194–204.

Francfort, H.-P. 2003. "La civilisation de l'Asie Centrale à l'âge du Bronze et à l'âge du Fer." In *Catalogue de l'Exposition De l'Indus à l'Oxus. Archéologie de l'Asie Centrale*, ed. O. Bopearachchi and C. Sachs in collaboration with N. Cayzac and F. Millet. Lattes: Association IMAGO – Musée de Lattes, pp. 29–59.

Frenez, D. 2017. "Manufacturing and Trade of Asian Elephant Ivory in Bronze Age Middle Asia: Evidence from Gonur Depe (Margiana, Turkmenistan)." *Archaeological Research in Asia* 15: 13–33.

Gubaev, A.G., G.A. Koshelenko, and M. Tosi (eds.) 1998. *The Archaeological Map of the Murghab Delta: Preliminary Reports, 1990–95*. Rome: IsIAO.

Hakemi, A. 1997a. "Kerman: The Original Place of Production of Chlorite Stone Objects in the 3rd MillenniumBC." *East and West* 47(1/4): 11–40.

Hakemi, A. 1997b. *Shahdad: Archaeological Excavations of a Bronze Age Center in Iran*. Rome: IsIAO.

Harrower, M.J., H. David-Cuny, S. Nathan, I. A. Dumitru, and S. Al-Jabri 2016. "First Discovery of Ancient Soft-Stone (Chlorite) Vessel Production in Arabia: Aqir al-Shamoos (Oman)." *Arabian Archaeology and Epigraphy* 27: 197–207.

Heydari, M., H. Fazeli Nashli, E. Cortesi, and M. Vidale 2015. "A Surface Collection at Chegerdak, a Bronze Age Centre in the Jazmurian Basin (South-Eastern Iran)." *Paléorient* 41(2): 133–155.

Hiebert, F.T. 1994. *Origins of the Bronze Age Civilization in Central Asia*. Cambridge: Cambridge University Press.

Hojamberdiev, M., P. Arifov, K. Tadjiev, and Y. XU 2010. "Characterization and Processing of Talc-Magnesite from the Zinelbulak Deposit." *Mining Science and Technology (China)* 20(3): 415–420.

Jarrige, J.-F. 2006. "Il Tesoro di Fullol." In *Afghanistan. I Tesori Ritrovati*, ed. P. Cambon. Torino: Umberto Allemandi and Co., pp. 25–32.

Kaniuth, K. 2010. "Long Distance Imports in the Bronze Age of Southern Central Asia: Recent Finds and Their Implications for Chronology and Trade." *Archäologische Mitteilungen aus Iran und Turan* 42: 3–22.

Kohl, P.L. 1975. "Carved Chlorite Vessels: A Trade in Finished Commodities in Mid-Third-Millennium." *Expedition*, 18(1): 18–31.

Kohl, P.L. 1976. "The Production of Chlorite and Its Distribution in the Early Third Millennium: The Integration Of Economic Theory to Archaeological Data." In *The Memorial Volume of the 6th International Congress of Iranian Art and Archaeological Research*, ed. M.Y. Kiani. Tehran: Iranian Centre for Archaeological Research, pp. 195–204.

Kohl, P.L. 1978. "The Balance of Trade in Southwestern Asia in the Mid-Third Millennium BC." *Current Anthropology* 19: 463–492.

Kohl, P.L. 1979. "The 'World Economy' of West Asia in the Third Millennium BC." In *South Asian Archaeology 1977*, ed. M. Taddei. Naples: Istituto Universitario Orientale, pp. 55–86.

Lamberg-Karlovsky, C.C. 1970. *Excavations at Tepe Yahya, Iran, 1967–1969, Progress Report I*. American School of Prehistoric Research, Bulletin 27. Cambridge, MA: Peabody Museum of Archaeology and Ethnology, Harvard University.

Lamberg-Karlovsky, C.C. 1988. "The 'Intercultural Style' Carved Vessels." *Iranica Antiqua* 23: 45–95.

Lamberg-Karlovsky, C.C. 2001. "Afterword: Excavations at Tepe Yahya: reconstructing the Past." In *Excavations at Tepe Yahya, Iran, 1967–1975*, ed. C.C. Lamberg-Karlovsky and D.T. Potts. American School of Prehistoric Research, Bulletin 27. Cambridge, MA: Peabody Museum of Archaeology and Ethnology, Harvard University, pp. 269–280.

Lamberg-Karlovsky, C.C. and D.T. Potts 2001. *Excavations at Tepe Yahya,Iran, 1967–1975*. American School of Prehistoric Research, Bulletin 27. Cambridge, MA: Peabody Museum of Archaeology and Ethnology, Harvard University.

Ligabue, G. and G. Rossi Osmida (eds.) 2007. *Sulla Via delle Oasi: Tesori Dell'Oriente Antico*. Trebaseleghe: Il Punto Edizioni.

Ligabue, G. and S. Salvatori 1988. *Battriana, una antica civiltà delle oasi dalle sabbie dell'Afghanistan*. Venice: Erizzo.

Madjidzadeh, Y. 2003. *Jiroft: The Earliest Oriental Civilization*. Tehran: Ministry of Culture and Islamic Guidance, Printing and Publishing Organization, Cultural Heritage Organization.

Martinetto, P., M. Anne, E. Dooryhée, and Ph. Walter 2000a. "Les formulations cosmétiques de l'Egypte ancienne retrouvées grace a la diffraction des rayons X." *Journal de Physique* IV, 10: 465–72.

Martinetto, P., M. Anne, E. Dooryhée, and Ph. Walter 2000b. "A Synchrotron X-ray Diffraction Study of Egyptian cosmetics." In *Radiation in Art and Archeometry*, ed. D.C. Creagh and D.A. Bradley. Amsterdam: Elsevier Science, pp. 297–316.

Miroschedji, P. 1973. "Vases et objets en stéatite susiens du Musée du Louvre." *Cahiers de la DAFI* 3: 9–79.

Namdar, D., R.J. Stacey, and St. J. Simpson 2009. "First Results on Thermally Induced Porosity in Chlorite Cooking Vessels from Merv (Turkmenistan) and Implications for the Formation and Preservation of Archaeological Lipid Residues." *Journal of Archaeological Science* 36, 11: 2507–2516.

Orris, G.J. and J.D. Bliss 2002. *Mines and Mineral Occurrences of Afghanistan*. United States Geological Survey Open-File Report 02-110. https://pubs.er.usgs.gov/publication/ofr02110.

Piran, S. (ed.) 2013. *Objects from the Jiroft Treasury: Soft Stone and Alabaster Objects (Recovered Collection) from the Halil River Basin*. Tehran: Pazineh.

Pottier, M.-H. 1984. *Matériel funéraire de la Bactriane méridionale de l'Age du Bronze*. Paris: Éditions recherche sur les Civilisations.

Potts, D.T. 2003. "A Soft-Stone Genre from Southeastern Iran: 'Zig-Zag' Bowls from Magan to Margiana." In *Culture Through Objects. Ancient Near Eastern Studies in Honour of P. R. S. Moorey*, ed. D. T. Potts, M. Roaf, and D. Stin. Oxford: Griffith Institute, pp. 77–89.

Potts, D.T. 2008. "An Umm an-Nar-Type Compartmented Soft-Stone Vessel from Gonur Depe, Turkmenistan." *Arabian Archaeology and Epigraphy* 18: 167–180.

Renaud, K.M. 2013. *The Mineral Industry of Afghanistan*. https://minerals.usgs.gov/minerals/pubs/country/2013/myb3-2013-af.pdf.

Salvatori, S. 2003. "Spunti per una definizione dei sistemi di relazione tra l'Iran orientale e l'Asia Centrale meridionale. I flaconi per cosmetici in clorite: Distribuzione e variabilita." In *Studi in onore di UmbertoScerrato per il suo settantacinquesimo compleanno*, ed. M.V. Fontana and B. Genito. Naples: Università degli Studi di Napoli 'L'Orientale,' pp. 721–734.

Salvatori, S. 2008. "Cultural Variability in the Bronze Age Oxus Civilisation and Its Relations with the Surrounding Regions of Central Asia and Iran." In *The Archaeological Map of the Murghab Delta II. The Bronze Age and Early Iron Age in the Margiana Lowlands: Facts and Methodological Proposal for a Redefinition of the Research Strategies*, ed. S. Salvatori and M. Tosi. BAR International Series 1806. Oxford: Archaeopress, pp. 75–98.

Sarianidi, V. 1998a. *Margiana and Protozoriastrism*. Athens: Kapon Editions.

Sarianidi, V.I. 1998b. *Myths of Ancient Bactria and Margiana on its Seals and Amulets*. Moscow: Pentagraphic.

Sarianidi, V. 2007. *Necropolis of Gonur*. Athens: Kapon Editions.

Sarianidi, V.I. and N.A. Dubova 2010. "Novye grobnitsy na territorii 'tsarskogo nekropolja' Gonura" [New graves in the area of the "Royal Necropolis" of Gonur]. In *Na puti otkrytija tsivilizatsii* [On the Track of uncovering a Civilization], vol. 3, Trudy Margianskoj Arkheologicheskoj Ekspeditsii, ed. P.M. Kozhin, M.F. Kosarev, and N.A. Dubova. St. Petersburg: Aletejja, pp. 144–171.

Sarianidi, V.I. and N. Dubova 2013. *Treasures of Ancient Margiana*. Aşgabat: Türkmen döwlet neşirýat gullugy.

Teufer, M. 2007. "Zu bronzezeitlichen Tulpendarstellungen in Zentralasien und im Iran." In *Scripta praehistorica in honorem Biba Teržan*, vol. 44, ed. M. Blečić, M. Črešnar, B. Hänsel, A. Hellmuth, E. Kaiser, and C. Metzner-Nebelsick. Ljubljana: Narodni muzej Slovenije, pp. 183–207.

Tosi, M. and R. Wardak 1972. The Fullol Hoard: A New Find from Bronze Age Afghanistan. *East and West* 22(1–2): 9–17.

Vahdati, A.A. and R. Biscione 2014. "Iran Italy Joint Excavations at Tepe Chalow, Jajarm Plain, North Khorasan Province." *Exhibition of Archaeological Finds 2013*. 13th Annual Symposium on the Iranian Archaeology, pp. 16–21.

Veresotskaja, G.E. 2016. "Issledovanie i restavratsija mozaichnogo dekora stenki lartsa-darokhranitel'nitsy kontsa III tys. n.e. iz tsarskoj grobnitsy 3230 Gonur-depe (Turkmenistan)" [Study and restoration of the mosaic decoration of the wall of the casket-tabernacle of the end of the 3rd millennium BC from the royal tomb 3230 at Gonur Depe]. In *Trudy Margianskoj Arkheologicheskoj Ekspeditsii*, vol. 6, ed. N.A. Dubova. Moscow: Staryj Sad, pp. 315–323.

Vidale, M. 2015. "Searching for Mythological Themes on The 'Jiroft' Chlorite Artefacts." *Iranica Antiqua*, L: 15–59.

Vidale, M. 2017. *Treasures from the Oxus. The Art and Civilization from Central Asia*. London and New York: I.B. Tauris.

Vidale, M., O. Craig, F. Desset, G. Guida, P. Bianchetti, G. Sidoti, M. Mariottini, and E. Battistella 2012. "A Chlorite Container Found on the Surface of Shahdad (Kerman, Iran) and Its Cosmetic Content." *Iran* 50: 27–44.

Vidale, M. and A. Lazzari (eds.) 2017. *Lapis Lazuli Bead Making at Shahr-i Sokhta*. Padova and Rome: Antilia, dBC, and ISMEO.

Vidale, M., G. Salviulo, F. Zorzi, and I. Mocchiutti 2016. "Cosmetics and Cosmetology at Shahr-I Sokhta." *Iran* 54(2): 1–24.

Wengrow, D. 2008. "Prehistories of Commodity Branding." *Current Anthropology* 49(1): 7–34.

Winkelmann, S. 2014. "'Trading Religious' from Bronze Age Iran to Bactria." In *Religions and Trade, Religious Formation, Transformation and Cross-Cultural Exchange Between East and West*, ed. P. Wick and V. Rabens. Leiden and Boston: Brill, pp. 199–232.

Ziolkowski, M.C. and A.S. Al-Sharqi 2006. "Dot-in-Circle: An Ethnoarchaeological Approach to Soft-Stone Vessel Decoration." *Arabian Archaeology and Epigraphy* 17: 152–162.

CHAPTER TEN

THE "ROYAL NECROPOLIS" AT GONUR DEPE

An attempt at systematization (plan, constructions, rituals)

Nadezhda A. Dubova

INTRODUCTION

With the opening of the excavations at Gonur Depe in 1974, extremely rich information was brought to light on the funeral rituals and on the tomb architecture in the Margush country. More than 5,000 tombs, from several cemeteries within or outside the city have been unearthed up to now. The Gonur Large Necropolis and partly other ones have been presented by S. Salvatori (1993, 1994, 1995) and in more in detail by V.I. Sarianidi (2001, 2006, 2007a, 2007b; Dubova 2004; Sarianidi and Dubova 2010a, 2010b; 2016; Dubova *et al.* 2018). However, not all the discoveries have been published yet.

Several types of burial construction are represented: shaft graves (i.e., pits with a niche in one of the walls) are the main one (55.2%), simple pit burials come next (32.3%), cists (i.e., rectangular pits surrounded by a mudbrick wall) are far behind (6.6%), followed by child burials in large ceramic vessels (2.3%). Underground roofed houses made of one or two rooms – also called chamber graves in the Large Necropolis, or hypogea in the "Royal Necropolis"[1] – with successive burials represent only 1.8%, and burnt pits with fired walls and bottom covered with a thick layer of sand or ash 1.45%. To these, six fractional burials and one *dakhma*[2] found by Sarianidi in the Gonur North palace must be added. Depending on the cemetery, the distribution of these types differs (Sarianidi 2001, 2007b; Sarianidi and Dubova 2016). When he described the Gonur Large Necropolis, Sarianidi wrote that the richest offerings were present in chamber tombs and in cists (5%), the poorest in simple pits (10%), while they were medium or rich in shaft graves (85%). He concluded that these differences reflected the probable social stratification of the Margush society (Sarianidi 2001: 35–36; 2007b: 124–125).

In 2004–2005, on the eastern bank of the largest Gonur water basin, a complex of unusual tombs containing both humans and animals was first unearthed. It included: (1) five underground constructions (hypogea) that looked like houses with several rooms and, also, a special courtyard in several cases (graves 3200, 3210, 3220, 3230, 3235); (2) two very large pits of ca. 6 m in diameter also called "ditches"

(3225, 3240); (3) some simple chamber tombs situated between them (3205, 3245, 3250); (4) one pit with several bones from a boar (3206); (5) two wells, each with two legs of a donkey lying on their tops (3330, 3331); and (6) to the north of these tombs, a ritual burial of a donkey lying next to a double hearth (3340).

The complex construction of these tombs, the presence of four-wheeled wagons, various human and animal sacrifices, many symbols of high social status, and numerous gold and silver funeral gifts though the graves had been grabbed several times showed that representatives of the elite or priests were buried there. Therefore, it was named the "Royal Necropolis" (Dubova 2004; Sarianidi 2006, 2007b).

Further discoveries were added to this complex in 2009: (7) one "ditch" (3900); (8) four hypogea (3880, 3905, 3910, 3915); (9) one rich shaft tomb (3870); and (10) two bulls' graves (3890, 3895) discovered to the south of the previous ones (Sarianidi and Dubova 2010a, 2010b). Finally, (11) three groups of dog burials were excavated in 2011 to the east, southeast, and south of the hypogea complex (Dubova 2012; Sataev and Dubova 2016).

Several publications concerning the very rich items that have been found in this "Royal Necropolis" have already been made (Dubova 2004; Sarianidi 2005: 204–116; 2006; 2007b: 184–185; 2008a: 216–240, figs. 70–178; 2008b: 148–225; Sarianidi and Dubova 2010a, 2010b). We present here only some of these artifacts. This chapter rather aims at a presentation of some of the major hypogea in order to present a preliminary systematization of the plan of the area, of the different constructions, and of the rituals that were performed in them. As will be pointed out, however, the severe looting suffered by these graves prevents any certainty.

THE PLANOGRAPHY OF THE "ROYAL NECROPOLIS"

The "Royal Necropolis" is situated on the southeastern part of Gonur Depe, close to a large, possibly residential house, partly destroyed by the bank of the large south basin. The preserved part of this building has thick walls and large well-planned rectangular rooms. These architectural features, together with some artifacts found there, gave Sarianidi reason to believe that this was the first building (possibly a palace), where the ruler lived during the construction of the central Gonur complex. One may suppose that the first aristocratic burials were made close to this first "royal" residence (Figures 10.1a and 10.1b).

The surface of the "Royal Necropolis" reaches 0.6 ha (100 m from north to south and 60 m from east to west). It includes 107 funerary constructions: 9 hypogea, 3 chamber tombs, 3 "ditches", 8 cist burials (two are cenotaphs, one contained a donkey, dismembered cattle bones were found in another), 8 shaft graves (3 cenotaphs, none with animal burials), as well as 62 pit graves and 3 child burials in jars. Human remains were found only in 28 pit graves (12 of which are adult individuals and 16 children). In 14 pit graves there were complete carcasses of animals and, in 10 others, only a few animal bones. Ten pits with accurate plastered walls were totally empty. These last pits were combined together with animal graves (mostly dog) and with other tombs where only some animal bones were recovered. Thirteen tombs on this territory are cenotaphs. Four of them are simple tombs including three pit burials (3251, 3876, and 3878) and one cist grave (3607) that contained real funeral offerings, while seven look more like memorial pits (i.e., small pits up to 1 m in

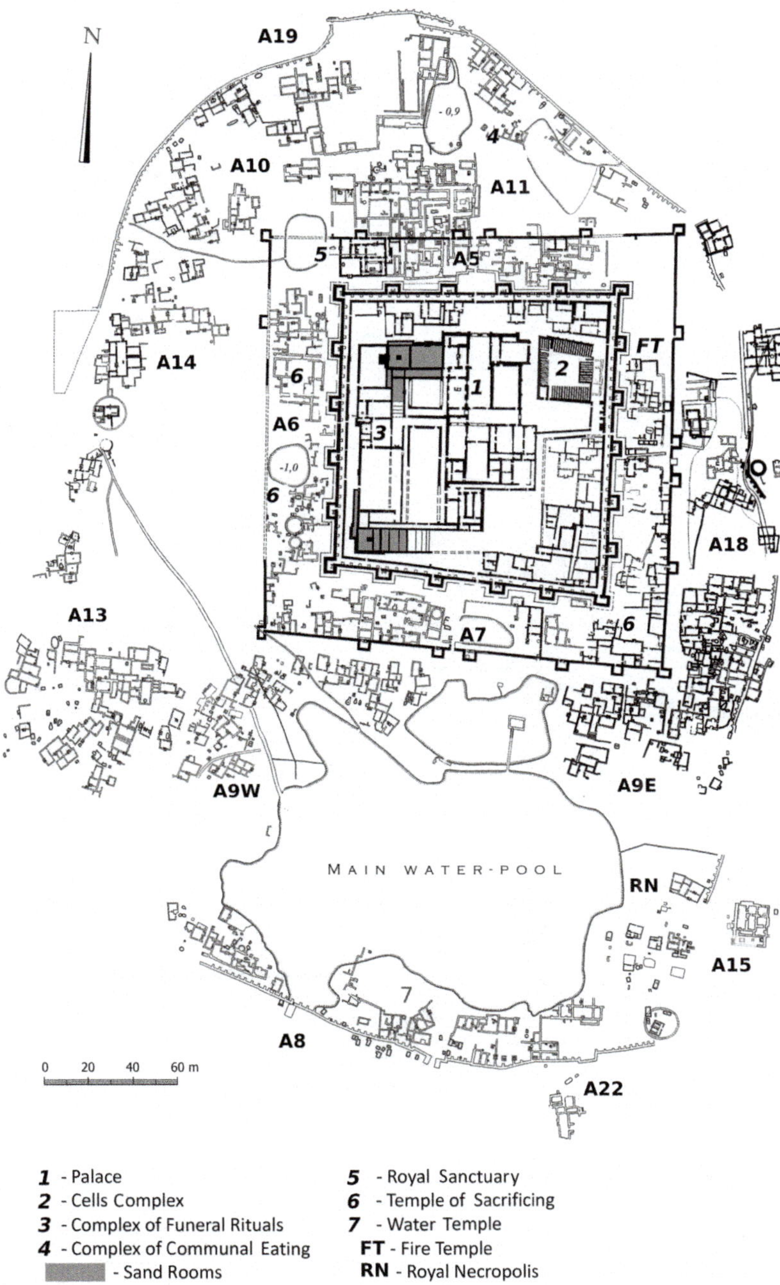

Figure 10.1a General plan of the central part of Gonur Depe North (Excavation areas A1 to A22) showing where the Royal Necropolis (RN) is situated, at its southeastern corner © Margiana Archaeological Expedition.

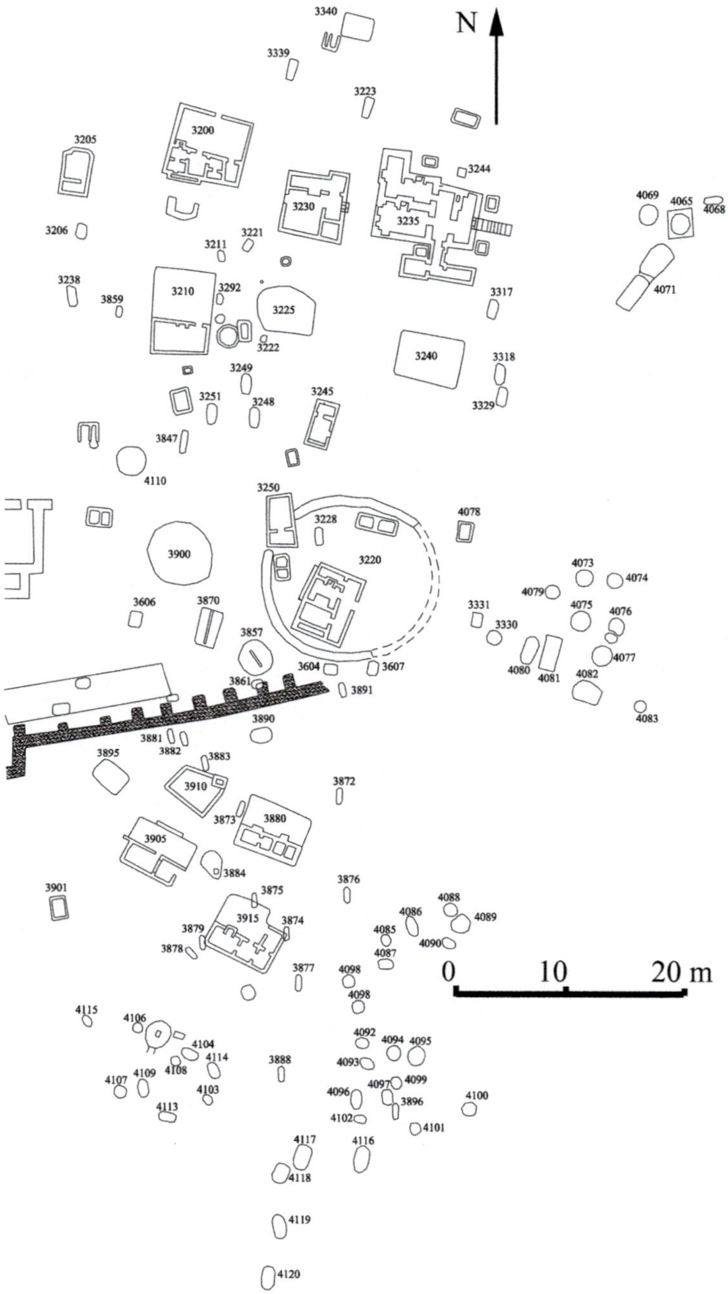

Figure 10.1b Plan of the "Royal Necropolis." In thick black, the encircling wall of the town, built after the Necropolis had been out of use. © Margiana Archaeological Expedition.

Figure 10.2 Tomb 3895, with the burial of a dismembered cow.
© Margiana Archaeological Expedition.

diameter containing only ceramic vessels), and two were empty pit graves. Most of the cenotaphs, memory pits, and human pit graves were not dug very deep, a fact that attests to their construction after that of the main burials, and it should be noticed that they never cut the hypogeums and other tombs.

Most of the elite burials were found inside the encircling wall of Gonur city, and only a few and the smaller ones (hypogea 3880, 3905, 3910, 3915; tombs 3889, 3890 and 3901) are on its outside. Since the encircling wall is standing on a thick rubbish layer, it can be said with certainty that it was built much later than most of the funeral constructions there. Furthermore, in this area, the wall does not have the oval form it has along most of its length, but presents sharp angles. It can therefore be assumed that it was constructed in successive portions during the second building phase of Gonur Depe (Sarianidi 2008a: 59).

A first and a very important observation is that this area is one that concentrates the greatest number of animal burials: 23, or 31.1% of their total amount at the site, have been found specifically in this territory (Dubova 2012: 109; 2015: 13) (for the animal burials, see also Chapter 12). Furthermore, a certain order in the disposition of these tombs can be noticed. The donkey ritual burial (3340) is located on the north of the "Royal Necropolis"; the burials of a bull (3890; see Chapter 12, fig. 7) and of a dismembered cow (3895, Figure 10.2) are outside the encircling wall near hypogeum 3910, and to the south of 3220 and the "ditch" 3900. A small square mudbrick construction (only 0.5 m on each side) containing the long bones of the legs of a boar (3206) was built close to the chamber tomb 3205. Two wells, on the top of which donkey's legs were lying (3330, 3331) were found to the east–southeast of hypogeum 3220. To the east of these two wells, one group of pits with dogs and other animal remains was discovered. A second group of dogs and other animals was

Figure 10.3 Collective pit burial 4110, containing a woman and seven children together with the full carcass of a puppy and dismembered animal remains. © Margiana Archaeological Expedition.

organized to the east of hypogeum 3235, and a third one to the south of hypogeum 3915 (Figure 10.1b).[3]

Collective pit burials are also a specific feature of the area. One of them (4110, see Figure 10.3), had not been looted. It was excavated in 2011, at 15 m to the northwest of "ditch" 3900. The pit (180 x 170 cm) contained the remains of a woman (25–30 years old) and of seven children (two aged 6–7 years old, two 9–11 years old, two 10–12 years old, and two 15–16 years old), together with a full carcass of a puppy and other incomplete animal remains. The human bodies were both in the center of the pit and along its walls. They were placed on their sides and with different orientation of the head. Four children were buried under the female remains and three above her. The height of the remains reached ca. 20 cm. On the bottom of the pit, traces of organic materials are probably those of a mat. At this same bottom level, the skeleton of a child (around 10 years old) was found near the western wall, the full skeleton of a puppy near the southwestern wall, the lower jaws of a another young person and that of an ovicaprine (sheep or goat) near the northeastern wall, and a donkey's metacarpus near the northern wall.

Near the city encircling wall, to the south of the mentioned "first palace," a very large rectangular pit (14 x 2m and ca.1.5 m deep),[4] filled with ground trash, was discovered. It contained several pits and no material but some animal remains. Near its southeastern angle, a round pit 2 m in diameter and more than 4.5 m deep (but not a well) has been excavated containing the complete skeleton of a puppy at –2.2 m from its upper edge. Some 5 m from the southwestern angle, just near the wall of the pit, another rectangular pit 180 x 100 cm (50 cm deep) was found containing a large standing ceramic vessel. At 3 m from it to the east, another round pit (1.5 m diameter) was excavated but was totally empty. Finally, in the western part of this large pit, the lower jaw of a donkey and the scattered bones of its skull were also found. The purpose of this large pit remains obscure.

The very center of the "Royal Necropolis" is occupied by hypogeum 3220 and "ditch" 3900. As mentioned above, these structures are surrounded immediately to the east by two wells with the legs of one donkey and graves of a group of dogs and other animals, to the south by the burials of a bull and a cow, and to the northwest by the collective burial 4110. Near this tomb 3220 are also two chamber graves (3250 and 3857) and the rich shaft tomb (3870).

The construction of hypogeum 3220 is unique: It is the only one that has a narrow encircling wall (two bricks wide and same in height, 12 m in diameter), partly destroyed on its northwestern part by the chamber grave 3250. This encircling wall forms a specific courtyard around the underground four-room house, situated in the southern part of the circle. On the ancient surface inside the circle were three altars made of mudbricks (one pear-shaped on the west side, and two rectangular ones joint together in one construction on the northern side). Only one later pit burial (3228) was made inside the circle.

To the north of hypogeum 3220 are four other similar constructions (3200, 3210, 3230, and 3235) and two "ditches" (3225 and 3240), while to the south (outside the main encircling wall of Gonur Depe) are four more hypogea (3880, 3905, 3910, and 3915) but no "ditches."

CONSTRUCTION AND ARRANGEMENT OF THE HYPOGEA

All the hypogea at the "Royal Necropolis" are underground constructions.[5] If we exclude no. 3220 because of its surrounding wall, two types of such tombs can be distinguished: (1) multiroom underground houses with a courtyard (3200, 3210, 3880, 3905, and 3915), and (2) more complex houses without a courtyard (3230 and 3235). The "ditches" were found close to the houses without a courtyard (3225 near 3230, and 3240 near 3235), while a similar "ditch" (3900) is close to hypogeum 3220.

Another hypogeum, 3910 (3.6 x 3 m), presents only the four outskirt walls and no courtyard, and it was totally empty, even of pottery vessels. It only contained on its northeastern corner a rectangular oven with hardly burnt walls. However, nothing shows that it could be a cenotaph. It is possibly either a special construction for memorial ceremonies or a completely looted tomb, although this latter interpretation is less probable.

All the courtyards contained remains of wagons, humans (possibly servants or slaves), and animals (as "escorting items"). The same kinds of finds were discovered in the "ditches," which leads us to consider that the "ditches" played the role of courtyards separated from the houses. In the first type of hypogeum, most of the houses have four rooms, including 3220. At excavation, hypogeum 3210 presented only one room, but traces on its walls show that it originally had at least three rooms and that the original partition walls between them had been demolished later and their traces carefully plastered with clay. Hypogeum 3235 has eight rooms.

Hypogeum 3230 (4.9 × 4.4 m) was built in mudbricks within a large pit 1.9 m deep (Sarianidi 2006: 180, fig. 32). It was looted three or four times in antiquity: In the upper layers, several beads of gold and semiprecious stones had been lost (see Figure 10.15). A thin (one-brick) partition wall, parallel to its northern wall, separates the main room. A double oven/chimney with slight traces of fire was built within this partition wall. Inside the tomb, the walls were covered with a thick layer of clay. Altogether, the "house" has three rooms, but another special small one is situated in its southeastern corner, with no entrance, and its floor 20 cm below that of the other rooms. This small room contained the skeletons of one woman aged 40–50 years old lying on her left side, head to the west; one man 50–60 years old on his right side, head to the west; and a young man 17–18 years old in supine position, head to the south. In the filling of the room many bone buttons covered with gold foil were found, as well as semiprecious beads.

In the main room outside this closed premise and near its western wall six individuals were buried, found in anatomical connection: one man and five adult women with rich offerings. Even after looting, gold beads, a bronze axe and seals, as well as ten flint arrowheads were discovered. A wooden container (or ostensory) decorated with mosaic inlays was found near the legs of this group of burials (see Chapter 11). Outside its northern wall, and near the entrance to the hypogeum, three more people were found, two women and one dwarf adult man (Dubova and Kufterin 2014) accompanied by several ceramic vessels. The remains of seven other individuals, three women and four men, were scattered at different places inside the "house" and, for some, along the walls due to the looting of the grave. At different places in all the rooms other funeral offerings were discovered (see pp. 356–361).

Some 5 m to the south of 3230 is the *"ditch" 3225* (4.95 × 3.50 m and 1.6 m deep), which contained the remains of ten people together with a four-wheeled wagon (near the west wall) and two camels (Sarianidi 2006: 178, fig. 31). It is a rare case of unviolated structure in this area. The adult camel had the central position and was lying on its left side, head to the east. Traces of a wooden rectangular seat on its back were still visible. A young camel was leaning against the eastern wall. The skeletons of one man (45–55 years old) and one woman (25–35 years old) were lying above the wheels. Two male skeletons were found in a kneeling position along the walls of the ditch: one along the northern, the other along the southern. One more man's skeleton was lying supine near the adult camel's head. Five human skulls without any other bones were found near the northern and southern walls and close to the animals' skeletons. No funeral offering was accompanying these burials.

At 2.5 m to the east of 3230 is *hypogeum 3235* (7.8 × 6.3 m, 1.55 m deep). It was looted five times. Seven stairs made in the eastern wall lead to a vestibule (Sarianidi 2006: 182, fig. 33). The doorway was tightly filled with bricks without mortar (as was the case for 3230), making it possible to re-enter the tomb easily. On the ancient surface, on both sides of the entrance and along the middle of the northern and southern walls, four rectangular fire altars were standing (120 × 100 cm) with slight traces of

Figure 10.4 Silver conical cup with a low-relief animal composition from hypogeum 3235. © Margiana Archaeological Expedition.

fire (Sarianidi and Dubova 2013: 206–207). The walls were made of mudbricks and covered with a thick layer of clay plaster.

In the western part of the vestibule, one passage leads to an inner hall. From there, one can enter into all the other rooms. To the south, a small corridor with a double hearth/chimney built in the wall leads to three premises containing only a few ceramic vessels and animal bones, possibly places intended for the household needs in the future life. To the northwest are two other rooms, and another is in the southwest.

Built-in ritual double hearths/chimneys were in all the west rooms. In the vestibule and the inner hall these in-wall constructions had only one hearth.

The richest offerings were found in the southwestern room that had four stepped-niches or "blind windows" (two in the western wall, one in the northern, and one in the southern). They consist of two long stone scepters (one intentionally broken, possibly sawn), many fragments of gold foil and different semiprecious beads. A hoard buried 85 cm deep under the floor was found unlooted and contained two gold, three silver, and two bronze vessels (Sarianidi 2008a: 168, fig. 80). One of the silver vessels is a conical vase with an exquisite animal composition in relief (Sarianidi 2005: figs. 109–112; 2008a: 171, figs. 81–82; Sataev and Sataeva 2015) (Figure 10.4). In the same room, three ivory rectangular slabs with geometric incisions were also discovered (Figure 10.5).

The lower part of the western wall, ca. 1 m high, had completely preserved the smooth surface of a clay coating. On the floor near this wall several hundreds of mosaic inlays were found, probably originally fixed on this clay coating. It can be assumed that the walls of this southwestern room were decorated with different mosaic compositions (Sarianidi 2006: 184; see also Chapter 11). Unfortunately, none of them was still *in situ*. Only in the northern room, small fragments were

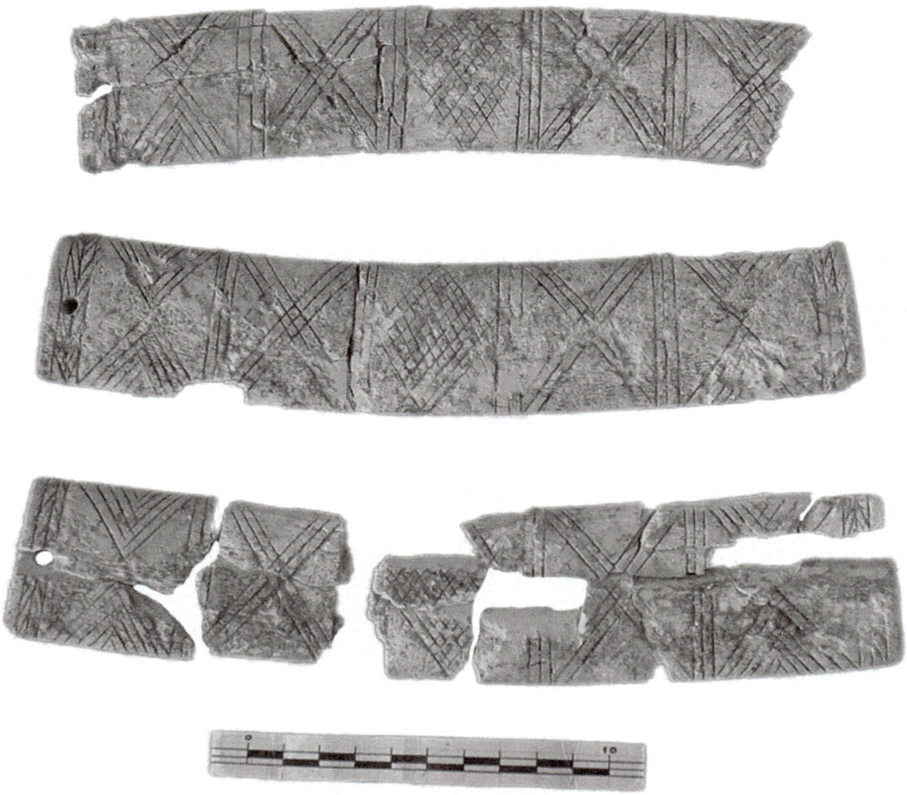

Figure 10.5 Three ivory plaques with geometric decorations from hypogeum 3235. © Margiana Archaeological Expedition.

still preserved on its northern wall (two fragments), as was the decoration of a wooden box found on the floor next to a silver vessel (Sarianidi 2008a: 204, fig. 114) (Figure 10.6).

Under the floor (at a depth of 0.40 m) of the southern part of the inner hall, human remains were excavated. They include three complete skeletons of young men aged 17–25 years old (all lying in a crouched position, head to the north) and those of four women aged 30–40 years old (in the same pose, but oriented to the south). No funeral inventory was found with them, except for a simple small bronze seal with a cross motif found on the chest of one of the women. One more complete female skeleton (16–17 years old) was lying on the narrow clay "bench" in the northern part of the vestibule, also without offerings. Above her, traces of a wooden box (75 × 20 cm) were visible. A last complete female skeleton (40–50 years old), was lying prone with her upper and lower limbs tightly folded up, squeezed between the sidewalls of the corridor leading to the "household" part of the tomb, next to the double hearth mentioned above, without offerings. To these complete skeletons must be added three human skulls (two of women and one of a man) discovered in the vestibule, and fragments of a man's long bones found on the top of the double hearth built between the southern and northern rooms. In all, 12 or 13 individuals

Figure 10.6 Mosaic inlays decoration *in situ* on the northern wall of hypogeum 3235, next to a silver vessel. © Margiana Archaeological Expedition.

(the connection between the long bones and the skull fragments is difficult and is not ascertained) have been unearthed in this hypogeum.

Except for the "household premises," it can be presumed that the western rooms were the "chambers for the funerary offerings." Because eight individuals (seven under the floor of the inner hall and one squeezed in the corridor) were buried in an unusual way, we may consider that they were servants. Scattered human remains found at the different places (mainly in the vestibule) may belong to the main deceased and may be the result of the looting of the tomb. This vestibule would then be the "burial chamber."

To this "house" can be associated a large rectangular pit, *"ditch"* 3240 (5.5 × 3.8 m, 1.95 m deep), situated at 6 m to its south. Unlike "ditch" 3225, this one had been robbed in antiquity. From the position of human and animal remains and of some artifacts (see Sarianidi 2006: 178–179), it is nevertheless possible to consider that in the eastern part of the pit a four-wheeled wagon together with two camels (one adult and one young), two big dogs, two adult sheep, and one calf were buried. The original position of the animals and the wagon cannot be determined with certainty. However, the bones of one of the dogs present traces of bronze oxides, which can only come from the bronze wheel rims of the wagon. This tends to show that the wheels (or the wagon itself) were lying on the floor in the southeastern part of the pit close to the southern wall, with one dog close to them. Unfortunately, the preservation of the human remains is not good enough to determine whether one or several individuals were near this wagon.

Other human remains are concentrated on the floor in the northwestern angle of the pit. Altogether, they represent seven men and four women, either still fully or partially articulated. Judging from the position of their limb bones, these people had been sacrificed either in the pit itself or not far from it. Only in a few cases their bodies were lying in a standard position, on the right side. Most of them were casually tossed one over the other.

In all, this ditch contained seven women (16–60 years old) and ten men (mainly 18–60 years old and one boy aged 10 years old). No funeral gifts were found, but it must be noted that even the wheel bronze rims had been retrieved by the robbers.

Hypogeum 3220 (5.3 × 3.7 m, 1.6 m deep), as mentioned above, is situated in the center of the "Royal Necropolis" (Sarianidi 2006: 169, fig. 17). It contains four rooms: a vestibule corridor along its eastern wall from which one could enter three rectangular premises in the western part. The entrance to the house was almost in the middle of the eastern wall, but was destroyed by robbers in antiquity. The upper part of the destroyed layer they left contained two stone miniature columns, fragments of two stone scepters/staffs, and sherds of ceramic vessels (Sarianidi 2005: 251).

A double hearth was built in the northern wall of the central room. On the floor in front of it, two female bodies (one without her skull) were lying on their right side, head to the west, and with bent legs. In front of the skeletons, near the southern wall of the same room was a table (120 × 70 cm) made of two layers of mudbricks and originally covered with a wooden board. A large niche (length 1.5 m, height 1.1 m) was at +50 cm above the floor level in the western wall of this room. In order to strengthen the wall, a second wall had been built outside, resting on the ancient surface. This supporting wall testifies that all the hypogea were roofed. In the other ones, no trace of any kind of roof has been found, leaving open the question of their shape. Given the preserved height of the walls (up to 2 m) and the presence of the supporting wall in tomb 3220, we can assume that they slightly rose above the level of the surrounding surface.[6] In contrast, "ditches" as well as all the pit graves had no roof and were filled with earth immediately after inhumation.

In the center of the northern wall of the 3220 northern room, a polished stone disk without a handle stood vertically. In this room, as in the southern one, many scattered human remains were discovered. In all, the skeletons of five men (18–50 years old) and one girl (12–13 years old) have been identified.

Near the southern wall of the southern room, a rectangular space (2 × 0.5 m) was separated with a low brick wall (preserved only on one brick height), on which two women's skulls were placed. The western part of this wall was completely preserved and reached the same height as the main wall of the tomb. So, one may consider that, at first, this space was a closed niche, also covered with the roof of the hypogeum. It may have been used for separate burials as was the case in 3230. In the eastern and western corners of this niche were small arched structures that we found empty (Figure 10.7).

The northwestern part of this southern room was filled with earth mixed with human bones and fragments of mosaics, either in separated inlays or in plates with compositions. The motifs exclusively consist of "hearts" (or pipal leaves) of different sizes with red painting (Figure 10.8). In the eastern part of the room, on the floor, a bronze axe and a compartmented seal were also discovered (Figures 10.9a and 10.9b). In the same room, an ivory spatula for cosmetics was found. The anthropological

Figure 10.7 Top view of hypogeum 3220 during cleaning. View from the northeast. © Margiana Archaeological Expedition.

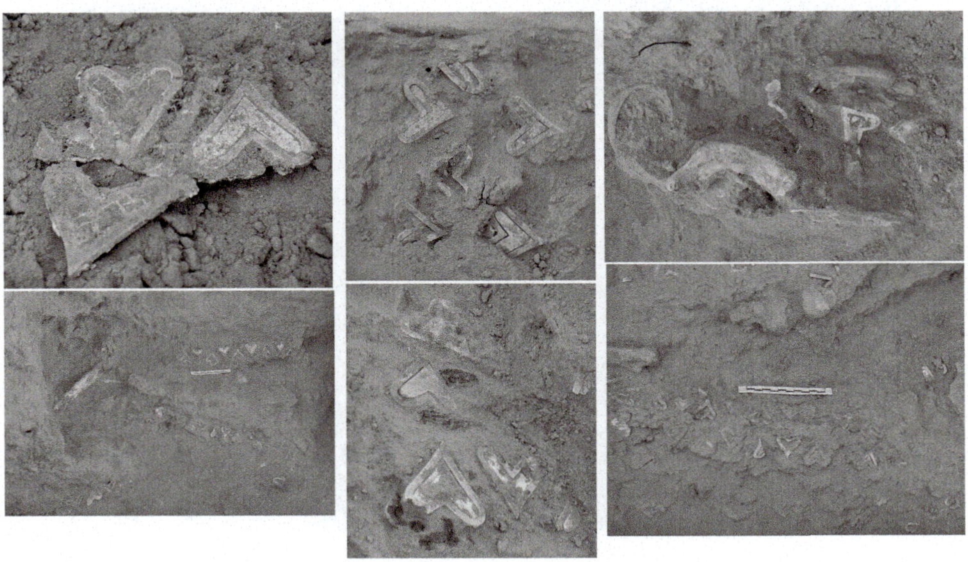

Figure 10.8 Different fragments of mosaic compositions from the southern room in hypogeum 3220. © Margiana Archaeological Expedition.

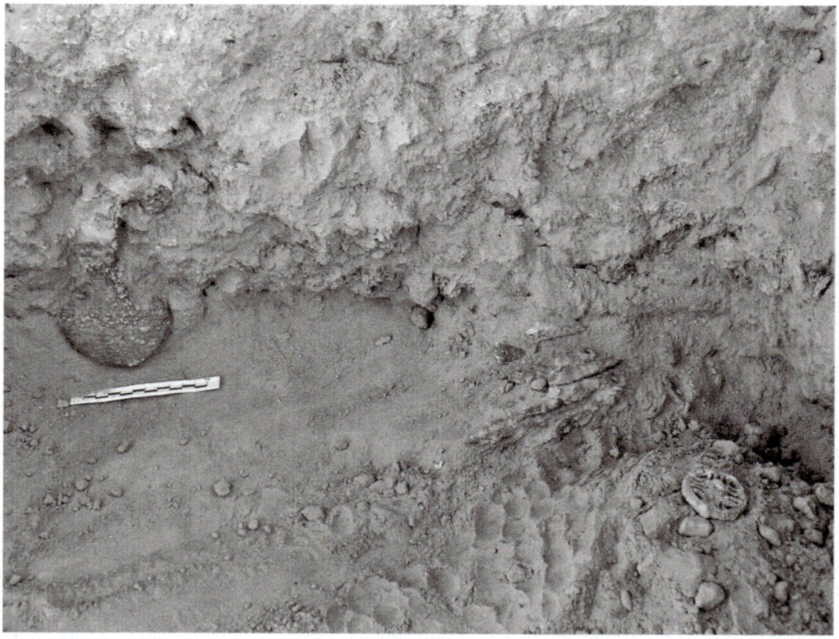

Figure 10.9a Ceremonial bronze axe and compartmented seal together with flint arrowheads *in situ* on the floor near the northern wall of the southern room in hypogeum 3220 during cleaning. © Margiana Archaeological Expedition.

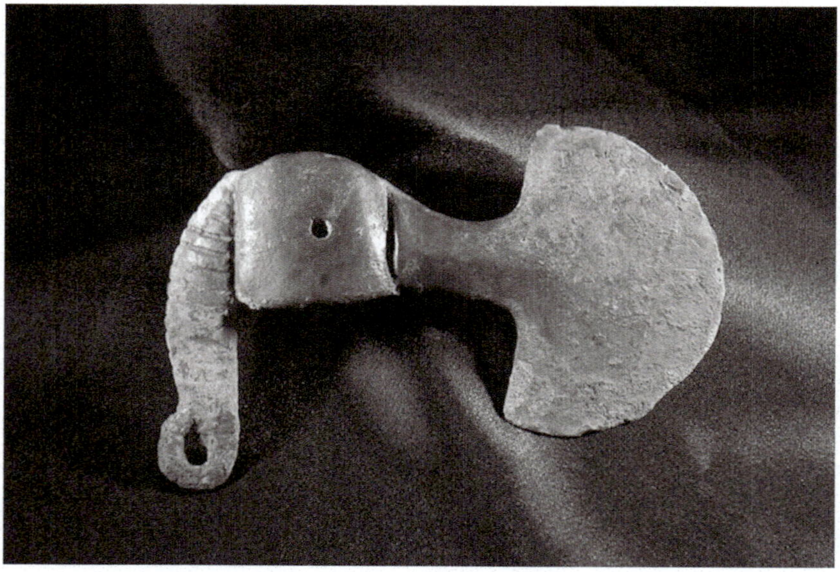

Figure 10.9b The bronze ceremonial axe from hypogeum 3220 after restoration. © Margiana Archaeological Expedition.

study of the bones excavated in this room show that it contained four women (18–60 years old) and six men (12–50 years old).

Under the floor of the southern part of the corridor a man was buried (35–40 years old), lying supine (Sarianidi 2008b: 341, fig. 100). His head was orientated to the northeast and rested on a stone sculpture of a ram (Sarianidi 2005: 233, fig. 93). Near his feet was a bronze knife, the handle of which was fastened with two bronze staples, and next to it, a lamb's rib. Near the knife, but dug deeper into the ground, a very rich hoard was discovered (Sarianidi 2008b: 334, figs. 87–88). It includes one gold jug with a narrow neck (weight 1 kg), a gold plate (Sarianidi 2008a: 161, fig. 73; 2008b: fig. 93) and 15 silver vessels (Figures 10.10a, 10.10b, and 10.10c). Among the last ones, especially noteworthy are a fine cylindrical vessel with two walking camels in relief (see the cover of this book; Sarianidi 2005: 236–237, figs. 95, 96; 2008a: 163, figs. 74–75; 2008b: 335–336, figs. 89–90), a large footed vessel made of two parts (each weighing over 4 kg), as well as a conical funnel-strainer. The hoard also contained eight bronze vessels. Above this man's burial a wooden "ostensory" decorated with "heart" inlays was found (Figure 10.11). Judging by the specific and rich headrest used in this burial,[7] and, also, by the very rich hoard, it can be assumed that this man was the main deceased in this hypogeum. By analogy with grave 3230, where three people were buried in a separate small chamber, it is possible that the niche in the southern room of tomb 3220 played a similar role and that the two skulls

Figure 10.10a Part of the silver vessels from the hypogeum 3220 hoard, after discovery. © Margiana Archaeological Expedition.

Figure 10.10b Silver spouted vessel from the hypogeum 3220 hoard, after restoration. © Margiana Archaeological Expedition.

Figure 10.10c Silver footed goblet from the hypogeum 3220 hoard, after restoration. © Margiana Archaeological Expedition.

lying on its low wall belonged to two other main deceased or were very close to the male one.

As mentioned above, at 2–3 m to the west from the "encircling wall" of hypogeum 3220, *"ditch" 3900* (5.8 × 4.9 m, ca. 2.5 m deep) was excavated (Sarianidi and Dubova 2010a, 2010b). Like "ditch" 3225, this one was fortunately undisturbed. It contained in its center a four-wheeled wagon that had been dropped on the floor from above 2 m so that one of its wheels was lying flat far from the other three.

Figure 10.11 Mosaic "ostensory" standing above the male burial in hypogeum 3220. © Margiana Archaeological Expedition.

On this wheel the remains of four men and one boy were found (two men 25–30 years old, one 30–40 years old, and one boy 10–12 years old). They were lying one above the other, not in the usual position. Their hands were stretched out above their heads and their legs had different orientations, showing that their bodies had been dragged from somewhere else without any respect. One male skeleton (30–40 years old) was found partly on the floor and partly along the southern wall as if he had slipped from the top. The second one (25–30 years old) was lying prone with his knees bent facing the first individual, near the northern wall. The third one (20–25 years old), whose legs were crossed over a ceramic vessel, lay under a large round bronze cauldron (around 1 m in diameter) in the northeastern part of the pit (Sarianidi and Dubova 2010a: 8, fig. 2; 9 figs. 5, 6). According to the archaeobotanist L.V. Sataeva (personal communication) black traces on the belly of this cauldron could be the remains of a leather bag or a cover into which the vessel was placed to carry it. Under this cauldron was a long stone scepter carefully sawn(?) in two parts, one of its ends being in a hoof shape (Sarianidi 2008a: 323, fig. 103). Another red stone scepter (ca. 1.5 m long), also carefully broken in two parts, had a spindle shape and silver bands on both ends (Figure 10.12). It was resting partly on the floor near the southern wall and partly on the body of a big dog, and was situated between the first male skeleton mentioned above, and the remains of a second dog lying on its right side (head to the north) near the eastern wall.

Figure 10.12 Eastern part of "ditch" 3900 with its finds *in situ*: a large bronze cauldron (in front), stone scepters (a black one under the cauldron and a red one in the background), a kneeling human skeleton (in the background), a dog skeleton (on the left) and cartwheels with bronze rims, during cleaning. © Margiana Archaeological Expedition.

Nearby, a bronze artifact made of seven balls put one on top of the other, with the smallest at the base (Sarianidi and Dubova 2010a: 8, fig. 3), was lying on the floor near the northeastern corner. Parallel to the vessel the second ceramic jar (40 cm diameter) of this burial was discovered.

In the northwestern part of this ditch, other human remains were found between the skeletons of two donkeys and two dogs (in the north) and two more big dogs lying one on top of the other (in the west). A fifth dog was lying on the edge of this large pit, maybe as a symbol of protection at the entrance to the grave.

Close to the wagon there were two adult camels together with traces of wood, possibly the shafts of the wagon. A bronze lamp or "brazier" was standing near the legs of one of the camels and a well-made bronze shovel was found under the dogs' skeletons near the western wall.

In all, seven humans, seven dogs (plus one on the edge), two camels, and two donkeys had been buried in this grave.

AN ATTEMPT AT SYSTEMATIZATION

From the excavations of these hypogea and their associated ditches, some recurrent features have been noticed.

In the houses without courtyards, some rooms contained tables and hearths together with ceramic items, so it is possible to consider that they had a "household" function. Some of the human remains were usually found in or near these premises and were possibly servants: either near the entrance (3230), under the floor (3235), or on the floor near the hearths (3220, 3230).

In the graves with a courtyard, servants were buried outside the "house." In these courtyards (3200, 3880, 3915) camels and dogs were also present. All the funeral gifts, including the rich ones, were stored in one (3200, 3210, 3220, 3880) or two to three (3230, 3235) special rooms.

Only in a very few cases is it possible to identify the main deceased (i.e., the person(s) to whom this construction was dedicated). It is the case for three persons in 3230, all situated in the small room without a door in the southeast corner of the grave, and for three others in 3220 (i.e., a man with the treasure near his legs in the corridor and two women inside the separated rectangular space in the southern room).[8] However, the adult man whose head was resting on the stone ram in 3220 can also be considered as the guardian of the rich hoard, because (as mentioned above) a bronze knife was placed between his legs and the set of gold, silver, and bronze vessels. In all the other tombs, and most clearly in 3200, only scattered human remains, usually concentrated in one room, have been found.

Three interpretations are possible:

1. The main individuals were at first placed in a special space like a *dakhma*(?) and their cleaned bones were transferred in the hypogea only after some time.
2. The elite individuals were buried in a special room of the tomb, and their bones were transferred to another unknown place after some time. This is supported by the existence of six secondary fractional burials at Gonur (as mentioned by Sarianidi 2007a: fig. 25), and by the fact that, in tomb 3200, two women in the eastern room are represented only by a few fragments of their bones (clavicles,

Table 10.1 Main characteristics of the hypogea and large "ditches" of the "Royal Necropolis" at Gonur Depe

Characteristics		*Hypogea*				
		3200	*3210*	*3220*	*3230*	*3235*
Dimensions (m)		6.5 × 5.6	6.5 × 4.5	5.3 × 3.7, diam. whole tomb 12 m	4.9 × 4.4	7.8 × 6.3
Depth of the graves (m)		1.1	1.2	1.6	1.6	1.55
Plundering		+	+	+	+	+
Number of (possible) main deceased		1M + 3F = 4	3F + 4M = 7	1M + 2F = 3	2M + 1F = 3	2M + 2F = 4
Number of buried (sacrificed) servants		2F + 1M = 3	4M	8F + 11M = 19	10F + 6M (incl. 1 dwarf) = 16	5F + 4M = 9
High social status symbols	Stone scepters	3	1	2	1	2
	Stone disks	1	2	1	1	1
	Miniature columns	2	2	2	1	-
	Four-wheeled wagons	1	–	–	–	–
Animals	Horse	1	–	–	–	–
	Donkey	–	–	–	–	–
	Camel	2	1	–	–	–
	Dog	1	1	–	–	–
	Calf	–	1	–	–	–
	Sheep	–	2	–	–	–
Mosaics		+	+	+	+	+
Gold foil and artifacts		+	+	+ (incl. 2 vessels)	+	+ (incl. 2 vessels)
Silver artifacts		+	+	+ (incl. 17 vessels)	–	+ (incl. 3 vessels)
Bronze vessels		–	–	5	–	3

				Large pits, "ditches"		
3880	*3905*	*3910*	*3915*	*3225*	*3240*	*3900*
5.6 × 3.8	5.4 × 4.1	3.6 × 3.0	5.9 × 5.0	4.95 × 3.5	5.5 × 3.8	5.8 × 4.9
1.6	1.55	1.3	1.55	1.6	1.95	2.5
+	+	+?	+	No	+	No
?	2? adult	–	2M + 1F = 3	–	–	–
3? 14–18 years old	1M + 1F + 1 (?8–9 years old) = 3	–	1M + 1F = 2	7M + 1F + 2? = 10	10M + 7F = 17	7M
3	–	–	1	–	–	2
1	–	–	–	–	–	–
4	–	–	1?	–	–	–
–	–	–	1?	1	1	1
–	–	–	–	–	–	–
–	–	–	–	–	–	2
1	1	–	3	2	2	2
1	-	–	1	–	2	7 + 1
–	–	–	–	–	1	–
–	–	–	–	–	2	–
+	+	–	+	–	–	–
+	+	–	+	–	–	–
+	+	–	+	–	–	+
–	–	–	–	–	–	–

(*continued*)

Table 10.1 Cont.

Characteristics	*Hypogea*				
	3200	*3210*	*3220*	*3230*	*3235*
Hided hoards	+	+	+	?	+
"Ostensories" (large boxes with mosaics)	–	? –	+	+	+
Ivory artifacts	+	+	+	+	+
Stone artifacts (except miniature columns and scepters)	Agate with gold caps; lapis, turquoise, carnelian beads	Set of processed small stones; carnelian beads	Sculpture of a ram; agate bead	Semiprecious beads (lapis, turquoise)	3 large agate beads with gold caps, 2 stone flowers; turquoise; cosmetic bottle
Bronze lamps "braziers"	2	–	–	–	–
Arrowheads	2 flint	50 flint	44 flint and 7 bone ones	10 flint	13 flint
Game boards	–	1	–	–	–
Other finds	Figurine of a Vegetation deity; 2 eagles; bronze seal; 6 ceramic vessels including 5 large ones	Silver signal trumpet; saddled (?) horse figurine; head of a composite statuette; >20 large ceramic vessels	Bronze axe and seal, marble hands from a composite statuette: bronze seal; nephrite pin	Bronze axe, two bronze seals, bronze knife; gold beads; >30 large ceramic vessels	Silver pin; silver vessel; wooden box with mosaics; silver funnel-strainer; pottery sherds

Note: M = male skeleton; F = female skeleton; ? = sex unknown.

				Large pits, "ditches"		
3880	*3905*	*3910*	*3915*	*3225*	*3240*	*3900*
–	–	–	–	–	–	–
+	–	–	–	–	–	–
+	–	–	+	–	–	–
Set of processed small stones	2 round balls (diam. 5 cm)	–	Fragments of unknown artifacts	–	–	–
–	–	–	–	–	–	1
16 flint	–	–	1	–	–	1
–	–	–	–	–	–	-
Large bronze vessel; 21 ceramic vessels/ including 4 large ones	>15 ceramic vessels	–	4 ceramic vessels; fragments of bronze artifacts	–	–	Bronze shovel; 2 ceramic vessels

pelvis, and jaw) and one man only by his clavicles. Only one complete female skeleton (possibly the last buried one) was lying at the entrance (Dubova 2004: 274–275). It cannot be excluded, that, similarly, the bones of earlier buried individuals were carefully taken out of the chamber tombs at the large Gonur necropolis (Sarianidi 2007b: 41–43).

3. These different types of burials depended on the social class of the deceased.

GRAVE GOODS

Symbols of power and ritual items like stone scepters, stone miniature columns and disks, mosaics, as well as gold, silver, stone, and ivory goods or their fragments have been discovered in all these tombs (Table 10.1). Most of the offerings unfortunately had been stolen and/or destroyed by robbers in antiquity, and the exact initial number of artifacts and their original position in the graves is unknown.

The sacrificed individuals (servants?) in the courtyards usually were not accompanied with any grave goods. However, one of the women in 3200 had a (simple) bronze pin on her dress and a bronze seal (with a crescent motif) lying near her right shoulder. Furthermore, among the bones of the servants(?) buried under the floor of tomb 3235 a large agate bead with gold caps was found, but it lay near skeletons that had been disturbed by robbers and could have fallen from their hands at that time.

Figure 10.13 Flint arrowheads *in situ* under the head of a male skeleton in hypogeum 3230. © Margiana Archaeological Expedition.

Figure 10.14a Gold and agate beads *in situ* on the neck of a female skeleton in hypogeum 3230. ©Margiana Archaeological Expedition.

Tomb 3230 is a special case, where, outside the small partly buried room in the southeast corner mentioned above, but very close to it, a male and five female skeletons have been excavated. Under the left temporal bone of the male skeleton, ten flint arrowheads were found (Figure 10.13). One female skeleton to the right of the male was wearing a necklace of gold beads and two other agate ones (Figures 10.14.a and 10.14b). A bronze ceremonial axe was near the vertex of her skull (top part of the temporal bones) (Sarianidi 2008a: 192; Sarianidi and Dubova 2013: 101). These six individuals, contrary to the others buried in the rooms of this hypogeum, were obviously not simple servants but possibly more closely related to the main deceased. A large amount of beads made of gold, lapis, and (one) turquoise from a necklace has been found in the upper mixed layer left by the robbers (Figure 10.15).

Contrary to shaft and pit burials, these hypogeums contained only a few pottery vessels. Moreover, many of them have a specific shape (i.e., large unstable jars of around 70–100 cm in height and a maximum diameter in the middle part around

Figure 10.14b The gold and agate beads necklace from hypogeum 3230, after restoration. © Margiana Archaeological Expedition.

30–40 cm) (Figure 10.16), which are typical of tombs 3200 (5 items), 3230 (30 items), and 3880 (8 items). These jars have striking similarity with some specific Indus valley jars (though usually black-slipped) (see, e.g., Méry 2000).

A large number of pots (21 items) were found in the 3880 courtyard. Cylindro-conical shapes and others have been discovered in tombs 3900, 3880, 3905 (Figure 10.17), and 3915. In grave 3200, a typical Bactria-Margiana Archaeological Complex (BMAC) conical vessel painted with clumsy red lines was next to two bronze "lamps" (Figure 10.18), one of which is very similar to that found in the Quetta treasure (Dubova 2004: 269, figs. 24, 25; Sarianidi 2005: 212), and the second similar to that from the large pit 3900 (Sarianidi and Dubova 2010a: 11, fig. 9). "Lamps" were only found in these two graves. A large round bronze vessel (ca. 50 cm diameter) with a shape similar to the cauldron from tomb 3900 was also found in the courtyard of tomb 3880.

Only one silver signal trumpet and one marble head from a female composite statuette have been discovered in the "Royal Necropolis." They come from grave 3210 (Sarianidi and Dubova 2013: 137). It is probably symbolic that only her head but not the complete statuette was carefully plastered inside the northern wall of the hypogeum, to the left of the entrance (Figure 10.19). However, other figurines come from these elite tombs, such as faience eagles in heraldic position (Figure 10.20), a composite "goddess of vegetation" (Figures 10.21a and 10.21b), a miniature gold

Figure 10.15 Turquoise, lapis, and gold beads left by the robbers in the upper levels of hypogeum 3230. Hypothetical reconstruction as a necklace. © Margiana Archaeological Expedition.

Figure 10.16 Three from the eight large jars discovered in hypogeum 3880. © Margiana Archaeological Expedition.

Figure 10.17 Part of other pottery vessels from hypogeum 3880. © Margiana Archaeological Expedition.

Figure 10.18 Two bronze lamps "braziers" and a conical cup with handmade red decoration *in situ* in hypogeum 3200. © Margiana Archaeological Expedition.

Figure 10.19 Head from a composite statuette found plastered in the northern wall of hypogeum 3210. © Margiana Archaeological Expedition.

Figure 10.20 One of the two eagles in gypsum and faience with a gold frame from hypogeum 3200. © Margiana Archaeological Expedition.

Figure 10.21a The "Vegetation" deity(?) *in situ* at the time of its discovery in hypogeum 3200 (gold, silver, gypsum, bronze). © Margiana Archaeological Expedition.

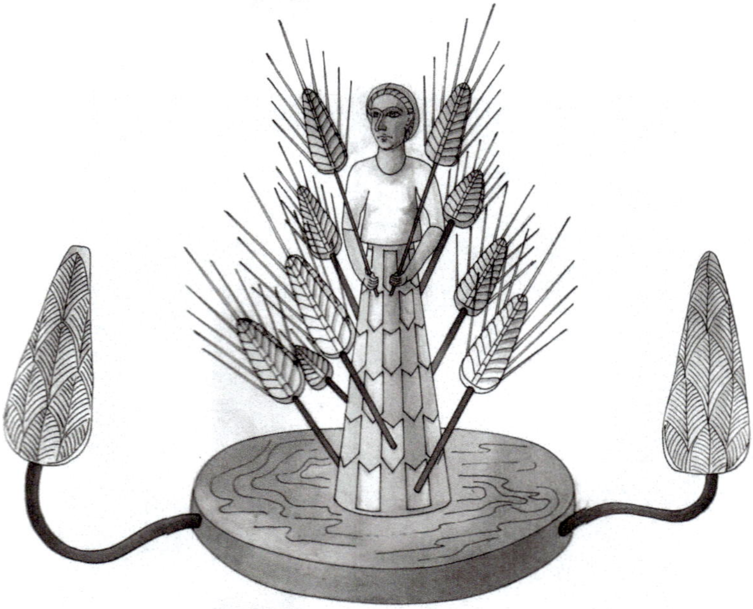

Figure 10.21b Reconstitution of the "Vegetation" deity(?). Drawing by A. Shajimova © Margiana Archaeological Expedition.

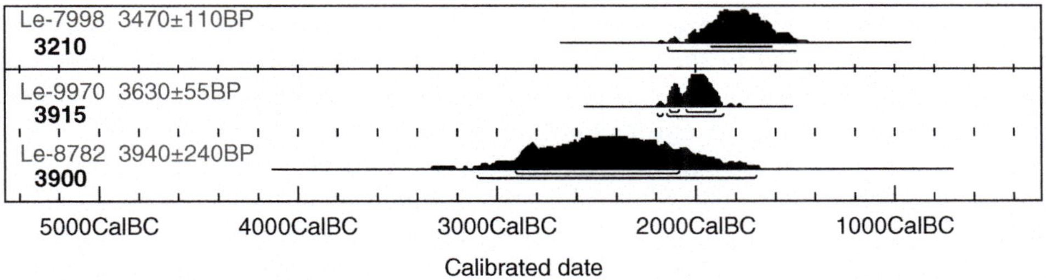

Figure 10.22 Three radiocarbon dates from the "Royal Necropolis". © Margiana Archaeological Expedition.

goat and a lion made of turquoise mounted on a pin found in tomb 3200 (Sarianidi 2005: 216, figs. 76, 77; 217, figs. 79; 218, fig. 80), as well as stone triangular trees discovered in tombs 3200, 3220, and 3230.

Many flint arrowheads have been found in all the hypogea except in 3905 and in "ditches" 3225, and 3240 (Table 10.1). The largest amount comes from tomb 3210 (55 items) and from tomb 3220 (44 items). In the last one, seven bone arrowheads have to be added. Only in one case (tomb 3230) were they found in a compact set, while in the other tombs, they were lying at different depths and scattered over the whole construction without any connection with the human bodies. We cannot exclude that the funerary ritual included archery inside the grave, like modern people who fire a rifle shot.

^{14}C DATES

Only three radiocarbon dates could be obtained from the Gonur "Royal Necropolis" (Figure 10.22): Ле-7998 from grave 3210, Ле-9970 from grave 3915, and Ле-8782 from grave 3900 (see the Appendix at the end of the volume). The earliest date, made on animal bones, is of little use because of its large spectrum. The latest date has also a large spectrum (tomb 3210), and the charcoal sample used for the dating comes from the "courtyard" and could be related not to the burial itself but to its later looting. Unfortunately, in general, there are only very few charcoal remains in these elite tombs. Taking into account the archaeological context of the site, we consider the dates between 2200 and 2000 BC for the whole "Royal Necropolis" to be adequate. We nevertheless hope to be able to make accelerator mass spectrometry (AMS) radiocarbon dates in the near future and to get better reliable information from them.

CONCLUSION AND ACKNOWLEDGMENTS

The aim of this chapter was not to fully describe the whole Gonur "Royal Necropolis" but only of some of it in order to attempt a systematization of the data. Some comparisons of the material have already been done by Sarianidi (2005, 2008a,

2008b). It has to be underlined again that these hypogea are the first to be discovered in the BMAC area and in the surrounding regions, and are also unique at Gonur Depe. The only place where analogies can be found is at Ur, where partly underground hypogeums with several chambers said to belong to the kings of the Third Dynasty (or the major priests) have been found close to the sacred area. They were originally lavishly decorated with gold, lapis lazuli, and were also surrounded by shaft graves containing simultaneous multiple burials of men, women, and children. Unfortunately, they had been totally looted, probably by the Elamites when they conquered the city (Moorey 1982: 163–174). Furthermore, in Shulgi's hypogeum room 1, remains of faience mosaics were found, probably coming from the wall (Moorey 1994: 175) (see Chapter 11).

The overall data coming from the rich Gonur cemetery shows, once more, that the culture brought to light presents an astonishing association of indigenous Central Asian traditional features together with influences from southwestern Asia, ancient Elam, and the Indus valley.

The works of the Margiana archaeological expedition were made effective and possible as a whole only thanks to the support of the Turkmenistan high administration, the efforts of many local workers, and scientists from different institutions and universities in Russia, Turkmenistan, and Finland, as well as to the financial support of the Russian Foundation for Basic Research, the Russian Humanitarian Foundation, the Greek Government, and the Pontic Greek societies. We are very grateful to all of them.

NOTES

1 Hypogea and chamber tombs are both house-like constructions but the latter are smaller and their architecture less complex than that of hypogea.
2 A *dakhma* is a place where the corpse of the deceased is placed for birds of prey or dogs to deflesh it according to the Zoroastrian ritual.
3 For a more detailed description of the tombs of dogs, see Chapter 12 and Sataev and Dubova (2016).
4 This pit has no number.
5 For a preliminary publication of tombs 3200, 3210, 3220, 3225, 3230, 3235, 3240, and their schemes, see Sarianidi (2005, 2006); tomb 3200 is also discussed in Dubova (2004). For tombs 3880 and 3900, see Sarianidi and Dubova (2010a, 2010b). The three tombs 3905, 3910, and 3915 are not yet published.
6 There is no evidence of a vaulted roof. Flat roofs are more likely.
7 A few stone headrests are known at Gonur but they are simple, without any specific decoration.
8 Only their skulls were in the niche, and their(?) long bones were scattered in the room.

REFERENCES

Dubova, N.A. 2004. "Mogil'nik i 'tsarskij nekropol' na beregakh Bol'shogo Bassejna Severnogo Gonura" [North Gonur graveyard and "Royal Necropolis" on the banks of the large basin]. In *U istokov tsivilizatsii* [Near the sources of civilizations], ed. M.F. Kosarev, P.M. Kozhin, and N.A. Dubova. Moscow: Staryj Sad, pp. 254–281.

Dubova, N.A. 2012. "Pogrebenija zhivotnykh v strane Margush" [Animal burials in Margush]. In *Trudy Margianskoj Arkheologicheskoj Ekspeditsii*, vol. 4, ed. V.I. Sarianidi. Moscow: Staryj Sad, pp. 101–139.

Dubova, N.A. 2015. "Animal Burials and Their Cults in Margiana." In *Defining the Sacred*, ed. N. Laneri. Oxford: Oxbow books, pp. 13–23.

Dubova, N.A. and V.V. Kufterin 2014. "Skelet karlika iz raskopok tsarskogo pogrebenija 3230 na Gonur Depe" [Skeleton of a dwarf from Gonur Depe royal grave 3230]. In *Trudy Margianskoj Arkheologicheskoj Ekspeditsii*, vol. 5, ed. V.I. Sarianidi. Moscow: Staryj Sad, pp. 166–171.

Dubova, N.A., R.M. Sataev, A.V. Fribus, S.P. Grushin, and V.V. Kufterin 2018. "Mogil'nik na jugo-zapade Gonur Depe (raskop 12). Soobshenie 1" [The graveyard on the south-western part of Gonur Depe (Area 12). Report 1]. In *Trudy Margianskoj Arkheologicheskoj Ekspeditsii*, vol. 7, ed. N.A. Dubova. Moscow: Staryj Sad, pp. 29–79.

Méry, S. 2000. "Les céramiques d'Oman et l'Asie moyenne. Une archéologie des échanges à l'Âge du Bronze." CRA Monographies. Paris: CNRS Editions.

Moorey, P.R.S. 1982. *Ur "of the Chaldees": A Revised and Updated Edition of Sir Leonard Woolley's Excavations at Ur*. Ithaca, NY: Cornell University Press.

Moorey, P.R.S. 1994. *Ancient Mesopotamian Materials and Industries: The Archaeological Evidence*. Oxford: Clarendon Press.

Salvatori, S. 1993. "The Discovery of the Graveyard of Gonur Depe (Murghab Delta, Turkmenistan): 1992 Campaign. Preliminary Report." *Rivista di Archeologia* XVII: 5–13.

Salvatori, S. 1994. "Excavations at the Namazga V Late Graveyard of Gonur 1 (Murghab Delta, Turkmenistan). Preliminary Report on the 1993 Field Season." *Rivista di Archeologia* XVIII: 14–39.

Salvatori, S. 1995. "The Gonur Depe 1 (Murghab Delta, Turkmenistan) Middle Bronze Age Graveyard: Preliminary Report on the 1994 Excavation Campaign." *Rivista di Archeologia* XVIX: 5–37.

Sarianidi, V.I. 2001. *Necropolis of Gonur and Iranian Paganism*. Moscow: Mir-Media.

Sarianidi, V.I. 2005. *Gonur Depe: Turkmenistan. City of Kings and Gods*. Ashgabat: Miras.

Sarianidi, V.I. 2006. "'Tsarskij nekropol' na Severnom Gonure" [The "Royal Necropolis" at Gonur North]. *Vestnik Drevnej Istorii* 2(257): 155–192.

Sarianidi, V.I. 2007a. "Dvortsovo-khramovyj kompleks Severnogo Gonura" [The palace-temple complex of Gonur North]. *Rossijskaja Arkheologija* 1: 49–63.

Sarianidi, V.I. 2007b. *Necropolis of Gonur*. Athens: Kapon editions.

Sarianidi, V.I. 2008a. *Margush: Mystery and True of the Great Culture*. Ashgabat: Türkmend övlethabarlary.

Sarianidi, V.I. 2008b. *Zoroastrianism: A New Motherland for an Old Religion*. Thessaloniki: Kyriakidis Brothers SA.

Sarianidi, V. and N. Dubova 2010a. "Nouvelles sépultures sur le territoire de la 'nécropole royale' de Gonur Depe." *Arts Asiatiques* 65: 5–26.

Sarianidi, V. and N. Dubova 2010b. "Novye grobnitsy na territorii tsarskogo nekropolja Gonura (predvaritel'noe soobshenie)" [New graves in the area of the "Royal Necropolis" at Gonur (preliminary report)]. In *Na puti otkrytija tsivilizatsii* [On the way of uncovering a civilization], ed. P.M. Kozhin, M.F. Kosarev, and N.A. Dubova. St. Petersburg: Aletejja, pp. 144–171.

Sarianidi, V.I. and N.A. Dubova 2013. *Treasures of Ancient Margiana: Photo Album*. Ashgabad: Türkmendövlethabarlary.

Sarianidi, V.I. and N.A. Dubova 2016. "Types of Graves at Gonur Depe Bronze Age Site in Turkmenistan." In *Proceedings of the 9th International Congress on the Archaeology of the Ancient Near East 2014, Basel*, vol. 3, ed. A.R. Stucky, O. Kaelin, and H.-P. Mathys. Wiesbaden: Harrassowitz Verlag, pp. 631–642.

Sataev, R.M. and N.A. Dubova 2016. "Pogrebenija sobak na tsarskom nekropole Gonura" [Dog burials at the Gonur "Royal Necropolis"]. In *Trudy Margianskoj Arkheologicheskoj Ekspeditsii*, vol. 6, ed. N.A. Dubova. Moscow: Staryj Sad, pp. 204–223.

Sataev, R.M. and R.M. Sataeva 2015. "Animalisticheskaja kompozitsija na kubke iz klada tsarskogo nekropolja Gonur-depe: analiz istochnika" [Animal composition on a goblet from a hoard in the "Royal Necropolis" at Gonur Depe: analysis of its origins]. In *Zapiski Istituta Istorii Material'noj Kul'tury RAN* 12: 103–117.

CHAPTER ELEVEN

POLYCHROME INLAYED AND PAINTED MOSAICS FROM GONUR DEPE (TURKMENISTAN)

Nadezhda A. Dubova

INTRODUCTION

Thanks to Victor Ivanovich Sarianidi's excavations at Gonur Depe, some of the finest examples of ancient art have been brought to light in Central Asia. Mosaic compositions are among the most remarkable discoveries of this art. They used to decorate the graves of the local elite (see Chapter 10).

No similar artwork has ever been found in any synchronous, later, or earlier historic monuments of Central Asia, Ancient Egypt, China, or the Indus valley, except perhaps in Mesopotamia.[1] The famous Uruk mosaics dating back to the end of the 4th millennium were made by pressing clay cones (8–10 cm long and 1.8 cm in diameter) into a wet plaster background. Their external surface formed the mosaic surface making geometric patterns like rosettes, triangles, or zigzags, generally painted in red, black, and white (Aruz and Wallenfels 2003: 18–19, figs. 4, 5). Inlays found in the ancient city of Mari dating from the 3rd millennium were manufactured from lapis lazuli, mother-of-pearl, and ivory (Parrot 1967: pl. XI; Aruz and Wallenfels 2003: figs. 96, 97, 100; Margueron 2007: 52, fig. 25).

The Gonur mosaics differ from these examples because, on the one hand, they combine two techniques – painting and inlays – and, on the other hand, because they include inserts made of specially prepared minerals. They are also unique in the scenes they depict. The only place with partly similar features is the frieze of warriors from the Early Dynastic II palace at Kish (2650–2550 BC) where the most important details are highlighted by the insertion of inlays in contrasting colors. The scene composition is hollowed out of a gray slate and then filled in with white limestone inserts (Aruz and Wallenfels 2003: fig. 48). No mosaic composition has been discovered at Mohenjo-daro, though a large number of separated pieces are attested, reminiscent, to a certain extent, of the Gonur examples (Marshall 1931: pl. CLV, *26–31*, *39–46*, *56*, *57*, *61*, *62*; CLVI, *12*).

Other kinds of mosaic compositions have also been found at Gonur, like game boards and containers. We present them in the next section, before going on to discuss the mosaics themselves.

THE GAME BOARDS

Several "game boards" made with inlays also have been discovered in several elite tombs at Gonur Depe (nos. 2900, 3130, 3210, 3310, 3623) (for some of them, see Chapter 10). These items and the technique of inlays used were rather widespread at that time. The pieces are made of ivory. The boards have a rectangular shape (40 × 19–20 cm) and are divided into three rows of roughly square sections (5 × 5–6 cm). The sections are separated by ivory insertions (i.e., four-sided sticks, 0.3–0.5 cm wide, 6 to 16 cm long, or possibly longer). The middle row contains eight square sections; at the junction of the third and fourth sections, ivory tesserae were inserted (Figures 11.1, 11.2a, and 11.2b).

In the tomb cist 2900, these inlays have a cross-type or four-petal shape decoration; in tomb 3210, resting lambs are depicted. In the other three graves (where, in fact, instead of a human body, sheep had been ritually buried, see Dubova 2008, 2012; Sarianidi and Dubova 2015), inlays were either absent or missing although paired empty spaces are attested. The shape of the game boards from Gonur is exactly the same as of those known at the Royal Cemetery of Ur (Woolley 1934: British Museum no. 120834), or Shahr-i Sokhta (Piperno and Salvatori 1983), but differs from those of the Susa and Khuzistan sites, Jiroft, Egypt, etc. (Dunn-Vaturi and Schädler 2006; Schädler and Dunn-Vaturi 2009; Rogersdotter 2011). Various stone, clay, and ivory

Figure 11.1 Game board from tomb 2900 during cleaning *in situ*. © Margiana Archaeological Expedition.

Figure 11.2a Game board from tomb 3210 during cleaning *in situ*.
© Margiana Archaeological Expedition.

dices were found in the tombs next to these boards, but these are also present in other graves where no game boards were found.

THE CONTAINERS OR "OSTENSORIES"

The first mosaic inlays were discovered in the Gonur North palace in the second half of the 1990s. Only very few were found there and they were not *in situ* (see pp. 376–377). At the same time, small wooden containers for round bronze mirrors decorated with inlays were found in two rather rich graves (nos. 194 and 199) at the Gonur Large Necropolis (see Sarianidi 2001: pl. 5, *11*, *12*; 2007: figs. 205, 206; 2008: fig. 144). These wooden boxes were decorated on the outside with geometric ornaments made with small, white stone tesserae 0.5 cm wide. Some of the squares and spaces between them were painted in red and black, creating a bright, colorful cross-like composition.

In 2004, other exquisite discoveries made during the excavations of the Royal Necropolis (see Chapter 10) made headline news. They consisted of skillfully executed multifigure compositions decorating the so-called "ostensories." These were large wooden boxes, probably for the disposal of valuable offerings as suggested by Sarianidi. They have been discovered in tombs 3230, 3220, and 3880 (Sarianidi 2008: figs. 100, 163, 165, 166; Sarianidi and Dubova 2013). Small boxes of different shapes and sizes,

Figure 11.2b Ivory insertion to the game board in the tomb 3210.
© Margiana Archaeological Expedition.

probably made of wood, were also found in tombs 3235, shaft tomb 3870, and possibly 3220 (see Sarianidi 2008: fig. 140; 2009: fig. 114 (Figures 11.3, 11.4, and 11.5).

THE MOSAIC PANELS

Most impressive are the painted and mosaic composition panels, all discovered in the "Royal Necropolis" (see Chapter 10). They were made either directly on the walls (their traces and fragments are preserved *in situ* only in tombs 3200 and 3235) (Sarianidi 2008: fig. 145) or on wooden panels that were later hung up on the walls (tomb 3220). Some of these have also been found on the walls of "house-like graves" or hypogea (tombs 3230, 3235), as well as in front of the facade of the "house of the dead" in the tombs with "courtyards" (nos. 3200, 3210) (Figures 11.6 and 11.7). Before discussing further the motifs and interpretation of the mosaics (see pp. 377–382), it is important to give some details about the way they were executed.

THE MANUFACTURING PROCESSES

During the conservation and restoration procedures, the technology used to make these mosaics was closely examined by Natalya Kovaleva and Galina Veresotskaja, experts from the State Research Institute for Restoration of the Ministry of Culture of

Figure 11.3 Side wall of the "ostensory" in hypogeum 3230 during cleaning *in situ*. © Margiana Archaeological Expedition.

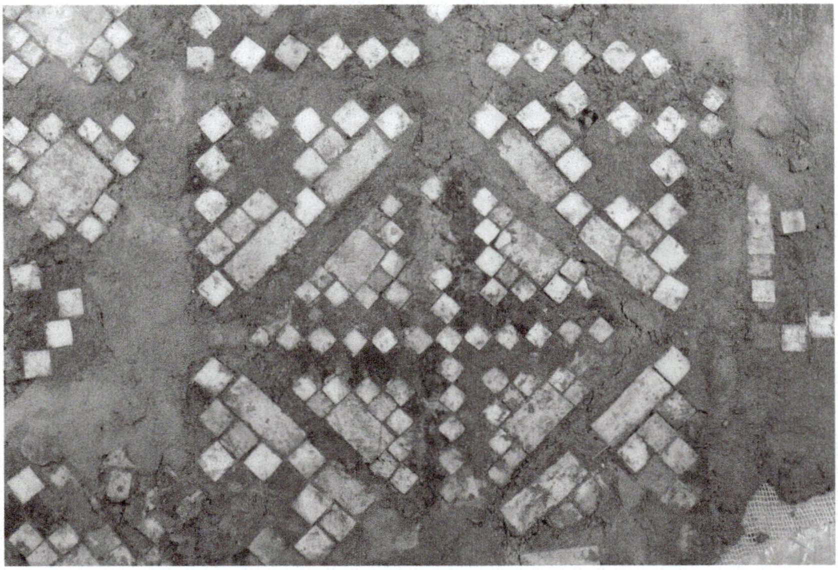

Figure 11.4 Side wall of the "ostensory" from hypogeum 3230 after restoration. Photo by K. Samurskij. © Margiana Archaeological Expedition.

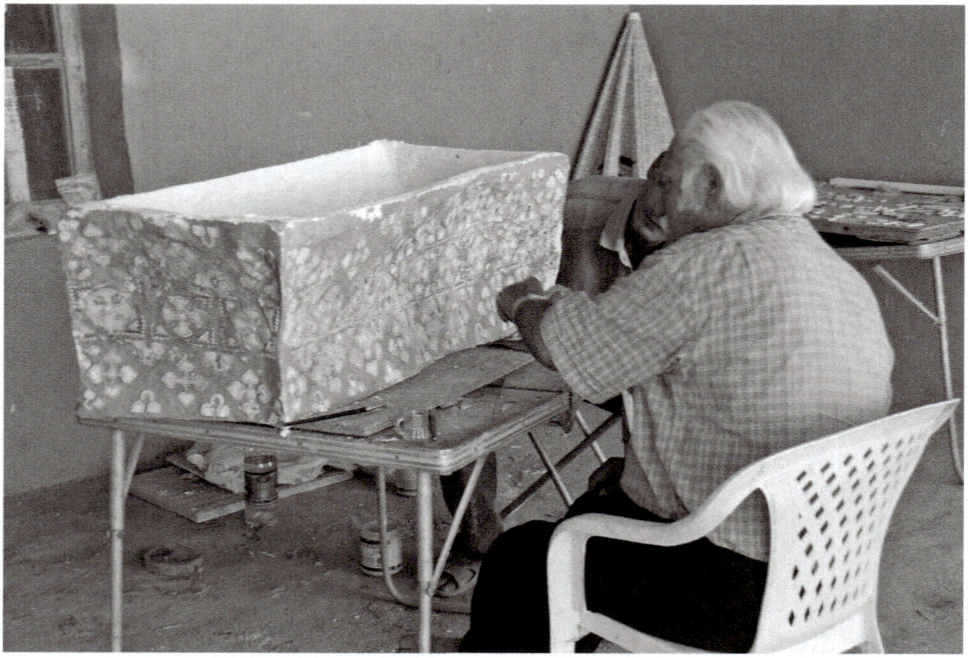

Figure 11.5 Gonur Depe, 2009. V.I. Sarianidi and restorer M. Rzakov during restoration of the "ostensory" from hypogeum 3880. © Margiana Archaeological Expedition.

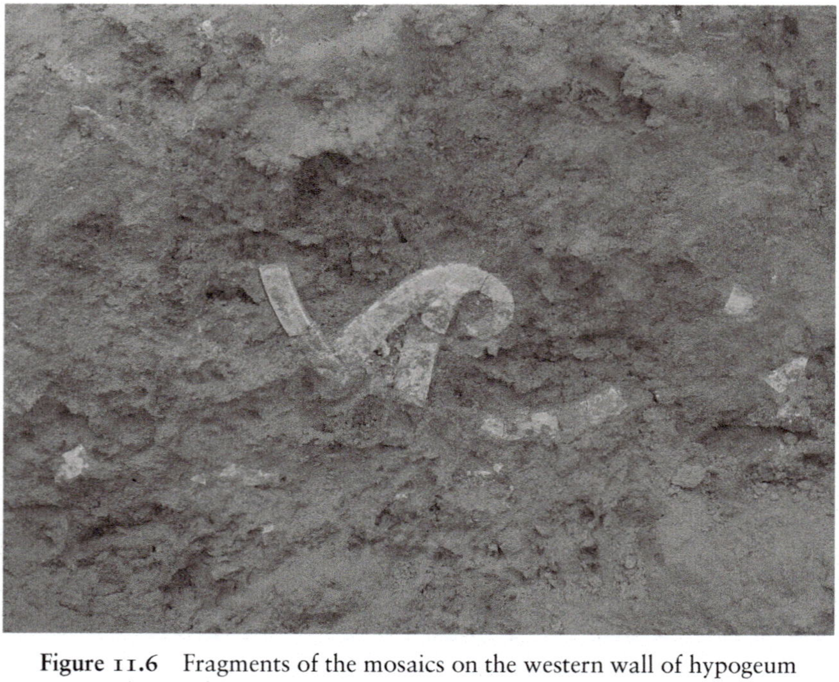

Figure 11.6 Fragments of the mosaics on the western wall of hypogeum 3200 during cleaning *in situ*. © Margiana Archaeological Expedition.

Figure 11.7 Fragment of the mosaics on the northern wall of hypogeum 3235 during cleaning *in situ*. © Margiana Archaeological Expedition.

Russia, together with their colleagues from various Russian (Natalja Chudnovskaja and V. Kireeva) and Turkmen organizations (Mukhammed Begliev, Ajdzhan Shajymova, Maksat Annanurov, and Annamurat Orazov).

Their study shows that the plaster used as a base layer was light gray, dense enough, and solid (thickness 0.7–0.8 cm). The stone inlays were fixed on the plaster mostly with a mastic material made from black carbonaceous pigment, quartz, and dihydrate and anhydride gypsum. A mixture of heat-treated natural resins was used as a binding agent. The components could not be identified due to their destruction during firing (Pisareva 2012; Kireeva 2014: 224–225; Veresotskaja 2016a: 318). Traces of this mixture survived on the reverse side of the inlays.

The motifs on the panels were made either directly (i.e., the stone tesserae were stuck face up directly on the plaster base), or indirectly (i.e., were first assembled face down on a temporary surface). N. Kovaleva considers that, in tomb 3230, the compositions were made in separate standard blocks (triangular or rectangular), isolated by wooden sticks (Kovaleva 2012: 180). In any case, after the composition was done, the spaces between the separate elements were filled with a red and black paint to underline the motifs. According to V. Kireeva (2014), a "black carbonic pigment" was the core element of the black paint or black-paste. As for the red paint, it was made of "red-brown clay minerals" (Kovaleva 2012: 180). G. Veresotskaja (2012: 187–188) proposes another possible method of manufacture for the inlayed compositions on the ostensory from tomb 3230. She believes that a wooden framework was used as the basis of the mosaic because the wooden strips (13–15 mm wide) were intertwined so as to make the main geometric pattern. According to Kireeva, the wooden base was covered with a solution of lime-sand coating that served as a preparation layer for the decoration. Several parts of the mosaic panels are well preserved in spite of the damage caused by robbers, and this seems at least partly due to their excellent preparation. The next step was the layout of the pattern: Specially made templates were pressed upon the surface into the still-wet layer of plaster, thus forming the main decoration grid. Thereafter the grid was filled in with the mosaic inlays.

Layers of the mastic material were applied on the surface. Apparently, due to the various materials used for the composition, each with its own physical and chemical properties, different kinds of mastic were used. We cannot exclude that a specific technological tradition unknown to us was at work in the region.

Once the tesserae were carefully stuck and leveled, the motifs were painted with various colors. A well-drawn line of this painting is still visible on the edge of the separate blocks. It is possible that the wooden framework slightly poked out above the base layer of the plaster and gypsum and added to the overall decoration (Veresotskaja 2012: 188).

The inlays were made of different minerals. Laboratory tests are still going on and the composition is not yet clear for all of them. Marble is attested, but the majority of the inlays are made of an artificial material. X-ray analysis of the inlays from tomb 3915 show that they are made of quartz up to 95%. The grains have a tabular shape and sharp edges. Their size varies around 10–20 microns, less often up to 50 microns, and they are chaotically placed in the siliceous matrix. In sections, quartz is characterized by irregular cracks, probably caused by changes in volume as it was transformed into various siliceous polytypes when heated. Judging from the mineral composition of the material, as well as from its structure and texture, a siliceous secondary rock (such as gaize, diatomite, and bergmehl) was used as raw material to make the inlays. Clearly, this raw material was heated, which changed its structure (Juminov 2012: 189–190; Veresotskaja 2016a).

This processing method (i.e., siliceous raw materials and heat treatment) has a long history in southwest Asia (Barthélémy de Saizieu 2003). Harappan seals were manufactured from elaborately fired limestone (Masson 1977: 149). The firing method aimed at facilitating the processing of minerals and has been used in the Indus valley since the Neolithic period (Vidale 1990, 2007; Lankton *et al.* 2003: 9). The same technique was used for the production of beads and drill bits (including those with picot edges) and was also known at Altyn Depe (Kircho and Kovnurko 2003).

At Gonur Depe, heated beads are frequent: for example, in the top layer on the western face of the site (Area 7), in tomb 3314 (Area 12), in tombs 3660 and 3716 (Area 16), etc. Without analysis, this material may often be confused with faience or gypsum. It was also used in the production of some seals (Sarianidi 2001: pl. 22, *1, 3, 4, 6–11, 13, 14*; 2002: 46, tab. II, *11–14*; 2007: 105, figs. 178, 179; Rossi Osmida 2011: 227, 228), leaf-shaped beads (in tomb 2827, see Sarianidi 2007: 98, fig. 149; 123, fig. 24), four-sided hairpins (Sarianidi 2007: 88, fig. 99), and many other items (Rossi Osmida 2011: 159).

Veresotskaja considers that there are traces of green glaze on the surface of some of the tesserae in the center of the carpet ornament of the ostensory from tomb 3230. Laboratory analyses made by A. Mazina and A. Sokolov in 2014 confirmed the presence of glaze (Veresotskaja 2016a: 316). However, the glazing layer formation

Figure 11.8 Beveled-cut edges on the back of the stone inlay from hypogeum 3230. © Margiana Archaeological Expedition.

process needs to be studied further, for there is no clear evidence that the tesserae were intentionally glazed. It is not completely excluded that this layer is due to later accidental contact with copper components.

During the restoration, it became clear that the spaces between the inlays were smooth, despite the fact that, in the same motif, the inserts significantly differed in thickness (0.25–0.5 cm; Kovaleva 2012: 181). This could have been achieved by fixing the tesserae on the plaster layer at a different depth. Furthermore, a great number of the edges of the inlays had been tapered before they had been bonded together in a block. According to T. Shaposhnikova's observations, the edges of most of the tesserae have a slope ranging from 45° to 85°, while fewer are cut at a right angle (90°) (Figure 11.8). According to the profile shape of the inlays, the type of junction between them was modified. Among the numerous stone artifacts from Gonur, A. Juminov found tools prepared and sharpened at the same angles as the tesserae of the mosaics found in the "royal tombs" (Sarianidi and Dubova 2013: fig. 17).

There is a long tradition in Greece, going back to the late Minoan period, of paved floors with river gravels and ornamented mosaics. The technique of cutting small pieces of stone (tesserae) slightly in cones is, however, said to have made its appearance only during the Hellenistic period and to *floruit* in Roman times (see, e.g., Bowersock 2006; Millz 2007). The mosaics discovered at Morgantina, Sicily (3rd century BC), for instance, are usually considered as the earliest examples of this technique. But the first attempts of figurative mosaics are known in the 3rd millennium (ca. 2500–2300 BC), on the famous standards from Ur (see, e.g., Margueron 1996) and from Mari, as well as on other panels (Couturaud 2014). The inlays, in these cases, are made of shell or limestone and incrusted on wooden panels covered with bitumen. At Ur, again, but during the Ur III period, mosaics are supposed to have covered the walls of Shulgi's hypogeum, but the severe looting of the tomb does not allow for their reconstruction (Moorey 1994: 175). The Gonur discoveries prove that this mosaic technique was also known in Margiana at the end of the 3rd millennium BC.

The way the individual tesserae were glued together to produce a whole composition is particularly interesting. In addition to the individual blocks used for geometric ornaments and the use of a template mentioned above, another peculiarity has been

Figure 11.9 Incised marks on the back of the inlays from hypogeum 3015. © Margiana Archaeological Expedition.

noticed; namely, special signs or symbols ("hearts" – possibly pipal leaves, crosses, circles, triangles, crossed circles, and almond-shaped symbols) that are visible on their reverse sides. Sarianidi was the first to notice these signs at the time of their discovery (Figure 11.9). This may imply that, to make the mosaics, the craftsmen were divided into different operation teams and that the composition was put together using the signs on the back.

THE LOCATION OF THE MOSAICS AT GONUR

The vast majority of these mosaic compositions were destroyed long ago, due to multiple lootings of the tombs in the Royal Necropolis. As a result, we are left today with thousands of individual pieces with geometric and figurative decoration (see Shaposhnikova 2012: figs. 6, 7) (Figure 11.10).

Apart from the finds in elite tombs of the Large Necropolis associated with containers for bronze mirrors mentioned above, a few compositions still in quite good condition have been brought to light in the hypogea. Those of the "fighting griffins," the "griffin in a cartouche," or the "dragons swallowing a goat" are obviously part of a large composition from tomb 3210 (Sarianidi 2005: figs. 67, 83, 85–86; 2006: fig. 11; 2008: figs. 147–159; 2009: figs. 117, 120–126). A "portrait" comes from the chamber grave 3245 (Sarianidi 2005: fig. 105; 2009: fig. 136; Kovaleva 2016: figs. 23, 24). The "fight between flying dragons and snakes" together with a geometric decoration come from the front facade of the "house-like" tomb. A plaquette with a geometric polychrome pattern hanging from the neck of a camel lying in the "courtyard" was found in tomb 3880 (Sarianidi and Dubova 2010: figs. 22, 24–25, 27, 29, 30; 2013: figs. 27–28; 30, 32, 36–38).

Figure 11.10 Four inlays in the shape of a dragon's head from hypogeum 3210. © Margiana Archaeological Expedition.

Several jewelry boxes were preserved. They have different shapes and are decorated with geometric ornaments (tomb 3200) (Dubova 2004: figs. 27, 29; Sarianidi 2006: fig. 4; Kovaleva 2016: figs. 18–22) and presumably also floral ones (tomb 3235) (Sarianidi 2009: fig. 114). Similarly, three large (0.8/1.0 x 0.6/0.8 x 0.6/0.8 m) "ostensories" have reached us: One, in tomb 3220 was decorated with "hearts" (or pipal leaves) (Sarianidi 2006: fig. 18; Sarianidi and Dubova 2013: fig. 10a); another, in tomb 3230, had a "carpet pattern" (Kovaleva 2012; Veresotskaja 2012, 2016a; Sarianidi and Dubova 2013: figs. 13, 14); the last and best-preserved one, from tomb 3880, is decorated with several rows of alternating crosses of different shapes (Sarianidi and Dubova 2010: figs. 17, 18; 2013: fig. 10c).

THE MOTIFS AND THEIR COMPARISONS

Today, most of the mosaics are stored and exhibited at the Museum of Fine Art of Turkmenistan (Ashgabat). The ostensory from tomb 3880 and some smaller compositions from burials 3870 and 3915 are stored in the Mary Regional History

Museum. N. Kovaleva and G. Veresotskaja are in charge of the restoration and study of these artifacts in the Museum of Fine Art. While analyzing the individual inlays from tomb 3210, G. Veresotskaja used the signs on their back. She was able to recover the lost shapes and sizes of the larger fragments and to complete in part the best-preserved composition of the "two griffins." Her study of the technological characteristics and of the way the inlays were assembled has revealed that the blocks with the wing feathers were made from two different kinds of material, one certainly mineral (probably limestone) and the other made of a less dense and light material, the analysis of which are in process. These feather blocks, made of various materials, are placed in a checker pattern, with spaces colored in red or black. These technological characteristics will allow a more accurate restoration of the complex composition scheme decorating this royal tomb. They also show the deep knowledge of those who made these mosaics. On the basis of the various techniques that have been identified, Veresotskaja has significantly improved the reconstruction of the griffins from tomb 3210. The first restorations dealt with two independent representations, the "eagles in a heraldic position" (see Sarianidi 2008: fig. 129) and the "fighting griffins" (Veresotskaja 2016b: fig. 23), which turned out to be two registers of the same composition (Figure 11.11).

The mosaics are still under study. Nevertheless, one cannot ignore the amazing variety of the color range in these pictures although only red, black, and white paints were used. This effect was achieved using different techniques in the position of the tesserae and in the choice of the filling between them. The compositions closely resemble some of the painted scenes of the so-called "Zimri-Lim's investiture" in the Amorite palace at Mari in Syria (Musée du Louvre, inventory no. AO19826).[2] This is particularly the case of the "griffin in a cartouche," exquisitely made by the Gonur craftsmen (Figure 11.12). In both the Gonur and Mari versions, the animal is placed in a frame surrounded by a dark border. The other representations in the investiture scene are framed in the same way. At Gonur, the black color was used in the final stage of the making of the mosaic, to outline or to draw the borders. The fantastic beast from Gonur Depe tomb 3210 has the torso of a lion, the head of a dragon, and spread wings. It is standing firm on its powerful legs. The head of the Syrian winged lion did not survive, but it may have worn horns. On the Gonur representation, the beast's head is crowned with black curved horns. It also has a black, slightly curved forward and narrow beard similar to that usually depicted on Egyptian priests' and pharaohs' faces. Other griffins wearing the same beards are depicted on another composition found in the same tomb at Gonur. The griffin's body is generally reminiscent of that of a lion's torso, but there are still noticeable differences: there is no mane, a distinctive feature of a male lion, and the lower part of the belly is colored in bright red. The back of the neck of the griffin is similar to that of a buffalo or a zebu since it is depicted with a hump rising immediately behind the head. Several blocks are made of stone tesserae: the head with its wide-open jaws, the legs with their underlined muscles, and the wings made of tiny elements in order to underline their volume. The Syrian beast's wings are also multicolored. But its torso is much less powerful, especially in the bottom part. In both pictures, the tails of the animals are raised up and tense. The Mari griffin's tail is rolled around a multicolored disk. Interestingly, in the Mari picture, the lion-griffin cartouche stands over that of a humpback buffalo or zebu with its front leg standing over mountains. These comparisons vividly illustrate

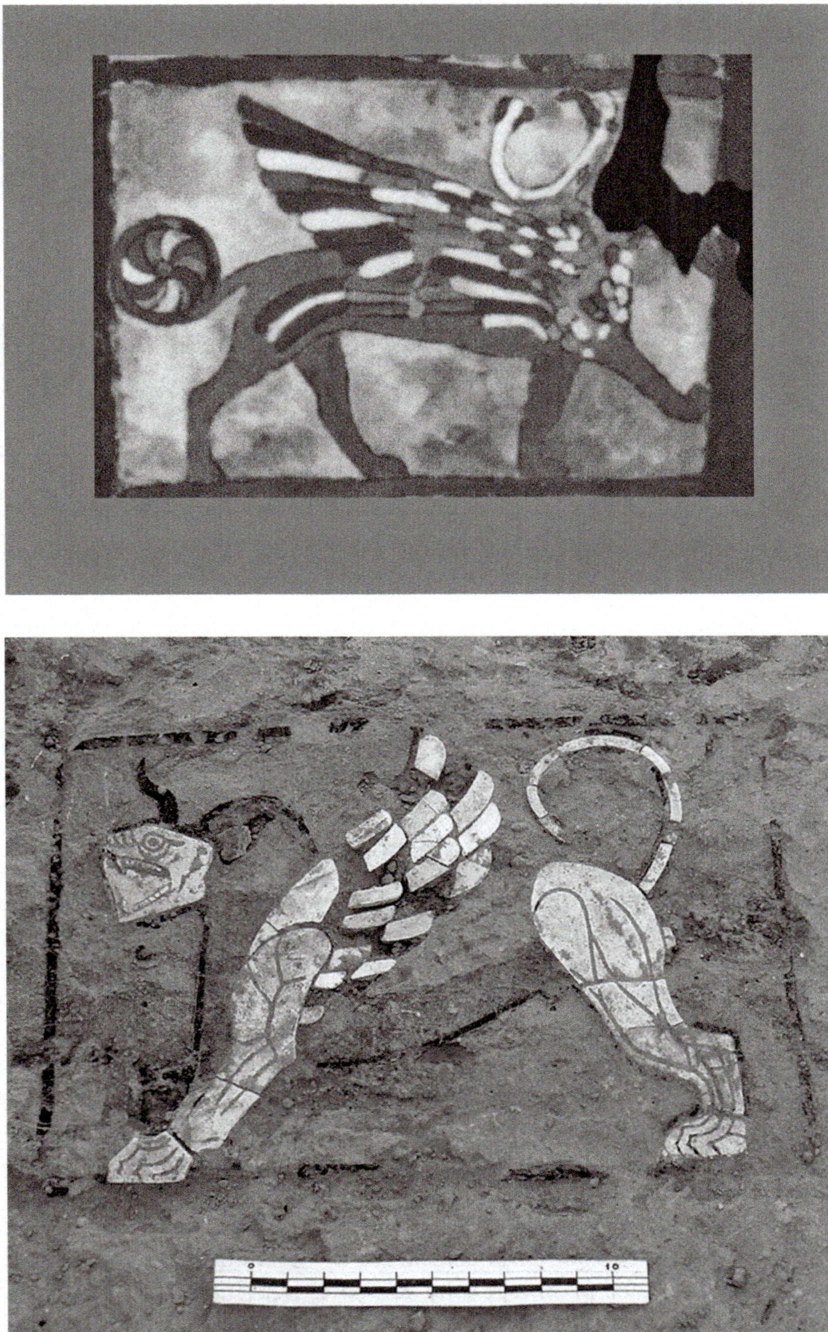

Figure 11.11 Comparison between the mosaics "griffin in a cartouche" from hypogeum 3210 at Gonur Depe (bottom © Margiana Archaeological Expedition) and a fragment of the "Zimri-Lim's investiture" painted scenes in the Amorite palace at Mari in Syria (top).

Figure 11.12 G. Veresotskaja's reconstruction of one mosaic scene from hypogeum 3210. © Margiana Archaeological Expedition.

the close cultural ties between the country of Margush and the Near East at dates close one to the other (see endnote 2).

Other comparisons can be advanced between the birds' figures (especially the heads) found at Gonur in hypogeum 3230 and the "small vulture" represented on a fragment of a gold vessel in the Fullol hoard of northeastern Afghanistan (Tosi and Wardak 1972: fig. 13).

As has already been mentioned, the Gonur mosaics are different from anything known before, as they combine two techniques – painting and inlays. Another remark is worthy of interest. Among the Gonur Depe finds, only one fragment represents a human face (burial 3245) (Figure 11.13), while complete human figures are typical in the third-millennium Mari mosaic compositions or on the famous standard of Ur, and they are also numerous in the early 2nd-millennium painted investiture scene mentioned above. However, many of the "tesserae" detached from the compositions in the Gonur tombs seem to represent elements of cloths (tombs 3210, 3230, 3235). There are also several pieces of large eyes (burial 3210; see, e.g., Shaposhnikova 2012: 173, 174). Apparently, human beings (or composite figures?) were represented in some of the lost compositions, and must have all been painted.

It is remarkable that only a few individual inlays have been discovered in the Gonur palace itself or in the other buildings considered by Sarianidi as a temple. Obviously, there, these inlays are not *in situ*. All the surviving figurative mosaic compositions come from the elite burials. However, even in the wealthiest burial chambers, only small mirror cases, game boards, and ostensories with mosaic decoration have been found. We can only assume their meaning.

A wooden box decorated inside with flowers inlays (Figure 11.14) was found in the shaft tomb 3870 built next to the rich tomb 3220 in the "Royal Necropolis" at Gonur. This looks more like the 3rd-millennium finds of Mari and Ur since there are

Figure 11.13 Mosaic inlay in the shape of a human portrait from tomb 3245 in the Gonur Depe "Royal Necropolis." Photo by K. Samurskij. © Margiana Archaeological Expedition.

Figure 11.14 Wooden box with mosaic inlays made of stone and ivory during cleaning *in situ* in chamber tomb 3870 at Gonur Depe. © Margiana Archaeological Expedition.

(almost) no painted elements. Four identical flowers are represented on the long sides of the box and two on the short sides.[3] All the flowers are identical and have four oval petals, the inner parts of which are decorated with drop-shaped ivory inlays; their external outline and central part are made of a white-colored mineral.[4] The rest of the sides are also decorated with inlays. The decoration consists of alternating stone and bone rectangular blocks from top to bottom, making six identical compositions where two right triangles made of a white mineral frame an isosceles triangle made of ivory standing on its summit. The edges of the whole composition, as its elements, are outlined in black. Some of the bone inlays making the frame of the flowers present traces of red paint.

To conclude this description, we would like to underline that the extraordinary mosaics discovered at Gonur, in tombs dated between 2300 and 2100 BC (Zajtseva *et al.* 2008) (or rather between 2150 and 1950, see the Appendix), demonstrate once more the high level of arts and crafts developed in the ancient Murghab oasis. We have just started the scientific analysis of the motifs and of the methods used to make these unique compositions that once decorated the walls of the elite burials and the ostensories. Remarkable results have already been achieved. Still, we are just at the beginning of our understanding of the symbolic or narrative meaning of the motifs. Hopefully, this will help us get an insight into the vision of the world and beliefs of these people. We hope that the uniqueness of the Gonur mosaics and the artistic mastery of their creators will also attract other researchers and specialists of the myths and legends of the ancient East, in order to reveal further the mysteries of this civilization, which remained hidden in the sands of the Kara Kum Desert for millennia.

NOTES

1 Faience inlays on walls are mentioned in the contemporaneous Shulgi hypogeum at Ur, but this building was so badly looted in antiquity that we do not know what it looked like (see Moorey 1994: 175).
2 The paintings found in the Amorite palace of Mari date most probably to an earlier period than the time of Zimri-Lim according to Margueron (2004: 509). He suggests either the 19th century or the beginning of the 18th century BC.
3 Unfortunately, the original decoration of the box was not preserved. The item displayed in the Mary Regional History Museum cannot be considered as a totally reliable reconstruction.
4 The technology used for the fabrication of this artifact has not yet been studied. We assume that the material was the same as in the other mosaics.

REFERENCES

Aruz, J. and R. Wallenfels (eds.) 2003. *Art of the First Cities: The Third Millennium BC from the Mediterranean to the Indus*. New York: Metropolitan Museum of Art.

Barthélémy de Saizieu, B. 2003. *Les parures de Mehrgarh: Perles et pendentifs du Néolithique pré-céramique à la période pré-Indus. Fouilles 1974–1985*. Paris: ERC.

Bowersock, G.W. 2006. *Mosaics as History: The Near East from Late Antiquity to Islam.* Cambridge, MA: Belknap Press.

Couturaud, B. 2014. "L'image et le contexte: nouvelle étude des panneaux figuratifs incrustés de Mari." *Syria: Archéologie, art et histoire* 91: 77–97.

Dubova, N.A. 2004. "Mogil'nik i tsarskij nekropol' na beregakh bol'shogo bassejna Severnogo Gonura" [The cemetery and "Royal Necropolis" on the banks of the large pool at Gonur North]. In *U istokov tsivilizatsii* [Near the sources of civilizations], ed. M.F. Kosarev, P.M. Kozhin, and N.A. Dubova. Moscow: Staryj Sad, pp. 254–281.

Dubova, N.A. 2008. "Ritual'noe zakhoronenie zhivotnykh na Gonur Depe" [Ritual animal burial at Gonur Depe]. In *Trudy Margianskoj Arkheologicheskoj Ekspeditsii*, vol. 2, ed. V.I. Sarianidi. Moscow: Staryj Sad, pp. 50–60.

Dubova, N.A. 2012. "Pogrebenija zhivotnykh v strane Margush" [Animal burials in the Margush country]. In *Trudy Margianskoj Arkheologicheskoj Ekspeditsii*, vol. 4, ed. V.I. Sarianidi. Moscow: Staryj Sad, pp. 101–139.

Dunn-Vaturi, A.-E. and U. Schädler 2006. "Nouvelles perspectives sur les jeux à la lumière de plateaux du Kerman." *Iranica Antiqua* XLI: 1–30.

Juminov, A.M. 2012. "Mozajchnye vstavki iz grobnitsy 3915" [Mosaics inlays from the royal sepulcher 3915]. In *Trudy Margianskoj Arkheologicheskoj Ekspeditsii*, vol. 4, ed. V.I. Sarianidi. Moscow: Staryj Sad, pp. 189–191.

Kircho, L.B. and G.M. Kovnurko 2003. "Prestizhno-kul'tovye predmety i ukrashenija Altyn-depe iz drevnikh iskusstvennykh mineralov (po dannym rentgenometricheskogo analiza)" [Prestigious cult artifacts and decoration from old artificial minerals at Altyn Depe (results from X-ray analysis)]. *Arkheologicheskie Vesti* 10: 108–114.

Kireeva, V.N. 2014. "Rezul'taty tekhniko-tekhnologicheskogo issledovanija mozaichnykh vstavok s Gonur-depe" [Results from the technical and technological studies of the Gonur Depe mosaic inlays]. In *Trudy Margianskoj Arkheologicheskoj Ekspeditsii*, vol. 5, ed. V.I. Sarianidi. Moscow: Staryj Sad, pp. 224–225.

Kovaleva, N.A. 2012. "Predvaritel'nye issledovanija i konservatsija dvukh fragmentov 'kovrovykh' mozaik iz Gonura" [Preliminary research and conservation of two fragments of "carpet" mosaic from Gonur]. In *Trudy Margianskoj Arkheologicheskoj Ekspeditsii*, vol. 4, ed. V.I. Sarianidi. Moscow: Staryj Sad, pp. 176–184.

Kovaleva, N.A. 2016. "Victor Sarianidi – Poslednij Krylatyj Lev Margiany" [Victor Sarianidi – the last winged lion of Margiana]. In *Trudy Margianskoj arkheologicheskoj ekspeditsii*, vol. 6, ed. N.A. Dubova. Moscow: Staryj Sad, pp. 108–124.

Lankton, J.W., J. Diamanti, and J.M. Kenoyer 2003. *A Bead Timeline*. Vol. I, *Prehistory to 1200 CE: A Resource for Identification, Classification and Dating*. Washington, DC: Beads Society of Greater Washington.

Margueron, J.-C. 1996. "'L'Étendard d'Ur': récit historique ou magique?" In *Collectanea Orientalia. Histoire, arts de l'espace et industrie de la terre. Études offertes en hommage à Agnès Spycket. CPOA* 3, ed. H. Gasche and B. Hrouda. Neuchâtel and Paris: Recherches et Publications, pp. 159–170.

Margueron, J.-C. 2004. *Mari, métropole de l'Euphrate au IIIe et au début du IIe millénaire av. J.-C.* Paris: Picard/Editions Recherche sur les Civilisations.

Margueron, J.-C. 2007. "Le royaume de Mari." *Studia Orontica* 3: 7–60.

Masson, V.M. 1977. "Pechati protoindijskogo tipa iz Altyn-depe" [Seals of proto-Indian type at Altyn Depe]. *Vestnik Drevnej Istorii* 4: 147–156.

Marshall, J. 1931. *Mohenjo-Daro and the Indus Civilisation*, vol. III. London: Arthus Probsthain.

Millz, T. 2007. *Iskusstvo mozaiki: Entsiklopedija* [Mosaics art: encyclopedia]. Moscow: Art-Rodnik.

Moorey, P.R.S. 1994. *Ancient Mesopotamian Materials and Industries: The Archaeological Evidence*. Oxford: Clarendon Press.

Parrot, A. 1967. *Les Temples d'Ishtarat et de Ninni-zaza*. Mission Archéologique de Mari, T. 3. Paris: Paul Geuthner.

Piperno, M. and S. Salvatori 1983. "Recent Results and New Perspectives from the Research at the Graveyard of Shahr-i Sokhta, Sistan, Iran." *Annali* 63: 172–191.

Pisareva, S.A. 2012. "Rezul'taty issledovanija sostava materialov arkheologicheskikh obraztsov iz raskopok na Gonur Depe" [Results of the studies on the materials of the archaeological samples from the Gonur Depe excavations]. In *Trudy Margianskoj Arkheologicheskoj Ekspeditsii*, vol. 2, ed. V.I. Sarianidi. Moscow: Staryj Sad, pp. 192–196.

Rogersdotter, E. 2011. *Gaming in Mohenjo-Daro as archaeology of Unities.* Gotarc. Ser. B. Gothenburg Archaeological Theses No 55. Gothenburg: University of Gothenburg.

Rossi Osmida, G. 2011. *Aji Kui Oasis.* Vol. II, *The Citadel of the Figurines.* Venice: Il Punto.

Sarianidi, V.I. 2001. *Necropolis of Gonur Depe and Iranian Paganism.* Moscow: Mir Medija.

Sarianidi, V.I. 2002. *Margush: Ancient Oriental Kingdom in the Old Delta of the Murghab River.* Ashgabat: Türkmendöwlethabarlar.

Sarianidi, V.I. 2005. *Gonur Depe, Turkmenistan: City of Kings and Gods.* Ashgabat: Türkmendöwlethabarlar.

Sarianidi, V.I. 2006. "'Tsarskij nekropol' na Severnom Gonure" [A "royal necropolis" at Gonur North]. *Vestnik Drevnej Istorii* 2: 155–192.

Sarianidi, V.I. 2007. "Dvortsovo-khramovyj kompleks Severnogo Gonura" [The palace-temple complex of Gonur North]. Rossijskaja Arkheologija 1: 49–63.

Sarianidi, V.I. 2008. *Zoroastrianism: A New Motherland for an Old Religion.* Thessaloniki: Kyriakidis Brothers.

Sarianidi, V.I. 2009. *Marguş, Türkmenistan: Mystery and Truth of the Great Culture.* Ashgab at: Türkmendöwlethabarlar.

Sarianidi, V. and N. Dubova N. 2010. "Nouvelles sépultures sur le territoire de la 'nécropole royale' de Gonur Depe." *Arts Asiatiques* 65: 5–26.

Sarianidi, V.I. and N.A. Dubova. 2013. "Mozaika III tys. do n.e. v Gonur Depe (Turkmenistan)" [Mozaics of the third millennium BC at Gonur Depe (Turkmenistan)]. In *Poslednij entsiklopedist: K jubileju so dnja rozhdenija B.A. Litvinskogo* [The last encyclopedist: volume in honor of the 90th anniversary of Boris Litvinskij], ed. G. Ju. Kolganova, A.A. Petrov, and S.V. Kullanda. Moscow: Oriental Institute, pp. 379–413.

Sarianidi, V.I. and N.A. Dubova. 2015. "Kul't zhivotnykh v Margiane" [Animal cult in Margiana]. In *Akademicheskoe vostokovedenie v Rossii i stranakh blizhnego zarubezh'ja (2007–2015): Arkheologija, istorija, kul'tura* [Academic oriental studies in Russia and the near foreign countries (2007–2015): archaeology, history, culture], ed. V.P. Nikonorov and V.A. Alekshin. St. Petersburg: OOO "Kontrast," pp. 521–558.

Schädler, U. and A.-E. Dunn-Vaturi 2009. "Board Games in Pre-Islamic Persia." *Encyclopædia Iranica.* www.iranicaonline.org/articles/board-games-in-pre-isl.

Shaposhnikova, T.A. 2012. "Mozaiki i rel'efnyj dekor sten tsarskogo nekropolja Gonur-Depe" [Mosaics and relief decoration of the walls of the "Royal Necropolis" at Gonur Depe]. In *Trudy Margianskoj Arkheologicheskoj Ekspeditsii*, vol. 4, ed. V.I. Sarianidi. Moscow: Staryj Sad, pp. 170–175.

Tosi, M. and R. Wardak 1972. "The Fullol Hoard: A New Find from Bronze Age Afghanistan." *East and West* 22(1–2): 9–17.

Veresotskaja, G.E. 2012. "Restavratsija fragmenta stenki lartsa-darokhranitel'nitsy i issledovanie tekhniko-tehnologicheskikh osobennostej materialov mozaiki kontsa III tys. do n.e. iz tsarskoj grobnitsy No. 3230 Gonur Depe (Turkmenistan)" [Restoration of the wall fragment of the casket ostensory, and study of the technical and technological characteristics of the end of third millennium BC mosaic materials from the royal tomb 3230 at Gonur Depe]. In *Trudy Margianskoj Arkheologicheskoj Ekspeditsii*, vol. 4, ed. V.I. Sarianidi. Moscow: Staryj Sad, pp. 185–188.

Veresotskaja, G.E. 2016a. "Issledovanie i restavratsija mozaichnogo dekora stenki lartsa-darokhranitel'nitsy kontsa III tys. do n.e. iz tsarskoj grobnitsy 3230 Gonur-depe (Turkmenistan)" [Research and restoration of mosaic wall decoration of the casket ostensory of the end of the 3rd millennium BC from the royal tomb of Gonur Depe

(Turkmenistan)]. In *Trudy Margianskoj Arkheologicheskoj Ekspeditsii*, vol. 6, ed. N.A. Dubova. Moscow: Staryj Sad, pp. 315–323.

Veresotskaja, G.E. 2016b. "Sarianidi Viktor Ivanovich: Korotkie vstrechi" [Sarianidi Victor Ivanovich: brief meetings]. In *Trudy Margianskoj arkheologicheskoj ekspeditsii*, vol. 6, ed. N.A. Dubova. Moscow: Staryj Sad, pp. 178–187.

Vidale, M. 1990. "Early Harappan Steatite, Faience and Paste Beads in a Necklace from Mehrgarh – Nausharo." *East and West* 39(1–4): 291–300.

Vidale, M. 2007. "Il Ruolo dei Bracciali nella Civilta della Valle dell'Indo (2600–1900 A.C.)." In *Sulla Via delle Oasi. Tesori dell'Oriente Antico*, ed. G. Ligabue and G. Rossi Osmida. Padova: Il Punto, pp. 242–249.

Woolley, C.L. 1934. *The Royal Cemetery: A Report on the Predynastic and Sargonid Graves Excavated Between 1926 and 1931. Ur Excavations.* 2 vols. London and Philadelphia: Joint Expedition of the British Museum and the University Museum, University of Pennsylvania.

Zajtseva G.I., N.A. Dubova, A.A. Sementsov, P. Reimer, J.-P. Mallory, and H. Jungner 2008. "Radiouglerodnaja khronologija pamjatnika Gonur-Depe" [Radiocarbon chronology of Gonur Depe]. In *Trudy Margianskoj Arkheologicheskoj Ekspeditsii*, vol. 2, ed. V.I. Sarianidi. Moscow: Staryj Sad, pp. 166–179.

CHAPTER TWELVE

ANIMAL BURIALS AT GONUR DEPE

Robert M. Sataev

INTRODUCTION

One of the characteristics of Gonur Depe is the presence of various ritual structures with animal remains. They have been found in human graves, in symbolic burials or cenotaphs, in "animal burials," and in sacrificial places (Sarianidi and Dubova 2008, 2015; Dubova 2012, 2015). Animal remains can be represented by individual bones ("meat offerings" in burials and in sacrificial places), entire skeletons or body parts (including skulls, limbs, horns), or charred bones after the cremation of whole carcasses.

Especially interesting are the so-called "animal burials" (i.e., archaeological structures where whole animal skeletons have been discovered). The main difference with the other forms of buried animal remains is that, in this case, whole animals have been intentionally buried. They have been discovered either in their natural anatomic order, or cut into pieces, or cremated. However, not all these structures are "animal burials" in a literal sense. Usually, the animals were not the main goal of the ritual but rather part of it. Nevertheless, the very fact that whole animals were used for rituals points at the significance and uniqueness of these structures and of the events linked with them. Furthermore, the use of whole animals demonstrates the wealth and high living standards of the ancient Gonur population who could sacrifice part of its valuable property.

ANIMAL BURIALS IN THE ANCIENT EAST

The use of whole animals in rituals is known at other places in the ancient East. Usually, animals are buried within human graves. At Mehrgarh, in the preceramic Neolithic period, burials that contained several goat skeletons with one individual have been discovered (Lechevallier *et al.* 1982). A camel burial was discovered at Mohenjo-daro (Meadow 1984). In a recent overview, J.L.R. Soldado (2016) has presented the different ways animal remains have been discovered at Near Eastern sites. He mentions several cases when a whole carcass has been found: animal burials inside human graves, animal burials associated with human graves, independent animal burials (i.e., not related to human graves), or animals in building foundations. Different animals were used in these rituals: equids (donkey, mule, onager), small cattle (sheep and goat), cattle, and pigs.

Animal burials related to human graves are the most widespread. G.M. Schwartz (2012) identifies four forms of using equids in ritual practice at the Syrian site of Umm el-Marra. In three of them the whole animal is concerned, in the last one, only parts of it. In the first case (type 1), the equids were placed between the tombs in a standing position. In the second (type 3), there was a pit burial for a group of animals. In the third one (type 4), equids were buried separately at different distances from the tombs and ritual complexes. All these animal burials somehow were connected with the human tombs.

According to B.A. Al-h (2012), the ritual of animal burials at the entrance of tombs probably has its origins in Mesopotamia, from where it spread during the Middle Bronze Age to Syria (see above) and Palestine (Jericho, Lachish, Tell Acco, Tell Harar, Tell El-Adjul) and finally to Egypt where a whole set of animal sacrifices have been discovered among the burials of two necropolises at Tell el'Dab'a (Avaris): donkeys (often in pairs) and rams and goats, placed either in the entrance pit, or in a separate pit at the tomb entrance (Bietak 1996; Forstner-Muller 2010). A ritual burial of a donkey with a metal horse-like harness dated to 1700/1650–1550 BC has been found at Tell Harar in Israel (Bar-Oz *et al.* 2013).

ANIMAL BURIALS IN CENTRAL ASIA

In southern Central Asia, to our knowledge, animal burials are exclusively related with the Bactria-Margiana Archaeological Complex (BMAC) period and have been found at several places.

At Dashly 1 (Afghanistan), the skeletons of young lambs surrounded by a large number of vessels were found in two burial pits (Sarianidi 1977: 55). In this case, the animals are supposed to replace a missing human being, since a lamb breast was placed in front of one of them as an afterlife meat offering. At Togolok 1 (Turkmenistan), in the eastern complex of rooms of the fortress, the burial of a "priestess" was excavated, according to Sarianidi (1990b: 164). The large burial pit contained two bulls and their driver, together with the main deceased and various gifts. At Togolok 24, in the cemetery, built graves contained lambs (Sarianidi 1990a: 52), and, in one case, a whole lamb skeleton was found outside the grave. At Sapallitepa (Uzbekistan), A. Askarov (1977: 42) has excavated four graves where there was no human remains but sheep and goat skeletons in the same positions as humans and surrounded by multiple clay vessels with food remains. At one of the cemeteries of Dzharkutan, in graves related to the Molali phase, two animal burials were also found (Askarov and Abdullaev 1983: 14). In another more or less contemporary cemetery, Buston VI, several burials with whole animal skeletons have been discovered, either with humans (12 cases) or without them (also 12 cases) (Avanesova 2016). The animals are mostly sheep (of both sex and various ages), but two contained dogs (one dog buried alone in grave 327, and two dogs with one individual in grave 79), and only in one case an animal (lamb) was buried without any associated gifts. At Buston VII, a burial with the cremated remains of a young woman also contained the whole skeleton of a lamb placed on its belly, with its legs bent, and its head oriented to the east (Avanesova and Kasparov 2015: 33). Finally, it is noteworthy that a camel burial was discovered at Mohenjo-daro (Meadow 1984), most certainly in relation with the BMAC.

As for Gonur Depe, in a previous study (Sataev 2008), we have already pointed out that there were three main types of animal burials: pieces of meat as a food offering, whole carcasses of animals accompanying the deceased, or full animals buried independently. We have also underlined that Gonur Depe was by far the place with the most numerous cases of animal burials ever found in Central Asia. In a more recent synthesis, E. Luneau (2014) came to similar conclusions and related the number of animal burials to the presence of an important elite at the site.

However, Gonur Depe is not only unique in the number but also in the diversity of the animal burial rituals. Altogether (excluding satellite sites), 64 ritual structures have been discovered at this site that involve whole animals, and their number was probably significantly higher, as shown by the study of destroyed structures that probably contained whole skeletons but for which we lack the evidence. In some structures, the presence of a whole animal skeleton could only be confirmed by the meticulous analysis of the skeletal remains.

As already mentioned, these animal burials do not constitute a separate, individual group of ritual structures, but are attributed to different kinds of rituals using animals.

Three main types have been identified, subdivided into groups, and will be reviewed here:

1. *Animals used for funeral and commemoration rituals*: They are placed in the graves along with various rituals. Two groups have been identified:
   (a) Whole animal(s) buried in graves of different construction types (pit, cist, chamber tomb) with specific arrangements (like a "household courtyard"), or buried outside the grave, usually at the entrance.
   (b) Whole animals buried within the burial construction in exact accordance with the human burial ritual.
2. *Individual burials of whole animals*: Here too, two groups can be distinguished:
   (a) Whole animal(s) buried in a specially prepared pit.
   (b) Dismembered parts of a whole animal buried in a specially prepared pit in a specific order.
3. *Burials of cremated whole animals*: Either the bones are collected and placed into containers (holes or vessels), or the skeleton is left at the place of cremation and covered with ash (the anatomical structure being then partially preserved).

These various forms of rituals, however, cannot always be clearly distinguished. It may be difficult, for example, to separate burials where the animals take the place of human beings (cenotaphs), from those where they are considered as burial gifts, and/or from individual animal graves. Thus, in the absence of human remains, animals from similar graves are sometimes considered as "substitutes," and in others as "companions." The first ones concern only the animals placed in graves where the same procedure as for humans was followed.

WHOLE ANIMALS USED FOR RITUALS LINKED WITH A GRAVE CONSTRUCTION (GROUP 1)

In most cases, at Gonur Depe, whole animals accompany the construction of graves (26 examples). These graves are of various types: pits, cists, and chamber tombs, and the range of animal species found in them is also quite varied: sheep, goat, cattle, pig, camel, donkey, and dog. In the simple burials (pits, cists) the most common animals are small ruminants, while in the more complex ones or "elite" tombs, cattle, dogs, donkeys, and camels are frequent.

All the animals attributed to these structures should probably be regarded as burial gifts for the afterlife. These can be divided into animate (animals, plants, people) and inanimate items (tools, utility artifacts, vehicles) (Smirnov 1997: 36, 73). The animal(s) placed in the grave or buried nearby is/are supposed to follow the deceased into his afterlife and to be part of his property (small ruminants and cattle symbolizing the "herd"; camel, donkey, horse for "transport"; dog as "guardian of the house and herd"). According to Smirnov's classification, the animals located within the grave, or within the main burial chamber in a multichambered structure (i.e., in close proximity to the deceased), are considered as "escorting items." When they are outside the pit or chamber, but within the burial construction (i.e., at the entrance), they should be viewed as "accompanying items." If further away, then they are classified as "neighboring items." In complex burial rituals of high social status (like at the Royal Cemetery of Ur, for instance), human beings can also be used as "living burial items," therefore it should be underlined that the animal burial must be considered in relation to the "main deceased" ("grave's owner"); that is, the person whose burial was the purpose of the ritual, and for whom the grave and/or accompanying structures were made.

Animals as "escorting items" (group 1a)

It has first to be underlined that burial constructions containing whole animal skeletons without human remains or without the "main deceased" have their own specific features in each area: They can vary in the construction type, in the choice of the animal species, or in the place and way the animals were positioned. Their common feature is the use of "living burial items."

In Area 16, one of the chambers of *tomb 3830*, built as a small above-ground double-room construction, contained one individual, while the other chamber contained one adult sheep (over 3.5 years old) and, under its belly, a lamb (3 months old). Together with the sheep, the chamber also contained ceramic vessels with remains of meat offering.

In Area 19, in the northern architectural building complex dated to period II, the most characteristic structure is *tomb 4155*, where there were no human remains but that contained the whole skeletons of a donkey, a dog, and a sheep at the entrance of the courtyard along the southern wall of the underground mudbrick cist. Although the grave was robbed in ancient times, and the initial order in which the animals had been laid inside the burial was partially disrupted, this can be reconstructed on the basis of the position of the bones. The donkey was first placed into the grave with its legs bent under the belly. The sheep was lying on its back at the front legs of the donkey, and, finally, above the head and legs of the donkey and above the sheep was placed a dog. The three animals symbolize "transport" (donkey), "herd" (sheep), and "guard" (dog).

In the same area, *tomb 4150* stands out by its construction and the number of animal species discovered. In the center of a nearly square pit/depression, between the western and eastern walls, was a trail of bones forming two main heaps – "western" and "eastern." Their careful analysis shows without any possible doubt that several whole animals were buried there: a donkey, a dog, a pig, two lambs, and two adult sheep. Furthermore, the partially preserved skeletal remains still in anatomical order allow us to reconstruct the positions of the animals at their burial. In the center of the pit was a donkey, with its head to the west and legs to the north. Two lambs were placed on top of each other at the back legs of the donkey (closer to the northern wall). Behind the back of the donkey, one on top of the other, were two sheep. A dog and a pig were placed at the front legs of the donkey, their heads facing the northern wall. This burial of a pig, with a more recent find, are the only two known cases of burials of these animals at Gonur Depe.

In this Area 19 of the settlement, it seems therefore that the ritual structures were related with the simultaneous ritual use of different animal species, at least a donkey, a dog, and a sheep.

Within the "Royal Necropolis" (Area 8, in the southeastern part of Gonur North) dated to period I (see Chapter 10), a significant number of burial constructions containing whole animal skeletons have been discovered. In the multichambered underground tombs or hypogea, also called "houses of the dead" (graves 3200, 3210, 3880, 3905, and 3915) either the remains of the individual(s) to whom these structures were devoted are lacking, or they were brought in only later on (but see Chapter 10).[1] Whatever the case, in these tombs, whole animal skeletons have also been discovered. It seems, thus, that these "houses of the dead" were not made as a burial site, but were more like some kind of ossuary. Moreover, according to their disposition, the human remains found in these tombs seem to be those of servants and probably played the role of "living burial items" like the animals (see Chapter 10). Since sometimes, like in tomb 3830, the animals are placed in the other rooms of the tomb, and, therefore, are not in the immediate proximity to the deceased, they should formally fall into the category of "accompanying items." However, considering the general purpose of such structure, both the animals and the people are supposed to accompany the "owner" of the tomb into afterlife even in his absence.

The most famous is the "royal" *tomb 3200* described in detail by N.A. Dubova (2004) (Figure 12.1). The tomb was built as an imitation of a residential dwelling with its neighboring "courtyard." A horse(?)[2] skeleton was found in the northeastern corner of the courtyard, next to the remains of a four-wheeled wagon, while a dog skeleton was found to the left of the courtyard entrance. At the eastern edge of the courtyard lay a whole camel skeleton with its legs facing outside. Remains of three human individuals were located next to it. In the northwestern corner of the courtyard was a chaotic pile of camel bones, probably from another whole skeleton.

In the same area, *tomb 3210* contained the remains of 11 human individuals.[3] According to Dubova (Chapter 10), four of them are the "main deceased." Skeletons of two sheep, one of a calf, a camel, and a dog were found with them. *Tomb 3225* consists of a round pit and contained the remains of ten individuals with a four-wheeled wagon and two camels (Figure 12.2). In *tomb 3240*, 17 individuals were discovered in a rectangular pit together with parts of wagon wheels. The wagon was possibly destroyed when the grave was robbed and the bronze wheel tires were stolen. A whole calf, two sheep, two camels, and two dogs were also found in the tomb. In *tomb 3280*, the skeletons of a sheep and two dogs were discovered in a

Figure 12.1 Gonur Depe "Royal Necropolis", hypogeum 3200 during excavation. General view from the southwest. © Margiana Archaeological Expedition.

rectangular pit together with the remains of two individuals. In *tomb 3880*, the whole skeletons of a dog and a camel were also found.

Another of these noteworthy structures is *tomb 3900*. It is a round-shaped pit, containing, among other items, a four-wheeled wagon and a large bronze cauldron together with seven individuals, seven dogs, two camels, and two donkeys (Figure 12.3). Judging from the position of the skeletons, the animals were slaughtered before they were thrown into the pit. It is evident that this burial was arranged for an individual with a high social status. As no remains of the "main deceased" were found, one can assume that this individual was buried in another place. In addition, on the surface of the burial, was the skeleton of a dog, tied and killed by a hit on the head. This dog burial was probably part of the rituals related to the same construction.

Finally, from this "Royal Necropolis" two more burials deserve special mention. *Tomb 3905* (of the type "house with a courtyard") contained the remains of two people in two separate chambers. None of them seems to be the "main deceased" of the tomb (see Chapter 10). In the southeastern corner of the northern chamber, the remains of a young camel were found on the floor. The skeleton is lying on its right side, head to the west, front limbs against the southern wall of the chamber, and back limbs in the corner. The neck of the animal is bent to the left, and the skull lies on its left side, its base pointing to the north. From the positions of the head and the body, it can be assumed that the animal was killed by cutting its throat when the head of the

Figure 12.2 Gonur Depe "Royal Necropolis", grave 3225 during excavation. General view from the west. © Margiana Archaeological Expedition.

Figure 12.3 Gonur Depe "Royal Necropolis", grave 3900 during excavation. General view from the north. © Margiana Archaeological Expedition.

camel was bent back and leftwards by the left arm of the person standing behind it (or possibly sitting on its back), while his right arm was cutting its throat. The animal had been killed on the spot where it was discovered. It is worth noting that immediately before death, the camel had been in its natural resting pose with its legs bent. The animal was then brought into the tomb and killed there.

In the multichambered *tomb 3915*, three heaps of animal bones belonging to two camels were discovered on the floor of the northernmost and largest chamber. The heaps are the result of the grabbing of the tomb when the thieves dismantled the skeletons, throwing the bones aside trying to reach valuable artifacts. The robbery happened sufficiently long after the burial had taken place, at least when the animals were already completely decayed. Considering the remains of a wooden arc-shaped piece and of numerous copper salts on the camel bones, one can assume that a four-wheeled wagon was present in the tomb. In addition, the whole skeleton of a dog was located along the northern wall of the chamber. It is possible that, when the tomb was re-dug for grabbing, the dog skeleton happened to be re-buried under the collapsed earth, which helped preserve it. Anyhow, one can assume that when the tomb was built, two camels harnessed with a wagon, and a dog, were buried there.

Animals as "substitutes" to humans (group 1b)

Graves where animals are buried like humans are also of various construction types. At Gonur, eight tombs are classified as such. Most of them are burial pits or cists and contain sheep skeletons but no human remains. Such burials usually contain

Figure 12.4 "Ritual animal burial" in Area 16 at Gonur Depe.
© Margiana Archaeological Expedition.

various artifacts (ceramic vessels, stone and metal items), and the animals can also wear jewelry. As a whole, such structures are usually interpreted as symbolic burials (cenotaphs). It is believed that, in this case, the animal symbolizes the deceased who is missing for unknown reasons. However, one cannot rule out that some of these structures may have played the role of sacrificial places (see p. 395). The sacrifice could be a tribute to a specific dead person, or meant to remove disease or death from a living person, or also could be a gift to supreme forces.

Among these structures one complex in Area 16 stands out by its scale. It contained four whole animals (a donkey and three sheep), together with a wealth of funeral offerings (Sarianidi and Dubova 2008; Dubova 2012). The burials are situated in a pit/depression, internally divided by walls of sun-dried bricks forming four separate chambers (*tombs 3597, 3621, 3622, 3623*) (Figures 12.4 and 12.5). The young sheep were buried in three parallel chambers (tombs 3621, 3622, 3623). In tomb 3622 the animal's skull was missing. Judging from the damage to its atlas, the head had been cut off by a flat tool with a sharp edge (possibly, a large knife). It seems that the head of the tied animal had been cut off while it was held back and received a straight chopping blow on the neck, which gave maximum damage to the right side of the front glenoid fossae of the atlas and the left occipital condyle. The head of the animal may have been buried separately elsewhere afterwards: The skull of a hornless sheep cut off in the exact same manner was buried intentionally in tomb 3622 excavated on

Figure 12.5 Gonur Depe grave 3597 (view from the south), with the burial of a donkey and three lambs (near its tail). © Margiana Archaeological Expedition.

the eastern edge of the large basin at Gonur Depe (Sataev 2008). Behind the southern wall of the three chambers with sheep, a donkey (tomb 3597) was found together with three lambs at its feet (aged up to 5 months). Originally, the southern wall had entrances into each of the sheep tombs, which were later sealed with mudbricks.

In an interview with the *Neutral Turkmenistan* newspaper in 2006, V.I. Sarianidi considered this structure as a sacrificial place, maybe dedicated to the *fravashi* or to the soul of ancestors, or gods of the other world. It had an undeniable special purpose as shown both by its monumental size and by its position on the right bank of a branch of the Murghab River opposite to the main necropolis.

It is also possible to view this complex as a "substitute," one of the forms of symbolic burial, when animals are buried to remove disease away from a human (the disease is "transferred" to the sacrificial animal), or to deliver it from a hex (Sataev 2008). Burials where the heads of sacrificial animals have been cut off are often considered as such (Danilov 1982). In the present case, considering the scale of the structure, it could have been related to a ritual of the "renewal of the king" or a person of high social status.

However, one cannot rule out either that the burials of the rams were cenotaphs, and that the animals symbolized missing persons (who, for example, had died far away). In this case, considering the monumental scale of the structure, these persons could have been representatives of the local elite, and the rams played the role

Figure 12.6 Gonur Depe cist grave 4140, with a young sheep in one of the rooms; three more sheep were discovered on the ground behind the southern wall of the tomb. © Margiana Archaeological Expedition.

of "the main deceased" while the donkey with the lambs were the "accompanying offerings."

In Area 9, *tomb 4140* deserves a separate discussion. It is a single chamber tomb made of sun-dried mudbricks, covered with a vaulted roof (Figure 12.6). Inside the construction, along its western wall, a young lamb was placed on top of three mudbricks. Skeletons of three other sheep of various ages were discovered on the floor behind the southern wall of the tomb, at the entrance. An adult sheep lay arch-bent on its left side with, at its feet, the skeleton of a lamb lying on the right side, its head facing the tomb entrance. Such double burials combining an adult and a young animal were a common practice at Gonur Depe and could be a symbolic wish for the herd to develop. Under the two skeletons, a third one was in anatomical disorder. This means that the body of the third animal had been buried significantly earlier and had decayed before the other two were placed on the floor in front of the tomb. When the fresh offerings were brought in, the bones of the previous one that probably hampered the access to the tomb were swept aside. Thus, one can assume that the tomb was used at least twice with a time gap in between.

Finally, within this category of "substitute" animal burials, there are also cases when human bones completely or partially cover up the decayed skeletons of small ruminants. In Area 5, *tomb 2900*,[4] for instance, the deceased was intentionally placed on top of a young sheep (Sarianidi 2007: 146–155). In two other burials (*tomb 3894*, Area 8; and *tomb 4449*, Area 12), both the position and the state of preservation of the animal bones (a sheep) show that it had been buried long before the human individual. In these cases, the human burial was clearly secondary, and the original burial was a cenotaph.

INDIVIDUAL ANIMAL BURIALS WITH WHOLE ANIMALS (GROUP 2A)

Some 27 of these graves have been found at Gonur Depe. They differ from the previous ones because the archaeological context shows that they cannot be directly related with human burials. However, one cannot rule out the possibility that such links existed. Thus, it is impossible to be always sure about the category of these burials. Most of the time, the individual animal burials are quite simple in their construction, usually simple pits, though cists are also attested, like *tomb 3790*, Area 16, which contained a whole young goat skeleton (less than 10 months old).

Bull burials

Among them, *tomb 3895* from the "Royal Necropolis" is particularly interesting. In a pit/depression a whole bull skeleton was excavated (Figure 12.7). The animal lay on its left side in a "running pose." Apparently, the bull had been killed by a hit on the head, dragged down into the pit, and then carefully laid out. The size of the bull is quite large for the Bronze Age (136 cm at wither). The specimen had relatively small

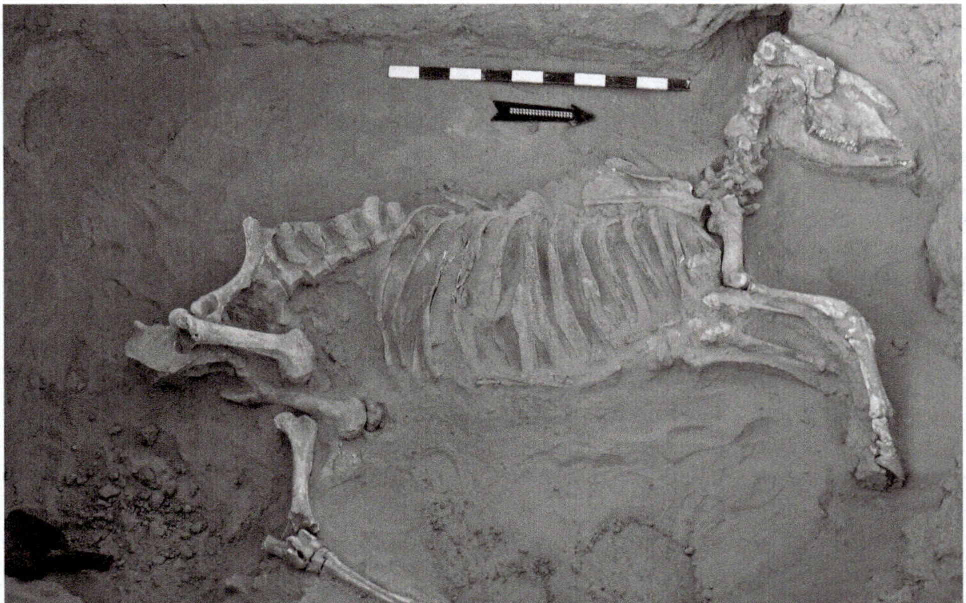

Figure 12.7 Gonur Depe grave 3895, with a complete bull burial. © Margiana Archaeological Expedition.

horns (length of the horn pith along its major curve 210 mm), growing backwards from the skull and then slightly curving upwards.

At Gonur Depe, bovids in rituals are significantly less common than small ruminants, and this may testify of their special status and of a hierarchy in sacrificial animals. Furthermore, bovids were significantly more valuable than sheep, since they could be used in different ways (milk, meat, pack or draft animals) and also since it took them longer to reach their mature size. Some consider that cult animals were the most significant species in the economy (Kuz'mina 1981). The image of a divine cow – symbol of abundance of all goods – is well represented in the Indo-Iranian epos (Lelekov 1979). Bull cult is connected with the ideas of male power and fertility (Bulatova 1988). In Hittite rituals, bovids were sacrificed for cult celebrations, and, depending on the sex of the deity, a cow or a bull was chosen. Bulls could also play the symbolic role of a ruler (Ardzinba 1982).

Thus, this burial could be considered as a sacrifice. It remains unclear, though, whether it was devoted to divine powers, or was a targeted tribute ("individual sacrifice"). It is possible that it is linked with some other significant burial, as it is located within the boundaries of the elite "Royal Necropolis."

Dog burials

Dog skeletons stand out among other individual animal graves. Usually dogs are buried in pits of various shapes. At the Gonur Large Necropolis, dog burials are found in pits previously fired. There is also a case of a dog buried in a cist (*tomb 3600*, Area 13). Both individual and multiple dog burials are attested at the site. Among the last ones, *tomb 3813* in Area 16 contained two dog skeletons of both sexes in a copulating pose. A group of dog burials discovered at the border of the "Royal Necropolis" raises questions. It is the case of *tomb 4065* at its eastern border, which contained seven dogs of various sex and age (Sataev and Dubova 2016) (Figures 12.8 and 12.9). A detailed study has shown that the dogs had died at various times and for different reasons, so the burial cannot be regarded as the result of a massive simultaneous animal sacrifice. Other dog burials found in this area contained from one to four animals. From the study of these specific dog burials we can conclude that all were made in so-called "empty graves" or "temporary graves," left empty after the human remains they had first contained had been removed.[5] The dogs were of both sex (but mostly females), of various ages (from 3 months to 6 years), and belonged to different morphological and functional species; they had died for various reasons, and, in some cases, were buried several hours (or days) after their death. Thus, the only common features in all these burials are their localization (at the border of the elite burial ground) and the use of "empty graves" for dog burials, while the arrangement within each burial is different in each case.

The motivation behind such individual dog burials in general, and in this elite area specifically, remains open for discussion. Sarianidi thought that the dog had an honored but also ambivalent place in the beliefs of the BMAC people (Sarianidi 2010). This is supported by the great number of individual dog burials at the site (whole skeletons or parts of them), and their use in burial rites. Several explanations have been advanced (Sataev and Dubova 2016): Either the possible special status of dogs in the religious beliefs demanded that dead animals be buried; or individual dog

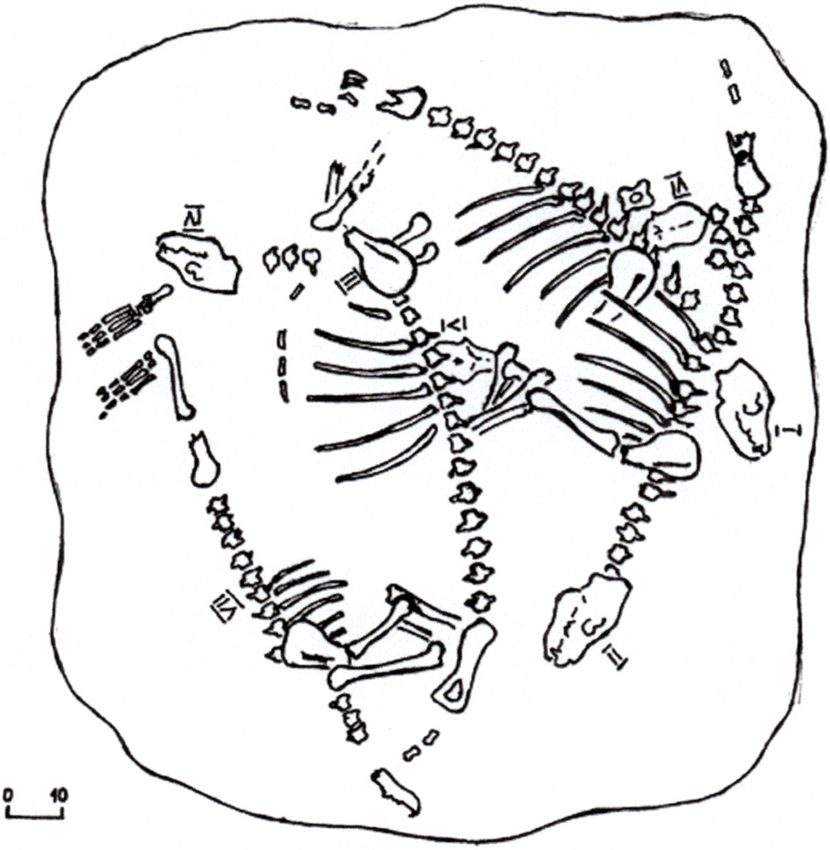

Figure 12.8 Gonur Depe "Royal Necropolis". Drawing of the upper layer of grave 4065, with the burial of seven dogs. © Margiana Archaeological Expedition.

burials could be part of some ritual, possibly a form of sacrifice adopted in the local cults. As for the dog burials at the edge of the "Royal Necropolis," they might have symbolized protection and watching over the remains of the representatives of the ruling elite at Gonur Depe.

DISMEMBERED ANIMAL SKELETONS (GROUP 2B)

Besides whole animals burials there is one case of an individual burial with a totally dismembered cattle skeleton at the "Royal Necropolis" *tomb 3890*. So far, it is the only case known at Gonur Depe, where the body was dismembered and its parts placed into the grave afterwards in a specific order. The grave is an irregular-shaped pit, slightly elongated from north to south (155 cm x 130 cm), and about 40 cm deep. No accompanying artifacts were found. The bones belong to a young cow (2.5–3 years old). The skull demonstrates intravital damage in the area of the nasal and joining frontal bones, so it seems that the animal was killed by a hit on the head with a heavy object.

Figure 12.9 Gonur Depe "Royal Necropolis", general view of the upper layer of grave 4065 during excavation. © Margiana Archaeological Expedition.

The cow had been carefully dissected at the articulations, without chopping or splitting, and the different parts were afterwards put into the grave in a specific sequence. The way the dissection was done indicates that the animal had been initially flayed (the distal limb segments remained with the body, hide intact) and gutted (ribs were separated along their joint surfaces). Probably, the meat was emaciated from the bones. Part of the rib cage is missing so that the buried skeleton numbers only 16 ribs, of which two front left ribs had not been separated so the joints with the vertebrae were preserved, while the four front right and the three last right and left ribs are missing. It is quite probable that these ribs were removed on purpose. The pelvis was split into two halves. The four whole limbs were separated. From the disposition of the animal skeletal elements in the grave, it is clear that they follow a specific order. The spinal parts were placed first, then the head, then the ribs. Simultaneously (or not long after), the left front and back limbs were placed into the pit.

The way the carcass was dissected and its parts disposed of in the pit speak of a complex ritual act. The skill and accuracy of the butchering indicate that the person who performed it had a good knowledge of the animal anatomy and a good experience of such work. After the animal had been brought towards the prepared pit and killed, it was butchered, flayed, and dismembered. Interior organs and meat were removed, and most probably, used for cooking ritual meals. Then, its bones were placed into the grave in a specific order, possibly that in which the pieces were cut off and cleared of the meat. We consider it probable that this burial was a true sacrifice, where the animal was devoted to divine powers.

The practice of burial of dissected animals may have been more common than is reflected in the archaeological finds. In the case described above, only the good state of preservation of the burial allowed its recognition and detailed study. It cannot be excluded that other skeleton parts found at Gonur Depe come from destroyed burials similar to this one.

CREMATION (GROUP 3)

The last form of recognized rituals using whole animals is cremation. In most cases, the bone remains from cremated animals discovered at different places in Gonur North were placed into especially prepared small pits ("sacrificial holes") or in ceramic vessels. The analysis of the charred bones found in the fills of these holes and vessels attest that they come from a whole sheep or goat (Sataev 2010; 2016: 134–137). The cremation probably took place in another place. After burning, the charred bones were gathered and brought to these holes or vessels.

In addition, whole charred animal skeletons have been found at Gonur, still lying where they had been cremated and buried under ashes. Thus, in Area 11, on the floor bordering *room 45* from the south (northern part of the "place of communal eating"), the remains of six cremated small ruminants (five sheep and one goat), varying in sex and age (from 6 months to 3.5 years) were found in a layer of ash. The skeletons were well preserved due to the fact that they had been purposely buried in the ash almost immediately after cremation. All the animals were orientated with their head to the north. The analysis of the disposition of the bones shows that the animals were placed on their bellies. Judging from the degree of cremation of the bones, the temperatures reached were above average and/or the procedure took a long time. The animals had been cremated simultaneously.

The scale of the event, with several animals cremated, is surprising. It could either be part of a ritual or the result of more mundane actions, like the burning of animals that had died of disease, but, nevertheless, the reason for it must have been quite extraordinary since it was very demanding from the material point of view (besides the animals, it required much wood or dung for fuel). It is possible to figure out the season when the event was performed. If we estimate that the age of the young specimen was 6 months, knowing that the birth time of small ruminants was probably late February–March as nowadays in southeastern Kara Kum, then the animals were burnt in late July–August. These are the hottest and driest months in the Kara Kum. A possible explanation for the ritual would then be that the channels of the ancient Murghab River that provided water to the city had dried up, especially if drought was general and affected also the Hindukush Mountains, sources of the river. The ritual would then be an address to the gods for water.

Remains of a cremated sheep (between 2 and 3 years old) were also discovered in an ashy layer in *room 7*, Area 17. Judging from the disposition of certain elements of the skeleton, the animal was left on the spot where it was burnt, and was covered with ash shortly after. The majority of the bones are burnt to a gray color (i.e., at medium temperature), but their carbonization is uneven, so the animal was probably burnt on an open ground covered with firewood, at a low and unstable temperature, and over a relatively short time. It does not seem that the sacrifice is connected to the building standing next to the burial, since the ash layer is below its floor level and wall foundations, and is significantly spread outside the walls of the room. Clearly, the building was constructed later over the ashes of the sacrificial place that were already partially covered with compacted fluvial silt at the moment of its construction. Despite the difference in scale of this event with the previous one, it shares some similarities with it and points at the popularity of such sacrificial rituals within the ancient population.

To sum up, there is clear evidence that the population of Gonur Depe widely practiced animal burial rituals, including whole animals. All kinds of domestic animals were used. They played either the role of living burial items, were substitutes for dead persons, or used for sacrifices. Thus, rituals involving the use of animals were an integral part of the religious beliefs of the Gonur residents.

NOTES

1 The human bodies are supposed to have been placed outside the graves for decomposition and their bones brought back afterwards.
2 This identification was proposed at the time of the excavations when no archaeozoologist was present and, due to the more frequent presence of donkeys at the site, doubts could be raised. However, we believe that an animal tooth from this tomb is undoubtedly from a horse.
3 For the radiocarbon dates of tombs 3210, 3240, 3900, and 3915, see the Appendix at the end of the book.
4 For the radiocarbon dates of tomb 2900, see the Appendix at the end of the book.
5 This is a deduction from the remains of small human bones left at the bottom of the graves, while the animal skeleton was complete.

REFERENCES

Al-h, B.A. 2012. "Avaris i Gonur" [Avaris and Gonur]. *Vestnik RUDN, ser. Vseobshchaja istorija* 1: 19–29

Ardzinba, V.A. 1982. *Ritualy i mify drevnej Anatolii* [Rituals and myths of ancient Anatolia]. Moscow: Nauka.

Askarov, A. 1977. *Drevnezemledel'cheskaja kul'tura epokhi bronzy juga Uzbekistana* [Ancient Bronze Age farmer culture of southern Uzbekistan]. Tashkent: Fan.

Askarov, A.A. and B.N. Abdullaev 1983. *Dzharkutan (K probleme protogorodskoj tsivilizatsii na juge Uzbekistana)* [Dzharkutan (on the question of a proto-urban civilization in southern Uzbekistan)]. Tashkent: Fan.

Avanesova, N.A. 2016. *Buston VI: nekropol' ognepoklonnikov dourbanisticheskoj Baktrii* [Buston VI: a necropolis of fire worshippers in proto-urban Bactria]. Samarkand: IICAS.

Avanesova, N.A. and A.R. Kasparov 2015. "Praktika vtorichnogo zakhoronenija nekropol'ja Buston VII" [The secondary burial practice at the Buston VII necropolis]. *Vestnik MICAI* 21: 27–54.

Bar-Oz, G., P. Nahshoni, H. Motro, and E.D. Oren 2013. "Symbolic Metal Bit and Saddlebag Fastenings in a Middle Bronze Age Donkey Burial," *PLoS One* 8(3): e58648. doi: 10.1371/journal.pone.0058648.

Bietak, M. 1996. *Avaris: The Capital of the Hyksos. Recent Excavations at Tell el'Daba'a.* London: British Museum Press.

Bulatova, A.G. 1988. *Traditsionnye prazdniki i obrjady narodov gornogo Dagestana v XIX–nachale XX veka* [Traditional holidays and customs of the people of the mountain Daghestan in the XIX–beginning of the XX centuries]. Leningrad: Nauka.

Danilov, S.V. 1982. "Ritual'nye zakhoronenija baranov v Zabajkal'e" [Ritual sheep burials in the Trans-Baikal region]. *Sovetskaja Arkheologija* 1: 229–233.

Dubova, N.A. 2004. "Mogil'nik i tsarskij nekropol' na beregah Bol'shogo bassejna Severnogo Gonura" [Graveyard and "Royal Necropolis" on the banks of North Gonur large pool]. In *U istokov tsivilizatsii* [Near the sources of civilizations]. Moscow: Staryj Sad, pp. 254–281.

Dubova, N.A. 2012. "Pogrebenija zhivotnykh v strane Margush" [Animal burials in Margush]. In *Trudy Margianskoj Arkheologicheskoj Ekspeditsii*, vol. 4. ed. V. Sarianidi. Moscow: Staryj Sad, pp. 101–139.

Dubova, N.A. 2015. "Animal Burials and Their Cults in Margiana." In *Defining the Sacred*, ed. N. Laneri. Oxford: Oxbow books, pp. 13–23.

Forstner-Muller, I. 2010. "Tombs and Burial Customs at Tell El-Dab'a during the late Middle Kingdom and the Second Intermediate Period." In *The Second Intermediate Period (Thirteenth–Seventeenth Dynasties): Current Research, Future Prospects – Orientalia Lovaniensia Analecta, 192*, ed. M. Marée. Leuven, Paris, and Walpole, MA: Peeters and Department of Eastern Studies, pp. 127–138.

Kuz'mina, E.E. 1981. "Slozhenie skotovodcheskogo khozjaistva v stepjakh Evrazii i rekonstruktsija sotsial'noj struktury obshchestva drevnikh pastusheskikh plemen" [The formation of the cattle-breeder economy in the Eurasian steppes and reconstruction of the social structure of the ancient pastoral tribes]. In *Materialy po khozjajstvu i obshchestvennomu stroju plemen Juzhnogo Urala* [Materials on the socioeconomical structure of southern Ural tribal groups], ed. N.A. Mazhitov. Ufa: BFAN SSSR, pp. 23–43.

Lechevallier, M., R.H. Meadow, and G. Quivron. 1982. "Depôts d'animaux dans les sépultures néolithiques de Mehrgarh, Pakistan." *Paléorient* 8: 99–106.

Lelekov, L.A. 1979. "Rannie formy iranskogo eposa" [Early forms of the Iranian epos]. *Narody Azii i Afriki* 3: 173–178.

Luneau, E. 2014. "Les dépôts funéraires animaliers de la civilisation de l'Oxus: diversité et singularité du rapport entre l'animal et la mort en Asie centrale méridionale à l'âge du Bronze (2300–1500 avant n.è.)." In *Rencontre autour de l'animal en contexte funéraire*, ed. I. Bede and M. Detante. Paris: Gaaf, pp. 195–209.

Meadow, R.H. 1984. "Camel Skeleton from Mohenjo-daro." In *Frontiers of the Indus Civilization*, ed. B.B. Lal and S.P. Gupta. Delhi: Books & Books, pp. 133–139.

Sarianidi, V.I. 1977. *Drevnie zemledel'tsy Afganistana. Materialy Sovetsko-Afganskoj ekspeditsii 1969 – 1974 gg* [Ancient farmers of Afghanistan. Materials of the Soviet–Afghan expedition 1969–1974 years]. Moscow: Nauka.

Sarianidi, V.I. 1990a. *Drevnosti strany Margush* [Ancient Margush]. Ashkhabad: Ylym.

Sarianidi, V.I. 1990b. "Sel'skij khram Togolok-1 v Margiane" [A rural temple at Togolok 1 in Margiana]. *Vestnik Drevnej Istorii* 2(193): 161–167.

Sarianidi, V.I. 2006. "Drevnjaja Margiana vzgljad iz XXI veka" [Ancient Margiana, a view from the 21st century]. Nejtral'nyj Turkmenistan, 25 September.

Sarianidi, V.I. 2007. *Necropolis of Gonur*. Athens: Kapon editions.

Sarianidi, V.I. 2010. *Long Before Zaratushtra*. Moscow: Staryj Sad.

Sarianidi, V.I. and N.A. Dubova. 2008. "Arkheologicheskie raboty na jugo-zapadnom kholme Gonur Depe" [Archaeological work on the southwestern hill of Gonur Depe (Area 16)]. In *Trudy Margianskoj Arkheologicheskoj Ekspeditsii*, vol. 2. ed. V. Sarianidi. Moscow: Staryj Sad, pp. 28–49.

Sarianidi, V.I. and N.A. Dubova. 2015. "Kul't zhivotnykh v Margiane" [Cult of animals in Margiana]. In *Akademicheskoe Vostokovedenie v Rossii i stranakh blizhnego zarubezhja*

(2007–2015): Arkheologija, Istorija, Kul'tura [Academic oriental studies in Russia and the neighbor countries (2007–2015): archaeology, history, culture], ed. V.P. Nikonorov and V.A. Alekshin. St. Petersburg: Contrast Ltd, pp. 521–558.

Sataev, R.M. 2008. "Zhivotnye v khozjajstve i dukhovnoj zhizni drevnego naselenija Gonur Depe" [Animals in the economy and spiritual life of the ancient Gonur Depe population]. In *Trudy Margianskoj Arkheologicheskoj Ekspeditsii*, vol. 2, ed. V. Sarianidi. Moscow: Staryj Sad, pp. 143–160.

Sataev, R.M. 2010. "Rekonstruktsija uslovij krematsii zhivotnykh iz zhertvennikov-lunok Gonur Depe" [Reconstruction of the animal cremation conditions in altar-pits at Gonur Depe]. In *Na puti otkrytija tsivilizatsii* [On the track of uncovering a civilization], ed. P.M. Kozhin, M.F. Kosarev, and N.A. Dubova. St. Petersburg: Aletejja, pp. 466–484.

Sataev, R.M. 2016. *Zhivotnye v kulture drevnej Margiany* [Animals in the culture of ancient Margiana]. Moscow: Staryj Sad.

Sataev, R.M. and N.A. Dubova. 2016. "Pogrebenija sobak na Tsarskom nekropole Gonura" [Dog burials in the Gonur "Royal Necropolis"]. In *Trudy Margianskoj Arkheologicheskoj Ekspeditsii*, vol. 6, ed. N. Dubova. Moscow: Staryj Sad, pp. 204–223.

Schwartz, G.M. 2012. "Era of the Living Dead: Funerary Praxis and Symbol in Third Millennium BC Syria." In *Constructing Funerary Rituals in the Ancient Near East*, ed. P. Pfälzner, H. Niehr, E. Pernicka, and A. Wissing. Wiesbaden: Harrassowitz, pp. 59–78.

Smirnov, A.Ju. 1997. *Labirint: Morfologija prednamerennogo pogrebenija: Issledovanija, teksty, slovar'* [Labyrinth: morphology of intentional burial: studies, texts, dictionary]. Moscow: Vostochnaja literatura RAN.

Soldado, J.L.R. 2016. *Structured Deposition of Animal Remains in the Fertile Crescent During the Bronze Age*. Oxford: Archaeopress.

CHAPTER THIRTEEN

FUNERARY RITUALS AND ARCHAEOTHANATOLOGICAL DATA FROM BMAC GRAVES AT ULUG DEPE (TURKMENISTAN) AND DZHARKUTAN (UZBEKISTAN)

Julio Bendezu-Sarmiento

INTRODUCTION

The Bronze Age cultures of Central Asia have long been viewed as marginal extensions of the more famous civilizations that surrounded them, particularly those of Mesopotamia and of the Iranian Plateau. However, discoveries from the past four decades have shown that this was, in fact, not the case. The Bronze Age of Central Asia, though lacking a writing system, has its own history and is an original complex civilization. The Bactria-Margiana Archaeological Complex (BMAC) refers to the apogee of a proto-urban culture with sites often fortified (Gonur Depe, Adji Kui, Namazga Depe, Ulug Depe, Altyn Depe, Kelleli, Taip, Togolok in Turkmenistan; Dashly in Afghanistan; Sapallitepa, Molalitepa, and Dzharkutan in Uzbekistan) and organized among dispersed oases along the Murghab and Amu Darya Rivers and areas around (Figure 13.1).[1] Alongside their local specificities, these dispersed regions maintained socioeconomic relationships with one another and shared common cultural elements (material elements, funerary practices, architecture) and symbolic ones (iconography and pantheon).

The archaeothanatological approach that is presented in this chapter provides part of the key to understanding the BMAC societies and, more broadly, Central Asia Bronze Age societies. The study of skeletal remains constitutes an essential source of information for the following thoughts and interpretations. Archaeothanatology is the study of human remains *in situ*. It combines the knowledge of human anatomy, the recording of the burial context, and an understanding of the taphonomic processes in order to recognize those that interacted with the corpse from the time of its burial to that of its excavation.

Two vast necropolises from the apogee of the Oxus Civilization have been excavated, at Gonur Depe and Dzharkutan. Thousands of other burials in smaller graveyards or scattered within the settlements like those from Ulug Depe have also been dug out. These graveyards have given us a considerable mass of information, and

Figure 13.1 Map of southern Central Asia with the distribution of Bronze Age archaeological sites.

we are now able to propose a synthesis of the mortuary norms and of the recurrences in the funerary practices of the whole BMAC population. We will base our study on the example of graves we excavated at two more or less contemporary sites, Ulug Depe and Dzharkutan.

STUDIES ON THE BRONZE AGE: ULUG DEPE AND DZHARKUTAN

For more than 15 years, two French archaeological expeditions have been leading excavations together with both a Turkmen (MAFTUR) and an Uzbek (MAFOuz-P) team on two major protohistoric sites of Central Asia: Ulug Depe[2] and Dzharkutan.[3]

The MAFTUR expedition began working at Ulug Depe in 2001. This site is located in the prefecture of Kahka, 175 km east of the Turkmen capital, Ashgabat (Figure 13.1), halfway between Namazga Depe (45 km to the west) and Altyn Depe (45 km to the east). Discovered in 1930 by A. Marushchenko, the site was partially excavated in the late 1960s by V.I. Sarianidi (Sarianidi and Kachuris 1968; Sarianidi 1969, 1972). From this short study, the historical importance of Ulug Depe as one of the most significant proto-urban sites in southern Central Asia could already be pointed out (Lecomte *et al.* 2002; Bendezu-Sarmiento and Lhuillier 2011, 2019; Lecomte 2011, 2013; Lhuillier *et al.* 2015; Lhuillier and Bendezu-Sarmiento 2018; Lhuillier 2019). Smaller than Namazga and Altyn Depe, Ulug stretches over an area of 13–15 ha at the top and 26 ha at the base, reaching a height of over 30 m. It testifies of the longest continuous stratigraphy of Central Asia, between the Early Chalcolithic and the Middle Iron Age.

Although no separated necropolis has yet been uncovered at the site, nearly 100 burials have been found in the habitation area (Bendezu-Sarmiento 2011, 2013). The first graves were excavated by Sarianidi between 1967 and 1970 (Sarianidi and Kachuris 1968; Sarianidi 1969; 1972), mainly in the northern zones of the site.

The site of Dzharkutan is located in southern Uzbekistan, 60 km to the north of the present-day city of Termez and about 10 km from the town of Sherabad, near the Afghani border (Figure 13.1). Situated on the left bank of the Bustansaj River, the city and the necropolis spread over multiple natural elevations in an area of ca. 50 ha. The chronology of the site covers most of the Bronze Age and overlaps into the Early Iron Age (Bendezu-Sarmiento and Mustafakulov 2013; Lhuillier *et al.* 2018; Lhuillier 2019).[4] The BMAC period is divided into several phases, successively the Dzharkutan, Kuzali, Molali, and Bustan, though there is debate on this succession (Kaniuth and Teufer 2001; *contra* Vinogradova 2001). At the beginning of the Middle Bronze Age, a simple settlement was founded (Bendezu-Sarmiento 2013; Bendezu-Sarmiento and Mustafakulov 2013; Bendezu-Sarmiento and Lhuillier 2019), evolving towards increasingly complex architecture to become a proto-urban site including a fortified "citadel," craft areas, and residential neighborhoods, as well as two monumental buildings known as the "temple" and the "palace."[5]

Our work began in 2006 with a prospection in the Surkhan Darya River valley, followed by an evaluation of the archaeological potential of the Dzharkutan site. In spring 2007, excavations started on necropolis 3 (Bendezu-Sarmiento and Mustafakulov 2008; Bendezu-Sarmiento 2013), and between 2009 and 2013 within the settlement, where other burials were excavated. In all, more than 1,500 graves have been uncovered by the successive Uzbeko-Soviet, German, and French missions at Dzharkutan, mainly in the several necropolises that cover three hills on ca. 2.5 ha (see esp. Askarov 1977; Askarov and Abdullaev 1983; Ionesov 1996, 2002; Shirinov and Baratov 1997; Bendezu-Sarmiento 2013; Teufer 2015).

PRINCIPAL CHARACTERISTICS OF THE FUNERARY PRACTICES

Our work on the Bronze Age stratum at both Ulug Depe and Dzharkutan helps document the relationships between the dead and the living, between residential and funerary zones.

During the Middle and Late Bronze Age, the majority of the dead were gathered in large necropolises situated around the edges of the cities (Bustan, Dzharkutan, Gonur Depe). But other burials were uncovered inside the settlements, in zones that were abandoned at the time or in spaces reserved for this purpose (Sapallitepa, Dzharkutan, Dashly, Altyn Depe, Namazga Depe).

A number of standards were followed in all these graves, whether in necropolises or settlements, throughout the different periods of the Bronze Age. Among them, for example, is the position of the dead in a lateral decubitus position on the right or left side, with folded limbs. While collective burial sites were more numerous during the Early Bronze Age, conversely individual graves in pits and, more rarely, burials in jars are the main types during the Middle and Late Bronze Age.

Among exceptional graves, very rich tombs are to be underlined. At Gonur Depe, in the "Royal Necropolis," hypogea decorated with paintings or mural mosaics constitute true houses for the dead with multiple inhumations including the main deceased and a number of "accompanying persons" associated with luxury artifacts, and often with carts yoked to draft animals (Sarianidi and Dubova 2011; see also Chapters 10 and 11). Animal graves (goats, sheep, camels, bulls, equines, or dogs) are also a particular feature of this civilization (see Chapter 12). These are not always sacrificed animals placed near human graves but are instead often specific burials of animals, perhaps with a substitutional function, accompanied by offerings that include prestigious objects (Luneau 2014).

A characteristic that has long been recognized involves women's graves that appear to be richer in both the quantity and quality of the artifacts placed in them as offerings (Askarov 1977; Alekshin 1986; Sarianidi 2001). This seems to indicate a specific social structure related to gender, with some objects – perhaps attributes linked to the roles or the powers of each gender – exclusively associated with women, and others with men (Luneau 2008, 2015).

Towards the end of the Bronze Age, a diversification of the funeral practices is noticeable, with evidence of manipulations on the bone remains linked with the reopening of the graves. At that time too, the first evidence of cremation appears (Avanesova 1995, 1996, 2013; Bendezu-Sarmiento 2004), revealing, according to some researchers, the influence of the northern steppe region of Central Asia in this southern zone.

PRIMARY INDIVIDUAL BURIALS AT ULUG DEPE

At Ulug Depe, the Bronze Age is characterized by a great homogeneity in the positions of primary inhumations, which are mostly individual (Figure 13.2), with subjects placed on their sides, with their limbs bent (especially the lower limbs) and their heads turned towards the north or northwest. These individual graves are found inside shallow pits, often filled in with soil containing chalk fragments and ashes.

The dislocation of certain skeletal elements (labile joints) of the hand, foot (phalanges), or thorax (cervical vertebrae) confirms the presence of a space that was left

Figure 13.2 Ulug Depe. Primary individual burials, sectors 1 and 17, graves 64, 91. © MAFTUR.

Figure 13.3 Ulug Depe. Two graves showing individuals with very tightly restrained positions of the upper and lower limbs, sector 1, graves 61, 66. © MAFTUR.

Figure 13.4 Ulug Depe. Perinatal and infant subjects, sector 1, graves 68, 69, 74. © MAFTUR.

temporarily empty during the decomposition process. This space was created due to the presence of a lightweight vegetal container that enveloped the deceased at the time of inhumation, traces of which have been discovered around bodies on several occasions. These "containers" often consisted of mats, but also sometimes of felt, placed at the bottom of the pit, making a mortuary shroud.

The extremely tight bend of the lower limbs of some subjects (Figure 13.3) suggests that, in addition to being wrapped in an organic material, they may also have been held in this position by ties that left no traces. Sometimes, this position can be explained by the small size of the pit, but it has nonetheless been shown that some pits were large enough to allow some parts of the body to shift around within an empty space, notably during the phases of putrefaction, dislocation, and decomposition.

Perinatal and infant subjects (Figure 13.4) seem to have been given the same treatment as adults (i.e., were enveloped in a vegetal "shroud" or placed inside a basket) (Bendezu-Sarmiento 2011). Their position is analogous with that of the adults, but their body orientation is more varied, and the burial pits of smaller dimensions.

Figure 13.5 Dzharkutan. Individual primary graves 1051 and 1045. © MAFOuz-P.

The perimortem subjects are never accompanied by grave goods. More rarely, older children might be provided with offerings. Moreover, the topographical situation of children's graves, usually placed in the center of a room, against a wall, or under the doorstep of a house (see Figure 13.4), leads us to believe that the houses were occupied at the time of these inhumations. In contrast, adult burials generally took place in more or less abandoned areas, with burial pits dug into or under entire wall segments of previous habitations.

INDIVIDUAL PRIMARY GRAVES AT DZHARKUTAN

At Dzharkutan, two types of pits, sometimes over 3 m deep, were uncovered in necropolises 3 and 4 (Askarov and Abdullaev 1983; Ionesov 1996, 2002; Shirinov and Baratov 1997; Bendezu-Sarmiento 2013): simple pits and sapping pits (*podboj*)[6] sealed with bricks (which implies that there was an empty space around the bodies during decomposition).

The subjects are placed in lateral decubitus, on the right or left side, with the upper and lower limbs bent (Figure 13.5). Although the axis of the body and the orientation of the head seem to show a preference for the northern axis (northeast, northwest), all the other directions are also present. On the whole, the majority of the graves that have been excavated belong to the Dzharkutan phase, while the Kuzali and Molali phases are more rarely represented and the Bustan phase is almost not attested (Luneau and Bendezu-Sarmiento 2013).

Among the rare positions, are some individuals placed on their backs with their knees bent, or burials in a seated position. These are all associated with the Kuzali and Molali phases. For the later Bustan phase, other burial practices existed, including cremations and "cenotaphs." These "cenotaphs," which contain no bodies, are difficult to identify as such. We excavated several rather deep pits comparable to those of regular burial pits, some of which contained only a few pottery vessels but most were empty, with only a few dark stains at the bottom, perhaps indicating the decomposition of seeds or other perishable goods, including bodies. These structures must certainly be attributed to a period between the Bronze and the Iron Age, a sort of intermediary stage before the total disappearance of every kind of necropolis or inhumation in the Early Iron Age (Bendezu-Sarmiento and Lhuillier 2015).

However, and unlike the neighboring necropolis of Bustan VI (Avanesova 2013), cremations have not yet been encountered in the Dzharkutan necropolises, with the possible exception of burnt and ground-up bones used in isolated bricks that closed off some of the deeper pits containing pottery vessels.

In addition to these large necropolises, a number of children's graves (often without any offerings) as well as adults' graves have been found inside the residential area of Dzharkutan and they nearly all date to the earliest occupation period of the site. As in the cemeteries, these individuals were often found in pits, positioned in lateral decubitus with their upper and lower limbs bent, and with offerings. These graves, especially the adults' ones, are situated in residential zones that were certainly abandoned at the time of the burials.

RELATIVELY RICH FUNERARY OFFERINGS

In the graves discovered at Ulug Depe, the offerings and decorative elements are abundant and include food presents, as well as various artifacts like pots, stone vases, beads made of semiprecious stones, pins, bracelets, and, in one case, a copper-alloy ring (Figure 13.6). Comparisons with the surrounding BMAC areas make it possible to firmly date this material to the Middle Bronze Age (see, in particular, Masson 1959, 1981; Askarov 1973, 1977; Askarov and Abdullaev 1983; Masson and Berezkin

Figure 13.6 Dzharkutan. Rich archaeological material remaining from looted graves (left to right): steatite vase, bronze flacon, detail of one of the ram's-head protomes on the bronze flacon. © MAFOuz-P.

2005; Sarianidi 2007; Bendezu-Sarmiento 2013; Teufer 2015). Sometimes accompanied by meat offerings (parts of bovine and/or ovine), the artifacts were placed at the feet of the deceased, or near the hips, head, or hands, and are much more frequent and opulent in the adults' graves than in those of younger subjects.

At Dzharkutan, the grave goods are also very luxurious (Askarov and Abdullaev 1983), with stone and metal objects (tools for working, toiletry artifacts), spindle whorls, flint arrowheads, and beads. The shape and style of some of the stone objects attest to short- and long-distance exchanges with southern Bactria and other areas. Clay pots are well represented, with sometimes dozens of vases accompanying the dead (see Figure 13.5). This material (open or covered jars, footed bowls in particular) was placed around the body and contained offerings such as seeds (cereals) and liquids (milk?), and quarters of mutton and beef (including ribs and legs). In the Late Bronze Age, the number of these offerings significantly diminishes and is often replaced towards the end of this period by miniature bronze items (knives, daggers) or clay ones (Luneau *et al.* 2014; see also Chapter 18).

The ceramic vessels, found in great number, pose a serious chronological problem since the presently used typo-chronology is based on earlier studies where the stratigraphic depth of the graves was considered as the main indicator. The graves that we have been able to excavate in the necropolises show that the situation is far more nuanced. Actually, ceramic material from the Late Bronze Age can be found very deep below the ground level (over 3 m), while graves from earlier periods can be situated at higher levels. This stratigraphic discrepancy can lead to errors and debate (Teufer 2005). In addition, graves containing homogeneous pottery from a single phase are quite rare, and the majority (except the earliest ones from the Dzharkutan phase) contain assemblages made from a mix of different phases that can be best explained by the secondary use of vessels from earlier graves (Luneau and Bendezu-Sarmiento 2013) (see p. 420).

SOME UNUSUAL FUNERARY STRUCTURES AT ULUG DEPE

The tomb of a "priestess"

In 2014, at Ulug Depe, a high platform made of *pakhsa* dated to the Middle Bronze Age was identified. On its northern part a sepulchral pit was dug that provided exceptional finds (Bendezu-Sarmiento and Lhuillier 2016).[7]

Unfortunately, the borders of the grave remain unclear, as water runoff caused the formation of a deep gully in this very area, to which must be added probable ancient looting. In the filling, and especially at the bottom of the pit, human remains were discovered and identified as belonging to a female individual. Qualitatively and quantitatively, this body is very poorly preserved, with only a few fragmentary elements of the spine, pelvis, and limbs found *in situ*. The subject seems to have been placed in right lateral decubitus (head to the east), the right leg flexed (the left one was no longer *in situ*). The left hand, partially connected, was lying on the hip bone of the same side.

The woman was wearing a belt made of small white limestone beads. Close to the body *in situ*, a second set of human bones, including a right hand, which may

Figure 13.7 Ulug Depe. "Priestess tomb," ceramic vessels characteristic of the NMG V period. © MAFTUR.

belong to the same individual, was also discovered in the filling of the pit. The hand was associated with limestone beads similar to the ones used for the belt, probably from a bracelet. Dozens of beads were discovered in both the filling of the pit and during the search of the gully that had cut through the grave. They present a wide variety of shapes and materials: gold, lapis lazuli, carnelian (including etched ones), agate, turquoise, fired mineral composite (frit or paste). The scattered bones were also associated with the imprint of a wickerwork basket and with 14 ceramic vessels characteristic of the Namazga (NMG) V period (Figure 13.7). A fifteenth small globular pot with painted decorations is of Indo-Iranian origin. Another important and unique discovery is that of eight alabaster vases (Figure 13.8), including types characteristic of Central Asia and others characteristic of the Iranian Plateau (for a detailed description, see Bendezu-Sarmiento and Lhuillier 2016).

The etched carnelian beads belong to the characteristic production of the Indus civilization, reflecting relations based on long-distance international trade as well as an established social hierarchy at Ulug Depe. Concerning the alabaster vases, it is as yet difficult to comment on their either unique or multiple origins: During the Bronze Age, southern Afghanistan and Baluchistan were known centers of alabaster production, while in southeastern Iran, Shahr-i Sokhta and Tepe Yahya were also major craft centers, especially thanks to their closeness to alabaster deposits (Casanova 1991: 51).

In addition to the beads, the alabaster vessels occupy a significant place in the inter-Iranian trade and illustrate once more the opulence of this exceptional funerary ensemble, and of the woman who was buried in this grave, confirming the important role women seem to have occupied in the Central Asia protohistoric society. This rich grave recalls the combination, encountered elsewhere, of high terraces and prestigious burials, such as the wealthy tomb of a "priest" at Altyn Depe (Masson 1981).

Figure 13.8 Ulug Depe. "Priestess tomb," alabaster vases. © MAFTUR.

A multiple burial structure

Among the atypical graves discovered at Ulug Depe, one deserves additional attention. It was excavated by Sarianidi in the 1960s. It is a multiple burial with a group of five individuals placed inside four closed, rectangular, brick-lined "cells" dated to the Middle Bronze Age (NMG V period), situated at the "southeast limit of the Depe" (Sarianidi and Kachuris 1968). In the juxtaposed cells (Figure 13.9), the subjects furthest to the south were lying on a stone pavement. Three of the graves were individual (graves 9, 10, and 13), while the fourth one contained two simultaneously buried individuals (graves 11 and 12). This last grave contained an adult accompanied by an immature placed near his/her legs, along with a number of ceramic vessels. The child's head was almost resting on its base, as if it had initially been placed on the gluteal muscles of the adult before its decomposition. Another indication of the simultaneous nature of these inhumations is visible in the immature's hand phalanges, which are above the adult's tarsus feet bones still in connection.

In all these graves, the subjects were positioned in right lateral decubitus with both their upper and lower limbs folded, and their heads turned towards the north-northwest. It has to be noticed that these three adults and two children were provided with exceptional funerary offerings compared with other Ulug Depe graves, as they included more than 50 ceramic vessels, and notably in the graves furthest to the east, stone vases and metallic objects including compartmented seals and a bronze ring. The pottery includes wheel-made vases like truncated conical or biconical bowls, and other vessels typical of the Middle and Late(?) Bronze Age (Figure 13.10).

For the earlier excavators, the presence of metallic seals was an indication that all these burials were related to women, which seems coherent with other observations made since then regarding the usual wealth of female graves and the gendered

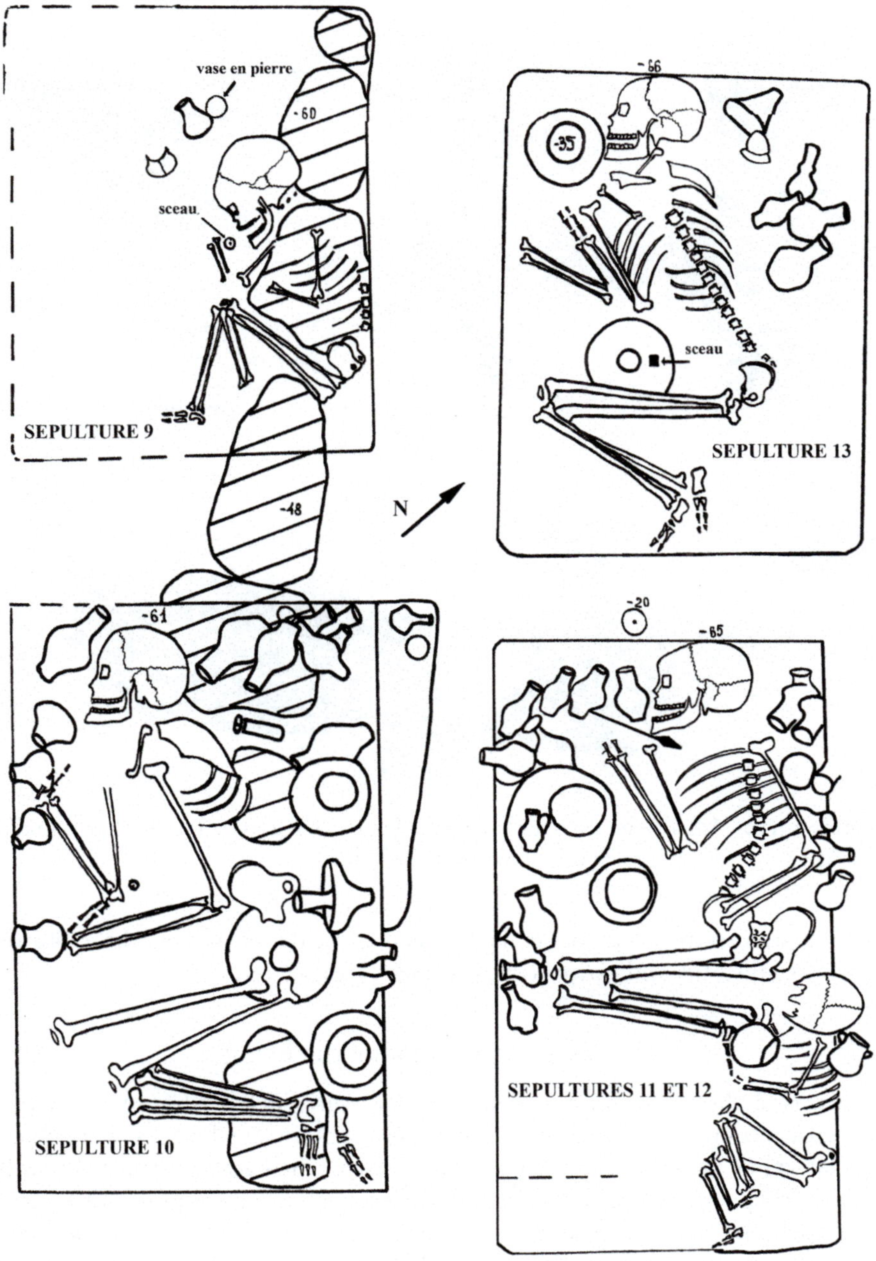

Figure 13.9 Ulug Depe. Female(?) graves in the form of cells. From V.I. Sarianidi; Bendezu-Sarmiento 2011: fig. 14.

Figure 13.10 Ulug Depe. Ceramics deposited in the cell graves. From V.I. Sarianidi; Bendezu-Sarmiento 2011: fig. 16.

repartition of certain types of artifacts, such as seals. However, the question of women's social status and power remains open (Luneau 2015).

Deceased infants among the living

The remains of about ten children have been discovered inside a house on excavation 1. This group contained immature and perinatal subjects aged 0 to 4 years placed in diverse lateral or dorsal decubitus positions, or with their backs against the walls of the house. They were close to two small alabaster concave columns with a groove on the upper part and two grinding stones arranged as to evoke the simulacrum of a human body stretched out on the ground (Figure 13.11) (Bendezu-Sarmiento 2006, 2011, 2013). A large jar without a bottom was found under one of the grinding stones, perhaps representing the head, the purpose of which remains unknown but that O. Lecomte (2013) has interpreted as symbolically intended for libations especially made for the children buried in this specific sepulchral zone.

No offerings or other accompanying artifacts were found in the grave itself. It should be recalled that the young were not recognized as full members of the social group in the Middle Bronze Age. This discovery is all the more interesting because

Figure 13.11 Ulug Depe. Human body's simulacrum of burial, sector 1. © MAFTUR.

we know that burials involving this age group are largely underrepresented and is important when considering the population growth during the Bronze Age.

THE LIVING AMONG THE DEAD AT DZHARKUTAN

Pre- and post-sepulchral practices in the Dzharkutan necropolises

In necropolis 3, soil stripping on a surface of over 1,000 m^2 brought to light the organization principles of a group of burial pits arranged in a circle and following a main northeast/southwest axis. Some "alinements" of mudbrick structures also stood out (i.e., walls made of header bricks of which three rows are preserved at best, slightly more than 1 m long and 50 cm wide), also oriented according to a northeast/southwest axis (Figure 13.12). These walls of medium height, built at the former ground level, served to show the surface location of graves or groups of tombs within the necropolis and to individualize them.

Less frequently, oval or circular traces of combustion (over 1 m in diameter) appear on the surface, as well as oval hearths not associated with any particular construction.

Figure 13.12 Dzharkutan necropolis 3. In the background, a brick structure found at the surface level; in the foreground, a set of offerings found near the surface. © MAFOuz-P.

The first type evokes what a wooden ring might leave after it burnt on the ground. In addition, we have discovered that, in these hearths, the inhabitants of Dzharkutan burnt seeds or grains, recognizable with the naked eye, perhaps with manure as a combustible. It remains difficult to link these (pre- or post-burial?) practices to any ritual that would have occurred directly around the burials (the inhumation or the closing or reopening of tombs?). Similar traces of fire are, however, well known later in the Late Bronze Age, in particular in the neighboring necropolis of Bustan, where N. Avanesova interprets them as ceremonial traces, perhaps connected to nomadic Andronovo populations who arrived from the northern steppes (Avanesova 1995, 2013).

At the same time, post-burial activity is noticeable in relation with culinary offerings associated with the graves and small vases sealed with a woven cloth containing local flowers, as shown by pollen analyses. These deposits or offerings were all situated between, against, or on top of the burial pits that were dug throughout the long history of the necropolis.

CONTEMPORANEOUS LOOTING OF THE BURIALS

Among the graves that we excavated, two out of three tombs had been looted in the past. For some of them, the evidence left by the robbers in formerly empty spaces

Figure 13.13 Dzharkutan. Looted grave 1001B. © MAFOuz-P.

testifies of a very early pillaging, occurring a short time after the inhumation. This is the case of tomb 1001B, for example, in which the upper part of the body, in lateral decubitus against the wall of the pit, was no longer in its original position (Figure 13.13). The entire right arm was in somewhat loose contact with the western wall. Based on the shifting of the body (in the areas of the hips, the spinal column, the head, and neck), it appears that this arm had been turned or thrown aside while it was still connected to the shoulder (Bendezu-Sarmiento and Grizeaud 2011).

However, despite the looting, some graves still contained fairly exceptional material such as beads made of metal, lapis lazuli, carnelian, or glazed faience, a very finely crafted steatite vase, a bronze flacon with three ram's-head *protomés* and bone inlay elements (see Figure 13.6). Their presence emphasizes the particular wealth of some of these tombs in comparison to others in the same necropolis.

Looting in the Dzharkutan necropolis is not exclusively a modern problem. The looted graves do not all appear to have been equally opulent initially, but the early date of interference with them proves that the looters were very likely the grave diggers themselves or at least that they had direct knowledge of the valuable items placed inside the graves. Reopening the graves and handling the bodies (buried deep below the ground level) was a collective effort and probably took a long time. These early incidences of systematic looting at less than 200 m from the settlement could suggest that the reopening of the graves occurred for practical or ritual purposes and was made by the families themselves or the community to recover objects that had been deposited with the deceased. Of course, other reasons are also possible, like looting by unrelated people in search of precious objects, or, occasionally and not on purpose, if a new burial was dug over an earlier one. In this case, we may imagine that the material recovered from the earlier tomb (of the Dzharkutan phase), in particular ceramic vessels that resist very well over time, was reused in the newly dug grave. This would explain the presence of different chronological items within a single grave (Luneau and Bendezu-Sarmiento 2013).

CONCLUSION

Studies of burial practices provide us with advanced perspectives for better understanding the Bronze Age societies between the 3rd and 2nd millennia. Clearly, however, questions remain about the enigmatic and complex matter of the origins, of the absolute chronology of the settlement, and of the evolution of the various local populations of Central Asia during the different phases of the Bronze Age. While the hypothesis of peaceful migratory movements of different cultures circulating north to south or east to west between the Middle East, the Caucasus, Siberia, or India is confirmed by the latest paleogenetic studies (see especially Narasimhan et al. 2019). It seems obvious that all the civilizations that surrounded the BMAC influenced, in one way or another, the internal historical evolution of its local population but also the whole of Central Asia.

NOTES

1 As explained in Chapter 1, the BMAC extends not only in Bactria, in the present-day south of Uzbekistan and north of Afghanistan, but also in Margiana in the zone of the endorheic delta of the Murghab River and in the foothills of the Kopet Dagh in Turkmenistan. Recent discoveries in northeastern Iran on the sites of Tepe Damghani (Francfort *et al.*, 2014) and Tepe Chalow (Vahdati and Biscione 2015; see Chapter 19) are also linked to this civilization.

2 Currently, the MAFTUR excavations are led by the author (CNRS/MNHN,UMR 7206) together with Mohamed Mamedow (Turkmen Direction for the Protection, Study, and Restoration of Historical and Cultural Monuments).

3 The work has been carried out in partnership with the Institute of Archaeology of Samarkand. The Franco-Uzbek Archaeological Mission: Protohistory (MAFOuz-P) was directed by the present author (CNRS-MNHN, UMR 7206), J. Lhuillier (CNRS, UMR 5133), S. Mustafakulov, and K. Rakhimov (Afrasiab Museum and Institute of Archaeology of Samarkand).

4 Dzharkutan was not occupied during the Early Bronze Age and no material from this period has so far been discovered in the region.

5 A number of excavation campaigns were undertaken between 1973 and 1990 on the large buildings of the settlement as well as on some residential quarters (Askarov 1977, 1980, 1981; Askarov and Shirinov 1991, 1993; Shirinov and Baratov 1997). Between 1994 and the early 2000s, the German-Uzkbek team focused on several residential mounds (Huff and Shajdullaev 1999; Huff 2000a, 2000b, 2001; Huff *et al.* 2001; Kaniuth 2006; Teufer 2015).

6 These are also called shaft graves at Gonur.

7 In 2014, trench 41 was opened on the northeastern flank of the Depe. The excavation, stretching over a surface of 170 m^2, revealed a massive horizontal terrace constructed with dense homogeneous blocks of *pakhsa*.

REFERENCES

Alekshin, V.A. 1986. *Sotsial'naja struktura i pogrebal'nyj obrjad drevnezemledel'cheskikh obshchestv* [Social structure and burial rituals of ancient farmers societies]. Leningrad: Nauka.

Askarov, A.A. 1973. *Sapallitepa*. Tashkent: Fan.

Askarov, A.A. 1977. *Drevnezemledel'cheskaja kul'tura bronzovogo veka Juzhnogo Uzbekistana* [The Bronze Age ancient farmers' culture of southern Uzbekistan]. Tashkent: Fan.

Askarov, A.A. 1980. "Raskopki poselenija Dzharkutan" [Excavations at the settlement of Dzharkutan]. *Arkheologicheskie Otkrytija* 1979: 445.

Askarov, A.A. 1981. "Southern Uzbekistan in the Second Millennium BC." In *The Bronze Age Civilization of Central Asia Recent Soviet Discoveries*, ed. Ph. L. Kohl. New York: M.E. Sharpe, pp. 256–279.

Askarov, A.A. and B.N. Abdullaev 1983. *Dzharkutan*. Tashkent: Fan.

Askarov, A.A. and T.S h. Shirinov 1991. "Le temple du feu de Dzharkutan, le plus ancien centre cultuel de la Bactriane septentrional." In *Histoire et cultes de l'Asie centrale préislamique*, ed. P. Bernard and F. Grenet. Paris: Editions du CNRS, pp. 129–142.

Askarov, A.A. and T. Sh. Shirinov 1993. *Rannaja gorodskaja kul'tura epokhi bronzy juga Srednej Azii* [The Early Bronze Age urban culture of southern Central Asia]. Samarkand: Fan.

Avanesova, N. 1995. "Bustan VI, une nécropole de l'âge du Bronze dans l'ancienne Bactriane (Ouzbékistan méridional): témoignages des cultes du feu." *Arts Asiatiques* 50: 31–46.

Avanesova, N. 1996. "Bustan VI: The Peculiarities of the Protobactrian Complex of Uzbekistan." *Unesco IASCCA Information Bulletin* 20: 141–155.

Avanesova, N. 2013. *Bustan VI: Nekropol' ognepoklonnikov dourbanisticheskoj Baktrii.* [Bustan VI: A necropolis of fire worshippers in pre-urban Bactria]. Samarkand: International Institute for Central Asian Studies.

Bendezu-Sarmiento, J. 2004. "Les structures funéraires à crémation de l'âge du bronze au Semirech'e (Kazakhstan). Etude anthropologique d'un ensemble d'échantillons provenant des nécropoles de Kul'saj I, de Kyzyl-Bulak (I, II) et d'Oj-Dzhajljau (III, VII)." *Paléorient* 30(2): 183–206.

Bendezu-Sarmiento, J. 2006. "Pratiques funéraires à Ulug-depe." In *Dossiers de l'Archéologie* 317: 28–33.

Bendezu-Sarmiento, J. 2011. "Pogrebal'nyj obrjad i chelovecheskie ostanki iz gorodishcha bronzovgo veka Ulug depe v Turkmenistane" [Burial ritual and human remains from the Bronze Age city of Ulug Depe in Turkmenistan]. In *Na puti otkrytija tsivilizatsii*, ed. P.M. Kozhin, M.F. Kosarev, and N.A. Dubova. St. Petersburg: Aletheia, pp. 513–535.

Bendezu-Sarmiento, J. 2013. "Archéologie funéraire et bio-anthropologie à Ulug dépé et Dzharkutan. Age du Bronze au Turkménistan et en Ouzbékistan." *Cahiers d'Asie centrale* 21–22: 501–532.

Bendezu-Sarmiento, J. and J.-J. Grizeaud 2011. "La pratique du pillage en Asie centrale dans les périodes anciennes." In *Halte au pillage*, ed. G. Compagnon. Paris: Éditions Errance, pp. 31–48.

Bendezu-Sarmiento, J. and J. Lhuillier 2011. "Iron Age in Turkmenistan: Ulug-depe in the Kopetdagh piedmont." In *Historical and Cultural Sites of Turkmenistan: Discoveries, Researches and Restoration for 20 Years of Independence*, ed. M.A. Mamedov. Ashgabat: Turkmen State Publishing Service, pp. 238–249.

Bendezu-Sarmiento, J. and J. Lhuillier 2015. "Sine Sepulchro Cultural Complex of Transoxiana (Between 1500 and the Middle of the 1st Millennium BC): Funerary Practices of the Iron Age in Southern Central Asia: Recent Work, Old Data, and New Hypotheses." *Archäologische Mitteilungen aus Iran und Turan* 45(2013): 283–317.

Bendezu-Sarmiento, J. and J. Lhuillier 2016. "Some More Exceptional Discoveries at Ulug-Depe." In *Trudy Margianskoj Arkheologicheskoj Ekspeditsii*, vol. 6, ed. N. Dubova. Moscow: Staryj Sad, pp. 522–528.

Bendezu-Sarmiento, J. and J. Lhuillier 2019. "Habitat and occupancy during the Bronze age in Central Asia: recent work at the sites of Ulug-depe (Turkmenistan) and Dzharkutan (Uzbekistan)." In *Urban cultures of Central Asia from the Bronze Age to the Karakhanids. Learnings and conclusions from new archaeological investigations and discoveries*, ed. Ch. Baumer and M. Novák. Wiesbaden: Harrassowitz Verlag, pp. 97–114.

Bendezu-Sarmiento, J. and S. Mustafakulov 2008. "Arkheo-antropologigicheskie issledovanija na mogil'nike Dzharkutan" [Archaeoanthropological research on a cemetery at Dzharkutan]. *Istorija Material'noj Kultury Uzbekistana* 36: 56–60.

Bendezu-Sarmiento, J. and S. Mustafakulov 2013. "Le site protohistorique de Dzharkutan durant les âges du Bronze et du Fer. Recherches de la Mission archéologique franco-ouzbèke-Protohistoire." *Cahiers d'Asie centrale* 21–22: 207–239.

Casanova, M. 1991. *La vaisselle d'albâtre de Mésopotamie, d'Iran et d'Asie centrale aux IIIe et IIe millénaires av. J.-C.* Paris: Editions Recherches sur les Civilisations.

Francfort, H.-P., A. Vahdati, J. Bendezu-Sarmiento, J. Lhuillier, E. Fouache, M. Tengberg, M. Mashkour, and Z. Shirazi 2014. "Preliminary Report on the Soundings at Tepe Damghani Sabzevar, Spring 2008." *Iranica Antiqua* XLIX: 111–158.

Huff, D. 2000a. "Djarkutan Archaeological Research on Tepe VI." *Istorija Material'noj Kultury Uzbekistana* 31: 58–69.

Huff, D. 2000b. "Tempel oder Palast? Dscharkutan – Monumentalbauten und Siedlung im bronzezeitlichen Baktrien." In *Archäologische Entdekkungen. Die Forschungen des Deutschen Archäologischen Instituts im 20. Jahrhundert*. Mainz: Ph. von Zabern, pp. 115–118.

Huff, D. 2001. "Djarkutan." *Arkheologicheskie issledovanija v Uzbekistane* 2000: 68–72.

Huff, D., Sh. Pidaev, and Sh. Shaidullaev 2001. "Uzbek-German Archaeological Researches in the Surkhan Darya Region." In *La Bactriane au carrefour des routes et des civilisations de l'Asie centrale: Termez et les villes de Bactriane-Tokharestan. Actes du colloque de Termez 1997*, ed. P. Leriche, Sh. Pidaev, M. Gelin, and K. Abdoullaev. Paris: Maisonneuve et Larose/IFEAC, pp. 219–233.

Huff, D. and Sh. Shajdullaev 1999. "Nekotorye rezul'taty rabot uzbeksko-germanskoj ekspeditsii na gorodishche Dzharkutan" [Some results from the Uzbek-German expedition work at the town of Dzharkutan]. *Istorija Material'noj Kultury Uzbekistana* 30: 19–26.

Ionesov, V.I. 1996. "K izucheniju pogrebal'nykh kompleksov Dzharkutana" [Analysis of the burial complexes from Dzharkutan]. *Istorija Material'noj Kultury Uzbekistana* 27: 17–28.

Ionesov, V.I. 2002. *The Struggle Between Life and Death in Proto-Bactrian Culture: Ritual and Conflict*. Mellen Studies in Anthropology, vol. 5. New York: Edwin Mellen Press.

Kaniuth, K. 2006. *Metallobjekte der Bronzezeit aus Nordbaktrien*. Mainz: Verlag Philipp von Zabern.

Kaniuth, K. and M. Teufer 2001. "Zur Sequenz des Gräberfeldes von Rannij Tulchar und seiner Bedeutung für die Chronologie des spätbronzezeitlichen Baktrien." *Archäologische Mitteilungen aus Iran und Turan* 33: 89–113.

Lecomte, O. 2011. "Ulug Depe 4000 Years of Evolution Between Plain and Desert." In *Historical and Cultural Sites of Turkmenistan. Discoveries, Researches and Restoration for 20 Years of Independence*, ed. M.A. Mamedow. Ashgabat: Turkmen State Publishing Service, pp. 221–237.

Lecomte, O. 2013. "Activités archéologiques françaises au Turkménistan." *Cahiers d'Asie centrale* 21–22: 165–190.

Lecomte O., H.-P. Francfort, R. Boucharlat, and M. Mamedow 2002. Recherches archéologiques récentes à Ulug-dépé (Turkménistan). *Paléorient* 28(2): 123–132.

Lhuillier, J. 2019. "The Settlement Pattern in Central Asia during the Early Iron Age." In *Urban cultures of Central Asia from the Bronze Age to the Karakhanids. Learnings and conclusions from new archaeological investigations and discoveries*, ed. Ch. Baumer and M. Novák. Wiesbaden: Harrassowitz Verlag, pp. 115–128.

Lhuillier, J., J. Bendezu-Sarmiento, and O. Lecomte 2015. "Ulug-Depe and an Overview of the Iron Age in Turkmenistan." *Journal of Iranian Archaeology* 4: 78–89.

Lhuillier, J. and J. Bendezu-Sarmiento 2018. "Central Asia and the interaction between the Iranian Plateau and the Steppes in Late First Millennium BC: case study from Ulug-depe in Turkmenistan". In *Ancient Chorasmia, Central Asia and the Steppes, Cultural relations and exchanges from the Achaemenids to the Arabs*, ed. M. Minardi and A. Ivanchik, Ancient Civilizations from Scythia to Siberia 24 (1-2): 331–353.

Lhuillier, J., J. Bendezu-Sarmiento, and S. Mustafakulov 2018. "The Early Iron Age Occupation in Southern Central Asia: Excavation at Dzharkutan in Uzbekistan." In *A Millennium of History: The Iron Age in Southern Central Asia (2nd and 1st millennia BC)*, ed. J. Lhuillier and N. Boroffka. Archäologie in Iran und Turan 17. Berlin: D. Reimer, pp. 31–50.

Luneau E. 2008. "Tombes féminines et pratiques funéraires en Asie centrale protohistorique. Quelques réflexions sur le 'statut social' des femmes dans la civilisation de l'Oxus." *Paléorient* 34(1): 131–157.

Luneau, E. 2014. "Les dépôts funéraires animaliers de la civilisation de l'Oxus: diversité et singularité du rapport entre l'animal et la mort en Asie centrale méridionale à l'âge du Bronze (2300–1500 avant n.e.)." In *Rencontres autour de l'animal en contexte funéraire. Actes de la rencontre de Saint-Germain-en-Laye 2012*, ed. I. Bede, C. Buquet-Marcon, M. Detante, and D. Perrier. Paris: Groupe d'anthropologie et d'archéologie funéraire, pp. 195–209.

Luneau, E. 2015. "Les relations de genre en Asie centrale protohistorique: redéfinition et discussion." *Les Nouvelles de l'Archéologie* 140: 28–34.

Luneau, E. and J. Bendezu-Sarmiento 2013. "Étude des céramiques de l'âge du bronze de la nécropole 3 de dzharkutan (Ouzbékistan): nouvelle approche typo-chronologique." *Cahiers d'Asie centrale* 21–22: 283–315.

Luneau, E., J. Bendezu-Sarmiento, and S. Mustafakulov 2014. "Ceramics and Chronology at Dzharkutan: A Revision of the Periodization of the Sapalli Culture (Uzbekistan c. 2100–1500 BC)." In *Proceedings of the 8th International Congress on the Archaeology of the Ancient Near East 2012, University of Warsaz*, vol. 2, ed. P. Bielinski, M. Gawlikowski, R. Kolinski, D. Lawecka, A. Soltisiak, and Z. Wygnanska. Wiesbaden: Harrassowitz Verlag, pp. 529–542.

Masson, V.M. 1959. *Drevnezemledel'cheskaja kul'tura Margiany* [The ancient farmers' culture of Margiana]. Materialy i Issledovanija po Arkheologii 73. Moscow: Nauka.

Masson, V.M. 1981. *Altyn-Depe*. Trudy JuTAKE XVIII. Leningrad: Nauka.

Masson, V.M. and Ju.E. Berezkin 2005. *Khronologija epokhi pozdnego eneolita – srednej bronzy Srednej Azii (pogrebenija Altyn-Depe)* [Chronology of the Late Chalcolithic–Middle Bronze Age in Central Asia (graves at Altyn Depe)], Trudy Institut Istorii Material'noj Kul'tury Rossijskaja Akademija Nauk, XVI. St. Petersburg: Nestor-Istorija.

Narasimhan V.M., N.J. Patterson, P. Moorjani, N. Rohland, R. Bernardos, S. Mallick, I. Lazaridis *et al.* 2019. "The Formation of Human Populations in South and Central Asia." *Science* 365 (6457), eaat 7487. doi: 10.1126/ science.aat7487.

Sarianidi, V.I. 1969. "Prodol'zhenie rabot na Ulug-depe" [Continuation of work at Ulug Depe]. *Arkheologicheskie Otkrytija* 1968: 434–435.

Sarianidi, V.I. 1972. "Raskopki 1970g. na Ulug-depe" [The 1970 excavations at Ulug Depe]. *Uspekhi sredneaziatskoj arkheologii* 1: 53–55.

Sarianidi, V.I. 2001. *Necropolis of Gonur and Iranian Paganism*. Moscow: Mir-Media.

Sarianidi, V.I. 2007. *Necropolis of Gonur*. Athens: Kapon Editions.

Sarianidi, V.I. and N.A. Dubova 2011. "Nouvelles sépultures sur le territoire de la 'nécropole royale' de Gonur-dépé." *Arts Asiatiques* 65: 5–26.

Sarianidi, V.I. and K.A. Kachuris 1968. "Raskopki na Ulug-depe" [Excavations at Ulug Depe]. *Arkheologicheskie Otkrytija* 1967: 342–345.

Shirinov, T. Sh. and S. Baratov 1997. "Bronzezeitliche Grabstätten der Nekropole Džarkutan 4c (Süd-Uzbekistan)." *Archäologische Mitteilungen aus Iran und Turan* 29: 65–120.

Teufer, M. 2005. "The Late Bronze Age Chronology of Southern Uzbekistan. A Reanalysis of the Funerary Evidence." In *South Asian Archaeology 2003*, ed. U. Franke-Vogt and H.-J. Weisshaar. Aachen: Linden Soft, pp. 199–209.

Teufer, M. 2015. *Spätbronzezeitliche Grabfunde aus Nordbaktrien und benachbarten Regionen. Studien zur Chronologie zwischen Aralsee und Persischem Golf. Keramik aus den Gräbern von Džarkutan*. Archäologie in Iran und Turan 13. Berlin: D. Reimer.

Vahdati, A. and R. Biscione 2015. *Preliminary Report on the 3rd Season of Joint Irano-Italian Archaeological Expedition at Tepe Chalow, Sankhast. Fall 2014*. Bojnord: Iranian Cultural Heritage, Handicrafts and Tourism Organization.

Vinogradova, N.M. 2001. "Südtadzhikistan in der Spätbronze – und Früheizenzeit." In *Migration und Kulturtransfer. Der Wandel vorder – und zentralasiatischer Kulturen im Umbruch vom 2. zum 1. vorchristlichen Jahrtausend. Akten des Internationalen Kolloquiums Berlin 1999*, ed. R. Eichmann and H. Parzinger. Kolloquien zur Vor- und Frühgeschichte, vol. 6. Bonn: Dr. Rudolf Habelt, pp. 199–219.

CHAPTER FOURTEEN

BIOARCHAEOLOGY OF THE BMAC POPULATION

A short review

Vladimir V. Kufterin

INTRODUCTION

During the many years of archaeological excavations in former Soviet Central Asia and the bordering regions, extensive anthropological data has been gathered, including reports on the Chalcolithic and Bronze Age periods (summaries in Ginzburg and Trofimova 1972: 48–98; Alekseev and Gokhman 1984: 19–27; Kijatkina 1987; Khodzhajov *et al.* 2011). A major part of this data originates from the sites now classified as the Bactria-Margiana Archaeological Culture (BMAC). Until recently, the main concern of anthropologists was the reconstruction of the anthropo-genetical types, and, therefore, most publications focused only on skull studies. This is also true for a number of early publications concerning the population of the surrounding territories – in particular Iran and the Indus valley (see, e.g., Dutta 1972; Cappieri 1973; Pardini and Sarvari-Negahban 1976). As a consequence, other significant anthropological aspects were either not addressed at all or given less attention in the scientific literature.

The goal of this chapter is to summarize the data on some specific aspects of the biological and social adaptation of the population of the BMAC area and of bordering territories, on the basis of a study of the pathological and stress markers. However, several limitations have to be underlined. First, the small amount of such data compels us to include a number of comparisons, and this explains why we will discuss material from remote sites, which, nevertheless, present the same general economic orientation as those of the BMAC (mainly based on agriculture). Second, often at multilayered sites, the anthropological data has been published as a whole, without possibilities to extract what concerns specifically the end of the 3rd and beginning of the 2nd millennium. This explains that we may speak of sites and periods apparently far from our theme.

Before going further into the matter, we should note that, unlike for the paleopathological data, the paleodemographical data of Central Asia, which also concerns the bioarchaeological reconstruction, has been recently partly gathered in T.K. Khodzhajov and A.V. Gromov's (2009) publication. Without focusing on demographic issues, we should underline that the use of paleodemographical indicators, such as the average age of the deceased (Table 14.1 and Figure 14.1), to reconstruct the interaction between ancient populations and the environment involves a number of

Table 14.1 Chalcolithic and Bronze Age settlements in Central Asia and adjacent regions with paleodemographical data. All belong to the Bronze Age except for Kara-Depe and Geoksjur which date to the Chalcolithic period and Altyn-Depe which dates to both the Chalcolithic and Bronze Age periods

Site	*Location*	N	*Reference(s)*
Gonur Depe	Southern Turkmenistan	2,461	Dubova and Rykushina 2007
Altyn Depe (Chalcolithic and Bronze Age)		132	Kijatkina 1987
Kara Depe, Geoksjur		36	Alekseev 1972
Parkhai II	Southwestern Turkmenistan	232	Kijatkina 1987
Sumbar		105	Khodzhajov and Gromov 2009
Sapallitepa	Southern Uzbekistan	92	Khodzhajov 1977
Dzharkutan		434	Alekseev *et al.* 1984
Buston VI		87	Dubova and Kufterin 2015
Tigrovaja Balka	Southern Tajikistan	52	Alekseev *et al.* 1986
Kokcha 3	Khorezm	54	Khodzhajov and Gromov 2009
Tasty-Butak	Western Kazakhstan	29	Alekseev 1972
Andronovo culture	Southern Siberia	32	Alekseev 1972

Note: N = number of individuals.

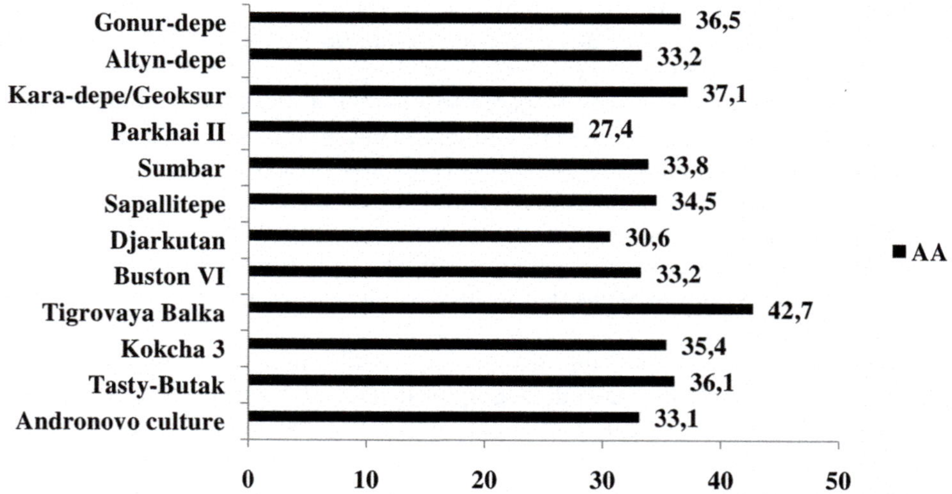

Figure 14.1 Average life expectancy of the adult population (AA) in the Chalcolithic and Bronze Age samples.

difficulties. In particular, there is no consistent pattern in the existing local variations of the demographic parameters, and thus they cannot be directly associated with a particular economic structure or "environmental niche" (Alekseev *et al.* 1986: 104). For example, the average life expectancy of the adult population in the "steppes" (Andronovo population of western Kazakhstan and southern Siberia) falls within the criteria limits common among the farming population of southern Central Asia. In addition, the paleodemographical data is to a certain extent a reconstruction (i.e., does not directly come out of the study of the material – the skeletal sample is just a prototype of a real population), and actually provides us with "models" only roughly reflecting the real demographic situation of different groups of the ancient population. Consequently, the paleopathological data gives more valuable information for the study of the adaptative process of the ancient population.

Up to now, there is very little evidence of the spread of any disease among the former population of the BMAC area. The data at hand is well representative for the population of Gonur Depe, but rather disparate for the inhabitants of the regions around, except for Harappa and Tepe Hissar. We have to mention here that we do not take into consideration the paleopathological data from Mesopotamia (territories of Syria and Iraq), since it has been recently gathered and systematized by A. Sołtysiak (2012). Thereafter in this study, we discuss and analyze material from the former Soviet territories in Central Asia, the Iranian Plateau, and the Indus valley (Table 14.2). The occurrence of pathological markers among the Andronovo communities from the Altai and Minusinsk basin (with a more cattle breeding orientation) serves as comparative data (Mednikova 2005; Tur and Rykun 2008; Svyatko 2014).

A highly descriptive interpretation of the Central Asian paleopathological discoveries is given in a few publications on the ancient and medieval population of Uzbekistan and Turkmenistan (Moldavskij and Khodzhajov 1983; Babakov *et al.* 1986; Babakov 1988; Khodzhaeva and Moldavskij 1990). In the light of modern methodological approaches to the anthropological sources, these publications are mainly of historiographical interest. Numerous case studies also remain beyond the scope of our discussion, since our aim is to characterize the adaptative process in the populations of the BMAC and the bordering territories using the most common paleopathological criteria. But even such a rough systematization of the data faces a number of problems, both objective and subjective in nature.

The first (objective) difficulties include: (1) the often bad or unsatisfactory conditions of preservation of the anthropological material; (2) the limitation due to skeletal selection (some series include only skulls or lack skeletons of children and teenagers); (3) the impossibility, in many cases, to reexamine the collections using modern approaches, because the remains have been re-buried; (4) the great variation in the number of samples from different chronological levels at a site, which obviously limits their direct comparison (e.g., the maximum number of selected samples for dental pathology analysis at Tepe Hissar I, II, and III concern, respectively, 18, 27, and 169 individuals).

Among the second (subjective) problems, we can mention: (1) incoherence in the methodological approaches used by different researchers: for example, the use of individuals for some of them or of dental number indicators for others (cf. Khodzhajov 1977; and Lorentz 2010); (2) discrepancies between different studies in the documentation of the pathological markers (the most illustrative in this regard

Table 14.2 BMAC settlements and sites from adjacent regions from which human remains were the subject of paleopathological examination

Site	*Location*	*Chronology*	*Reference(s)*
Gonur Depe	Murghab oasis (southern Turkmenistan)	2300–1500 BC	Babakov *et al.* 2001; Rykushina *et al.* 2003; Kufterin 2012, 2016b
Altyn Depe (Chalcolithic and Bronze Age)	Tedjen oasis (southern Turkmenistan)	3650–1850 BC	Kijatkina 1987; Kufterin 2016a
Sapallitepa	Sherabad oasis (southern Uzbekistan)	First half of the 2nd millennium BC	Khodzhajov 1977
Buston VI	Sherabad oasis (southern Uzbekistan)	End of the first half of the 2nd millennium BC	Dubova and Kufterin 2015; Rykushina 2015
Tepe Hissar (Chalcolithic and Bronze Age)	Damghan region (northeastern Iran)	4300–1800 BC	Krogman 1940; Rathbun 1980, 1984; Afshar 2014; Speakman 2017
Dinkha Tepe[a]	Western Azerbaijan (northwestern Iran)	1900–1350 BC(?)	Rathbun, 1980, 1984; Tavassoli 1999; Tavassoli *et al.* 2016: tab. 6; Wang 2012
Hasanlu (all periods)[a]	Western Azerbaijan (northwestern Iran)	6th millennium–800 BC	Rathbun 1980, 1984; Tavassoli 1999; Tavassoli *et al.* 2016: tab. 6; Wang 2012
Shahr-i Sokhta[a]	Western Seistan (southeastern Iran)	3000–2200 BC	Lorentz 2007, 2008, 2010
Mehrgarh (Chalcolithic)	Baluchistan (southwestern Pakistan)	4500–3200 BC	Lukacs and Minderman 1992
Mohenjo-daro[a]	Sindh (southeastern Pakistan)	2600–1400 BC	Kennedy 1984; Lovell 2016
Harappa (all the cemeteries and graves together)	Punjab (central Pakistan)	2300–1750 BC	Kennedy 1984; Hemphill *et al.* 1991; Lukacs 1992; Lovell 1994, 1997, 2014, 2016; Robbins Schug *et al.* 2012, 2013
Lothal[a]	Gujarat (western India)	2300–1200 BC	Sarkar 1985

Table 14.2 Cont.

Site	Location	Chronology	Reference(s)
Sarai Khola	Taxila region (northwestern Pakistan)	200–100 BC	Bernhardt 1969; Lukacs *et al.* 1989
Timargarha	Swat valley (northwestern Pakistan)	1400–850 BC	Bernhardt 1967; Lukacs *et al.* 1989

Note: Sites from which the material is not discussed in this chapter.

is the data on the prevalence rate of anemia among the Indus valley population, see Lovell 2016: 176); (3) manipulations with the various selected samples from the same site, which often lead to contradictory results (compare the data on Tepe Hissar from Afshar 2014 and Speakman 2017); (4) different sets of characteristics and/or levels of characterization in the study reports of the different researchers; (5) emphasis on the craniological analysis related mainly with ethnic and genetic issues and the fragmentary presentation of concurrent data (e.g., the Sialk publication provides only information on the prevalence of carious lesions; see Vallois 1939).

Despite these caveats, we have tried to summarize the available data on the status of the population's health during the Chalcolithic, Bronze, and Early Iron Age, with an emphasis on the BMAC settlements. As noted above, the paleopathological data on the populations of the surrounding areas (Iranian Plateau, Indus valley, and the "steppe world" or Andronovo culture) serves as comparative data.

The prevalence of certain dental diseases among the adult population of the compared groups is shown by the figures given in Table 14.3. The most noticeable feature is an almost complete absence of dental caries among the representatives of the Andronovo culture, while it prevails among the farming population (the Sapallitepa data is not entirely convincing in the general scheme). From the available information, we can observe a chronological trend: Dental caries increase over time – in the population of the later periods, the percentage of pathology is generally higher (i.e., data from Sarai Khola and Timargarha in northern Pakistan; from Mehrgarh and Harappa in southern Pakistan; Buston VI in Uzbekistan). The prevalence of the dental abscesses does not demonstrate any consistent pattern, except for abnormally low frequencies at Sapallitepa and abnormally high at Tepe Hissar. Almost all the examined individuals from the Andronovo population have dental calculus. This is also true among the farming population (the use of grinding stones increases the abrasive properties of cooked products), but it is a little less noticeable in the Harappa sample, which may indicate some specific cooking technology. The abnormally low rate of the antemortem tooth loss (AMTL) in the Altyn Depe sample can be explained by two factors: the small volume of the sample and the prevalence of the younger subgroup (up to 35 years) in it. The enamel hypoplasia stress marker represents the most controversial data. This is due to discrepancies in the ways the markers were documented in the different studies (data on the Large Necropolis and

Table 14.3 Frequency of dental disease in some Chalcolithic, Bronze, and Early Iron Age groups (individual account, both sexes together)

Site/culture	*Reference*	*Dental caries*			*Dental abscesses*		
		N	*n*	%	N	*n*	%
Gonur (Large Necropolis)	Rykushina *et al.* 2003	403	133	33.0	–	–	–
Gonur (latest graves in the "ruins")[a]	This chapter	177	59	33.3	194	32	16.5
Altyn Depe (Chalcolithic and Bronze Age)	Kufterin 2016a	22	7	31.8	20	3	15.0
Sapallitepa	Khodzhajov 1977	42	6	14.3	157	6	3.8
Buston VI	Rykushina 2015	48	23	47.9	82	10	12.2
Tepe Hissar I–III	Speakman 2017	71	26	36.6	71	17	24.0
Tepe Hissar III	Afshar 2014	163	57	35.0	169	102	60.4
Harappa	Lukacs 1992	39	17	43.6	38	7	18.4
Mehrgarh (Chalcolithic)	Lukacs and Minderman 1992	69	16	23.2	–	–	–
Sarai Khola	Lukacs *et al.* 1989	36	19	53.0	35	5	14.3
Timargarha	Lukacs *et al.* 1989	83	29	34.9	–	–	–
Andronovo culture (Altai)	Tur and Rykun 2008	83	2	2.4	83	23	27.5
Andronovo culture (Minusinsk basin)	Svyatko 2014	19	0	0	19	3	15.8

Site/culture	*Calculus*			*AMTL*			*EH*		
	N	*n*	%	N	*n*	%	N	*n*	%
Gonur Depe (Large Necrop.)	410	235	57.3	–	–	–	376	210	55.8
Gonur Depe ("ruins")[a]	177	122	68.9	206	77	37.4	188	24	12.8
Altyn Depe (Chalcolithic and Bronze Age)	22	15	68.2	20	1	5.0	22	3	13.6
Sapallitepa	–	–	–	–	–	–	–	–	–
Buston VI	52	42	80.8	82	22	26.8	41	32	78.0
Tepe Hissar I–III	–	–	–	–	–	–	71	36	50.7
Tepe Hissar III	164	89	54.3	169	80	47.3	128	97	75.8
Harappa (altogether)	40	17	42.5	41	13	31.7	36	26	72.2
Mehrgarh (Chalcolithic)	–	–	–	34	10	29.4	–	–	–
Sarai Khola	–	–	–	35	14	40.0	36	9	25.0

Table 14.3 Cont.

Site/culture	*Calculus*			*AMTL*			*EH*		
	N	*n*	%	N	*n*	%	N	*n*	%
Timargarha	–	–	–	50	16	32.0	–	–	–
Andronovo culture (Altai)	82	?	≈100	82	24	29.3	82	59	72.0
Andronovo culture (Minusinsk basin)	19	19	100.0	19	6	31.6	19	14	73.7

Note: AMTL = antemortem tooth loss; EH = Enamel hypoplasia; N = number of individuals; n = number of cases (affected individuals).

[a] Period III (latest graves in the ruins at Gonur Depe).

late burials at Gonur Depe); in case of understatement of the occurrence rates, we deal with cases when only distinct front-teeth enamel defects have been documented.

From the data obtained on the dental diseases among the sedentary population of the BMAC area and of the bordering territories, similarities in the occurrence of jaw and teeth pathologies (with a few exceptions), are significant. The situation is markedly different for the "steppe" groups. Despite the complex causes for dental caries, we tend to associate their high occurrence among the BMAC, Iranian Plateau, and Indus valley populations with their diet based on a farming economy. The prevalence of dental calculus and the almost total absence of dental caries is a particular feature of the Andronovo series. These results lead us to suppose that, despite minor differences in the frequency of occurrence of these markers, the causes of the antemortem tooth loss and enamel hypoplasia were quite different in the sedentary farming societies and in the steppe pastoral groups.

Information on the pathological and stress-marker occurrences on the skulls and the postcranial skeletons is presented in Table 14.4. The relatively low frequency of the osteological anemia marker – *cribra orbitalia* – in the samples from Harappa and from the late burials of Gonur (those dug in the "ruins") is worthy of interest. On the one hand, it may indicate a higher "quality of life" for these groups, but, on the other hand, it may reflect the differences in the landscape and ecological living conditions of the various groups of the ancient population. It should be recalled that other figures (i.e., 25%) have been given for the *porotic hyperostosis* occurrence in the Indus valley sites (Lovell 2016: 176; see also Kennedy 1984; Lovell 1997). The incidence of traumatic injuries in farming groups is relatively low. Some higher figures in the cranial trauma at Harappa and the postcranial trauma at Gonur may reflect a higher level of interpersonal aggression, related to a high population density. It is interesting that traumatic injuries are more common among women than men in the Gonur group (Kufterin 2016b). The frequency of osteoarthritis of the articulations and spine appears to be quite high among the farming populations while, on the contrary, there is very little osteological evidence of infection (in particular limb *periostitis*). The latter fact should not be considered as a proof of a low risk of infection within the proto-urban settlements. It is known that a number of common but critical infectious

Table 14.4 Frequency (%) of stress indicators in some Bronze Age groups

Site	*CO*	*Trauma (cranial)*	*Trauma (post-cranial)*	*OA (joints)*	*OA (spine)*	*Periostitis*
Gonur Depe (latest graves in the "ruins")	11.1	2.8	13.8	33.2	44.3	4.5
Altyn Depe (Chalcolithic and Bronze Age)	22.7	10.0	–	–	–	–
Sapallitepa	–	3.2	0.6	–	–	–
Buston VI	–	2.4	7.0	–	23.3	2.3
Harappa	6.9	15.5	1.9	Relatively high		6.1
Andronovo culture (Altai)	24.2	22.0	–	–	–	–

Note: CO = Cribra orbitalia; OA = osteoarthritis.

Table 14.5 "Correspondence analysis" results: loads of marks for vectors I and II

Indicators	*I*	*II*
Dental caries (Dcar)	–0.18	0.43
Calculus (Dcal)	0.25	0.02
Antemortem tooth loss (AMTL)	–0.28	–0.15
Enamel hypoplasia (LEH)	–0.20	–0.13
Cribra orbitalia (CO)	0.37	0.03
Cranial trauma (T)	0.21	–0.34

diseases are quickly followed by death without leaving osteological manifestations. Apparently, we face here the phenomenon of an "osteological paradox": Bones without evidence of inflammatory diseases may not be those of healthy individuals.

Table 14.5 and Figure 14.2 show the results of a formalized comparison (through correspondence analysis) of six groups (Gonur Depe, Altyn Depe, Buston VI, Tepe Hissar, Harappa, and the Andronovo sites in the Altai) and six pathological markers (dental caries, dental calculus, tooth loss, enamel hypoplasia, *cribra orbitalia* – for the Buston VI series the absent value was substituted by an average value – and skull injuries). Figure 14.2 shows a coherent highly isolated position of the Andronovo series, which is characterized by a high positive loading on vector I and high negative loading on vector II. The specific characteristics of the Andronovo populations include a high frequency of cranial injuries, evidence of dental calculus, and anemia (the latter feature, however, is not common for the pastoral populations of Eurasia,

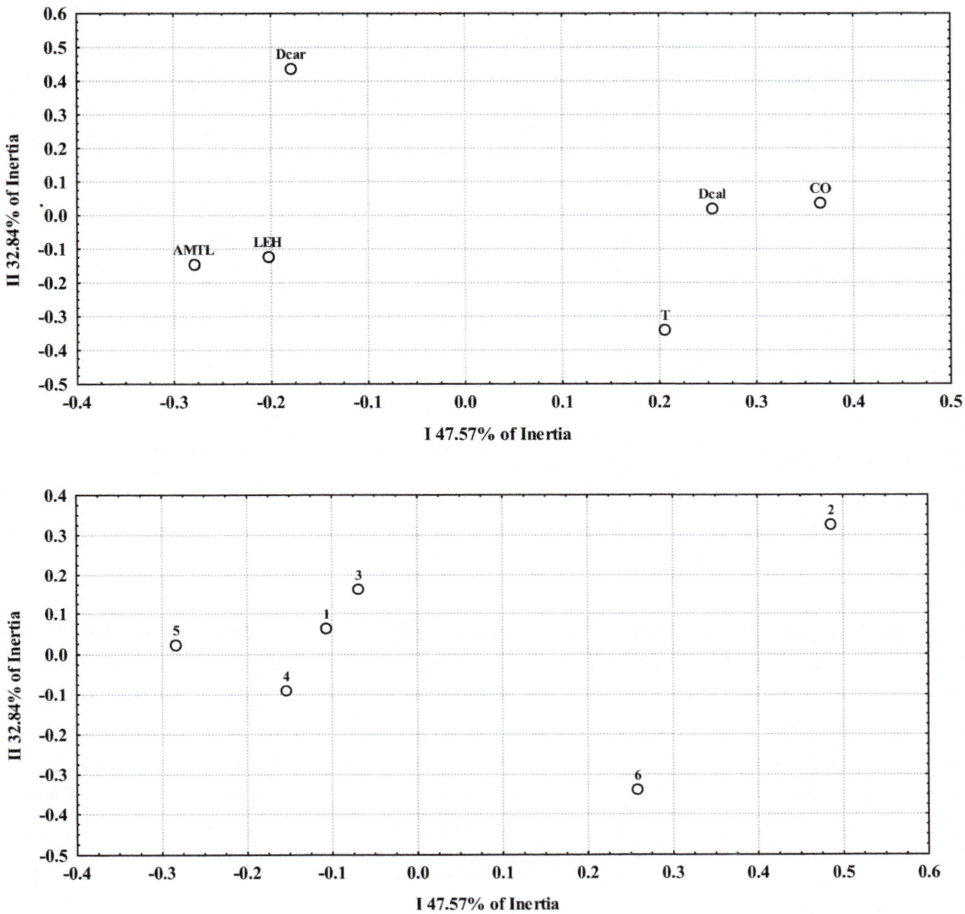

Figure 14.2 "Correspondence analysis" results. Layout of features and groups in the range of I and II vectors (47.6% and 32.8% of inertia): 1: Gonur-depe; 2: Altyn Depe; 3: Buston VI; 4: Tepe Hissar III; 5: Harappa; 6: Andronovo culture (Altai).

see, e.g., Mednikova 2005: 279), with a total absence of dental caries. The isolation of the Altyn Depe series probably represents a statistical artifact due to the very small number of samples. The groups from Bactria (Buston VI) and Margiana (Gonur Depe) are the closest ones, as are also the series from Iran (Tepe Hissar) and Bronze Age Harappa. This underlines the similarity between the pathological profiles and the biosocial adaptation of the ancient farming population in regions with different landscapes and ecological environments.

The results of the study are presented in Table 14.6. In all, these figures correspond to those from an "idealized model of subsistence systems correlated with demography and health" for farming populations, as proposed by J. Lukacs (2007: 291). As mentioned above the high infection risk does not have clear osteological evidence in the materials from the majority of the BMAC sites or from the bordering territories. Judging from the paleopathological characteristics, the biosocial adaptation of the

Table 14.6 Summary of results: frequency of diseases and stress indicators among the Chalcolithic and Bronze Age populations

Indicators	*Central Asia oases*	*Iranian Plateau*	*Indus valley*	*Western and southern Siberia (Andronovo culture)*
Dental caries	**/ ***	***	***	*
Calculus	***	**	**	***
Enamel hypoplasia	*/**/***	*/ **/***	**/***	***
Porotic hyperostosis	*/ **	**	*/ **	*/**
Joint disease	**/ ***	***	***	?
Infectious disease	*	*	*/ **	*/ **
Trauma	*/ **	**	**	***

Note: * = low; ** = average; *** = relatively high.

farming and pastoral population were significantly distinct from one another. At the same time, the populations of the Central Asia oases (BMAC), the Iranian Plateau, and the Indus valley present obvious similarities in pathological status. It infers that the local variations can be caused, on the one hand, by the special nature of adaptation to the different environmental conditions and, on the other hand, by the differences in the methodological approaches used by the researchers in their studies.

REFERENCES

Afshar, Z. 2014. "Mobility and Economic Transition in the 5th to the 2nd Millennium BC in the Population of the Central Iranian Plateau – Tepe Hissar." PhD thesis, Durham University, Durham, UK.

Alekseev, V.P. 1972. "Paleodemografija SSSR" [Paleodemography of USSR]. *Sovetskaja Arkheologija* 1: 3–21.

Alekseev, V.P. and I.I. Gokhman 1984. *Antropologija Aziatskoj chasti SSSR* [Anthropology of the Asian part of USSR]. Moscow: Nauka.

Alekseev, V.P., T.P. Kijatkina, and T.K. Khodzhajov 1986. "Paleoantropologija Srednej Azii epokhi neolita i bronzy" [Paleoanthropology of Central Asia during the Neolithic and Bronze Age]. In *Materialy k etnicheskoj istorii naselenija Srednej Azii* [Materials for the ethnic history of the population of Central Asia], ed. V.P. Alekseev. Tashkent: Fan, pp. 100–131.

Alekseev, V.P., T.K. Khodzhajov, and H. Khalilov 1984. *Naselenie verhov'ev Amudar'i po dannym paleoantropologii* [Population from the upper Amu Darya (anthropological data)]. Tashkent: Fan.

Babakov, O. 1988. *Srednevekovoe naselenie Turkmenistana (po materialam antropologii)* [Medieval population of Turkmenistan (anthropological data)]. Ashhabad: Ylym.

Babakov, O., H. Almazov, and P. Tangrykuliev 1986. "Zabolevanija zubov i cheljustej srednevekovogo naselenija Turkmenii" [Dental and jaw diseases of the medieval population of Turkmenistan]. *Zdravookhranenie Turkmenistana* 4: 22–25.

Babakov, O., G.V. Rykushina, N.A. Dubova, S.V. Vasiliev, A.P. Pestrjakov, and T.K. Khodzhajov 2001. "Human Skeletal Remains from the Necropolis of Gonur Depe." In *Necropolis of Gonur and Iranian Paganism*, ed. V.I. Sarianidi. Moscow: Mir-Media, pp. 219–240.

Bernhardt, W. 1967. "Human Skeletal Remains from the Cemetery of Timargarha." *Ancient Pakistan* 3: 291–407.

Bernhardt, W. 1969. "Human Skeletal Remains from the Prehistoric Cemetery of Sarai Khola." *Pakistan Archaeology* 6: 100–115.

Cappieri, M. 1973. *The Iranians of the Copper/Bronze Ages*. Miami, FL: Coconut Grove, pp. 1–19.

Dubova, N.A. and V.V. Kufterin 2015. *Antropologija naselenija Juzhnogo Uzbekistana epokhi pozdnej bronzy (po materialam nekropolja Buston VI)*. Etnicheskaja antropologija Srednej Azii, 4 [Anthropology of the Late Bronze Age population of southern Uzbekistan (materials of the Buston VI necropolis). Ethnic anthropology of Central Asia, 4]. Moscow: Staryj Sad.

Dubova, N.A. and G.V. Rykushina 2007. "Paleodemografija Gonur-depe" [Paleodemography of Gonur Depe]. In *Chelovek v kul'turnoj i prirodnoj srede* [Humans in cultural and natural environment], ed. T.I. Alekseeva. Moscow: Nauka, pp. 309–319.

Dutta, P. 1972. "The Bronze Age Harappans: A Re-Examination of the Skulls in the Context of a Population Concept." *American Journal of Physical Anthropology* 36: 391–396.

Ginzburg, V.V. and T.A. Trofimova 1972. *Paleoantropologija Srednej Azii* [Paleoanthropology of Central Asia]. Moscow: Nauka.

Hemphill, B.E., J.R. Lukacs, and K.A.R. Kennedy 1991. "Biological Adaptations and Affinities of Bronze Age Harappans." In *Harappa Excavations 1986–1990: A Multidisciplinary Approach to Third-millennium Urbanism*, ed. R.H. Meadow. Madison, WI: Prehistory Press, pp. 137–182.

Kennedy, K.A.R. 1984. "Trauma and Disease in the Ancient Harappans." In *Frontiers of the Indus Civilization*, ed. B.B. Lal and S.P. Gupta. New Delhi: Books and Books, pp. 425–436.

Khodzhajova, G.K. and M.I. Moldavskij 1990. "Materialy k paleopatologii Uzbekistana" [Data for the paleopathology of Uzbekistan]. *Istorija Material'noj Kul'tury Uzbekistana* 24: 220–228.

Khodzhajov, T.K. 1977. *Antropologicheskij sostav naselenija epokhi bronzy Sapallitepa* [Anthropological composition of the Bronze Age population of Sapallitepa]. Tashkent: Fan.

Khodzhajov, T.K. and A.V. Gromov 2009. Paleodemografija Srednej Azii [Paleodemography of Central Asia]. Moscow: Institute of Ethnology and Anthropology, Russian Academy of Sciences.

Khodzhajov, T.K., S.I. Mustafakulov, and G.K. Khodzhajova 2011. *Paleoantropologija juga Srednej Azii epokhi eneolita i bronzy* [Paleoanthropology of southern Central Asia in the Chalcolithic and Bronze Age]. Moscow: Institute of Ethnology and Anthropology, Russian Academy of Sciences.

Kijatkina, T.P. 1987. *Paleoantropologija zapadnykh rajonov Central'noj Azii epokhi bronzy* [Paleoanthropology of the western regions of Central Asia]. Dushanbe: Donish.

Krogman, W.M. 1940. "The Skeletal and Dental Pathology of an Early Iranian Site." *Bulletin of the History of Medicine* 8.1: 28–48.

Kufterin, V.V. 2012. "Antropo-ekologija i osobennosti biosocial'noj adaptatsii drevnego naselenija juga Srednej Azii" [Anthropoecology and peculiarities of the biological adaptation of the ancient population of southern Central Asia]. PhD thesis, Institute of Ecology of the Volga basin, Tol'jatti, Russia.

Kufterin, V.V. 2016a. "Eshche raz o kraniologicheskih materialakh iz Altyn-depe (Turkmenistan): paleoekologicheskie aspekty issledovanija" [Once more on the craniological materials from Altyn Depe: paleoecological aspects]. *Ekologija drevnikh i traditsionnykh obschestv* 5(1): 38–42.

Kufterin, V.V. 2016b. "Osobennosti biologicheskoj i social'noj adaptatsii naselenija strany Margush" [Peculiarities of the biological and social adaptation of the population of Margush]. In *Trudy Margianskoj Arkheologicheskoj Ekspeditsii*, vol. 6, ed. N. Dubova. Moscow: Staryj Sad, pp. 272–294.
Lorentz, K.O. 2007. "Teeth as Tools: Health and Disease at Shahr-e Sokhte." In *Proceedings of the 9th Annual International Symposium on Iranian Archaeology. Archaeological Reports*, 7. Tehran: ICAR, pp. 321–332.
Lorentz, K.O. 2008. "Activity-Induced Patterns of Dental Abrasion and Antemortem Tooth Loss at 3rd Millennium BC Shahr-e Sokhte (Iran)." *Iran* XLVI: 319–327.
Lorentz, K.O. 2010. "Hair, Bone and Teeth: Reconstructing Life from Human Remains at 3rd Millennium BC Shahr-i Sokhta (Sistan, Iran)." In *Proceedings of the 6th International Congress of the Archaeology of the Ancient Near East*, vol. 1, ed. P. Matthiae, F. Pinnock, L. Nigro, and N. Marchetti. Wiesbaden: Harrassowitz, pp. 391–408.
Lovell, N.C. 1994. "Spinal Arthritis and Physical Stress at Bronze Age Harappa." *American Journal of Physical Anthropology* 93(2): 149–164.
Lovell, N.C. 1997. "Anaemia in the Ancient Indus Valley." *International Journal of Osteoarchaeology* 7(2): 115–123.
Lovell, N.C. 2014. "Additional Data on Trauma at Harappa." *International Journal of Paleopathology* 6(1): 1–4.
Lovell, N.C. 2016. "Bioarchaeology of the Indus Valley Civilization: Biological Affinities, Paleopathology and Chemical Analysis." In *A Companion to South Asia in the Past*, ed. G. Robbins Schug and S.R. Walimbe. Chichester: John Wiley & Sons, pp. 169–186.
Lukacs, J.R. 1992. "Dental Paleopathology and Agricultural Intensification in South Asia: New Evidence from Bronze Age Harappa." *American Journal of Physical Anthropology* 87: 133–150.
Lukacs, J.R. 2007. "Interpreting Biological Diversity in South Asian Prehistory: Early Holocene Population Affinities and Subsistence Adaptations." In *The Evolution and History of Human Populations in South Asia: Inter-Disciplinary Studies in Archaeology, Biological Anthropology, Linguistics and Genetics*, ed. M.D. Petraglia and B. Allchin. Dordrecht: Springer, pp. 271–296.
Lukacs, J.R. and L. Minderman 1992. "Dental Pathology and Agricultural Intensification from Neolithic to Chalcolithic Periods at Mehrgarh (Baluchistan, Pakistan)." In *South Asian Archaeology 1989*, ed. C. Jarrige, J.P. Gerry, and R.H. Meadow. Madison, WI: Prehistory Press, pp. 167–179.
Lukacs, J.R., M. Schultz, and B.E. Hemphill 1989. "Dental Pathology and Dietary Patterns in Iron Age Northern Pakistan." In *South Asian Archaeology 1985*, ed. K. Frifelt and P. Sorensen. London: Curzon Press, pp. 475–496.
Mednikova, M.B. 2005. "Paleoekologija Tsentral'noj Azii po dannym antropologii" [Paleoecology of Central Asia according to the anthropological data]. In *Antropoekologija Tsentral'noj Azii* [Anthropoecology of Central Asia], ed. T.I. Alekseeva. Moscow: Nauchnyj Mir, pp. 256–289.
Moldavskij, M.I. and T.K. Khodzhajov 1983. "Paleopatologicheskie issledovanija v Uzbekistane" [Paleopathological research in Uzbekistan]. In *Problemy sovremennoj antropologii* [Problems of modern anthropology], ed. B.A. Nikitjuk and L.I. Tegako. Minsk: Nauka i Tekhnika, pp. 53–54.
Pardini, E. and A.A. Sarvari-Negahban 1976. "Craniologia degli Inumati di Shahr-i-Sokhta (Sistan, Iran). Studio Preliminare." *Estratto anticipato dall'Archivio per l'Antropologia e la Ethnologia* CVI: 1–49.
Rathbun, T.A. 1980. "Patterns of Pathology Among Metal Age Iranian and Mesopotamian Populations." *American Journal of Physical Anthropology* 52: 269.

Rathbun, T.A. 1984. "Skeletal Pathology from the Paleolithic through the Metal Ages in Iran and Iraq." In *Paleopathology at the Origins of Agriculture*, ed. M.N. Cohen and G.J. Armelagos. New York: Academic Press, pp. 137–167.

Robbins Schug, G., K. Blevins, B. Cox, K. Gray, and V. Mushrif-Tripathy 2013. "Infection, Disease and Biosocial Processes at the End of the Indus Civilization." *PLoS ONE* 8(12): 1–20.

Robbins Schug, G., K. Gray, V. Mushrif-Tripathy, and A.R. Sankhyan 2012. "A Peaceful Realm? Trauma and Social Differentiation at Harappa." *International Journal of Paleopathology* 2(2): 136–147.

Rykushina, G.V. 2015. "Odontologicheskaja kharakteristika naselenija `epokhi pozdnej bronzy, ostavivshego pamjatnik Buston VI" [Odontological characteristics of the Late Bronze population at Buston VI]. In N.A. Dubova and V.V. Kufterin *Antropologija naselenija Juzhnogo Uzbekistana epokhi pozdnej bronzy (po materialam nekropolja Buston VI)* [Anthropology of the Late Bronze Age population of southern Uzbekistan (materials of the Buston VI necropolis)]. Moscow: Staryj Sad, pp. 133–142.

Rykushina, G.V., N.A. Dubova, and N.A. Suvorova 2003. "Odontologicheskaja kharakteristika drevnego naselenija Turkmenistana (po materialam mogil'nika epokhi bronzy Gonur-depe)" [Odontological characteristics of the ancient population of Turkmenistan (from the Bronze Age cemetery Gonur Depe)]. In *Nauka o cheloveke i obshchestve: itogi, problemy, perspektivy* [Human and social sciences: results, problems and perspectives]. Moscow: Institute of Ethnology and Anthropology Russian Academy of Sciences, pp. 130–140.

Sarkar, S.S. 1985. "Human Skeletal Remains from Lothal." In *Lothal – A Harappan Port Town: Memoirs of the Archaeological Survey of India 1955–62*, vol. 2, ed. S.R. Rao. New Delhi: Archaeological Survey of India, pp. 269–304.

Sołtysiak, A. 2012. "Paleopathology in Mesopotamia. A Short Overview." *Światowit* X(LI/A): 91–108.

Speakman, A. 2017. "Sex Differences in Dental Lesions at Tepe Hissar During Periods of Stress." PhD thesis, University of Delaware, Newark.

Svyatko, S.V. 2014. "Dental Palaeopathological Analysis of the Eneolithic–Early Iron Age Populations from the Minusinsk Basin, Southern Siberia: Palaeodietary Implications." *Archaeology, Ethnology and Anthropology of Eurasia* 42(2): 143–156.

Tavassoli, M.M. 1999. *Palaeoanthropology of Iran*. Rawalpindi: S.T. Printers and Publishers.

Tavassoli, M.M., A. Sajjadi, S. Barani-Nia, and M. Tavassoli 2016. "A Bronze Age Grave Discovered at Tepe Foroudgah, Iran: An Anthropological and Palaeopathological Study." *Mediterranean Journal of Social Sciences* 7(6): 259–269.

Tur, S.S. and M.P. Rykun 2008. "Naselenie andronovskoj kul'tury Altaja po dannym bioarkheologicheskogo issledovanija" [Population of the Altai Andronovo culture according to the studies of the bioarchaeological data]. *Izvestija Altajskogo gosudarstvennogo Universiteta. Ser. Istorija* 4–2(60): 191–198.

Vallois, H.V. 1939. "Les ossements humains de Sialk." In *Fouilles de Sialk près de Kaskan 1933, 1934, 1937*, ed. R. Ghirshman. Paris: Librairie Orientaliste Paul Geuthner, vol. 2, pp. 113–192.

Wang, A. 2012. "Linear Enamel Hypoplasias in Hasanlu: A Survey." PhD thesis, University of Pennsylvania, Philadelphia.

CHAPTER FIFTEEN

ANIMAL EXPLOITATION AT GONUR DEPE

Robert M. Sataev

INTRODUCTION

The Oxus Civilization is an ancient farming civilization with a complex economy including irrigated agriculture, household and distant pastureland cattle breeding, as well as the manufacture of tools and handicrafts. This is well demonstrated by the material excavated at the proto-urban center of Gonur Depe – the largest and the best explored settlement.

Finds made during the excavations have underlined its cultural and trade relations with other centers of the Middle East and the Indus valley (Harappan Civilization). At some of these foreign sites like Shahr-i Sokhta, Tepe Yahya, Mehrgarh, Mohenjo-daro, Harappa, and Shortughai, studies made on the animal bone material have given valuable information on the various species and their use. Though most of these studies concern periods that precede Gonur, we feel it is necessary to review them before we present our study on the fauna from Gonur.

COMPARATIVE FAUNA DATA[1]

For a long time, the data on animal remains from Shahr-i Sokhta was considered as preliminary (Caloi *et al.* 1978; Bökönyi 1985). A number of studies were devoted to discuss specific species or animal groups (Caloi 1978; Compagnoni 1978a, 1978b; Caloi and Compagnoni 1981). Only later, a synthetic article based on the analysis of 15,141 bones (including mammals, birds, and reptiles) finally provided a clearer insight into the species composition and relative ratios of the animal remains excavated at the site (Bökönyi and Bartosiewicz 2000: 127). However, this data concerned only periods I–III and not the last period IV. Among the animal remains from period III (ca. 2500–2200 BC according to Salvatori and Tosi 2005), over 78% of the bones belong to small domestic or wild ungulates (*Capra*, *Ovis*, and *Gazella*). Cattle bones account for 21%, and those of other species are extremely rare (Bökönyi and Bartosiewicz 2000: 128, tab. 3). Bones of pigs, horses, and donkeys, especially, are represented only by singular finds.

At Tepe Yahya, the data on the animal remains is fairly limited (15,500 bones were collected, see Meadow 1987: 890). Their study concerns period IVC (ca. 3100–2800 BC) and earlier ones, and is presented in a diagram providing only a general presentation by taxons (Meadow 1987: 891, fig. 1). Among the domesticated species, the material includes remains of ovicaprines (*Ovisaries*, *Capra hircus*), cattle (*Bos sp.*), and dogs (*Canis lupus familiaris*) (Meadow 1987: 889). Period IVC is dominated by bones of small ungulates, while only 6% of the bone remains represent large cattle (Meadow 1987: 891). Paleozoological data from Tepe Yahya later periods synchronous with the Oxus Civilization (periods IVA and IVB) is as yet not published.

Zooarchaeological material from the excavations at Mehrgarh in Baluchistan, at the edge of the Indus valley, covers all its very long occupation – from the 7th to the beginning of the 2nd millennium BC (Meadow 1981, 1987, 1989a, 1989b). Data on the species identification from period VII (ca. 2900–2500 BC according to Jarrige *et al.* 2011), and on their ratio, is published with various ranges of details in several articles. The domestic species are represented by sheep, goat, and cattle. On average, for phases A–B of period VII, the ovicaprines represent 63%, and cattle over 33% (Meadow 1984a: 317, fig. 1). In the material from the early levels attributed to M.VIIA, ovicaprines (*Caprine*) represent 61%, and cattle 36% (Meadow 1989a: 174, fig. 6). In various areas of the site, in the layers of M.VIIB, ovicaprines vary between 38% and 70%, and cattle between 27% and 49%. In the last M.VIIC layers, ovicaprines are over 46%, and cattle 40% (Meadow 1984a: 317, fig. 1). Thus, in the course of Mehrgarh VII, the role of cattle, represented by the zebu humped ox (*Bos indicus*), increased (Meadow 1984c: 37).

Mohenjo-daro in the Indus valley was the first site in the region where R.B.S. Sewell and B.S. Guha studied the animal remains (Sewell and Guha 1931). But this study did not take into consideration the inner periodization of the site. The authors have identified 37 species of invertebrates (sponges, corals, shellfish) and vertebrates. Domestic species are represented by sheep, goat, zebu (*Bos indicus*), buffalo (*Bubalus bubalis*), domestic pig (*Sus scrofa domesticus*), domestic horse (*Equus caballus*), camel (*Camelus dromedarius*), and dog (Sewell and Guha 1931; Wheeler 1953). Most of the bones were related to zebus. However, some of the species definitions need to be revised. Thus, Sewell and Guha themselves expressed their doubt that the horse bones really belonged to a domesticated form. As for R.H. Meadow, he supposes that the camel bones excavated at the site belong to a double-humped species (Meadow 1984b: 136).

The material from the old Harappa excavations includes remains of 26 species of vertebrates (Prashad 1936). Mammals are represented by 19 species, 10 of which are domestic, including sheep, goat, zebu, buffalo, domestic pig, camel (*Camelus dromedarius*), domestic horse (*Equus caballus*), donkey (*Equus asinus*), and domestic cat (Prashad 1936). Bulls (*Bos indicus*) at Harappa are represented by two forms: large, massive, and small short-horned (Prashad 1936). According to Meadow, the number of cattle bones in some Harappa levels may exceed 50%, and at other Harappan sites, it even reaches 75% (Meadow 1989b: 68) while ovicaprines altogether only represent 25% (Meadow 1989b: 67, fig. 6.1). In the late layers of Harappa, B. Nath (1962) has identified the remains of domestic horse, which, according to him, are the first evidence that true horses were present in the region. B. Prashad (1936) regards the cat bones found at Harappa as the first representative of domestic cat on the subcontinent. The camel bones, as in the case of Mohenjo-daro, most probably belong to a Bactrian camel.

The animal species composition and their ratios at other sites of the Indus valley attributed to the Harappan civilization, such as Chanhu-daro, Jalilpur, Kot Diji, Rahman Dheri, Gumla, Amri, Rupar, or Bara, does not differ much from what is known from the material of Mohenjo-daro and Harappa (Majumdar 1934: 129–130; Mackay 1943: 270–273; Nath 1962, 1968; Dani 1970–1971; Fairservis 1975: 184; Allchin and Allchin 1982: 151; Meadow 1987–1988). The main feature of the Harappan animal husbandry is the domination of cattle.

Within the Indus Civilization, the Shortughai settlement, located in the upper part of the Amu Darya River, far beyond its motherland, holds a special place. It is considered as one of the Harappan trading posts through which commercial relations between the Indus valley and the Bactria-Margiana Archaeological Complex (BMAC) were carried out, at least during its early periods I–II. Osteological material from the excavations includes 3,973 mammal bones (Desse 1989: 188). From the 2,077 mammal bones that can be defined up to the species and genus, 2,031 bones (97.8%) belong to domestic species (Desse 1989: 190). These are represented by sheep (*Ovis aries*), goat (*Capra hircus*), bull (*Bos taurus*, *Bos indicus*), and buffalo (*Bubalus bubalis*). Most of the bones (83.8%) are those of ovicaprines, among which the majority consists of sheep. Cattle (bull and buffalo) accounts for only 13.9%. The role of cattle in the household of Shortughai was thus far less important in comparison with the settlements located within the main Harappan area. From the results of the zooarchaeological study, the composition of the domestic species at this site was very limited.

THE GONUR FAUNA DATA

Despite the relatively long history of research in the ancient delta of the Murghab, only a few studies have been done on the fauna excavated there (Ermolova 1986; Hiebert 1993; Hiebert and Moore 1993; Moore 1993, 1994; Moore *et al.* 1994). The long excavations made at Gonur Depe have led to the recovery of a large number of fauna remains that now allows an evaluation of the animals in the economy (Sataev 2008, 2016). A detailed analysis was made in order to determine specific uses of the animals. Furthermore, Gonur is exceptional because of its high content of such species as domestic pig (*Sus scrofa domesticus*), camel (*Camelus bactrianus*), domestic donkey (*Equus asinus asinus*), or domestic horse (*Equus caballus*), which are either totally absent or extremely rare at the above mentioned sites and that are undoubtedly domesticated as shown by the archaeological context of their discovery. In general, considering the scale of the excavations and the amount of faunal material, the data obtained at Gonur Depe can be extrapolated to the whole territory of ancient Margiana.

The main problem impeding our studies is the low content of animal bones in the cultural layers of the site. Compared to the numerous animal remains related to ritual practices (burials, cenotaphs, ritual places, etc., see Chapter 12), the finds of bones discarded in kitchen midden are relatively rare at Gonur Depe. Furthermore there is always the risk that some of them come from destroyed (or unrecognized) ritual events. Thus, the bones that are used as sources of information about the household exploitation of animals come from midden layers, short-time trash pits, floors of rooms, and spaces between them ("courtyards," "squares").

The archaeobotanical and zoological data shows that the climate at that time was not significantly different from the actual one, and that, within the oases, it was even more favorable for agriculture (Sataev and Sataeva 2010, 2012a, 2012b, 2013).

Bones of wild species are rare (Moore 1994; Sataev 2008, 2016; Sataev and Sataeva 2010, 2014). Among these the goose (*Ancer sp.*) and duck (*Anas sp.*), bustard (*Otis tarda*), hare (*Lepus tolai*), wild ram (*Ovis ammon orientalis*), Persian gazelle (*Gazella subgutturosa*), boar (*Sus scrofa*), and koulan (*Equus hemionus*) were probably intentionally hunted. However, one cannot rule out that some other discovered wild species like the middle-Asian tortoise (*Agrionemys horsfieldi*), kurgan eagle (*Buteo rufinus*), jackal (*Canis aureus*), wolf (*Vulpes vulpes*), or fox (*Vulpes corsac*) might also have been used for their hides, feathers, shells (notably because tortoise shells were found in burials), fangs, etc. Among the finds, worked antlers of red deer (*Cervus elaphus*) and a tooth of a brown bear (*Ursus arctos*) should also be noted. As other bones from these two last species are not present at the site, they must have been imported. Deer antlers were used for bone cutting (together with flint blades), while the bear tooth probably originates from a hide with its head (or hunter trophy). In general, hunting did not play a significant role in the economy of the Gonur inhabitants. It was probably occasional (except for antelope hunting), and was apparently practiced only by the elite if one is to rely on the presence of weapons exclusively in such graves.

Eight species of domestic animals are represented: sheep (*Ovis aries*), goat (*Capra hircus*), cattle (*Bos taurus taurus*), domestic pig (*Sus scrofa domesticus*), double-humped camel (*Camelus bactrianus*), donkey *(Equus asinus asinus*), horse *(Equus caballus*), and dog (*Canis lupus familiaris*) (Sataev 2016). It is quite probable that the residents of Gonur also knew domestic cats (*Felis silvestris catus*). Bones of this species have not been found yet at the site, but an easily recognizable clay figurine depicting a cat with a groove around its neck has been discovered (Figure 15.1). It is possible that the groove around the neck symbolizes a "collar." Possibly, cats were kept as elite, decorative animals. Based on this figurine, the cats of Gonur were probably short-faced, short-legged, and fluffy, different from the long-faced, long-legged, and short-haired Egyptian cats. It should also be noted that, despite the fact that wild cat species can be found around the site nowadays, none has been discovered in the excavations.

Figure 15.1 Cat terracotta figurine from Gonur Depe, room 46, Area 19. © Margiana Archaeological Expedition.

Size and appearance

Our data on the size and appearance of the animals comes mostly from the study of whole skeletons or large skeleton parts found in burials and other ritual structures. For the species found only as separated bones, the reconstruction of their size is based on their bone measurements.

The average height at withers of the male sheep (*Ovis aries*) is estimated at 70–75 cm, and that of the females is 60–65 cm. During the earlier Chalcolithic period at Ilgynly Depe (Turkmenistan), the sheep size was between 66 and 80 cm (Kasparov 2006: 68), so it seems that Bronze Age sheep were a little smaller than Chalcolithic ones. At Shahr-i Sokhta, their size is estimated at 65–70 cm, a height close to that of the Gonur sheep (Bökönyi and Bartosiewicz 2000: 136). Among the sheep, there were hornless specimens and others with small "bumpy" horns. Some animals also combined features of domestic and wild animals and were possible hybrids. Among the clay zoomorphic figurines of sheep found at the site, some have specific fat tails that may indicate that this kind of sheep was raised at Gonur.

Goats are more uniform in appearance than sheep. Hornless specimens have not been discovered. Domestic goats were a little smaller than sheep and reached around 65 cm in height. At Shahr-i Sokhta the size of the goats is estimated to 74.5 cm, but bones of wild animals (*Capra aegagrus*) may have been mixed with the remains of domestic animals (Bökönyi and Bartosiewicz 2000: 136).

All the studied skeletons of bovids, discovered at Gonur, are from *Bos Taurus*, some of which were castrated bullocks. Although certified bones of zebu (*Bos indicus*) have not been found yet at Gonur, we cannot exclude that this species was known since three clay figurines of "bulls" with humps (Figure 15.2) have been discovered at the site (Sataev 2016) and since zebus are depicted on stamps from Margiana and Bactria (Sarianidi and Dubova 2013: 167). Furthermore, A.K.

Figure 15.2 Zebu clay figurine from Gonur Depe upper layer, Area 17. © Margiana Archaeological Expedition.

Kasparov has mentioned a bull figurine with a characteristic hump found at Ilgynly Depe (Turkmenistan) in the Chalcolithic Jalangach levels (3700–3300 BC) (Kasparov 2006: 56, fig. 10; 57).

The bovids were relatively large. Height at withers for cows was no less than 120 cm, for bulls 130 cm, and for bullocks 135 cm. In southern Turkmenistan, during the Chalcolithic period (4th millennium BC), the bovid height is estimated at 122–124 cm (Kasparov 2006: 58). At Shahr-i Sokhta the sizes are similar to those of the Gonur animals, with a height at withers estimated at 130–140 cm (Bökönyi and Bartosiewicz 2000: 136). The cattle from Gonur was also not smaller than that from settlements in the steppe and forest-steppe areas of the Late Bronze Age, estimated in the Trans-Ural northern forest-steppe zone at 122 cm, and in the southern area at 125 cm (Kosintsev 1989: 91, tab. 3). The closest sizes to the Gonur ones come from the Bronze Age site of Arkaim (Trans-Ural steppe zone), with 115 cm for cows, 130 cm for bulls, and 135 cm for bullocks (Kosintsev 2000: 31).

It seems that the cattle raised at Gonur Depe had short horns, which may have resulted from a purposeful selection of animals in that direction. Representations, on seals and other artifacts, of short-horned bovids are also found at various sites of ancient Margiana (Sarianidi 2008: figs. 102, 169).

Pig (*Sus scrofa domesticus*) remains, although they are not very frequent, have been regularly found throughout the whole area of the site. Besides kitchen waste, pig bones come from meal offerings in burials and from other ritual structures (even more frequently than in kitchen waste), and pig astragali were used as amulets(?) together with those of sheep and goats. Although no whole pig skeleton has been found at Gonur, the bone measurements point towards an average height of about 65 cm. This size is smaller than that known in Bronze Age steppe settlements.

Camel (*Camelus bactrianus*) bones belong to the double-humped species. Based on the studies of camel full skeletons and of separate bones, one can assume that their height at withers (without the humps) was around 165 cm on average. Their appearance can be well imagined by looking at the realistic image of two walking camels depicted on a cylindrical silver vessel found in one of the graves of the "Royal Necropolis" (Sarianidi 2005: 237, fig. 96; see the cover of this book) or on other representations. They are heavy built, with high humps (especially the front one), calluses on their front legs, thick forelocks, and fur hanging from the bottom of their necks (Figure 15.3). All these features are characteristic for the domestic animals. Wild camels, according to the description given by A.G. Bannikov (1975: 63–64), are light, slim, long-legged, and have a narrow body. They have low, steep, cone-like humps, their hair is relatively short, males have shorter forelocks on their heads, and there are no calluses on the front feet. Thus, one can say, that the camels of Gonur had long been domesticated (see also Chapter 26).

Numerous finds of donkey (*Equus asinus asinus*) bones and whole skeletons of this animal speak of its common use at Gonur. Their height is estimated 115 cm at withers. One must note that apparently, compared to modern times or the Middle Ages, these animals were more respectfully treated, and to some extent were deemed as sacred since their remains are found in a whole range of ritual situations. Possibly, such an attitude is reflected in the later Zoroastrian tradition when a three-legged donkey is put at the center of the world, serving certain divine functions: "Three-legged donkey is in the center of the Warkash sea, and all water, poured on corpses,

Figure 15.3 Drawing of camels represented in relief on a silver cylindrical vessel from Gonur Depe, hypogeum 3220. © Margiana Archaeological Expedition.

menstrual excreta and other waste, when they reach the three-legged donkey – it purifies and blesses all of that with its gaze" (Chunakova 1997: 25).

Eight bones undoubtedly belonging to the domestic horse (*Equus caballus*) have been found at Gonur Depe. Judging by the paucity of such bones, this animal was a rare, elite species at the site. The most interesting and informative find is the horse skull found in 2006 in the southern corner of room 149 at Gonur North Area 9. Its skull and teeth specificities demonstrate that it is a horse and not of kulan/half-ass (*Equus hemionus*) or a donkey (*Equus asinus asinus*), more common at the site. The age of the animal, determined by the condition of its cheek teeth, was over 5 years. Judging by the size of the skull and its other bones, its height at withers is estimated around 136–144 cm. During the Chalcolithic and Early Bronze Age all the bones of wild equids discovered in Central Asia belong exclusively to kulan/half-ass (*Equus hemionus*) (Kasparov 2006: 33, 37). The first other known bones of domestic horses appear in Late Bronze Age sites. N.M. Ermolova has underlined the existence of undeniable horse bones in the Namazga (NMG) VI levels of the Namazga Depe multilayer settlement, as well as at Tekkem Depe (first construction period) (Ermolova 1986: 116). She also identified one horse bone at the BMAC settlement of Taip in the Murghab delta (Ermolova 1986: 116). Another horse bone mentioned at Kelleli is doubtful according to her and may have belonged to a large kulan. Outside Turkmenistan, remains of domestic horses have also been found at the Bronze Age settlement of Kangurttut, in the upper reach of the Tajrsu River (Tajikistan) where they are dated to the late 2nd–early 1st millennium BC (as determined by V.P. Danil'chenko and E.E. Antipina) (Vinogradova *et al.* 2008: 189–190). All the above-mentioned sites being later than Gonur, one can therefore assume that the domestic

Figure 15.4 Dog steatite figurine from Gonur palace, room 220.
© Margiana Archaeological Expedition.

horse bones from this site probably mark the very beginning of its use in southern Central Asia (on horses in Central Asia, see also Chapter 26).

Dog (*Canis familiaris*) remains are numerous at Gonur. A few come from the cultural levels but many come from various ritual structures and are mostly represented by whole skeletons. Dogs were used as a guardian of households and herds, and possibly also raised as a fighting animal. Three morphological types ("breeds") of dogs are represented: large "fighting" dogs (around 80 cm at withers), medium-sized "shepherd" dogs (around 50 cm at shoulder), and small dogs (up to 40 cm at shoulder) (Figure 15.4). Large dogs have the appearance of those depicted on the Hittites' stelae with their long tail. In contrast, the tails of the "shepherd" dogs were cut in most cases. The state of the teeth of these shepherd dogs indicates that they ate hard food, probably the bones of the animals eaten by humans. The function of the small dogs is yet unclear. Considering the fact that an almost complete skeleton of a small dog was found in an elite burial (burial 3210 in the territory of the "Royal Necropolis"), and that it was entirely covered with minute white beads probably sewn over a "blanket" that has disappeared, we cannot exclude that they were decorative "pets" (Dubova 2012: 129) (Figure 15.5). They may also have played the role of mouse catchers, since domestic cats are not numerous at Gonur (though small forms of wild felines undoubtedly inhabited the area in the Bronze Age).

Meat use

Proof of the use of animals for meat comes mainly from kitchen waste, found in three different areas at Gonur North.[2] The total number of bones that can be identified up to a certain species or higher taxon, is estimated at 2,311 samples altogether (Table 15.1). The three areas are located to the north (Area 19) and east (Area 18) (architectural-building complexes of period II[3]), as well as to the south (Area

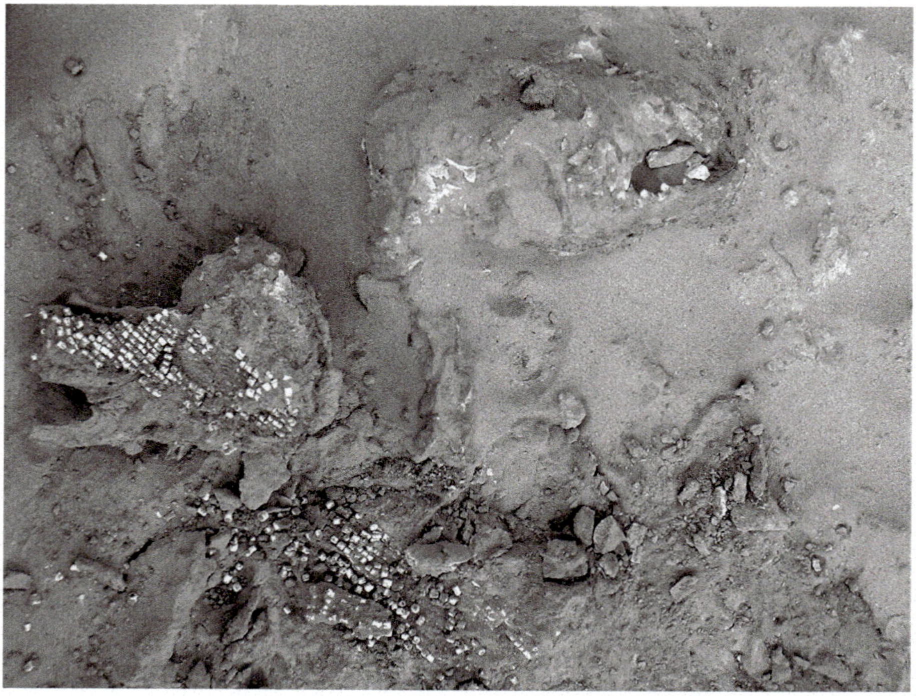

Figure 15.5 Small dog burial *in situ* in Gonur Depe, hypogeum 3210. The animal was covered by a now gone "blanket" entirely sewn with minute white beads. © Margiana Archaeological Expedition.

23) (period III). Animal remains from kitchen-household origin in other areas of the site and other periods are extremely rare. The overall species composition (including bones from periods I and IV) matches that established for the areas where the representative collection comes from.

Seven species of domestic animals come from these three midden deposits. When considered together, large cattle bones represent between 13.1 and 29.4%, ovicaprines between 61.5 and 81.7% (of which 52.9 to 79.5% are sheep bones), domestic pig between 1.8 and 3.0%, double-humped camel between 0.8 and 2.9%, donkey between 0.9 and 10.9%, and dog between 0.4 and 5.8%.

The majority of the eaten animal bones comes from cattle and ovicaprines (sheep and goat), which were the main meat sources. Despite the more numerous bones of ovicaprines in the three areas, the volume of meat produced implies that the share of cattle meat was equal in the diet to that of ovicaprines. The share of the other domestic species, like pigs and camels, is significantly smaller. Donkeys, horses, or dogs were apparently not part of the diet as shown by the condition of their preserved bones that probably arrived into kitchen waste by accident, either from destroyed ritual structures, or from animal burials.

The analysis of the skeleton parts and the age of butchering of the animals demonstrates that, depending on the areas and periods, the meat of cattle and ovicaprines was used differently.

Table 15.1 Characteristics of the zooarchaeological material from Areas 18, 19, and 23 at Gonur Depe

Zoological identification	*Structures*					
	Area 18		*Area 19*		*Area 23*	
	n	%	*n*	%	*n*	%
Domestic mammals						
Cattle (*Bos taurus taurus*)	175	29.4	124	16.2	108	13.1
Individual small cattle	242	61.5	330	63.2	562	81.7
Sheep (*Ovis aries*)	97		81		89	
Goat (*Capra hircus*)	27		72		23	
Pig (*Sus scrofa domesticus*)	11	1.8	23	3.0	16	1.9
Bactrian camel (*Camelus bactrianus*)	17	2.9	6	0.8	16	1.9
Donkey (*Equus asinus asinus*)	16	2.7	84	10.9	8	0.9
Dog (*Canis familiaris*)	10	1.7	44	5.8	3	0.4
Wild mammals						
Hare (*Lepus tolai*)	1	–	–	–	–	–
Goitered gazelle (*Gazella subgutturosa*)	3		15		12	
Half-ass (*Equus hemionus*)	3		–		–	
Fox corsac (*Vulpes corsac*)	–		1		–	
Brown bear (*Ursus arctos*)	1				–	
Unidentified	31		234		91	–
Total mammals	634		1,014		928	

(*continued*)

Table 15.1 Cont.

Zoological identification	*Structures*					
	Area 18		*Area 19*		*Area 23*	
	n	%	*n*	%	*n*	%
Birds						
Goose (*Ancer sp.*)	2	–	–	–	-	–
Bustard (*Otis tarda*)	1		–		–	
Unidentified	1		–		1	
Reptiles						
Tortoise (*Agrionemys horsfieldi*)	1	–	–	–	1	–
Gray monitor (*Varanus griseus*)	–	–	–	–	1	–

Ovicaprines

The material from Area 19 is largely dominated by bones of animals aged between 10 months and 1.5 years, while animals older than 2 years are relatively rare. In addition, the distribution of bones from different parts of the skeleton is more even in this area than in others. Thus, one can assume that the population of this part of Gonur North during period II raised ovicaprines both for meat and, possibly, for secondary products (milk ones, wool).

In Area 18 (period II) and Area 23 (period III) another picture emerges. Both areas present the same use of ovicaprines, where the animals were most commonly butchered between 10 months and 2 years of age (over 60%) and much less afterwards, so that at 3.5 years old over 80% of these animals had been eliminated but butchering continued. The further decrease of butchering is probably due to the deteriorating culinary properties of the meat from these old adult specimens. Males are slightly more common among the slaughtered animals (53.3–60%), but the share of females, forming the staple of the breeding stock, is also significant (40–46.7%). Thus, one can assume that the inhabitants of these "quarters" had no preferences for the age or sex of the slaughtered animals. At the same time, the analysis of the bone composition shows that the bones come from the most valuable parts for meat. Thus, the slaughter of the ovicaprines and their primary butchering had taken place elsewhere and the residents of these areas (or the personnel working there) received parts of animals probably in the form of already butchered meat elements or offerings.

Cattle

At Gonur North, in the northern sector, the peak of the cattle-age slaughter was between 1.5 and 2.5 years, and slowly diminished afterwards. The osteological material represents all the different parts of the animal, so that cattle was probably butchered by the residents of this building complex themselves. Thus, one can assume that the inhabitants of Gonur North used these animals exclusively for meat.

The material from the eastern sector of Gonur North demonstrates that the animals were most intensely slaughtered between 1.5 and 2.5 years old, probably because calf meat was not prized and they raised the animals until they had reached enough weight. After that age, the amount of slaughters significantly decreased to rise again after the animals were 5 years old, probably because they had lost by then their working and productive qualities. The different parts of the skeleton are represented quite evenly so that all the animal parts were used for food. In terms of quantity, the distal limb segments are quite numerous (18.8%), which points at butchering within the residential area. However, cattle was raised at Gonur not only for meat, but also for milk and as draft animals.

The analysis of the cattle bones from the southern Area 23 (period III) shows that the slaughter of the animals did not depend on their age, and that only the most valuable parts were used as food. Just as in the case of ovicaprines, it may indicate that the inhabitants of this complex did not breed the cattle themselves, but received the meat from outside.

Pig

The bone material from Gonur concerns all the age groups, from newborn to old specimens, although most of them were under 2 years (i.e., were slaughtered when they were young). Although pig was less common than sheep, goats, and cattle, domestic pigs were raised, and pig was one of the staple sources of meat. It should be underlined that pig is not an especially exotic species in arid zones. At Tell Khazna I in northeastern Syria, for instance, during the Early Dynastic I period (beginning of the 3rd millennium BC) the percentage of domestic pig sometimes reaches 44% (Antipina and Lebedeva 2008: 381). Considering that the share of pig in the meat diet at Gonur was small, several hypotheses can be advanced. This species could be a relict brought to Gonur by one of the groups who came from the foothills of the Kopet Dagh and was an element of their traditional culture, or its meat was used only on certain occasions, for example during burial and remembrance rituals. However, it has to be underlined that domestic pig seems to appear in southern Turkmenistan only during the BMAC (Oxus Civilization) since its presence earlier is doubtful. A.I. Shevchenko (1960) and Ermolova (1979), for instance, consider pigs from Chalcolithic sites as wild, and V.I. Tzalkin (1970) as domestic, while Kasparov (2006: 42) accepts the possibility that pigs were kept enclosed in Chalcolithic settlements of southern Turkmenistan, but believes that they were wild.

Double-humped camel

The use of double-humped camel for meat is uncertain. Camel remains, just as those of pig, are found practically in all the areas and all periods at Gonur, but their percentage in kitchen-household waste is extremely low. At the same time, as this zooarchaeological material includes purposefully split and charred bones of camels of various ages, one can assume that some of the animals, rejected for certain reasons, could have been slaughtered for meat. Thus, we cannot exclude that camel meat was eaten on certain special occasions.

To sum up, one can state that the inhabitants living in the vicinity of the Gonur North palace-temple complex (possibly, the personnel that served the administrative and religious elite) either kept a limited amount of animals for meat, or received their meat from the exterior (i.e., from "rural" satellite settlements). The animals kept in households were probably mainly used for milk and as a working force (draft or pack animals).

Other animal use

Animals were not only used as meat sources. Ovicaprines were also bred for wool, as confirmed by the presence of textiles made of animal wool, and of special tools like bone combs for wool.

Cattle was probably also bred for its physical force, for milk, and for dung to manure the barren soils. In the arid conditions in which this BMAC farming culture existed, cattle raising was vital for its development. Bullocks were intensively used as draft animals even before the domestication of camel and horse. G.N. Lisitsina considered that plowing in Turkmenia had started in the second half the 3rd millennium because of Bronze Age finds of miniature terracotta models of chariots with bull and camel heads applied on their front sides, which prove their use as draft animals in agriculture (Lisitsyna 1978: 213). At Ulug Depe in southern Turkmenistan, "some figurines of bulls, probably for rituals, have painted harnesses" (Lisitsina 1978: 55).[4] Moreover, a fragment of a clay "bull" figurine with a distinct groove around its neck possibly depicting also part of a harness was found at Gonur (Figure 15.6).

The suggestion that the population of Gonur milked its cattle is directly supported by the discovery of remains of a millet porridge cooked with milk in a vessel from burial 3901 (Sataev and Sataeva 2012a: 160). It is possible that some of the undecorated bone tubes, which, according to V.I. Sarianidi (2002: 151) were used to drink hallucinogenic drinks, were also used to increase milk output during milking operations, by blowing air into the animal's vagina (a practice widely used by pastoralists).

Undoubtedly, the inhabitants used hides of cattle and ovicaprines to make clothes and shoes, despite the fact that no such archaeological remains are preserved. From Gonur, for example, come anthropomorphic statuettes dressed in *kaunakes* – a wide dress sown from hides or a fabric imitating hide patches using the so-called *filikli* technique, a variety of nodular wool weaving (Tsareva 2009, 2012; for these statuettes see Chapter 7). Clay models of shoes have also been discovered (Sarianidi 2008: 233, fig. 143), most probably originally made of leather.

Camels and donkeys were used as draft and pack animals and were often harnessed into four-wheeled wagons. Camel skeletons have been discovered in several elite

Figure 15.6 Bull clay figurine with a distinct groove around its neck, from Gonur Depe, room 12, Area 18. © Margiana Archaeological Expedition.

burials of the "Royal Necropolis" (in the courtyards of the so-called "houses of the dead") and in pits/ditches (burials 3225, 3240, and 3900) (see Chapters 10 and 12) together with four-wheeled wagons. It is also possible that camels and donkeys were harnessed together to draw carts, depending on the weight of the cargo and the distance of the journey, as shown by the remains of two camels and two donkeys found in pit 3900 together with one wagon.

After eating, the bones of the slaughtered animals were also used as technical raw materials, mostly for the manufacture of various artifacts including tools, weapons, jewelry, or cult items. Tools include awls, needles, and combs. Awls are the most common bone tools discovered at Gonur Depe. Bone arrowheads were probably used for hunting. Bone jewelry includes "hairpins" and beads. Cult artifacts concern decorated "tubes" for ritual drinking and "ceremonial axes." Another specialized category of bone items includes "astragals" – worked or not (i.e., worn on one or both sides) from small ungulates (sheep, goat, pig). Astragals of ovicaprines were found in significant quantities at Gonur either in rooms, in burials, or in other structures. Besides singular finds, astragals were also found in sets. Their functional purpose is still unclear and one can assume that they had multiple purposes: utility (leather polishers), sacral-symbolic (amulets, charms), and for games (gambling dice).

Bone items also include numerous horn antlers from ungulates, mostly Persian gazelle. These horns testify to intentional break-off (removal) and/or preliminary processing. The studied finds fall into several groups reflecting the first stages and/or different stages of use. As the "cult of horns" is one of the constituents of the religious tradition at ancient farming centers (Antonova 1984), it is possible that these horns may have had a double function: The horn itself was used for crafting (raw material for bone carving), while the pith was preserved for rituals, which could explain their frequent presence in burials. As other bones of Persian gazelles on the site are rare, it is possible that these animals (males) were mainly hunted for their horns and that the animals were butchered and eaten directly at the hunting posts outside the major settlement. This may also indicate the existence of professional groups of hunters leaving the sites for expeditions of various length.

To sum up, though few bone artifacts have been identified at Gonur Depe, their use as tools show that bone had not lost all its value as a technical raw material, despite the more frequent use of metal tools. This was especially the case for leather or textile activities.

In conclusion, cattle breeding was an important part of the economy of the BMAC society, as was the use of specific species, like Bactrian camels, donkeys, or bullocks as draft or pack animals. The burials of some of them in rich graves show that they were highly prized. The very limited number of animal bones found in middens tends to support Sarianidi's hypothesis that Gonur was not a living place but rather an administrative and religious specific center.

NOTES

1 The absolute dates of these sites have constantly been modified. The dates given here are the last admitted ones. Unfortunately, the comparative fauna mostly predates the BMAC period except for Shortughai.

2 The material coming from the midden deposits excavated by F. Hiebert is not included here.

3 According to the existing stratigraphic data and radiocarbon dating (Sarianidi 2007; Zajtseva *et al.* 2008), four periods have been recognized in the proto-urban center. They are successively attributed to: (I) first construction period of the Gonur palace, from its foundation to the "huge fire" (ca. 2300–2000 BC); (II) second construction period of the palace, from the beginning of its restoration after the fire until the beginning of its interior reorganization (ca. 2000–1800 BC); (III) third construction period of the palace, from the beginning of its interior reorganization until the end of its use as the residence of the rulers (ca. 1800–1600 BC); (IV) from the ruling elite's departure to the complete abandonment of the settlement (1600–1500 BC).

4 At that time, the author dated them to the 4th millennium. They more probably date to the 3rd millennium.

REFERENCES

Allchin, B. and F.R. Allchin 1982. *The Rise of Civilization in India and Pakistan*. Cambridge: Cambridge University Press.

Antipina, E.E. and E.Ju. Lebedeva 2008. "Produktsija zemledelija i skotovodstva na Tell Hazne I" [Farming and cattle-breeding production at Tell Hazna]. In *Arkheologija Kavkaza i Blizhnego Vostoka* [Archaeology of the Caucasus and the Near East], ed. N.Ja. Merpert and S.N. Korenevskij. Moscow: TAUS, pp. 374–389.

Antonova, E.V. 1984. *Ocherki kul'tury drevnikh zemledel'tsev Perednej i Srednej Azii* [Essays on the ancient farmers culture of southwestern and Middle Asia]. Moscow: Nauka.

Bannikov, E.G. 1975. "Dikij verbljud havtagaj" [Wild Havtagai camel (*Camelus ferus*)]. Priroda 2: 63–70.

Bökönyi, S. 1985. "Preliminary Results of a Thorough Evaluation of the Mammal Bone Material from Shahr-i Sokhta." *East & West* 35(4): 426–429.

Bökönyi, S and L. Bartosiewicz 2000. "A Review of Animal Remains from Shahr-i Sokhta (Eastern Iran)." In *Archaeozoology of the Near East. IV. Proceedings of the Fourth*

International Symposium on the Archaeozoology of Southwestern Asia and Adjacent Areas, ed. M. Mashkour, A.M. Choyke, H. Buitenhuis, and F. Poplin. Groningen: Rijksuniversiteit, pp. 116–152.

Caloi, L. 1978. "The Bone Remains of Small Wild Carnivores from Shahr-i Sokhta." In *Approaches to Faunal Analysis in the Middle East*, ed. R.H. Meadow and M.A. Zeder. Peabody Museum Bulletin 2. Cambridge, MA: Harvard University Press, pp. 129–132.

Caloi, L. and B. Compagnoni 1981. "Preliminary Remarks on the Bovine Remains at the Site of Shahr-i Sokhta (Iranian Sistan) (3200–1800 BC)." In *South Asian Archaeology 1979*, ed. H. Härtel. Berlin: D. Reimer Verlag, pp. 181–190.

Caloi, L., B. Compagnoni, and M. Tosi 1978. "Preliminary Remarks on the Faunal Remains from Shahr-i Sokhta." In *Approaches to Faunal Analysis in the Middle East*, ed. R.H. Meadow and M.A. Zeder. Peabody Museum Bulletin 2. Cambridge, MA: Harvard University Press, pp. 87–90.

Chunakova, O.M. 1997. *Zoroastrijskie teksty. Suzhdenie Dukha razuma (Dadestan-i menog-i hrad). Sotvorenie osnovy (Bundahishn)* [Zoroastrian texts. Judgments of the spirit of reason. Creation of the foundation]. Moscow: Vostochnaja Literatura.

Compagnoni, B. 1978a. "The Bone Remains of *Equus hemionus* from Shahr-i Sokhta." In *Approaches to Faunal Analysis in the Middle East*, ed. R.H. Meadow and M.A. Zeder. Peabody Museum Bulletin 2. Cambridge, MA: Harvard University Press, pp. 105–118.

Compagnoni, B. 1978b. "The Bone Remains of Gasella subgutturosa from Shahr-i Sokhta." In *Approaches to Faunal Analysis in the Middle Eastn*, ed. R.H. Meadow and M.A. Zeder. Peabody Museum Bulletin 2. Cambridge, MA: Harvard University Press, pp. 119–128.

Dani, A.H. 1970–1971. "Excavation in the Gomal Valley." *Ancient Pakistan* 5: 1–177.

Desse, J. 1989. "Etude archéozoologique." In *Fouilles de Shortughaï. Recherches sur l'Asie centrale protohistorique*, vol. I, ed. H.-P. Francfort. Paris: de Boccard, pp. 187–206.

Dubova, N.A. 2012. "Pogrebenija zhivotnykh v strane Margush" [Animal burials in Margush]. In *Trudy Margianskoj Arkheologicheskoj Ekspeditsii*, vol. 4, ed. V.I. Sarianidi. Moscow: Staryj Sad, pp. 101–139.

Ermolova, N.M. 1979. "K kharakteristike mlekopitajuschikh Turkmenii bronzovogo veka (po osteologicheskomu materialu iz raskopok Altyn-depe v 1975 g.)" [Characteristics of mammals of Bronze Age Turkmenistan (based on the osteological materials from the Altyn Depe 1975 excavations)]. *Karakumskie drevnosti* VIII: 113–116.

Ermolova, N.M. 1986 "Materialy k izucheniju skotovodstva i okhoty v Tsentral'noj Azii v epokhu eneolita i bronzy" [Materials for the study of cattle breeding and hunting in Central Asia during the Chalcolithic and Bronze Ages]. In *Drevnie tsivilizatsii Vostoka* [The Ancient Orient civilizations], ed. V.M. Masson. Tashkent: Fan, pp. 110–117.

Fairservis, W.A. 1975. *The Roots of Ancient India*. Chicago: University of Chicago Press.

Hiebert, F.T. 1993. "Excavations of Domestic Quarters from Gonur Depe (North)." *Unesco IASCCA Information Bulletin* 19: 78–95.

Hiebert, F.T. and K.M. Moore 1993. "New Stratigraphic Excavations at Gonur Depe (North)." *Unesco IASCCA Information Bulletin* 19: 96–108.

Jarrige, J.-F., A. Didier, and G. Quivron 2011. "Shahr-i Sokhta and the Chronology of the Indo-Iranian regions." *Paléorient* 37(2): 7–34.

Kasparov, A.K. 2006. *Skotovodstvo i okhota epokhi neolita-paleometalla v Juzhnom Turkmenistane (razvitie strategii ispol'zovanija zhivotnykh resursov)* [Cattle breeding and hunting during the Neolithic–Chalcolithic periods in southern Turkmenistan (development of animal resource management strategies)]. St. Petersburg: Evropejskij Dom.

Kosintsev, P.A. 1989. "Okhota i skotovodstvo u naselenija lesostepnogo Zaural'ja v epokhu bronzy" [Hunting and cattle breeding of the Bronze Age population in the Trans-Ural steppes]. In *Stanovlenie i razvitie proizvodjaschego khozjajstva na Urale*, ed. V.S. Korshunov. Sverdlovsk: UrO AN SSSR, pp. 84–104.

Kosintsev, P.A. 2000. "Kostnye ostatki zhivotnykh iz ukreplennogo poselenija Arkaim" [Animal bone remains from the Arkaim fortified settlement]. In *Arkheologicheskij istochnik i modelirovanie drevnikh tekhnologij. Trudy muzeja-zapovednika Arkaim* [Archaeological source and the ancient technologies modeling. Transactions of Arkaim Museum Reserve], ed. G.B. Zdanovich. Cheljabinsk: Poligraficheskoe ob'edinenie "Kniga," pp. 17–44.

Lisitsyna, G.N. 1978. *Stanovlenie oroshaemogo zemledelija v Juzhnoj Turkmenii. Opyt istoricheskogo analiza materialov kompleksnykh issledovanij na juge SSSR i Blizhnem Vostoke* [Formation of irrigated agriculture in southern Turkmenistan: historical analysis of the materials from a combination of investigations in southern USSR and the Near East]. Moscow: Nauka.

Mackay, E.J.H. 1943. *Chanhu-daro Excavations 1935–1936*. American Oriental series, 20. New Haven, CT: American Oriental Society.

Majumdar, N.G. 1934. *Explorations in Sind, 1929–1930*. Memoirs of the Archaeological Survey of India, 48. New Delhi: Archaeological Survey of India.

Meadow, R.H. 1981. "Early Domestication in South Asia: A First Report of the Faunal Remains from Mehrgarh, Pakistan." In *South Asian Archaeology 1979*, ed. H. Härtel. Berlin: D. Reimer Verlag, pp. 143–179.

Meadow, R.H. 1984a. "Animal Domestication in the Middle East: A View from the Eastern Margin." In *Animals and Archaeology*. Vol. 3, *Early Herders and Their Flocks*, ed. J. Clutton-Brock and C. Grigson. BAR International Series, 202. Oxford: Archaeopress, pp. 309–337.

Meadow, R.H. 1984b. "A Camel Skeleton from Mohenjo-daro." In *Frontiers of the Indus Civilization*, ed. B.B. Lal and S.P. Gupta. Delhi: Books & Books, pp. 133–139.

Meadow, R.H. 1984c. "Notes on the Faunal Remains from Mehrgarh, Pakistan, with a Focus on Cattle (Bos)." In *South Asian Archaeology 1981*, ed. B. Allchin. Oriental Publications No. 34. Cambridge: Cambridge University Press, pp. 34–40.

Meadow, R.H. 1987. "Faunal Exploitation Patterns in Eastern Iran and Baluchistan: A Review of Recent Investigations." In *Orientalia Josephi Tucci Memoriae Dicata*, 2, ed. G. Gnoli and G. Tucci. Rome: Istituto Italiano per il Medio et Estremo Oriente, pp. 881–961.

Meadow, R.H. 1987–1988. "The Faunal Remains from Jalilpur." *Pakistan Archaeology* 23: 204–220.

Meadow, R.H. 1989a. "Continuity and Change in the Agriculture of the Greater Indus Valley: The Palaeoethnobotanical and Zooarchaeological Evidence." In *Old Problems and New Perspectives in the Archaeology of South Asia*, ed. J.M. Kenoyer. Wisconsin Archaeological Reports, 2. Madison: University of Wisconsin, pp. 61–74.

Meadow, R.H. 1989b. "A Note on the Distribution of Faunal Remains During the Later Period at Mehrgarh (Baluchistan, Pakistan)." In *South Asian Archaeology 1985*, ed. K. Frifelt and P. Sørensen. London: Curzon Press, pp. 167–175.

Moore, K.M. 1993. "Animals Use at Bronze Age Gonur Depe." *Unesco IASCCA Information Bulletin* 19: 164–176.

Moore K.M. 1994. "Zhivotnye v khozjajstve Gonur-depe epokhi bronzy" [Animals in the household of Bronze Age Gonur Depe]. *Unesco IASCCA Information Bulletin* 19: 200–216.

Moore, K.M., N.F. Miller, F.T. Hiebert, and R.H. Meadow 1994. "Agriculture and Herding in the Early Oasis Settlements of Oxus Civilization." *Antiquity* 68(259): 418–427.

Nath, B. 1962. "Remains of the Horse and Indian Elephant from the Prehistoric Site of Harappa (West Pakistan)." In *Proceedings of the First All-India Congress of Zoologists, 1959: Part 2, Scientific Papers*, ed. R.V. Seshaiya and B.S. Chauhan. Calcutta: Zoological Society of India, pp. 1–14.

Nath, B. 1968. "Animal Remains from Rupar and Bara sites, Ambala District, E. Punjab, India." *India Museum Bulletin* 3(1–2): 69–116.

Prashad, B. 1936. *Animal Remains from Harappa*. Memoirs of the Archaeological Survey of India, 51. Delhi: Archaeological Survey of India.

Salvatori, S. and M. Tosi 2005. "Shahr-i Sokhta Revised Sequence." In *South Asian Archaeology 2001*, ed. C. Jarrige and V. Lefèvre. Paris: Editions Recherche sur les Civilisations, pp. 281–292.
Sarianidi, V.I. 2002. *Margush: Ancient Oriental Kingdom of the Old Delta of Murghab River.* Ashgabad: Türkmendöwlethabarlary – TDH.
Sarianidi, V.I. 2005. *Gonur Depe: Turkmenistan. City of Kings and Gods.* Ashgabad: Miras.
Sarianidi, V.I. 2007. "Dvortsovo-khramovyj kompleks Severnogo Gonura" [Palace-temple complex of the North Gonur]. *Rossijskaja Arkheologija* 1: 49–63.
Sarianidi, V.I. 2008. *Margush: Mystery and True of the Great Culture.* Ashgabad: Türkmendö wlethabarlary – TDH.
Sarianidi, V. and N. Dubova. 2013. *Treasures of Ancient Margiana: Photo Album.* Asgabat: Türkmendöwletneşirýatgullugy. Türkmendöwlethabarlary – TDH.
Sataev, R.M. 2008. "Zhivotnye v khozjajstve i dukhovnoj zhizni drevnego naselenija Gonur-depe" [Animals in the economic and spiritual life of Gonur Depe ancient population]. In *Trudy Margianskoj Arkheologicheskoj Ekspeditsii*, vol. 2, ed. V.I. Sarianidi. Moscow: Staryj Sad, pp. 143–160.
Sataev, R.M. 2016. *Zhivotnye v kul'ture Drevnej Margiany* [Animals in the culture of ancient Margiana]. Moscow: Staryj Sad.
Sataev, R.M. and L.V. Sataeva 2010. "Problema rekonstruktsii prirodnykh uslovij drevnego Gonurskogo oazisa" [Problem of the reconstruction of the natural conditions in the ancient Gonur oasis]. In *Dinamika ekosistem v golocene* [Dynamics of ecosystems in the Holocene], ed. N.G. Smirnov. Ekaterinburg-Cheljabinsk: Rifej, pp. 190–193.
Sataev, R.M. and L.V. Sataeva 2012a. "Arkheozoologicheskie i arkheobotanicheskie issledovanija na Gonur-depe v 2010 g" [Archaeozoological and archaeobotanical studies at Gonur Depe]. In *Trudy Margianskoj Arkheologicheskoj Ekspeditsii*, vol. 4, ed. V.I. Sarianidi. Moscow: Staryj Sad, pp. 57–61.
Sataev, R.M. and L.V. Sataeva 2012b. "Krupnyj rogatyj skot v khozjaistve Drevnej Margiany" [Cattle in the economy of ancient Margiana]. In *Trudy Margianskoj Arkheologicheskoj Ekspeditsii*, vol. 4, ed. V.I. Sarianidi. Moscow: Staryj Sad, pp. 154–158.
Sataev, R.M. and L.V. Sataeva 2013. "Istoricheskie predposylki razvedenija krupnogo rogatogo skota v aridnykh uslovijakh juga Srednej Azii" [Historical backgrounds of cattle breeding in the arid conditions of southern Central Asia]. *Vestnik Bashirskogo Gosudarstvennogo Agrarnogo Universiteta* 1(25): 62–65.
Sataev, R.M. and L.V. Sataeva 2014. "Results of Archaeozoological and Archaeobotanical Research at the Bronze Age Gonur Depe Site (Turkmenistan)." In *Proceedings of the 8th International Congress on the Archaeology of the Ancient Near East 2012*, vol. 3, ed. P. Bieliński, M. Gawlikowski, R. Koliński, D. Ławecka, A. Sołtysiak, and Z. Wygnańska. Wiesbaden: Harrassowitz Verlag, pp. 369–372.
Sewell, R.B.S. and B.S. Guha 1931. "Zoological Remains." In *Mohenjo-daro and the Indus Civilization*, ed. J. Marshall, vol. II. London: Arthur Probsthain, pp. 649–673.
Shevchenko, A.I. 1960. "K istorii domashnikh zhivotnykh Juzhnogo Turkmenistana" [History of the domestic animals of southern Turkmenistan]. In *Trudy Juzhno-Turkmenistanskoj Arkheologicheskoj kompleksnoj ekspeditsii*, vol. 10. Ashkhabad: Akademii Nauk Turkmenskoj SSSR, pp. 464–477.
Tsalkin, V.I. 1970. "Drevnejshie domashnie zhivotnye Srednej Azii" [The oldest domestic animals of the Middle Asia]. *Bjulleten' Moskovskogo Obshchestva Ispytatelej Prirody. Otdelenie biologii* [Bulletin of the Moscow Society of Naturalists. Biological Section] LXXV(1): 145–149.
Tsareva, E.G. 2009. "Where and When Knotted Rugs Began to Be Produced." In *Filikl, Anatolia XX Century: Moreschini Collection*, ed. P. Barbetti, A. Boralevi, A. Levi, and E. Tsareva. Florence: Angelo Pontecoroli editore, pp. 48–55.

Tsareva, E.G. 2012. "Filikli: k istorii uzelkovogo tkachestva v Tsentral'noj Azii (po materialam Mesopotamii i Turkmenistana)" [Filikli: history of knotted weaving in Central Asia (from Mesopotamian and Turkmenistan data)]. In *Trudy Margianskoj Arkheologicheskoj Ekspeditsii*, vol. 4, ed. V.I. Sarianidi. Moscow: Staryj Sad, pp. 240–249.

Vinogradova, N.M., V.A. Ranov, and T.G. Filimonova 2008. *Pamjatniki Kangurttuta v Jugo-zapadnom Tadzhikistane (epokha neolita i bronzovyj vek)* [The settlements of Kangurttut in southwestern Tajikistan (Neolithic and Bronze Age)]. Moscow: Institut vostokovedenija RAN.

Wheeler, R.E.M. 1953. *The Indian Civilization*. Cambridge: Cambridge University Press.

Zajtseva, G.I., N.A. Dubova, A.A. Sementsov, P. Reimar, J. Mallory, and H. Yungner 2008. "Radiouglerodnaja khronologija pamjatnika Gonur Depe" [Radiocarbon chronology of the Gonur Depe site]. In *Trudy Margianskoj Arkheologicheskoj Ekspeditsii*, vol. 2, ed. V.I. Sarianidi. Moscow: Staryj Sad, pp. 166–179.

CHAPTER SIXTEEN

LIFE IN THE COUNTRYSIDE

The rural archaeology of the Sapalli culture

Kai Kaniuth

INTRODUCTION

This chapter describes the rural facet of Late Bronze Age life in the Sapalli culture. It is based on recent survey work and the results of excavations conducted between 2007 and 2010 at the site of Tilla Bulak in the Pashkhurt plain on the western border of the Surkhan Darya valley (Figure 16.1).

The Sapalli culture, the local northern Bactrian variant of the Oxus Civilization,[1] flourished from the 20th to the 15th century BC.[2] According to ceramic parallels, it is contemporaneous with the Turkmen Namazga (NMG) VI phase and with all – except the earliest – Bronze Age remains in the Murghab delta.[3] In Uzbekistan, it can be divided into two phases, referred to here as Late Bronze (LB) I and LB II. The Late Bronze Age is clearly separated from the following Early Iron Age (the Jaz I/Kuchuk cultures with their distinctive handmade painted wares), which developed from ca. 1450 BC onwards. This terminology requires some explanation: My division between the Middle and Late Bronze Ages follows the initial classification of pottery from Namazga Depe (Ganjalin 1956; Kuftin 1956), which considered the NMG V period as Middle Bronze Age, and the subsequent NMG VI period as Late Bronze Age, without any chronological overlap. Subsequent redefinitions of the Middle Bronze Age have focused rather on social attributes and cultural-historical aspects, such as settlement structure (urban/nonurban), and thereby confused the matter. By using the original relative chronological concept, in no way do we embrace accompanying notions of absolute chronology, ethnogenesis, or migratory history. Furthermore, there is still some terminological disagreement about the final phase of the Late Bronze Age in southern Uzbekistan. This concerns the period between ca. 1700–1500 BC, which includes the successive phases of Kuzali, Molali, and Bustan according to Soviet and Uzbek scholars; it is referred to as LB II by most German archaeologists, and as "Bronze Final" in recent French publications (e.g., Luneau 2014). Its inclusion within the Oxus Civilization was endorsed by H.-P. Francfort, although he considered the dates of 2300–1500 BC "larges et symboliques" (Francfort 2005: 258) and later on consistently used the higher date of 1750 BC for its end (Francfort 2009; Francfort and Tremblay 2010).

The first Late Bronze Age sites of southern Uzbekistan were discovered in the late 1960s, when researchers of the Fine Arts Institute from Tashkent discovered the site of Molalitepa (Beljaeva and Khakimov 1973) and a team from the newly founded Samarkand Institute of Archaeology started excavations at Sapallitepa (Al'baum 1969; Askarov 1971, 1973).[4] A few years later, in 1973, A. Askarov began working at Dzharkutan, the largest Sapalli culture settlement (Askarov 1977; Askarov and Abdullaev 1983). Excavations there subsequently continued in cooperation with German (Huff 1995, 2000) and French scholars (Bendezu-Sarmiento and Mustafakulov 2008, 2013; see also Chapter 13). In the Pashkhurt plain, initial surveys did not discover any Bronze Age remains (Rtveladze and Khakimov 1973; Rtveladze 1974), and the first artifacts dated to the 2nd millennium were published only in 2006 (Mokroborodov 2006).

SETTLEMENT PATTERNS

In Margiana, the core region of the Oxus Civilization, several hundred Bronze Age sites are known along the deltaic fan of the Murghab River (Sarianidi 1990; Gubaev *et al.* 1998; Salvatori and Tosi 2008; Markofsky 2010). This settlement zone remained largely intact thanks to the subsequent retraction of the Murghab delta, which kept the sites from disturbance. Along the river branches, subsistence was based on irrigation agriculture (Miller 1999; Castiglione and Cottini 2002; Sataev and Sataeva 2012) and mixed animal husbandry (Moore 1993; Sataev and Sataeva 2014). Irrigation was presumably organized by local landlords, resident in fortified *qalas* at the heart of the riverine oases.[5] Francfort's early postulate of Bronze Age urban settlement in Bactria and Margiana (Francfort 1984, 1989, 2005) and V.I. Sarianidi's assumption of a highly stratified social system (Sarianidi 1990, 1998, 2002, 2005, 2008; see also Chapters 10 and 11) have been justified through the stunning results of the excavations at Gonur Depe. Here, the central *qala* with its double enceinte merely makes up the monumental center of a larger agglomeration comprising manufacturing zones, domestic architecture, and prestige burial grounds, all enclosed by a city wall. With around 50 ha, this is the largest Murghab site, and it may have been the center of a tiered settlement system (Salvatori 1998: 57–65; see also Chapter 5).

In northern Afghanistan (southern Bactria), the settlement structure along the Amu Darya's tributaries seems to resemble that of Margiana, judging by survey work in the 1970s (Kohl 1984: 162 map 18, summing up Sarianidi 1977: figs. 5, 7, 8).

In the Surkhan Darya valley, the natural conditions differ. The deeply incised Surkhan River requires substantial hydraulic efforts in order to be used for irrigation. Consequently, all Bronze Age settlements are concentrated along the wider western half of the plain, where smaller tributaries – the Sherabad Darya, Ulanbulaksaj, Khalkadzarsaj, and Sangardak – could be tapped more easily for runoff irrigation. When considering the Surkhan Darya plain's settlement pattern, it must be borne in mind that since the 1960s the region underwent a thorough land reshuffle to prepare for intensive agriculture and cotton growing. This process resulted in a massive loss of archaeological data and irrevocably changes our perception of the Sapalli culture living conditions. Only for one irrigation zone, the Bustansaj oasis, do we have an estimate of 20 km^2 of arable land (Shirinov 2002: 155).

At Dzharkutan, what remains of the site measures some 40 ha, half of which constitutes part of a single settled zone with several monumental compounds and production areas, the rest being devoted to several necropolises. A glimpse into

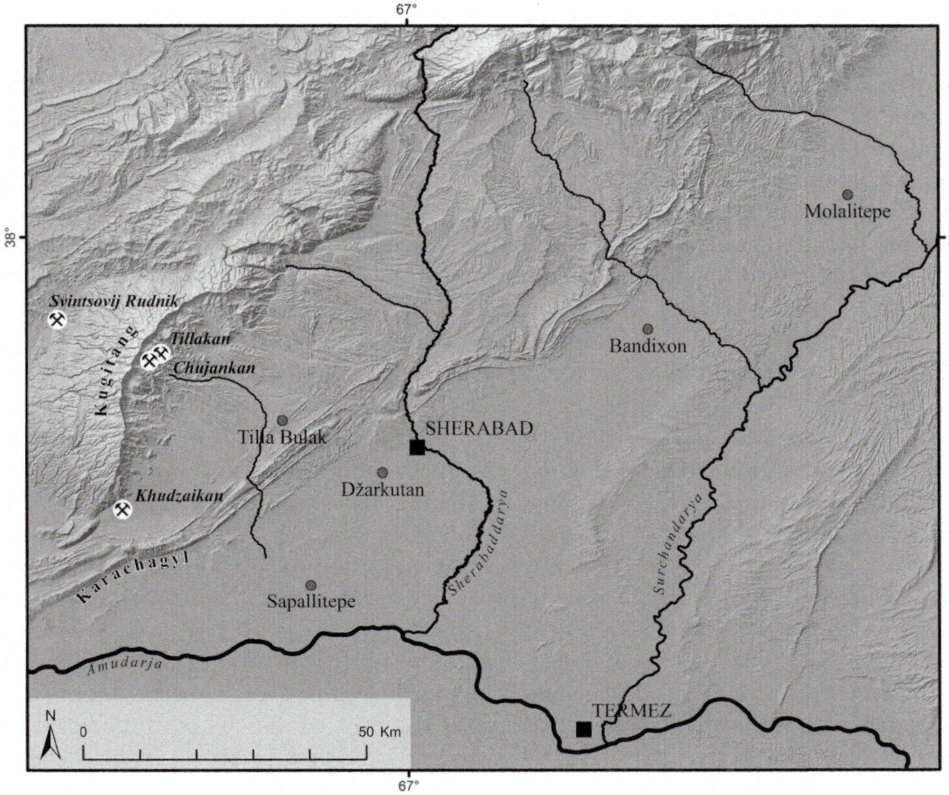

Figure 16.1 Surkhan Darya region map with major Late Bronze Age sites (map by St. E. Metz; base map: NASA SRTM GL-1).

the as yet unpublished data from a geophysical prospection carried out in 2012 (Mustafakulov *et al.* 2014: fig. 8) shows a dense occupation between the excavated *qalas* of the citadel, the "temple" (Tepe 6) and the remains on Tepes 4 (Kaniuth 2014), 5 and 8. Whether this should qualify Dzharkutan as an "urban" site is open to discussion, but the occupation is more substantial than commonly thought. At Sapallitepa, the known remains comprise a square *qala* of 82 × 82 m, but soundings in the neighboring cotton fields encountered less well-preserved settlement remains. It is not much to build an argument upon, but leaves open the possibility that the Sapallitepa settled zone was larger than the 3–4 ha usually accorded. Further north, the settlements of Molalitepa and Ochamajlitepe measure some 8–10 ha each, but have not been thoroughly investigated. Judging by the two excavated sites of Dzharkutan and Sapallitepa, therefore, the settlements of the Sapalli culture were made up of larger agglomerations with at least one central *qala* and surrounding unfortified housing. City walls have been detected with certainty only at Gonur.[6] At Dzharkutan, a very thick wall opened on Tepe 7 may, however, be part of a similar structure (Askarov and Shirinov 1993: fig. 68, *II*).

Even though the Bronze Age settlement pattern in the Surkhan Darya valley differed from that of Margiana and southern Bactria, until recently the question remained if adaptions to more marginal zones existed, and if so, what form they took. When the site of Tilla Bulak was discovered by chance in the summer of 2006 in a region previously thought to be entirely devoid of pre-Hellenistic occupation, the possibility arose of investigating such a settlement area.

The Pashkhurt plain is a small, sparsely settled intermontane zone (between 700 and 1,200 m asl), 30 km long in a north–south direction, and perched between the Karachagyl chain (elevation ca. 900–1,000 m asl) to the east and the Kugitang Mountains (up to 3,000 m asl) to the west. It has only a single major watercourse, fed by minor tributaries flowing from the Kugitang range. This Dabilsaj River changes its name to Ulanbulaksaj after its descent into the Surkhan Darya plain at the village of Gaz, and, in these early times, continued to flow past the site of Sapallitepa.[7]

The Kugitang mountain chain consists of Mesozoic (Jurassic) limestone and intrusive (Paleozoic?) granite, while the Karachagyl chain consists of folded Mesozoic (Jurassic, Cretaceous) and Neogene sediments and sedimentary rocks (mudrock). The depression between them is filled with Quaternary sediments of gravel and loess. The rich mineral resources of the Kugitang may have constituted an incentive for settling this intermontane zone. On both sides of the mountain, salt,[8] copper, tin, lead, and iron[9] are readily available, but no pre-medieval exploitation has yet been documented (Ionin and Shafranov 1937; Raevskij and Fiveg 1973; Pruger 1980; Ruzanov and Burjakov 1997; Sverchkov 2009).

When Tilla Bulak was discovered, it was the first prehistoric settlement known in the valley.[10] Since then, unsystematic prospection[11] discovered several more sites in the eastern half of the Pashkhurt valley, the majority of which are of medieval or modern date (Kaniuth 2010b; Dvurechenskaja *et al.* 2014). After 2015, an Uzbek–Czech research project of the universities of Termez and Prague has investigated the region in a more systematic fashion.[12] It has demonstrated that the northern part of the Pashkhurt plain was densely inhabited for most of the 2nd millennium BC, with both Sapalli culture and Jaz I sites well represented (Danielisová *et al.* 2009; Stančo *et al.* 2014; Augustinovà *et al.* 2015; Stančo 2016; Augustinovà, Stančo, Damasêk, Khamidov *et al.* 2017; Augustinovà, Stančo, Damasêk, Mrva *et al.* 2017; Stančo *et al.* 2017).

LATE BRONZE AGE SITES IN THE PASHKHURT REGION

Altogether, almost two dozen Sapalli culture sites have so far been identified in the Pashkhurt region (Figure 16.2 and Table 16.1). In its northern part, around the villages of Karabag, Zarabag, and Maydan, ten Bronze Age sites have been documented by the Uzbek–Czech expedition (Augustinovà, Stančo, Damasêk, Khamidov *et al.* 2017; Augustinovà, Stančo, Damasêk, Mrva *et al.* 2017).[13] The major ones are the following.

Bobolangar, ca. 30 × 35 m, is situated 15 m above a spring in the southern part of the Zarabag oasis. Due to the good preservation of the survey pottery it was initially considered to be a cemetery (Augustinovà, Stančo, Damasêk, Mrva *et al.* 2017: 125–128). Following excavation of three soundings at Bobolangar, it became clear that we are dealing with a stratified settlement involving several occupation, destruction, and rebuilding events. The published pottery dates to the very earliest phase of the Sapalli culture (LB Ia) (Kysela *et al.* 2018: 164–173). Tulkitepa is also located in the

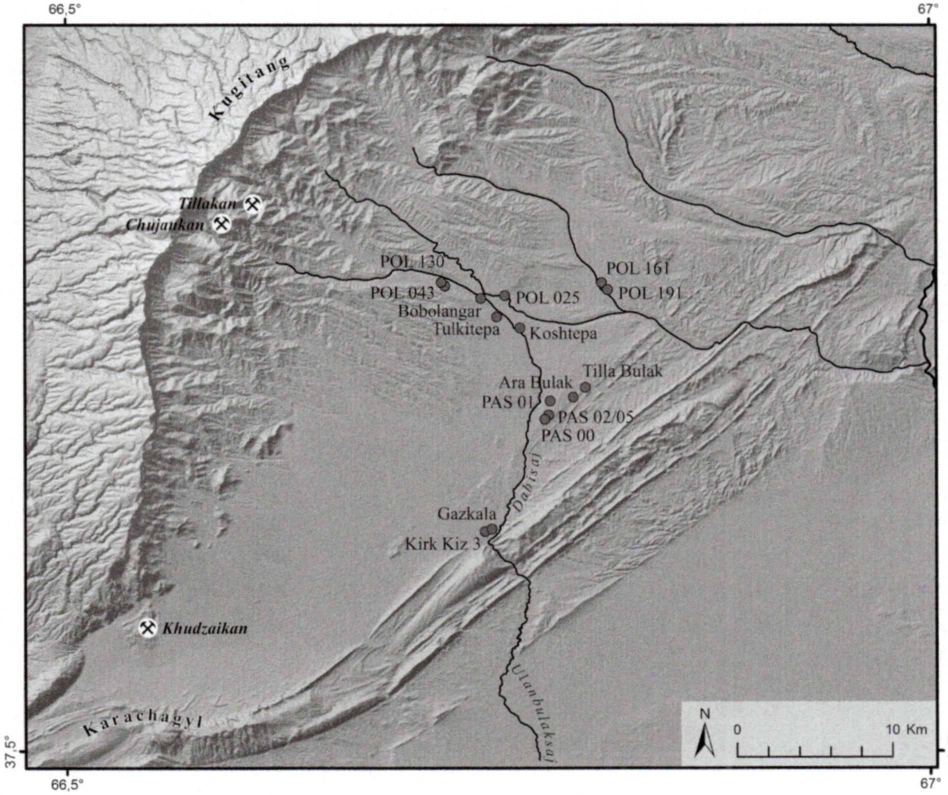

Figure 16.2 Pashkhurt valley map with known Late Bronze Age sites (map by St. E. Metz; base map: NASA SRTM GL-1).

Zarabag oasis and is likewise ca. 50 × 50 m large. Half of the collected pottery (20 of 38 dateable sherds) is LB in date, indicating a quite substantial Bronze Age occupation (Augustinovà, Stančo, Damasêk, Mrva *et al.* 2017). Koshtepa (T1) (Augustinová *et al.* 2015), 1.2 km south of Zarabag, is an oval hill 2 m high and 50 m wide at its largest point. Several well-preserved vessels were recovered here. Out of 16 dateable sherds, five go back to the Late Bronze Age. A very recent addition (Stančo *et al.* 2017) is the site of Yultepa, halfway between the Pashkhurt and Surkhan Darya valleys and cut by the modern road. If its extent, as seen from satellite imagery, should turn out to relate exclusively to a Late Bronze Age occupation, this place, with 2 ha or more, would be the largest intermontane Sapalli culture site.

In the central part of the Pashkhurt plain, just east of the eponymous village, two Late Bronze Age settlements were discovered on the westernmost ridge of the Karachagyl folding zone: Tilla Bulak (0.4 ha, see pp. 471–479) and, 800 m to the south, Ara Bulak (ca. 0.3 ha). Both date to the LB I phase. A second ceramic distribution zone was documented on the crest of the elevated left bank of the Dabilsaj, on the edge of the modern town. In two instances (PAS 00, PAS 01), whole vessels of LB II date were recorded (Mokroborodov 2006; Kaniuth 2010b), suggesting that this place was used for burials, without visible contemporary settlements.

Table 16.1 Late Bronze Age sites in the Pashkhurt valley

Name	*Alternative name(s)*	*Coordinates E*	*(WGS 84) N*	*Village*	*Investigation*
Koshtepa	POL_023, KuPi_001	66.76292°	37.74502°	Zarabag	Survey
POL 025	POL_025	66.75412°	37.76380°	Zarabag	Survey
Bobolangar	POL_039, KuPi_008	66.74039°	37.76207°	Zarabag	Survey
POL 043	POL_043	66.71900°	37.76951°	Zarabag	Survey
Tulkitepa	POL_085, KuPi_011	66.74943°	37.75139°	Zarabag	Survey
POL 130	POL_130	66.71865°	37.77035°	Zarabag	Survey

Description	*Size*	*Quantity of dateable LBA (vs. all) finds*	*Date*	*References*	*Notes*
Tepe	0.2 ha (?) Stančo *et al.* 2017: tab. 2)	5 (15)	LB	Augustinová *et al.* 2015; Augustinová, Stančo, Damašek, Mrva *et al.* 2017	1 sherd illustrated in Augustinová *et al.* 2015: fig. 1, 1
Isolated finds	–	5 (7)	LB	Augustinová, Stančo, Damašek, Mrva *et al.* 2017	–
Tepe or cemetery(?)	0.1 ha	97 (438)	LB	Augustinová, Stančo, Damašek, Mrva *et al.* 2017; Kysela *et al.* 2018	11 sherds illustrated in Augustinová, Stančo, Damašek, Mrva *et al.* 2017: fig. 20; 23 pottery items illustrated in Kysela *et al.* 2018: figs. 13, 14, and pl. 4, 2
Isolated finds	–	3 (4)	LB	Augustinová, Stančo, Damašek, Mrva *et al.* 2017	–
Tepe	0.2 ha/2 ha (Stančo *et al.* 2017: tab. 2)	20 (38)	LB	Augustinová, Stančo, Damašek, Mrva *et al.* 2017	–
Isolated finds	–	3 (20)	LB	Augustinová, Stančo, Damašek, Mrva *et al.* 2017	–

(*continued*)

Table 16.1 Cont.

Name	*Alternative name(s)*	*Coordinates* E	*(WGS 84)* N	*Village*	*Investigation*
POL 161	POL_161	66.81022°	37.77121°	Karabag	Survey
POL 191	POL_191	66.81368°	37.76716°	Karabag	Survey
Kalapushtepa	POL 218, KuPi 065	66.86056°	37.73242°	Maydan	Survey
Yultepa	ShD 254	66.922049°	37.744933°	Maydan	Survey
Tilla Bulak	PAS_TB01	66.80075°	37.71067°	Pashkhurt	Excavation

Description	*Size*	*Quantity of dateable LBA (vs. all) finds*	*Date*	*References*	*Notes*
Isolated finds	–	4 (9)	LB	Augustinová, Stančo, Damašek, Mrva *et al.* 2017	–
Isolated finds	–	3 (10)	LB	Augustinová, Stančo, Damašek, Mrva *et al.* 2017	–
Tepe	0.1 ha	5 (6)	LB	Augustinová, Stančo, Damašek, Khamidov *et al.* 2017	5 LBA sherds reported
Tepe	2 ha	?	LB	Stančo *et al.* 2017	Several dozen LBA sherds were observed by the author during a visit to the site in 2017; exact figures from the survey are not yet available
Tepe (settlement)	0.5 ha	ca. 60,000	LB I	Kaniuth 2007, 2009, 2010b, 2011, 2016	180 sherds and vessels have been published in Kaniuth 2007: figs. 13–14; 2009: figs. 7–12; Kaniuth 2010b: figs. 8–11; 2011: figs. 11, 12, 20; 2016: fig. 4

(*continued*)

Table 16.1 Cont.

Name	*Alternative name(s)*	*Coordinates* E	*(WGS 84)* N	*Village*	*Investigation*
Ara Bulak	PAS_AB01	66.79370°	37.70506°	Pashkhurt	Excavation
PAS 01	PAS_01	66.78030°	37.70290°	Pashkhurt	Survey (unsystematic)
PAS 02/05	PAS_ 02.03.04.05	66.77940°	37.69480°	Pashkhurt	Survey (unsystematic)
PAS 00	Unnamed (Bezimjannoe)	66.77° (?)	37.69° (?)	Pashkhurt	Survey (unsystematic)
Kirkkiz 3	Kyrkkyz-Ata?	66.74253°	37.62783°	Gaz	Survey (unsystematic)

Description	*Size*	*Quantity of dateable LBA (vs. all) finds*	*Date*	*References*	*Notes*
Tepe (settlement)	0.3 ha	74	LB I	Kaniuth 2010b	55 sherds illustrated in Kaniuth 2010b: figs. 43–45
Burials(?)	0.2 ha (sherd distr.)	32	LB II	Kaniuth 2010b	3 vessels in Kaniuth 2010b: fig. 40; 11 vessels this chapter, Figure 16.10
Burials(?)	0.6 ha (sherd distr.)	8	LB II	Kaniuth in press	–
Burials(?)	?	12	LB II	Mokroborodov 2006	12 items illustrated in Mokroborodov 2006: fig. 1
Tepe or cemetery(?)	0.2 ha (sherd distr.)	21	LB I	Dvurechenskaya *et al.* 2014 (Kyrkkyz-Ata); Kaniuth, this chapter, Figure 16.3	9 sherds this chapter, Figure 16.3; 12 sherds (from "Kyrkyz-Ata", Dvurechenskaja *et al.* 2014; according to A. Augustinová (pers. comm.) this may be the spot where the LB I vessels kept in the Gaz school may come from

(*continued*)

Table 16.1 Cont.

Name	*Alternative name(s)*	*Coordinates* E	*(WGS 84)* N	*Village*	*Investigation*
Gazkala	Goz Kala Buloq/Kirkkiz (Augustinovà, Stančo, Damasêk, Khamidov *et al.* 2017: 149)	66.74642°	37.62922°	Gaz	Excavation
Goz Dagana 1	POL 240.241, KuPi 074	66.74696°	37.62178°	Gaz	Survey
Kyzylbay 1/2	POL 237.238, KuPi 073	66.7057°	37.6101°	Gaz	Survey
Kyzylbay 3	POL 239, KuPi 072	66.704231°	37.608542°	Gaz	Survey

Note: LBA = Late Bronze Age.

Description	*Size*	*Quantity of dateable LBA (vs. all) finds*	*Date*	*References*	*Notes*
Tepe (settlement)	0.7 ha	40	LB I	Dvurechenskaja *et al.* 2014; Rukavishnikova *et al.* 2015	37 sherds illustrated in Dvurechenskaja *et al.* 2014: fig. 3; 3 sherds published in Rukavishnikova *et al.* 2015: fig. 4
Tepe	100m²	8 (62)	–	Augustinová, Stančo, Damašek, Khamidov *et al.* 2017; Stančo *et al.* 2017;	8 LBA sherds reported
Tepe	1.420m² (K.2) 1.100 m² (K.1)	4 (85)	–	Augustinová, Stančo, Damašek, Khamidov *et al.* 2017; Stančo *et al.* 2017	4 LBA sherds reported by Augustinová, Stančo, Damašek, Khamidov *et al.* 2017; 3 of these come from the small tepe (POL 238), 1 from the lower terrace (POL 237)
Tepe	730m²	12 (89)	–	Augustinová, Stančo, Damašek, Khamidov *et al.* 2017; Stančo *et al.* 2017	4 sherds illustrated in Augustinová, Stančo, Damašek, Khamidov *et al.* 2017: figs. 11, *1*; 12, *1–3*

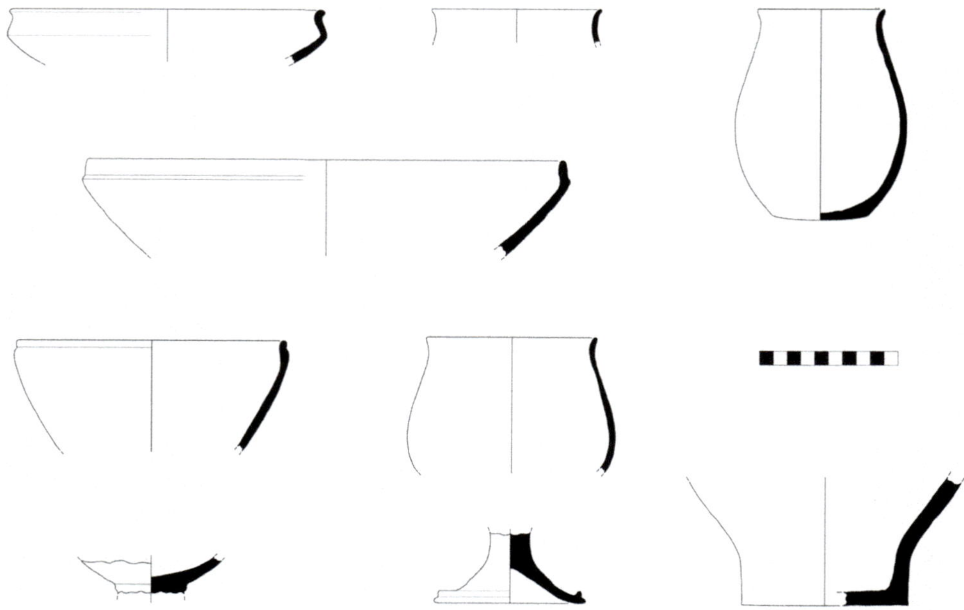

Figure 16.3 Late Bronze Age pottery from Kirkkiz 3.

Near the village of Gaz, the most important Bronze Age site appears to be Gazkala,[14] a dominant rocky spur with terracing on its western side. The (LB I) pottery distribution is said to extend for several hundred meters up to the banks of the Dabilsaj (Dvurechenskaja *et al.* 2014). In its immediate environs, west of Gazkala, at least one further concentration of LB I ceramics, labeled Kirkkiz 3, was discovered by the Tilla Bulak team (Figure 16.3).[15] The pottery scatter did not exceed 0.25 ha, but it may well be related to the Gazkala settlement 400 m away. Possibly, Kirkkiz 3 marks the location of the settlements' necropolis. At the site of Dzharkutan, for instance, the burial grounds (Dzharkutan 4a, b, c, and the Bustan group) are also separated from the living place by small streams. Gazkala and Kirkkiz 3 are separated by modern fields and by the remains of the sprawling late antique and early medieval settlement of Gaz-Ata (also named Gaztepa), which may, of course, have covered up a previous Bronze Age occupation. In any case, the Gaz region contained Late Bronze Age occupational remains distributed over an area of no less than 20 ha.

Some 4 km south of Gaz, another settled zone existed at Kyzylbay. Late Bronze Age ceramics stem from a tepe (POL 239) and two localities several hundred meters upstream (POL 237, 238). All the documented sherd scatters do not add up to more than 0.3 ha. Once the current survey program has covered the southern end of the Pashkhurt valley, around the villages of Aqtash and Chorvoq, this picture will most likely be complemented by further sites.

The overall impression then, is of fairly substantial, but spatially delimited habitational zones, clustered in or close to contemporary villages near springs, all answering the same primary requirement of water availability. The proximity to the mountains and the resulting dynamic relief of the landscape offer concomitantly very specific opportunities and limits to an economic valorization. We can currently only guess as to what degree the mineral resources of the Pashkhurt region were exploited. Clearly, however, the limits of easily arable and irrigable land must have restricted the sustainable population, and consequently the settlement sizes. The settlement structure therefore differs much from that of the (comparatively) densely clustered and populated centers of Margiana, where the major sites exhibit such a high degree of architectural elaboration. Since Margiana's smaller hamlets are still waiting to be studied in detail, and since only one is known in Bactria (Tilla Bulak, see the next section), we are at a loss for drawing any conclusions about regional specific residential or subsistence patterns.

TILLA BULAK

Of the Late Bronze Age sites in the Pashkhurt region, only Tilla Bulak (the "Gold Spring") has been the target of a major field expedition, from 2007 until 2010.[16] Geomorphological investigations showed that the site had been founded in a particularly favorable environmental niche. Very conscious choices were made in its placement, taking into account visibility, water availability, but also conditions for building: The settlement is located on a drainage divide, and, therefore, on permanently dry ground. At the same time, it controlled a pocket of arable land to the east, easily watered by small channels fed from a perennial ("golden") spring at the foot of the site (Makki in Kaniuth 2010b). By contrast, only livestock could be kept to the west of Tilla Bulak. Irrigation farming there requires either the tapping of groundwater resources or the construction of an elaborate canal system to bring water over larger distances.[17]

Tilla Bulak saw two major occupational phases, with radiocarbon (^{14}C) dates falling between the 20th and 18th century BC[18] (see also the Appendix at the end of this volume). The first phase ended in a fire sometime in the late 19th century BC. The settlement was then rapidly rebuilt, only to be left once and for all in the mid-18th century BC after the relatively short reoccupation of phase 2. Initially, settlers occupied a small, steep hill of half a hectare in size and laid the ground for dense mudbrick architecture covering the entire elevation by terracing operations (Figure 16.4).

Contrary to what is regularly encountered in other larger Bactria-Margiana Archaeological Complex (BMAC) settlements, not much energy was spent on defensive architecture. Walls of different thickness at Tilla Bulak probably indicate terracing rather than protective measures, and a single round corner bastion within the settlement was, likewise, of limited value against military threats. To the northwest, access to the site was channeled by a 3 m wide artificial ditch. A 6 m wide opening remained between the moat heads, again without any visible fortification.

A central lane led from this entrance up to the top of the hill, where it ended in a blind alley just before the topmost compound (rooms 1 and 2). The households consisted of two to four rooms (i.e., about 30–60 m^2).[19] At least one larger room in each household contained a fireplace with a chimney (Figure 16.5).

In some places, mud plaster still stuck to the walls, and floors consisted of beaten earth. Entrances from the lanes were in several instances marked by exterior stepping stones, which had to be raised as waste accumulated in the alleys. No hierarchy in the rooms or among the households can be postulated on the basis of either size,

Figure 16.4 Tilla Bulak. Plan of phase 1 (plan by M. Gruber, taken from Kaniuth 2016: fig. 3). The major access route through the village is marked by arrows.

Figure 16.5 Chimney of Tilla Bulak, phase 1.

embellishments, or small finds, although a larger proportion of "prestigious" objects were discovered in the upper parts of the village.

The first settlement was destroyed at the end of the 19th century BC by a conflagration of unknown cause. This phase 1 was documented only in parts of the site, but yielded much valuable information on Bronze Age life since a lot of material had been preserved *in situ* under the debris. Phase 2 architecture was built immediately above the earlier walls in many cases, but the settlement pattern seems less dense. Several households were now surrounded with open spaces, others were abandoned and filled with rubble and refuse.

From individual rooms of phase 1, the *in situ* inventories that have been recovered consist of both fine tableware and cooking ware, grinding stones, and storage vessels (some with volumes of up to 175 l). The botanical finds show that the inhabitants of Tilla Bulak were engaged in mixed farming. Among cereals, barley (*H. vulgare*) and hulled wheat (*T. aestivum*, *T. spelta*) are predominant, each making up about 30% of the macro remains recovered during flotation.[20] Rye (*S. secale*), oat (*Avena*), millet (*P. miliaceum*), peas (*P. sativum*), and beans (*Fabaceae/Vicia faba*) make up the remainder. The question is still open as to the identification of wild versus domesticated grape (*V. vinifera*), which is abundantly attested (Peters in Kaniuth 2009). Animal bones were exclusively recovered from refuse contexts. As can be expected, sheep and goat dominate the assemblage, but cattle also contributed significantly to the diet, while the proportion of pig is negligible (Sachs in Kaniuth 2010b). This alimentation, based on half a dozen staples, was supplemented through hunting, as shown by finds of deer, bird, and even a few bear bones, and possibly gathering, as suggested by cherry kernels. The processing of food was a household affair, judging by the even distribution of grinding stones.

Pottery is by far the predominant find category. Some 60,000 sherds have been recovered, 6,400 of which are diagnostic (rims, bases, and 142 archaeologically complete shapes). Technologically, they are indistinguishable from the contemporary material of Sapallitepa and the early levels of Dzharkutan. The clay of the wheel-thrown fine wares (90% of all the sherds) is well prepared, with very few visible inclusions. The firing conditions (800–850°C) were tightly controlled and both reducing and oxidizing atmospheres used. The proportion of sherds with gray surface color is about 1% by number, and considerably less by weight. Given the high quality of this tableware throughout the Sapalli culture area, and the lack of kilns at Tilla Bulak, the question arose of its local production. A first series of XRF analyses on 40 sherds from Tilla Bulak showed that practically all the tableware had been produced from the same clay source, strongly suggesting that this pottery was locally manufactured (Daszkiewicz and Schneider in Kaniuth 2007; Kaniuth *et al.* in press).[21] Typologically, very few forms from Tilla Bulak are unattested at other Sapalli culture sites. The frequency of conspicuous vessel classes, such as footed bowls (5–6%) or goblets (ca. 1%), more or less agrees with that in graves from the LB I period and sheds an interesting light on the relationship between burial and settlement assemblages.[22] Cooking ware, a handmade and heavily shell-tempered pottery, completely absent from graves, is much more important (up to 25%) in Tilla Bulak primary contexts than published evidence from elsewhere would lead us to expect. This probably reflects the poorer preservation conditions in secondary contexts at Dzharkutan and Sapallitepa. An example of an *in situ* room inventory is reproduced in Figure 16.6, and the associated pottery in Figure 16.7.

A range of household utensils were discovered at the site: terracotta and stone spindle whorls, flint and bone arrowheads, bone awls, astragali, and stone grinding tools. Grooved stone balls, possibly loom counterweights, may indicate local weaving. Though very few items of jewelry would be expected from a rural settlement context, we did find some beads of lapis lazuli, turquoise, and agate, some copper or bronze dress pins and one limestone eagle-shaped pendant (see Figure 16.9). This latter piece is interesting for its wide distribution across southern Middle Asia, southwestern Iran, and up to western Syria, where one similar find is known at Ebla.[23] The eagle-shaped pendant bears on its back a number of drill holes, leading over to the large group of typical BMAC "amulets," mostly flat disk-shaped stones with a geometric motif on one or on both sides. At Sapallitepa and Dzharkutan, the majority of these amulets consisted of gypsum beads strung as part of necklaces, but larger examples exist (Askarov 1973: 95, fig. 47, pl. 30; 1977: pl. 43, 45). Similar pieces from the Afghan art market (Sarianidi 1998: 200–233, nos. 1060–1233) and from Margiana (Sarianidi 1998: 304f, nos. 1669–1670) are more often made of colored stones. M. Teufer (2015: 51) has pointed to the widely shared canon of images and interpreted it as part of an interregional system of communication (see also Chapter 6).

Figure 16.6 Tilla Bulak, phase 1. Context TB-852 with *in situ* inventory covered by mudbrick collapse. Note the *in situ* grinding stones in the northeastern corner of the room. Phase 2 walls are lightly shaded.

Figure 16.7 Tilla Bulak. Room inventory 3145 from context TB-852, phase 1.

When such flat-surfaced objects have a lug handle (mostly on the larger pieces over 2 cm in diameter), they are usually described as seals, a term that is questionable. Sealing, apparently, was mostly restricted to pottery vessels, which leaves a wide range of interpretative possibilities and places them in the same context as other pot or potter's marks applied before firing.[24] It is interesting to note that here is no evidence for the use of both seals and pottery marks after LB I. Sealing on clay stoppers or bullae is very rare and does not seem to have played a role in substantial trade activities, but rather for strictly local storage practices.[25] The huge number of metal

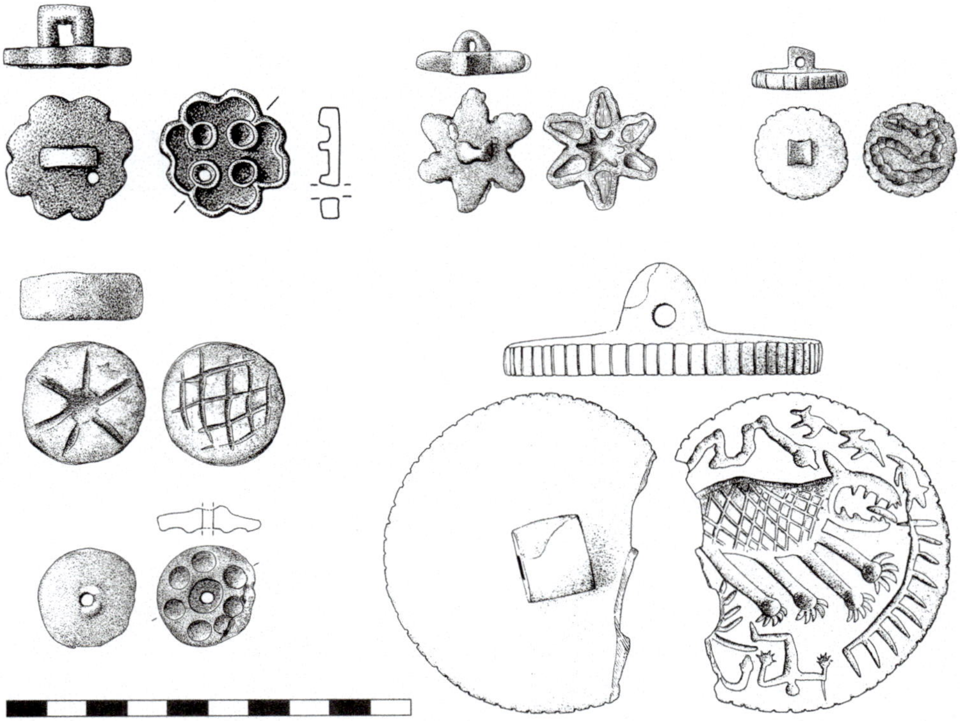

Figure 16.8 Tilla Bulak. LB I seals of stone, metal, and alabaster (all drawings by C. Wolff).

compartmented seals known in the BMAC area (Baghestani 1997; Sarianidi 1998; see also Chapter 8) is undoubtedly indicative of an awareness of the importance of administrative practices, but it does not in itself constitute proof of its application.

At Tilla Bulak, the only evidence for seal use was that of compartmented seals that had been impressed on pottery before firing (Kaniuth 2011: fig. 17), even though we paid particular attention to burnt or unburnt clay fragments during excavations and dry-sieved part of the excavated sediments. Given this negative evidence for the administrative use of seals, the variety of items considered as such discovered at Tilla Bulak is noteworthy, especially in a rural settlement context (Figure 16.8). From small circular stones with drilled compartments to casted metal compartmented seals proper, almost the entire range of shapes is encountered at the site. Furthermore, for the first time wooden inlays in compartmented seals and a wooden seal have been documented (Kaniuth 2010b). Filling the compartments of a metal seal—and thereby making the resulting impression less distinguishable—again indicates that aesthetic considerations superseded functional ones.

One particularly large, but unfortunately damaged, piece cut from a translucent white stone, possibly alabaster, found in a phase 2 refuse pit replicates the common narrative element of dominant dragons in Middle Asian glyptic art (Sarianidi 1986; Francfort 1992): A large quadruped with a flame(?) emanating from its jaw dominates a tumbling human figure. The same scene is depicted on a piece from Togolok 21 (Sarianidi 1998: 292, no. 1621). The size of our object (8 cm in diameter)

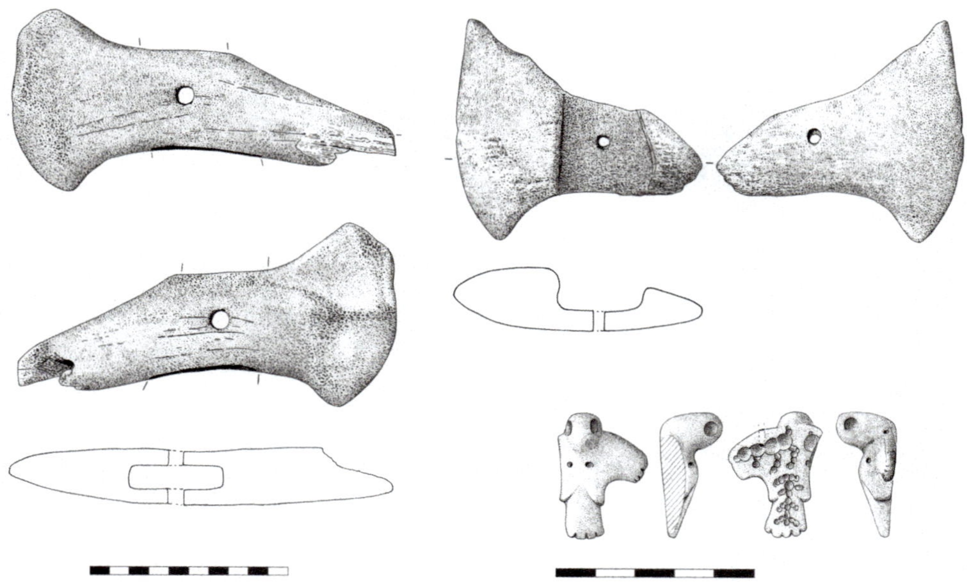

Figure 16.9 Tilla Bulak. LB I axes and eagle pendant from phases 1 and 2.

is out of proportion compared to all other contemporary "seals" and certainly to that of known impressions. Here again, an emblematic function appears as the primary motive for its creation. In the Oxus Civilization, and also in the wider region of eastern Iran, there is a clear association between seals and female burials (Baghestani 1997: 149–152; Kaniuth 2006: 74ff.). Contrary to previous assertions, this custom is not consistently observed at Gonur, where of all clearly sexed single inhumations with seals one in five is male (Sarianidi 2007: app. 1).

The necropolis of Tilla Bulak remains to be found. In the Sapalli culture the dead were buried in separate graveyards, very rarely in settlements, and, if so, then most often in abandoned houses.[26] Of the ten burials from Tilla Bulak, seven date to the Bronze Age, and all stratigraphically postdate the phase 2 occupation. Only one contained the body of an adult, the others are child burials in jars.[27] The lack of adult burials also explains the dearth of metal finds, a frequent status-related burial gift. However, another category of prestige items was encountered; namely, antler axes with curved blade, sloping neck, and lateral riveting (Figure 16.9). At Gonur and Sapallitepa, bone (antler?) and metal examples appear only in well-furnished male graves (Askarov 1973; Kaniuth 2006; Sarianidi 2007: 53).[28]

The occupation at Tilla Bulak ended sometime before the mid-18th century BC (at, or just after, the transition from subphase LB Ia to LB Ib). The reasons for this abandonment are unknown, especially since the street levels in the upper part of the site demonstrate clearly that we are missing several phases towards the end of the occupation, destroyed by erosion. Some ephemeral walls have been grouped as phase 3, but beyond their general LB I date, not much can be said of this final occupation.

The shift of settlements from LB I to LB II corresponds to a wide scale pattern, visible also at Sapallitepa, where life stops with the end of LB Ia, at Molali, which starts

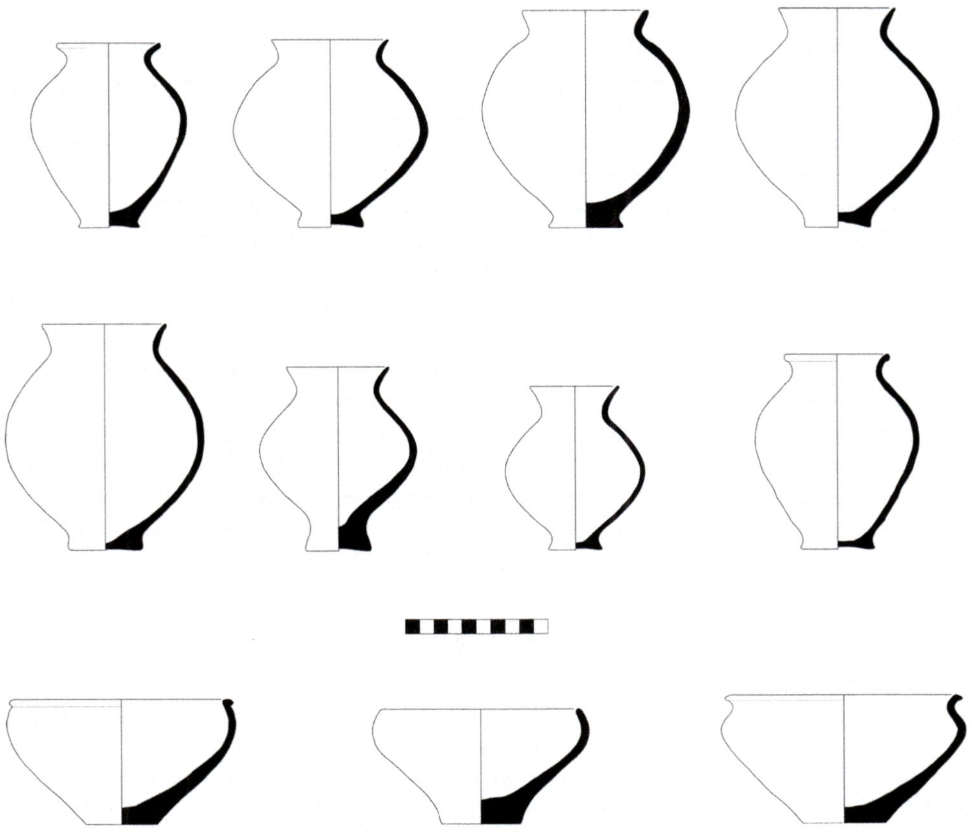

Figure 16.10 Pashkhurt. LB II pottery collected at PAS 01.

with LB II, and at Gaztepa, which is only occupied during LB I, etc. The only known settlement in southern Uzbekistan that offers a long occupation sequence covering all the Late Bronze Age periods is the large site of Dzharkutan, which apparently continued to offer suitable living conditions.

This shift in location affected also the Pashkhurt valley. Tilla Bulak was abandoned, but LB II pottery has been discovered at the southeastern limits of the modern village of Pashkhurt (PAS 00, PAS 01; see Figure 16.2). Since the material found (vessels) is very well preserved, we are probably dealing with the remains of a burial ground. The pottery of the LB II period is less uniform typologically than in the previous period and the Pashkhurt vessels are no exception. Even with the small sample available (Figure 16.10), the divergence from known assemblages at Dzharkutan or Molali is marked.

CONCLUSION

The dense occupation of the Pashkhurt valley during the Late Bronze Age (and also the Early Iron Age) comes as a surprise only when neglecting the massive impact of modern centrally managed agriculture in the Surkhan Darya valley proper. A landscape dotted with small rural hamlets was probably the rule there, too, and only the largest agglomerations have survived in some form to this day. Given the small size of the Pashkhurt settlements (no more than 0.5 ha, with the possible exception of Yultepa), the presence of a material culture of high quality, including prestige items identical to those encountered in the largest sites of the Sapalli culture is notable. Apparently, in these hamlets, the inhabitants' way of life and needs for ostentatious display did not differ significantly from that of the (proto-)urban sites. The major difference is directly related to their smaller size: Large architectural compounds within these hamlets have not yet been found. The rural adaption of the Sapalli culture must, until more sites have been investigated, be considered one of small, unfortified villages where social distinction was not communicated through architectural form, but probably existed as shown by status-related artifacts. Whether the availability of metals in the surrounding mountains was an incentive for settling the region is not proven by any workshop at Tilla Bulak, but the readily accessible salt deposits of Khodzaikan were probably a coveted resource. It remains to be explained why all excavated sites in the area are contemporary with the very beginning of the Late Bronze Age in southern Uzbekistan. Whatever the reasons for the choice of locations, after less than 150 years all evidence for settlement ceases in the Pashkhurt valley.

NOTES

1 For the history of the terms Oxus Civilization and Bactria-Margiana Archaeological Complex (BMAC), see Chapter 1. Both terms have been employed too freely to be used as a meaningful frame of reference (see also Salvatori 2016, whose criticisms I share, if not his conclusions). For the purposes of this chapter they are considered as synonymous.

2 Relevant radiocarbon dates were assembled by Görsdorf and Huff (2001) and Kaniuth (2013). See also the Appendix to this volume and for the BMAC general chronology, see Chapter 1.

3 Compare the pottery for northern Bactria in Askarov (1977: fig. 17), Askarov and Abdullaev (1983: fig. 7), and Teufer (2015: fig. 10); for Margiana in Sarianidi (1990; 2007: 52–67) and Udeumuradov (1993); and for southern Bactria in Francfort (1989: figs. 30–42).

4 These discoveries happened at the same time when Soviet scholars started working in northern Afghanistan, and discovered a flourishing Bronze Age civilization (Kruglikova and Moustamandy 1970; Kruglikova and Sarianidi 1971).

5 The assumption, that settlement was organized in separate oases, has been called into question by Markofsky (2010: 250), who noted a "significant occupational continuity" and concomitant "lack of clearly bounded and isolated regions."

6 But see also the recently excavated Saridzar in southern Tajikistan (Koch *et al.* 2013; Teufer 2014; see also Chapter 5).

7 The Ulanbulaksaj originally discharged into the Amu Darya, but is now completely diverted for irrigation purposes.

8 The famous Khodzaikon salt deposit near the village of Aktash has an output of rock salt estimated at several thousand tons annually. Caves in the mountain served for treatment of respiratory diseases during Soviet times, and concentrated brine can be tapped in the shafts to this day. Salt is also produced on the western (Turkmen) side of the Kugitang Mountains, near the village of Svintsovij Rudnik ("lead mine" – see note 9). It is currently impossible to determine any ancient work at the site of Khodzaikon, although E. Pruger (quoted in Sverchkov 2009: 152) considers the possibility of exploitation in the 1st century AD.

9 All Russian-language sources for pre-modern mining activities in the region are conveniently, if succinctly, summarized by Sverchkov (2009). Of particular relevance are the polymetallic deposits of Chujankan and Tillokan with mineralizations of lead, iron, copper, and tin, but the author remains wary of any Bronze Age exploitation. In the western Kugitang iron and lead are the predominant metals (Sverchkov 2009: 152f).

10 Earlier survey work was summed up in Rtveladze (1974: 76ff) and Arshavskaja *et al.* (1982: 134ff).

11 Surveys were made by the Tilla Bulak team between 2008 and 2010, and by members of the Tokharistan expedition (Uzbek Academy of Sciences) and the Middle Asian expedition (Institute of Archaeology of the Russian Academy of Sciences).

12 For methodology and selection procedures see the introductory notes to Augustinová *et al.* (2015), Augustinovà, Stančo, Damasêk, Khamidov *et al.* (2017), Augustinovà, Stančo, Damasêk, Mrva *et al.* (2017), and Stančo (2016).

13 Some 23 Bronze Age sites are listed in the survey report. Here, we only take into account those sites that produced at least three datable Bronze Age sherds. Given the small proportion of Late Bronze Age attributions within the collected lots and the general similarity of Late Bronze Age and Early Iron Age wheel-made pottery, it seems advisable to remain cautious in these cases until excavations are made. This also takes into account occasional misattributions: The rim sherds in Augustinová *et al.* (2015: fig. 6.2) – one of two Late Bronze Age sherds reported from Kurgantepa (POL_142; KuPi_055) – and Augustinová *et al.* (2015: fig. 7.3) (POL_161) are so far not attested within the Sapalli culture repertoire of shapes.

14 Also named Chakmaktosh ("flintstone") in the survey report. A preliminary excavation report has been published with details on the stone architecture, some pottery, and finds (Rukavishnikova *et al.* 2015: fig. 4). Unfortunately, these do not contain standard BMAC pottery.

15 No sites in the Gaz area were previously referred to in the literature, but Kirkkiz 3 could be identical with the find spot of Bronze Age pottery at the northern outskirts of Gaz village (mentioned by Dvurechenskaja *et al.* 2014: 72ff).

16 The excavations were conducted by a joint expedition of the Institute for Art Historical Research, Tashkent, and Ludwig-Maximilians-University, Munich, headed by the author.

17 An attempt to irrigate was made in recent times, but the inherent instability of the water management system lead to an abortion of all efforts.

18 For absolute chronology, see the ^{14}C dates given in Kaniuth (2016).

19 Rooms of 8–25 m^2 were the norm at Dzharkutan, Tepe 5, one of the living quarters of the settlement (Shirinov 2002: 92).

20 This figure is somewhat distorted by a deposit within a single *tannur* (TB-271), which alone yielded 2,500 fragments of *T. spelta*. Actually, the balance may therefore lean towards *H. vulgare*, which was the main agricultural product at Dzharkutan, Sapallitepa, and also in the Murghab delta sites (Miller 1993, 1999; Shirinov 2002: 155, with further literature).

21 A further, as yet unpublished, series of analyses that compare the Tilla Bulak sherds with material from Ara Bulak, Dzharkutan, and Sapallitepa may lead to a reinterpretation of these findings.

22 Type quantities from the Dzarkutan Tepe 6 "temple" monumental building are 10% for footed bowls and 2% for footed goblets (Shirinov 2002: 66, tab. 4), possibly reflecting the specific function of this place.

23 For a discussion of the distribution and chronological relevance of this artifact, see Ascalone (2007) and Kaniuth (2010a).

24 The variety of potters' marks from the Sapalli culture is illustrated by Askarov (1973: pl. 31) and Askarov and Shirinov (1993). Geometric motifs (single or double strokes, crosses, semicircles or full circles, wavy lines, and combinations thereof) are regularly found on

medium and large storage vessels. Pots stands, which could be turned over and used as molds for forming pithos bases, show an even wider range, particularly of zoomorphic designs (deer, birds, wheels, swastikas, and spoked wheels) (see Chapter 26). Many of these potters' marks from the Pashkhurt valley are comparable, but one is unique so far, depicting a bow and arrow, and a standard or tree of life (Kaniuth 2007).

25 The only evidence for the sealing of bullae or containers comes from Gonur (Sarianidi 1998: 316–319 nos. 1745–1762; see also Chapter 8), while a single sealing on clay is known from Dzharkutan (Kaniuth 2010a: 15f, fig. 13). For the Sapalli culture seal impressions on pottery, see also Mustafakulov *et al.* (2012: fig. 3) and Bendezu-Sarmiento and Mustafakulov (2013: fig. 10). Bonora *et al.* (2014) give a very helpful overview of Chalcolithic tokens and sealings from Turkmenistan. Simple administrative practices have a long history in the Kopet Dagh, but the same cannot currently be said for southern Uzbekistan. A more optimistic view of sealing practices and their role in interregional trade in Bronze Age Middle Asia is given by Lyonnet (2005: 196, with further literature).

26 This is the case at both Dzharkutan Tepe 5 and Sapallitepa, where intrusive burials have wrongly been considered to be intramural (i.e., constructed at a time when the rooms were still inhabited) by their excavators, see Teufer (2015).

27 For a full list, see Gruber *et al.* (2012: 346).

28 But see again the Gonur necropolis for exceptional cases: Axes of bone and bronze were associated with females in graves 220 and 2406 (Sarianidi 2007: app. 1).

REFERENCES

Al'baum, L.I. 1969. "Pamjatnik epokhy bronzy na territorii Surkhandary" [A Bronze Age site in the Surkhandarya region]. *Obshchestvennye Nauki Uzbekistana* 5: 46–47.

Arshavskaja, Z.A., E.V. Rtveladze, and Z.A. Khakimov 1982. *Srednevekovye pamjatniki Surkhandary* [Medieval sites in Surkhandarya]. Tashkent: Izdatel'stvo literatury i iskusstva imeni Gafura Gulama.

Ascalone, E. 2007. "Archaeological Links Between Syria and Iran: Stratigraphical Correlation and Regional Connections." In *Proceedings of the International Colloquium From Relative Chronology to Absolute Chronology, Rome 2001*, ed. P. Matthiae. Rome: Bardi, pp. 33–70.

Askarov, A.A. 1971. "Poselenie drevnykh zemledeltsev na juge Uzbekistana" [A settlement of ancient farmers in southern Uzbekistan]. *Obshchestvennye nauki Uzbekistana* 8: 40–43.

Askarov, A.A. 1973. *Sapallitepa*. Tashkent: FAN.

Askarov, A.A. 1977. *Drevenezemledel'cheskaja kul'tura bronzovogo veka juzhnogo Uzbekistana* [The ancient Bronze Age farmers' culture of southern Uzbekistan]. Tashkent: FAN.

Askarov, A.A. and B. Abdullaev 1983. *Dzharkutan*. Tashkent: FAN.

Askarov, A.A. and T. Shirinov 1993. *Rannaja gorodskaja kul'tura epokhi bronzy juga Srednej Azii* [The Bronze Age early urban culture of southern Middle Asia]. Samarkand: Institute of Archaeology.

Augustinová, A., L. Stančo, L. Damašek, O. Khamidov, T. Kolmačka, and S. Shaydullaev, 2017. "Archaeological Survey in the Oases of Maydon and Goz in the Piedmont of the Kugitang Mountains (South Uzbekistan): Preliminary Report for Season 2017." *Studia Hercynia* 21(2): 139–152.

Augustinová, A., L. Stančo, L. Damašek, M. Mrva, and S. Shaydullaev 2017. "Archaeological Survey of the Oases of Zarabag, Karabag and Kampyrtepa in the Piedmont of the Kugitang Mountains, South Uzbekistan: Preliminary Report for the Season 2016." *Studia Hercynia* 21: 104–148.

Augustinová, A., L. Stančo, S. Shaydullaev, and M. Mrva. 2015. "Archaeological Survey in the Zarabag Micro Oasis South Uzbekistan: Preliminary Report on the Season 2015." *Studia Hercynia* 19: 262–281.

Baghestani, S. 1997. *Metallene Compartimentsiegel aus Ost-Iran, Zentralasien und Nord-China*. Rahden: Leidorf.

Beljaeva, T.V. and Z.A. Khakimov 1973. "Drevnebaktrijskie pamjatniki Mirshade" [The ancient Bactrian sites of Mirshade]. In *Iz istorii antichnoj kul'tury Uzbekistana* [From the history of ancient culture in Uzbekistan], ed. G.A. Pugachenkova. Tashkent: Izdatel'stvo literatury i iskusstva imeni Gafura Gulama, pp. 35–51.

Bendezu-Sarmiento, J. and S. Mustafakulov 2008. "Arkheoanthropologicheskie trudy frantsusko-uzbekskoj ekspeditsii na Dzharkutanskom nekropole Surkhandarja/Travaux archéo-anthropologiques sur la nécropole de Dzharkutan, Surkhan-Daria." *Istorija Material'noj Kultury Uzbekistana* 36: 75–87.

Bendezu-Sarmiento, J. and S. Mustafakulov 2013. "Le site proto-urbain de Dzharkutan durant les âges du bronze et du fer. Recherches de la Mission archéologique franco-ouzbèke-Protohistoire." *Cahiers d'Asie centrale* 21–22: 207–236.

Bonora, G., M. Vidale, M. Mariottini, and G. Guida 2014. "On the Use of Tokens and Seals along the Kopet Dagh Piedmont, Turkmenistan ca. 6000–3000 BCE." *Paléorient* 40: 55–71.

Castiglione, E. and M. Cottini 2002. "Archaeobotanical Analyses on Materials from the Necropolis and City of Gonur." In *Margiana. Gonur-Depe Necropolis*, ed. G. Rossi Osmida. Trebaseleghe: Il Punto, pp. 180–195.

Danielisová, A., L. Stančo, and A. Shaydullaev 2009. "Preliminary Report of Archaeological Survey in Sherabad District, South Uzbekistan in 2009." *Studia Hercynia* 14: 67–90.

Dvurechenskaja, N.D., V.V. Mokroborodov, and D.V. Rukavishnikova 2014. "Marshrutnye issledovanija na juge Uzbekistana v 2013 godu" [Regional survey in southern Uzbekistan]. *Kratkie Soobshchenija Instituta Arkheologii* 236: 69–81.

Francfort, H.-P. 1984. "The Early Periods of Shortugai (Harappan) and the Western Bactrian Culture of Dashly." In *South Asian Archaeology 1981*, ed. B. Allchin. Cambridge: Cambridge University Press, pp. 170–175.

Francfort, H.-P. 1989. *Fouilles de Shortughaï. Recherches sur l'Asie centrale protohistorique*. Paris: de Boccard.

Francfort, H.-P. 1992. "Dungeons and Dragons: Reflections on the System of Iconography in Protohistoric Bactria and Margiana." In *South Asian Archaeology Studies*, ed. G. Possehl. New York: International Science Publisher, pp. 179–208.

Francfort, H.-P. 2005. "La civilisation de l'Oxus et les Indo-Iraniens et Indo-Aryens en Asie centrale." In *Aryas, Aryens et Iraniens an Asie Centrale*, ed. G. Fussmann, J. Kellens, H.-P. Francfort, and X. Tremblay. Paris: De Boccard, pp. 253–332.

Francfort, H.-P. 2009. "L'âge du bronze en Asie centrale: La civilisation de l'Oxus." *Anthropology of the Middle East* 4: 91–111.

Francfort, H.-P. and X. Tremblay 2010. "Marhashi et la civilisation de l'Oxus." *Iranica Antiqua* 45: 51–224.

Ganjalin, A. 1956. "K stratigrafii Namazga-tepe" [About the stratigraphy of Namazga Depe]. In Trudy Instituta Istorii Arkheologii i Etnografii Akademii Nauk Turkmenistanskoj SSR, vol. 2, ed. O.K. Kuliev. Moscow: Nauka, pp. 37–66.

Görsdorf, J. and D. Huff 2001. "^{14}C-Datierungen von Materialien aus der Grabung Dzarkutan, Uzbekistan." *Archäologische Mitteilungen aus Iran und Turan* 33: 75–87.

Gruber, M., J. Il'yasov, and K. Kaniuth 2012. "A Decorated Ivory Belt from Tilla Bulak, Southern Uzbekistan." *Ancient Civilizations from Scythia to Siberia* 18: 339–375.

Gubaev, A., G. Koshelenko, and M. Tosi (eds.) 1998. *The Archaeological Map of the Murghab Delta: Preliminary Reports 1990–95*. Rome: Istituto Italiano per il Antico Oriente.

Huff, D. 1995. "Fragen zur Siedlungsgeschichte von Dzarkutan." In *Archaeologia Iranica et Orientalis: Miscellanea in honorem Louis Vanden Berghe*, ed. L. de Meyer and E. Haerinck. Ghent: Peeters, pp. 22–26.

Huff, D. 2000. "Tempel oder Palast? Dscharkutan – Monumentalbauten und Siedlung im bronzezeitlichen Baktrien." In *Archäologische Entdeckungen. Die Forschungen des Deutschen Archäologischen Instituts im 20. Jahrhundert*, ed. H. Kyrieleis. Mainz: Zabern, pp. 115–118.

Ionin, N.B. and S.A. Shafranov 1937. "Tillokan i Chujankan – mestorozhdenija zheleza i polimetallov" [Tillokan and Chujankan – iron and polymetallic ore deposits]. In *Tadzhiksko-Pamirskaja ekspeditsija 1935 g*, ed. A.I Kovalerov and D.I. Sherbakov. Moscow: Akademija Nauk SSSR. pp. 499–525.

Kaniuth, K. 2006. *Metallobjekte der Bronzezeit aus Nordbaktrien*. Mainz: Zabern.

Kaniuth, K. 2007. "Tilla Bulak 2007 – Vorbericht zur ersten Kampagne." *Archäologische Mitteilungen aus Iran und Turan* 39: 31–47.

Kaniuth, K. 2009. "Tilla Bulak 2008 – Vorbericht zur zweiten Kampagne." *Archäologische Mitteilungen aus Iran und Turan* 41: 75–96.

Kaniuth, K. 2010a. "Long Distance Imports in the Bronze Age of Southern Central Asia: Recent Finds and Their Implications for Chronology and Trade." *Archäologische Mitteilungen aus Iran und Turan* 42: 3–22.

Kaniuth, K. 2010b. "Tilla Bulak 2009 – Vorbericht zur dritten Kampagne." *Archäologische Mitteilungen aus Iran und Turan* 42: 129–163.

Kaniuth, K. 2011. "Tilla Bulak 2010 – Vorbericht über die vierte Grabungskampagne." *Archäologische Mitteilungen aus Iran und Turan* 43: 261–282.

Kaniuth, K. 2013. "A New Late Bronze Age Site in Southern Uzbekistan." In *South Asian Archaeology 2007*, ed. D. Frenez and M. Tosi. Oxford: Archaeopress, pp. 151–155.

Kaniuth, K. 2014. "Die Bestattungen des Hügels 4 von Dzarkutan." *Archäologische Mitteilungen aus Iran und Turan* 46: 95–121.

Kaniuth, K. 2016. "The Late Bronze Age Settlement of Tilla Bulak (Uzbekistan) – A Summary of Four Years' Work." In *South Asian Archaeology and Art 2012*. Vol. I, *Man and Environment in Prehistoric and Protohistoric South Asia: New Perspectives*, ed. V. Lefèvre, A. Didier, and B. Mutin. Turnhout: Brepols, pp. 117–128.

Kaniuth, K., B. Ögüt, and C. Salgado Ceballos in press. "The Late Bronze Age Pottery from Tilla Bulak." In *The Bronze Age in Central Asia and Khorassan*, ed. N. Boroffka, E. Luneau, and M. Teufer.

Koch, J., J. Fassbinder, and M. Teufer 2013. "Archäologie am Fuße des Pamirgebirges. Mit Magnetometer und Kappameter im Süden Tadschikistans." *Denkmalpflege Informationen* 154: 68–70.

Kohl, P. 1984. *Central Asia: Palaeolithic Beginnings to the Iron Age*. Paris: ERC.

Kruglikova, I.T. and C. Moustamandy 1970. "Resultats préliminaires des travaux de l'expédition archéologique Afghano-Soviétique en 1969." *Afghanistan* 23: 84–97.

Kruglikova, I.T. and V.I. Sarianidi 1971. "Drevnjaja Baktrija v svete novykh arkheologicheskikh otkrytij" [Ancient Bactria in the light of recent archaeological discoveries]. *Sovetskaja Arkheologija* 4: 154–177.

Kuftin, B.A. 1956. "Polevoj otchet o rabote XIV otrjada JuTAKE po izucheniju kul'tury pervobytno-obshchinnikh poselenij epokhi medi i bronzy v 1952 g" [Field report on the work of the XIVth JuTAKE brigade in 1952 for the study of the first farmer settlements during the Chalcolithic and Bronze Age]. In *Pamjatniki kul'tury kammenogo i bronzovogo veka Jurzhnogo Turkmenistana* [Sites of the Stone and Bronze Age in southern Turkmenistan], Trudy Juzhno-Turkmenistanskoj Arkheologicheskoj Kompleksnoj Ekspeditsii, vol. 7, ed. V.M. Masson. Moscow: Nauka, pp. 260–290.

Kysela, J., A. Augustinová, and R. Kinaston 2018. "Preliminary Reports on the Excavations at Burgut Kurgan and Bobolangar in 2017. *Studia Hercynia* 22: 158–182.

Luneau, E. 2014. *La fin de la civilisation de l'Oxus. Transformations et recompositions des sociétés de l'âge du Bronze final en Asie centrale méridionale 1800–1500/1400 avant n. è.* Paris: De Boccard.

Lyonnet, B. 2005. "Another Possible Interpretation of the Bactro-Margiana Culture (BMAC) of Central Asia: the Tin Trade." In *South Asian Archaeology 2001*, ed. C. Jarrige and V. Lefèvre. Paris: Éditions Recherche sur les Civilisations, pp. 191–200.

Markofsky, S.B. 2010. "Illuminating the Black Sands: Survey and Settlement in the Bronze Age Murghab Delta, Turkmenistan." PhD thesis, University College London.

Miller, N.F. 1993. "Preliminary Archaeobotanical Results from the 1989 Excavation at the Central Asian Site of Gonur Depe, Turkmenistan." *Unesco IASCCA Information Bulletin* 19: 149–163.

Miller, N.F. 1999. "Agricultural Development in Western Central Asia in the Chalcolithic and Bronze Ages." *Vegetational History and Archaeobotany* 8: 13–19.

Mokroborodov, V.V. 2006. "Keramika epokhi bronzy iz kishlaka Pashkhurt" [Bronze Age pottery from the Pashkhurt village]. In *Arkheologicheskie issledovanija v Uzbekistane 2004–2005 gody*." Samarkand: Institut Arkheologii, pp. 160–164.

Moore, K. 1993. "Animal Use at Bronze Age Gonur Depe." *Unesco IASCCA Information Bulletin* 19: 164–185.

Mustafakulov, S.M., J. Bendezu-Sarmiento, U. Rakhmanov, M. Khasanov, J. Lhuillier, and E. Luneau 2012. "Polevye raboty Uzbeksko-Frantsuzkoj MAFOuz ekspeditsii na Dzharkutane v 2009–2011 gg" [French–Uzbek MAFOuz expedition fieldwork at Dzharkutan in 2009–2011]. In *Arkheologicheskie issledovanija v Uzbekistane 2010–2011 gg*. Samarkand: Institut Arkheologii, pp. 203–213.

Mustafakulov, S.M., J. Bendezu-Sarmiento, U. Rakhmanov, J. Lhuillier, K. Rakhimov, and D.S. Sarmiento 2014. "Rezultaty rabot na pamjatnike Dzarkutan v 2012 godu" [Work results at Dzharkutan in 2012]. In *Arkheologicheskie issledovanija v Uzbekistane 2012 g*. Samarkand: Institut Arkheologii, pp. 137–144.

Pruger, E.B. 1980. "Drevnie gornopromyshlennye obekty Juga Uzbekistana" [Ancient mining sites of southern Uzbekistan]. *Materialy po istorii, istoriografii i arkheologii. Sbornik nauchnikh trudov Tashkentskogo gosudarstvennogo universiteta* 630: 27–34.

Raevskij, V.I. and M.P. Fiveg 1973. *Mestorozhdenija kalijnikh solej SSSR: Metody ikh poiskov i razvedki* [Deposits of potassium salts of the USSR: methods of their search and exploration]. Leningrad: Nedra.

Rtveladze, E.V. 1974. "Razvedochnoe izuchenie baktrijskikh pamjatnikov na juge Uzbekistana" [Survey study of Bactrian sites in southern Uzbekistan]. In *Drevnjaja Baktrija. Predvaritel'niye soobshchenija ob arkheologicheskikh rabotakh na juge Uzbekistana* [Ancient Bactria, preliminary report on the archaeological works in southern Uzbekistan], ed. V.M. Masson. Leningrad: Nauka, pp. 74–85.

Rtveladze, E.V. and Z.V. Khakimov 1973. "Marshrutnye issledovanija pamjatnikov Severnoj Baktrii" [Regional survey of sites in northern Bactria]. In *Iz istorii antichnoj kultury Uzbekistana* [From the history of ancient culture in Uzbekistan], ed. G.A. Pugachenkova. Tashkent: Izdatel'stvo Gafura Gulama, pp. 10–34.

Rukavishnikova, I.V., V.V. Mokroborodov, A.N. Gorin, and A.A. Gladchenkov 2015. "Gazkala – novij pamjatnik Baktrijsko-Margianskogo arkheologicheskogo kompleksa" [Gazkala – a new site of the Bactria-Margiana Archaeological Complex]. *Kratkie Soobshchenija Instituta Arkheologii* 239: 72--87.

Ruzanov, V.D. and Ju.F. Burjakov 1997. "Drevnye rudniki i pamjatniki metallurgii v gorakh Kugitangtau" [Old mines and metallurgical sites in the Kugitangtau Mountains]. *Ozbekiston Moddiy Madaniyati Tarixi* 28: 175–178.

Salvatori, S. 1998. "Margiana Archaeological Map: The Bronze Age Settlement Pattern." In *The Archaeological Map of the Murghab Delta: Preliminary Reports 1990–95*, ed. A. Gubaev, G. Koshelenko, and M. Tosi. Rome: Istituto Italiano per la Archeologia Orientale, pp. 57–65.

Salvatori, S. 2016. "Bactria-Margiana Archaeological Complex: How Terminology Hides Historical Processes." In *Trudy Margianskoj Arkheologicheskoj Ekspeditsii*, vol. 6, ed. N. Dubova. Moscow: Staryj Sad, pp. 449–460.

Salvatori, S. and M. Tosi (eds.) 2008. *The Archaeological Map of the Murghab Delta. Studies and Reports II: The Bronze Age and Early Iron Age in the Margiana Lowlands – Facts and Methodological Proposals for a Redefinition of the Research Strategies.* Oxford: Archaeopress.

Sarianidi, V.I. 1977. *Drevnie Zemledeltsy Afghanistana. Materiali Sovetskoj-Afghanskoj Ekspeditsii 1969–1974* [Ancient agriculturalists of Afghanistan: materials of the Soviet–Afghan expedition 1969–1974]. Moscow: Nauka.

Sarianidi, V.I. 1986. "Zmen i drakony v gliptike Baktrii i Margiany" [Snake and dragons in the glyptic of Bactria and Margiana]. In *Vostochnyj Turkestan i Srednjaja Azija v sisteme kul'tur drevnego i srednevekovogo vostoka* [Eastern Turkestan and Central Asia in the oriental culture system of the ancient and Middle Ages], ed. B.A. Litvinskij. Moscow: Nauka, pp. 66–71.

Sarianidi, V.I. 1990. *Drevnosti strany Margush* [Ancient Margush]. Ashkhabad: Ylym.

Sarianidi, V.I. 1998. *Myths of Ancient Bactria and Margiana on Its Seals and Amulets.* Moscow: Pentagraphic.

Sarianidi, V.I. 2002. *Margush: Ancient Oriental Kingdom in the Delta of the Murghab River.* Ashgabat: Türkmendöwlethabarlary.

Sarianidi, V.I. 2005. *Gonurdepe: City of Kings and Gods.* Ashgabat: Miras.

Sarianidi, V.I. 2007. *Necropolis of Gonur.* Athens: Kapon.

Sarianidi, V.I. 2008. *Margush, Turkmenistan: Mystery and Truth of the Great Culture.* Ashgabat: Türkmendöwlethabarlary.

Sataev, R.M. and L.V. Sataeva 2012. "Archaeozoological and Archaeobotanical Studies at Gonur Depe in 2010." In *Trudy Margianskoj Arkheologicheskoj Ekspeditsii*, vol. 4, ed. V.I. Sarianidi. Moscow: Staryj Sad, pp. 57–61.

Sataev, R.M. and L.V. Sataeva 2014. "Results of Archaeozoological and Archaeobotanical Studies of Gonur Depe in 2012–2013." In *Trudy Margianskoj Arkheologicheskoj Ekspeditsii*, vol. 5, ed. V.I. Sarianidi. Moscow: Staryj Sad, pp. 177–184.

Shirinov, T. 2002. "Die frühurbane Kultur der Bronzezeit im südlichen Mittelasien. Die vorgeschichtliche Siedlung Dzarkutan." *Archäologische Mitteilungen aus Iran und Turan* 34: 1–170

Stančo, L. 2016. "Archaeological Survey in the Surroundings of Kayrit, South Uzbekistan: Preliminary Report for Season 2015." *Studia Hercynia* 20: 73–85.

Stančo, L., S. Shaydullaev, J. Bendezu-Sarmiento, A. Pažout, and H. Vondrová 2014. "Kayrit Burial Site, South Uzbekistan: Preliminary Report for Season 2014." *Studia Hercynia* 18: 31–41.

Stančo, L., S. Shaydullaev, A. Shaydullaev, A. Augustinová, and J. Havlik 2017. "Archaeological Survey in the Eastern Kugitang Piedmonts (South Uzbekistan): Preliminary Report for Seasons 2016 and 2017." *Studia Hercynia* 21(2): 121–138.

Sverchkov, L. 2009. "A History of Research on Ancient Mining in Uzbekistan." *Archäologische Mitteilungen aus Iran und Turan* 41: 141–164.

Teufer, M. 2014. "Rezultaty raskopok na pamjatnikakh Saridzhar 2 i Karim Berdy v Juzhnom Tadzhikistane" [Results of excavations at the sites Saridzhar 2 and Karimberdy in southern Tajikistan]. *Arkheologicheskie Raboty v Tadzhikistane* 37: 120–179.

Teufer, M. 2015. *Spätbronzezeitliche Grabfunde aus Nordbaktrien und benachbarten Regionen. Studien zur Chronologie zwischen Aralsee und Persischem Golf.* Berlin: Reimer.

Udeumuradov, V.N. 1993. *Altyn-Depe i Margiana: Svjazi, khronologija, proiskhozhdenie.* Ashkhabad: Ylym.

CHAPTER SEVENTEEN

WHO INTERACTED WITH WHOM? Redefining the interaction between BMAC people and mobile pastoralists in Bronze Age southern Turkmenistan

Barbara Cerasetti[*]

INTRODUCTION

In recent years, the interest of researchers working in Central Asia and, more specifically, in the territory of the inner delta of the Murghab River in southern Turkmenistan, has increasingly focused on the understanding of the relationship between the two main archaeological cultures that have been identified in the region during the Bronze Age, one mobile and the other sedentary. Between the Middle Bronze Age (2400–1950 BCE) and the beginning of the Iron Age (1300–900 BCE), this phenomenon shaped the social patterns and settlement processes that will develop in the later historical periods. Focusing on this interrelation and not considering that it is, at best, a marginal element to the better studied Bactria-Margiana Archaeological Complex (BMAC) phenomenon, highlights many unclear aspects of several Bronze Age archaeological overviews (Rouse and Cerasetti 2014, 2018; Cerasetti *et al.* 2018, 2019, Cerasetti *et al.* in press; Spengler *et al.* 2016; Rouse *et al.* 2019). Certainly, a lot has been made, but much more needs to be done. We are like brave sailors in a sea of still-unknown knowledge!

When the editors, friends, and colleagues asked me to write a contribution on the relationship between the two cultures, I smiled. Not because they asked me something unusual or impossible, but because this is one of the most unknown subjects in the Murghab region and, at the same time and for this reason, one of the most exciting.

THE BACKGROUND

The first steps in the analysis of this complex phenomenon date back to 1990, at the beginning of *The Archaeological Map of the Murghab Delta Project* (Gubaev *et al.* 1998; Salvatori *et al.* 2008). Ever since the pioneering intensive surface survey made in the northeastern section of the wide alluvial fan, the presence of the so-called Andronovo or steppe culture in campsites located on the sand dunes among BMAC settlements or close to them, has been clearly brought to light (see Cerasetti 1998).

This culture is characterized by a typical gray-brown handmade pottery with incised geometrical decoration (Incised Coarse Ware – ICW). On the basis of the extensive literature written on this subject, at that time mainly by Soviet and post-Soviet authors (Askarov 1962; Grjaznov 1966; Guljamov *et al.* 1966; Itina 1977; Gening *et al.* 1992; Kuz'mina 1994), we thought that, by the end of the Bronze Age and due to an important and constant process of aridity, mobile populations had left the steppes of Kazakhstan to enter portions of the inner Murghab delta where the former agriculture land was converted into an uncultivated pasture area (Cattani 2008b; Cattani *et al.* 2008; Cerasetti 2012).

Our interest in what we generically call "nomadic/steppe culture" – actually as we shall see a melting pot of many different cultural phases and of different populations – was strengthened by V.I. Sarianidi's (1975) article.[1] In this pioneering study, Sarianidi reported a basic concept of purely pastoral sites alongside with semi-nomadic occupation levels inside BMAC settlements, which was confirmed later by further excavations. From the very beginning of the study of the interaction between mobile and sedentary cultures, Sarianidi had already defined their coexistence as peaceful, made of the mutual exchange of daily products for produced ones, as well as of the exchange of ideas and lifestyles. Thus, we were convinced that mobile pastoralists and BMAC farmers coexisted peacefully in the vast region of the Murghab alluvial fan, the territory of which was divided. On one side were the fields still available for agriculture and, on the other, those only suitable for pastoralism because of the drying up of the climate.

FURTHER RESEARCH

To ensure this vision, we decided to investigate different sites in more detail.

We excavated the pastoral sites of Chopantam (Cattani 2008a; Cerasetti *et al.* 2018; Rouse and Cerasetti 2018) and Ojakly (Rouse and Cerasetti 2014; Cerasetti *et al.* 2018; Rouse and Cerasetti 2018), located respectively at about 6 km southeast of Takhirbaj Depe and 12 km northeast of Gonur North. And we also excavated at the BMAC sites of Adji Kui 1 (Spengler *et al.* 2016; Cerasetti *et al.* 2018; Rouse and Cerasetti 2018) and Togolok 1 (Rouse and Cerasetti 2018; Cerasetti *et al.* 2019; Cerasetti *et al.* in press), situated respectively in the northwestern section and in the center region of the northeastern part of the alluvial fan (Figure 17.1).

What emerged was much more than just a peaceful coexistence. Central Asian specialists in mobile and semi-mobile pastoral cultures who deal with the problems of exchange and integration between farmers and herders, usually consider that these two groups are totally different populations, each with its own way of life, ideas, and material culture.

But why shouldn't it be the same population differentiated into groups employed in various activities? Why not, indeed, the same individuals within the same ethnic group who interchange roles, farmer and shepherd, according to the needs of their community? Investigations in urban contexts validated the possibility of individual members of the BMAC population being employed in various daily tasks.

It is highly possible that in the Murghab area lived various groups of pastoralists with different ethnic backgrounds, more or less regular or seasonal mobility, minor or major agricultural knowledge, as well as specific craft specializations. The various

Figure 17.1 The alluvial fan of the Murghab River (southern Turkmenistan). Main Bronze Age archaeological sites along ancient fluvial ridges. Basemaps: CORONA KH-4B (May 1972); Landsat 5TM (June 1987) (GIS elaboration by R. Arciero).

archaeological contexts that have been investigated so far clearly show that the situation is extremely complex and that there is no conclusive opinion.

Today, the inhabitants of the villages established on the edge of the desert in the Mary Region (i.e., more or less the present alluvial fan of the Murghab River) change their role regularly from shepherd to farmer in accordance with the social demands. Every week or fortnight, the adult male members turn from guarding the herds grazing in the northernmost territories of the fan to land cultivation within their own *sovkhoz*. We interviewed them as well as those shuttling between the desert and the village to supply the shepherds with basic foodstuffs and to supervise the movement of the herds – currently on off-road vehicles and previously on horseback or camel. Their answers were illuminating for us as archaeologists trying to interpret at least part of the internal working organization of this population that we can safely define as "agropastoralists" (Arciero and Forni 2018).

So far, on the basis of the previous literature and current excavations, we can say that, during the Bronze Age, the Murghab region was characterized by the presence of two ethnic pastoral groups who seem to have had a similar lifestyle and material culture (Cattani 2008a; Rouse and Cerasetti 2014, 2018; Cerasetti *et al.* 2018, 2019). One of the two groups inhabited the BMAC urban centers and used undecorated handmade pottery, while the other was located in close proximity to the BMAC settlements, or in the intra-site areas, and used ICW pottery. This clear but important distinction hampered the progress of our studies and the real understanding of the complex Bronze Age population's scenario in the Murghab region. It seems that the unique feature that really differentiates the two groups is the use of a different type of pottery. In both contexts handmade pottery has always been found in combination with BMAC wheel-made pottery, on the surface as well as within the cultural layers from well controlled excavations, with the predominance of one or the other.

In the Murghab region, pastoralists are attested as early as the second half of the Middle Bronze Age (ca. 2210–1960 BCE). Evidence comes from the excavations made in three trenches just outside the defensive walls of the Bronze Age site of Adji Kui 1 (Spengler *et al.* 2016; Cerasetti *et al.* 2018; Rouse and Cerasetti 2018). There, the coexistence of the BMAC people living in the "citadel," as defined by G. Rossi Osmida (2003, 2007), with a pastoral population located on the edge of the town is clearly attested. Unfortunately, the quality of the excavations conducted by Rossi Osmida at this site has obscured further traces anywhere else on the site, and therefore the possibility to better outline within the settlement the presence of this pastoral group and understand its connection with the BMAC inhabitants has disappeared. What we can say, however, is that the cultural phase we have investigated in all of the three trenches is related undoubtedly to mobile pastoralists and concerns their living structures, material culture, and organic remains. The semi-subterranean sunken dwelling found in the trench immediately behind the defensive walls of the citadel, and the numerous postholes drilled into artificial platforms in the other two trenches lead us to presume the presence of light structures such as tents, together with fireplaces and storage pits, and leave no doubt on the cultural belonging of the investigated context. Moreover, the significant presence of millet in a fireplace of the westernmost trench, well dated by radiocarbon analysis that place this occupation in the Middle Bronze Age (3708 ± 45 BP, i.e., 2210–1960 BCE), is exceptional evidence for an ancient pastoral existence in a sedentary context (Cerasetti *et al.* 2018).

A similar image comes from the more recent phases of the Bronze Age. A careful and proper stratigraphic control of the excavations made at the site of Togolok 1 produced excellent and quite unexpected results as well. To understand their importance, we must add here that one of the main causes for our difficulty in answering the questions on the interaction/integration between mobile and sedentary populations during the Bronze Age relies upon the very negative old practice of removing from BMAC contexts any occupation remains of populations involved in herding tasks. This is an even greater handicap, of course, if these cultures are indeed two sides of the same coin.

At Togolok 1,[2] the pastoral phase seems to conclude the lifetime of the settlement after a period of abandonment. This gave us unique information that we had never obtained before. The living structures of the settlement after its abandonment are very similar to those excavated at Adji Kui 1: postholes probably supporting poles for light structures, storage pits, fireplaces with abundant remains of meals, and a massive presence of ovicaprid dung that let us suppose the existence of a shelter for the herds. All these features tend to indicate that pastoralists involved in the daily task of looking over grazing herds were living there.

However, these "pastoralists" handled terracotta figurines (Forni 2015, 2017) and owned chlorite seals (Forni 2017–2018), consumed beef and pork, and supplemented their diet with legumes and grains (for all these material classes, see Cerasetti *et al.* 2019; Cerasetti *et al.* in press). This very significant data makes us consider these people rather as "agropastoralists." The urban setting of Togolok 1 is finally suggesting the real settlement dynamics during the Bronze Age.

In the purely nomadic sites of Ojakly[3] and Chopantam that we also excavated, tents or sunken dwellings, fireplaces with meal remains, and storage pits with pot fragments remind us of those in urban contexts. In Chopantam, for instance, a significant amount of BMAC wheel-made pottery is present together with ICW handmade pottery, and the practice of agriculture is definitely attested along with herding. At Ojakly, a craft area with a pottery kiln points at a partial stabilization of this population and is the only one of the Andronovo culture found in the Murghab alluvial fan. It seems that only pottery, with its incised decoration, makes this cultural scenery different from that found in Adji Kui 1 and Togolok 1.

TO CONCLUDE: QUESTIONS

At the current state of research, there is no evidence of ICW pottery with a secure stratigraphic provenience within BMAC urban contexts. Once more, the same questions arise: What can we, or should we, argue on this basis? Are we dealing with different ethnic groups (i.e., those with handmade pottery without decoration at Adji Kui 1 and Togolok 1, and those with ICW pottery at Ojakly and Chopantam), but both of pastoral culture? Should we suppose that pastoral groups in the urban contexts, characterized by undecorated handmade pottery, are one of the expressions of the BMAC culture, while the population of ICW pottery – who inhabited campsites or villages on a seasonal basis close to BMAC settlements, or in the intra-site areas – belongs to a different pastoral culture coming from another area?

The idea that the presence/absence of ICW pottery might have been an important distinctive feature in the overall framework of the Bronze Age settlement in the Murghab region is intriguing. I can already hear my colleague and friend Lynne M. Rouse, my partner in this adventure in the Kara Kum Desert, who, like Talking Cricket in Pinocchio's fable, keeps my feet firmly on the ground:

> Things are more complex than they seem and we are only at the beginning of understanding. Pottery-making and pottery use should not be the only way to characterize the population groups. Defining "ethnic groups" and their interaction depends on many different kinds of data and to restrict it to pottery ignores all the other scientific analyses that we have gathered during the last several years. These early hypotheses will be challenged as work continues in the Murghab, while revisions, new ideas, and better understandings are the ultimate goal!

She said, reading a preliminary version of this chapter … and she is right!

Thus, it is not our intention to simplify the complex issue of the mobile/sedentary coexistence, interaction, and integration, or identity and diversification of the two cultures. For now, we prefer not to be seduced by the fascinating speculations that these early results suggest, and to go back to the only possible attitude (i.e., to base our hypothesis on the data at hand).

However, recent innovating studies of the ancient genome made on the human remains found during excavations can contribute to clarifying the genetic formation of the inhabitants from mobile pastoralist sites as well as from large BMAC urban settlements and provide more information about their possible origin within the larger framework of migration that characterized Central Asia during the Bronze Age. The A-DNA analysis might be the right way to shed some light on this integration process and on what appears to become a single social identity of two separate populations. A southward spread of genetic ancestry from the Eurasian steppe has been documented and correlates with the archaeologically known expansion of pastoralist sites from the steppe to Central Asia in the Middle to Late Bronze Age (Narasimhan *et al.* 2019). These steppe communities mixed genetically with BMAC peoples, whom they encountered in Central Asian regions. Local interactions between the occupants of BMAC urban centers and the mobile pastoral communities in their surroundings must have been a key aspect of the political and economic integration of the BMAC system.

We consider that one of the more crucial issues is to go deeper and to better define the complex phenomenon of the Bronze Age population in southern Turkmenistan in order to understand fully the following historical periods. The comparative research made by our colleagues in the neighboring regions of Central Asia is particularly important in that matter (Mace *et al.* 1993; Salzman 2002; Frachetti 2008, 2012; Brite 2013; Frachetti and Maksudov 2014; Brite *et al.* 2017, Rouse 2020 to name just a few; see also Chapter 21). Unfortunately, in Turkmenistan up to now only slight attention has been given to the pastoral components. This has left aside a critical element for the understanding of the complex BMAC society, which still escapes us although much research has been made on its wealthy features.

It is not possible to walk with one leg!

NOTES

* There are many thanks to be made, especially to colleagues who are dealing with this problem in Central Asia and who have given me numerous ideas for this chapter. Thanks to my inflexible and infallible editors, B. Lyonnet and N. Dubova, whose dedication has made it possible to carry out this work. A special thanks to Lynne M. Rouse, my partner in this great adventure and continuous term of comparison; to my senior collaborators Luca Forni and Roberto Arciero, without whose help and enthusiasm this work could not go on; and to colleagues of the Ministry of Culture of Turkmenistan, who have always shared our love for this research. Another special thanks to our sponsors the Ministry of Foreign Affairs and International Cooperation (MAECI), University of Bologna, Washington University in St. Louis, ISMEO-The International Association of Mediterranean and Oriental Studies (Rome), Deutsches Archäologisches Institut (DAI), University of Naples "L'Orientale," and to the Italian Embassy of Turkmenistan, without whose support I could never overcome the thousand obstacles that every vast project like this entails.

1 In this article, Sarianidi mentions the presence in Bactria of handmade pottery that is different to that of the Andronovo people, and Teufer (Chapter 25, this volume) also mentions handmade pottery from Bactria related to the Vakhsh culture.

2 The site is made of several mounds. In the 1970s, V.I. Sarianidi made soundings on two of them up to virgin soil and on two potter's kiln areas associated with it (11 kilns in total). A long description is given in Sarianidi (1990: 34–45). He mentions three main periods of construction, the presence of many women terracotta figurines, a cult vessel with applied animals on its walls and rim, a sherd with the imprint of a cross-shaped seal, a stone amulet, a fragment of a cylinder seal, and stone artifacts including lapis, carnelian, and metal ones.

He also mentions in the first two upper *jarus* (i.e., around 1 m deep) of the main mound the presence of a large number of handmade vessels without any decoration, one of which has the same shape as usual wheel-made pottery, and others considered to be cooking ware (but not all). These handmade ceramics are still present but in much lesser quantities in the third *jarus* below, where the remains of a large wall were found, said to probably belong to a monumental building (editors' note).

3 For the absolute dates of Ojakly see the Appendix at the end of the volume.

REFERENCES

Arciero, R. and L. Forni 2018. "La prima urbanizzazione in Turkmenistan: coesistenza tra nomadi e sedentari nel delta interno del fiume Murghab – risultati preliminari delle indagini archeologiche ed etnografiche presso il sito di Togolok 1." In *Armenia, Caucaso e Asia Centrale*, ed. A. Ferrari, E. Pupulin, M. Ruffilli, and V. Tomelleri. Venice: Edizioni Ca'Foscari, pp. 11–38.

Askarov, A. 1962. "Pamjatniki andronovskoj kul'tury v nizov'jakh Zarafshana" [Andronovo-culture sites in the lower Zeravshan]. *Istorija Material'noj Kul'tury Uzbekistana* 3: 28–41.

Brite, E.B. 2013. "Mobility in the Central Asian Oasis States: Archaeological Evidence from Antique Khorezm." *Bulletin of International Institute for Central Asian Studies* 18: 3–25.

Brite, E.B., G. Khozhaniyazov, J.M. Marston, M.N. Cleary, and F.J. Kidd 2017. "Kara-Tepe, Karakalpakstan: Agropastoralism in a Central Eurasian Oasis in the 4th/5th Century A.D. Transition." *Journal of Field Archaeology* 42(6): 514–529.

Cattani, M. 2008a. "Excavations at Sites No. 1211 and No. 1219 (Final Bronze Age)." In *The Bronze Age and Early Iron Age in the Margiana Lowlands: Facts And Methodological Proposals for a Redefinition of the Research Strategies*, ed. S. Salvatori, M. Tosi, and B. Cerasetti. BAR International Series 1806. Oxford: Archaeopress, pp. 119–132.

Cattani, M. 2008b. "The Final Phase of the Bronze Age and the 'Andronovo Question' in Margiana." In *The Bronze Age and Early Iron Age in the Margiana Lowlands: Facts And Methodological Proposals for a Redefinition of the Research Strategies*, ed. S. Salvatori, M. Tosi, and B. Cerasetti. BAR International Series 1806. Oxford: Archaeopress, pp. 133–151.

Cattani, M., B. Cerasetti, S. Salvatori, and M. Tosi 2008. "The Murghab Delta in Central Asia 1990–2001: The GIS from Research Resource to a Reasoning Tool for the Study of Settlement Change in Long-Term Fluctuations." In *The Bronze Age and Early Iron Age in the Margiana Lowlands: Facts and Methodological Proposals for a Redefinition of the Research Strategies*, ed. S. Salvatori, M. Tosi, and B. Cerasetti. BAR International Series 1806. Oxford: Archaeopress, pp. 39–45.

Cerasetti, B. 1998. "Preliminary Report on Ornamental Elements of 'Incised Coarse Ware,'" In *The Archaeological Map of the Murghab Delta: Preliminary Reports 1990–95*, ed. A. Gubaev, G. Koshelenko, and M. Tosi. Reports and Memoirs. Series Minor, III. Rome: IsIAO, pp. 67–74.

Cerasetti, B. 2012. "Remote Sensing and Survey of the Murghab Alluvial Fan, Southern Turkmenistan: The Coexistence of Nomadic Herders and Sedentary Farmers in the Late Bronze Age and Early Iron Age." In *Proceedings of the 7th International Congress on the Archaeology of the Ancient Near East*, vol. 1, ed. R. Matthews and J. Curtis. Wiesbaden: Harassowitz, pp. 539–558.

Cerasetti, B., R. Arciero, M. Carra, A. Curci, J. De Grossi Mazzorin, L. Forni, E. Luneau, L.M. Rouse, and R.N. Spengler III 2019. "Bronze and Iron Age Urbanization in Turkmenistan: Preliminary Results from the Excavation of Togolok 1 on the Murghab Alluvial Fan." In *Urban Cultures of Central Asia from the Bronze Age to the Karakhanids: Learnings and Conclusions from New Archaeological Investigations and Discoveries*, ed. C. Baumer and M. Novák. Wiesbaden: Harrassowitz, pp. 63–72.

Cerasetti, B., R. Arciero, L. Forni, L.M. Rouse, M. Carra, A. Curci, J. De Grossi Mazzorin, E. Luneau, and R.N. Spengler III in press. "Interactions Between Sedentary and Mobile Peoples in the Bronze Age of Southern Central Asia: Excavations at Togolok 1 in the Murghab Region." In *Farmers, Traders and Herders: The Bronze Age in Central Asia and Khorasan*, ed. E. Luneau, M. Teufer, and N. Boroffka. Archäologie in Iran und Turan. Berlin: Dietrich Reimer.

Cerasetti, B., L.M. Rouse, and I. de Nigris 2018. "The Influence of Late Bronze Age Sedentary–Mobile Interactions on the Iron Age: Mobile Pastoral Occupation Sites in the Murghab Alluvial Fan, Turkmenistan." In *A Millennium of History: The Iron Age in Central Asia (2nd and 1st millennia BC)*. Archäologie in Iran und Turan 17/Mémoires de la Délegation Archéologique Française en Afghanistan XXXV. Berlin: Dietrich Reimer, pp. 17–30.

Forni, L. 2015. "Le figurine antropomorfe e zoomorfe in terracotta dell'Età del Bronzo dalla regione del fiume Murghab (Turkmenistan)." Master's thesis, University of Bologna, Bologna.

Forni, L. 2017. "Bronze Age Terracotta Anthropomorphic and Zoomorphic Figurines from the Murghab Region (Turkmenistan): New Evidence and Interpretations." *Bologna: OCNUS, Quaderni della Scuola di Specializzazione in Beni Archeologici* 25: 9–20.

Forni, L. 2017–2018. "I sigilli dell'età del Bronzo dalla regione del fiume Murghab (Turkmenistan). Nuove evidenze dai materiali della Missione Archeologica Italo-Turkmena tra il 1990 e il 2018." PhD thesis, Inter-University School of Specialization in Archaeological Heritage of the Universities of Trieste, Udine, and Ca'Foscari.

Frachetti, M.D. 2008. *Pastoralist Landscapes and Social Interaction in Bronze Age Eurasia*. Berkeley: University of California Press.

Frachetti, M.D. 2012. "Multiregional Emergence of Mobile Pastoralism and Non-uniform Institutional Complexity across Eurasia." *Current Anthropology* 53: 2–38.

Frachetti, M.D. and F. Maksudov 2014. "The Landscape of Ancient Mobile Pastoralism in the Highlands of Southeastern Uzbekistan, 2000 B.C.–A.D. 1400." *Journal* of *Field Archaeology* 39: 195–212.

Gening, V.F., G.B. Zdanovich, and V.V. Gening 1992. *Sintashta. Arkheologicheskie pamjatniki ariiskikh plemen Uralo-Kazakhstanskikh stepej* [Sintashta: archaeological sites of Aryan groups from the Uralo-Kazakhstan steppes]. Cheljabinsk: Juzhno-Ural'skoe knizhnoe izd.

Grjaznov, M.P. (ed.) 1966. *Andronovskaja kul'tura* [The Andronovo culture]. Moscow: Nauka.

Gubaev, A., G. Koshelenko, and M. Tosi (eds.) 1998. *The Archaeological Map of the Murghab Delta: Preliminary Reports 1990–95*. Rome: IsIAO.

Guljamov, Ja.G., U. Islamov, and A. Askarov 1966. Pervobytnaja kul'tura i vozniknovenie oroshaemogo zemledelija v nizov'jakh Zarafshana [Indigenous culture and origin of farming irrigation in the lower Zeravshan valley]. Tashkent: Fan.

Itina, M.A. 1977. *Istorija stepnykh plemen Juzhnogo Priaral'ja* [History of the steppe groups of the southern Aral area]. Moscow: Nauka.

Kuz'mina, E.E. 1994. *Otkuda prishly Indoarii? Material'naja kul'tura plemen andronovskoj obshchnosti i proiskhozhdenie indoirantsev* [Where from came the Indo-aryans? Material culture of the Andronovo-culture community and origin of the Indo-Iranians]. Moscow: MGP Kalina, Rossijskaja Akademija Nauk.

Mace, R., D.M. Anderson, T. Bierschenk, L. Cronk, I. Köhler-Rollefson, W. Lancaster, F. Lancaster, P.D. Little, E.A. Morris, and J. Rossignol 1993. "Transitions Between Cultivation and Pastoralism in Sub-Saharan Africa." *Current Anthropology* 34(4): 363–382.

Narasimhan, V.M., N.J. Patterson, P. Moorjani, N. Rohland, R. Bernardos, S. Mallick, I. Lazaridis *et al.* 2019. "The Formation of Human Populations in South and Central Asia." *Science*, 365(6457): n.p. doi: 10.1126/science.aat7487.

Rossi Osmida, G. (ed.) 2003. *Central Asia Cultural Values*. Padova: Il Punto.
Rossi Osmida, G. 2007. *Adji Kui Oasis: La cittadella delle statuette, I*. Padova: Il Punto.
Rouse, L.M. 2020. Silent partners: archaeological insights on mobility, interaction and civilization in Central Asia's past. Central Asian Survey. DOI: 10.1080/02634937.2020.1769024
Rouse, L.M. and B. Cerasetti 2014. "Ojakly: A Late Bronze Age Mobile Pastoral Occupation Site in the Murghab Delta, Turkmenistan." *Journal of Field Archaeology* 39(1): 32–50.
Rouse, L.M. and B. Cerasetti 2018. "Mixing Metaphors: Sedentary–Mobile Interactions and Local–Global Connections in Prehistoric Turkmenistan." *Antiquity* 92(363): 674–689.
Rouse, L.M., K.M. Grillo, R. Piermartiri, E. Rotandaro, H. Cogo-Moreira, G.M. Bargossi, and B. Cerasetti 2019. "Not Just 'Nomadic Jars': The Late Bronze Age Ceramic Assemblage from the Mobile Pastoralists Site of Ojakly, Murghab Region, Turkmenistan." *Archaeological Research in Asia* 18: 100–119.
Salvatori, S., M. Tosi, and B. Cerasetti (eds.) 2008. *The Bronze Age and Early Iron Age in the Margiana Lowlands: Facts and Methodological Proposal for a Redefinition of the Research Strategies. The Archaeological Map of the Murghab Delta*. BAR International Series 1806. Oxford: Archaeopress.
Salzman, Ph.C. 2002. "Pastoral Nomads: Some General Observations Based on Research in Iran." *Journal* of *Anthropological Research* 58: 245–264.
Sarianidi, V.I. 1975. "Stepnye plemena epokhi bronzy v Margiane" [Bronze Age steppe groups in Margiana]. *Sovetskaja Arkheologija* 2: 20–29.
Sarianidi, V.I. 1990. *Drevnosti strany Margush* [Ancient Margush]. Ashkhabad: Ylym.
Spengler, R., I. de Nigris, B. Cerasetti, M. Carra, and L.M. Rouse 2016. "The Breadth of Dietary Economy in Bronze Age Central Asia: Case Study from Adji Kui 1 in the Murghab Region of Turkmenistan." *Journal of Archaeological Science* 22: 372–381.

CHAPTER EIGHTEEN

THE END OF THE OXUS CIVILIZATION

Elise Luneau

INTRODUCTION

The last centuries of existence of the Oxus Civilization have been described from the beginning as a time of decline without major material and cultural expressions (Masson and Sarianidi 1972: 137). Although poorly known due to a lack of research, this three-century period was until recently viewed as the "collapse" of the previous complex society that had spread in southern Central Asia, and was considered as unattractive for research (Luneau 2014). In addition, due to wide changes that occur, it is often perceived as a break in the cultural continuity of the Oxus Civilization.

This period may, however, be regarded otherwise (i.e., as a significant phase of transformation within a more or less preserved cultural frame), with the development of different models of social complexity, an optimization in the exploitation of the territory in connection with diverse economic systems and lifestyles, interactions between supposedly sedentary and mobile populations, and the gestation of peculiar beliefs and new symbolic practices.

This chapter presents a concise synthesis of the state of knowledge on the end of the Oxus Civilization. It brings new lights on diverse issues about the interpretation of this particular moment of sociocultural evolution and on the causes of transformations.

AN UNCLEAR CHRONOLOGY AND A PROBLEMATIC TERMINOLOGY

Central Asian prehistoric archaeology still faces many problems related to the chronology with several discrepancies and inconsistencies. First, the Bronze Age culture is not consensually defined, although the Bactria-Margiana Archaeological Complex (BMAC) and Oxus Civilization appear to be increasingly considered as synonymous. If the term "BMAC" was first used to underscore the rather strong homogeneity over Bactria and Margiana during the so-called peak phase of this cultural entity (Sarianidi 1974; Hiebert 1994a; Kohl 2007; Salvatori 2016), it tends now to be used for the designation of this culture till the middle of the second millennium BC. The

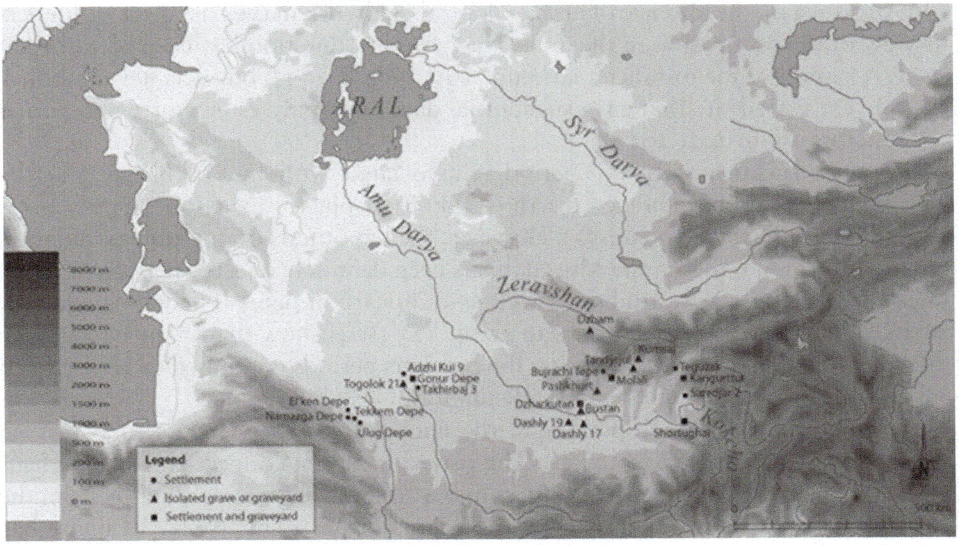

Figure 18.1 Map of the main sites related to the Final Bronze Age in southern Central Asia.

term "Oxus Civilization" has been created to better inscribe this entity as a phenomenon similar to what happened in Middle Asian societies (Francfort 1984; Hiebert 1994b; Lamberg-Karlovsky 1994, 2013; Salvatori 2008). At once, it emphasizes the processes and evolution of the archaeological features through time in the 3rd and 2nd millennium BC.

The latest phase of the Oxus Civilization that is of our concern here is approximately positioned between 1800/1700 BC and 1500/1400 BC, when the Early Iron Age starts, according to radiocarbon dates from different areas of its distribution (for the main sites related to this period, see Figure 18.1; for the ^{14}C dates see Luneau 2014: 272–277; see also the Appendix at the end of the volume).

A review of the available data underlines the imprecision of the regional periodization as well as great disparities in surveys and excavations. New programs of surveys have been recently carried out,[1] but many sites are still poorly or not yet excavated. In general, archaeological layers dated to this period have mainly been investigated in the southern parts of Uzbekistan and Tajikistan, and are still barely known in the other areas.

In southern Uzbekistan, this period, called here the Final Bronze Age (Luneau *et al.* 2014) or Late Bronze Age II (Teufer 2005; Kaniuth 2006) (see Chapter 16) has been primarily excavated at the sites of Dzharkutan, Bustan, and Molali (Askarov 1977; Avanesova 2013), but is also identified at many other poorly or not yet excavated small settlements (e.g., Bujrachi Depe, Pashkhurt, etc.). It is mainly recognized through specific pottery complexes (named "Kuzali," "Molali," and "Bustan"), which, despite apparent local variation, have also been used to characterize the sequence of southern Tajikistan, where for a long time the majority of the Bronze Age sites recorded were related to this period (Kangurttut, Teguzak, Kumsaj, Tandyrjul, etc.) (Vinogradova 2004).

In Afghanistan, the absence of archaeological fieldwork in the past decades hinders our view of the occupation in the area. The period in question has been identified at Dashly 17 and 19, for instance (Sarianidi 1977: 69, fig. 31), but was never studied. The only excavated site that gave substantial data and analyses is Shortughai period IV (Francfort 1989: 242; 2016).

In central Turkmenistan, the period is associated with the Takhirbaj 3 settlement, which serves as reference for the "Takhirbaj phase" (Masson 1959: 12–17; Sarianidi 1990; Hiebert 1994a: 71). However, the hypothesis of one-time occupation of many Bronze Age sites in the Murghab has been definitely rejected (Salvatori 2008; Luneau 2014: 192). New studies as well as radiocarbon dates (Hiebert 1994a: 71; Rossi Osmida 2007: 118–120; Luneau 2014: 123–124) show that many of the large settlements that had developed during the previous periods continued to be occupied well into the first half of the second millennium BCE.

In the foothills of the Kopet Dagh, the end of the BMAC refers to Namazga (NMG) VI (Shchetenko 1972, 2000). The period has been connected with an abandonment or decrease of the urban centers and a "migration" of populations from the Kopet Dagh towards the Murghab area (Sarianidi 1990). This NMG VI phase has been identified at numerous other sites in the area (El'ken Depe, Tekkem Depe, Ulug Depe, etc.) but lacks a good characterization of its material culture. So far, the NMG VI layers of the Turkmen foothills have still not been precisely published, and it is thus unclear to which exact part of the BMAC they are related.

In northern Iran, although a permanent occupation related to the Oxus Civilization has been recently highlighted by new investigations (Vahdati and Biscione forthcoming; see also Chapter 19) and is clearly connected with the mature Oxus period in southern Central Asia, its continuation into the Final Bronze Age is not yet ascertained (Luneau 2018).

The chronology is thus imprecise, mostly due to regional differences, the complexity of chronological subdivisions, and methodological issues. It often relies on older studies, which sometimes proved to be invalid and that create serious confusion. It is also mostly based on few clear stratigraphic studies and on series created from the distribution of the funerary material. The sequences previously defined, as well as the connections between the different regions, are still a large matter of debate in southern Uzbekistan and Tajikistan (Teufer 2005, 2015; Luneau *et al.* 2014), as well as in Turkmenistan (P'jankova 1993; Lamberg-Karlovsky 2003; 2013; Salvatori 2016) and Iran (Luneau 2018).

In addition, discrepancies in the terminology adopted by different scholars, like the use of the same term for different periods, disturb the perception (Luneau 2014: 46–47; Salvatori 2016). The final phase of the Oxus Civilization has been named as being "Final Bronze Age" but is also designated as "Late Bronze Age" (see Chapter 16). This last term serves as well to designate a larger chronological time period (Teufer 2005; Kaniuth 2006; Stančo 2016), which sometimes relates to the whole BMAC cultural complex. Lastly, the term "Final Bronze Age" was also previously used to designate cultures dated to the second half of the second millennium BC (Chust, Jaz I) because of the (quasi) absence of iron items, the imprecise nature of the transition, and the better harmonization with the Near Eastern chronologies. These cultures and periods are, however, now conventionally attributed to the "Early Iron Age."

In order to highlight the evolution and phenomena of transition in the long run, we will use here the terms "Final Bronze Age" to designate the final phase of the Oxus Civilization between ca. 1750/1400 BC, "Middle/Late Bronze Age" to refer to the previous phases of this cultural entity between ca. 2300 and 1750 BC, and "Early Iron Age" to name the next "Jaz I" cultural phase (ca. 1450–1000/900 BC).

ARCHAEOLOGICAL FEATURES AND ECO-SOCIOCULTURAL TRANSFORMATIONS

At least 100 sites are said to be related to the final phase of the Oxus Civilization. However, more than half have not been studied at all and only a few have been published. The inconsistent knowledge on their periodization for several of them interferes with an understanding of the sociocultural evolution. In addition, the information provided, often coming from old excavations, lacks precision and systematization. Therefore, many aspects related to the society and many quantitative and qualitative details on the material culture are still unknown. Data is particularly lacking on the socioeconomic features (profile of the subsistence economy, quantification of artifacts, studies of the technological processes, etc.) as well on the social structures (settlement pattern and organization, etc.). These gaps significantly impede, as of now, the possibility of an elaborated and precise analysis of the period as well as evaluation of the transformations.

Settlement pattern and architecture

The sites concern both settlements and graveyards approximately in similar number (Figure 18.1), with significant regional divergences apparently resulting from the history of research. They are positioned in different ecological environments, mostly on the river terraces and in the alluvial plains, but some are also higher in the foothills. The surface of occupation, smaller than in the previous period, varies greatly between the settlements, from ca. 22.5 ha (e.g., Takhirbaj 3 in Turkmenistan) to ca. 2 ha (e.g., Teguzak in Tajikistan). We have only very few hints at what the architecture looked like. Monumental buildings are rare and could represent a reuse of earlier buildings rather than new constructions. Continuities have been identified in the building techniques. Like previously, the household structures are built of unbaked mudbricks or *pisé*, although they tend to become simpler (Francfort 1989: 59). New types with stone fundaments, smaller and lighter, seem to appear especially in the foothills, like in Tajikistan, with terraced installations on the slopes, sometimes with semi-opened structures. Other light structures, such as fences of branches or bulrushes have also been identified. They fit in with the ecological environment, and perhaps are related to a more pastoral economy.

Previous studies have mostly highlighted important changes in the settlement pattern with the previous period. A trend towards a decrease in size of the sites and a lower dimensional differentiation is indeed perceptible in all the areas. This shift in the settlement pattern has been interpreted as the replacement of a territorial ranking system by one without prevailing centers (Salvatori 2008). However, the absence of hierarchical settlement structuration in the Final Bronze Age cannot be ascertained according to the available data, especially because of the chronological issues previously mentioned as well as the current impossibility to evaluate the space inhabited in many multiperiod sites during the final phase. This evolution may be further balanced by the continuous appearance and disappearance of sites during the whole period. Actually, the population dynamics do not appear to be interrupted or restrained during the final period.

Subsistence economy

Such a settlement pattern underlies various economic strategies, combining agriculture, animal husbandry, and a small amount of hunting, but the specificities of the subsistence scheme between the diverse population groups are still not well known. Archaeobotanical and archaeozoological data come from only two sites, one located in the alluvial plain (Shortughai, period IV, see G. Willcox and J. Desse in Francfort 1989: 175–206) and the other in the foothills (Kangurttut, see Vinogradova 2004: 225–226, 248–253, 257–266). At the first site, agriculture still relies on the main cultivated species known previously, i.e. one type of barley, wheat, millet and lentil. Different species of pulses and other plants may have been harvested from the wild. Farmers continued to use irrigation canals already operating in the previous periods. Regarding animal husbandry, the range of species and the proportion between animals persist compared to previous periods. Cattle (bovid and water buffalo) and caprine (sheep and goat) are predominant, and the ages of slaughter of the animals point at a husbandry strategy not merely for meat consumption. At the second site, in the foothills, the flora spectrum reveals a higher diversity of species of cultivated barley and wheat, as well as millet and, perhaps, flax. The presence of bovine, sheep and goat, donkey, horse, and dog is attested. The bovine age of slaughter tends to show that milk production did not play a major role compared to meat production and to the draft or pack force of the animals. To this short description has to be added the presence of horse bones at several other sites, such as Bustan 6, Takhirbaj 3, Namazga Depe, and Tekkem Depe.

Thus, the socioeconomic profile seems not to have radically changed since the mature Oxus period. However, the landscape management, especially in respect to water, is still largely unknown, and new dynamics of human/environment interaction can be expected. Irrigation may indeed have decreased or been maintained, but certainly did not develop more (Lyonnet 1997: 76–78; Rouse and Cerasetti 2015). At Shortughai, some changes have been detected in comparison with the earlier periods, such as the possible disappearance of *Cyperaceae* (viewed as a potential indicator of irrigation), the rise of barley in comparison to wheat, and the more frequent use of millet (better adapted to droughts) (Willcox in Francfort 1989: 175–183). These changes may indicate a decrease in irrigation in parallel with a more common practice of dry agriculture, as also suggested by the frequent pair *Panicum/Setaria*. However, additional dry agriculture may also have been a complement to increase the bulk of the harvest.

A permanency over the whole period of occupation in the animal exploitation according to the range of species, the proportions of caprine, bovine, and wild animals, as well as the peculiar practices of slaughter has been suggested in the northern Afghan site (Desse in Francfort 1989: 190–204). If the amount of settlements had arisen in the foothills, it would have necessitated the development of specific socioeconomic practices, but the data is presently insufficient to evaluate the possible increase of herding in the different ecological niches.[2]

The data brought to light tends to point at the permanency of a mostly sedentary population with a complex agropastoral economy. It is also possible that part of the population was involved in a more mobile way of life, with seasonal movements related to stockbreeding, particularly in the foothill areas where lighter dwelling structures may have been preferred.

Material culture

The material culture consists of pottery vessels, as well as metal, stone, unbaked clay, and bone artifacts. Various implements also attest to work with wool, skins, textile, and mineral raw material. The ceramic production is varied with several types of fine, medium and coarse wares identified so far, showing a regional distribution. The range of shapes is large and shifted significantly compared to the previous periods (Figure 18.2). Different techniques were used: handmade shaping by coiling, or from a lump of clay, or over a lump of clay in a cloth; wheel-throwing; and wheel-coiling (i.e., a mixed technique of coiling and wheel at different steps of the *chaîne opératoire*). Like previously, pottery complexes are mostly dominated by the wheel-made or wheel-coiled pottery, but recent excavations at Saredjar 2 in Tajikistan have brought to light a ceramic corpus where 85% is handmade pottery (Teufer *et al.* 2014) so that pottery practices need to be reevaluated at the local scale. The pottery related to the Final Bronze Age is mostly said to be red-burnished (Hiebert 1994a: 71), or with a red slip (P'jankova 1993; Pilipko 2010: 232–233), or with incised decorations (curved and straight lines) (Sarianidi 1977: 25). However, first, exhaustive studies at the Dzharkutan necropolis and the settlement of Molali in southern Uzbekistan show that only 20–30% of the vessels are decorated, mostly by polishing, slip, and incisions (Luneau 2014: 114–145; Luneau et al. forthcoming). Second, these types of decoration also existed in the previous periods. Even considering regional variation, these features cannot be used as the only criteria of identification for the Final Bronze Age pottery without the precise description of the different pottery complexes through time.

Similarly, for the other materials the types of artifacts significantly changed at the beginning of the Final Bronze Age. The corpus of metal items is characterized by a new phenomenon: the appearance of miniature objects deposited in the graves, especially in southern Uzbekistan and southern Tajikistan, which represent approximately four-fifths of the whole metallic assemblage (Figure 18.3, *1*). Although unequally distributed (Figure 18.4), they are present in one-seventh of all the burials recorded for this period, and are connected to all kinds of graves (Luneau 2014: 150–151). Except for these miniature objects, few other metal artifacts have been discovered in the settlements and are related mainly to ornaments, but a few tools and weapons are also present. The great majority (around 90%) is copper-based. Archaeometrical analyses currently attest to the common use of tin bronze (half of the samples in the most eastern areas) in addition to copper alloys.

The stone production mostly concerns implements and weapons (axes, arrowheads, working stones, casting molds for metallurgy, etc.). Ornaments are few, mainly beads. Unlike the previous period, precious and semiprecious stones are rare, even in the

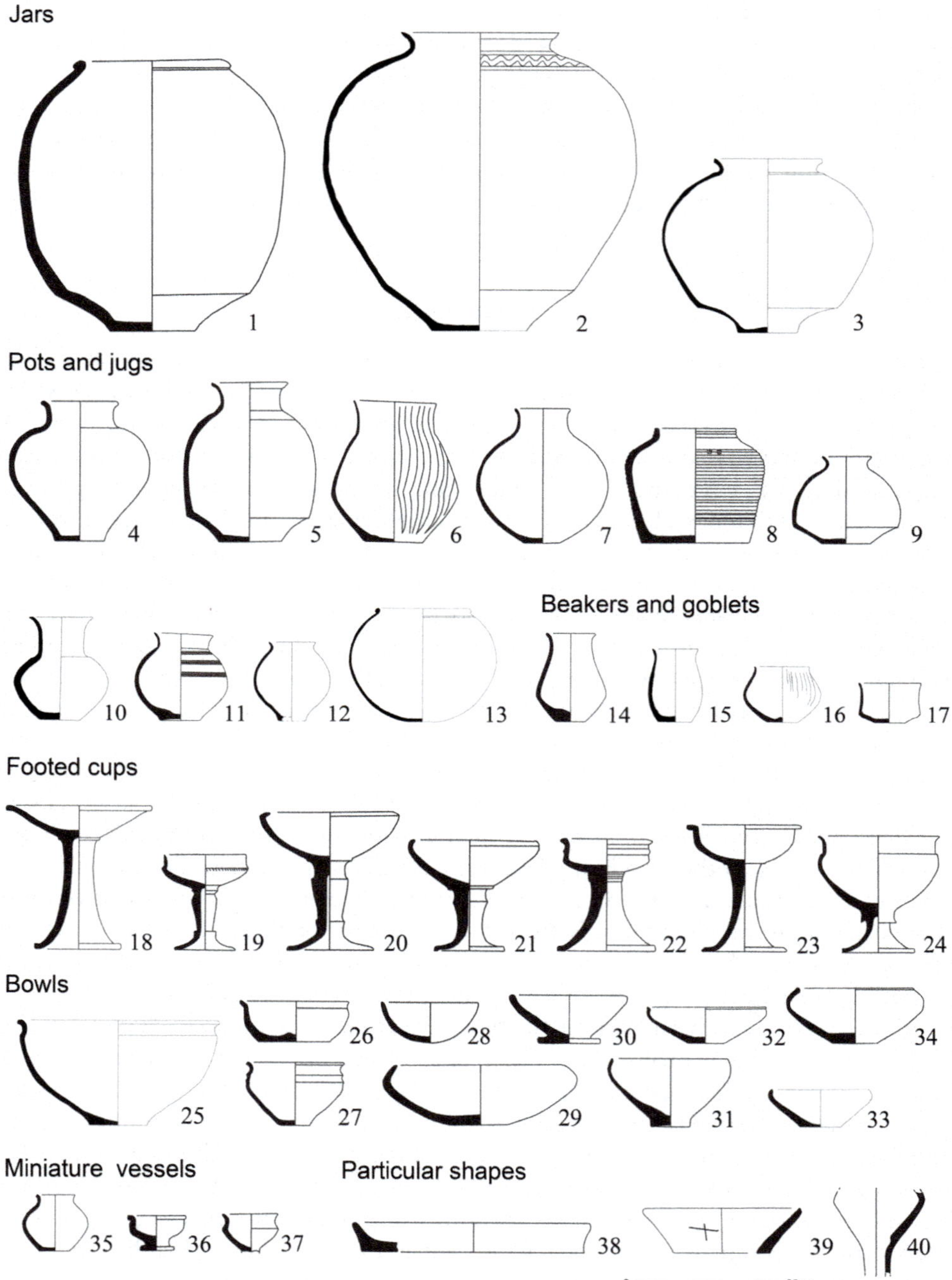

Figure 18.2 Examples of pottery types related to the Final Bronze Age. 1. Dzharkutan 4A (after Askarov and Abdullaev 1983: 117, fig. LII, 5); 2–3. Molali (©Deutsches Archaölogisches Institut, Eurasien-Abteilung); 4. Tandyrjul (after Vinogradova 2004: 131, fig. 9, 5); 5. Dzharkutan 4A (after Askarov and Abdullaev 1983: 116, fig. L, 3); 6. Nurek (after P'jankova 1979: 84, fig. 2, 3); 7. Takhirbaj 3 (after Sarianidi 1990: 266, fig. LXI, 23); 8. Bustan 6 (after Avanesova 1997: 171, fig. 14, 1); 9. Molali (after Beljaeva and Khakimov

graves. Beads with an axe shape (made of lapis lazuli or an artificial material) seem characteristic of the period in the eastern sites (Figure 18.3, 2). The same can be said of various unbaked clay objects unknown before (figurines, spindle whorls, miniature cones or spoons, etc.) (Figure 18.3, 3). Bone awls and tubes have also been recorded.

Changes compared to the previous periods are clear both in the quantity of artifacts discovered and in their style (Luneau 2014: 165–169). Although the study lacks a detailed comparative study from reliable stratigraphic layers between the different periods (such as that done at Shortughai, see Francfort 1988), the examination of the data from funerary and settlement contexts suggests a rise in the amount of pottery in contrast with a significant decrease in the quantity of artifacts made of other materials. A strong decline in the number of tools, weapons, and ornaments in metal is indeed clearly attested until they almost disappear (at least in the graves). The miniature metal objects that had only appeared around 1800/1700 BC, also progressively decrease throughout the Final Bronze Age and vanish around 1500/1400 BC.

Besides these quantitative changes, others dealing with the shapes and styles are obvious. The ceramic assemblage is affected by a higher variability, possibly in relation with functional and organizational shifts in the pottery production, though the production techniques are the same (clay sources, paste composition, fashioning, firing, etc.) (Luneau et al. forthcoming). Concerning metallurgy, the development of the techniques of bronze metallurgy is attested in southern Uzbekistan. No diachronic evaluation has been made in the other areas. Old analyses also point to possible modifications of the ore supplies throughout the Bronze Age (Ruzanov 1999), raising different interpretations (depletion of sources, geographical preference, changes of commercial partners). Globally, a general trend to less time investment and less skill in the fabrication process is sometimes noticeable but still needs to be evaluated properly. Cultural changes are noticed in the size of items, especially for the metal items (miniaturization), which are more likely connected to symbolic changes than to a raw material shortage,[3] but questions remain on the signification of this practice

Figure 18.2 (Cont.)

1973: 41, fig. 8, 1); 10–11. Dzharkutan 3 (©MAFOUZ-Protohistory); 12–13. Molali (©Deutsches Archaölogisches Institut, Eurasien-Abteilung); 14. Bustan 6 (after Avanesova 2007: 86, fig. 16, II, 5); 15–16. Dzharkutan 3 (©MAFOUZ-Protohistory); 17. Takhirbaj 3 (after Masson 1959: 177, fig. IX, 6); 18. Bustan 6 (after Avanesova 1997: 161, fig. 7, 1); 19. Dzharkutan 4A (after Askarov and Abdullaev 1983: 115, fig. XLVIII, 11); 20. Dzharkutan 4A (after Askarov 1977: 222, fig. LIX, 3); 21. Dzharkutan 3 (©MAFOUZ-Protohistory); 22. Takhirbaj 3 (after Masson 1959: 177, fig. IX, 3); 23. Dzharkutan 4A (after Askarov and Abdullaev 1983: 115, fig. XLVIII, 9); 24. Dzharkutan 4A (after Askarov and Abdullaev 1983: 115, fig. XLVIII, 3); 25. Molali (©Deutsches Archaölogisches Institut, Eurasien-Abteilung); 26. Kangurttut (after Vinogradova *et al.* 2008: 351, 70, 3); 27. Tandyrjul (after Vinogradova 2004: 131, fig.9, 9); 28. Tandyrjul (after Vinogradova 1980: 67, fig.3, 13); 29. Takhirbaj 3 (after Masson 1959: fig. VII, 8); 30. Kangurttut (after Vinogradova *et al.* 2008: 350, 60, 6); 31. Shortughaï (after Francfort 1989: pl. 56, 4); 32–33. Dzharkutan 3 (©MAFOUZ-Protohistory); 34. Kangurttut (after Vinogradova *et al.* 2008: 346, 27, 2); 35. Takhirbaj 3 (after Sarianidi 1990: 266, fig. LXI, 15); 36. Kangurttut (after Vinogradova *et al.* 2008: 350, 61, 2); 37. Bustan 6 (after Avanesova 2007: 70, fig.6, II, 1); 38. Takhirbaj 3 (after Masson 1959: 174, fig. VI, 16); 39. Namazga Depe/Vyshka (after Khlopina 1972: 208, fig.7); 40. Kangurttut (after Vinogradova *et al.* 2008: 324, fig. 29a, 25).

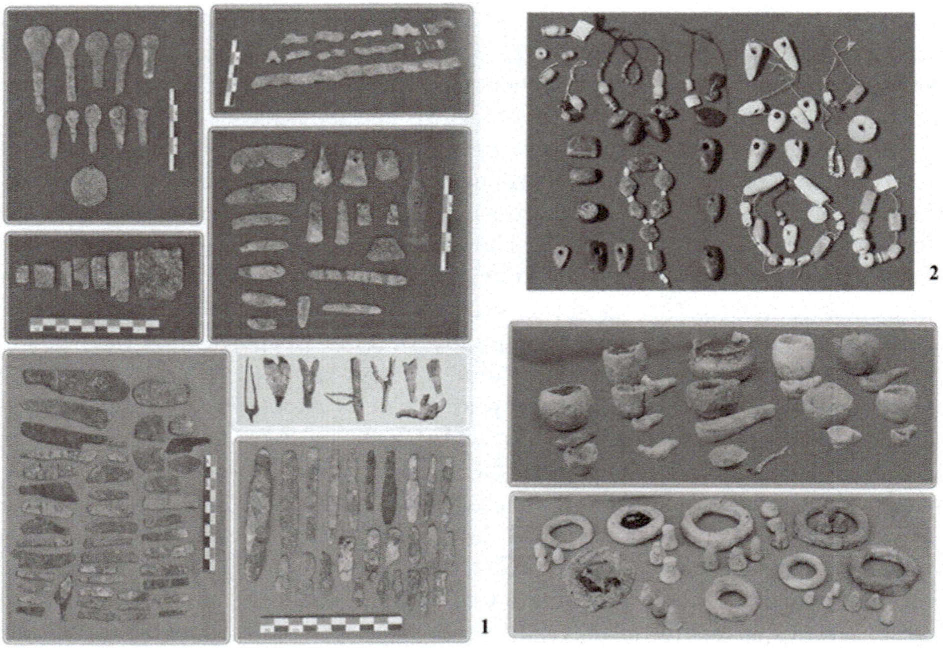

Figure 18.3 Examples of specific artifacts related to the Oxus Civilization from Uzbekistan in the Final Bronze Age. 1. Miniature artefacts in copper alloy (Bustan 6, after Avanesova 2013: 560, fig. XVI); 2. Beads, especially axe-shape (Bustan 6, after Avanesova 2013: 558, fig. XIV); 3. Small items in unbaked clay (Bustan 6, after Avanesova 2013: 553, fig. IX).

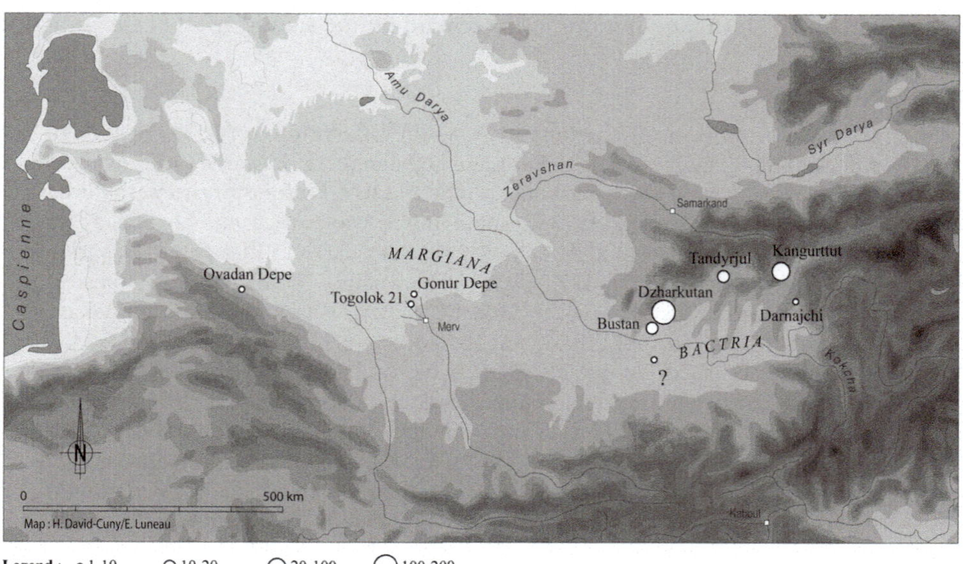

Figure 18.4 Geographic and quantitative distribution of the miniature metallic items at the Final Bronze Age sites of the Oxus Civilization.

and on the choices of the artifacts that were miniaturized or that have disappeared. Some new types of objects also emerge in the assemblages, such as the small articles in unbaked clay (see Figure 18.3, *3*) or specific items related to the northern Central Asian traditions (Figure 18.6). Nevertheless, there is, in addition, a disappearance or sharp rarefaction of numerous types of artifacts considered previously as male prestige goods (ceremonial axes, stone miniature columns, discs and staff, etc.) and female ones (composite figurines, seals, ornaments, cosmetic items, etc.), especially those items requiring long-distance acquisition and long technological operations. The disappearance of the most prestigious and valuable items goes together with a decrease in the volume of exchanges between the cultures of the Near East/Middle Asia (raw materials, prestige goods from the neighboring societies, etc.; Luneau 2015).

By contrast, one of the new signals for the period is the rise of elements linked with the northern "steppe" Central Asian cultural tradition (Luneau 2014: 151; 2017): one-seventh of the quantity of artifacts in metal and stone (e.g., daggers, molds, and beads; see Figure 18.6),[4] millet, domestic stone structures, and funerary stone constructions or cremation, for instance (see Table 18.1). The presence of artifacts stylistically connected with northern Central Asia possibly testifies to a switch in the direction of the exchange network and of intensified relations with the "Andronovo" populations. Since these populations were already located at the fringe of the Oxus Civilization area (Zeravshan, see Chapter 24), in northern Central Asia (Kazakhstan, Kyrgyrzstan, see Chapter 26), but also in southern Central Asia itself (see Chapter 17), the degree of extension and the modalities of the exchanges need to be further examined. This factor surely depended upon the territorial organization of the different communities. In all, these disparate parameters still need to be better investigated and analyzed to further assess a reformulation of the exchanges.

Burial practices

Funerary practices are extremely diverse and include different types of surface and underground structures, as well as different types of burials and body treatments (Luneau 2014: 154–160). Primary burials (in grave pits or in jars), secondary burials, cremations, burials without bodies (often called cenotaphs), human burials with animals, and animal-only burials have been recorded so far (Figure 18.5, *1–7*). Except for cremation, all were already attested during the mature phase of the Oxus Civilization. Primary and individual burials are the majority. The practice of burials without human bodies, sometimes replaced by unbaked clay figurines (Figure 18.5, *6*), is a significant and specific component of the final phase of the Oxus Civilization (almost 21% of the graves recorded) but is very unevenly distributed, according to the data collected so far (Figure 18.7). Contrary to the previous period, there is less differentiation in the quality and quantity of the funerary goods between the individuals (between 0 to 15 ceramics and 0 to 14 objects not counting beads). Meat offerings were found in half of the graves. Various evidence of fire and possible post-sepulchral rituals with the deposition of a few ceramic pots (Figure 18.5, *8*) are also common in some graveyards.

Together with the changes in the burial structures – aboveground (appearance of big clay blocks or large stone deposits) or underground (disappearance of mudbrick graves, appearance of large stone fillings) – and with those in the quantitative and

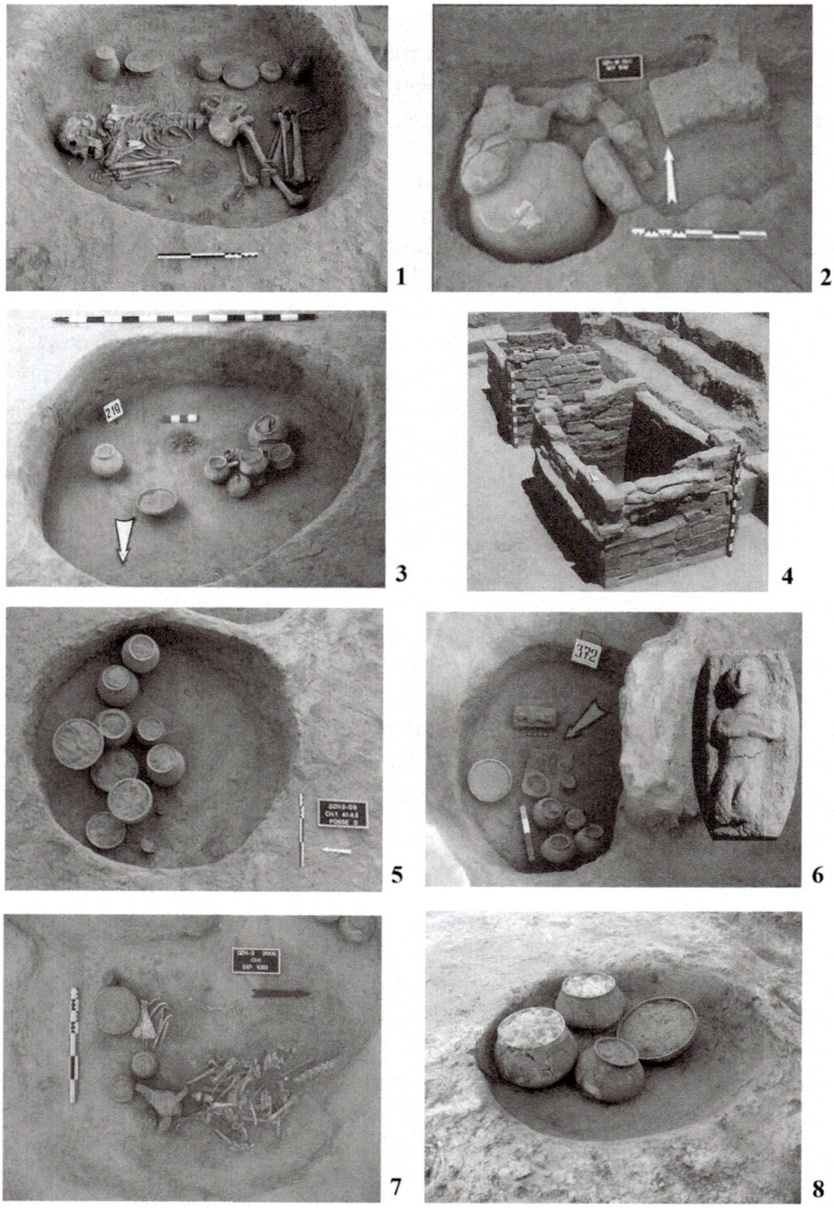

Figure 18.5 Types of graves and funerary practices related to the Oxus Civilization from Uzbekistan in the Final Bronze Age. 1. Primary inhumations in the ground (Dzharkutan, ©MAFOUZ-Protohistoire); 2. Primary inhumation in jar (Dzharkutan, ©MAFOUZ-Protohistoire); 3. Cremation (Bustan 6, after Avanesova 2013: 547, fig. III); 4. Cremation structures (Bustan 6, after Avanesova 2013: 550, fig. VI); 5. Burial without body (Dzharkutan, ©MAFOUZ-Protohistoire); 6. Burial without body and unbaked clay figurine (Bustan 6, after Avanesova 2013: 548, fig. IV); 7. Animal burial (Dzharkutan, ©MAFOUZ-Protohistoire); 8. Offering pit with few vessels at the surface of the graves (Dzharkutan, ©MAFOUZ-Protohistoire).

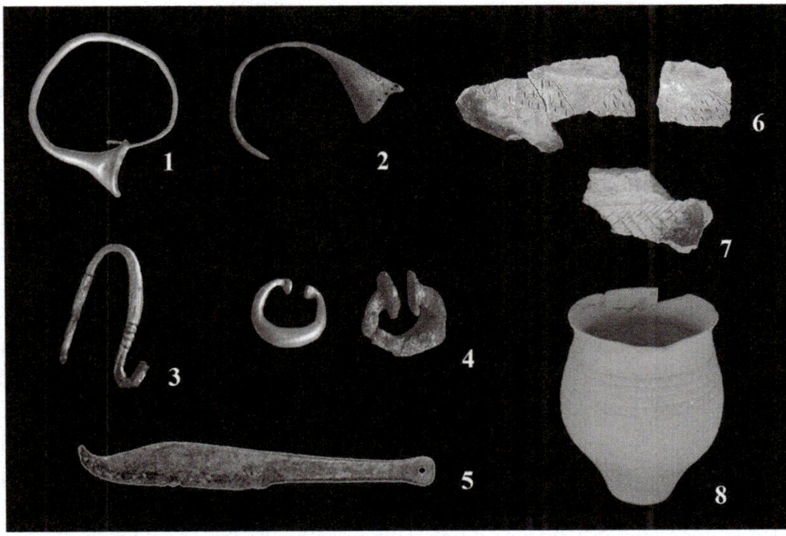

Figure 18.6 Examples of imported artifacts and/or of influences related to the northern steppe central Asian cultures in the Oxus Civilization. 1–4. Earring and pendants in gold and copper alloy from Bustan 6 (after Avanesova 2013: 558, fig. XIV); 5. Knife in copper alloy from Nurek (after Zejmal' 1985: fig. 28); 6–7. Wheel-coiled and handmade pottery of coarse ware with "Andronovo"-like decorations (Molali, ©Deutsches Archäologisches Institute, Eurasien-Abteilung); 8. Wheel-made pottery of fine ware with "Andronovo"-like shape and decorations (Kutimov 2013).

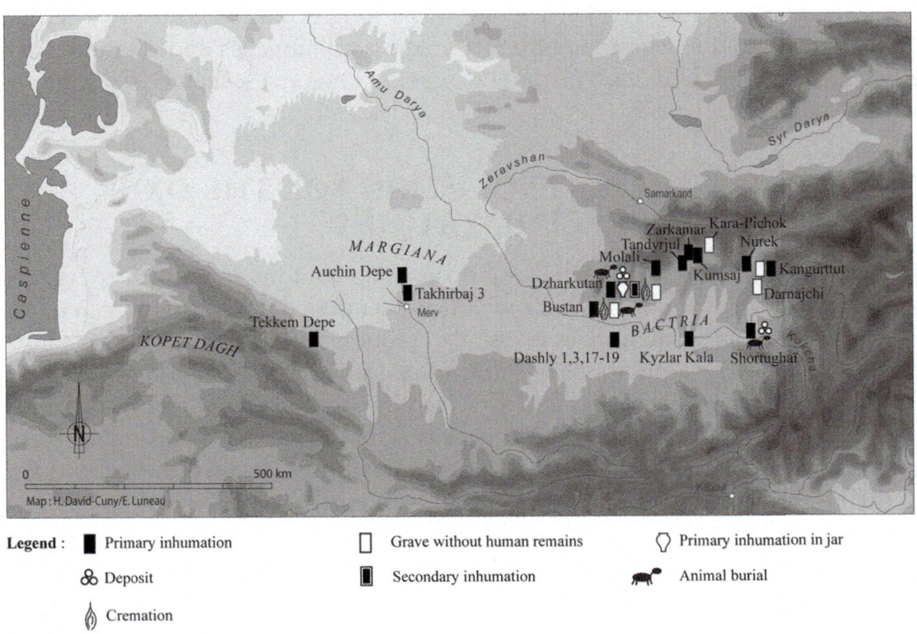

Figure 18.7 Geographic distribution of the main types of burial practices at the Final Bronze Age sites of the Oxus Civilization.

qualitative homogenization of the grave goods or in the miniaturization of the artifacts, a larger variety in the funerary treatment is obvious and attests significant symbolic mutations in comparison with the previous period. It is particularly visible in the eastern areas (southern Uzbekistan and Tajikistan) with a high increase of burials without a human body and the appearance of cremation while, at the same time, the number of burials with human remains decreases. This major shift in the ideological and symbolic beliefs of the populations is likewise attested by the disappearance of most of the figurative representations (except few animal and anthropomorphic figurines) on the whole territory.

A turning point is also observed in the differentiation of male and female graves, suggesting a possible modification in the social status and functions of men and women in comparison with the previous periods (Luneau 2008). The current archaeological data reveals a reduction in the wealth discrepancy between the different burials, suggesting a more equal distribution of goods between the deceased. But this may also point to another funeral strategy and conception of death, in which graves are not necessarily supposed to indicate the social status of the deceased and cannot be used alone as a sign of a lesser social stratification between individuals.

Regionalization

Although the disparity in research may be misleading, it is during the latest phase of the Oxus Civilization that regional differences (stylistic and technological features in the material culture, burial practices in particular) seem to be the most vivid, a possible testimony to decentralization and greater local autonomy.

Finally, diverse and complex mutations are characteristic for the final phase of the Oxus Civilization (Table 18.1). They attest to a long evolution over three centuries at least, but the missing chronological information still impedes understanding of the general and detailed dynamics of these processes. Ongoing research is still unravelling these complex changes.

WHICH SOCIAL EVOLUTION FOR THE OXUS CIVILIZATION?

The evolution between the Middle/Late Bronze Age and the Final Bronze Age testifies to various types of changes, as previously mentioned. Some phenomena ensure continuity, either as permanencies or as an evolution within the previous traditions, whereas others are real mutations or innovations. Some of these mutations have been viewed as declines, which are often perceived as an anomaly within a progressive model of evolution. Consequently, all the transformations were considered in a global perspective of "collapse" of the BMAC/Oxus Civilization. Nevertheless, we argue that the evolution actually reflects a more ambiguous situation.

The term "collapse" is initially and commonly defined in a broader view as the fall or disintegration of complex societies quite suddenly, within a rather rapid time period (Tainter 1988; Yoffee and Cowgill 1988). It is used for describing specific societal transformations and transitions of societies, but the precision of this definition and the evaluation of collapse are highly debated in relation to mechanisms and parameters – such as demographic decline, political fragmentation, joint loss in

Table 18.1 The Oxus Civilization final phase: continuities and changes

	Continuities	*Evolution/breaks*
Settlement pattern		• Reduction in size of the settlements • Change in the territorial arrangement
Territory	• Northeastern Iran • Turkmenistan • Northern Afghanistan • Southern Uzbekistan • Southern Tajikistan	
Ecology	• Plains and foothills	
Architecture	• Mudbrick architecture	• Cessation of the construction of monumental buildings with continuity of occupation • Simplification and reduction in size of the architecture • Architecture with stone fundaments • Seasonal habitats?
Ceramic production	• Paste composition • Fabrication techniques • Decorations	• Morphology • Apparent regionalization of the assemblages • New decorations • Less investment
Metal production	• Some shapes • Fabrication techniques (types of alloys)	• Apparition and predominance of miniature items • Few non-miniature items • Augmentation of items related to the northern Central Asian traditions • Augmentation of 15% of the bronze metallurgy in southern Uzbekistan and Tajikistan
Lithic production	• Some shapes	• New shapes

(*continued*)

Table 18.1 Cont.

	Continuities	*Evolution/breaks*
Burial practices	• Some types of structures • Primary inhumations (individual or double, in jar) • Animal burials • Burials without bodies • Deposits of funerary goods and food (pieces of meat)	• Disappearance of the monumental graves • New type of surface structures • Diversification • Appearance of cremation • Appearance of secondary burials? • Increase of burials without bodies • Increase of post-burial practices
Elites and symbolic	–	• Disappearance of the prestige and luxury items • Disappearance of most of the figurative expressions, especially concerning the elites and the mythology
Exchange network	–	• Apparent limitation of exchanges • Sharp slowdown of exchanges with Middle Asia • Higher contacts with the populations related to the "Andronovo" community
Subsistence economy	• Maintenance of the irrigation network • Similar storage mode	• Increased use of millet? • Possible evolution of the subsistence economy with more pastoralism
Way of life	–	• Development of mobility?

economic, social, and political complexity, or cultural break (Tainter 1988; Yoffee and Cowgill 1988; Johnson 2017; Middleton 2017). New theoretical developments – such as resilience, adaptation and sustainability, multiple factors, slow collapse, variability of scales in space and time, etc. (McAnany and Yoffee 2010; Butzer and Endfield 2012; Tainter 2016; Storey and Storey 2017) – also highlight how the concept of collapse is historically constructed on values centered on evolutionism (unilinear evolution) and ethnocentrism (notion of social progress or civilization) and demonstrate the complexity and variety of evolutionary processes. The multilinear evolution of societies is now well accepted in archaeological research together with the recognition of isolated, punctual declines that need to be evaluated in a constantly changing general context. Considering the abilities of adaptation and transformations of all the social groups in order to be perpetuated, whatever their complexity level, it is difficult to state on the collapse of past societies, and in that way, to pull away judgments on their failure or success (McAnany and Yoffee 2010).

Causes linked with societal collapse have often been debated in terms of opposition between several single causes. These can be internal and related to resource depletion, demographic pressure, warfare, or societal dysfunctions. They also can be external and due to climate and environmental changes or to interference from foreign populations. Recently, several works have criticized these single-factor theories and highlighted the dynamic interaction of all interconnected systems within the society in transformation (Butzer and Endfield 2012; Johnson 2017: 4–5). Besides, authors also particularly advocate the role of the "human element" (Johnson 2017) or the social structures (McAnany and Yoffee 2010; Butzer 2012) and their capacity or not for resiliency and adaptation to changes. The investigation of multiple factors – resulting from the intersection of various social aspects and systems constituting a specific society – appears as a more circumspect and fruitful approach for analyzing these complex processes.

Hence, a "qualitative" loss towards more simplicity and less investment may be underlined in the Oxus Civilization (smaller sites, absence of new monumental buildings, dimensional standardization, architecture simplification, less investment in the pottery production, etc.). However, these phenomena connected with the disappearance of the urban structures also fall within a context of a deep modification of the lifestyles, and of new sociopolitical conceptions and/or organization at different scales, between sites, groups of people, and individuals. The question arises of the choice and/or the necessity of an alternative option as the driving force for changes (Biscione and Tosi 1979; Tainter 1988: 120–121; Wilcox 2010). This new form of social organization could also represent an optimized (or simply differently organized) exploitation of the territory (Francfort 1989: 386; 2009), in which more mobile populations, perhaps recently arrived, may have had a significant role.

In regard to the material culture, without evidence suggesting the lack of raw material (metal, precious stones, etc.), their reduced use is possibly due to a change in the trade networks as mentioned above or to the loss of a technical know-how, but the disappearance of several types of artifacts, especially those considered as social status markers, can also be interpreted as significant changes in social behaviors. Social status may have been exhibited in other ways than with these specific artifacts requiring a long investment and deposited in the graves. By contrast, the miniaturization of specific objects, previously well known in their full size, may at the same time reflect a resilience and the desire of a particular group to maintain some traditions (here in specific types of items), even in an altered form.

Besides, although the level of commercialization and craft production still needs to be examined with detailed data from new excavations, the apparent reduction of the exchanges and in the craft specialization may not necessarily be linked with an economic decrease as advocated by supporters of the commercial development model (Smith 2004), but may suggest the rise of a new economic system.

The diversification and changes in the funerary practices together with the disappearance of representations, especially the mythological ones, may also correspond to the presence of other symbolic and religious considerations, perhaps linked to diverse populations. The previous symbolic world may not have carried the same significance into the Final Bronze Age. Such a perspective indeed could be prompted with the appearance of proto-Zoroastrianism, often linked with the migration of Indo-Aryan/Indo-Iranian populations (Sarianidi 2001; Kuz'mina 2007). There is no compelling evidence to support a tight correlation between population, religion, language, and archaeological features (Lyonnet 1994; Lamberg-Karlovsky 2002; Francfort 2005). However, several groups of populations, with diverse burial practices – among which possible speakers of Indo-Aryan/Indo-Iranian languages that are attested in the region during the next period seem to coexist during the Final Bronze Age. The presence of a proto-Zoroastrian component is possible but not demonstrated, and does not appear to be predominant according to the classical Zoroastrian practices.

The problem, then, lies in the interpretation of these phenomena. To what extent do the archaeological records reveal deep structural changes of the society and not only superficial modifications? Are the material expressions only symbols, unconnected with the deep social structures of this society, or do they express the essential value of the Oxus Civilization, so that their disappearance means the demise of sociocultural features as well? Anyhow, the multiple transformations in the Final Bronze Age indicate a major and significant evolution in the historical trajectory of southern Central Asia.

This process has been interpreted as one of "de-urbanization," a transition from an "urban phase" (the mature Oxus Civilization phase) to a "post-urban phase" (the Final Bronze Age and subsequent Early Iron Age) mainly through changes in economic parameters (Francfort 1988, 1989). The mutations are illustrated by a less urban society with a settlement pattern less concentrated, more dispersed, and more broadly distributed in the different ecological areas. It cannot be explained only by the abandonment of the urban structures. On the one hand, the strong transformations visible in the disappearance of the elite material (no prestige goods and no social indicators for prestige and power, demise of the iconographical supports and representations, absence of new monumental buildings), in the settlement pattern (uniformity, possible more local concentration), in the exchanges (strong reduction of exchanges with the Near East/Middle Asia), and in the burial practices (disappearance of high investment structures, uniformity, decline in the quantity and quality of the funerary goods, diversification of burials and body treatments, shift with respect to gender) apparently express a significant shift in the modes and/or forms of legitimate power. On the other hand, the changes of the funerary practices (decrease of inhumations along with the increase of various practices) and the mutation in part of the material culture (disappearance of specific items, emergence of new objects) point at a deep modification in the ways of thinking and in the symbolic expressions, perhaps related to the loss of influence of the dominant beliefs and values in favor of new ones. The combination of several social, material, and mental changes allows us to consider this mutation as a multifaceted, overall, and deep evolution of the whole sociocultural system.

At the transition between the Late Bronze Age and the Final Bronze Age, these transformations signal a major disjuncture, especially considering political integration: the end of the previous governance arrangement, i.e. a system ruled by distinct elites, based on a long-distance network of exchanges with Middle Asian societies, an urban model of settlement pattern, particular burial practices connected to its members' social ranks, and diverse ostentation strategies characterized by the display of specific material and symbolic representations. Thereafter, even if the territorial and cultural framework appears more or less preserved, the society experienced a social and institutional reorganization. It looks apparently less hierarchized and more rural, with a less centralized and perhaps more autonomous settlement pattern, possibly based on new landscape socioeconomic strategies (role of herding and of mobility?), integrated in a more local network of exchanges, along with beliefs apparently less materially assertive and more diversified. More than before, the cohabitation of different values, practices, and human groups seems to play a crucial role in this development. The difficulty is to assign its material culture to each group, to demonstrate the mixes, and to understand the subsequent evolution in this complex and untangled multicultural society. Likewise, it is for now still hard to differentiate the specific practices, roles, and symbols of the (potentially new) group of rulers, as well as the (potentially new) social norms shared by the members of the Final Bronze Age society.

In any case, the decline of specific features on the one hand, and the permanencies and innovations on the other, cannot, in the end, be interpreted as testimonies of a simplification of the Oxus Civilization. The archaeological records indicate rather a change in the sociopolitical system (end of the "proto-state"?) associated with a diversification, probably reflecting more a significant transformation in the social structure than a loss in social complexity. The definition of such a structure, whether ranked, hierarchical, heterarchical, transegalitarian, or non-uniform, is a difficult issue. Moreover, it is hard to characterize this evolution in the sociopolitical system since that of the previous phase is as yet not well defined.

Furthermore, precisions on the speed of the changes have still to be clarified to go further in the definition of the evolution. It is obvious that we observe the "end of something" (Middleton 2017: 87), but the rapidity of the changes can be discussed. It may well be that a political breakdown, possibly ideological and structural, rapidly occurred around 1800/1700 BC, but the disappearance of the elements and dimensions of the culture (Great Tradition) associated with the Oxus Civilization was not fast. Beyond the breaks in elite culture, sociocultural continuities are still visible through time (see Table 18.1). Issues in the definition of the processes of collapse particularly originate from the "relation of a political unit with a cultural unit" (Middleton 2017: 97). Rapid collapse of the political and socioeconomic system may go alongside longer cultural and ideological transition. In addition, Johnson's (2017) view of a "transition to a different way of life" according to the "nonrapid nature of the overall changes" matches with the transformations of the Oxus Civilization. However, the concept appears too vague and insufficient to precisely describe the evolution. Should we speak about a political fall (Tainter 1988; Storey and Storey 2017)? Or about a slow collapse (Storey and Storey 2017) of the large and complex cultural system defined as the Oxus Civilization?

WHY THESE CHANGES IN THE BMAC/OXUS CIVILIZATION?

To explain these major changes, assumptions suggested a demographic increase (Khlopina 1972; Mousavi 2008), environmental modifications (Dolukhanov 1981; Kuz'mina and Ljapin 1984; Gentelle 2001), socioeconomic disruption (Lyonnet

2005), or the impact of the "Andronovo" populations (Kuz'mina 1976, 2007). None of these single factors is for now sufficient to explain the major changes but recent data improves our perception of the possible causes.

Considering a demographic evolution, although there are not yet reliable demographic studies, signs do not tie in with any drastic positive or negative demographic change that would have led to a population depletion or pressure at the beginning of the 2nd millennium BC. No bioarchaeological studies indicate severe pathologies or epidemics. Recent work by the MAFOUZ-Protohistoire carried out on the necropolis of Dzharkutan reveals an approximately similar number of individuals in the graves related to the Late Bronze Age and those related to the Final Bronze Age, each being equivalent in time, ca. 300 years long (Luneau 2014: 209–210). Compared to the reduction in size of the settlements, the increasing number of smaller graveyards during the last phase of the Oxus Civilization rather states a redistribution of the population in more scattered groups of smaller size.

The examination of the current environmental data reveals the general inadequacy between climate variations and the sociocultural phenomena (Luneau 2019). However, the impact of local shifts in relation with the great variability of the Central Asian ecology, the processes of human/environment co-interaction, as well as the synchronism between different timescale events are still a matter of research. Beyond the ecological change, anthropological processes must also be better determined, since conditions for adaptation were possible especially with the exploitation of the highlands. The frequent use of millet, for instance, may be viewed as an increase of dry farming as a response to aridity. But some proxies used to evaluate climate changes may correspond to human choices due to other considerations, material or ideological (Marro 2009). In that case, the rise in the use of millet may be connected with an adaptation of the farming practices to a decrease of irrigation but also to other factors, like a better adaptation to some ecological areas or a specific relationship with the northern Central Asian populations who are the possible vector of diffusion of this crop (Spengler *et al.* 2014).

The impact of the populations traditionally identified as "Andronovo" has indeed long been suggested (especially in connection with the supposed Indo-Aryan or Indo-Iranian migrations), since the society seems more multicultural (or less dominated by a single conspicuous group) during the latest phase of the Oxus Civilization. The Oxus Civilization's relationships with these groups has been perceived in varied ways: bellicose (Kuz'mina 1976), peaceful through trade (Sarianidi 1975; Vinogradova and Kuz'mina 1996), through processes of contact/avoidance (Lamberg-Karlovsky 2002), or through integration/assimilation into the BMAC populations (Masson and Sarianidi 1972: 154; Masson 2002; P'jankova 2002; Kohl 2007: 193; Kuz'mina 2007: 211–291; Avanesova *et al.* 2010; Cerasetti *et al.* 2019; see also Chapter 17). The variety of assumptions reveals the difficulty in understanding the interactions between the diverse groups – partly mobile and sedentary, herders and farmers at different degrees – due to (still) very little material evidence and the nuanced and dynamic dimensions of the social boundaries challenging our heuristic abilities (Luneau 2017; Rouse and Cerasetti 2018; Rouse 2020). Moreover, the ethnocentric dominant view of permanent competition between human groups (Servigne and Chapelle 2017) often conceals the fact that diverse relationship dynamics may have been set up, such as collaboration especially under difficult conditions, as illustrated by case studies in past Central Asia (Stride *et al.* 2009).

The "Andronovo" groups were largely in contact with the Oxus populations, especially for trade. Exchanges of raw material, artifacts, and practices testify to direct contacts and negotiations, perhaps based on an economic and cultural specialization (Chernykh 1992; Vinogradova and Kuz'mina 1996; Avanesova 2002; Boroffka *et al.* 2002; Hiebert and Moore 2004; Lyonnet 2005; Kuz'mina 2007; Cattani 2008; Rouse and Cerasetti 2014; see also Chapter 26). Nevertheless, the volume of exchanges appears limited, and models of exchanges must be tested by future research. Besides, the transfers and appropriations so far identified are in the techno-cultural, stylistic, or ideological spheres (Luneau 2017). In the Oxus world, changes have been suspected in the pottery production (decoration, and perhaps rise of handmade shaping), in metallurgy (increase of tin bronze, use of "steppe"-like artifacts), and in the burial practices (cremation, stone arrangements). Conversely, technological features related to the Oxus traditions (double chamber kilns, pottery molds, and wheel-shaping) have been discovered at "Andronovo" campsites, and some of the major southern burial practices (pit with funeral chamber) have been adopted by the northern "newcomers." A social proximity has been materially identified with the presence of Oxus pottery at "Andronovo" campsites (Gonur-N, Ojakly, Chopantam) and conversely with that of a few "steppe"-like artifacts at Oxus sites; with the juxtaposition of tombs attributed to both communities at some cemeteries (Kumsaj, Tandyrjul); with the association of material stylistically related to the northern Central Asian traditions in Oxus graves (Tandyrjul); and with the presence of burials containing material typical of the "Andronovo" populations at Oxus sites (Takhirbaj 3). In all, according to the limited archaeological evidence we have so far, our perception of the interactions between the two communities mostly falls within a framework of coexistence with reciprocal influences but with the general maintenance of spatial and social distances at most places. Only at a few sites (mostly in the eastern areas located at the fringe of the Oxus Civilization) do the sporadic acculturation processes tend to highlight possible integration.

Yet, research on ancient DNA has recently brought out the significant role played by northern populations in the genetic history of Central Asia through important population mix (Narasimhan *et al.* 2019). However, the precise connecting features between groups, local implantation, chronological range, and patterns of their movements are still scarcely defined. The challenge currently rests on the recognition of these mixing processes between human groups, both on an anthropological/cultural (nature and degree of interactions) and chronological (succession of events) perspective.

Indeed, it is quite likely that this change in the population of the area (higher influence of northern populations) disturbed older social norms and institutions. But many aspects of this issue are still unclear. Was the infiltration of populations fully or partly responsible for the changes, or did internal upheavals of the Oxus Civilization create favorable conditions for its wider infiltration in southern Central Asia? Can we identify a real migration or were these movements related to the socioeconomic behavior of these populations? The chronological range for the movement of the Bronze Age northern populations into southern Central Asia is still uncertain, due to the low number of dated campsites and the confusion of the stratigraphic sequence at the Oxus sites. There is, indeed, a rise in features related to the northern populations, but it may just as well result from the development of singular human groups already present on the territory.

In parallel, the sequence of events in the Near East/Middle Asia suggests that drastic changes in the "interregional" trade network could be responsible for the demise of the Oxus Middle/Late Bronze Age system (Luneau 2015), if one is to consider the dependence of the Oxus Civilization on the commercial exchanges especially through the tin trade (Lyonnet 2005; Steinkeller 2016). Economic difficulties may have created unstable political conditions with a loss of legitimacy of the ruling elites and the weakening of its control over the territory. In addition, and/or consequently, this degradation could have stimulated a shift in dominance to the benefit of other population groups, in terms of political or economic leadership. The Oxus populations may have needed to redirect the exchanges towards new partners, whereas the northern Central Asian populations could have engaged the dependency of the Oxus populations on the tin trade. As secondary elites, they could have gained more influence, slowly undermining the sociopolitical, ideological, and cultural institutions still existing, and finally creating the conditions for the emergence of a new society.

Other internal causes, such as disruption of the economic structures, internal conflicts for power, or sociopolitical dysfunction, are difficult to determine but often recurring in collapse. This perspective is indeed challenging methodologically, entailing the development of new concepts and tools, such as modeling, with the aim to identify the actors (ruling people, secondary elites, and the major population), the nature of power, the ways of exercising power, and the modalities of political changes.

Lastly, a structural approach examining the capacity of the BMAC system for adaptation (fragility/stability) should also be endeavored to better understand the attempts or lack of attempts to react to changes.

Anyhow, a multicausal approach makes sense to better tackle the social changes in the Oxus Civilization. Multiple events appear to have occurred simultaneously. Hence, several factors (environment, interactions with other populations, economy, etc.) may be "critically important for sustainability, but such interrelationships are filtered through a web of complex social responses" (Butzer and Endfield 2012: 3631). Since alternative causes may explain the same evidence, the combination of external and internal movements may be a reasonable research pathway. The challenge is thus to acquire missing data, to investigate the timeline of events (cause–effect relationship), and to acutely investigate the intersection between all the data sets to evaluate the importance and role of each factor.

THE SUCCESSION OF THE OXUS CIVILIZATION AND THE MECHANISMS OF CULTURAL CHANGES

The changes observed in the material culture attributed to the Final Bronze Age led some scholars to point at a break in the cultural continuity of the BMAC/Oxus Civilization (Hiebert 1994a: 71–72). The coexistence of various cultural systems (emergence of new cultural phenomena along with the maintenance of the previous values) is most likely. In the 300 years this last phase approximately lasted, however, a more or less overall cultural permanency and maintenance of some of the underlying social structures allows us to consider that transformations occurred in a relatively preserved BMAC frame. In a wide historical perspective, just as the BMAC is the result of a long "formative" phase with the setting up of sociocultural features that only became characteristic in the subsequent mature period, its end did not happen hastily. It is only around 1500/1400 BC that the last cultural features related to the Oxus Civilization vanished, at the time of emergence of another cultural complex known as the "handmade painted ware cultures"[5] of the Early Iron Age, characterized by the homogenization of new specific features over Central Asia despite regional peculiarities. This transition between the Final Bronze Age and the Early Iron Age may be seen as the end of the cultural framework of the Oxus Civilization.

Nevertheless, similarly, the sociocultural profile observed in the next period of the Early Iron Age has more in common with the society of the final phase of the Oxus Civilization than with that of the mature period. In accordance with the ideas of resilience and regeneration (Schwartz and Nichols 2006; McAnany and Yoffee 2010), continuities through collapse occur in most cases and the new traditions are to some extent connected with the earlier society. As a matter of fact, even though the transitional processes are not well understood, breaks and continuities are evident between the Final Bronze Age and the Early Iron Age in southern Central Asia (Luneau 2014: 217–223). Alongside innovations, some of the deep transformations experienced by the Oxus Civilization since 1800/1700 BC and the continuous and slow movement of sociocultural reconfiguration between all the Bronze Age groups seem to intensify in the mid-2nd millennium BC and to form the roots of the Early Iron Age society. This transition has often been explained by a migration of mobile groups coming from the north who replaced the local population. But since these groups were present earlier in the beginning of the 2nd millennium BC, the modifications cannot rely only on their further migration. Certainly, the local evolution of the Oxus and "Andronovo" groups in the Final Bronze Age, although still poorly understood, played an important role and appears as a driving force for constructive changes. As a result, the intricate dynamic of "interpenetration" or "interweaving" of the different cultures and human groups, already at work during the Final Bronze Age, led to a deeper transformation of the whole society. More investigations on this long-term evolution should better describe and identify the mechanisms and vectors of these cultural changes in a context of constant sociocultural interaction.

CONCLUSION

Changes in the settlement pattern with the abandonment and the reduction in size of the major sites has long been the basis for a "collapse" interpretation of the last centuries of the Oxus Civilization. Yet, the complexity of the evolution must be acknowledged in regard with the multiple, arrhythmic, unequal, and nonlinear changes that occurred at various levels, in different fields, and at the same time. Transformations most likely took place over decades, or maybe centuries through various complex processes. The sociocultural evolution in the Final Bronze Age may better be described as a combination of resiliencies, adaptations, innovations, and/or declines, relying on a new structuration without major homogeneity and on a possibly different cohabitation (or new configuration) between the various human groups compared to the previous phase. The precise examination of the sociocultural interactions and symbiosis is crucial for the identification of variable sociocultural and political dynamics – necessarily complex, pragmatic, unsteady, fluctuating, and adaptive – between supposedly mobile/sedentary populations in order to explain the end of the Oxus Civilization.

Our challenge now is to develop multidisciplinary research, crossing material culture, practices, and genetics, collecting new datasets, and implementing simulation of environmental change and modeling of social response (Butzer 2012; Angourakis *et al.* 2014) for correlating models and data, as well as for testing working hypotheses. Many aspects still need specific investigation, including the subsistence systems, territory management and settlement patterns, rituals, sociopolitical structuration, regional variability, and the "Andronovo" groups, as well as a precise study in the measurement of declines considering qualitative and quantitative approaches and in the perception of the eco–social interactions. In parallel, we must urgently deal with the chronological uncertainties that seriously impede our understanding of both the social continuities/discontinuities and pace of changes, and the causal factors. In such an intricate society as that known at the end of the Oxus Civilization, made of different sociocultural groups, every element interacts, affects, and modifies the other, and can alter or strengthen pre-existing economic, social, and political behavior patterns.

In all, contrary to the dominant disruptive perspective of a collapse, the viewpoint of "constant evolution" (Winock 2007) suggests that the final phase of the Oxus Civilization was also a productive and rich period of sociocultural interaction, symbiosis, and innovation, which was highly significant in the development of the so-called mobile/sedentary relationships through Central Asian history. It also encourages us to reconsider the idea of "cultures" and "civilizations" as immutable and indivisible, and, as M. Wilcox (2010: 125) puts it, to interrupt "the logical links between what archaeologists interpret as an 'abandonment,' the ready-made parable of 'failure' and the social deaths of invented groups" in order to better understand the factors and logics of sociocultural change in human history.

NOTES

1 Ongoing surveys in Uzbekistan, Tajikistan, and northern Iran bring new sites related to the Bronze Age (Stančo 2016; Kutimov 2017; Vahdati 2018; see also Chapters 16, 19, 23, and 25).

2 If the present state of knowledge suggests an increase of the settlement in the foothills of Tajikistan and southern Uzbekistan, recent surveys in southwestern Uzbekistan have revealed a strong presence of Bronze Age populations before the sole Final Bronze Age (Kaniuth 2010; Stančo 2016; see also Chapter 16). A significant study of the settlement pattern in the various ecological areas is now urgently needed to better understand the population movements.

3 The existence of miniaturized ceramics (Luneau *et al.* 2014) is a step in this direction.

4 The proportion increases up to one-fifth according to the types of metal items (Luneau 2014: 151).

5 For Western academic review of the handmade painted ware cultures, see for instance Kohl (1984), Francfort (1989, 2001), Lyonnet (1997: 83–119), Bonora and Vidale (2008), Boroffka (2009), and Lhuillier (2013).

REFERENCES

Angourakis, A., B. Rondelli, S. Stride, X. Rubio-Campillo, A.L. Balbo, A. Torrano, V. Martinez, M. Madella, and J.M. Gurt 2014. "Land Use Patterns in Central Asia. Step 1: The Musical Chairs Model." *Journal of Archaeological Method and Theory* 2(2): 405–425.

Askarov, A.A. 1977. *Drevnezemledel'cheskaja kul'tura epokhi bronzy juga Uzbekistana* [Ancient Bronze Age farming culture in southern Uzbekistan]. Tashkent: Fan.

Askarov, A.A and V.N. Abdullaev 1983. *Dzharkutan*. Tashkent: Fan.

Avanesova, N.A. 1997. "Spätbronzezeitliche Kulturkontakte in der baktrischen Flussoase nach den Befunden der Nekropole Bustan 6." *Archäologische Mitteilungen in Iran und Turan* 29: 147–178.

Avanesova, N.A. 2002. "Novye materialy epokhi bronzy Zerafshanskoj doliny." [New Bronze Age material from the Zeravshan valley]. *Arkheologicheskie Issledovanija v Uzbekistane* 2001: 19–26.

Avanesova, N.A. 2007. "Relative und absolute Chronologie agrarischer Fundorte der Spätbronzezeit in Südwest-Tadzikistan." *Archäologische Mitteilungen in Iran und Turan* 39: 61–89.

Avanesova, N.A. 2013. *Buston VI: nekropol' ognepoklonnikov dourbanisticheskoj Baktrii* [Buston VI: a fire worshipper necropolis in proto-urban Bactria]. Samarkand: MITSAI.

Avanesova, N.A., N.A. Dubova, and V.V. Kufterin 2010. "Skeletal Remains from Bustan VI: A Sapalli Culture Cemetery in Uzbekistan." *Archaeology Ethnology & Anthropology of Eurasia* 38(1): 118–137.

Beljaeva, T.V. and Z.A. Khakimov. 1973. Drevnebaktrijskie pamjatniki Mirshade [The sites of the ancient Bactria at Mirshade]. In *Iz istorii antichnoj kul'tury Uzbekistana* [On the history of the ancient culture of Uzbekistan], ed. G.A. Pugachenkova. Tashkent: Izd-vo literatury i iskusstvo, pp. 35–51.

Biscione, R. and M. Tosi 1979. *Protostoria degli stati turanici. Aspetti dell'evoluzione urbana e forme d'insediamento nel popolamento dell'Asia centrale nell'Età del Bronzo (2500–100 a.c.) alla luce dei dati archeologici*. Napoli: Supplemento no. 20 agli Annali.

Bonora, G.L. and M. Vidale 2008. "An Aspect of the Early Iron Age (Yaz I) Period in Margiana: Ceramic Production at Site No. 999." In *The Bronze Age and Early Iron Age in the Margiana Lowlands: Facts and Methodological Proposals for a Redefinition of the Research Strategies*. The Archaeological Map of the Murghab Delta, Studies and Reports, vol. II, ed. S. Salvatori and M. Tosi. Oxford: Archaeopress, pp. 155–196.

Boroffka, N. 2009. "Siedlungsgeschichte in Nordbaktrien – Bandichan zwischen Spätbronzezeit und Frühmittelalter." In *Alexander der Grosse und die Öffnung der Welt*, ed. S. Hansen, A. Wieczorek, and M.Tellenbach. Mannheim: Verlag Schnell, pp. 135–143.

Boroffka, N., J. Cierny, J. Lutz, H. Parzinger, E. Pernicka, and G.Weisgerber 2002. "Bronze Age Tin from Central Asia: Preliminary Notes." In *Ancient Interactions: East and West in Eurasia*, ed. K. Boyle, C. Renfrew, and M. Levine. Oxford: Oxbow Books, pp. 135–159.

Butzer, K.W. 2012. "Collapse, Environment, and Society." *Proceedings of the National Academy of Sciences* 109(10): 3632–3639.

Butzer, K.W. and G.H. Endfield 2012. "Critical Perspectives on Historical Collapse." *Proceedings of the National Academy of Sciences* 109(10): 3628–3631.

Cattani, M. 2008. "The Final Phase of the Bronze Age and the 'Andronovo Question' in Margiana." In *The Bronze Age and Early Iron Age in the Margiana Lowlands: Facts and Methodological Proposals for a Redefinition of the Research Strategies. The Archaeological Map of the Murghab Delta Studies and Reports*, vol. II, ed. S. Salvatori and M. Tosi. Oxford: Archaeopress, pp. 133–151.

Cerasetti, B., R. Arciero, M. Carra, A. Curci, J. De Grossi Mazzorin, L. Forni, E. Luneau, L.M. Rouse, and R.N. Spengler 2019. "Bronze and Iron Age Urbanisation in

Turkmenistan: Preliminary Results from the Excavation of Togolok 1 on the Murghab Alluvial Fan." In *Urban Cultures of Central Asia from the Bronze Age to the Karakhanids*, ed. C. Baumer and M. Novák. Wiesbaden: Harrassowitz, pp. 63–72.

Chernykh, E.N. 1992. *Ancient Metallurgy in the USSR: The Early Metal Age*. Cambridge: Cambridge University Press.

Dolukhanov, P.M. 1981. "The Ecological Prerequisites for Early Farming in Southern Turkmenia." In *The Bronze Age Civilization of Central Asia. Recent Soviet Discoveries*, ed. P.L. Kohl. New York: M.E. Sharpe, pp. 359–385.

Francfort, H.-P. 1984. "The Early Periods of Shortughaï (Harappan) and the Western Bactrian Culture of Dashly." In *South Asian Archaeology 1981*, ed. B. Allchin. Berlin: Dietrich Reimer Verlag, pp. 170–175.

Francfort, H.-P. 1988. "A propos de l'urbanisation du site de Shortughaï (Afghanistan): Une approche archéologique des transformations de l'économie de production." *Bulletin du Centre Genevois d'Anthropologie* 1: 15–34.

Francfort, H.-P. 1989. *Fouilles de Shortughaï: recherches sur l'Asie centrale protohistorique*. Paris: De Boccard.

Francfort, H.-P. 2001. "The Cultures with Painted Ceramics of South Central Asia and Their Relations with the Northeastern Steppe Zone (Late 2nd–Early 1st Millennium B.C.)." In *Migration und Kulturtransfer*, ed. R. Eichmann and H. Parzinger. Bonn: Dr. Rudolf Habelt, pp. 221–235.

Francfort, H.-P. 2005. "La civilisation de l'Oxus et les Indo-Iraniens et Indo-Aryens." In *Aryas, Aryens et Iraniens en Asie Centrale*, ed. G. Fussman, J. Kellens, H.-P. Francfort, and X. Tremblay. Collège de France, Publications de l'Institut de Civilisation Indienne, vol. 72. Paris: De Boccard, pp. 253–328.

Francfort, H.-P. 2009. "L'âge du Bronze en Asie centrale. La civilisation de l'Oxus." *Anthropology of the Middle East* 4(1): 91–111.

Francfort, H.-P. 2016. "How the Twins Met: Indus and Oxus Bronze Age Civilizations in Eastern Bactria. Shortughaï Revisited Forty Years Later." In *Trudy Margianskoj Arkheologicheskoj Ekspeditsii*, vol. 6, ed. N.A. Dubova. Moscow: Starij Sad, pp. 461–475.

Gentelle, P. 2001. "Irrigations antiques en Bactriane du nord: l'image de satellite, la prospection archéologique, les inférences historiques." In *La Bactriane au carrefour des routes et des civilisations d'Asie centrale. Termez et les villes de Bactriane-Tokharestan*, ed. P. Leriche, C. Pidaev, M. Gelin, and K. Abdoullaev. Paris: Maisonneuve et Larose/IFEAC, pp. 163–172.

Hiebert, F.T. 1994a. *Origins of the Bronze Age Oasis Civilization in Central Asia*. Cambridge, MA: Peabody Museum.

Hiebert, F.T. 1994b. "Production Evidence for the Origins of the Oxus Civilization." *Antiquity* 68(259): 372–387.

Hiebert, F.T. and K.M. Moore 2004. "A Small Steppe Site Near Gonur." In *U Istokov Tsivilizatsii* [At the origins of civilization], ed. M.F. Kosarev, P.M. Kozhin, and N.A. Dubova. Moscou: Staryj Sad, pp. 294–302.

Johnson, S. 2017. *Why Did Ancient Civilizations Fail?* New York: Routledge.

Kaniuth, K. 2006. *Metallobjekte der Bronzezeit aus Nordbaktrien*. Mainz: Verlag Philipp von Zabern.

Kaniuth, K. 2010. "Tilla Bulak 2009 – Vorbericht zur dritten Kampagne." *Archäologische Mitteilungen aus Iran und Turan* 42: 129–163.

Khlopina, L.I. 1972. "Southern Turkmenia in the Late Bronze Age." *East and West* 22(3–4): 199–214.

Kohl, P.L. 1984. *Central Asia: Paleolithic Beginnings to the Iron age*. Paris: Editions Recherche sur les Civilisations.

Kohl, P.L. 2007. *The Making of Bronze Age Eurasia*. Cambridge: Cambridge University Press.

Kutimov, J. 2013. "Novye issledovanija mogil'nika epokhi pozdnej bronzy Kumsaj (Tadzhikistan)" [New research on the Late Bronze Age cemetery of Kumsaj (Tadjikistan)]. *Arkheologicheskie Vesti* 19: 75–91.

Kutimov, J. 2017. "Novye arkheologicheskie materialy epokhi bronzy v zapadnoj chasti Gissarskoj doliny Tadjikistana" [New Bronze Age archaeological material in the western part of the Hissar plain in Tadjikistan]. *Arkheologicheskie Vesti* 23: 115–131.

Kuz'mina, E.E. 1976. "The 'Bactrian Mirage' and the Archaeological Reality: On the Problem of the Formation of the North Bactrian Culture." *East and West* 26(1–2): 111–131.

Kuz'mina, E.E. 2007. *The Origin of the Indo-Iranians*. Leiden: Brill.

Kuz'mina, E.E. and A.A. Ljapin 1984. "Novye nakhodki stepnoj keramiki na Murgabe" [New steppe pottery discoveries in the Murghab]. In *Problemy arkheologii Turkmenistana*, ed. V.M. Masson. Ashkhabad: Ylym, pp. 7–22.

Lamberg-Karlovsky, C.C. 1994. "The Oxus Civilization: The Bronze Age of Central Asia." *Antiquity* 68(259): 353–354.

Lamberg-Karlovsky, C.C. 2002. "Archaeology and Language: The Indo-Iranians." *Current Anthropology* 43(1): 63–88.

Lamberg-Karlovsky, C.C. 2003. "Civilization, State or Tribe? Bactria and Margiana in the Bronze Age." *Review of Archaeology* 24(1): 11–19.

Lamberg-Karlovsky, C.C. 2013. "The Oxus Civilization." *Cuadernos de Prehistoria y Arqueología de la Universidad Autónoma de Madrid* 39: 21–63.

Lhuillier, J. 2013. *Les cultures à céramique modelée peinte en Asie centrale méridionale: Dynamiques socioculturelles à l'âge du Fer ancien (1500–1000 av. n.è.)*. Paris: De Boccard.

Luneau, E. 2008. "Tombes féminines et pratiques funéraires en Asie centrale protohistorique. Quelques réflexions sur le 'statut social' des femmes de la civilisation de l'Oxus." *Paléorient* 34(1): 131–157.

Luneau, E. 2014. *La fin de la civilisation de l'Oxus: Transformations et recompositions des sociétés de l'âge du Bronze final en Asie centrale méridionale*. Paris: Editions De Boccard.

Luneau, E. 2015. "The Fall of the Oxus Civilization and the Role of Exchanges with the Neighboring Societies During the First Half of the Second Millennium B.C." In *South Asian Archaeology and Art 2012: Man and Environment in Prehistoric and Protohistoric South Asia – New Perspectives*, ed. V. Lefèvre, A. Didier, and B. Mutin. Turnhout: Brepols, pp. 169–183.

Luneau, E. 2017. "Transfers and Interactions Between North and South in Central Asia during the Bronze Age." In *Interaction in the Himalayas and Central Asia*, ed. E. Allinger, F. Grenet, C. Jahoda, M.-K. Lang, and A. Vergati. Vienna: Austrian Academy of Sciences Press, pp. 13–27.

Luneau, E. 2018. "Further Insight on the Cultural Dynamics Between North-Eastern Iran and Central Asia in the Bronze Age Through the Study of Pottery." In *Proceedings of the Symposium of Iranian Young Archaeologists 2015*, vol. 3, ed. M.H. Azizi Kharanaghi, M. Khanipour, and R. Naseri. Tehran: Iranology Foundation, pp. 103–123.

Luneau, E. 2019. "Climate Change and the Rise and Fall of the Oxus Civilization in Southern Central Asia." In *Socio-Environmental Dynamics Along the Historic Silk Road*, ed. L.E. Yang, H.-R. Bork, X. Fang, and S. Mischke. New York: Springer Nature, pp. 275–298.

Luneau, E., J. Bendezu-Sarmiento, and S. Mustafakulov 2014. "Ceramics and Chronology at Dzharkutan: A Revision of the Periodization of the Sapalli Culture (Uzbekistan, c. 2100–1500 BC)." In *Proceedings of the 8th International Congress on the Archaeology of the Ancient Near Eas*, vol. 2, ed. P. Bielinski, M. Gawlikowski, R. Kolinski, D. Ławecka, A. Soltysiak, and Z. Wygnanska. Wiesbaden: Harrassowitz, pp. 529–542.

Luneau, E., U. Rakhmanov, N. Boroffka, and L. Sverchkov forthcoming. "Pottery from Molali, Uzbekistan: A Final Bronze Age Pottery Complex of the Oxus Civilization." In *Farmers,*

Traders and Herders: The Bronze Age in Central Asia and Khorasan, Archäologie in Iran und Turan. Berlin: Dietrich Reimer.

Lyonnet, B. 1994. "Central Asia, the Indo-Aryans and the Iranians: Some Reassessments from Recent Archaeological Data." In *South Asian Archaeology 1993*, ed. A. Parpola and P. Koskikalio. Helsinki: Suomalainen Tiedeakatemia, pp. 425–434.

Lyonnet, B. 1997. *Prospections archéologiques en Bactriane orientale (1974–1978)*. Vol. 2, *Céramique et peuplement du Chalcolithique à la conquête arabe*, ed. J.-C. Gardin. Paris: Editions Recherche sur les Civilisations.

Lyonnet, B. 2005. "Another Possible Interpretation of the Bactria-Margiana Culture (BMAC) of Central Asia: The Tin Trade." In *South Asian Archaeology 2001*, ed. C. Jarrige and V. Lefevre Paris: Éditions Recherche sur les Civilisations, pp. 191–200.

McAnany, P.A. and N. Yoffee 2010. *Questioning Collapse: Human Resilience, Ecological Vulnerability, and the Aftermath of Empire*. Cambridge: Cambridge University Press.

Marro, C. 2009. "Réflexions autour de l'hypothèse d'une 'crise' en Mésopotamie à la fin du IIIème millénaire av. n.è." *Scienze dell'antichita. Storia Archeologia Antropologia* 15: 131–142.

Masson, V.M. 1959. *Drevnezemledel'cheskaja kul'tura Margiany* [Ancient farming culture of Margiana]. Moscou: Nauka.

Masson, V.M. 2002. "Cultures of the Steppe Bronze Age and Urban Civilizations in the South of Central Asia." In *Complex Societies of Central Eurasia from the 3rd to the 1st Millennium BC: Regional Specifics in Light of Global Models*, vol. II, ed. K. Jones-Bley and D.G. Zdanovich. Washington, DC: Institute for the Study of Man, pp. 547–557.

Masson, V.M. and V.I. Sarianidi 1972. *Central Asia: Turkmenia Before the Achaemenids*. London: Thames & Hudson.

Middleton, G.D. 2017. "The Show Must Go On: Collapse, Resilience, and Transformation in 21st Century Archaeology." *Reviews in Anthropology* 46(2–3): 78–105.

Mousavi, A. 2008. "Late Bronze Age in North-Eastern Iran: An Alternative Approach to Persisting Problems." *Iran* XLVI: 105–120.

Narasimhan, V.M., N. Patterson, P. Moorjani, N. Rohland, R. Bernardos, S. Mallick, I. Lazaridis *et al.* 2019. "The Formation of Human Populations in South and Central Asia." *Science* 365(6457): n.p. doi: 10.1126/science.aat7487.

Pilipko, V.N. 2010. "Dushakskoe selishche (juznyj Turkmenistan)" [The Dushak settlement (southern Turkmenistan)]. In *Na puti otkritija tsivilizatsii* [Discovering a civilization], ed. P.M. Kozhin, M.F. Kosarev, and N.A. Dubova. St. Petersburg: Aletheia, pp. 226–236.

P'jankova, L.T. 1979. "Otchet o rabote nurekskogo arkheologicheskogo otrjada" [Report of the Nurek archaeological team]. *Arkheologicheskie Raboty v Tadzhikistane* XIV: 78–92.

P'jankova, L.T. 1993. "Pottery of Margiana and Bactria in the Bronze Age." *International Association for the Study of the Cultures of Central Asia Information Bulletin* 19: 109–127.

P'jankova, L.T. 2002. "South Tajikistan: Synthesis of Settled and Steppe Cultures at the End of the Bronze Age." In *Complex Societies of Central Eurasia from the 3rd to the 1st Millennium BC: Regional Specifics in Light of Global Models*, vol. II, ed. K. Jones-Bley and D.G. Zdanovich Washington, DC: Institute for the Study of Man, pp. 558–573.

Rossi Osmida, G. 2007. *Adji Kui Oasis: La citadella delle statuette*. Venice: Il Punto Edizioni.

Rouse, L.M. 2020. "Silent partners: archaeological insights on mobility, interaction and civilization in Central Asia's past." *Central Asian Survey* 39(3): 398–419.

Rouse, L.M. and B. Cerasetti 2014. "Ojakly: A Late Bronze Age Mobile Pastoralist Site in the Murghab Region, Turkmenistan." *Journal of Field Archaeology* 39(1): 32–50.

Rouse, L.M. and B. Cerasetti 2015. "Micro-Dynamics and Macro-Patterns: Exploring New Archaeological Data for the Late Holocene Human–Water Relationship in the Murghab Alluvial Fan, Turkmenistan." *Quaternary International* 437: 20–34.

Rouse L.M. and B. Cerasetti 2018. "Mixing Metaphors: Sedentary–Mobile Interactions and Local–Global Connections In Prehistoric Turkmenistan." *Antiquity* 92(363): 674–689.

Ruzanov, V.D. 1999. "K voprosu o istokakh olovjanistykh bronz v metallobrabatyvajushchikh proizvodstvakh plemen juga Srednej Azii v epokhu bronzy" [On the sources of tin bronzes in the southern Central Asia Bronze Age metallurgical production]. *Isledovanie Material'noj Kul'tury Uzbekistana* 30: 37–40.

Salvatori, S. 2008. "The Margiana Settlement Pattern from the Middle Bronze Age to the Parthian-Sasanian: A Contribution to the Study of Complexity." In *The Bronze Age and Early Iron Age in the Margiana Lowlands: Facts and Methodological Proposals for a Redefinition of the Research Strategies. The Archaeological Map of the Murghab Delta Studies and Reports*, vol. II, ed. S. Salvatori and M. Tosi. Oxford: Archaeopress, pp. 57–74.

Salvatori, S. 2016. "Bactria-Margiana Archaeological Complex: How Terminology Hides Historical Processes." In *Trudy Margianskoj Arkheologicheskoj Ekspeditsii*, vol. 6, ed. N.A. Dubova. Moscow: Starij Sad, pp. 449–460.

Sarianidi, V.I. 1974. "Bactria v epokhu bronzy." *Sovetskaja Arkheologija* 4: 49–71.

Sarianidi, V.I. 1975. "Stepnye plemena epokhi bronzy v Margiane" [Bronze Age steppe groups in Margiana]. *Sovetskaja Arkheologija* 2: 20–29.

Sarianidi, V.I. 1977. *Drevnie zemledel'tsy Afganistana* [Ancient farmers of Aghanistan]. Moscou: Nauka.

Sarianidi, V.I. 1990. *Drevnosti strany Margush* [Ancient Margush]. Ashkhabad: Ylym.

Sarianidi, V.I. 2001. "The Indo-Iranian Problem in the Light of the Latest Excavations in Margiana." *Studia Orientalia* 94: 417–442.

Schwartz, G.M. and J.J. Nichols. 2006. *After Collapse: Regeneration of Complex Societies*. Tucson: University of Arizona Press.

Servigne, P. and G. Chapelle 2017. *L'entraide: L'autre loi de la jungle*. Paris: Les Liens qui Libèrent.

Shchetenko, A.J. 1972. "Raskopki 'vyshki' Namazga-Depe" [Excavations of the "tower" at Namazga Depe]. *Uspekhi Sredneaziatskoj Arkheologii* 1: 52–53.

Shchetenko, A.J. 2000. "K probleme periodizatsii kul'tury Namazga VI" [On the question of the Namazga VI culture periodization]. In *Vsaimodejstvie kul'tur i tsivilizatsij* [Interactions of cultures and civilizations], ed. J.E. Berezkin. St. Petersburg: Institut Istorii Material'noj Kul'tury Rossijskaja Akademija Nauk, pp. 127–141.

Smith, M.E. 2004. "The Archaeology of Ancient State Economies." *Annual Review of Anthropology* 33: 73–102.

Spengler, R., M. Frachetti, P. Doumani, L. Rouse, B. Cerasetti, E. Bullion, and A. Mar'yashev 2014. "Early Agriculture and Crop Transmission Among Bronze Age Mobile Pastoralists of Central Eurasia." *Proceedings of the Royal Society B:* Biological Sciences 281(1783): n.p. http://doi.org/10.1098/rspb.2013.3382.

Stančo, L. 2016. "Archaeological Survey in the Surroundings of Kayrit (South Uzbekistan): Preliminary Report for Season 2015." *Studia Hercynia* XX(2): 73–85.

Steinkeller, P. 2016. "The Role of Iran in the Inter-Regional Exchange of Metals: Tin, Copper, Silver and Gold in the Second Half of the Third Millennium BC." In *Ancient Iran: New Perspectives from Archaeology and Cuneiform Studies*, ed. K. Maekawa. Kyoto: Kyoto University, Center for Eurasian Cultural Studies, pp. 127–150.

Storey, R. and G.R. Storey 2017. *Rome and the Classic Maya: Comparing the Slow Collapse of Civilizations*. New York: Routledge.

Stride, S., B. Rondelli, and S. Mantellini 2009. "Canals Versus Horses: Political Power in the Oasis of Samarkand." *World Archaeology* 41: 73–87.

Tainter, J.A. 1988. *The Collapse of Complex Societies*. Cambridge: Cambridge University Press.

Tainter, J.A. 2016. "Why Collapse Is so Difficult to Understand." In *Beyond Collapse: Archaeological Perspectives on Resilience, Revitalization and Transformation in Complex Societies*, ed. R.K. Faulseit. Carbondale: Southern Illinois University Press, pp. 27–39.

Teufer, M. 2005. "The Late Bronze Age Chronology of Southern Uzbekistan. A Reanalysis of the Funerary Evidence." In *South Asian Archaeology 2003*, ed. U. Franke-Vogt and H.-J. Weisshaar. Aachen: Linden Soft, pp. 199–209.

Teufer, M. 2015. *Spätbronzezeitliche Grabfunde aus Nordbaktrien und benachbarten Regionen: Studien zur Chronologie zwischen Aralsee und Persischem Golf*. Archäologie in Iran und Turan 13. Berlin: Dietrich Reimer.

Teufer, M., N. Vinogradova, and Y. Kutimov 2014. "Der Übergang von der Spätbronze- zur Früheisenzeit in Mittelasien." *Das Altertum* 59: 109–148.

Vahdati, A.A. 2018. "Preliminary Report on the Archaeological Survey in Esfarayen county, Northeastern, Iran." *Archäologische Mitteilungen aus Iran und Turan* 47: 253–279.

Vahdati, A.A. and R. Biscione forthcoming. "The Bronze Age Cemetery of Chalow: A Brief Report on Three Seasons of Excavations." In *Farmers, Traders and Herders: The Bronze Age in Central Asia and Khorasan*. Archäologie in Iran und Turan. Berlin: Dietrich Reimer.

Vinogradova, N.M. 1980. "Otchet o raskopkakh mogil'nika Tandyrjul v 1975 g" [Report on the excavations of the graveyard Tandyrjul in 1975]. *Arkheologischeskie Raboty Tadzhikistana* XV(1975): 63–74.

Vinogradova, N.M. 2004. *Jugo-zapadnyj Tadzhikistan v epokhu pozdnej bronzy* [Southwestern Tadjikistan in the Late Bronze Age]. Moscow: Institut Vostokovedenija – Rossijskaja Akademija Nauk.

Vinogradova, N.M. and E.E. Kuz'mina 1996. "Contacts Between the Steppe and Agricultural Tribes of Central Asia in the Bronze Age." In *Between Lapis and Jade: Ancient Cultures of Central Asia, Anthropology and Archaeology of Eurasia*, ed. F.T. Hiebert and N. Di Cosmo. New York: Sharpe, pp. 29–54.

Vinogradova, N.M., Ranov V.A., and Filimonova T.G. 2008. *Pamjatniki Kangurttuta v jugo-zapadnom Tadzhikistane (epokha neolita i bronzovyj vek)* [The settlements of Kangurttut in southwestern Tajikistan (Neolithic and Bronze Age)]. Moscow: Institut Vostokovedenija Rossija Akademija Nauk.

Wilcox, M. 2010. "Marketing Conquest and the Vanishing Indian: An Indigenous Response to Jared Diamond's Archaeology of the American Southwest." In *Questioning Collapse: Human Resilience, Ecological Vulnerability, and the Aftermath of Empire*, ed. P.A. McAnany and N. Yoffee. Cambridge: Cambridge University Press, pp. 113–141.

Winock, M. 2007. "… 'Mutation,' répond l'historien." *L'Histoire* 316: 74–75.

Yoffee, N. and G.L. Cowgill 1988. *The Collapse of Ancient States and Civilizations*. Tucson and London: University of Arizona Press.

Zejmal', E.V. 1985. *Drevnosti Tadzhikistana. Katalog vystavki [Ancient Tajikistan. Exhibition Catalogue]*. Dushanbe: Donish.

PART III

THE "SURROUNDING AREAS"

CHAPTER NINETEEN

THE BMAC PRESENCE IN EASTERN IRAN

State of affairs in December 2018 – towards the Greater Khorasan Civilization?

Raffaele Biscione and Ali A. Vahdati

INTRODUCTION

One of the most interesting phenomenon of the Bactria-Margiana Archaeological Complex (BMAC) is its diffusion in a significant part of Middle Asia as shown by the presence of various classes and types of artifacts that probably indicate several modalities of expansion. In order to study this phenomenon it is important to precisely identify and determine the four different areas that are concerned within this vast region (Biscione and Vahdati, forthcoming). Before we proceed to their presentation, it must be underlined here that the authors prefer to adopt the term "Greater Khorasan Civilization," or GKC, to that of BMAC, for the reasons explained below.

The formative area

There is no doubt that the BMAC/GKC has its roots in the Namazga culture, therefore it can be expedient to consider as the "formative area" the region covered by the Namazga (NMG) periods I–IV, from the Early Chalcolithic to the Early Bronze Age (Figures 19.1 and 19.2) (see also Chapter 4 and *contra* Chapters 1 and 20). As can be seen from Figure 19.1,[1] the formative area embraces a good deal of northeastern Iran (Biscione and Vahdati 2012: 356; Vahdati 1394/2015). It includes: the upper Atrek valley from Shirvan to Quchan (Venco Ricciardi 1980); the areas of Darreh Gaz (Kohl and Heskel 1980: 163–169; Yusefi Zoshk and Bāghizadeh 1391/2011) and Mashhad (Bāsafā and Rahmati 1393/2014; Habibi *et al.* 2015); the western part of the region of Nishapur (Hiebert and Dyson 2002: 117–122);[2] the area of Sabzevar (Vahdati and Francfort 1389/2010; Francfort *et al.* 2014); and the region of Torbat-e Jam, southeast of Mashhad, where the site of Tepe Mokhar, on the eastern outskirts of the city, has given GKC pottery and ceramics of period NMG IV and perhaps III.[3]

The recent discoveries of Tepe Mokhar and Tepe Chupan, west of Bardaskan, and of Tepe Nāy, east of Kashmar (Figure 19.3),[4] the last two with local pottery similar

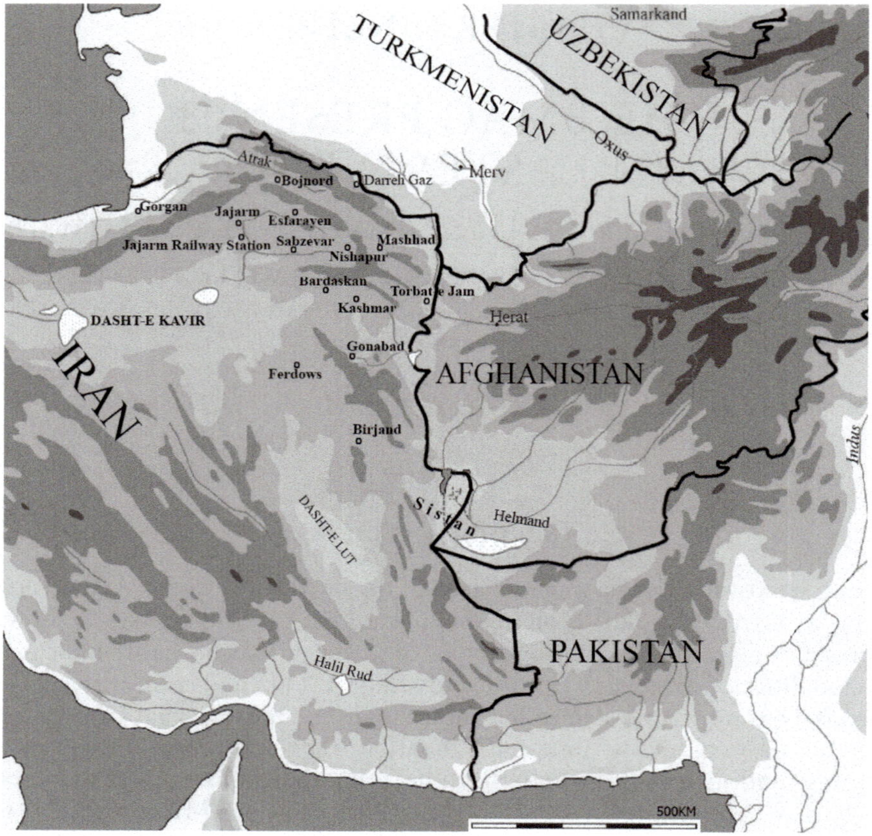

Figure 19.1 Geographical features and modern cities mentioned in the chapter. (Elaboration by the authors based on maps downloaded from Maps-for-free, https://maps-for-free.com/.)

to that known at Tepe Borj (Garajian 2012: 59–62), show that the southern border of the formative region follows approximately the line Sabzevar–western Nishapur plain–Torbat-e Jam–Mashhad region (Figures 19.1 and 19.2). Thus, the surface occupied by the Namazga culture on the Iranian Plateau is indeed larger than that known in southern Turkmenistan.

The core area

The BMAC/GKC extends over most of the "formative area" but also far beyond it and includes Margiana (if not significantly settled during the Early Bronze Age) and Bactria.[5] In this vast area, the zone where the BMAC/GKC is only preceded by the Namazga culture or where it represents the first occupation, either because the regions were not inhabited or because earlier sites are unknown,[6] can be considered as the "core area" (see Figure 19.2). In Iran, it seems to coincide with the presently known border of the "formative area."

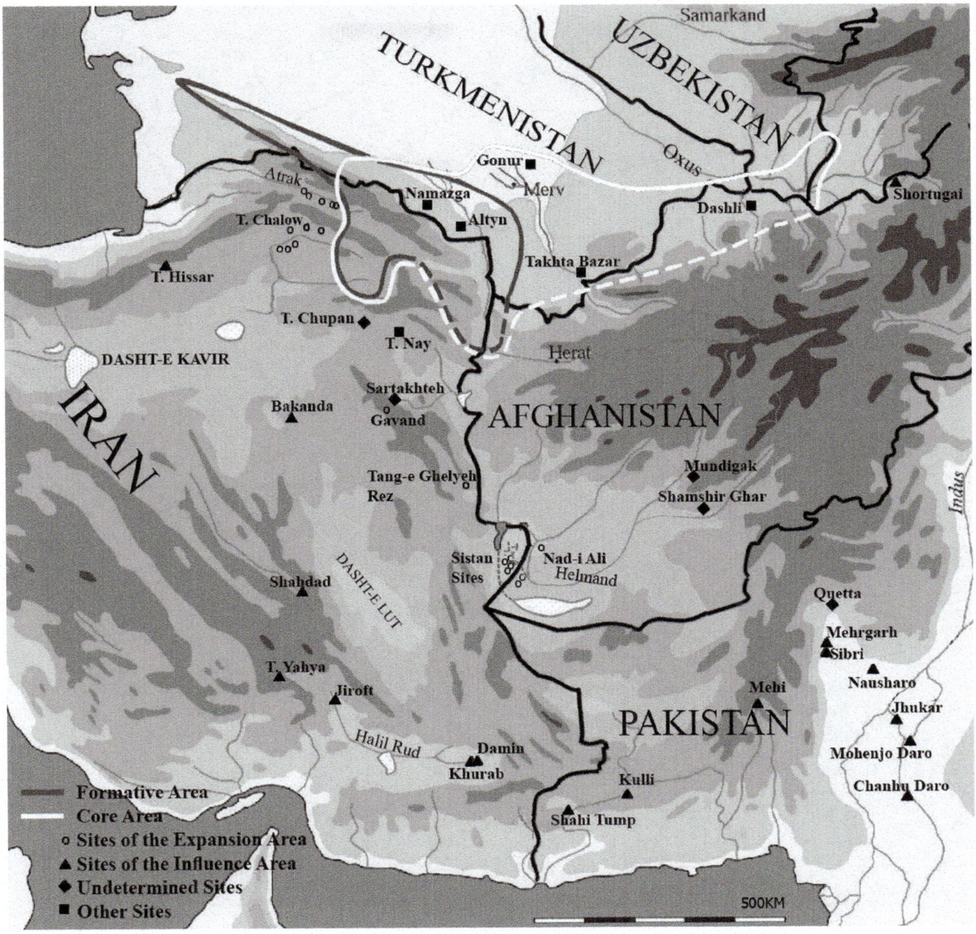

Figure 19.2 "Formative" and "core" areas of the BMAC/ GKC and archaeological sites with BMAC/ GKC materials determining expansion and influence areas. (Elaboration by the authors based on maps downloaded from Maps-for-free, https://maps-for-free.com/.)

The expansion area

In recent years, outside the "core area," a number of cases in which the BMAC/ GKC replaced the previously existing cultures/civilizations have been identified. The regions concerned by this replacement can be defined as the "expansion area" and include parts of eastern Iran, beyond the borders of the "core area." The sites where this expansion is clear (see Figure 19.2) are presented in the following sections, but the borders of this area are still fuzzy and uncertain. Part of this expansion went westwards, to the region of Bojnord in the Middle Atrek valley, and of Jajarm–Esfarayen in the valley of the northern Kal-e Shur, both areas located in northern Khorasan. Another (greater) part is located southwards, in the southern part of Khorasan and in the Iranian and Afghan Seistan (see also Chapter 20).

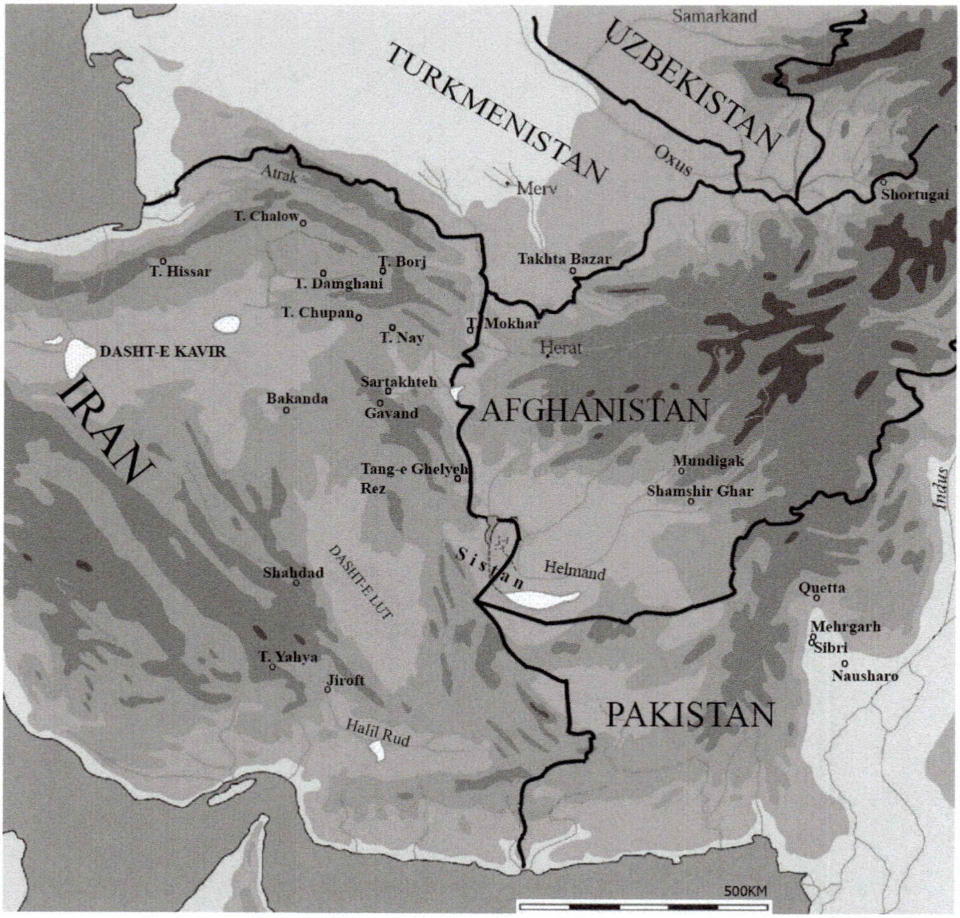

Figure 19.3 Archaeological sites mentioned in the chapter. (Elaboration by the authors based on maps downloaded from Maps-for-free, https://maps-for-free.com/.)

The influence area

There is also a significant BMAC/GKC presence in regions occupied by other cultures. This presence ranges from a coexistence where both material cultures are represented, to a sporadic presence of BMAC/GKC luxury objects. In the opinion of the authors, this could be defined as the "influence area." It includes sites already known (see Figure 19.2), like the massive BMAC/GKC presence, with a substantial amount of pottery[7] found to the east of the core area, at Shortughai (Francfort 1989: 65–70), and to the southeast, at Mehrgarh (Santoni 1988; Jarrige 1989), Nausharo (Jarrige 1988), and Sibri (Santoni 1984, 1988) (see also Chapters 20 and 21).[8] Similarly, in Iran, it includes the well-known cases of Shahdad, Tepe Yahya, Jiroft, Khurab (see Chapter 20), and Tepe Hissar, where its presence is essentially represented by luxury objects (like at Jiroft). In a number of other sites the BMAC/GKC evidence is limited to a few or isolated objects (Biscione and Vahdati, forthcoming).

This chapter focuses on the expansion area and on the sites of the influence area not yet published or difficult for Western scholars to access. Table 19.1 provides information on the type of site, settlement, or graveyard we are dealing with.

NORTHERN KHORASAN

In the present-day province of northern Khorasan (i.e., immediately outside the western part of the core area), the expansion involved a limited area around Bojnord and the northern Kal-e Shur River valley, in the plain going from Jajarm to the Jajarm railway station and Esfarayen (see Figure 19.1). In these two zones, the previous culture of the Hissar/Gorgan sites disappeared and was replaced by BMAC/GKC settlements and graveyards (see Table 19.1). It is interesting to note that, in this part of the expansion area, the Yaz I pottery is largely widespread during the second half of the 2nd millennium, like all over the core area (Biscione and Vahdati 2012: 356–358; Dānā and Hojabri 1393/2014: 308; Vahdati 2014; 1394/2015: 47–50; 2015: 266–268, 2016; 2018; Vahdati and Meier 2019; Vahdati and Biscione, forthcoming).

South of the Sabzevar–Nishapur–Mashhad line, data is scarce: We have the information coming from Tepe Mokhar on the outskirts of Torbat-e Jam (see above), and the authors have been told that BMAC/GKC pottery was found at Gonabad, but it has not been possible to visit the site and check the pottery, so these last occurrences remain uncertain for the moment.[9]

SOUTHERN KHORASAN

Slightly more to the south, at Gavand (Figure 19.3), 15 km north–northeast of the city of Ferdows in the province of southern Khorasan, at least one looted tomb contained typical BMAC/GKC artifacts now confiscated (Figures 19.4 and 19.5). The pins and bracelets suggest that the burial ritual was similar to that known at T. Chalow, where almost every skeleton wore at least one bracelet, or more generally two, and one or more pins at the shoulder (Biscione and Vahdati 2011: 237; Vahdati *et al.* 2018). This burial feature, however, is not attested in southern Central Asia.

After the discovery of the looted materials, in 2015 Mohammad Farjami, archaeologist of the southern Khorasan region, opened 31 test trenches to examine the stratigraphy and extension of the archaeological deposits of the flat site where the objects had been reportedly found (Farjami 1394/2016; 1395/2017: 474–475, 478 fig. 4). Since the site was heavily damaged by intensive agriculture, no traces of structures have been discovered and the sherds collected are from mixed layers dated to the Bronze Age (BMAC/GKC), the Historic, and Islamic periods. According to the digger, the BMAC/GKC material is part of the burial goods of a Bronze Age graveyard destroyed by modern agricultural activities.[10]

Farjami recently gave us a list of other sites in southern Khorasan where BMAC/GKC material has been found; namely, Sartakhteh, Bakandā, and Tang-e Ghelye Rez (see Figure 19.3). Among them, Sartakhteh is the most recently discovered. It is located near the village of Bārāz, 23 km west of Qa'en, and some of the surface pottery was collected by Mostanser Qolinejad, archaeologist of the southern Khorasan region. It contained BMAC/GKC sherds, probably from graves, together with local pottery. Without excavations it is not possible to say whether these graves

Table 19.1 BMAC/GKC materials on sites in eastern Iran and on other sites mentioned in the chapter

Site name	*Country/region*	*Fig. no.*	*Settlements*	*Graveyards*	*References*
Formative and core areas					
Nishapur P (Yusefabad? Kalateh Shuri?)	Iran, Khorasan Razavi	–	+	–	Bāsafā and Rezāi 1395/2017
Shahrak-e Firuzeh	Iran, Khorasan Razavi	–	–	+	Bāsafā and Rahmati 1391/2011
Tepe Mokhar	Iran, Khorasan Razavi	–	+	–	Unpublished (see endnote 3)
Tepe Naderi	Iran, Khorasan Razavi	–	+	–	Habibi *et al.* 2015: 831
Tepe Tup Derakht	Iran, Khorasan Razavi	–	+	–	Visited by the authors
Tepe Qasemabad	Iran, Khorasan Razavi	–	+?	–	Pottery examined by A. Vahdati
Tepe Damghani	Iran, Khorasan Razavi	19.3	+	–	Vahdati and Francfort 1389/2010: 29; Francfort *et al.* 2014: 130
Tepe Ferizi	Iran, Khorasan Razavi	–	+	–	Vahdati and Francfort 1389/2010: 26; Francfort *et al.* 2014: 124
Expansion area					
Tepe Eshgh	Iran, northern Khorasan	19.2	–	+ (1 grave)	Vahdati 2014
Tepe Yavar	Iran, northern Khorasan	19.2	–	+?	Vahdati 2014: 22
Tepe Jangi	Iran, northern Khorasan	19.2	–	–	Vahdati 2015: 267
Jayran Tepe	Iran, northern Khorasan	19.2	+	–	Vahdati 2016
Tepe Rafteh	Iran, northern Khorasan	19.2	+	–	Vahdati 2011: 184
Tepe Chalow	Iran, northern Khorasan	19.2, 19.3	+	+	Biscone and Vahdati 2011: 237
Fakhraniyeh (Tepe Rig)	Iran, northern Khorasan	19.2	+	–	Visited by the authors
Tepe Shafiabad	Iran, northern Khorasan	19.2	+	–	Visited by the authors
Gavand	Iran, southern Khorasan	19.2, 19.3	–	+ (1 grave)	Farjami 1395/2017

Tang-e Ghelye Rez	Iran, southern Khorasan	19.2, 19.3	–	+	Sorush and Yusefi 1393/2014
Seistan sites	Iran, Seistan	19.2	+	–	Dales 1977
Nad-i Ali	Afghanistan, Seistan	19.2	+	–	Besenval and Francfort 1994: 5
Gardan Reg 2	Afghanistan, Seistan	19.2	+	–	Unpublished
Gardan Reg 3	Afghanistan, Seistan	19.2	+	–	Unpublished
Gardan Reg 7	Afghanistan, Seistan	19.2	+	–	*
Influence area					
Hissar	Iran, Semnan Ostan	19.2, 19.3	–	+	Schmidt 1937: 171–175
Bakanda	Iran, southern Khorasan	19.2, 19.3	–	+	Unpublished
Jiroft	Iran, Kerman Ostan	19.2, 19.3	–	+	Francfort 2019
Shahdad	Iran, Kerman Ostan	19.2, 19.3	–	+	Hiebert and Lamberg-Karlowsky 1992
Tepe Yahya	Iran, Kerman Ostan	19.2, 19.3	+		Hiebert and Lamberg-Karlowsky 1992: 6
Khurab	Iran, Baluchistan	19.2		+ (1 grave)	Hiebert and Lamberg-Karlowsky 1992: 6–7
Shortughai	Afghanistan, Takhar	19.2, 19.3	+	–	Francfort 1989
Mehrgarh VIII	Pakistan, Kachi plain	19.2, 19.3	–	+	Santoni 1988; Jarrige 1989
Nausharo	Pakistan, Kachi plain	19.2, 19.3	+	–	Jarrige 1989
Sibri	Pakistan, Kachi plain	19.2, 19.3	+	+	Santoni 1984, 1988
Undetermined					
Tepe Chupan	Iran, Khorasan Razavi	19.2, 19.3	–	+	Unpublished
Sartakhteh	Iran, southern Khorasan	19.2, 19.3	–	+	Unpublished

(*continued*)

Table 19.1 Cont.

Site name	*Country/region*	*Fig. no.*	*Settlements*	*Graveyards*	*References*
Quetta	Pakistan, Baluchistan	19.2, 19.3	–	+ (1 grave)	Jarrige and Hassan 1989
Mundigak	Afghanistan, Kandahar	19.2, 19.3	–	–	Balali 2010
Shamshir Ghar	Afghanistan, Kandahar	19.2, 19.3	–	+ (1/more?)	Francfort and Tremblay 2010: 107–108
Sites without GKC/BMAC materials					
Tepe Borj	Iran, Khorasan Razavi	19.3	+	–	–
Tepe Nāy	Iran, Khorasan Razavi	19.2, 19.3	+	–	–
Korond	Iran, southern Khorasan	19.2, 19.3	–	+	–
Shahr-i Sokhta	Iran, Seistan	19.2	+	+	–
Tepe Taleb Khan 1	Iran, Seistan	19.2	+	–	–
Tepe Taleb Khan 2	Iran, Seistan	19.2	+	–	–

Note: + = settlement or graveyard; – = no settlement or graveyard.

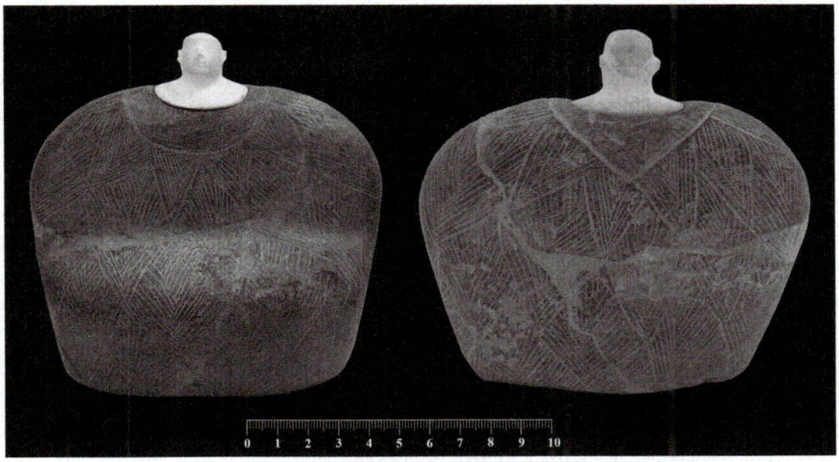

Figure 19.4 A composite figurine from the pillaged grave at Gavand (Photo © M. Farjami, courtesy of the author).

Figure 19.5 Other materials from the pillaged grave at Gavand (Photo © M. Farjami, courtesy of the author).

are contemporary to the settlement or later, and whether their grave goods are entirely related to the BMAC/GKC or mixed. Therefore, in Figure 19.2, this site is considered as "undetermined."

Bakandā (see Figure 19.2),[11] near the village of Pir-e Hājāt, lies 70 km north–northeast of Tabas and 90 km west–northwest of Boshruyeh. It is a graveyard where burials present both local and BMAC/GKC pottery (Farjami 1395/2017: 475).

These two sites have been discovered by chance during small rural construction activities. In both cases, evidence of graves was observed. The culture they are related to is not yet clear, showing analogies but not total similarities with that known at

Tepe Borj (Garajian 1387/2007) and at other sites of the eastern part of the Nishapur plain (see p. 544, note 2).

In mid-2017, R. Heydari, a student in archaeology, worked at Tepe Chupan, located about 25 km west–northwest of Bardaskan (see Figures 19.1 and 19.2), a site already known to the local office of the Cultural Heritage, Tourism and Handicraft Organization. Heydari and his colleague S. Babazadeh[12] have identified BMAC/GKC pottery on the surface. The authors were able to visit the site on December 11, 2017 and found that the great majority of the pottery was local but that, on the surface, there were some clusters of BMAC/GKC sherds, probably indicating eroded graves. Like at Sartakhteh, a surface survey cannot determine whether these graves are contemporary to the settlement or later, and whether their grave goods are entirely BMAC/GKC or mixed. Therefore, in Figure 19.2, Tepe Chupan is also considered as "undetermined." Furthermore, a cursory check of the surface pottery left on the site showed that the shape of some of the BMAC/GKC sherds was somehow atypical and did not fully correspond to those of the core area.[13]

The southernmost find-spot in southern Khorasan is a graveyard at Tang-e Ghelye Rez, near the town of Razeh (see Figure 19.3), located not far from the Afghan border, about 102 km east–southeast of Birjand. The necropolis has been excavated and preliminary published (Sorush and Yusefi 1393/2014).[14] The material is definitely BMAC/GKC, without any local element (Figures 19.6 and 19.7). Obviously, in the Birjand region there could be other BMAC/GKC sites.[15] The archaeological exploration of southern Khorasan and of the southern part of Khorasan-e Razavi has just started and it is very likely that, in the near future, new data will enrich our knowledge of the BMAC/GKC phenomenon.[16]

SEISTAN

Seistan is doubtless the region of the "expansion area" where the presence of the BMAC/GKC is the most significant. The first information came when G. Dales, then at the University of Pennsylvania Museum of Archaeology and Anthropology, made his survey in Afghan Seistan (1968–1971) and studied the Shela Rud region

Figure 19.6 Pottery from the Tang-e Ghelyeh Rez necropolis (1) (Photo © S. Yusefi, courtesy of the author).

Figure 19.7 Pottery from the Tang-e Ghelyeh Rez necropolis (2) (Photo © S. Yusefi, courtesy of the author).

(Figure 19.8). He discovered a group of what we now know as BMAC/GKC prestige objects (i.e., "miniature columns" and "stone disks" with or without a handle) (Figure 19.9). These artifacts came from one or more unidentified graves in the vicinity and were brought as offerings to the modern tomb of a local holy man in the shrine of Godar-i Shah, but they were evidently much more ancient (Dales 1972: 31–35, figs. 16–19; 1977). Until the mid-1980s archaeologists could not yet assess and understand the significance of such discoveries and the only known parallels were similar finds from Hissar IIIC. Only later did it become obvious that they were part of a complex of related BMAC/GKC luxury objects discovered over a very wide area from the border of Mesopotamia to the Indus valley and from Margiana to the southern coasts of the Persian Gulf. Unfortunately, the results of Dales's survey in the Shela Rud and in the nearby Gardan Reg areas have been published only in a very preliminary way and no other information is available (Dales and Flam 1969; Dales 1972, 1977, 1993).

Later on, R. Besenval and H.P. Francfort underlined the presence at Nad-i Ali (Afghanistan) (Figure 19.8) of a BMAC/GKC jar with a molded bottom and noticed that some of the published pottery of Dales's survey in the Gardan Reg area could be considered as BMAC/GKC (Besenval and Francfort 1994: 5, 10–11).

The documentation of Dales's survey, kept in the archives of the University of Pennsylvania Museum of Archaeology and Anthropology,[17] shows that the pottery is similar to that of periods Shahr-i Sokhta II–IV, BMAC/GKC, Nad-i Ali I-II (Iron Age and Achaemenid), Parthian-Sasanian, and Islamic. There is a good amount of BMAC/GKC fragments (Figures 19.10, 19.11, and 19.12) and some of them show peculiarities not present in the pottery of the "core area," like, for instance, incised wavy lines outside of bowls. All this pottery can be dated exclusively to the flourishing period of the BMAC/GKC, with no fragment later (for this period) than 1800/1700 BC. This pottery was found on at least three out of the possibly 12 sites identified and studied

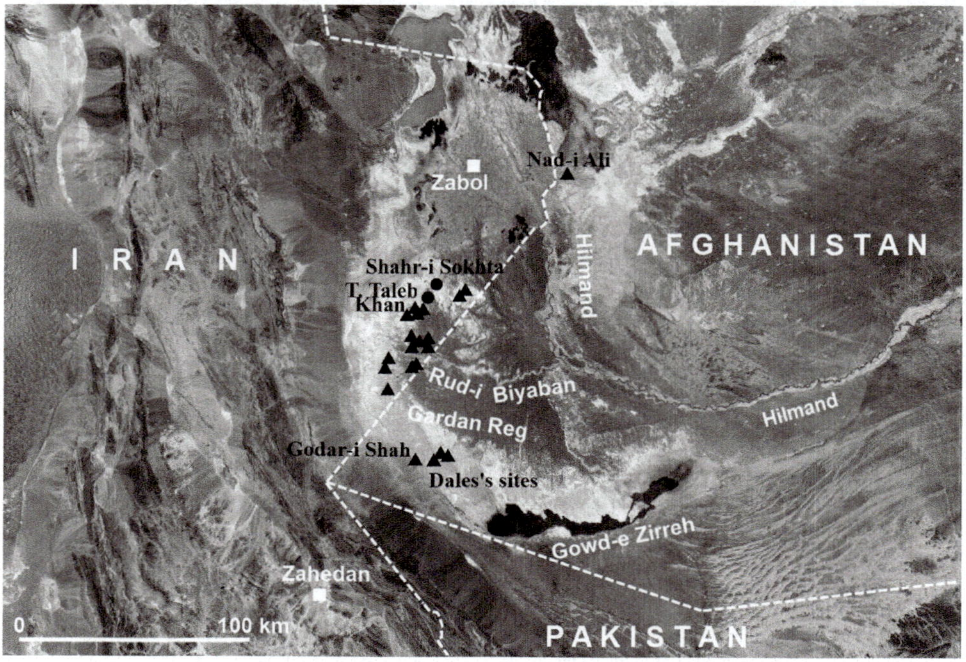

Figure 19.8 Iranian and Afghan Seistan, with the main geographical features and archaeological sites mentioned in the chapter. The black triangles are sites with BMAC/GKC pottery. (Elaboration by the authors based on Landsat false color imagery.)

Figure 19.9 "Miniature columns" and stone disks at the shrine of Godar-i Shah, Afghan Seistan (Photo © University of Pennsylvania Museum of Archaeology and Anthropology, Philadelphia, courtesy of the Museum).

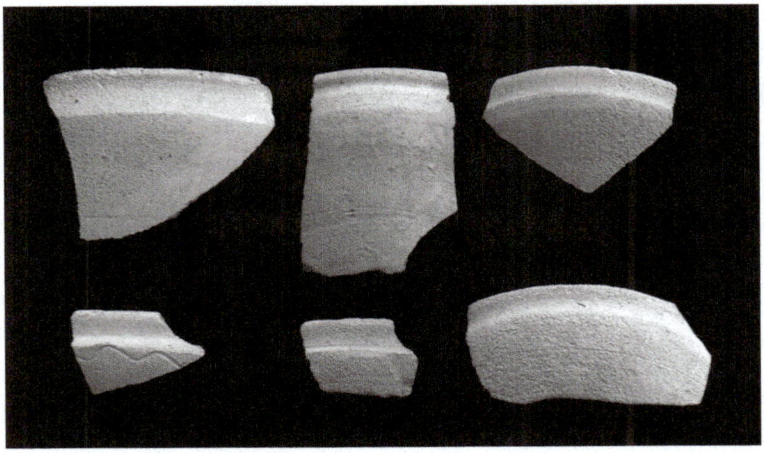

Figure 19.10 BMAC/GKC pottery from the Dales survey in Afghan Seistan (1) (Photo © University of Pennsylvania Museum of Archaeology and Anthropology, Philadelphia, courtesy of the Museum).

Figure 19.11 BMAC/GKC pottery from the Dales survey in Afghan Seistan (2) (Photo © University of Pennsylvania Museum of Archaeology and Anthropology, Philadelphia, courtesy of the Museum).

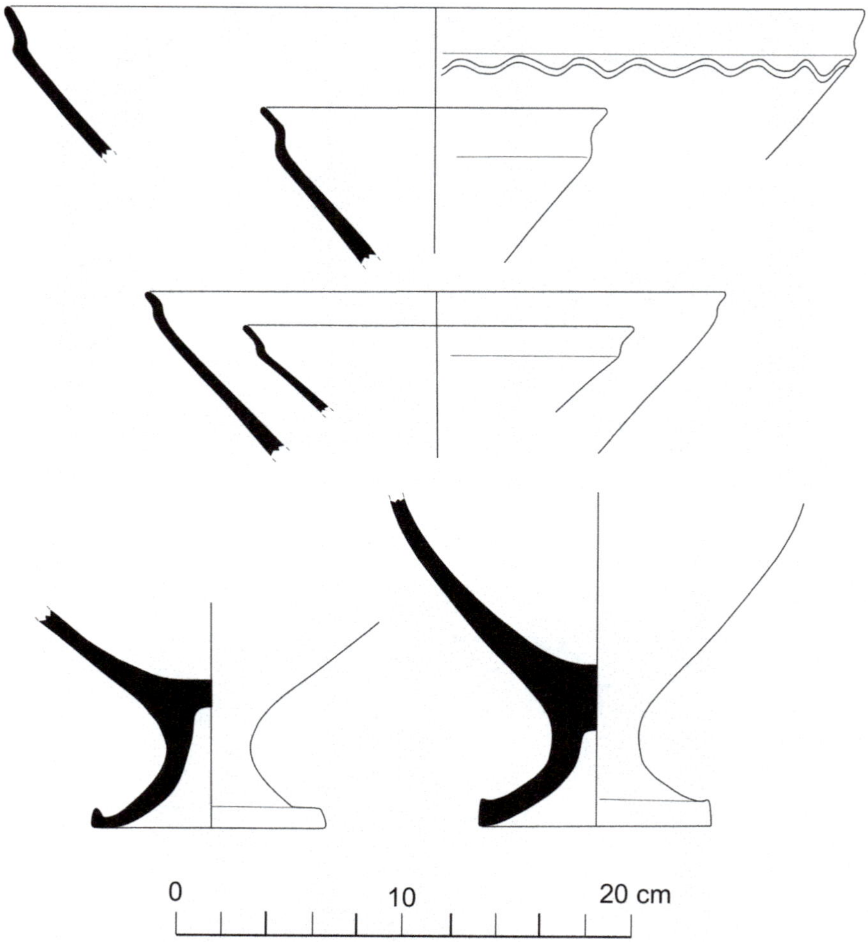

Figure 19.12 BMAC/GKC pottery from the Dales survey in Afghan Seistan (3). (Drawing by Mr. A. Mancini for the authors.)

by Dales in the Gardan Reg–Shela Rud area;[18] one of them contained pottery of Shahr-i Sokhta (SiS) I–III–IV and BMAC/GKC; another of SiS III and BMAC/GKC; and the last one was a new foundation that had only BMAC/GKG potsherds.

This kind of pottery had never been noticed earlier in the Iranian Seistan, but recent finds change the picture. In 2009–2010 the University of Zahedan, under the direction of professors S.R. Mousavi Haji and R. Mehrafarin, carried out a systematic survey over the whole region. It led to the discovery of over 1,200 sites covering all the periods.[19] A cursory examination of the results of the survey revealed 17 settlements with very diagnostic BMAC/GKC pottery and 31 other sites where such pottery was probably represented. These sites are distributed on the various branches of the Rud-i Biyaban, but apparently with a higher concentration in its southern part, which corresponds to the 3rd–early 2nd millennium delta of the Helmand River. Of the 17 sites with diagnostic BMAC/GKC pottery (see Figure 19.8), 11 were settled in the SiS III and BMAC/GKC periods, 2 contain period SiS IV and BMAC/GKC pottery, 1 has SiS III–IV and BMAC/GKC sherds, and 3 were new foundations.

DISCUSSION

This BMAC/GKC presence raises, again, the question of the end of the Shahr-i Sokhta sequence and of the abandonment of the Rud-i Biyaban. On three sites of the delta, there is an association of BMAC/GKC and SiS IV pottery on the surface but neither at Shahr-i Sokhta nor at Tepe Taleb Khan 1 and 2, the other excavated sites dated to period SiS IV,[20] was a single fragment of BMAC/GKC pottery ever found although some shapes are rather similar. This suggests that the sequence of Shahr-i Sokhta had already ended and that the delta of the Rud-i Biyaban was already abandoned when the BMAC/GKC pottery arrived in Seistan.[21] This moment cannot yet be determined on the basis of the pottery date, but allowing BMAC/GKC a reasonable amount of time to reach Seistan and to develop a settlement system with at least 20 sites, including the ones found by Dales, we can roughly estimate that this cannot be later than 2000 BC. This moment also marks the *terminus ante quem* for the end of the Helmand Civilization. As period SiS IV (which includes phase 1, a gap, and phase 0) is generally dated 2200–1700 BC (Salvatori and Tosi 2005: 288–290, figs. 12–13), this means either telescoping its three moments in a maximum of 200 years, or making the beginning of period IV earlier than 2200.

The finds in Seistan corroborate a BMAC/GKC presence at Nad-i Ali (Besenval and Francfort 1994: 5) and at Shamshir Ghar (Francfort and Tremblay 2010: 107–108), and this pottery can be found also at Mundigak (Balali 2010). Certainly, then, the middle valley of the Helmand (Figure 19.2) was also opened to BMAC/GKC, but it is not possible to verify whether it was part of the "expansion" or of the "influence" areas. This also changes our view on the remarkable BMAC/GKC presence at Quetta, Sibri, Mehrgahr, and Nausharo (Figures 19.2 and 19.3), which cannot be considered as isolated occurrences but are parts of a much larger phenomenon (see also Chapters 20 and 21).

It has to be underlined that, in the "expansion area," the presence of BMAC/GKC is marked both by graveyards and by settlements distributed in two distant areas (see Table 19.1). On the one hand, it concerns discoveries made in northern Khorasan in the area immediately west and south of Bojnord, in the Kal-e Shur valleys, among which is T. Chalow, and those near the Jajarm railway station. On the other hand, it concerns the sites found in the Iranian and Afghan Seistan. There is no trace of settlements in between these two areas, except graves, at Bakandā, Tang-e Ghelye Rez, and probably at Gavand, Sartakhteh, and Tepe Chupan (Figures 19.1 and 19.2).

The "influence area" of the BMAC/GKC (Figure 19.2) has been recently studied (e.g., Biscione and Vahdati, 2011, 2012, forthcoming), and we will not list its occurrences again. The differences between the "expansion" and "influence" areas are as evident as they are strong and they point at different reasons for the presence of BMAC/GKC artifacts.

In the "influence area" the presence of isolated prestige objects, or groups of them, testifies to a movement of goods, indicating connections with the "core area," most probably generated by trade. These occurrences of single objects are generally located on a belt far from the core area (Biscione and Vahdati, forthcoming), extending from southern Mesopotamia to the Indus valley and beyond (Lothal, in Gujarat), and encompassing also the southern shore of the Persian Gulf.

In the same "influence area," groups of BMAC/GKC pottery and elite objects found in local contexts are evidence of stronger connections with the "core area." Isolated burials with grave goods mostly or fully BMAC/GKC indicate the presence of individuals originating from the core area who moved to other places for a variety of reasons (trade, specialized knowledge, and abilities needed abroad, etc.), as already stressed, for instance, by Hiebert and Lamberg-Karlovsky (1992: 3, 6) or Thornton (2013: 195). The sites with groups of BMAC/GKC pottery and objects are distributed in an oblique band going from the Gorgan plain to the Halil Rud valley, closer to the "expansion area" and to the "core area" than the outer band of isolated objects (Biscione and Vahdati, forthcoming).

The sites of the "expansion area" (i.e., settlements and graveyards with fully BMAC/GKC material culture), are disposed in a line oriented approximately northwest–southeast from Bojnord and Jajarm to Seistan and they indicate the presence of a significant number of people with strong connections to the "core area" (Figure 19.2). It is unlikely that these sites were trading stations because they are too numerous. In regions like the Bojnord and the northern Kal-e Shur valleys or Seistan they testify to a real occupation of the territory and have replaced the previous local traditions that had lasted for centuries before, like the Helmand Civilization in Seistan and the Hissar/Gorgan culture in northern Khorasan. This movement seems to have touched abandoned areas, like Seistan (see above) and the northern Kal-e Shur valley (in this last region there is a gap between the early 3rd millennium occupation and the arrival of the BMAC/GKC), or previously unoccupied areas, like the region of the Jajarm railway station.

Data about Mundigak, Shamshir Ghar, and Nad-i Ali and their territory is too scant to allow placing the sites in the "expansion" or the "influence" area, but Shortughai, Nausharo, Sibri, and Mehrgarh are definitely in the "influence area," as is shown by the association of BMAC/GKC and other material cultures. The "expansion area" is therefore sandwiched inside the "influence area."

It is not fortuitous if Seistan is the area where the BMAC/GKC presence is the strongest and if Mehrgarh, Nausharo, Shortughai, and Sibri offer a very large amount of BMAC/GKC pottery. Earlier, during periods SiS I/Mundigak III, Seistan and Mundigak had also been reached by strong Central Asian influences as shown by the NMG III pottery found at both sites (Biscione 1973; see also Chapter 1). Similarly, the Quetta valley shows the same Central Asian pottery (Fairservis 1956: figs. 414–415) and Mehrgarh, Nausharo, and Sibri are close to it and are on the natural route connecting it – and ultimately Afghanistan – with the Indus valley (see Chapters 20 and 21). Apparently, the expansion of the BMAC/GKC followed the same routes as those of the NMG III expansion a thousand years earlier.

The geographical location of the "core" and of the "expansion" areas raises another question (see also Chapter 1). There is no consensus about the denomination of the civilization subject of this study. The most widely used is "Bactria-Margiana Archaeological Complex" (BMAC) but, while this term stresses the places where this civilization has its most splendid remains, it neglects the "formative area" and the other parts of the "core" area. The term "Oxus Civilization," is, again, too limited, referring only to northern and southern Bactria. The old name "Namazga VI" ties it to an even more restricted geographical area and to a specific aspect of the Namazga sequence. If we consider the "core area" and its extension – from Sabzevar to Tajikistan and from the northernmost part of the Murghab delta to Nishapur and to Takhta Bazar (on the Murghab River 30 km north of the Afghan border) (Figures 19.1 and 19.2) – it is clear that it actually matches most of the Greater Khorasan at the times of the Sasanian and Early Islamic periods (Ibn al-Faqih al-Hamedani 1996: 615; Rante 2015: 9–17). This area is located on the itinerary of the Great Khorasan Road (or the High Road in Zagros), in later periods called the "Silk Road" (i.e., the trade route that brought lapis lazuli, raw ores, and other commodities from Central Asia and the Iranian Plateau to the West from the 4th millennium BC onwards). Greater Khorasan always had a specific individuality that, after the late 3rd and early 2nd millennium, continued throughout the Iron Age with the diffusion of the Yaz I pottery (Biscione 1981: 210–211, 212; Lhuillier 2013: 127), through the Achaemenid period (Cattenat and Gardin 1977: 230–235 figs. 2–4), and, somehow, still existed until the Early Islamic period (Rante 2015: 21–22). For these reasons the authors think it more advisable to use the name "Greater Khorasan Civilization" (GKC).

Neither the beginning nor the end of the GKC expansion can be determined with certainty. At the moment, on the basis of the scant amount of pottery gathered and studied, the only thing that can be said is that the GKC pottery of the "expansion area" does not show any of the characteristics of the later Bronze Age Central Asian sequence (i.e., the Takhirbaj/Molali and Bustan stages). It can therefore be assumed that the GKC expansion dates to the flourishing period, between 2300/2200 and 1800/1700 BC, and that it disappears after that. This, in turn, underlines a gap in the Bojnord area and of the Kal-e Shur valley sequences, since the two regions are unoccupied from 1800/1700 to 1400 until the appearance of the Yaz I pottery.[22]

The lack of documentation and of stratigraphic studies does not allow, for the moment, a more precise statement (see Chapter 18).

With the exception of the northern Kal-e Shur valley, the Bojnord area, and possibly Seistan,[23] much of the GKC expansion took place in marginal areas with poor water resources. Thus, until recently, it was considered that life could only be sustained there with the use of qanats or very deep wells (i.e., beyond the technological possibilities of protohistoric times). This is the case, for instance, in the Jajarm station area, a remarkably dry zone settled for the first time in the GKC period, or in the other areas of southern Khorasan. This fact may suggest that either the GKC water management and water exploitation techniques were more efficient than those of the previous cultures, or that the GKC economy and society needs for water were not high, unless we suppose that the growing population pressure was so high that they had to use every possible area, even those short of water. Unfortunately, no data allows us to choose among these different hypotheses or to propose any other plausible one (but see Chapter 21). The fact remains that much of the GKC expansion in Iran took place in very dry, peri-desertic areas.

What we can say, however, even at the actual preliminary stage of the studies, is that the GKC expansion was pacific because it happened into unsettled or abandoned areas. Furthermore, the association in the same graves of both local and GKC pottery like at Bakandā (see above), would be improbable in the case of warlike, violent relationships. This cohabitation is visible not only in the "influence area," but also in the "expansion area": In one of the graves at T. Chalow, both gray ware and GKC vases were found together (Vahdati and Biscione, forthcoming). At a moment so strongly connected to cultural, social, religious, and group identity, the coexistence of two different traditions shows clearly that none was trying to destroy the other and that they were equally accepted. This is a good demonstration that the relationships between the newcomers and the local people were pacific and that most probably there were individuals with mixed heritage fully accepted in the social context.

The reasons for the GKC expansion are as yet unknown (on this theme see also Chapters 1 and 20). Some possible explanations were presented above (e.g., occupation of unsettled or abandoned regions and exploitation of marginal areas) but we do not yet have direct evidence for the reasons of such a large phenomenon. Similarly, its end and the abandonment of these areas together with the fall of the GKC are also unclear. It is tempting to connect them to the drought period that began ca. the mid-3rd millennium (Gupta *et al.* 2006; Madella and Fuller 2006; Dixit *et al.* 2014; Fouache *et al.* 2015: 27). This could give a wonderful and perfect explanation to the crisis that began around 2200 BC at Shahr-i Sokhta and Tepe Hissar, leading eventually to the abandonment by 1800/1700 BC of almost every proto-urban center east of the Tehran-Bandar Abbas line, from Yahya to Hissar, or from Shahdad to Harappa and Mohenjo-daro, including also the great centers of GKC. Such an explanation is definitely tempting, but evidence is still scarce. We hope that new research will help in solving this problem and the others connected with the Greater Khorasan Civilization.

NOTES

1 The presence of NMG IV pottery in Margiana is scarcely documented and the question is still open, therefore the writers for the moment prefer not to include this region in the formative area.

2 The research carried out up to now in the Nishapur plain revealed that there are two or three sites with pottery of the Namazga culture, one of them being the mound called Nishapur P (Hiebert and Dyson 2002) to be identified with Yusefabad according to O. Garajian, or Kalateh Shuri according to H. Basafa. Both sites lie west of present-day Nishapur and were briefly excavated, respectively by Garajian and Basafa (Bāsafā and Rezāi 1395/2017). If Nishapur P is not to be identified with them, it means that there are three sites with Namazga pottery in this plain. The other sites east of Nishapur have local Chalcolithic/Early Bronze Age pottery, including Tepe Borj (Garajian 1387/2007: 59–62). Garajian and Basafa kindly showed us their excavations and materials on December 12–13, 2017, and exposed their ideas about the Nishapur plain in the pre-Achaemenid periods. The presence of a strong cultural border dividing an open plain without natural barriers into two different cultural traditions practically without interrelations is a very peculiar fact, still to be studied and exactly defined. A possible reason might be the natural, easy road linking the western Nishapur plain to the area of Quchan in the upper Atrek valley.

3 The data about the region of Torbat-e Jam, still unpublished, were presented in Mashhad on December 17, 2018 by M. Toghra'i, archaeologist of the Provincial Office of the Cultural Heritage, Tourism and Handicraft Organization of the Khorasan-e Razavi province.

4 The authors were able to visit the sites on December 11 and 12, 2017 thanks to the kindness of Javad Rajabi, of the Provincial Office of the Cultural Heritage, Tourism and Handicraft Organization of the Khorasan-e Razavi province, branch of Bardashkan, and of Reza Yusefi, director of the branch of Kashmar of the Provincial Office of the Cultural Heritage, Tourism and Handicraft Organization of the Khorasan-e Razavi province.

5 We exclude from it eastern Bactria with Shortughai, since this area was earlier settled with a culture related to Mundigak and Baluchistan (Lyonnet 1981, 1988).

6 Bactria, both northern and southern, has been surveyed and it is reasonable to think that, if a significant number of pre-BMAC/GKC settlements existed in the area of diffusion of that civilization, their traces would have been most probably detected.

7 The BMAC/GKC Quetta hoard, coming from a grave (Jarrige and Hassan 1989) was not excavated by archaeologists, therefore the accompanying pottery was not retrieved and it is not possible to determine whether it is a full BMAC/GKC burial or whether BMAC/GKC luxury objects were associated with local pottery.

8 It is not always possible to attribute some sites precisely to the expansion or to the influence area, but in this chapter our reasoning concerns areas rather than individual sites.

9 The authors checked the Gonabad area and visited the region of Torbat-e Jam in March 2019, too recently to enter the results in this chapter, but the results are outside the scope of this chapter (limited to 2018).

10 On October 23, 2015, Mohammad Farjami, whom we warmly thank, took us to the site and kindly showed us the pottery and the other materials from his excavations.

11 On October 23, 2015, Farjami kindly showed us the materials, gave us relevant information, and supplied us with photographs.

12 Heydari kindly supplied us with relevant information about the site on December 12, 2017 and is publishing with Babazadeh the data of the surface survey.

13 The same can be said of Shahrak-e Firuzeh, immediately west of the city of Nishapur (Bāsafā and Rahmati 1391/2011), a BMAC/GKC necropolis without local pottery most probably in the core area. Some of the vases in the local museum exhibition are remarkably atypical and do not fully match with the shapes of the core area.

14 The final publication is being prepared by S. Yusefi, archaeologist of the southern Khorasan region, who kindly supplied us with significant information and photographs in 2014 and 2015.

15 In 1978, R. Biscione identified BMAC/GKC globular jars with molded bottom in an antique dealer's shop at Birjand. The man said that he did not know their provenance and did not allow taking pictures of them, so that no report could be made. It cannot be excluded that the objects come not from Tang-e Ghelyeh Rez but from some other necropolis.

16 In 2015 the authors (Biscione and Vahdati, forthcoming) listed, among the sites with BMAC/GKC pottery, that of Korond, 27 km west–northwest of the city of Boshruyeh. The necropolis excavated there by Farjami contains pottery with very simple and not diagnostic shapes that originally was thought to be BMAC/GKC. Recently, however, Farjami (personal communication, December 2017) attributed this necropolis to the historical periods, most probably the Parthian.

17 R. Biscione was able to study the documentation of this survey in September 2008 thanks to the kindness of Dr. Richard Hodges, then director of the museum, and to the invaluable help of Professor Holly Pittman, to whom we extend our warmest thanks. The pottery gathered during the survey is kept in the reserves of the Kabul Archaeological Museum but was badly damaged by a rocket during the war (B. Lyonnet, personal communication), while the graphic and photographic documentation are stored in the archives of the university museum. The results of the study will be published soon.

18 Unfortunately, no map with the number and the position of the sites was found in the archives of the university museum. One was kindly supplied by Barbara Dales, G. Dales's widow, and had three sites marked on it, but without numbers or names. On the basis of the existing documentation the total number of the sites remains unknown.

19 The results of the survey are yet unpublished as they are the subject of dissertations for students at Zahedan University, but thanks to the kind cooperation of professors S.R. Mousavi Haji and R. Mehrafarin it was possible to consult the reports with the registration of the sites and the graphic documentation of the finds.

20 The pottery of period SiS IV has never been fully published. The only extensive publication (Biscione 1979) did not have the scope of a total typology but, as the shapes are very standardized with few variations, the article actually gives a full typology of the SiS IV pottery. This was confirmed by an unpublished dissertation (Giacummo 1987). Tepe Taleb Khan 1, close to Shahr-i Sokhta, was excavated in 2001–2003 for Zabol University by Dr. S.M.S. Sajjadi, director of the Shahr-i Sokhta excavations. Zabol University continued the work in the following years. Dr. Sajjadi, to whom we extend our warmest thanks, allowed R. Biscione to examine the pottery from his excavations. Tepe Taleb Khan 2, very close by, is currently being excavated by Drs. Kavosh and Naseri, of Zabol University, who kindly showed R. Biscione the pottery of their excavation and supplied important information. Further information was given by Professor M. Vidale, who is collaborating on the publication of the site. Our warmest thanks go to professors Kavosh, Naseri, and Vidale.

21 It is unlikely, in the authors' opinion, that BMAC/GKC and SiS IV coexisted. In the Nishapur plain during the Chalcolithic period there is clear evidence of the coexistence of local and Namazga pottery, with a cultural border between the two clearly separated areas (see p. 544, note 2), but, in the Late Bronze Age, BMAC/GKC and local pottery were found associated at Tepe Borj (Garajian 1387/2007). In Seistan, neither Shahr-i Sokhta nor Tepe Taleb Khan 1–2 show any trace of BMAC/GKC pottery, and it is not possible to draw a border between sites with SiS IV pottery and those with BMAC/GKC ceramics, because they are sometimes located side by side. If there was a period of coexistence, the incidence of association of the two material cultures on the surface of the sites would have been much higher, and, most probably, some BMAC/GKC sherd would have been found in the excavations.

22 The core area fits with the distribution of the Yaz I pottery, but data on the Iranian part of the core area is still too scant to ascertain whether or not the Takhirbaj/Molali and Bustan phases are attested there. In Seistan there are no traces of the Yaz I pottery and there seems to be a gap between the end of the GKC and the beginning of the Nad-i Ali sequence around 1000/800 BC.

23 The case of Seistan is not clear. The distribution of the GKC sites suggests that they exploited different and smaller channels of the Rud-i Biyaban delta than those used at the time of SiS period IV. This could mean that those channels were abandoned because of droughts, and that the GKC settlers arrived in a more arid environment. The date of the drying up of the Rud-i Biyaban, due to a shifting of the Helmand delta possibly combined with a severe drought, is not yet known. It can only be said that it happened during the 2nd millennium since, at the beginning of the 1st millennium, the site of Nad-i Ali is located far from the Rud-i Biyaban and only very few Nad-i Ali I-II sherds (Iron Age and Achaemenid) have been found in Iranian Seistan or in the Gardan Reg area. This means that, by the early 1st millennium BC, the Helmand delta had already shifted somewhere else and that the Rud-i Biyaban was dry. But again this is just a *terminus ante quem*.

REFERENCES

Balali, N. 2010. "Bronze Age in Afghanistan." Poster presented at the 20th EASAA conference, Vienna 2010, www.univie.ac.at/easaa-vienna-2010/images/post_proc/easaa_pp_balil.pdf.

Bāsafā, H. and M. Rahmati 1391/2011. "Gozāresh-e moqadddamāti-ye kāvosh-e bāstān shenākhti dar mohavvateh-ye Shahrak-e Firuzeh (bahār 1388 va 1389)" [Preliminary report of the archaeological excavation at Shahrak-e Firuzeh, Spring 2009 and 2010]. In *Be koushesh-e J. Kianfar, and P. Istakhri: Mazdaknāmeh (5), Yādbud- e Panjomin Sālgardehdar Gozashteh Mazdak Kiānfar* [In memory (fifth anniversary) of Mazdak Kianfar's death], ed. J. Kianfar and P. Istakhri. Tehran: Pazhuheshgāh-e Miraz-e Farhangi va Gardeshkari, pp. 613–623.

Bāsafā, H. and M. Rahmati 1393/2014. "Gamānehzani be manzur-e tadqiq-e 'arseh-ye esteqrār pish az tārikhi Chenārān" [Sounding for delimitation of Chenārān prehistoric settlement]. In *Maqallehā-ye Kutāh Davāzdahomin Gerdehamāi-ye Sālāne-ye Bāstān Shenāsi-ye Irān, 29 tā 31 Ordibehesht 1393* [Brief articles 12th Annual Symposium on the Iranian Archaeology, 19–21 May, 2014]. Tehran: Pazhuheshgāh-e Miraz-e Farhangi va Gardeshkari, pp. 78–79.

Bāsafā, H. and M.H. Rezāi 1395/2017. "Avvalinfasl-e kāvush dar mohavvateh-ye Kalāteh Shuri (Khorāsān-e Razāvi, Nishābur)" [First season of archaeological excavation at Kalāteh Shuri (Khorāsān-e Razāvi, Nishapur]. In *Gozāreshhā Pānzdahomin Gerdehamāi Sālānehi Bāstān Shenāsi-ye Irān, 15 tā 17 Esfand Māh 1395* [Proceedings of the 15th Annual Symposium on the Iranian Archaeology, 5–7 March, 2017]. Tehran: Pazhuheshgah-e Miraz-e Farhangiva Gardeshkari, pp. 28–31.

Besenval, R. and H.P. Francfort 1994. "The Nad-i Ali Surkh Dagh: A Bronze Age Monumental Platform in Central Asia." In *From Sumer to Meluhha, Contributions to the Archaeology of South and West Asia in Memory of George F. Dales Jr.*, ed. J.M. Kenoyer. Madison: University of Wisconsin Press, pp. 3–14.

Biscione, R. 1973. "Dynamics of an Early South Asian Civilization: First Period of Shahr-i Sokhta and Its Connections with Southern Turkmenistan." In *South Asian Archaeology*, ed. N. Hammond. London: Duckworth & Co., pp. 105–118.

Biscione, R. 1979. "The Burnt Building of Period Shahr-i Sokhta IV: An Attempt of Functional Analysis from the Distribution of Pottery Types." In *Iranica*, ed. G. Gnoli and A. Rossi. Napoli: Istituto Universitario Orientale, pp. 291–306, pls. 1–4 and figs. 1–22.

Biscione, R. 1981. "Centre and Periphery in Late Protohistoric Turan: the Settlement Pattern." In *South Asian Archaeology 1979*, ed. H. Härtel. Berlin: D. Reimer, pp. 201–213.

Biscione, R. and A.A. Vahdati 2011. "Excavations at Tepe Chalow, Northern Khorasan, Iran." *Studi Micenei ed Egeo-Anatolici* 53: 236–241.

Biscione, R. and A.A. Vahdati 2012. "The Iranian–Italian Archaeological Mission, Season 2012: The Identification of Cultural Areas." *Studi Micenei ed Egeo-Anatolici* 54: 355–360.

Biscione, R. and A.A. Vahdati forthcoming. "The Diffusion of Bactria-Margiana Archaeological Complex in Eastern Iran: New Evidence and Considerations." In *Farmers, Traders and Herders: The Bronze Age in Central Asia and Khorāsān*, ed. E. Luneau, M. Teufer, and N. Boroffka. Archäologie in Iran und Turan. Berlin: Dietrich Reimer.

Cattenat, A. and J.-C. Gardin 1977. "Diffusion comparée de quelques genres de poterie caractéristiques de la période achéménide sur le Plateau Iranien et en Asie Centrale." In *Le Plateau Iranien et l'Asie Centrale des origines à la conquête islamique, leurs relations à la lumière des documents archéologiques*, ed. J. Deshayes. Paris: CNRS Editions, pp. 225–248.

Dales, G.F. 1972. "Prehistoric Researches in Southern Afghan Seistan." *Afghanistan* 4(25): 14–40.

Dales, G.F. 1977. "Hissar IIIC Stone Objects in Afghan Sistan." In *Mountains and Lowlands, Essays in the Archaeology of Greater Mesopotamia*, ed. L. Levine and T. Cuyler Young. Malibu, CA: Undena, pp. 17–27.

Dales, G.F. 1993. "A Line in the Sand, Explorations in Afghan Seistan." In *South Asian Archaeology Studies*, ed. G. Possehl. New York: Oxford & IBH Publishing, pp. 19–32.

Dales, G.F and L. Flam 1969. "On Tracking the Woolly Kullis and the Like." *Expedition* 12(1): 15–23.

Dana, M. and A. Hojabri 1393/2014. "Olguhā-ye estekhāri mohavvatehā-ye pish az tārikhi-e Jājarm, Khorāsān-e Shomali" [Settlement pattern of pre-historic sites in Jajarm plain, northern Khorasan]. *Majmu'e-ye Maqāllāt-e Hamāyesh-e Beinolmellali-e Bāstānshenāsān-e Javān* [Proceedings of the International Conference of Young Archaeologists], ed. K.A. Niknami, M.H. Azizi Kharanaghi, M. Khanipour, and R. Naseri. Tehran: University of Tehran Press, pp. 295–310.

Dixit, Y., D.A. Hodell, and C.A. Petrie 2014. "Abrupt Weakening of the Summer Monsoon in Northwest India ~4100 yr ago." *Geology* 42(4): 339–342.

Fairservis, W.A. Jr. 1956. *Excavations in the Quetta Valley, West Pakistan.* New York: American Museum of Natural History.

Farjami, M. 1394/2016. "Now yāftehā-ye 'Asr-e Mefragh dar Khorāsān-e Jonubi (Motāle'e-ye moredi: Mohavate-ye Gavand-e Bāghestān-e Ferdows)" [Bronze Age recent discoveries in southern Khorasan (case study: Gavand site near Bāghestan, Ferdows)]. *Dovvomin Hamāyesh-e Melli-e Bāstān Shenāsi-ye Iran*, Jeld-e dovvom [Proceedings of the 2nd National Conference on the Iranian Archaeology, vol. 2], ed. H. Hashemi. Birjand: Nashr-e Chehar Derakht, pp. 231–257.

Farjami, M. 1395/2017. "Ta'in'Arseva Harim-e Gavand-e Bāghestan-e Ferdows" [Sounding for delimitation of Gavand, near Bāghestan, Ferdows]. *Panzdahomin Gerdehamāi-ye Sālāne-ye Bāstān Shenasi-ye Iran, 15–17 Esfand Māh 1395* [Proceedings of the 15th Annual Symposium on the Iranian Archaeology, March 5–7, 2017].Tehran: Pazhuheshgah-e Miraz-e Farhangi va Gardeshkari, pp. 474–479.

Fouache, E., H.P. Francfort, C. Cosandey, and C. Adle 2015. "La crise d'aridité climatique de la fin du 3e millénaire av. J.-C., à la lumière des contextes géomorphologiques de 3 sites d'Iran Oriental (Bam, Tepe Damghani, Jiroft)." In *Greater Khorasan (Studies in the History and Culture of the Middle East)*, ed. R. Rante. Berlin: De Gruyter, pp. 27–38.

Francfort, H.P. 1989. *Fouilles de Shortughai, Recherches sur L'Asie Centrale Protohistorique.* Paris: de Boccard.

Francfort, H.P. 2019. "Iran and Central Asia: The Grand Route of Khorasan (Great Khorasan Road) During the Third Millennium BC and the 'Dark Stone' Artefacts." In *The Iranian Plateau During the Bronze Age: Development of Urbanisation, Production, and Trade*, ed. J.W. Meyer, E. Vila, M. Mashkour, M. Casanova, and R. Vallet. Lyon: Mom Editions, pp. 247–266.

Francfort, H.P and X. Tremblay 2010. "Marhaši et la civilisation de l'Oxus." *Iranica Antiqua* 45: 51–224.

Francfort, H.P, A.A. Vahdati, J. Bendezu-Sarmiento, J. Lhuillier, E. Fouache, M. Tengberg, M. Mashkour, and Z. Shirazi 2014. "Preliminary Report on the Soundings at Tappeh Damghani, spring 2008." *Iranica Antiqua* 49: 111–158.

Garajian, O. 1387/2007. "Lāyenegāri dar gamāneh yek-e TappehBorj, Nishabur: rouykardi tafsiri" [Stratigraphy in trench 1 of TepeBorj, Nishapur: an interpretive approach]. *Majale-ye payām-e Bāstānshenās* 5(10): 1–18.

Garajian, O. 2012. "Patterns of Change During the Transitional Process from Chalcolithic Cultures to the Bronze Age in Northeastern Iran, Based on Pottery Studies." In *Nāmvarnāmeh: Studies in Honour of Massoud Azarnoush*, ed. H. Fahimi and K. Alizadeh. Tehran: Iran Negar, pp. 55–68.

Giacummo, M. 1987. "La ceramica del periodo IV del Palazzo Bruciato di Shahr-i Sokhta." Unpublished PhD thesis, Istituto Universitario Orientale, Naples.

Gupta, A.K., D.M. Anderson, D.N. Pandey, and A.K. Singhvi 2006. "Adaptation and Human Migration, and Evidence of Agriculture Coincident with Changes in the Indian Summer Monsoon During the Holocene." *Current Science* 90(8): 1082–1090.

Habibi, Kh., B. Shaik Baikloo Islam, and A.A. Eshqi 2015. "Evidences from Cultures of the Central Asia in Naderi Tepe, Mashad Plain, in Northeastern Iran." *MAGNT Research Report* 3(1): 826–844.

Hiebert, F.T. and R.H. Dyson 2002. "Prehistoric Nishapur and the Frontier Between Central Asia and Iran." *Iranica Antiqua* 37: 113–149.

Hiebert, F.T. and C.C. Lamberg-Karlowsky 1992. "Central Asia and the Indo-Iranian Borderland." *Iran* 30: 1–15 and pls. 1–3.

Ibn al-Faqih al-Hamedani, A. 1996. *Kitab al-buldan*, ed. Y. al-Hadi. Beirut: A'lam al-kutub Publications.

Jarrige, J.F. 1988. "Le complexe culturel de Mehrgarh (Période VIII) et de Sibri. Le 'Trésor' de Quetta." In *Les cités oubliées de l'Indus. Archéologie du Pakistan. Catalogue de l'exposition du Musée National des Arts Asiatiques Guimet*, ed. J.-F. Jarrige. Paris: Association Française d'Action Artistique, pp. 111–128.

Jarrige, J.-F. 1989. "Excavation at Nausharo 1987–88." *Pakistan Archaeology* 24: 21–67.

Jarrige, J.-F. and M.U. Hassan 1989. "Funerary Complexes in Baluchistan at the End of the Third Millennium in the Light of Recent Discoveries at Mehrgarh and Quetta." In *South Asian Archaeology 1985*, ed. K. Frifelt and P. Sørensen. London: Curzon Press, pp. 150–166.

Kohl, P.L. and D. Heskel 1980. "Archaeological Reconnaissance in the Darrreh Gaz Plain: A Short Report." Iran 18: 160–172.

Lhuillier, J. 2013. "Les cultures 'à céramique modelée peinte' en Asie Centrale: un aperçu de l'assemblage céramique de la deuxième moitié du 2e millénaire av. J.C." *Iranica Antiqua* 48: 103–146.

Lyonnet, B. 1981. "Etablissements chalcolithiques dans le Nord-Est de l'Afghanistan: leurs rapports avec les civilisations du bassin de l'Indus." *Paléorient* 7(2): 57–74.

Lyonnet, B. 1988. "Les relations de la Bactriane orientale avec le monde indo-baluche à partir des données céramologiques, du 3e au 2e millénaire avant notre ère." In *L'Asie centrale et ses rapports avec les civilisations orientales des origines à l'Age du Fer. Actes du colloque franco-soviétique, Paris 1985*, ed. J.-C. Gardin. Paris: de Boccard, pp. 143–151.

Madella, M. and D.Q. Fuller 2006. "Palaeoecology and the Harappan Civilisation of South Asia: A Reconsideration." *Quaternary Science Reviews* 25: 1283–1301.

Rante, R. 2015. "'Khorasan Proper' and 'Greater Khorasan' within a Politico-Cultural Framework." In *Greater Khorasan (Studies in the History and Culture of the Middle East)*, ed. R. Rante. Berlin: De Gruyter, pp. 9–26.

Salvatori, S. and M. Tosi 2005. "Shahr-i Sokhta revised sequence." In *South Asian Archaeology 2001*, ed. C. Jarrige and V. Lefèvre. Paris: Éditions Recherche sur les Civilisations, pp. 281–292.

Santoni, M. 1984. "Sibri and the Southern Cemetery of Mehrgarh: Third Millennium Connections Between the Northern Kachi Plain (Pakistan) and Central Asia." In *South Asian Archaeology 1981*, ed. B. Allchin. Cambridge: Cambridge University Press, pp. 52–60.

Santoni, M. 1988. "Aspects matériels des cultures de Sibri et de Mehrgarh VIII (plaine de Kachi, Baluchistan, Pakistan) à la fin du troisième et au début du deuxième millénaires." In *L'Asie Centrale et ses Rapports avec les Civilisations Orientales des Origines à l'Age du Fer*. Paris: de Boccard, pp. 135–141.

Schmidt, E.F. 1937. Excavations at Tepe Hissar, Damghan. Philadelphia: University of Pennsylvania Press.

Sorush, M.R. and S. Yusefi 1393/2014. "Mohavvateh-ye Razeh, shāhedi az esteqrahā-ye hezāreh-ye sevom tā dorān-e tārikhi dar Khorasān-e Junubi" [Razeh: evidence of 3rd millennium to historical period settlements in southern Khorasan]. In *Maqallehā-ye Kutā-e Davāzdahomin Gerdehamāi Sālāne-ye Bāstān Shenāsi-ye Irān, 29 tā 31 Ordibehesht 1393* [Brief articles on 12th Annual Symposium on the Iranian Archaeology, May 19–21, 2014], Tehran: Pazhuheshgah-e Miras-e Farhangi va Gardeshkari, pp. 271–273.

Thornton, C. 2013. "The Bronze Age in Northern Iran." In *The Oxford Handbook of Ancient Iran*, ed. D.T. Potts. Oxford: Oxford University Press, pp. 179–202.

Vahdati, A.A. 2011. "A Preliminary Report on a Newly Discovered Petroglyphic Complex Near Jorbat, the Plain of Jajarm, Northeastern Iran." *Paléorient* 37: 177–187.

Vahdati, A.A. 2014. "A BMAC Grave from Bojnord, North-Eastern Iran." *Iran* 52: 19–27.

Vahdati, A.A. 2015. "Asr-e Mefragh va Āhan dar Khorāsān (3000 tā 500 p.m.)" [Bronze and Iron Ages in Khorasan (3000–500 BCE)]. In *Gozari bar Bāstān Shenāsi-e Khorāsān* [A review on the archaeology of Khorasan], ed. M. Labbāf Khaniki. Tehran: Sazman-e Miras Farhangi va Gardeshgari (ICHTO), pp. 39–49.

Vahdati, A.A. 2015. "Preliminary Report on the Archaeological Survey in Esfarāyen County, Northeastern, Iran." *Archäologische Mitteilungen aus Iran und Turan* 47: 261–287.

Vahdati, A.A. 2016. "A Preliminary Report on the First Season of Excavation at Jayran Tepe in the Plain of Esfarayen, Northeastern Iran." *Iranica Antiqua* 51: 85–101.

Vahdati, A.A. 2018. "The Early Iron Age in Northern Khorāsan." In *A Millennium of History: The Iron Age in Central Asia (2nd and 1st Millennia BC)*, ed. J. Lhuillier and N. Boroffka. Archäologie in Iran und Turan, 17/Mémoires de la Délégation Archéologique Française en Afghanistan, XXXV. Berlin: D. Reimer, pp. 51–66.

Vahdati, A.A. and R. Biscione forthcoming. "The Bronze Age Cemetery of Chalow: a Brief Report on Three Seasons of Excavations." In *Farmers, Traders and Herders: The Bronze Age in Central Asia and Khorāsān*, ed. E. Luneau, M. Teufer, and N. Boroffka. Archäologie in Iran und Turan. Berlin: Dietrich Reimer.

Vahdati, A.A. and H.-P. Francfort 1389/2010. "Gozāresh moghadamāti gamāneh zani dar Tappeh Dāmghāni, Sabzavār, bahār 1387" [Preliminary report on the sounding at Tepe Damghani, Sabzevar, spring 2008], *Majalle-ye Bāstān Shenāsi va Tārikh* 24(2): 17–36 (with English abstract).

Vahdati, A.A., R. Biscione, M. Tengberg, and M. Mashkour 2018. "Kavosh dar tappeh Chalow: shavahedi az majmueh-ye bastan shenakhti-ye balkhi-marvi (BMAC) dar dasht-e Jajarm" [Excavation at Tepe Chalow: some evidence of the "Bactria-Margiana Archaeological Complex" (BMAC) in the plain of Jajarm, northeastern Iran]. *Archaeology (Journal of the Iranian Center for Archaeological Research)* 1(1): 70.

Vahdati, A.A. and D.M.P. Meier 2019. "Objects of the 'Oxus Civilization' and of the 'Jiroft Culture' from the Mofakham Museum (Bojnord, North Khorasan Province). In *Tales of the Three Worlds: Archaeology and Beyond – Asia, Italy, Africa. A Tribute to Sandro Salvatori*, ed. D. Usai, S. Tuzzato, and M. Vidale. Oxford: Archaeopress Archaeology, pp. 98–112.

Venco Ricciardi, R. 1980. "Archaeological Survey in the Upper Atrek Valley (Khorassan, Iran), Preliminary Report." *Mesopotamia* 15: 51–72.

Yusefi Zoshk, R. and S. Bāghizadeh 1391/2011. "Kārbord-e sistemhā-ye ettelā'āt-e joghrāfyāi (GIS) dar tahlilolgu-ye esteqrāri motāl'eh moredi mohavvatehā-ye dasht-e Dargaz as dore-ye Nowsangi tā pāyān-e 'Asr-e Āhan IV" [Application of geographic information systems (GIS) to the analysis of settlement patterns (case study: sites of Darreh Gaz plain from the Neolithic to the end of Iron Age IV)]. *Nāmeh-ye Bāstān Shenāsi* 2: 7–24 (with English summary).

CHAPTER TWENTY

THE RELATIONSHIP BETWEEN THE OXUS CIVILIZATION AND THE INDO-IRANIAN BORDERLANDS

Benjamin Mutin and C.C. Lamberg-Karlovsky

INTRODUCTION

Between ca. 3000 and 1500 BC, the vast area from the Mediterranean Sea to the Indus valley and the Persian Gulf to southern Central Asia saw the emergence, in some places, of the earliest major Bronze Age civilizations, with sites conforming to cities, major improvements in all types of crafts, greater cultural integration encompassing vaster expanses, and an overall complexification of the society ruled by elites. As said in Chapter 1, this period is also defined as the "Middle Asian Interaction Sphere" (Possehl 2007), a term that designates the trade and, more broadly, the multiple interactions that took place from around the mid-third to the early to mid-second millennia BC between various cultural complexes in this area (some we know were kingdoms thanks to the Mesopotamian texts) such as Meluhha (Indus Civilization), Marhashi (probably the Halil Rud Civilization), Magan (Oman), Dilmun (northern part of the Persian Gulf), Elam (southwestern Iran or more broadly Iran), Sumer in Mesopotamia, and the Oxus Civilization in Central Asia. The scope of this interaction was even larger, extending at least from the eastern Mediterranean to China. Although this chapter focuses on the relationship between the Oxus Civilization and the Indo-Iranian borderlands (i.e., southeastern Iran, southern Afghanistan, and Pakistan), it is important to keep in mind that this relationship was incorporated within – and cannot be understood without an understanding of – these much wider dynamics, which included movements of peoples, animals, goods, styles, technologies, and beliefs, and resulted from various forms of interactions including political alliances, migrations/colonizations, conquests/military expeditions, exchange/trade, and imitation/emulation. For instance, it would certainly be misleading to ignore that the art of the Oxus Civilization partly relates to that of Elam and Mesopotamia (see Chapters 1 and 7).

The topic of this chapter may be addressed through three much-debated aspects: (1) the origins of the Oxus Civilization; (2) the nature and interpretations of the Oxus-related material evidence outside of Central Asia; and (3) the purported impacts of the Oxus Civilization on the Indo-Iranian borderlands: the spread of violence, language, and religion. This chapter certainly is not the place to answer all these questions (see also Chapter 1); such an enterprise would take many lifetimes and much additional effort in the field. It is nonetheless a good opportunity to review the material evidence at hand and provide additional comments on these aspects in light of recent discoveries and analyses.

MATERIAL EVIDENCE

Southern parallels for the material culture recorded in the Oxus Civilization are distributed across Iran and Pakistan from Susa in Khuzistan (Amiet 1986) to the Indus valley, hundreds of kilometers distant from the main sites and core of this civilization. More distant stylistic similarities are even detected between the Oxus Civilization and Syrian and Anatolian assemblages (Sarianidi 1998; see also Francfort 2005b: 31; Amiet 2007c: 76). Sites with Oxus-related remains are recorded in northern Iran such as at Tepe Hissar (Schmidt 1937; Dyson and Howard 1989) and in numerous locations in the Indo-Iranian borderlands such as: Shahdad (Hakemi 1997), Khinaman (Curtis 1988; Maxwell-Hyslop 1988), Tepe Yahya (Hiebert and Lamberg-Karlovsky 1992; Lamberg-Karlovsky and Potts 2001), the Halil Rud basin (Madjidzadeh 2003; Akbarzadeh and Piran 2013), and Khurab in the Bampur valley (Stein 1937) in southeastern Iran; Nad-i Ali, Godar-i Shah, and in the Gardan Reg in southwestern Afghanistan (Besenval and Francfort 1994); and Mehi, Kulli (Stein [1931] 1982), Quetta, Sibri, Mehrgarh, Nausharo (Jarrige 1985; Jarrige and Hassan 1989; Jarrige 2016), Harappa, and Mohenjo-daro (During-Caspers 1994a, 1994b; Franke 2010; Vidale 2017: 56) in Pakistan (Figure 20.1). These sites are usually considered outliers of the main Oxus sphere, whereas other sites in Iran are considered true Oxus settlements, especially in the northeastern Khorasan province,

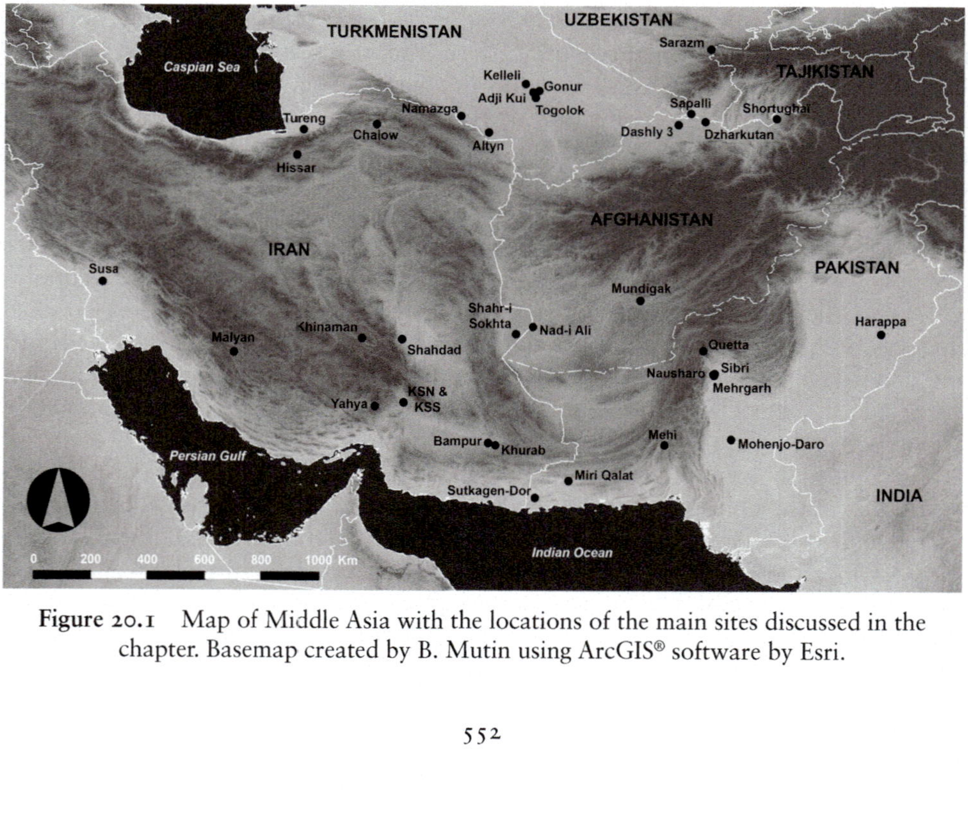

Figure 20.1 Map of Middle Asia with the locations of the main sites discussed in the chapter. Basemap created by B. Mutin using ArcGIS® software by Esri.

where the Oxus cemetery of Tepe Chalow and additional Oxus sites are located (Biscione and Vahdati 2011; Vahdati and Biscione 2013; and Chapter 19).

While the evidence for relationships between the Oxus Civilization and the Indo-Iranian borderlands may hardly be debated, its dating and interpretation have varied (see also Chapter 1). *Contra* V.I. Sarianidi's first dating of the mature phase of this civilization (so-called Gonur and Togolok periods) to between ca. 1500 and 1000 BC (Sarianidi 1998: 76–78), with new radiocarbon analyses and detailed comparative examinations of stratigraphic sequences and material culture in southern Central Asia most scholars agree on a dating between ca. 2200/2100 and 1750 BC or between 2300 and 1700 BC (see Lamberg-Karlovsky 2003: 13; Francfort 2007: 104; Francfort and Tremblay 2010: 104; Lamberg-Karlovsky 2012: 60), including, later on, Sarianidi (2004; see Lamberg-Karlovsky 2012: 72; see also Chapter 1 and the Appendix at the end of the volume). There are, however, disagreements in detail as to the chronological subdivision of the Oxus Civilization and whether it was contemporaneous with the Namazga (NMG) period V, its later part, and/or NMG period VI (compare Francfort 2005a: 258–260n13 with Hiebert 1994a: 165–178; 1994b: 374 tab. 1; P'jankova 1994: 363), while a most important issue remains that the stratigraphic sequences of most major Oxus sites excavated in Central Asia have never been documented (see, e.g., Lamberg-Karlovsky 2007). According to F.T. Hiebert, the cultural contacts between the Oxus and its southern neighbors occurred between 2000/1900 and1750 BC (Hiebert 1994a: 163–164, 177–178, 177 fig. 10.8), and H.-P. Francfort dates the climax of this civilization, including major artistic developments and long-distance relationships, to between ca. 2200 and 1750 BC, and the end or decrease of these relationships after ca. 1700 BC (Francfort 2005a: 260, 262). S. Salvatori places its beginning in the second half of the third millennium BC, ca. 2400/2300 BC (Salvatori 2000: 101–103, 136, 137 fig. 16; 2016: 450), and emphasizes a drastic change around 1900 BC, which manifested in the collapse of settlements and foundation of new ones with different architectural features (independent, smaller fortresses representing a fragmented power), the end of the use of copper-alloy compartmented seals, and a marked reduction in "international" trade (Salvatori 2016: 454–455). Most importantly, he criticizes the use of the labels "Oxus Civilization" and "Bactria-Margiana Archaeological Complex (BMAC)"; to him, these terms are most often imprecisely used and tend to mask significant historical developments that occurred between the Middle and Late Bronze Age, including the above-mentioned discontinuity around 1900 BC (Salvatori 2016: 455–456). This is important to keep in mind, although the present chapter is not designed to address this topic further. In any case, it is today established that the Oxus Civilization was at least partly contemporaneous with most major Bronze Age civilizations of Middle Asia and included within the above-mentioned Middle Asian Interaction Sphere, and not later as was first suggested by Sarianidi (1998: 78).

As for the interpretation, combined with Sarianidi's low dating of the Oxus Civilization, the Oxus-related material evidence on the southeastern Iranian Plateau – which he therefore thought was older – provided him with support for his central hypothesis on the origin of this civilization. He believed that this civilization emerged as the result of two main migrations of Indo-Aryan tribes beginning in the Syro-Anatolian regions: one crossing northern Iran and ending in Margiana and Bactria as well as up to northwestern China, and the other one crossing southern Iran and ending in Khorasan and Pakistani Baluchistan (Sarianidi 1998: 158, 163 fig. 75; 2009: 42–43; see also Hiebert and Lamberg-Karlovsky 1992: 3; Francfort 2005a: 265; Lamberg-Karlovsky 2012: 60). P. Amiet sees an impetus from what he defines as the "trans-élamite" culture in eastern Iran in the development of this civilization. He emphasizes the potential role of "trans-élamite" merchants and seminomadic artisans in this development and their preeminent position as an elite community in southern Central Asia (Amiet 1986: 138, 206–207, 212–214; 2007b: 51). J.-F. Jarrige stresses the presence of many objects typical of the Oxus Civilization south of the Hindukush before it emerged in Central Asia; he believes in particular in "a Western Harappan origin for some of the most characteristic pottery shapes of ... the Bronze Age culture of Bactria and Margiana" (C. Jarrige *et al.* 1995: 95). It is today most commonly admitted that the origins of the Oxus Civilization are mostly local and that the material evidence in the southern regions reflect a later diffusion from Central Asia (see Lamberg-Karlovsky 2012: 60–61). This does not completely discard the possibility that, in a context of intense intercultural interaction, cultures in eastern Iran and Pakistan affected aspects of the formation and evolution of the Oxus Civilization. Also, the question of the southern relationship and its impact on the formation of the Oxus Civilization is a question that, timewise, requires both looking at possible immediate antecedents in the southern regions as well as more distant contributions. Furthermore, it is also important to retell that, although it is unlikely that the Oxus Civilization had a direct origin in southern Iran and Pakistani Baluchistan (see pp. 567–570), the question certainly remains partly open as far as the numerous territories located further north, just south of Margiana and Bactria in Iranian Khorasan, Afghanistan, and northern Pakistan, are concerned, since these territories are still poorly known (Lamberg-Karlovsky 2003: 12; see also Chapter 19).

Lastly, a difficulty in interpreting the material evidence in southeastern Iran, southern Afghanistan, and Pakistan is that it is varied both quantitatively and qualitatively. The above-mentioned chronological issues do not help. This configuration still opens the door to a variety of interpretations as to its nature, from just stylistic influences to state expansionism, and as to the directions, southwards or northwards, and dating of the stimuli. In the end, it appears that various types of processes, both intercultural and transcultural, were at play in this relationship and took place at different times, which now require better definition and chronological resolution.

Shahdad

Over 700 km south of the Oxus cemetery at Tepe Chalow, located on the western edge of the Lut Desert, Shahdad undoubtedly is at present the site holding the most connections to the Oxus Civilization on the southeastern Iranian Plateau. This site, however, is not a southern replica of Gonur Depe; it has evident, strong connections to the Oxus Civilization and at the same time seems embedded within the main Middle Bronze Age tradition, or its late form, typical of southeastern Iran. The site consists of burial grounds, areas with domestic architecture, and a metal production center, extending over 100–150 ha (Salvatori and Vidale 1982). Salvatori and Tosi (1997: 126) put it this way:

> The artefacts' distribution suggested that at its climax, before 2000 BC, the Bronze Age settlement might have extended between 50 and 100 ha ... Cemeteries,

> orchards and suburban craft activity areas, spread at close range around the town might double this figure for the total occupational area to some 200 ha.

Parallels for Shahdad's material culture at Tepe Yahya periods IVB–IVA (see pp. 558–560) place its occupation between ca. 2300 and 1700 BC (Hiebert and Lamberg-Karlovsky 1992: 8), and Salvatori and Tosi (1997: 123) believe that "the graves excavated by Hakemi can be chronologically distributed over a time span of some seven hundred years, most likely between 2500 and 1800 BC." A. Hakemi, in the same publication, places the Bronze Age occupation at Shahdad between ca. 2700–2450 and 1500 BC and within the local chronological periods Tak. IV1–Tak. II1 (Hakemi 1997: 75–82). Of particular importance is an inscription recorded on a ceramic, written in Elamite Linear dating to ca. 2100 BC (Hakemi 1997: 183 obj. no. 0193) while a single ^{14}C date from Site D provided 2050+/–80 cal. BC (Hakemi 1997: 112–113). While Hakemi's proposed dating appears too long, S. Paklayan, who reanalyzed the material culture from Shahdad, concludes that the bulk of the assemblage can be dated to between ca. 2300/2200 and 1900/1800 (Paklayan 2004: 176).

Connections to the Oxus Civilization are numerous and various, particularly in the burials. The artifacts at Shahdad with parallels in this civilization include stone objects and containers, metal axes and seals, and ceramics (Hiebert and Lamberg-Karlovsky 1992: 8, pls. IId–III; Lamberg-Karlovsky and Hiebert 1992; Sarianidi 1998: 138–139, 139 fig. 71; Kohl 2007: 227–228), although it is important to note that many of them are also reported from additional distant areas across Iran and the Persian Gulf (see Chapter 22). Some are included in what Amiet names the "trans-élamite" culture, a culture centered on southeastern Iran, defined mostly by the assemblages from Tepe Yahya, Shahdad, and the Halil Rud basin, and related to Elam, but characterized by a distinctive style of stone and metal objects and iconography (Amiet 1986: 138, 170, 195–204, 206–207, 212–214; 1994: 1; 2007b: 51; 2007c: 71–72, 77–79). Putting aside his above-mentioned model on the origin of the Oxus Civilization, a stylistic influence from southeastern Iran and more broadly Elam on certain types of objects – not all – recorded in southern Central Asia may hardly be denied, while typical Central Asian objects are observed at Shahdad.

Similar types of stone objects and containers in chlorite/steatite, alabaster, calcite, and limestone are recorded both at Shahdad and at various sites in Central Asia including Gonur Depe, Togolok, and Adji Kui (e.g., columns: Hakemi 1997: 625 Fr.1–3; containers: Hakemi 1997: 56 figs. 32–33, 605–611, 694; perfume jars or vials: Hakemi 1997: 58 fig. 36, 618–620; quadrangular boxes: Hakemi 1997: 57 figs. 34–35, 611–615; see parallels in Central Asia in Hiebert 1994a: 149–150 figs. 9.10–9.11; Sarianidi 1998: 50–51 figs. 18–19, 20, 53; Rossi Osmida 2007: 167 fig. 17; Vidale 2017: 44–51, 58–62). As far as the chlorite/steatite material is concerned (see Chapter 9), two styles, defined as "trans-élamite," are distinguished: one, older, termed "Intercultural Style" or "*série ancienne*"[1] and a recent one termed "*série récente*" (Miroschedji 1973; Kohl 2001). At Shahdad, the "Intercultural Style" is virtually absent (in contrast to Tepe Yahya; see pp. 558–560), while this site is connected to the Oxus Civilization through the "*série récente*" style including containers with typical incised dot-in-circle motifs (see Francfort and Tremblay 2010: 99, 133–135). These are recorded from the Persian Gulf to southern Central Asia and are also sometimes labeled "Gulf style" (e.g., Thornton 2012: 605).

Regarding the metallic objects, D. Meier (2015), who recently reassessed almost all the metallic remains from Shahdad held by the National Museum of Iran in Tehran (504 artifacts including containers, axes, blades, adzes, points, rings, mirrors, and needles/pins; Meier 2015: 213), observes that most of these objects have greater similarities in the Oxus Civilization, particularly at Gonur Depe and Adji Kui, in the sense that they share the same types and variety (see Amiet 1986: 196–204; see also Sarianidi 1998: 58–60 figs. 23–25 for a selection of metal objects from Margiana). He also suggests a parallel "intended continuation of an arsenical bronze metallurgical tradition at a time when tin bronze metallurgy was already widespread and in common use in the neighboring Bronze Age cultures" (Meier 2015: 245, 246–252). The use of arsenical copper is demonstrated by compositional analysis of bronze objects from Shahdad, Khinaman, Susa, and Bactria (Salvatori 1995: 41; see also Lamberg-Karlovsky 1996: 210). One may emphasize here the iconic two shaft-hole axe heads from grave 44 in cemetery B and grave 36 in cemetery A at Shahdad (Hakemi 1997: 207 no. 0393, 195 no. 0302) with parallels at Khinaman (Curtis 1988: 102 fig. 2) and in Bactria and Margiana (Pittman 1984: 65–78; Amiet 1986: 196, 314–316 figs. 165–170; Hiebert 1994a: 162 fig. 9.26 no. 6; Salvatori and Tosi 1997: 129; Vidale 2017: 27–30).[2] M. Vidale and P. Steinkeller agree with Amiet (1986: 156, 197) who believes that these axes as well as hammers

> were badges of acquired authority that the kings of the polity known in Mesopotamia as Shimashki gave to their high-status officials … These axes and hammers … were evidently awarded by the rulers of Elam to dignitaries in recognition of particular merits.
>
> (Vidale 2017: 28)

This was in practice during both the Shimashki and Sukkalmah dynasties (Steinkeller 2014: 699), between the 21st and 15th centuries BC. Metal trumpets (Hakemi 1997: 635 figs. Gn. 1 and Gn. 2) also are objects from Shahdad shared with Oxus sites in Central Asia as well as Tepe Hissar (Lawergren 2003). The case of the copper dishes with embossed decoration found at Shahdad (Hakemi 1997: 645–648 Gs. 4–7) deserves mention. They are often assigned a Bactrian origin (see Vidale 2017: 31–32), although the only excavated examples come from Shahdad and Tepe Hissar (Pittman 1984: 24, 27), while an additional one is reported from illegal excavation supposedly in the Halil Rud basin (see Vidale 2017: 31; see also below p. 562 on the origin of this material). The metallic compartmented seals from Shahdad (Hakemi 1997: 659–660) are overall comparable to specimens from Central Asia and share certain motifs with the latter (e.g., Salvatori and Tosi 1997: 128), although more numerous and the most elaborated seals are in the latter region (e.g., Pittman 1984: 53–57; Amiet 1986: 197; Sarianidi 1998: 59 fig. 24, 66 fig. 29). The practice of stamping ceramic vessels using compartmented seals, observed at Shahdad (Hakemi 1997: 672–678, 679–690) and Tepe Yahya period IVA (Pittman 2001: 266–267 figs. 10.60–10.63), is recorded in Central Asia (Pittman 2001: 239; see also Hiebert 1994a: 60 fig. 4.32, 61).

In addition to the above-mentioned objects, the Oxus material assemblage shares significant iconographic elements with Shahdad and, more broadly, "trans-élamite" southeastern Iran, Iran, and its Near Eastern neighbors (see Amiet 1986: 195–204;

2007c: 76; Francfort 2007). Several examples from Shahdad deserve mention. The well-known Oxus goddess (see also Chapter 7), whose stone composite figurines are recorded at Gonur Depe, Adji Kui, and Togolok as well as Nishapur in northeastern Iran (see Vidale 2017: 98–166), "appears in the glyptic art of south eastern Iran from Shahdad [cylinder-seals: Hakemi 1997: 661 figs. Ib.2 and Ib.4]" (Pittman 1984: 51) and is recorded further west.[3] A stylistic relationship is also noted between Shahdad and Central Asia as far as toreutics is concerned (Francfort 2005b: 28). Amiet points out similarities in the way human figures are represented on the famous Shahdad's copper "standard" (Hakemi 1997: 649) and on a silver cup from Bactria (Amiet 1986: 203, 295 fig. 126, 328–329 fig. 202).

Besides the objects, significant parallels are observed between typical fireplaces excavated at Shahdad, Gonur Depe, and sites in the Adji Kui oasis (Adji Kui 9) as well as Susa in levels dating to the Shimashki and Sukkalmah dynasties (Meier 2015: 155–167). Certainly, a detailed comparative analysis of the other architectural aspects at Shahdad would be productive. Additional architectural comparisons that may be noted here are in the "distinctive stepped niches and sagging lintels reminiscent of the designs on the carved Intercultural Style chlorite vessels" (Kohl 2007: 219, 220 fig. 5.23), reported from Tepe Hissar and Altyn Tepe. The sagging lintels are at Shahdad observed on clay house models (Hakemi 1997: 63 fig. 39, 692 figs. Ja.1–2) and through the shape and incised motifs of chlorite/steatite boxes and vials (Hakemi 1997: 469 obj. no. 3523, 515 obj. no. 4077, 614–615 figs. Fk.10–11, 707 fig. Wa.4) found in burials (see also Aruz and Wallenfels 2003: 332 no. 228: a copper pin reportedly from eastern Iran or Central Asia, of "trans-élamite" style, representing a couple inside a house with a sagging lintel). The stepped niche, or crenellated pattern, a motif typical of Central Asia (e.g., Aruz and Wallenfels 2003: 354–355 no. 248, 355 fig. 89, 358–363 nos. 251–254) is observed at Shahdad on painted motifs on a clay house model, mat, and seal (Hakemi 1997: 63–64 figs. 40–41, 320 obj. no. 1791).

Lastly, it should be mentioned that the variety of burial types at Shahdad, from simple pits to more sophisticated structures with architecture and inner partition (Hakemi 1997: 47, 74), matches the variety and descriptions of some of the types recorded in Central Asia. For instance, although simple pits are the most common type of burials at Shahdad, tombs with mudbrick or chineh walls are also observed. These could be of the shaft and lateral chamber or mudbrick chamber types reported, for instance, from Gonur Depe (Sarianidi 1998: 67–76; see a recapitulation in Vidale 2017: 40–42; see also Chapter 10) and Shahr-i Sokhta in Seistan in older periods (Piperno and Salvatori 2007: 17). The mudbrick chambers from Gonur Depe also are paralleled in Elam such as at Susa (Francfort 2005a: 276). To this inventory should be added the presence of cenotaphs, reported from Dzharkutan, Gonur Depe, Togolok, Shahdad, and Khurab, as well as Sibri, Mehrgarh, and Quetta in Pakistan (Hiebert and Lamberg-Karlovsky 1992: 3; Hiebert 1994a: 128, 129 fig. 7.25; see also below p. 565).

Again, a part of these aspects that connect Shahdad to the Oxus Civilization are also recorded at other places in Middle Asia and the Persian Gulf, and this site is not solely connected to Central Asia. Alongside the pottery vessels that link it to the Oxus Civilization (e.g., Hiebert and Lamberg-Karlovsky 1992: pl. IIIc–d; Hakemi 1997: 408 obj. no. 2996), a solid connection to the distinct, southeastern Iranian tradition is manifested in the numerous painted ceramics found at this site (Hakemi 1997: 53–55 figs. 26–31, 571–573, 584–603). Although they do not all relate directly to the Bronze Age painted pottery tradition from the Halil Rud basin such as at Konar Sandal South and Tepe Yahya IVB (Madjidzadeh 2003; Madjidzadeh and Pittman 2008; Lamberg-Karlovsky and Potts 2001), which are partly older, they are closer to it as well as to ceramics recorded in the Persian Gulf during the Umm an-Nar period (see Méry 2000) than to any other traditions present at that time and before in Central Asia and the Middle East. Considering their chronological position and stylistic differences, some specimens may be interpreted as a late form of this tradition. These painted ceramics are recorded in Shahdad Tak. III2–III1 periods (ca. 2450–2200 and 2200–1900 BC), while plain, buff ware with sometimes "engraved geometric designs" characterizes the following Tak. II2 period (ca. 1900–1700 BC).[4] The latter are reminiscent of Tepe Yahya period IVA assemblage (see p. 559) as well as materials from Central Asia. The above-noted Linear Elamite inscription observed on one of these plain vessels provides an additional connection between Shahdad, and more broadly Elam, and Central Asia where similar inscriptions are recorded (see Francfort 2005a: 284). Conversely, the clay human statues recovered from burials at Shahdad are an additional example of a practice rarely observed in Central Asia (Hakemi 1997: 298 no. 1453, 299–300 no. 1479, 333 no. 1988, 345 no. 2145, 347 no. 2162, 353 no. 2229, 371 no. 2520, 375 no. 2553, 379 no. 2599, 382 no. 2629, 392 no. 2750, 428 no. 3160, 444 no. 3322, 446 no. 3346, 451 no. 3392, 457 no. 3432, 469 no. 3520, 541 no. 4300, 549 no. 4402).[5] Hakemi writes that these statues are recorded in both Tak. III2 and Tak. III1 periods, although most are observed in the former (Hakemi 1997: 79).

Tepe Yahya

Objects interpreted as Oxus-related material from Tepe Yahya in the Soghun plain are usually reported as being quantitatively very limited and coming from phases IVB1 and IVA4 contexts dating to the late third and early second millennia BC (Hiebert and Lamberg-Karlovsky 1992: 6). These include a complete truncated ceramic with a typical outturned rim recorded in phase IVA4 (Hiebert and Lamberg-Karlovsky 1992: pl. IIa; Lamberg-Karlovsky and Potts 2001: 149 fig. 5.4d), additional rim fragments of seemingly same form from phases IVB4–1 (Lamberg-Karlovsky and Potts 2001: 145, 146, 149 fig. 5.3) and phases IVC1–IVB5 (Lamberg-Karlovsky and Potts 2001: 63 fig. 2.5, 114), and a copper seal (Hiebert and Lamberg-Karlovsky 1992: pl. IIb). The seal, however, was in fact found in a Proto-Elamite jar inside the IVC period Proto-Elamite building dating to around 3000 BC (Lamberg-Karlovsky 1984; Lamberg-Karlovsky and Potts 2001: 13, 47 fig. 1.50; Mutin 2013b: 159, 429 fig. 4.25: 5). Instead of being connected to the Oxus specimens, it more consistently parallels Late Chalcolithic artifacts (discussed later in the chapter) found in Kech-Makran (Mutin 2020). Very minimal relationship with Central Asia is observed as far as period IVB chlorite/steatite assemblage is concerned, although Tepe Yahya (phases IVB4–1) is known for having a "chlorite working area … still the only manufacturing site for carved chlorite artifacts so far identified and scientifically excavated" (Vidale 2015: 17; see also Kohl 2001). The style of carved chlorite/steatite artifacts produced in this period is named "Intercultural Style" for the reason that it is recorded in assemblages from various archaeological cultures across Iran, the Persian Gulf, and

Mesopotamia (Kohl 2001: 215),[6] while just one lid fragment of "Intercultural Style," collected on the surface of Gonur Depe, bearing what seems to be an incised "hut-pot" motif, attests to its presence in Central Asia (Hiebert and Lamberg-Karlovsky 1992: 6, 8 fig. 3). The most rational explanation for this situation is that this style is older than most chlorite/steatite objects in the Oxus Civilization as well as at Shahdad (Kohl 2001: 210, 212), which mostly relate to the later "*série récente*" style. A terminus ante quem for the "Intercultural Style" in the early Akkadian period, or ca. 2300/2200 BC, is now also largely agreed upon (see Francfort and Tremblay 2010: 99; Vidale 2015: 16–20).[7] Most of the cylinder seals from Tepe Yahya period IVB (and one from period IVA) have parallels at Shahdad and are characteristic of the distinctive style observed at these two sites and named "trans-élamite" by Amiet. Like at Shahdad, these seals include representations of the above-mentioned goddess (see Pittman 2001: 237–238). Only two stamp seals from period IVB are connected to Shahdad as well as to Central Asia (Pittman 2001: 239, 266 fig. 10.58[8]).

Compared to Shahdad, Tepe Yahya thus appears connected to the Oxus Civilization through a limited series of objects. This remark, however, applies to period IVB, and, while it is now evident that this period relates to the older "Intercultural Style" period characteristic of the Halil Rud Civilization, the question of the cultural association of period IVA, the subsequent settlement at Tepe Yahya, is still not entirely resolved. The dating of period IVA has varied[9] and still is not totally fixed, although current data tend to show that this period lies somewhere in the first half of the second millennium BC. The single reliable ^{14}C date (Beta 6473) from period IVA is between the 18th and 15th centuries BC,[10] while two from phase IVB1, the architectural phase just prior to period IVA, are between the 22nd and the 18th/17th centuries BC (an additional one from phase IVB1 is between the 19th and the 13th centuries BC).[11] Period IVA ceramic assemblage mostly consists of plain vessels, which is in stark contrast with that of period IVB, characterized by fine, painted material (see Lamberg-Karlovsky and Potts 2001). Many of them have incised marks, located on or near the bottom on the exterior surface. These ceramics are reminiscent of the above-mentioned plain vessels from Shahdad as well as, more remotely, the plain vessels from the Oxus Civilization. In any case, they best compare to those, while no antecedent is observed on the south-eastern Iranian Plateau, at a minimum not in the previous period as observed at Tepe Yahya IVB and Konar Sandal South; Tepe Yahya IVA ceramics are plain, likewise the Oxus and many Shahdad vessels, and the stamped and incised marks they bear are also observed at this site and in Central Asia, although they are not common in the latter (Hiebert 1994a: 59–61; Hakemi 1997: 665–688; Pittman 2001: 239).[12] In Shahdad's chronology, Tepe Yahya period IVA ceramics seem to relate more to period Tak. II (ca. 1900–1500 BC), when "the red ware is replaced by a kind of coarse buff ware and the painted ware disappeared" (Hakemi 1997: 118). In Margiana, the incised and stamped marks are recorded only between ca. 2000 and 1500 BC (Margiana periods 2 and 3; Hiebert 1994a: 59–61), while at Tepe Yahya the incised marks and the practice of stamping ceramics are limited to period IVA.

Chlorite/steatite production considerably decreases in period IVA (Kohl 2001: 211), but it is in this period that the best, albeit limited, parallels to Shahdad and, ultimately, Central Asia are observed. These parallels include objects in the "*série récente*" style, characterized at Tepe Yahya by bowls with incised dot-in-circle motifs or zigzag bands and tall goblets with incised concentric triangles and chevrons (Lamberg-Karlovsky 1988: 52–53, 78 fig. 3h–l, 80 fig. 4z–dd). Ph. Kohl mentions undecorated bell-shaped bowls (found in both periods IVB–IVA) and flat-based cups as parallels, but he also notes that the above-mentioned boxes, hut house models, and vials decorated with dot-in-circle motifs from Shahdad are either rare or absent at Tepe Yahya (Kohl 2001: 212, 221; see also Hakemi 1997: 605–607, 609–611; Sarianidi 1998: 49–51 figs. 17–19; see above the parallels noted in Central Asia). There are also "Intercultural Style" objects in period IVA, although there is a possibility that these are intrusive or heirlooms from period IVB (Lamberg-Karlovsky 1988: 52–53; see also Kohl 2001: 214). The heirloom hypothesis may also explain the presence of a few "Intercultural Style" objects at Shahdad (Kohl 2001: 212–213). The possibility remains that certain forms and decorations of period IVB continued to be produced in the subsequent period (see Kohl 2001: 221–222, 226–227), but the heirloom and intrusion hypotheses and the existence of a chronological discrepancy between Shahdad/Tepe Yahya IVA and Tepe Yahya IVB/"Intercultural Style" seem to us more rational (also given the fact that the problematic examples are relatively rare). Following this, one observes that, after a time when chlorite products and raw material from Kerman were exchanged through a trade system centered in the Persian Gulf, a shift occurred whereby the former now became more connected than before to Central Asia at the time of the Oxus Civilization (see Kohl 2001: 228).

Certainly, in spite of an increasing general consensus on Tepe Yahya's chronology, the precise dating of periods IVB–IVA remains a crucial question (see Lamberg-Karlovsky 2001) that would greatly help to characterize these cultural dynamics and the reasons for the apparent change in material culture at both this site and Shahdad around 2000 BC (e.g., plain vs. painted ceramics and possibly "*série récente*" vs. "*série ancienne*"/"Intercultural Style"). An interruption is visible in the tradition seen before, and the reasons for this could be: a chronological gap between the traditions represented at Tepe Yahya periods IVB and IVA; a military conquest or successive military campaigns – those led by Sargon, Rimush, and Naram-Sin, granted that Tepe Yahya IVB was part of the kingdom named Marhashi (Lamberg-Karlovsky 2001: 280); a climate change (see Francfort 2005a: 303–304; Fouache *et al.* 2013); a population replacement; new stylistic influences; and/or technological changes. In any case, in these changing times and in the new settlement at Tepe Yahya IVA, connections with Central Asia and Shahdad are evident. Like at Shahdad, Tepe Yahya IVA dwellers may be from the Oxus Civilization, from a remote Central Asian tradition, or, at least, they seem to have incorporated elements of this civilization within the context of southeastern Iran while developing new relationships with culturally distinct neighbors in Fars. The current reassessment of Tepe Yahya IVA material assemblage and a better dating of this period would certainly help narrow down the range of possible reconstructions.

Halil Rud basin

The Halil Rud basin, ca. 90 km east of Tepe Yahya, was densely populated during the third millennium BC. About 170 sites dating to this period are recorded in this region, including 70 sites with evidence of metallurgical activities and the major sites of Konar

Sandal South and Konar Sandal North (Madjidzadeh and Pittman 2008). Because of these discoveries, in addition to the settlements at Tepe Yahya and Shahdad at some point, the Halil Rud basin is considered the center of an important Bronze Age polity, named the Halil Rud (or Jiroft) Civilization and characterized by distinctive chlorite/steatite and ceramic assemblages, relating to those at Tepe Yahya IVB ("Intercultural Style"/"*série ancienne*" and painted ceramics) (see Pittman 2013a). Most scholars believe that this civilization is Marhashi, the above-mentioned kingdom listed in Mesopotamian texts (Steinkeller 1982, 2014), although Francfort and X. Tremblay (2010) argue that Marhashi designates the Oxus Civilization (on this name theme see also Chapter 2). Shahdad has relationships to the Halil Rud basin, especially through its painted ceramics, although, as noted above, this site appears mostly more recent than Konar Sandal South, based on current available data in Madjidzadeh and Pittman (2008). We also noted above that Tepe Yahya IVA and the Oxus Civilization too are mostly later than the "Intercultural Style" as well as the deposits from Konar Sandal South characterized by typical painted ceramics. We shall, however, await complete publication of the excavations at this site and perhaps new excavation before drawing too definitive conclusions. The "citadel" and surrounding wall uncovered at this site are dated to ca. 2400–2300 BC, but C.P. Thornton (2012: 604) notes that the "latest surface material around the mound corresponds with the earliest material so far identified from [Konar Sandal North]," dating to the early second millennium BC (see pp. 561–562). Additionally, parallels are noted at Shahdad as well as Central Asia, Shahr-i Sokhta, and Bampur to the southeast for a dozen metal compartmented seals from Konar Sandal South (Madjidzadeh and Pittman 2008: 95–97, 96 fig. 28),[13] although one may argue that these seals also are reminiscent of older specimens recorded in Kech-Makran, southwestern Pakistan, and at Bampur (see pp. 568–569). Cylinder-seal impressions from this site are also compared to Shahdad's glyptic. An illustrated example represents a scene with two seated figures facing each other (Madjidzadeh and Pittman 2008: 97, 99, 97 fig. 29b, right part of the impression), which is not very different from scenes present on an open-worked pin in the Louvre Museum said to be from eastern Iran or Central Asia, a lapis lazuli disk also supposedly from south-eastern Iran, and a compartmented seal from Bactria (Amiet 2007a: 173, 222 fig. 89; Aruz and Wallenfels 2003: 332 no. 228; see also Shahdad's "standard" in Hakemi 1997: 649). Additional seal impressions "can be closely paralleled with seals found at Tepe Yahya and Shahdad, showing the pantheon of southeastern Iran" (Madjidzadeh and Pittman 2008: 99, fig. 31a, b, c, d). Most of the (few) cylinder seals found at Gonur Depe are said to come from the Halil Rud basin (Pittman 2014). Lastly, like at Shahdad and in Central Asia, a Linear Elamite inscription (on a baked brick) was found at Konar Sandal South (Madjidzadeh and Pittman 2008: 81). The stylistic connections observed in the glyptic may indicate the continuation of an older form of art at Shahdad and in Central Asia, as Amiet argues in general for the "trans-élamite" art that he defines as "archaic" (Amiet 1986: 197, 199). The Linear Elamite inscription, in contrast, represents a solid chronological marker.

Konar Sandal North, the second major site excavated recently in the Halil Rud basin, is thought to be more recent. Provisionally and based on the very few ceramics published so far, Francfort estimates that this site is not older than ca. 1500 BC (Francfort and Tremblay 2010: 99). Certain types of pottery, however, are considered reminiscent of the Oxus assemblage (Madjidzadeh and Pittman 2008: 94–95, 95 fig. 27). Also, compartmented seal impressions with close parallels at Shahdad are recorded on jars (Madjidzadeh and Pittman 2008: 100 fig. 33). Additionally, M. Vidale reports the presence of chlorite/steatite fragments of "*série récente*" style, the style recorded at Shahdad and in Central Asia, on the surface of this site (Vidale 2015: 19). Thornton confirms the presence of such material in the earliest levels at Konar Sandal North, including chlorite/steatite vessels bearing the typical dot-in-circle motif. Interestingly, an imposing stepped mudbrick platform (300 x 300 x? 6.5 m) was built on top of these levels. The ceramic assemblage associated with it has no ready parallels, but the bricks used in the construction compare to those employed in the last phase of the "citadel" at Konar Sandal South, and, overall, both bricks and ceramics point to the early second millennium BC, with the possibility that the two structures were used in the same period (Thornton 2012: 604–605; see also Madjidzadeh and Pittman 2008: 88–89). If the date of around 2000 BC for this platform is confirmed, then it could be linked to the similar Bronze Age monumental constructions observed in the broader region at Altyn Tepe, Ulug Depe, and Tureng Tepe as well as Nad-i-Ali (see Francfort 2005a: 281; see also below p. 563).

The Halil Rud basin, more specifically the Jiroft region, is also considered the point of origin of a large amount of Bronze Age ceramic, stone, and metal artifacts seized in Iran around 2001. The origin of these objects is, however, still debated (see Muscarella 2005). Amiet, for instance, suggests that some of them are from Shahdad (Amiet 2007c: 71). While certain doubts remain as to the provenience of these antiquities, Vidale is quite certain of their origin being in the Jiroft region including Mahtoutabad, a looted Bronze Age cemetery located near Konar Sandal South (Vidale 2015: 16–17). The material collected includes many "Intercultural Style" chlorite/steatite objects, which are older than the Oxus Civilization, although a "*série récente*" vial, stylistically connected to Shahdad and Central Asia, is observed (Francfort and Tremblay 2010: 98). Besides this material, the series of stone vessels, containers, and miniature columns seized (Akbarzadeh and Piran 2013: 50–59) have clear parallels at Shahdad (Hakemi 1997: 608–611) and in Central Asia (e.g., from Togolok: Hiebert 1994a: 156 fig. 9.21, 156 fig. 9.21; Gonur Depe: Rossi Osmida 2007: 167 fig. 16, 171 figs. 26–27; Vidale 2017: 44 fig. 41; Ulug Depe: Amiet 1986: 308 fig. 150; "Bactriane": Amiet 1986: 311 figs. 157–158; additional ones, from private collections, said to be from Central Asia, are illustrated in Vidale 2017: 48–50 figs. 44–46, 59–61 figs. 51–52, although caution should be exercised as to the origin and authenticity of some of this material).

Lastly, accepting the view that the Halil Rud basin was the core of Marhashi, Vidale interprets a scene depicted on a silver goblet from a private collection, purportedly from Central Asia, as a representation of

> a ruler of Marhashi wounded and handcuffed in a ceremonial Bactrian procession, demonstrating that [the contacts between the Oxus Civilization and southeastern Iran] were far from peaceful and that military expeditions, at least in form of raids, could cross the Iranian Plateau.
>
> (Vidale 2017: 44, 66–74; again, caution is required as to the origin and authenticity of this material)

Seistan

The Nad-i Ali "Surkh Dagh" mound in the lower Helmand valley, Afghan Seistan, is known for its monumental brick platform, likely measuring ca. 50 x 50 x 35–40 m.[14] This site was originally dated by R. Ghirshman and G.F. Dales to between the Median-Achaemenid and Parthian/Kushan to perhaps Sasanian periods until R. Besenval and Francfort (1994) reexamined the archives of these scholars and dated the platform to the Bronze Age. They observed that a jar found on top of the platform was typical Oxus material, while architectural features of this construction including brick sizes are consistent with those of other similar Bronze Age monuments, particularly at Tepe Hissar and Tureng Tepe. Additional ceramic fragments collected by W.A. Fairservis confirm the presence of a Bronze Age settlement at Nad-i Ali (Besenval and Francfort 1994: 10). Beyond this site, a series of artifacts – ceramics as well as stone vessels, an amulet, disks, "handbags," and columns – collected by Dales in the Gardan Reg and at Godar-i Shah, as well as by N. Hammon in the former group of sites, connect to the Oxus and Indus civilizations while also showing relationships to southeastern Iranian Bronze Age cultural assemblages. In the Gardan Reg, these objects were associated with substantial remains of copper smelting and pottery making (Besenval and Francfort 1994: 10–11). The platform and other finds in Afghan Seistan substantiate with the evidence recorded in Kerman and Pakistan the existence of strong relationships between the Oxus Civilization and southern territories located hundreds of kilometers distant. This data might even reflect a stronger connection to the Central Asian civilization than in Kerman.

Across the border, in Iranian Seistan, Shahr-i Sokhta is famous for its expansive settlement dating to the third millennium BC. This site was occupied mostly before the Oxus Civilization emerged save perhaps for its last period, period IV, dating to between the late third and early second millennia BC (Salvatori and Tosi 2005). The chronology of this site is, however, still the topic of vivid discussion (see Cortesi *et al.* 2008; Jarrige, Didier, and Quivron 2011), and while Salvatori and Tosi (2005: 290) mention the presence of Oxus-related vessels in period IV, R. Biscione (1979: 294–295) emphasizes the lack of connection with Central Asia at that time. Hopefully, current fieldwork at this site, which has exposed new impressive constructions with massive walls with pilasters dating to period IV, will continue providing new information as to this period. So far, there is no mention in these recent excavations of Oxus-related material (Sajjadi and Moradi 2015). However, we learn from Biscione and A.A. Vahdati (see Chapter 19) that Oxus-related sites are recorded in Iranian Seistan.

Khurab

Khurab is located in the Bampur valley in the Seistan and Baluchistan province, approximately 400 km distant from Tepe Yahya and Shahr-i Sokhta. This site, excavated by M.A. Stein (1937), yielded burials of different dates from several trenches including some with typical third-millennium BC fine, painted ceramic vessels of types recorded from Kerman to Kech-Makran in southwestern Pakistan (see Wright 1989; Didier 2013), as well as some with Oxus-related material. In the latter group, Hiebert and Lamberg-Karlovsky (1992: 6) emphasize three tombs from trenches L, E, and D. They believe these are tombs of individuals from Central Asia because their ceramics have parallels in this area and none in the context of the Indo-Iranian borderlands. They observe that burial Li in trench L contains ceramics relating to the Oxus Civilization with more specific parallels at Dashly Tepe 3 (southern Bactria) dating to 1900–1700 BC (Stein 1937: 121, fig. 42, pl. VI: L.i.276, pl. XV: L.i.278–279; Hiebert and Lamberg-Karlovsky 1992: 6, 7 fig. 2). In this burial, however, Stein also reported one painted bowl, decorated with "wing" or "palm" motifs, a decoration typical of assemblages dating to the third millennium BC in southwestern Pakistan and Iranian Seistan and Baluchistan province (Stein 1937: pl. XV: L.i.291).[15] This vessel type is dated to the first half of the third millennium BC further east in Kech-Makran and even limited to the period between 2800 and 2600 BC (Kech-Makran period IIIb) according to A. Didier (2013). This certainly is in contradiction with the above-mentioned suggested dating to the early second millennium BC, unless one considers the unlikely possibility that this fine, brittle ceramic became an heirloom and did not break for centuries. This possibility would not however appear entirely unlikely if one were to envision that this ceramic remained untouched within one of the older graves present at Khurab and retrieved from that grave to be reused in burial Li. From trench E is a burial with 32 ceramics, two metal bowls, an alabaster jar, an agate bead with gold ferrules at both ends, and a shaft-hole axe decorated with a seated camel (Stein 1937: 121, fig. 41, pl. VI: E.i.252, pl. XVI: E.i.253–254, 256, pl. XVIII: E.i.251, 258, pl. XXXIV: 1, E.i.252, pl. XXXII: 15, E.i.254). The axe is composed of a copper-arsenic alloy with impurities consisting mostly of antimony and lead (Lamberg-Karlovsky 1969: 164–165). The ceramics are described as mostly unpainted by Stein (1937: 121) and were paralleled later with Oxus vessels (Hiebert and Lamberg-Karlovsky 1992: 6). Lastly, a burial from trench D also yielded unpainted ceramics, comparable to vessels from Central Asia (Hiebert and Lamberg-Karlovsky 1992: 6), particularly two conical cups (Stein 1937: 120, fig. 39, pl. XV: D.245–246, pl. XXXIV: 2, D.245, 4, D.246). However, again, painted ceramics are also recorded in this burial (Stein 1937: pl. XV: D.244, 249–250, pl. XXXIII: 6, D.250, 8, D.244). While two of them (D.249–250) tend to more specifically connect to the painted assemblage from Shahdad, one jar (D.244) belongs to a general type found from Kerman to Kech-Makran and the Oman Peninsula and usually dated to the third millennium BC (Méry 2000; Didier 2013). In Kech-Makran, such vessel types and canisters with similar painted decorations are dated to between ca. 2800 and 2300 BC (Jarrige, Didier, and Quivron 2011: 29; Didier 2013); that is, before the Oxus Civilization. A copper or bronze bowl as well as preserved bread rolls and loaves in a ceramic were also found in this burial (Stein 1937: 120–121, pl. XVIII: D.247). This practice of using vessels with bread offerings at Khurab may be compared to an example from a burial at Sapalli Depe (Bactria), where a vessel contained grain or a porridge (Hiebert and Lamberg-Karlovsky 1992: 6; see also Askarov 1973: 46–47). From Khurab, Kohl (2001: 221) mentions a tall chlorite/steatite goblet with an incised decoration of "*série récente*" style (Stein 1937: pl. VI: F.i.263[16]); it parallels above-mentioned examples from Shahdad and Tepe Yahya IVA.

Kachi plain

In 2012, at the South Asian Archaeology and Art Conference in Paris, J.-F. Jarrige (2016) addressed in a keynote lecture the question of the Oxus presence in Pakistan, with a focus on the discoveries relating to this civilization he and his team made in the Kachi plain. He had coined the term "Mehrgarh VIII/Sibri/Dauda Damb cultural complex" after the sites with such finds he had investigated in this plain (see also Chapter 21). The site of Mehrgarh, known for having the longest sequence of pre- and protohistoric archaeological deposits in Pakistan is, in its period VIII, characterized by burials and cenotaphs with a material culture very similar to that typical of the Oxus Civilization. Although these graves were plundered, Jarrige points out a series of Oxus-related objects they were able to recover, including complete ceramics, a stone rod, a gold and turquoise pendant, fine copper/bronze objects including one that resembles the so-called typical *harpés* from Bactria, and one faience vessel (Jarrige 2016: 22, 31–32 figs. 9–11; these finds relating to the Oxus Civilization at Mehrgarh as well as those from Quetta, Dauda Damb, Nausharo, and Sibri discussed below are also detailed and illustrated in Jarrige 1985, 1991; C. Jarrige *et al.* 1995: 41–50, 88–95, see the reports on the fourth to eighth seasons; see also Santoni 1984, 1988). The team located additional graves with similar material culture in the same area at Dauda Damb, just as this site was being bulldozed for road construction. The Oxus-related objects from Dauda Damb include many ceramics similar to those from Mehrgarh VIII, compartmented metal seals, and miniature faience rams (Jarrige 2016: 20, 22, 29 figs. 4–5). A few kilometers south of Mehrgarh, Jarrige and his team also excavated the site of Nausharo. There, they identified, together in the same archaeological deposits assigned to Nausharo period IV, artifacts relating to the third phase of the Indus Civilization dating to between 2100 and 1900 BC, material relating to the Kulli cultural complex, whose core is located within southern Pakistan, as well as a few Oxus-related ceramics (Jarrige 1994; 2016: 20). Near Nausharo lies Sibri, which was briefly excavated also by Jarrige and his team. In contrast to the above-mentioned Oxus-related elements recorded at Nausharo and in the graves from Mehrgarh and Dauda Damb, Sibri yielded mudbrick architectural structures over two archaeological phases. These vestiges include rooms with niches and buttresses, which Jarrige links to the architecture known in Central Asia around 2000 BC, as well as ovens and pottery kilns with other installations and remains relating to pottery making. The numerous finds include cylinder seals, amulets, compartmented seals, and human figurines. Some are clearly connected to the Oxus Civilization, such as two cylinder seals, although a few also relate to the Indus Civilization, including an amulet with two Indus signs (Santoni 1984, 1988; Jarrige 1991; Jarrige 2016: 20, 22–23, 27–28 figs. 1–3). Jarrige points out the erosion that affected the graves from Mehrgarh VIII and Dauda Damb as well as the Sibri settlement, an erosion that is, to him, probably responsible for the lack of permanent settlements with such material culture in the Kachi plain region. Sibri is, however, not alone, as "traces of [such] settlements [were identified] near the graves of Dauda Damb, at Damboli, north of Sibri, and at Shah Ghair" (Jarrige and Quivron 2008: 69, 71; Jarrige 2016: 21–22, quote at 21).

Quetta

An additional grave relating to the Oxus Civilization was found around 110 km northwest of Mehrgarh, precisely where the Serena Hotel today stands in Quetta City, as a swimming pool was being dug in this hotel. This grave yielded abundant jewelry and other types of artifacts in gold and various types of semiprecious stones, copper objects including an incense burner ("brazier"),[17] ivory artifacts, and alabaster vessels, as well as stone-grooved disks, shafts/rods, and miniature grooved columns. A steatite composite figurine that seems to represent the above-mentioned Oxus goddess observed at Gonur Depe, Adji Kui, and Togolok as well as Nishapur was also recorded in this grave. It is important to recall that this grave had already been plundered when these objects were rescued and that probably other graves of the same type, "as rich as some of the graves of the 'Royal Necropolis' at Gonur," remain to be excavated in this area (Jarrige 2016: 21–22, 30–31 figs. 7–8, quote at 22).

Mohenjo-daro and Harappa

In addition to the above-mentioned Oxus-related objects found within a material assemblage mostly relating to the Indus Civilization at Nausharo, other Oxus-related materials are recorded at the two iconic Indus sites of Mohenjo-daro and Harappa (see also Chapter 21). They consist of seals and sealing impressions (During-Caspers 1994a, 1994b; Franke 2010; Vidale 2017: 51) as well as the well-known "priest-king" statue at the former, one that had long symbolized the Indus Civilization and that is nowadays recognized by most scholars as an item of Central Asian origin (see Vidale 2017: 56). The relationship between the Oxus and the Indus civilizations is not limited to the introduction of these objects into the Indian subcontinent, since typical Indus-related objects are also recorded in Central Asia. These include:

> steatite stamp seals with animal images ... transport and storage jars ... precious beads in carnelian, jasper and other stones, small stone composite animal sculptures and ivory gaming pieces presumably imported from the Indus Valley [at Gonur Depe as well as seals at Altyn Depe].
>
> (Vidale 2017: 51; see also Chapters 1 and 21)

Kulli, Mehi, and Nindowari

The Kulli cultural complex, set within the geographic expanse of the Indus Civilization in southern Pakistan, is best known through three sites: Kulli, Mehi, and Nindowari. We noted above the discovery of materials relating to this cultural complex together in the same archaeological deposits with Indus-related and Oxus-related objects at Nausharo. Jarrige has also pointed out "the occurrence in the final stage of the Kulli culture at Nindowari of pottery related to the same cultural assemblage as several graves from Mehi with grave goods similar to those of the [above-mentioned Mehrgarh] VIII graveyard" (Jarrige 2016: 24; see also Jarrige, Jarrige *et al.* 2011). The graves from Mehi had been excavated by Stein and indeed contain materials characteristic of the Oxus Civilization, including typical plain ceramics (Stein [1931] 1982: 154–163, pls. XXVII–XXXII; plain ceramics, e.g., pl. XXX: Mehi.III.6.2 and Mehi.III.6.3). H. Pittman emphasizes one copper-alloy mirror from this site, which, "because of the large number of examples associated with Central Asia, [she believes] is likely [to have been] brought from the north" (Pittman 1984: 39; probably the one illustrated in Stein 1931: pl. XXXII: Mehi. II.2.1a). Additionally, a miniature stone column similar to the above-mentioned specimens recorded in Central Asia and eastern Iran is reported by Stein from Kulli (Stein 1931: pl. XXIII: Kul.1.x.1).

Further west, in-between Mehi and the core of the Kulli cultural complex in Pakistan and the above-mentioned Iranian sites of Tepe Bampur and Khurab, no remains relating to the Oxus Civilization have been recovered so far in the Kech-Makran region of southwestern Pakistan. None has been identified at the third millennium BC sites recorded in this region, including at Miri Qalat where an Indus occupation was excavated, and a gap is then apparent in the ancient occupation of this region after the Indus Civilization disappeared (see Besenval and Sanlaville 1990; Besenval 1997).

INTERPRETATIONS AND CONCLUDING REMARKS

The major issue that hinders a clear appreciation of this material evidence is the lack of a well-defined chronology including the very chronology of the major Oxus sites in Central Asia. In southeastern Iran, the above examples from Khurab show significant discrepancies in the dating of some materials. We may add the much-discussed chronology of Shahr-i Sokhta, the long-debated dating of Tepe Yahya IVB and heirloom problem, the current lack of available data from Konar Sandal North, and so on. This issue is enhanced by the intensive lootings that have affected sites in the Oxus Civilization, eastern Iran, southern Afghanistan, and Pakistan, and we are left to define their material cultures partly based on objects from museums and private collections with no provenience, claimed as coming vaguely from eastern Iran, Central Asia, or both. Certainly, the general chrono-cultural framework of Bronze Age southeastern Iran and Pakistan is established, but more precise dates for its periods, excavated layers, and material cultures would greatly help us understand the processes involved in the relationship between the Indo-Iranian borderlands and the Oxus Civilization. At present, two aspects of this relationship, distinct in time and direction, are distinguished: one relates to the origins of the Oxus Civilization and the other to a later expansion of this civilization.

Origins of the Oxus Civilization

Contra Sarianidi's hypothesis that sees a Syro-Anatolian origin in the development of the Oxus Civilization (Sarianidi 1998: 158, 163 fig. 75), and Amiet (1986: 138, 206–207, 212–214; 2007b: 51) who puts forth the concept of an impetus from "trans-élamite" merchants and seminomadic artisans from eastern Iran, as well as Steinkeller (2014: 701) who, partly agreeing with Amiet on an origin in Iran, considers the Oxus Civilization as an "offshoot" of eastern Iran and that the two "shared a similar culture," it is most commonly accepted that the roots of this civilization are in southern Central Asia, or, perhaps, in the regions just south of it in eastern Iran and Afghanistan (Hiebert 1994a: 165, 174–175; 1994b; Lamberg-Karlovsky 2002: 70, 72; Francfort 2005a: 258–259; Anthony 2007: 421; Lamberg-Karlovsky 2012: 60–61; Salvatori 2016: 450–451; Vidale 2017: 18; see also Chapters 1, 4, and 19). Many aspects including distinctive material characteristics such as in the stone, metal, and ceramic industries as well as the distant relationships that are observed in the Oxus Civilization have their best prior parallels in the cultural assemblages and technologies in southern Central Asia as well as in the networks of interregional interaction between ca. 3000 and 2500/2200 BC (see Hiebert 1994a: 164–171; P'jankova 1994; Francfort 2005a: 254, 258–260; 2007: 120–121; see also Chapter 1). The Oxus Civilization, however, certainly did not develop in complete isolation, and some form of synchronism with its close and distant neighbors is visible (see Kohl 2007: 217–218): "Its material culture is *sui generis* while its aesthetic products and technological achievements are the equal of any Bronze Age community in the Near East" (Lamberg-Karlovsky 2012: 67–68). At least, the relationship with Iran is undeniable in several other aspects of its material culture including the iconography and metal and stone objects with close parallels with/within what Amiet has called the "trans-élamite" culture and more broadly with the Elamite world in Iran. Jarrige (C. Jarrige et al. 1995: 95, quote at 95; see Jarrige 1985) observes that

> many of the objects which are considered characteristic of the Bronze Age in Bactria and Margiana were already present south of the Hindu Kush as early as the first half of the 3rd millennium as, for example, with the miniature columns (Shahr-i Sokhta), alabaster lamps (Mundigak, Nausharo ID), handled weights (Mundigak), or copper compartmented stamp seals sometimes with the design of an animal (Nausharo IC and other sites).

He also believes that some of the most characteristic ceramic forms of the Oxus Civilization have their origins within the western border of the Indus valley (C. Jarrige *et al.* 1995). As noted above, the networks of interregional interaction observed between ca. 3000 and 2500/2200 BC are emphasized as a key aspect in the formative period of this civilization, and it now appears that it is probably best to consider the formation of the Oxus Civilization as a multifocal process of *longue durée*. In other words, although it is of Central Asian origin, it also incorporates elements from distant cultures and chronological periods, including from the archaeological complexes located in the Indo-Iranian borderlands between the Late Chalcolithic-Middle Bronze Age. As Jarrige puts it (C. Jarrige *et al.* 1995: 95): "The Bronze Age culture of Bactria and Margiana represents, without doubt, a brilliant combination of indisputably original features with many borrowings from earlier traditions of various regions on both sides of the Hindu Kush." The existence of extensive interactions in the Late Chalcolithic–Early Bronze Age within the Indo-Iranian borderlands and between these territories and southern Central Asia is noted as are the easternmost records of the Proto-Elamite culture (Hiebert 1994a: 168; Francfort 2005a: 259–260; see also Chapter 1). These interactions are observed through the distribution of raw materials and similar artifacts, technologies, and pottery designs across these territories, and such interactions certainly were not limited to eastern Middle Asia but consisted of a chain of intercultural and transcultural relationships that encompassed many more cultural spheres or polities between at least Mesopotamia, the Indus valley, and Central Asia. Their processes may have included various forms of political alliances, migrations/colonizations, conquests/military expeditions, exchange/trade, and imitation/emulation (see Mutin 2013b: 206–214). New data has been collected on this interaction on the southeastern Iranian Plateau around 3000 BC as far as the ceramics are concerned (Besenval 1997; Mutin 2013a; Mutin *et al.* 2017), and it also now appears increasingly evident that the iconic Bronze Age metallic compartmented seals, usually regarded as a Central Asian creation (Francfort 2005a: 259), most likely originated in this area between ca. 3500 and 3000 BC (see Chapter 8). This does not mean that the roots of the Oxus Civilization are to be sought in this area, but that this typical and most sophisticated aspect of this civilization probably ultimately derives from there (Mutin 2020; see also Besenval 1997: 15, 18).

As for the "trans-élamite" and Elamite aspects of the Oxus Civilization, Amiet recognizes that they only concern stone and metal luxury goods (Amiet 1986: 195). Although this evidence is too limited to establish that the Oxus Civilization derived from the "trans-élamite" culture, the existence of relationships between the two may hardly be disputed. These relationships affected certain aspects – but not all – of this civilization. Among the objects we reviewed above, the metal axes, interpreted as symbols of authority, may have been given to seal political or economic alliances. With regard to this, Francfort (2005a: 291–292; 2007: 121) evokes the possibility that "une suzeraineté iranienne 'élamite' sur la Civilisation de l'Oxus [existât] à certains moments" [an Iranian "Elamite" suzerainty existed at some points over the Oxus Civilization]. In any case, that some form of diplomatic and/or commercial relationship existed is not in contradiction with the abundant and varied evidence of raw materials imported and transformed into objects in Central Asia (alabaster, chlorite/steatite, lapis lazuli, precious metals, copper, and seashells; see Hiebert and Lamberg-Karlovsky 1992: 5; Hiebert 1994a: 176). The ubiquitous, transcultural, nature of certain prestige objects, such as these axes, seem to tell us that the elites in the Indo-Iranian borderlands and southern Central Asia, as well as far afield to the west, "were developing power legitimization" at a large scale (see Jarrige 1991: 103; Salvatori and Tosi 1997: 131). Records of writing are extremely rare around 2000 BC both in eastern Iran and the Oxus Civilization, and, in this context, it may rationally be conceived that the Linear Elamite, possibly a sort of *scriptura franca* (Vidale 2017: 100), was employed for specific occasions, perhaps in selected commercial or diplomatic liaisons.

The art of the Oxus Civilization, also best represented on high-status items, incorporates conventions, compositions, and certain motifs and themes derived from Iran and more broadly from the Middle East and eastern Mediterranean. The Western influences on Oxus art cannot be taken as evidence for the roots of this civilization to be in the West, but "pour aboutir au résultat constaté, il faut une tradition locale solide, ainsi que des liens multiples, intenses et de longue durée avec le Moyen-Orient" [to reach the observed result, a substantial local tradition is necessary, as well as multiple, intensive and long-term relations with the Middle East] (Francfort 2007: 117). Its artists adapted these aspects while combining them with distinctive local traits on distinctive Central Asian objects, creating an original artistic tradition (Amiet 1986: 198–199; 2007c: 76). In Francfort's words (2007: 121): "l'art ... de la Civilisation de l'Oxus ... imprime sa marque originale sur des concepts qui étaient partagés dans le concert des civilisations de l'époque et il fait montre de sa créativité" [the art ... of the Oxus Civilization ... stamps its original brand over the concepts shared within the contemporary civilizations, and thus reveals its creativity]. In relation to this topic, the Oxus pantheon represented by this art is different from that of the Middle East but relates, to some extent, to that of Elam (see Chapters 6 and 7). Moreover, as an apparent syncretism, its symbolic system is "a cultural stratigraphy of old shamanic Central Asian pre-agrarian conceptions, parts of Elamite mythology and iconic language, plus Indian and Syro-Hittite elements, local mythic creations or innovations" (Francfort 1994: 416; see also 2007: 112–113). This agrees with the facts noted by Hiebert (1994b: 376, 384, 386) that metal and stone resources were imported and objects in these materials were locally produced in Central Asia.

The picture that emerges is that the Oxus Civilization originated and evolved as the result of *longue durée* and multifocal – mostly local, but also intercultural and transcultural – processes with various origins, directions, and impacts depending on the sphere of activity considered.[18] This is not surprising as it seems to be a universal characteristic of the development and continuous evolution of cultures that are not in complete isolation. As noted above, the Oxus Civilization was part of a larger Middle Asian sphere of interactions that included movements of peoples, animals, goods, styles, technologies, and beliefs and resulted from various forms of relationships including political alliances, migrations/colonizations, conquests/military expeditions, exchange/trade, and imitation/emulation. In this context, the relationship with the Indo-Iranian borderlands certainly had an impact on the Oxus Civilization,[19] and this impact should be discussed within the limits of its material manifestations (see Chapter 8); it appears that its art, the style of which was most likely commissioned[20] or at least valued by elites as far as the artifacts recovered are concerned, partly relates to Iran, including, compatibly, certain status items possibly representing political/commercial alliance and power legitimization – an aspect that seems to echo some of the political relations documented in the Mesopotamian texts during that time (see p. 580) – while an array of other materials and representations are clearly Central Asian.

Oxus-related evidence in the Indo-Iranian borderlands: arrows or a circle?

Oxus-related evidence in the Indo-Iranian borderlands is varied and probably reflects different situations. Hiebert and Lamberg-Karlovsky (1992: 3) observe that

> at Khurab, Shahdad, and Khinaman(?), there are burials of several periods which appear to include tombs that (1) contain material exclusively of Iranian type, (2) contain material of Iranian type with an object or two of Central Asian origin, and (3) contain, in the rare instance, material exclusively of Central Asian derivation.

The burials from Shahdad and Khurab (as well as those from Pakistan) of the third category are interpreted as tombs of individuals from Central Asia. But who were these individuals and what were their motives: Were they dignitaries, tradesmen, raiders, individuals looking for new lands? (On this question, see also Chapter 21.)

The clearest and most substantial excavated Oxus-related evidence in southeastern Iran is at Shahdad, where it is seen through various types of objects, technology, iconography, architecture, and possibly burial practices. Salvatori and Tosi (1997: 122) conclude that

> the connections of the Bactrian materials with Shahdad were many and striking for the degree of similarity expressed in the iconography as well as the technologies

> used to manufacture the most sophisticated objects in metal and stone, from seals to ornaments and tools. Trade cannot be considered as [the] only factor of explanation.[21] Not just economical exchanges connected Kerman with Bactria and the Murghab, but a shared ideolog[y] that pervaded the system to determine political structures and religious beliefs.

Salvatori sees Shahdad as a market place, a colony, "at the intersection point of two cultural spheres" and believes that "the presence of Bactria-Margiana tradesmen at Shahdad is highly probable and justified by the geographical position of the Kerman region, a true hinge ... at the eastern border of the Elamite world" (Salvatori 1995: 50; 2016: 454). He provides the same hypothesis regarding the Oxus-related evidence in Pakistan (Salvatori 1995: 51). Amiet (1986: 206–207, 212–213) also emphasizes an economic motive in a model involving seminomadic artisans and merchants. Jarrige highlights the large amount and wealth of vestiges recovered in Pakistan relating to the Mehrgarh VIII/Sibri/Dauda Damb cultural complex and the permanent character of the settlement at Sibri, leaving to him no doubt that "a complex society [relating to the Oxus Civilization], including elites, craftsmen and farmers as well as probably peddlers ... lived and interacted with the local population at the western margin of the Indus valley" (Jarrige 2016: 25). Hiebert (1994a: 177–178) suggests that "the interaction with southern Iran (Khurab), northern Iran (Hissar), and South Asia (Mehrgarh) is an expansion in quest of resources." An alternative model sees in this southern evidence the result of state expansionism, a territorial expansion or colonization, in which "the political subordination of a neighboring region was the paramount concern, while interest in their resources would have been a secondary and subsequent concern" (Lamberg-Karlovsky 1996: 214; see also Hiebert and Lamberg-Karlovsky 1992: 3–4; Hiebert 1994b: 386; Salvatori 1995: 50–51 *contra*). The expansion of the Oxus Civilization brings to mind a rather common phenomenon of Bronze Age civilizations in the Old World. The Uruk expansion was not alone. Expansionism was a common process within numerous relatively contemporaneous civilizations: the Kura-Araxes, the Proto-Elamite, and the Egyptian, to be followed by the Indus and Oxus expansions. The expansionist process, save for the Uruk expansion, has been poorly explored and the commonality of the process left unexamined. G. Algaze (2005) has argued that trade, even imperialist/colonialist motivations, inspired the Uruk expansion. It is unlikely that Algaze's "cause" for the Uruk expansion explains all of the above expansions. Conditions motivating cultural expansion may differ and such conditions should be a matter of research. The results of their expansions, however, appear to be the same. As each cultural expansion moves into regions of indigenous and distinctive cultures, they are not sustained for more than a century or two. One must imagine that, as the two distinctive cultures confronted each other, a process of acculturation and/or segregation ensued. In the end, the distinctive culture of the expansionist culture all but disappeared from the continuous indigenous settlements. What motivated the expansionist movements and what caused their collapse remain without explanation. It is not unreasonable to suggest that different causes brought about expansionism: warfare, climate change, trade, or combinations of these and more. The causes for expansionism remain largely unexplored while the results of expansion seem to be uniformly shared; namely, the eventual disappearance of expansionist settlements, whether as separate settlements within foreign lands or as communities embedded within indigenous settlements. Advocates of expansion share certain universal theoretical postulates. Algaze (2008: xv) goes so far as to suggest that the Uruk expansion "may be conceptualized as unwittingly creating the world's earliest world system: [It] united previously independent regions and polities into an overarching system of asymmetrical relationships of interdependency that were principally, but not solely, economic in nature." In fact, *none* of the conceptualized evidence for cultural expansions, enumerated below, is evident in the Chalcolithic and Bronze Age periods of the Near East, whether it be the Uruk or the Oxus expansions: (1) the core dominates the periphery, be it by organizational efficiency, military means, or ideological agency; (2) the core exploits the periphery by asymmetric trade (i.e., the extraction of valuable resources from the periphery by exporting cheap goods from the core); (3) the politics of the periphery, indeed its increased complexity, are structured by the core's organization of trade and exchange.

Another question emerges as to who controlled this expansion, be it commercial, military, or of any other nature. The structure of the Oxus Civilization is not precisely known (see also Chapter 1). Does it consist of interacting city-states, a single state, competing tribes? (Hiebert and Lamberg-Karlovsky 1992; Lamberg-Karlovsky 2012: 70–71). In the absence of texts, a reconstruction of the economic and political structure of the Oxus Civilization remains elusive. Nevertheless, ethnographic and historical texts of later periods offer insights into the Central Asia cultural world. The architectural signature of the Oxus Civilization is the large fortified settlement whose entire complex can cover several acres. In the later periods this type of structure can be found throughout the Central Asian world extending into Iran and Afghanistan. Typically, it is referred to as a *qala* and is inhabited by a political leader referred to as the *khan*, who rules over his *khanate*. Within the Oxus Civilization, these impressive structures are situated within an average distance ranging between 50 and 75 km. The distance from Gonur to Togolok, both with truly impressive *qalas*, is 57 km. Perhaps the heavy fortification systems surrounding the Oxus *qala* attests to persistent tribal warfare, well documented in the ethnographic and historic literature (Vámbéry 1873; O'Donovan 1882; Schrader 1893). We believe that the architectural signature of the *qala* descends in an unbroken line from the Oxus Civilization to the nineteenth century. The *qala* is both home to the *khan*, his subordinates, and a center of production and storage. Ceramics, metal artifacts, a wide variety of stone artifacts from statuary to bowls as well as jewelry and utilitarian objects of gold and silver were manufactured in the *qala*. Irrigation in the Murghab oases would have been a central concern requiring *corvée* labor to dig, clean, and manage the irrigated fields for the agricultural production on the lands owned by the tribal lineage (see Lamberg-Karlovsky 1994; 2003: 14; 2012: 70–71). If the structure of the Oxus Civilization was based on a *qala* within a *khanate* model, then was the southward expansion the result of separate enterprises led by each *khanate*, or the result of a collaborative effort? How did this operate in conflicting times, particularly after ca. 2000/1900 BC, when a greater fragmentation characterized by "smaller fortresses settled by minor, independent seats of power" is envisioned (Vidale 2017: 84; see also Salvatori 2016)? It also is important to note that no settlement has been found in the Indo-Iranian borderlands that could compare to a *qala*, suggesting a dissimilar socioeconomic organization where local cultural patterns dominated the social order.

When speaking of the Oxus expansion, this apparent absence of *qala* in the Indo-Iranian borderlands must be recalled, as must be the fact that Oxus-related evidence in Iran and Pakistan is in many cases of the limited, luxury, or specific utility type (i.e., miniature trumpets at Tepe Hissar, grooved stone columns from the "haute terrasse" at Tureng Tepe and Shah Tepe), while composite stylized female statues have been recovered from a number of sites in Khorasan as have compartmented bronze seals (Pittman 2013b; Thornton 2013). In these cases, the evidence "suggests that wealthy and influential immigrants from Margiana likely migrated into northeastern Iran around 2000 BC, where they seem to have integrated into local communities in ways that we do not yet understand" (Thornton 2013: 195). The same interpretation may be expressed as far as the graves from Quetta and Dauda Damb are concerned. However, a more speculative view on the nature of the Oxus-related evidence in Iran may be expressed. While the above proposals may be represented as arrows or lines pointing to the south, connecting southern Central Asia to dots in the Indo-Iranian borderlands (as represented in Hiebert 1994a: 177 fig. 10.8; Francfort 2007: 105 fig. 3), a circle, which would represent a large cultural sphere, could perhaps be used to schematize some of these north–south dynamics. Salvatori's interpretation of Shahdad "at the intersection point of two cultural spheres" implies that, more than Shahdad being a colony, the Oxus cultural sphere extended considerably much southwards, down to this site (see also Chapter 19). The above recapped material assemblage from this site also raises the question as to how one should interpret Shahdad; should it be considered an Oxus settlement acculturated to the southeastern Iranian tradition, or a southeastern Iranian settlement with Oxus-related elements – how should one consider the above-mentioned, indeed, striking similarities between the Oxus Civilization and Shahdad? The same idea, extended further east, of a cultural integration of the Afghan–Iranian borderland with Central Asia is expressed by Besenval and Francfort, who also cautiously emphasize the potential role of this area in the formation and development of the Oxus Civilization (Besenval and Francfort 1994: 11, 13). A possibility indeed exists that Shahdad, sites in Seistan, and the evidence recorded in the Quetta-Kachi area in Pakistan, all roughly along the 30th parallel north, marked at some point the southern limit of a much larger and more compact Oxus cultural sphere.[22] This cannot be proven at present because of the apparent lack of data in most territories between southern Central Asia and southeastern Iran and southern Afghanistan. It should however be noted that, in addition to Tepe Chalow in Khorasan, numerous new Oxus sites and materials are reported further south in this province and in Seistan (see Chapter 19). In western Afghanistan, a survey in 2004–2006 in Herat province (Herat city is ca. 400 km north of Nad-i Ali) failed to identify a substantial Oxus-related settlement; only one site is recorded near the Turkmen border. However, the German team that conducted this survey points out its limited extent and expects more to be found in this province. This hypothesis is supported by the numerous Oxus-related objects in the Herat Museum reportedly from the northern part of Herat province and possibly from the neighboring Badghis province, including typical metal and stone objects (Franke 2016).

As for the areas south of Shahdad, a current dearth of data also hinders a good appreciation of the situation. Tepe Yahya IVA certainly is presently the most substantial source, but the study of this period is not completed (it is still in progress), there is not enough reliable absolute dates, and, although its material assemblage is close to that of Shahdad, it remains difficult to conclude precisely on this relationship since the chronology is not fixed and the contexts in which it was collected are mostly dissimilar. Further east, we shall wait and see to what extent the earliest levels at Konar Sandal North, including the promising mudbrick platform, and the latest deposits at Konar Sandal South connect to Tepe Yahya IVA, Shahdad, the platform at Nad-i Ali and additional sites in Seistan, and Central Asia. Further south, the tombs at Khurab with Oxus-related vessels appear relatively isolated in this context. It is probable that these tombs are those of individuals from Central Asia, but this evidence offers a configuration different from that observed at Shahdad and Tepe Yahya IVA, and the reason of the presence of these individuals in this geographical context is unclear; were they tradesmen, nomadic artisans, dignitaries sent on a political mission, individuals ensuring the liaison on the route between southern Central Asia and the Persian Gulf? An issue with the Oxus-related materials from Khurab remains the dating of some typical local painted ceramics found in association with them, thought to be much older. We do not learn more from Tepe Bampur, in the same area, where Oxus-related material is lacking,[23] but columns are reported further south from "elite burials ... in a recently discovered cemetery near Nikshar" (Vidale 2017: 44). In Pakistan, we noted the lack of Oxus-related data in the Kech-Makran region of southwestern Pakistan, while Oxus-related evidence within the Kulli cultural complex, further east, seems at present less abundant than that recorded to the north in the Kachi plain and at Quetta.

Lastly, an aspect seemingly in contradiction with the model that sees a large Oxus cultural sphere encompassing some of the territories to the south is that the Oxus-related elements recorded in these areas are usually considered intrusive (e.g., Hiebert 1994a: 163–164). Are they always? It certainly is easy to notice a difference between the plain ceramic tradition that characterizes the Oxus Civilization and the typical fine, painted vessels recorded from Kerman to Kech-Makran, including the Jaz-e Murian in Iranian Baluchistan, as well as Shahr-i Sokhta in the third millennium BC. Conversely, it is difficult to distinguish between "foreign" and "local" in the second millennium BC, because most major sites in this vast region apparently are abandoned, save for Tepe Yahya, and no site has been excavated that could provide a clear and complete understanding of this millennium; the "local" assemblage and its evolution throughout the second millennium BC is not well known. However, an important observation may be noted: At Tepe Yahya IVA, a new plain ceramic tradition dominates four architectural levels with rooms with plastered walls and floors as well as chimneys, whereas painted material has completely disappeared. Period IVA is also a period of settlement expansion at Tepe Yahya and of higher settlement density in the Soghun valley (Lamberg-Karlovsky 1977). Was this settlement contemporary with Shahdad, was it partly more recent? Regardless, the evidence at this site and potentially the monumental platform in the Halil Rud basin may reflect a more anchored and perhaps more enduring Oxus-related or "Oxus-related-related" presence in southeastern Iran, south of Shahdad, than just the short-lived intrusion of a limited series of objects. Although a break is apparent in the Indo-Iranian borderlands' material cultures after ca. 2000 BC, current evidence in southeastern Iran in the first half of the second millennium BC does not conform with the classic model, frequently offered to explain this break, of Central Asian tribes raiding and disrupting their southern neighbors. In Pakistan, Jarrige (1985: 114–115) and

M. Santoni (1984) come to the same conclusion, particularly as far as the Oxus-related permanent settlement at Sibri is concerned, and, as noted above, the former emphasizes the large amount and wealth of vestiges recovered in Pakistan that relate to the Oxus Civilization (Jarrige 2016: 23–25). In contrast to current data in southeastern Iran, however, Jarrige observes a transitional phase at Nausharo, from the end of period III through period IV, during which ceramic components that later become typical of Mehrgarh VIII begin to evolve from the previous Indus-related ceramic repertoire (C. Jarrige *et al.* 1995: 89–90, 95). Also, although he tends to express a more direct connection between the Mehrgarh VIII/Sibri/Dauda Damb cultural complex and the Oxus Civilization in his more recent contribution on this topic (Jarrige 2016), he also recalls the local aspects of the material in this cultural complex: "We are dealing here with a relatively original culture which was too complex to have been simply imported in bulk from South Central Asia or from eastern Iran, despite connections with these regions" (Jarrige 1991: 97; see also C. Jarrige *et al.* 1995: 89).

Regardless of how to interpret the origins of this culture, a fact remains that is similar to the situation observed in southeastern Iran that there is an evident shift between the material culture typical of the Indus Civilization and that of the Mehrgarh VIII/Sibri/Dauda Damb and Quetta cultural complex; in particular, painted decorations on ceramic vessels are no longer observed (C. Jarrige *et al.* 1995: 91). As noted above, the associations of Indus-related, Kulli-related, and Oxus-related artifacts observed in the archaeological deposits from Nausharo, as well as evidence from other sites on the western border of the Indus valley system, led Jarrige to think that some of the most characteristic ceramic forms of the Oxus Civilization have their origins in this area (C. Jarrige *et al.* 1995: 95; see also Jarrige 1985). Although this seems to be a little bit of a stretch at the present state of knowledge and not all scholars agree on this interpretation, a fact remains that these associations show that these interactions between Central Asia and the Indian subcontinent seem to have occurred, not at the end but at a climax of the Indus Civilization, a time of great prosperity during which major architectural extensions took place at Mohenjo-daro and Harappa and extensive exchange networks existed between the Indus valley and Mesopotamia, including along the Persian Gulf (Jarrige 2016: 24). This certainly disagrees with the above-mentioned model of Central Asian tribes raiding and disrupting their southern neighbors that is commonly offered to explain the apparent break in the Indo-Iranian borderlands' material cultures after ca. 2000 BC.

The Oxus Civilization in the Indo-Iranian borderlands: violence, language, and religion?

An important, most debated, and controversial aspect associated with the Oxus Civilization and its spread to the south is the introduction of Indo-Iranian languages onto the Iranian Plateau and in South Asia, Persian in Iran, and Sanskrit speakers in India (see also Chapter 1). Indeed, linguists think that they were spoken in these territories by ca. 1500 BC (see Lamberg-Karlovsky 2002: 72), and archaeologists have attempted to find the archaeological cultures involved in this diffusion. Sarianidi (e.g., 1998) believed in the idea that the Oxus communities were Indo-Iranians and practiced a form of proto-Zoroastrianism. He interpreted the numerous Oxus archaeological remains he excavated in light of the Rig Veda and Avesta. His interpretations find no or minimal support in the archaeological records (see, e.g., Lamberg-Karlovsky 2007, 2012).[24] On the topic of the spread of Indo-Iranian languages, J.P. Mallory and D.Q. Adams (1997: 73) take the lead. In their words:

> [During] the Indo-Iranian migrations from the steppe lands south into the historical seats of the Iranians and Indo-Aryans ... these steppe cultures were transformed as they passed through a membrane of Central Asian urbanism. The fact that typical steppe wares are found on BMAC sites and that intrusive BMAC material is subsequently found further to the south in Iran, Afghanistan, Nepal, India and Pakistan, may suggest then the subsequent movement of Indo-Iranian speakers after they adopted the culture of the BMAC.

In this view the steppe nomads are Indo-Iranians who assimilate with the Oxus communities in Bactria/Margiana. The Oxus communities then adopt an Indo-Iranian language that, in their southern migration, bifurcates into the related languages of Persian and Sanskrit. M. Witzel (2013) is among many others prepared to see the Oxus Civilization as Indo-Iranian (see Hiebert and Lamberg-Karlovsky 1992: 1, 10; but see also Lamberg-Karlovsky 2002 and comments; Anthony 2007). The identification of the Indo-Iranian homeland, their continental wide migrations and chronology, has been an industry of speculation and unresolved debate for more than a century. Recently, A. Parpola (2015: 71, 76) similarly wrote that

> an elite of Aryan-speaking pastoralist warriors from the steppes took over the leadership in the BMAC ... The takeover ... could have occurred relatively peacefully ... it could have given the BMAC a more effective hierarchical leadership, which in turn would provide a plausible explanation for the BMAC's increased dynamism and aggressiveness in the later part of its "urban" phase.

Francfort (2005a: 271, 297–304) argues for a later steppic arrival with Indo-Iranian or Indo-Aryan speakers in southern Central Asia between ca. 1700 and 1400 BC, but he also notes the probability that certain elites in the Oxus Civilization were Indo-European, Indo-Iranian, or Indo-Aryan speakers beforehand. These southward dynamics – by the Oxus Civilization or by steppic components (Andronovo and others) assimilated to the Oxus Civilization – identified by the archaeology are chronologically and geographically consistent with this southward linguistic diffusion (see Renfrew's comment in Lamberg-Karlovsky 2002: 82). Also, that new beliefs and languages, or aspects of them, circulated with peoples and materials is self-evident, and this occurred probably not just as a single episode during the first half of the second millennium BC, but most likely as a continuous process between the Middle and Late Bronze Age (and beyond). The question, however, remains open in this particular case as there is a considerable period of time between the early to mid-second millennium BC here considered archaeological cultures and the actual Iranian speakers recorded in the first millennium BC. An important issue also remains that archaeology is ill-equipped to search for linguistic evidence in the material record and to trace movements of peoples who assimilate to local cultures (Lamberg-Karlovsky 2002: 74–75; see also Francfort 2005a; Anthony 2007 *contra*).

Another aspect, often thought to be concomitant with the diffusion of the Oxus Civilization to the south, is violence, the breakdown of the Bronze Age polities in the Indo-Iranian borderlands (see Lamberg-Karlovsky 1996: 242–243).[25] In the above-mentioned Parpola's reconstruction lies the idea that steppic populations are responsible for the Oxus Civilization's "increased dynamism and aggressiveness in the later part of its 'urban' phase." This scholar thinks that the steppic takeover of the leadership in the Oxus Civilization was a peaceful process. First, one may argue that the dynamism of this civilization is evident before the "later part of its 'urban' phase" as exemplified by the numerous and varied types of relationships reviewed here. In Pakistan, we noted above that the material evidence for an interaction between the Oxus and Indus civilizations dates to the climax of the latter and does not seem to coincide with its breakdown. Second, the fact that both communities in the steppe and in the Oxus Civilization have their settlements fortified, particularly after ca. 2000 BC, suggests conflicting, not peaceful, times between the two and within this civilization (Lamberg-Karlovsky 2002: 73; see also Vidale 2017: 8, 20; but see Chapter 17 *contra*).[26] The "aggressiveness" depicted by Parpola, regardless of whether it resulted from the introduction of a steppic component, is echoed by Vidale (2017: 44, 72–74) in his interpretation of the above-mentioned scene on a silver goblet of alleged Central Asian origin; he believes this scene represents a handcuffed ruler of Marhashi and demonstrates the existence of military expeditions led by Oxus individuals in southeastern Iran. Granted that some degree of contemporaneity existed between the Oxus Civilization, the Indus Civilization, the Halil Rud Civilization (possibly Marhashi), and other polities not discussed here (represented at Tepe Hissar and Tureng Tepe for instance as well as in the steppe and further west in Iran, the Persian Gulf, and Mesopotamia), as well as some form of expansionism, evident for instance through the distribution of Oxus-related and Indus-related (at Shortughaï, Sutkagen-Dor, Sotka-Koh, and Miri Qalat) settlements, possibly outposts, and individuals outside of their respective heartlands, it certainly is highly possible that competition and conflict existed for the control of resources, territories, or for any other reason. Increasing contact leads to tension and conflict. The expansion of empires, whether the later French, German, British, or Russian ones, had the inevitable result of increasing conflict between nations. This is exemplified by the Anglo-French wars, the Anglo-Spanish wars, the Anglo-American wars, as well as the various alliances of European nations fighting with or against Russia, and so on. It seems reasonable to say that every experience of globalized efforts leads to conflict and war. Similarly, the Egyptian–Hittite, Assyrian–Babylonian, Babylonian–Elamite, Elam/Shimashki–Ur III empire conflicts or wars took place in the context of one party expanding into the realm of another. We hear echoes of this for instance in the alliances of Marhashi with Shulgi, of Shimashki joining the Elamites to bring down the Ur III empire, and of the Ur III empire campaigning in the Persian Gulf. Throughout the third millennium BC, Mesopotamia was in a state of near constant warfare as indicated by the texts. Yet, there is scant evidence for war/conflict indicated in the archaeological record, although it is correct to admit that such evidence is usually not detected archaeologically (Lamberg-Karlovsky 1994: 403).

It also is apparent that the relationship between the Oxus Civilization and the Indo-Iranian borderlands was likely not based solely on violence. We noted above the possibilities that alliances existed, signified by specific status items, and that the southward spread may have led to the settlement of more endured Oxus-related or "Oxus-related-related" occupations in both southeastern Iran, Afghanistan, and Pakistan, although these possibilities remain hypothetical. One may add that it is difficult to conceive that this civilization had the capacity of maintaining networks of exchanges, settlements, and individuals over such a considerable area and at places located hundreds of kilometers distant from its heartland (be it in southern Central Asia and/or Khorasan-Afghanistan) in a context of constant hostility. Would the Oxus-related individuals at Khurab have received proper burial, and specifically so in a cemetery with a strong presence of the local tradition? If clashes of cultures occurred (see Lamberg-Karlovsky 1996: 241–242), which seems to be a universal, inevitable constant, we believe that they did not characterize the totality of the relationship between Central Asia and the Indo-Iranian borderlands including during the Oxus expansion. In Pakistan, Jarrige seems to envision a more peaceful reconstruction in which

> distinct social and cultural organizations … coexisted on the western margins of the Indus valley between about 2100 and 1900 BC. They include the Indus civilization in its third major period, the Kulli culture in its second major period and the cultural complex related to the Oxus Civilisation.
>
> (Jarrige 2016: 24; see also Jarrige, Jarrige, and Quivron. 2011: 196–199)

Trade certainly had a role in bringing together these various cultural entities in Pakistan as well as those in southeastern Iran, Afghanistan, and beyond (Lyonnet 2005; see also Chapter 1). On this topic, Vidale (2017: 56) writes:

> The increasing interaction of the Oxus with the Indus worlds, in the last centuries of the third millennium, was part of a more generalized intensification of cultural contacts and interregional links. Some scholars have recognized in this process the opening of a prehistoric "Silk Road," controlled at its western end by Oxus merchants and political leaders. Studies and discoveries point to the existence of important copper- and tin-mining districts in the territory of present Uzbekistan (the Zeravshan valley) and central Kazakhstan, while the traditional notion of the presence of valuable tin deposits in Afghanistan exploited in the third millennium has recently been questioned. Thus, one of the reasons of the sudden explosion of the Oxus polities might have been the control of the tin trade, that, judging from the available data on copper alloys, superseded the traditional arsenic alloys and became suddenly crucial in the last two centuries of the third millennium.

Although the economic motive aspect in, or the totality of, this statement may be correct, one should nonetheless recall here the above-mentioned continuation of an arsenical bronze tradition at Shahdad, Khinaman, Susa, and in Bactria (Salvatori 1995: 41; see also Lamberg-Karlovsky 1996: 210; Meier 2015: 245, 246–252). Regardless, the picture resulting from this interregional interaction, be it originally instigated by an economic motive or not, is, as envisioned by Jarrige (2016: 25) on the western margins of the Indus valley, that various categories of populations including elites, craftsmen, and farmers, and not only isolated traders, from various cultural groups including ones relating to the Oxus Civilization, lived and interacted in this area as well as in southeastern Iran.

It is presently difficult to understand when, where, and for how long within the span of the Oxus Civilization its relationship with the cultural groups present in the Indo-Iranian borderlands was cooperative or coercive. While it remains possible that raids existed and destabilized the civilizations in this region (Hiebert 1994a: 178), the above remarks and current evidence tend to also show a different picture. Tentatively, the evidence reviewed here may illustrate a case of gradual cultural assimilation and replacement whereby Central Asian aspects became progressively dominant further south, at least in southeastern Iran: (1) Central Asian aspects and aspects of the third millennium BC southeastern Iranian tradition are associated at Shahdad; (2) the southeastern Iranian tradition disappears (though probably not in all its aspects) while Central Asian aspects continue to exist at Shahdad and Tepe Yahya IVA. The Central Asian component then disappears for some unknown reasons together with the end of these occupations and the remaining Bronze Age sites in the Indo-Iranian borderlands. This reconstruction is also not in contradiction with part of the data recorded in the Kachi plain in Pakistan. One may assume that this assimilation was favored by centuries of north–south interactions and reciprocal influences such as recorded at least since the Late Chalcolithic and throughout the Middle Bronze Age. It remains to determine how the rest of the evidence reviewed here fits in this coarse and tentative reconstruction, to verify it with each type of available material evidence, and certainly to get a greater chronological control.

One last observation on the topic of this cultural interaction and integration deserves mention. Most of our evidence for cultural interaction as well as for chronology building and interpretations regarding economic, political, and social structures are derived from material remains: ceramic types and styles, seals and sealings, architecture, metal products, statuary, etc. Yet, each of the above, and particularly the complex iconography of the "Intercultural Style," distributed from Mari to Mohenjo-daro, the seals and sealings, and the male and female statuary, are all widely distributed. The recent publication of *Idols: The Power of Images* (Caubet 2019) provides a remarkable array of the male and female statuary of the Oxus Civilization. Stylistic sharing over a wide geographical expanse in similar artifact production, whether it be ceramics, elaborately carved stone bowls, or seals speaks to a non-material sharing of complex belief systems and ideologies, which are doubtlessly reverberated in sacred rituals. When these objects disappear from the archaeological record, we can assume that the belief systems and activities associated with them are dramatically transformed or disappear from their cognitive and behavioral world. It is only in the past few decades that we have come to appreciate that at least from the third millennium BC, from Mesopotamia to the Indus valley and from the Arabian Peninsula to the Oxus Civilization in Central Asia, we are looking at a world of expansive communication. With the unfolding of this *longue durée* we must ask what brought about the rise and dissolution of this very first expansive interaction, one that unfolded for several centuries on near global scale.

The "name game"

Lastly, do Mesopotamian texts, which refer to numerous neighbors situated to their East, make reference to the Oxus Civilization? Authors differ on what can be called the "name game" (see also Chapter 2). D.T. Potts (2008), in considering the archaeological and textual evidence, takes the place referred to as Shimashki in the texts to be the Oxus Civilization. Francfort and Tremblay (2010) disagree. They identify this civilization as the kingdom of Marhashi, while Steinkeller sees it as the region referred to as Tukrish and situates Shimashki on the eastern flanks of the Zagros Mountains, and not in Central Asia. In southeastern Iran, *contra* Francfort and Tremblay, most scholars believe that the territories in the Kerman province down to the Bampur area are identified in the Mesopotamian texts as Marhashi. Potts (2002) has reviewed, as documented in Mesopotamian texts, the "gifts" exchanged between Marhashi and Mesopotamia. In the 18th year of the reign of Shulgi, at the height of the Ur III empire, 22nd to 21st centuries, his daughter married the king of Marhashi. Such a diplomatic royal wedding would have strengthened the political and economic ties between the Ur III empire and the kingdom of Marhashi. Potts demonstrates that exchanges between Marhashi and Mesopotamia included the exchange of animals, military assistance, and shared festivals. The wedding of Shulgi's daughter to the king of Marhashi would have been contemporary with the Oxus presence in the region of southeastern Iran and underscores the wide-ranging contacts that brought different territorial kingdoms into contact. Potts (2008) also emphasizes the numerous relationships noted in the Mesopotamian texts between Shimashki, Elam, and the Ur III empire, including the fact that, early in the reign of Ibbi-Sîn in the mid-20th century, a Shimashkian king, Yabrat/Ebarat, and his followers took control of Susa and Anshan in southwestern Iran. If Potts's view on the location of Shimashki is correct, then this would mean that, in the mid-20th century, an Oxus king took control of Susa, where incidentally numerous Oxus-related materials are recorded. This reconstruction certainly is attractive; however, as noted above, not everybody agrees on this, including Steinkeller, who identifies the Oxus Civilization as Tukrish and places Shimashki in the eastern Zagros. In any case, although opinions differ as to the textual identity of the Oxus Civilization, there can be little doubt of its extensive contacts with Mesopotamia, Iran, the Indus, and the Arabian Peninsula, so well documented by the early pioneers in recognizing its importance (Amiet 1986; T. Potts 1994) and, most importantly, its discoverer Victor Sarianidi.

CONCLUSION

Although the Oxus Civilization certainly emerged mostly as the result of a southern Central Asian process, the cultures in the Indo-Iranian borderlands contributed, directly or remotely, to shape some of its aspects in its formative and mature phases. It also appears increasingly evident, based on combined evidence from Shahdad, Tepe Yahya, the Halil Rud basin, Seistan, and Khurab (with reservations for some of these until more data are made available), in addition to new discoveries recently reported from Khorasan and those made in Pakistan that the impact of the Oxus Civilization on its southern neighbors was not limited to the mere distribution of discrete Oxus elements within distinct cultural polities. It seems to have been more substantial and anchored and perhaps endured longer than previously thought, regardless of the much-debated hypothesis that the diffusion of new proto-Zoroastrian beliefs and Indo-Iranian languages were associated with it, a hypothesis that remains possible but that is presently difficult to prove.

NOTES

1 It is now agreed upon that this style originated in southeastern Iran and more specifically in Kerman. Some scholars also term this style "Halil Rud" (Vidale and Frenez 2015: 145) or "Jiroft" (Thornton 2012: 604) with reference to the location where numerous examples were recovered or are reported as coming from (see also Vidale 2015: 16).

2 This type of axe is also recorded in Elamite Iran, at Susa as an actual object and on glyptic representations, as also is a type of hammer (Amiet 1966: 407 fig. 307, 258 fig. 187; Pittman 1984: 65–78; Steinkeller 2014: 699–700 figs. 19–21; Vidale 2017: 27–30).

3 She also appears "on a silver vessel said to have been found in the province of Fars, in south-western Iran, assigned to the latter half of the third millennium, around the time of the Akkadian empire. The motif occurs later on Iranian cylinder seals. A number of clay impressions of cylinder seals with this same female figure have been found at Tall-i Malyan, ancient Anshan, in Fars; and at Susa, in Khuzistan. These are dated by stratigraphic and epigraphic evidence to the dynasty of the Sukkalmah (c.1900–1450? BC)." (Pittman 1984: 51). Amiet (2007c: 72–73, 74–75) defines the examples from Shahdad and Tepe Yahya as "trans-élamite" adaptations of Akkadian iconography, adaptations that also spread to Central Asia. D.T. Potts sees in this motif a Central Asian influence on Anshan (Potts 2008: 186–187).

4 These are Hakemi's (1997) dates, and we noted above that these may need revision (see, e.g., Paklayan 2004: 176, 177, tab. 7.1).

5 Francfort (1994: 407; 2005b: 41n164), however, mentions the presence of clay statues at Gonur Depe, Susa, and in Tajikistan and Uzbekistan.

6 As noted above, different, new labels are today offered as it has become increasingly evident since recent fieldwork conducted in the Jiroft area that this style is characteristic of the Halil Rud (Jiroft) Civilization in the Halil Rud basin and adjacent sectors in Kerman, including Tepe Yahya IVB.

7 The less likely possibility remains that "this observed difference is also cultural or spatial. The carved 'Intercultural-Style' vessels have a more southerly (marine-oriented?) distribution than the Shahdad chlorites" (Kohl 2001: 212).

8 This seal has a stepped cruciform outline that is reminiscent of designs in Central Asia, although H. Pittman does not explicitly mention this. However, she mentions its relation to period IVA impressions and points out parallels in Bactria for the second one (not illustrated).

9 For example, 2500–2200 BC (Lamberg-Karlovsky 1970: 5; 1971: 87; 1972: 89); 2400–1800 BC (Lamberg-Karlovsky 1977: 35, 43); from ca. 2200 or 2100 BC to ca. 1800 or to ca. 1600/1500 BC (Potts 1980: 579); 1700–1400 BC (Beale and Lamberg-Karlovsky 1986: 11); 1900–1550 BC cal. or 1700–1400 BC uncal. (Prickett 1986: 409 tab. 3.1, 459).

10 Beta 6473, recalibrated using OxCal (v.13): 1661–1502 (68.2%) and 1744–1433 cal. BC (95.4%).

11 Beta 6470, recalibrated using OxCal (v.13): 2122–1885 (68.2%) and 2195–1767 cal. BC (95.4%); TF 1140, recalibrated using OxCal (v.13): 2110–1746 (68.2%) and 2203–1627 cal. BC (95.4%); and Beta 6474, recalibrated using OxCal (v.13): 1683–1401 (68.2%) and 1875–1221 cal. BC (95.5%).

12 The ceramic assemblage from period IVA also incorporates a distinctive burnished ware. This ware is sometimes slipped and usually has a series of parallel short, punctuated lines on the rim on the exterior surface. Incised marks are also sometimes present. Period

IVA also yielded painted buff ceramics similar to types recorded in Fars in the second millennium BC.

13 A more recent study of the seals and sealings from Konar Sandal South has been published (see Pittman 2018).

14 Its height, "more than 35 m" or "about 40 m," is known, but its length and width are estimated based on the size of the mound (200 x 50 m) and the hypothesis that the platform was squarish (see Besenval and Francfort 1994: 8, 11).

15 Pl. XV: L.i.280 is painted and does not seem to relate to the Oxus Civilization either, although the photograph illustrated by Stein is not clear enough to ascertain this.

16 Alabaster vessels in the photograph illustrated by Stein in the same plate (pl. XV: L.i.280) also compare to materials from Shahdad.

17 A similar object at Gonur is considered to be a lamp (see Chapter 10) (editors' note).

18 Therefore, concepts such as Kohl's "shared social fields" (2008) and P. Bourdieu's (1992) "champ social" certainly are useful to isolate and approach aspects of this complexity.

19 The seal impressions found at Konar Sandal South have led Pittman (2014) to consider the Halil Rud basin, or southeastern Iran more generally, as the origin for much of the Oxus Civilization pantheon and iconography.

20 It is also possible that artists traveled (Francfort 2005a: 292).

21 It should be noted here that a number of pyrotechnological structures at Shahdad interpreted earlier as smelting furnaces are in fact most probably domestic structures (Meier 2015: 252–253).

22 Oxus-related material is also reported from Shamshir Ghar in the region of Kandahar, southern Afghanistan (Francfort and Tremblay 2010: 107–108).

23 One metal compartmented seal is recorded, although its context is older (period IV) and, given the existence of much older specimens in Kech-Makran, it should now be admitted that this type of object in this chronological context does not necessarily relate to Central Asia only. A rectangular stone seal, found in a post-period VI context, resembles examples from Shahdad and is chronologically more compatible (see de Cardi 1970: 326 fig. 47 nos. 15–16, 327–329).

24 A recent analysis by Vidale of the "Balafrés" is in line with Sarianidi's proposal. These chlorite/steatite and limestone statues are usually assigned a Central Asian or Iranian origin and are believed to be Oxus objects, although it is unclear where they come from since none of them (16 are known in total) have been recovered from archaeological excavation. In his analysis, Vidale (2017: 200, see 168–200) concludes that "the *balafré* represented an early Indo-Iranian ancestor of the Rgvedic Indra as dragon-slayer, by means of a synoptic visual narrative" (see Francfort 1994 for a different interpretation).

25 Alternative, "classic" hypotheses have been offered to explain the end of the Bronze Age civilizations in the Indo-Iranian borderlands, including the drought that began in the late third millennium BC in the Middle East and Central Asia (see Fouache *et al.* 2013, 2015), an overexploitation of the environment, and internal sociopolitical changes (see Lamberg-Karlovsky 1994: 404; Francfort 2005a: 303–304).

26 It also is unclear what type of "more effective hierarchical relationship" the elite of Aryan-speaking pastoralist warriors from the steppes who purportedly took over the Oxus Civilization brought to this civilization.

REFERENCES

Akbarzadeh, D. and S. Piran 2013. *Objects from the Jiroft Treasury, Tehran: Soft Stone and Alabaster Objects (Recovered Collection) from the Halil River Basin – National Museum of Iran, with a Report by Dr. Y. Majidzadeh*. Tehran: Pazineh Press.

Algaze, G. 2005. *The Uruk System: The Dynamics of Expansion of Early Mesopotamia*, 2nd ed. Chicago: University of Chicago Press.
Algaze, G. 2008. *Ancient Mesopotamia at the Dawn of Civilization. The Evolution of an Urban Landscape*. Chicago: University of Chicago Press.
Amiet, P. 1966. *Elam*. Auvers-sur-Oise: Archée.
Amiet, P. 1986. *L'âge des échanges inter-iraniens: 3500–1700 avant J.-C.* Paris: Éditions de la Réunion des musées nationaux.
Amiet, P. 1994. "Un sceau trans-élamite à Suse." *Revue d'assyriologie et d'archéologie orientale* 88(1): 1–4.
Amiet, P. 2007a. "Antiquities of Bactria and Outer Iran in the Louvre Collection." In *Bactria: An Ancient Civilization from the Sands of Afghanistan* (Studies and Documents 3), ed. G. Ligabue and S. Salvatori. Venice: Erizzo, pp. 159–180.
Amiet, P. 2007b. "Elam et trans-Elam. À propos de sceaux-cylindres de la collection du Dr. Serge Rabenou." *Revue d'assyriologie et d'archéologie orientale* 101(1): 51–58.
Amiet, P. 2007c. "L'âge des échanges inter-iraniens: 3500–1700 avant J.-C." In *Sulla Via delle Oasi: Tesori dell'Oriente Antico*, ed. G. Ligabue and G. Rossi Osmida. Trebaseleghe: Il Punto Edizioni, pp. 64–87.
Anthony, D.W. 2007. *The Horse, the Wheel, and Language: How Bronze Age Riders from the Eurasian Steppes Shaped the Modern World*. Princeton, NJ: Princeton University Press.
Aruz, A. and R. Wallenfels (eds.) 2003. *Art of the First Cites: The Third Millennium BC from the Mediterranean to the Indus*. New York: The Metropolitan Museum of Art.
Askarov, A. 1973. *Sapallitepa*. Tashkent: FAN.
Beale, T.W. and C.C. Lamberg-Karlovsky 1986. *Excavations at Tepe Yahya, Iran. 1967–1975. The Early Periods* (American School of Prehistoric Research Bulletin 38). Cambridge, MA: Peabody Museum of Archaeology and Ethnology, Harvard University.
Besenval, R. 1997. "Entre le Sud-Est iranien et la plaine de l'Indus: le Kech-Makran. Recherches archéologiques sur le peuplement ancien d'une marche des confins indo-iraniens." *Arts Asiatiques* 52: 5–36.
Besenval, R. and H.-P. Francfort 1994. "The Nad-i Ali 'Surkh Dagh': A Bronze Age monumental platform in Central Asia?" In *From Sumer to Meluhha: Contributions to the Archaeology of South and West Asia in Memory of George F. Dales Jr.* (Wisconsin Archaeological Reports 3), ed. J.M. Kenoyer. Madison: Prehistory Press, pp. 3–14.
Besenval, R. and P. Sanlaville 1990. "Cartography of Ancient Settlements in Central Southern Pakistani Makran: New Data." *Mesopotamia* 25: 79–146.
Biscione, R. 1979. "The Burnt Building of Period Shahr-i Sokhta IV: An Attempt of Functional Analysis from the Distribution of Pottery Types." *IRANICA, Seminario di studi asiatici Series Minor* 10: 291–306.
Biscione, R. and A.A. Vahdati 2011. "Excavations at Tepe Chalow, Northern Khorasan, Iran." *Studi Micenei ed Egeo-Anatolici* 53: 236–241.
Bourdieu, P. 1992. *Les règles de l'art. Genèse et structure du champ littéraire*. Paris: Seuil.
Cardi, B. de 1970. *Excavations at Bampur, a Third Millennium Settlement in Persian Baluchistan, 1966* (Anthropological Papers of the American Museum of Natural History 51(3)). New York: American Museum of Natural History, pp. 237–355.
Caubet, A. 2019. *Idols: The Power of Images*. Milan: Skira.
Cortesi, E., M. Tosi, A. Lazzari, and M. Vidale 2008. "Cultural Relationships Beyond the Iranian Plateau: The Helmand Civilization, Baluchistan and the Indus Valley in the 3rd Millennium BC." *Paléorient* 34(2): 5–35.
Curtis, J. 1988. "A Reconsideration of the Cemetery at Khinaman, South-East Iran." *Iranica Antiqua* 23: 97–128.

Didier, A. 2013. *La production céramique du Makran (Pakistan) à l'âge du Bronze Ancien (2800–2500 av. J.-C.)* (Mémoires des Missions Archéologiques Françaises en Asie Centrale et en Asie Moyenne 14, Série Indus-Baluchistan). Paris: de Boccard.

During-Caspers, E.C.L. 1994a. "Further Evidence for 'Central Asian' Material from the Arabian Gulf." *Journal of the Economic and Social History of the Orient* 37: 33–53.

During-Caspers, E.C.L. 1994b. "Widening Horizons: Contacts Between Central Asia (the Murghabo-Bactrian Culture) and the Indus Valley Civilization Towards the Close of the Third and the Early Centuries of the Second Millennium B.C." *Annali dell'Istituto Universitario Orientale di Napoli* 54: 171–197.

Dyson, R.H. Jr. and S.M. Howard 1989. *Tappeh Hesar. Reports of the Restudy Project, 1976* (Monograhie de Mesopotamia 2). Florence: Case editrice Le Lettere.

Fouache, E., H.-P. Francfort, C. Cosandey, and C. Adle 2015. "La crise d'aridité climatique de la fin du 3ème millénaire av. J.-C., à la lumière des contextes géomorphologiques de trois sites d'Iran Oriental (Bam, Tepe Damghani, Jiroft)." In *Greater Khorasan*, ed. R. Rante (Studies in the History and Culture of the Middle East). Berlin: De Gruyter, pp. 27–38.

Fouache, E., H.-P. Francfort, C. Cosandey, C. Adle, J. Bendezu-Sarmiento, and A.A. Vahdati 2013. "Les régions de Bam et de Sabzevar (Iran): une évolution dans l'implantation des sites archéologiques et dans la gestion des ressources en eau compatible avec l'hypothèse d'une aridification croissante du climat entre 2500–1900 BC." *Cahiers d'Asie centrale* 21–22: 559–572.

Francfort, H.-P. 1994. "The Central Asian Dimension of the Symbolic System in Bactria and Margiana." *Antiquity* 68(259): 406–418.

Francfort, H.-P. 2005a. "La civilisation de l'Oxus et les Indo-Iraniens et Indo-Aryens." In *Aryas, Aryens et Iraniens en Asie Centrale* (Collège de France, Publications de l'Institut de Civilisation Indienne 72), ed. G. Fussman, J. Kellens, H.-P. Francfort, and X. Tremblay. Paris: de Boccard, pp. 253–328.

Francfort, H.-P. 2005b. "Observations sur la toreutique de la civilisation de l'Oxus." In *Afghanistan: Ancient carrefour entre L'Est et L'Ouest*, ed. O. Bopearachchi and M.-F. Boussac. Turnhout: Brepols, pp. 21–64.

Francfort, H.-P. 2007. "L'art de la Civilisation de l'Oxus." In *Sulla Via delle Oasi: Tesori dell'Oriente Antico*, ed. G. Ligabue and G. Rossi Osmida. Trebaseleghe: Il Punto Edizioni, pp. 102–127.

Francfort, H.-P. and X. Tremblay 2010. "Marhaši et la Civilisation de l'Oxus." *Iranica Antiqua* 45: 51–224.

Franke, U. 2010. "From the Oxus to the Indus: Two Compartmented Seals from Mohenjo-daro Pakistan." In *Corpus of Indus Seals and Inscriptions*. Vol. 3.1, *Supplement to Mohenjo-daro and Harappa*, ed. A. Parpola, B.M. Pande, and P. Koskikallio. Helsinki: Suomalainen Tiedeakademia, pp. 17–43.

Franke, U. 2016. "Prehistory." In *Herat Through Time: The Collections of the Herat Museum and Archive*. Vol. 3, *Ancient Herat*, ed. U. Franke and M. Müller-Wiener. Berlin: Staatliche Museen zu Berlin – Preußischer Kulturbesitz, pp. 21–58.

Hakemi, A. 1997. *Shahdad: Archaeological Excavations of a Bronze Age Center in Iran* (IsMEO Reports and Memoirs 27). Rome: Oxford and IBH.

Hiebert, F.T. 1994a. *Origins of the Bronze Age Oasis Civilization in Central Asia* (American School of Prehistoric Research Bulletin 42). Cambridge, MA: Peabody Museum of Archaeology and Ethnology, Harvard University.

Hiebert, F.T. 1994b. "Production Evidence for the Origins of the Oxus Civilization." *Antiquity* 68(259): 373–387.

Hiebert, F.T. and C.C. Lamberg-Karlovsky 1992. "Central Asia and the Indo-Iranian Borderlands." *Iran* 30: 1–15.

Jarrige, C., J.-F. Jarrige, R.H. Meadow, and G. Quivron 1995. *Mehrgarh Field Reports 1974–1985: From Neolithic Times to the Indus Civilization*. Karachi: Department of Culture and Tourism, Government of Sindh, Pakistan, in Collaboration with the French Ministry of Foreign Affairs.

Jarrige, J.-F. 1985. "Les relations entre l'Asie centrale méridionale, le Baluchistan et la vallée de l'Indus à la fin du 3e et au début du 2e millénaire." In *L'Archéologie de la Bactriane ancienne*, ed. J.-C. Gardin. Paris: Editions du CNRS, pp. 105–118.

Jarrige, J.-F. 1991. "The Cultural Complex of Mehrgarh (Period VIII) and Sibri: The 'Quetta Hoard.'" In *Forgotten Cities on the Indus: Early Civilization in Pakistan from the 8th to the End of the 2nd Millennium BC.*, ed. M. Jansen, M. Mulloy, and G. Urban. Mainz: Verlag Philipp von Zabern, pp. 94–103.

Jarrige, J.-F. 1994. "The Final Phase of the Indus Occupation at Nausharo and Its Connection with the Following Cultural Complex of Mehrgarh VIII." In *South Asian Archaeology 1993*, vol. 1, ed. A. Parpola and P. Koskikallio. Helsinki: Suomalainen Tiedeakatemia, pp. 295–313.

Jarrige, J.-F. 2016. "Key-Note Lecture: Indus-Oxus Civilisations: More Thoughts." In *South Asian Archaeology and Art 2012*. Vol. 1, *Man and Environment in Prehistoric and Protohistoric South Asia: New Perspectives*, ed. V. Lefèvre, A. Didier, and B. Mutin. Turnhout: Brepols, pp.19–33.

Jarrige, J.-F., A. Didier, and G. Quivron 2011. "Shahr-i Sokhta and the Chronology of the Indo-Iranian Regions." *Paléorient* 37(2): 7–34.

Jarrige, J.-F. and M.U. Hassan 1989. "Funerary Complexes in Baluchistan at the End of the Third Millennium in the Light of Recent Discoveries at Mehrgarh and Quetta." In *South Asian Archaeology 1985*, ed. K. Frifelt and P. Sorensen. London: Curzon Press, pp. 150–166.

Jarrige, J.-F., C. Jarrige, and G. Quivron 2011. *Nindowari (Pakistan): The Kulli Culture, Its Origin and Its Relations with the Indus Civilisation*. Paris: Ginkgo.

Jarrige, J.-F. and G. Quivron 2008. "The Indus Valley and the Indo-Iranian Borderlands at the End of the 3rd Millennium and the Beginning of the 2nd Millennium BC." In *South Asian Archaeology 1999*, ed. E.M. Raven. Groningen: Brill, pp. 61–83.

Kohl, Ph. 2001. "Reflections on the Production of Chlorite at Tepe Yahya: 25 Years Later." In *Excavations at Tepe Yahya, Iran 1967–1975: The Third Millennium* (American School of Prehistoric Research Bulletin 45), ed. C.C. Lamberg-Karlovsky and D.T. Potts. Cambridge, MA: Peabody Museum of Archaeology and Ethnology, Harvard University, pp. 209–230.

Kohl, Ph. 2007. *The Making of Bronze Age Eurasia*. New York: Cambridge University Press.

Kohl, Ph. 2008. "Shared Social Fields: Evolutionary Convergence in Prehistory and Contemporary Practice." *American Anthropologist* 110(4): 495–506.

Lamberg-Karlovsky, C.C. 1969. "Further Notes on the Shaft-Hole Pick-Axe from Khurab, Makran." *Iran* 7: 162–168.

Lamberg-Karlovsky, C.C. 1970. *Excavations at Tepe Yahya, Iran, 1967–1969* (American School of Prehistoric Research Bulletin 27). Cambridge, MA: Peabody Museum of Archaeology and Ethnology, Harvard University.

Lamberg-Karlovsky, C.C. 1971. "The Proto-Elamite Settlement at Tepe Yaḥya." *Iran* 9: 87–96.

Lamberg-Karlovsky, C.C. 1972. "Tepe Yaḥya 1971: Mesopotamia and the Indo-Iranian Borderlands." *Iran* 10: 89–100.

Lamberg-Karlovsky, C.C. 1977. "Foreign Relations in the Third Millennium at Tepe Yahya." In *Le Plateau Iranien et L'Asie Centrale des origines à la conquête islamique*, ed. J. Deshayes. Paris: Éditions du CNRS, pp. 33–43.

Lamberg-Karlovsky, C.C. 1984. "An Idea or Pot Luck." In *Frontiers of the Indus Civilization: Sir Mortimer Wheeler Commemoration Volume*, ed. B.B. Lal and S.P. Gupta. New Delhi: Books and Books, pp. 347–351.

Lamberg-Karlovsky, C.C. 1988. "The 'Intercultural Style' Carved Vessels." *Iranica Antiqua* 23: 45–95.

Lamberg-Karlovsky, C.C. 1994. "The Bronze Age khanates of Central Asia." *Antiquity* 68(259): 398–405.

Lamberg-Karlovsky, C.C. 1996. *Beyond the Tigris and Euphrates: Bronze Age Civilizations* (Beer-Sheva 9). Jerusalem: Ben-Gurion University of the Negev Press.

Lamberg-Karlovsky, C.C. 2001. "Afterword. Excavations at Tepe Yahya: Reconstructing the Past." In *Excavations at Tepe Yahya, Iran 1967–1975: The Third Millennium* (American School of Prehistoric Research Bulletin 45), ed. C.C. Lamberg-Karlovsky and D.T. Potts. Cambridge, MA: Peabody Museum of Archaeology and Ethnology, Harvard University, pp. 269–280.

Lamberg-Karlovsky, C.C. 2002. "Archaeology and Language: The Indo-Iranians." *Current Anthropology* 43(1): 63–88.

Lamberg-Karlovsky, C.C. 2003. "Civilization, State, or Tribes? Bactria and Margiana in the Bronze Age." *Review of Archaeology* 24(1): 11–19.

Lamberg-Karlovsky, C.C. 2007. "The BMAC: Pivot of the Four Quarters: Temples, Palaces." In *Sulla Via delle Oasi: Tesori dell'Oriente Antico*, ed. G. Ligabue and G. Rossi Osmida. Trebaseleghe: Il Punto Edizioni, pp. 88–102.

Lamberg-Karlovsky, C.C. 2012. "The Oxus Civilization (aka the Bactria-Margiana Archaeological Complex)." *Review of Archaeology* 30: 59–75.

Lamberg-Karlovsky, C.C. and F.T. Hiebert 1992. "The Relations of the Finds from Shahdad to Those of Sites in Central Asia." *Journal of the Ancient Near Eastern Society* 21: 135–140.

Lamberg-Karlovsky, C.C. and D.T. Potts (eds.) 2001. *Excavations at Tepe Yahya, Iran 1967–1975: The Third Millennium* (American School of Prehistoric Research Bulletin 45). Cambridge, MA: Peabody Museum of Archaeology and Ethnology, Harvard University.

Lawergren, B. 2003. "Oxus Trumpets, ca. 2200–1800 BC: Material Overview, Usage, Societal Role, and Catalog." *Iranica Antiqua* 38: 41–118.

Lyonnet, B. 2005. "Another Possible Interpretation of the Bactro-Margiana Culture (BMAC) of Central Asia: The Tin Trade." In *South Asian Archaeology 2001*, ed. C. Jarrige and V. Lefèvre. Paris: Editions Recherche sur les Civilisations, pp. 191–200.

Madjidzadeh, Y. 2003. *Jiroft: The Earliest Oriental Civilization.* Tehran: Ministry of Culture and Islamic Guidance, Iran Cultural Heritage Organization Research Center Printing and Publishing.

Madjidzadeh, Y. and H. Pittman 2008. "Excavations at Konar Sandal in the Region of Jiroft in the Halil Basin: First Preliminary Report (2002–2008)." *Iran* 46: 69–103.

Mallory, J.P. and D.Q. Adams (eds.) 1997. *Encyclopedia of Indo-European Culture.* London and Chicago: Fitzroy Dearborn.

Maxwell-Hyslop, K.R. 1988. "A Comment on the Finds from Khinaman." *Iranica Antiqua* 23: 129–138.

Meier, D.M.P. 2015. "Traces of Protohistorical Metallurgical Activities During the 3rd Millennium BC in Eastern Iran with a Special Focus on the Case of Shahdad in the Dasht-eh Lut (Kerman province)." PhD dissertation, Freie Universität, Berlin.

Méry, S. 2000. *Les céramiques d'Oman et l'Asie moyenne: Une archéologie des échanges à l'Âge du Bronze* (CRA Monographies 23). Paris: Éditions du CNRS.

Miroschedji, P. de 1973. "Vases et objets en stéatite susiens du musée du Louvre." *Cahiers de la Délégation Archéologique Française en Iran* 3: 9–80.

Muscarella, O.W. 2005. "Jiroft and 'Jiroft-Aratta': A Review Article of Yousef Madjidzadeh, Jiroft – the Earliest Oriental Civilization." *Bulletin of the Asia Institute* 15: 173–198.

Mutin, B. 2013a. "Ceramic Traditions and Interactions on the Southeastern Iranian Plateau during the Fourth Millennium BC." In *Ancient Iran and Its Neighbours: Local Developments and Long-Range Interactions in the 4th Millennium BC*, ed. C.A. Petrie (British Institute of Persian Studies Archaeological Monographs Series III). Oxford: Oxbow Books, pp. 253–275.

Mutin, B. 2013b. *The Proto-Elamite Settlement and Its Neighbors: Tepe Yahya Period IVC* (American School of Prehistoric Research Monograph Series, Harvard University). Oxford and Oakville: Oxbow Books.

Mutin, B. 2020. "Of Intercultural Interaction and Innovation: A Case-Study Based on Late Chalcolithic Copper Objects from Pakistan and Their Parallels in Iran and Central Asia." *Archäologische Mitteilungen aus Iran und Turan* 49: 101–121.

Mutin, B., L. Minc, C.C. Lamberg-Karlovsky, and M. Tosi 2017. "Regional and Long-Distance Exchanges of an Emblematic 'Prestige' Ceramic in the Indo-Iranian Borderlands. Results of Neutron Activation Analysis." *Paléorient* 43(1): 141–163.

O'Donovan, E. 1882. *The Merv Oasis*. London: Smith, Elder & Co.

Paklayan, S. 2004. "An Analysis of the Site of Shahdad (Khabis), Kerman, Iran: Internal Chronology and External Relations during the Third and Early Second Millennia BC." BA thesis, University of Sydney, Sydney.

Parpola, A. 2015. *The Roots of Hinduism. The Early Aryans and the Indus Civilization*. New York: Oxford University Press.

Piperno, M. and S. Salvatori 2007. *The Shahr-i Sokhta Graveyard (Sistan, Iran). Excavation Campaigns 1972–1978* (IsIAO Reports and Memoirs New Series 6). Rome: IsIAO.

Pittman, H. 1984. *Art of the Bronze Age: Southeastern Iran, Western Central Asia, and the Indus Valley*. New York: The Metropolitan Museum of Art.

Pittman, H. 2001. "Glyptic Art of Period IV." In *Excavations at Tepe Yahya, Iran 1967–1975. The Third Millennium* (American School of Prehistoric Research Bulletin 45), ed. C.C. Lamberg-Karlovsky and D.T. Potts. Cambridge, MA: Peabody Museum of Archaeology and Ethnology, Harvard University, pp. 231–268.

Pittman, H. 2013a. "Eastern Iran in the Early Bronze Age." In *The Oxford Handbook of Ancient Iran*, ed. D.T. Potts. Oxford: Oxford University Press, 304–324.

Pittman, H. 2013b. "Seals and Sealings in the Sumerian World." In *The Sumerian World*, ed. H. Crawford. London: Routledge, pp. 319–344.

Pittman, H. 2014. "Anchoring Intuition in Evidence: A Continuing Discussion of Cylinder Seals from Southeastern Iran." In *Edith Porada zum 100: Geburstag. A Centenary Volume*, ed. E. Bleibtreu and H.U. Steymans (Orbis Biblicus et Orientalis 268). Fribourg: Academic Press/Göttingen: Vandenhoeck & Ruprecht, pp. 375–393.

Pittman, H. 2018. "Administrative Role of Seal Imagery in the Early Bronze Age: Mesopotamian and Iranian Traders on the Plateau." In *Seals and Sealings in the Ancient World: Case Studies from the Near East, Egypt, the Aegean, and South Asia*, ed. M. Ameri, S.K. Costello, G. Jamison, and S.J. Scott. Cambridge: Cambridge University Press, pp. 13–35.

Possehl, G.L. 2007. "The Middle Asian Interaction Sphere. Trade and Contact in the 3rd Millennium BC." *Expedition* 49(1): 40–42.

Potts, D.T. 1980. "Tradition and Transformation: Tepe Yahya and the Iranian Plateau during the Third Millennium B.C." PhD dissertation, Harvard University, Cambridge, MA.

Potts, D.T. 2002. "Total Prestation in Marhashi-Ur Relations." *Iranica Antiqua* 37: 343–357.

Potts, D.T. 2008. "Puzur-Inshushinak and the Oxus Civilization (BMAC): Reflections on Shimashki and the Geo-Political Landscape of Iran and Central Asia in the Ur III Period." *Zeitschrift für Assyriologie* 98: 165–194.

Potts, T. 1994. *Mesopotamia and the East: An Archaeological and Historical Study of Foreign Relations ca. 3400–2000 BC* (Oxford University Committee for Archaeology Monograph 37). Oakville, CT: Oxbow Books.

Prickett, M.E. 1986. "Man, Land and Ware: Settlement Distribution and the Development of Irrigation Agriculture in the Upper Rud-i Gushk Drainage, Southeastern-Iran." PhD dissertation, Harvard University, Cambridge, MA.

P'jankova, L.T. 1994. "Central Asia in the Bronze Age: Sedentary and Nomadic Cultures." *Antiquity* 68(259): 355–372.

Rossi Osmida, G. 2007. "Margiana e Via delle Oasi." In *Sulla Via delle Oasi: Tesori dell'Oriente Antico*, ed. G. Ligabue and G. Rossi Osmida. Trebaseleghe: Il Punto Edizioni, pp. 154–177.

Sajjadi, S.M.S. and H. Moradi. 2015. "Shahr-i Sokhta 2014–2015 Excavations. The new Results in Areas 1, 20, 26 and 28." *Archeologia Aerea. Studi di Aerotopografia Archeologica* 9(15): 149–167.

Salvatori, S. 1995. "Protohistoric Margiana: On a Recent Contribution." *Rivista di Archeologia* 19: 38–55.

Salvatori, S. 2000. "Bactria and Margiana Seals: A New Assessment of Their Chronological Position and a Typological Survey." *East and West* 5(1/4): 97–145.

Salvatori, S. 2016. "Bactria-Margiana Archaeological Complex: How Terminology Hides Historical Processes." In *Trudy Margianskoj Arkheologicheskoj Ekspeditsii*, vol. 6, ed. N.A. Dubova. Moscow: Staryj Sad, pp. 449–460.

Salvatori, S. and M. Tosi 1997. "Postscriptum: Some Reflections on Shahdad and Its Place in the Bronze Age of Middle Asia." In *Shahdad: Archaeological Excavations of a Bronze Age Center in Iran* (IsMEO Reports and Memoirs 27), ed. A. Hakemi. Rome: Oxford and IBH, pp. 121–132.

Salvatori, S. and M. Tosi 2005. "Shahr-i Sokhta Revised Sequence." In *South Asian Archaeology 2001*, ed. C. Jarrige and V. Lefèvre. Paris: Editions Recherche sur les Civilisations, pp. 281–292.

Salvatori, S. and M. Vidale 1982. "A Brief Surface Survey of the Protohistoric Site of Shahdad (Kerman, Iran)." *Rivista di Archeologia* 19: 5–10.

Santoni, M. 1984. "Sibri and the South Cemetery of Mehrgarh: Third Millennium Connections Between the Northern Kachi Plain (Pakistan) and Central Asia." In *South Asian Archaeology 1981*, ed. B. Allchin. Cambridge: Cambridge University Press, pp. 52–60.

Santoni, M. 1988. "Aspects matériels des cultures de Sibri et Mehrgarh VIII (plaine de Kachi, Baluchistan, Pakistan) à la fin du troisième et au début du deuxième millénaires." In *L'asie centrale et ses rapports avec les Civilisations orientales des Origines à l'Age du Fer* (Mémoires de la Mission Archéologique Française en Asie Centrale 1), ed. J.-C. Gardin. Paris: de Boccard, pp. 135–141.

Sarianidi, V.I. 1998. *Margiana and Protozoroastrism*. Athens: Kapon.

Sarianidi, V.I. 2004. "Strana Margush otkryvaet svoi tainy dvortsovo-kul'tovyi ansambl' severnogo Gonura" [The Margush country opens its North Gonur secret palace-temple monument]. In *U istokov tsivilizatsii* [At the origins of civilization], ed. M.-F. Kosarev, P.M. Kozhin, and N.A. Dubova. Moscow: Staryj Sad, pp. 229–253.

Sarianidi, V.I. 2009. *Margush: Mystery and Truth of the Great Culture*. Ashgabat: Türkmen döwlet nesirgy'at gullugy.

Schmidt, E.F. 1937. *Excavations at Tepe Hissar Damghan*. Philadelphia: University of Pennsylvania Press.

Schrader, O. 1893. *Turkestan*. London: Charles Griffin & Sons.

Stein, M.A. [1931] 1982. *An Archaeological Tour in Gedrosia* (Memoirs of the Archaeological Survey of India 43). New Delhi: Cosmo Publications.

Stein, M.A. 1937. *Archaeological Reconnaissances in the North-Western India and South-Eastern Iran*. London: Macmillan.

Steinkeller, P. 1982. "The Question of Marhashi: A Contribution to the Historical Geography of Iran in the Third Millennium B.C." *Zeitschrift für Assyriologie* 72: 237–265.

Steinkeller, P. 2014. "Marhashi and Beyond: The Jiroft Civilization in a Historical Perspective." In *"My Life Is Like the Summer Rose": Maurizio Tosi e l'Archeologia come Modo di Vivere – Papers in Honour of Maurizio Tosi for His 70th Birthday* (BAR International Series 2690), ed. C.C. Lamberg-Karlovsky and B. Genito. Oxford: Archeopress, pp. 691–707.

Thornton, C.P. 2012. "Iran." In *A Companion to the Archaeology of the Ancient Near East*, ed. D.T. Potts. Malden, MA: Wiley-Blackwell, pp. 596–606.
Thornton, C.P. 2013. "The Bronze Age in Northeastern Iran." In *The Oxford Handbook of Ancient Iran*, ed. D.T. Potts. Oxford: Oxford University Press, pp. 179–203.
Vahdati, A.A. and R. Biscione 2013. "Iran–Italy Joint Excavation at Tepe Chalow in Jajarm Plain, North Khorasan Province." *Catalogue of the Exhibition of Archaeological Finds 2013: 13th Annual Symposium on the Iranian Archaeology*. Tehran: National Museum of Iran, pp. 16–23.
Vámbéry, A. 1873. *History of Bokhara: From the Earliest Period Down to the Present*. London: Henry King & Co.
Vidale, M. 2015. "Searching for Mythological Themes on the 'Jiroft' Chlorite Artefacts." *Iranica Antiqua* 50: 15–59.
Vidale, M. 2017. *Treasures from the Oxus*. London and New York: I.B. Tauris & Co.
Vidale, M. and D. Frenez 2015. "Indus Components in the Iconography of a White Marble Cylinder Seal from Konar Sandal South (Kerman, Iran)." *South Asian Studies* 31(1): 144–154.
Witzel, M. 2013. "Iranian Migration." In *The Oxford Handbook of Ancient Iran*, ed. D.T. Potts. Oxford: Oxford University Press, pp. 423–441.
Wright, R.P. 1989. "New Perspectives on Third Millennium Painted Grey Wares." In *South Asian Archaeology 1985*, ed. K. Frifelt and P. Sorensen. London: Curzon Press, pp. 137–149.

CHAPTER TWENTY-ONE

INTERACTION BETWEEN THE WORLDS OF SOUTH ASIA AND CENTRAL ASIA

Shereen Ratnagar

INTRODUCTION

From the Kopet Dagh piedmont, Margiana, and the Oxus region, to central Afghanistan, then to Seistan and the Kandahar area, and further east to the Kachi plain and the realm of the Harappan civilization in the Indus valley, lies the uneven topography of a geographically and culturally diverse region (Figure 21.1). Uninhabited areas separate clusters of early sites, and we can detect multiple axes or loops of cultural interaction.

While the narrow strip of the Kopet Dagh piedmont, where the Anau/Namazga culture developed, had assured groundwater and the flow of several medium but dependable rivers, in Margiana and South Bactria, where the fortified centers of multiple chiefdoms were established (Lamberg-Karlovsky 1994), several small and branching rivers with endoreic drainage were flowing. In South Asia, at the heart of the Harappan culture, the formidable and annually flooding Indus River system initiating at great heights in the Himalaya, and its agricultural expanses lying on an unconfined aquifer, were regularly renewed with fertile alluvium. Yet similar sequences from the Neolithic to the Bronze Age were founded on much the same domesticates in both civilizations, except for specific species, like pulses in the Indus region or the camel in Central Asia.

Harappan settlement occurs south of the latitudes of Harappa and Rakhi Garhi but not on the fertile Potwar plateau of Pakistan with its good winter rainfall. In rocky Kutch, which has no perennial river, fortified Indus sites dependent on ground water are widely dispersed, and very few settlements occur in Saurashtra. Finally, there were important Harappan outposts in places as distant as Makran and north-eastern Afghanistan. The signature of this culture comprises mass-produced baked bricks for urban buildings, fortified citadels-cum-lower towns, small standardized cubical weights of chert, a logographic writing mainly on small but expertly carved stamp seals, flash-flood harvesting in Kutch, and a complex pyrotechnology for the use of steatite. There is evidence that state formation had encompassed a large area (Ratnagar 2015). Harappan cities were smaller than Mesopotamian Uruk and Ur, and dispersed, even though the area of the civilization is huge, but the total settled area is appreciable when compared with Bronze Age southern Central Asia.

Unlike Mesopotamia and Egypt of the third millennium BC, in southern and Central Asia the Bronze Age civilizations were short-lived and lacked grandiose images of persons with political power.

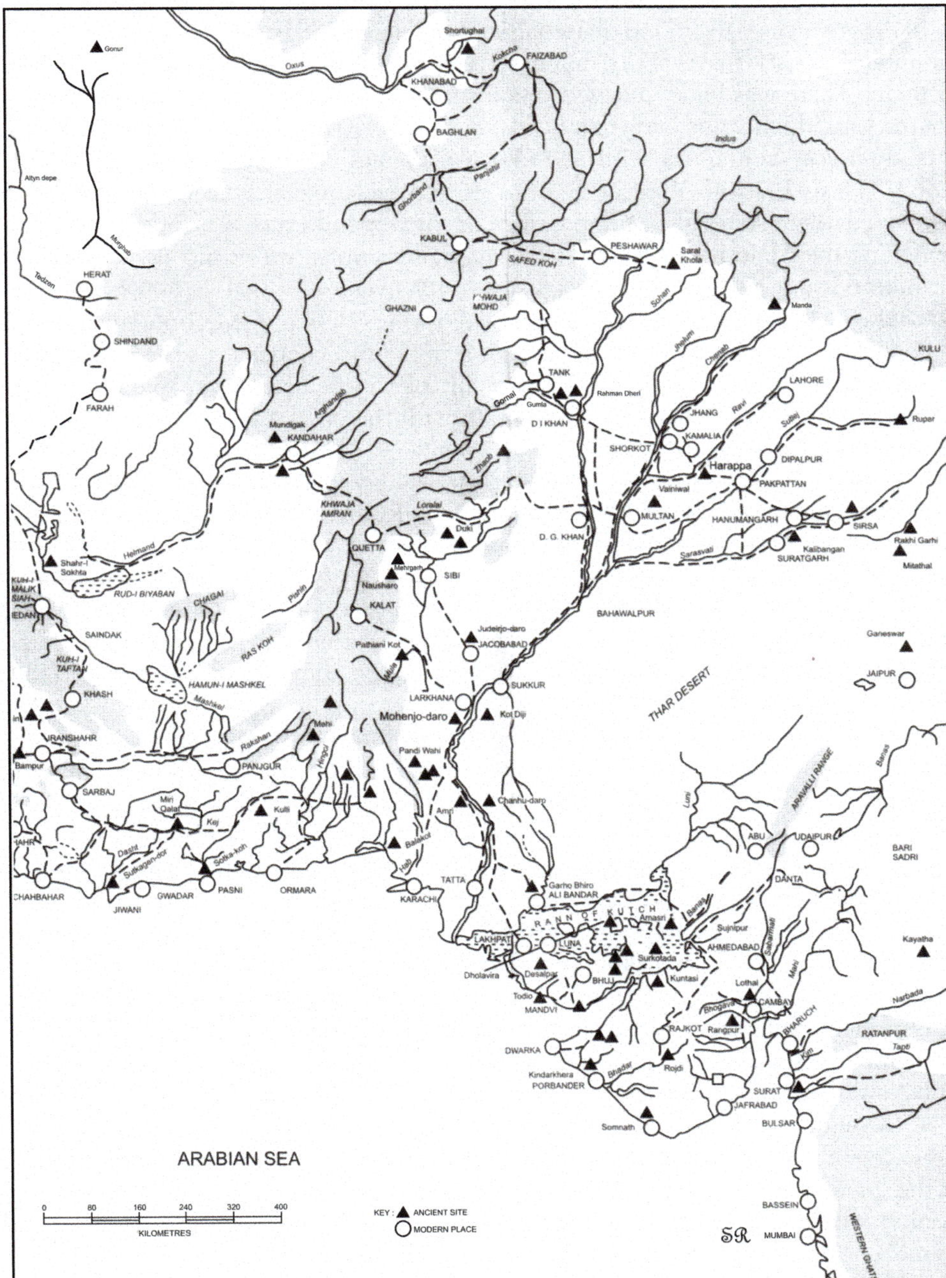

Figure 21.1 The area of South Asia and Central Asia.

DIFFERENT MODES OF INTERACTIONS

The survey of material culture similarities, linkages, and influences across the whole region, reveals that the interconnections between the Harappan world and the Bactria-Margiana Archaeological Complex (BMAC) were of an order different from those between the Harappan civilization and the southern regions to Mesopotamia.

In the latter case, all kinds of materials and artifacts were carried: the produce of high mountains, coasts, river valleys, and deserts, either by pack animals, wheeled vehicles, or boats. There was bulky diorite stone for large statues, but also miniature monkey figurines and high-value carnelian beads. Fine materials such as ivory and intricately carved chlorite containers with mythological themes contrasted with ship building wood. The societies involved in the crafting and exchange of such wares were either developed urban states like Mesopotamia, or tribal polities such as copper-rich Oman.

Behind the Central and South Asian interconnections, we see no signs of the use of military power to procure special stones for oversized royal statues (as was the case when Manishtushu of Akkad conquered Magan/Oman). Mesopotamian rulers imported buffalos, bred only in India, as is attested by cylinder seals (Boehmer 1975) depicting Mesopotamian heroes in worship of the animal. More significant is a cylinder seal whose owner was a Sumerian with the official designation of eme.bal Meluhha, or the official translator of the language of Meluhha (South Asia) (Gelb 1968; Collon 1996: 215, fig. 12). These seals, together with one from Tell Asmar with the elephant, rhinoceros, and fish-eating crocodile, are testimony to Mesopotamian knowledge of the world outside and, more significantly, to the institutionalization of communication with foreigners.

Where Central and South Asian links are concerned, there is nothing comparable to such state-organized trade.[1] The linkages are certainly early. At Mehrgarh, lapis lazuli, seashells, and turquoise beads are reported in some burials of the Neolithic period I graves (Jarrige 2008: 146),[2] but there are no Harappan cubical weights in Central Asia, and the latter area lacks indigenous forms of weights or script.

In the Kopet Dagh piedmont zone, the odd etched carnelian beads occur in small number at Altyn Depe, in one of the richest burials (252) and in a collective vault (Kohl 1984: 130–131 and pl. 12c) – strung with agate and gold beads in one case, and in a waist girdle with carnelian, turquoise, agate, onyx, and crystal beads in another. A few etched carnelian beads are also known at Gonur (Sarianidi 2007: 115–116), and five others were found in an intrusive burial at Ulug Depe, in association with gold, turquoise, agate, and lapis (Bendezu-Sarmiento and Lhuillier 2016: 527). At Altyn Depe (Kohl 1984: 133; Masson 1988: pl. XXII.1a and 1b) there were two seals of Harappan origin, one from excavation 9 with two Harappan signs said to be authentic Harappan writing, and the second, from a "priest's grave" in excavation 7, with a swastika on its obverse.

In the endoreic delta of the Murghab, an inscribed Harappan-type seal with an elephant emblem comes from Gonur (Sarianidi 2005: 258, fig. 114) (Figure 21.2). At the same time, a few open-work round stamp seals from Bactria showing a seated monkey with a long tail have no counterparts in the Harappan world (Francfort and Tremblay 2010: 143, fig. 16) – perhaps an argument for the existence of monkeys up to the northeastern flanks of the Hindukush (Francfort and Tremblay 2010: 144–145; *contra* Ratnagar 2004: 203).

Ivory of South Asian origin was found in hoards and burials in the uppermost levels of Altyn Depe. There was raw ivory cached away inside walls, as well as rods with a rectangular section and incised geometric motifs that Masson (1992: 242) interpreted as "divination sticks" of Harappan origin. Several such batons have also been noted by Frenez (2017) in grave deposits at Gonur, as well as a number of other ivory artifacts (see Chapter 7 and 22). Still at Altyn, ivory beads interspersed with lapis beads that had been stitched onto the clothes of a deceased man have been discovered. Interestingly, the exceptional burial dug into the summit in the northern part of Ulug Depe mentioned above contained high-value alabaster containers, and also beads of precious stone, but no seals or ivory.

Faience bangles are also mentioned in Margiana, as well as the pipal leaf motif (see, e.g., Francfort and Tremblay 2010: 110). With regard to craftsmanship, a calcite statue of a seated ram found in a grave at Gonur Depe (Sarianidi 2005: 233, fig. 93), size unknown, is evocative of the Indus style.

Were single etched carnelian beads mere strays, or did their rarity indicate the high value of an exotic (South Asian) technology? In Mesopotamia one such bead was part of the Ur treasure found in Mari, a gift from one royal house to the deity of another. Such beads have been excavated in elite living quarters at Altyn Depe, but do not occur in the jewelry hoards cached away in the last days of Mohenjo-daro. Yet the special technique of treating the surface of hard carnelian makes us infer that the beads were valued as much at home as abroad, and not exclusively for export: There were unfinished etched carnelian beads at Chanhu-daro (Mackay 1943: 199) and about 14 clay imitations of them at Mohenjo-daro (Mackay 1938: 506).

Figure 21.2 Indus seal with Indus script found at Gonur North Area, 9 room 19. © Margiana Archaeological Expedition, courtesy of N. Dubova.

Ivory may tell its own story. In Harappan towns ivory was used for everyday items like thin incised cylinders, combs, rods, containers, furniture fittings, clasps for boxes, and even a jar stopper. Mackay (1938: 579) noted that "so much was ivory used [in Mohenjo-daro] that bone takes a very subordinate place." In contrast, judging from the archaeological contexts, it was fairly precious in Turkmenistan. An unusual find, however, occurred in the late levels of Mohenjo-daro (Mackay 1938: 117). Two elephant tusks (one of these more than a meter long) lay in close proximity to the unburied skeletons of nine persons, five of them children. One of the children wore an ivory comb and Mackay notes that it was likely that the group were local residents because of a known Harappan type of bronze bangle on one skeleton. Thus, Mackay (1938: 118) suggests that a group of ivory carvers may have been attempting to flee the city at a time of strife. Such an event in the last days of Mohenjo-daro would be roughly contemporary with the contexts in which ivory was used/hidden away in Altyn Depe. Kohl (1984: 133) points out that the strongest similarities between the material culture of Altyn Depe and that of Harappan sites "seem to come from the very latest levels" at Altyn. The excavators of Ulug Depe say much the same – Namazga (NMG) V pottery occurs in association with the funerary finery there. In several tombs at Gonur North, ivory objects were deposited: Frenez (2017) has noted Harappan-style combs, batons, and flat round or rectangular pieces with incised patterns, as well as a fragment of ivory tusk.

Perhaps, then, the Harappan-type seals and ivory in Turkmenistan indicate that some people (craftsmen?) migrated towards the northwest from the Indus region during a time beset with political unrest (see Ratnagar 2000: 33–45). Admittedly this is a shaky hypothesis because (a) no ivory has been reported at the Harappan settlement of Shortughai near the Oxus (though we could speculate that refugees would have avoided a Harappan-administered outpost), and (b) the burials with etched carnelian beads have also other diverse ornaments and artifacts that are not traceable to South Asia.

THE PROCUREMENT OF RAW MATERIAL

Lapis lazuli

In contrast to the lapis lazuli bead almost 12 cm long and inscribed with a royal dedication that was found at Mari (Cholidis 2003: 143) and the lapis spouted cup almost 7 cm high in an Ur royal burial (Collins 2003: 117), only small beads and small items of this stone have been found in Central Asia and South Asia. But the fact that lapis lazuli is rare at Mohenjo-daro does not necessarily mean that it was unimportant. The same applies to finished lapis ornaments at Shahr-i Sokhta and the numerical preponderance (9:1) of waste lapis flakes there. In both cases lapis was imported. And it was an item important enough to warrant a collecting outpost at Shortughai, only 2.5 ha in size, on the loess near the confluence of the Kokcha River with the Amu Darya. On the upper Kokcha are the major lapis quarries, and pebbles of it could have been retrieved from the riverbed as shown by one such find at Shortughai (Francfort and Tremblay 2010: 128fn252). In more than one research paper, Jean-Francois Jarrige reported having seen lapis lazuli in the Chagai highlands, north of Dalbandin town. However, he pointed out (Jarrige 1996: 869) that the Badakhshan lapis was probably known in Bronze Age South Asia as a stone with a better hue than the Chagai one (see also Casanova 1995: 16). Incidentally, plentiful marble onyx is found in the Chagai district as a whole (Ciarla 1981: 48) and this stone was used for the making of vessels at Shahr-i Sokhta. Recently, Law (2014) has made a logical and carefully researched argument that the Chagai source is geologically unlikely and has no documentary mention in geological or travel literature. Local assertions about the occurrence of lapis could come from a confusion with copper ore. Nevertheless, he leaves the question of the source open.

Lyonnet (1977) has made a reasoned argument that Shortughai does not stand directly on the road between Turkmenistan and the upper Indus valley, and that the Indus Civilization was not the result of a population influx. Thus, the *raison d'être* of Shortughai was trade in a foreign country.[3] On excavation (Francfort 1989), the site yielded in its earliest habitation levels a material culture with Harappan ceramic forms and bricklaying conventions, a rhinoceros on a seal with Harappan script, a couple of sherds with scratched signs of the script, five etched carnelian beds, a gold bead enclosing an axial tube, a long (7.7 cm) carnelian elongated barrel bead, and bangles of the same marine shell as used in South Asian sites. There were also four lapis beads and many more waste flakes. While there were no cubical chert weights and no masses of chert blades for harvest work as in the Indus and Kutch, and while the bone tools and camel figurine (Francfort 1989: 161, pl. 82.1) are not recognizably Harappan, the other items belong to the Indus repertoire. This was clearly an implanted habitation.

Turquoise

Turquoise is reported in a few headbands at Neolithic Mehrgarh (see note 2) (Sellier 1991: 76), threaded with beads of steatite, lapis, and terracotta. In period III (ca. 4800–3500 BC), there are indications that small beads of lapis, turquoise, etc. were made locally (Sellier 1991: 77). In use in Turkmenistan from the NMG II period (ca. 4000–3500 BC), the nearest source was across the Kopet Dagh, in Iran's Kuh-i Binalud near Nishapur.[4]

At Shahr-i Sokhta, a very small number of turquoise ornaments occur together with lapis lazuli (Piperno 1976, 1979; Sajjadi 2003). Could Seistan have been the source of the Mehrgarh stones, perhaps even of the entire finery worn by the dead? No exact chronological synchronism of the two sites and periods is possible, but one would certainly expect people to have been moving with their animal herds along the Kandahar–Quetta–Bolan Pass–Kachi route. In the Early Indus period there was a little lapis lazuli coming to the sites nearby from hills-to-plains routes: Mehrgarh, Rahman Dheri (the Gomal Pass), Sheri Khan Tarakai (the Tochi Pass), Jalilpur, Amri, and Pandi Wahi. For the Mature Harappan period, lapis has been reported at several sites, in the form of a few irregular-shaped beads (it is not a stone with a conchoidal fracture and will not flake in a predictable way). At Chanhu-daro, for instance, Mackay found 15 lapis beads compared to 14 of jasper, 16 of shell, and 16 of copper/bronze. There were also two unfinished beads of lapis, which "prove[d] that this stone was actually worked at Chanhu-daro" (Mackay 1943: 209). One may argue that the exotica found in Mohenjo-daro and at Altyn or Gonur were a matter of random gifts or incidental exchange, yet valued objects could have been given as gifts at marriages, or as diplomatic gifts between elite families for reasons of prestige or political alliance, or even as tribute or homage.[5] Only four beads but dozens of waste pieces of the lapis stone are reported at Harappan Shortughai, as stated above. Importantly, in the Harappan period contemporary with the NMG V culture, Shortughai had no artifacts from Turkmenistan or Margiana unless one is to refer to the clay figurine that appears to represent a camel (Francfort 1989: 167, pl. XCI.5). Relevant to the setting up of Shortughai is the discovery of a scatter of waste lapis lazuli on mound HD4 on the coast of Oman at Ras al Hadd, near a site with sherds of Harappan transportation jars (see Ratnagar 2004: 189), a clear testimony to its Harappan export towards the west by sea.

Thus far incidental exchange, organized trade, royal gifts, and migrations (of people with their craftsmanship) have been considered as explanations for the South and Central Asian connections. Indeed, in tandem with trade is the possibility of migrations of groups (in one direction), large or small. As suggested, worked and unworked ivory and etched carnelian beads could have been part of the baggage of migrants to Turkmenistan, at a time when the world was in upheaval. Kohl (1984: 133) wrote that "contact was strongest at the eve of the collapse of the site (or even later, if the hoards were intrusive)." Harappan artifacts occur in disparate contexts in Turkmenistan and Margiana. Not all the Harappan seals occurred in burials as at Altyn Depe. The splendid inscribed elephant seal of Harappan type from Gonur Depe mentioned above comes from outside the palace, area 9, room 19 (with a "trans-Elamite-like" cylinder seal discovered in another room, see Chapter 8). One cannot help wonder if the elephant symbol was an old connection with ivory carving in South Asia. As far as I know, elephant seals have been found only at Harappa, Mohenjo-daro, and Chanhu-daro.

Several stone columns were found in the "Gonur (Large) Necropolis" (Kohl 2007: 196–201; Sarianidi and Dubova 2010; Lamberg-Karlovsky 2013) and in graves at Altyn Depe and Ulug Depe on the Kopet Dagh piedmont, as also on house floors at Altyn. Such artifacts appear to indicate movements from northwest to southeast. Such polished stone artifacts have been found in Baluchistan and the Kachi plain. Scepters, columns, and disks occurred together in a substantial hoard (perhaps a votive funerary deposit) at Quetta (Jarrige and Usman Hasan 1989). Materials from graves and cenotaphs at Mehrgarh period VIII include a fine pottery seen in the Quetta hoard with bowls and cups on stands, a long marble staff, a bronze pin with a figured head, and a cosmetic container (Santoni 1984). A short distance south is Sibri, a much eroded habitation site with a stone column, bronze compartmented seals, and a socketed bronze adze-axe. There were also a large number of schematic human figurines with violin-shaped bodies and stumps for arms (Santoni 1984).

Jarrige made remarks on the variety of modes of disposal of the dead, which could attest to the arrival of new groups – not necessarily settlers but visitors or prospectors – whose burials were interspersed among the graves and memorials of other migrants. Yet no marble miniature column or grooved disk was found in Mohenjo-daro or Chanhu-daro. At Chanhu-daro there is only a fluted bronze cosmetic flacon, in a house from Jhukar levels (Mackay 1943: 177, pl. LXXVI.32), comparable to that at Mehrgarh VIII and others at Gonur. At Mohenjo-daro the rare marble that was used was coarse grained and no skills are evident in shaping vessels out of it (Ratnagar 2004: 151). The argument of Jarrige and Usman Hasan (1989) and Hiebert and Lamberg-Karlovsky (1992: 3–4) for migrants is thus convincing (see also Chapter 20).

Many of the intercultural artifacts are without context or correlations. True, tanged bronze mirrors occur all over our region and also at Harappan sites, mainly in the burials. If these were of high tin bronze and therefore had reflecting surfaces, was the spread of this artifact type connected with metallurgical technology, or was it part of shamanic paraphernalia for "seeing"? Does the occurrence of a single-socketed adze-axe at Mohenjo-daro (and one at Sibri) reveal actual contact with Gonur Depe, where the same advanced type of artifact occurs? A few animal-headed pins occur in Margiana (Dashly) and at Mohenjo-daro, but the resemblance is generic. A truly striking occurrence is an eagle with outstretched wings on a stepped seal from Harappa (Vats 1940: pl. XCI.255), matched in Bactria. One would again assume that such artifacts were personal possessions, and would thus think of travelers or migrants into South Asia. We do not know if the destruction of Mundigak IV.1 was connected with such movements, perhaps of single travelers or prospectors.

Seals are distributed across our region and beyond (Ferioli *et al.* 1979; Salvatori 2000; see also Chapter 8). They are made of bronze, stone, and terracotta. Open work is common in Margiana, and stepped outlines are known in all the relevant regions. I have found it impossible to discern distribution patterns among the various types; not all illustrations give the dimensions, and a great number come from the antiquities market. The visual content of some of the seals, too, is minimal, and similarities may be fortuitous. A raised reverse, or a perforated boss on many of these, indicate that they were pressed onto wet surfaces. At post-Harappan Pirak, several seals are about 4.0 cm in diameter, with prominent perforated bosses (see Jarrige and Santoni 1979: fig. 97). While there may be a good match in Tepe Hissar or at Mundigak (Jarrige and Santoni 1979: fig. 96), the designs themselves are elementary and the similarity may be a coincidence. Pirak gives us a rare context that is not a funerary one. In period II (1300–1100 BC), a complex of rooms with storage jars, finely painted pottery, and unbaked clay figurines have yielded a number of round seal impressions with a cross design on unbaked clay. Also found in the vicinity are six terracotta seals. So the repeated stamping of these artifacts for accounts/records/authentication is clear (Jarrige and Santoni 1979: 368, 371, 380). At the same time, the borderlands area lacked writing: At Gonur only several irregular clay lumps scored with lines across and along were found (see Lamberg-Karlovsky 2013: 49, fig. 17a). For their part, some cross-shaped and animal-outline stone and metal seals of very small dimensions, 0.5 or 2.0 cm across, could have been tokens of herding agreements or "entry passes" for herds in agricultural domains, or the clientship of a chief.

PASTORALISM AND ITS RELEVANCE

To make the case for the role of pastoralists in the economy of Mehrgarh, I have argued previously (Ratnagar 1987) that the site saw its development as a regional center as it managed relationships between nomads and villagers: it is not located on the best agricultural land; from the beginning there was pottery that matched that of the Quetta valley, Zhob, Loralai, etc.; and the Kachi lies on the annual winter migration route of nomads from the Kalat region, the Marri hills, and the Quetta plateau – open to ingress from three directions (see also Audouze and Jarrige 1991). For periodic access to winter water, grazing, and exchange partners in an agricultural village, the good offices of a village elder would be important. Ties of kinship or marriage would make it easier for shepherds to return each year to the Kachi. A herdsman may give token gifts to the elder of a settled village in order to express his goodwill and his endeavor not to let his flocks harm the sown fields; he could have in-laws in the village. This argument, however, was based on the premise of separation between farming and stockbreeding.

Actually, Khazanov (2009) makes a distinction between pastoralists of the Chalcolithic and Bronze Ages and those who rode horses and/or camels in the Iron Age and later. The latter were true nomads, but in earlier periods, pastoralists would be in special need of integration into the rural economy. Yet this distinction may not always work: Many pastoralists who walk along their migrations cover long distances. For instance, the Bakkarwal annual circuit with sheep, goats, buffalo, and horses – all movement is on foot – is more than 500 km (Suri 2014: 55–56). Also, Chetan Singh (2003) has written about Himalayan pastoralists placing their goods on sturdy sheep bred to carry loads. In any case, the early settlement at Shahr-i Sokhta, a Late Chalcolithic Bronze Age settlement, has produced evidence for the husbandry of the Bactrian camel, far to the south of the zone of its original domestication. Long-distance pastoral nomadism was part of the interaction across this world.

Lyonnet (2009) suggests the likelihood that some Syrian settlements in the third millennium were not towns in the accepted sense, but gathering places for herdsmen, their rituals, and grain storage or periodic meetings. This suggests the possibility that the collective granaries of period II (ca. 5500–4800 BC) at Mehrgarh (Quivron 1991; Samzun 1991) indicate a village facility for both herders and crop growers – an integrated community – up to the Chalcolithic period in the Kachi. Thereafter, individual house storage jars replaced collective storage buildings.

Mehrgarh was not incorporated into the Harappan realm and at Harappan Nausharo nearby, there is no lapis. Traffic down the Bolan Pass ceased at some point: After period IV.3, Mundigak lay deserted for a long time (Jarrige 1996: 864), and many settlements along the Bolan routes taken by sheep herders saw a hiatus of occupation. I have mentioned an intervention by the Harappan authorities in the movement of lapis lazuli (and alluvial gold, possibly) through the establishment of Shortughai near the Oxus–Kokcha confluence, to organize collection of the blue stone and also of placer gold from the Kokcha sands, either directly, in season, or from the local people. The motivation behind what I see as the closure of the Bolan frontier could have been the special value of lapis lazuli. It was treasured in Mesopotamian palaces, worn by goddesses at the neck, inscribed with the deeds of Gilgamesh, presented by a king to a king, and was the material of some exquisite craftsmanship. The long narrative of Enmerkar and the En of Arata shows that the Sumerians had no knowledge of the land of lapis, no experience of the high mountains. If left to nomads to handle, the supply of this beautiful stone could be unreliable, and it could lose its value if by chance large amounts became available and it was worn by just anyone. But after the Mature Harappan period or at the time of its demise (Santoni 1984; Jarrige 1991), movement across the Baluchistan highlands is again discernible with the establishment of Pirak (Jarrige and Santoni 1979).

At the end of the Harappan period, ca. 1700 BC, besides Pirak itself, other sites had Pirak pottery, such as a mound near Pathani Damb, Nausharo on its surface, and coeval sites on the Quetta plateau, or to the north of Kalat, and at Dabar Kot in Loralai. As Jarrige (1985: 44–45) points out, "pottery in the Pirak style has been found not only along the whole western side of Kachi but also on the Quetta plateau." It appears that pastoral movements into the Kachi had resumed after some centuries and it is not surprising that about four or five lapis lazuli beads were found in period II levels at Pirak in contexts that also had evidence for the Bactrian camel and for horse riding. Equally important is Durkhan near Dadhar, also in the north Kachi, where surface exploration revealed an entire lapis industry associated with Pirak III related gray pottery (Jarrige and Santoni 1979: 354).

Things could have begun to trickle into the plains again, but Pirak, occupied between 1700 and 700 BC, is testimony to migration from Central Asia. Located at 150 m elevation below the Bolan Pass and within reach of the Nari and Bolan Rivers, this 9 ha settlement presents a change from the Harappan culture (Jarrige 1997).[6] The crops, for instance, included not only barley and wheat but also rice and *jowar* (sorghum). Long chert blades used in harvest work in the Harappan realm were not found here, instead there were blades with deep and regular notches and hundreds of other kinds of flake tools. In period III, house walls had deep niches regularly spaced, an ingenious arrangement not seen at any Harappan site. About 20 terracotta seals with convex, perforated backs were found that, with their simple designs, have counterparts at Shahr-i Sokhta and also at sites on the Murghab. In period III, after 1100 BC, the use of iron began at Pirak. Horse rider figurines and skeletal remains speak eloquently for immigration down this ancient highway from the steppe. There were also painted figurines and skeletal remains of the Bactrian camel. Scores of unbaked clay figurines of the camel were found in all three levels. A bronze tube with pointed end and multiple perforations in periods II and III at Pirak has parallels at Tillja Tepe, but its function remains unknown.

As regards the Helmand civilization, in a report on the excavations at Deh Morasi Ghundai southwest of Kandahar, L. Dupree (1963) contrasted this site with Mundigak in the hills northwest of the Arghandab (Casal 1961; C. Jarrige and Tosi 1981; Barthélémy de Saizieu *et al.* 1993). He noted the lack of clear evidence at Deh Morasi for wheat and for regular house architecture, and the limited range of its metal and flaked stone assemblage compared with Mundigak. He suggested that while the latter was a peasant village, Deh Morasi was its nomadic pastoral adjunct.

One should consider also the founding of the culturally related Shahr-i Sokhta, around 3200 BC (Biscione 1973, 1984), in an arid landscape (Tate 1910; Tosi 1968, 1969; Piperno and Tosi 1975), partly at least by people who were migrants from southern Turkmenistan and the Tedjen delta. It appears that interactions with the north were maintained in periods II and III: Other than the making of marble-onyx vessels (Ciarla 1981), there was an industry of cutting and paring blocks of lapis lazuli for export; there were graves in which infant kids or lambs were interred with the dead (Sajjadi 2003); and the dung, bones, and hair of the Bactrian camel are evidenced (Compagnoni and Tosi 1977). Moreover, of the sealings found at Shahr-i Sokhta not all were affixed to packages, nor were they part of written communications: some appear to have been placed over closed doors in houses.

Pastoral routes lead along the Helmand to Kandahar and thence Quetta and the Kachi plain. It would not be fanciful to ascribe the procurement of lapis lazuli at Shahr-i Sokhta, the center in that period (soon after 3000 BC) of the lapis work, to the movements of pastoralists on a route coming south to Seistan. In a travelogue, A. Toynbee (1961: 58) described a "chain of mountain-ringed pastures" that linked the Oxus with the Helmand. On this route he saw a series of "green amphitheaters encircled by mountains." These pasture "bowls" were "full of" browsing camels, sheep, goats, and their herdsmen in May. From the Hari Rud River, his route left the green meadows and took a right bend to Mazar-i Sharif and then Kunduz, where there were hundreds of animals on their way to Badakhshan and the Oxus.

Lapis cutting in Seistan came to an end before Shahr-i Sokhta period IV. As stated, the Harappan polity appears to have stopped traffic down the Bolan Pass: There were no cultural exchanges with Mundigak and no settlements to interact with on the Quetta plateau, for instance. The transitional Harappan phase (or period ID) at Nausharo (Jarrige *et al.* 2011: 27; *contra* Cortesi *et al.* 2008) lacks elements of Quetta or Kandahar ceramics, has no Nal polychrome, and no parallels with Baluchi female figurines either. Importantly, there was no lapis lazuli at Mundigak IV.2 or Shahr-i Sokhta IV (Piperno 1973: 120).

It cannot be a coincidence that, contemporary with this long pause, Shortughai appeared, established *de novo* near the confluence of the Kokcha and Oxus,[7] an appropriate location for the collection of lapis lazuli and gold, possibly tin as well. The narrative of Wood (1841) who traveled in the area specifies that the largest lapis mines were located on the upper Kokcha, upstream of Jurm, where the river flows in a narrow (200 yards) and steep gorge. Wood (1841: 255–265) gave a brief account of the process of quarrying. On the route taken by travelers to the lapis quarries lies Faizabad, the principal town of Badakhshan. It is a node not only on the routes of flocks to the Shewa plateau and its fine summer grazing at 3,400 m asl, but also situated on the Kokcha whose upper reaches flow past the lapis mines.

G. Herrmann (1968) was part of an Oxford expedition to Badakhshan and gave some more details about the Sar-i Sang quarries, with a useful map. A subsequent French expedition to the area in 1964, 1967, and 1971 (Wyart *et al.* 1981) specified that the quarry/mine was accessible only between June and November. Importantly, the route to Sar-i Sang on foot and pony mentioned by these sources is Kunduz-Khanabad-Taluqan, a long way to Faizabad, and then upstream along the Kokcha banks to Jurm.

Stech and Piggott (1986), on the basis of the excavated rolled pebble found at Shortughai, added the issue of stream tin to their discussion of the source of lapis in Badakhshan. They state, however, that stream tin occurs in the Zeravshan valley to the north of the Badakhshan Mountains. Was placer tin available, together with pieces of lapis, in the bed of the Kokcha-Oxus near Shortughai? They do not make a specific mention of alluvial gold, but a recent Afghan government publication states that gold panning is a traditional skill of the villages on the lower Kokcha and is practiced between June and September (Afghanistan Water Resources Authority 2013: 48).

It is sometimes forgotten that Dupree (1967) reported high-altitude caves with occupation dated to the Paleolithic, Neolithic, and also with "Nomadic Bronze Age assemblages" (with bronze socketed points, bracelets, and carnelian beads). A few of these are located south of Mazar-i Sharif and west of Kunduz, and the Dara-i Kur cave is in Badakhshan – about 40km from Shortughai – with a Paleolithic, Neolithic (goat cult?),[8] Iron Age, and medieval sequence.

Lyonnet (1981) reported about six chalcolithic sites, pre-Shortughai, in the plain of Taluqan, on the much traveled route descending from Faizabad to Kunduz and connected with the procurement of lapis lazuli from the upper Kokcha, which several herdsmen take today. At these sites potsherds with analogues at Mundigak III and early Indus Amri IA–B were noticed. The plain would have been settled in the early third millennium and, according to Lyonnet, this possibly signals a population movement from the Helmand valley.

Balland and de Benoist (1982: 127–128) found that, in recent times, routes entered the Hindukush from Turkmenistan, and also that hundreds of families wintered in the Kunduz plain, leaving for the Shewa plateau in the summer. Pastoralist routes via the Kokcha and Faizabad are repeatedly mentioned in the ethnography of the Afghan herding population. It was noted that "thousands of nomads pass through Faizabad" with their flocks (Schurmann 1962: 99–102, 401–402; see also Glatzer and Casimir 1983: 308–309; Kreutzmann and Schutte 2011: 100–101, 103).[9] The Shewa highland affords rich summer pasture for flocks from the Takhar, Kunduz, and Baghlan provinces, as well as for local animals of the Shugni people. In the summer of 2003, some 150,000 sheep were grazing here. Although numbers would have been smaller in the third millennium, the point is the striking coincidence of the climb to the lapis quarries and of the shepherds' mountain trails.[10]

A theoretical exploration of pastoralism is now necessary. We need to know why movements of trinkets or valuables or more bulky items may be ascribed to high-altitude pastoralists: What makes them traders? David Clarke (1978: 328, 345–51) has observed that the nomadic pastoralist techno-complex is a "larger and looser" entity than any named archaeological culture.

The theory set out by Khazanov (1994) is as follows. Pastoral nomads need to keep flocks of a minimum size over and above the core herd of male and female breeding and lactating sheep and goats. With an increasing flock, or because of pressures in the home base, herders need to move seasonally. Families with, say, 400 sheep must move with the changing seasons, as no natural grassland can support them all through the year (see Kerven *et al.* 2004).[11] Second, while families move together in clans or descent groups, and claims to pasture are exercised jointly, it is the individual family that owns its animals. It is free to cull its herd and sell animals or their produce. Thus, individual nomads negotiate their sales of wool or sheep milk or cheese independently at markets or villages (also Barth 1960). Third, living on animals and their produce involves many risks. Several animals may die at one time because of snow drifts or infection in winter on the mountains (Kerven *et al.* 2004). The interval between two arrivals at a pasture may be too short for the vegetation to have regrown. There is a fine line between a healthy expansion of a herd vis-à-vis overstocking and depletion, not to mention the fluctuations in labor requirements. As for the trading activities of pastoralists, newborn animals may be in demand in spring, as is wheat for the daily bread. Glatzer and Casimir (1983: 321, 323) call the nomads of western Afghanistan "part of a larger economic system of animal husbandry, agriculture, handicrafts, and trade" (Glatzer and Casimir 1983: 308). Khazanov shows that specialization coupled with instability of production, means that it is the nomad who is the party interested in trade. It is the pastoralist who must take the active role and negotiate his way into village society around him, without waiting for traders to come to him. Even though it may be the sedentary people who gain, it is the nomads who most need to trade (Khazanov 1994: 202–212).

CONCLUSION

To casual exchanges, elite gifts, organized trade, and the migrations of craftsmen, I have added the relevance of the pastoral economy. The migration of groups, tribes, or clans may have been occasioned by the colonization and later abandonment of Shahr-i Sokhta, the destructions at Mundigak IV.1, Gumla, and Kot Diji, the population shifts from Turkmenistan to Margiana and Bactria, and the possible strife and desertion at Mohenjo-daro. Major institutional intervention by South Asians in the trade to Mesopotamia, with the setting up of settlements not only at Shortughai but also at the all-weather sea station of Sutkagen-Dor in Makran, did not work in the long run, perhaps because of the sheer distances involved, perhaps also because such institutions of craftsmanship as had flowered in Mesopotamia did not endure to keep demand for lapis and other products alive.

NOTES

1 Archaeologists have found similarities between a headless seated stone statue (L 950) from Mohenjo-daro, and an embossed silver cup from Bactria with seated men (see Ardeleanu-Jansen 1991: 167–169). I doubt, however, the cultural content in this seeming similarity. Only three men on the silver vessel have their right hand on the right knee, their garments are patterned but not so the plain cloth that covers statue L 950, and it is only the "priest king" of Mohenjo-daro who wears an embroidered shawl. More importantly, the silver container has no provenience – it was purchased in the Kabul bazaar (Ardeleanu-Jansen 1991: 167).

2 Jarrige (2008: 146) writes that in the aceramic Neolithic period at Mehrgarh, burials were dug into the remains of abandoned houses with decaying mud walls. Nine graveyards alternated with nine building levels. But there is no clear indication of the levels from which particular graves were cut down, into the abandoned debris of houses, as he admits (Jarrige 2008: 139). It is not known for how long the site was used as a graveyard each time before burial activity ceased, and the ground was leveled for fresh house construction. The Neolithic I material at the site forms some 7 m of deposit and one wonders just how early were the burials, with their exotic bits and pieces, given that the tops of the grave cuts were not detected. Aside from all this is the caveat Jarrige (2008: 150) lodges that a dense mass of roots sprawls within and across the Neolithic mound to a depth of 7 m, which could have affected the radiocarbon dating.

3 Such a trading post does not have to be located in the immediate vicinity of a mine or quarry. More importantly, it is argued that Shortughai's location made it possible for the colonists to practice their traditional subsistence activities. It has been noted (Francfort 1989: 269) that all the pottery of the early period at Shortughai was made at the site or in its vicinity.

4 In the NMG V period (ca. 2500 to 2000 BC), turquoise was found in a few rich graves at Altyn Depe.

5 At Kunal, a Haryana site mistakenly identified as Harappan, a rich hoard of small stone beads and a few silver and gold ornaments were found – an elite prestation? There were 2,806 steatite beads, but 5,690 beads of lapis (Khatri and Acharya 1997).

6 Geographically it stands on the arid Kachi plain in contradiction to the alluvial plain of the Indus, the boundary between the two falling near Jacobabad. Its location is also between the plains and the mountains of Baluchistan, on the route to Quetta.
7 Francfort is convinced that there would have been some settlement in the area in the previous period.
8 It is now clear that the "Neolithic goat cult" of Dara-i Kur has in fact to be dated to the BMAC.
9 Many pastoralists spend the winter (September to April) in the lowlands of western Afghanistan south of the Hari Rud and then move northeast. Because of dense vegetation growth, the lake Shiwa pastures are prized and flocks may graze there for three months; Pashtuns of the Kunduz district move to this lake at an altitude above 3,000 m, via Faizabad. Uzbek families in Kunduz also drive their flocks up to the high altitudes of Badakhshan. Thus, Kreutzmann and Schutte (2011: 103) estimate that 1.2 million cattle, sheep, goats, horse, and camels are driven along this route and the entire migration takes more than four weeks, with halts at say 31 stages for rest, water, some stored fodder and free grazing, and also animal shelter at night.
10 Özyarkent (2013) argues that pastoral production need not rely on all-year mobility and that there could have been changes of residence by animal herders on a seasonal basis, as mining was also an activity in some areas.
11 Fodder cannot be laid in for more than 50 animals. As Singh (2003: 52–54) has shown, it is a long and tedious procedure to allow natural grass to dry on a hill, then cut it, dry it some more, and then transfer it to a village store.

REFERENCES

Afghanistan Water Resources Authority. 2013. *Lower Kokcha Irrigation Project*. Kabul: Afghanistan Water Resources Authority.

Ardeleanu-Jansen, A. 1991. "The Sculptural Art of the Harappan Culture." In *Forgotten Cities on the Indus*, ed. M. Jansen, M. Mulloy, and G. Urban. Mainz: P von Zabern, pp. 167–178.

Audouze, F. and C. Jarrige 1991. "Nomadic Pastoralists and Sedentary Agriculturists in the Kachi Plain." *Studies in History* 7(2): 232–254.

Balland, D. and A. de Benoist 1982. "Nomades et Semi-Nomades Baluč en Afghanistan." *Revue Géographie de l'Est* 22(1–2): 117–144.

Barth, F. 1960. *Nomads of South Persia*. Oslo: University of Oslo.

Barthélémy de Saizieu, B., A. Bouquillon, and A. Duval 1993. "Les parures en pierre de Mundigak." *Paléorient* 19(2): 65–94.

Bendezu-Sarmiento, J. and J. Lhuillier 2016. "Some More Exceptional Discoveries at Ulug-Depe." In *Trudy Margianskoj Arkheologicheskoj Ekspeditsii*, vol. 6, ed. N. Dubova. Moscow: Staryj Sad, pp. 522–528.

Biscione, R. 1973. "Dynamics of an Early South Asian Urbanization." In *South Asian Archaeology*, ed. N. Hammond. London: Duckworth, pp. 105–118.

Biscione, R. 1984. "Baluchistan Presence in the Ceramic Assemblage of Period I at Shahr-i Sokhta." In *South Asian Archaeology 1981*, ed. B. Allchin. Cambridge: Cambridge University Press, pp. 69–80.

Boehmer, R.M. 1975. "Das Aufreten des Wasserbuffels in Mesopotamien und seine sumerische Beseichnung." *Zeitschrift fur Assyriologie* 64: 1–12.

Casal, J.-M. 1961. *Fouilles de Mundigak*. Paris: Librairie Klincksieck.

Casanova, M. 1995. "Le lapis-lazuli dans l'Orient Antique." In *Les Pierres Précieuses de l'Orient ancien*, ed. F. Tallon. Paris: Musée du Louvre, pp. 15–20.

Cholidis, N. 2003. "The Treasure of Ur from Mari." In *Art of the First Cities: The Third Millennium BC from the Mediterranean to the Indus*, ed. J. Aruz and R. Wallenfels. New York: Metropolitan Museum, pp. 139–147.

Ciarla, R. 1981. "Study of the Manufacture of Stone Vases at Shahr-i Sokhta." In *South Asian Archaeology 1979*, ed. H. Härtel. Berlin: D. Reimer, pp. 45–63.

Clarke, D.L. 1978. *Analytical Archaeology*, 2nd rev. ed. New York: Columbia University Press.

Collins, P. 2003. "The Tomb of Puabi." In *Art of the First Cities: The Third Millennium BC from the Mediterranean to the Indus*, ed. J. Aruz with R. Wallenfels. New York: Metropolitan Museum, pp. 108–132.

Collon, D. 1996. "Mesopotamia and the Indus: The Evidence of the Seals." In *The Indian Ocean in Antiquity*, ed. J. Reade. London: Kegan Paul, pp. 209–225.

Compagnoni, B. and M. Tosi 1977. "The Camel." In *Approaches to Faunal Analysis in the Middle East*, ed. R.H. Meadow and M. Zeder. Peabody Museum Bulletin 2. Cambridge, MA: Harvard University Press, pp. 91–103.

Cortesi, E., M. Tosi, A. Lazzari, and M. Vidale 2008. "Cultural Relationships Beyond the Iranian Plateau." *Paléorient* 34(2): 5–35.

Dupree, L. 1963. *Deh Morasi Ghundai*. New York: American Museum of Natural History.

Dupree, L. 1967. "The Prehistoric Period of Afghanistan." *Afghanistan* 20(3): 8–28.

Ferioli, P., E. Fiandra, and S.Tusa 1979. "Stamp Seals and the Functional Analysis of their Sealings at Shahr-i Sokhta II–III." In *South Asian Archaeology 1975*, ed. J.A. van Lohuizen de Leeuw. Leiden: E.J. Brill, pp. 7–27.

Francfort, H.-P. 1989. *Fouilles de Shortughai*. Paris: Diffusion de Boccard.

Francfort, H.-P. and X. Tremblay 2010. "Marhasi et la Civilisation de l'Oxus." *Iranica Antiqua* 45: 51–151.

Frenez, D. 2017. "Manufacturing and Trade of Asian Elephant Ivory. Evidence from Gonur Depe." *Archaeological Reseach in Asia*. http://doi.org/10.1016/j.ara.2017.08.002.

Gelb, I.J. 1968. "The Word for Dragoman in the Ancient Near East." *Glossa* 2: 93–104.

Glatzer, B. and M.J. Casimir 1983. "Herds and Households Among Pashtun Nomads." *Ethnology* 22(4): 307–325.

Herrmann, G. 1968. "Lapis Lazuli: The Early Phases of it Trade." *Iraq* 30: 21–57.

Hiebert, F.T. and C.C. Lamberg-Karlovsky 1992. "Central Asia and the Indo-Iranian Borderlands." *Iran* 30: 1–15.

Jarrige, C. and M. Tosi 1981. "The Natural Resources of Mundigak." In *South Asian Archaeology 1979*, ed. H. Härtel. Berlin: D. Reimer, pp. 115–142.

Jarrige, J.-F. 1985. "Continuity and Change in the North Kachi Plain." In *South Asian Archaeology 1983*, ed. J. Scotsmans and M. Taddei. Naples: Istituto Universitario Orientale, pp. 35–68.

Jarrige, J.-F. 1991. "Mehrgarh: Its Place in the Development of Ancient Cultures in Pakistan." In *Forgotten Cities on the Indus*, ed. M. Jansen, M. Mulloy, and G. Urban. Mainz: P. von Zabern, pp. 34–50.

Jarrige, J.F. 1996. "Les Fouilles de Nausharo." *Comptes-rendus des Séances de l'Académie des Inscriptions et Belles-Lettres* 140(3): 821–878.

Jarrige, J.-F. 1997. "From Nausharo to Pirak: Continuity and Change in the Kachi/Bolan Region from the 3rd to the 2nd Millennium BC." In *South Asian Archaeology 1995*, ed. R. and B. Allchin. London: Curzon Press, pp. 11–32.

Jarrige, J.-F. 2008. "Mehrgarh Neolithic." *Pragdhara* 18: 135–154.

Jarrige, J.-F., A. Didier, and G. Quivron 2011. "Shahr-i Sokhta and the Chronology of the Indo-Iranian Regions." *Paléorient* 37(2): 7–34.

Jarrige, J.-F. and M. Santoni 1979. *Fouilles de Pirak*. Paris: Diffusion de Bocard.

Jarrige, J.-F and M. Usman Hasan 1989. "Funerary Complexes in Baluchistan at the End of the Third Millennium." In *South Asian Archaeology 1985*, ed. K. Frifelt and P. Sorensen. London: Curzon Press, pp. 150–166.

Kerven, C., I.I. Alimaev, R. Behnke, G. Davidson, L. Franchois, N. Malmakov, E. Mathijs, A. Smailov, S. Temirbekov, and I. Wright 2004. "Retraction and Expansion of Flock Mobility in Central Asia: Costs and Consequences." *African Journal of Range and Forage Science* 21(3): 159–169.

Khatri, J.S. and M. Acharya 1997. "Kunal: The Earliest Pre-Harappan Settlement." In *Facets of Indian Civilization*, ed. J.P. Joshi. Delhi: Aryan Books, pp. 88–91.

Khazanov, A. 1994. *Nomads and the Outside World*, 2nd ed. Madison: University of Wisconsin.

Khazanov, A. 2009. "Specific Characteristics of Chalcolithic and Bronze Age Pastoralism in the Near East." In *Nomads, Tribes, and the State in the Ancient Near East*, ed. J. Szuchman. Chicago: Oriental Institute, pp. 119–127.

Kohl, P.L. 1984. *Central Asia: Palaeolithic Beginnings to the Iron Age*. Paris: Editions Recherche sur les Civilisations.

Kohl, P.L. 2007. *The Making of Bronze Age Eurasia*. Cambridge: Cambridge University Press.

Kreutzmann, H. and S. Schutte 2011. "Contested Commons: Multiple Insecurities of Pastoralists in North-Eastern Afghanistan." *Erdkunde* 65(2): 99–119.

Lamberg-Karlovsky, C.C. 1994. "The Bronze Age Khanates of Central Asia." *Antiquity* 68: 398–405.

Lamberg-Karlovsky, C.C. 2013. "The Oxus Civilization." *Cuadernos de Prehistoria y Arqueología, Universidad Autónoma de Madrid (CuPAUAM)* 39: 21–63.

Law, R. 2014. "Evaluating Potential Lapis lazuli Sources." In *"My life Is Like the Summer Rose": Maurizio Tosi e l'Archeologia come modo di vivere – Papers in Honour of Maurizio Tosi for His 70th Birthday*, ed. C.C. Lamberg-Karlovsky and B. Genito. BAR International Series 2690. Oxford: Archeopress, pp. 419–430.

Lyonnet, B. 1977. "Découvertes de sites de l'Age du Bronze dans le N.E. de l'Afghanistan: leurs rapports avec la civilisation de l'Indus." *Annali* 37: 19–35.

Lyonnet, B. 1981. "Etablissements chalcolithiques dans le Nord-Est de l'Afghanistan." *Paléorient* 7(2): 57–74.

Lyonnet, B. 2009. "Who Lived in the Third-Millennium 'Round Cities' of Northern Syria?" In *Nomads, Tribes, and the State in the Ancient Near East*, ed. J. Szuchman. Chicago: Oriental Institute, pp. 179–200.

Mackay, E. 1938. *Further Excavations at Mohenjo-daro*. Delhi: Govt Press.

Mackay, E. 1943. *Chanhu-daro Excavations*. New Haven, CT: American Oriental Society.

Masson, V.M. 1988. *Altyn-Depe*. Philadelphia, PA: University Museum.

Masson, V.M. 1992. "The Bronze Age in Khurasan and Transoxiana." In *History and Civilizations of Central Asia*, vol. I, ed. A.H. Dani and V.M. Masson. Paris: UNESCO, pp. 225–245.

Özyarkent, H. 2013. "Animal Herder-Miners of the Andronovo Culture." *Metalla* 20(2): 5–16.

Piperno, M. 1973. "Microdrilling at Shahr-i Sokhta." In *South Asian Archaeology 1971*, ed. N. Hammond. London: Duckworth, pp. 119–139.

Piperno, M. 1976. "Grave 77 at Shahr-i Sokhta." *East and West* 26: 9–12.

Piperno, M. 1979. "Socio-Economic Implications from the Graveyard at Shahr-i Sokhta." In *South Asian Archaeology 1977*, ed. M. Taddei. Naples: Istituto Universitario Orientale, pp. 123–39.

Piperno, M. and M. Tosi 1975. "The Graveyard of Shahr-i Sokhta." *Expedition*: 186–197.

Quivron, G. 1991. "The Neolithic Settlement at Mehrgarh." In *Forgotten Cities on the Indus*, ed. M. Jansen, M. Mulloy, and G. Urban. Mainz: Philipp von Zabern, pp. 59–65.

Ratnagar, S. 1987. "Pastoralists in the Prehistory of Baluchistan." *Studies in History* 3(2): 137–154.
Ratnagar, S. 2000. *The End of the Great Harappan Tradition*. Delhi: Manohar.
Ratnagar, S. 2004. *Trading Encounters from the Euphrates to the Indus*. Delhi: Oxford University Press.
Ratnagar, S. 2015. *Harappan Archaeology: Early State Perspectives*. Delhi: Primus.
Sajjadi, S.M. 2003. "Excavations at Shahr-i Sokhta: Excavations of the Graveyard 1977–2000." *Iran* 41: 21–97.
Salvatori, S. 2000. "Bactria and Margiana Seals." *East and West* 50: 97–145.
Samzun, A. 1991. "The Early Chalcolithic: Mehrgarh Period III." In *Forgotten Cities on the Indus*, ed. M. Jansen, M. Mulloy, and G. Urban. Mainz: Philipp von Zabern, pp. 66–72.
Santoni, M. 1984. "Sibri and the South Cemetery of Mehrgarh." In *South Asian Archaeology 1981*, ed. B. Allchin. Cambridge: Cambridge University Press, pp. 52–60.
Sarianidi, V.I. 2005. *GonurDepe, Turkmenistan, City of Kings and Gods*. Ashgabat: Miras.
Sarianidi, V.I. 2007. *Necropolis of Gonur*. Athens: Kapon Editions.
Sarianidi, V. and N.A. Dubova 2010. "Nouvelles sépultures sur le territoire de la nécropole royale de Gonur Depe." *Arts Asiatiques* 65: 5–26.
Schurmann, H.F. 1962. *The Mongols of Afghanistan*. The Hague: Mouton.
Sellier, P. 1991. "Mehrgarh: Funerary Rites." In *Forgotten Cities on the Indus*, ed. M. Jansen, M. Mulloy, and G. Urban. Mainz: Philipp von Zabern, pp. 75–86.
Singh, Ch. 2003. "Between Two Worlds: Trader Pastoralists of Kinnaur." In *Mobile and Marginalized Peoples: Perspectives from the Past*, ed. R.C. Heredia and S. Ratnagar. Delhi: Manohar, pp. 35–64.
Stech, T. and V.C. Pigott 1986. "The Metals Trade in Southwest Asia in the Third Millennium BC." *Iraq* 48: 39–64.
Suri, K. 2014. "Impact of Armed Conflict on Seasonal Migration Practices of Gujjar and Bakkarwal Tribes in Jammu and Kashmir." *International Organization of Scientific Research (IOSR) Journal of Humanities and Social Science* 19(2): 54–62.
Tate, G.P. 1910. *Seistan: A Memoir*. Calcutta: Govt of India.
Tosi, M. 1968. "Excavations at Shahr-i Sokhta, a Chalcolithic Settlement." *East and West* 18: 9–66.
Tosi, M. 1969. "Excavations at Shahr-i Sokhta, the Second Campaign." *East and West* 19: 283–386.
Toynbee, A. 1961. *From the Oxus to the Jumna*. Oxford: Oxford University Press.
Vats, M.S. 1940. *Excavations at Harappa*. Delhi: Govt of India.
Wood, J. 1841. *Personal Narrative of a Journey to the Source of the River Oxus*. London: John Murray.
Wyart, J., P. Bariand, and J. Filippi 1981. "Lapis Lazuli from Sar-e-Sang, Badakhshan." *Gems and Gemology* 17: 184–190.

CHAPTER TWENTY-TWO

THE OXUS CIVILIZATION/BMAC AND ITS INTERACTION WITH THE ARABIAN GULF

A review of the evidence

*Pierre Lombard**

INTRODUCTION

The relationship between the Arab/Persian Gulf cultures and the Oxus Civilization at the end of the 3rd and in the first quarter of the 2nd millennium BC is a recurring theme in the archaeology of the Arabian Peninsula.

As is long known, the Gulf has played a major role in the circulation of different raw materials or goods that the Mesopotamians were searching for abroad (Oppenheim 1954; Leemans 1960). During the Akkadian period, the Gulf was the main artery for the earlier trade between Magan/Oman and Mesopotamia, which passed through Dilmun (Potts 1990; Begemann *et al.* 2010). It is also at that time that the first mention is made in the cuneiform texts of Meluhha/the Indus area (see Laursen and Steinkeller 2017). Magan is usually considered as covering the actual territory of the United Arab Emirates (UAE) and a large part of the Oman Sultanate, but several authors have also proposed that a portion of the Persian coast would be part of it (Possehl 2002; Potts 2016). The definition of Dilmun may have changed across time but, for the period of our concern, it included Bahrain together with a still poorly defined part of the eastern Arabian Peninsula including the Tarut (Saudi Arabia) and Failaka (Kuwait) islands (Figure 22.1).

At Dilmun or Magan, beside finds attributed with certainty to Mesopotamia or the Indus, a growing number of other discoveries presenting features foreign to them has led us to consider that the Gulf was involved in a larger cultural sphere, that of Amiet's (1986) "Iran extérieur" or Possehl's (2002) "Middle Asian Interaction Sphere" that included Central Asia (see also the map in Possehl 2007: 41). These reflections were initiated in the 1970s, but they mostly developed during the 1980s together with the rise of excavations in the Gulf. They owe much to the work of E.C.L. (Inez) During Caspers (1989, 1992, 1994a, 1994b, 1994c, 1996), and many articles related to her work and to these questions have been published in a book dedicated to her memory (Olijdam and Spoor 2008). Other specialists also occasionally participated in this discussion (Tosi 1974; Potts 1983, 1993, 2008; Amiet 1986: 171–180; Masson 1988; Olijdam 2008; Weeks 2016). This research focused mainly on the inventory of the artifacts potentially reflecting such interrelations, or concerned a particular case.

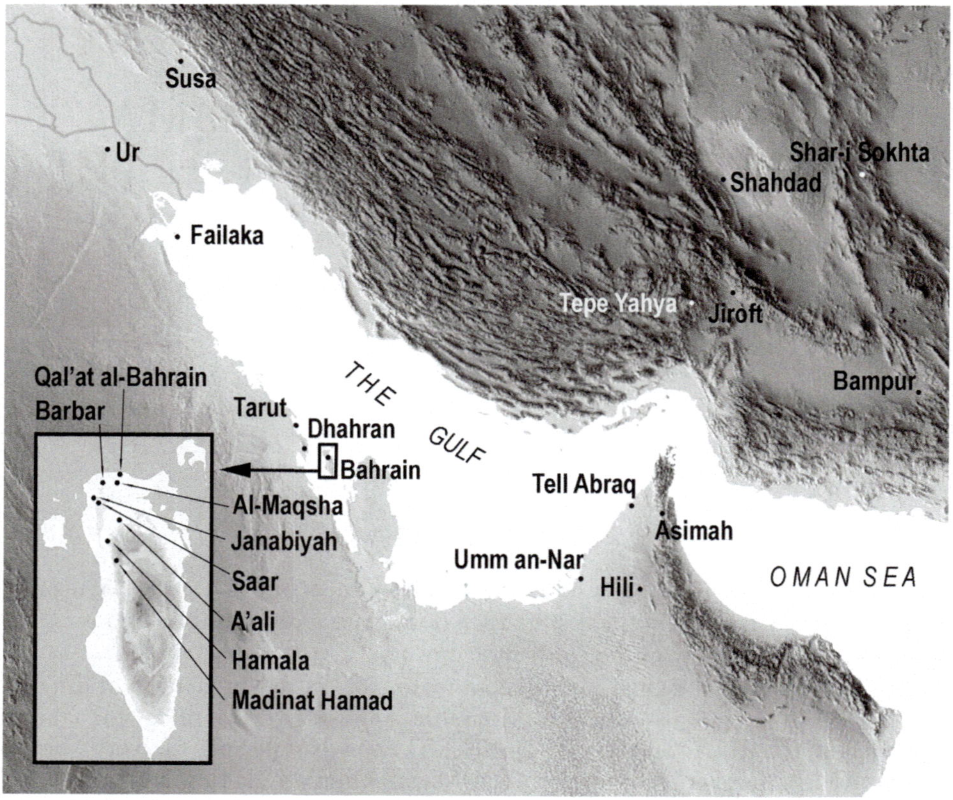

Figure 22.1 Map of the Arabian Gulf area, with archaeological sites mentioned in the chapter (cartography UMR 5133-Archéorient, CNRS, Lyon/ P. Lombard).

The debate is extensive, sometimes contradictory over time, and often stained with inaccuracies about the exact provenance or the precise dating of the discussed artifacts. Several recent discoveries give us an opportunity to reconsider and question the exact nature of the interactions that existed between Central Asia and the Gulf.

A FEW DIRECT RELATIONS

We have first to admit that, altogether, the artifacts testifying of connections between the Gulf and the Oxus area are rare though unquestionable and present in both areas. Moreover, as we will see, it is not always obvious to distinguish the items that are really foreign from those that were made locally but under a deep foreign influence.

Exports from the Gulf discovered in Central Asia

As far as we know, only two artifacts undoubtedly coming from the Gulf have been found in Central Asia. They are made of soft stone and illustrate the characteristic production of Dilmun and Magan.

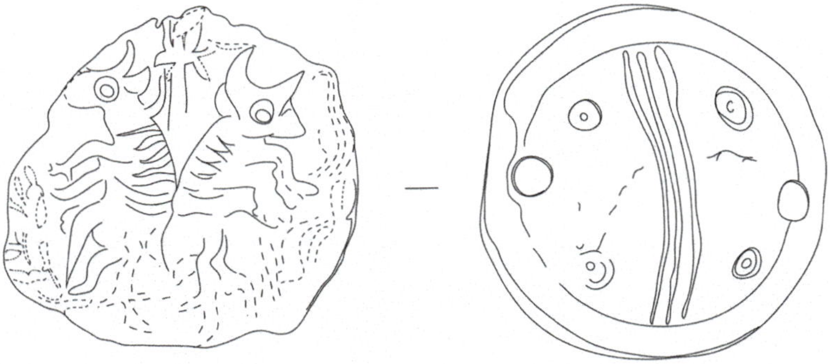

Figure 22.2 Dilmun stamp seal from unknown provenience, Afghanistan(?) (Drawing: J. Marchand, after Sarianidi 1986: 231).

The first is a typical Dilmun stamp seal, as shown by its morphology, style of engraving, and iconography (Figure 22.2). Unfortunately, its provenience is uncertain. It was published laconically by V.I. Sarianidi (1986: 231, right margin), who said it was from Afghanistan without further precision. The rather early date of the publication shows that it probably came from the Kabul Bazar where, at that time, the material was mostly that from the looted graves in Bactria, but we have no way to prove it. Only a drawing and an imprint of its reverse side is published, revealing a rather complex iconography but the drawing is not totally clear. It represents two bovines in standing position, placed back to back, accompanied by symbols difficult to identify. This discovery was reexamined later by E. Olijdam (2008: 282, fig. 14), who compared it with a so-called "Dilmun" stamp seal from the now underwater Harappan port of Dwarka (Gujarat), already noticed by H. Crawford (1998: 90–91, and fig. 5.7). Though the motif on the Dwarka piece could effectively recall the Dilmun iconographic style, Olijdam's view is quite surprising, as this latter seal, in our view, cannot be attributed to the Dilmun production because of its square shape that rather relates it with that of the Late Harappan. Moreover, and as stressed by Olijdam, if double or multiple animal patterns are present in the Dilmun iconographic repertoire, the disposition of animals back to back is unusual and the animals are more frequently represented with crossed bodies or necks (see, e.g., Al-Sindi 1999: 262–264). Nevertheless, even if no clear *comparendum* exists for this seal supposed to come from Afghanistan, its general style and the characteristic organization of its back – a wide perforated boss decorated with three parallel lines running at right angles to the perforation and with four dot-in-circle motifs placed on both sides of the lines – definitely point towards a chronological attribution to Qal'at al-Bahrain City IIb–c (ca. 1950–1800 BC[1]); that is, a very intensive period of the Gulf trade.

The second artifact comes from a woman's grave of the "Royal Necropolis" (grave 3245)[2] at Gonur North.[3] This grave is a simple two-chambered tomb situated between the larger hypogea, and, by chance, was never looted (Sarianidi 2005: 253–254; Potts 2008; Salvatori 2010; for the Gonur "Royal Necropolis," see Chapter 10) (Figure 22.3). The object is a more or less rectangular chlorite container with two compartments, decorated on the exterior long sides with two rows of incised double-dotted circles (Figure 22.4) (Sarianidi 2005: fig. 103; Potts 2008: fig. 20; Salvatori

Figure 22.3 Grave 3245 from the Gonur "Royal Necropolis" (© Margiana Archaeological Expedition).

Figure 22.4 Compartmented chlorite vessel from grave 3245, Gonur "Royal Necropolis" (© Margiana Archaeological Expedition).

Figure 22.5 Some of the burial offerings from tomb 3245, Gonur "Royal Necropolis" (© Margiana Archaeological Expedition).

2010: 247 and fig. 5; on chlorite containers, see also Chapter 9). Together with this container, in the smaller room dedicated to the offerings, were several other artifacts (Figure 22.5): another chlorite larger casket with four compartments, only decorated on its lid with rows of deep grooves along the sides, and a snake coiled at right angles in the center; a small cylindrical and undecorated chlorite container; four long jadeite pins or needles; a round fluted alabaster vessel; a plain cylindrical alabaster vessel with flat horizontal rim; an ivory cosmetic palette representing a winged griffin swallowing a calf (see Chapter 7, Figure 7.1), and other ivory items with incised dotted circles; part of an unworked elephant tusk; a painted inlay representing a woman's face in profile; a large copper/bronze bowl and another deeper metal vessel with a wide horizontal rim; a triangular copper/bronze spatula with a coiled handle; and a marine shell belonging to the *Cardium* genus (Figure 22.6). The woman in the larger room was wearing a faience bracelet on each arm and a necklace of tiny gold, lapis, and other semiprecious stone beads. Several pottery vessels accompanied these offerings (see Figure 22.3).

While several of these artifacts have close parallels with finds from the famous Quetta hoard found by chance in Baluchistan and undoubtedly related to the Bactria-Margiana Archaeological Complex (BMAC) (Jarrige and Hassan 1989), many of them also point at close connections with different southern areas: (1) the Indus for the ivory items or the faience bracelets; (2) southeastern Iran/Baluchistan(?) for the alabaster vessels, and the four-compartmented chlorite box with the snake motif on the lid; and (3) the Gulf area for the marine shell and the two-compartmented container with incised dotted circles.

This last artifact can clearly be associated with the classical "*série récente*" of chlorite objects. These were first identified in Iran (at Susa and Tepe Yahya) but mainly

Figure 22.6 *Cardium* shell from grave 3245, Gonur "Royal Necropolis" (© Margiana Archaeological Expedition).

produced in the Gulf (Vogt 1985: 31–32; David 1996: 34–38; see also Chapter 9), notably during the Late Umm an-Nar period in the Oman Peninsula where this vessel was most probably made as shown by the presence of several similar items there. To the examples of Tell Abraq mentioned by Potts (2008: 174, fig. 21; 175, figs. 22–25), one must add the five similarly compartmented specimens from Tomb A at Hili North (David 2002: 179, fig. 9, *1–6*), and the eight specimens from Tomb N at Hili Garden (David 2002: 182, fig. 13, *1–3*). D.T. Potts (2008: 176, fig. 29) also mentions only one such vessel from Saar at Bahrain, but many more such containers (with or without compartments) have actually been found in the cemeteries of this settlement. Beside the few specimens published by M. Ibrahim (1982: 154, figs. 3–4; 215, pl. 53, 3–4), at least four unpublished containers, found notably by a Jordanian expedition in the so-called "honeycomb cemetery" at Saar in 1998, are kept in the finds of the Bahrain National Museum. The average dates of the Late Umm an-Nar period, ca. 2300–1950 BC (Görsdorf and Vogt 2001: 3–4) fit well with the occupation of Gonur North grave 3245 (2200–1957, 2 σ cal. BC) (see the Appendix at the end of the volume and Salvatori 2010: 248).

Imports from Central Asia found in the Gulf

The artifacts testifying of relations with the BMAC found in the Gulf are more numerous, but, as mentioned earlier, it is not always easy to separate those that are clear imports from those that are BMAC related and could have been made locally.

Seals and seal impressions

A clear import from the BMAC was discovered in 1985 at the Madinat Hamad cemetery (sector BS2, tumulus 107) on Bahrain Island (Figure 22.7). It is a stone step-sided, perforated, square/rhomboidal stamp seal or amulet (BNM A21327). One side depicts a winged leaping lion with a large opened maw and tail turned up on its rump. The other side shows two squatting(?) wolves placed top to tail one above the other

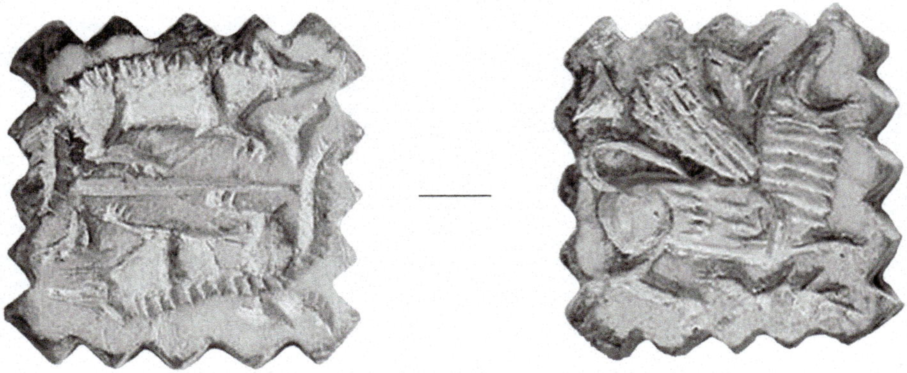

Figure 22.7 BMAC stamp seal or amulet from the Madinat Hamad cemetery (Bahrain) (courtesy of the Bahrain National Museum).

with a horizontal line in between. On both sides, other elements (geometric designs or not clearly identified animals – snake?) are visible (During Caspers 1994c: 106–107 n32; 1996: 52; Crawford and Al-Sindi 1995: 2, fig. 1–2; Sarianidi 1998: fig. 941, *1–2*; Salvatori 2000: 130 (but listed with wrong provenience); Lombard and Boksmati-Fattouh 2017: 68).

This stamp seal belongs to a type frequently found in the BMAC (type 19a in Salvatori 2000: 130–131). Most of the known specimens unfortunately come from the antique market or are housed in private collections (e.g., Sarianidi 1981: 221–255; 1998: cat. 944, 947–957, 1518), but some come from Margiana where the localization is secure (mostly Gonur South, or Togolok 21, i.e., BMAC period II) (Sarianidi 1998: cat. 1622–1631, 1784, etc.). The distinctive shape of the Bahrain stamp seal (Salvatori 2000's "stepped lozenge") is a recurrent one, either in soft stone (fired steatite) or copper (Ligabue and Salvatori 1988: fig. 18; Sarianidi 1998: 164–165, cat. 888, 890; 192–193, cat. 1014; 296–297, cat 1635; Salvatori 2000: figs. 14, 2 and 15, 4; Sarianidi 2002: 267; Vidale 2017: 85–86, figs. 68–71; see also Chapter 8, Figures 8.1, 8.5, and 8.6). The winged lion with opened maw, frequently accompanied by snakes, is a traditional theme of the BMAC glyptic according to Sarianidi (1981: 250; 1998: 34–36) while the pattern of the opposite squatting wolves seems rather uncommon (cf. Sarianidi 1998: 71, cat. 107 for a different, but close representation). Up to now, such a find is unique in the Arabian Gulf and clearly an import.

A second stamp seal comes from Tell F6 at Failaka (Kuwait) (Kjaerum 1983, no. 335) and could be related to type 19b in Salvatori (2000: 132). It is engraved on both sides and is badly eroded but the chest of a kneeling antelope is visible on one of them.

Still from Failaka comes finally another testimony of the Gulf–BMAC interrelations that Olijdam's (2008) careful analysis makes particularly convincing. It concerns a cylinder-seal impression on a fragment of a thick-walled jar discovered in 1988, unfortunately in a disturbed context in the "temple tower" unearthed by the French mission on Tell F6 of the island (Pic 1990: 139, no. 26; Calvet 1996; 2005: 54) (Figure 22.8). At first, it was compared with the Late Mittani glyptic and given a rather late date but M. Pic and Y. Calvet pointed out that seal impressions on jars

Figure 22.8 Cylinder-seal impression on a pottery jar fragment from Tell F6, Failaka, Kuwait (Drawing: J. Marchand, after Calvet 1996: 60, fig. 3).

before firing are very rare in the Gulf, and that the clay and temper of this fragment – buff with coarse mineral inclusions – excluded local or even regional production in the Gulf or southern Mesopotamia (Calvet 1996: 59). Despite Olijdam's statement, which could give the impression of a widespread practice in the BMAC area, the sealing on jars before firing is not very frequent in Central Asia, even more so with a cylinder seal, which is not a local type and started to develop only in the BMAC period II (Salvatori 2008: 80; see also Chapter 8). Reported cases of cylinder-seal impressions (together with that of a stamp seal in one case) have been brought to light in Margiana at Gonur South (Hiebert 1994: 60–61) and at Taip Depe 1 (Kohl 1984: fig. 15c) on the shoulder of large storage jars ca. 1 m in height and over 0.70 m in diameter. In addition to these examples mentioned by Olijdam, we can add the more recent discovery of two large fragments of jars with seal impressions (though not from a cylinder seal) coming from northern Bactria, at the citadel of Dzharkutan, in an area interpreted as a possible storage building (Bendezu-Sarmiento and Mustafakulov 2013: 220). This last site has provided several other finds that also place it within the large exchange network including the Indus valley (Bendezu-Sarmiento and Mustafakulov 2013: 220). At Gonur, the southern part of the settlement is dated to period II, as emphasized by Olijdam (2008: 271) who reproduces this seal impression (Olijdam 2008: 272, figs. 4a–b) as well as several fragments from Taip 1 (Olijdam 2008: 273–274, figs. 5a–c, 6–7). Their technical features are close to those of the Failaka sherd.

One of Olijdam's major arguments in favor of a BMAC interrelation for the Failaka impression concerns its iconography (Olijdam 2008: 275–281), which he develops with much detail. The cylinder seal that was used is characterized by its large size (45 mm high x 15 mm in diameter), which is somewhat exceptional when compared to the average dimensions of the few cylinder seals used in the Gulf (ca. 17–31 mm high x ca. 10–14 mm in diameter). Its impression depicts a figurative scene where two birds of prey with their bodies in profile, outstretched wings and tails, seize the backs of long-tailed and squatting quadrupeds (calves?) in their talons. Two additional motifs complete the field: a rosette and a crescent, as well as a coiled snake with a triangular head, which appear respectively between the wings of the two birds of prey. This scene is repeated twice (12.5 cm long). Such an iconographic theme is totally alien to Dilmun and, according to Olijdam (2008: 275–276), predatory birds with unfolded wings attacking an animal remain exceptional in iconography of the late 3rd and 2nd millennium in the whole Near East. Conversely, such a scene is frequent in the Oxus/BMAC repertoire, as well as the snake motif (associated with it or not). The detailed execution of some features of the birds of prey (eagles?), as the detailed anatomy of their unfolded wings, for example, strongly suggest a first-hand knowledge of these animals by the seal cutter, an observation that could strengthen a Bactrian origin for the seal associated with the Failaka impression. More generally, such a scene where animals are engaged in an antagonist relationship could reflect the ideological hierarchical system that seems to characterize the BMAC iconography (Olijdam 2008: 276–277; cf. also Francfort 1994: 406ff; 413: tab. 2).

Soft-stone containers

Several such chlorite artifacts, often small in size, discovered in the Gulf are clearly foreign to the local production and related to types produced by the Oxus Civilization.

This is the case for a small squarish vial discovered in the 1960s on the site of Al-Rufay'a, Tarut island (Saudi Arabia) (Figure 22.9). Though wrongly attributed to the Hellenistic phase in the original publication, its dating to the Bronze Age and Central Asian origin is not in any doubt today (Zarins 1978: 70, no. 584; see also Chapter 9). We owe its correct identification to S. Salvatori who convincingly related it to his own type E (Salvatori 2003: 724–725, tab.1 – wrongly labeled "F"– and fig. 1; 2008: 83, tab. 6.1 and fig. 6.2) while comparing this small container with examples from Bactria (see, e.g., Pottier 1984: 152–153, fig. 20).

Two other stone vessels have been found at either end of the Gulf and may have been imported from Bactria. One was discovered in the northern part of Bahrain in an Early Dilmun grave at Al-Maqsha (Lombard 1999: 97, no. 95; Potts 2008: 171, fig. 8). The other in the Al-Ain area (Abu Dhabi, UAE) in Tomb A at Hili North (Cleuziou and Vogt 1985: fig. 4.5; Vogt 1985: pl. 28, *11–12*; During Caspers 1994a: 37, 41, fig. 7; 1996: 49; David 2002: 181 and fig. 11; Potts 2008: 170–171,

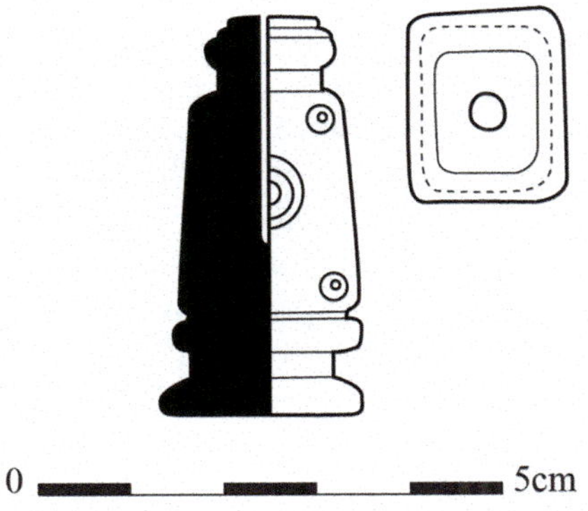

Figure 22.9 Cosmetic vial from Al-Rufay'a (Tarut island, Saudi Arabia) (Drawing: J. Marchand, after Zarins 1978: 70, no. 584).

Figure 22.10 Cubic-shaped chlorite bottle/vial from Al-Maqsha (Bahrain) (courtesy of the Bahrain National Museum).

Figure 22.11 Cubic-shaped chlorite bottle/vial from Hili North (Abu Dhabi, UAE) (Drawing: J. Marchand, after David 2002: fig. 11, 7).

figs. 9–10). Both of them belong to the same type; namely, a small, cosmetic(?), cubic-shaped bottle, with a constricted circular rim (cf. type B2 in Salvatori 2003: 724, fig. 1; 2008: 83). The first specimen (Figure 22.10) is decorated on its four sides with a combination of three incised dotted circles, while the second one (Figure 22.11) displays a simpler pattern (a unique dotted circle on each side). These two miniature vials are clearly distinguishable from the usual production of chlorite vessels from the Oman Peninsula, sometimes exported to Bahrain. While the latter are generally characterized by very visible interior marks of carving and chiseling, these two cosmetic bottles reveal internal traces of drilling and lathe turning. It is clear that these are alien to the Gulf sector and therefore imported. It is on this basis that S. Cleuziou and B. Vogt (1985: 255–257) first suggested an import from Central Asia, later validated by During Caspers (1996: 49; see also Chapter 9).

Ivory

The site of Tell Abraq in the UAE has delivered several ivory combs, a rather rare artifact in the Middle East, most of them being discovered in the Indus valley in Mature Harappan contexts and a few others in the BMAC area. Two of these artifacts were found in 1993 and 1998 in the Late Umm an-Nar collective tomb situated about 10 m west of the fortress (Potts 1993; 1994: 620–622, fig. 53, 6–8; 2000: 126–127; During Caspers 1996: 51–52 and pl. IV). They display a specific incised decoration that Potts clearly associates with the Central Asian iconography. The patterns consist of two long-stemmed flowers (interpreted as tulips), with three petals and wavy leaves; one may note that the petals of this flower are shown closed on the second specimen. On the most complete comb (Figure 22.12), the composition also includes three dotted circles. This "tulip" pattern observed at Tell Abraq never appears on the similar Indus combs. Potts (1994: 593) compared them with a similar pair of long-stemmed flowers reproduced on a group of soft-stone vessels coming from the Kabul antique market and first published by M.-H. Pottier (1984: 84, pl. 20; 150; see also Chapter 9). The convergence noted by Potts is actually striking, and is also validated by M. Teufer (2007: 203) who devoted a thorough study on this particular iconographic theme. The flower species *Tulipa montana* Lindl., the natural range of which expands from South Turkmenistan to Iran – thus covering ancient Bactria and Margiana – could have directly inspired the decoration of the Tell Abraq combs.

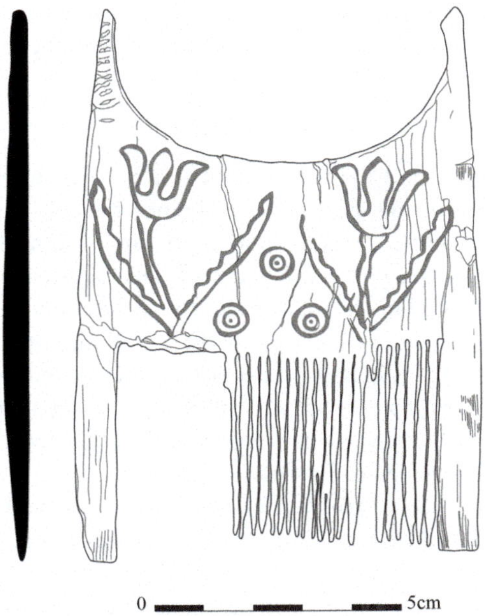

Figure 22.12 Ivory comb, with incised decoration from Tell Abraq (Sharjah, UAE) (Drawing: J. Marchand, after Potts 2008: 172, fig. 13).

A DISPUTED CASE: THE DEPOSITS FROM THE BARBAR TEMPLE II AT BAHRAIN

Other artifacts have regularly entered the debate on the relationship between the Oxus and the Arabian Gulf region.

As is known, the Dilmun culture has mainly been characterized by the emblematic excavations made at Bahrain on the Barbar Temple with its different phases: they have deeply marked the historiography of the regional research on this culture. The diggers have proposed to date this building between ca. 2100–2050 BC (Temple I), 2050–1900 BC (Temple II), and 1900–1800 BC (Temple III), but radiocarbon dates now rather push these dates several decades earlier (see the Appendix at the end of the volume).[4]

This architectural complex, discovered and excavated between 1954 and 1961 by the Danish Moesgaard Museum expedition, under the successive direction of P.V. Glob, P. Mortensen, and H.H. Andersen, has been the subject of several stratigraphic re-interpretations and chronological revisions in the last 30 years. The comprehensive and final publication (Andersen and Højlund 2003a, 2003b) describes several phases of a building organized around a central elevated platform with two neighboring altars equipped with drains, and connected by a processional staircase to a (sacred?) pool built around a freshwater spring to the west. This central structure is connected to the east via an oval-shaped courtyard, probably used as a place for slaughtering sacrificed animals. Both Temples I and II clearly illustrate this tripartite plan, which seems to evolve and become simpler in Temple III, very poorly preserved unfortunately. The various artifacts discussed further have all been found below the floor level of the central platform, some of them grouped together in a structure described as a "pit for offerings." We will see how their connection with Central Asia proposed by several scholars has changed through time and now seems rather weak.

The copper bull's head

The Early Dilmun bull's head from Barbar has now become the most famous and iconic piece of the Dilmun material culture (Figure 22.13). It was found ca. 50 cm below the floor of the central, trapezoidal platform of Temple II, together with a deposit of rectangular copper plates inlayed with several rows of nails (Andersen and Højlund 2003a: 109, 268). This head, apparently molded in one piece (Andersen and Højlund 2003a: 265–266, fig. 698; cf. also Aruz and Wallenfels 2003: 311, cat. 206), is about 18.10 x 15 cm (max. height and width of the curved horns).

As early as 1974, M. Tosi (1974: 167) had proposed to compare this piece with another bull's head discovered by V.M. Masson in 1972 in the collective "priestly sepulcher" of excavation 7 at Altyn Depe (Masson 1973: 114; 1976: 16–18 and pl. Ia; 1988: 64–71 and pl. XXIII; Kohl 1984: 130–131; cf. also Aruz and Wallenfels 2003: 356, cat. 249). In so doing he refuted During Caspers's first studies (During Caspers 1971: 222) where she had proposed "a Sumerian ED I period prototype" for this representation. However, his proposal has never really convinced us since these two animal heads differ so widely in many of their stylistic elements. In Masson's final publication on Altyn Depe, he mentions at length the Barbar's discovery, which he also considers to be the closest analogue to the Altyn Depe head (Masson 1988: 75–76), and also highlights

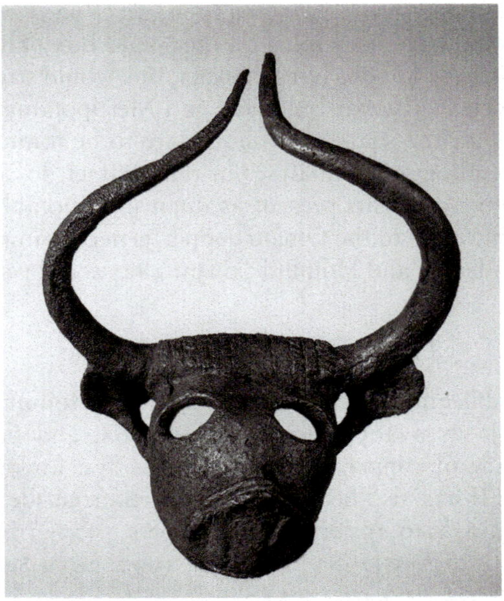

Figure 22.13 Copper bull's head from the Barbar Temple (Bahrain) (courtesy of the Bahrain National Museum).

their differences. Indeed, they are significant. If the dimensions of the two artifacts are close, the materials differ: The piece from Altyn Depe is in gold, with silver horns and turquoise inlays for the eyes and on the front while the Barbar head is in copper. The points of stylistic convergence are also not as obvious, as said peremptory by Tosi (1974) or more cautiously suggested by Masson (1988). Moreover, the Barbar copper head presents very peculiar features that in no case appear on that of the piece from Altyn Depe. This concerns notably the shape of the horns, or that of the ears, and the place and way they are incorporated to the head: at Barbar they are molded in one piece with the support while at Altyn Depe they are attached on it. Barbar's depiction also has a kind of curly fringe on top of the animal head that is absent at Altyn Depe and it does not reproduce the specific eyebrow treatment observed on the Margiana bull's head either. The only common elements in the two pieces, clearly distinguishing them from Mesopotamian models, is the flatness of the animal's muzzle in each, the rendering of the nose ("as if it were a pig's snout," cf. During Caspers 1971: 222), as well as the specific frontal position of the eyes, most probably also inlaid at Barbar, while they are usually placed in lateral position on the Mesopotamian Early Dynastic examples.

Rather than a direct relationship between Turkmenistan and Bahrain in this particular case, it seems then wiser to consider that both pieces are the result of two local productions both influenced by their pertaining to a common Middle Asian Interaction Sphere network, an opinion already suggested by During Caspers (1976: 32). The pair of gold pendants featuring bulls, found in a probable grave exposed during public works in Quetta (Pakistan) and associated with Mergharh VIII pottery (Jarrige 1988: 124, cat. 153–154), could also belong to the same multicultural network at the beginning of the 2nd millennium BC. One will recall, finally, an earlier suggestion made by D. Barnett (1969: 101 fn23) who thought that the Barbar bull's head could have been fixed on the sound box of a musical instrument, like the Ur Royal Cemetery famous lyres; if exact, this would strengthen the existence of connections between the Barbar bull head and Mesopotamia where undoubtedly the best comparisons for the shape of its horns are to be found (Zettler and Horne 1998: 63, cat. 9). Metallographic analyses of this artifact do confirm a rather pure copper with a few other elements present as impurities, notably arsenic and nickel, which indicate a profile close to the Omani copper, generally imported as bun-shaped ingots to Dilmun (Andersen and Højlund 2003a: 282; 2003b: 19).

The "copper" mirror's handle

Despite its laconic publication in H.H. Andersen and F. Højlund's final monography of the Barbar Temple excavations (2003a: 262, 264, and fig. 696), a male figurine, apparently made of copper, and found in the "pit for offerings" in the platform floor of Temple II has also been the topic of recurrent debates for many years (Glob 1954: 148, fig. 6; Rao 1970; During Caspers 1973; 1992: 10 and pl. 7a–b; 1994a: 43–44; 1996: 51; Potts 1983: 131–132; 1990: 205n 84; 2008: 168; cf. also Aruz and Wallenfels 2003: 309, cat. 205).

The piece features a standing nude male character with shaved head, eyes staring forward, and two hands clasped at the breast in a traditional Mesopotamian attitude of worship (Figure 22.14). Due to the particular device present below the feet of

Figure 22.14 Copper mirror's handle from the Barbar Temple (Bahrain) (courtesy of the Bahrain National Museum).

this male representation – a fan-shaped base cast in one piece with the figurine, with two rivets/holes – this piece has rightly been interpreted as a mirror's handle by Rao (1970: 218), who compared it with a copper/bronze mirror from Mehi (southern Baluchistan), discovered by Sir Aurel Stein in the late 1920s. We have to admit that Rao's proposition, validated a few years later by During Caspers (1973) was not surprising in the general context of the rarity of copper mirrors in the ancient Near East (outside Egypt) at that time. Ten years later, while reviewing several aspects of the archaeological and historical research on Barbar, Potts (1983: 131–132) proposed another perspective. Taking into consideration the growing material evidence of mirrors with anthropomorphic handles found in burial contexts in the BMAC area, for instance at Sapallitepa in grave 41 (Askarov 1973: pl. 25, *14* and pl. 32, *14*), or from looting in Bactria (Amiet 1977: 115–116, fig. 19; 1986: 195, 199, 201, and figs. 162–163; Sarianidi 1977: 78–79 and fig. 40, 2;[5] Pottier 1984: 39–40 and figs. 268–273; Kaniuth 2006: 66–68 and figs. 10–11; Sarianidi 2008: fig. 165 left; Benoit 2010: fig. 4), he suggested that this type of artifact probably originated in Bactria, and that a direct link could exist between this production and the Barbar specimen. Interestingly, at the time of the first discoveries, Sarianidi (2008) had considered that these handled mirrors may have originated in Iran since similar ones were known in Luristan. No such artifact has been found up to now in Margiana (N. Dubova, personal communication; cf. also Salvatori 2008: 81–82, for a general discussion about Bactrian mirrors).

If the Barbar piece effectively displays a certain stylistic convergence with the Bactrian examples, some important differences are also obvious. In the case of the Bactrian mirrors, the anthropomorphic handle generally reproduces a (very) stylized female silhouette whose head, absent, is in fact replaced by the functional, circular part of the mirror, reflecting the face of the user. The fact that the Barbar mirror's handle is actually arranged in the opposite position means it can no longer play the same role. Moreover, much more than a simple sketch of human representation as on the Bactrian pieces, this handle depicts a perfect figurine in the round, with careful modeling and precise iconographic details. Here we should recall During Caspers's (1973: 129) first comments where she highlighted the proximity of this figurine with the Sumerian stylistic canons of the Early Dynastic III and Early Akkadian statuary. As for the previous – and contemporaneous – bull's head discussed above, such a relationship still appears wiser today. The Barbar mirror handle should thus be considered as a new example of a typical "multicultural production" from Dilmun. However, we must note that the metallographic analysis of this piece revealed an unusually high percentage of tin (over 10%), which would make it, in this case, a real "bronze" (Andersen and Højlund 2003a: 282). This composition, which remains exceptional for Bahrain's Early Dilmun phase, is however strangely not confirmed in the table of metal analyses reproduced in the volume of appendices by these same authors (Andersen and Højlund 2003b: 19).

The alabaster vessels

The last group of artifacts worthy of discussion found in the same central platform of Bahrain Barbar Temple II is a series of alabaster vases (four complete and numerous fragments). They were discovered in a clearly disturbed context since

Figure 22.15 Alabaster/calcite vessels from the Barbar Temple (Bahrain) (courtesy of the Bahrain National Museum).

fragments of one of them were lying in different sectors (Andersen and Højlund 2003a: 91–94; 109, fig. 129) (Figure 22.15). Analyses have since then pointed out that some of these vessels are not made of alabaster (gypsum), but of very pure calcite, as is the case for the majority of the so-called "alabaster" vessels from the ancient Near East (Casanova 1991: 15). The provenance of these containers, of various shapes, has been the subject of many comments: Egypt, southern Babylonia, eastern Iran (Glob 1954: 153 n8; 1959: figs. 2–4; Mortensen 1971: 391, fig. 7; 394; Howard Carter 1972: 22 fn19; Casanova 1991: 34 fn38, 44 fn81, 72, 88; 2003), as well as Central Asia. This last potential source was notably suggested by H. Crawford and K. Al-Sindi (1995: 3) but without convincing arguments and it has been reiterated more cautiously by Potts (1983: 129–131). In fact, Bahrain National Museum (BNM) items A19059 and A19057 (cf. Figure 22.15, first and second from left) have their counterparts in southern Mesopotamia, mainly at Ur, as already mentioned a long time ago by P. Glob (1954, 1959) and P. Mortensen (1971) (see, e.g., the proximity between BNM A19059/Danish ref. 517.YL; and Woolley 1934: pl. 178c, pl. 241, *5*) and as completed by T. Howard Carter (1972). This kind of vessel actually seems to be quite frequent according to M. Casanova (1991, types IVa and IVc: 33–34 fn37 and 39; fig. 3/pl. 3, *41*, *46*; and fig. 4/pl.4, *48–49*), especially in Egypt (Aston 1994: 80, fig. 9) and in Iran. In this last region several specimens have been recovered from the clandestine digging at Jiroft (Majidzadeh 2003: 145, 146) and complete their well-known occurrence at Susa, Shahdad, or Shahr-i Sokhta II–III. They are also known in Central Asia where Casanova, however, limits her comparisons to a few cases at Mundigak IV (Casal 1961: fig. 134, 22) and in southern Bactria (Amiet 1977: fig. 7, *1*, *3*; Pottier 1984: pl. LXIV, *356–357*; LXV, *361*; LXVI, *369*), mentioning also examples from Ulug Depe and Altyn Depe but without more details (Casanova 2003: 284). We can add several additional cases coming from BMAC contexts: from the Quetta hoard already mentioned (Jarrige and Hassan 1989: fig. 2, three different items); from the same grave 3245 at Gonur "Royal Necropolis" discussed above where one of the alabaster vessels appears close to this type (see Figure 22.5) and is moreover copied in metal; or from recent antique market finds (Vidale 2017: 60–61, fig. 52). These discoveries confirm that this shape was known both in western and in Central Asia.

The shape of the third Barbar vessel (alabaster *pyxis* BNM A19058/Danish ref. 517.BX, cf. Figure 22.15, third from left) appears much more original and has no convincing parallels: The only comparison suggested by Glob (1954: n9) with a lidded jar from Ur (Woolley 1934: 100, pl. 250) appears inadequate today due to the external shape of the Ur piece and the attachment system of its lid, while another suggested by Casanova (2003: 284) with a pentagonal *pyxis* from Mundigak (cf. Casal 1961: 234, fig. 134, *15*) seems even less proper, in our view.

Nevertheless, if there is no doubt that these alabaster/calcite vessels are found within all this vast geographical area, we have no real clue, up to now, as to where they were actually produced, though eastern Iran/Baluchistan might be the best production center due to the abundance of finds there. The BMAC area can hardly be considered as another center of production: The alabaster containers found there are either limited in size (Amiet 1977: 97, fig. 7; Vidale 2017), or have very specific shapes (Jarrige and Hassan 1989: fig. 2; Sarianidi 2005: 243, fig. 102; Bendezu-Sarmiento and Lhuillier 2016: 526, figs. 4–5; see also Chapter 13, Figure 13.8).

To sum up, as proposed by Casanova (2003: 288) in her final study of the Barbar vases, an eastern Iranian origin is most likely for this material.

CLAY AND METAL PEDESTALED GOBLETS FROM THE GULF: A PARTICULAR CASE

The last category of material to be discussed here has, since the early 1970s, attracted the attention of During Caspers and led her to more closely examine a possible Central Asian influence on the Gulf area.

Pedestaled goblets of various sizes and slightly different shapes, generally in clay but also in copper/bronze for a few examples, are regularly reported in old and ongoing excavations in the Gulf area, and more generally in the eastern Arabian Peninsula. They have been discussed on several occasions by various scholars (During Caspers 1989: 19–20; 1992: 6–9; 1994a: 34–40; 1996: 49; Lombard 1999: 96, 99; Potts 2000: 126–127; 2008: 169, 173–174; Aruz and Wallenfels 2003: 312).

Clay pedestal goblets

We will first focus on the most distinctive of these goblets (i.e., the pottery ones). They are rather numerous on Bahrain Island, where they have been discovered at A'ali in 1962 (Højlund 2007: 55–56, figs. 69–71), Qal'at al-Bahrain during the late 1960s (Højlund and Andersen 1994: 93, figs. 194–196), Hamala North in 1968 (During Caspers 1972; 1972–1974; 1980: fn24; 1994a: 34–35 and fig. 2), Saar in 1978 (Ibrahim 1982: 32, fig. 38, *1–3*, pl. 49, *1–3*; Lombard 1999: 64, cat. 35–36; 96, cat. 92), Janabiyah (2018 unpublished rescue excavation), and at a site without location (During Caspers 1994a: 34 and fig. 1, center – wrongly labeled as coming from Saar). All these vessels, recovered from the traditional burial mounds of the Early Dilmun phase of Bahrain are dated (still conventionally as mentioned in note 4) between ca. 1950 and 1800 BC.

In the eastern province of Saudi Arabia, three specimens come from the Dhahran tumuli attributed also to the Early Dilmun phase (Zarins 1989: fig. 13, *6*, *12*, *18*) while one was recovered in 1998 from the Late Umm an-Nar grave at Tell Abraq, UAE (Potts 2000: 127; 2008: 173–174, figs. 17–18).

It is also possible, finally, that a few fragments from Failaka (Kuwait) belong to a similar type of vessel (Højlund 1987: 31–32, fig. 80).

When they are entirely preserved or give a complete profile, these pieces all display a similar overall shape (i.e., a round, ovoid, or tulip-shaped goblet with a simple beaded rim on top of a small- to medium-height foot). The numerous specimens from Bahrain show that they were wheel-turned and made in a rather fine red-brick or gray ware, generally sand-tempered, sometimes burnished on the surface. Their fabric, which sometimes slightly differs from one vase to the other, is unquestionably distinct from the usual local one, and points to their probable importation. Unfortunately, no clay analysis has ever been done on any of them to try to establish where these vases could have been produced.

A comparison of their shape with the productions from Central Asia shows obvious parallels and can easily explain During Caspers's first analysis. This is, for instance, the case for the goblet in fine gray ware from Saar tumulus S100 (Ibrahim 1982: fig. 38, *3*), with its characteristic ring in relief at the junction between its hollow pedestal and the container (Figure 22.16, *1*). Its shape is similar to vases from the cenotaphs of Sibri/Mehrgarh VIII, long ago recognized as being connected with the Oxus Civilization/BMAC (Santoni 1984: 52–54, fig. 8, *1*; Jarrige 1988: cat. 131; Jarrige *et al.* 1995: 356, fig. 7, 27), or from Nindowari III (a site related to the contemporary Kulli B culture, see Jarrige *et al.* 2011: figs. 11, 18). Another quite large specimen from Bahrain (height ca. 17.3 cm), found in 2018 in a still unpublished tumulus at the Janabiyah site, is made in a red fine fabric alien to Bahrain (Figure 22.16, *3*), and can be confidently compared to quite similar shapes from Togolok 21 (P'jankova 1989: fig. 6), Gonur 1 (Sarianidi 1990: pl. XIII, *7*; 2007: 58, type 3.8 according to Udeumuradov), Dashly 1 and 3 (Sarianidi 1976: fig. 18, *4*; fig. 33, *9*; fig. 34, *23*; 1977: 64, fig. 26, *1*), or Sapallitepa (Askarov 1973: tab. 15, *5*, *10–12*; tab. 16, *2*, *3*, *9–10*, *12–13*). Such *comparenda* also convincingly apply to a last Bahraini goblet, from a not located Early Dilmun burial mound (Figure 22.16, 2). The large "chalice" from Tell Abraq in the UAE (Figure 22.16, 4) has a different profile (Potts 2008: figs. 17–18), but is nevertheless also close to examples from Bactrian sites (Sarianidi 1976: fig. 25, *1*) and more so to a shape from Tepe Yahya VIB4-2 (Potts 2001: fig. 3.19 bottom left).

Though their shape makes them rather fragile, we cannot exclude that such undecorated pieces constituted true imports from the BMAC or BMAC-related region. However, this kind of vessel is not limited to Central Asia: they are present, though far less represented, for instance at Tepe Yahya in level IVB6 (Potts 2001: fig. 3.14i–j) and IVB4–2 (Potts 2001: fig. 3.19), and are also known made of soft stone in the Jiroft area (Perrot and Madjidzadeh 2005: 126, fig. 3e; 127, pl. 1a–e).

Beside the plain pottery vessels discussed above, several of the goblets discovered in the Dilmun area display a painted geometric or naturalistic decoration (usually black or dark red, either on a red or yellow slip or directly on the red clay). It goes from

Figure 22.16 Unpainted pedestal-pottery vessels from the Arabian Gulf. 1: Saar (Bahrain) (Drawing: J. Marchand, after Ibrahim 1982: fig. 38, 3); 2: unlocated site (Bahrain); 3: Janabiyah (Bahrain) (courtesy of the Bahrain National Museum); 4: Tell Abraq (Sharjah, UAE) (from Potts 2008: fig. 18).

simple groups of vertical and horizontal lines (Dhahran tumuli, Figure 22.17, *1–3*), to herringbone or crosshatched linear patterns (Saar, Figure 22.17, *5–6*), crosshatched upturned triangles (A'ali, Figure 22.17, *7–8*), crosshatched "butterflies" associated with clusters of geometrical leaves (Hamala North, Figure 22.17, *4*), or frieze of caprids associated with hatched bands and triangles (Qal'at al-Bahrain, Figure 22.17, *9*, and possibly Failaka). Painted decoration is extremely rare on the overall Bactria or Margiana pottery production, and only represented by simple horizontal dark bands on a red background probably under the influence of the Indus Civilization. In contrast, painted pottery vessels are quite frequent in the Dilmun material culture, either imported from the east (Baluchistan, southeastern Iran, or Indus) or locally made, and sometimes illustrate similar patterns. It can thus be proposed that these goblets are hybrid productions where a Central Asian shape is painted in the local fashion.

Finally, a unique vase from Bahrain, discovered during the 1988 season of excavations at Saar (tumulus 140, cf. Lombard ed. 1999: 99, cat. 98; Aruz and Wallenfels 2003: 312, cat. 207) deserves attention. This masterpiece from the Bahrain National Museum (A7050) belongs to the same type of goblets discussed above but is characterized by a very fine red clay, a concave strongly carinated shape – ending with a flat bottom – of the upper bowl, as well as by a peculiar geometric bichrome

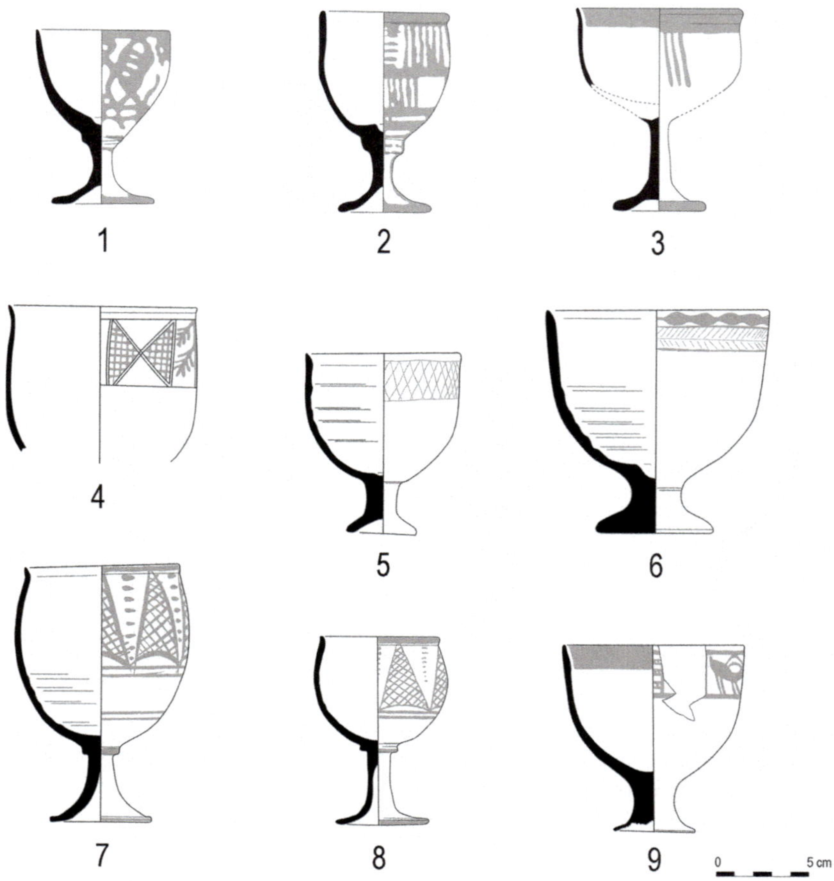

Figure 22.17 Painted pedestal-pottery vessels from the Arabian Gulf. 1–3: Dhahran (Saudi Arabia); 4: Hamala North (Bahrain); 5–6: Saar (Bahrain); 7–8: A'ali (Bahrain); 9: Qal'at al-Bahrain (Drawing: J. Marchand, after Zarins 1989: fig. 13, 6, 12, 18; During Caspers 1994b: fig. 2; Ibrahim 1982: fig. 38, 1, 2; Højlund 2007: figs. 70–71; Højlund and Andersen 1994: figs. 194–196).

painted decoration (Figure 22.18). The profile of this vase does recall types from several BMAC sites, like Dashly 3 (Sarianidi 1976: fig. 37, *4–9*; fig. 40, *13*) or the Gonur Large Necropolis (Sarianidi 2007: 65, fig. 29, top right), or from related sites like Khurab in Baluchistan (Stein 1937: pl. XXXII, *19*), although all these vases have a less pronounced carination at the base of the container. Its painted decoration, nevertheless, prohibits consideration of it being any import from Central Asia and points rather at production from Baluchistan where such dense and bichrome motifs are frequent in the Kulli B culture (Jarrige *et al.* 2011: fig. 30). Actually, the closer contemporary parallels to our vase are alabaster vessels found also in BMAC funerary contexts at the Gonur Large Necropolis (Sarianidi 2007: 112, fig. 196), at Ulug Depe (Bendezu-Sarmiento and Lhuillier 2016: 526, fig. 4, *5–7*), in the Quetta hoard (Jarrige and Hassan 1989: fig. 2), or in the looted graves from Bactria (Pottier 1984:

Figure 22.18 Painted pedestal-pottery vessel from Saar (Bahrain) (courtesy of the Bahrain National Museum).

fig. 29, *217*, *218*). Another close shape also coming from a Gonur grave is made of chlorite and is heavily decorated, but the bowl is not as deep as in our case (Sarianidi 2007: 112, fig. 198).

We also have to mention here that, beside comparisons with BMAC or related sites, our vessel also finds parallels with earlier pedestalled vases, like the chalice found at Mehrgarh VII in gray ware with a red slip considered to be a possible import from northeastern Iran (Jarrige 1988: 108, fig. 115), or another similar gray chalice coming from a grave at Parkhai 2 in the Sumbar valley, Turkmenistan (Khlopin 2002: 303, fig. 11) contemporary with the Namazga (NMG) IV period (ca. 2500 BC). These last examples are, however, much too early for the Gulf vessels.

Copper/bronze pedestal goblets

Interestingly, pedestal goblets are also attested in copper/bronze and a few of these were also found in the Arabian Gulf. Most of them come from Bahrain. They are also known in rich graves of the BMAC area, usually in silver (Sarianidi 2008: 271). Their shape is very similar to the pottery goblets and we do not know which copied the other.

At Bahrain, the largest pedestal goblet (height ca. 18 cm and diameter ca. 21 cm) was found during amateur excavations in the tumulus of Hamala North in 1968 (During Caspers 1992: 8–9, and fig. 3b): it has a globular bowl with slightly everted rim, and a large hollow pedestal (Figure 22.19, *1*). A quite similar vessel but of smaller size (Figure 22.19, 2) was recovered from the "Alignment A" grave 2 at Asimah, Ras al-Khaimah Emirate; its excavator does not exclude an eventual importation from Central Asia, and validates During Caspers's analysis of the Hamala North specimen

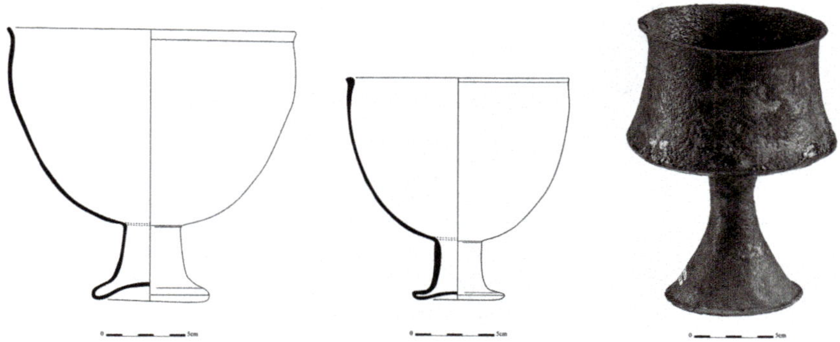

1 2 3

Figure 22.19 Copper pedestal-pottery vessels from the Arabian Gulf. 1: Hamalah North (Bahrain); 2: Asimah (Ras al-Khaimah, UAE). (Drawing: J. Marchand, after During Caspers 1992: fig. 3b; Vogt 1994: fig. 54, 3). 3: Madinat Hamad (Bahrain) (courtesy of the Bahrain National Museum).

(Vogt 1994: 129, fig. 54, 3; pl. 18c). The last vessel worthy of mention here is a very close metal version of the pedestalled beaker A7050 from Bahrain National Museum, discussed above (A 21213 found in Madinat Hamad, Figure 22.19, 3); both pieces could thus share the same provenience. Unfortunately, no metallographic analyses were carried out on these copper/bronze items up to now.

CONCLUSION

This short study underlines a rather limited manifestation of a direct circulation of artifacts between the BMAC and the Gulf. Nevertheless, this observation should not underestimate the existence of more or less regular but indirect contact between these two regions during the first quarter of the 2nd millennium BC. If most of the BMAC-related artifacts come from graves, none of them contains exclusively this material as is the case in several tombs of eastern Iran or Baluchistan (see Chapters 19 and 20). In the Gulf, these Central Asia products are always mixed with local ones and we can therefore dismiss the idea of the presence there of a BMAC group of people. Artifacts, techniques, or artistic traditions peculiar to each of the two areas have been pointed out. In the Gulf, this is especially visible in the introduction of some of the new pottery shapes. Most of the time, however, real imports seem exceptional and we are mostly dealing with hybrid products where a strong assimilation of the foreign tradition leads to a local production with adapted features.

Some artifacts certainly came through the Indus valley (cf. Potts 2008, the "Meluhhan trajectory"), which was related with the BMAC at one end and had developed special maritime connections of long duration with the Gulf at the other. Goods may also have come from southeastern Iran or Baluchistan/Magan (i.e., on a longer pedestrian road), but little is known of the Iranian coast except for the port of Liyan/Bushir. It is highly probable that these artifacts hide a far more important circulation of other goods that have now disappeared, such as spices, textiles, semiprecious stones, metals including tin, etc. (cf. notably Kohl and Lyonnet 2008: 37–40; Kaniuth 2010).

The great number of new excavations in the Gulf, now all precisely dated and associated with analyses of all kinds, should help in bringing other data based on firmer ground to answer these questions.

NOTES

* I wish to express my deepest gratitude to both editors for their patience and support. I especially warmly thank Bertille Lyonnet for her wise advice, corrections, and constant help in finding the necessary Central Asia references I was lacking, and for her encouragement when needed.

1 The terminal date proposed here is later than that from the radiocarbon dates (see the Appendix at the end of the volume), but these are still scarce and many other arguments that we cannot develop here speak in favor of such a later date.

2 I am extremely grateful to N. Dubova for generously lending me the pictures of all the artifacts found in this grave.

3 This location is confirmed *contra* Potts (2008: 176) who had expressed some reservation about it.

4 Only two radiocarbon dates are available for the the same level at the Barbar Temple (see the Appendix at the end of this volume), and the chronology of this building, drastically revised in 1983, still mainly relies on the ceramic sequence that was established only by comparing selected stratified assemblages from the site with sequences from Danish excavations at Qal'at al-Bahrain and on Tell F3 and F6 at Failaka in Kuwait. The final publication of the Barbar Temple (Andersen and Højlund 2003a, 2003b) remains silent on the detailed chronology of each of its architectural stages, and recent, average dates curiously appear only on two synthetic chronological tables, which are considered as the main references today (Højlund 2007: 12, fig. 3; Laursen 2017: 382–383, fig. 511).

5 There is probably an error in the legend of fig. 40 where it says that the mirror comes from Dashly 3, while the text mentions that it comes from plundered graves and was found in the Kabul bazar.

REFERENCES

Al-Sindi, K. 1999. *Dilmun Seals, Part One*. Bahrain: Ministry of Cabinet Affairs and Information and Bahrain National Museum.

Amiet, P. 1977. "La Bactriane proto-historique." *Syria* 54: 89–121.

Amiet, P. 1986. *L'âge des échanges inter-iraniens, 3500–1700 avant J.-C.* Paris: Réunion des musées nationaux.

Andersen, H.H. and F. Højlund 2003a. *The Barbar Temples*. Højbjerg: Jutland Archaeological society.

Andersen, H.H. and F. Højlund 2003b. *The Barbar Temples: Appendices*. Højbjerg: Jutland Archaeological society.

Askarov, A. 1973. *Sapallitepa*. Tashkent: Fan.

Aston, B.G. 1994. *Ancient Egyptian Stone Vessels: Materials and Forms*. Studien zur Archäologie und Geschichte Altägyptens, Band 5. Heidelberg: Heidelberg Orientverlag.

Aruz, J. and R. Wallenfels (eds.) 2003. *Art of the First Cities: The Third Millennium BC from the Mediterranean to the Indus*. New York: Metropolitan Museum of Art.

Barnett, R.D. 1969. "New Facts About Musical Instruments from Ur." *Iraq* 31(2): 96–103.

Begemann, F., A. Hauptmann, S. Schmitt-Stecker, and G. Weisgerber 2010. "Lead Isotope and Chemical Signature of Copper from Oman and Its Occurrence in Mesopotamia and Sites on the Arabian Gulf Coast." *Arabian Archaeology and Epigraphy* 21: 135–169.

Bendezu-Sarmiento, J. and J. Lhuillier 2016. "Some More Exceptional Discoveries at Ulug-depe." In *Trudy Margianskoj Arkheologicheskoj Ekspeditsii*, vol. 6, ed. N. Dubova. Moscow: Staryj Sad, pp. 522–528.

Bendezu-Sarmiento, J. and S. Mustafakulov 2013. "Le site proto-urbain de Dzharkutan Durant les âges du bronze et du fer. Recherches de la mission archéologique franco-ouzbèke-Protohistoire." *Cahiers d'Asie centrale* 21–22: 207–236.

Benoit, A. 2010. "Some Thoughts on the Cosmetology of the Near East During the Third Millennium B.C.: Containers and Contents." In *Trudy Margianskoj Arkheologicheskoj Ekspeditsii*, vol. 3, ed. P.M. Kozhin, M.F. Kosarev, and N.A. Dubova. St. Petersburg: Aletejja, pp. 312–322.

Calvet, Y. 1996. "Une empreinte de cylindre-sceau de Failaka (Koweit)." In *Collectanea Orientalia: Histoire, arts de l'espace et industrie de la terre. Études offertes en hommage à Agnès Spycket*, ed. H. Gasche and B. Hrouda. Neuchâtel-Paris: Recherches et Publications, pp. 57–64.

Calvet, Y. 2005. "Agarum, une île de la civilisation de Dilmoun." In *L'île de Failaka. Archéologie du Koweït*, ed. G. Galliano. Lyon: Musée des Beaux-Arts, pp. 41–61.

Casal, J.-M. 1961. *Fouilles de Mundigak*, 2 vols. Mémoires de la Délégation Archéologique Française en Afghanistan, t. XVII. Paris: Klincksieck.

Casanova, M. 1991. *La vaisselle d'albâtre de Mésopotamie, d'Iran et d'Asie centrale aux IIIe et IIe millénaires av. J.-C.* Paris: Éditions Recherche sur les Civilisations.

Casanova, M. 2003. "Alabaster and Calcite Vessels." In *The Barbar Temples. Appendices.* ed. H.H. Andersen and F. Højlund. Højbjerg: Jutland Archaeological society.

Cleuziou, S. and B. Vogt 1985. "Tomb A at Hili North (United Arab Emirates) and Its Material Connections to Southeast Iran and the Greater Indus Valley." In *South Asian Archaeology 1983*, ed. J. Schotsmans and M. Taddei. Naples: Istituto Universitario Orientale, pp. 255–257.

Crawford, H. 1998. *Dilmun and Its Gulf Neighbours*. Cambridge: Cambridge University Press.

Crawford, H. and K. Al-Sindi 1995. "A Seal in the Collections of the National Museum, Bahrain." *Arabian Archaeology and Epigraphy* 6: 1–4.

David, H. 1996. "Styles and Evolution: Soft Stone Vessels During the Bronze Age in the Oman Peninsula." *Proceedings of the Seminar for Arabian Studies* 26: 31–46.

David, H. 2002. "Soft Stone Vessels from Umm an-Nar Tombs at Hili (UAE): a comparison." *Proceedings of the Seminar for Arabian Studies* 32: 175–185.

During Caspers, E.C.L. 1971. "The Bull's Head from Barbar Temple II, Bahrain: A Contact with Early Dynastic Sumer." *East and West* 21: 217–223.

During Caspers, E.C.L. 1972. "The Bahrain Tumuli." *Proceedings of the Fifth Seminar for Arabian Studies*: 9–19.

During Caspers, E.C.L. 1972–1974. "The Bahrain Tumuli." *Persica* VI: 131–156.

During Caspers, E.C.L. 1973. "Sumer and Kulli Meet at Dilmun in the Arabian Gulf." *Archiv für Orientforschung* 24: 128–132.

During Caspers, E.C.L. 1976. "Cultural Concepts in the Arabian Gulf and the Indian Ocean: Transmissions in the Third Millennium and their Significance." *Proceedings of the Seminar for Arabian Studies* 6: 8–39.

During Caspers, E.C.L. 1980. *The Bahrain Tumuli: A Catalogue of Two Important Collections.* Leiden: Nederland Historish-Archeologich Instituut te Istanbul.

During Caspers, E.C.L. 1989. "Some Remarks on Oman." *Proceedings of the Seminar for Arabian Studies* 19: 13–31.

During Caspers, E.C.L. 1992. "Intercultural/Mercantile Contacts Between the Arabian Gulf and South Asia at the Close of the Third Millennium BC." *Proceedings of the Seminar for Arabian Studies* 22: 3–28.

During Caspers, E.C.L. 1994a. "Further Evidence for 'Central Asian' materials from the Arabian Gulf." *Journal of the Economic and Social History of the Orient* 37: 33–53.

During Caspers, E.CL. 1994b. "Non-Indus Glyptics in a Harappan Context." *Iranica Antiqua* 29: 83–106.

During Caspers, E.C.L. 1994c. "Triangular Stamps Seals from the Arabian Gulf, Once Again." *Proceedings of the Seminar for Arabian Studies* 24: 97–114.

During Caspers, E.C.L. 1996. "Local MBAC Materials in the Arabian Gulf and Their Manufacturers." *Proceedings of the Seminar for Arabian Studies* 26: 47–64.

Francfort, H.-P. 1994. "The Central Asian Dimension of the Symbolic System in Bactria and Margiana." *Antiquity* 68(259): 406–418.

Glob, P.V. 1954. "Temples at Barbar." *Kuml* 4: 142–153.

Glob, P.V. 1959. "Alabaster Vases from the Bahrain Temple." *Kuml* 51: 138–145.

Görsdorf, J. and B. Vogt. 2001. "Asimah: Neue ^{14}C-Datierung zum späten 3: vorchristlichen Jahrtausend auf der Halbinsel Oman." *Arabian Archaeology and Epigraphy* 12: 1–9.

Hiebert, F.T. 1994. *Origins of the Bronze Age Oasis Civilization in Central Asia*. American School of Prehistoric Research, Bulletin 42. Cambridge, MA: Peabody Museum of Archaeology and Ethnology, Harvard University.

Højlund, F. 1987. *Failaka/Dilmun: The Second Millennium Settlements*. Vol. 2, *The Bronze Age Pottery*. Jutland Archaeological Society Publications XVII. 2. Højbjerg: Jutland Archaeological Society.

Højlund, F. 2007. *The Burial Mounds of Bahrain: Social Complexity in Early Dilmun*. Højbjerg: Jutland Archaeological Society.

Højlund, F. and H.H. Andersen. 1994. *Qal'at al-Bahrain*. Vol. 1, *The Northern City Wall and the Islamic Fortress*. Højbjerg: Jutland Archaeological Society.

Howard Carter, T. 1972. "The Johns Hopkins University Reconnaissance Expedition to the Arab-Iranian Gulf." *Bulletin of the American Schools of Oriental Research* 207: 6–40.

Ibrahim, M. 1982. *Excavations of the Arab Expedition at Saar el-Jisr, Bahrain*. Bahrain: Ministry of Information.

Jarrige, J.-F. (ed.) 1988. *Les cités oubliées de l'Indus. Archéologie du Pakistan*. Paris: Association française d'action artistique.

Jarrige, J.-F. and M.U. Hassan 1989. "Funerary Complexes in Baluchistan at the End of the Third Millennium in the Light of Recent Discoveries at Mehrgarh and Quetta." In *South Asian Archaeology 1985*, ed. K. Frifelt and P. Sorensen. Copenhagen: Curzon Press, pp. 150–166.

Jarrige, J.-F., C. Jarrige, R. Meadow, and G. Quivron 1995. *Mehrgarh, Field Reports 1974–1985 from Neolithic Times to the Indus Civilization: The Reports of Eleven Seasons of Excavations in Kachi District, Balochistan, by the French Archaeological Mission to Pakistan*. Karachi: Department of Culture and Tourism.

Jarrige, J.-F., G. Quivron, and C. Jarrige 2011. *Nindowari, Pakistan: La culture de Kulli. Ses origines et ses relations avec la Civilisation de l'Indus*. Paris: Gingko.

Kaniuth, K. 2006. *Metallobjekte der Bronzezeit aus Nordbaktrien*. Archäologie in Iran und Turan, 6. Mainz: Philipp von Zabern.

Kaniuth, K. 2010. "Long Distance Imports in the Bronze Age of Southern Central Asia: Recent Finds and Their Implications for Chronology and Trade." *Archäologische Mitteilungen aus Iran und Turan* 42: 3–22.

Khlopin, I.N. 2002. *Epokha Bronzy Jugo-Zapadnogo Turkmenistana*. St. Petersburg: Institut Istorii Material'noj Kul'tury.

Kjærum, P. 1983. *Failaka/Dilmun: The Second Millennium Settlements*. Vol. 1, *The Stamp and Cylinder Seals*. JASP XVII. Aarhus: Jutland Archaeological Society.

Kohl, P.L. 1984. *Central Asia: Palaeolithic Beginnings to the Iron Age*. Paris: Recherche sur les Civilisations.

Kohl, P.L. and B. Lyonnet 2008. "By Land and By Sea: The Circulation of Materials and Peoples, ca. 3500–1800 BC." In *Intercultural Relations Between South and Southwest Asia: Studies in Commemoration of E.C.L. During Caspers (1934–1996)*, ed. E. Olijdam and R.H. Spoor. BAR International Series 1826. Oxford: Archaeopress, pp. 29–42

Laursen, S. 2017. *The Royal Mounds of A'ali in Bahrain: The Emergence of Kingship in Early Dilmun*. Aarhus: Jurland Archaeological Society.

Laursen, S. and P. Steinkeller 2017. *Babylonia, the Gulf Region, and the Indus: Archaeological and Textual Evidence for Contact in the Third and Early Second Millennia B.C.* Mesopotamian Civilization 21. Winona Lake, IN: Eisenbrauns.

Leemans, W.F. 1960. *Foreign Trade in the Old Babylonian Period*. Leiden: Brill.

Ligabue, G. and S. Salvatori 1988. *Battriana, una antica civiltà delle oasi dalle sabbie dell'Afghanistan*. Venice: Erizzo.

Lombard, P. (ed.) 1999. *Bahrain, the Civilisation of the Two Seas: From Dilmun to Tylos*. Paris and Gent: Institut du Monde Arabe and Snoek-Ducaju & Zoon.

Lombard, P. and N. Boksmati-Fattouh 2017. "Dilmun i gorizonty mezhdunarodnoj torgovli" [The power of trade]. In *In the Land of Dilmun Where the Sun Rises: Archaeological Treasures from the Bahrain National Museum (3rd–1st Millennia B.C.)*, ed. N. Koslova. St. Petersburg: State Hermitage Publishers, pp. 56–67.

Majidzadeh, Y. 2003. *Jiroft: The Earliest Oriental Civilization*. Teheran: Ministry of Culture and Islamic Guidance.

Masson, V.M. 1973. "Raskopki zhretseskej grobnitsy v Turkmenii." *Priroda* 3: 114.

Masson, V.M. 1976. "Altin-Depe and the Bull Cult." *Antiquity* 50: 14–19.

Masson, V.M. 1988. *Altyn-Depe*. University Museum Monograph, 55. Philadelphia: University Museum, University of Pennsylvania.

Mortensen, P. 1971. "On the Date of the Temple at Barbar in Bahrain." *Kuml* 1970: 385–398.

Olijdam, E. 2008. "A Possible Central Asian Origin for the Seal-Impressed Jar from the 'Temple Tower' at Failaka." In *Intercultural Relations Between South and Southwest Asia: Studies in Commemoration of E.C.L. During Caspers (1934–1996)*, ed. E. Olijdam and R.H. Spoor. BAR International Series 1826. Oxford: Archaeopress, pp. 268–287.

Olijdam, E. and R.H. Spoor (eds.) 2008. *Intercultural Relations Between South and Southwest Asia: Studies in Commemoration of E.C.L. During Caspers (1934–1996)*, ed. E. Olijdam and R.H. Spoor. BAR International Series 1826. Oxford: Archaeopress.

Oppenheim, A.L. 1954. "Seafaring Merchants of Ur." Journal of American Oriental Society 74: 6–17.

Perrot, J. and Y. Madjidzadeh 2005. "L'iconographie des vases et objets en chlorite de Jiroft (Iran)." *Paléorient* 31(2): 23–152.

Pic, M. 1990. "Quelques éléments de glyptique." In *Failaka. Fouilles Françaises 1986–1988*, ed. Y. Calvet and J. Gachet. TMO 18. Lyon: Maison de l'Orient, pp. 125–140.

Possehl, G.L. 2002. *The Indus Civilization: A Contemporary Perspective*. Walnut Creek, CA: Altamira.

Possehl, G.L. 2007. "The Middle Asian Interaction Sphere. Trade and Contact in the 3rd Millennium BC." *Expedition* 49(1): 40–42.

Pottier, M.-H. 1984. *Matériel funéraire de la Bactriane méridionale de l'Age du Bronze*. Paris: Recherche sur les Civilisations.

Potts, D.T. 1983. "Barbar miscellanies." In *Dilmun. New Studies in the Archaeology and Early History of Bahrain*, ed. D.T. Potts. Berliner Beiträge zum Vorderen Orient, Bd. 2. Berlin: Dietrich Reimer, pp. 127–129.

Potts, D.T. 1990. *The Arabian Gulf in Antiquity. Vol. I, From Prehistory to the Fall of the Achaemenid Empire*. Oxford: Clarendon Press.

Potts, D.T. 1993. "A New Bactrian Find from Southeastern Arabia." *Antiquity* 67: 591–596.

Potts, D.T. 1994. "South and Central Asian Elements at Tell Abraq (Emirate of Umm al-Qaiwain, United Arab Emirates), c. 2200 BC–AD 400." In *South Asian Archaeology 1993*, ed. A. Parpola and P. Koskikallio. Helsinki: Suomalainen Tiedeakatemia, pp. 615–628.

Potts, D.T. 2000. *Ancient Magan: The Secrets of Tell Abraq*. London: Trident.

Potts, D.T. 2001. *Excavations at Tepe Yahya, Iran, 1967–1975: The Third Millennium*, ed. C.C. Lamberg-Karlovsky. American School of Prehistoric Research, Bulletin 45. Cambridge, MA: Harvard University.

Potts, D.T. 2008. "An Umm an-Nar-Type Compartmented Soft-Stone Vessel from Gonur Depe, Turkmenistan." *Arabian Archaeology and Epigraphy* 18: 167–180.

Potts, D.T. 2016. *People of the Lower Sea: Magan and Its Neighbours. The Archaeological Heritage of Oman*, vol. 3. Mascate: Ministry of Heritage and Culture, Sultanate of Oman.

P'jankova, L. 1989. "Pottery Complexes of Bronze Age Margiana (Gonur and Togolok 21)." *International Association for the Study of the Cultures of Central Asia Information Bulletin* 16: 27–54.

Rao, M.S.N. 1970. "A Bronze Mirror Handle from the Barbar Temple, Bahrain: A Further Link with the Kulli Culture – South Baluchistan)." *Kuml* 1969: 218–220.

Salvatori, S. 2000. "Bactria and Margiana Seals: A New Assessment of Their Chronological Position and a Typological Survey." *East and West* 50: 97–145.

Salvatori, S. 2003. "Spunti per una definizione dei sistemi di relazione tra l'Iran orientale e l'Asia Centrale meridionale. I flaconi per cosmetici clorite: Distribuzione e variabilità." In *Studi in onore di Umberto Scerrato per il suo settantacinquesimo compleanno*, ed. M.V. Fontana and B. Genito. Naples: Università degli Studi di Napoli 'L'Orientale,' pp. 721–734.

Salvatori, S. 2008. "Cultural Variability in the Bronze Age Oxus Civilisation and Its Relations with the Surrounding Regions of Central Asia and Iran." In *The Archaeological Map of the Murghab Delta II: The Bronze Age and Early Iron Age in the Margiana Lowlands – Facts and Methodological Proposal for a Redefinition of the Research Strategies*, ed. S. Salvatori and M. Tosi. BAR International Series 1806. Oxford: Archaeopress, pp. 75–98.

Salvatori, S. 2010. "Thinking Around Grave 3245 in the 'Royal Graveyard' of Gonur (Murghab Delta, Turkmenistan)." In *Na puti otkrytija tsivilizatsii* [On the track of uncovering a civ-ilization]. *Trudy Margianskoj Arkheologicheskoj Ekspeditsii*, vol. 3, ed. P.M. Kozhin, M.F. Kosarev, and N.A. Dubova. St. Petersburg: Aletejja, pp. 244–257.

Santoni, M. 1984. "Sibri and the South Cemetery of Mehrgarh: Third Millennium Connections Between the Northern Kachi Plain (Pakistan) and Central Asia." In *South Asian Archaeology*, ed. B. Allchin. Cambridge: Cambridge University Press, pp. 52–60.

Sarianidi, V.I. 1976. "Issledovanie pamjatnikov Dašlinskogo oazica." *Drevnajaja Baktrija* 1: 21–86.

Sarianidi, V.I. 1977. *Drevnij zemledel'tsi Afghanistana*. Moscow: Nauka.

Sarianidi, V.I. 1981. "Seal-Amulets of the Murghab Style." In *The Bronze Age civilization of Central Asia*, ed. P. Kohl. New York: M.E. Sharpe, pp. 221–255.

Sarianidi, V.I. 1986. *Die Kunst des Alten Afghanistan: Architektur, Keramik, Siegel, Kunstwerke aus Stein und Metall*. Leipzig: E.A. Seemann.

Sarianidi, V.I. 1990. *Drevnosti Strany Margush*. Ashkhabad: Akademija Nauk Turkmenskoj SSR.

Sarianidi, V. 1998. *Myths of Ancient Bactria and Margiana on Its Seals and Amulets*. Moscow: Pentagraphic Ltd.

Sarianidi, V.I. 2002. *Margush: Ancient Oriental Kingdom in the Old Delta of the Murghab River*. Ashgabad: Türkmen dövlet habarlary.

Sarianidi, V.I. 2005. *Gonurdepe, Türkmenistan: City of Kings and Gods*. Aşgabat: Miras.

Sarianidi, V.I. 2007. *Necropolis of Gonur*. Athens: Kapon Editions.

Sarianidi, V.I. 2008. *Margush, Turkmenistan: Mystery and Truth of the Great Culture.* Aşgabat: Türkmen döwlet neşirýat gullugy.

Stein, A. 1937. *Archaeological Reconnaissances in North-Western India and South-Eastern Īrān.* London: Macmillan.

Teufer, M. 2007. "Zu bronzezeitlichen Tulpendarstellungen in Zentralasien und im Iran." In *Scripta Praehistorica in honorem Biba Teržan*, ed. M. Blečić, M. Črešnar, B. Hänsel, M. Hellmuth, E. Kaiser, and C. Metzner-Nebelsick. Ljubljana: Narodni muzej Slovenije, pp. 183–207.

Tosi, M. 1974. "Some Data for the Study of Prehistoric Cultural Areas on the Persian Gulf." *Proceedings of the Seminar for Arabian Studies* 4: 145–171.

Vidale, M. 2017. *Treasures from the Oxus: The Art and Civilization of Central Asia.* London: I.B. Tauris & Co.

Vogt, B. 1985. "The Umm an-Nar Tomb A at Hili North: A Preliminary Report on Three Seasons of Excavation, 1982–1984." *Archaeology in the United Arab Emirates* IV: 20–37.

Vogt, B. 1994. *Asimah: An Account of a Two-Month Rescue Excavation in the Mountains of Ras al-Khaimah, United Arab Emirates.* Ras al-Khaimah: Department of Antiquities and Museums.

Weeks, L. 2016. "Iran and the Bronze Age Metals Trade in the Persian Gulf." *International Journal of the Society of Iranian Archaeologists* 2(3): 13–25.

Woolley, C.L. 1934. *Ur Excavations II: The Royal Cemetery.* London and Philadelphia: Trustees of the Two Museums.

Zarins, J. 1978. "Steatite Vessels in the Riyadh Museum." *Atlal* 2: 65–93.

Zarins, J. 1989. "Eastern Saudi Arabia and External Relations: Selected Ceramic, Steatite and Textual Evidence, 3500–1900 B.C." In *South Asian Archaeology 1985*, ed. K. Frifelt and P. Sørensen. Scandinavian Institute of Asian Studies, Occasional Papers 4. London: Curzon Press, pp. 74–103.

Zettler, R.L. and L. Horne (eds.) 1998. *Treasures from the Royal Tombs of Ur.* Philadelphia: University of Pennsylvania Museum of Archaeology and Anthropology.

CHAPTER TWENTY-THREE

THE FORMATION OF THE OXUS CIVILIZATION/BMAC IN SOUTHWESTERN TAJIKISTAN

Natal'ja M. Vinogradova

INTRODUCTION

Historically, from the middle of the 3rd to the beginning of the 2nd millennia BC, southern Central Asia was occupied by settled farming cultures and city-state civilizations, but the territory of southern Tajikistan was considered as a distant and different borderland. Recent discoveries of earlier Middle Bronze Age sites on this territory, however, provide ground for a new approach on the issue of the cultural formation of this area.

CHALCOLITHIC (NMG III) SITES

As yet, no sites of this period have been discovered in southwestern Tajikistan. However, the site of Sarazm is known to the north, in the Zeravshan valley (Isakov 1991) and testifies of relationships with South Turkmenistan. Typical painted Geoksjur ceramics (Namazga (NMG) III period) have been discovered at Sarazm (mainly periods II–III) in the upper Zeravshan (Isakov 1991; Lyonnet 1996; see also Chapter 4). A few similar fragments discovered in the Murghab delta (Masimov 1979: 72–77; 2000: 25–26) may point at the route of the Geoksjur migrants (Masson 2006: 57). The end of these contacts is dated either at the beginning of the 3rd millennium (Kircho 2010: 40), or around 2500 BC (Lyonnet 1996: 60–61, 65–66). Beside the northern route mentioned above, a southern one – the lapis lazuli road – is also known via Serakhs and the Tedjen-Hari Rud: It had two branches, one to northern Afghanistan, and the other to southern Afghanistan and southeastern Iran (Kircho 2010: 40).

The interactions that these sites underline between different areas of Central Asia at the end of the 4th/early 3rd millennium, and the trade routes and population movements they imply could also have concerned southern Tajikistan.

EARLY BRONZE AGE AND BEGINNING OF MIDDLE BRONZE AGE SITES (NMG IV AND EARLY V)

Until recently, in the region to the east of the Kopet Dagh area, except for the site of Zamanbaba in the lower Zeravshan (Guljamov *et al.* 1966), only fortuitous finds pointed at the presence of NMG IV/early NMG V sites and no discovery had

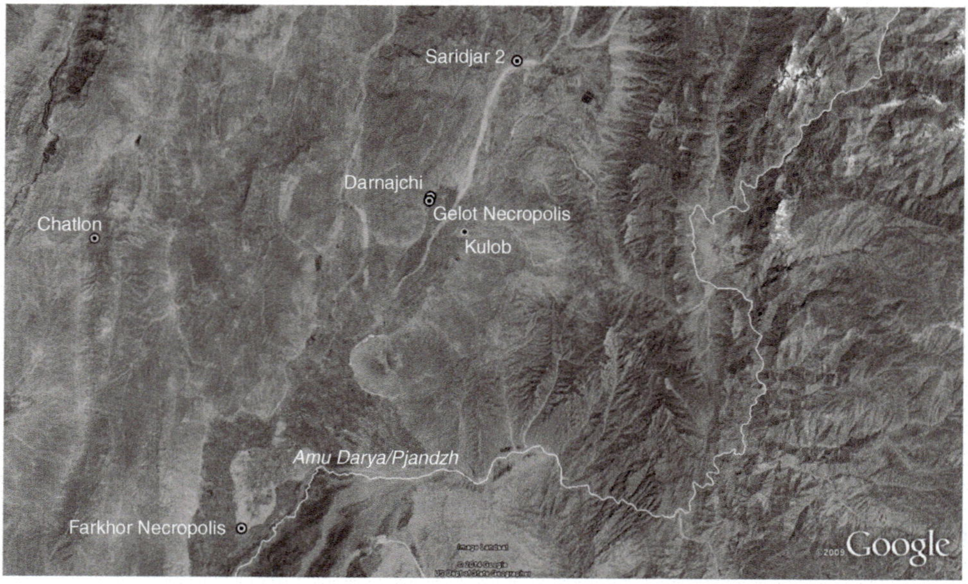

Figure 23.1 Map. Localization of the Farkhor, Darnajchi, and Gelot cemeteries.

been made *in situ* (Vinogradova 2013: 281–283). This has now changed with the excavations of cemeteries at Farkhor, Gelot, and Darnajchi (Figure 23.1) and these discoveries considerably alter our point of view on the formation of the sedentary/farming and pastoralist/steppe cultures of this area.

The Farkhor cemetery

The Farkhor cemetery is, at present, the earliest known cemetery in southern Tajikistan as shown by the material from the graves, but ^{14}C dates are not yet available.[1] It is situated on the southern part of the Urtoboz loess hill, Chiltanbobo area, in the suburbs of Parkhar.[2] This loess hill separates the Rivers Pjanj and Kyzylsu. Numerous artifacts related to the Early and Middle Bronze Age were discovered while digging contemporary graves: marble "weights," stone "rods," onyx vessels, clay "house" models, silver bracelets, and ceramic vessels (Figure 23.2) (Vinogradova 2011: 84–98; 2015: 21–27; Bobomulloev *et al.* 2014: 39–47; 2015: 47–66; 2017: 78–127). Up to now, in the zones yet free of modern tombs covering the slopes of the hill, seven areas have been excavated and 50 burials discovered. The burial constructions consist either of pits with lateral niche (*podboj*)/catacomb-like or of simple pits (Bobomulloev *et al.* 2014: 40–41; 2017: 78, 120, fig. 22a). The skeletons are in a flexed position, men mostly lying on their right side, and women on their left, but in some cases also on their right. There is no strict orientation though the most common direction is to the northwest. Funerary gifts are positioned around the head, neck, arms, pelvis, legs, and behind the back. They are of various kinds and made of different materials and they can be related either to sedentary/farming communities or to the pastoralist/steppe culture.

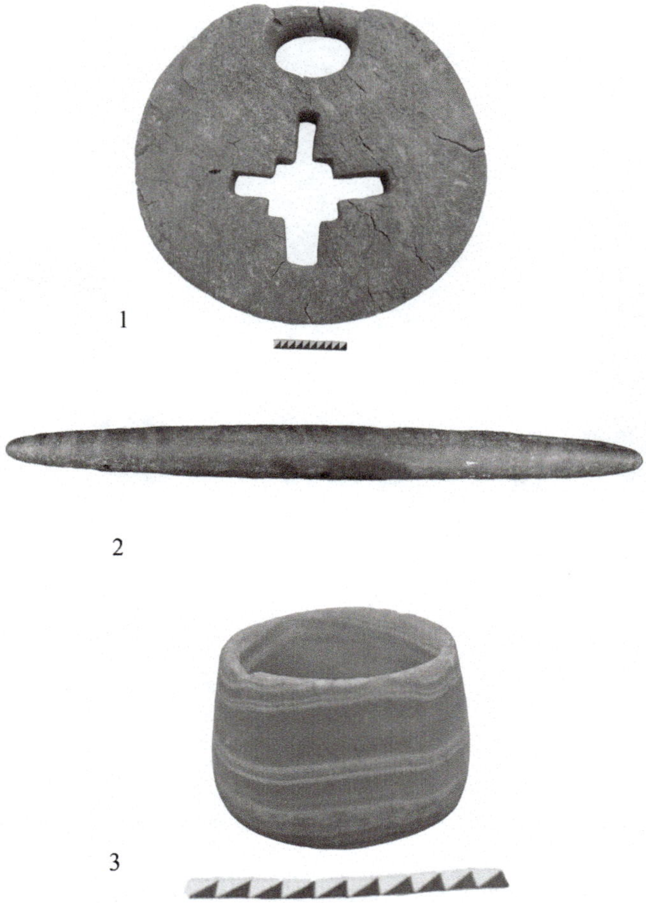

Figure 23.2 Farkhor cemetery. Chance finds: (1) stone weight; (2) stone "rod"; (3) onyx vessel.

Funerary artifacts related to farming communities

The pottery in these burials includes both wheel-made and handmade vessels, with white- or red-slipped exterior surfaces. Three pots from burials 32 and 33 are especially noteworthy: the first is jar-shaped, rather large and has an unusual rim, everted with an inner projection for a lid, and with 18 regularly spaced perforations all around (Figure 23.3, 2); the second is a pot with a funnel-shaped neck (Figure 23.3, *3*) very similar in shape with one from Shahdad necropolis A, burial 104 (Hakemi 1997: 260, no. 0962), as well as with another, but made of gypsum, from Altyn Depe, NMG III layer (burials 295–296) (Kircho 2005: tab. 59, 2); the third one (Figure 23.4, *1*), from burial 33, is biconical and painted with horizontal black stripes along the neck and on the middle part of the body. There are no close parallels to its shape, but a similar painted motif is known at Altyn Depe period NMG IV (Kircho 2005: 350, fig. 4, *34*; 351, fig. 5, *21*) and at Tepe Hissar IIIB (Schmidt 1937: pl. XXXIX, H 2212).

Figure 23.3 Farkhor cemetery. Area 6A, grave 32, funerary material.

Several graves contained almost-rectangular clay "boxes" with inner partitions (Figure 23.4, 2–2*a*). P. Amiet considered them to be models of houses (Amiet 1977: 96, fig. 6). Similar artifacts come from the looted graves of southern Bactria (Sarianidi 1979: 26, 27), and a few fragments from Shortughai III and IV (Francfort 1989: pl. 62, 11–13). Others have been found in the Zamanbaba culture (lower Zeravshan valley), which is dated to the NMG IV–V period on the basis of the figurines, beads, metal items, and painted vessels imported from southern Central Asia found together with the local material (Guljamov *et al.* 1966: 148, 163).

Funeral offerings at Farkhor also include two silver vessels. One of them presents three large ribs (Figure 23.3, 4) and is comparable with vases from looted graves in northern Afghanistan (Salvatori 1989: 210, fig. 74).

Two nearly cylindrical-shaped stone vessels made of onyx were found in burial 32 (Figure 23.3, *1*). According to L.B. Kircho, such stone vessels of varying height and proportions are the predominant type during the NMG IV period (Kircho 2014: 67). A stepped cross-shaped stone vessel was found in the same burial (Bobomulloev *et al.* 2017: 123, fig. 24b, *10*) (Figure 23.3, *6*). Similar vases made of marble-like stone are frequent during the Early and Middle Bronze Age in Afghanistan, Baluchistan, northern Iran, and Mesopotamia (Casanova 1991). The rich burials 18 and 19 contained "stone lamps with their lids."[3] Their sides are decorated with three dotted circles incised into each of the four sides (Figures 23.5 *in situ*; 23.6, *6*; 23.7, *5*). Very close

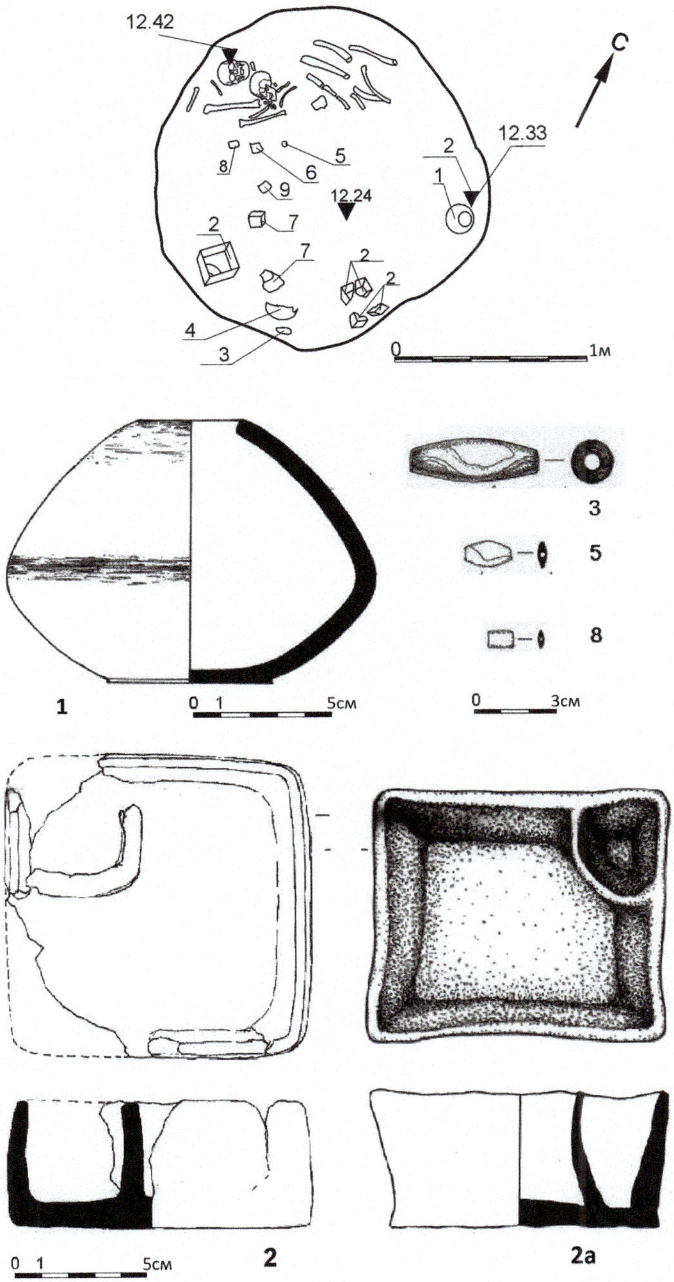

Figure 23.4 Farkhor cemetery. Area 6, grave 33 and its funerary material. On top, plan; below, funerary material: 1, pottery vessel decorated with black paint; 2, 2a, terracotta "house model"; 3, 5, 8, stone beads (+ shown only on the plan: 4, stone flake; 6, pebbles; 7, pottery sherds).

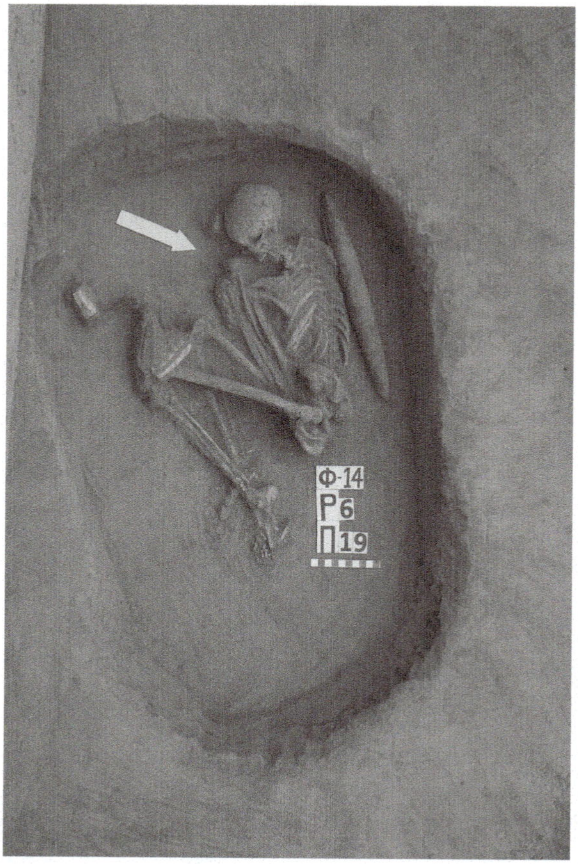

Figure 23.5 Farkhor cemetery. Area 6A, grave 19, photo.

analogies can be found at Shahr-i Sokhta (Sajjadi 2003: 83, fig. 40, type 1) and at Altyn Depe (burial 828, dated to the early NMG IV period) (Kircho 2005: 350, fig. 4, 23). According to Kircho (2005: 452), at Altyn Depe lamps were found exclusively in burials and cultural layers of the Early Bronze Age (NMG IV period), and only one example from burial 893 dates to the beginning of the Middle Bronze Age (NMG V).

Flat stone "weights" with a handle and cross-shaped perforations are chance finds (Bobomulloev *et al.* 2015: 48, figs. 1, 2) (see Figure 23.2, *1*). Similar artifacts, with the same or other decoration, come from looted graves in Afghanistan (Pottier 1984: 175, fig. 41, no. 291; Winkelmann 1997: 207, fig. 12; Sarianidi 1998: fig. 7, *7*) and from Ulug Depe in Turkmenistan (Sarianidi 2001: 69, fig. 36). Flat weights are characteristic for the second half of the 3rd millennium (Kircho 2007: 201). Stone "rods," however, were found *in situ* for the first time in Tajikistan (Figures 23.5 and 23.6, *1*). Their length ranges from 60 to 90 cm and they weigh around 6.0 kg. Other rods out of context had been discovered in the Gelot cemetery (Vinogradova *et al.* 2009: 41) and in the Kyzylsu River valley (Vinogradova *et al.* 2003: 107, fig. 5, *3*). At Altyn Depe, "rods" (or "staff" or "scepter") come from the "sanctuary," and from burial 362 (Kircho 2005: 415). These items are also known at Tepe Hissar IIIB (Schmidt 1937: 222, pl. LXIV, H 2894) and in burial complexes of Baluchistan (Kircho 2005: 415).

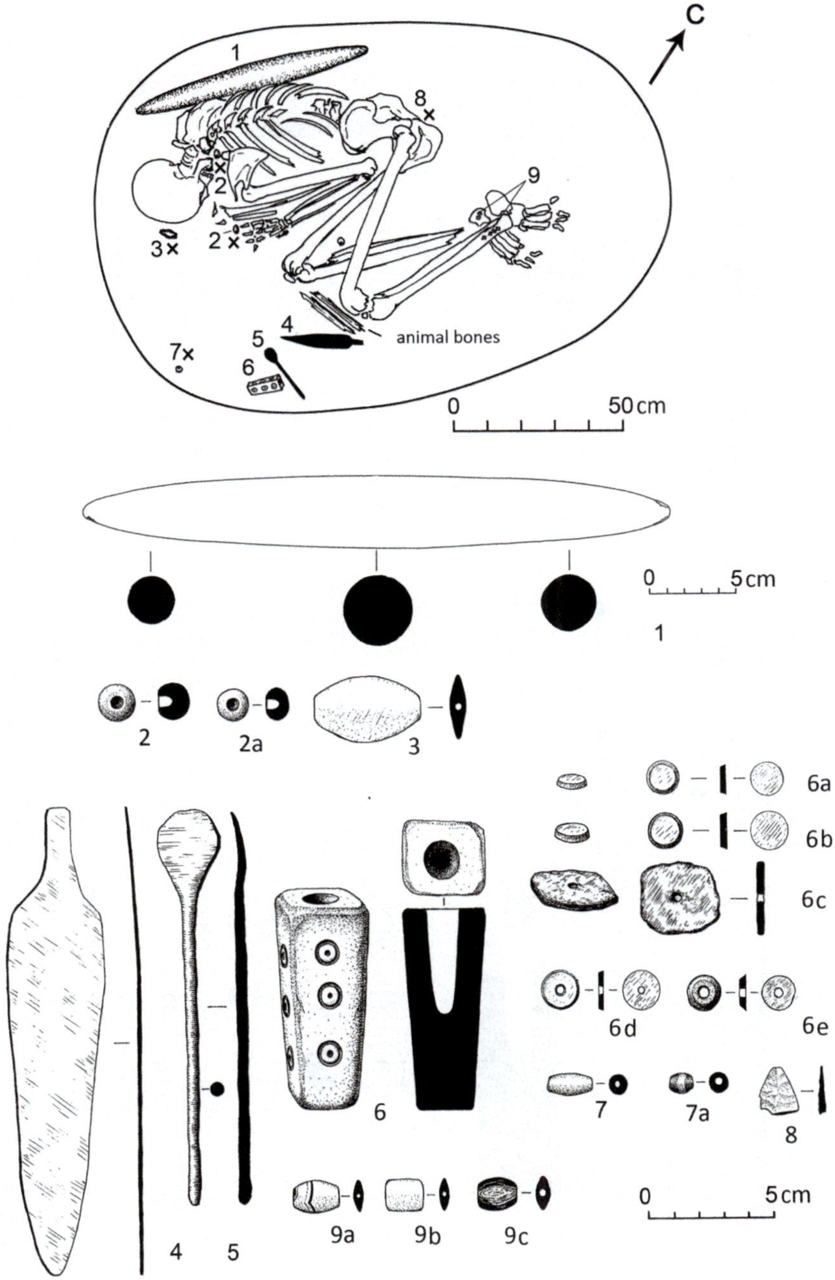

Figure 23.6 Farkhor cemetery. Area 6A, grave 19, drawing with funerary inventory. On top, plan; below, funerary material: 1, stone rod; 2,2a top, limestone finials; 3, agate bead; 4, bronze knife; 5, bronze cosmetic spatula; 6, stone lamp; 6a–6c, lamp stone lids; 7, sandstone with calcite bead; 8, flint arrowhead; 9a, 9c, agate beads; 9b, turquoise bead.

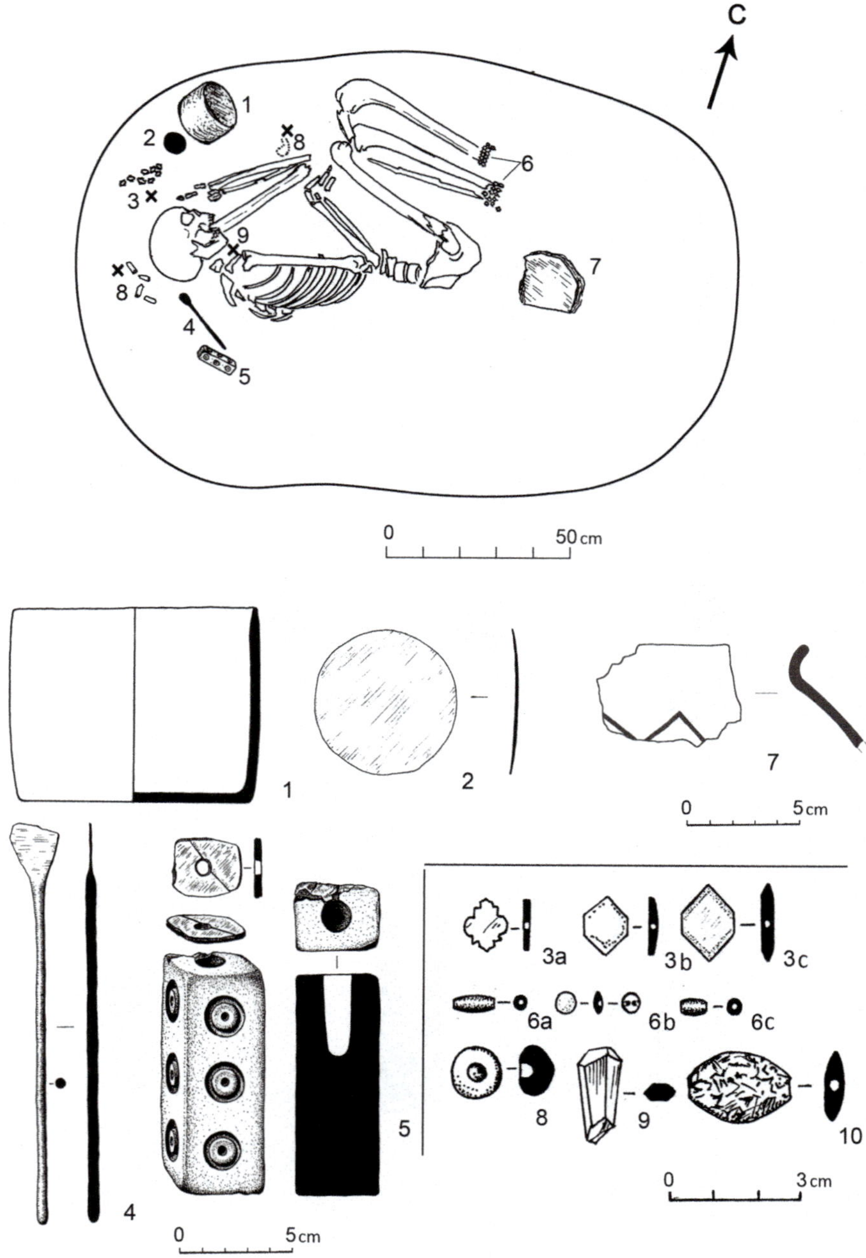

Figure 23.7 Farkhor cemetery. Area 6A, grave 18 and its funerary inventory. On top, plan; below, funerary inventory: 1, ceramic vessel; 2, bronze mirror; 3a, b, c, plaster beads; 4, bronze cosmetic spatula; 5a, stone lamp with lid; 6a, agate bead; 6b, turquoise bead; 6c, gold bead; 7, pottery sherd with ornament; 8, stone finial; 9, rock-crystal fragment; 10, stone pendant.

Several metal cosmetic pins with heads in the form of spatulae of various shapes also come from these graves (Figures 23.6, *5* and 23.7, *4*).[4] Such pins are frequent among the artifacts of the farming societies and were used for a long time, from the end of 4th to the beginning of the 2nd millennium (Alekshin 2010: 34–43). Among them, one with a round head and edges carved into stepped rhombs is noteworthy (Bobomulloev *et al.* 2017: 106, fig. 20, *1*) and finds close parallels at Altyn Depe (tomb 828, early NMG IV; see Kircho 2005: 350, fig. 4, *5*), at Sarazm (Isakov 1991: fig. 23, *2*), and at Mundigak period IV, 1 (Casal 1961: fig. 140, *19*). Another interesting silver pin with a cross-shaped head must also be noted (see Figure 23.3, *5*).

The shape of some of the stone beads is significant both from a cultural and chronological point of view. Cross-shaped lapis lazuli and alabaster beads (square with stepped edges) (Figure 23.7, *3a*) (Vinogradova and Bobomulloev 2016: 543, fig. 9, *3*; 545, fig. 11, *3a*) are comparable to cross-shaped beads frequent in southern Turkmenistan, at Altyn Depe (Kircho 2005: 391, fig. 22), Parkhai II (Khlopin 2002: 63, tab. 94, *4*), as well as in the looted graves from northern Afghanistan (Amiet 1977: 102, fig. 12; Sarianidi 1979: 25, fig. 1). As imports from southern areas, they are attested in the Zeravshan valley (Uzbekistan) both in Zamanbaba burials (Guljamov *et al.* 1966: 154, fig. 48; tab. VIII, XX) and at the Zhukov cult complex (Avanesova and Jurakulova 2008: 26, fig. 4, *2*, *4*).

The analogies mentioned above clearly testify to contact between the population buried at Farkhor and other areas of southern Central Asia (Altyn Depe, Ulug Depe), Iran (Hissar IIIB–C, Shahdad), and Afghanistan (looted graves) at an early time. All these artifacts date to the Early and beginning of the Middle Bronze Age (NMG IV–V periods) according to the chronological sequence of southern Turkmenistan.

Funerary artifacts related to steppe culture communities

However, the funerary material of some graves (tombs 2, 4, 12, 14, 22) is different and presents a different circle of analogies. The pottery is handmade with round and flat bottoms, sometimes finished on the potter's wheel. The whole surface is covered with white or red slip, but some gray-ware and black polished pots are also attested. They are decorated with a burnished "tree-like" or herringbone pattern (Figure 23.8, *2–3*). Handmade pottery with such a decoration is well known among the Vakhsh culture sites in the valley of the Kyzylsu, especially at the Darnajchi cemetery (Vinogradova *et al.* 2013: 115, fig. 14, *1*, *2*; fig. 30, *1*, *2*; see also Chapter 25). The polished "tree"/ herringbone pattern is characteristic for the Astrabad culture of northeastern Iran at Tureng Tepe III, Shah Tepe II, Hissar IIIC (Stankevich 1978: 24). Gray-ware pots with the same ornament are also known at Dashly 1 in northern Afghanistan (Sarianidi 1976: 37, fig. 24).

One handmade pot (Figure 23.9, *2*) from tomb 12, associated with a fragment of a bronze knife and sheep bones, is especially interesting. Its closest analogies, both for the shape and the temper of the clay, are among the Zamanbaba cemetery group VI pottery (Guljamov *et al.* 1966: pl. XIVB, *1*, *7*). Less close parallels can also be found in the handmade pottery from the Rannij Tulkhar cemetery in the Bishkent valley (Mandel'shtam 1968: 151, tab. X, *8* – pits with slope). This type of pot is mainly common among the steppe cultures where it is dated to the Early and Middle Bronze as confirmed by a radiocarbon date on human bones from a grave (no. 2) in another

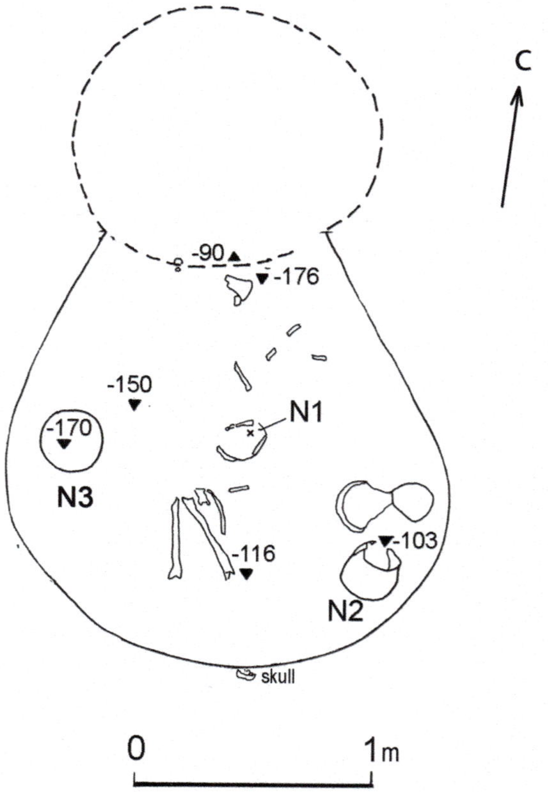

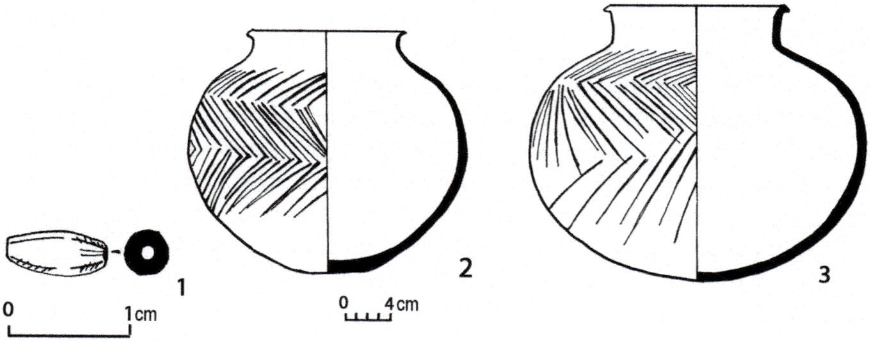

Figure 23.8 Farkhor cemetery. Area 1, grave 4 and funerary inventory. On top, plan; below, funerary inventory: 1, lapis lazuli bead; 2–3, "tree-like" polished pots.

nearby cemetery, Darnajchi (see p. 649–650) (Table 23.1). In this last cemetery, two graves (burials 2 and 3) have produced the same funerary material as that at Farkhor, while the others are chronologically later.

According to the anthropological data, all the human remains from the Farkhor burials (except tomb 12), are characterized by their common gracile and undeniable European (Mediterranean type) appearance.[5]

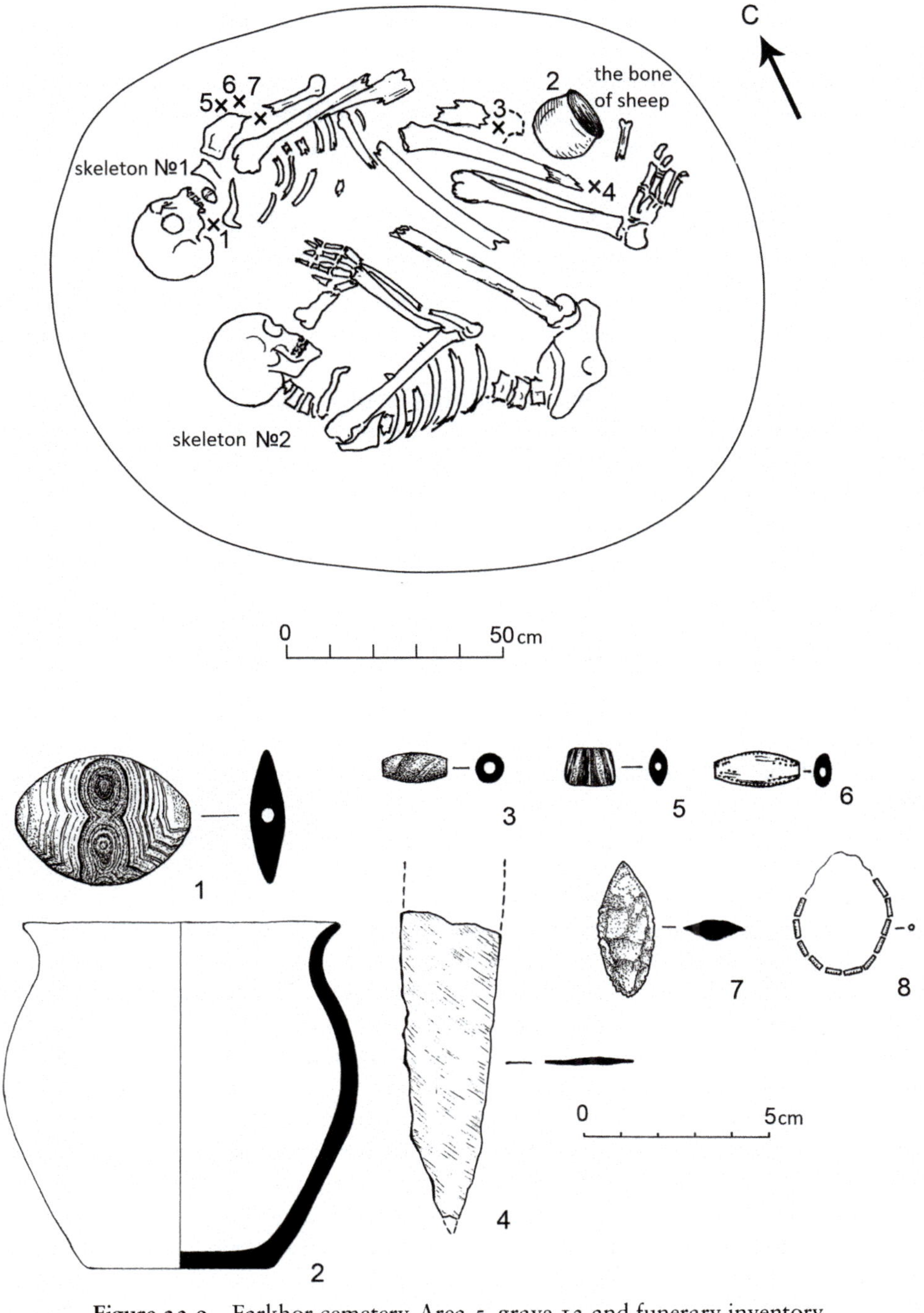

Figure 23.9 Farkhor cemetery. Area 5, grave 12 and funerary inventory. On top, plan; below, funerary inventory: 1, agate pendant; 2, pot; 3, 5, 6, 8, grinded mineral beads; 4, fragment of a bronze knife; 7, flint arrowhead.

To sum up, some ceramic shapes found among the funeral offerings of the Farkhor cemetery were unknown before and are the first testimonies of the Vakhsh culture in southern Tajikistan (Bobomulloev *et al.* 2015: 61; see also Chapter 25). The Farkhor and some of the Darnajchi burial finds testify to relationships with groups of the steppe Bronze Age cultures (Zamanbaba cemetery, Rannij Tulkhar – pit with slope).

MIDDLE BRONZE AND EARLY LATE BRONZE AGE SITES (NMG V AND VI)

The other burial complexes in southern Tajikistan, Gelot, and Darnajchi date to the mature phase of the Oxus Civilization or Bactria-Margiana Archaeological Complex (BMAC) (NMG V/early NMG VI).

Both are located within 6 km to the northwest of Kuljab, on the second terrace of the Jakhsu River, along the loess hills bordering the river valley from the northwest. Several Bronze Age cemeteries have been found on the northern slope of the hills (Vinogradova *et al.* 2009: 35–72; 2010: 389ff; Vinogradova and Kutimov 2018). They have been used for a long time, both by farming sedentary communities and by the Vakhsh culture population, which is traditionally considered as a steppe pastoralist culture (see Chapter 25).

Gelot burials related to farming communities

These graves are widely scattered, and are sometimes over 2 m deep, consisting either of pits or catacombs. The entrance to the burial chamber is sealed with lumps of clay or with bricks. The bodies are flexed on the left (females) or right (males) side, without any systematic orientation. The funerary offerings include from 5 to 11 vessels, and, in grave 3, Area 4, a biconical stone spindle whorl. A rich burial (grave 2, Area 6) provided more material (see below). Animal meat offerings (sheep) are present inside the graves, and two sheep burials (graves no. 5 and 6, Area 6) are also attested; the animals were lying on the right or left side, heads to the west without grave goods.

All the pots from these tombs (except for one wall fragment in burial 3) are wheel-made, and almost all are covered with a dark-red slip on the exterior (and interior for bowls). Some vessels have a white slip. The majority of the pots are not polished. The most specific shapes are stemmed goblets, solid stemmed bowls, flat-base goblets,[6] large open bowls, jugs, spouted pots, etc. (see Figures 23.10 and 23.11). Comparisons place this pottery material within the Sapalli and Dzharkutan phases of the Sapalli culture (i.e., late NMG V/early NMG VI). Similar pots are present in burials at Sapallitepa and Dashly 1, in NMG V layers at Altyn Depe, Gonur, or in the Kelleli phase of Margiana. Other pots have close parallels at Dzharkutan and Dashly 3 (Vinogradova *et al.* 2010: 402–405).[7] Analogies are also obvious with the pottery from the cemeteries in northeastern Iran, Chalow (Biscione and Vahdati 2011: 240, fig. 6), Bojnord (Vahdati 2014: 24, figs. 7, 8), and Tepe Damghani (Francfort *et al.* 2014: 131, fig. 5).

The rich woman's burial 2 was found in Area 6 (Figures 23.12 and 23.13). Besides 11 vessels (Figure 23.14), it contained a rich inventory including a bronze pin and a bronze mirror,[8] lapis lazuli beads and one gold bead, a fragment of a pink marble stamp seal or amulet, and a marble anthropomorphic figurine (Figures 23.16 and 23.17). The seal probably depicts a four-petal rosette (Figure 23.15, 5), a well-known

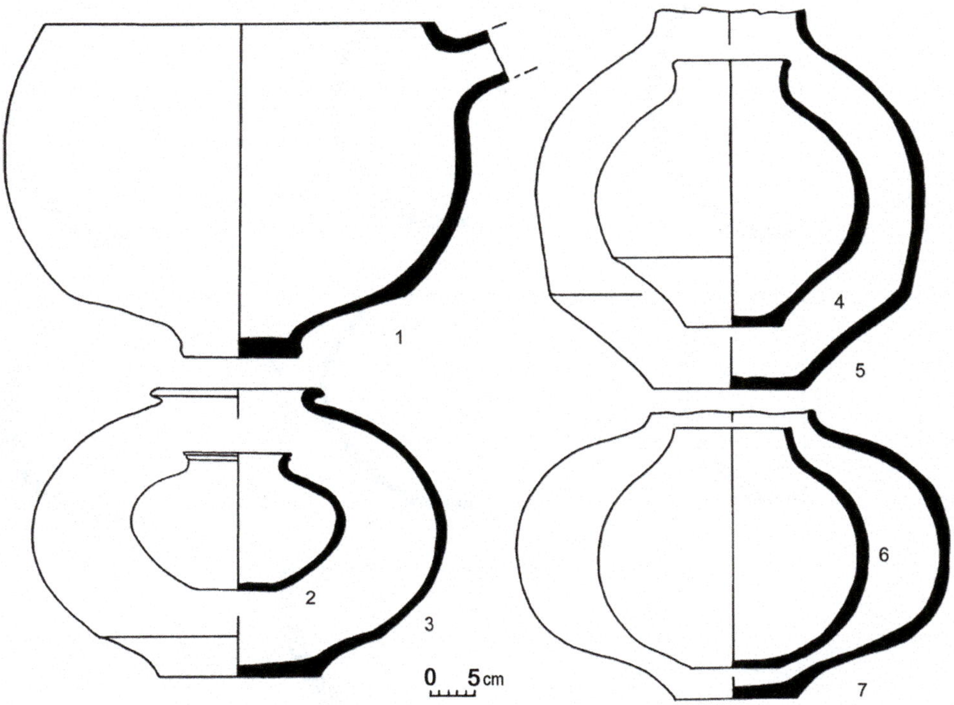

Figure 23.10 Gelot cemetery. Pottery from various graves: 1, 2, Area 4, grave 2; 3, 5, Area 5, grave 1; 4, Area 6, grave 2; 6, 7, Area 4, grave 3.

motif among metal seals at Altyn Depe (Masson 1981: tab. XVI, *8*), Shahr-i Sokhta (Salvatori 2000: 115, fig. 11, *10* – type 8), or among serpentine amulets in Margiana (Sarianidi 1991: 222). The most significant discovery made in this burial, however, is the stone human figurine found under the bone of her right foot heel (Figures 23.13 *in situ*, 23.16, and 23.17) (Vinogradova and Kutimov 2009: 102–116; Vinogradova and Winkelmann 2016: 361–377).[9] The figurine (13.2 cm high) most probably depicts a man[10] and was lying on the back, face up. The lower part of the statuette was carved in the form of a truncated cone. Its head is round-shaped with a visible straight nose, chin, and "smiling" lips. The eyes are large and almond-shaped. The hands rest on the waist, thumbs slightly raised up, touching each other, fingers in closed fists. The hair is straight, underlined with thin incised lines, ending in a braid, and rolled up into a bun at the back of the head. The position of the arms probably points at a praying position.

At first sight, this figurine has no direct analogies in the BMAC. Its style is reminiscent rather of the Mesopotamian sculptures of the Early Dynastic period (broad shoulders, position of the arms, "smiling lips," and large almond-shaped eyes), like those of Mari (Braun-Holzinger 1997: tab. 24a, d; 30a, b). However, some features link it to some representations from Bactria and Margiana, like the head of a silver pin found at Gonur Depe (Sarianidi 1995: 78, fig. 3) that depicts a woman wearing a *kaunakes* and siting on a chair. Her face differs from the Gelot figure, but there are similarities in the large almond-shaped eyes, the ellipsoidal ear, and the hairdo that is also modeled with thin incised lines and gathered in a bun at the back of the

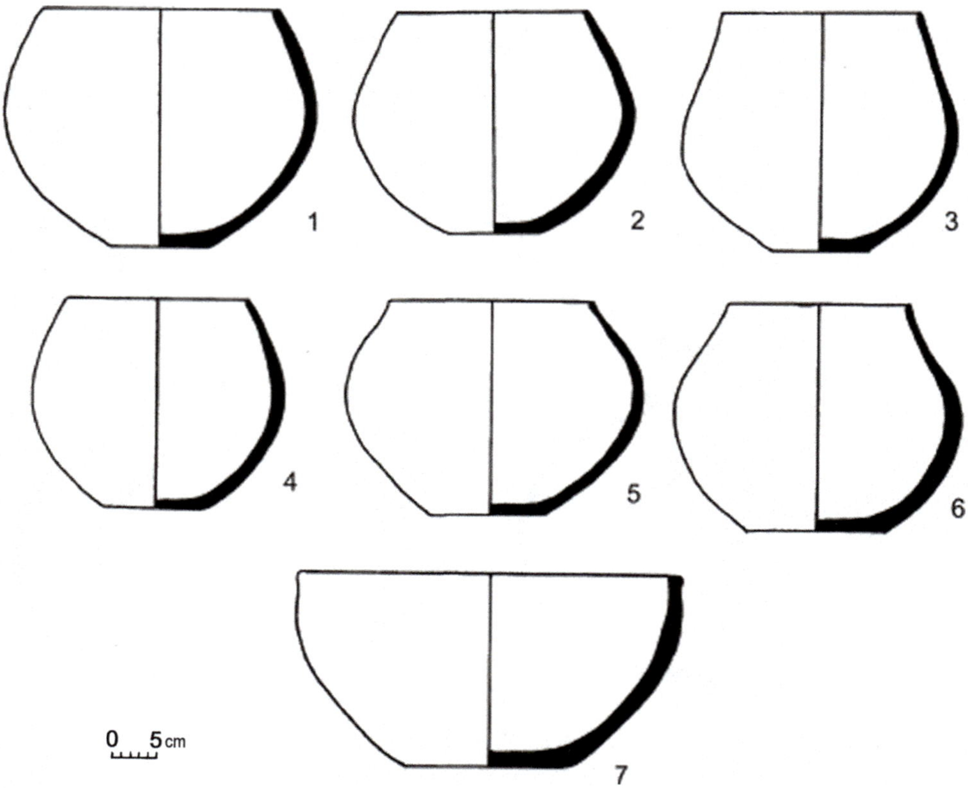

Figure 23.11 Gelot cemetery. Pottery from various graves: 1–3, 5, 7, Area 6, grave 2; 4, Area 4, grave 3; 6, Area 5, grave 1.

head. Such a hairstyle is sometimes visible on Bactrian composite statuettes (Pottier 1984: pl. XLI, 303).

Some similarities in the treatment of the eyes can also be found in the unique "priest-king" statue found at Mohenjo-daro (Possehl 2002: figs. 6.6 and 6.7) and in the head from Mundigak IV.3 (Casal 1961: pl. XLIV).

Other parallels to this statuette come from eastern Iran, at Shahdad, where clay statues in praying poses were found in Necropolis A. They are, however, larger in size (from 45 to 97 cm) (Hakemi and Sajjadi 1990: 147) and represent portraits of the deceased. Some sculptures are modeled under the influence of Mesopotamian plastics (broad shoulders, wide arms that either rest on the chest or on the waist). These specific ways of modeling the torsos of the statues connect the Gelot statuette with the sculptures from Shahdad (Hakemi 1997: nos. 1479, 3432, 3520). According to S. Winkelmann, the cylinder seals from Shahdad Necropolis A date from the Early Dynastic period to the beginning of the Akkadian period. We tend to consider that the Gelot figurine was made in Bactria from local stone but under Mesopotamian influence. H.-P. Francfort informed us of a similar figurine found in a Hellenistic context at Ai Khanoum (not far from Gelot, on the left bank of the Darya-i Panj) (Francfort 2015: 41–51).

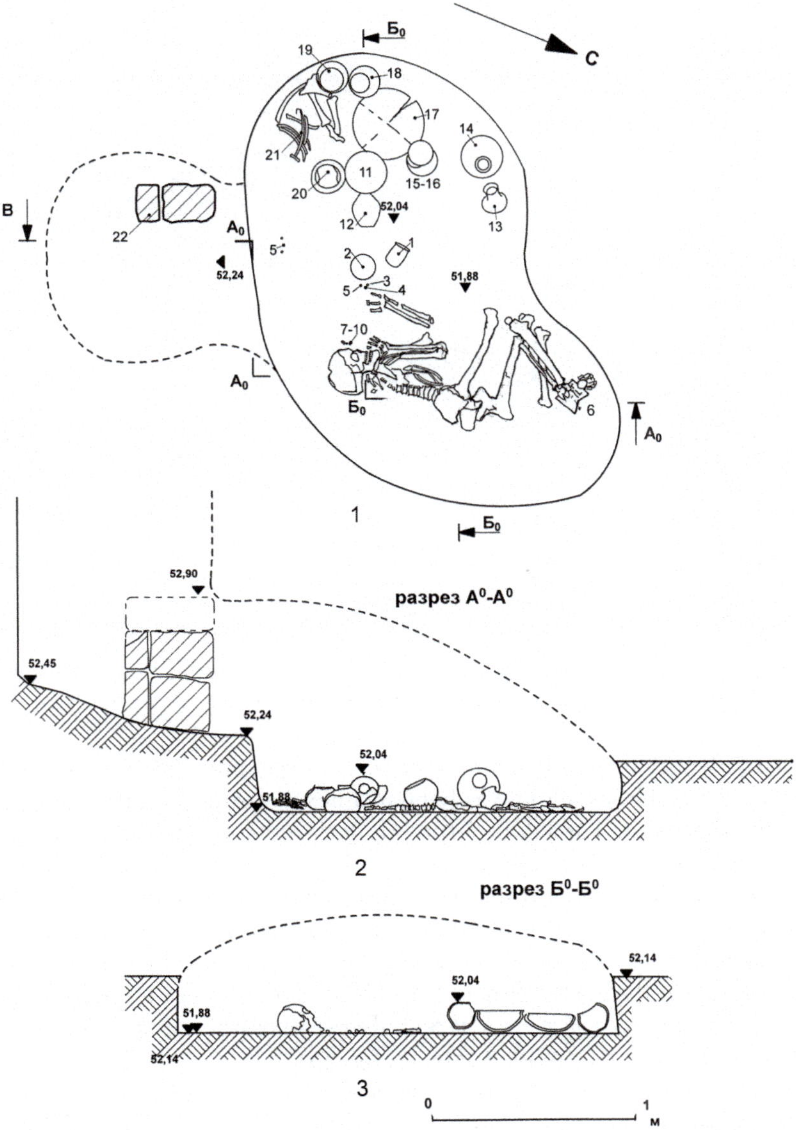

Figure 23.12 Gelot cemetery. Area 6, grave 2. Plan and section of the grave. *1*, *11–20*, pottery; *2*, bronze mirror; *3*, tip of a spatula; *4*, broken seal; *5*, fragments of copper artifacts; *6*, marble statuette; *7–10*, beads; *21*, sheep bones; *22*, bricks.

Darnajchi burials related to farming communities

In 2010–2012, excavations started in five areas near the small village of Darnajchi on loess hills along the Vose-Khovaling road. Only two of the graves are undoubtedly attributed to sedentary farmers thanks to their pottery: burial 5 in Area 1 (Figure 23.18) and burial 5 in Area 2 (Figure 23.19), but other wheel-turned vessels were present in destroyed tombs (Figure 23.20, *1–5*). Only with tomb 5, Area 2, was

Figure 23.13 Gelot cemetery. Area 6, grave 2. Photo, *in situ*. Notice the stone figurine under the foot of the deceased.

it possible to identify the construction type (i.e., a catacomb with a one-step entrance pit) (Figure 23.19). The skeleton was lying in a flexed position on the right side, head towards the northwest. Three wheel-turned, dark-red slipped vessels were found in the grave: a jug with a cross-like incision, a deep bowl, a stemmed-goblet (broken stem), as well as a handmade round-bottomed bowl that was also dark-red slipped. All the shapes, except for the round-bottomed bowl, have direct analogies with the ceramic material from the rich burial 2, Area 6 at Gelot. The incised cross mark on pottery is typical for the NMG V/early NMG VI period: at Altyn Depe (Udeumuradov 1993 levels 0–2: 52, fig. 19, *7–13*), at Sapallitepa (Askarov 1973: 169, tab. 31, *16–19*), and at Dashly 1, 3 (Sarianidi 1976: 77, fig. 48).

To sum up, the wheel-turned pottery material from Gelot and Darnajchi is close to that from the late NMG V/early NMG VI period (or BMAC) of Central Asia.

Burial complexes of the Vakhsh culture at Gelot and Darnajchi

At these two cemeteries, other burials contained a different material and are related to the Vakhsh culture (see Chapter 25). The graves are located on the same slopes of the loess hills, next to the previous ones. Nothing is visible on the surface. They do not overlap each other, being separated by a significant space (from 2 to 6 m). The burial pits are the same as the previous ones, either consisting of a simple pit or a pit with a niche (*podboj*) or a catacomb. The chamber entrance is sealed with lumps of clay. Differences are in the extreme flexed position of the individual (heels touching the pelvis). In most cases, the heads are oriented to the north or northwest. In male

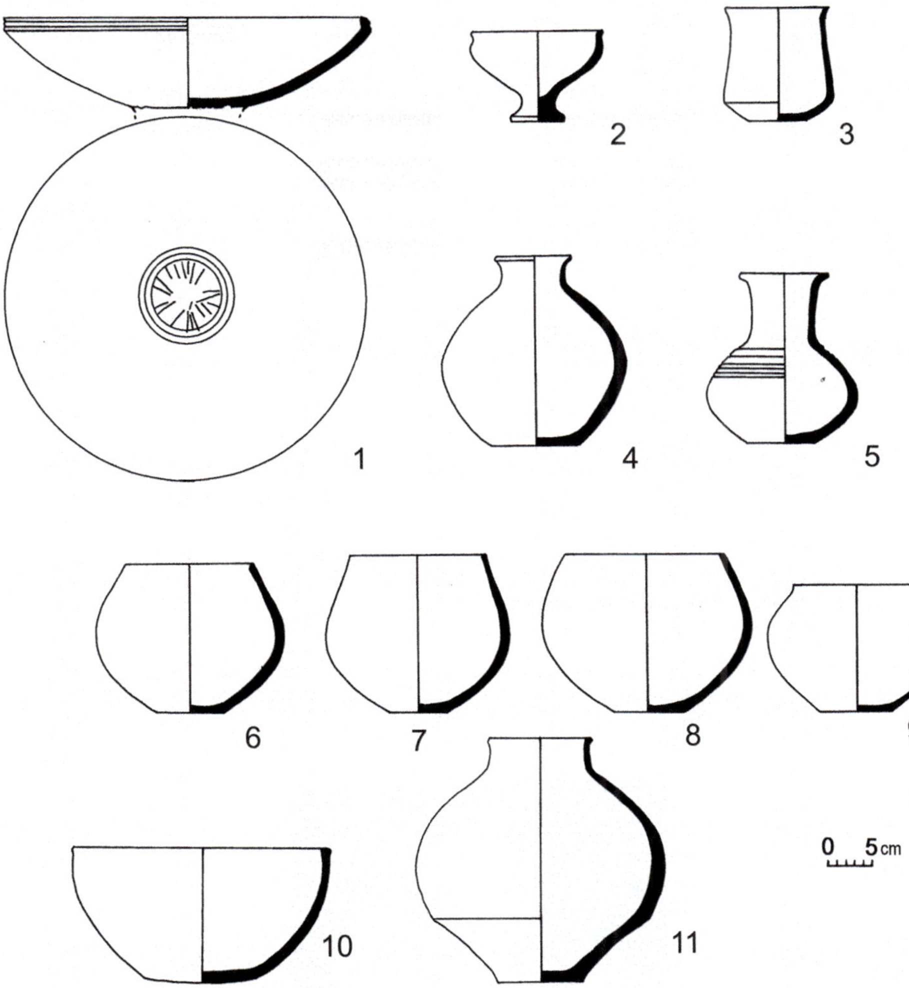

Figure 23.14 Gelot cemetery. Area 6, grave 2. Funerary material: the pottery vessels.

burials the skeleton lies on the right side, in female burials on the left. The hands are in front of the face, on the chest, or the waist. At Gelot, double burials and secondary burials with dismembered skeletons were also discovered: only the skull and the long bones of the arms and legs were found. Sometimes bones of sheep or wild animals (deer) are present in the graves. The inventory is quite standard: one to four vessels; beads made of paste, carnelian, or lapis lazuli; pestles for grinding color; in rare cases, knives and bone items; finally, stone "rods" were found in looted graves.

The Vakhsh culture pottery is both wheel-turned (40%) and handmade. The wheel-turned pottery consists of tulip-shaped "goblets" on a disk-shaped solid bottom, or goblets with a biconical body, ring-based bowls, deep bowls with a conical lower part, large dishes, or spherical pots sometimes with a disk-shaped base (Figures 23.21).

Handmade vessels are sometimes finished on the wheel and are mostly represented by round-bottomed shapes. Remains of red or white slip, with a "tree-like" polished

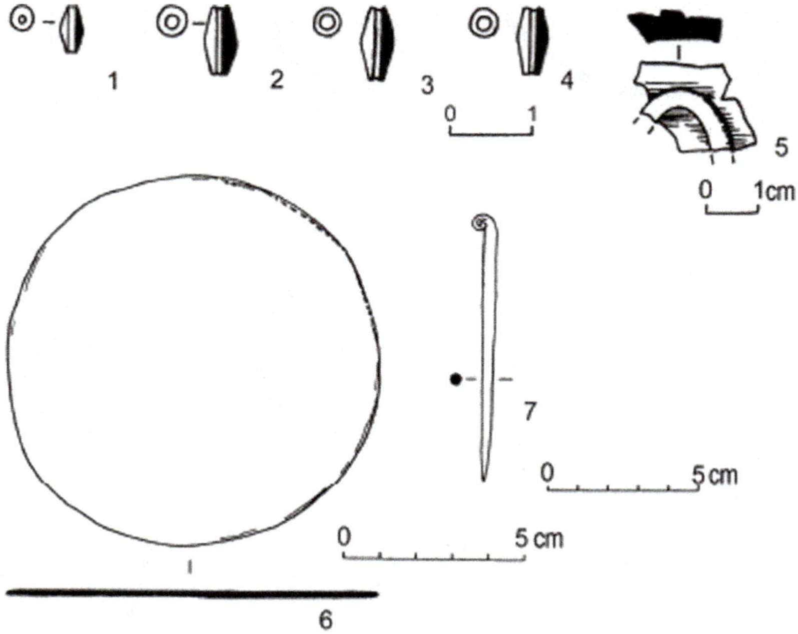

Figure 23.15 Gelot cemetery. Area 6, grave 2. Funerary material: 1, gold bead; 2, 3, 4, lapis lazuli beads; 5, fragment of a stone seal; 6, bronze mirror; 7, bronze pin.

Figure 23.16 Gelot cemetery. Area 6, grave 2. Stone figurine (front view).

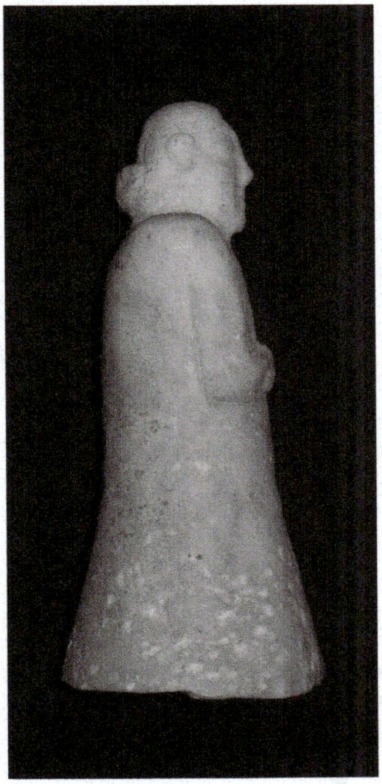

Figure 23.17 Gelot cemetery. Area 6, grave 2. Stone figurine (side view).

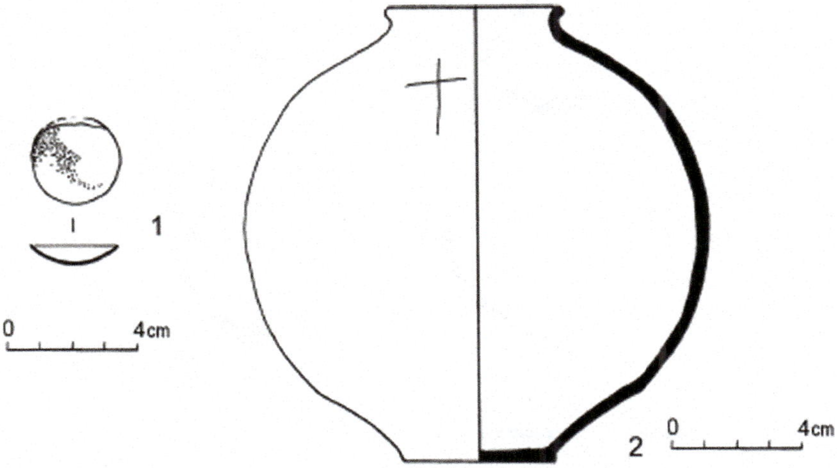

Figure 23.18 Darnajchi cemetery. Area 1, grave 5 funerary inventory. 1, miniature bronze bowl; 2, ceramic vessel.

decoration are visible on some pots (Figure 23.22, *1–2*). Some shapes can be considered as tableware, like the tulip-shaped "goblets" on a ring or a solid disk base (found only at Gelot), jugs, deep bowls with a conical lower part or with a round bottom, and spherical pots with a slightly round bottom (Figure 23.23). Pear-shaped vessels are also attested. At Darnajchi, the most characteristic types are round-bottomed spherical or pear-shaped pots (Figure 23.20).

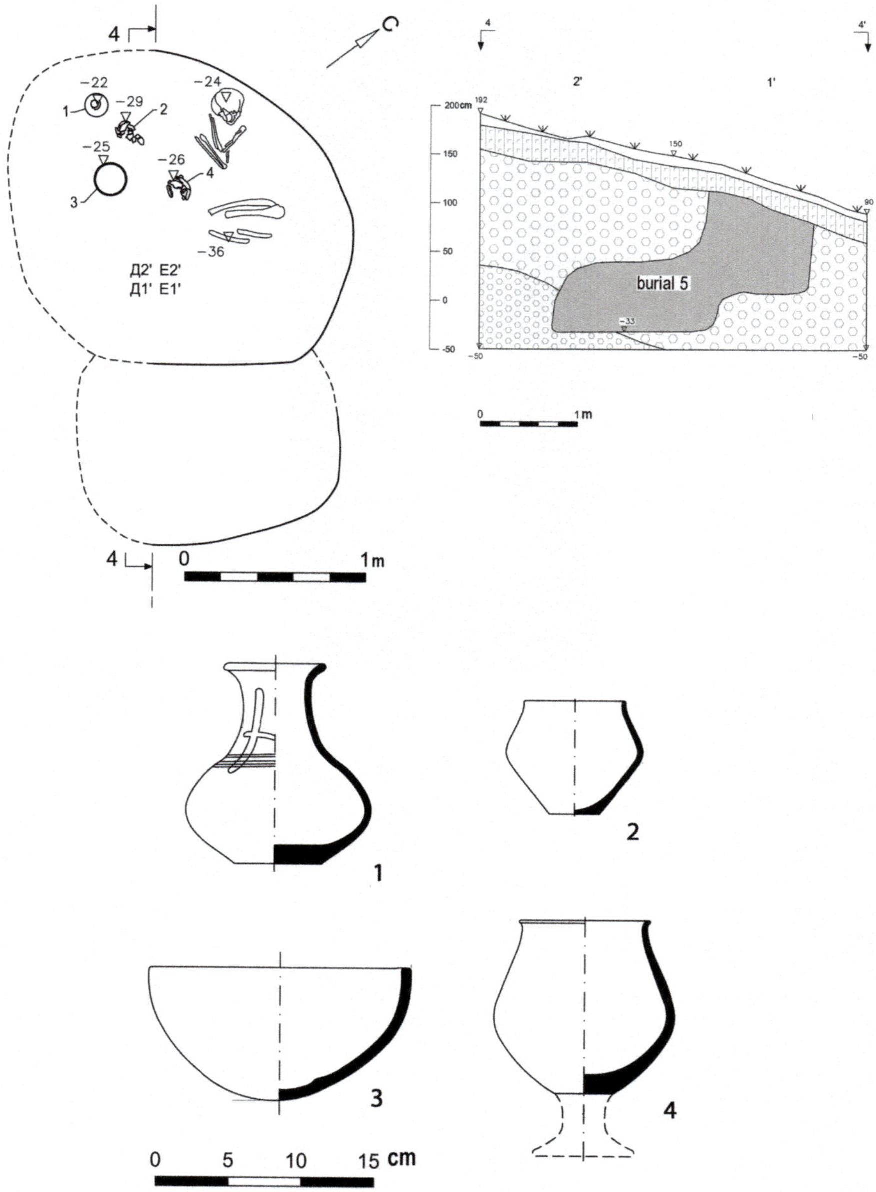

Figure 23.19 Darnajchi cemetery. Area 2, grave 2 and funerary inventory. On top, plan and section of the grave; below, funerary material: 1–4, ceramic vessels.

Both the wheel-turned and handmade pottery from Gelot finds analogies in the Vakhsh and Bishkent cultures. Tulip-like or egg-shaped "goblets" are one of the most typical forms of vessels of these cultures, with a solid disk-shaped base when wheel-turned and with a ring base when handmade. Very close pottery is present in the Obkuh cemetery in the Kyzylsu valley (P'jankova 2003: 227, fig. 15, *7*, *8*). Two egg-shaped wheel-turned vessels have been discovered in burials at Dashly 17 and Dashly

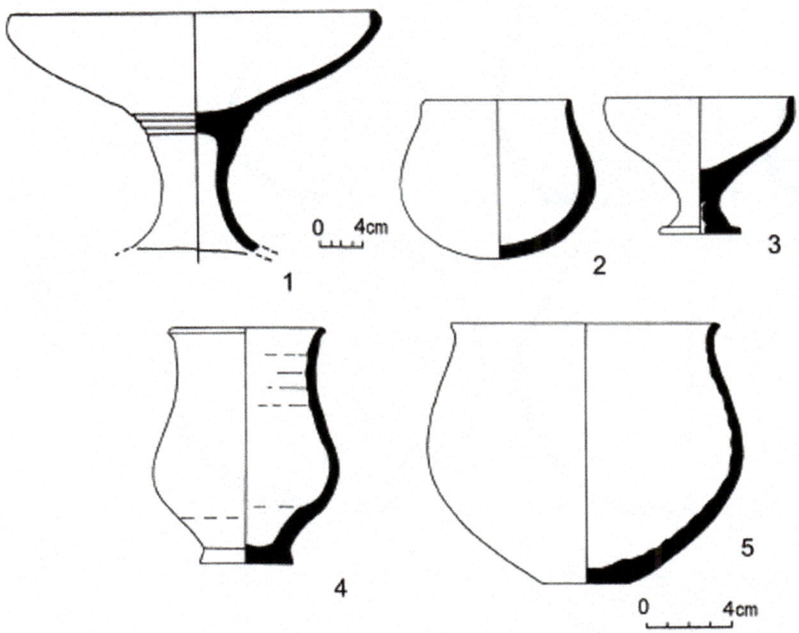

Figure 23.20 Darnajchi cemetery. Pottery from destroyed graves.

18 in northern Afghanistan (Sarianidi 1977: 69, fig. 31, *8*, *9*). Other close comparisons exist with the pottery from the cemeteries at Rannij Tulkhar (Mandel'shtam 1968), Makoni Mor (P'jankova *et al.* 2009), and Tigrovaja Balka (P'jankova 1989).

However, it should be underlined that pot jars (*banki*), so typical of the late phase of the Vakhsh and Bishkent culture, are not present here and that handmade pots with a polished tree-like/herringbone pattern all over their surface are known at the earlier Farkhor cemetery as already mentioned.

The bronze items found in the graves from Gelot include a pin awl and a double-blade knife with a spatula-like blade. In addition two bone items, an arrow, and a hollow tube were also found. The arrow has a cone-like elongated shape with a beveled tang. Four similar arrows come from kurgan 24 at Makoni Mor (P'jankova *et al.* 2009: 114, fig. 20, *1–4*). The bone tube was cut on both ends with a knife, and polished.

The graves include numerous beads made of gold, agate, carnelian, lapis lazuli, turquoise, and paste, and are of various shapes: cylindrical, barrel-like, drop-like, or six-faceted. Drop-like and six-faceted beads are found only in Darnajchi. Agate or turquoise beads have a wide range of analogies among both the farming and steppe cultures of Central Asia. Six-faceted and drop-like lapis lazuli beads from Darnajchi are similar to those found at Shahdad (Hakemi 1997: 656, Hb2), while at Sarazm this shape is attested in carnelian (Razzokov 2008: fig. 33, *1–4*). Drop-like beads are known in Iran at Hissar III (Schmidt 1937: pl. LXIX, H3546) and Shahdad (Hakemi 1997: 656, Hb13). The great number of large or small beads in the burials of the Vakhsh culture at Gelot and Darnajchi points indirectly to their local manufacture

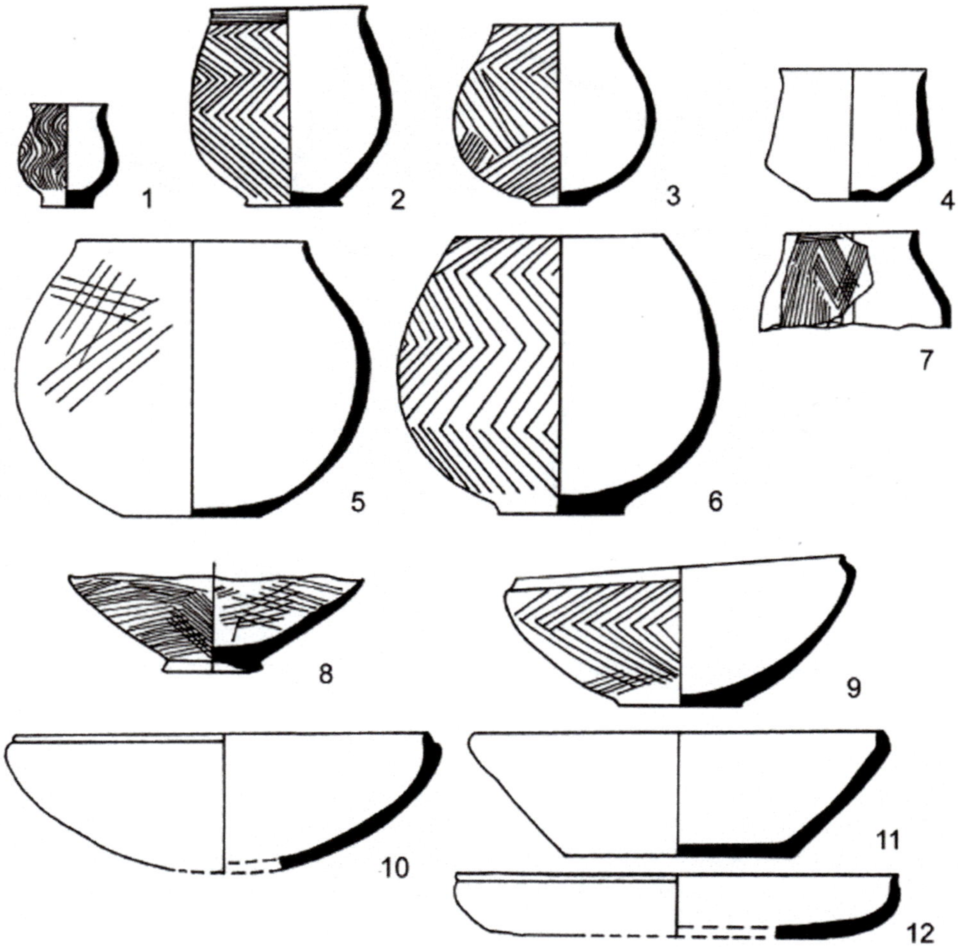

Figure 23.21 Gelot cemetery. Vakhsh culture wheel-turned pottery.

and to close cultural and trade or exchange relationships with the farming communities of that area.

Three radiocarbon dates on human bones from burials at Gelot date them to the end of the 3rd/beginning of the 2nd millennium BC (Teufer *et al.* 2014: 110–119) (Table 23.1). These dates significantly shift backwards the relative chronology of the Gelot burials that had been built upon pottery typological comparisons with cultures of the neighboring regions (Vinogradova *et al.* 2012: 139–145; 2013: 94–96; see also the Appendix at the end of this volume). In addition, the Vakhsh pottery from Darnajchi (round-bottomed vessels with the polished herringbone pattern), rare shapes of beads (multifaceted and drop-like), and analogies with artifacts from Farkhor, Hissar III, and Shahdad suggest that the Vakhsh burials there are older than the Gelot ones. As we have seen, at least two of the graves (burials 2 and 3 from Area 2) are contemporary with the Farkhor cemetery (see Table 23.1).

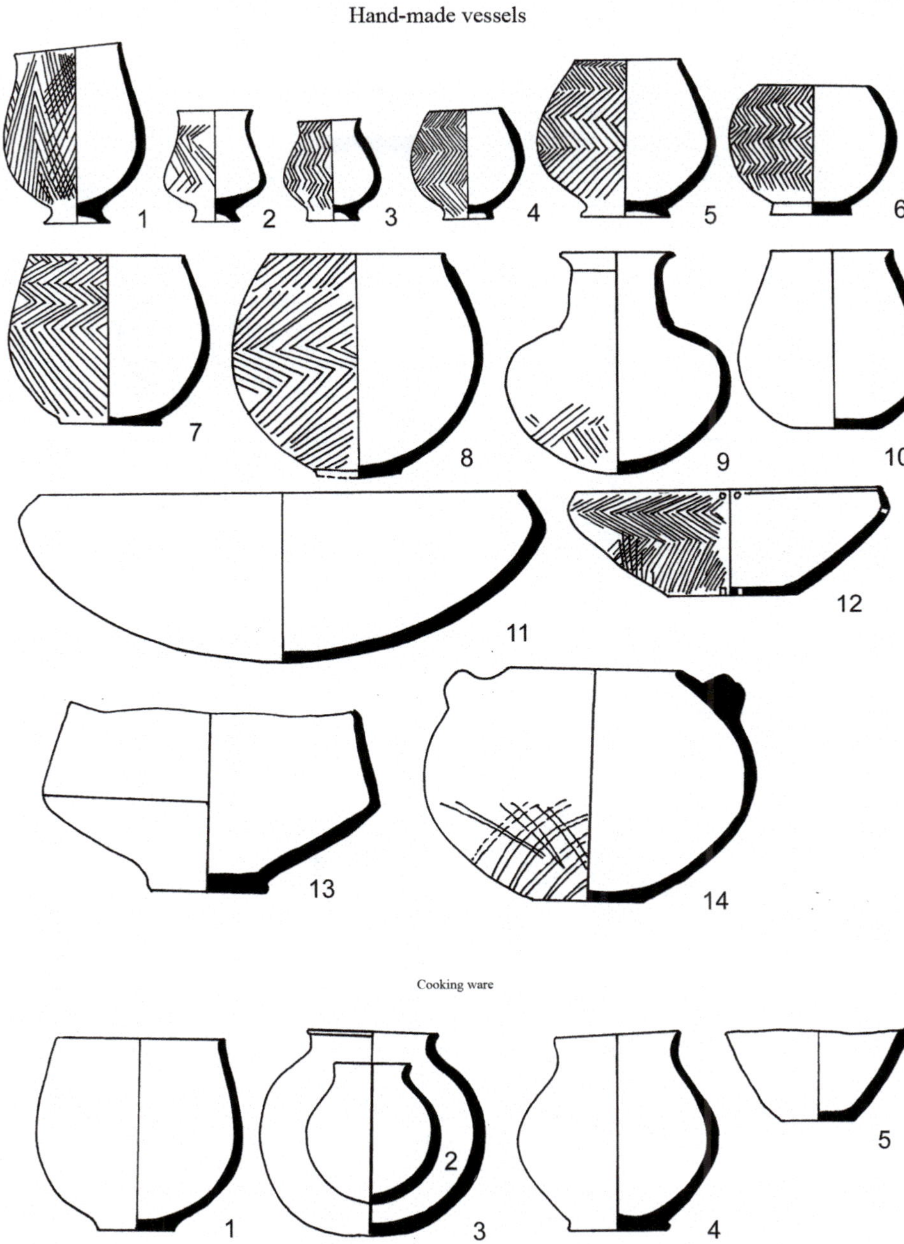

Figure 23.22 Gelot cemetery. Vakhsh culture handmade and cooking pottery.

CONCLUSION

To sum up, archaeological investigations at the Farkhor, Gelot, and Darnajchi cemeteries allow us to reevaluate the ethnocultural processes that took place in southern Tajikistan at the end of the 3rd/early 2nd millennium BC. The new data from Farkhor,

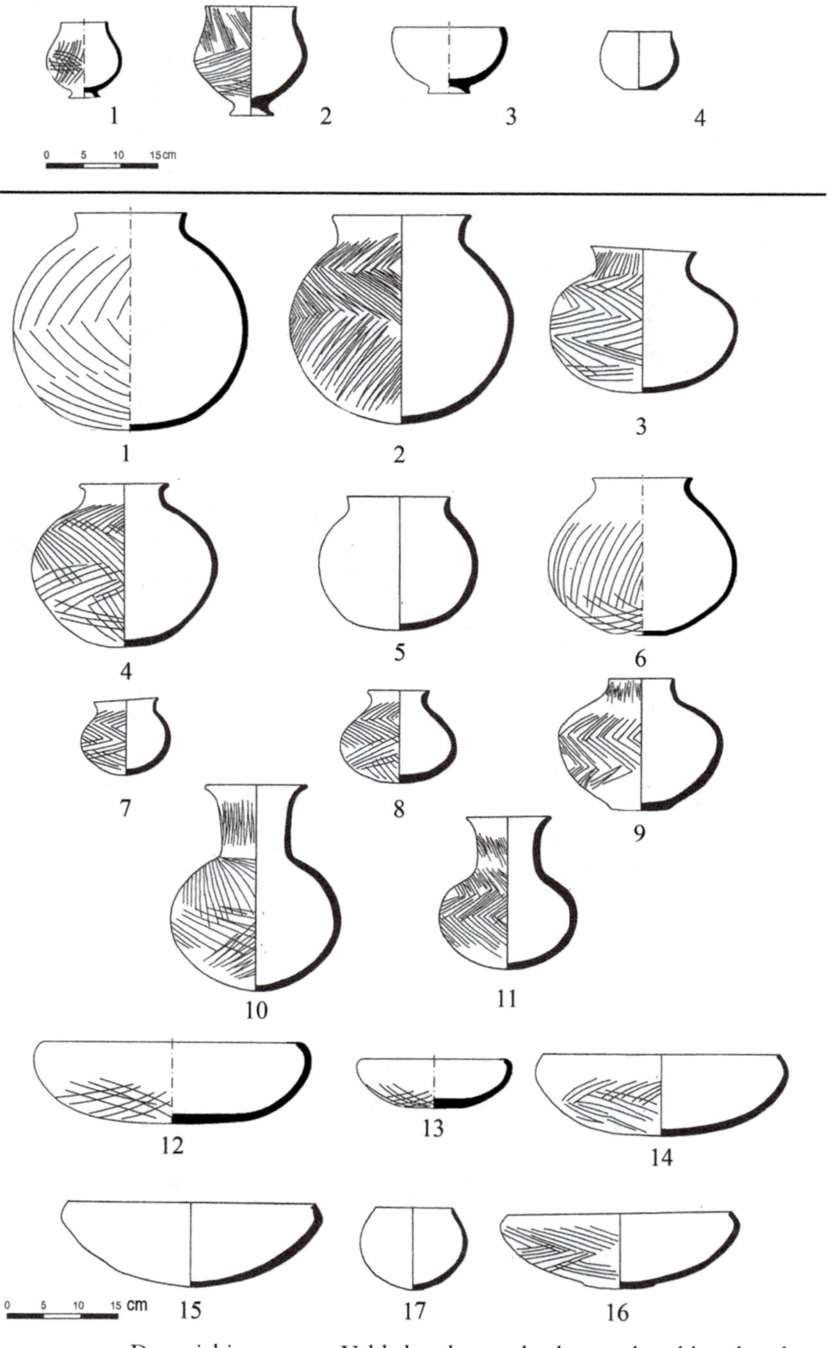

Figure 23.23 Darnajchi cemetery. Vakhsh culture wheel-turned and handmade pottery: 1–4 (top), wheel-turned; 1–16 (bottom), handmade.

Table 23.1 Radiocarbon dates from the Gelot and Darnajchi burials (OxCal v4 1, 4 Bronk Ramsey *et al.* 2010; r5 atmospheric data from Reimer *et al.* 2009)

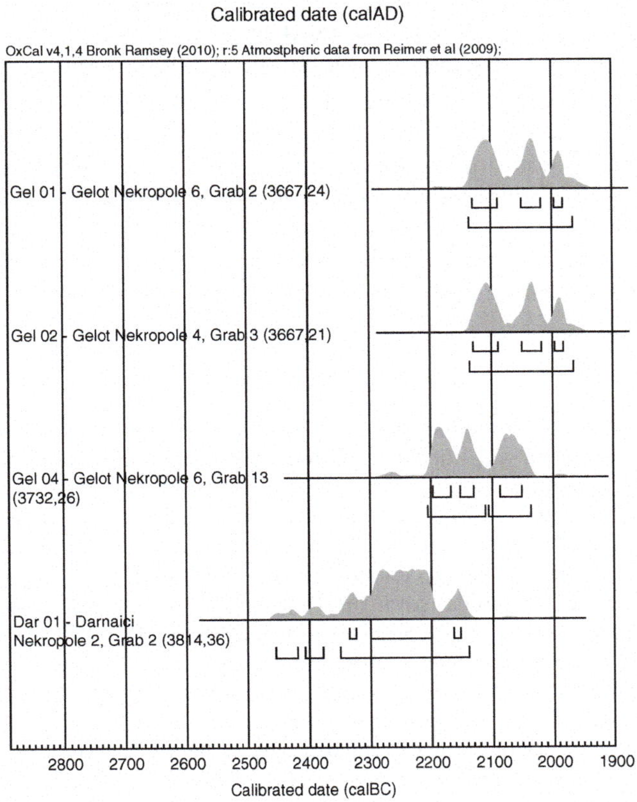

Labornr. MAMS	Probenname	Kennung	C14 Alter	±	13C	Cal 1 sigma	Cal 2 sigma	C:N	%C	%coll
15064	Gel 01 - Gelot Nekropole 6, Grab 2	Gel 01	3667	24	–41,8	calBC 2128–1981	calBC 2135–1965	3,9	18,6	0,3
15065	Gel 02 - Gelot Nekropole 4, Grab 3	Gel 02	3667	21	–14,8	calBC 2126–1982	calBC 2134–1966	3,3	18,1	1,1
15067	Gel 04 - Gelot Nekropole 6, Grab 13	Gel 04	3732	26	–43,1	calBC 2197–2049	calBC 2203–2036	0,0	0,0	0,2
15072	Dar 01 - Darnaici Nekropole 2, Grab 2	Dar 01	3814	36	–37,8	calBC 2333–2153	calBC 2456–2140	3,4	2,5	0,2
15073	KB 01 - Karim Berdy Grab 1	KB 01	1585	20	–25,1	cal AD 430–533	cal AD 425–537	3,2	34,5	4,5

and possibly from Kangurttut 2, point at the presence of a local Bactrian culture already during the second half of the 3rd millennium.

The analysis of the funeral offerings from the Farkhor, Gelot, and Darnajchi cemeteries attests to relationships between their population and the farming communities of southern Central Asia (Altyn Depe, Gonur Depe, Sapallitepa, Dzharkutan), Iran (Hissar IIIC, Shahdad Necropolis A, Tepe Chalow, and other northeastern sites), and Afghanistan (Dashly 1, 3). Chronologically, it correlates them with the late Akkadian and Ur III periods in Mesopotamia.

Contrary to what was previously thought, this region presents all the prerequisites for the formation of the Oxus or BMAC culture. Francfort has recently suggested that the eastern Bactria culture be renamed as the "Pjandzh culture" (Francfort 2016: 471).

This material culture results from the combination of a local Bactrian tradition with deep Mesopotamian and Elamite influences. The art of ancient Bactria, especially in toreutics and glyptics, developed under a strong Mesopotamian influence, maybe reflecting the cultural desires of the local Bactrian society. The Gelot stone statuette discovery confirms Amiet's considerations about the influence of Elam ("second Sumer") on the development of the Bactrian civilization (Amiet 1990: 135). According to him, under the leadership of southern Mesopotamia and southwestern Iran, a cultural unity (*koinè*) constituted Greater Iran in the second half of the 3rd millennium (Amiet 2004: 5–25; 2007: 64–87). It stretched from the south of the Persian Gulf and southeast of Kerman to the Helmand valley, South Turkmenistan, Afghanistan, and northeastern Iran. This *koinè* developed until the early 2nd millennium BC with the BMAC/Oxus Civilization in Bactria and Margiana (Amiet 1986: 172–204).

NOTES

1 In spring 2016, S. Bobomulloev discovered another Early Bronze burial ground, Kangurttut 2 (Akrami 2017: 38–53) not far from Shulyptu village, opposite to the Kangurttut Late Bronze Age settlement.
2 The town of Parkhar is the administrative center of the eponymous district, Khatlon Region. It is located in the valley of the Panj River, 2 km from the Afghan border.
3 Analyses make them cosmetic containers rather than lamps (see Chapters 4 and 9).
4 According to K.B. Kalinina, from the Department of Restoration and Conservation at the Hermitage State Museum (St. Petersburg), the base of the alloy is copper, and impurities have not been determined. Bronze items from Farkhor have also been studied by D. Steineger from the German Archaeological Institute (Berlin). The analyses show that all the objects are made of arsenical bronze (from 1 to 5% As).
5 Most of the anthropological investigations were conducted by N.A. Dubova from the Institute of Ethnology and Anthropology of the Russian Academy of Science.
6 The name is conventional, as the shape has no stem. It is small-sized and has thin walls.
7 H.-P. Francfort (1989: 368) considers the material from Dashly 1 as earlier than that from Dashly 3 and synchronizes Dashly 1 with the Sapalli period, and Dashly 3 with that of Dzharkutan.
8 Spectral analyses show that the pin is made of copper with a minor lead content (1.4%). Another fragment of a bronze object is a copper alloy with arsenic (0.8%) and lead (1%). The first analyses were performed by A. Egorkov at the laboratory of archaeological technology of the Institute of the History of Material Culture (IIMK), St. Petersburg. The second was performed by A. Dubrovin at the laboratory of the Institute of Restoration, Moscow.
9 Spectral analysis made by the Central Laboratory of Ecology Department (Dushanbe) show that the figurine is made of a soft white stone – anhydrite (variety of gypsum), attested in the Neogene rock formations of the Afghan–Tajik depression. The weight of the figurine is 555 g.
10 According to S. Winkelmann, however, the figurine represents a woman (Vinogradova and Winkelmann 2016: 361–377).

REFERENCES

Akrami, Z.I. (ed.) 2017. *Kangurttut. Materialy arkheologicheskoj ekspeditsii 2015–2016 gg* [Kangurttut: materials of the archaeological expedition 2015–2016], vol. 1. Dushanbe: Donish.

Alekshin, V.A. 2010. "O funktsional'nom naznachenii terrakotovykh 'relikvariev' Altyn-depe" [About the function of the terracotta "reliquaries"]. *Zapiski Instituta Istorii Material'noj Kultury* 5: 105–116.

Amiet, P. 1977. "Bactriane proto-historique." *Syria* 54(1–2): 89–121.

Amiet, P. 1986. *L'âge des échanges inter-iraniens 3500–1700 avant J.-C.* Paris: Édition de la Réunion des musées nationaux.

Amiet, P. 1990. "Antiquities of Bactria and Outer Iran in the Louvre Collection." In *Bactria: An Ancient Oasis Civilization from the Sands of Afghanistan*, ed. G. Ligabue and S. Salvatori. Venice: Erizzo, pp. 157–180.

Amiet, P. 2004. "L'élargissement des échanges inter-iraniens. Essai de mise à jour." *Akkadica* 125: 5–25.

Amiet, P. 2007. "L'âge des échanges inter-iraniens 3500–1700 avant J.-C." In *Sulla Via delle Oasi: Tesori dell'Oriente Antico*, ed. G. Ligabue and G. Rossi Osmida. Padova: Il Punto: 64–87.

Askarov, A.A. 1973. *Sapallitepa*. Tashkent: Fan.

Avanesova, N.A. and Dzhurakulova, D.M. 2008. "Drevnie nomady Zeravshana" [Ancient nomads of Zeravshan]. In *Kul'tura nomadov Tsentral'noj Azii* [Culture of Central Asian nomads]. Samarkand: IICAS, pp. 13–33.

Biscione, R. and A.A. Vahdati 2011. "Excavation at Tepe Chalow, Northern Khorasan, Iran." *Studi Micenei ed Egeo-Anatolici* 53: 236–241.

Bobomulloev, S., N.M. Vinogradova, and B. Bobomulloev 2014. "Nekropol' Farhor: novyj pamjatnik epokhi srednej bronzy Tadzhikistana" [Farkhor necropolis: a new Middle Bronze Age site in Tajikistan]. *Izvestija Akademii Nauk Respubliki Tadzhikistan. Otdelenie obshchestvennykh Nauk* 3(235): 39–47.

Bobomulloev, S., N.M. Vinogradova, and B. Bobomulloev 2015. "Raskopki mogil'nika srednego bronzovogo veka Farhor na juge Tadzhikistana" [Excavations of the Middle Bronze Age cemetery at Farkhor in southern Tajikistan]. *Zapiski Instituta Istorii Material'noj Kultury* 5: 47–66.

Bobomulloev, S., N. Vinogradova, and B. Bobomulloev (with contributions by M. Teufer and G. Lombardo) 2017. "Preliminary Results of Research of the Burial Ground of Farkhor, a Middle Bronze Age Monument in the South of Tajikistan." In *Materialy Farkhorskoj ekspeditsii*, vol. I, ed. S. Bobomulloev. Dushanbe: Institute of History, Archaeology and Ethnography of Academy of Sciences of Republic of Tajikistan, pp. 78–127.

Braun-Holzinger, E.A. 1977. *Frühdynastische Beterstatuetten*. Abhandlungen der Deutschen Orient Geselschaft, 19. Berlin: Gebr. Mann.

Bronk Ramsey, C., M. Dee, S. Lee, T. Nakagawa, and R. Staff 2010. "Developments in the calibration and modelling of radiocarbon dates." Radiocarbon, 52(3): 953–961.

Casal, J.M., 1961. *Fouilles de Mundigak*. Paris: Klincksieck.

Casanova, M. 1991. *La vaisselle d'albâtre de Mésopotamie, d'Iran et d'Asie Centrale aux IIIe et IIe millénaires avant J.-C.* Paris: Éditions Recherche sur les Civilisations.

Francfort, H.-P. 1989. *Fouilles de Shortughaï. Recherches sur l'Asie Centrale protohistorique.* Paris: de Boccard.

Francfort, H.-P. 2015. "Une idole protohistorique d'Aï Khanoum hellénistique (Bactriane, Afghanistan)." In *De Samarcande à Istanbul: étapes orientales*, vol. II, ed. V. Schiltz. Paris: CNRS editions, pp. 41–51.

Francfort, H.-P. 2016. "How the Twins Met: Indus and Oxus Bronze Age Civilizations in Eastern Bactria. Shortughaï Revisited Forty Years Later." In *Trudy Margianskoj Arkheologicheskoj Ekspeditsii*, vol. 6, ed. N. Dubova. Moscow: Staryj Sad, pp. 461–475.

Francfort, H.-P., A.A. Vahdati, J. Bendezu-Sarmiento, J. Lhuillier, E. Fouache, M.Tengberg, M. Mashkour, and Z. Shirazi 2014. "Preliminary Report on the Soundings at Tepe Damghani Sazevar, Spring 2008." *Iranica Antiqua* 49: 111–143.

Guljamov Ya.G., U. Islamov, and A. Askarov 1966. *Pervobytnaja kul'tura i vozniknovenie oroshaemogo zemledelija v nizov'jakh Zerafshana* [Indigenous culture and origin of irrigated agriculture in the lower Zeravshan]. Tashkent: Fan.

Hakemi, A. 1997. *Shahdad: Archaeological Excavations of a Bronze Age Center in Iran.* Rome: IsIAO.

Hakemi, A. and S. Sajjadi 1990. "Shahdad Excavations in the Context of the Oases Civilization Bactria." *An Ancient Oasis Civilization from the Sands of Afghanistan*, ed. G. Ligabue and S. Salvatori. Venice: Erizzo, pp. 141–157.

Isakov, A.I. 1991. *Sarazm.* Dushanbe: Donish.

Khlopin, I.N. 2002. *Epokha bronzy Jugo-Zapadnogo Turkmenistana* [Bronze Age of south-western Turkmenistan]. St. Petersburg: Peterburgskoe Vostokovedenie.

Kircho, L.B. 2005. "Pogrebal'nyj inventar' Altyn-depe" [Funerary offerings at Altyn Depe]. In *Khronologija epokhi pozdnego eneolita – srednej bronzy Srednej Azii (Pogrebenija Altyn Depe)* [Chronology of the Late Chalolithic–Middle Bronze Age of Central Asia (burials of Altyn Depe)], ed. V.M. Masson and Ju.E. Berezkin. St. Petersburg: Nestor-Istorija, pp. 347–421.

Kircho, L.B. 2007. "Drevnie svjazi naselenija Juzhnogo Turkmenistana i doliny Zeravshana (Nachalo formirovanija torgovykh putej v Srednej Azii)" [Ancient relationships between southern Turkmenistan and the Zeravshan valley (first steps of the trade routes in Central Asia)]. *Zapiski Instituta Istorii Material'noj Kultury* 2: 193–208.

Kircho, L.B. 2010. "K voprosu ob ekonomicheskoj osnove rastsveta Margianskoj tsivilizatsii" [Question on the economic basis for the development of the Margiana civilization]. In *Na puti otkrytija tsivilizatsii* [On the track of uncovering of civilization], ed. P.M. Kozhin, M.F. Kosarev, and N.A. Dubova. St. Petersburg: Aletejja, pp. 38–43.

Kircho, L.B. 2014. "Izdelija iz kamnja i istochniki syr'ja v period rannej bronzy (po dannym Altyn-depe, juzhnyj Turkmenistan)" [Stone artifacts and sources of raw materials in the Early Bronze Age (from the Altyn Depe data, southern Turkmenistan)]. In *Arkheologija drevnikh obshchestv Evrazii: khronologija, kul'turogenez, religioznye vozzrenija* [Archaeology of the ancient Eurasian societies: chronology, culture genesis, religious beliefs]. St. Petersburg: Art-Express, pp. 62–85.

Lyonnet, B. (with A. Isakov and N. Avanesova) 1996. *Sarazm (Tadjikistan). Céramiques (Chalcolithique et Bronze ancien).* Paris: de Boccard.

Mandel'shtam, A.M. 1968. *Pamjatniki epokhi bronzy v Juzhnom Tadzhikistane* [Bronze Age sites in southern Tajikistan]. Moscow: Nauka.

Masimov, I.S. 1979. "K voprosu ob osvoenii nizovij Murgaba drevnezemledel'cheskimi plemenami" [The lower Murghab development by ancient farming communities]. *Izvestija Akademii Nauk Turkmenskoj SSR. Serija Obshchestvennykh Nauk* 2: 72–77.

Masimov, I.S. 2000. "Istoki drevnemurgabskoj tsivilizatsii" [Origins of the ancient Murghab civilization]. In *Kul'turnoe nasledie Turkmenistana (glubinnye istoki i sovremennye perspektivy)* [Cultural heritage of Turkmenistan (ancient sources and modern perspectives)]. Materialy k mezhdunarodnoj nauchnoj konferentsii. St Petersburg: Evropejskij Dom, pp. 25–26.

Masson, V.M. 1981. *Altyn-depe.* Trudy JuTAKE, XVIII. Leningrad: Nauka.

Masson, V.M. 2006. *Kul'turogenez Drevnej Tsentral'noj Azii* [Culture genesis of ancient Central Asia]. St. Petersburg: Izdatel'stvo filologicheskogo fakulteta SpbGU.

Possehl, G.L. 2002. *The Indus Civilization: A Contemporary Perspective.* Lanham, MD: Rowman & Littlefield.

Pottier, M.-H. 1984. *Matériel funéraire de la Bactriane Méridionale de l'âge du bronze.* Paris: Editions Recherche sur les Civilisations.

P'jankova, L.T. 1989. *Drevnie skotovody Juzhnogo Tadzhikistana* [Ancient cattle breeders of southern Tajikistan]. Dushanbe: Donish.

P'jankova, L.T. 2003. "Raskopki pogrebenij Vakhshskoj kul'tury v mogil'nike Obkukh (Moskovskij rajon) v 2002 godu" [Excavations of Vakhsh culture burials in the Obkukh cemetery (Moscow district) in 2002 year]. *Arkheologicheskie Raboty v Tadzhikistane* 28: 194–228.

P'jankova, L.T., B.A. Litvinskij, S. Bobomulloev, K. Kaniuth, and M. Teufer. 2009. "Das bronzezeitliche Gräberfeld von Makonimor, Tadžikistan." *Archäologische Mitteilungen aus Iran und Turan* 41: 97–140.

Razzokov, A. 2008. *Sarazm (Orudija truda i khozjajstvo po eksperimental'no-trasologicheskim dannym)* [Sarazm (Tools and economy according to experimental use-wear data)]. Dushanbe: Ejod.

Reimer, P.J., M.G.L. Baillie, E. Bard, A. Bayliss, J. Warren Beck, P.G. Blackwell, C. Bronk Ramsey *et al.* 2009. "IntCal09 and Marine09 radiocarbon age calibration curves, 0–50,000 years cal BP." *Radiocarbon* 51: 1111–1150.

Sajjadi, S.M.S. 2003. "Excavations at Shahr-i Sokhta. First Preliminary Report of the Excavations of Graveyard 1997–2000." *Iran* 41: 21–97.

Salvatori, S. 1989. "Early Bactrian Objects in Private Collections." In *An Ancient Oasis Civilization from the Sands of Afghanistan*, ed. G. Ligabue and S. Salvatori. Venice: Erizzo, pp. 181–247.

Salvatori, S. 2000. "Bactria and Margiana Seals: A New Assessment of Their Chronological Position and Typological Survey." *East and West* 50(1–4): 97–145.

Sarianidi, V.I. 1976. "Issledovanie pamjatnikov Dashlinskogo oazisa" [Investigations on the sites of the Dashly oasis]. In *Drevnjaja Baktrija (Materialy Sovetsko-Afganskoj ekspeditsii 1969–1973 gg.)* [Ancient Bactria (materials of the Soviet–Afghan expedition 1969–1973)], ed. I. Kruglikova. Moscow: Nauka, pp. 21–86.

Sarianidi, V.I. 1977. *Drevnie zemledel'tsy Afganistana (Materialy Sovetsko-Afganskoj ekspeditsii 1969–1974 gg.)* [Ancient farmers of Afghanistan (materials of the Soviet–Afghan expedition 1969–1974 years)]. Moscow: Nauka.

Sarianidi, V.I. 1979. "K voprosu o kul'ture Zamanbaba" [About the Zamanbaba culture]. In *Etnografija, arkheologija Srednej Azii* [Ethnography and archaeology of Central Asia]. Moscow: Nauka, pp. 23–28.

Sarianidi, V. 1991. *Die Kunst des alten Afghanistan*. Leipzig: Wiley-VCH.

Sarianidi, V.I. 1995. "Sostavnye statuetki Baktrii i Margiany" [Composite statuettes of Bactria and Margiana]. *Vestnik Drevnej Istorii* 1: 75–79.

Sarianidi, V. 1998. *Myths of Ancient Bactria and Margiana on Its Seals and Amulets*. Moscow: PentaGraphic, Ltd.

Sarianidi, V.I. 2001. *Necropolis of Gonur and Iranian Paganism*. Moscow: Mir-media.

Schmidt, E.F. 1937. *Excavations at Tepe Hissar, Damghan*. Philadelphia: University Museum, University of Pennsylvania Press.

Stankevich, I.L. 1978. "Keramika Juzhnoj Turkmenii i Irana v bronzovom veke" [Ceramics of southern Turkmenistan and Iran in the Bronze Age]. In *Drevnost' i srednevekov'e narodov Srednej Azii* [Antiquity and Middle Age populations of Central Asia], ed. B.G. Gafurov and B.A. Litvinskij. Moscow: Nauka, pp. 17–31.

Teufer, M., N.M. Vinogradova, and Yu. G. Kutimov 2014. "K voprosu ob absoljutnoj datirovke pogrebenij bronzovogo veka s nekropolej Gelot i Darnajchi" [On the absolute chronology of the Bronze Age burials in the Gelot and Darnaichi cemeteries]. *Arkheologicheskie Raboty v Tadzhikistane* 37: 110–119.

Udeumuradov, B.N. 1993. *Altyn-depe i Margiana: svjazi, khronologija, proiskhozhdenie* [Altyn Depe and Margiana: relationships, chronology and origin]. Ashkhabad: Ylym.

Vahdati, A.A. 2014. "A BMAC Grave from Bojnord, North-Eastern Iran." *Iran* 52: 19–27.

Vinogradova, N.M. 2011. "Pamjatniki epokhi srednej bronzy v Jugo-Zapadnom Tadzhikistane" [Middle Bronze Age sites in southwestern Tajikistan]. *Vestnik Drevnej Istorii* 4: 84–98.

Vinogradova, N. 2013. "Will Sites from the Middle Bronze Age Be Discovered on the Territory of Southern Tajikistan?" In *South Asian Archaeology 2007*, vol. 1, ed. D. Frenez and M. Tosi. BAR International Series 2454. Oxford: Archaeopress, pp. 281–283.

Vinogradova, N.M. 2015. "Preliminary Results of the Excavations at the Necropolis of Farkhor, a Middle Bronze Age in Southern Tajikistan." *Routes de l'Orient. Revue d'archéologie de l'Orient ancien* 2: 21–27.

Vinogradova, N.M. and S. Bobomulloev 2016. "Predvaritel'nye rezul'taty issledovanij mogil'nika Farkhor – pamjatnika epokhi bronzy na juge Tadzhikistana, vesnoj 2014 g" [Preliminary results of the investigations at the Farkhor Bronze Age cemetery in southern Tajikistan]. In *Trudy Margianskoj Arkheologicheskoj Ekspeditsii*, vol. 6, ed. N.A. Dubova. Moscow: Staryj Sad, pp. 534–552.

Vinogradova, N.M., T. Götzel't, and L.T. P'jankova 2003. "Arkheologicheskaja razvedka v bassejne reki Kyzylsu (Juzhnyj Tadzhikistan)" [Archaeological reconnaissance in the Kyzylsu River basin (southern Tajikistan)]. *Rossijskaja Arkheologija* 1: 103–115.

Vinogradova, N.M. and Yu. G. Kutimov 2009. "Bogatoe zemledel'cheskoe pogrebenie epokhi bronzy v Gelote" [A rich Bronze Age sedentary burial at Gelot]. In *Problemy kul'turogeneza i kul'turnogo nasledija* [Problems of culture genesis and cultural heritage], ed. V. Masson. St. Petersburg: Info ol, pp. 102–116.

Vinogradova, N.M. and Yu. Kutimov 2018. *Pogrebal'nye pamjatniki epokhi bronzy v Juzhnom Tajikistane (mogil'niki Gelota i Darnajchi)*. Moscow: Nauka.

Vinogradova, N.M., Yu. G. Kutimov, and G. Lombardo 2009. "Arkheologicheskie issledovanija otrjada po izucheniju pamjatnikov epokhi bronzy v Khatlonskoj oblasti v 2007 godu" [Archaeological investigations on Bronze Age sites in the Khatlon district in 2007]. *Arkheologicheskie Raboty v Tadzhikistane* 33: 35–66.

Vinogradova, N.M., Yu.G. Kutimov, G. Lombardo, and M. Teufer 2012. "Arkheologicheskie issledovanija mogil'nika Gelot v 2009 godu" [Archaeological investigations at the Gelot cemetery in 2009]. *Arkheologicheskie Raboty v Tadzhikistane* 35: 136–166.

Vinogradova, N.M., Yu. G. Kutimov, G. Lombardo, and M. Teufer 2013. "Arkheologicheskie issledovanija mogil'nikov Gelot i Darnajchi v 2010 godu" [Archaeological investigations at the Gelot and Darnaichi cemeteries in 2010]. *Arkheologicheskie Raboty v Tadzhikistane* 36: 91–118.

Vinogradova, N.M., Yu. G. Kutimov, and M. Teufer 2010. "Pogrebenija epokhi bronzy v okrestnostjakh kishlaka Gelot na juge Tadzhikistana" [Bronze Age burials in the outskirts of Gelot in southern Tajikistan]. In *Na puti otkrytija tsivilizatsii* [On the track of uncovering of civilization], ed. P.M. Kozhin, M.F. Kosarev, and N.A. Dubova. St. Petersburg: Aletejja, pp. 389–411.

Vinogradova, N. and S. Winkelmann 2016. "The Rich Grave in the Burial Ground of Gelot (South Tajikistan)." In *South Asian Archaeology and Art 2010: Contextualizing Material Culture in South and Central Asia in Pre-Modern Times*, ed. V. Widorn, U. Franke, and P. Latschenberger. Turnhout, Belgium: Brepols Publishers, pp. 361–377.

Winkelmann, S. 1997. "Gedanken zur Herkunft und Verbreitung iranischer und mittelasiatischer Gewichte." *Mitteilungen der Deutschen Orient-Gesellschaft zu Berlin* 129: 187–224.

CHAPTER TWENTY-FOUR

THE ZERAVSHAN REGIONAL VARIANT OF THE BACTRIA-MARGIANA ARCHAEOLOGICAL COMPLEX

Interaction between two cultural worlds

Nona A. Avanesova

INTRODUCTION

Many books and articles have been written on the cultural attribution of the highly developed civilization referred to as the Bactria-Margiana Archaeological Complex (BMAC) (Sarianidi 1977, 1990) or the Oxus Civilization (Francfort 1994, 1997, 2005), but several aspects are still a subject of debate. They concern primarily the problems of its genesis, geographical position, chronology, specific ritual practices, and, partially, methodology. This situation is explained by an acute deficit of synthetical works on specific regions, where all the material already at hand would be analyzed and accessible for scientific use. This remark concerns the Zeravshan Bronze Age discoveries, and this chapter aims at filling this gap.

Here we will determine the cultural specificities of burial complexes and rare settlements in a delimited area, analyzing all the currently available material in the same way. These sites testify of evident synthetical features issued from two cultural systems – based on farming on the one hand, and cattle breeding on the other hand, in which metalworking played an important economical role. Within the BMAC framework, the graves and settlements of the Zeravshan valley are considered as a new independent cultural formation, but it significantly broadens the geographical horizon of this phenomenon. This original historico-cultural situation lies mainly in the complex integration of cultural processes that led to the formation of multicultural groups. A key factor in the formation of this original Zeravshan culturo-geographical area, and that determined its historical role at that time, is the presence of important mineral resources and its location in the middle of the Eurasian continent.

GEOGRAPHICAL SETTING AND POTENTIAL RESOURCES

The Zeravshan River valley is located between the Turkestan-Aktau and Zeravshan mountain ranges (to the east), and the Karshi steppe and Kyzyl Kum desert (to the west). The valley stretches along an east–west axis, and consists of a flat plain and its terraces above the current bed of the river. Nowadays, the eastern mountainous part belongs to Tajikistan, and its western, piedmont, and plain part – to Uzbekistan (Baratov 1977: 8). This territory is a contact zone between the Eurasian steppe populations and those of the southern Central Asia farming oases. The development of the economic activities of the area rests primarily on the river itself that created an ecological niche with rich alluvial soils in a plain reaching 40–60 km wide in its middle and lower parts, a fact that would certainly have attracted its first known farmers of the 4th millennium, at Sarazm. The formation of this ancient culture is almost contemporaneous to those known further west. Vast and rich pasture lands are also present in the plain and in the piedmont areas along its tributaries. They represented a stable feeding basis for cattle breeding throughout the year and may explain the Zhukov sanctuary (Avanesova 2013). One of the main reasons for the active development of the region was the availability of numerous mineral ore deposits (copper, lead, tin, silver, and gold) as well as semiprecious stones (lapis lazuli, turquoise, chalcedony), accessible from the river valley. The proximity of water, mineral resources, and forests (a thick forest-gallery or *tugai* existed almost along the whole length of the river, while the mountain massifs were also covered with *archa*/jenniper forests) promoted the development of metalworking. During the Chalcolithic and Bronze Age, the region benefitted from its own ore-mining deposits. The geo-ecological wealth of the Zeravshan is visible in different aspects of its economy: farming, cattle breeding, mining, and metalworking. Thus, the river served as the backbone for a specific unified economic area, though the natural conditions in the territories of upper, middle, and lower Zeravshan are very different. The different ecological niches created by various landscape and climatic specificities led the inhabitants to have different activities.

A CONTACT ZONE WITH SEVERAL CULTURAL COMPLEXES

However, although a chronological succession is traceable from the beginning of the production economy to the appearance of the first urban centers, it is nevertheless difficult to distinguish any common genetic roots among the various cultural remains. At each historical stage, features of contamination and interaction between different cultural traditions are visible and explain the remarkable individuality of the region. Its geographical location and favorable environment played a crucial role in the structural organization of these cultures. All the ancient formations known in the Zeravshan valley (Sarazm, Zamanbaba, Tugai, Zhukov, Sazagan, etc.) appear rather suddenly, without clear formative prototypes, but share a number of features, reflecting the particularities of the historical situation. The overall population density of the region has varied, its maximum is said to have been reached in the 2nd millennium BC.

During the Paleometal period (Chalcolithic and Bronze Age), the Zeravshan area was one where cultural integration was the most intense in Central Asia, mixing different cultural traditions. The local basis for this process was the Sazagan-Kel'teminar substratum. Significant changes happened by the middle of the 4th millennium with the settlement of Geoksjur-type farming groups among the indigenous population at Sarazm (Isakov 1991a). The local Neolithic hunter-fisherman population of the valley probably adopted the new and more productive methods of these newcomers. At about the same time and later on, other groups came from the steppes, from the Afanasievo and Jamnaja cultural horizons, and brought with them their own economic model (Avanesova 2012b). Thus, in the Zeravshan valley, from ca. the second half of the 4th to the middle of the 2nd millennium, several cultural complexes, each with its own economical orientation, have been identified:

1. Sarazm, which developed on a complex farming and cattle breeding economy, combined with metalworking.
2. Early mobile, small cattle breeders (primarily based upon sheep raising) of the Afanas'evo-Jamnaja type (Zhukov cultic complex, Sarazm circular tomb(?), dispersed seasonal sites like Siab-2, Ljavljakan, and Ajakagitma).
3. The Zamanbaba culture, with an economic structure based on pastoral cattle breeding together with riverside agricultural and hunting activities.
4. Kel'teminar groups continuing their traditional way of subsistence appropriation, albeit with partial transformations.
5. Specialized industrial complexes dedicated to mining and the metalworking production cycle (based upon local mineral resources) but also with elements of settled cattle breeding (Tugai, Medomi, Aktashi, etc.). Within them production areas directly linked to metal-bearing deposits exploited by Andronovo groups (Karnab, Ajjakkuduk, Bajsojchanek, Auminzatau, Mushiston, etc.) can be distinguished.
6. Complex Andronovo pastoral communities of cattle breeders (settlements and cemeteries of Chakka, Gudzhajli, Muminobad, Dashtikozy, Sajgus, etc.).

Another well-identified cultural group is visible in complexes close both to the early BMAC circle (i.e., with pottery of the early stages of the Sapalli culture, etc.), and to pre-Andronovo groups (horse disk-shaped cheek-pieces, metal items, etc.). Its material comes from isolated burials (Siab, Zardcha Khalifa, Dhzam, Sazagansaj, Aksaj, Tusunsaj) and from the Tugai settlement, which will be discussed later in the chapter.

Specific sources enable the modeling of a multibranch economical system with traces of specialized production aiming at intra-regional and trans-regional cultural relations. The formation of economic groups was driven by natural conditions and migration processes. The parallel development of various economic types and the synthesis of two or more cultural systems are one of the specific features of the Zeravshan sites during the Paleometal Age. From a culturo-chronological point of view, the materials are diverse. Close contact and the infiltration of separate groups of farmers and cattle breeders were driven by their dependence on the same metallurgical centers.

MINERAL RESOURCES AND METAL EXPLOITATION

The impressive feature of the region is the presence, besides mining exploitation, of the settlements of ancient metallurgists, where they smelted copper, tin, and, possibly, gold. In most cases, ore processing took place in the vicinity of its sources. Traces of metal production have been documented at settlements, short-time stations, places of ancient smelting works, manufacturing areas (Sarazm, Tugai, Ljavljakan,

Karnab, Aktashty, Tym, Madami, Beshbulak, Chakka, Navbag, etc.). The number of sites attributed to metallurgy and metalworks has increased due to geological-archaeological studies made between 1961 and 1975 by the "Ancient Mine Openings" project of the Uzbekistan Ministry of Geology. A great number of these sites (Bronze Age miners and metalworkers' seasonal camps) have not yet been published, or consist of hardly accessible reports (Lopatin *et al.* 1964; Kireev *et al.* 1969; Maksudov *et al.* 1975). The sites are localized in areas near streams, or on non-submersible terraces between two catchment basins where the link is clear between shepherding and the location of mineral resources. The character of these complexes (thin cultural layer with little material, small area) points at their non-permanent nature (with the exception of Sarazm), with short-time camps of various populations on their routes to mineral sources.

The Zeravshan valley contains significant mineral and ore resources (see also Chapter 28). Here one can find deposits and occurrences of gold, silver, copper, lead, zinc, tin, arsenic, bismuth, and iron ore. They are mostly attributed to the main mountainous formations – Nurotau, Mal'guzar, Karatjube, Ziaetdin, Zirabulak. The basic raw resources also include non-ore forms of mineral resources (feldspar, talc, graphite, etc.), stones for coating (marble, granite, gypsum, etc.), and for ornaments (turquoise, chalcedony, onyx, etc.) (Popov and Tyljaganov 1976). Many of these places attest to ancient exploitations. Thus, in its work in the Nuratau Mountains, the "Ancient Mine Openings" project has identified a great number of gold and turquoise deposits with earlier traces of exploitation; ancient traces of mining of cinnabar, copper, and turquoise have been found in the Malguzar Mountains; tin and silver-arsenic ores in the Ziaetdin-Zirabulak Mountains; and copper and turquoise in the Karatube Mountains. It has to be underlined here that the majority of these ancient mining works, slag heaps, remains of smelting furnaces, and mining settlements functioned during the Bronze Age and were operated by Catacomb-Srubno-Andronovo prospectors and metallurgists (Avanesova *et al.* 2005: 17–27; Avanesova 2012a: 3–35). New research enables us to assume that the region was involved in an expansion process because of its ore resources. From our current knowledge and the extent of the gathered materials, we are now able to make a general interpretation of the ongoing activities in this region.

The majority of the complexes are represented by short duration and reoccupied stations, scatters (the cultural layers having been destroyed by later anthropic factors), and production sites related to mining work. Traces of underground activities (tunnels) and specific remains of the daily lives of Bronze Age metalworkers in the Ziaetdin-Zarabulak Mountains speak for themselves (Ruzanov 2000; Partsinger and Boroffka 2008; Avanesova 2012a: 3–35). According to geologists' reports, ancient smelting sites are attested all over the Zeravshan area (Maksudov *et al.* 1975). Unfortunately, none of the settlements has been studied yet, except for Sarazm and Tugai, which give only a fragmentary knowledge of their character. However, despite our small database, we have a complete picture of all the stages of the technological development of metalworking in the area: miners, metallurgists, blacksmiths, foundrymen. The character of the sites is determined by the presence of metal ore, metallurgic slags, smelting furnaces (or their remains), stone ore-mining tools, crucible fragments, molds, sintered (fritted) elements, metal prills, etc. The cultural background of these sites is illustrated by their material remains (pottery etc.), which characterize groups of shepherds and farmers of different periods and with different cultures. Metalworking, in which the polyethnic steppe group led a major mediating role, was oriented not only to answer local needs, but also to meet massive barter exchanges with areas devoid of metal ores.

CHARACTERISTICS OF THE ARCHAEOLOGICAL REMAINS

Our aim here is to introduce into the scientific discussion the materials of the Tugai settlement and of isolated burials that enter into the BMAC sphere like Siab, Dzham, Zardcha Khalifa, Sazagansaj, Aksaj, and Tusunsaj.[1] It is important to note that these complexes are all situated close to river systems. They present an amazing homogeneity, with clear integration of both the Anau/Sapalli and Sintashta/Petrovka cultural elements, which makes them so specific. Due to the limitation of this chapter, we do not pretend to be exhaustive and will give here only brief but the main information. After a short presentation of the context of the discoveries, the different categories of the material found will be analyzed in more detail.

Siab is located within the Samarkand city limits, on the left bank of the Siab River along the Afrasiab mound, at 100 m from the Khodzha Donijar mausoleum. It was found at +4 m above the Siab actual water level and ca. –15 m from the modern surface of the loess Afrasiab Tepe. The burial chamber (1.75 x 1.55 m) was oriented northwest–southeast. A poorly preserved 20–25-year-old female skeleton (Dubova 2008) was found in a crouching position on her right side, with her head oriented to the northwest. Traces of ocher were visible under her bones. The burial artifacts consisted of two pottery vessels placed at her head, a semiprecious bead necklace (turquoise, lapis lazuli, agate, carnelian), and metal temple pendants (headwear elements), bracelets, and rings (Figure 24.1) (Avanesova 1991: 72, fig. 60, 2, 3; 2001: 63, fig. 4).

Dzham is close to the current village of the same name (60 km southwest of Samarkand). Six different places showed remains of a Paleometal period occupation.[2] Most of them are related to the Dhzam stream (Avanesova *et al.* 2005: 12–33). In 1994–1998, other burial remains were discovered at the foot of the natural Galasherik hill, destroyed by land plowing. Four burials were identified. Three of them are related to the Sapalli culture, while the last one is characteristic of the Andronovo (Fedorovo) culture. From our observations they all consist of simple pits, each containing an individual in a crouching position on his/her right or left side. Two burials inventories included distinct features of both pre-Andronovo cultures (pottery, single-edged blade knife), and those from the southern oases (pottery, mirrors, pin, cosmetic spatula) (Figure 24.2).

Sazagansaj is outside the eastern border of the Sazagan village, 26 km southwest of Samarkand. It was discovered in 1995 during a sand quarry within the old Sazagansaj stream.[3] Two burials have been documented: one almost completely preserved, the other one only partially (Avanesova 2010: 337–339).

The graves consist of almost rectangular pits (2.1 x 1.65 x 0.6 m), with their long sides north–south orientated. Grave 1 contained the remains of a 40–50-year-old woman, probably (judging from the number of animal bones) accompanied by a whole sheep carcass (for the rest of the material, including a clay nozzle, see p. 686) (Figure 24.3). In grave 2, the undamaged remains of a 25–30-year-old male were

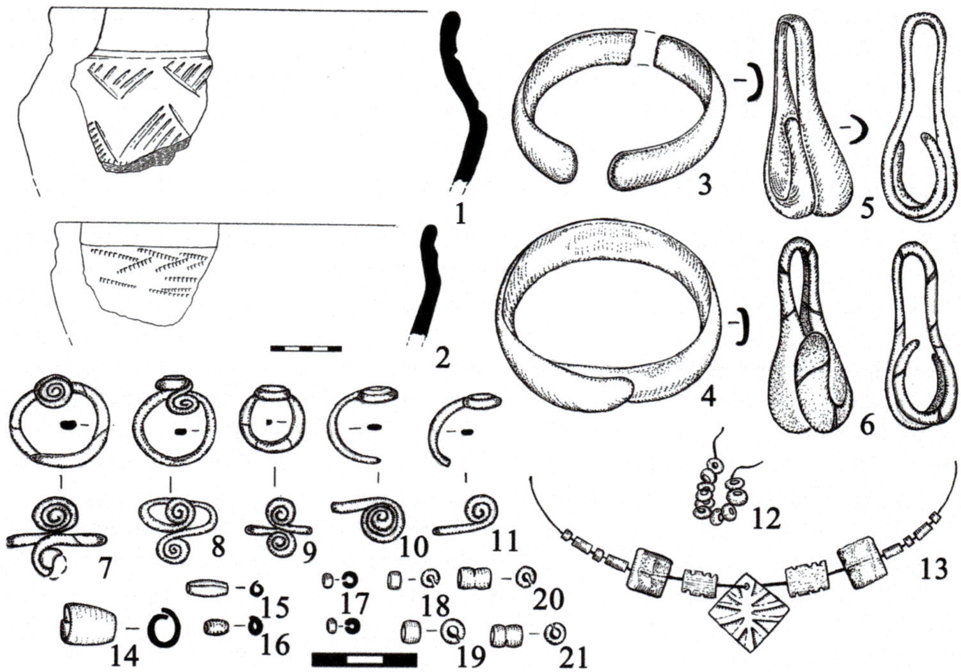

Figure 24.1 Siab burial. From N. Avanesova's excavations in 1976. Funerary inventory: 1–2: pottery; 3–12, 14–17: bronze artifacts; 13, 18–19: stone artifacts; 20–21: paste (heated material) items. Collection stored in the Archaeological Museum of Samarkand University (here after AMUS).

found, lying on his right side in a crouching position, head to the north with a deviation to the northwest. The set of items placed next to the deceased were preserved and consisted of an ornamented handmade vessel near the head, and a single-edged blade knife next to it. Wheel-turned vessels were standing behind the deceased's back. In the bowl unprocessed pieces of semiprecious stones were found – turquoise, lapis lazuli, aquamarine. Two horse cheek-pieces and remains of food offerings with a double-edged blade knife were lying on his chest. At his feet were stone tools, a pile of defective beads and prepared carnelian and lapis lazuli blanks, bronze and stone jewelry and arrowheads (Avanesova 2010: 344–345, figs. 2, 3) (Figure 24.4).

To sum up, the material acquired from these extraordinary burials shows clear features of cultural synthesis between the proto-urban of southern Central Asia and the northern pre-Andronovo Eurasian steppe.

Zardcha Khalifa is situated 5 km west of the modern town of Pendzhikent, 13 km east of Sarazm, a famous Chalcolithic/Early Bronze Age settlement. Salvage excavations were made there in 1986 by R. Razzokov and, within an area of 6 m^2, brought to light a human skeleton crouching on his right side, head facing the northwest, and 20 cm to the west of his skull, the remains of a mountain ram lying on its right side. In the burial pit, the following items were discovered: three ceramic and three bronze vessels (two of the latter are votive); bronze artifacts including two horse bits, single-edged and double-edged blade knives, a cosmetic spatula

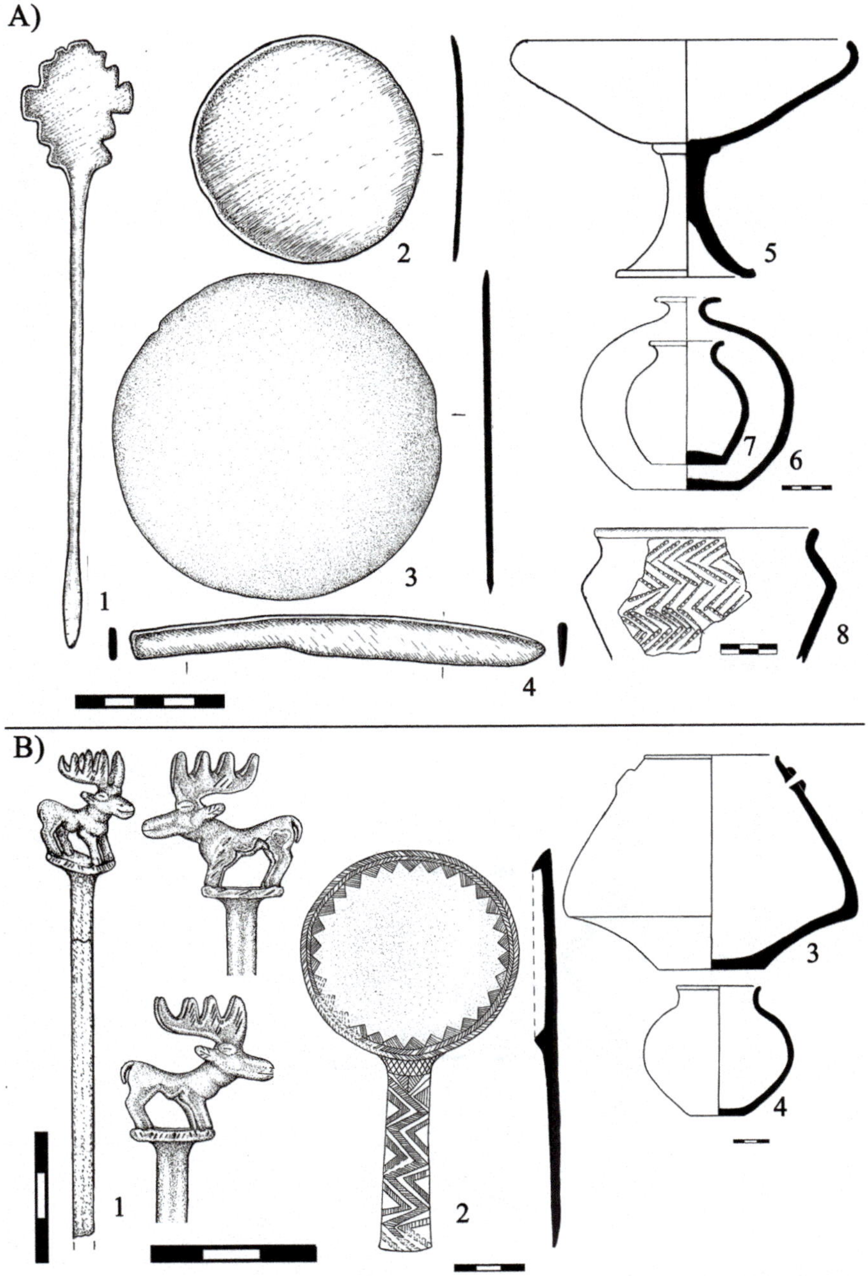

Figure 24.2 Dzham burials. Funerary inventory: A) tomb 1: 1–4: bronze items; 5–8: pottery. B) tomb 2: 1–2: bronze items; 3–4: pottery. Collection stored in AMUS funds.

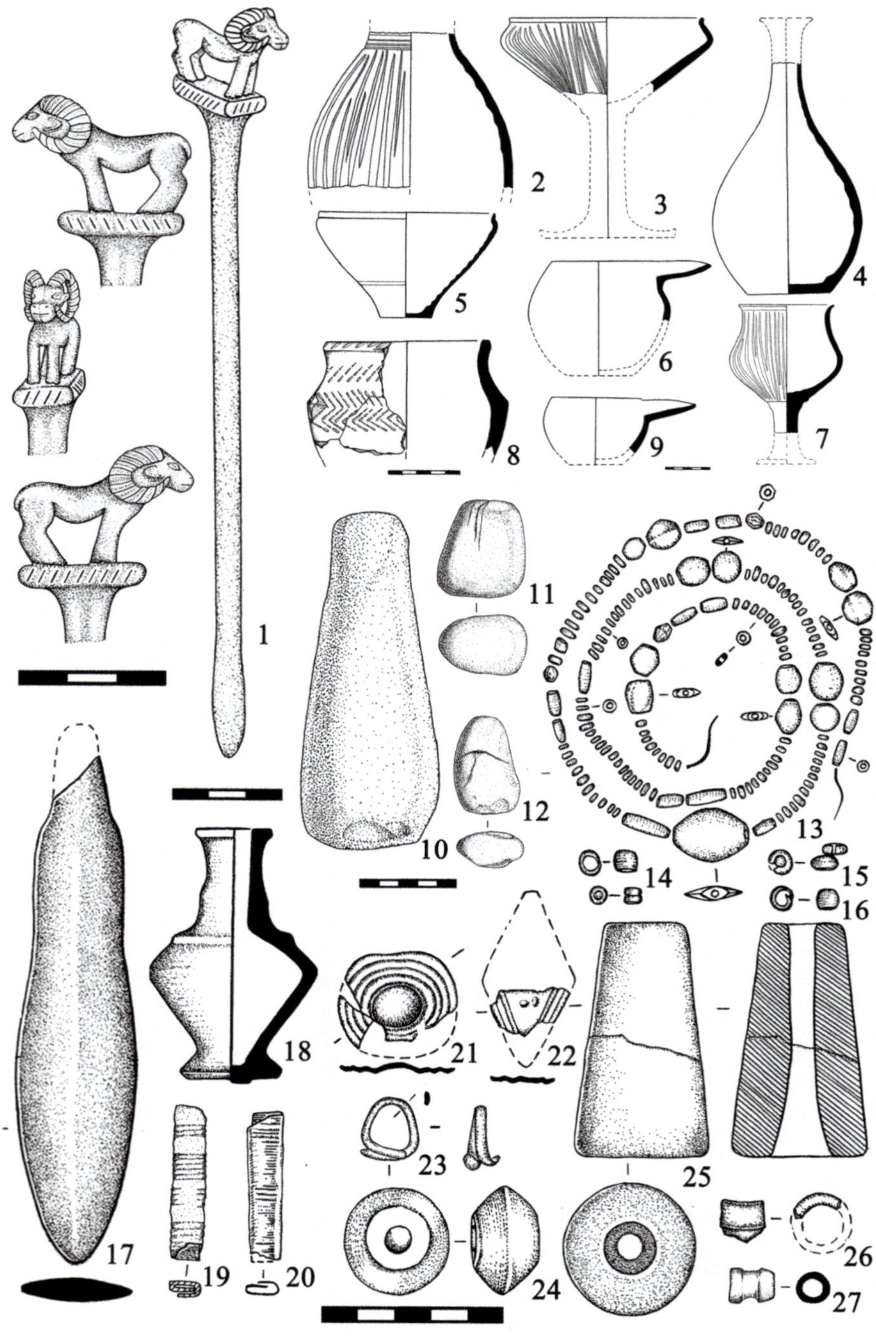

Figure 24.3 Sazagansaj. Tomb 1. Funerary inventory: 1, 14–23: bronze artifacts; 2–9: pottery; 10–13, 24: stone artifacts; 25: clay artifacts; 26–27: bone artifacts. Collection stored in AMUS funds.

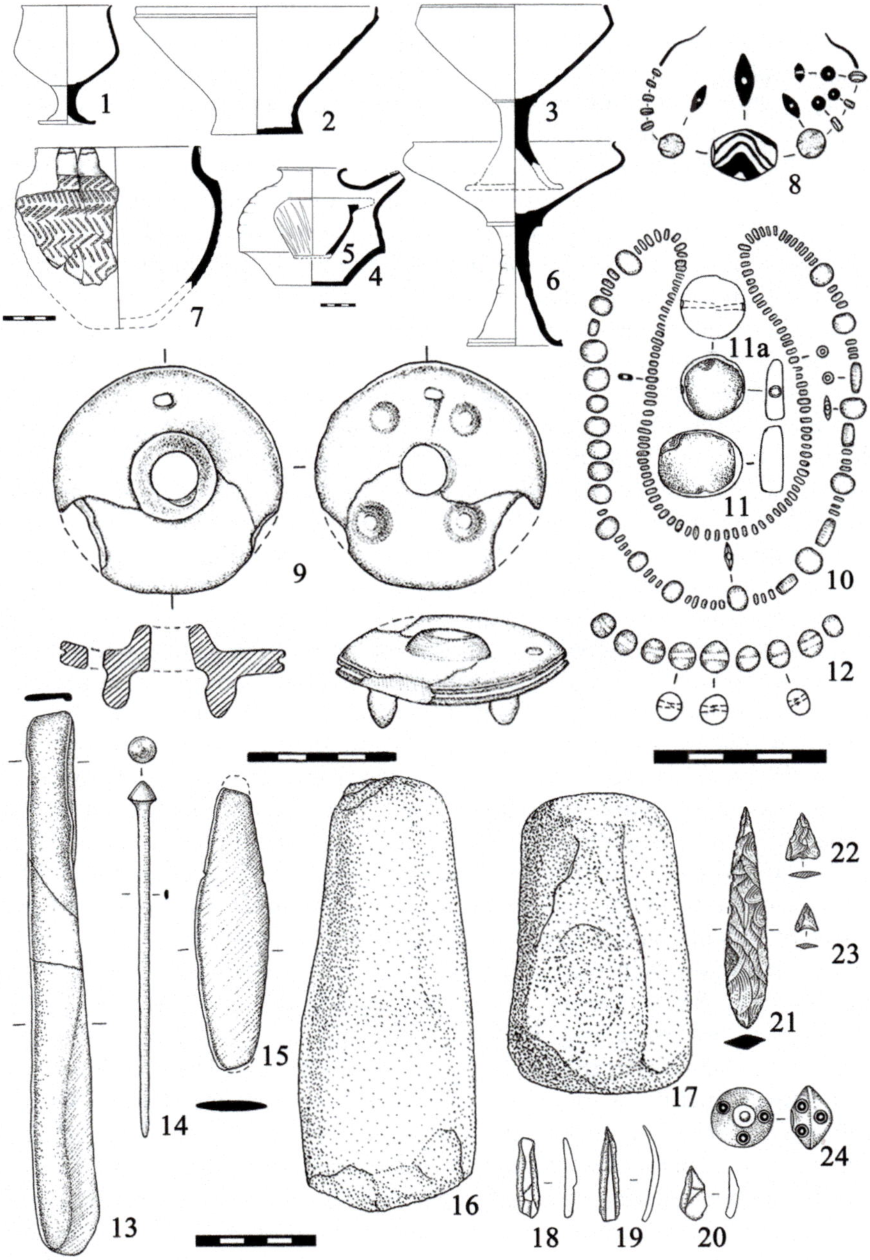

Figure 24.4 Sazagansaj. Tomb 2. Funerary inventory: 1–7: pottery; 9: horn; 13–15: bronze artifacts; 8, 10–12, 16–24: stone artifacts. Collection stored in AMUS funds.

with a horse-shaped head, patches for repairing wooden vessels; gold clips, small beads, and a spherical-shaped "applied element"; four horn horse cheek-pieces; and a phallomorphic pestle (Bobomulloev 1999: 307–313) (Figure 24.5). The burial inventory composition points at syncretism between steppe and sedentary southern cultures (see also Chapter 26).

Aksaj is 32 km southwest of Samarkand, 1.5 km above the village of Aksaj. It was discovered in the spring of 1992 on the left slope of the third terrace (according to E.D. Mamedov) during preparation of the area for agricultural use. Fragments of two pottery vessels, a stone seal, a bronze bracelet, and beads were given to the archaeological management units (AMUS). In the autumn of the same year, we investigated the place where this material was coming from and concluded that it was related to a pit burial. In the sounding (2 x 1.5 m) we made, three whole vessels, a bronze pin with a sickle-shaped head, two stone arrowheads, fragments of a stone votive vessel, a turquoise pendant, and beads were found (Figure 24.6). In general, the pottery and jewelry correspond to the standards of the early Sapalli culture.

Tusunsaj is 48 km northeast of Samarkand. It was discovered in 1985 during archaeological work at Safar Tepe near the Tusunsaj water reservoir by G. Vafaev, an employee at the Institute of Archaeology of Uzbekistan. In 1987 he brought to the AMUS four Bronze Age pottery vessels found during agricultural works near the village of Beshkal. Vafaev also mentioned the discovery there of a bronze pin with a dog-shaped figured head (Burjakov 1987).[4] Undoubtedly, these finds came from a destroyed burial. The pots are clearly related with the Sapalli culture, while the zoomorphic bronze pin falls into the category of the rare art artifacts from BMAC sites (Figure 24.7).

Tugai is a miners' and metalworkers' settlement at the northwest border of a village of the same name, 18 km east of Samarkand. It was discovered in 1986 by drill men of the Samarkand hydrogeological service during their survey of the Zeravshan River left-bank terraces. The site is located at the edge of the first terrace, over 6 m above the current river level. Salvage excavations were performed by the Department of Archaeology of Samarkand University under the author's direction. In an area of 248 m^2, half-buried constructions for living and storage as well as for production purposes with traces of metalworking activities have been excavated (Avanesova 1995: 82–86; 2015: 47–62). Within the half-buried household, the cultural layer does not exceed 0.9 m, and, on its outside, 0.2–0.4 m. This very thin cultural layer indicates that the site was a seasonal dwelling. The remains of a semi-buried heating structure, a smelting furnace with its firing installation, have especially drawn our attention. Besides the obvious metalworking complex, copper prills and fragments of crucible with a slagged surface were discovered. On a 64 m^2 area around the copper smelting structure separate piles of oxidized ores with various types of copper mineralization, scattered slag pieces, dumps of ashes and charcoal, charred animal bones, and stone tools for the fragmentation and crushing of the ore were also found. There is no doubt that the copper smelting structure was related to the smelting of burnt and roughly prepared ores thanks to the influx of an air flow as shown by the discovery of nozzle fragments. The metal was probably further refined into crucibles.[5]

The cultural and chronological interpretation of Tugai is primarily based on the pottery found at the site and the comparisons made with the techno-morphological characteristics of that from Sintashta-Petrovka sites and from Sarazm (Avanesova

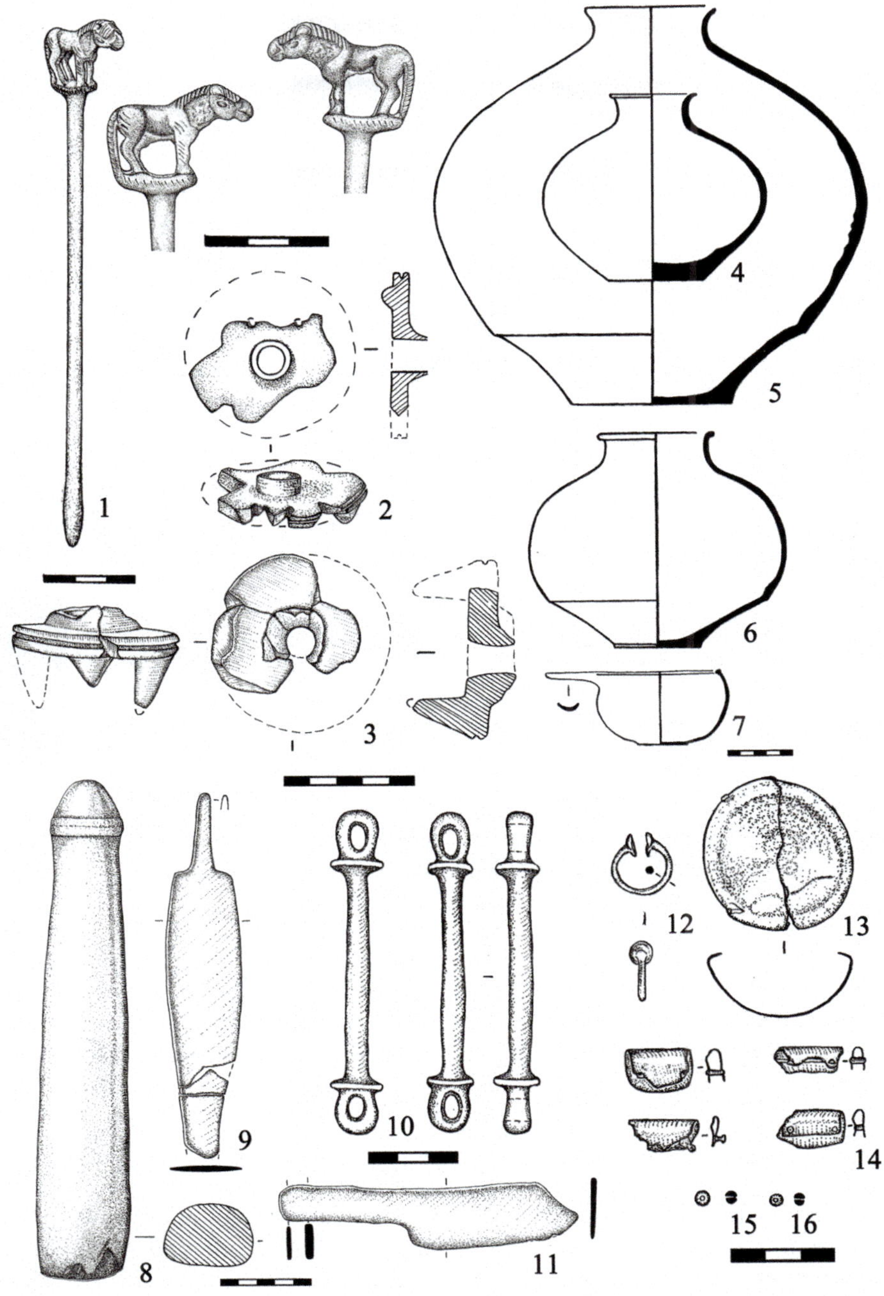

Figure 24.5 Zardcha Khalifa. Funerary inventory: 1: bronze items; 2–3: horn; 4–6: pottery; 8, 15, 16: stone items; 7, 9–11, 14: bronze items; 12–13: gold items. Collection stored at the A. Rudaki Museum, Penjikent (Tadjikistan).

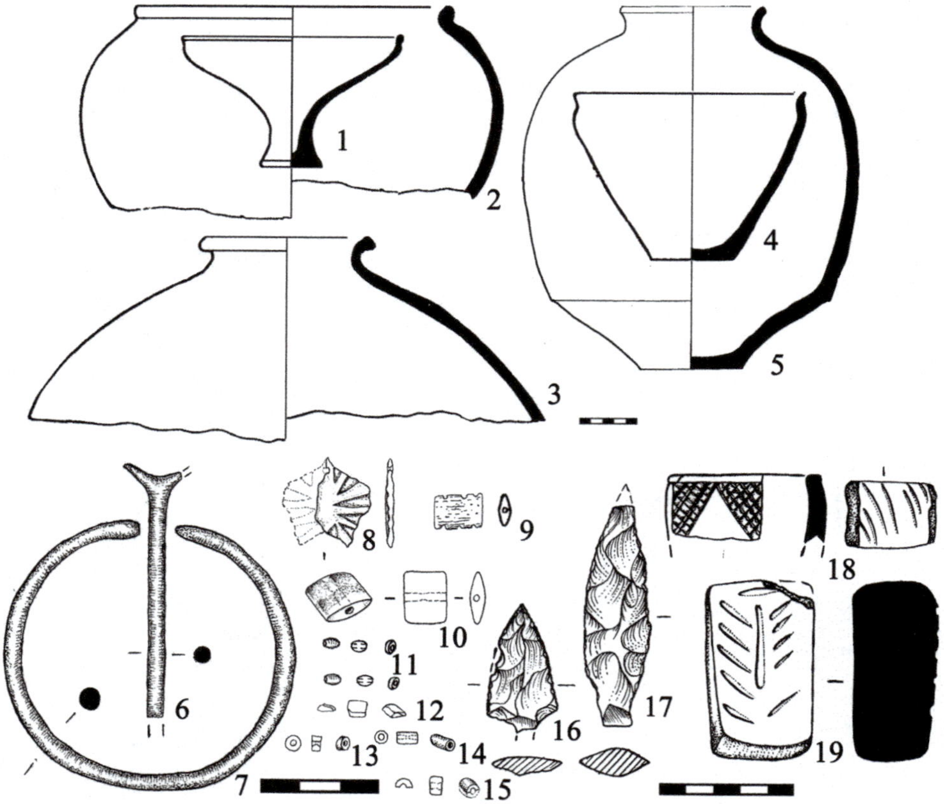

Figure 24.6 Aksaj. Funerary inventory: 1–5: pottery; 6–7: bronze items; 8, 10–19: stone items; 9: paste (heated material) items. Collection stored at AMUS.

1996), as well as certain types of jewelry close to the South Turkmenistan artifacts (Figure 24.8). The collected artifacts give a clear view of their syncretic character, with wide analogies in the southern Urals and in southern Central Asia. All these sites clearly point at a whole system of relations uniting the steppes and the ancient southern farmers' world.

COMPARATIVE ANALYSIS OF THE SITE'S INVENTORY

The material acquired from the above-mentioned excavations testifies to a quite distinct cultural and chronological unity. It has provided unique and rather unusual multicultural material. The collection features artifacts of varying shapes and material with clear signs of symbiosis. The limited framework of the present chapter does not allow a detailed description of the whole block of material culture, but we will underline the general cultural and chronological significance of the artifacts. Pottery is, up to now, the most determining factor as far as chronology and cultural background are concerned.

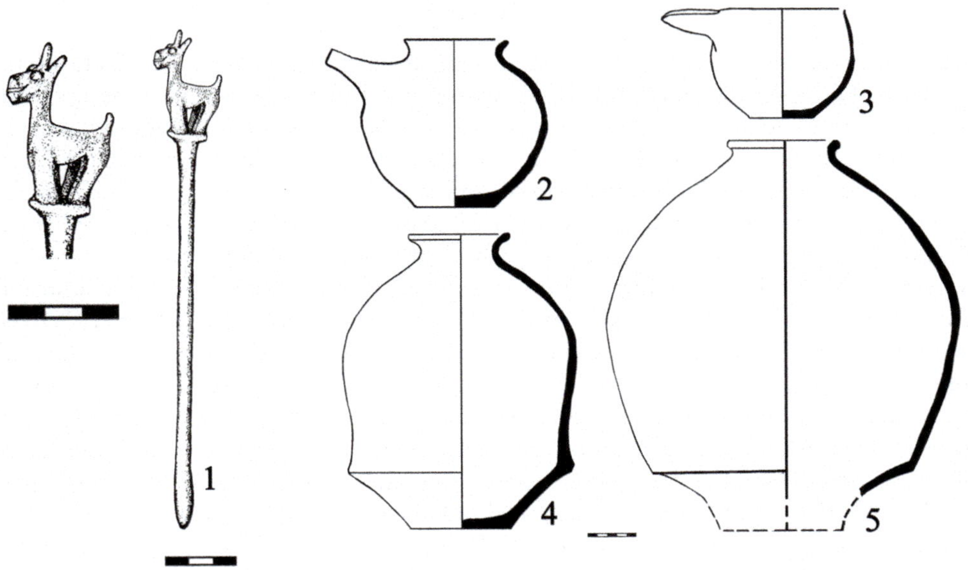

Figure 24.7 Tusunsaj. Funerary inventory: 1: bronze items; 2–5: pottery.

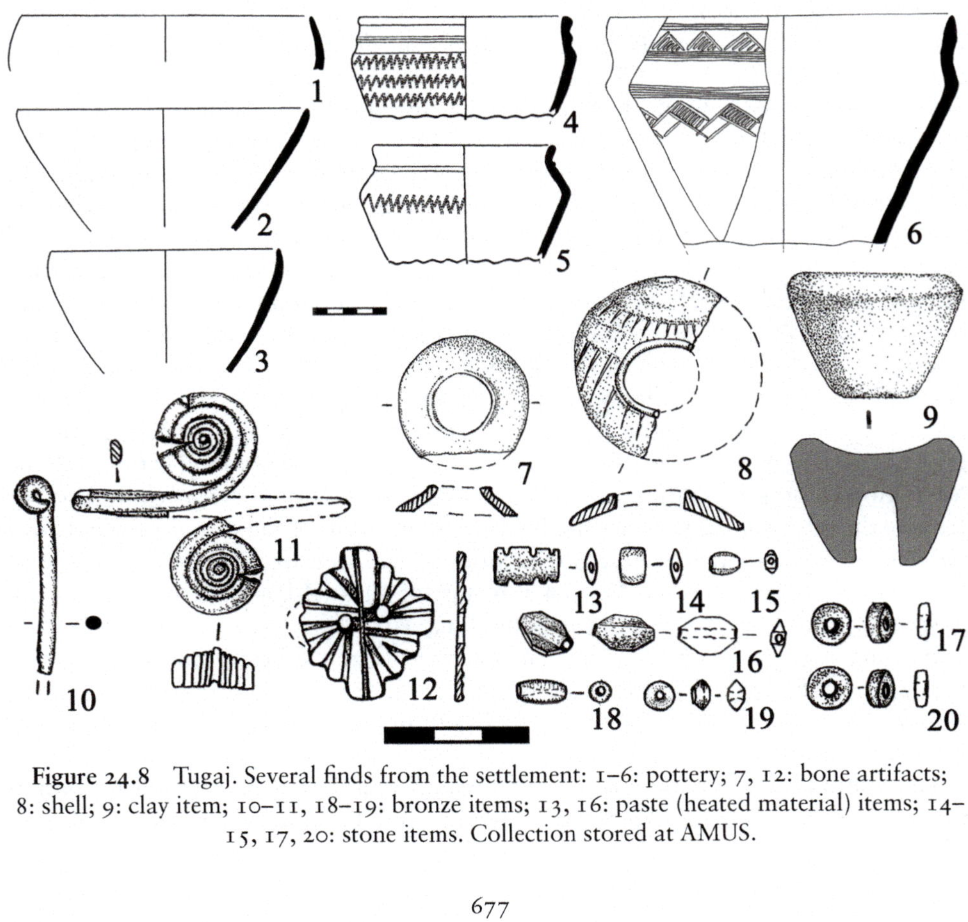

Figure 24.8 Tugaj. Several finds from the settlement: 1–6: pottery; 7, 12: bone artifacts; 8: shell; 9: clay item; 10–11, 18–19: bronze items; 13, 16: paste (heated material) items; 14–15, 17, 20: stone items. Collection stored at AMUS.

Pottery

The pottery complex clearly demonstrates the peculiar character of the Zeravshan remains. The collection includes vessels with great differences in their function and in their fabric technology. The majority (29 vessels) are made on a potter's wheel, and only five are handmade and decorated. The wheel-turned pots are of high quality (clay without visible impurities) and well fired. The surface of the vessels is pinky-red, greenish-white, grayish-black; many of them have vertical polished bands (some very shiny), sometimes on both the inner and outer sides. Traces of slip are still visible. The color section is reddish or gray, firing is even. The usual assortment includes opened vases and stemmed goblets, spouted "teapots," bowls with a pouring spout, conical bowls, narrow-neck bottles, small jars, and jugs. It mostly consists of tableware (see Figures 24.2a, *5–7*; 24.2b, *3–4*; 24.3, *3–4*; 24.4, *6–11*; 24.5, *1–6*; 24.6, *4–6*; 24.7, *1–5*; 24.8, *2–5*). The pots coming from the burials provide interesting and very important comparisons from a cultural and historical point of view. They present a wide circle of analogies among BMAC sites, but especially with the Sapalli culture. Their technological and typological features point more precisely to the Sapalli and early Dzharkutan phases (Askarov 1973; Askarov and Abdullaev 1983; Rakhmonov 1987; Askarov and Shirinov 1993). A significant number of pottery shapes are also similar to those from Margiana (P'jankova 1990, 1994; Udeumuradov 1993).

The handmade vessels found together with wheel-turned pottery are quite unusual. Their specificities rest mainly in their shapes and decoration methods (Avanesova 2010: 342–343).

Pot-shaped vessels with an emphasized exterior rib are comparable by many parameters with Poltavkino ceramics (see Figure 24.2a, *8*; 24.3, *8*; 24.4, *7*). The similarities can be seen in: (1) the technique and methods of decoration (frequency of stamped impressions, depth of the impressions giving a ribbed aspect to the surface), and the themes and character of the covering (comb-impressed "fishbone" and zigzags covering most of the vessel surface); (2) teeth-like impressions – bent, thin, narrow, made with a shell; (3) clay composition: clay + sand + chamotte + organics and clay + crushed shells + chamotte + organics; (4) a tendency towards a sharp carination on the upper part of the shoulder. All the above-mentioned features are characteristic of the late phases of the Poltavkino pottery at sites from the lower Volga and Caspian Sea regions (Kachalova 1962: 37–39; 1983: 4–17; 2001: 36–38, 51–52, figs. 1, 4, 7; Salugina 1994: 177). However, despite these similarities with the Poltavkino ceramics tradition, there is no complete identity. It is especially true for the way the necks of the pots are made, the flat-cut finish of the rim, the proportions and thickness of the pots. These last-mentioned features are very close to those of the Potapovka-Sintashta and Petrovka groups of sites (Zdanovich and Zdanovich 1980: fig. 2, *7*, *10*; 3, *1*, *3*, *7*, *10*; Potemkina 1985: fig. 75, *1*, *4*, *9*, *11*; Gening *et al.* 1992: fig. 55, *4*; 63, *2*, *4*; 95, *16*, *17*; 121, *1*, *2*; Vasil'ev *et al.* 1994: fig. 21, *1*, *3*, *5*; 1995: fig. 3, *8*; Batalov *et al.* 1996: fig. 18, *8*; 23, *7*; D.G. Zdanovich 2002; Vinogradov 2003: fig. 8, *6*; 27, *7*; 90, *2*; Maljutina and Zdanovich 2004: fig. 6, *3*; Epimakhov 2005: fig. 22, *1*; 34, *3*; 96, *6*).

The cultural and genetic role the Poltavkino population had in the formation processes of the Petrovka-Sintashta, Arkaim, and Potapovka-Pokrovka cultures, which are all very close, has often been considered (Smirov and Kuz'mina 1977: 51; Potemkina 1990: 127; Vasil'ev *et al.* 1995: 16; Zdanovich 1997: 54; Kachalova 2001: 55; Maljutina and Zdanovich 2003: 125). This is also confirmed by the pottery vessels from the Siab and Tugai burials. A special article has already been dedicated to them (Avanesova 2015: 53–59), and we will only give here our general observations. These vessels are pot- and jar-like. According to their morphological features, one type may be declined in several shapes. These include biconical and carinated pots (see Figures 24.1, *1–2*; 24.8, *4–6*). All the pots are decorated with a limited number of simple, monotonous patterns, mainly crosshatched triangles, horizontal zigzags, or a repetition of the "rocker pattern." The pots are made using a mold at their base (i.e., using old broken pots, often wrapped in a cloth to shape new ones). The clay is tempered with sand, crushed shells, and talc fragments.

To sum up, the ceramic complex is close (but not identical) to that of Petrovka of the northern and central Kazakhstan, Tobol'sk area, southern Urals, and Trans-Urals regions. The discussed pottery combines almost all the elements described by G.B. Zdanovich and other authors for vessels of the Petrovka culture (Zdanovich and Zdanovich 1980: 187–189; G.B. Zdanovich 1988: 100–101, figs. 3, 7, 17, 18; 2002: 42; Vinogradov 2003: 246–247, etc.). This is proof of their simultaneous appearance and allows us to assume that steppe communities appeared in Central Asia earlier than originally thought.

Our studies once again underline the multicomponent character of the Petrovka culture. According to the archaeological research made in the Volga and the Urals, the Petrovka culture is a combination of various substrates of Jamnaja, Poltavkino, Abashevo, and Catacomb traditions. By their decorative, morphological, and technological features, the Zeravshan pottery complexes are close to those of monocultural sites in the South Urals and Kazakhstan like Petrovka, Novyj Kumak, Verkhnjaja Alabuga, Kamyshnoe I, Stepnoe, Kulevchi III, and Krivoe ozero (Smirnov and Kuz'mina 1977: 20, fig. 6; Zdanovich and Zdanovich 1980: 188, fig. 3; Potemkina 1985: fig. 75, 9; fig. 38, 4, 7; Vinogradov 1982: fig. 2; 2003: 246–248). These links are confirmed when comparing other items than ceramics.

The closest similarity is found in northern Kazakhstan and the Urals region, where the main marker of decoration is the technique of the "walking comb."

In the Tugai complex described above, part of the ceramics contrasts sharply with the Petrovka pottery but finds exact analogies among the Sarazm unpainted pottery (see Figure 24.8, *1–3*). The shapes and technology used to make these vessels accords with Sarazm period IV pottery as described by A. Isakov (1991a: 88–91, tab. VIII).[6] The majority of the vessels have red polished slipped walls (one fragment is black-gray polished) and are represented by conical or hemispherical cups, bowls, and possibly goblets (according to Isakov's terminology). The finish of the vessels is mostly similar. They were made by coiling and finished on a wheel. The clay is significantly sand tempered.[7] The surface treatment is typical for the Sarazm complex with polishing over a black or red slip on both sides. The small size of our collection of Sarazm-type ceramics may point at a brief period of coexistence between Tugai and Sarazm that are only 27 km apart. Of course, such a close proximity of the two sites certainly affected them and this allows us to synchronize them at least partially.

Bronze knives

Bronze[8] knives are represented by two variants (i.e., blades with one or two cutting edges). One knife (from Sazagan) consists of a long tangless blade with a single cutting edge. The handle part is not distinct from the blade but blunt and lowered down. The cutting edge is sharpened by hammering, the tip of the knife is round. The described item is considered as a multifunctional tool, used as a knife, a sickle, and/or a chopper (see Figure 24.4, *13*). It is typologically closest to the Petrovka artifacts of northern Kazakhstan (Petrovka II settlement) and southern Urals (Kulevchi III, Kinzarskoe settlements) (Avanesova 1991: 26, fig. 30, *1*, *2*; fig. 31, *1*, 2). This can be completed by finds of identical knives in the burial grounds of Sintashta (C1, in the barrow surmounting burial 14, see Gening *et al.* 1992: 274, fig. 153, *20*) and Kamennij Ambar-5 (Kurgan 2, tomb 8, see Epimakhov 2005: 151, fig. 30, *10*). These knives are relatively large in size (37–22 cm). Similar-shaped tangless blade tools, where the handle is in the continuation of the blade, are found in the Turbino and Potapovka burial grounds (Bader 1964: 83, fig. 77; Vasil'ev *et al.* 1994: 55, fig. 51, *8*) and in the First Maslovka Abashevo settlement (Prjakhin 1976: 145, fig. 27, *19*). The above-mentioned knives, despite their general morphological similarity, differ because of their double-edged blades.

The single-edged blade knives from the Dzham and Zardcha Khalifa burials are handle-casted knives (see Figures 24.2a, *4* and 24.5, *11*). Here, the transition between the handle and the blade differs from the previous one. By their morphological features they are close to similar items found in Petrovka settlements of the southern Urals (Kulevcho III and Ust' Suerskoe) and northern Kazakhstan (Petrovka II, Konezavod III) (Evdokimov 1976: 510; Vinogradov 1982: 3–12; Zdanovich 1983: 41–50, tab. 10; Potemkina 1985: 53, *1*, *2*; Avanesova 1991: 26–28, tab. 30, *6*; 31, *4*, *8*). It should be noted that the same kinds of finds have been discovered in Andronovo burials of western Siberia at Elovka II and Chernozer'e I (Matjushenko 1973: 8–9; Gening and Stefanova 1994: 13, tab. 6).

Double-edged blade knives are represented by two different types – dagger and knife. The dagger has a leaf-shaped double-edged blade, a slightly raised central rib, an almost unpronounced cross-guard, and a distinct shaft. The central axis of the blade is slightly thickened giving it a lens-like profile. The shaft has a narrow rectangular profile (Figure 24.3, *17*). In the general scheme of evolution of the double-edged blade dagger-knives with rib, the Sazagan dagger represents the first stage (Avanesova 1991: fig. 22, *3*, *4*). New materials from Sintashta-like sites (the C1 kurgan and pit burial at Sintashta, Kamennij Ambar, Verkhnjaja Alabuga) complete and assure these assumptions (Potemkina 1985: fig. 80, *12*; 82, *18*; Gening *et al.* 1992: fig. 140, *1*; 148, *19*; 152, *6*; Epimakhov 2005: fig. 25, *1*; 48, *20*). Variations in this knife shape are attested in syncretic (Abashevo-Srubnaja, Poltavkino-Srubnaja) burial grounds of the Volga–Ural interfluve like Natal'ino-2 (Malov 1980: 153; 1993: tab. 2, *19*; 3, *25*) and Solnechnoe (Agapov *et al.* 1983: fig. 6, *4*). Not less interesting is the similarity (weak median rib) with the knife from Sapallitepa (Askarov 1977: tab. 30, *6*) (Figure 24.9, *5*).

The other Sazagan knife with a short leaf-like blade and a wide tang smoothly issued from the blade is also worthy of interest. It has a flat lens-like section (see Figure 24.4, *15*). Knives of this kind represent the second type (B) of early double-edged tools in

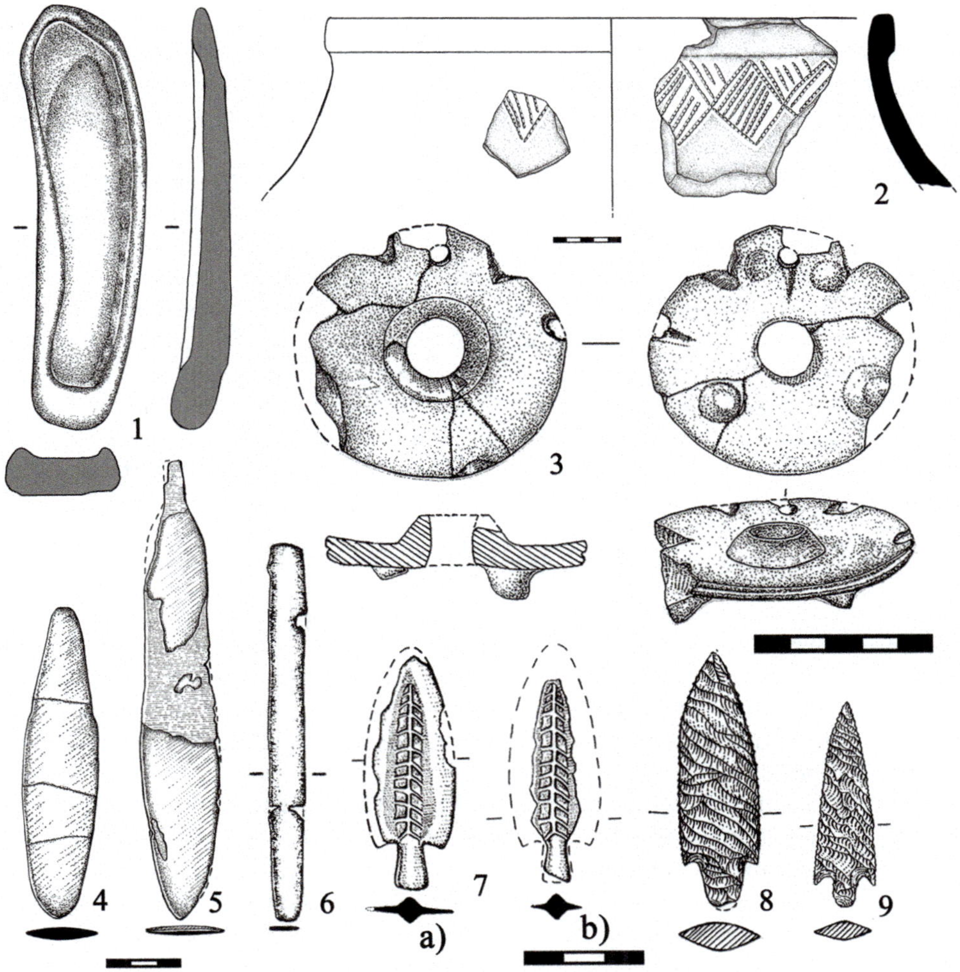

Figure 24.9 Pre-Andronovo artifacts from the early layers of the Sapalli culture: 1: stone mold: Dzharkutan, citadel, room 5; 2: pre-Andronovo pottery: Dzharkutan, temple, corridor 3, reliquary; 3: horn cheek-piece: Dzharkutan, temple, subtower premise of the eastern wall; 4: metal plate knife: Dzharkutan, citadel, room 4; 5: metal knife with overlapping notches: Sapallitepa, room 15; 6: single-blade metal knife: Dzharkutan, residential quarter, hill 5; 7a): petioled metal tip: Sapallitepa, grave 82 (field drawing by O.M. Grjaznov); 7b): The same tip after electro-mechanic cleaning; 8: stone arrowhead: Dzharkutan, citadel; 9: stone arrowhead: Dzharkutan, temple, outside the northern part of the wall.

the development of dagger-knives (Avanesova 1991: 25, figs. 22–27). Their closest typological similarities are in the knives from Sintashta complexes (Gening *et al.* 1992: fig. 51, *13*; 88, *5*; 96, *14*; 140, *4*; 146, *1*, *2*; 148, *9*; Batalov *et al.* 1996: fig. 6, *2*), and they share close characteristics with a similar item from a Potapovka burial ground (Vasil'ev *et al.* 1994: fig. 46, *6*). Finally, there is a closer (geographically) analogy with the double-edged blade knife discovered in the citadel of Dzharkutan (excavation VI, room 4, Dzharkutan period), which serves as yet more evidence for the early relationships between the steppe and the ancient farmers' population (Figure 24.9, *4*). In general, the Sazagan artifacts demonstrate close typological similarities with the South Ural knives of the Eurasian metallurgic province early phase (Chernykh 2007: 74, fig. 5, *3*).

The knife from Zardcha Khalifa is different from the above-mentioned knives (Figure 24.5, *9*). It has a double-edged blade and a tang, a leaf-like elongated shape with a slightly raised rib. The blade section is flat on one side, and slightly swollen on the other. The tang has a sharpened tip and is rectangular in section. The blade is severely worn out. Typologically close knives were present in the Middle Bronze Age layers at Altyn Depe (Kircho 2001: 76, fig. 3), at Dashly-3 (Sarianidi 1977: 76–77, fig. 36), and at Shortughai, mound B, period IV (Francfort 1989: vol. II: pl. 78, 2; pl. XXXIX, 2). They are known at Sapallitepa (Askarov 1977: 72, tab. 30, 7), at Sarazm IV (Isakov 1991b: fig. 77, 1–2), as well as in graves of the Farkhor cemetery (Bobomulloev *et al.* 2016: 27, fig. 18, *4*). In general, double-edged, tanged, flat-section leaf-like blades were common in the second half of the 3rd–2nd millennia BC (Ruzanov 2013: 122).

Bronze jewelry

Bronze[9] jewelry is an interesting category of artifacts adding more elements to the relations between Sapallitepa and their neighbors. It serves as an important ethnographic and chronological indicator. The overall number is rather low, and is represented by single, badly preserved items like clothes' accessories (buckle, tube, beads, pin), head decorations (pendant, trinket, hairpin), and toilet artifacts (mirrors, cosmetic spatulas).

The items included a pendant in the form of an oval spiral with one and a half revolutions and inward-oriented ends (without widening), made of flattened thick wire (from Sazagan, see Figure 24.3, *23*). Such a jewel with several variations has a large circle of typological parallels (from Transylvania to Altai) since the Early Bronze Age. Almost all of them are made of a gutter-like metal strip (Avanesova 1991: 53–54, figs. 43, 47). Wire pendants are most commonly associated with the Jamnaja and Balanovskaja cultures (Bader and Khalikov 1976: tab. 50; Bogdanov 2004: fig. 56, *10*, *11*). Isolated examples are known in the Abashevo culture (Kuz'mina 2002: 157, figs. 1–6), the Sintashta burial ground (C1, tomb 14, see Gening *et al.* 1992: fig. 148, *8*; 153, 2), and in kurgan 24 of Bol'shekaraganskij necropolis (Batalov *et al.* 1996: 83; fig. 19, 2).

Bronze decorative items

Among the other bronze artifacts, four items are in the form of a rod with a round bulge at one end and a decorated end on the other. They are usually referred to in the literature as pins, hairclips (if the end is sharp), hairpins, cosmetic rods, cosmetic sticks (if the end is drop-shaped), or, finally, votive scepters (in cases of animal-figure heads). It is difficult to make a distinction, as publications often lack their precise provenience and good illustrations, except for the undisputed cases when small cosmetic "pins" were found inside cosmetic bottles.[10]

Hairpins with figured zoomorphic head are represented by round-section rods ending on one side into an elongated smooth bulge, and on the other side into a hammered rectangular horizontal platform on which a full standing animal, ram, horse, deer, or dog is soldered (see Figures 24.2, *1*; 24.3, *1*; 24.5, *1*; 24.7, *1*).

Such similar animal figures are known in the geographically close and contemporary BMAC cultural context. Among them, three examples come from Sapalli and Dzharkutan. Their heads (moufflon, sheep) are made according to the Bactrian tradition of volumetric sculpture (Askarov 1977: tab. XLl, *1*; LVI, *5*, *6*). Similar artistic style artifacts are present in the Gonur Necropolis (Sarianidi 2002: 113; 2007: fig. 102, 103, 105), or in plundered burials and accidental finds of Bactria (Sarianidi 1977: fig. 43; 2002: 112). These artifacts, we think, are different manifestations of the same Bactrian artistic tradition. It must be added that this "animal style" in the BMAC applied arts is unevenly distributed throughout Bactrian history. Most finds are attributed to protohistoric Bactria, with hairpins and cosmetic rods with figured heads extensively used at the end of the 3rd and beginning of the 2nd millennium BC. Their existence thereafter is limited to the first half of the 2nd millennium BC, after which they totally disappear.

Among these finds, a cosmetic spatula with a drop-shaped bulge at one end and a flat cross-shaped stepped rhomb on the other discovered in the Dhzam burial is significant (see Figure 24.2a, *1*). Similar objects are frequently found inside metal, stone, or ceramic cosmetic bottles. Its closest analogies are at Altyn Depe in the Early Bronze Namazga (NMG) IV period (Kircho 2001: 77, fig. 6); in the graves of the Farkhor graveyard in Tajikistan (Bobomulloev *et al.* 2016: 29, fig. 20, *1*, 24, 4); in the destroyed burials of the Shagym burial ground in Ferghana (Amanbaeva *et al.* 2006: 256–265, fig. 2, *1*, 2); at Sarazm (occasional find, Isakov 1991a: fig. 23, 2); in plundered graves of the Dashly oasis (Sarianidi 1977: color insertion 5) and in an NMG IV grave in northeastern Iran (Francfort *et al.* 2014: pl. 7). Their map distribution, unlike hairpins with zoomorphic heads, expands over the northeastern area of BMAC, covering the eastern part of the Ferghana valley (Kyrgyzstan) and southern Tajikistan at the end of the 3rd and beginning of the 2nd millennium BC.

The mirror with a solid-cast handle found at Dzham cannot be left unnoticed. The back of the disc and the handle are decorated with a composition of geometric figures related to the sun cycle (see Figure 24.2b, 2). The decoration is made by a combination of alternated groups of elements (i.e., triangles and crosshatched bands). From its morphological and technological features, the Dzham mirror is a typical Bronze Age Bactrian mirror (Askarov 1977: 73, tab. XXXVII; Sarianidi 1981: 289, fig. 21), but the incised decoration complies with the cultural markers of the Abashevo-Sintashta pottery. It is thus the result of a cultural synthesis where cattle breeding (ornamentation) and farming (shape) communities coexist. It is remarkable that the same phenomenon is visible in the materials of the Sazagansaj, Zardcha Khalifa, and Siab burials, and at the Tugai settlement.

The above-mentioned pins, cosmetic spatula, and mirrors are very common finds among the BMAC assortment of decorative items.

Horse cheek-pieces

A special category in our inventory concerns the horse cheek-pieces (*psalia*) made of horn, because they are clear sociocultural and chronological markers (see also Chapter 26). Seven similar items have been discovered (in various conditions of preservation): two pairs come from Zardcha Khalifa burials, one pair from Sazagansaj, and another from Dzharkutan (see Figures 24.4, *9*; 24.5, *2–3*; 24.9, *3*). The description given here is based on the best-preserved ones (Figure 24.4, *9*).

The object is disk-shaped, shield-like with four solid spikes and has one central and one additional hole. The central hole is framed with a protruding truncated-conical bulge, forming a hub. A groove runs along the exterior rim of the disk, surrounding it in relief. The shield is slightly concave. The dimensions are the following: overall diameter 7.4 x 7.5 cm, thickness 0.7 cm, diameter of the hub inner hole 1.4–1.6 cm, height of the hub protruding part 0.7 cm, and the smaller hole 0.8 x 0.4 cm. The longest preserved more or less biconical spike is 1.5 cm, the smallest is 1.2 cm. The psalium is made of a deer antler, the main trunk of which was sawn.[11] Use-wear analysis has shown that the item was made using a set of different technical methods – drilling, splitting, scraping, hewing, sawing, polishing with abrasive. In general, the manufacturing features of the Sazagan cheek-pieces are the same as those observed on the Dzharkutan ones (Figure 24.9, *3*). They were most probably made according to the Volga–Urals tradition, as shown by the great degree of similarity they have with cheek-pieces from the Potapovka-Sintashta culture (Gening *et al.* 1992: fig. 57, *8*; Vasil'ev *et al.* 1994: fig. 28, *15*; 33, *1*). It is useful here to carefully consider the similarities in design between the protruding hub of the central hole and the groove on the side and those on similar items from Kamennij Ambar-5 burial ground, as noticed by A.N. Usachuk (2005: 185–186; see also Epimakhov 2005: fig. 21, *1*, *2*). Other very interesting parallels complement these analogies and concern the modeled bulges around the hub of the cheek-pieces from the Bol'shekaraganskij burial ground (Batalov *et al.* 1996: fig. 17, *10*; 18, *4*), the Sintashta large pit burials cemetery (Gening *et al.* 1992: fig. 75, *1*, *2*; 79, *9*, *10*), the Solntse graveyard (Epimakhov 1996: fig. 12, *1*), and Krivoe Ozero (Vinogradov 2003: fig. 98). All these cheek-pieces are attributed to the Sintashta period. Such similarities among distant regions can only be followed in relative cultures and they attest to the existence of trans-Eurasian routes. The cheek-piece finds in the Central Asian interfluve are the markers of groups from the Sintashta-Petrovka cultural circle migrating from the territory of the Volga–Urals region. Their cultural interactions with the southern farmers' world go as far south as the Sapalli culture (northern Bactria) (Figure 24.9).

The similarity in construction of the Central Asian cheek-pieces with their northern examples is also clear in their internal standardization. The degree of morphological similarity is extremely high. All are made of horn, shield-like, disk-shaped with solid spikes, with two or three different-sized holes, with decorations of the central hole and butt surfaces of the shield, and with overall proportions up to 8 cm. These features testify of a well-established tradition of *psalia*-making. However, the Central Asian artifacts of horse-riding equipment demonstrate differences in the level of certain features that allows us to distinguish them as a variant and, among other things, to synchronize them chronologically. In the evolution of this part of the horse harnesses, the Central Asian cheek-pieces represent one of the most archaic types in Eurasia. Without trying to prove an earlier date for the shield-like *psalia*, it has to be noticed that they point at a chronological priority (Poltavka and Sapalli culture pottery) compared to those of the Sintashta and Potapovka circle. The discovery of a new region of shield-like cheek-pieces once again raises the question of the origins of horse-drawn chariots and their role in the ancient East. However, this is a topic for another specific study. Let us just remark here that burials with cheek-pieces can be interpreted as the burials of a chariot-riding aristocracy, initially participating in barter–exchange operations as representatives of their family clans and who later have significantly monopolized such trade (see also Chapter 26). The Sazagansaj burials demonstrate such a phenomenon with high probability.

Stone inventory

The stone inventory concerns an assortment of small and large items. The majority of the former are beads, arrowheads, flint blades; the latter are represented by grinding stones, pestles, and anvils.

Among the flint items, made of blades and flakes, one must note a piercer and a drill, executed with a few retouches on the face and more near the very tip. These are small tools up to 5 cm long, made of a brown or dark-green flint. Use-wear analyses show that these tools were used for cutting, drilling, and preparation work, and were widely used for stone jewelry craft production (see Figure 24.4, *18–20*).

Arrowheads are of two types: tangless elongated leaf-like shape and arrowheads with a notch at the base. The leaf-like arrowhead is made of a large, pinkish-gray flint flake with even sides and a carefully carved tip. It is made on both sides by thin, flattened, faceted pressure flaking. Its relatively narrow tip has vertical, graceful proportions, its base is round (Figure 24.4, *21*). Traces of reuse are preserved – the tip of the blade is worn out, partially chipped and polished out. The character of the use-wear features shows that it was used as a drill for stone. A similar pattern of arrowhead reuse has been documented by different authors: G.F. Korobkova, when studying the Kel'teminar arrowheads (Korobkova 1969: 109, 113, 120) and those from Altyn Depe (Korobkova 2001: 147–148), or T.Sh. Shirinov during the analysis of those from Sapallitepa (Shirinov 1986: 39). The same is said about those from grave 1200 of a master stonecutter in the Gonur Large Necropolis (Sarianidi 2001: 71). Analysis of the functional purpose of the Sazagansaj burial stone items points at their use in certain technological operations for bead making.

Two other arrowheads have a triangular squat shape and a concave base with pointed ends. The widest part is at the base. The two ends were formed by the central notch and are slightly pronounced. The arrowhead section is almost lens-like. These arrowheads are very similar to those from the developed stage of the Poltavkino culture (II Berezhkovskij graveyard) in the lower Volga (Sinitsyn 1959: 119, fig. 38, *5*; Kachalova 2001: 43, fig. 4, 42) and to some examples from Catacomb sites (Bratchenko 1976: 55, fig. 25, 22, 23; 99; 55, 7; Prjakhin 1982: figs. 11–12; Subbotin 2000: 374, fig. 11, 24, *25*).

An important category of stone finds are beads made of various semiprecious stones. Beads are one of the most common items in burials. They decorate clothes, headwear, and footwear. In the Zeravshan collection, they are made of lapis lazuli, turquoise, carnelian, agate, and limestone. Most of this stone jewelry and decorative material was extracted in the Central Asian interfluve and in protohistoric Bactria.

Lapis lazuli beads are the majority of all the jewelry found in the Sazagan burials. They have different shapes: cylindrical (short and elongated), tubular, barrel-shaped, ring-shaped small beads, flattened disk-shaped, rhombic, truncated-oval, and nearly square. Their closest analogies are at Sapallitepa, Gonur, and Altyn Depe (Askarov 1973: 95, tab. 30; Kircho 2005: 390–400; Sarianidi, 2007: 115).[12]

In our collection, one object deserves special interest. It is a clay nozzle (i.e., a pipe with an internal conical empty channel from Sazagan burial 1). It must be related to a copper-smelting furnace and is made of clay tempered with gypsum powder and fine sand. The surface is well leveled out, it has a reddish-brown color. The clay core is brownish-black. The cone-shaped hole for air intake is covered with a grayish coating. It is 8.2 cm long, the diameter of the conical air-blowing hole is 2 cm on one side and 0.8 cm on the other (see Figure 24.3, *25*).

This find of a nozzle together with a functional set of tools for smelting and the processing of metal is extremely rare in BMAC sites. As mentioned above, fragments of such items have been found at the Tugai smelting place. Another unusual feature is that the "metallurgist" was a woman. Similar burials are documented in the Sintashta-type cemetery of Kamennij Ambar-5 (Epimakhov 2005: 150, 156).

Until recently, burials of metal- and foundry workers were known only in some cattle-breeder cultures of Eastern Europe (Bochkarev 1974: 48–53) starting with the middle of the 3rd millennium. Considering the chronology and the comparisons, the most significant parallel to our nozzle is found at the Poltavkino Kalinovsky burial ground (lower Volga region) (Shilov 1959: 15; fig. 2, 4, *5*; 17, fig. 5, 6–9; Kachalova 1983: 9, tab. 3, *41*, 42, 47–48, *53–54*). There are also matches in Sintashta burial grounds, at Solntse II (Epimakhov 1996: 38, fig. 11, *11*, *12*), Sintashta III (Gening *et al.* 1992: 336; fig. 195, *6*, *7*), and a stone nozzle was found in a cemetery near the Berezovaja Mountains (Khaljapin 2001: 424, fig. 3, *11*). Similar items were also found in the Arkaim settlement (Zdanovich, 1995: 32, fig. 6, 26, 27).[13]

Bronze ingots and a piece of copper ore represent a special category of artifact in this Sazagan burial 1. Undoubtedly, the grave indicates the deceased's specialization, connected with metal production. Burial 2 inventory also demonstrates clear evidence of the deceased's crafting activities in stone cutting, as shown by the accompanying working tools and raw materials – uncut pieces of lapis lazuli and turquoise. Their processing is also proved by faulty lapis lazuli and carnelian beads and by a half-finished and unpolished oval carnelian bead without drilled hole. The high status of this master jeweler in beads and precious stones is further demonstrated by the presence of horn cheek-pieces in the grave. The closest analogy to such a burial type can be found in the Gonur Large Necropolis (Sarianidi 2001: 71–72). Burials reflecting the professional status of the deceased are also known at Sapallitepa, and include potters, carpenters, and weavers (Askarov 1977: 153).

Finds associated with craft production raise an important topic. The extraordinary Sazagan burials reflect the social ranks of the deceased, rarely encountered at BMAC sites. The appearance of these complexes in the Zeravshan valley can probably be explained by the work of the deceased on the nearby turquoise and copper-tin deposits (Avanesova *et al.* 2005: 17, 18, 24). The high level of manufacturing (employing professional craftsmen) must have had a solid economic basis. It is assumed from studies made on some artifacts that stone craft production based on local resources existed in protohistoric Bactria and Sogdiana. The raw material for jewelry was mined both in the Zeravshan and in the Amu Darya basins. Turquoise, lapis lazuli, and probably metal were products of active exchanges with the inhabitants of the southern oases who imported both raw material and ready items. The tools discovered in the burials ensure almost the whole *chaîne opératoire* for mining and smelting activities and for stonecutting production (jewelry), which probably were separate specialized activities. Their expansion into the Zeravshan region probably had a transitory character, as shown by these isolated discoveries.

CULTURO-HISTORICAL AND CHRONOLOGICAL POSITION OF THE ZERAVSHAN SITES

The different objects found in the sites of our study are considered as resulting from the activities of interrelated population groups. The composition of the sets of artifacts, as well as the rituals of this cultural phenomenon, leave no room to doubt that a new, very localized ancient farming province, close to that of the BMAC, has been discovered in the Zeravshan area. Its atypical character is due to its contact with the steppe population of Eurasia. As we have seen, the sites present clear parallels with the Amu Darya regional complexes of the BMAC in its Sapalli variant. The Zeravshan sites are at the fringe of its northern territory, to the north and northeast of the Sapalli culture. This territory was certainly partially included into the BMAC formation area, but its development stopped due to the interference of steppe communities. One can assume that this process was due not only to a demographic expansion among the farmers and to a sudden degradation of the climatic conditions, but to deeper socioeconomic changes both among the farmers and the steppe population.

The rich ore resources of the region and its trade connections were certainly the main motivation for the arrival of new populations. It is possible that the introduction of metalworking in the region reflects the firm establishment of cattle breeding and farming communities. The Zeravshan population, being located between two cultural worlds (cattle breeders in the north and farmers in the south) gathered cultural features from both worlds. At the meeting point of the two cultural areas, the Zeravshan valley testifies of active contact between them. The mechanism behind its development can be traced by looking at the raw material resources of the Zeravshan and the Amu Darya basins: Active trade exchanges relied on turquoise, lapis lazuli, and metal. It is known that the consumers' value of turquoise and lapis lazuli equaled that of metal (Stech 1999; Lyonnet 2005). Traces of turquoise local mining and processing are known since the Late Neolithic period (Vinogradov *et al.* 1965: 127–130). However, the exchanges were, of course, not limited to the goods mentioned above and there were other trade items as well (e.g., salt, cloth fabrics) not visible anymore in the archaeological finds. It is very likely, for example, that silk, discovered at Sapallitepa (Askarov 1977: 158), was traded from protohistoric Bactria to other areas. Judging by the geological data (reflected in a number of unpublished reports), the Zeravshan ore deposits contain the two metals required for bronze making (i.e., copper and tin) (Lopatin *et al.* 1964; Kireev *et al.* 1969). Copper deposits are known across the whole Zeravshan area. Tin is found as cassiterite, a polymetallic ore with mineralization of tin (Litvinskij 1950, 1954; Ruzanov 1979, 1999). Many deposits (Ajjakkuduk, Auminzatau, Basojchanek, Karnab, Mushiston, etc.) testify to ancient activities and metal production (Boroffka *et al.* 2000; Ruzanov 2000; Partsinger and Boroffka 2008). There are examples of pre-Andronovo and Andronovo ceramic finds in ancient mines and metal processing sites, including gold (Maksudov *et al.* 1975: 107, figs. 11–12).

The extent of our knowledge of the Paleometal Age of the Zeravshan region unfortunately does not allow a more detailed vision of the character, extent, and volume of interrelations of the reviewed communities. Cultural interactions undoubtedly directly depended on the intensity of direct contacts. The Zeravshan valley was the center of an interregional system of trade and exchange, culturally connected both to steppe groups and to southern Central Asia. The direction followed by these connections is visible in the materials of Sapallitepa and Sintashta. The particular features of the latter can easily be identified in the system of the Eurasian province. The finds of typical steppe products such as shield-like cheek-pieces, laminar knives, bronze tanged-arrowheads, casting-molds for sickles, stone arrowheads from jasper-like flint stone, and, finally, pottery fragments at Sapallitepe and in the early layers of Dzharkutan (Khuff and Shajdullaev 1999: 19–26. fig. 4, 7; Shajdullaev 2000: 36, fig. 17, *1*) testify to an early presence of steppe communities in protohistoric Bactria (see also Chapter 17) (Figure 24.9).

It is possible to assume that besides goods, people also moved to the southern oases (as shown by the introduction of their own pottery and by the shapes of casting molds), while probably only goods were supplied to the north (Avanesova 1987: 53–55). The new direction of cultural connections was tied to the market of craft production, as is well demonstrated by the Sazagansaj material. Who were buried there is a question that is difficult to answer. The burial inventory reflects a very complex situation: From the anthropological studies, it is possible that the deceased were from Margiana since the craniometric characteristics are close to those from the Gonur Depe Large Necropolis (Kufterin 2010: 504–512). The anthropological data from Zardcha Khalifa, on the other hand, provides evidence for an origin in the steppe areas of the southern Urals (Razzokov 2006: 201–202). The originality and uniqueness of the Zeravshan sites leaves a wide field for speculations about their ethno-cultural origin.

Another issue has yet to be solved; namely, the chronological position of the Zeravshan burials. The author considers that, on the one hand, the entering of BMAC-type sites in the Zeravshan valley places them into the chronological frame of the ancient farmer cultures of Central Asia, and, on the other hand, that they are synchronous with the late Poltavkino sites of the Volga region and the Sintashta-type sites of the southern Trans-Urals. In broad chrono-cultural terms this puts them at the end of the NMG V and beginning of NMG VI in the southern Central Asia chronology, and with pre-Andronovo and pre-Srubnaja cultures in the steppe chronology. The traditional chronology would attribute them to the early 2nd millennium BC, but calibrated radiocarbon dates place them at the transition between the 3rd and 2nd millennium BC (see the Appendix at the end of the volume). From the correlation made between the chronological position of the different artifacts it is possible to date them between the 21st and the 19th centuries BC.

CONCLUSION

The protohistoric remains discovered in the Zeravshan valley represent an historical phenomenon that appeared at the junction of two culturo-economic areas.

The BMAC expansion, due to interior integration processes and stimulation from exterior impulses, led to the creation of new territorial groups (involving the Zeravshan region) with their own specific features. The originality of the BMAC Zeravshan group is very clearly demonstrated by the presence of artifacts of different cultures in the same complex, with a prevalence of the Sapalli substrate (or one close to it with a farming basis). It is this specificity (coexistence of two cultural worlds) that allows us to speak of a Zeravshan variant of the BMAC, a particular cultural formation. This, in turn, gives another possible way to analyze the BMAC: The perspectives opened by its intercultural relations with the steppe world are of vital importance for their synchronization and for understanding the interaction mechanisms (see also Chapter 26). The presence of BMAC-type sites in the Central Asian interfluve authorizes assumptions about the possible existence not only of a trade route, but also of ethno-cultural infiltration within the frame of the Sintashta-Petrovka community that has a multicomponent character. As a whole, the number of the Zeravshan sites attests to a cultural and historical unity of the BMAC with protohistoric Sogdiana, but with local specificities. Further research and studies on the proto-urban culture of the Zeravshan valley are clearly necessary.

NOTES

1 With the exception of Zardcha Khalifa, all of them have been discovered thanks to the students and employees of the archaeology department at the University of Samarkand during exploration and conservation surveys initiated by the author. Due to natural and anthropic factors, the majority were either threatened or destroyed at the time of their excavations. Many of these remains are not visible on the actual surface but are buried under several meters of alluvial and silty deposits made by the Zeravshan and its tributaries, or by small isolated streams. Burial remains can be worked out thanks to the offering artifacts, human bone fragments, and the stratified position of these item clusters.

2 These discoveries are due to the efforts of a single volunteer, enthusiast, and devoted local historian from the nearby village – Abdumanon Jerkulov. The author is extremely grateful to him for the collection he gathered and presented to the AMUS. In spring 1998, a survey was made at the location of the burial complexes on top of a natural elevation commonly known as Galasherik by researchers from the archaeological department of Samarkand University (including the author) and from the Institute of Archaeology of Uzbekistan (Academy of Science), led by A.E. Berdimuradov. Excavations were made in the northern part of the hill, which significantly overlooks the surrounding area.

3 Archaeological investigations were made by O.E. Ibragimov, professor at the archaeology department of Samarkand University and resident of Sazagan. In 1996, the author made further investigation and cleanings within the quarry. The finds from the two burials have been first presented to the author in order to determine their cultural background, and she made drawings of the whole complex. Later she was granted the excavation diary, the field drawings, and pictures for publication. Currently all the material is stored in AMUS.

4 The figure of the pin is courtesy of Ju.F. Burjakov, for which I am deeply grateful.

5 For metal analyses from the Tugai finds, see Avanesova (2015).

6 For another presentation of this pottery, see B. Lyonnet (1996: types I.1 to I.10 and pp. 63–65).

7 The temper composition was studied visually along fresh breaks and on the surface of the ceramics.

8 Unfortunately, no chemical analyses were made on these artifacts and their exact composition is as yet unknown.

9 See note 8.

10 Not speculating on their functional purpose, we consider them as jewelry and conventionally call them pins or hairpins, when the rod has a smooth bulge at the end.

11 Determination made by B.Kh. Batyrov, paleozoologist at Samarkand University.

12 Unfortunately, with rare exceptions, publications do not provide detailed description of the beads. This impedes research on the regional distribution of the various types and on their chronological attribution. The study of specific morphological types could be the basis for verification or better precision in the dates of the burials with stone beads.

13 A review of nozzle finds in settlements and of the frequency of their occurrence in burials within the Eurasian metallurgical province can be found in Val'kov and Kuz'minykh (2000: 73–83).

REFERENCES

Agapov, S.A., I.B. Vasil'ev, O.V. Kuz'mina, and A.P. Semenova 1983. "Srubnaja kul'tura lesostepnogo Povolzhja" [The Srubnaja culture of the Volga forest-steppe region]. In *Kul'tury bronzovogo veka Vostochnoj Evropy* [Bronze Age cultures of eastern Europe], ed. S.G. Basin. Kujbyshev: Kujbyshevskij gosudarstvennyj pedagogicheskij universitet, pp. 6–58.

Amanbaeva, B.E., A.E. Rogozhinskij, and D. Merfi 2006. "Mogilnik Shagym – novyj pamjatnik epokhi bronzy Vostochnoj Fergany" [Shagym cemetery – a new Bronze Age site in eastern Ferghana]. *Arkheologicheskie Issledovanija v Uzbekistane 2004–2005* 5: 256–265.

Askarov, A.A. 1973. *Sapallitepa*. Tashkent: Fan.

Askarov, A.A. 1977. *Drevnezemledel'cheskaja kultura juga Uzbekistana* [Ancient Bronze Age farming culture of southern Uzbekistan]. Tashkent: Fan.

Askarov, A.A. and B.N. Abdullaev 1983. *Dzharkutan*. Tashkent: Fan.

Askarov, A.A. and T.Sh. Shirinov 1993. *Rannaja gorodskaja kul'tura epokhi bronzy juga Srednej Azii* [Early urban Bronze Age culture of southern Central Asia]. Samarkand: Fan.

Avanesova, N.A. 1987. "Stepi Evrazii i oazisy Juga: problema kulturnyh vzaimodejstvij v epokhu bronzy" [Eurasian steppes and southern oases: problem of cultural interactions in the Bronze Age]. In *Interactions of the Nomadic Cultures and Ancient Civilizations: Abstracts of the Soviet–French Symposium on the Archaeology of Central Asia and Neighboring Regions*, ed. V.M. Masson. Alma-Ata: Nauka, pp. 53–55.

Avanesova, N.A. 1991. *Kul'tura pastusheskikh plemen epokhi bronzy aziatskoj chasti SSSR* [Bronze Age pastoral groups culture in the Asian part of the USSR]. Tashkent: Fan.

Avanesova, N.A. 1995. "Novoe o proniknovenii pastusheskikh plemen bronzovogo veka v zemledel'cheskie oazisy na juge Uzbekistana" [New data on the penetration of the Bronze Age pastoral tribes into the farming oases of southern Uzbekistan]. In *Izuchenie kul'turnykh vzaimodejstvij i novye arkheologicheskie otkrytija* [Study of the cultural interactions and new archaeological discoveries]. Materialy plenuma Instituta Istorii Material'noj Kul'tury, 22. St. Petersburg: Institute for the History of Material Culture, pp. 82–86.

Avanessova, N.A. 1996. "Pasteurs et agriculteurs de la vallée du Zeravshan (Ouzbékistan) au début de l'Age du Bronze: relations et influences mutuelles." In *Sarazm (Tadjikistan), céramiques (Chalcolithique et Bronze Ancien)*, ed. B. Lyonnet (in collaboration with A. Isakov). Paris: de Boccard, pp. 117–131.

Avanesova, N.A. 2001. "U istokov urbanisticheskogo Afrosiaba" [At the origins of the city of Afrosiab]. *Istorija Material'noj Kul'tury Uzbekistana* 32: 57–72.

Avanesova, N.A. 2010. "Zeravshanskaja kul'turnaja provintsija Bactrijsko-Margianskoj tsivilizatsii" [The Zeravshan cultural province of the Bactria-Margiana civilization]. In *Na puti otkrytija tsivilizatsii* [On the track of uncovering a civilization]. *Trudy Margianskoj Arkheologicheskoj Ekspeditsii*, vol. 3, ed. M.F. Kosarev, P.M. Kozhin, and N.A. Dubova. St. Petersburg: Aletejja, pp. 334–364.

Avanesova, N.A. 2012a. "Drevnie gornjaki Zeravshana" [Ancient miners of Zeravshan]. *Arkheologija Uzbekistana* 1(4): 3–35.

Avanesova, N.A. 2012b. "Svjatilishe rannikh nomadov Zeravshana" [Sanctuary of the early nomads of Zeravshan]. In *Afanas'evskij sbornik*, vol. 2 [Afanas'evo collection of articles, vol. 2], ed. N.F. Stepanova. Barnaul: Azbuka, pp. 8–27.

Avanesova, N.A. 2013. "Zhukov, un sanctuaire énéolithique dans la vallée du Zeravshan (Ouzbékistan)." *Paléorient* 39(2): 85–108.

Avanesova, N.A. 2015. "Keramika poselenija gornjakov-metallurgov Zeravshana" [Ceramics of the miners-metallurgists of a Zeravshan settlement]. *Istorija Material'noj Kul'tury Uzbekistana* 39: 47–62.

Avanesova, N.A., Sh. B. Shajdullaev, and A. Erkulov 2005. "K voprosu o kul'turnoj prinadlezhnosti dzhamskikh drevnostej epokhi paleometalla" [On the cultural attribution of the Dzham antiquities to the Paleometal period]." In *Tsivilizatsii skotovodov i zemledel'tsev Tsentralnoj Azii* [Civilization of cattle breeders and farmers of Central Asia]. Samarkand-Bishkek: International Institute of the Central Asian Investigations, pp. 12–33

Bader, O.N. 1964. *Drevnejshie metallurgi Priural'ja* [The oldest metallurgists of Cisural]. Moscow: Nauka.

Bader, O.N. and A.Kh. Khalikov 1976. *Pamjatniki balanovskoj kul'tury* [Sites of the Balanovo culture]. Svod arheologicheskikh istochnikov, B 1–25. Moscow: Nauka.

Baratov, P.B. 1977. *Prirodnye resursy Zeravshanskoj doliny i ikh ispolzovanie* [Natural resources of Zeravshan valley and their use]. Tashkent: Fan.

Batalov, S.G., S.A. Grigor'ev, and G.B. Zdanovich 1996. "Pogrebal'nye kompleksy epokhi bronzy Bol'shekaraganskogo mogil'nika (publikatsija resul'tatov arkheologicheskikh raskopok 1988 goda)" [Funerary complexes of Bol'shekaraganskij graveyard (publication of the 1988 archaeological excavations)]. In *Materialy po arkheologii i etnografii Juzhnogo Urala: Trudy muzeja-zapovednika "Arkaim"* [Materials on the archaeology and ethnography of southern Ural: transactions of the Museum-Reserve "Arkaim"], ed. A.D. Tairov. Cheljabinsk: Kamennyj pojas, pp. 64–88.

Bobomulloev, S. 1999. "Raskopki grobnitsy bronzovogo veka na verkhnem Zeravshane" [Excavations of a Bronze Age sepulcher in the upper Zeravshan]. *Stratum Plus* 2: 307–313.

Bobomulloev, S., N.M. Vinogradova, and B. Bobomulloev 2016. "Predvaritel'nye rezul'taty issledovanij mogil'nika Farkhor: pamjatnika epokhi srednej bronzy na juge Tadzhikistana" [Preliminary results of the investigations at the Farkhor cemetery: a Middle Bronze age site of southern Tajikistan]. *Vestnik MICAI* 24: 9–52.

Bochkarev, V.S. 1974. "Pogrebenija litejshikov epokhi bronzy (metodologicheskij peresmotr)" [Casting smiths burials of the Bronze Age (methodological revision)]. *Problemy Arkheologii* 2: 48–53.

Bogdanov, S.V. 2004. *Epokha medi stepnogo Priural'ja* [The Chalcolithic period in the Cisural steppe]. Ekaterinburg: Ural'skoje otdelenie RAS.

Boroffka, N., M. Bubnova, G. Vajsgerber, I. Luts, G. Partsinger, E. Pernitska, D. Starshinin, Ja. Tserny, and Ju. Jakubov 2000. "Issledovanija v rajone drevnikh razrabotok olova v Severnom Tadzhikistane (Mushiston i Takfon)" [Research in the ancient tin-mining area in northern Tajikistan (Mushiston and Takfon)]. *Arheologicheskie Raboty v Tadzhikistane* 27: 71–118.

Bratchenko, S.N. 1976. *Nizhnee Podonje v epokhu srednej bronzy* [Lower Don in the Middle Bronze Age]. Kiev: Naukova dumka.

Burjakov, Ju.F. 1987. "Chto rasskazal Tusunsaj" [What Tursunsaj told]. *Leninskij put' Newspaper* 149–14707: 3.
Chernykh, E.N. 2007. *Kargaly. Tom V, Kargaly: fenomen i paradoksy razvitija; Kargaly v sisteme metallurgicheskikh provintsij; Potaennaja (sakral'naja) zhizn' arkhaichnykh gornjakov i metallurgov* [Kargaly. Vol. V, Kargaly: a phenomenom and paradox of development; Kargaly in the system of the metallurgical provinces; the secret (sacred) life of ancient miners and metallurgists]. Moscow: Nauka.
Dubova, N.A. 2008. "Eshche raz o rannikh skotovodakh i zemledel'tsakh Srednej Azii. Antropologija Siabskogo pogrebenija" [Once more on the early cattle breeders and farmers of Central Asia: anthropology of the Siab burial]. *Aktual'nye voprosy antropologii* 3: 166–172.
Epimakhov, A.V. 1996. "Kurgannyj mogil'nik Solntse II: nekropol' ukreplennogo poselenija Ust'e epokhi srednej bronzy" [The kurgan cemetery of Solntse II: necropolis of the Ust'e fortified Middle Bronze Age settlement]. In *Materialy po arkheologii i etnografii Juzhnogo Urala: Trudy muzeja-zapovednika "Arkaim"* [Materials on the archaeology and ethnography of southern Ural: Transactions of the Museum-Reserve "Arkaim"], ed. A.D. Toirov. Cheljabinsk: Kamennyj pojas, pp. 22–42.
Epimakhov, A.V. 2005. *Rannie kompleksnye obshchestva severa tsentral'noj Evrazii (po materialam mogil'nika Kamennij Ambar-5)* [Early complex societies of northern central Eurasia (from the material of the Kamennij Ambar-5 graveyard)]. Book 1. Cheljabinsk: Cheljabinskij dom pechati.
Evdokimov, V.V. 1976. "Raboty v Kustanajskoj oblasti" [Works in the Kustanaj district]. *Arkheologicheskie Otkrytija* 1975: 510–512.
Francfort, H.-P. 1989. *Fouilles de Shortugaï. Recherches sur l'Asie centrale protohistorique.* Paris: de Boccard.
Francfort, H.-P. 1994. "The Central Asian Dimension of the Symbolic in Bactria and Margiana." *Antiquity* 68: 411–420.
Frankfor, A.-P. 1997. "Pechati Oksa: raznoobraznie form i izmenjaemost' funktsii" [Oxus seals: diversity and variability of functions]. *Vestnik Drevnej Istorii* 4: 60–68.
Francfort, H.-P. 2005. "La civilisation de l'Oxus et les indo-iraniens et indo-aryens en Asie Centrale." In *Āryas, aryens iraniens en Asie centrale*, ed. G. Fussman. Publications de l'Institut de Civilisation Indienne, vol. 72. Paris: Collège de France, pp. 183–192.
Francfort, H.-P., A. Vahdati, J. Bendezu-Sarmiento, J. Lhuillier, E. Fouache, M. Tengberg, M. Mashkour, and Z. Shirazi 2014. "Preliminary Report on the Soundings at Tepe Damghani Sabzevar, Spring 2008." *Iranica Antiqua* 49: 111–158.
Gening, V.F., G.B. Zdanovich, and V.V. Gening 1992. *Sintashta.* Cheljabinsk: Juzhno-Uralskoe knizhnoe izdatelstvo.
Gening, V.F. and N.K. Stefanova 1994. *Chernoozer'e I: mogil'nik epokhi bronzy srednego Priirtysh'ja* [Chernoozer'e I: a Middle Bronze Age graveyard of the Middle Irtysh]. Ekaterinburg: Ural'skij universitet.
Isakov, A. 1991a. *Sarazm.* Dushanbe: Donish.
Isakov, A. 1991b. *Verkhov'ja Zeravshana v epokhu eneolita i bronzy (K probleme mnogoochagovogo razvitija Srednej Azii v rannezemledel'cheskuju epokhu)* [Upper Zeravshan in the Chalcolithic and Bronze Age (questions on the multifocal development of Central Asia in the early farming period)]. Avtoreferat Dissertatsii. Leningrad: IIMK.
Kachalova, N.K. 1962. "K voprosu o pamjatnikakh poltavkinskogo tipa" [About the Poltavkino-type sites]. *Arkheologicheskij sbornik Gosudarstvennogo Hermitazha* 5: 31–49.
Kachalova, N.K. 1983. "O lokal'nykh razlichijakh poltavkinskoj kul'turno-istoricheskoj obshchnosti" [On the local differences of the Poltavkino cultural-historical community]. *Arkheologicheskij sbornik Gosudarstvennogo Hermitazha* 24: 4–19.

Kachalova, N.K. 2001. "Otnositel'naja khronologija poltavkinskikh pamjatnikov" [Relative chronology of the Poltavkino sites]. *Arkheologicheskij sbornik Gosudarstvennogo Hermitazha* 35: 32–58.

Khaljapin, M.V. 2001. "Pervyj beskurgannyj mogil'nik sintashtinskoj kul'tury v Stepnom Priural'e" [The first cemetery without kurgans of the Sintashta culture in the steppe region of Cisurals]. In *Bronzovyj vek Vostochnoj Evropy: kharakteristika kul'tur, khronologija i periodizatsija. Materialy mezhdunarodnoj nauchnoj konferentsii "K stoletiju periodizatsii V.A. Gorodtsova bronzovogo veka juzhnoj poloviny Vostochnoj Evropy"* [Bronze Age of Eastern Europe: characteristics of cultures, chronology and periodization. Materials of the International Scientific Conference "To the century of V. Gorodtsov's periodization of the Bronze Age of the southern half of eastern Europe"]. Samara: Samarskij universitet, pp. 417–425.

Khuff, D. and Sh. Shajdullaev 1999. "Nekotorye rezul'taty rabot Uzbeksko-Germanskoj ekspeditsii na gorodishche Dzharkutan" [Several results from the Uzbek–German expedition works at the Dzharkutan settlement]. *Istorija Material'noj Kul'tury Uzbekistana* 30: 19–26.

Kircho, L.B. 2001. "Metallicheskie izdelija Altyn-depe" [Altyn Depe metal artifacts]. In *Osobennosti proizvodstva poselenija Altyn-depe v epokhu paleometalla* [Production specificities at Altyn Depe during the Paleometal Age], ed. V.M. Masson and L.B. Kircho. Materialy Juzhno-Turkmenistanskoj arkheologicheskoj kompleksnoj ekspeditsii, 5. St. Petersburg: Institut Istorii Material'noj Kul'tury RAN, pp. 60–71.

Kircho, L.B. 2005. "Pogrebal'nyj inventar' Altyn-depe" [Altyn Depe funerary inventories]. In *Khronologija epokhi pozdnego eneolita – srednej bronzy Srednej Azii (pogrebenija Altyn-depe)* [Chronology of the Late Chalcolithic and Middle Bronze Age of Central Asia (Altyn Depe burials)], ed. V.M. Masson and Ju.E. Berezkin. Trudy Instituta Istorii Material'noj Kul'tury RAN, T. XVI. St. Petersburg: Nestor Istorija, pp. 347–421.

Kireev, V.I., E.B. Pruger, and B.I. Bahmetova 1969. *Otchet po teme: "Revizija i izuchenie drevnikh gornykh vyrabotok na territorii Uzbekistana za 1964–1967gg."* [Report on the theme: "revision and research of the ancient mine workings on the territory of Uzbekistan in 1964–1967"]. Tashkent: Rukopisnye fondy arkhiva "Samarkandgeologija" [Manuscript fund of the archive "Samarkand geology"]. Inv. no. 935.

Korobkova, G.F. 1969. *Orudija truda i khozjajstvo neoliticheskikh plemen Srednej Azii* [Tools and economy of the Neolithic population of Central Asia]. Materialy i Issledovanija po Arkheologii SSSR, 158. Moscow: Nauka.

Korobkova, G.F. 2001. "Metodika izuchenija kamennykh, kostjanykh i drugikh izdelij iz nemetallicheskogo syr'ja epokhi paleometalla" [Methods of study of Paleometal Age stone, bone and other products from non-metallic raw materials]. In *Osobennosti proizvodstva poselenija Altyn-depe v epokhu paleometalla* [Specificities of production at Altyn Depe in the Paleometal Age], ed. V.M. Masson and L.B. Kircho. Materialy Juzhno-Turkmenistanskoj arkheologicheskoj kompleksnoj ekspeditsii, 5. St. Petersburg: Institut Istorii Material'noj Kul'tury RAN, pp. 142–212.

Kufterin, V.V. 2010. "Kraniologicheskij material epokhi bronzy iz Sazagana pod Samarkandom" [Bronze Age craniological material from Sazagan near Samarkand]. In *Na puti otkrytija tsivilizatsii* [On the track of uncovering a civilization]. Trudy Margianskoj Arkheologicheskoj Ekspeditsii, vol. 3, ed. M.F. Kosarev, P.M. Kozhin, and N.A. Dubova. St. Petersburg: Aletejja, pp. 504–512.

Kuz'mina, O.V. 2002. "Ukrashenija abashevskoj kul'tury" [Abashevo culture ornaments]. In *Problemy arkheologii Evrazii. K 80-letiju N.Ja. Merperta* [Problems of archaeology of Eurasia: for the 80 years of N.Ja. Merpert], ed. R.M. Munchaev. Moscow: Nauka, pp. 157–180.

Litvinskij, B.A. 1950. "K istorii dobychi olova v Uzbekistane" [On the history of tin mining in Uzbekistan]. In *Trudy Sredneaziatskogo gosuniversiteta* [Transactions of Central Asian

University]. New series, XI, Gumanitarnye nauki, book 3. Tashkent: Sredneaziatskij gosuniversitet, pp. 51–68.

Litvinskij, B.A. 1954. *Drevnejshie stranitsy istorii gornogo dela Tadzhikistana i drugikh respublik Srednej Azii* [The oldest pages in the history of mining in Tajikistan and the other republics of Central Asia]. Nauchno-populjarnaja biblioteka Instituta istorii, arkheologii i etnografii [Popular science library of the Institute of History, Archaeology and Ethnography], 19. Stalinabad: Institut Istorii, Arkheologii i Etnografii.

Lopatin, S.V., E.B. Pruger, and A.G. Ermilov 1964. *Karta razmeshchenija drevnikh vyrabotok na territorii Uzbekistana. Tom I,* Objasnitel'naja zapiska [A map of the distribution of ancient minings on the territory of Uzbekistan. Vol. 1, Explanatory notice]. Tashkent: Rukopisnye fondy arkhiva "Samarkandgeologija" [Manuscript fund of the archive "Samarkand geology"]. Inv. no. 659.

Lyonnet, B. (with A. Isakov) 1996. *Sarazm (Tadjikistan), céramiques (Chalcolithique et Bronze Ancien)*. Paris: de Boccard.

Lyonnet, B. 2005. "Another Possible Interpretation of the Bactro-Margiana Culture (BMAC) of Central Asia: The Tin Trade." In *South Asian Archaeology 2001*. Vol. I, *Prehistory*, ed. C. Jarrige and V. Lefevre. Paris: Editions Recherche sur les Civilisations, pp. 191–200.

Maksudov, I.Z., V.I. Kireev, L.P. Shevtsova, and E.B. Pruger 1975. *Otchet po poiskam, revizii i otsenke ob'ektov blagorodnykh metallov po sledam drevnikh gornykh rabot v Tsentral'nykh Kyzylkumakh, Kuratinskom i Chatkalo-Kuraminskom regionah (po rabotam za 1974–1975gg.)* [Report on the search, revision and evaluation of precious metals from the remains of ancient mining operations in the Central Kyzyl-Kum, Kurata and Chatkalo-Kurama regions (1974–1975)]. Tashkent: Rukopisnye fondy arkhiva "Samarkandgeologija" [Manuscript fund of the archive "Samarkand geology"]. Inv. no. 1639.

Maljutina, T.S. and G.B. Zdanovich 2003. "Keramika ukreplennogo poselenija Arkaim" [Ceramics of the Arkaim fortified settlement]. In *Drevnjaja keramika: problemy i perspektivy kompleksnogo podkhoda* [Ancient ceramics: problem and perspectives of interdisciplinary approach], ed. D.G. Zdanovich. Cheljabinsk: Cheljabinskij gosuniversitet, pp. 99–131.

Maljutina, T.S. and G.B. Zdanovich 2004. "Keramika Arkaima: opyt tipologii" [Arkaim ceramics: a typological attempt]. *Rossijskaja Arkheologija* 4: 67–82.

Malov, N.M. 1980. "Issledovanie pamjatnikov epokhi medi i bronzy v Saratovskom Povolzh'e" [Research on the Copper and Bronze Age sites in the Saratov province on the Volga River]. *Arkheologicheskie Otkrytija 1979*: 153.

Malov, N.M. 1993. "Kurgannye mogil'niki, poselenija, sluchajnye nakhodki i klad Volgo-Ural'skogo mezhdurech'ja" [Kurgan cemeteries, settlements, occasional finds and hoard in the Volgo–Ural interfluve]. In *Pamjatniki srubnoj kul'tury. Volgo-Ural'skoe mezhdurech'e*. Svod arkheologicheskikh istochnikov, no. 1–10, T.1. Saratov: Saratovskij universitet, pp. 18–19.

Matjushenko, V.I. 1973. "Andronovskaja kul'tura na Verkhnej Obi" [Andronovo culture on the upper Ob']. In *Iz istorii Sibiri* [From the history of Siberia], vol. 11, ed. V.I. Moshinskaja. Tomsk: Izdatel'stvo Tomskogo universiteta, pp. 3–12.

Partsinger, G. and N. Boroffka 2008. "Poselenie metallurgov epokhi bronzy v Karnab-Sichkonchi (Uzbekistan)" [Bronze Age metallurgists' settlement at Karnab-Sichkonchi (Uzbekistan)]. *Izvestija laboratorii arkheologicheskikh issledovanij Kemerovskogo universiteta* [News of the laboratory of archaeological researches of Kemerovo State University], vol. 20. Pervobytnaja arkheologija. Chelovek i iskusstvo. Sbornik nauchnykh trudov, posvjashchennyj 70-letiju so dnja rozhdenija Ja.A. Shera [Prehistoric archaeology. Man and art. Collection of articles dedicated to the 70 years of the birthday of Ja.A. Sher]. Novosibirsk: Izdatel'stvo Instituta Arkheologii i Etnografii SO RAN, pp. 163–167.

Popov, V.I. and Kh. T. Tyljaganov (eds.) 1976. *Mineral'no-syr'evye resursy Uzbekistana* [Mineral resources of Uzbekistan]. Part 1. Tashkent: Fan.

Potemkina, T.M. 1985. *Bronzovyj vek lesostepnogo Pritobol'ja* [The Bronze Age in the Tobol River forest-steppe region]. Moscow: Nauka.

Potemkina, T.M. 1990. "O nekotorykh spornykh voprosakh rannej i srednej bronzy Volgo-Ural'skogo regiona" [On the debated questions of the Early and Middle Bronze Age in the Volga–Ural region]. *Sovetskaja Arkheologija* 1: 118–130.

Prjakhin, A.D. 1976. *Poselenija abashevskoj obshchnosti* [Settlements of the Abashevo community]. Voronezh: Voronezhskij universitet.

Prjakhin, A.D. 1982. *Poselenija katakombnogo vremeni lesostepnogo Podon'ja* [Settlements of the Catacomb period in Don River forest-steppe region]. Voronezh: Voronezhskij universitet.

P'jankova, L.T. 1990. "Keramicheskie kompleksy Margiany epokhi bronzy (Gonur i Togolok-21)" [Ceramic complexes of Bronze Age Margiana (Gonur and Togolok-21]. In *Unesco IASCCA Informatsionnyj bjulleten'* 16, ed. S.I. Potabenko, L.A. Chvyr', and A.V. Nagornaja. Moscow: Nauka, pp. 24–44.

P'jankova, L.T. 1994. "Keramika Margiany i Baktrii epokhi bronzy" [Ceramics of Margiana and Bactria]. *Rossijskij etnograf* 21: 135–156.

Rakhmonov, U. 1987. "Keramicheskoe proizvodstvo epokhi bronzy Juzhnogo Uzbekistana" [Ceramic production in the Bronze Age of southern Uzbekistan]. Avtoreferat dissertatsii. Samarkandskij gosudarstvennyj universitet, Samarkand.

Razzokov, A. 2006. "Sarazm – rannij gorod Ariany" [Sarazm – an early city in Ariana]. In *Arijskaja tsivilizatsija v kontekste evroaziatskikh kul'tur* [Aryan civilization in the context of the Eurasian cultures], ed. A. Radzhabov, R. Mukimov. P. Dzhamshedov, and Kh. Nazarov. Dushanbe: Donish, pp. 198–206.

Ruzanov, V.D. 1979. "O nekotoryh drevnikh olovorudnykh istochnikakh na territorii Uzbekistana" [About some ancient tin-mine sources on the territory of Uzbekistan]. *Istorija Material'noj Kul'tury Uzbekistana* 15: 98–104.

Ruzanov, V.D. 1999. "K voprosu o istokakh olovjanistykh bronz v metalloobrabatyvajushikh proizvodstvakh plemen juga Srednej Azii v epokhu bronzy" [About tin bronze sources in the Bronze Age southern Central Asian metalworking industry]. *Istorija Material'noj Kul'tury Uzbekistana* 30: 37–40.

Ruzanov, V.D. 2000. "Karnabskoe kassiteritovoe mestorozhdenie: istochnik olova epokhi bronzy" [Karnab cassiterite deposit: a source of tin in the Bronze Age]. *Istorija Material'noj Kul'tury Uzbekistana* 31: 54–57.

Ruzanov, V.D. 2013. *Metalloobrabotka na juge Srednej Azii v epokhu bronzy* [Bronze Age metalworking in southern Uzbekistan]. Samarkand: Institute of Archaeology Ja.G. Guljamov.

Salugina, N.P. 1994. "Tekhnologicheskoe issledovanie keramiki Potapovskogo mogil'nika." Prilozhenie 2 [Technological research on the ceramics of the Potapovka cemetery. Supplement 2]. In *Potapovskij kurgannyj mogil'nik indoiranskih plemen na Volge*, ed. I.B. Vasil'ev, P.F. Kuznetsov, and A.P. Semenova. Samara: Samarskij universitet, pp.173–186.

Sarianidi, V.I. 1977. *Drevnie zemledel'tsy Afganistana* [Ancient farmers of Afghanistan]. Moscow: Nauka.

Sarianidi, V.I. 1981. "Zerkala drevnej Baktrii" [Mirrors of ancient Bactria]. *Sovetskaja Arkheologija* 1: 289–293.

Sarianidi, V.I. 1990. *Drevnosti strany Margush* [Ancient Margush]. Ashhabad: Ylym.

Sarianidi, V.I. 2001. *The Gonur Necropolis and Iranian paganism.* Moscow: Mir-media.

Sarianidi, V.I. 2002. *Margush: Ancient Oriental Kingdom in the Old Delta of Murghab River.* Ashhabad: Türkmendöwlethabarlary.

Sarianidi, V.I. 2007. *Necropolis of Gonur.* Athens: Kapon Edition.

Shajdullaev, Sh.B. 2000. *Severnaja Baktrija v epokhu rannego zheleznogo veka* [Northern Bactria in the Early Iron Age]. Tashkent: Fan.

Shilov, V.P. 1959. "O drevnej metallurgii i metalloobrabotke v Nizhnem Povolzh'e" [On ancient metallurgy and metalworking in the lower Volga]. In *Materialy i Issledovanija po Arkheologii SSSR*, vol. 60, ed. E.N. Krupnov. Moscow: Nauka, pp. 11–38.

Shirinov, T.Sh. 1986. *Orudija proizvodstva i oruzhie epokhi bronzy Sredneaziatskogo mezhdurech'ja (Po dannym eksperimental'no-trassologicheskogo izuchenija)* [Bronze Age tools and weapons of the Central Asian interfluve (according to data from experimental use-wear analyses)]. Tashkent: Fan.

Sinitsyn, I.V. 1959. "Arkheologicheskie issledovanija Zavolzhskogo otrjada (1951–1953 gg.)" [Archaeological research of the Zavolzhskij team (1951–1953)]. In *Materialy i Issledovanija po Arkheologii SSSR*, vol. 60, ed. E.N. Krupnov. Pamjatniki Nizhnego Povolzh'ja [The lower Volga sites]. T. 1 Moscow: Nauka, pp. 56–129.

Smirnov, K.F. and E.E. Kuz'mina 1977. *Proiskhozhdenie indoirantsev v svete novejshikh arkheologicheskikh otkrytij* [Origins of the Indo-Iranians in the light of the last archaeological discoveries]. Moscow: Nauka.

Stech, T. 1999. "Aspects of Early Metallurgy in Mesopotamia and Anatolia." In *The Archaeometallurgy of the Asian Old World*, ed. V.C. Pigott. Philadelphia: University of Pennsylvania Museum of Archaeology and Anthropology, pp. 59–71.

Subbotin, L.V. 2000. "Severo-Zapadnoe Prichernomor'e v epokhu rannej bronzy" [Northwestern Black Sea coast in the Early Bonze Age]. *Stratum plus* 2: 350–387.

Udeumuradov, B.N. 1993. *Altyn-depe i Margiana: svjazi, khronologija, proiskhozhdenie* [Altyn Depe and Margiana: contacts, chronology, origin]. Ashgabat: Ylym.

Usachuk, A.N. 2005. "Kamennoambarskie psalii (trasologicheskij analiz)" [Kamennij Ambar cheek-pieces (use-wear analysis)]. In *Rannie kompleksnye obshchestva Severa Tsentral'noj Azii (Po materialam mogil'nika Kamennij Ambar-5), Kniga 1* [Early complex communities of northern Central Asia (from the materials of the Kamennij Ambar-5 cemetery), book 1], ed. A.V. Epimakhov. Cheljabinsk: Cheljabinskij dom pechati, pp. 179–189.

Val'kov, D.V. and S.V. Kuz'minykh 2000. "Sopla evrazijskoj metallurgicheskoj provintsii. (K probleme odnoj arkheologicheskoj zagadki)" [Nozzles of the Eurasian metallurgical province (on one archaeological puzzling problem)]. In *Problemy izuchenija eneolita i bronzovogo veka Juzhnogo Urala. Sbornik nauchnykh trudov* [Problems in the study of the Chalcolithic and Bronze Age of southern Ural: collection of scientific articles], ed. V.V. Tkachev. Orsk: Uralskoe otdelenie RAS, pp. 73–83.

Vasil'ev, I.B., P.F. Kuznetsov, and A.P. Semenova 1994. *Potapovskij kurgannyj mogil'nik indoiranskikh plemen na Volge* [The Potapovka kurgan cemetery of Indo-Iranian tribes on the Volga]. Samara: Samarskij universitet.

Vasil'ev, I.B., P.F. Kuznetsov, and A.P. Semenova 1995. "Pamjatniki potapovskogo tipa v lesostenom Povolzh'e" [Monuments of Potapovka type in the forest-steppe part of Volga]. In *Drevnie indoiranskie kultury Volgo-Uralja* [Ancient Indo-Iranian cultures of the Volga–Ural region], ed. I.B. Vasil'ev. Samara: Samarskij universitet, pp. 5–37.

Vinogradov, N.B. 1982. "Kulevichi III: pamjatnik petrovskogo tipa v Juzhnom Zaural'e" [Kulevichi III: a Petrovka-type site in southern Trans-Ural]. *Kratkie Soobshchenija Instituta Arkheologii* 169: 94–99.

Vinogradov, N.B. 2003. *Mogil'nik bronzovogo veka Krivoe Ozero v Juzhnom Zaural'e* [The Bronze Age graveyard of Krivoe ozero in southern Trans-Ural]. Cheljabinsk: Cheljabinskij pedagogicheskij institut.

Vinogradov, A.V., S.V. Lopatin, and E.D. Mamedov 1965. "Kyzylkumskaja birjuza. (Iz istorii dobychi i obrabotki)" [Kyzylkum turquoise (history of mining and processing)]. *Sovetskaja Etnografija* 2: 114–134.

Zdanovich, D.G. 2002. *Arkaim: nekropol' (po materialam kurgana 25 Bol'shekaraganskogo mogil'nika)* [Arkaim: necropolis (from the materials of kurgan 25 of the Bol'shekaraganskij graveyard)]. Cheljabinsk: Juzhno-Ural'skoe knizhnoe izdatel'stvo.

Zdanovich, G.B. 1983. "Osnovnye kharakteristiki petrovskikh kompleksov Uralo-Kazakhstanskikh stepej (K voprosu o vydelenii petrovskoj kul'tury)" [Main characteristics of the Petrovka complexes in the Ural-Kazakhstan steppes (on the question of the identification of the Petrovka culture)]. In *Bronzovyj vek stepnoj polosy Uralo-Irtyshskogo mezhdurech'ja* [The Bronze Age of the steppe belt of the Ural–Irtysh interfluve], ed. G.B. Zdanovich. Cheljabinsk: Bashkirskij gosudarstvennyj universitet, pp. 48–68.

Zdanovich, G.B. 1988. *Bronzovyj vek Uralo-Kazakhstanskikh stepej* [The Bronze Age of the Ural-Kazakhstan steppes]. Sverdlovsk: Ural'skogo universiteta.

Zdanovich, G.B. 1995. "Arkaim: Arii na Urale ili nesostojavshajasja tsivilizatsija" [Arkaim: Arians in the Urals or an unfulfilled civilization]. In *Arkaim: issledovanija, poiski, otkrytija* [Arkaim: research, studies, discoveries], ed. G.B. Zdanovich. Cheljabinsk: Kamennyj pojas, pp. 21–42.

Zdanovich, G.B. 1997. "Arkaim: kul'turnyj kompleks epokhi srednej bronzy Juzhnogo Zaural'ja" [Arkaim: a Middle Bronze Age cultural complex of southern Ural]. *Rossijskaja Arkheologija* 2: 47–62.

Zdanovich, G.B. 2002. "Uralo-Kazakhstanskie stepi v epokhu srednej bronzy" [The Ural-Kazakhstan steppes in the Middle Bronze Age]. Avtoreferat Dissertatsii. Cheljabinskij gosudarstvennyj universitet, Cheljabinsk.

Zdanovich, G.B. and S.Ja. Zdanovich 1980. "Mogil'nik epokhi bronzy u sela Petrovka" [A Bronze Age cemetery near the Petrovka village]. *Sovetskaja Arkheologija* 3: 183–192.

CHAPTER TWENTY-FIVE

THE "CLASSICAL VAKHSH CULTURE"

A Bronze Age culture of the 3rd and early 2nd millennium BC in southern Tajikistan

Mike Teufer[*]

INTRODUCTION

After the discovery made by Vadim A. Ranov and Boris A. Litvinskij in the beginning of the 1950s of the Bronze Age kurgan burial grounds at Vakhsh I, Tigrovaja Balka, Ojkul', and Dzharkul' in the lower Vakhsh valley, systematic research on these cemeteries started from the 1960s onwards (Figure 25.1). It became evident that the graves of several of these necropolises not only shared similar structures, but also had very similar sets of grave goods. This led Litvinskij to define a common archaeological culture – the Vakhsh culture. The Makoni Mor cemetery, located further east in the lower Kyzylsu, was studied between 1962 and 1963 and was also considered by Litvinskij as belonging to the Vakhsh culture (Litvinskij 1964: 158).

Two features are characteristic of these five cemeteries: burials under kurgans (see Figure 25.2a, *1*; 25.2b, *1*, *2*), and similar ceramic material (see Figures 25.3–25.12).

However, only the burial grounds of Tigrovaja Balka (P'jankova 1986) and Dzharkul' (Teufer *et al.* 2015) in the lower Vakhsh valley have been fully published. Kai Kaniuth and the present author recently recorded the material coming from the cemetery of Makoni Mor and published it on the basis of Ljudmila T. P'jankova's preliminary studies and with the help and support of Saidmurod Bobomulloev and B.A. Litvinskij (P'jankova *et al.* 2009). The cemeteries of Vakhsh I and Ojkul' are currently being prepared for publication by the author.

The Vakhsh culture has been considered up until now as a phenomenon of the Late Bronze Age. This assumption was based on its chronological attribution through analogies of pottery types with the neighboring regions. At first, this seemed highly probable due to the state of research on the Sapalli-Dzharkutan culture in the Surkhan Darya region (southern Uzbekistan). It led to the widespread assumption that the Vakhsh culture occurred during the latest phase of the Late Bronze Age (see Francfort 1981: 198–202; P'jankova 1982: 36; 1986: 56–58; 1989: 92–101; Kaniuth and Teufer 2001: 100–104, 112; Vinogradova 2004: 89–92; P'jankova *et al.* 2009; Luneau *et al.* 2011: 283–312; Teufer 2015: 236–253), which, based on its absolute dating, meant more or less the period between the 17th and the 15th centuryies BC.

A number of different theories regarding the origin of the Vakhsh culture have been proposed, many of which are coupled with migration hypotheses. Litvinskij,

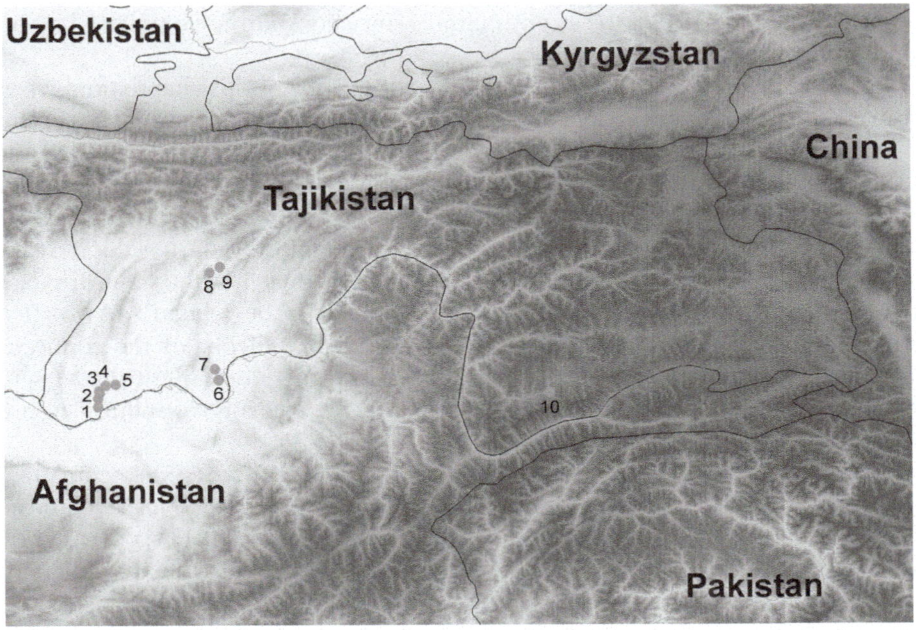

Figure 25.1 Distribution map of the "classical Vakhsh culture." 1: Amudaria; 2: Dzharkul'; 3: Ojkul'; 4: Tigrovaja Balka; 5: Vakhsh I; 6: Makonimor; 7: Akhtam Sakhoba; 8: Krugsaj; 9: Teguzak; 10: Jushbok.

for instance, supposed that the bearers of the Vakhsh culture were immigrants from Turkmenistan, and that influences from the steppe cultures were recognizable when they arrived in southern Tajikistan (Litvinskij 1973: 9–13). The latter proposal has been further developed by Kuz'mina who underlined that the Vakhsh culture was likely due to a considerable participation of populations from the steppe region. Until today, the relation between the Vakhsh culture and the Bishkent culture is debated. Kuz'mina refers to them together as the Bishkent culture (Kuz'mina 1972a: 119 fn16; 1972b: 138–139, 141; 1976: 121; 2007: 275, 277). Mandel'shtam had previously considered a possible steppe influence for the Bishkent culture, for which only the settlement of Rannij Tulkhar is known up to now. Even more interesting are the connections he recognized with Zamanbaba in the lower Zeravshan valley (Mandel'shtam 1968: 99, 135–136). This led Askarov to conclude that the cultures of Bishkent, Zamanbaba, and the younger phase of Sapalli (Molali period) all shared a common genesis. Despite differences in anthropological contexts, Askarov sees the Bishkent and Vakhsh cultures as one and the same (Askarov 1977: 115; 1981). According to P'jankova, there are similarities between the pottery of the Bishkent and Vakhsh cultures. She believed them to be two aspects of the same, but influenced to different degrees by the oases cultures. Thus, the Bishkent culture had likely been influenced more by the Sapalli culture from northern Bactria, and the Vakhsh culture by the Dashly culture from southern Bactria (P'jankova 1986: 74). As for Sarianidi, he considered the Vakhsh culture to have branched off from southern Bactria and questioned any influence of the steppe cultures since no Andronovo pottery had been found in Vakhsh burials (Sarianidi 1977: 147–148).[1]

Migration theories, however, are not within the scope of this chapter and we will not discuss them in further detail. Considering the recent excavations in southern Tajikistan, we should focus the research more on the local developments and less on the possible migrations. Cemeteries like Farkhor (Bobomulloev *et al.* 2017; see Chapter 23) and Kangurttut II (Bobomulloev and Bobomulloev 2017) now provide evidence for a beginning of the Bronze Age in Tajikistan during the transition from Namazga (NMG) IV to V. The Vakhsh culture is thus likely to have originated much earlier than originally thought. This has also to be related to two of the statements mentioned above: First, Mandelstam's early theory of the connection between the Zamanbaba and Bishkent cultures and, therefore, in a broader sense, with the Vakhsh culture, and Sarianidi's denial of the Andronovo influence on the Vakhsh culture.

Despite contradicting opinions (Sarianidi 1979; Askarov 1981), Zamanbaba is related with NMG V and occasionally also with NMG IV.[2] This is confirmed by the connections we can now establish with the more recent burials at Farkhor (Bobomulloev *et al.* 2017: 78). With a relatively secure dating in the 3rd millennium BC, this allows for a possible earlier beginning of the Vakhsh culture. A similar opinion had already been suggested by Lyonnet through the comparisons she made of the Vakhsh pottery with much older assemblages (see Lyonnet 1997: 65, 74 fn61, 80).

Thanks to Sarianidi, it is also possible to depart from a research tradition that has brought more strain than benefits on our comprehension of the Vakhsh culture. There are no indications of Andronovo features in the Vakhsh culture. Finds from the Andronovo culture first appear in southern Tajikistan only at the end of the Bronze Age, during a period that I have labeled as Late Bronze Age IIb (Teufer 2014: 129, fig. 25, 3–7; 2015: 232–236); that is, after the Vakhsh culture had already ended – a fact on which I will come back to later on.

Northern – or rather northeastern – influences can be recognized at best in the construction of kurgans (see p. 728). Yet, this feature has nothing to do with the Andronovo culture, as none of the Andronovo burials found until now in southern Tajikistan (Kumsaj, Tujun) shows evidence of such constructions.

Thus, the relationship between the Bishkent and Vakhsh cultures is seen less as an expression of different migration processes than as a model of cultural intervals, discussed in more detail below. With respect to the recent research in southern Tajikistan, the chronological model for Rannij Tulkhar developed by Kaniuth and Teufer should be reconsidered (Kaniuth and Teufer 2001). This, however, also lies outside the scope of this chapter.

This chapter will focus on the cemeteries from the lower Vakhsh valley, which were considered by Litvinskij as being part of the Vakhsh culture and can now be dated on the basis of the latest excavation results from southern Tajikistan (cf. Chapter 23). We will see that these burial grounds are clearly at least partially contemporary with the Bactria-Margiana Archaeological Complex (BMAC).

CHARACTERISTICS OF THE "CLASSICAL VAKHSH CULTURE"

The "Vakhsh culture" is a pure construct of an archaeological culture created by Litvinskij. As mentioned in the introductory remarks, the term "Vakhsh culture" is used inconsistently, especially in terms of its combination with the Bishkent culture. Alongside this problem, the general use of an "archaeological culture" is a theme hotly debated in current research (cf. Roberts and Vander-Linden 2011: 1–21). However, since we will use the term "Vakhsh culture," we should first clarify what in this context it means and what it does not. We take as a basis the definition of archaeological cultures as constructs of modern science. I draw here upon the reflections of the German cultural scientist Klaus Hansen, who described a culture as an expression of standardization (Hansen 1995: 114–120). I will also use the terms "cultural resource" and "distance" ("*écart*") proposed by the French philosopher and sinologist François Jullien (2017: 35–43). The following premises are to be assumed:

1. Standardization is part of human coping strategies.
2. Standardization can be found in archaeological classes of finds (pottery, bronze materials, etc.) and contexts (grave complexes, etc.).
3. The definition of an archaeological culture is a constructive process that combines different standardizations that are spatially and temporally close. The combined standardization of an archaeological culture is thereby the result of a subjective selection. The users of these standardization forms are representative of a so-called "*Abstraktionskollektiv*."[3] This "*Abstraktionskollektiv*," however, exists only because certain standardizations are ascribed to it.
4. Standardization constitutes the repertoire of cultural resources. These are theoretically exchangeable by means of communicative processes. Which of these cultural resources will be used depends on the agents (Jullien 2017).

This last point is quite significant for the Vakhsh culture in southern Tajikistan. It is there that we find the kurgans that are a characteristic of the five cemeteries mentioned in the introductory remarks to the chapter. But Vakhsh pottery can also be found in graves that do not have a kurgan construction, as at Gelot and Darnajchi (see Chapter 23), Obkukh (P'jankova 2003), and some graves at Rannij Tulkhar (Mandel'shtam 1968: 8–52). In this case there is an interrelationship that Jullien (2017) described as "*l'écart*" or as "*l'entre*." The concept of identity does not play a role here and it is difficult to know to what extent the people understood themselves as being part of a larger community of individuals buried in kurgans or of an even larger group, that of the users of the Vakhsh pottery.

Henceforth, when only the cemeteries with kurgan burials – for which the term "classical Vakhsh culture" has been introduced – are considered, it is meant to be seen as an "*Abstraktionskollektiv*" with two standardizations (pottery and kurgans) for which the question of cultural identity or even ethnic attribution is not addressed and, due to methodological reasons, almost should not be addressed (Teufer 2012: 271–276).

The term "classical Vakhsh culture" refers mainly to graves due to the scarcity of settlements. The two standardizations are the pottery shapes and the grave structures (kurgans). To this definition of "classical Vakhsh culture" we must also include the above-mentioned four cemeteries in the lower Vakhsh valley (Tigrovaja Balka, Vakhsh I, Ojkul', Dzharkul'), Makoni Mor in the lower Kyzylsu, the so-called Amudarja graves (P'jankova 1991: 165–176), the kurgans from Teguzak (P'jankova 1993: 227–241), Akhtam Sakhoba (Mukhitdinov 1975: 75) and Juzhbok (Babaev 1980: 75–87), and perhaps the Krugsaj grave 3.[4] Thus, the distribution of the "classical Vakhsh culture" spans from the Vakhsh River in the west to the western Pamir in the east (see Figure 25.1).

The standards of the "classical Vakhsh culture" will be briefly presented in the following section.

The grave structures of the "classical Vakhsh culture"

The characteristic feature of the "classical Vakhsh culture" is the burial in catacombs or pits under flat mounds (height between 0.1 m and 1.10 m). The kurgans often feature one or two stone circles. The mound above the grave is made of loess and stones (Figure 25.2a, *1*; 25.2b, *1*), but there are also mounds made of pure loess deposits, as at Teguzak. If stone circles are present, then the mound rests above them or is made within the outer circle. A specific circular form made of hearths distinguishes two kurgans at Tigrovaja Balka (Kurgans 7 and 8; P'jankova 1986: 14–15). The mounds are commonly round and occasionally rectangular. In general, the diameter of the flat mounds can be over 20 m in diameter for the bigger kurgans, and 5 to 6 m for the smaller ones.

The graves are dug into the natural ground soil (Figure 25.2b, *2*). In most of the cases there is only one grave under the mounds (Figure 25.2a, *2*), although cases of two graves are also attested (Figure 25.2b, *3*), and a few cases of multiple burials. The graves are not always positioned at the center under the mound. Usually they are catacomb-like (i.e., they consist of an entrance shaft and a chamber), but there are also cases of simple pits. In the chambers, the deceased are buried in a flexed position. Occasionally, so-called cenotaphs, or graves without a body, have been found.[5]

The pottery of the "classical Vakhsh culture"

Vessels are present as grave goods in many but not in all the graves of the "classical Vakhsh culture" (Figures 25.3–25.12). The most extensive pottery assemblages can reach six vessels, but these are rather scarce. In most cases, there are one or two vessels in the graves.

Pottery of the "classical Vakhsh culture" is normally handmade. Within this group, a subgroup is characterized by a coarse mineral-tempered clay. This grayish-black to dark-brown pottery is traditionally designated to kitchenware (P'jankova 1989: 57). This pottery – especially when decorated with a raised plastic band running along the shoulder (see Figure 25.8, *24*) – is often related to "steppe pottery" – in these contexts used as a general term for the Andronovo cultures[6] (P'jankova 1986: 50; 1989: 58), and more specifically the Federovo variant (Kuz'mina 2007: 228, 278). However, like Sarianidi, we disagree with this relation. We are confirmed in our position by the fact that this kitchenware is also present at Gelot, grave 3 in necropolis 4, dated to the 21st century BC on the basis of ^{14}C dates (see pp. 714–715) – the rest of the inventory

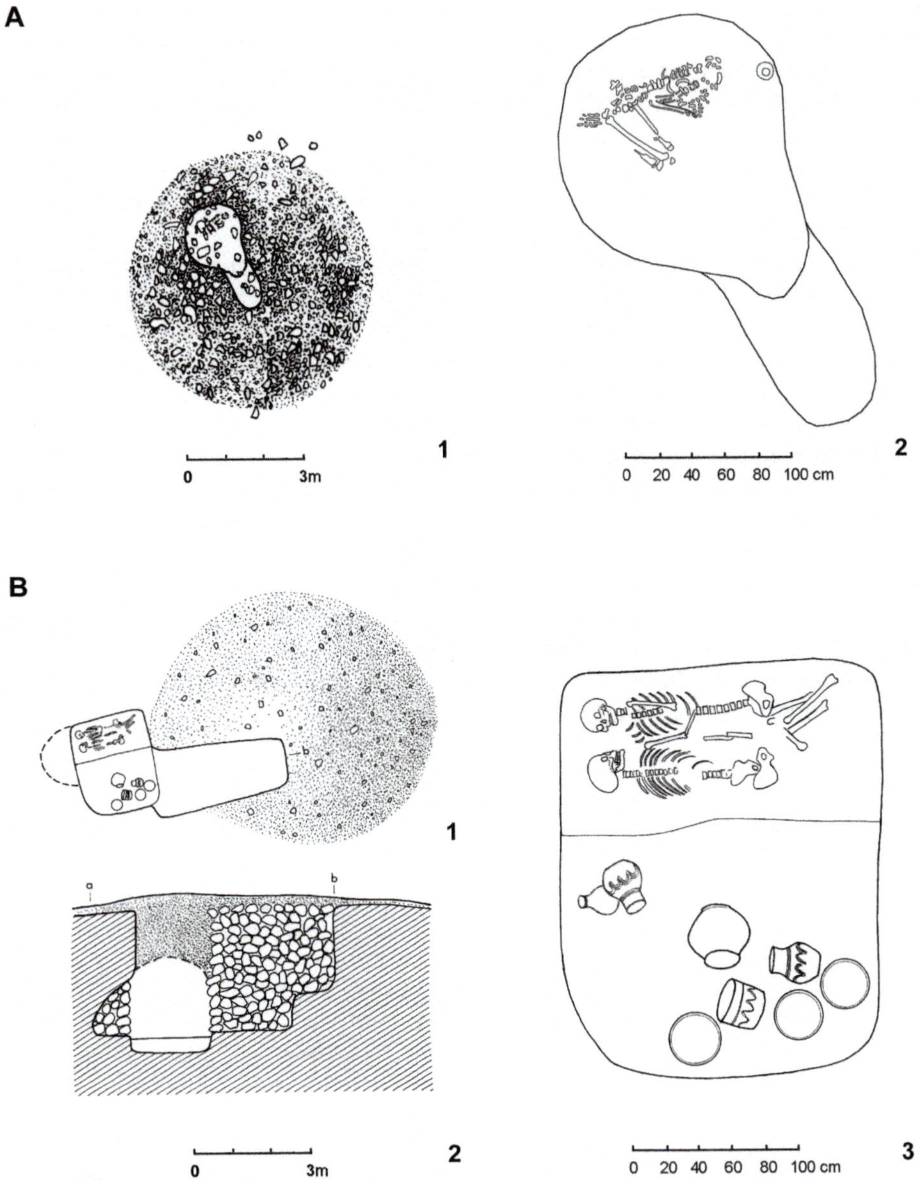

Figure 25.2 Kurgans of the "classical Vakhsh culture." A. Dzharkul', Kurgan 38; B. Makonimor, Kurgan 32.

being exclusively made of wheel-turned pottery and, as for all the other graves of this necropolis, without any connections with the so-called "steppe culture." This pottery fabric has rather to be derived from its practical use as kitchenware, leaving soot and smolder residue, which effectively differentiates it from the "tableware." This kitchenware is also known in settlements of the Sapalli Culture (cf. Kaniuth 2007: 39–40). Certainly, vessels with a raised plastic decoration around the shoulder can be found in numerous steppe cultures of the second half of the 3rd or early 2nd millennium BC that either pre-date or are contemporary with the Andronovo cultures: Worth mentioning here are the Samus (Parzinger 2006: 282, fig. 92, *20*), Elunino (Kirjushin 2002: fig. 82, *3*; 89, *1*, *5*, *24*; 95, *3*, *4*, *5*; 100, *1*; 106, *1*), Krotovo (Parzinger 2006: 286, fig. 94, *7.3*), and Chermurchek (Kovalev and Erdenebaatar 2009: 155–160, fig. 3, *10–11*) cultures. In my opinion, however, convincing parallels to the Vakhsh culture pottery can neither be found in the cultures mentioned above, nor in the Andronovo cultures.

In some cases, wheel-thrown pottery is also attested. However, its percentage among grave good pottery is low, as at Tigrovaja Balka – only 29.4% (P'jankova 1986: 41). The exact ratio between wheel-thrown pottery and handmade pottery is difficult to determine for each cemetery because it is not always evident to know to which grave the materials belong.[7] Still, it is possible to determine that handmade pottery is predominant and that wheel-thrown pottery comprises barely more than a third of the material found in all burial grounds.

The range of pottery shapes in the "classical Vakhsh culture" is limited to beakers, jugs, pots, and bowls. This is completed by a single find of a pedestal bowl at Makoni Mor and a goblet at Tigrovaja-Balka and at Kumsaj, respectively. Many publications have already dealt with this pottery (e.g., P'jankova 1974: 176–180; 1978: 30–64; 1986: 41–50; 1989: 57–77; P'jankova *et al.* 2009: 104–108; Luneau *et al.* 2011: 286–297; Teufer 2015: 236–243; Teufer *et al.* 2015: 104–105).

The following classification for the "classical Vakhsh culture" includes five groups of shapes containing 23 types.

I. Beakers and pots

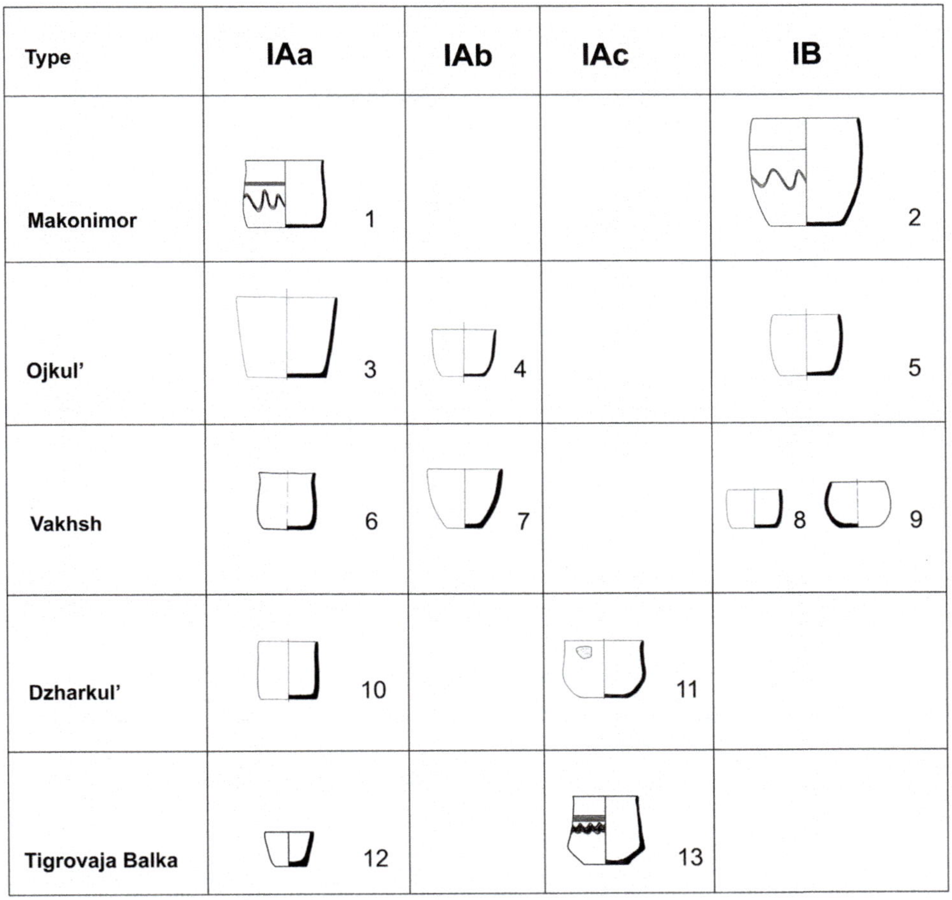

Figure 25.3 Pottery types of the "classical Vakhsh culture": beakers.

IAa Straight-sided beaker (Figure 25.3, *1, 3, 6, 10, 12*).
IAb Straight-sided beaker with slightly everted lip (Figure 25.3, *4, 7*).
IAc Beaker with elongated concave upper half and steep carinated lower half (Figure 25.3, *11, 13*).
IB Slightly globular beaker (Figure 25.3, *2, 5, 8, 9*).

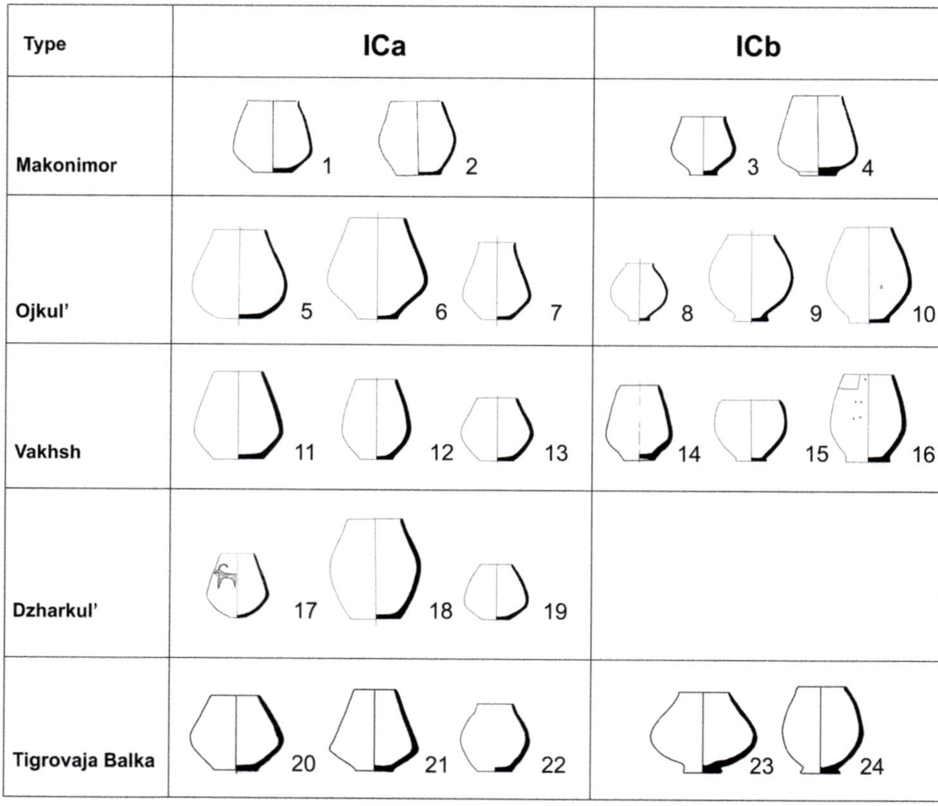

Figure 25.4 Pottery types of the "classical Vakhsh culture": beakers and pots.

ICa Carinated beakers and pots with or without slightly raised foot (Figure 25.4, *1*, *2*, *5–7*, *11–13*, *17–22*).

ICb Carinated beakers and pots with clear base ring (Figure 25.4, *3*, *4*, *8–10*, *14–16*, *23*, *24*).

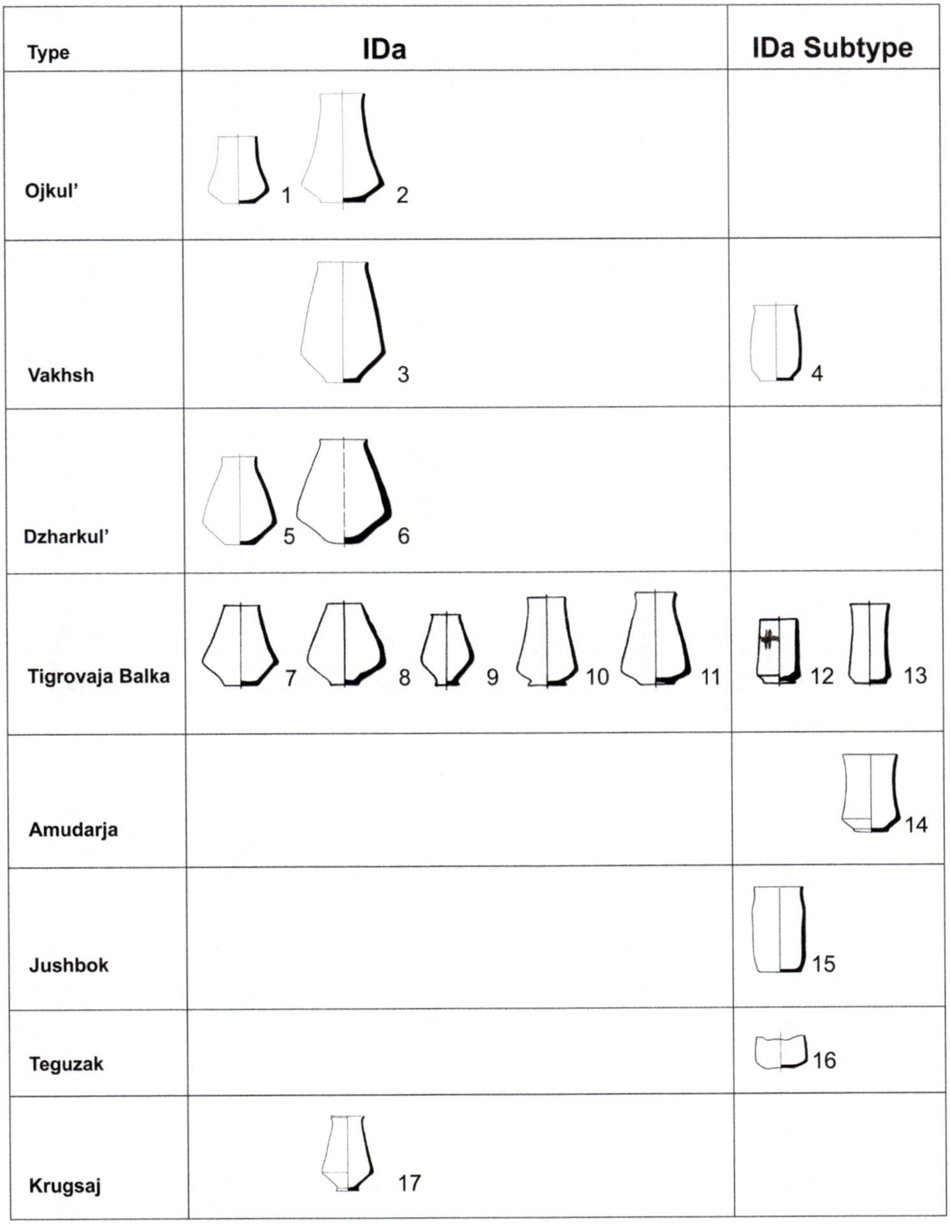

Figure 25.5 Pottery types of the “classical Vakhsh culture”: beakers.

IDa Biconical elongated beaker (Figure 25.5, *1–3*, *5–11*, *17*). Infrequently occurring subtypes can include an elongated beaker with straight or concave walls (Figure 25.5, 4, *12–16*).

IDb Elongated beaker (Figure 25.6, *1*, *10–14*).

IDc Elongated beaker with slightly or strongly everted rim and slightly globular lower half (Figure 25.6, 2–9, *15–17*).

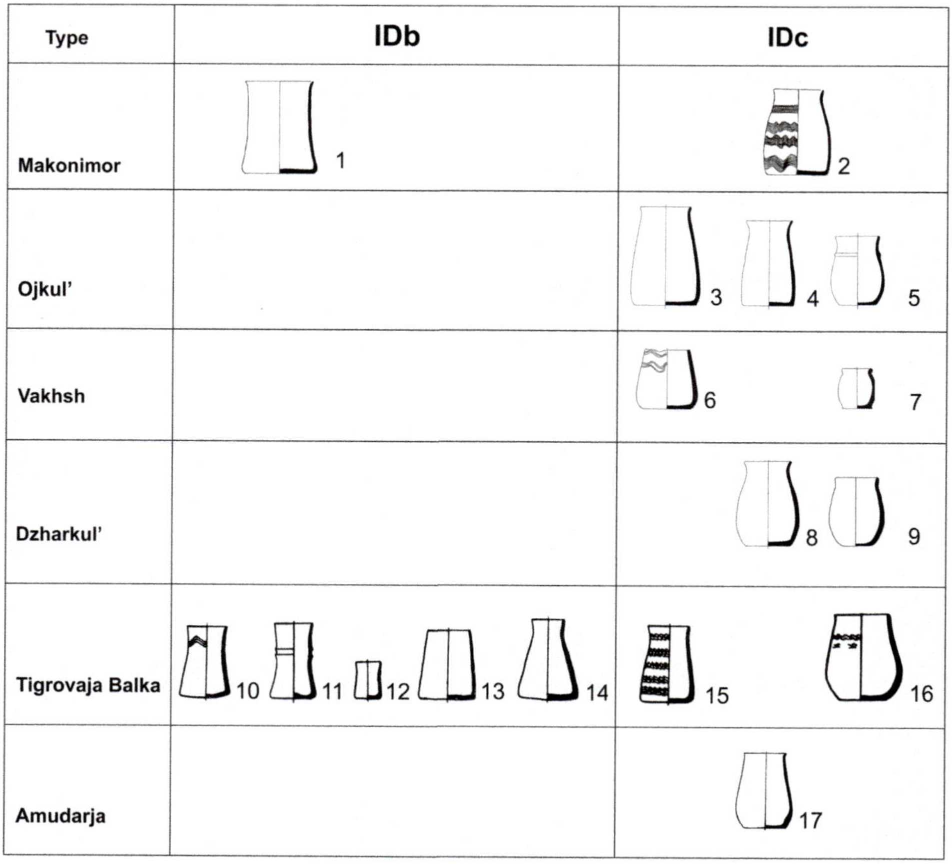

Figure 25.6 Pottery types of the "classical Vakhsh culture": beakers.

II. Jugs

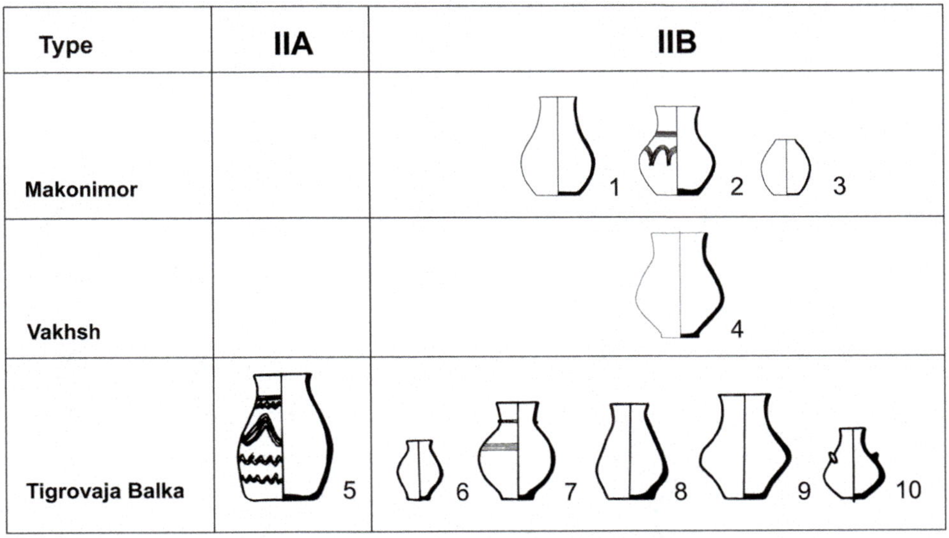

Figure 25.7 Pottery types of the "classical Vakhsh culture": jugs.

IIA Jugs without strong profile (Figure 25.7, *5*).
IIB Jugs with strong profile (Figure 25.7, *1–4*, *6–10*).

III. Globular pots

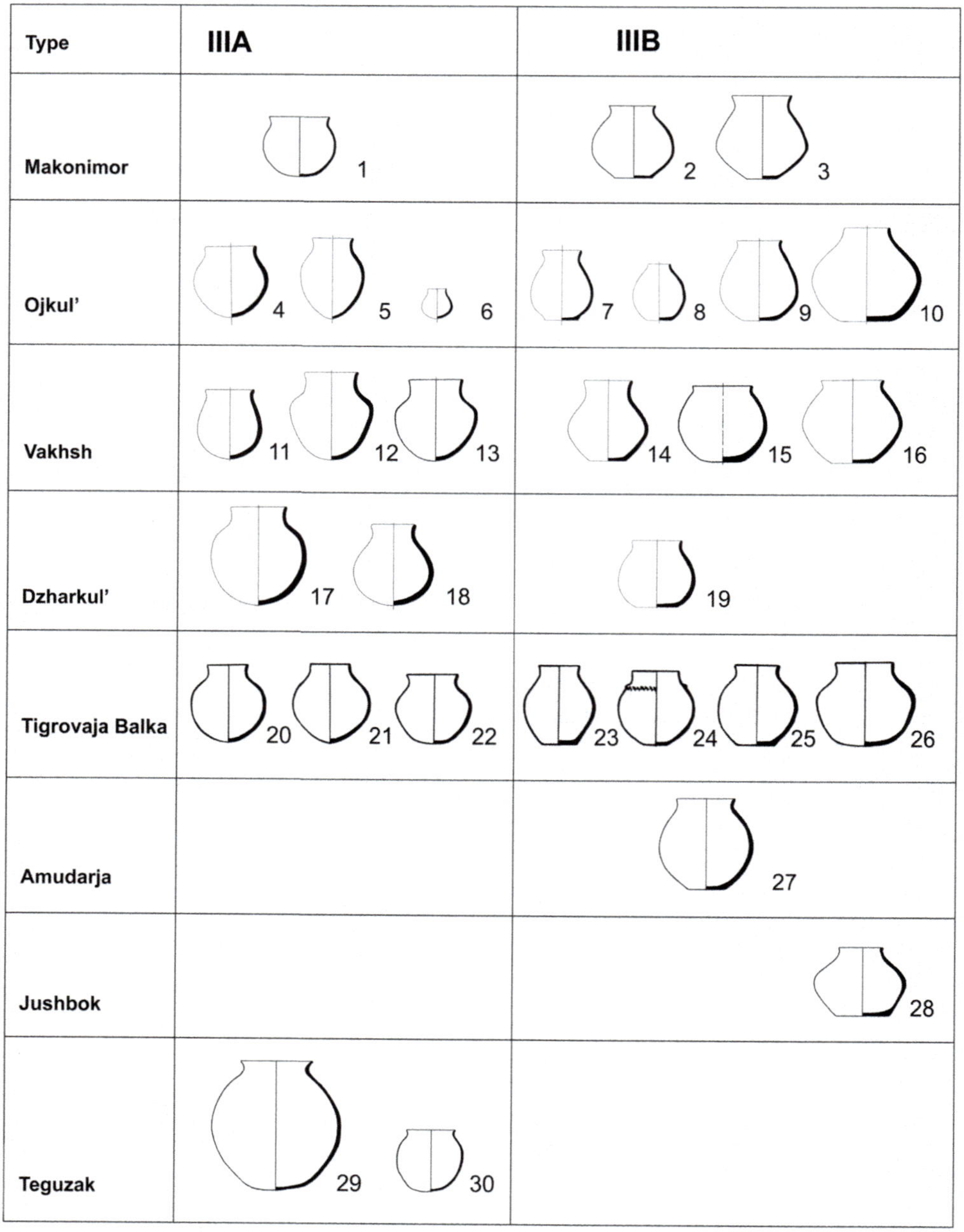

Figure 25.8 Pottery types of the "classical Vakhsh culture": globular pots.

IIIA Round-bottomed globular pots (Figure 25.8, *1*, *4–6*, *11–13*, *17*, *18*, *20–22*, *29*, *30*).

IIIB Flat-bottomed globular pots (Figure 25.8, *2*, *3*, *7–10*, *14–16*, *19*, *23–28*).

IV. Bowls

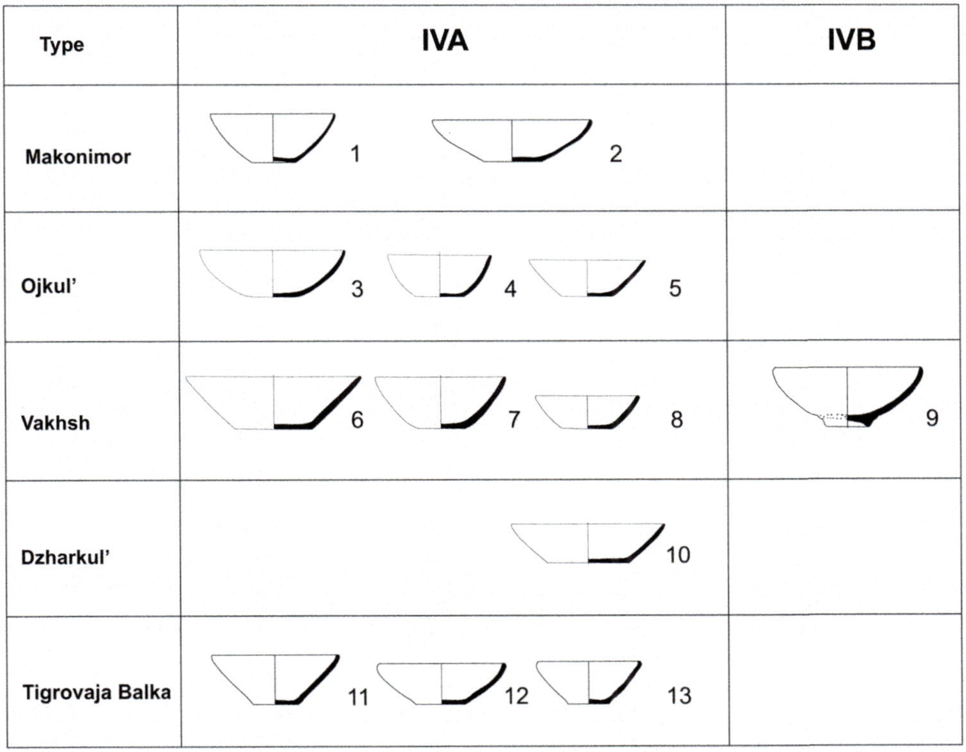

Figure 25.9 Pottery types of the "classical Vakhsh culture": bowls.

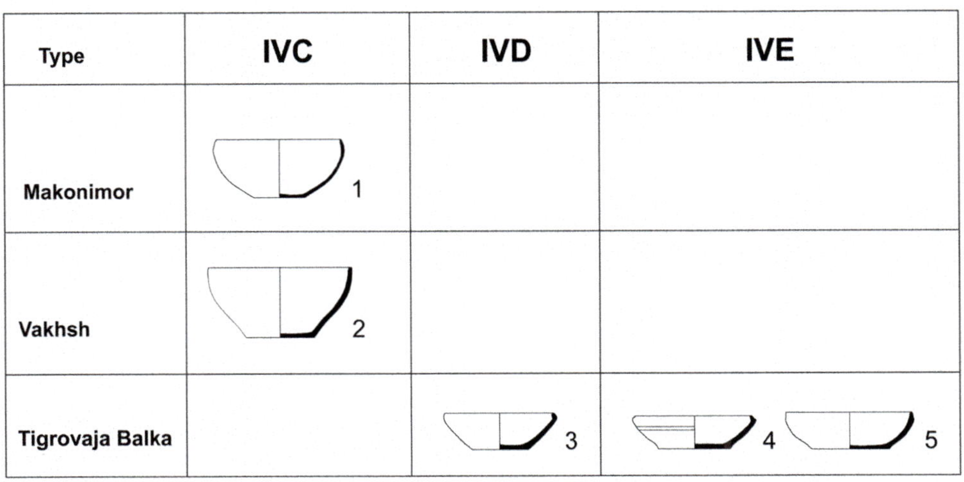

Figure 25.10 Pottery types of the "classical Vakhsh culture": bowls.

Type	IVF	IVG
Makonimor	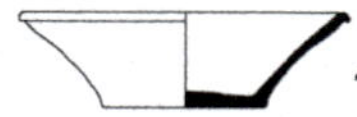 1	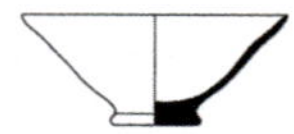 2
Tigrovaja Balka	 3	
Krugsaj		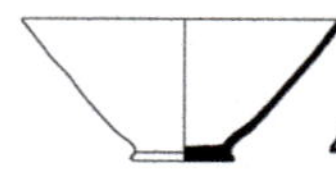 4

Figure 25.11 Pottery types of the "classical Vakhsh culture": bowls.

IVA Bowls with straight or slightly convex profiles without strong rim or with very slightly inverted rim (Figure 25.9, *1–8*, *10–13*).
IVB Slightly convex bowl with raised base (Figure 25.9, *9*).
IVC Bowls with very rounded profile and slightly narrower bottom half (Figure 25.10, *1*, *2*).
IVD Straight-walled bowl with inverted rim (Figure 25.10, *3*).
IVE Shallow bowl with slightly concave, s-shaped rim (Figure 25.10, *4*, *5*).
IVF Bowl with concave lower profile and strongly everted rim (Figure 25.11, *1*, *3*).
IVG Bowl with slanted profile and raised, flat base (Figure 25.11, *2*, *4*).

V. Goblets and pedestal bowls

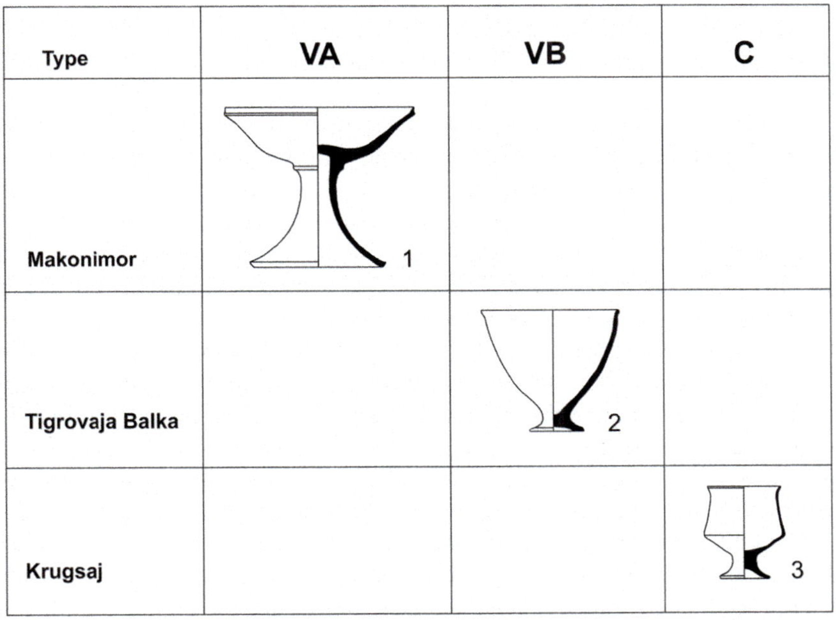

Figure 25.12 Pottery types of the "classical Vakhsh culture": goblets and pedestal bowls.

There are three single vessels in this group, making three different types. One is the pedestal bowl with ridged rim, similar to type IVE bowls. This type is designated as VA (Figure 25.12, *1*). The second type is a goblet composed of a large beaker on a short foot designated as type VB (Figure 25.12, *2*). The third is a goblet the archetype of which is known in the Sapalli-Dzharkutan and Dashly culture, which I have used elsewhere and designated as type C (Figure 25.12, *3*) (Teufer 2015: 35; 156, figs. 13 and 76). This designation will be maintained. This vessel from the "classical Vakhsh culture" is handmade, while the specimens from the Sapalli-Dzharkutan and Dashly cultures were wheel-thrown. The other examples (types VA and VB) were likewise wheel-thrown.

THE CHRONOLOGY OF THE "CLASSICAL VAKHSH CULTURE"

The recent research made at Farkhor, Gelot, and Darnajchi (see Chapter 23) has also substantial consequences for the chronological position of the "classical Vakhsh culture."

A combined table (Table 25.1) has been made for the grave finds from the neighboring cemeteries of Gelot and Darnajchi, and lists the individual graves (first column on the left) along with their grave goods. A seriation for them was excluded[8] because of the low number of burials: The small amount of statistically relevant vessels prevents the generation of a thorough separate type in almost each case. To illustrate this point further, I decided to work on a typology. The vessels in the individual columns corresponding to a "type" actually represent a wide spectrum of features. This applies especially to the goblets (fifth column from the left), where, instead of assigning numbers to the inventory, I rather chose to provide images of these vessels that externalize this vast spectrum of features. In the combined table, the vessels made by hand are represented by half-shaded icons and the wheel-thrown ones by non-shaded icons. This table highlights a diagonal that – due to the reasons just mentioned – at best allows us to extrapolate a development tendency. This is, in fact, confirmed by three radiocarbon dates that underline a progressive development (right-hand column of the table). The oldest date is between 2333 and 2153 (cal. BC 1σ), or 2456 and 2140 (cal. BC 2σ). Two other dates are located at the bottom end of the table, one of which lies between 2128 and 1981 (cal. BC 1σ), or between 2135 and 1965 (cal. BC 2σ), the second between 2126 and 1982 (cal. BC 1σ), or between 2134 and 1966 (cal. BC 2σ)[9] (see also the Appendix at the end of this volume). This gives us an absolute timespan between the 23rd and 22nd centuries BC and the 21st century BC. The individual grave finds are placed within this interval, allowing for the following description from a rough pottery development based on the simplest, mostly handmade larger pots and storage vessels, or bowls and jug-like containers, goblets, and carinated vessels, up to exclusively wheel-thrown bowls, pots, and storage vessels.

A probable use of the Gelot and Darnajchi cemeteries from the 23rd to 22nd centuries to the 21st century BC can be deduced from the ^{14}C dates alone. But these cemeteries continued to be used up to the beginning of the 2nd millennium BC as shown by the youngest grave in the sequence (i.e., grave 2 from necropolis 4 at Gelot, bottom-most row on Table 25.1). In this grave, the ceramic inventory was exclusively made of wheel-thrown pottery, including a goblet (Figure 25.13, 2); a globular pot with a spout, a broad rim, and narrow base (Figure 25.13, 4); a storage vessel (Figure 25.13, *1*); and a globular pot (Figure 25.13, *3*). Parallels for many of the vessels can be found in the Sapalli culture: the globular pot with spout is present at Tilla Bulak (Kaniuth 2009: fig. 12; 2010: fig. 9) (Figure 25.13, *8*, *12*); the globular pot is known at Dzharkutan, necropolis 3, grave 28 (Teufer 2015: pl. 226, *6*) (Figure 25.13, *7*); and the goblet at Dzharkutan, necropolis 3, grave 18 (Teufer 2015: pl. 215, *5*) (Figure 25.13, *11*). The parallels mentioned with the Sapalli culture all date to stage I of the Late Bronze Age, and therefore already to the early 2nd millennium, as will be developed in more detail later in the chapter. Similar goblets can be found in southern Bactria (see p. 721) (Figure 25.13, *5*, *6*, *9*, *10*).

Table 25.1 Sequence of the graves from Gelot and Darnajchi. D – Darnajchi; G – Gelot; N – Necropolis; G – Grave. Vessels without shading: wheel-thrown vessels; vessels with shading: handmade vessels

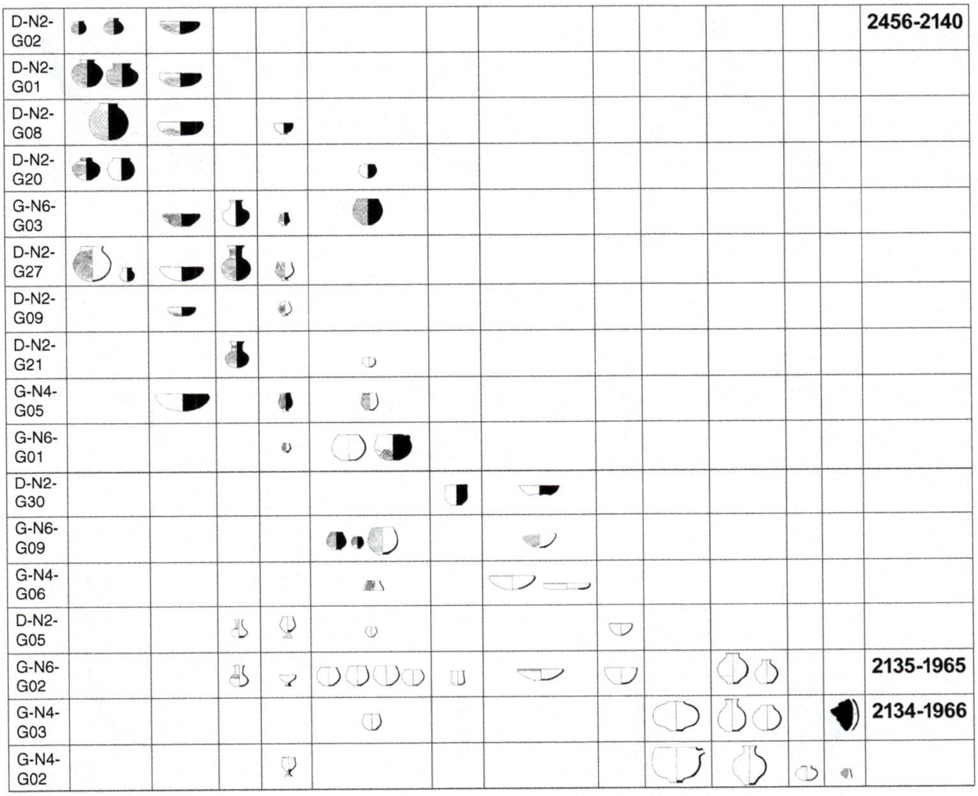

Carinated biconical vessels are known during the Gelot–Darnajchi sequence, identified by ^{14}C dates (Table 25.1, sixth column from the left), and can be related with the ICa and ICb types from the Vakhsh burials in the lower Vakhsh valley (see Figure 25.21). This underlines the relevance of the Gelot–Darnajchi sequence for the chronology of these Vakhsh burials. Furthermore, grave 2 at Darnajchi necropolis 2, is dated to the beginning of the Gelot–Darnajchi sequence. This grave contains a pottery spectrum with solid parallels in the "classical Vakhsh culture," including two handmade, round-bottomed pots (type IIIA) (Figure 25.14, *2*, *3*) and another handmade bowl with inverted rim (type IVD) (Figure 25.14, *1*). The same kind of round-bottomed pot is known at sites of the "classical Vakhsh culture" (Figure 25.8, *1*, *4–6*, *11–13*, *17*, *18*, *20–22*, *29*, *30*). The bowl finds comparisons at Tigrovaja Balka (Figure 25.10, *3*). As mentioned above, a ^{14}C date is provided for this burial.

A further grave with typical Vakhsh pottery is grave 13 from Gelot, necropolis 6, not listed in Table 25.1, dated to 2197–2049 cal. BC (1σ) or 2203–2036 cal. BC (2σ). It contained a round-bottomed pot of type IIIA (Figure 25.15, *3*), a flat-bottomed pot of type IIIB (Figure 25.15, *1*), a bowl of type IVG (Figure 25.15, *3*), and a carinated

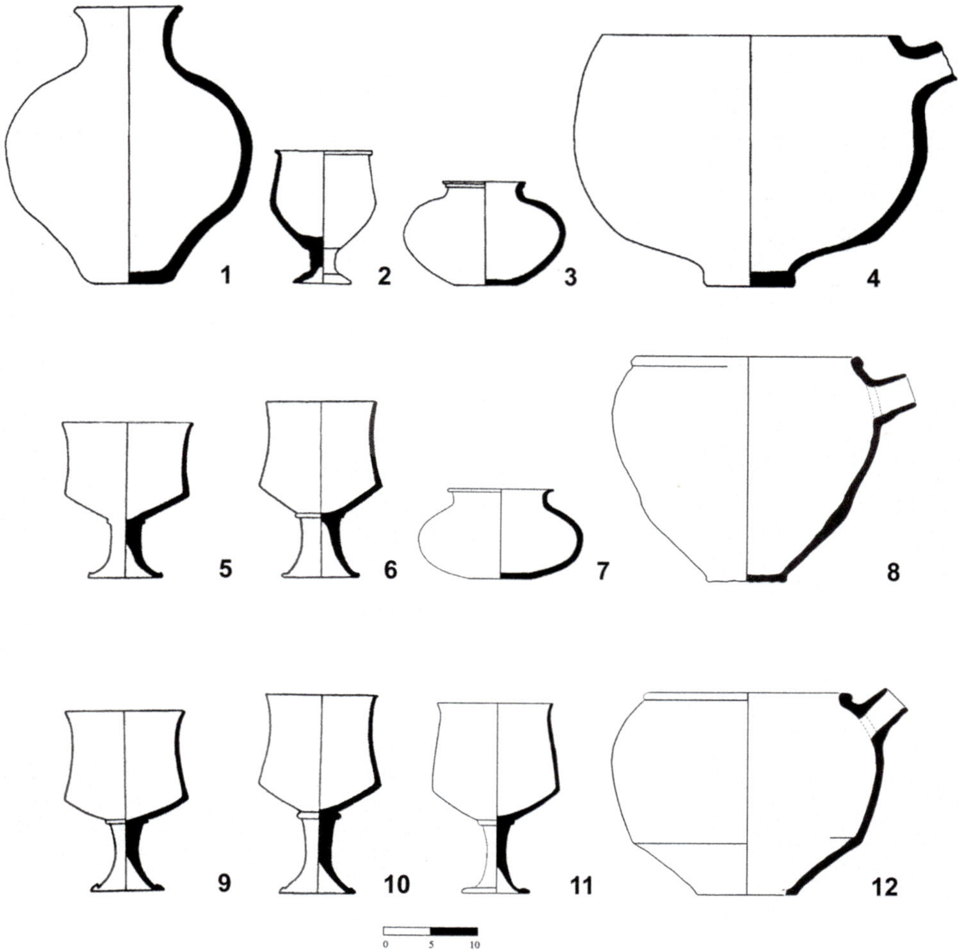

Figure 25.13 Pottery inventory from grave 2 from Gelot, necropolis 2, and its parallels from the Sapalli-Dzharkutan and Dashly cultures. 1–4: Gelot, Necropolis 2; 5: Dashly 3, grave 34; 6: Dashly 3, grave 44; 7: Dzharkutan N3, grave 28; 8, 12: Tilla Bulak; 9: Dashly 3, grave 47; 10: Dashly 3, grave 53; 11: Dzharkutan N3, grave 18 (8 after Kaniuth 2009; 12 after Kaniuth 2010; 5. 6. 9. 10 after Sarianidi 1976).

beaker with or without a slightly raised foot of type ICa (Figure 25.15, 2). All these vessel types have parallels in the pottery finds of the kurgans from the "classical Vakhsh culture."

It is therefore evident that the Vakhsh culture is more likely a phenomenon of the 3rd millennium than of the 2nd millennium BC.

However, unlike the "classical Vakhsh graves," the tombs at Gelot and Darnajchi have no visible aboveground, kurgan-like construction and should therefore be distinguished from them. The underground part of the tomb is nevertheless similar – namely, a simple pit (Gelot, necropolis 6, grave 13) or catacomb (Darnajchi, necropolis 2, grave 2).

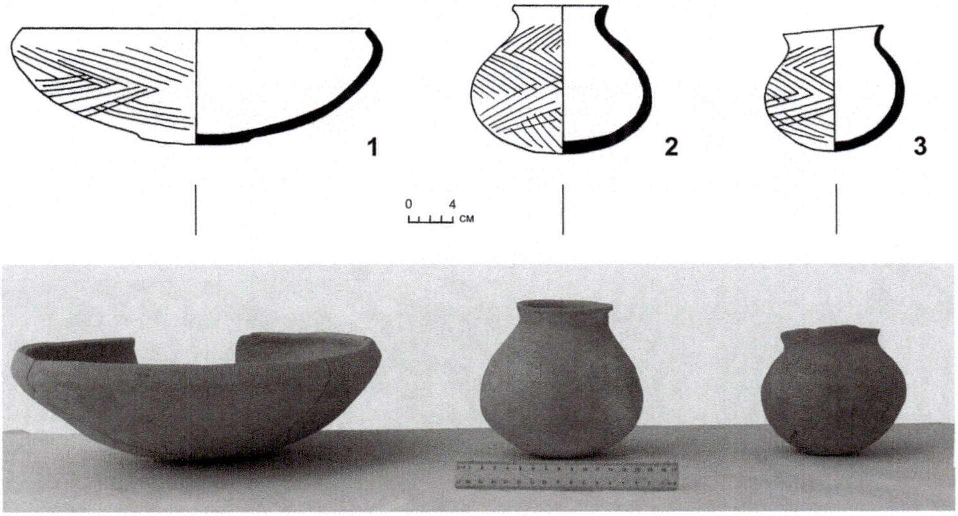

Figure 25.14 Ceramic inventory from Darnajchi, necropolis 2, grave 2.

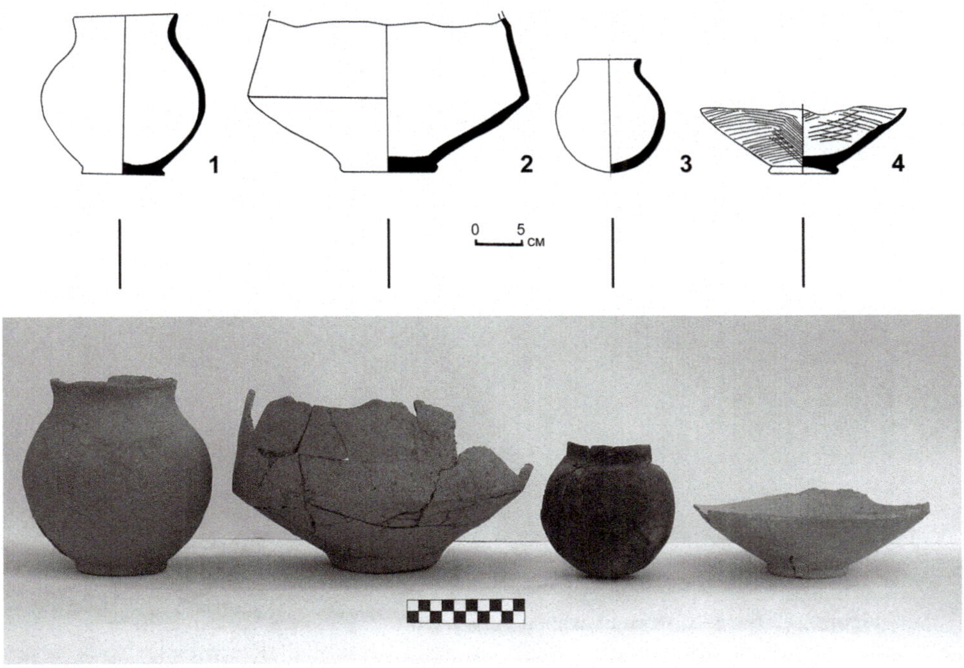

Figure 25.15 Ceramic inventory from Gelot, necropolis 6, grave 13.

The excavations at Farkhor (cf. Chapter 23) offer further evidence for the early dating of the Vakhsh culture. Grave 18 was not only rich in metal and stone objects (Figure 25.16, *5–14*), but also contained two pots, among which a straight-walled beaker of type IAa (Figure 25.16, *4*), typical of the "classical Vakhsh culture"

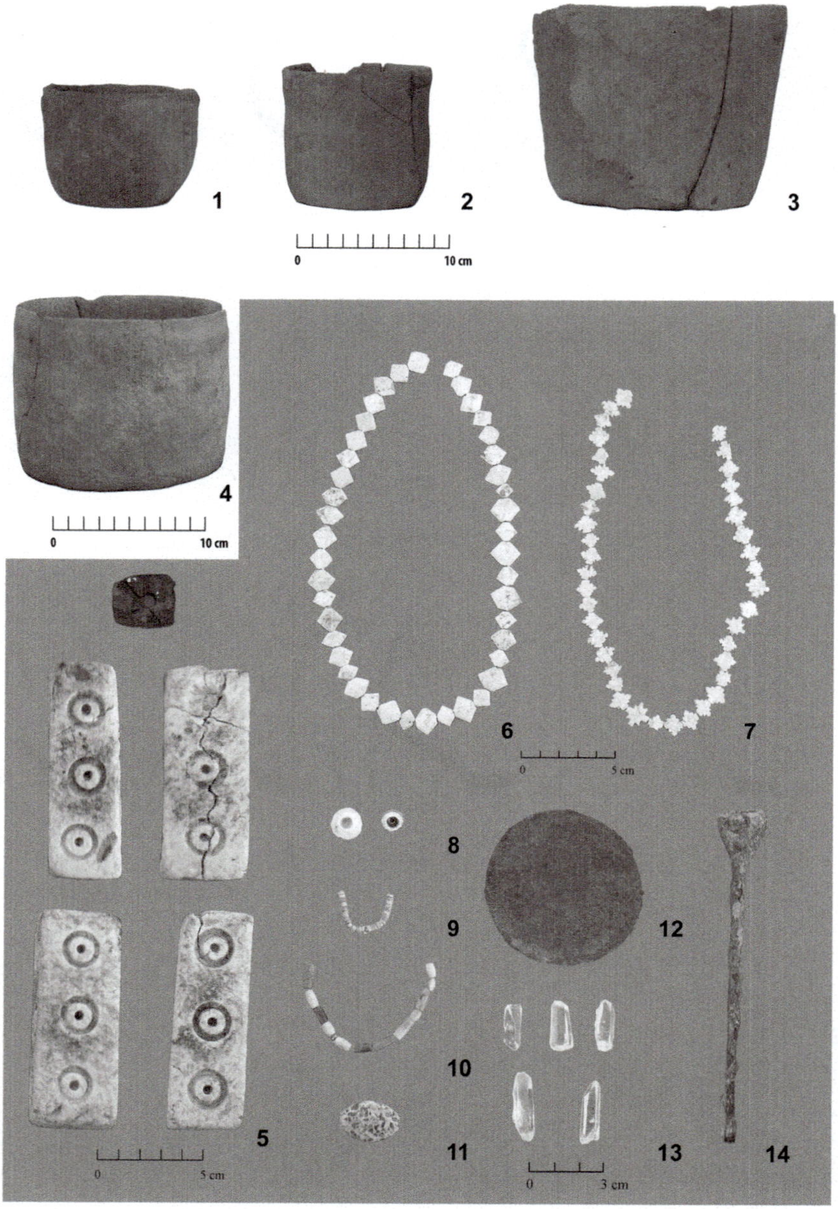

Figure 25.16 1–4, straight-walled cups from the "classical Vakhsh culture": 1: Ojkul', Kurgan 4; 2: Ojkul', Kurgan 46; 3: Ojkul', Kurgan 35; 4–14, inventory from Farkhor, necropolis 6A, grave 18.

(Figures 25.3, *1*, *3*, *6*, *10*, *12*; 25.16, *1–3*). Grave 18 should be dated to the transition from Early (NMG IV) to Middle Bronze Age (NMG V), as mentioned above (Teufer *et al.* 2015: 112–113).

It is now possible to establish a sequence for the cemeteries of the lower Vakhsh valley based on the ^{14}C dates from Gelot and Darnajchi and the parallels with

Farkhor. The sequence is founded on the already published graves from Tigrovaja-Balka and Dzharkul' and on the material from the burial grounds from Ojkul' and Vakhsh (publication in preparation). For this, we also use a combined table (Table 25.2). Seriation was, again, excluded as not only the number of graves but also the number of types that can be used is insufficient. This meant we could record only part of the types in the table. The sequence progresses from the top left to the bottom right. Rows within the table indicate individual graves. Because two of the cemeteries have not yet been published, the inventory of the individual graves has no number. Again, the handmade vessels are represented by half-shaded icons and the wheel-thrown ones by non-shaded icons. The beginning of the sequence shows straight-walled vessels of type IAa (Figure 25.16, *1*, *2*, *3*), which are similar to those at Farkhor (Figure 25.16, *4*) (cf. Teufer *et al.* 2015: 138, fig. 24, *1*) and Zamanbaba (Guljamov *et al.* 1966: pl. 15, *9*). The spectrum of shapes then consists of a few carinated vessels (types ICa and ICb), elongated beakers (type IDb), round-bottomed (type IIIA) and flat-bottomed pots (type IIIB), as well as simple bowls (type IVA). There are clear comparisons with the pottery material from Zamanbaba; this applies especially to the shapes of the pots.[10] At the end of the sequence, biconical elongated beakers (type IDa) and elongated beakers with slightly or strongly everted rim and slightly globular lower half (type IDc) are present. The carinated vessels (types ICa and ICb) (Table 25.2, cols. 3 and 4) in particular demonstrate connections with the Gelot–Darnajchi sequence (Table 25.1, col. 6). Here we can see a change from carinated vessels (types ICa and ICb) in favor of biconical, elongated beakers (type IDa). This could possibly indicate a gradual change in form between the two types ICa and IDa. The elongated beakers with slightly or strongly everted rim and slightly globular lower half (type IDc) and the carinated vessels (types ICa and ICb) are mutually exclusive. The elongated beakers (type IDb) are the forerunners of the elongated beakers with slightly or strongly everted rim and slightly globular lower half (type IDc). This underlines a dichotomy that runs through the table via types IIIA and IIIB as well as type IVA – which perhaps appeared later and took the place of the carinated vessels of type ICb.

An older spectrum includes types IAa, ICa, ICb, and IDb. According to the abovementioned parallels with grave 18 at Farkhor, type IAa could have appeared already in the middle of the 3rd millennium (transition NMG IV–V). As we have already said, analogies for types ICa and ICb can be found in the Gelot–Darnajchi sequence, pointing at a time period between the 23rd/22nd to the 21st centuries BC. A younger spectrum includes types IDa and IDc. It is not yet possible to determine the chronological transition more accurately. The absolute date of grave 13 at Gelot (2197–2049 cal. BC 1σ or 2203–2036 cal. BC 2σ), the inventory of which tends to point to the older section, nevertheless shows clearly that it extends until the end of the 3rd millennium.

Types IDa, IDb, and IDc are not found at Gelot and Darnajchi. Except for vessels of types ICa and ICb (see Figure 25.21), there are otherwise barely any overlap in the spectrum of pottery types of the Vakhsh burials and the burials from Gelot and Darnajchi. It is very likely that this is due to the fact that the Vakhsh burials are a local phenomenon while the graves from Gelot and Darnajchi display a stronger influence from the western oases. The wheel-thrown pottery – strongly reminiscent of the Sapalli-Dzarkutan and Dashly cultures – is especially dominant at the end of

Table 25.2 Sequence of the Vakhsh graves from the lower Vakhsh valley. Dz – Dzharkul'; Oj – Oikul'; TB – Tigrovaja Balka; Va – Vakhsh; K – Kurgan. Vessels without shading: wheel-thrown vessels; vessels with shading: handmade vessels

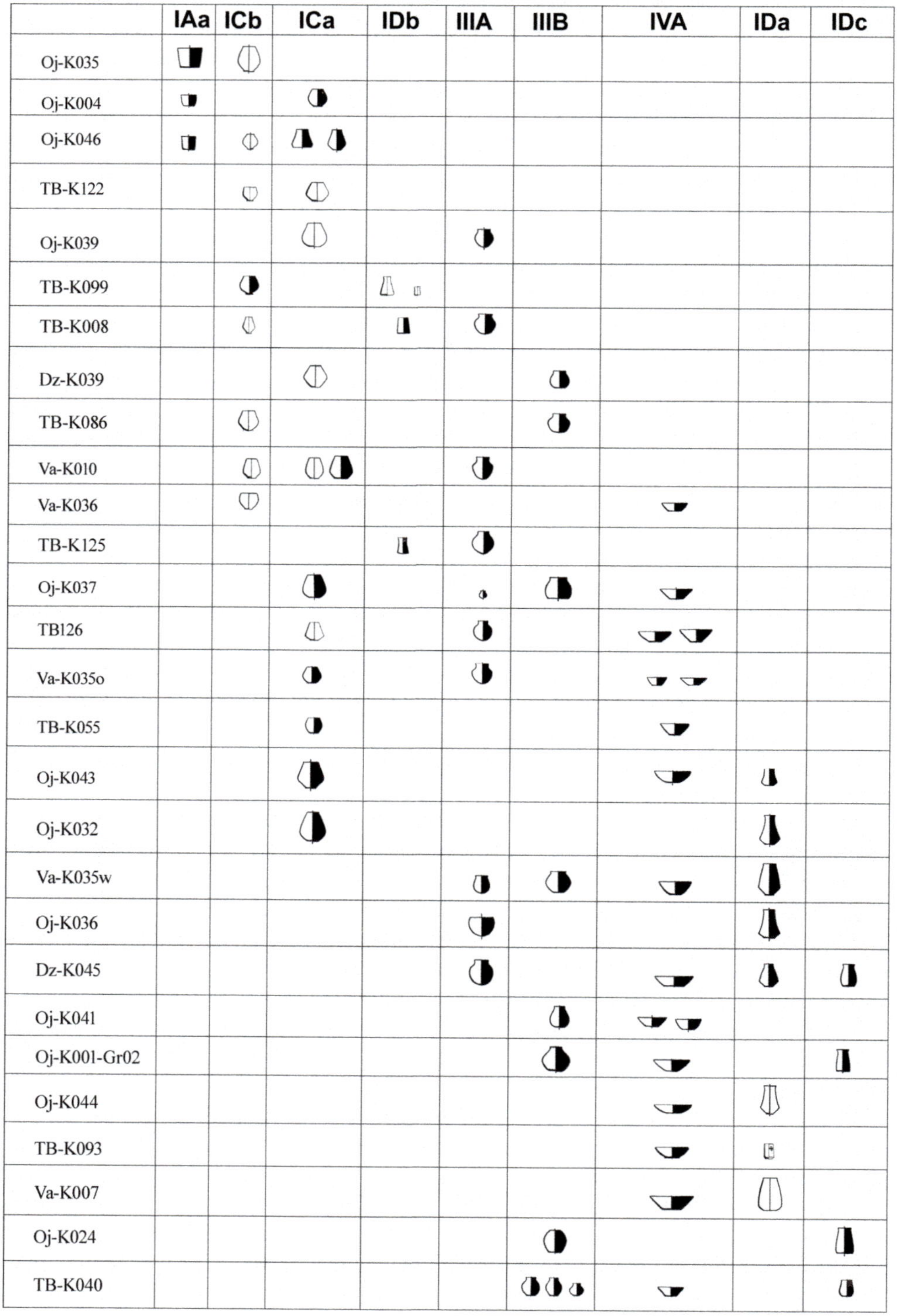

the Gelot–Darnajchi sequence as shown by the youngest grave of the sequence (grave 2 from necropolis 4 in Gelot) (Figure 25.13, *1–4*).

Here, I would like to come back again to a goblet discovered in this last grave (Figure 25.13, *2*), as it can possibly help to determine the end of the Vakhsh burial sequence in the lower Vakhsh valley. This goblet can be ascribed to a type also found in northern and southern Bactria, which I have classified as type C. This type is known in grave 18 of necropolis 3 at Dzharkutan (Figure 25.13, *11*). In the sequence made for northern Bactria created with the use of a combined table (Teufer 2015: 665–667), this grave is placed in the Late Bronze Age Ib phase (Teufer 2015: 666). Here, we are thus dealing with the younger phase of an older stage of the Late Bronze Age (LtB I) in northern Bactria (for more on this period, cf. Teufer 2015: 53–54). In absolute chronology, this stage encompasses the approximate timespan of 2000 to 1700 BC (cf. the absolute chronology in Kaniuth 2006: 48).

Other goblets of this type are known in southern Bactria at Dashly 3, graves 22, 23, 31, 34, 44, 47, and 53 (Sarianidi 1976: fig. 37, *4*, *9*; 39, *20*; 40, *13*; 43, *19*; 45, *10*; 46, *1*) (Figure 25.13, *5*, *6*, *9*, *10*). Thanks to a combined table, a grave sequence for southern Bactria was established, in which these graves belong to groups 2 and 3 (Teufer 2015: 670). Group 2 is related with phase Ib in northern Bactria, and group 3 dates to the transition from Ib to IIa in the same region.[11] Thus, the end of the occurrence of these goblets should be in the 17th century BC, if we assume a transition from Late Bronze Age Ib to IIa to have taken place during the 17th century BC (cf. Kaniuth 2006: 48).

To determine the end of the Vakhsh burial sequence in the lower Vakhsh valley, the most importance find is the discovery of one of these goblets in grave 3 at Kumsaj, which may possibly be ascribed to the "classical Vakhsh culture" (Figures 25.12, *3*; 25.17, *2*). This grave also contained a biconical elongated beaker (type IDa) (Figure 25.17, *3*), which points to the end of the sequence of the Vakhsh culture.

Biconical elongated beakers of type IDa are also present in both northern and southern Bactria (beaker type A2; Teufer 2015: 36, fig. 14; 157, fig. 77) and date to phase IIa, or group 3 (Teufer 2015: 666, 670). However, the carination of the beakers (beaker type A2) is not as strong on those from northern Bactria as it is on the biconical elongated ones of type IDa (Figure 25.18).

We must therefore assume an end for the Vakhsh culture in the lower Vakhsh valley at the beginning of Late Bronze Age IIa. Grave 3 from Kumsaj in southern Tajikistan contains a typical assemblage of vessels from this transition period from Ib to IIa during the 17th century BC.

This late end is also related to the find of a bronze dagger in Kurgan 7 at Tigrovaja Balka (Figure 25.19, *1*). The edges of the handle are extended in the cross-section. This may explain why P'jankova connected this dagger with those from the Near East and the Caucasus, or with the so-called "daggers of Near Eastern type," which date to the second half of the 2nd millennium BC (P'jankova 1989: 81–82). In fact, there are significant differences between these daggers – also referred to as "*Randgriffdolche*" (Figure 25.19, *3–12*)[12] – and the specimen from Tigrovaja Balka, as P'jankova also notes. Daggers with wide handle edges already appear in the Near East during the Old Babylonian period, as clearly shown by a specimen from Tall as-Sulaima, but that also has additional rivet holes (Müller-Karpe 2004: pl. 46, *771*)

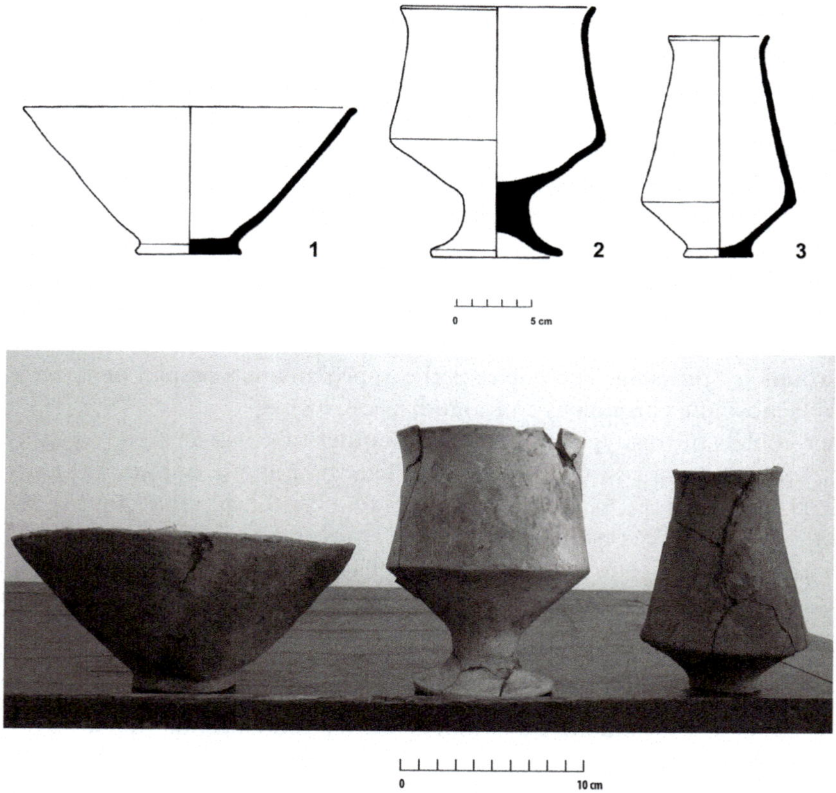

Figure 25.17 Pottery inventory from Krugsaj, grave 3.

(Figure 25.19, 2).[13] A dating of the dagger from Kurgan 7 at Tigrovaja Balka around the 18th/17th century BC is therefore quite possible.

Altogether, we can determine the beginning of the Vakhsh culture to have occurred in the middle of the 3rd millennium due to the connection given by the straight-walled beaker from grave 18 at Farkhor. The end falls presumably in the 17th century BC based on the connections with grave 3 from Kumsaj and its correlation with the sequence of northern and southern Bactria. Both radiocarbon dates from the graves containing Vakhsh pottery at Gelot and Darnajchi can be placed in the second half of the 3rd millennium (Figure 25.20). The vessels of types ICa and ICb are also associated with the Gelot–Darnajchi grave sequence of the older stage of the Vakhsh culture (23rd/22nd–21st centuries BC).

Older radiocarbon dates for the Vakhsh culture of the lower Vakhsh valley come from Tigrovaja Balka. These dates were, so far, considered not usable due to their large range (P'jankova 1986: 58; Kaniuth 2006: 51). This was due to the fact that none of the dates belonged to the presumed time frame of the Late Bronze Age. A reevaluation of the chronology of the Vakhsh culture shows that only one of the five dates (LE-1407: 1070–830 cal. BC 2 σ) is divergent. Three of the dates, despite their very wide range, fall into the 3rd millennium with fluctuations even into the 4th millennium (GIN-2524: 2570–2200 cal. BC 1σ; 2900–2000 cal. BC 2σ; GIN 2525: 3650–2700 cal. BC 1σ; 4000–2200 cal. BC 2σ; GIN 2526: 2850–2400 cal. BC

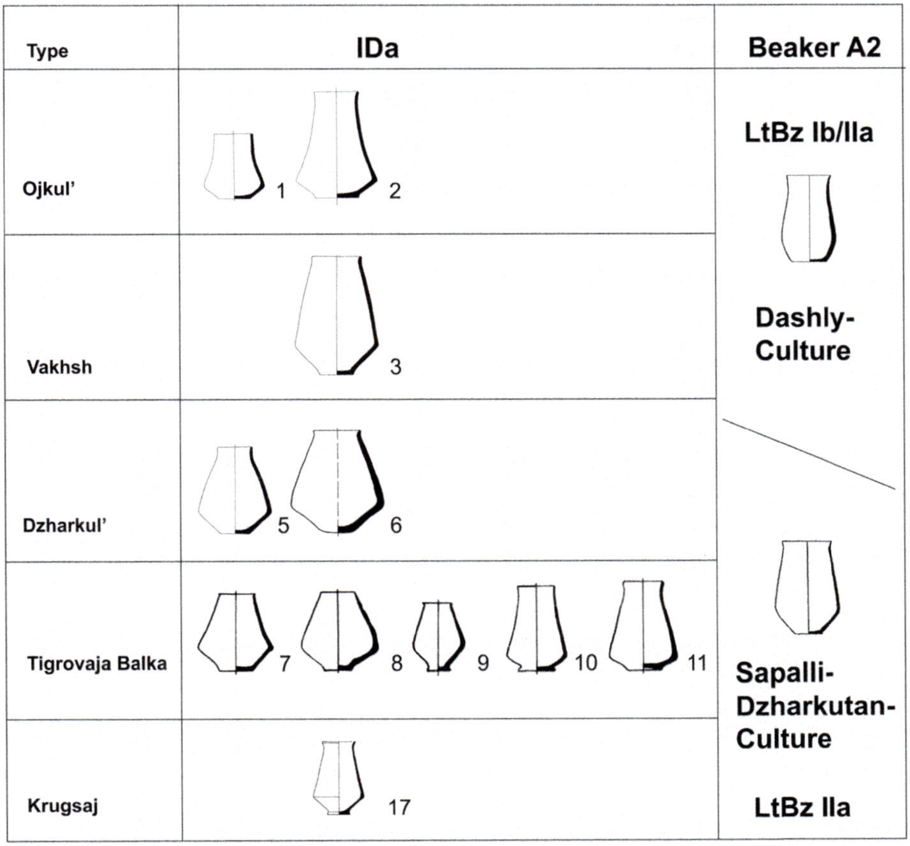

Figure 25.18 Beakers of type IDa and similar beaker of type A2 from the Sapalli-Dzharkutan and Dashly cultures.

1σ; 2900–2300 cal. BC 2σ). These three samples were taken from the hearths that surround Kurgan 8 at Tigrovaja Balka. A fourth sample comes from a hearth around Kurgan 7 in the same cemetery and dates between 1730–1520 cal. BC 1σ or 1860–1590 cal. BC 2σ[14] and therefore belongs to the LBA Ib/LBA IIa (Teufer 2015: 52–57; 256, tab. 43) (see the Appendix at the end of the volume). The speculated date mentioned above for the dagger from Kurgan 7 would be well in accordance with this ^{14}C date.

Assuming that the Vakhsh culture starts perhaps with NMG V and lasts until the LBA Ib – a period between ca. the 25th and the 17th centuries – there would be a correspondence with these ^{14}C dates, although this would only be general. The majority of the dates are not against an earlier start of the Vakhsh culture.

Therefore, it is not surprising that, at Tashguzor, so far the only settlement with Vakhsh pottery, vessel shapes have been found that can be compared with those from the Zamanbaba settlement.[15] This site, and the graves around it, can securely be dated to the 3rd millennium, as was stated above. The stone weight found in it (pit house 2 of Area 6 from Tashguzor; Vinogradova 1999: fig. 5, *11*) points also more likely to the 3rd millennium.

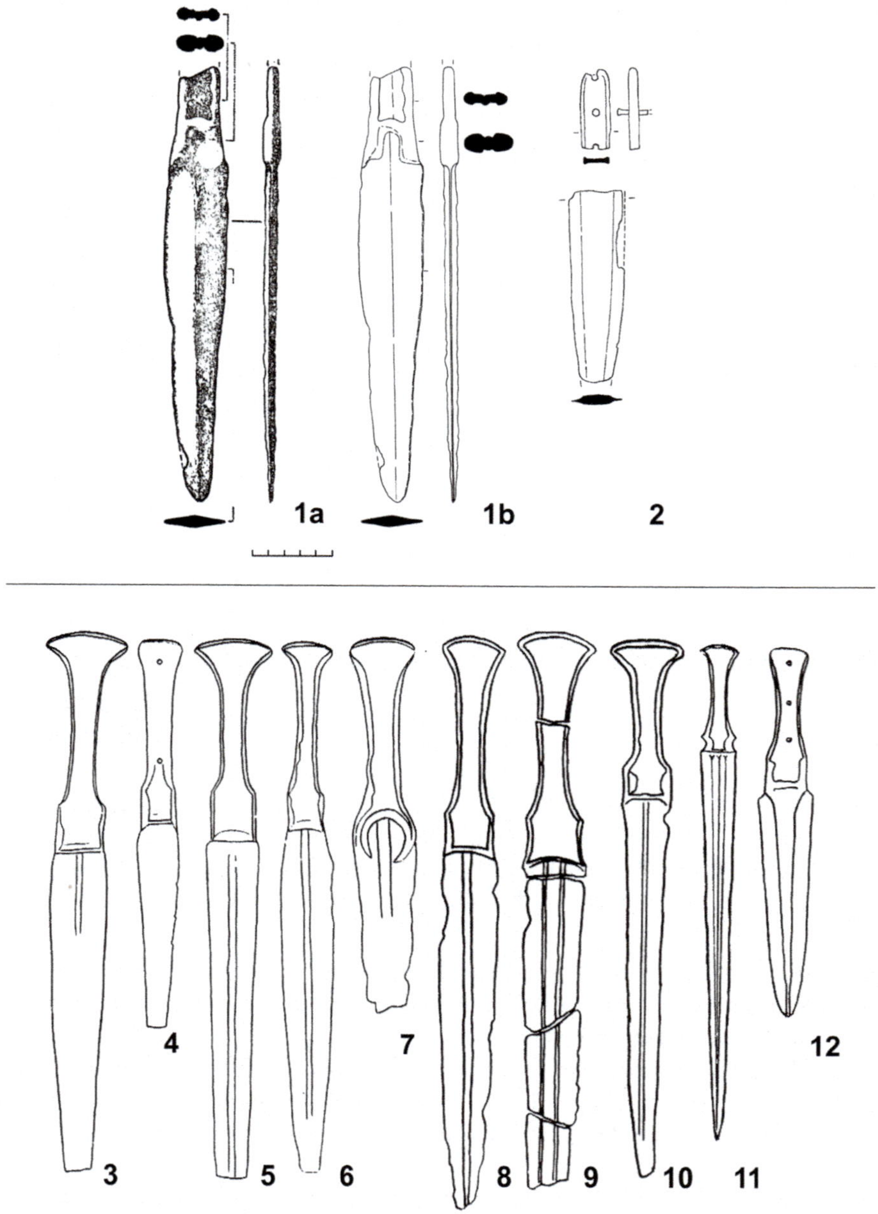

Figure 25.19 Daggers: 1: Tigrovaja Balka, Kurgan 7; 2: Tall as-Sulaima; 3–12: Collection of "Randgriffdolche" (after Motzenbäcker 1996) – 3: Faskau; 4: Kumbulte; 5: Bogvi; 6: Kyzyl Vank; 7: Kedabek; 8–10: "Armenia"; 11: Artik; 12: Arich.

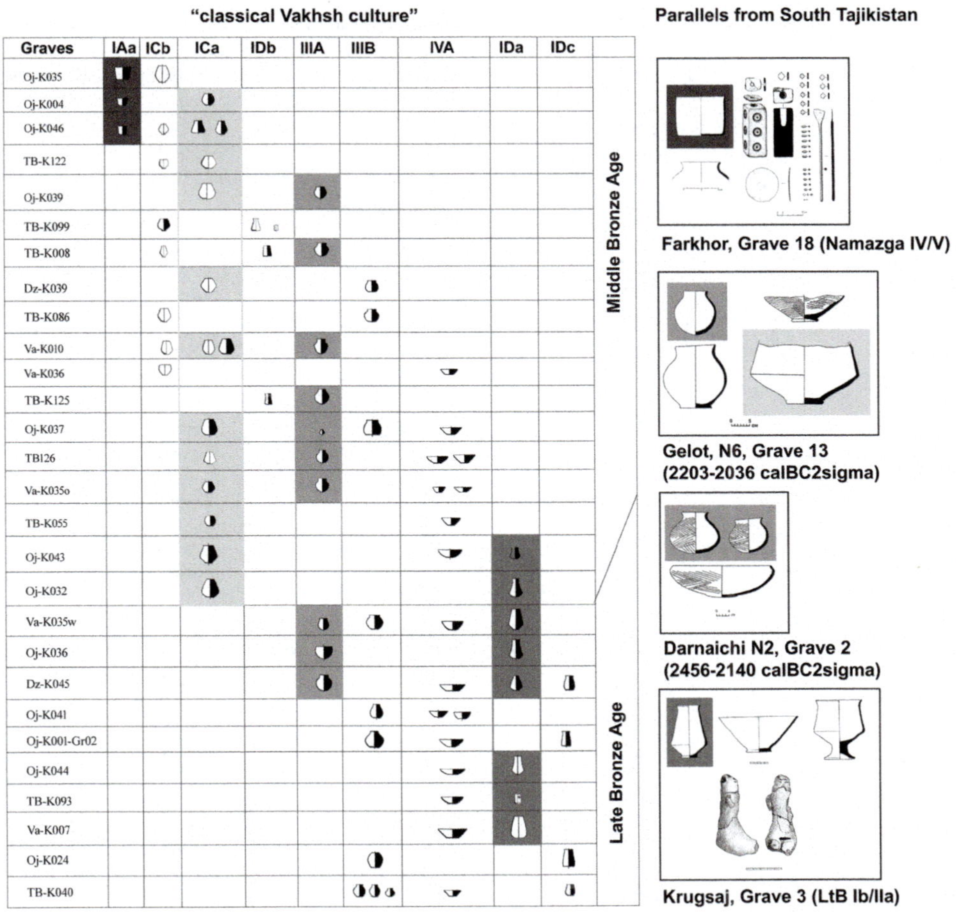

Figure 25.20 Sequence of the Vakhsh graves from the lower Vakhsh valley and dated grave finds from southern Tajikistan.

CONNECTIONS BETWEEN THE "CLASSICAL VAKHSH CULTURE" AND THE OASES CULTURES OF WESTERN CENTRAL ASIA

Relations between the Vakhsh pottery and some of the pottery shapes of the Middle Bronze Age in Margiana are also obvious. This applies especially to the carinated vessels of type ICa and ICb. They are closely related to the beakers that I have classified as type 3 from Margiana, which date to the beginning of the Bronze Age sequence there (Figure 25.21).[16]

The presence of vessel shapes in the Herat Museum with clear parallels to those of the Vakhsh culture in northern Afghanistan (Franke 2016: 54–55, cat. nos. 114–116, 118) and the discovery of Vakhsh pottery in the Dashly oasis (Sarianidi 1977: 69, fig. 31, 3, 8, 9) underlines a network connecting the oases cultures of western Central Asia with southern Tajikistan. Here it is striking that the Margiana biconical beakers

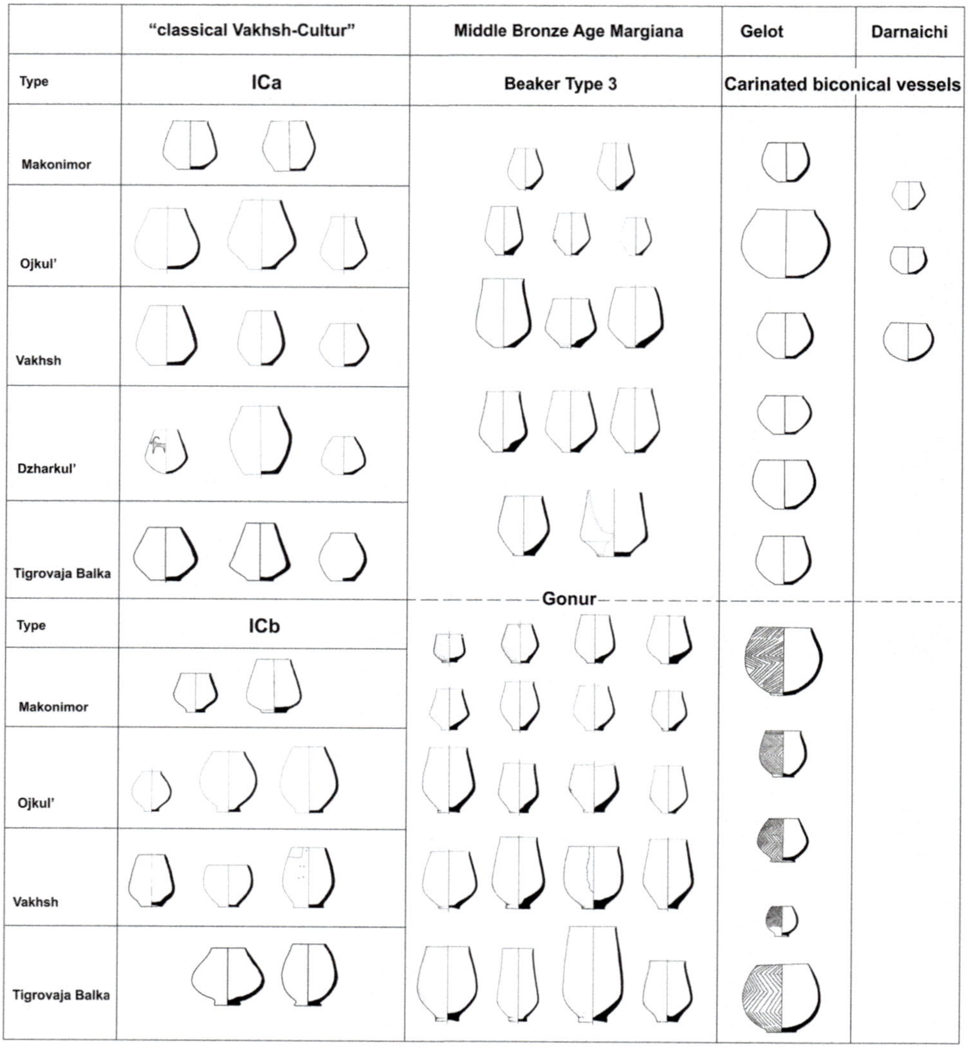

Figure 25.21 Older beakers of types ICa and ICb compared to the carinated biconical vessels from Gelot and Darnajchi and the beakers of type 3 from the Middle Bronze Age in Margiana (the examples of beaker type 3 from Margiana originate from the Gonur Necropolis, excavated by Salvatori: grave 1, 2, 5, 9, 12, 18, 21, 40, 42, 72, 107, 300, 84, 91, 404, 410, 412, 414, 417, 419, 427, 430, 432, 500 [Salvatori 1993, 1994, 1995]).

(type 3) are typologically closer to the "classical Vakhsh culture" beakers (types ICa and ICb) than to those from Gelot and Darnajchi (Figure 25.21).

As for the younger vessel types (types IDa and IDc), we have already referred to similarities between the northern and southern Bactria beakers (type A2) and the biconical elongated beakers of type IDa (Figure 25.18). Typological analogies can also be found in the oases cultures of western Central Asia for the second vessel type of the younger phase – beaker type IDc. These vessels occur at the Gonur Large

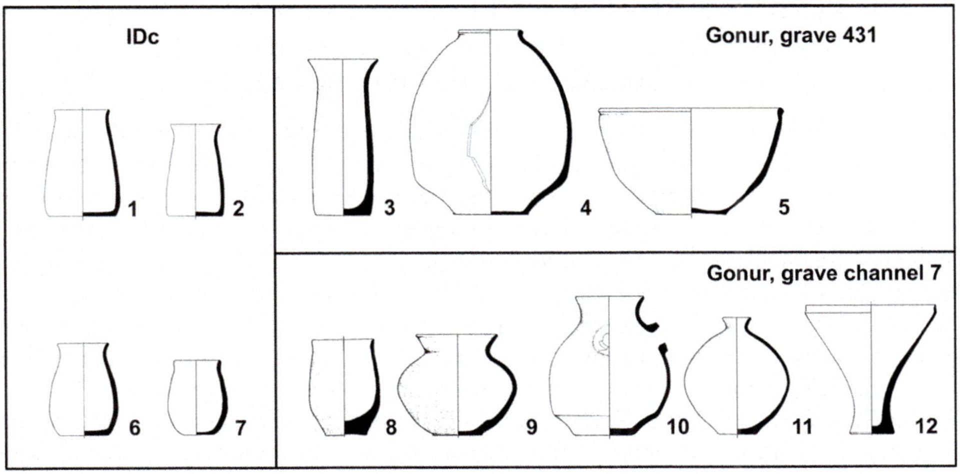

Figure 25.22 Younger beakers of type IDc and the pottery inventory from Gonur, grave 431 and grave channel 7: 1 – Ojkul', Kurgan 24; 2 – Ojkul', Kurgan 1; 3–5 – Gonur, grave 431; 6 – Dzharkul', Kurgan 45; 7 – Dzharkul', Kurgan 21; 8–12 – Gonur, grave channel 7 (3–5, 8–12 after Salvatori).

Necropolis in graves excavated by Salvatori (1993, 1994, 1995). One of them, grave 431, contained a large ovoid jar, a truncated conical basin, and a high subcylindrical beaker with flared, everted rim (Figure 25.22, *3–5*) in addition to two other vessel sherds. The beaker (Figure 25.22, *3*) – which I have classified elsewhere as a beaker type 2 (Teufer 2015: 93, fig. 37) – shows certain similarities to the beakers of type IDc. The second grave – or "grave channel 7" – at Gonur contains a globular pot (Figure 25.22, *9*), a spouted vessel ("teapot") (Figure 25.22, *10*), a globular jar (Figure 25.22, *11*), a truncated conical bowl with a high subcylindrical concave foot (Figure 25.22, *12*), as well as a pear-shaped beaker (Figure 25.22, *8*). This last beaker is comparable to the elongated beaker with slightly or strongly everted rim and slightly globular lower half (type IDc). While grave 431 dates to the transition from the Middle to the Late Bronze Age (Teufer 2015: 669, suppl. 5), grave channel 7 points to the early Late Bronze Age, indicated especially by the spouted vessel.[17] Grave channel 7 at Gonur therefore clearly shows that the "classical Vakhsh culture" traces down to the Late Bronze Age.

These connections between the oases cultures of western Central Asia area and southern Tajikistan are also visible in other artifacts (bronze and stone objects, statuettes) coming from cemeteries that are not considered as belonging to "classical Vakhsh culture" (like Farkhor, Kangurttut II, and Gelot) because of the absence of kurgan-like construction (see Chapter 23). In the "classical Vakhsh culture," these influences are limited to the occurrence of bronze mirrors at Tigrovaja Balka that are spatially and chronologically widespread (Kaniuth 2006: 72–73).

These cemeteries without kurgans are themselves related through their pottery (type ICa/ICb) to those of the "classical Vakhsh culture." The above-described model of the cultural "*écart*" is clearly visible here. This also clarifies the question of the grave constructions, which will be discussed in more detail next.

DIFFERENT BURIAL COLLECTIVES IN SOUTHERN TAJIKISTAN IN THE 3RD MILLENNIUM: THE PARTICULARITY OF THE KURGANS

From the dates presented above of Gelot and Darnajchi (22nd/23rd century BC until the Late Bronze Age Stage Ib of the 2nd millennium) and those of the Vakhsh sequence (from the second half of the 3rd to the early 2nd millennium BC), a chronological overlap can be clearly seen with the carinated biconical vessels of types ICa and ICb. Otherwise, chronological overlaps are few. Significant differences can be found in the burial structure. It should be assumed that there were two "*Abstraktionskollektive*" in southern Tajikistan in the 3rd millennium, each of them burying their dead in different ways. One buried their dead in simple underground graves in pits or catacombs, the other built a mound made of stone and loess (kurgans) above the same underground graves.

Outside of the "classical Vakhsh culture," these kurgan graves are not attested anywhere in Central Asia between the Caspian Sea and the Pamir during the 3rd and early 2nd millennium.

Contemporary parallels are only found in Kazakhstan, Xinjiang and southern Siberia. We will first consider the kurgans that are similar to those of the "classical Vakhsh culture" in their cultural standardization; namely, the kurgans from Saensaj in northeastern Xinjiang (China) (Kovalev 2015: 293–306), which have a low mound and are enclosed by a stone circle. The tomb chamber is dug into the naturally elevated ground. It is significant that they contain bronze artifacts such as bronze mirrors (Kovalev 2015: 299, figs. 1 and 5) or bronze spatulas (Kovalev 2015: 306, fig. 8, 2) and stone beads (Kovalev 2015: 306, fig. 8, 4), which find their closest parallels in Central Asia.[18] Bronze mirrors can also be found in the "classical Vakhsh culture" graves (P'jankova 1986: fig. 73, *10*, *11*, *12*). A particularity of some of the tombs at Saensaj is the presence of four-wheeled wagons. According to radiocarbon dates, they date back to the second half of the 3rd millennium (Kovalev 2015: 297). They are currently the best parallels to the kurgans of the Vakhsh culture despite certain divergences in their structures.[19]

Still in the Altai region, in the area of the Karakol culture, low kurgans 18 to 20 m in diameter enclosed by stone circles are present (Kubarev 2009: fig. 3, *20*). These correspond in size to the larger kurgans of the Vakhsh culture. But the deceased individuals do not lie in catacombs but in stone cists, the stone slabs of which bear figurative decorations.

Finally, in the area of the Okunevo culture, structures are attested that could be considered as kurgans and, aside from pits and stone cists, some cases of catacombs are also known (Teufer *et al.* 2015: 115).

CONCLUSION

New research on the Bronze Age in southern Tajikistan not only makes it possible to shift with certainty the beginning of the Bronze Age in this region to the 3rd millennium, but also to show that several types of burial rituals were taking place. One of them is the construction of kurgans, considered as the most characteristic feature of the "classical Vakhsh culture" and distinguishing it from the contemporary western and southwestern neighboring oases cultures of the Bronze Age. This includes the

BMAC, whose cultural hegemony at the beginning of the 2nd millennium BC is traceable to the edges of Pamir. The "classical Vakhsh culture" has, instead, certain affinities at least in burials with the Bronze Age communities located in the Altai and Xinjiang.

Thus, it is possible that the "classical Vakhsh culture" played an intermediary role in the transfer of technology and products – bronze metallurgy and cultivation of wheat/millet – between the Central Asian oases cultures of the 3rd/early 2nd millennium and western China.[20]

NOTES

* I would like to thank Luna Watkins and Hannah Gilb for their translation of this chapter. I additionally thank Bertille Lyonnet for her redaction of a previous version, which motivated the development of certain parts.

1 For a more detailed account of the research history with corresponding literature, see P'jankova *et al.* (2009: 134–136).

2 For example, cf. Teufer (2003: 127–128), with a short excursus of the research history in fn39.

3 For the notion "*Abstraktionskollektiv*," cf. Hansen (2009: 30–31).

4 Unfortunately, there is no image of this grave. The description mentions a diameter of 4 m for the graves from Krugsaj. Grave 3 is covered by stone slabs (cf. Jusupov 1991: 160–161).

5 For further accounts of the grave structures and burial rites, cf. P'jankova (1986: 12–26; 1989: 16–32), P'jankova *et al.* (2009: 102–103), Teufer *et al.* (2015: 102–103).

6 For the use of the plural of Andronovo cultures, cf. Varfolomeev and Evdokimov (2013: 289–305).

7 Compare with P'jankova *et al.* (2009: 103–104) and Teufer *et. al.* (2015: 103–104).

8 This method is explained in numerous introductions of archaeological textbooks; for example, see Giligny (2005: 140–158). For a more detailed explanation of this method, first used in 1899, see Eggert (2001: 201–221). This method of grave analysis is virtually never implemented in modern research of Bronze Age Central Asia.

9 For the radiocarbon dating, cf. Teufer *et al.* (2014a: 116–118; 2014b: 112–113).

10 For the flat-bottomed pots from Zamanbaba cf. Guljamov *et al.* (1966: pl. 6, *1*, *2*; 11, *1*, *2*; 14, *5*); for the round-bottomed pots, cf. Guljamov *et al.* (1966: pl. 10, *B1*; 11, 4; 13, *1*).

11 For more on this period in southern Bactria, see Teufer (2015: 157–159).

12 For more on these daggers, see Motzenbäcker (1996: 42–48) and Reinhold (2007: 34–36).

13 For the dating, see Helwing (in press).

14 This concerns sample LE 715. The second sample from Tigrovaja Balka, Kurgan 7, in Kaniuth's table (Kaniuth 2006: 51, tab. 13 Tigrovaja Balka sample B) is probably based on a misunderstanding of Francfort's statement (Francfort 1981: 201). Francfort referred to sample LE-715.

15 For example a storage vessel with an elongated neck and flat bottom from Tashguzor (P'jankova 1999: fig. 1, *1*; 5, *1*) and Zamanbaba (Guljamov *et al.* 1966: pl. 9, *1*).

16 This description of the Middle Bronze Age is based on the completely published graves excavated by Salvatori from the Gonur necropolis (Salvatori 1993, 1994, 1995). For the chronological classification of the beakers type 3, cf. Teufer (2015: 669, suppl. 5).

17 The same spouted vessels are considered to be a characteristic type of the Late Bronze Age phase Ia; cf. Teufer (2015: 55, fig. 21).

18 The bronze object can also be compared with Farkhor in southern Tajikistan, cf. Bobomulloev *et al.* (2017: fig. 10, *5*; 14, *1*; 15, *1*; 16, 4; 18, *5*; 26, *1*); for the stone beads cf.

Bobomulloev *et al.* (2017: 126 third row from the top, figure on the right; 127 second row from the top, central figure; 128 second row from the top, figure on the right; 130 second row from the top, figure on the right).

19 For more details on the connections with China, see Teufer (forthcoming).

20 Such a transmission is, so far, especially presumed for the Eurasian steppe or rather the so-called "Inner Asian mountain corridor" (cf. Spengler *et al.* 2014; Linduff 2018: 35–60).

REFERENCES

Askarov, A.A. 1977. *Drevnezemledel'cheskaja kul'tura epokhi bronzy juga Uzbekistana* [Early Bronze Age farming culture of southern Uzbekistan]. Tashkent: Fan.

Askarov, A.A. 1981. "K peredatirovke kul'turi Zamanbaba" [A new chronology of the Zamanbaba culture]. In *Kul'tura i iskusstvo drevnego Khorezma* [Culture and art in ancient Khwarezm], ed. M.A. Itina. Moscow: Nauka, pp. 99–110.

Babaev, A.D. 1980. "Mogil'nik Juzhbok II: pamjatnik epokhi bronzy na Zapadnom Pamire" [The graveyard of Juzhbok II: a Bronze Age archaeological site in western Pamir]. *Arkheologicheskie Raboty v Tadzhikistane* 15: 75–87.

Bobomulloev S. and B. Bobomulloev 2017. "Kangurttut 2: novyj pamjatnik epokhi srednej bronzy na juge Tadzhikistana" [Kangurttut 2: a new Middle Bronze Age site in southern Tajikistan]. In *Kangurttut: Materialy arkheologicheskoj ekspeditsii 2015–2016 gg*, I [Kangurttut: material from the archaeological expedition, 2015–2016, vol. I], ed. Z.I. Akramov. Dushanbe: Donish, pp. 38–53.

Bobomulloev, S., N.M. Vinogradova, and B. Bobomulloev (with contributions by M. Teufer and G. Lombardo) 2017. "Preliminary Results of Research of the Burial Ground of Farkhor, the Middle Bronze Age Monument in the South of Tajikistan." In *Materialy Farkhorskoj arkheologicheskoj ekspeditsii*, vol. I, ed. T. Khudzhageldiev. Dushanbe: Donish, pp. 68–114.

Eggert, M.K.H. 2001. *Prähistorische Archäologie. Konzepte und Methoden*. Tübingen, Basel: A. Francke.

Francfort, H.-P. 1981. "The Late Periods of Shortughai and the Problem of the Bishkent Culture (Middle and Late Bronze Age in Bactria)." In *South Asian Archaeology 1979*, ed. H. Härtel. Berlin: Dietrich Reimer, pp. 191–202.

Franke, U. 2016. "Prehistory." In *Herat Through Time: The Collections of the Herat Museum and Archive. Ancient Herat*, vol. 3, ed. U. Franke and M. Müller-Wiener. Berlin: Staatliche Museen zu Berlin – Preußischer Kulturbesitz, pp. 21–57.

Giligny, F. 2005. "De la fouille à l'interprétation: le traitement des données." In *Guide des méthodes de l'archéologie*, ed. J.-P. Demoule, F. Giligny, A. Lehoërff, and A. Schnapp. Paris: La Découverte, pp. 129–186.

Guljamov, Ju.G., U. Islamov, and A.A. Askarov 1966. *Pervobytnaja kul'tura i vozniknovenie oroshaemogo zemledelija v nizov'jakh Zarafshana* [Indigenous culture and the origin of irrigated agriculture in the lower Zerafshan]. Tashkent: Fan.

Hansen, K.P. 1995. *Kultur und Kulturwissenschaft: Eine Einführung*. Tübingen, Basel: Francke.

Hansen, K.P. 2009. *Kultur, Kollektiv, Nation*. Passau: Stutz.

Helwing, B. in press. "Klassifikation, Typologie und Datierung der untersuchten Metallobjekte." In *Die Metallindustrie Mesopotamiens von den Anfängen bis zum 2. Jahrtausend v. Chr. – Textband*, ed. H. Hauptmann and E. Pernicka. Rahden/Westf.: Marie Leidorf GmbH.

Jullien, F. 2017. *Es gibt keine kulturelle Identität*. Berlin: Suhrkamp.

Jusupov, A.Kh. 1991. "Arkheologicheskie issledovanija Bajpazinskogo i Vakhshskogo arkheologicheskikh otrjadov v 1983 g" [Archaeological research of the Bajpazin and Vakhsh archaeological sections, 1983]. *Arkheologicheskie Raboty v Tadzhikistane* 23: 151–164.

Kaniuth, K. 2006. *Metallobjekte der Bronzezeit aus Nordbaktrien*. Archäologie in Iran und Turan 6. Mainz: Philipp von Zabern.
Kaniuth, K. 2007. "Tilla Bulak 2007 – Vorbericht zur ersten Kampagne." *Archäologische Mitteilungen aus Iran und Turan* 39: 31–47.
Kaniuth, K. 2009. "Tilla Bulak 2008 – Vorbericht zur zweiten Kampagne." *Archäologische Mitteilungen aus Iran und Turan* 41: 75–96.
Kaniuth, K. 2010. "Tilla Bulak 2009 – Vorbericht zur dritten Kampagne." *Archäologische Mitteilungen aus Iran und Turan* 42: 129–163.
Kaniuth, K. and M. Teufer 2001. "Zur Sequenz des Gräberfeldes von Rannij Tulchar und seine Bedeutung für die Chronologie des spätbronzezeitlichen Baktrien." *Archäologische Mitteilungen aus Iran und Turan* 33: 89–113.
Kirjushin, Ju.F. 2002. *Eneolit i rannaja bronza juga Zapadnij Sibiri* [The Eneolithic and Early Bronze Age in southwestern Siberia]. Barnaul: Altajskogo Universiteta.
Kovalev, A.A. 2015. "Saen'saj: novaja kul'tura bronzovo veka vostochnoevropejskogo proiskhozhdenija na Tian'-shan" [Saen'sai: a new Bronze Age culture of eastern European descent in Tian Shan]. In *Drevnejshie evropejtsy v serdtse Azii: Chemurcheksij kul'turnyj fenomen II: Rezul'taty issledovanii v tsentral'noj chasti Mongol'skogo Altaja i v istokakh Kobdo; pamjatniki Sin'tsziana i okrainnykh zemel* [Earliest Europeans in the heart of Asia: the Chemurchek cultural phenomenon II. Excavations from the central Mongolian Altai and the Khovd River; sites and finds in Xinjiang and in outlying regions], ed. A.A. Kovalev. St. Petersburg: Sankt-Peterburgskij gosudarstvennyj muzej-institut sem'i Rerikhov, razrabotka i sostavlenie serii, pp. 293–306.
Kovalev, A.A. and D. Erdenebaatar 2009. "Discovery of New Cultures of the Bronze Age in Mongolia According to the Data Obtained by the International Central Asian Archaeological Expedition." In *Current Archaeological Research in Mongolia: Papers from the First International Conference on "Archaeological Research in Mongolia," Ulaanbaatar 2007*, ed. J. Bemmann, H. Parzinger, E. Pohl, and D. Tseveendorzh. Bonn: Universität Bonn Institut für Vor- u. Frühgeschichtliche Archäologie, pp. 149–170.
Kubarev, V.D. 2009. *Pamjatniki karakol'skoj kul'tury Altaja* [Karakol' culture sites, Altai]. Novosibirsk: Institute of Archaeology and Ethnography Press.
Kuz'mina, E.E. 1972a. "Kul'tura Swata i ee svjazi s severnoj Baktriej" [The Swat culture and its connections with northern Bactria]. *Kratkie Soobshchenija o dokladakh i polevykh issledovanijakh Instituta Arkheologii Akademii Nauk SSSR* 132: 116–121.
Kuz'mina, E.E. 1972b. "K voprosu o formirovanii kul'tury severnoj Baktrii" [Question on the cultural genesis in northern Bactria]. *Vestnik Drevnej Istorii* 1: 131–147.
Kuz'mina, E.E. 1976. "The 'Bactrian Mirage' and the Archaeological Reality." *East and West* 26(1–2): 113–131.
Kuz'mina, E.E. 2007. *The Origin of the Indo-Iranians*. Leiden/Boston: Brill.
Linduff, K.M. 2018. "Technoscapes and the Materialization of Ideas in Metal on the Inner Asian Frontier." In *Ancient China and Its Eurasian Neighbors: Artifacts, Identity and Death in the Frontier, 3000–700 BCE*, ed. K. M. Linduff, Yan Sun, Wei Cao, and Yuanqing Liu. Cambridge: Cambridge University Press, pp. 35–71.
Litvinskij, B.A. 1964. "Tadzhikistan i Indija." [Tajikistan and India]. In *Indija v drevnosti* [Ancient India], ed. V.V. Struve and G.M. Bongard-Levin. Moscow: Nauka, pp. 143–165.
Litvinskij, B.A. 1973. "Arkheologicheskie raboty v Tadzhikistane v 1962–1970 gg. (nekotorye itogi)" [Archaeological research in Tajikistan from 1962 to 1970: some results]. *Arkheologicheskie Raboty v Tadzhikistane* 10: 5–41.
Luneau, E., N.M. Vinogradova, L. P'jankova, and Kh.Ju. Mukhtidinov 2011. "Productions céramiques de la culture du Vakhsh (Tadjikistan âge du Bronze): de nouvelles perspectives chrono-culturelles." *Archäologische Mitteilungen aus Iran und Turan* 43: 283–312.

Lyonnet, B. 1997. *Prospections archéologiques en Bactriane orientale (1974–1978). Sous la direction de Jean-Claude Gardin.* Vol. 2, *Céramique et peuplement du chalcolithique à la conquête arabe.* Paris: Editions Recherche sur les Civilisations.

Mandel'shtam, A.M. 1968. *Pamjatniki epokhi bronzy v juzhnom Tadzhikistane* [Bronze Age sites in southern Tajikistan]. Materialy i issledovanija po Arkheologii SSSR 145. Leningrad: Nauka.

Motzenbäcker, I. 1996. *Sammlung Kossnierska. Der Digorische Formenkreis der kaukasischen Bronzezeit.* Berlin: Staatliche Museen zu Berlin – Preußischer Kulturbesitz.

Mukhitdinov, Ju. 1975. "Izuchenie pogrebal'nykh pamjatnikov v Parkharskom rajone Tadzhikistana" [Studies on burial sites in the Parkhar region]. *Uspekhi sredneaziatskoj arkheologii* 3: 75.

Müller-Karpe, M. 2004. "Katalog I: Untersuchte Metallobjekte aus Mesopotamien." In *Die Metallindustrie Mesopotamiens von den Anfängen bis zum 2. Jahrtausend v. Chr.*, ed. H. Hauptmann and E. Pernicka. Rahden (Westf.): Marie Leidorf GmbH, pp. 1–89.

Parzinger, H. 2006. *Die frühen Völker Eurasiens. Vom Neolithikum bis zum Mittelalter.* Munich: C.H. Beck.

P'jankova, L.T. 1974. "Mogil'nik epokhi bronzy Tigrovaja Balka" [The Bronze Age cemetery of Tigrovaja Balka]. *Sovetskaja Arkheologija* 3: 165–180.

P'jankova, L.T. 1978. "Keramika iz Tigrovaja Balka (bronzovyj vek)" [Bronze Age pottery from Tigrovaja Balka]. *Material'naja Kul'tura Tadzhikistana* 3: 29–68.

P'jankova, L.T. 1982. "South-Western Tajikistan in the Bronze Age." *Unesco IASCCA Information Bulletin* 1: 35–46.

P'jankova, L.T. 1986. *Jungbronzezeitliche Gräberfelder im Vachš-Tal, Süd-Tadžikistan.* Materialien zur Allgemeinen und Vergleichenden Archäologie 36. München: C.H. Beck.

P'jankova, L.T. 1989. *Drevnie skotovody Juzhnogo Tadzhikistana* [Early pastoralists in southern Tajikistan]. Dushanbe: Donish.

P'jankova, L.T. 1991. "Novyj mogil'nik Vakhsh Kul'tury (raskopki 1983g.)" [A new cemetery of the Vakhsh culture: excavations from 1983]. *Arkheologicheskie Raboty v Tadzhikistane* 23: 165–176.

P'jankova, L.T. 1993. "Otchet o rabotakh Vakhshskogo arkheologicheskogo otrjada v 1984g" [Report on the research of the Vakhsh archaeological section: 1984]. *Arkheologicheskie Raboty v Tadzhikistane* 24: 227–241.

P'jankova, L.T. 1999. "Keramik aus der Siedlung Tašguzor." *Archäologische Mitteilungen aus Iran und Turan* 31: 33–67.

P'jankova, L.T. 2003. "Raskopki pogrebenii vakhshskoj kul'tury v mogil'nike Obkukh (Moskovskij rajon) v 2002 g." [Excavations of Vakhsh culture graves in the Obkukh cemetery (Moskovski District), 2002]. *Arkheologicheskie Raboty v Tadzhikistane* 28: 194–228.

P'jankova, L.T., B.A. Litvinskij, S. Bobomulloev, K. Kaniuth, and M. Teufer 2009. "Das bronzezeitliche Gräberfeld von Makoni Mor, Tadžikistan." *Archäologische Mitteilungen aus Iran und Turan* 41: 97–140.

Reinhold, S. 2007. *Die Spätbronze- und frühe Eisenzeit im Kaukasus: Materielle Kultur, Chronologie und überregionale Beziehungen.* Bonn: Habelt.

Roberts, B.W. and M. Vander-Linden 2011. "Investigating Archaeological Cultures: Material Culture, Variability, and Transmission." In *Investigating Archaeological Cultures. Material Culture, Variability, and Transmission*, ed. B.W. Roberts and M. Vander Linden. New York: Springer, pp. 1–21.

Salvatori, S. 1993. "The Discovery of the Graveyard of Gonur-Depe 1 (Murghab Delta, Turkmenistan): 1992 Campaign, Preliminary Report." *Rivista di Archeologia* 17: 5–43.

Salvatori, S. 1994. "Excavations at the Namazga V Late Graveyard of Gonur 1 (Murghab Delta, Turkmenistan). Preliminary Report on the 1993 Field Season." *Rivista di Archeologia* 18: 14–39.

Salvatori, S. 1995. "Gonur-Depe 1 (Margiana, Turkmenistan): The Middle Bronze Age Graveyard. Preliminary Report on the 1994 Excavation Campaign." *Rivista di Archeologia* 19: 5–37.

Sarianidi, V.I. 1976. "Issledovanija pamjatnikov Dashlinskogo oazisa" [Research on sites of the Dashly oasis]. In *Drevnjaja Baktrija. Materialy sovetsko-afganskoj ekspeditsii 1969–1973 gg* [Ancient Bactria: material from the Soviet–Afghan Expedition, 1969–1973], ed. I.T. Kruglikova. Moscow: Nauka, pp. 21–86.

Sarianidi, V.I. 1977. *Drevnie zemledel'tsy Afganistana.* Materialy sovetsko-afganskoj ekspeditsii 1969–1974 gg [Early farmers in Afghanistan: material from the Soviet–Afghan expedition, 1969–1974]. Moscow: Nauka.

Sarianidi, V.I. 1979. "K voprosu o kul'ture Zamanbaba" [Question on the Zamanbaba culture]. In *Etnografija i Arkheologija Srednej Azii* [Ethnography and archaeology in Central Asia], ed. A.V. Vinogradov. Moscow: Nauka, pp. 23–28.

Spengler, R.N., M. Frachetti, P. Doumani, L. Rouse, B. Cerasetti, E. Bullion, and A. Mar'yashev 2014. "Early Agriculture and Crop Transmission Among Bronze Age Mobile Pastoralists of Central Eurasia." *Proceedings of the Royal Society B: Biological Sciences* 281(1783): 20133382. http://dx.doi.org/10.1098/rspb.2013.3382.

Teufer, M. 2003. "Kulturkontakte zwischen Sogdien und Baktrien am Beginn der Spätbronzezeit. Dargestellt am Grab von Zardča Chalifa." *Mitteilungen der Gesellschaft für Anthropologie, Ethnologie und Urgeschichte* 24: 121–140.

Teufer, M. 2012. "Der Streitwagen: Eine 'indo-iranische' Erfindung? Zum Problem der Verbindung von Sprachwissenschaft und Archäologie." *Archäologische Mitteilungen aus Iran und Turan* 44: 271–312.

Teufer, M. 2014. "Rezul'taty raskopok na pamjatnikakh Saridzhar 2 i Karim Berdy v Juzhnom Tadzhikistane" [Excavation results from the sites of Saridjar 2 and Karim Berdy in southern Tajikistan]. *Arkheologicheskie Raboty v Tadzhikistane* 37: 120–179.

Teufer, M. 2015. *Spätbronzezeitliche Grabfunde aus Nordbaktrien und benachbarten Regionen. Studien zur Chronologie zwischen Aralsee und Persischem Golf.* Archäologie in Iran und Turan 13. Berlin: Dietrich Reimer.

Teufer, M. forthcoming. "New Research on the Bronze Age in Eastern Middle Asia and the Connections to Western China."

Teufer, M., S. Bobomulloev, and L.T. P'jankova 2015. "Das bronzezeitliche Gräberfeld von Džarkul und seine Einbindung in das Chronologiesystem des 3. Jahrtausends v. Chr. in Südtadžikistan." *Mitteilungen der Gesellschaft für Anthropologie, Ethnologie und Urgeschichte* 36: 101–140.

Teufer, M., N.M. Vinogradova, and Ju.G. Kutimov 2014a. "Der Übergang von der Spätbronze- zur Früheisenzeit in Mittelasien." *Das Altertum* 56(2): 109–148.

Teufer, M., Vinogradova N.V., and Ju.G. Kutimov 2014b. "K voprosu ob absoljutnoj datirovke pogrebenii bronzovogo veka s nekropolej Gelot i Darnajchi" [Question on the absolute chronology of the Gelot and Darnaichi necropolises]. *Arkheologicheskie Raboty v Tadzhikistane* 37: 110–119.

Varfolomeev, V. and V. Evdokimov 2013. "Die Andronovo-Kulturen." In *Unbekanntes Kasachstan. Archäologie im Herzen Asiens I. Katalog der Ausstellung des Deutschen Bergbau-Museums Bochum vom 26. Januar bis zum 30. Juni 2013*, ed. Th. Stöllner and Z. Samašev. Bochum: Deutsches Bergbau-Museum Bochum, pp. 289–305.

Vinogradova, N.M. 1999. "Die spätbronzezeitliche Siedlung Tašguzor." *Archäologische Mitteilungen aus Iran und Turan* 31: 33–48.

Vinogradova, N.M. 2004. *Jugo-Zapadnyj Tadzhikistan v epokhu pozdnej bronzy* [Southwestern Tajikistan in the Late Bronze Age]. Moscow: Institut Vostokovedenija RAN.

CHAPTER TWENTY-SIX

THE OXUS CIVILIZATION AND THE NORTHERN STEPPES

Gian Luca Bonora

INTRODUCTION

The steppe communities of the Bronze Age have been traditionally classified under the broad term "Andronovo culture." They share similar design elements and forms of material culture, burial traditions, as well as pastoralism as their primary economic activity.[1] Regional variations such as Petrovka, Alakul' and Fedorovo, and Sargary and Alekseevka spread out across the Central Eurasian steppes: from the Urals to the borders of western Mongolia and as far south as the foothill terraces and oases of eastern Tajikistan, Uzbekistan, north Afghanistan, and Margiana in South Turkmenistan (P'jankova 1994; Kuz'mina 2007, 2008a). The material culture and economic nature of this intricate patchwork of mobile communities arguably began to emerge in locally independent contexts as early as the middle of the 3rd millennium (Frachetti 2008).

Evidence of sustained contacts and interactive processes between the Central Eurasian steppes and southern Central Asia, where agricultural economy prevailed, is so far unclear and scant (Figure 26.1). The reasons are twofold: First, archaeological finds only partially document the cultural, linguistic, and biological interactions that certainly took place in prehistory; and, second, few specialists have studied the archaeological evidence of early interaction between these two geographical and cultural worlds.

During the Neolithic and Chalcolithic periods, ca. 7th–4th millennia, the human communities of the two expanses were isolated and unaware of their cultural developments and achievements. The relationships increasingly intensified from the late 4th to early 3rd millennium and can be considered as well established during the 2nd millennium BC. Their reciprocal isolation reversed through the occupation of the desert oases surrounding the Kyzyl Kum and Kara Kum, which would not have been possible without the domestication of the Bactrian camel. The occupation of the desert oases may be considered as an extension of the southern agricultural world northwards, thus promoting the development of extensive pastoralism in north–south directions, and forging economic and cultural bonds between the oasis communities and the steppe pastoralists. A continental synergic network, implying the reciprocal diffusion of knowledge, skills, and innovations in an area encompassing the Ural region, southern Siberia, eastern Turkestan, southwest Asia, the Indo-Iranian borderland and the Indian subcontinent was then created. It even impacted the Bronze Age empires of the Near East and Mediterranean Sea, millennia before the advent of the Silk Road (Potts 1999). The principal link between the geographical regions of the ancient world was the circulation of raw materials, finished goods, and innovative ideas and technologies: Each geocultural region contributed significantly to the growth and ideological structure of the Eurasian sociopolitical entities. The prime movers that drove and shaped this convoluted inter-regional trade were the voracious demands and requests of both the Near and Middle Asiatic cities and the Eurasian steppe communities for metal, animals, slaves, semiprecious and other ornamental stones, woods, textiles, leather goods, and exotic artifacts to increase power and feed appetites for high status and prestige (see Chapter 1).

CULTURAL CHRONOLOGICAL ASSESSMENTS

North–south or vice versa movements were most likely protracted processes occurring over millennia and centuries and not sudden events. Moreover, they were largely peaceful in character, involving gradual bidirectional assimilation. From a chronological point of view, the precise beginning of these slow but progressive interactions east of the Caspian Sea has been a matter of heated discussion among scholars (see Chapter 17). This is partly due to the fact that archaeological evidence is open to various interpretation, and partly because the chronological framework for the prehistoric steppe cultures was, and still is, somewhat uncertain, based mainly on cross-cultural analogies and formal comparisons.

Central and eastern Kazakhstan as well as the Minusinsk basin along the middle course of the Enisej River and the Altai Mountains in south Siberia were occupied from the mid-4th millennium by the Afanasievo culture, whose communities were characterized by mobility associated with cattle breeding economic activity. The Pontic area, west of the Urals, were occupied by the Repin (from ca. 3700 BC) followed by the Jamnaja or Pit-Grave culture (from ca. 3400 BC). The material culture of the Jamnaja and Afanasievo was very similar and ancient DNA research seems to have confirmed that the two groups were genetically almost indistinguishable (Haak *et al.* 2015). The Catacomb culture followed the Jamnaja. It is documented in the 3rd millennium in the Ukrainian and southern Russian steppes, while the valleys of the Volga and Kama Rivers north of Samara and the southern Ural hills were occupied in the late 3rd millennium by the Abashevo culture, which played a significant role in the origin of the Sintashta-Arkaim culture. The geographical space between the Pontic region and the Altai Mountains, from the semidesert and desert areas of the Kara Kum and Kyzyl Kum to the deltas of the Amu Darya, Syr Darya, and Zeravshan Rivers, was occupied by the Neolithic and Chalcolithic Kel'teminar culture with communities engaged into a mobile hunting, gathering, and fishing subsistence way of life. The earliest evidence of the Kel'teminar culture (Dar'jasaj phase) is attested already in the late 7th to early 6th millennium, while the latest (Ajakagytma, Dzhambas Kala 5 and 11 and some sites of the Ljavljakan complex) are dated to the late 3rd to the beginning of the 2nd millennium (Brunet 2005). Around the 17th or 16th century BC at least, the alluvial plains of ancient Chorasmia and the Bukhara oasis were inhabited by communities of the Tazabag'jab culture. The archaeological data shows that it combined features of the Kel'teminar

Figure 26.1 Map of Central Eurasia with the location of the main archaeological sites mentioned in the chapter (in italic) and the most important modern cities (in roman, bold) (ESRI software).

substratum, typical elements of the Srubnaja culture in its South Urals variant as well as pottery types and decorative patterns of the Alakul' and Fedorovo traditions. The Tazabag'jab populations had a mixed economy of cattle breeding and farming. Their pastoralist economy together with irrigation systems distinguish them from the other Bronze Age communities of the Eurasian steppes.

The cultural complexes of the late 3rd millennium BC of the northern Eurasian steppes are today fragmented into numerous archaeological cultures and their related local variants (*facies*) bearing different names depending on individual scholars and their subjective perception of the archaeological facts. However, there seems to be a general agreement regarding the periodization of the late Middle Bronze Age, or Sintashta-Arkaim period (2100–1800 BC), and the following Petrovka period (1900–1750 BC),[2] which marks the beginning of the Late Bronze Age and the start of the Andronovo culture. Sintashta was coeval to the Potapovka and the Late Catacomb culture, while Petrovka was coeval to the Pokrovka; Potapovka and Pokrovka spread out along the middle Volga (see also Chapter 24).

The next phase is represented by two cultures, Alakul' and Fedorovo, whose area of diffusion overlaps that of the Andronovo complex. To distinguish them is extremely problematic because they share many features. As similar ceramic styles occurred in the same sites and burial grounds, and even in the same semi-subterranean pit house and grave, it is nearly impossible to accurately interpret the formation and evolution of the steppe cultures in the first half of the 2nd millennium BC. Moreover, different scholars have fragmented both cultures into numerous *facies*, some of which are characterized by an earlier and later phase. For the Alakul' culture, for instance, the Cheljabinsk, Ujsk-Uvel', Magnitogorsk, Tobol'sk, northern Kazakhstan (split into the Petropavlovsk and Kokshetav *facies*), Sol'-Iletzk, and western and central Kazakhstan are mentioned. For the Fedorovo, the Urals, north Kazakhstan, central Kazakhstan, Semirech'e, upper Ob', middle and upper Enisej River valley, Irtysh valley near Pavlodar, eastern Kazakhstan, Tian Shan, and Pamir are known. The outcome is a very complex mosaic of large and small, intersecting and overlapping "tesserae" that make it troublesome to see and to interpret unequivocally the whole picture.

The Srubnaja or Timber Grave culture was spread in the eastern part of the Pontic Caspian area in the second quarter of the 2nd millennium BC and it is interpreted as the direct result of the evolution of three earlier cultures: Late Catacomb, Poltavka, and Potapovka. The Kozhumberdy and Amangel'dy cultures, respectively in western and northern Kazakhstan, and the Nura along the Ishim River valley pertain to the mid-2nd millennium BC. The whole region of Sarjarka (or central Kazakhstan) was occupied by the Sargary and Alekseevka cultures. The latter produced unique pottery, known as *valikovaja keramika*, characterized by an applied band on the shoulders of most of the vessels. The Sargary-Alekseevka complex can be further divided in the *facies* of Atasu, between Karaganda and Dzhezkazgan, and Tautary along the slopes of the Karatau Mountains. The Begazy-Dandybaj is today regarded as a local variant of the Sargary culture, and the large stone-built hypogeums of Begazy, Aksu Ajuly 2, Buguly 3, Sangru (or Sangyru) 1, Ajbas Darasy, Karazhartas, and Dandybaj are interpreted as monumental burials of the local Sargary elite who controlled and managed the rich local resources (copper ores of Dzhezkazgan and Kenkazgan and huge pastures).

New calibrated radiocarbon dates of steppe materials have been recently achieved, which have extended the old and obsolete periodization leading to alternative "high" chronologies (i.e., 300–500 years earlier than the traditional chronologies established in Soviet times). Although these dates are few and have not been accepted by all scholars, they more properly correspond to the relative chronology. Moreover, they can be confidently linked to artifacts from southern Central Asia, which in turn, in the best scenario, can be related to historical chronologies established for the ancient Near and Middle East. In fact, research and fieldwork in southern Turkmenistan and

Margiana, Uzbekistan, and Tajikistan suggest a "South Asian line of synchronization," based upon clusters of radiocarbon dates from sites across the Indo-Iranian borderland. This could be the foundation of an acknowledged chronology of the Eurasian steppe in the Bronze Age, from the Caspian Sea to the Tian Shan Mountain and beyond.

From all the gathered data, and though it cannot yet be in agreement with all of it, a chronological correlation between these different cultures is provisionally proposed (Table 26.1).

EARLIEST EVIDENCE AT THE TURN OF THE 3RD MILLENNIUM: SARAZM AND ZHUKOV

The issue of the stratigraphic position of all the anthropic layers and finds from Sarazm, which consequently affects their periodization, has yet to be solved. However, the rich Afanasievo-like burial from excavation 4 attributed to Sarazm period I and dated to the mid-4th millennium (Lyonnet 1996: 33, 35 (type III.11) and 47), although it raises several interpretative problems,[3] is the earliest evidence of interactive processes east of the Caspian Sea between the cattle herders of the Eurasian steppe and the proto-urban communities of western Central Asia. The round funerary structure, interpreted as a typical Afanasievo stone fence with a stone stele set in its center, contained five graves and a rich assemblage of stone beads, but no steppe ceramics. Burial 1 was located along the southeastern edge of the stone fence and contained a 6–7-year-old child adorned with a necklace of about 200 lapis lazuli beads and 22 in silver. In the center of the stone fence, a 20-year-old couple was buried accompanied by a subadult individual. The woman wore rich clothing adorned with beads of lapis lazuli, turquoise, and carnelian, and her head and arms were covered with 49 pierced beads of gold and 24 of silver (Isakov 1992; 1994: 5–6). Two seashell bracelets, manufactured from *Turbinella pyrum*, attest to her high social status and indicate long-distance trade with coastal sites in Baluchistan (Besenval 1987).[4] The cultural importance of this burial is very significant: Not only does it document one of the earliest interactions between north and south in Central Asia, but also between the northeastern and the southern frontiers of Middle Asia.

The earliest steppe ceramics, represented by Late Kel'teminar pottery, were found at Sarazm in its phase II (according to Isakov 1991: fig. 4, *9–10*; 26, *20*), dated at the turn of the 3rd millennium, or in phase III (according to Lyonnet 1996), dated to the first centuries of the 3rd millennium. Turquoise workshops belonging to the same cultural period, Late Kel'teminar, have been found at Kaptarnikum, Beshbulak 3, and Ljavljakan in the Kyzyl Kum Desert east of the Ajdar Kol' and north of the modern city of Nurota, at the northern edge of the middle Zeravshan valley (Vinogradov 1972, 1981; Vinogradov and Mamedov 1975).

Turquoise from the Kyzyl Kum, north of the Zeravshan valley, as well as lapis lazuli from the Badakhshan Mountains of north Afghanistan, were then traded southward through Altyn Depe across Iran (Shahr-i Sokhta, Tepe Hissar), into Mesopotamia and the Indus valley. Turquoise, lapis lazuli, and other semiprecious stones were regarded as prized trade commodities requested by the southern communities, who exchanged them with other valuables, mainly cereals and textiles.

Table 26.1 The approximate position of the Namazga (NMG) V and VI periods in relation to the chronological framework for the Middle and Late Bronze Age cultures across the Eurasian steppes

Chronology	*Trans-Urals steppe region*	*Western Kazakhstan*	*North and Central Kazakhstan*	*South Kazakhstan and Semirech'e*	*Upper and lower Zeravshan*	*Southern Central Asia*	*Margiana*
2500	Jamnaja		Botai		Sarazm III/ Zamanbaba (early phase)	EBA/NMG IV period	
2400	Jamnaja		Botai		Sarazm III/ Zamanbaba (early phase)	NMG V/MBA/ proto-urban phase	
2300	Jamnaja/ Poltavka		Elunino		Sarazm IV/Soch weight	NMG V/MBA/ proto-urban phase	Gonur North/ Kelleli phase
2200	Jamnaja/ Poltavka		Elunino/ Krotovo		Soch weight	NMG V/MBA/ proto-urban phase	Gonur North ("Royal Necropolis")/Adji Kui 9
2100	Jamnaja/ Sejma-Turbino		Elunino/ Krotovo/Sejma Turbino		Khaka hoard	NMG V/MBA/ proto-urban phase	Gonur North/ Gonur South phase I/Adji Kui 1 and 9
2000	Sintashta-Srubnaja	Sintashta	Sejma Turbino		Tugai/Siab grave	NMG VI/LBA (Taip/Sapalli)	Gonur North/ Gonur South phase I/Taip

(continued)

Table 26.1 Cont.

Chronology	*Trans-Urals steppe region*	*Western Kazakhstan*	*North and Central Kazakhstan*	*South Kazakhstan and Semirech'e*	*Upper and lower Zeravshan*	*Southern Central Asia*	*Margiana*
1900	Sintashta/ Petrovka Srubnaja-Alakul'	Sintashta/ Petrovka (Early Alakul')	Petrovka (beginning of the Andronovo period)	Fedorovo (Begash 1B)	Zardcha Khalifa	NMG VI/ LBA (Gonur South/Sapalli/ Dzharkutan)	Gonur South phase II
1800	Petrovka Srubnaja-Alakul'	Petrovka/Alakul'	Petrovka/ Alakul'/ Fedorovo	Fedorovo (Begash 1B)		NMG VI/LBA (Togolok/ Dzharkutan)	Togolok/ Gonur North camp site
1700	Alakul'	Alakul'	Alakul'/ Fedorovo	Fedorovo (Begash 1B)	Zamanbaba (late phase)	NMG VI/LBA (Togolok – Kuzali)	Togolok/ Ojakly/ Gonur North camp site
1600	Alakul'/ Sargary	Kozhumberdy	Fedorovo/ Amangel'dy and Nura cultures/ Sargary (Atasu)	Atasu/ Dandybaj (Begash 2)	Dasht-i Kozi/ Tazabag'jab sites/Kajrak Kum	NMG VI/FBA (Takhirbaj 3/ Molali and Bustan)	Takhirbaj 3/Ojakly
1500	Sargary	Kozhumberdy	Sargary (Begazy-Dandybaj) Alekseevka	Atasu/ Dandybaj (Begash 2)	Karnab/ Tazabag'jab sites	NMG VI/FBA (Takhirbaj 3/ Molali and Bustan)	Takhirbaj 3/sites 1211–1219

1400	Sargary	Kozhumberdy	Sargary (Begazy Dandybaj) Alekseevka/ Tagisken South	Atasu/ Dandybaj (Begash 2)		EIA (Jaz I)	EIA (Jaz I)
1300	Sargary/Irmen		Sargary (Begazy Dandybaj) Alekseevka/ Tagisken South	Atasu/ Dandybaj (Begash 2)		EIA (Jaz I)	EIA (Jaz I)
1200	Sargary/Irmen		Dongal	Izmailovka		EIA (Jaz I)	EIA (Jaz I)
1100	Late Irmen		Dongal (end of the Andronovo period)	Izmailovka		EIA (Jaz I)	EIA (Jaz I)
1000	Late Irmen		Tasmola/Arjan and Tunnug 1			EIA (Jaz II)	EIA (Jaz II)

Note: EIA = Early Iron Age; FBA = Final Bronze Age; LBA = Late Bronze Age; MBA = Middle Bronze Age; NMG = Namazga.
Source: from Görsdorf *et al.* 1998, 2001; Zajtseva *et al.* 2004; Kircho and Popov 2005: 528–539; Anthony 2007; Epimakov 2007, 2016; Frachetti and Maryashev 2007; Kuz'mina 2007: 467–476; Hanks *et al.* 2007; Korjakova and Epimakhov 2007; Salvatori 2008; Panjushkina *et al.* 2008, 2010; Svjatko *et al.* 2009; Molodin *et al.* 2011, 2012, 2014; Epimakhov and Krause 2013; Doumani *et al.* 2015; Teufer 2015; Grigor'ev 2016; Tkachev 2016.

Despite the scarcity and subjectivity of these finds, they seem to suggest that during the second half of the 4th and early 3rd millennium, the mobile tribes of the steppe and the settled communities of the south began to interact. On the other hand, the complete absence of Central Asian materials at that time in the steppe is more likely linked to archaeological chance and contingent situations connected to a lack of specific knowledge of exogenous materials, than to the actual historical situation.

The Afanasievo-like burial of the Sarazm "princess" can no longer be considered as an isolated instance of such contact at an early time. The isolated discovery of some Afanasievo ceramic sherds in Margiana, near Gonur (Sarianidi 1990: 203), is now completed by the Zhukov "sanctuary" (Avanesova 2013), located some 16 km downstream from Sarazm and east of Samarkand in the Zeravshan valley and characterized by the wealth and variety of its assemblage. It is yet another extremely important witness to early north–south contacts. This therefore confirms that the region around Samarkand was a dreamland of sorts for northern cattle herders (see also Chapter 24). In the late 4th to early 3rd millennium it became the meeting point of diverse Eurasian cultures: the Kel'teminar Central Asian Neolithic culture, the Pit-Grave (Jamnaja) culture that spread across the Eurasian steppes, and the Afanasievo culture of north Kazakhstan and southern Siberia. The area had great abundance of resources: metal deposits, water, varied pastures, wild animals, semiprecious stones, and wood as building material. Due to the complexity and variety of the Zhukov finds, it is impossible to establish a precise and concrete cultural attribution, or to determine which archaeological culture this "sanctuary" belongs to. However, the predominant Afanasievo cultural component in the pottery assemblage seems to indicate a hypothetical movement of an Afanasievo group through Central Asia in search for new resources and pastures. Nevertheless, the aforementioned Kel'teminar component, as well as some cultural traits belonging to the Pit-Grave culture of the Cis-Urals steppes, cannot be ignored.[5] These cultural features document the intricate synthesis of the various cultural groups that presumably participated in the earliest settlement of northern cattle herders in the Zeravshan valley.

REGARDING THE MIDDLE ZERAVSHAN VALLEY

The Zeravshan River valley was not only a vast expanse of pastureland and the chief marketplace for buying and selling turquoise and other ornamental and semiprecious stones, it also contained polymetallic deposits of copper, lead, silver, and especially tin: The salient vector behind the rapid development of bronze metallurgy in the late prehistory (see Chapter 24). The most probable sources of tin in the 3rd and 2nd millennium were in western and northern Afghanistan, but, above all, along the Zeravshan (Besenval 1988; see Chapter 28, and also Chapter 29 for another point of view). Modern mineral surveyors found tin ore in the former, although no ancient mines have been identified there so far, whereas the oldest exploited tin mines in the ancient world were discovered in the latter, near the sites of Sarazm and Zhukov. These archaeological sites represent the physical gateway through which steppe cultures, with their horses and chariots (discussed in more detail later in the chapter), first arrived at the northern fringes of Central Asia. Their southward movement to the Zeravshan was not an accidental by-product of herding patterns but instead was intentional, motivated by the desire for trade, prestige, and riches. Thus, it may be conjectured that steppe herders from the north, possibly bearers of the Afanasievo and/or Kel'teminar culture, might have begun to purposely explore the vast southern Kazakh expanses and plains towards the Syr Darya, Amu Darya, and Zeravshan River valleys ca. 3000 BC or perhaps earlier. Social rivalry between steppe chiefs was the main motivation behind the quest for new sources of prestige valuables: The settlements at Sarazm and Zhukov were a logical consequence of this "hoard hunt." Interactions with Central Asia commenced as an extension of rivalries between tribal chiefs and subsequently transformed into a relationship that fundamentally altered enmity, metal production, and ritual competition among steppe cultures.

Why has evidence of initial contact been discovered in this valley and not elsewhere? The abundant natural resources of the Zeravshan valley have already been discussed. Moreover, it is worth mentioning that this territory is the southernmost natural extension of the Eurasian steppe and pasturelands, as well as being adjacent to the high-evolved proto-urban centers of western Central Asia. The valley, as a gravitational force equidistant from two geographical and cultural areas, was crossed by a trade route connecting remote sources of lapis lazuli, turquoise, and other semiprecious stones to the most renowned workshops and marketplaces of the ancient Middle Asia world. Last but not least, one did not have to cross the Kyzyl Kum and Kara Kum Deserts to reach Zeravshan: thus domesticated camels were not needed. Both deserts served as an effective cultural barrier until roughly the 3rd millennium, at which time the domestication of the Bactrian camel began somewhere in Central Asia or Inner Mongolia (see pp. 750–752). The establishment of the earliest cattle herders along the piedmont plain of the Kopet Dagh mountain range would have been impossible or extremely difficult had it not been for the aforementioned domesticated "ship of the desert."

The middle Zeravshan valley was, on the contrary, easily accessible. A path wound along the Amu Darya and Syr Darya, then along the northern side of the mountain pasturelands of southern Kazakhstan, subsequently across the well-watered Tashkent plain and then through the small "hunger steppe" around Jizzah, finally entering the large river valley in front of Samarkand. Prior to camel domestication, voyaging through Central Asia would have entailed cabotage: following large river valleys, floodplains, and foothills where men and animals could find water and fodder for survival during their journey. From this point of view, the destination of Sarazm and Zhukov were a sort of waypoint, in much the same way as Samarkand was when the Silk Road trade flourished between east and west, north and south across the Eurasian continent.

A PETROVKA COLONY IN THE MIDDLE ZERAVSHAN VALLEY

Following the chronological scheme of the archaeological discoveries, from the earliest to the latest, the next chapter is represented by early Andronovo ceramics discovered at Tugai, another site in the middle Zeravshan valley, close to Zhukov between Samarkand and Sarazm. These ceramics had predominant Petrovka and secondary Abashevo elements, and were found in association with wheel-made red- and black-polished ware typical of the Sarazm IV period. This site demonstrates that more substantial relations had been established by the middle of the 3rd millennium BC according to the dating of Sarazm IV through the ceramic comparisons (Avanesova 1996, 2015; see also Chapter 24) or much more probably between late 3rd and early 2nd millennium BC if one is to follow the latest dating for the Petrovka culture.[6] At Tugai, a comparatively large quantity of steppe pottery was found: Some 22 pots are made with the paddle-and-anvil technique over a textile-impressed form and are tempered with both crushed shell, typical of the Petrovka technique of manufacturing, and talc or steatite minerals, typical of the Abashevo culture, suggesting that the original homeland of these steppe pots was located in northeastern Kazakhstan and/or along the southern Urals region. Extensive evidence of copper smelting and metalworking (crucibles with copper slag, two copper-smelting furnaces, slag, etc.) was also found (Avanesova 2015). Tugai may have been a base or colony used by steppe traders or perhaps a temporary settlement of northern entrepreneurs responsible for the propagation of the chariot into the civilized world of the south, and accompanied by valuable horses and supplies of copper (see pp. 752–753).

The evidence of copper working at Tugai is of outstanding importance. Other Andronovo-related groups, whose numbers were increasing, later moved south into Central Asia, at the beginning of the 2nd millennium BC, where they exploited the tin mines of the Zeravshan valley (Alimov *et al.* 1998; Parzinger und Boroffka 2003). Two Bronze Age tin mines with evidence of copper working were excavated between 1997 and 1999 (see also Chapter 28). The largest, located on the lower Zeravshan at Karnab, Lapas, and Changali (Uzbekistan), about 170 km west of Sarazm, was rich in cassiterite ore with a moderate tin content (ordinarily about 3%, although some samples yielded as much as 22%). The pottery decorated with comb-stamped, incised, and grooved patterns, and the radiocarbon dates, which range from 1600 to 1000 BC, show that the Karnab mine was operated by people from the northern steppes, connected with the Andronovo horizon (Boroffka *et al.* 2002; Garner 2013). In the Bronze Age miners' settlement of Sichkonchi,[7] north of Lapas, steppe pottery of the Andronovo-Tazabag'jab culture was mainly found (Garner 2013: 295, cat. nos. 131–135).[8] Moreover, the deeper layers of Karnab mine 5/5 yielded Andronovo-Tazabag'jab pottery and a well-polished diorite pestle-scepter.[9] The shape of this artifact is reminiscent of others documented both across the Eurasian steppe (from Romania to the Enisej River valley and to the northern slope of the Trans-Ili Alatau Mountains) and in southern Central Asia (South Turkmenistan and eastern Iran),[10] mainly in the 2nd millennium BC (Boroffka and Sava 1998; Garner 2013: 171–175; fig. 127; cat. no. 1097).

The other mining complex at Mushiston in the upper Zeravshan (Tajikistan), just 40 km east of Sarazm (Parzinger und Boroffka 2003; Garner 2013; see also Chapter 28), is characterized by stannite (tin sulfide), cassiterite, and copper ores with a very high tin content (maximum 34%). According to 11 radiocarbon dates from wood beams, Mushiston was exploited in three different phases between 2400 to 1300 BC. The latest is characterized by several pottery sherds of the Andronovo-Tazabag'jab culture (Garner 2013: 295–297, cat. nos. 136–148). Thus, it is the oldest exploited ore mine in Central Asia known to date.

Other evidence from this key region are dated to the second quarter and the middle of the 2nd millennium (1700–1400 BC). They are worth describing, as they highlight the mixed composition of the ethno-cultural communities participating in the syncretic population process. A very rich female burial was found near Urgut, southeast of

Samarkand and not far from the border between Uzbekistan and Tajikistan (Masson 1966). The funerary inventory consisted of a vessel with a carved steppe decorative pattern, a bronze circular mirror, four bracelets, a bell-shaped hanging earring also of steppe origin, a ring, and more than 1,000 metal beads. Near Muminobad, also southeast of Samarkand, an individual was buried with a biconical vessel with a typical Srubnaja decorative pattern and numerous personal ornaments, among them trumpet-shaped earrings typical of the Fedorovo artistic circle (Lev 1966). A Srubnaja incised vessel, characteristic Andronovo bracelets, and the same type of trumpet-shaped earring were found in the Chakka cemetery north of Samarkand (Kuz'mina 2008a: 299). The burials of Siab and Sajguz, as well as the finds from Sazagan 2, about 20 km south of Samarkand, where Fedorovo ceramics and trumpet-shaped earrings were recovered (Dzhurakulov and Avanesova 1984), could be considered further proof (see Chapter 24).

ZARDCHA KHALIFA

The grave of Zardcha Khalifa (Bobomulloev 1997; Bostongukhar 1998; see also Chapter 24), close to Sarazm, provides early evidence of two of the most important discoveries in the human history, which occurred in the Eurasian steppe and then propagated southward: domestic horses and, probably, the chariot. Dated to the beginning of the 2nd millennium BC, the burial contained a male buried in a contracted pose. The tomb contained also the skeleton of a ram, three wheel-made Bactria-Margiana Archaeological Complex (BMAC) (Dzharkutan phase) ceramic vessels and a trough-spouted BMAC-type bronze vessel, as well as fragments of two others metal vessels. Also found in the tomb were a Fedorovo gold earring (an open ring with disks at both ends), a bronze pin decorated by a horse cast at its top, similar to the Sejma-Turbino knives and scepters with horse figurines on the hilts, a stone phallus-like scepter/pestle, as well as two sets of Sintashta horse tack consisting of two bronze bar bits with looped ends and four circular bone cheek-pieces (*psalia*). The association between the bar bits and the four cheek-pieces seem to imply that this was the tack for a chariot team. The individual buried may well have been the owner of a couple of horses, and perhaps also of a chariot, imported from the steppe (see pp. 752–753) and acquired perhaps as a diplomatic gift or by trade. Thus, this burial illustrates early process of hybridization: The horseman buried at Zardcha Khalifa was of mixed cultural origins, relating both to the steppe and the sown.

Sintasha-type cheek-pieces have also been found in the Sazagan grave 2 (Avanesova 2010; see also Chapter 24) in close association with BMAC ceramics of the Sapalli phase (1950–1780 BC). The same association of southern vessels and steppe artifacts (an Andronovo double-bladed arrowhead with a protruding stem, Fedorovo trumpet-shaped earrings, bracelets, and beads) also occurred in the chance finds from Dzham, southwest of Samarkand (Avanesova *et al.* 2001; see also Chapter 24), as well as in the Dasht-i Kozi cemetery, ca. 50 km upstream from Sarazm, dated to the middle of the 2nd millennium BC. Here, in burials no. 8 and 9, Late Fedorovo and Molali-type pots were found. Of the 12 skulls analyzed, three were male individuals and belonged to the Andronovo anthropological type (Isakov and Potemkina 1989).

Similar anthropological evidence comes from the skulls found in the cemeteries of Patma-saj and Karalemata-saj, located in the Bol'shye Balkhany Desert of southwestern Turkmenistan both characterized by Late Srubnaja and Andronovo ceramic types (Ginzburg 1959). These burial grounds, together with the isolated grave of Gyzylgy-kum and the cemeteries of Parau I and II (Mandel'shtam 1966a, 1966b; Vinogradova and Kuz'mina 1986), in the foothills of the Kopet Dagh west of Kyzyl Arvat, represent the latest southward movements of steppe populations in prehistorical times. This took place during the Takhirbaj phase, or the Final Bronze Age in Margiana (ca. mid-2nd millennium BC), contemporary to the Andronovo grave from Takhirbaj 3 and the steppe ceramic sherds from Anau, Tekkem Depe, El'ken Depe, and many sites in Margiana (Cattani 2008; see also Chapter 17). The southernmost attestations of steppe culture materials in Turkmenistan are found in the Tedjen (Kuin-kuju) and the Murghab (Imambaba, Takhta Bazar, and Kashan) valleys (Kuz'mina and Ljapin 1984; Vinogradova and Kuz'mina 1986).

In Bactria, the southernmost presence of the steppe culture is documented in the oases of Farukhabad, and to the east, at Shortugai in periods III and IV (Francfort 1989: 101). The mountainous areas of Central Asia were also settled by various groups of Andronovo tribes as considerable infiltration of steppe populations started with the second quarter of the 2nd millennium BC. The evidence based on material from southern Tajikistan allows the assertion that active contacts between different tribes and communities were going on in the region, together with some form of contact with the Oxus Civilization that is documented in almost all the sites (Vinogradova *et al.* 2008; see Chapters 18 and 25). The pottery of the Late Andronovo steppe variety represents a considerable percentage of the total inventory in sites of the first phase of the Late Bronze Age (ca. 1600 BC), such as Kangurttut and Teguzak, as well as in those of the second phase (ca. 1500 BC), such as Kumsaj and Tandyrjul, and it was associated with wheel-thrown vessels made of light clay typical of the BMAC tradition (Teufer 2015; see also Chapter 25). In contrast to pottery, the metal artifacts found in sites such as the Rannij Tulkhar cemetery, dated between the first and the second phase of the Late Bronze Age (ca. 1600–1500 BC), are all typical of the steppe Bronze Age. In this burial ground, posthumous trepanation was recorded in several cases as was cremation, a trait that also occurred in graves of the Fedorovo variant of the Andronovo culture (P'jankova 1981). Very few Bronze Age settlements presently have been discovered in southern Tajikistan, except for Teguzak and Kangurttut that are both nestled in a steep hilly landscape of mountainous regions not suitable for the practice of intensive agriculture. Considering that the architecture in both sites is dominated by stone constructions and mudbrick structures without marked evidence of substantial floor levels, they may represent seasonal encampments, perhaps periodically reoccupied as highlighted by their mixed appearance in the ceramic assemblage (P'jankova 1981).

Also in the far south of Tajikistan, in the Vakhsh and Bishkent valleys, the environmental conditions were not suitable for dry farming. The inhabitants focused on herding throughout the 2nd millennium BC. The Vakhsh valley culture (see Chapter 25) was characterized by burial in tumuli containing handmade vessels with carved decoration as well as plain wheel-made pots, while Bishkent valley burials involved the cremation of the dismembered skeleton in stone cists and catacomb graves with steppe-like materials, BMAC artifacts as well as finds from the Vakhsh culture. Such a ritual use of fire was characteristic of the Bishkent tombs (P'jankova 1994).

THE COMPLEX QUESTION OF THE DOMESTICATION OF THE HORSE

The Surtanda, Tersek, and Botai culture sites in the Tobol-Ishim drainage basin on the north Kazakh steppe date to the mid and second half of the 4th millennium BC. The second phase of occupation of Botai is characterized by more than 150 households relying primarily on horses for meat. Botai alone yielded over 300,000 animal bones, 99% of which were from the horse (Zajbert 1993; Levine *et al.* 2003). Research carried out by S. Olsen and V.V. Zajbert provides increasing evidence to support the argument for incipient horse domestication and horse herding at Botai, where equestrian horse hunters kept a few horses for riding and hunted numerous indigenous wild exemplars. Botai is rich in bone artifacts, not found in coeval prehistoric sites, such as horse mandibles that functioned as thong smoothers. This tool, easily recognizable on account of its broad notch with microscopic striation running perpendicular to the edge, was used in the manufacture of rawhide thongs (lassos, bridles, reins, whips, and hobbles), indispensable in an equestrian economy. Furthermore, large concentrations of horse manure, indicative of horse herding, penning, and the possible use of horse dung as insulating material in house construction were also discovered at Botai. This steppe custom is still persistent in most of north and central Kazakhstan.[11] Another crucial element is the discovery of organic lipid residue on 5,500-year-old Botai pottery sherds. The analyses performed by different scholars led to the identification of hydrogen isotope deuterium, indicating mare's milk (Dudd *et al.* 2003; Outram *et al.* 2009). Since it is impossible to milk a wild mare, this data, in conjunction with the evidence of organic bit wear leaving a characteristic mark on the enamel of the horse premolars,[12] is strongly supportive of early domestication and riding at Botai (Levine *et al.* 2003). Bit wear was also discovered at a later date at several other sites, like Sergeivka (2800–2600 BC), where significant bevels were recorded in 30% of horse premolars, Utjevka (2000–1850 BC) where bevels of 6 and 5 mm were encountered on a horse stallion with accompanying decoration, and in Petrovka-Alakul' culture sites (1900–1600 BC) where a bevel of about 4 mm was identified (Anthony and Brown 2000: 83–84).

Substantial evidence for horse domestication and herding has also been documented in the Eurasian steppe rock art, beginning from the late 3rd or early 2nd millennium BC. Unlike earlier periods (Paleolithic and Chalcolithic), when horses were primarily depicted in hunting scenes or alone, the Bronze Age petroglyphs show both horses being hunted and already domesticated. Engravings and carvings of the latter have been discovered at the Bronze Age site of Bajkonur in central Kazakhstan, depicting horses with straps attached to the nostrils or mouth (Novozhenov 2002: 42, tabs. 31, 36). Nearly 200 zoomorphic petroglyphs have been identified at Terekty Aulie, some 70 km northeast of Zhekazgan in central Kazakhstan. Some 90% of them depict horses alone, but one shows a man standing on a horse, another a man holding a horse with a rope, and a third one represents two riders (Samashev *et al.* 2013: figs. 70, 71, and 76). Other examples come from the coeval rock-art site at Eshkiolmes, on the right bank of the Koksu River in the Almaty region of southeast Kazakhstan. There, one petroglyph depicts a chariot drawn by a mounted horse, a second represents a chariot with a driver drawn by two horses, while a third shows a horse held by a strap. Some horses are held by grooms in a fourth engraving, and yet another represents a man holding a horse by a bridle in one hand and grasping an axe in the other to slaughter it (Mar'jashev *et al.* 1998: figs. 71, 72, 85–87). At Arpauzen, on the northern slope of the Karatau Mountains of southern Kazakhstan, a group of bird hunters composed of archers, camels, and a man holding two horses by straps is depicted (Kadyrbaev and Mar'jashev 2007: fig. 12). The same subject is represented at the site of Kul'zhabasy, located in the Chu-Ili River valley (Bajpakov and Mar'jashev 2004). One of the most widespread depictions in all the rock-art sites along the slopes of the Karatau mountain range is that of a chariot drawn by a pair of horses. These representations clearly highlight the long and complex transitional period when the steppe communities progressively abandoned the practice of hunting wild horses and began harnessing and mounting them. The same evidence is documented on Bronze Age petroglyphs in Mongolia (Jacobson *et al.* 2001: nos. 983–987 and 1001).

Wild horses were native to the area from the northern Eurasian steppe until the region of Sarjarka in central Kazakhstan. In the Near and Middle East as well as Central Asia the native equids were onagers. Thus, any horse artifacts and remains found in the Bronze Age sites of these latter regions must be interpreted as evidence of trading from the steppes far off to the north.

The earliest images of men riding equids in the Near East cannot be dated before 2000 BC. The Akkadian seal impression from Kish (2400–2300 BC) shows the coarse depiction of a rider on an indeterminate equid: it may be a donkey or a donkey–onager hybrid (Buchanan 1966). A few dozen terracotta figurines of cavaliers from Syria and Mesopotamia dating to ca. 2000 BC are equally ambiguous (Badre 1980: 63, 99, 116–117, pl. LXVII). The earliest known unequivocal representation of a horse rider in the Near East is on a Ur III seal impression of Abbakalla, the person in charge of the payments for the king Shu-Sin, dated to 2037–2029 BC according to the middle chronology, and to 1972–1964 BC according to the short chronology (Owen 1991). The man on this seal impression is depicted precariously riding bareback on an equid with a flat mane and flowing tail, two characteristics of the horse (Drews 2004: 32–34). A slightly later Mesopotamian plaque of the early 2nd millennium BC of unknown provenance shows a horse and a rider sitting on its hindquarters. The rider controls the horse by means of a nose ring and his only security is provided by a girth that he grips with his left hand, while with the right hand he holds reins (Moorey 1970: pl. XIIIa; Drews 2004: fig. 3, 6). A very important Near Eastern image of a horse without rider comes from a seal impression of Ishar-Beli, most likely a high officer of Tar'am Agade, daughter of Naram-Sin (who reigned ca. 2254–2218 BC, according to the middle chronology, and ca. 2190–2154 BC, according to the short chronology), found at Urkesh in northern Syria (Buccellati and Kelly-Buccellati 2002). This seal depicts a divine presentation scene wherein a god leads another god before a third divinity, who sits on a throne. The latter appears to be feeding a prancing equid,[13] its mane and tail plaited, while the god who is being presented carries a young such equid (a colt?). A buffalo seems to look at the scene. In this seal, as in other Akkadian seals, animals such as lions, buffaloes, and horses are possibly symbolic representations of the domination (or projected domination) of the kings represented in the seals. Therefore, it is perhaps possible to advance the hypothesis that horses(?) would represent the northern steppes, while the buffalo is the symbol of the south or southeast. The elegance of the plaited mane and tail of the equid leads us to believe that it is a horse, that such aesthetic practices were customary among royalty, and that horses were entirely domestic (i.e., docile), and accepted the care of their outer appearance.

Domestic horses in southern Central Asia

In southern Central Asia, some evidence of horses comes from Gonur Depe (see also Chapters 10, 12, and 15). A horse skeleton was said to have been found in the "Royal Necropolis" grave 3200 located on the northern edge of Gonur North and associated with period I dated to ca. 2200 BC (Sarianidi 2005: 210, fig. 72; 2008; see Chapter 12 on the horse identification, and Chapter 1 and the Appendix at the end of the volume for the date). The excavation of this brick-lined burial pit also brought to light a four-wheeled wagon with wheels made of wooden boards bound with bronze rims. Moreover, the burial also contained the osteological remains of two camels, a dog, and a decapitated foal. The bodies of ten human adults, who appear to have been sacrificed before they were buried, were also discovered within the burial (see Chapter 10). The so-called "tomb of a warrior," no. 2380, comes from the Gonur Large Necropolis; it yielded a bronze mace head depicting a horse *protomé* with ears pointed forward, clearly modeled eyes and nostrils, and a perfectly combed mane (Sarianidi 2007: 127, fig. 3). Not far from this "warrior's" tomb, a stallion burial (no. 3310) contained only parts of its body: It lacked its head, rump, and tail. It was interpreted as a cult burial of a domestic horse (Sarianidi 2008: 327). Beside the discovery of a jaw belonging to a young horse near the other "warrior's" burial (no. 2900) and a horse head buried in trench no. 9, an entire stallion was buried in excavation 16 (Sarianidi 2008: 327, fig. 183). A stone statuette from burial no. 3210 seems to depict a horse with a saddle (Sarianidi 2005: fig. 90). In another grave pit or ditch from the "Royal Necropolis" (no. 3225), which contained numerous human skeletons, possibly slaves or servants, the identification of two horses is no longer confirmed and they are now thought to be camels. A four-wheeled wagon[14] with copper-alloyed rim wheels was dismantled before burial (Sarianidi 2005: fig. 99; 2008: 173–207, 323–329, fig. 180; for a more precise description of the finds see Chapter 10).

To sum up, the earliest evidence of horses in Central Asia is represented by the finds from rich tombs in the Gonur "Royal Necropolis" and Large Necropolis dated to the turn of the 3rd millennium, and the grave of Zardcha Khalifa, dated slightly later. Two other artifacts, a horse-head image adorning a crested copper axe of the Oxus Civilization type, and an Oxus-style seal depicting a man riding a galloping equid that looks very much like a horse (Sarianidi 1986: 41, 44, fig. 8, 10), unfortunately, have no exact provenance. The first one, now housed in the Louvre Museum in Paris, was purchased in the art market (Amiet 1986: fig. 167) and the second was probably plundered from a Late Bronze Age burial ground in Bactria (North Afghanistan).

These Central Asian finds, as well as the aforementioned ones from the Near and Middle East, suggest that domestic horses began to appear at the end of the 3rd millennium. Given their simultaneous appearance across Iran and Mesopotamia, and the position of the Oxus sites between the steppes and the southern civilizations, horses were probably a highly prized trade commodity, also used in funerary rituals and ceremonies, as a pack animal, to pull wagons and carts, and as an artistic subject to adorn prized implements. Terracotta statuettes of horses pulling wagons and carts (Kircho 2009) were also manufactured. There is no evidence that horses were used as a food source in Central Asia prehistory (for Gonur, see Chapter 15).

THE QUESTION OF THE DOMESTICATION OF THE CAMEL IN EURASIA

If the question of the domestication of the horse is complex, the history behind that of the camel (*Camelus bactrianus*) in Eurasia is still an enigma, owing to the relative paucity of well-documented and dated faunal assemblages from archaeological sites. Both camels and dromedaries have been used as they are able to withstand prohibitive conditions and inhospitable environments. In ancient times these species were not subject to a significant artificial selection that changed the morphology of some of their physical features, as was the case with sheep and goats, and, to a lesser extent, bulls and cows (but see *contra* Chapter 15). It is therefore very difficult, if not impossible, to determine when and where the Bactrian camel was first domesticated on the basis of the bone finds. Therefore, the earliest evidence of domestication is only indirectly ascertained, by the finds of terracotta and stone figurines, or of camel dung and wool in some specific and favorable archaeological contexts. These finds seem to suggest that the process of domestication of the Bactrian camel was underway at the turn of the 3rd millennium.

Two-humped camelid faunal remains are said to have been found at 7th millennium Neolithic sites of Inner Mongolia and northeast Xinjiang. These bone finds are seriously considered as remains of domestic Bactrian camels (Peters and von den Driesch 1997: 661). The hunting of wild Bactrian camels, and probably the exploitation of some tamed camels, was practiced at Ajakagytma, at the edge of the Kyzyl Kum Desert in Uzbekistan, where camel bones constitute 50–85% of the total osteological remains during the Neolithic period, between 5500 and 3500 BC (Szymczak and Khudzanazarov 2006).

According to archaeozoological research, it appears that the *Camelus ferus* natural habitat stretched from the great bend of the Yellow River in northwestern China through Mongolia to central and southern Kazakhstan.[15] No evidence has been found to suggest that its original range included those parts of southern Central Asia and Iran where some of the earliest Bactrian camel remains have been found. It is highly probable that wild Bactrian camels were being domesticated as beasts of burden, as well as for their milk and fine neck wool somewhere in Inner Mongolia (Potts 2005; *contra* Trinks *et al.* 2012, who suggest a single process of domestication taking place in the southwestern regions of Central Asia rather than Mongolia or East Asia). Therefore, if zoologists are correct in believing that the *bactrianus* natural habitat did not extend west of central and southern Kazakhstan, the BMAC regions, as well as eastern Iran, Baluchistan, and the Indus valley are simply geographical areas into which it was introduced from further northeast.

As a matter of fact, osteological remains of *Camelus bactrianus*, interpreted as domestic by the excavators, were brought to light in southern Turkmenistan: at Chong Depe and Anau North, in levels of the late 4th millennium; at Shor Depe and Khapuz Depe in Early Bronze Age contexts of the first half of the 3rd millennium; at

Altyn Depe and Namazga Depe in Middle Bronze Age layers dated to the second half of the same millennium (Duerst 1908; Masson and Sarianidi 1972); and at Gonur Depe in several rich animal burials (see Chapters 10 and 12).

Bones of *bactrianus* were found at Shah Tepe (mid-3rd millennium), in the Gorgan plain of northeastern Iran (Amschler 1940), while at Sialk Tepe, near Kashan on the rim of the central Iranian desert, the zoomorphic representation on a sherd from period III.4 has been interpreted as evidence of *bactrianus* in the late 4th millennium (Ghirshmann 1938). The excavations at Shahr-i Sokhta in Iranian Seistan have brought to light some interesting remains in reference to the *bactrianus* species: bones, wool, and dung residues (Compagnoni and Tosi 1975). The dung was found in a biconical vessel, which was nearly filled to the rim. The vessel was almost completely buried in the floor of room CLXVI, attributed to phase 7 of the chrono-stratigraphical sequence of Shahr-i Sokhta and dated around 2700 BC by a series of calibrated radiometric determinations (Salvatori and Tosi 2005). Therefore, this evidence leads us to believe that, although perhaps not fully domesticated, the *Camelus bactrianus* was used not only for its meat, but also for its fleece and droppings, an excellent fuel source, as far back as the early or the first half of the 3rd millennium. All these finds may be interpreted as embryonic forms of domestication. Moreover, the scarce percentage of bones found in the Iranian and South Turkmenistan settlements should not be construed as evidence against the domestication of the animal, as the camel is used as a form of transport outside the inhabited center, being associated with nomadism. It is also worth mentioning that excavations in the residential areas of archaeological sites along the caravan routes are notoriously scarce in camelid finds. N.M. Ermolova's studies on the fauna found in the large medieval center at Shehr Islam, inhabited uninterruptedly in the 9th–14th centuries AD, or at the ancient Merv, one of the main urban centers in mediaeval Central Asia along the Silk Road, indicate that the remains of *Camelus* do not even reach 1% (Ermolova 1970: 215).

Figurines of two-humped camelids have been found in the Kachi plain sites of Mehrgarh (from the site surface) and Pirak (from contexts dating to the 2nd and 1st millennia BC), at Zayak, in an inter-mountain valley of Baluchistan between Kharan and Nal (2nd millennium BC), at Shahr-i Sokhta (3rd millennium) and at Altyn Depe (second half of the 3rd millennium) (Bonora 2006). The evidence from this latter proto-urban site in southern Turkmenistan, which also includes terracotta models of wheeled carts drawn by Bactrian camel (Kircho 2009; see also Chapter 4), highlights the use of the Bactrian camel as a draft animal, which seems to be more or less contemporary with its domestication.

By the late 3rd and early 2nd millennium the Bactrian camel is documented in the iconography of numerous copper stamp seals of Oxus production (Sarianidi 1998b: 71–73, nos. 108–111) and on soft-stone stamp seals from excavations at Togolok 1 (burial 10), Togolok 21 (southern court), and Gonur South (room 592) in Margiana (Sarianidi 1998b: 297, nos. 1634–1635). Gonur North has also yielded a silver goblet depicting a camel procession in relief, from the "royal" burial 3220 (Sarianidi 2005: figs. 95–96; see also the cover of this book), as well as a theriomorphic ceramic vessel in the shape of a Bactrian camel and a graffito of a Bactrian camel carved on a ceramic drain pipe (Sarianidi 1998a: figs. 14–15). A clay bulla with a cylinder seal impression from the temenos at Togolok 1 shows a human and a caprid between two Bactrian camels (Sarianidi 1998a: fig. 28.1).

In conclusion, although the data is discontinuous, the aforementioned finds seem to indicate that the earliest proof of camel domestication occurred in Inner Mongolia during the Neolithic period. This was followed by a westward propagation, with evidence appearing in Turkmenistan (mid-4th millennium), Margiana and Bactria (mid-3rd millennium), and the surrounding areas to the south (Pirak, in Pakistani Baluchistan, and Mohenjo-daro, along the Indus River) as we move into the late 3rd and early 2nd millennium BC (Meadow 1984, 1993). The full domestication of the camel, that is, used as a means of transport, as a draft animal, and as a source of secondary products (hair, milk), marks the establishment of north–south pastoralism across the desert oases, involving the use of desert lands as pasture. This kind of pastoralism promoted consistent bidirectional interaction, between the pastoralist world of the northern steppe and the agricultural world of southern Central Asia. As previously stated, this system was well established in the final centuries of the 3rd millennium.

CHARIOTS

Between 2200 and 1900 BC, the early steppe sites of the Sintashta, Arkaim, and Petrovka culture, on the Ural-Tobol steppe, were situated in regions rich in copper and gold ores, such as Kargaly and Maichailovo. Excavations at these sites document progress in furnace construction with chambered and dome-shaped hearths, permitting firing at higher temperatures and the production of first arsenical and ultimately tin bronze. Moreover, excavations of the burial grounds located near the sites carried out in the 1970s and 1980s uncovered no fewer than 16 early spoke-wheeled chariots from nine cemeteries; that is, the discovery of the first wheeled vehicles designed for speed, an innovation that changed land transport forever (Novozhenov 2012). Radiocarbon dates from five Sintashta culture graves containing the impressions of spoked wheels, with probability distributions that fall predominantly before the turn of the 2nd millennium BC, show that these chariots were the oldest known anywhere (Anthony 2007: tab. 15.1). In conjunction with funerary inventory (weaponry and the sacrificial head and forelegs of horses), chariots were buried, as shown by the outlines of vehicle structures, wheel rims, and spokes in the earth. These mortuary offerings are a testament to a 1,000-year-long steppe tradition: More than 100 cart and wagon burials, or graves furnished with various types of wheeled vehicles, have been discovered, some of them also containing the remains of a covered structure (Izbitser 1998). The chariot axle length was virtually the same as that of the earlier steppe wagons, thus making the case for a local evolution rather than a foreign appropriation and borrowing. Only recently have steppe societies, rather than Near Eastern ones, been credited with the invention of the chariot.

The finds from Kamennij Ambar 5, Sintashta, Berlyk, Solntze 2, Ulubaj, Satan, and Krivoe Ozero, provide the earliest incontrovertible record of rapid horse transport. Weighing less than 35 kg, or one-twentieth the weight of the ancient wagons (Piggott 1992: 17–18), these horse-drawn chariots were a new breed of vehicle, infinitely lighter and swifter than the older solid-wheeled conveyances. The site of Krivoe Ozero, located some 130 km north of Sintashta, yielded the discovery of two horse skulls buried with a chariot. Four disc-shaped bone cheek-pieces, radiocarbon dated from 2136 to 1904 BC with an average of 2026 BC, lay next to the horse skulls, amid projectile points and a bronze axe and dagger (Anthony and Vinogradov 1995: 38). Horse bones in association with similar disk-shaped cheek-pieces have been found at sites across the Volga, Don, and Donetz regions (Kuz'mina 2000). In fact, it appears that a widespread experimentation with different types of bridling was well under way at this time. Moreover, advances in bridling may not have been limited to the chariot: It is highly probable that actual bits for riding had also evolved by this time.

The early bridle type had one strap on the horse's nose fastened by simple cheek-pieces. Subsequent designs featured a cheek-strap that was held in place by more intricate cheek-pieces with multiple small holes. The concurrent use of shield- (with or without tenons) and rod-like cheek-pieces indicates a continued quest for the most efficient form (Dietz 2003; Kuz'mina 2007, 2008b).

Since the mid-3rd millennium, solid-wheeled battlewagons drawn by teams of donkeys or donkey–onager hybrids had been deployed in Mesopotamian warfare. As attested by seal impressions from *kārum* Kanesh II (Kültepe), a four-spoke-wheeled war chariot, driven by a single figure holding a battle-axe and drawn by two equids of indeterminate type (horses or native asses or onagers) controlled by lines attached to nose rings, was in use in Anatolia ca. 1950–1850 BC. Not only did the light horse-drawn chariot from the northern steppes render the clumsy battlewagons of the 3rd millennium immediately obsolete, but the vital northern innovation of harnessing with bridles, bits, and reins, soon replaced the archaic nose rings and nose bands for equid control throughout the Near and Middle East (Drews 2004: 49–51).

In summary, the earliest spoke-wheeled chariots originate from the steppes, where they were used mainly for hunting and ceremonial purposes. They were subsequently introduced to the Near East possibly through the BMAC sites, in concurrence with steppe horses and studded disk cheek-pieces. By the mid-2nd millennium BC the light horse chariot also diffused rapidly westward to the Aegean and central Europe and south to India, extending east to China later in the millennium (Pare 1992; Parpola 1999).

THE OCCUPATION OF THE DESERT OASES

As previously stated, the occupation of the desert oases (Margiana, lower Zeravshan or Makhandarya, lower Kashkadarya, and Akchadarya) could be considered as an extension of the southern agricultural world northwards. This included the development of extensive pastoralism in north–south directions, thus placing the oases in close economic and cultural contact with the complex agropastoral cultures that developed on the southern margins of the Andronovo world. This process began more or less in the 3rd millennium, in close connection with the domestication of the Bactrian camel throughout Eurasia and the southward propagation of horses and chariots from the northern steppes. Small sites pertaining to the Zamanbaba culture, located in the lower Zeravshan River valley, along the dry course of the Makhandarya, and along the lowermost course of the Kashkadarya, were first occupied in an earlier (Kel'teminar) period,[16] then subsequently abandoned and reoccupied (Kuz'mina 1958; Vinogradov 1960, 1981; Guljamov *et al.* 1966).

The last phase can be chronologically dated to the second quarter of the 2nd millennium BC. The villages were permanently occupied, characterized by semi-subterranean pit houses, typical dwellings of the Kel'teminar and northern steppe tradition, and were sustained by irrigation agriculture practiced in the larger oasis of the river delta. About 6% of the ceramics were wheel-made, which chronologically overlaps with handmade sherds, clearly similar to those pertaining to the Tazabag'jab culture of the southern Akchadarya delta in Khorezm. This great similarity could certainly indicate an exchange of materials or signify that bearers of the Tazabag'jab culture spread or migrated from the northwest to the southeast in the early to mid-2nd millennium BC.[17]

The diversion of deltaic branches for irrigation purposes, the construction of irrigation canals, and the discovery of grinding stones in the anthropic layers of the dwellings are well attested in the Tazabag'jab sites along the deltaic plains of the lower Akchadarya, in the eastern part of the lower Amu Darya (Andrianov [1969] 2016; Itina 1977). This local context of permanent agricultural villages, chronologically attributed to the second quarter of the 2nd millennium BC, may reflect strong influence or even some initial colonization from the south, where the practice of agriculture with artificial irrigation was known and implemented as early as the beginning of the 4th millennium, if not earlier (but, concerning the BMAC, see *contra* Chapter 1).[18] However, apart from the practice of irrigation agriculture, the Tazabag'jab culture south of the Aral Sea exhibits none of the cultural complexity attested south in the Oxus Civilization sites. The Tazabag'jab residential structures, the pottery, and small finds are those typical of the steppe pastoral tradition suggesting clear links to the steppes farther north.

The Zamanbaba and the Tazabag'jab cultures may thus be viewed as the diverse outcomes, at different chronological periods, of the protracted processes of southward movement into the desert oasis, and the consequent assimilation and rapid transformation of the earlier Neolithic Kel'teminar culture.

BMAC MATERIALS IN THE NORTHERN STEPPES

Thus far, we have examined and described the movement of goods, technologies, and ideas from north to south, as well as the beginning of the occupation of the desert oasis characterized by permanent villages of pit houses supported by irrigation agriculture. We will now analyze the south–north direction, bearing in mind that, so far, the Neolithic and Chalcolithic periods suggest an absence of reciprocal awareness.

The earliest evidence of Bronze Age finds of southern Central Asian origin in the northern steppe are documented in sites of the Sintashta-Arkaim and Petrovka culture. The early mobile explorers who introduced steppe horses, light chariots, and various accoutrements southward, buried their dead and established colonies in the middle Zeravshan valley, were interested in exploiting new resources and gaining prized artifacts to increase their economic and social status or rank. They succeeded in entering the trade network established between the 4th and the 3rd millennium by farmers scattered across the Iranian Plateau and along the Kopet Dagh Mountains of Turkmenistan, connecting sites like Sarazm, Mundigak, Shortughai, Shahr-i Sokhta, Konar Sandal A and B (Jiroft), Shahdad, Tepe Yahya, and Tepe Hissar to resources in the Hindukush Mountains (Vidale and Lazzari 2017). Lapis lazuli, turquoise, and other commodities were traded, bought, and sold across Middle and Central Asia and set the stage for enduring economic ties not only between the Indus and the Near East but, according to recent archaeological discoveries in the regions of Kazakhstan and southern Siberia, north into the steppe as well.

The stepped pyramid design, also known as stepped cross or crenellation patterns, was a basic element in the decorative artwork of the southern areas. It was commonly used on pottery, woven textiles, jewelry, metalwork at the end of the 4th/early 3rd millennium at Namazga, Sarazm, and Helmand sites (Mundigak, Shahr-i Sokhta, and Rud-i Biyaban), and even recognized in a wall painting from the early 3rd millennium palace at Tal-i Malyan in Fars. The same design has been identified on 5% of the ceramic decoration of the Sintashta and Potapovka ceramic tradition (Figure 26.2). This is possibly among the earliest evidence of southern influence and movement of goods to the north. It subsequently became a standard design element also on Petrovka and Andronovo pottery (Vinogradov 1995). It has been suggested that the design was conveyed to the northern steppes on textiles,[19] possibly one of the commodities exchanged for metal. The sudden shift to large-scale copper production in the area (Chernykh 2004), which began about 2100–2000 BC in the earliest Sintashta-Arkaim settlements, must have been stimulated by a sharp increase in demand, most likely from southern Central Asia, as well as the Near and Middle East. The mudbrick fortified Oxus towns were the intermediaries between the Sintashta-Arkaim miners of the northern steppes and an already established and limitless market where copper was one of the most valuable merchandises. To this regard, we can underline a noteworthy text discovered at Ur, dated to the reign of Rim-Sin of Larsa (1822–1763 BC), recording the receipt of nearly 20 tons of copper in a single shipment, most of it allocated to a single trader (Muhly 1995). Zimri-Lim, king of the powerful city-state of Mari in northern Syria between 1776 and 1761 BC, distributed gifts totaling more than 410 kg of tin to his allies during a political trip in his eight-year reign (Dossin 1970).

Other evidence of the relationship between the vast regions of the Central Eurasian steppe and Central Asia are represented by a thick grayish-brown lead wire, made of two braided strands and weighing 8.7 g, found among the metal objects at the Sintashta culture settlement of Kujsak (Maljutina and Zdanovich 1995: 103; Zaykov *et al.* 1999). The analyzed microprobe shows that the specimen consists of practically pure lead. From this analysis and the morphology of the objects, it has been concluded that the material was part of a rod-like ingot, in its foundry form. The lead was probably obtained from pure galena, the typical source for this metal in ancient times. The find has been interpreted as a possible import from the south, especially from the Zeravshan valley (Anthony 2007, 2009), because lead had never before appeared in the northern steppes as a pure metal but was common at Sarazm where a single ingot of lead weighing 10 kg was found (Besenval 1987: 455). However, objects made of lead and litharge (lead oxide, PbO), improperly called "ingots," as well as bits of galena ore, were also found along the eastern foothill plain north of the Kopet Dagh, both at Ilgynly Depe (4th millennium) and Altyn Depe (second half of the 3rd millennium). In both cases, they were discovered on the site surface (Salvatori *et al.* 2002, 2009). These finds suggest refining processes of lead and/or silver.

Some semiprecious stone beads of BMAC origin, such as lapis lazuli and turquoise, as well as faience beads, made of a glaze over a clay core and considered as imports from the south, were found in several southern Urals and Siberia sites, such

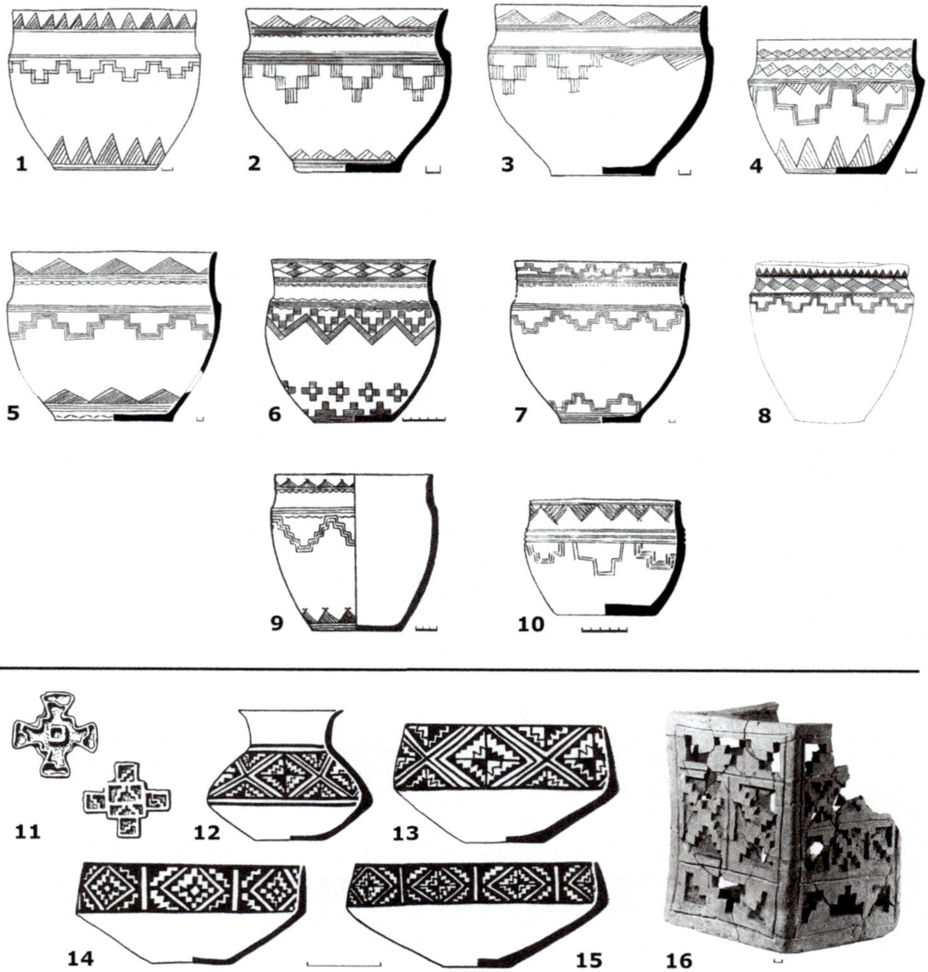

Figure 26.2 Stepped pyramid decorative patterns (crenellation motifs) on steppe pottery and on southern Central Asian pottery. 1: Alakul' barrow 8, grave 1; 2: Alakul' barrow 13, grave 1; 3: Alakul' barrow 13, grave 1; 4: Chernjaki I; 5: Alakul' barrow 13, grave 23; 6: Isakovo; 7: burial ground near Cheljabinsk; 8: Chernjaki I; 9: Tzarev kurgan; 10: Efimovka; 11: Bronze Age seals from the surface of Altyn-depe; 12: Painted pottery from burial 296, excavation 1 of Altyn-depe; 13: Carved terracotta box from room 188, excavation 2 of Altyn-depe (Masson 1988; Kuz'mina 2008).

as Sintashta (Kuz'mina 1997; 2001: 20), Alabuga, Ushkatta, Keembaj, Krivoe Ozero, Ural-Saj, Aksajman, Borovoe, Nurtai, Rostovka, and Sopka 2 (Matjushchenko and Sinitzyna 1988: 7; Vinogradov 2003: 239; Kuz'mina 2007). All these finds are chronologically attributed to the Sintashta period, between late 3rd and early 2nd millennium BC. A mirror belonging to the same period, characterized by an Oxus (Bactrian) type of handle, was found in the Sintashta culture burial mound of Krasnoe Znamja (Sungatov and Safin 1995; Kuz'mina 2001: 20).

In the 2nd millennium BC settlement of Shagalaly II, between Kokshetau and Zerinda, in the Akmola region of northern Kazakhstan, T.S. Maljutina brought to light imported pottery of Oxus origin, or imitation, in stratigraphic association with Fedorovo and Alekseevka pottery (Maljutina 1987, 1991). Subsequent research revealed that the earliest level of this site belongs to the Petrovka and Alakul' period, the second to the Nura *facies* of the Fedorovo period, and the third to the Sargary-Alekseevka culture (Bonora and Sakenov 2014). The Oxus-like pottery (predominantly of the Togolok and Takhirbaj phases in Margiana, or Kuzali and Molali in Bactria) was stratified in the first 0.3–0.4 m of anthropic deposit in close association with predominantly Alekseevka culture sherds and, to a lesser extent, with those of the Nura variant. The percentage of Central Asian pottery in the upper layers is between 5% and 15% of the whole ceramic corpus, while in the small semi-subterranean dwelling number 4 excavated by the author in 2013 and 2014, the percentage of wheel-thrown ceramic sherds of Oxus origin is about 20%. In 2013, fragments of a wheel-made truncated cone-shaped support/stand were found, known as *podstavska*[20] (Figures 26.3, 26.4, and 26.5). Other scholars (Hiebert 1994: 57) interpret this ceramic object as a chuck or mold; that is, a tool used by potters in the shaping of the large conical molded-base vessels typical of the BMAC and later periods in southern Central Asia. The support is said to have been placed with its larger mouth facing upwards on the potter's wheel to start the forming sequence

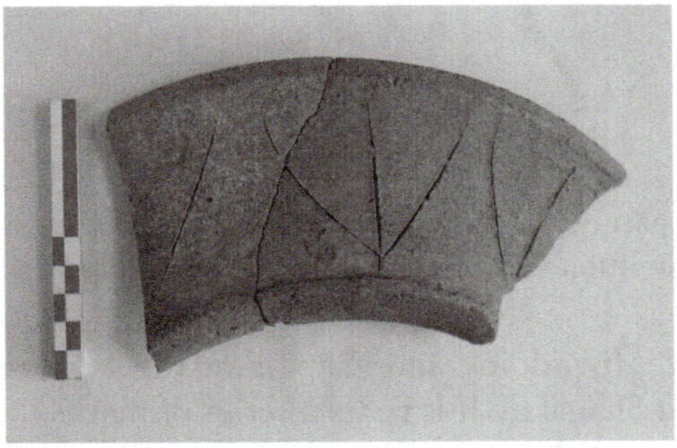

Figure 26.3 Pottery *podstavska* (stand) from Shagalaly II, with potters' marks on its outer surface (photo by the author).

Figure 26.4 Plain inner surface of the pottery *podstavska* (stand) from Shagalaly II (photo by the author).

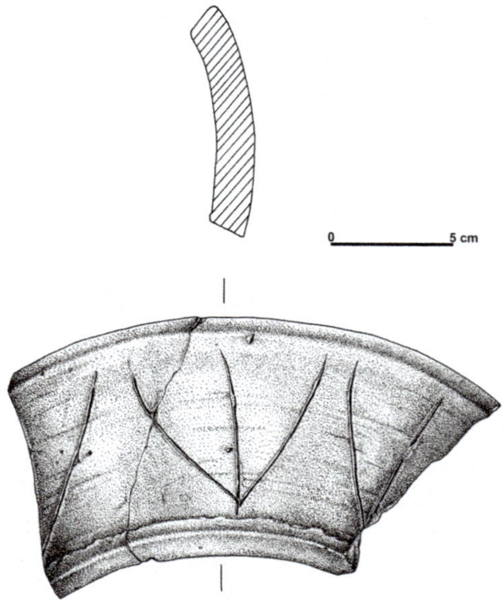

Figure 26.5 Pottery *podstavska* (stand) from Shagalaly II (drawing by the author).

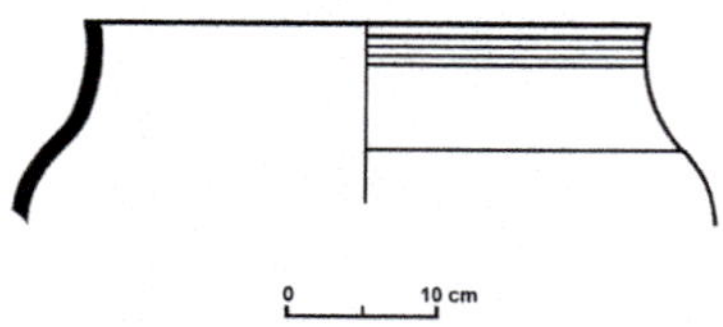

Figure 26.6 Upper part of the body of the large wheel-made jar found within pit 6, in stratigraphic association with the pottery *podstavska* (drawing by D. Baytyleu).

of jars of variable size, by adding and shaping at the same time series of slabs or superimposed coils. At Shagalaly II it bears a series of marks incised on its outer surface (Bonora forthcoming). Three marks are clearly visible, possibly the repetition of the same pattern. The central mark is complete and depicts a trident with a long stem emerging from the central prong. An oblique line is all that remains of the mark on the right side, while on the left are one straight and one oblique line. The complete carved sign is widely attested across the repertories of the Indo-Iranian borderlands during the 4th, 3rd, and 2nd millennia BC. The *podstavska* was found near a large wheel-made jar with carinated part of the body (Figure 26.6).

More than ten *podstavska* (Figure 26.7), in association with other wheel-thrown ceramic sherds[21] and with Sargary and Alekseevka pottery, were discovered at sites (Zharkovo 3, Rublevo 6, Kalinovka 2, and Burly 3) of the Kulunda and Khabary plains in the steppe Altai region of the Barnaul district in southern Siberia (Kirjushin *et al.* 2013: fig. 12; Kirjushin *et al.* 2014; Sitnikov 2015). This evidence has led some archaeologists to create the distinctive Burly *facies* of the Late Bronze Age in the steppe Altai region. Most of these pot stands, or molds, were incised with potters' marks as the Shagalaly II case and most of the pot stands found in the Oxus Civilization core regions.

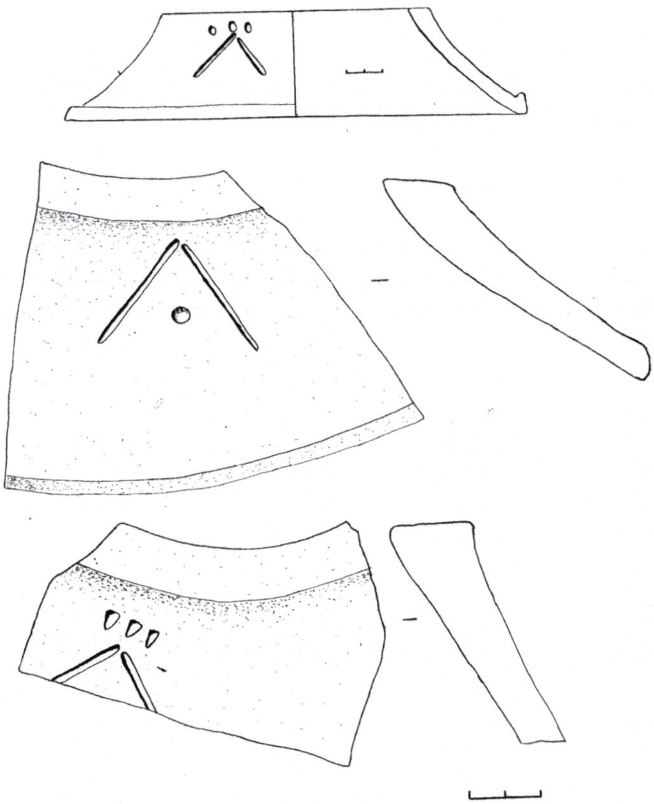

Figure 26.7 Finds of pottery *podstavska* (stand) with incised potters' marks from Burly 3, steppe Altai region of southern Siberia (drawing by S. Sitnikov).

Without addressing the question of whether a *podstavska* is a support for large jars or a potter's tool for the manufacture of large carinated vessels, it is worth mentioning that potters' marks on the surface or base of a ceramic container are almost universally accepted as elements of meta-writing. Indeed, they represent an embryonic system of intelligible symbols for communication as well as, wrong or right, stepping stones in the emergence of literacy. It seems clear that the cattle breeder communities in the Shagalaly and Burly valleys not only imported different wheel-thrown pottery types (and their content?) from the south but that they also understood, accepted, and used an exogenous system of communication, possibly of ancient southern Central Asian origin.

A baked-brick hypogeum dated back to the 14th–13th centuries BC was excavated by V.K. Mertz in the graveyard of Karaoba, along the Irtysh River valley, south of the Burly archaeological *facies* (Mertz 2013, 2016). It contained some wheel-made pottery sherds belonging to large jars of the late Oxus Civilization production in close association with numerous Begazy-Dandybaj complete vessels (Bejsenov *et al.* 2014: 179, fig. 3, *10–11*). This funerary structure is the earliest and easternmost baked-brick hypogeum excavated thus far in the steppes of Eurasia.

Other finds of wheel-made Oxus pottery have been documented at Begazy-Dandybaj sites in central Kazakhstan: Myrzhyk (Kadyrbaev and Kurmankulov 1992: 213), Kent, Domalaktas, Bajshura, and the Tasyrbaj burial ground (Varfolomeev 2013: 175–177; Bejsenov *et al.* 2014: 157, fig. 82, *1–4*, *6*, *7*, *14*, and photo 36, *1–4*). Similar finds are known in the Inner Syr Darya River delta, in the Tagisken 2 burial ground, where ceramic vessels of southern Central Asian production stratigraphically are associated with funerary hypogeums built in baked bricks (Bonora 2018).

The presence of late Oxus wheel-thrown pottery at different sites in the Eurasian steppes, as well as the close cultural links between Karaoba along the Irtysh, sites of the Begazy-Dandybaj culture in central Kazakhstan and the necropolis of southern Tagisken in the inner delta of Syr Darya, seem to indicate that the use of baked bricks is of Near Eastern and/or Indus Civilization origin and may have spread across the Eurasian steppe through Oxus sites. However, this hypothesis is advanced with great caution and needs more substantial evidence. In fact, although present in Mesopotamian architecture from the Early Dynastic period (ca. 2900–2350 BC), and in numerous Indus Civilization sites such as Jalilpur (already around 2800 BC), Harappa, Mohenjo-daro, Kot Diji, Ganweriwala, Rakhigarhi, and Lothal (Mughal 1970; Jansen 1985), durable baked bricks are not attested in large quantities in monumental and residential structures across the Iranian Plateau and southern Central Asia at the turn of the 2nd millennium BC. In the BMAC area, some rooms of the southern residential complex at Adji Kui 1 belonging to the upper architectural layer and dated to the early 2nd millennium and excavated in 2003 revealed a sector built entirely of 165 baked bricks of excellent craftsmanship (Orazov 2016: 228–230). Despite this isolated piece of evidence and the lack of architectural structures in baked bricks across the Iranian Plateau, it would appear plausible that this innovation during the 2nd millennium spread northward. This was most likely on account of transhumance routes to the summer pastures of the Irtysh valley and southern slopes of Altai, and to the winter pastures of the peri-fluvial ecological niches of the Inner Syr Darya delta. Central Kazakhstan was crossed over by most of the transhumance paths in late spring and early autumn (Bonora 2018).

A lock-shaped undecorated "handbag," or weight (Figure 26.8), made of grayish granite[22] found by chance in 1984 in the Lenin village of the Kordaj district, at the border between Kazakhstan and Kyrgyzstan, is displayed today in a showcase of the Historical Museum at Zhambyl (formerly, Taraz). Four other stone weights, two of the same type as the Kordaj lock-shaped exemplar, one shaped as a flat disk and the fourth shaped as an elongated parallelepiped bag with long handle,[23] were found also by chance in the Semirech'e region of southeastern Kazakhstan and today are stored in the funds of the Central State Museum at Almaty. These discoveries may be contextualized with the several mining and metallurgical centers of Kazakhstan such as Kendyktas, Balkhash, Dzhezkazgan, and Karaganda, or Kalba-Narym (on metallurgy in Kazakhstan, see also Chapter 28). The Kendyktas center is in the Kordaj and Shu districts of the Zhambyl Region. It includes more than 20 copper ore outcrops with remarkable traces of ancient mining activities, the largest of which are located in Shatyrkol', Zhajysan, Ungurlu, and near the mediaeval city of Kordaj. While the Balkhash, Dzhezkazgan, and Karaganda centers, located in central Kazakhstan, north of the Balkhash Lake, are characterized by numerous and very large copper

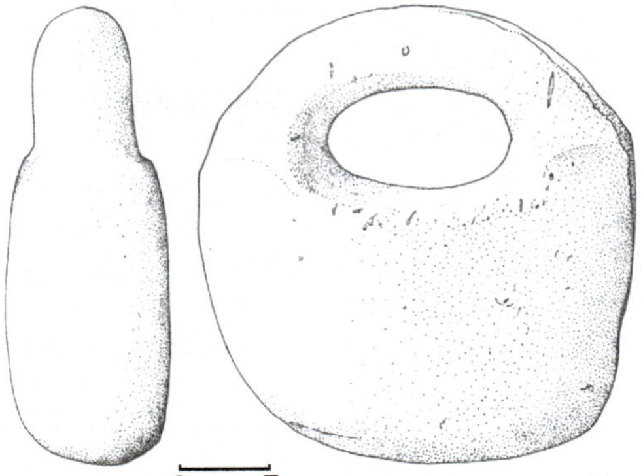

Figure 26.8 Lock-shaped "handbag" from the Lenin village of the Kurdaj district, at the border between Kazakhstan and Kyrgyzstan (drawing by M. Vidale).

ore outcrops (Margulan 2001), the mining and metallurgical center of Kalba-Narym was exploited in the Bronze Age for the presence of lead, copper, and gold ore (Garner 2013).

These "handbags" or weights from southern Kazakhstan, similar but not equal to some exemplars from Sarazm, Bactria, Tajikistan, and southern Turkmenistan (Altyn Depe, Kara Depe, Anau, and Tepe Hissar), are a typical artifact of the Oxus Civilization core area. The rounded shape of most of the south Kazakhstan items, their large handle heavily attached to the body and the absence of geometric and figurative decorations distinguish them from the specimens from southeastern Iran. Equal proportions between the handle and the body, a thin and elegant handle, and, above all, complex carved decorations on both faces of the body depicting geometric, zoomorphic, and figurative subjects with inserts for semi-precious stones are typical for the southeastern Iranian production (Majidzadeh 2004).[24] A chronological attribution to the second half of the late 3rd millennium is put forth here for these southern Kazakhstan handbags chance finds. They may testify the earliest attempts made by southern Central Asian surveyors to explore and exploit the southern and central Kazakhstan steppes in search for metal and other resources.

Double spiral-headed pin heads were widespread in Near and Middle Asian contexts beginning as early as the 4th millennium. The earliest Central Asian finds come from the Chalcolithic and Bronze Age burial ground of Parkhai II in south-western Turkmenistan and are dated to the mid-4th millennium (Khlopin 1997). In this cemetery and in many other Central Asian sites, such artifacts are documented during the whole 3rd as well as in the first half of the 2nd millennium BC. Double-spiral headed pins represent another item that spread from south to north, into the steppes, at the turn of the 2nd millennium. However, the mobile communities did not passively accept this metal artifact. On the contrary, they were subject to a process of transformation and individualization, which was not the case among the communities of the Indo-Iranian borderland. In fact, the specimen from the Balykty burial ground, in central Kazakhstan (Tkachev 2002, vol. II: 28; Usmanova 2010: 158, fig. 163, no. 2) has a rather rough appearance, suggesting a local production conceived by a local craftsman willing to create an artifact only seen elsewhere or described in a concise manner. The discovery of a mold fragment to make a double-spiral pin or pendant at the Tazabag'jab site of Kokcha 15A (near the Aral Sea), dated back to the middle of the 2nd millennium (Itina 1977: 132–133, fig. 68, no. 1), is an additional confirmation of their local production by the mobile cattle breeding communities of the Eurasian steppe. The pin from the Petropavlovsk graveyard excavated in the early 19th century documents a completely new type of double-spiral pin, undoubtedly of local manufacture. This pin was crafted to recreate an artifact of *longue durée*, characterized by peculiar features to differentiate it from all the others. The volutes do not begin directly from the long and pointed stem, but from a "shield" between the stem and the spirals (Orazbaev 1958: 261, fig. 30, *1*; Akishev 1959: tab. VII; Kuz'mina 1966: 79; 2007: fig. 33, on the right, no number; Agapov and Kadyrbaev 1979: 61, figs. 7–8). The two specimens from Ilekshar, dated to the mid-2nd millennium BC (Bajpakov *et al.* 2012; Bisembaev and Dujsengali 2013: 93 and 103), display another feature where the stem of both pins is twisted and not plain as in the other cases. This testifies of a local re-invention or local re-elaboration in the Eurasian steppe of a shape common in the west and southwest since the 4th millennium and that remained unchanged for almost two millennia. Lastly, the pin from the Tagisken South graveyard is the one of the latest from a chronological point of view (Itina and Jablonskij 2001: fig. 123, *15*). Dated back to the second half of the 2nd millennium, it documents a new type of pin, different from the previous ones, where the top is shaped in the form of double glasses, suggesting that the early shape was completely modified and then forgotten.

The diffusion of double-spiral-headed pins across northern and southern Eurasia demonstrates that objects, and the idea behind them, propagate between and through cultures due to various mechanisms and for various reasons, and that they are often transformed. Perhaps this explains why the spirals and their variations are so geographically and culturally scattered. Yet, once in the Eurasian steppe, the pins undoubtedly took on a local and not a Central Asian shape, and possibly also another meaning. The foreign idea was translated into a local medium, functioning in a local setting with its specific and local significance. Characteristic material culture forms and materials are transportable, but their meanings can be re-shaped and re-applied to fit diverse sociopolitical settings and a different array of participants (Appadurai 1986; Urban 2001). Thus, while the double-spiral-headed pins may have indeed originally been first produced in southwestern Central Asia (after Europe and the Near East), once they found their way to the Eurasian steppe they took on an indigenous shape and meaning, with each case different from the others. This same concept may also be true for the pot stands from Shagalaly II and the Burly culture sites, as well as for other classes of artifacts that moved from south to north and vice versa.

The mobile groups of specialized pastoralists and herders, the merchants and traders, pilgrims and travelers, moving from region to region and leaving tokens and traces behind as they moved onward, are the key to this interaction and plausible mechanism for the dispersal of artifacts from the southern sown towards the northern steppe, and vice versa.

CONCLUSION: THE CONNECTING ROLE OF THE MOBILE GROUPS

Regular interaction between Oxus Civilization communities and nomadic pastoral populations explains quite logically how Oxus craftsmen may have acquired their raw materials, because the semiprecious stones and metals they used had to be imported from outside their core area.

Oxus craftsmen were highly skilled in transforming copper, tin, arsenic, silver, and gold, as well as semiprecious stones into finished goods ready to be exchanged. The goods also communicated a clear ideological message of human mastership over nature and its resources and of dominance and mercantile superiority over the other mobile and settled communities. Control over the endorheic hydrographic systems and the construction of artificial canals and water reservoirs for irrigation agriculture allowed them to control the natural cycle of harvest, and the fortified citadels, with one or two defensive walls and towers, enabled them to withstand and control their economic rivals coming from close or distant lands. From an ideological point of view, all these achievements (artificial irrigation, fortification, and great artistic skills in manufacturing goods) were represented in the numerous Oxus seals found in Margiana and Bactria where muscular men are depicted clutching snakes and fighting against other animals and monsters. The muscular man is the representation of the strength, mastery, and dominance of these Oxus communities, while the serpents, as well as the other animals, real and legendary, symbolize both natural resources and trade competitors. However, within this economic system established between the steppe and the sown, the northern mobile pastoralists were not regarded as enemies but as allies playing a fundamental role in the development and success of the Oxus culture. They controlled, managed, and increased the trade of some raw materials that were crucial to the ideological representation of the Oxus culture. According to this perspective, the mobile groups of the northern steppe were not peripheral communities. They were essential and indispensable to the ideological structure and sociopolitical welfare of the Oxus Civilization.

NOTES

1 From an environmental point of view, the Eurasian steppe is a patchwork of local ecological niches. It is the home of a variety of natural resources, and of pasture lands that are present everywhere.

2 Petrovka is an eastern offshoot of the Sintashta-Arkaim culture. The material culture and funerary rituals of both cultures are very similar. However, Petrovka pottery sherds have been found stratified above Sintashta-Arkaim layers at several sites (Kuz'mina 2001: 9; Vinogradov 2003), showing that Petrovka arose from Sintashta-Arkaim and should be considered as the latter of the two.

3 From a stratigraphic point of view, the burial is below the earliest architectural layer of excavation 4, so it may have existed prior to the first building installations.

4 This assessment is supported by a few pottery sherds from Baluchistan found in period I layers of Sarazm (see Lyonnet 1996: 56–57).

5 The very intricate question of the origin of the Afanasievo culture and of its cultural connections with the Repin and Pit-Grave cultures, which suggests that "a process of migration and return migration that installed the early Afanas'evo culture continued across

the north Kazakh steppe during the Yamnaya period" (Anthony 2007: 307), is outside the scope of this chapter. This series of migrations have been recently questioned by the most recent radiocarbon datings, which seem to indicate that Afanasievo is older than Jamnaja (Brunet 2013: 86).

6 The last occupation period at Sarazm is dated by ^{14}C dates to the third quarter of the 3rd millennium (see the Appendix at the end of this volume). However, all the pottery finds are well related to other cultures in Baluchistan, the Kopet Dagh area, or northeast Iran, and point at a date around the middle of the 3rd millennium (see Lyonnet 1996).

7 Four radiocarbon dates from Sichkonchi lie between the 18th and the beginning of the 14th centuries BC (Parzinger und Boroffka 2003).

8 H. Parzinger and N. Boroffka (2003: abb. 51, *12*; abb. 71, *4*) mention a stone vessel in breccia, possibly from southern Turkmenistan or southwestward to eastern Iran, as well as some sherds that they consider as imports from Bactria (Parzinger und Boroffka 2003: 302). However, this statement is not supported by Garner (2013).

9 Type III, according to the classification by Boroffka and Sava (1998).

10 A great similarity can be recognized particularly between the Karnab artifact and the pestle or scepter from Anau South, although the latter is fragmentary (Warner 1980). Anau South is dated to the Bronze Age. The pestle-scepter found by chance in 1912 along the Malaja Almatinkaja stream of Almaty, along the northern slope of the Trans-Ili Alatau, differs from the Karnab exemplar only for its mushroom-shaped top (Samashev *et al.* 2005: 34).

11 Author's personal observation at the village of Koshkar and Shagalaly, south of Kokshetau, summer 2013 and 2014.

12 With regard to the Botai horses, 26% of the mature measurable second premolars (P2) examined had significant bevel measurements.

13 An onager or a hybrid mare, according to Buccellati and Kelly-Buccellati (2002).

14 The remains of the wheels of another wagon or cart were found in pit burial no. 3240, plundered by robbers. In association there were at least 15 human skeletons, two camels, one calf, two dogs, and presumably two sheep (Sarianidi 2008: 179, 198–200, fig. 87; see Chapter 10).

15 A single bone find belonging to the species *Camel bactrianus* species is documented from the site of the Tolstov camp, dated to the 5th–4th millennium, in the lower Akchadarya (Vinogradov 1981: 140–141).

16 The earlier period is also documented in the cemetery of Zamanbaba where Early Bronze Age and/or Middle Bronze Age materials were brought to light: a painted biconical vessel, consisting of two annular lines, and fragments of gray-ware footed vessels.

17 Tazabag'jab ceramics have also been reported near Tashkent and near Ashkhabad, in the Late Bronze Age levels of Namazga and Tekkem Depe, as well as in numerous sites of the Murghab delta (see Chapter 17).

18 Three parallel canals merged nearly at right angles from an ancient deltaic river have been detected at Geoksjur 1 (4000–3500 BC), on the clay and silt plains of the Tedzhen delta (Lizitzyna 1981). The irrigation system also included side canals branching off from the three main ones and supplied water to the fields. A triangular cross-section channel, dated to the late 6th–early 5th millennium, was discovered at Tepe Pardis, in the Tehran plains. It may be the earliest example of artificial water management in Iran and the surrounding areas (Gillmore *et al.* 2009).

19 This hypothesis, put forward by P.L. Kohl, is conceivable but difficult to prove considering the poor state of conservation of organic materials in the southern Urals sites. Ceramics cannot be considered as a suitable candidate because the pottery manufactured in the towns of southern Central Asia is unpainted during the Middle Bronze Age.

20 The *podstavska* has a lower diameter of 40 cm and an upper diameter of 24 cm. The thickness is comprised between 1.6 and 1.8 cm.

21 The percentage of the wheel-made pottery in these sites does not exceed 5% of the whole ceramic corpus.
22 The "handbag" is 25 cm high, 24 cm long, and 6 cm thick. No information regarding the discoverer and other artifacts associated with it have been published.
23 Two other elongated parallelepiped weights or handbags are reported in the Regional Museum of History of Ferghana (Uzbekistan). The author has not seen them yet.
24 The beautiful lock-shaped handbag or amulet from Soch, in Ferghana, characterized by fine figurative zoomorphic decoration and inserts containing colored stones, has very close similarities with artifacts from southeastern Iran and seems to document that groups from southern Iran and eastern Baluchistan were involved in the process of searching for resources and their exploitation at the end of the 3rd millennium BC (see Chapter 9, *contra* Chapter 8).

REFERENCES

Agapov, P. and M.K. Kadyrbaev 1979. *Sokrovishcha drevnego Kazakhstana* [Treasures of ancient Kazakhstan]. Alma-ata: Zhalin.

Akishev, K.A. 1959. "Pamjatniki stariny Severnogo Kazakhstana" [Ancient sites of northern Kazakhstan]. *Trudy Instituta Istorii Arkheologii i Etnografii* 7: 3–31.

Alimov, K., N. Boroffka, M. Bubnova, J. Burjakov, J. Cierny, J. Jakubov, J. Lutz *et al.* 1998. "Prähistorischer Zinnbergbau in Mittelasien. Vorbericht der ersten kampagne 1997." *Eurasia Antiqua* 4: 137–199.

Amiet, P. 1986. *L'Âge des échanges inter-iraniens, 3500–1700 avant J.C.* Paris: Editions de la Réunion des musées nationaux.

Amschler, J.W. 1940. "Tierreste der Ausgrabungen von dem 'Grossen Königshügel' Shah Tepe, in Nord-Iran." In *Reports from the Scientific Expedition to the North-Western Provinces of China Under the Leadership of Dr. Sven Hedin: The Sino-Swedish Expedition*. No. 9, Archaeology 4. Stockholm: Elanders Boktryckeri Aktiebolag, pp. 35–129.

Andrianov, B.V. [1969] 2016. *Ancient Irrigation Systems of the Aral Sea Area: The History, Origin, and Development of Irrigated Agriculture*, ed. S. Mantellini with the collaboration of C.C. Lamberg-Karlovsky and Maurizio Tosi. Oxford and Philadelphia: Oxbow Books.

Anthony, D.W. 2007. *The Horse, the Wheel and Language: How Bronze Age Raiders from the Eurasian Steppe Shaped the Modern World*. Princeton, NJ and Oxford: Princeton University Press.

Anthony, D.W. 2009. "The Sintashta Genesis: The Roles of Climate Change, Warfare, and Long-Distance Trade." In *Social Complexity in Prehistoric Eurasia. Monuments, Metals, and Mobility*, ed. K.M. Hanks and B.K. Linduff. New York: Cambridge University Press, pp. 47–73.

Anthony, D.W. and D.R. Brown 2000. "Eneolithic Horse Exploitation in the Eurasian Steppes: Diet, Ritual, and Riding." *Antiquity* 74: 75–86.

Anthony, D.W. and N.B. Vinogradov 1995. "The Birth of Chariot." *Archaeology* 48(2): 36 – 41.

Appadurai, A., 1986. *The Social Life of Things: Commodities in Cultural Perspective*. Cambridge: Cambridge University Press.

Avanesova, N.A. 1996. "Pasteurs et agriculteurs de la vallée du Zeravshan (Ouzbékistan) au début de l'âge du Bronze: relations et influences mutuelles." In *Sarazm (Tadjikistan) céramiques (Chalcolithique et Bronze Ancien)*, ed. B. Lyonnet. Paris: de Boccard, pp. 117–124.

Avanesova, N.A. 2010. "Zeravshanskaja kul'turnaja provintsija Baktrijsko-Margianskoj Tsivilizatsii" [The Zeravshan cultural province of the Bactria-Margiana Civilization]. In *Na puti otkrytija tsivilizatsii* [On the track of uncovering a civilization]. Trudy Margianskoj

Arkheologicheskoj Ekspeditsii, vol. 3, ed. P.M. Kozhin, M.F. Kosarev, and N.A. Dubova. St. Petersburg: Aletejja, pp. 334–364.

Avanesova, N.A. 2013. "Zhukov, un 'sanctuaire' énéolithique d'anciens nomades dans la vallée du Zeravshan (Ouzbékistan)." *Paléorient* 39(2): 85–108.

Avanesova, N.A. 2015. "Keramika poselenija gornjakov-metallurgov Zarafshana." *Istorija Material'noj Kul'tury Uzbekistana* 39: 47–62.

Avanesova, N., Sh. Shajdullaev, and A. Erkulov 2001. "Dzham – ein neuer bronzezeitlicher Fundort in der Sogdiana." *Archäologische Mitteilungen aus Iran und Turan* 33: 63–73.

Badre, L. 1980. *Les figurines anthropomorphes en terre cuite a l'âge du Bronze en Syrie.* Paris: Librairie orientaliste Paul Geuthner.

Bajpakov, K.M., I.V. Erofeeva, V.A. Inochkin, S.S. Krivobokova, and M.N. Sdykov 2012. *Ural'sk: drevnyj i sovremennyj: Ocherki istorii* [Uralsk: ancient and modern: historical notes]. Ural'sk: Zapadno-Kazakhstanskij oblastnoj tsentr istorii i arkheologii.

Bajpakov, K.M. and A.N. Mar'jashev 2004. *Petroglify v gorakh Kul'zhabasy* [Petroglyphs in the Kul'zhabasy Mountains]. Almaty: Institut Arkheologii im. A.Kh. Margulana.

Bejsenov, A.Z., V.V. Varfolomeev, and A.E. Kasenalin 2014. *Pamjatniki Begazy-Dandybaevskoj Kul'tury Tsentral'nogo Kazakhstana* [Begazy-Dandybay culture sites of Central Kazakhstan]. Almaty: Institut Arkheologii im. A.Kh. Margulana.

Bejsenov, A.Z., V.V. Varfolomeev, V.K. Mertz, and I.V. Mertz 2014. "Raskopki mogil'nika Karaoba v 2013 g" [The excavation of the Karaoba burial ground in 2013]. In *Dialog kul'tur Evrazii v arkheologii Kazakhstana* [Dialogue of Eurasian cultures in the Kazakhstan archaeology]. Astana: Sarjarka, pp. 173–186.

Besenval, R. 1987. "Découvertes récentes à Sarazm (R.S.S.du Tadjikistan): attestation des relations au IIIe millénaire, entre l'Asie Centrale, l'Iran du Nord-Est et le Baluchistan." *Comptes Rendus de l'Académie des Inscription et Belles-Lettres*, April–June: 441–456.

Besenval, R. 1988. "L'étain dans l'Asie centrale protohistorique: une source possible pour les métallurgies moyen-orientales. Examen de l'hypothèse." In *L'Asie centrale et ses rapports avec les civilisations orientales, des origines à l'âge du fer*, ed. J.-C. Gardin. Paris: de Boccard, pp. 229–235.

Bisembaev, A.A. and M.N. Dujsengali 2013. *Drevnosti aktjubinskogo kraja* [Antiquities of Aktobe region]. Aktobe: Universitet baspa ortalygy.

Bobomulloev, S. 1997. "Ein bronzezeitliches Grab aus Zardča Chalifa bei Pendžikent (Zeravšan-Tal)." *Archäologische Mitteilungen aus Iran und Turan* 29: 121–134.

Bonora, G.L. 2006. "Prospettive archeologiche nella ricostruzione linguistica del Baluchistan protostorico." Unpublished PhD thesis, Università di Napoli "L'Orientale," Naples.

Bonora, G.L. 2018. "A General Revision of the Chronology of the Tagisken North Burial Ground." *Ancient Civilizations from Scythia to Siberia* 24: 307–330.

Bonora, G.L. forthcoming. "A Podstavska Incised with a Trident-like Potter's Mark from the Bronze Age site of Šagalaly II (Akmola Region, Kazakhstan) and its Links with Middle Asian Sites." In *Farmers, Traders and Herders: The Bronze Age in Central Asia and Khorasan*, ed. E. Luneau, M. Teufer, and N. Boroffka. Archäologie in Iran und Turan. Berlin: Dietrich Reimer.

Bonora, G.L. and S.K. Sakenov 2014. "A Topographical and Stratigraphical Periplous Around the Bronze Age Settlement of Shagalaly II (Pavlovka), Akmolinsk Region, Kazakhstan." In *Materialy Mezhdunarodnoj nauchno-prakticheskoj konferentsii "Margulanovskie Chtenija-2014"* [Proceedings of the Scientific-Practical Conference "Margulan Lectures 2014"]. Almaty-Pavlodar: EKO, pp. 107–131.

Boroffka, N., J. Cierny, J. Lutz, H. Parzinger, E. Pernicka, and G. Weisgerber 2002. "Bronze Age Tin from Central Asia: Preliminary Notes." In *Ancient Interactions: East and West in Eurasia*, ed. K. Boyle, C. Renfrew, and M. Levine. Cambridge: McDonald Institute for Archaeological Research, pp. 135–160.

Boroffka, N. and E. Sava 1998. "Zu Den steinernen 'Zeptern/Stoessel-Zeptern,' 'Miniatursaeulen' und 'Phalli' del Bronzezeit Eurasiens." *Archäologische Mitteilungen aus Iran und Turan* 30: 17–113.

Bostongukhar, S. 1998. *Verkhov'ja Zeravshana vo vtorom tys. do n.e.* [The upper Zeravshan valley in the 2nd millennium BC]. Dushanbe: FAN.

Brunet, F. 2005. "Pour une nouvelle étude de la culture néolitique de Kel'teminar, Ouzbékistan." *Paléorient* 31(2): 87–106.

Brunet, F. (with N.A. Avanesova) 2013. "Zhukov, un 'sanctuaire' énéolithique d'anciens nomades dans la vallée du Zeravshan (Ouzbékistan)." *Paléorient* 39(2): 86.

Buccellati, G. and M. Kelly-Buccellati 2002. "Tar'am-Agade, Daughter of Naram-Sin, at Urkesh." In *Of Pots and Plans: Papers on the Archaeology and History of Mesopotamia and Syria Presented to David Oates in Honour of His 75th Birthday*, ed. L. Al-Gailani Werr, J. Curtis, H. Martin, A. McMahon, J. Oates, and J. Reade. London: Nabu Publications, pp. 11–31.

Buchanan, B. 1966. *Catalogue of Ancient Near Eastern Seals in the Ashmolean Museum.* Vol. 1, *Cylinder seals.* Oxford: Clarendon.

Cattani, M. 2008. "The Final Phase of the Bronze Age and the 'Andronovo Question' in Margiana." In *The Bronze Age and Early Iron Age in the Margiana Lowlands: Facts and Methodological Proposals for a Redefinition of the Research Strategies*, ed. S. Salvatori and M. Tosi with B. Cerasetti. BAR International Series 1806. Oxford: Archaeopress, pp. 133–151.

Chernykh, E.N. 2004. "Kargaly: The Largest and Most Ancient Metallurgical Complex on the Border of Europe and Asia." In *Metallurgy in Ancient Eastern Eurasia, from the Urals to the Yellow River*, ed. K.M. Linduff. Lewiston: Mellen Press, pp. 223–237.

Compagnoni, M. and M. Tosi 1975. "The Camel: Its Distribution and State of Domestication in the Middle East During the Third Millennium BC in Light of Finds from Shahr-i Sokhta." In *Approaches to Faunal Analysis in the Middle East*, ed. R.H. Meadow and M.A. Zeder. Peabody Museum Bulletin 2. Cambridge, MA: Peabody Museum editions, pp. 91–103.

Dietz, U.L. 2003. "Horseback Riding: Man's Access to Speed?" In *Prehistoric Steppe Adaptation and the Horse*, ed. M. Levine, C. Renfrew, and K. Boyle. Cambridge: McDonald Institute for Archaeological Research, pp. 189–199.

Dossin, G. 1970. "La route de l'étain en Mésopotamie au temps de Zimri-Lim." *Revue d'Assyriologie* 64: 97–106.

Doumani, P., M. Frachetti, R. Beardmore, T. Schmaus, R. Spengler, and A. Mar'yashev 2015. "Burial Ritual, Agriculture, and Craft Production Among Bronze Age Pastoralists at Tasbas (Kazakhstan)." *Archaeological Research in Asia* 1–2: 17–32. https://doi.org/10.1016/j.ara.2015.01.001.

Drews, R. 2004. *Early Riders: The Beginning of Mounted Warfare in Asia and Europe.* New York: Routledge.

Dudd, S.N., R.P. Evershed, and M. Levine 2003. "Organic Residue Analysis of Lipids in Potsherds from the Early Neolithic Settlement of Botai, Kazakhstan." In *Prehistoric Steppe Adaptation and the Horse*, ed. M. Levine, C. Renfrew, and K. Boyle. Cambridge: McDonald Institute for Archaeological Research, pp. 45–54.

Duerst, J.U. 1908. "Animal Remains from the Excavation of Anau and the Horse of Anau in Its Relation to the History and the Races of Domesticated Horses." In *Explorations in Turkestan: Expedition of 1904 – Prehistoric Civilizations of Anau. Origins, Growth and Influence of Environment*, ed. R. Pumpelly. Washington DC: Carnegie Institution of Washington, pp. 341–442.

Dzhurakulov, M.D. and N.A. Avanesova 1984. "Novye issledovanija po Sazaganskomu poselenija" [New research on the Sazagan settlement]. *Istorija Material'noj Kul'tury Uzbekistana* 19: 32–40.

Epimakhov, A.V. 2007. "Otnositel'naja i absoljutnaja khronologija sintashtinskikh pamjatnikov v svete radiokarbonnykh datirovok" [Relative and absolute chronology of the Sintashta sites in the light of radiocarbon datings]. *Problemy istorii, filologii, kul'tury* 17: 402–421.

Epimakhov, A.V. 2016. "K voprosu o radiouglerodnoj argumentatsii rannej datirovki alakul'skikh drevnostej" [On the ^{14}C argumentation for the early dating of the Alakul' antiquities]. *Vestnik Arkheologii, Antropologii i etnografii, Elektronnyj zhurnal* 3(34): 60–67.

Epimakhov, A. and R. Krause 2013. "Relative and Absolute Chronology of the Settlement Kamennyi Ambar." In *Multidisciplinary Investigations of the Bronze Age Settlements in the Southern Trans-Urals (Russia)*, ed. R. Krause and L. Koryakova. Bonn: Habelt, pp. 129–143.

Ermolova, N.M. 1970. "Novye materialy po izucheniju ostatkov mlekopitajushchikh iz drevnikh poselenii Turkmenii" [New materials for the study of mammal remains from ancient Turkmenian sites]. *Karakumskie Drevnosti* 3: 205–232.

Frachetti, M.D. 2008. "Variability and Dynamic Landscapes of Mobile Pastoralism in Ethnography and Prehistory." In *The Archaeology of Mobility. Old World and New World Nomadism*, ed. W. Barnard and H., Wendrich. Los Angeles, CA: Cotsen Institute of Archaeology, pp. 366–396.

Frachetti, M. and A. Maryashev 2007. "Long-Term Occupation and Seasonal Settlement of East Eurasian Pastoralists at Begash, Kazakhstan." *Journal of Field Archaeology* 32: 221–242. https://doi.org/10.1179/009346907791071520.

Francfort, H.-P. 1989. *Fouilles de Shortugai. Recherches sur l'Asie Centrale protohistorique.* Paris: de Boccard.

Garner, J. 2013. *Das Zinn Der Bronzezeit in Mittelasien II. Die Montanarchäologischen Forschungen an Den Zinnlagerstätten.* Berlin-Bochum: Deutsches Archäologisches Institut Eurasien-Abteilung und Deutsches Bergbau-Museum Bochum.

Ghirshmann, R. 1938. *Fouilles de Sialk, près de Kashan*, 2 vols. Paris: Geuthner.

Gillmore, G.K., R.A.E. Coningham, H. Fazeli, R.L. Young, M. Magshoudi, C.M. Batt, and G. Rushworth 2009. "Irrigation on the Tehran Plain, Iran: Tepe Pardis – the Site of a Possible Neolithic Irrigation Feature?" *Catena* 78: 285–300.

Ginzburg, V.V. 1959. "Materialy k antropologii naselenija Juzhnoj Turkmenii v epokhu pozdnej bronzy" [Anthropological materials of the South Turkmenia population in the Late Bronze Age]. In Trudy JuTAKE *IX*, ed. M.E. Masson. Ashgabad: Nauka, pp. 102–106.

Görsdorf, J., H. Parzinger, and A. Nagler 2001. "New Radiocarbon Dates of the North Asian Steppe Zone and Its Consequences for the Chronology." *Radiocarbon* 43(2B): 1115–1120. https://doi.org/10.1017/S003382220004176X.

Görsdorf, J., H. Parzinger, A Nagler, and N. Leont'ev 1998. "Neue ^{14}C Datierungen für die Sibirische Steppe and ihre Konsequenzen für die regionale Bronzezeitchronologie." *Eurasia Antiqua* 4: 73–80.

Grigor'ev, S.A. 2016. "Problema khronologii i proiskhozhdenija alakul'skoj kul'tury v svete novykh raskopok v Juzhnom Zaural'e" [The question of chronology and origin of the Alakul' culture according to the excavations in the south Trans-Urals region]. *Vestnik Arkheologii, Antropologii i etnografii, Elektronnyj zhurnal* 3(34): 44–53.

Guljamov, Ya.G., U. Islamov, and A. Askarov 1966. *Pervobytnaja kul'tura i vozniknovenie oroshaemogo zemledelija v nizov'jakh Zarafshana* [Indigenous culture and formation of irrigated agriculture in the lower Zeravshan]. Tashkent: Fan.

Haak, W., I. Lazaridis, N. Patterson, N. Rohland, S. Mallick, B. Llamas, G. Brandt *et al.* 2015. "Massive Migration from the Steppe was a Source for Indo-European Languages in Europe." *Nature* 522(7555): 207–211.

Hanks, B., A. Epimakhov, and C. Renfrew 2007. "Towards a Refined Chronology for the Bronze Age of the Southern Urals, Russia." *Antiquity* 81(312): 353–367. https://doi.org/10.1017/S0003598X00095235.

Hiebert, F.T. 1994. *Origins of the Bronze Age Oasis Civilizations of Central Asia*. Bulletin of the American School of Prehistoric Research 42. Cambridge, MA: Peabody Museum of Archaeology and Ethnology.

Isakov, A.I. 1991. *Sarazm*. Dushanbe: Donish.

Isakov, A.I. 1992. "Bogatoe zhenskoe pogrebenie iz Sarazma (Tadzhikistan)" [A rich woman burial from Sarazm (Tajikistan)]. *Arkheologicheskie Vesti* 1: 64–75.

Isakov, A.I. 1994. "Sarazm: An Agricultural Center of Ancient Sogdiana." *Bulletin of the Asia Institute* 8: 1–12.

Isakov, A.I. and T.M. Potemkina 1989. "Mogil'nik plemen epokhi bronzy v Tadzhikistane" [A Bronze Age tribal burial ground in Tadjikistan]. *Sovetskaja Arkheologija* 1: 145–167.

Itina, M.A. 1977. *Istorija stepnykh plemen juzhnogo Priaral'ja (Vtorogo – nachalo pervogo tysjachlet'ja do n.e.)* [History of the steppe tribes of the southern Aral Sea region (2nd–beginning of the 1st millennium BC)]. Moscow: Nauka.

Itina, M.A. and L.T. Jablonskij 2001. *Mavzolej severnogo Tagiskena, Pozdnij bronzovyj vek Nizhnej Syrdar'i* [The mausoleums of Tagisken North: the Late Bronze Age of the lower Syr Darya]. Moscow: Vostochnaja Literatura RAN.

Izbitser, J. 1998. "Wheeled Vehicles and the Homeland of the Indo-Europeans." In *Proceedings of the Seventh Annual UCLA Indo-European Conference*, ed. A della Volpe and E.C. Polomé. Washington, DC: Institute for the Study of Man, pp. 144–149.

Jacobson, E., V.D. Kubarev, D. Tseevendorj, G. Tepfer, and J. Meacham 2001. *Répertoire des pétroglyphes d'Asie centrale. Fascicule No. 6:* Mongolie du nord-ouest, Tsagaan Salaa/Baga Oigor. Mémoires de la Mission Archéologique en Asie Centrale, vol. 6. Paris: de Boccard.

Jansen, M. 1985. "Mohenjo-Daro, City of the Indus Valley." *Endeavour* 9: 161–169.

Kadyrbaev, M.K. and Zh. Kurmankulov 1992. *Kul'tura drevnikh skotovodov i metallurgov Sary-Arki (po materialam severnoj Betpak-Daly)* [The culture of the ancient herdsmen and metallurgists of Sary-Arka (from the materials of northern Betpak-Dale)]. Alma-ata: Gylym.

Kadyrbaev, M.K. and A.N. Mar'jashev 2007. *Petroglify khrebta Karatau* [Petroglyphs of the Karatau Mountains]. Almaty: Institut Arkheologii im. A.Kh. Margulana.

Khlopin, I.N. 1997 *Eneolit jugo-zapadnogo Turkmenistana* [The Aeneolithic of the south-western Turkmenistan]. St. Petersburg: Evropejskij Dom.

Kircho, L.B. 2009. "The Earliest Wheeled Transport in Southwestern Central Asia: New Finds from Altyn Depe." *Archaeology, Ethnology and Anthropology of Eurasia* 37(1): 25–33. doi: 10.1016/j.aeae.2009.05.003.

Kircho, L.B. and S.G. Popov 2005. "K voprosu o radiouglerodnoj khronologii arkheologicheskikh pamjatnikov Srednej Azii, V-II tys. do n.e." [On the radiocarbon chronology of the Central Asia archaeological sites, 5th–2nd millennium BC]. In *Khronologija epokhi Pozdnego Eneolita–Srednej Bronzy Srednej Azii (Pogrebenija Altyn-depe)* [Chronology of the Late Chalcolithic–Middle Bronze Age of Central Asia (Burials of Altyn Depe)], ed. V.M. Masson and Ju.E. Berezkin. St. Petersburg: Nestor-Istorija, pp. 528–539.

Kirjushin, Ju.F., M.A. Demin, D.V. Papin, A.A. Rednikov, and A.S. Fedoruk 2014. "Arkheologicheskie issledovanija na poselenii Burla-3 v 2013–2014 gg." [Archaeological research at Burly 3 in 2013 and 2014]. In *Polevye issledovanija v Priirtysh'e, Verkhnem Priob'e i na Altae 2014 g* [Field research along the Irtysh, in the upper Ob' and in the Altai in 2014]. Barnaul: AltGPU, pp. 49–56.

Kirjushin, Ju.F., D.V. Papin, A.S. Fedoruk, A.A. Rednikov, and O.S. Fedoruk 2013. "Problema izuchenija pamjatnikov 'burkinskogo tipa' epokhi pozdnej bronzy na territorii stepnogo Altaja" [Questions on the Late Bronze Age "Burly-type" sites in the steppe Altai region]. *Problemy arkheologii, etnografii, antropologii Sibiri i sopredel'nykh territorij* 19: 212–215.

Korjakova, L. and A. Epimakhov 2007. *The Urals and Western Siberia in the Bronze and Iron Ages*. New York: Cambridge University Press.

Kuz'mina, E.E. 1958. "Mogil'nik Zaman-Baba" [The burial ground of Zaman-Baba]. *Sovetskaja Etnografija* 2: 24–33.

Kuz'mina, E.E. 1966. "Metallicheskie izdelija eneolita i bronzovogo veka v Srednej Azii" [Metal artifacts of the Copper and Bronze Age in Central Asia]. In *Arkheologija SSSR svod arkheologicheskikh istochnikov*, issue B4-9, ed. V.I. Moshinskaja. Moscow: Nauka.

Kuz'mina, E.E. 1997. "The Cultural Connections Between the Shepherds of the Steppes and South Central Asia, Afghanistan and India in the Bronze Age." In *South Asian Archaeology 1995*. ed. R. Allchin and B. Allchin. New Delhi: Science Publishers and IBH Publishing Co., pp. 279–289.

Kuz'mina, E.E. 2000. "The Eurasian Steppe: The Transition from Early Urbanism to Nomadism." In *Kurgans, Ritual Sites, and Settlements: Eurasian Bronze and Iron Age*, ed. J. Davis-Kimball, E.M. Murphy, L. Koryakova, and L.T. Yablonsky. BAR International Series, 890, Oxford: Archaeopress, pp. 118–123.

Kuz'mina, E.E. 2001. "The First Migration Wave of Indo-Iranians to the South." *Journal of Indo-European Studies* 29(1–2): 1–40.

Kuz'mina, E.E. 2007. *The Origin of the Indo-Iranians*. Leiden: Brill.

Kuz'mina, E.E. 2008a. *Klassifikatsija i periodizatsija pamjatnikov Andronovskoj kul'turnoj obshchnosti* [Classification and periodization of the Andronovo culture sites]. Aktobe: Print A.

Kuz'mina, E.E. 2008b. *The Prehistory of the Silk Road*, ed. V.M. Mair. Philadelphia: University of Pennsylvania Press.

Kuz'mina, E.E. and A.A. Ljapin 1984. "Novye nakhodki stepnoj keramiki epokhi bronzy na Murgabe" [New finds of Bronze Age steppe pottery along the Murghab]. In *Problemy arkheologii Turkmenistana* [Questions of archaeology in Turkmenistan], ed. V.M. Masson. Ashgabad: Ylym, pp. 6–22.

Lamberg-Karlovsky, C.C. 2002. "Archaeology and Language: The Indo-Iranians." *Current Anthropology* 43(1): 63–88.

Lev, D.N. 1966. "Pogrebenie bronzovoj epokhi bliz g. Samarkanda" [A burial of the Bronze Age near Samarkand]. *Kratkie Soobshchenija Instituta Arkheologii* 108: 101–105.

Levine, M.A., C. Renfrew, and K. Boyle (eds.) 2003. *Prehistoric Steppe Adaptation and the Horse*. Cambridge: McDonald Institute for Archaeological Research.

Lizitzyna, G.N. 1981. "The History of Irrigation Agriculture in Southern Turkmenia." In *The Bronze Age Civilisation of Central Asia: Recent Soviet Discoveries*, ed. P.L. Kohl. New York: M.E. Sharpe, pp. 350–358.

Lyonnet, B. 1996 (with A. Isakov and N. Avanessova). *Sarazm (Tadjikistan) céramiques (Chalcolithique et Bronze Ancien)*. Paris: de Boccard.

Majidzadeh, Y. 2004. *Jiroft Earliest Oriental Civilization*. Tehran: Organization of the Ministry of Culture and Islamic Guidance.

Maljutina, T.S. 1987. *Poselenie i mogil'nik Pavlovka. Otchet. Polevye arkheologicheskie issledovanija UKAE v Kazakhstane v 1982–1983 gg* [The Pavlovka settlement and burial ground report: field archaeological research of the Uralo-Kazakhstan archaeological expedition in Kazakhstan (1982–1983)]. Cheljabinsk: Juzhnij Uralskij Gosudarstvennij Universitet, pp. 3–26.

Maljutina, T.S. 1991. "Stratigraficheskaja pozitsija materialov Fedorovskoj kul'tury na mnogoslojnykh poselenijakh kazakhstanskikh stepej" [The stratigraphical position of the Fedorovo culture materials in the multilayered sites of the Kazakhstan steppes]. In *Drevnosti vostochno-evropejskoj lesostepi: Mezhvuzovskij sbornik nauchnykh trudov* [Ancient eastern European forest-steppe: inter-university collection of scientific works]. Samara: Samarskij gosudarstvennij pedagogicheskij institut imeni V.V. Kujbysheva, pp. 141–162.

Maljutina, T.S. and G.V. Zdanovich 1995. "Kujsak – ukreplennoe poselenie protogorodskoj tsivilizatsii juzhnogo Zaural'ja" [Kujsak: a fortified settlement of the proto-urban civilization of south Trans-Urals]. In *Rossija i Vostok: Problemy vzaimodejstvja, Chast' 5, blok 1: Kul'tury Eneolita-Bronzy stepnoj Evrazii* [Russia and the East: problems of interaction, part 5, block 1: the Chalcolithic and Bronze Age culture of the Eurasian steppe]. Cheljabinsk: Juzhnij Uralskij Gosudarstvennij Universitet, pp. 100–106.

Mandel'shtam, A.M. 1966a. "Pamjatniki stepnogo kruga epokhi bronzy na juge Srednej Azii" [Bronze Age steppe sites in southern Central Asia]. In *Srednjaja Azija v epokhu kamnja i bronzy* [Central Asia in the Stone and Bronze Age]. Leningrad: Nauka, pp. 239–259.

Mandel'shtam, A.M. 1966b. "Pogrebenija srubnoj kul'tury v Juzhnoj Turkmenii" [Srubnaja culture burials in South Turkmenii], *Kratkie Soobshchenija Instituta Arkheologii* 108: 105–108.

Margulan, A.Kh. 2001. Sochinenija *Tom 2: Sarjarka. Gornoe Delo i metallurgija v epokhu bronzy. Dzhezkazgan – drevniy e srednevekovyy metallurgicheskiy tzentr (gorodishche Milykuduk)* [Works Vol. 2. Sarjarka. Bronze Age mining and metallurgy. Dzhezkazgan: ancient and mediaeval metallurgical center (town of Milykuduk)]. Almaty: Dayk Press.

Mar'jashev, A.N., A.A. Gorjachev, and S.A. Potapov 1998. *Kazakhstan I: choix de pétroglyphes du Semirech'e. Répertoire des Pétroglyphes d'Asie centrale.* Fascicule no. 5. Mémoires de la Mission Archéologique Française en Asie centrale, vol. 5. Paris: de Boccard.

Masson, V.M. 1966. "Kul'tura Zaman-Baba" [The Zaman-Baba culture]. In *Srednjaja Azija v epokhu kamnja i bronzy* [Central Asia in the Stone and Bronze Age], ed. V.M. Masson. Moscow: Nauka, pp. 208–212.

Masson, V.M. and V.I. Sarianidi 1972. *Central Asia Before the Achaemenids.* New York: Praeger Editions.

Matjushchenko, V.I. and G.V. Sinitzyna 1988. *Mogil'nik u derevnja Rostovka vblizi Omska* [The burial ground near the village of Rostovka, around Omsk]. Tomsk: Tomskogo gosudarstvennogo Universiteta.

Meadow, R.H. 1984. "A Camel Skeleton from Mohenjo-daro." In *Frontiers of the Indus Civilization: Sir Mortimer Wheeler Commemoration Volume*, ed. B.B. Lal and S.P. Gupta. New Delhi: Books & Books on behalf of the Indian Archaeological Society and the Indian History & Culture Society, pp. 133–140.

Meadow, R.H. 1993. "Animal Domestication in the Middle East: A Revised View from the Eastern Margin." In *Harappan Civilization: a Recent Perspective*, 2nd ed., ed. G.L. Possehl. Delhi: Oxford & IBH and the American Institute of Indian Studies, pp. 295–320.

Mertz, V.K. 2013. "Begazinskij kompleks iz Karaoba (Semipalatinskoe Priirtysh'e)" [The Begazy Complex of Karaoba (in the Irtysh valley of Semipalatinsk)]. In *Begazy-Dandybaevskaja Kul'tura Stepnoj Evrazii. Sbornik Nauchnykh Statej, Posvjashchennyj 65-Letiju Zh. Kurmankulova* [The Begazy-Dandybay culture of the Eurasian steppe: collection of scientific works celebrating the 65th birthday of Zh. Kurmankulov]. Almaty: Begazy-Tasmola, pp. 270–282.

Mertz, V.K. 2016. "Fenomen Begazy-Dandybaja v Priirtysh'e" [The Begazy-Dandybaj phenomenon in the region of the Irtysh River]. *Madeni Mura* 4: 78–87.

Molodin, V., A.V. Epimakhov, and Zh.V. Marchenko 2014. "Radiouglerodnaja khronologija epokhi bronzy Urala i juga Zapadnoj Sibiri: printsipy i podkhody, dostizhenija i problemy" [Radiocarbon chronology of the Bronze Age in the Urals and southwestern Siberia: principles and approaches, results and problems]. *Arkheologija i Etnografija* 13(3): 136–167.

Molodin, V., Z. Marchenko, Y. Kuzmin, A. Grishin, M. van-Strydonck, and L. Orlova 2012. "C-14 Chronology of Burial Grounds of the Andronovo Period (Middle Bronze Age) in the Baraba Forest Steppe, Western Siberia." *Radiocarbon* 54 (3–4): 737–747. https://doi.org/10.1017/S0033822200047391.

Molodin, V., L. Mylynikova, O. Novikova, I. Durakov, L. Koveleva, N. Efrmova, and A. Soloviev 2011. "Periodization of Bronze Age Cultures in the Ob-Irtysh Forest-Steppe." *Archaeology, Ethnology and Anthropology of Eurasia* 39: 40–56. https://doi.org/10.1016/j.aeae.2011.11.003.

Moorey, P.R.S. 1970. "Pictorial Evidence for the History of Horse-Riding in Mesopotamia Before the Kassites." *Iraq* 32: 36–50. https://doi.org/10.2307/4199890.

Mughal, M.R. 1970. "The Early Harappan Period in the Greater Indus Valley and Baluchistan." Unpublished PhD thesis, University of Pennsylvania, Philadelphia.

Muhly, J.D. 1995. "Mining and Metalwork in Ancient Western Asia." In *Civilizations of the Ancient Near East*, vol. 3, ed. J.M. Sasson, J. Baines, G. Beckman, and R. Rubinson. New York: Scribner's, pp. 1501–1519.

Novozhenov, V.A. 2002. *Petroglify Sary-Arki* [Petroglyphs of Sary-Arka]. Almaty: Institut Arkheologii im. A.Kh. Margulana.

Novozhenov, V.A. 2012. *Chudo kommunikatsii i drevnejshij kolesnyj transport Evrazii* [The miracle of communication and the earliest wheeled transport in Eurasia]. Moscow: TAUS.

Orazbaev, A.M. 1958. "Severnyj Kazakhstan v epokhu bronzy" [North Kazakhstan in the Bronze Age]. *Trudy Instituta Istorii Akademii Nauk KazSSR* 5: 216–294.

Orazov, A.T. 2016. "Epokha bronzy v Margiane: stroitel'nye materialy i konstruktsii" [The Bronze Age in Margiana: building materials and constructions]." In *Trudy Margianskoy Arkheologicheskoy Ekspeditzii*, vol. 6, ed. N.A. Dubova. Moscow: Staryj Sad, pp. 224–231.

Outram A.K., N.A. Stear, R. Bendrey, S. Olsen, A. Kasparov, V.F. Zaibert, N. Thorpeand, and R.P. Evershed 2009. "The Earliest Horse Harnessing and Milking." *Science* 323: 1332–1335. https://doi: 10.1126/science.1168594.

Owen, D.I. 1991. "The 'First Equestrian': An Ur III Glyptic Scene." *Acta Sumerica* 13: 259–273.

Panyushkina, I., C. Chang, A. Clemens, and N. Bykov 2010. "First Tree-Ring Chronology from Andronovo Archaeological Timbers of Bronze Age in Central Asia." *Dendrochronologia* 28: 13–21. https://doi.org/10.1016/j.dendro.2008.10.001.

Panyushkina, I., B. Mills, E. Usmanova, and C. Li 2008. "Calendar Age of Lisakovsky Timbers Attributed to Andronovo Community of Bronze Age in Eurasia." *Radiocarbon* 50: 459–466. https://doi.org/10.1017/S0033822200053558.

Pare, C.F.E. 1992. *Wagons and Wagon Graves of the Early Iron Age in Central Europe.* Oxford University Committee for Archaeology Monograph 35. Oxford: Oxford University Committee for Archaeology.

Parpola, A. 1999. "The Formation of the Aryan Branch of Indo-European." In *Archaeology and Language*. Vol. 3, *Artifacts, Languages, and Texts*, ed. R. Blench and M. Spriggs. London: Routledge, pp. 180–207.

Parzinger, H. and N. Boroffka 2003. *Das Zinn in der Bronzezeit in Mittelasien I: Die siedlungsarchäologischen Forschungen im Umfeld der Zinnlagerstätten.* Archäologie in Iran und Turan 5. Mainz: Von Zabern.

Peters, J. and A. von den Driesch 1997. "The Two-Humped Camel (Camelus bactrianus): New light On Its Distribution, Management and Medical Treatment in the Past." *Journal of Zoology* 242: 651–679.

Piggott, S. 1992. *Wagon, Chariot and Carriage: Symbol and Status in the History of Transport.* New York: Thames & Hudson.

Potts, D.T. 1999. *The Archaeology of Elam.* Cambridge: Cambridge University Press.

Potts, D.T. 2005. "Bactrian Camels and Bactrian-Dromedary Hybrids." *Silk Road* 3(1): 49–58.

P'jankova, L.T. 1981. "Bronze Age Settlements of Southern Tadjikistan." In *The Bronze Age Civilization of Central Asia. Recent Soviet Discoveries*, ed. P.L. Kohl. New York: Armonk, pp. 287–310.

P'jankova, L. 1994. "Central Asia in the Bronze Age: Sedentary and Nomadic Cultures." *Antiquity* 68(259): 355–372.

Salvatori, S. 2008. "Cultural Variability in the Bronze Age Oxus Civilization and Its Relations with the Surrounding Regions of Central Asia and Iran." In *The Bronze Age and Early Iron Age in the Margiana Lowlands: Facts and Methodological Proposals for a Redefinition of the Research Strategies*, ed. S. Salvatori and M. Tosi with the editorial collaboration of B. Cerasetti. BAR International Series 1806. Oxford: Archaeopress, pp. 75–98.

Salvatori, S. and M. Tosi 2005. "Shahr-i Sokhta Revised Sequence." In *South Asian Archaeology 2001*. Vol. I, *Prehistory*, ed. C. Jarrige and V. Lefèvre. Paris: Éditions Recherche sur les Civilisations, pp. 281–292.

Salvatori, S., M. Vidale, G. Guida, and G. Gigante 2002. "A Glimpse on Copper and Lead Metalworking at Altyn-depe (Turkmenistan) in the 3rd Millennium BC." *Ancient Civilizations from Scythia to Siberia* 8(1–2): 69–106.

Salvatori, S., M. Vidale, G. Guida, and E. Masioli 2009. "Ilgynly-Depe (Turkmenistan) and the 4th Millennium BC Metallurgy of Central Asia." *Paléorient* 35(1): 47–67. https://doi:10.3406/paleo.2009.5277.

Samashev, Z., D. Baytileu, and Zh. Kurmankulov 2013. *Petroglify Terekty Aulie* [Petroglyphs of Terekty Aulie]. Astana: Institut Arkheologii im. A.Kh. Margulana.

Samashev, Z., F. Grigor'ev, and G. Zhumabekova 2005. *Drevnosti Almaty* [Ancient Almaty]. Almaty: KazIzdat KT.

Sarianidi, V.I. 1986. "Mezopotamija i Baktria vo vtorom tys. do n.e." [Mesopotamia and Bactria in the 2nd millennium BC]. *Sovetskaja Arkheologija* 2: 34–46.

Sarianidi, V.I. 1990. *Drevnosti Strany Margush* [Ancient Margush]. Ashgabat: Ylym.

Sarianidi, V.I. 1998a. *Margiana and Protozoroastrism*. Athens: Kapon Editions.

Sarianidi, V.I. 1998b. *Myths of Ancient Bactria and Margiana on Its Seals and Amulets*. Moscow: Pentagraphic.

Sarianidi, V.I. 2005. *Gonurdepe: City of Kings and Gods*. Ashgabat: Miras.

Sarianidi, V.I. 2007. *Necropolis of Gonur*. Athens: Kapon Editions.

Sarianidi, V.I. 2008. *Margush: Mystery and Truth of the Great Culture*. Ashgabat: Turkmend-owlethabarlary.

Sitnikov, S.M. 2015. *Kul'tura sargarinsko-alekseevskogo naselenija lesostepnogo i stepnogo Altaja* [The culture of the Sargary-Alekseevka communities in the Altai forest-steppe and steppe]. Barnaul: AltGPU.

Sungatov, F.A. and F.F. Safin 1995. "Issledovanija kurgannykh mogil'nikov v Zaural'e v 1991g" [Study of barrow burial grounds in the Trans-Ural regions in 1991]. In *Nasledie vekov. Okhrana i izuchenie pamjatnikov arkheologii v Bashkortostane* [The legacy of the centuries: preservation and study of archaeological sites in Bashkortostan]. Ufa: Natsional'nyj Muzej, pp. 58–64.

Svyatko, S.V., J.P. Mallory, E.M. Murphy, A.V. Polyakov, P.J. Reimer, and R.J. Schulting 2009. "New Radiocarbon Dates and a Review of the Chronology of Prehistoric Populations from the Minusinsk Basin, Southern Siberia, Russia." *Radiocarbon* 51(1): 243–273. https://doi.org/10.1017/S0033822200033798.

Szymczak, K. and M. Khudzhanazarov 2006. *Exploring the Neolithic of the Kyzyl-Kums: Ayakahytma, "the Site" and Other Collection*. Warsaw: Institute of Archaeology, Warsaw University.

Teufer, M. 2015. *Spätbronzezeitliche Grabfunde aus Nordbaktrien und benachbarten Regionen. Studien zur Chronologie zwischen Aralsee und Persischem Golf*. Berlin: Dietrich Reimer Verlag GmbH.

Tkachev, V.V. 2002. *Tsentral'nyj Kazakhstan v epokhu Bronzy* [Central Kazakhstan in the Bronze Age]. Tjumen': TjumGNGU.

Tkachev, V.V. 2016. "Radiouglerodnaja khronologija kuzhumberdynskoj kul'turnoj gruppy na zapadnoj periferii alakul'skogo areale" [The radiocarbon chronology of the Kozhumberdy culture group of sites in the western periphery of the Alakul' area]. *Vestnik Arkheologii, Antropologii i etnografii, Elektronnyj zhurnal* 3(34): 68–77.

Trinks, A., P. Burger, N. Benecke, and J. Burger 2012. "Ancient DNA Reveals Domestication Process: The Case of the Two-Humped Camel." In *Camels in Asia and North Africa: Interdisciplinary Perspectives on Their Past and Present Significance*, ed. E.-M. Knoll and P. Burger. Österreichische Akademie der Wissenschaften Philosophisch-Historische Klasse. Veröffentlichungen zur Sozialanthropologie 18. Vienne: Österreichische Akademie der Wissenschaften, pp. 79–86.

Urban, G. 2001. *Metaculture: How Culture Moves Through the World*. Public Worlds. Minneapolis: University of Minnesota Press.

Usmanova, E.R. 2010. *Kostjum zhenshchiny epokhi bronzy Kazakhstana. Opyt rekonstruktsii* [Female dress of the Bronze Age Kazakhstan: attempt of reconstruction]. Karaganda: Lisakovsk.

Varfolomeev, V.V. 2013. "Keramika superstratnogo oblika iz pamjatnikov Begazy-Dandybaevskoj kul'tury" [Pottery of superstratum appearance from the Begazy-Dandybay culture sites]. In *Begazy-Dandybaevskaja kul'tura Stepnoj Evrazii. Sbornik Nauchnykh statej, posvjashchennyj 65-letiju Zh. Kurmankulova* [The Begazy-Dandybay culture of the Eurasian steppe: collection of scientific articles dedicated to the 65th birthday of Zh. Kurmankulov]. Almaty: TOO NITZIA "Begazy-Tasmola," pp. 167–197.

Vidale, M. and A. Lazzari (eds.) 2017. *Lapis Lazuli Bead Making at Shahr-i Sokhta. Interpreting Craft Production in a Urban Community of the 3rd Millennium BC*. Rome: IsMEO and Università degli Studi di Padova.

Vinogradov, A.V. 1960. "Novye neoliticheskie nakhodki Khorezmskoj ekspeditsii AN SSSR 1957 g." [New Neolithic finds of the Khorezm expedition (USSR Academy of Science) in 1957]. In *Polevye issledovanija Khorezmskoj ekspeditsii v 1957 g. Materialy Khorezmskoj ekspeditsii, tom 4* [Field research of the Khorezm expedition in 1957: materials of the Khorezm expedition, vol. 4], ed. S.P. Tolstov. Moscow: Nauka, pp. 63–81.

Vinogradov, A.V. 1972. "Kyzylkumskie juveliry" [Jewelers of the Kyzylkum]. *Uspekhi sredneaziatskoj arkheologii* 2: 43–45.

Vinogradov, A.V. 1981. *Drevnie okhotniki i rybolovy sredneaziatskoj mezhdurech'ja. Trudy Khorezmskoj arkheologo-etnograficheskoj ekspeditsii* [The ancient hunters and fishermen of the Central Asia interfluve: works of the Khorezm archaeological and ethnographical expedition]. Moscow: Nauka.

Vinogradov, A.V. and E.D. Mamedov 1975. *Pervobytnyj Ljavljakan* [Prehistoric Ljavljakan]. Materialy Khorezmskoj arkheologo-etnograficheskoj ekspeditsii. Moscow: Nauka.

Vinogradov, N.B. 1995. "Juzhnye motivy v keramicheskikh kompleksakh epokhi bronzy v Juzhnom Zaural'e" [Southern motives in the pottery complexes of the Bronze Age of south-eastern Urals]. In *Konvergentsija i divergentsija v razvitii kul'tur epokhi eneolita-bronzy Srednej i Vostochnoj Evropy* [Convergence and divergence in the cultures development during the Chalcolithic and Bronze Age in Central and Eastern Europe], ed. S. Bochkarev, L.B. Kircho, V.M. Masson, and M.B. Rysin. St. Petersburg: Institute for the History of Material Culture of the Russian Academy of Sciences, pp. 71–74.

Vinogradov, N.B. 2003. *Mogil'nik bronzovogo veka Krivoe Ozero v Juzhnom Zaural'e* [The Bronze Age burial ground of Krivoe Ozero in southern Trans-Urals]. Cheljabinsk: Juzhno-Uralskoe Knizhnoe Izdatel'stvo.

Vinogradova, N.M. and E.E. Kuz'mina 1986. "Kontakty stepnykh i zemledel'cheskikh plemen Sredney Azii v epokhu Bronzy" [Contacts of steppe and farming tribes of Central Asia in the Bronze Age]. In *Vostochnyj Turkestan i Srednjaja Azija v sisteme kul'tur drevnego i srednevekovogo Vostoka* [Eastern Turkestan and Central Asia in the cultural system of the ancient and mediaeval east], ed. B.A. Litvinskij. Moscow: Nauka, pp. 126–151.

Vinogradova, N.M., V.A. Ranov, and T.G. Filimonova 2008. *Pamjatniki Kangurttuta v Jugo-Zapadnom Tadzhikistane (Epokha Neolita i Bronzovogo Veka)* [The Kangurttut sites in southwestern Tajikistan (Neolithic and Bronze Age)]. Moscow: Institut Vostokoedenija RAN.

Warner, L. 1980. "Stone Implements and Skeletons excavated in Anau." In *Explorations in Turkestan: Expedition of 1904. Prehistoric Civilizations of Anau: Origins, Growth and Influence of Environment*, ed. R. Pumpelly. Washington DC: Carnegie Institution of Washington, pp. 477–484.

Zajbert, V.V. 1993. *Eneolit Uralo-Irtyshskogo mezhdurech'ja* [The Chalcolithic of the Ural–Irtysh interfluve]. Petropavlovsk: Republic of Kazakhstan Academy of Science.

Zajtseva, G.I., B. van Geel, N. Bokovenko, K.V. Chugunov, V.A. Dergachev, V.G. Dirksen, M.A. Koulkova *et al.* 2004. "Chronology and Possible Links Between Climatic and Cultural Change During the First Millennium BC in Southern Siberia and Central Asia." *Radiocarbon* 46(1): 259–276. https://doi.org/10.1017/S0033822200039576.

PART IV

METALS AND METAL DEPOSITS

CHAPTER TWENTY-SEVEN

ARCHAEOMETALLURGICAL STUDIES ON BMAC ARTIFACTS

Steffen Kraus

INTRODUCTION

Copper minerals like malachite and azurite as well as native copper were already used in the Near East for the production of jewelry during the 9th/8th millennium (Pernicka 1990: 29; Roberts *et al.* 2009: 1013–1014). In Central Asia, the earliest metal finds are isolated discoveries from the Anau IA layers of Kaushut, Monjukly Depe, Chakmakly Depe, and Anau Depe, which date the beginning of metalworking in this region to the 6th/5th millennium (Kuz'mina 1966: 86; Terekhova 1981: 315–316). The first typologically determinable metal objects appeared during the Early Chalcolithic (Namazga (NMG) I) and increased in number in the following NMG II and III periods (4th/3rd millennium). These were mainly small objects such as pins and awls, while larger objects have only been occasionally found, such as the axe from Jalangach Depe or the gouge from Kara Depe (Kuz'mina 1966: 86–88; Chernykh 1992: 29–31). The most recent find is a dagger made of pure copper from the early layers of Dashly Depe in Turkmenistan (Boroffka and Kurbanov 2015; Kraus and Yagshimuradov 2015). While in the NMG I–III periods mainly pure copper with natural admixtures of arsenic, antimony, silver, and lead was processed, during NMG IV (first half of the 3rd millennium) copper alloys with arsenic, lead, and tin were increasingly used: However, whereas copper-arsenic alloys were the dominating alloys, tin bronze alloys were scarce (Ruzanov 1999: 104; Salvatori *et al.* 2002: 87).

With the appearance, at the end of the 3rd millennium, of the large proto-urban centers at the foothills of the Kopet Dagh, an increase in metal processing is evident. More than 1,000 copper and bronze objects are known. The most intensively investigated objects are those from Sapalli, Dzharkutan, Dashly Tepe, and Gonur Depe, which show typological parallels with metal artifacts from Afghanistan, Iran, and the Indus valley (Chernykh 1992; Kaniuth 2006; Ruzanov 2013; Kraus 2016, Kraus *et al.* in prep.). In this context, Chernykh (1992: 179) has differentiated two production centers: Bactria, said to be a metallurgical center with mining and smelting; and Margiana, considered as a center for just metal processing. At that time, in Bactria, tin bronze objects occur more frequently, while they are rarely found in Margiana, where according to Chernykh (1992: 179–182) mainly pure copper and occasionally copper alloys containing arsenic and/or lead were used, suggesting that polymetallic deposits were exploited. However, recent analyses of more than 850 metal artifacts from Gonur Depe in Margiana show that copper alloys with arsenic and/or lead represent more than 50% of the total metal inventory, and that pure copper and tin bronze were also used (Kraus 2015: 202; 2016: 259).

At the end of the NMG V period (at the beginning of the 2nd millennium BC) and in the following Late Bronze Age (NMG VI), the urban centers in the foothills of the Kopet Dagh such as Altyn Depe or Namazga Depe were abandoned and replaced by smaller settlements. This is said to have led to a massive migration eastwards towards Margiana and Bactria, where new settlement centers were then established (Sarianidi 1981: 188–189; Kohl 1984: 135–137). A change is also evident in the metal finds. A close connection to the Kopet Dagh motherland is still visible, but the influence of the Late Bronze Age Sapalli culture of northern Bactria is considerably stronger, which is why the term Bactria-Margiana Archaeological Complex (BMAC) became widely accepted. According to Ruzanov, the Sapalli culture mainly used copper-tin alloys (Ruzanov 1999: 104), but also arsenic and lead bronze as well as pure copper, which led him to consider an import of the metal from southern Central Asia (Iran/Afghanistan) (Ruzanov 1982: 10). More recent analyses of metal objects from the Sapalli culture showed similar results, with arsenic bronze preferably used, followed by tin bronze (Kaniuth 2006: 166). The first evidence of the use of iron appears in the middle of the 2nd millennium BC, which, however, becomes the dominant metal only during the 1st millennium BC (Garner 2013: 10).

AIMS AND METHODS

Since the 1960s, a considerable amount of research on prehistoric Central Asian metal artifacts has been carried out by several scholars, and numerous metal analyses have been published (e.g., Chernykh 1962, 1976; Kuz'mina 1966; Terekhova 1974, 1990; Galibin 1983, 1997, 2002; Isakov 1991; Bostongukhar 1998; Ruzanov 1999, 2013; Ravich 2004, 2008; Kaniuth 2006; Vinogradova *et al.* 2010; Erb-Satullo 2011; Vertman and Dubova 2013, 2014). The results of the archaeometallurgical investigations on over 1,000 metal artifacts from different sites within the territory of the BMAC (Figure 27.1) presented here have been achieved as part of the interdisciplinary project "ROXIANA" – archaeological research on the metallic and pottery assemblages from the Oxus basin to the Indus valley during protohistory, funded by the French National Research Agency (ANR) and the Deutsche Forschungsgemeinschaft (DFG). The aim of the project was a comprehensive study of the Central Asia metal finds and ceramics dating from the Chalcolithic to the Early Iron Age (4th–1st millennium BC).

The primary aim of the archaeometallurgical research was the identification of the alloys used for the metal objects. Furthermore, investigations were made to determine if there was a relation between specific alloys and the various typological groups in order to highlight possible choices in the raw material. Typologically the artifacts provide evidence of trade and cultural contacts with the Bronze Age Indus valley culture in the south, as well as with the steppe cultures in the north (Pittman 1984: 89; Chernykh 1992: 176; Parzinger 2006: 433). The contact with the steppe cultures is related to an increase in the occurrence of tin bronze. In this context, the provenance of the raw materials used for metal production might provide information on the metal distribution and cultural exchanges. Therefore, a selection of objects was

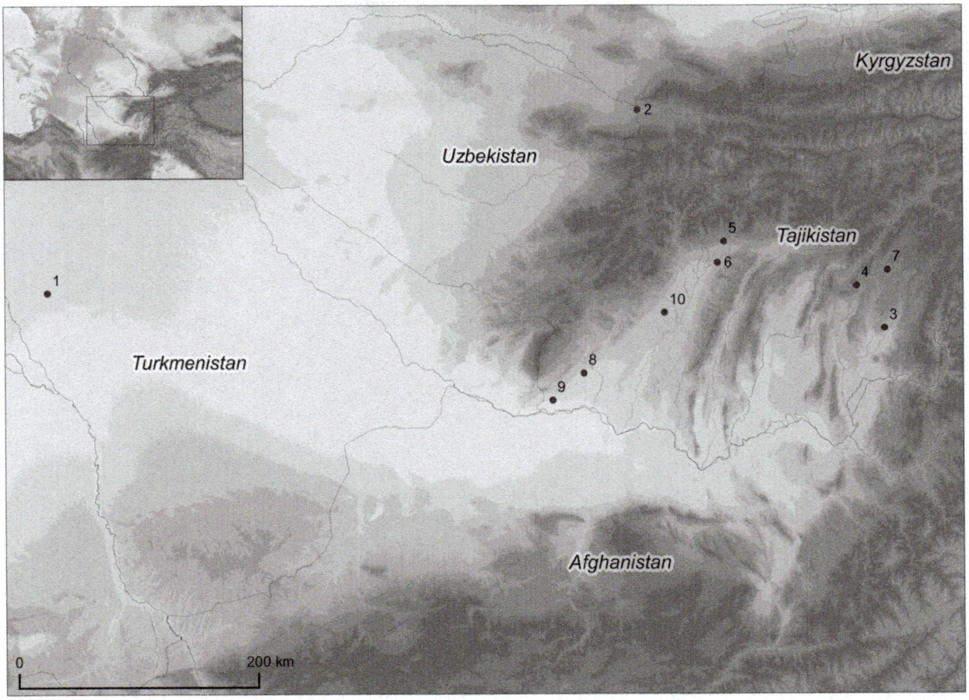

Figure 27.1 General map showing BMAC sites investigated within the ROXIANA project: 1: Gonur Depe; 2: Sarazm; 3: Gelot; 4: Kangurttut; 5: Kumsaj; 6: Tandyrjul; 7: Tujun; 8: Dzharkutan; 9: Sapallitepa; 10: Molalitepa (S. Kraus).

analyzed for their lead isotope signatures in addition to the determination of their major and trace elements. The chemical composition of the objects was determined by energy dispersive X-ray fluorescence spectrometry (ED-XRF) and neutron activation analysis (NAA). The determination of the lead isotope ratios was carried out using a multi-collector mass spectrometer with inductively coupled plasma (MC-ICP-MS). All the analyses were performed at the laboratories of the Curt-Engelhorn Centre of Archaeometry gGmbH (CEZA), Mannheim, Germany.[1]

RESULTS AND DISCUSSION

According to their chemical composition, the studied metal finds were assigned to different types of metal. Apart from precious metal finds made of gold and silver, mainly nonferrous metals were investigated. These were mainly copper-based alloys as well as a few lead objects.

Precious metal artifacts

All the investigated gold objects are pieces of jewelry that were found in burials at Gonur Depe. The chemical analyses revealed at least two types of gold metal on the basis of their silver content (Figure 27.2): objects containing between

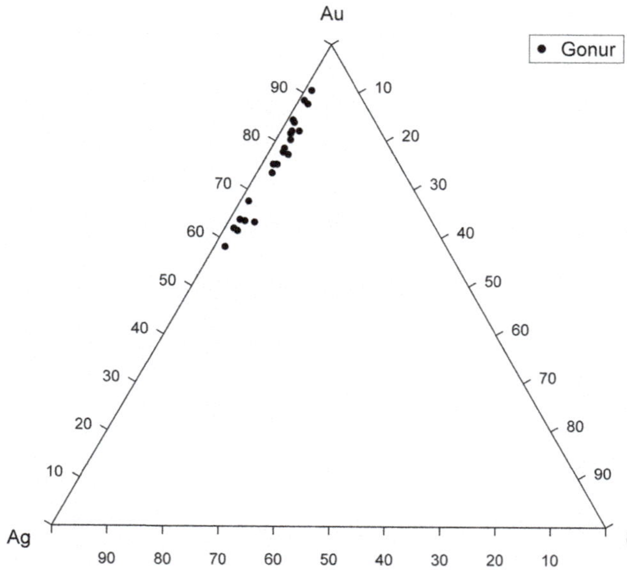

Figure 27.2 Ternary diagram of analyzed gold objects from Gonur Depe showing two types of gold metal classified by their silver content (S. Kraus).

10 and 25% silver on the one hand, and, on the other, objects with relatively high concentrations of more than 30% silver. In addition, all the gold objects contain between 0.5 and 4.5% copper. These results correspond to the recent studies of Juminov and Dubova on gold foils and on some golden beads from the "Royal Necropolis" of Gonur Depe, which contain between 1.4 and 4.0% copper (Juminov and Dubova 2014: 201). High copper concentrations indicate an intentional alloying with copper, as natural gold contains only 100 mg/kg copper (Schmiderer 2008: 90). Earlier gold analyses of a few objects from the Sumbar necropolis, Turkmenistan, also revealed copper concentrations of up to 6.8%, indicating an intentional alloying (Galibin 1983: 227).

Most of the studied silver objects come from Gonur Depe, and they mostly consist of jewelry but also some vessels. In addition, a few artifacts from Uzbekistan and Tajikistan were analyzed: a miniature vessel from Dzharkutan, a pendant from Tujun, as well as decorative objects from Molalitepa. The majority of the silver objects from Gonur as well as the artifacts from Molali were heavily corroded. Therefore, their classification on the basis of their chemical composition seemed problematic. Moreover, it has to be underlined that almost all the silver artifacts had significant copper concentrations (up to 20%) indicating a systematic addition of copper in most cases. Only a small vessel from Gonur was made of pure silver. The miniature vessel from Dzharkutan and the pendant from Tujun also consisted of pure silver containing up to 0.6% lead, which is typical for cupelled silver. Evidence for the use of copper-containing silver have already been advanced by the previous studies of E.E. Kuz'mina (1966: 104), V.A. Galibin (1983: 227), and V.D. Ruzanov (Isakov and Ruzanov 2008: 227; Ruzanov 2013: 316–326, 337).

Lead artifacts

The investigated lead items include several ring ingots and a vessel from Gonur Depe as well as an ingot from Sarazm, an earlier site dated to the Late Chalcolithic/Early Bronze Age. With the exception of the vessel from Gonur Depe, which contains about 1% copper, all the other objects consist of pure lead. Their chemical analyses revealed that all the detected elements are natural impurities. However, some significant differences in their composition are also evident. Very low contents of arsenic and antimony, partially below 1 µg/g, seem to be characteristic for the lead from Gonur Depe, while significantly higher concentrations of arsenic and antimony have been found in the lead from Sarazm. Furthermore, all the lead samples show relatively high silver content, indicating that the lead was not desilvered, since the threshold value for desilvering lead is of about 0.01% Ag in the Late Bronze Age (Pernicka 1987: 684; Begemann and Schmitt-Strecker 2001: 161).

Considering the results of the lead isotope analyses (Figure 27.3), it turns out that the lead objects from Gonur Depe are scattered over a wide range and form two groups. The artifacts of the first group have similar signatures to the lead ores from Nakhlak, a deposit in the region of Anārak in Central Iran. However, the objects of the second group rather suggest an origin further east, in the Karamazar Mountains of Uzbekistan, especially in the deposits of Lashkerek, where presumably the lead found at Sarazm also comes from.

Copper and bronze artifacts

From our analyses, the copper-based metal artifacts from Bronze Age Central Asia have been assigned to at least five different alloy types. The major alloying elements

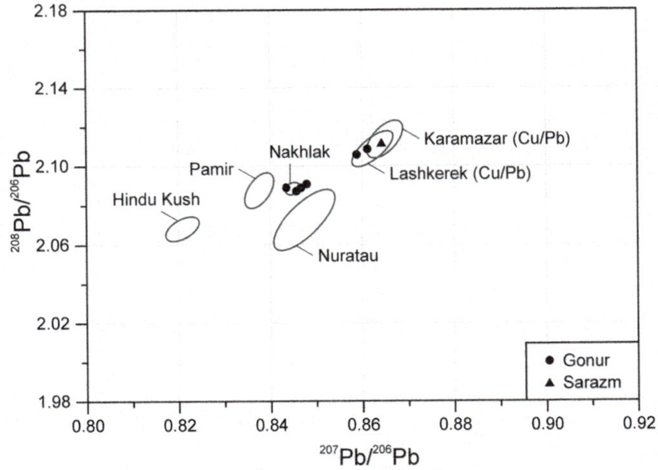

Figure 27.3 Lead isotope diagram comparing the lead artifacts from Gonur Depe and Sarazm with potential lead ore deposits in Central Asia (ore data taken from Brill *et al.* 1997; Chiaradia 2006; Pavlova and Borisenko 2009; Pernicka *et al.* 2011; Merkel 2016). The standard deviation is smaller than the symbols (S. Kraus).

are arsenic, lead, and tin as well as antimony and zinc. Often, arsenic and lead occur together at relatively high concentrations. In these cases, the dominant alloying constituent determined the designation of the alloy. Furthermore, unalloyed copper was still in use.

Sarazm, Tajikistan (4th/3rd millennium)

The earliest copper objects analyzed in this project were discovered at the ancient settlement of Sarazm, located in the upper Zeravshan valley in northern Tajikistan. Radiocarbon dates provide evidence that the settlement was inhabited from the Late Chalcolithic until the end of the Early Bronze Age (middle of the 4th to the middle of the 3rd millennium BC) (Isakov *et al.* 1987: 90–91; Isakov and Ruzanov 2008: 225). Therefore, the investigation of the metal finds from this site is of particular interest, as they represent a very early phase of the development of metallurgical production in Tajikistan and in the entire eastern zone of Central Asia. Among the investigated finds were mainly weapons such as daggers, arrowheads, and spearheads, but also some spatulas as well as a cross-edged ("*kreuzschneidige*") axe, all of which are very well-preserved.

The chemical analyses show that nearly all the finds consist of unalloyed copper (Figure 27.4). Only two daggers contain up to 2% arsenic and one spatula about 2% lead. Earlier studies carried out with XRF and spectral analyses gave similar results (Isakov *et al.* 1987: 91–92; Isakov 1991: 117–136; Isakov and Ruzanov 2008: 227–228). It is conceivable that the copper was extracted from oxidic ores and, in some cases, from sulphidic ores of the Zeravshan region (Isakov *et al.* 1987: 101–102; Isakov and Ruzanov 2008: 232). The low tin concentration in the artifacts is particularly remarkable as the settlement is situated close to the copper/tin deposit of Mushiston (Berger *et al.* in prep.). The early dating of the settlement in the 4th and first half of the 3rd millennium BC, however, is a possible explanation since tin bronze was not yet in use.

The results of the lead isotope analyses show that all the artifacts from Sarazm have similar lead isotope signatures (Figure 27.5), which is in contrast to the large spread of their element patterns. Presumably, different ore deposits within a geochemical region were exploited. Compared with lead isotope data of copper ore deposits,

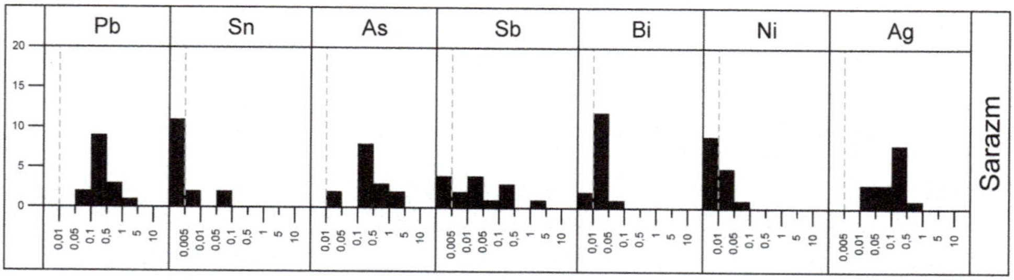

Figure 27.4 Frequency distribution of selected element concentrations in the analyzed metal objects from Sarazm measured with ED-XRF. All values in mass percent with the first value indicating the limit of detection (marked by the dashed line) (S. Kraus).

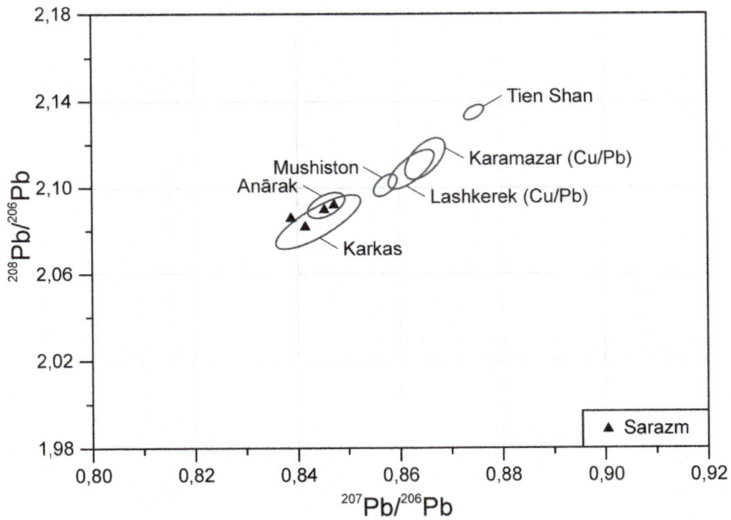

Figure 27.5 Lead isotope diagram comparing the copper-based metal artifacts from Sarazm with potential copper ore deposits in Central Asia (ore data taken from Chiaradia *et al.* 2006; Pernicka *et al.* 2011; Merkel 2016; Berger *et al.* in prep.). The standard deviation is smaller than the symbols (S. Kraus).

they match with the ores from Central Iran. But since these copper ores are rich in arsenic, the copper used at Sarazm did not come from that region. On the other hand, we saw that the ores from the nearby deposits of Mushiston were not exploited, so the origin of the metal from Sarazm remains yet unclear.

Gonur Depe, Turkmenistan (2200–1700 BC)

The majority of the finds studied during the project come from Gonur Depe, including the contemporaneous sites of Gonur 20 and Gonur 21. Nearly 870 copper-based metal objects have been analyzed for their chemical composition, the majority (around 720 objects) with portable XRF. Located in the northeastern region of the Murghab delta, Gonur Depe was the largest Middle Bronze Age settlement in Margiana and southeastern Turkmenistan, covering ca. 50 ha (Kohl 1984: 148; Sarianidi 1990: 14; 2006: 34; 2009: 40). According to radiocarbon data, the site is dated from 2300/2250 BC until the middle of the 2nd millennium BC (Hiebert 1994: 78–80; Sarianidi 2006: 36; Zajtseva *et al.* 2008; see the Appendix at the end of this volume).

The analyses carried out with portable XRF (Figure 27.6) revealed that almost two-thirds of the objects are made of Cu-As alloys. Similarly, earlier studies performed by N.N. Terekhova (1990: 182–192) and recent investigations made by V.D. Ruzanov (2013: 304–305) or E.G. Vertman and N.A. Dubova (2013: 7; 2014: 191–196) demonstrate the dominance of Cu-As in the metallic artifacts from Gonur Depe. Arsenical copper appeared in the 4th millennium in a vast area ranging from southern Iran to central Europe. In southern Central Asia the preference for arsenical copper is evident at the beginning of the 3rd millennium (Erb-Satullo 2011: 161). At Gonur Depe especially, tools and weapons were made of Cu-As alloys with high contents

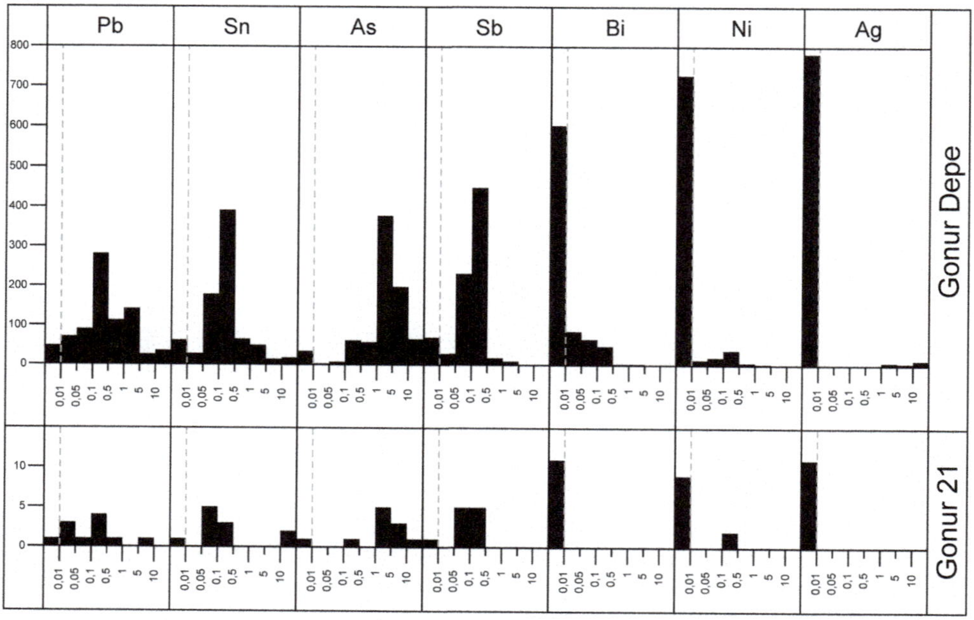

Figure 27.6 Frequency distribution of selected element concentrations in the analyzed metal objects from Gonur Depe and Gonur 21 measured with portable XRF. All values in mass percent with the first value indicating the limit of detection (marked by the dashed line) (S. Kraus).

of arsenic, which provide better casting properties and increase the hardness of the metal, an advantage for tool or weapon use. Lead, on the other hand, is insoluble in copper in solid state and weakens the mechanical stability of the alloy. Therefore, Cu-As alloys containing lead as well as Cu-Pb alloys were preferably used for cosmetic instruments (mostly flacons and spatulas) and for some jewelry artifacts (mainly pins and bracelets). Only occasionally were tools, weapons, and vessels made of copper alloys containing lead. Tin bronze resembles arsenical copper in its physical properties but has a different color. It is harder than unalloyed copper and also easier to cast than pure copper (Pernicka 1998: 136). Frequently tin bronze artifacts also contained arsenic and/or lead. When Cu-As alloys were mostly used for tools and weapons, the analyses show that tin bronze was used for the majority of the ceremonial axes, for some bowls, as well as for some mirrors. One mirror also attests to a gold content in the percent range that indicates gilding. However, unalloyed copper was still in use and is present in all the typological groups. From our results, no preferred alloy for a specific group is discernible. Furthermore, some objects, mainly classified as ingots, have important iron contents, up to 30% Fe. A low iron content may be due to impurities on the surface but a high content suggests the use of raw copper, and underlines a local metalworking at Gonur Depe.[2]

About 100 copper-based metal finds from Gonur Depe were sampled for further analyses at the CEZA laboratories. Overall, the measurements in the laboratory confirmed the results of measurements with portable XRF performed at the museum (Figure 27.7), but differences between the analytical results due to corrosion have also

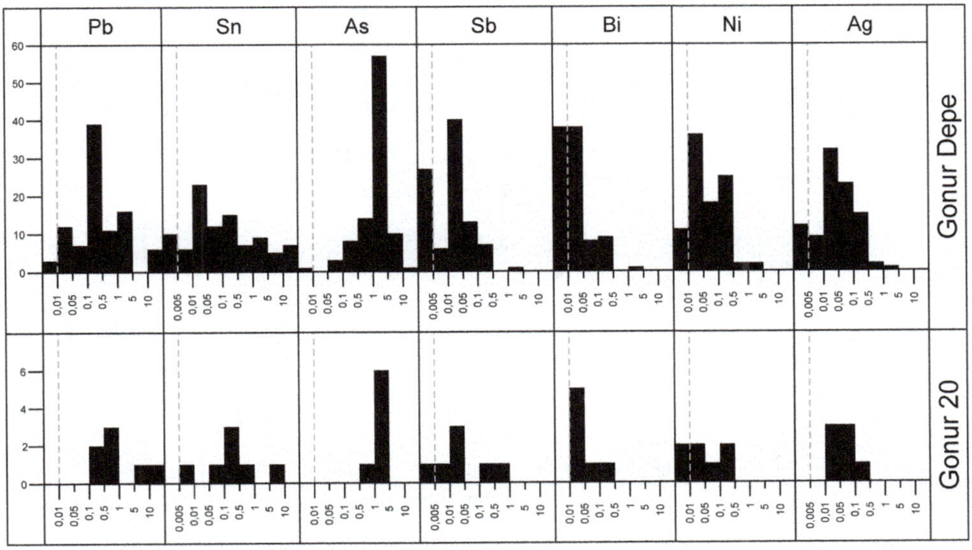

Figure 27.7 Frequency distribution of selected element concentrations in the analyzed metal objects from Gonur Depe and Gonur 20 measured with ED-XRF. All values in mass percent with the first value indicating the limit of detection (marked by the dashed line) (S. Kraus).

been observed. Normally, corroded tin bronzes show higher contents of tin at the surface, since copper tends to become depleted on the surface, while tin forms insoluble compounds and is therefore enriched at the surface. According to the measurements performed in the laboratories, however, some objects had a lower tin content at the surface than in the metal, which may be explained by pXRF measurements at areas where a secondary corrosion layer (mainly malachite) was formed. In addition, slightly higher iron contents were observed in the measurements with pXRF, which certainly did not originate from the metal itself but from corrosion products with the iron from the soil. The results show that arsenical copper dominates the metal assemblage of Gonur Depe. Furthermore, it turned out that nearly half of the tin bronze and lead bronze objects also contained arsenic. The enhanced arsenic content could be due to the smelting of arsenic-rich copper ores (e.g., tennantite). Such ores are known from the deposits near Arisman in Central Iran (Berthoud *et al.* 1982: 42; Pernicka *et al.* 2011: 634–640). Finally, an unspecified object should be mentioned that, in addition to lead, had an unusually high bismuth content. The only known objects with a similar composition have been found at Altyn Depe (Ruzanov 2013: 302), Togolok (Ruzanov 2013: 313), and Gonur Depe (Ruzanov 2013: 304; Vertman and Dubova 2013: 7; 2014: 191–196).

On the basis of their lead isotope ratios, the copper-based metal artifacts from Gonur Depe have been divided into two groups (Figure 27.8). Three tin bronzes have very similar lead isotope ratios but are significantly different from the other samples. Comparing the trace element contents, especially nickel, arsenic, antimony, and silver as relevant elements for the determination of the origin of the metal, the three objects generally present significantly less of them, indicating a different source for their raw material. Comparisons with lead isotope ratios of potential copper

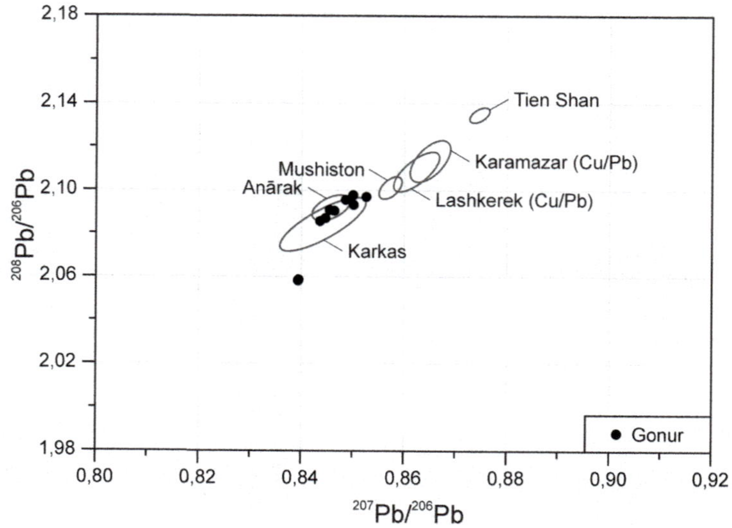

Figure 27.8 Lead isotope diagram comparing the copper-based metal artifacts from Gonur Depe with potential copper ore deposits in Central Asia (ore data taken from Chiaradia *et al.* 2006; Pernicka *et al.* 2011; Merkel 2016; Berger *et al.* in prep.). The standard deviation is smaller than the symbols (S. Kraus).

deposits in Central Asia are currently only possible with a few data, and they do not show any correlation with the three objects. Similarly, the lead isotope ratios of the second group do not correlate with the available data from Central Asian copper ore deposits, but with the signatures of deposits in Central Iran (Figure 27.8), especially with the arsenic-rich copper deposits of the Anārak and Karkas region near Arisman. This points to long-distance contacts of Gonur Depe and emphasizes its central role as a junction of early trade routes along the later Silk Road. However, we should keep in mind that the data available for comparison is far from being exhaustive, especially for the Central Asian deposits. Therefore, the question on the provenance of metals at Gonur Depe cannot as yet be definitively answered.[3]

Late Bronze Age sites in Tajikistan

The majority of the metal artifacts analyzed from Tajikistan come from Late Bronze Age[4] sites. From Gelot (2nd millennium BC) (see also Chapter 23), two completely corroded objects – a mirror and an unspecified object – have been studied. The latter originally consisted of unalloyed copper, while the mirror is made of tin bronze with a relatively high tin content (11% Sn), which is probably due to enrichment by corrosion. Previous analyses of a pin and an undetermined object from Gelot indicate the use of copper with an increased lead content but without tin (Vinogradova *et al.* 2010: 407).

At Kangurttut (20th–15th century BC, see Vinogradova 2004b: 95), according to the analyses (Figure 27.9), unalloyed copper dominates the metal assemblage, a result in contrast with earlier studies, where tin bronze was considered as the dominant metal (Ravich 2004: 224; 2008: 424; Vinogradova 2004a: 58; Ruzanov

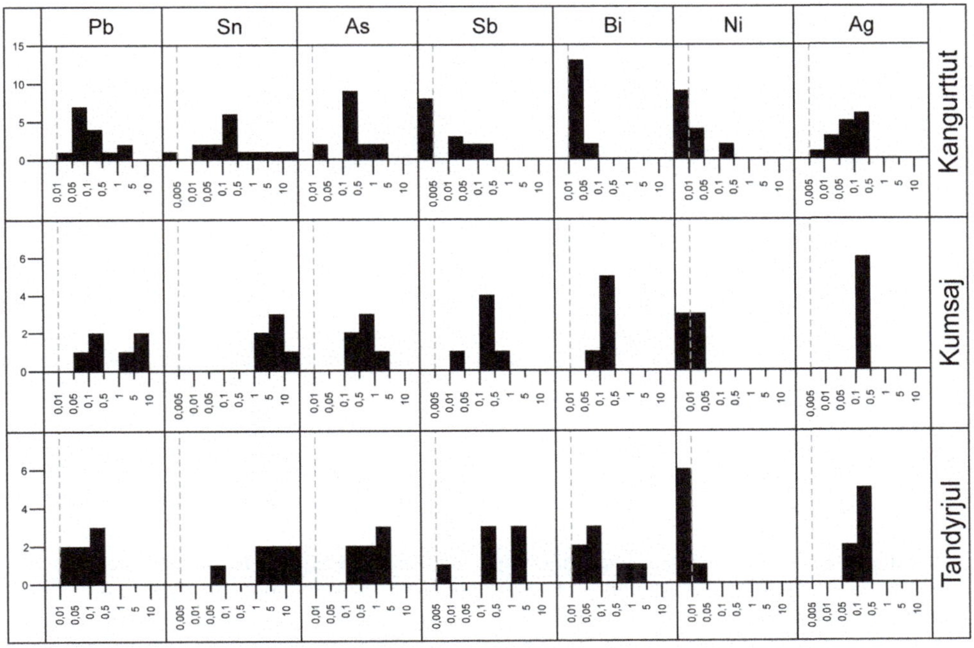

Figure 27.9 Frequency distribution of selected element concentrations in the analyzed metal objects from Bronze Age sites in Tajikistan measured with ED-XRF. All values in mass percent with the first value indicating the limit of detection (marked by the dashed line) (S. Kraus).

2013: 45–46). Only three objects consist of tin bronze with a tin content of up to 12% and we therefore consider the collection of samples currently studied as not representative. Finally, among the artifacts we analyzed is an object of unknown function, which according to the measurements contains 12.4% zinc. Only very few prehistoric artifacts made of copper-zinc alloys are known and they raise the question of whether they were intentional or accidental products (Biswas 1993; Thornton and Ehlers 2003; Thornton 2007; Fan *et al.* 2012). However, the high zinc content of this object might be questionable since it consists of heavily corroded material.

From Kumsaj (1750–1500 BC, see Luneau 2014), some of the investigated metal artifacts, such as four beads from a necklace, had already been analyzed (Ravich 2004), but, due to different designations, a direct assignment was not possible. All the analyzed artifacts consist of tin bronze (Figure 27.9). A ring and two bracelets also contain 4 to 10% lead. Although only a few objects have been analyzed, the results correspond with previous studies, where the appearance of leaded tin bronzes has already been demonstrated (Ravich 2004: 224). In this context it is noteworthy that, in the nearby and contemporaneous site of Tandyrjul, almost no leaded tin bronze was found (Figure 27.9). The metal artifacts consist mainly of tin bronze and of one in arsenical copper. Indeed, earlier studies on the metallic material from this site had shown that lead was only rarely used as an alloying component (Ruzanov 2004: 199; 2013: 335). Instead, some artifacts have increased contents of antimony and arsenic, an enrichment that could be explained by their very bad state of preservation and

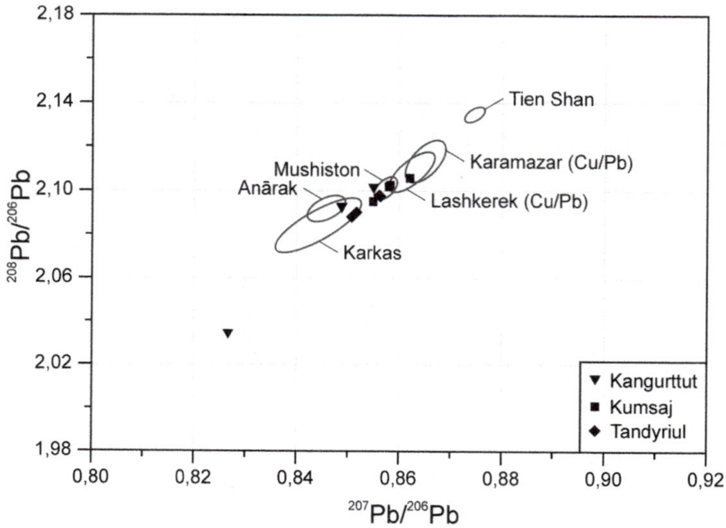

Figure 27.10 Lead isotope diagram comparing the copper-based metal artifacts from Bronze Age sites in Tajikistan with potential copper ore deposits in Central Asia (ore data taken from Chiaradia *et al.* 2006; Pernicka *et al.* 2011; Merkel 2016; Berger *et al.* in prep.). The standard deviation is smaller than the symbols (S. Kraus).

the corrosion of the copper originally produced from fahlores like tetrahedrite and tennantite.

The results of the lead isotope analyses show that most of the tin bronzes from Kumsaj have similar signatures to the copper/tin ores from Mushiston (Figure 27.10). However, comparing the trace elements, the artifacts agree only marginally with these ores. Furthermore, there are other possible deposits 20 km north of Kumsaj in the Hissar Mountains, where copper, lead, and tin ores occur together (Novikov and Radililovskij 1990). As far as Tandyrjul is concerned, some artifacts present similar lead isotope ratios as the ore from Mushiston. Due to the fact that all the artifacts are tin bronzes it is assumed that copper and tin ores were smelted together in a single process. In addition, it is worth mentioning that the copper-zinc object from Kangurttut has a completely different lead isotope ratio. Its high content of radiogenic lead indicates an origin of the metal in the deposits of the southern Tian Shan (Chiaradia *et al.* 2006: 420).

Late Bronze Age sites in Uzbekistan[5]

The Late Bronze Age settlement of Dzharkutan is stratigraphically divided into two phases (Kaniuth 2006: 49): Late Bronze Age I (LBA I: 20th–18th century BC) and Late Bronze Age II (LBA II: 17th–15th century BC).[6] According to the chemical analyses, a significant difference between the two phases can be observed (Figure 27.11). In the earlier phase (LBA I) copper-arsenic alloys dominate the metal assemblage, while only a few objects are made of unalloyed copper and tin bronze. Some artifacts, mainly flacons and spatulas, contain lead as additional alloying element. The preference of lead bronze for flacons at Dzharkutan has already been demonstrated in

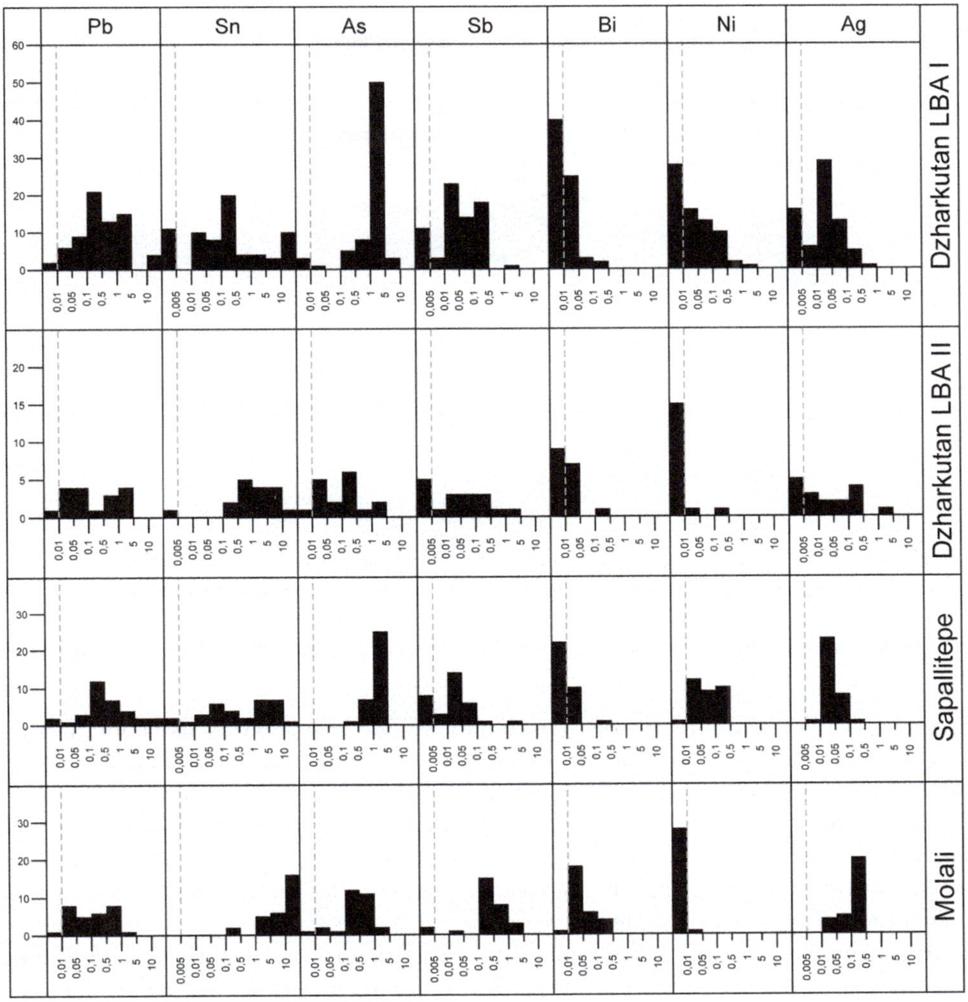

Figure 27.11 Frequency distribution of selected element concentrations in the analyzed metal objects from Dzharkutan measured with ED-XRF. All values in mass percent with the first value indicating the limit of detection (marked by the dashed line) (S. Kraus).

previous and recent studies (Kaniuth 2006: 165–167; Ruzanov 2013: 327). In the later phase (LBA II), unalloyed copper and tin bronze with only low arsenic contents are the predominant metal indicating a change of raw materials, and one can assume that this is related to the beginning of the exploitation of the copper ore deposits in the upper Zeravshan valley (Kaniuth 2006: 176).

The results from Sapalli, where all the samples date to the early phase of the LBA, reveal similar trends as those seen at Dzharkutan (Figure 27.11). Copper-arsenic alloys dominate the metal inventory, but no artifact from unalloyed copper has been found. Earlier studies carried out with XRF and spectral analyses had already shown that copper-arsenic alloys and tin bronze were the major types of metal at Sapallitepa, while unalloyed copper was only rarely used (Kaniuth 2006: 161–171; Ruzanov

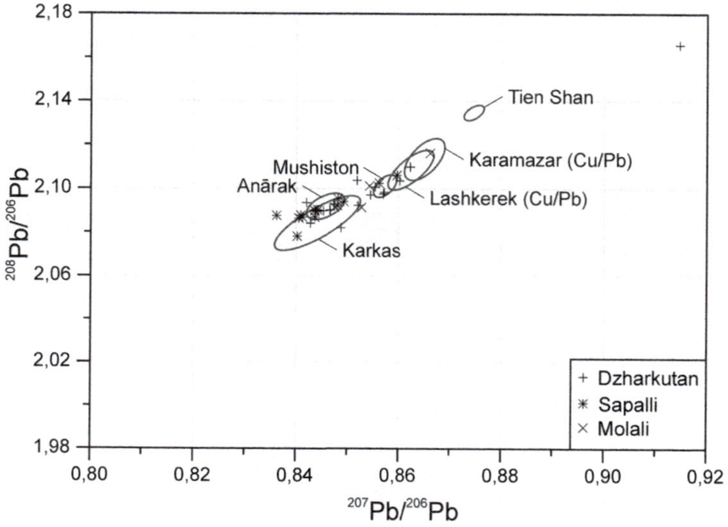

Figure 27.12 Lead isotope diagram comparing the copper based metal artifacts from Bronze Age sites in Uzbekistan with potential copper ore deposits in Central Asia (ore data taken from Chiaradia *et al.* 2006; Pernicka *et al.* 2011; Merkel 2016; Berger *et al.* in prep.). The standard deviation is smaller than the symbols (S. Kraus).

2013: 321–324). Since the majority of the investigated tin bronzes also contain high arsenic, it is to be assumed that the metallurgy of LBA I is based on arsenic copper ores. Finally, similarly to the LBA II phase at Dzharkutan, the metal assemblage of the contemporaneous Molali complex (Kaniuth 2006: 53) consists predominantly of unalloyed copper and of tin bronze. However, some artifacts, mostly chain links, are made of tin bronze with slightly increased contents of antimony, silver, and arsenic, indicating fahlores as raw material.

According to the lead isotope analyses, the artifacts from southern Uzbekistan show a high variability (Figure 27.12). On the one hand, artifacts have the same signature as the ores from Central Iran, but, on the other hand, some objects show a correlation with ores from Central Asian deposits. The question arises, especially for Dzharkutan, whether the LBA I and LBA II phases can be isotopically differentiated, as is the case with the alloys. For such an interpretation, however, a more precise dating of the artifacts is necessary. In addition, there is a dramatic lack of available data for the Central Asian copper ore deposits, although several of them present traces of prehistoric mining (see Chapter 28). Therefore, it is currently not possible to suggest any provenance for the raw material. To overcome this problem, more research on the copper deposits of Central Asia is needed, especially in view of their possible prehistoric exploitation. It resulted in a project, led by the author and funded by the German Research Foundation (DFG), in which copper ores from various deposits in Uzbekistan are investigated to obtain new data on the prehistoric exploitation of copper ore deposits in Central Asia. The preliminary results emphasize the importance of the Central Asian copper ore deposits, in particular those of the Karamazar/Kuraminsk and Chatkal Mountains, as supplier for raw materials at least since the Middle Bronze Age (Kraus 2018, 2020).

CONCLUSION

Though the majority of the metal finds consist of corroded material, our investigations have reached some conclusions concerning the types of alloys and the metal groups of the BMAC. In most cases, our results are concordant with previous analyses.

The earliest finds investigated come from Sarazm, a Late Chalcolithic and Early Bronze Age settlement, and were all produced from unalloyed copper. In spite of the small distance to the tin ore deposit of Mushiston, their tin content is very low, which can be explained by the early date of the site, when tin bronze was not yet common. Unalloyed copper continued to be used during the Middle Bronze Age and the early phase of the Late Bronze Age, as evidenced by the investigations at Gonur Depe and Dzharkutan. The predominant metal during that time, however, was copper-arsenic alloys due to their improved casting and mechanical properties compared to unalloyed copper. Furthermore, flacons and other cosmetic utensils additionally contain lead or consist of copper-lead alloys. In the later phase of the Late Bronze Age arsenical copper was replaced by unalloyed copper and by tin bronze with a low arsenic content, as demonstrated by the investigations at Dzharkutan and Molali. It is possible that either other ore sources were used due to a change of trade routes or that the technology of enrichment with arsenic was no longer applied. The Late Bronze Age sites in southwest Tajikistan were clearly dominated by tin bronze, which is explained by the contact of the local population with the steppe nomads from the north (see Chapters 25 and 26). The unusually high number of unalloyed copper artifacts at the site of Kangurttut might be due to a biased selection of the analyzed objects, since previous studies demonstrate indeed a domination of tin bronze.

The discussion of possible ore sources based on lead isotope ratios was, however, limited to only a few data from deposits of Central Asia and Iran. Therefore provenance analyses for most of the metal artifacts are not possible. However, for the copper-based metal artifacts from Gonur Depe, on the basis of actual knowledge, it is possible that the arsenic-rich copper ores of the central Iranian Plateau were used for the production of the raw metal and then traded to Gonur Depe where it was further processed. Furthermore, the lead isotope data also demonstrate a possible bronze production from the Mushiston ores at Tandyrjul. Because only a few data are available for comparison, it would be most useful for prospective research to investigate the geochemical characterization of ore deposits exploited in prehistorical times in Central Asia. In this context the preliminary results of an ongoing project, led by the author and funded by the DFG, show that the Central Asian copper ore deposits were probably exploited during the Chalcolithic, and certainly since the beginning of the Bronze Age. Although it is not yet possible to identify individual deposits as raw material suppliers for specific sites, the potential of the Central Asian deposits was undoubtedly known by the BMAC.

NOTES

1 At Gonur Depe (Turkmenistan), the determination of the chemical composition of about 850 metal objects was conducted during a field campaign in 2013 with a portable ED-XRF spectrometer (Kraus 2016: 257). It allowed a fast, nondestructive, multi-element analysis of the surface of the artifacts (for the discussion on the significance of such analyses, see Giumlia-Mair 2005).

2 Within the project, three other finds of unknown function were analyzed. It turned out that these artifacts mainly contained iron, arsenic, and copper. According to their macroscopic properties and their chemical compositions, these objects were identified as speiss fragments indicating an early production of arsenical copper at Gonur Depe (Kraus *et al.* in prep.). In this context, the crucibles found at the site should be mentioned (Dubova 2008: 94; Papakhristu 2008: 213) since they were most probably used for the enrichment of copper with arsenic.

3 Recent investigations also suggest potential copper deposits in Kazakhstan (pers. comm. A. Gontscharov) (see Chapter 28).

4 The ^{14}C dates put this cemetery at the turn of the 3rd to 2nd millennium BC (see the Appendix at the end of the volume).

5 The discussion of the chemical analyses of the metal finds from Sapalli and Dzharkutan also includes some as yet unpublished results kindly provided by E. Pernicka and J. Lutz, who carried out the investigations on the ores in a previous project.

6 On the Late Bronze Age denomination by some authors for Sapalli and Dzharkutan, see Chapter 1.

REFERENCES

Begemann, F. and S. Schmitt-Strecker 2001. "Zur Herkunft sardischer Unterwasser-Bleifunde." *SKYLLIS – Zeitschrift für Unterwasserarchäologie* 4(2): 160–165.

Berger, D., G. Brügmann, J. Lutz, and E. Pernicka in preparation. "Mušiston, Tadjikistan, a Tin Source for the Central Asian Bronze Age from Tin Isotopes" (working title).

Berthoud, T., C. Volfovsky, M. Menu, L.P. Hurtel, and S. Cleuziou 1982. "Cuivres et alliages en Iran, Afghanistan, Oman au cours des IVe et IIIe millénaires." *Paléorient* 8(2): 39–54.

Biswas, A.K. 1993. "The Primacy of India in Ancient Brass and Zinc Metallurgy." *Indian Journal of History of Science* 28(4): 309–330.

Boroffka, N. and A. Kurbanov 2015. "New Data on the Neolithic and Eneolithic of Central Asia: Dashly Depe (Turkmenistan)." *Miras: Ylmy-Köpçülikleýin Žurnal* 1(57): 38–55.

Bostongukhar, S. 1998. *Verkhov'ja Zarafshana vo II tysjachletii do n.e.* [The upper Zeravshan valley in the 2nd millennium BC]. Dushanbe: Institut Istorii, Arkheologii i Etnografii AN Respubliki Tadzhikistan.

Brill, R.H., C. Felker-Dennis, H. Shirahata, and E.C. Joel 1997. "Lead Isotope Analyses of Some Chinese and Central Asian Pigments." In *Conservation of Ancient Sites on the Silk Road: Proceedings of an International Conference on the Conservation of Grotto Sites*, ed. N. Agnew. Los Angeles, CA: J. Paul Getty Trust, pp. 369–378.

Chernykh, E.N. 1962. "Nekotorye rezul'taty izuchenija metalla anauskoj kul'tury" [Some results from the analysis of the Anau culture metals]. *Kratkie Soobshchenija* 91: 30–38.

Chernykh, E.N. 1976. *Drevnjaja metalloobrabotka na Jugo-zapade SSSR* [Ancient metalwork in southwest SSSR]. Moscow: Nauka.

Chernykh, E.N. 1992. *Ancient Metallurgy in the USSR: The Early Metal Age.* New Studies in Archaeology. Cambridge: Cambridge University Press.

Chiaradia, M., D. Konopelko, R. Seltmann, and R.A. Cliff 2006. "Lead Isotope Variations Across Terrane Boundaries of the Tien Shan and Chinese Altay." *Mineralium Deposita* 41: 411–428.

Dubova, N.A. 2008. "Masterskaja po proizvodstvu splavov na osnove medi Severnogo Gonura (zapadnaja chast' raskopa 9)" [A workshop for smelting copper in Gonur North (western part of excavations 9)]. In *Trudy Margianskoj Arkheologicheskoj Ekspeditsii*, vol. 2, ed. V.I. Sarianidi. Moscow: Staryj Sad, pp. 94–104.

Erb-Satullo, N.L. 2011. "Alloys from Anau: The Manipulation of Metallic Properties in 3rd Millennium BC Southern Central Asia." In *Materials Issues in Art and Archaeology IX*, ed. P.B. Vandiver, W. Li, J.L. Ruvalcaba Sil, C.L. Reedy, and L.D. Frame. Materials Research Society Proceedings, 1319. Cambridge: Cambridge University Press, pp. 157–163.

Fan, X., G. Harbottle, Q. Gao, W. Zhou, Q. Gong, H. Wang, X. Yu, and C. Wang 2012. "Brass Before Bronze? Early Copper-Alloy Metallurgy in China." *Journal of Analytical Atomic Spectrometry* 27(5): 821–826.

Galibin, V.A. 1983. "Spektralnyj analiz nakhodok iz Sumbarskikh mogil'nikov" [Spectral analyses of the finds from the Sumbar cemeteries]. In *Jugo-zapadnaja Turkmenija v epokhu pozdnej bronzy: po materialam Sumbarskikh mogil'nikov* [Southwest Turkmenistan in the Late Bronze Age from the finds of the cemeteries in the Sumbar valley], ed. I.N. Khlopin. Leningrad: Nauka, pp. 224–234.

Galibin, V.A. 1997. "Metall rannego i razvitogo eneolita iz mogil'nika Parkhai II" [Metal of the Early and Middle Chalcolithic period from the cemetery of Parkhai II]. In *Eneolit Jugo-Zapadnogo Turkmenistana* [The Chalcolithic period of southwest Turkmenistan], ed. I.N. Khlopin. St. Petersburg: Evropejskij Dom, pp. 296–298.

Galibin, V.A. 2002. "Metall iz mogil'nika Parkai II" [Metal from the Parkhai II cemetery]. In *Epokha bronzy Jugo-Zapadnogo Turkmenistana* [The Bronze Age of southwest Turkmenistan], ed. I.N. Khlopin. St. Petersburg: Peterburgskoe vostokovedenie, pp. 195–199.

Garner, J. 2013. *Das Zinn der Bronzezeit in Mittelasien II: Die montanarchäologischen Forschungen an den Zinnlagerstätten.* Archäologie in Iran und Turan, 12. Berlin: Deutsches Archäologisches Institut-Deutsches Bergbau Museum.

Giumlia-Mair, A. 2005. "On Surface Analysis and Archaeometallurgy." *Nuclear Instruments and Methods in Physics Research B* 239: 35–43.

Hiebert, F.T. 1994. *Origins of the Bronze Age Oasis Civilization in Central Asia.* American School of Prehistoric Research Bulletin, 42. Cambridge, MA: Peabody Museum of Archaeology and Ethnology, Harvard University.

Isakov, A.I. 1991. *Sarazm. K voprosu stanovlenija rannezemledel'cheskoj kul'tury Zeravshanskoj doliny (raskopki 1977–1983 gg)* [Sarazm: questions on the evolution of the early farmer culture in the Zeravshan valley (1977–1983 excavations)]. Dushanbe: Donish.

Isakov, A., P.L. Kohl, C.C. Lamberg-Karlovsky, and R. Maddin 1987. "Metallurgical Analysis from Sarazm, Tadjikistan SSSR." *Archaeometry* 29(1): 90–102.

Isakov, A.I. and V.D. Ruzanov 2008. "Rezul'taty spektral'nykh issledovanij metalla poselenija Sarazm" [Results from spectral analyses on the metals of Sarazm]. In *Trudy Margianskoj Arkheologicheskoj Ekspeditsii*, vol. 2, ed. V.I. Sarianidi. Moscow: Staryj Sad, pp. 225–233.

Juminov, A.M. and N.A. Dubova 2014. "Rezul'taty mikrozondovogo analiza zolotykh izdelij iz Gonur Depe" [Results from micro-probe analysis of gold artifacts from Gonur Depe]. In *Trudy Margianskoj Arkheologicheskoj Ekspeditsii*, vol. 5, ed. V.I. Sarianidi. Moscow: Staryj Sad, pp. 197–203.

Kaniuth, K. 2006. *Metallobjekte der Bronzezeit aus Nord-Baktrien.* Archäologie in Iran und Turan, 6. Mainz am Rhein: Philipp von Zabern.

Kohl, P.L. 1984. *Central Asia: Palaeolithic Beginnings to the Iron Age.* Paris: Éditions Recherche sur les Civilisations.

Kraus, S. 2015. "Untersuchungen an prähistorischen Metallfunden aus Gonur Depe, Turkmenistan." In *Archäometrie und Denkmalpflege 2015: Jahrestagung an der Johannes-Gutenberg-Universität Mainz*, ed. T. Gluhak, S. Greiff, K. Kraus, and M. Prange. Metalla, Sonderheft, 7. Bochum: Deutsches Bergbau-Museum, pp. 201–203.

Kraus, S. 2016. "Metallurgical Investigations in Gonur Depe, Turkmenistan." In *Trudy Margianskoj Arkheologicheskoj Ekspeditsii*, vol. 6, ed. N.A. Dubova. Moscow: Staryj Sad, pp. 257–264.

Kraus, S. 2018. Prähistorischer Bergbau und Metallurgie in Usbekistan – Erste Ergebnisse. In *Archäometrie und Denkmalpflege 2018*, DESY-PROC, ed. L. Glaser. Hamburg: Deutsches Elektronen-Synchrotron, pp. 79–82.

Kraus, S. 2020. "Prehistoric Mining and Metallurgy in Uzbekistan: An Introduction." In *Proceedings of the 11th International Congress on the Archaeology of the Ancient Near East 3–7 April 2018, Munich. Vol. 2, Field Reports, Islamic Archaeology*, ed. A. Otto, M. Herles, K. Kaniuth, L. Korn, and A. Heidenreich. Wiesbaden: Harrassowitz, pp. 197–206.

Kraus, S., T. Schifer, and E. Pernicka in preparation. "The Archaeometallurgical Studies of the ROXIANA Project" (working title).

Kraus, S. and G. Yagshymuradov 2015. "XRF Analysis of a Chalcolithic Dagger from Dashly Depe (Turkmenistan)." *Miras: Ylmy-Köpçülikleýin Žurnal* 1(57): 56–57.

Kuz'mina, E.E. 1966. "Metallicheskie izdelija eneolita i bronzovogo veka v Srednej Azii" [Metal artifacts from the Chalcolithic and Bronze Age Central Asia]. In *Arkheologija SSSR svod arkheologicheskikh istochnikov*, issue B4-9, ed. V.I. Moshinskaja. Moscow: Nauka.

Luneau, E. 2014. *La fin de la Civilisation de l'Oxus. Transformations et recompositions des sociétés de l'Age du Bronze final en Asie centrale méridionale (1800–1500/1400 avant n.è.).* Paris: de Boccard.

Merkel, S.W. 2016. *Silver and the Silver Economy at Hedeby.* Der Anschnitt, Beiheft, 33. Bochum: Deutsches Bergbau-Museum.

Novikov, V.P. and V.V. Radililovskij 1990. "K istorii gornorudnykh promylov i metallurgii drevnego Gissara" [History of the mines industry and of metallurgy in ancient Hissar]. *Ochenki po Istorii Geologii Tadzhikistana* 1: 54–103.

Papakhristu, O.A. 2008. "K rekonstruktsii tiglej dlja plavki splavov na osnove medi iz masterskoj" [Reconstruction of crucibles for producing alloys on a copper base found in a workshop]. In *Trudy Margianskoj Arkheologicheskoj Ekspeditsii*, vol. 2, ed. V.I. Sarianidi. Moscow: Staryj Sad, pp. 196–224.

Parzinger, H. 2006. *Die frühen Völker Eurasiens: Vom Neolithikum bis zum Mittelalter.* München: Verlag C.H. Beck.

Pavlova, G.G. and A.S. Borisenko 2009. "The Age of Ag-Sb Deposits of Central Asia and Their Correlation with Other Types of Ore Systems and Magmatism." *Ore Geology Reviews* 35(2): 164–185.

Pernicka, E. 1987. "Erzlagerstätten in der Ägäis und ihre Ausbeutung im Altertum: Geochemische Untersuchungen zur Herkunftsbestimmung archäologischer Metallobjekte." *Jahrbuch des Römisch-Germanischen Zentralmuseums* 34: 607–714.

Pernicka, E. 1990. "Gewinnung und Verbreitung der Metalle in Prähistorischer Zeit." *Jahrbuch des Römisch-Germanischen Zentralmuseums* 37(1): 21–129.

Pernicka, E. 1998. "Die Ausbreitung der Zinnbronze im 3. Jahrtausend." In *Mensch und Umwelt in der Bronzezeit Europas*, ed. B. Hänsel. Kiel: Oetker-Voges, pp. 135–147.

Pernicka, E., M. Momenzadeh, A. Vatandoust, K. Adam, M. Böhme, Z. Hezarkhani, N. Nezafati, M. Schreiner, and B. Winterholler 2011. "Archaeometallurgical Research on the Western Central Iranian Plateau." In *Early Mining and Metallurgy on the Western Central Iranian Plateau: The First Five Years of Work*, ed. A. Vatandoust, H. Parzinger, and B. Helwing. Archäologie in Iran und Turan, 9. Mainz am Rhein: Philipp von Zabern, pp. 633–687.

Pittman, H. 1984. *Art of the Bronze Age: Southeastern Iran, Western Central Asia, and the Indus Valley.* New York: Metropolitan Museum of Art.

Ravich, I.G. 2004. "Osobennosti sostava i tekhnologii izgotovlenija metallicheskikh izdelij, najdennikh na poselenii Kangurttut" [Composition and technological characteristics of the metal artifacts from Kangurttut]. In *Jugo-Zapadnij Tadzhikistan v epokhu pozdnej bronzy* [Southwest Tadjikistan in the Late Bronze Age], ed. N.M. Vinogradova. Moscow: Institut Vostokovedenija, Academy of Sciences, pp. 200–224.

Ravich, I.G. 2008. "Arkheometallografija izdelij, najdennykh na poselenii i v mogil'nike Kangurttut" [Archaeometalography of artifacts found at Kangurttut (settlement and cemetery)]. In *Pamjatniki Kangurttuta v jugo-zapadnom Tadzhikistane (epokha neolita i bronzovyj vek)* [The sites of Kangurttut in southwest Tadjikistan (Neolithic and Bronze Age)], ed. N.M. Vinogradova, V.A. Ranov, and T.G. Filimonova. Moscow: Institut Vostokovedenija, Academy of Sciences, pp. 412–428.

Roberts, B.W., C.P. Thornton, and V.C. Pigott 2009. "Development of Metallurgy in Eurasia," *Antiquity* 83: 1012–1022.

Ruzanov, V.D. 1982. "Istorija drevnej metallurgii i gornogo dela Uzbekistana v epokhu bronzy i rannego zheleza" [History of ancient metallurgy and mining of Uzbekistan in the Bronze Age and Early Iron Age]. Abstract of PhD thesis, Moscow State University, Moscow.

Ruzanov, V.D. 1999. "Zum frühen Auftreten der Zinnbronze in Mittelasien." In *The Beginnings of Metallurgy*, ed. A. Hauptmann, E. Pernicka, T. Rehren, and Ü. Yalçın. Der Anschnitt, Beiheft, 9. Bochum: Deutsches Bergbau-Museum, pp. 103–105.

Ruzanov, V.D. 2004. "Spektral'nij analiz izdelij iz mogil'nika Tandyr-jul'" [Spectral analysis of artifacts from the Tandyr-jul' cemetery]. In *Jugo-Zapadnij Tadzhikistan v epokhu pozdnej bronzy* [Southwest Tadjikistan in the Late Bronze Age], ed. N.M. Vinogradova. Moscow: Institut Vostokovedenija, Academy of Sciences, p. 199.

Ruzanov, V.D. 2013. *Metalloobrabotka na juge Srednej Azii v epokhu bronzy* [Metalwork in southern Central Asia during the Bronze Age]. Samarkand: Institut Arkheologii Akademii Nauk Respubliki Uzbekistan.

Salvatori, S., M. Vidale, G. Guida, and G. Gigante 2002. "A Glimpse on Copper and Lead Metalworking at Altyn-Depe (Turkmenistan) in the 3rd Millennium BC." *Ancient Civilizations from Scythia to Siberia* 8(1–2): 69–106.

Sarianidi, V.I. 1981. "Margiana in the Bronze Age." In *The Bronze Age Civilization of Central Asia: Recent Soviet Discoveries*, ed. P.L. Kohl. Armonk, NY: M.E. Sharpe, pp. 165–193.

Sarianidi, V.I. 1990. *Drevnosti strany Margush* [Ancient Margush]. Ashkhabad: Ylym.

Sarianidi, V.I. 2006. *Gonur-Depe: City of Kings and Gods*. Aşgabat: Miras.

Schmiderer, A. 2008. "Geochemische Charakterisierung von Goldvorkommen in Europa." Unpublished PhD thesis, Martin-Luther-Universität, Halle-Wittenberg.

Terekhova, N.N. 1974. "Tekhnologija izgotovlenija pervykh metallicheskikh orudij u drevnejshikh zemledel'tsev Juzhnoj Turkmenii" [Technology of production of the ancient farmers' first metal tools in southern Turkmenia]. *Sovetskaja Arkheologija* 1: 69–78.

Terekhova, N.N. 1981. "The History of Metalworking Production Among the Ancient Agriculturalists of Southern Turkmenia." In *The Bronze Age Civilization of Central Asia: Recent Soviet Discoveries*, ed. P.L. Kohl. Armonk, NY: M.E. Sharpe, pp. 313–324.

Terekhova, N.N. 1990. "Obrabotka metallov v drevnej Margiane" [Metalworks in Ancient Margiana]. In *Drevnosti strany Margush* [Ancient Margush], ed. V.I. Sarianidi. Ashkhabad: Ylym, pp. 177–202.

Thornton, C.P. 2007. "Of Brass and Bronze in Prehistoric Southwest Asia." In *Metals and Mines: Studies in Archaeometallurgy*, ed. S. La Niece, D. Hook, and P.T. Craddock. London: Archetype Publications, pp. 123–135.

Thornton, C.P. and C.B. Ehlers 2003. "Early Brass in the Ancient Near East." *Institute of Archaeo-Metallurgical Studies* 23: 3–8.

Vertman, E.G. and N.A. Dubova 2013. "Rekonstruktsija khimicheskogo sostava metalla pamjatnika bronzogo veka Gonur Depe (Turkmenistan) po dannym analiza metodom mass-spectrometrii s induktivno svjazannoj plazmoj" [Chemical composition of the metals from the Bronze Age settlement of Gonur Depe (Turkmenistan) according to ICP-MS analyses]. *Vestnik Tomskogo Gosudarstvennogo Universiteta, Istorija* 4(24): 5–9.

Vertman, E.G. and N.A. Dubova 2014. "Analiz metalla Gonur Depe metodom mass-spektrometrii s induktivno-svjazannoj plazmoj" [ICP-MS analyses of metal from Gonur Depe]. In *Trudy Margianskoj Arkheologicheskoj Ekspeditsii*, vol. 5, ed. V.I. Sarianidi. Moscow: Staryj Sad, pp. 188–203.

Vinogradova, N.M. 2004a. *Jugo-Zapadnij Tadzhikistan v epokhu pozdnej bronzy* [Southwest Tadjikistan in the Late Bronze Age]. Moscow: Institut Vostokovedenija, Academy of Sciences.

Vinogradova, N.M. 2004b. "The Settlement and the Cemetery of Kangurttut among the Late Bronze Age Sites of North-Eastern Bactria (South Tajikistan)." *Annali (dell'Università degli studi di Napoli "L'Orientale")* 64: 75–97.

Vinogradova, N.M., Ju.G. Kutimov, and M. Teufer 2010. "Pogrebenija epokhi bronzy v okrestnostjakh kishlaka Gelot na juge Tadzhikistana" [Bronze Age burials near Gelot, southern Tadjikistan]. In *Na puti otkrytija tsivilizatsii* [On the track of uncovering a civilization], *Trudy Margianskoj Arkheologicheskoj Ekspeditsii*, vol. 3, ed. P.M. Kozhin, M.F. Kosarev, and N.A. Dubova. St. Petersburg: Aletejja, pp. 389–411.

Zajtseva, G.I., N.A. Dubova, A.A. Sementsov, P. Reimer, J. Mallory, and H. Jungner 2008. "Radiouglerodnaja khronologija pamjatnika Gonur Depe" [Radiocarbon chronology of the Gonur Depe site]. In *Trudy Margianskoj Arkheologicheskoj Ekspeditsii*, vol. 2, ed. V. Sarianidi. Moscow: Staryj Sad, pp. 166–179.

CHAPTER TWENTY-EIGHT

METAL SOURCES (TIN AND COPPER) AND THE BMAC

Jennifer Garner

INTRODUCTION

Among the wealth of finds that have been discovered in Bactria-Margiana Archaeological Complex (BMAC) contexts, mainly in graves, metal objects such as ceremonial axes, pins, cosmetic items, etc., with ornaments are frequent. It remains unclear, however, where exactly the BMAC obtained its metal resources (tin and copper). The lowland territories where the major BMAC centers are located (the plain of Bactra and the Murghab delta) are devoid of this kind of raw material, and, currently, no known mines within the mountains nearby can be attributed to the BMAC. This makes it more likely that the suppliers of this kind of raw material were the northern surrounding nomadic steppe populations. The Andronovo people who lived in the steppes to the north of the BMAC are well known for their mining and metallurgical activities (Kuz'mina 1991; 1994: 137–146; 2007: 85–99; Chernykh 1992: 193–194, 212; Parzinger 2006: 357). Furthermore, the Andronovo metal inventory is distinctive for its high-quality tin bronzes (with 3–10% tin) (Chernykh 1992: 213). Some people have also advanced that one of the reasons for their nomadic lifestyle was the search for metal resources (Parzinger and Boroffka 2003: 272).[1]

The Andronovo culture is one of semi-mobile herders. It has a wide distribution from the Urals across Kazakhstan to Siberia, and southwards down to Chorasmia, Sogdiana, and even to the peripheries of Bactria and southern Turkmenistan. It is composed of various groups, each having their own name and specific local characteristics (see Chapter 26). However, discussion is still ongoing regarding their chronological contemporaneity or succession (Table 28.1). The southernmost group, Tazabag'jab, is centered upon the lower Zeravshan and lower Amu Darya valleys close to the Aral Sea (Itina 1977). There are many archaeological examples of a close relationship between the Andronovo groups and the BMAC. These include, for example, the grave of Zardcha Khalifa with offerings typical of the BMAC but found in the Zeravshan valley inhabited by Andronovo groups (Bobomulloev 1997; Isakov 1998: 56–60; Teufer 2003; see also Chapters 24 and 26), and numerous cemeteries and a few settlements in southwest Tajikistan (northeast Bactria) where the local Bishkent-Vakhsh culture has a mixed inventory: The pottery is both wheel-turned (BMAC sphere) and handmade (Andronovo) (Vinogradova 2001, 2004, 2008: 213–216; see also Chapters 23 and 25). Moreover, the analysis of the metal finds of the

Table 28.1 The different Andronovo groups, their geographical and chronological position (after Parzinger and Boroffka 2003: 280 fig. 1) (see also Chapter 26)

Period	*Zeravshan valley*	*Khorezm*	*Ferghana valley*	*Northern steppe/ forest zone (Kazakhstan)*	*Bactria*	*South Turkmenistan*
1850–1500 BC	Andronovo-Tazabag'jab	Andronovo-Tazabag'jab Kokcha 3	Andronovo-Tazabag'jab Kajrak Kum variant	Andronovo-Fedorovo Semirech'e and Tian-Shan Semirech'e variant	Shortugai III/IV Dashly 3 Sapalli I/II BMAC Bishkent-Vakhsh	NMG VI BMAC Togolok

latter reveal typical Andronovo high-quality tin bronzes with 10% tin but with 5–10% lead, which is untypical for this culture (Vinogradova 2004: 82; but see *contra* Chapter 27). In southern Tajikistan, Uzbekistan,[2] and Margiana (Turkmenistan)[3] there are yet more sites with evidence of contact between nonurban pastoral groups and BMAC (see also Chapter 17).

The Tazabag'jab group could well have played an intermediary role in the trade that developed between the northern steppe groups and the southern oasis cultures (Kuz'mina 1994: 243; Vinogradova 2001: 215; 2008: 213, 216–217). This trade may have mainly been an exchange of farming or craft products from the southern oasis for raw metal resources (copper and tin) or semifinished and finished metal items from the Andronovo area (Lyonnet 2005). This could be an explanation for the settlement of steppe populations at the periphery of the southern oasis (Hiebert 1994: 69–70; Kohl 2007: 192–195, 201, 205–206; Kuz'mina 2008: 73–76; Vinogradova 2008: 217; Cerasetti 2012: 546–547; Rouse and Cerasetti 2014: 47).

STATE OF MINING ARCHAEOLOGICAL RESEARCH IN CENTRAL ASIA

Investigations of "old mines" in Central Asia was undertaken during the Soviet period in the first half of the 20th century as part of large-scale interdisciplinary exploration campaigns of the numerous ore deposits and mineral veins. The focus at this time was to map the ores, to investigate their composition, and to estimate the reserves for future industrial exploitation. These eco-geological expeditions were occasionally accompanied by archaeologists who recorded both archaeological sites and the remains of the old mining works (Sverchkov 2009: 141–142). However, most of the newly discovered old mines were described by geologists, a situation that M.E. Masson strongly criticized when he called for a larger archaeological presence in the field (Masson 1930: 38).

This geological background has been used as the basis for later archaeological investigations, mainly in the form of surveys or small exploratory excavations, but invariably restricted to those medieval mines for which there were Chinese or Arabic written sources. Prehistoric mining was very seldom a topic of research, and the publications gave only a vague idea of the actual extent of the mining districts.[4] Therefore individual metallogenic complexes were described in a very generalized way, without providing any detailed maps for the mines, and with finds either sporadically recorded, or not at all. For example, neither P.S. Nazarov, a mining engineer who found bronze tools within copper mines at Naukat (Ferghana, Uzbekistan) in 1907, nor A.P. Agafonov, a geologist who found a bronze axe within the "Karasu" section of Uchkurgan in 1948 (Islamov 1976: 38), record the exact position of their finds, or properly publish their discoveries. Furthermore, invariably the dating of the earliest mines within most of the metallogenic districts of Central Asia is based upon the dating of surface finds such as pottery, or perhaps on the association of these mines with neighboring settlements, but not on radiocarbon dates.

One of the problems encountered in the search for the old mines at the time of the BMAC is that most metal analyses have been done on objects found within settlements or graves, but rarely on the semifinished products or metallurgical waste, such as slag or crucibles.[5] Characterization of the ore types from the different deposits is also usually absent. V. Ruzanov was one of the few people to have worked on this topic. He was of the opinion that the identification of prehistoric raw material can only now be achieved through geochemical analysis, given that most of the old mining remains have since been covered up or erased by medieval and modern mining activities (Ruzanov 1980: 61–62). He is also one of the few scientists to attempt an attribution of the different mining districts to specific cultures through a comparison of the chemical composition of the ores and the objects (Ruzanov 1979, 1980, 1982, 1999a). Most of the other metallurgical analyses were made simply to provide tables of compositions for the appendices of the archaeological publication(s).

Following the collapse of USSR, new impulses in mining archaeological research were initiated through a range of international research projects.[6] These projects have mainly been concerned with attempts to understand how the mining zone was used; that is, the organized exploitation of these deposits, and the identification of ores by means of lead isotope analysis. However, the question of trade routes between the mining zone and its surroundings, and the role of the nearby settlements, was rarely touched upon.

To sum up, the use of archaeometallurgical evaluation techniques (including phase diagrams, cluster analysis, and statistics) or supra-regional comparative studies are uncommon within this field of research. Instead we are left with hypotheses.

TIN DEPOSITS AND MINING IN CENTRAL ASIA: AN OVERVIEW

At around 2500 BC the production of tin bronze artifacts started to become a commonplace in the Near East, particularly in Mesopotamia. Today more and more questions arise about the origins of this metal, since neither copper nor tin deposits are to be found there (Muhly 1973; Crawford 1974). However, there are copper deposits not far away, for example in Anatolia, the Caucasus, Oman, and Iran (see Chapter 29). Nonetheless, the origins of the tin used have yet to be deciphered. Studying the geology is not of much help because the established tin deposits within Europe or in East Asia

(Malaysia and Thailand) either do not have traces of prehistoric mining or have no connections with the Near East during the 3rd and 2nd millennium BC (Penhallurick 1986: 51–54). Analytical provenance methods such as stable lead isotopes do not work on tin, because tin contains either no lead or only insignificant trace amounts (Begemann and Schmitt-Strecker 2009: 2, 22–23). The study of cuneiform texts delivers few hints as to the source of the metals (Lyonnet 2005: 193–195).

Within the former Soviet Union, large tin deposits in the steppes of Central Asia associated with traces of Bronze Age mining dating to the Andronovo and Begazy Dandybaj periods (2nd millennium BC) have long been known (Masson 1930, 1935, 1953; Chernikov 1949, 1951, 1960; Litvinskij 1950; Margulan 1972: 73; 1979; 2001). However, tin was also being produced here earlier, during the Early Bronze Age, since the first tin bronze artifacts appear within the Sejma-Turbino circle at the end of 3rd millennium BC (Chernykh 1992: 215–226; Parzinger 1997: 223–225; Kuz'mina 2007: 253; Pigott 2018).

An overview of tin sources and their exploitation has been provided by C. Roden (1985) and R.D. Penhallurick (1986). Their reviews begin with Europe, where the best-known sources are the placer deposits of Cornwall (England), which have been considered as a potential tin supplier for the Aegean Bronze Age (Laing 1968; Muhly 1973: 410–412; 1985: 287–288; Piggott 1977; Dayton 2003: 165; Gülçur 2005: 15). However, these European deposits would not have been the focus of the BMAC supply. This is also true for the tin ore sources in the Zagros Mountains in Iran, like Deh Hosein, which are closely associated with the so-called Luristan bronzes (Begemann *et al.* 2008; Nezafati *et al.* 2008; Nezafati *et al.* 2009; but see Chapter 29). These mines date to the Early Iron Age (ca. 1300/1250 to 700/650 BC), and thus are not contemporaneous with the BMAC.[7]

We will focus here upon Central Asia and nearby areas both geographically and chronologically close to the BMAC.

India

Since the place name "Meluhha" mentioned in Mesopotamian texts as a provider of copper and tin is identified as the Indus valley, the question arises as to whether the source of these metals could lie somewhere there. An inscription from Lagash, for instance, mentions that copper, silver, tin, lapis lazuli, and carnelian were imported from Meluhha (Maddin *et al.* 1977; Maxwell-Hyslop 1982: 32). Could the BMAC have obtained tin along the same route, since its connections with the Indus valley are well known? (Crawford 1974: 244; Dales 1977: 72–75; Kohl 2005; Lyonnet 2005: 195). In the Indus valley itself, however, there are no known tin sources. The nearest ones lie further east, for example at Nurango in Bihar, Tosham in the Bastar region, and the Chattisgarh or Koraput regions of Orissa. However, nothing there hints at a prehistoric exploitation (Hedge 1978; Babu 2003: 174–175, 179–180). Therefore, Penhallurick suggests that Meluhha acts more as a distribution center, similar to Dilmun (located at the Persian Gulf), and not as a tin source itself (Penhallurick 1986: 21; Helwing 2009: 210).

With regards to a possible tin source for the Harappa culture, N. Kochhar *et al.* (1999) point to a polymetallic ore deposit that contains cassiterite located within the Tusham region (Bhiwani district of Haryana). However, the concentration of tin in this ore is between 0.05 and 1.5%, which is rather low. Finds of pre-industrial smelting furnaces and slag proves that there was metallurgy in this area, yet the authors do not mention what type of ore was being smelted, and the date of these furnaces is not stated.

Nevertheless, it is interesting to note that the first tin bronzes appeared in the Indus valley during the 3rd millennium BC at Mohenjo-daro, where they made up 6% of the site inventory. During the Harappan period the amount of tin bronze increases to around 23%, but then decreases in post-Harappan times (Stech and Pigott 1986: 43–44). This can be seen as an argument either for the import of tin during the Harappan period or for the rapid exhaustion of a much smaller ore source.

Afghanistan

Further to the north are the placer tin deposits in modern-day Afghanistan, an area that was culturally connected to the Indus valley (Sarianidi 1971: 34–35; Crawford 1974; Dales 1977; Maxwell-Hyslop 1982; Francfort 1989; Kohl 2005) (Figure 28.1, mining center IV and V).

Potential tin resources in Afghanistan, where the famous lapis lazuli sources of Badakhshan are also located, have been mentioned long ago (Berthoud *et al.* 1978; Muhly 1985: 281; Stech and Pigott 1986; Pigott 1999: 118; see also Chapter 29). Tin and lapis lazuli, together with other minerals such as native gold or nephrite, can be found throughout the Near East. This points to the manifold trade relationship existing between Mesopotamia, Iran, and Afghanistan (Alimov *et al.* 1999: 142). The use of tin bronze at Mundigak in southern Afghanistan is assumed to have started around the first half of the 3rd millennium BC (Lamberg-Karlovsky 1967: 147; Moorey 1994: 299). At Gardan Reg (Seistan, southwestern Afghanistan) large scatters of slag have been found next to pottery kilns associated with painted ceramics from the Mundigak IV period, pointing to possible copper and bronze production during the Early Bronze Age (Dales 1971: 158; 1972: 19–22; 1992: 23, 25–26; Weisgerber 2004a: 197–198). Although the current state of research regarding early metal production in Afghanistan is still very fragmentary due to the contemporary political situation, there exists a strong possibility that local metal resources were already being exploited in the 3rd millennium BC, with the extraction of lapis lazuli going back to at least the 4th millennium BC.[8] During the 1970s, Soviet geologists discovered 135 different tin sources in Afghanistan (Weisgerber 2004b: 198), all of which were alluvial. Two main ore-bearing zones have been identified. The first extends from the south of Kandahar to the Badakhshan province, while the second is located farther to the southwest within the area of Seistan, and extends northwards towards Herat. These deposits consist of cassiterite with an average tin content of 5–6%, which are often found together with other metal ores such as copper, lead, and gold, alongside minerals such as chlorite (Stech and Pigott 1986: 44; Moorey 1994: 299). There are also isolated tin deposits within the provinces of Badakhshan, Maydan, Parwan, and Oruzgan, as well as in Herat (Misgaran) and Farah. In addition, there are many smaller occurrences of tin mineralization alongside other ores containing traces of tin (Stech and Pigott 1986: 44–45). Many "old" copper mines and slag heaps containing fragments of stone tools used for ore preparation were found close to the tin ore deposits, and this points to pre-Islamic mining activity. At Misgaran in Herat province for example, a polymetallic copper deposit (malachite

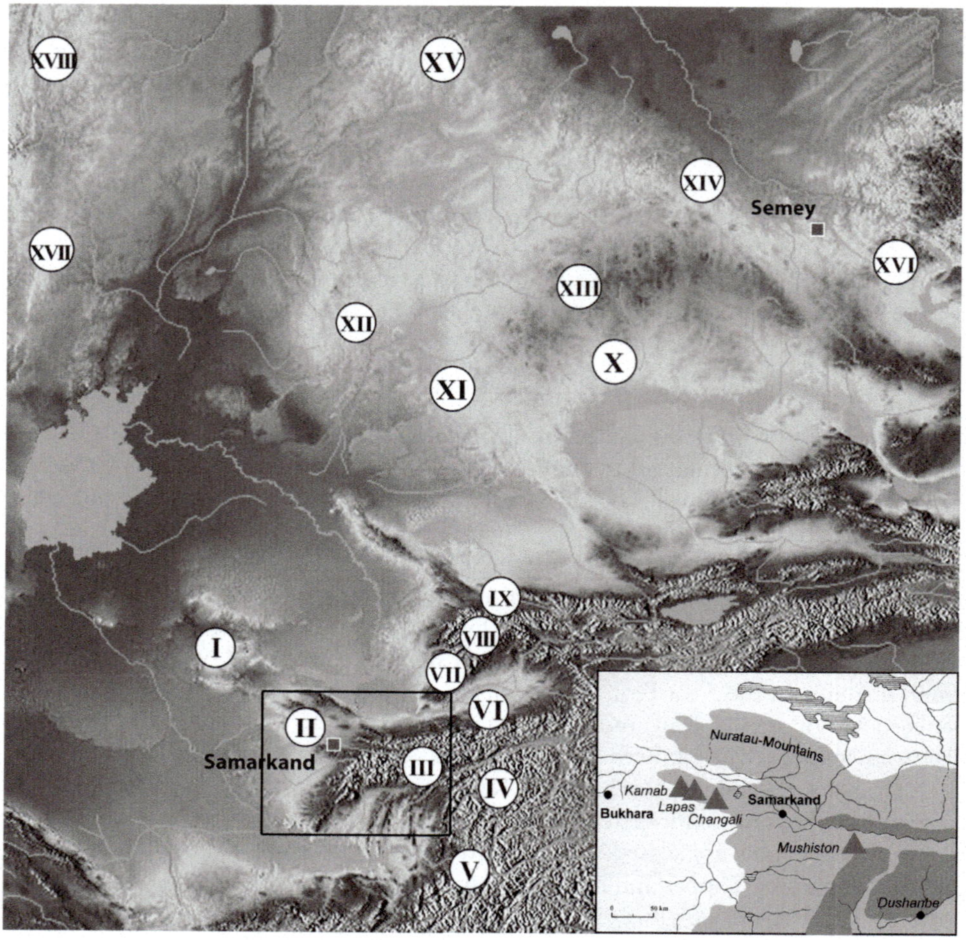

Figure 28.1 Map of Central Asia, showing the different mining and metallurgy centres, especially the tin mines of the Zeravshan valley. I: Kyzylkum Desert; II: Sogdian; III: Hissar Mountains; IV: Gorno-Badakhshan Region; V: Afghanistan; VI: Ferghana; VII: Shash-Ilak Region; VIII: Chatkal Mountains; IX: Talas; X: Balkhash Region; XI: Northern Betpak-dala; XII: Dzhezkazgan Region; XIII: Karaganda-Karkaralinsk Region; XIV: Bajanaul Region; XV: Northkazakhstan (Kokshetau); XVI: Eastkazakhstan (Kalba-Narym); XVII: Westkazakhstan (South-Urals); XVIII: Southeastern Urals-Region (map by https://maps-for-free.com/).

and copper-arsenic ores together with lead and zinc ores, etc.) is associated with alluvial cassiterite deposits (Berthoud *et al.* 1978: 2–3; Stech and Pigott 1986: 45). There, small mining galleries and open-cast mining pits dug to a depth of 4 m in the (copper?) ore vein testify to the early nature of this working, while the presence of grooved stone hammers is a strong argument for prehistoric mining (Weisgerber 2004b: 192). The tin occurrence at Shindand may also be related to the bronzes found nearby at the site of Mundigak (Dayton 2003: 167).

Southern Central Asia

Tin deposits with evidence of Bronze Age exploitation are known from the area to the north of Afghanistan in Central Asia, both within the steppes and mountain ranges of what is today Uzbekistan, Tajikistan, Kyrgyzstan, and Kazakhstan.

Important deposits are to be found within the mountain ranges to the south of the Zeravshan valley, and in the Hissar Mountains up towards the Pamir and Tian-Shan. These then extend further to the northeast up into Kyrgyzstan, the Atasu Mountains (central Kazakhstan), and into the Kalba-Narym Massif and the Altai Mountains (eastern Kazakhstan). In these areas there are hundreds of identified tin mines, some of which show evidence of Bronze Age exploitation (Chernikov 1949, 1951, 1960).[9]

The tin-mining district south of the Zeravshan valley, between Bukhara and Samarkand (Karnab, Lapas, Changali), was first investigated and described by B.A. Litvinskij in 1946 as part of a geological exploration expedition (Litvinskij 1950, 1954). In 1997–1999 international research[10] started in this area. Several different tin-mining deposits were excavated: at Karnab and Lapas, between Samarkand and Bukhara; at Changali near Kattakurgan; and at Mushiston, not far from Pendjikent and the prehistoric settlement of Sarazm (Figure 28.1, mining center II and III). At Karnab the mines reached a length of 1,500 m and were 500 m wide (Figure 28.2).

Unfortunately, the area was heavily disturbed by hundreds of meters of Soviet prospection trenches running perpendicular to the west–east ore veins. Geologically the area is composed of metamorphosed Cretaceous sediments with a massive granitic intrusion. The quartz veins stand almost vertically within the granite body and are between 0.4 m and 1.2 m thick. Cassiterite (SnO_2) is found primarily within these

Figure 28.2 The open-cast tin mines of Karnab, Uzbekistan (photo by DBM).

Figure 28.3 View of rock pillars with remains of an ore vein. Note the clear traces of fire setting on the roof (photo by DBM).

quartz veins. The mining took place within open-cast trenches that followed the veins to an unknown depth. The water table was reached after 9 m, but in one of the mines, situated upon a hill, it had been possible to mine to a depth of 17 m. All of the mines were extremely narrow and irregular in shape, taking the form of the ore veins. Evidence of climbing access routes such as footholds or steps cut into the rock were visible, as well as rock pillars left for support (Figure 28.3). To crack the rock, the ancient miners used fire in combination with heavy stone hammers made of local limestone (Figure 28.4). Fire setting takes advantage of the fact that the heated rock cracks from the tensions arising out of the expansion and contraction of the various different minerals inside of it. Following this action, the cracked rock was hammered away. A fire-setting experiment in Karnab showed that it was possible to crack the much harder granite and quartz just using a limestone hammer (Garner 2013: 62–111).

Figure 28.4 Thousands of stone hammer tools found inside the mines at Karnab (photo by DBM).

The settlement of Karnab-Sichkonchi that lay close to the mine was also excavated (Boroffka *et al.* 2002: 149–153; Parzinger and Boroffka 2003). Similar artifacts were found in both the settlement and the mine, and tin metallurgy was clearly identified. A round furnace (0.75 m in diameter x 0.15 m preserved height) was found inside an oval-shaped structure (2.5 m x 4 m diameter) that was partly sunk into the ground. Analysis of the soil revealed a high tin content, so a metallurgical use, perhaps for tin smelting, seems likely. Within the middle of this structure, but in a later phase of this feature, were found two cast-drops of tin bronze (Parzinger and Boroffka 2003: 50, fig. 20a, fig. 51, *8*, *10*). Furthermore, two similar furnaces were found that may have been used in the same way. Finds of grinding stones, stone hammers, and concentrations of crushed quartz containing cassiterite all point to tin processing (Parzinger and Boroffka 2003: 50–51, fig. 20a–b, fig. 51, *9*). Associated finds of camel bones and pottery from the BMAC (Dzharkutan phase) all testify to the wide-ranging connections between these mines and the region of Bactria (Parzinger and Boroffka 2003: 58 and fn25, fig. 26, *1*, fig. 71.1a, b). Ceramic typologies and radiocarbon dates date these mines in Karnab, Lapas, Changali, and the settlement to the first half of 2nd millennium BC, and link them with the Andronovo-Tazabag'jab culture (Parzinger and Boroffka 2003: 71–73; Garner 2013: 232, fig. 154).[11]

Further east within the same mountain range overlooking the Zeravshan valley lies another large site in northwestern Tajikistan on the slope of the Mushiston Mountain at ca. 3,000 m above sea level (Figure 28.5). There, a prehistoric underground mine with galleries had been cut by modern mining at a depth of 35 m below the surface. The excavations proved that two different types of galleries could be distinguished.

Figure 28.5 View of the Mushiston deposit in Tajikistan (photo by DBM).

The first type consisted of small single chambers, only a few meters wide, which could be elongated to become an *adit*. The second type took the form of a huge labyrinth, by far exceeding the scale of the first one (Figure 28.6).

The mine could not be documented completely due to the presence of several collapses. The constant temperature (6°C) underground had preserved many of the wooden props, and one was still standing upright *in situ*. At Mushiston the main tin mineral is stannite ($CuFeSnS_4$), which occurs together with malachite and azurite in the oxidation zone. Another mineral called mushistonite ($(Cu, Zn, Fe)Sn(OH)_6$) was also discovered here, and this was named after the site. Indeed the special feature of the Mushiston ore deposit is that here both tin ore and copper ores occur together. Smelting experiments carried out in Freiberg, Germany, showed that this ore could easily be smelted and in one step produce a tin bronze.[12] At Mushiston stone hammers and bone tools were found together with Andronovo-Tazabag'jab ceramics. Radiocarbon dates showed that the mining started at the end of the 3rd millennium BC and stopped in the first half of the 2nd millennium BC (Garner 2013: 234–235, fig. 154).[13]

Additional tin ore deposits in southern Central Asia have barely been investigated. This is the case of those in the Artshamaidan River basin and near Takfon in central Tajikistan (Ruzanov 1999b: 103).[14] Other polymetallic deposits containing tin were exploited in medieval times, yet it would require further research to look for evidence of older exploitation. This is probably the case for the Kugitangtau Mountains west of Surkhan Darya (southern Uzbekistan); the northern Ferghana valley at "Tugrysaj" (Kuraminsk Mountains) (Ruzanov and Burjakov 1997; Ruzanov 1999b: 103); and the southern Chatkal Mountains in Uzbekistan, for example in the lowlands of the Kok-Su River where investigations were carried out in two of the four areas of the

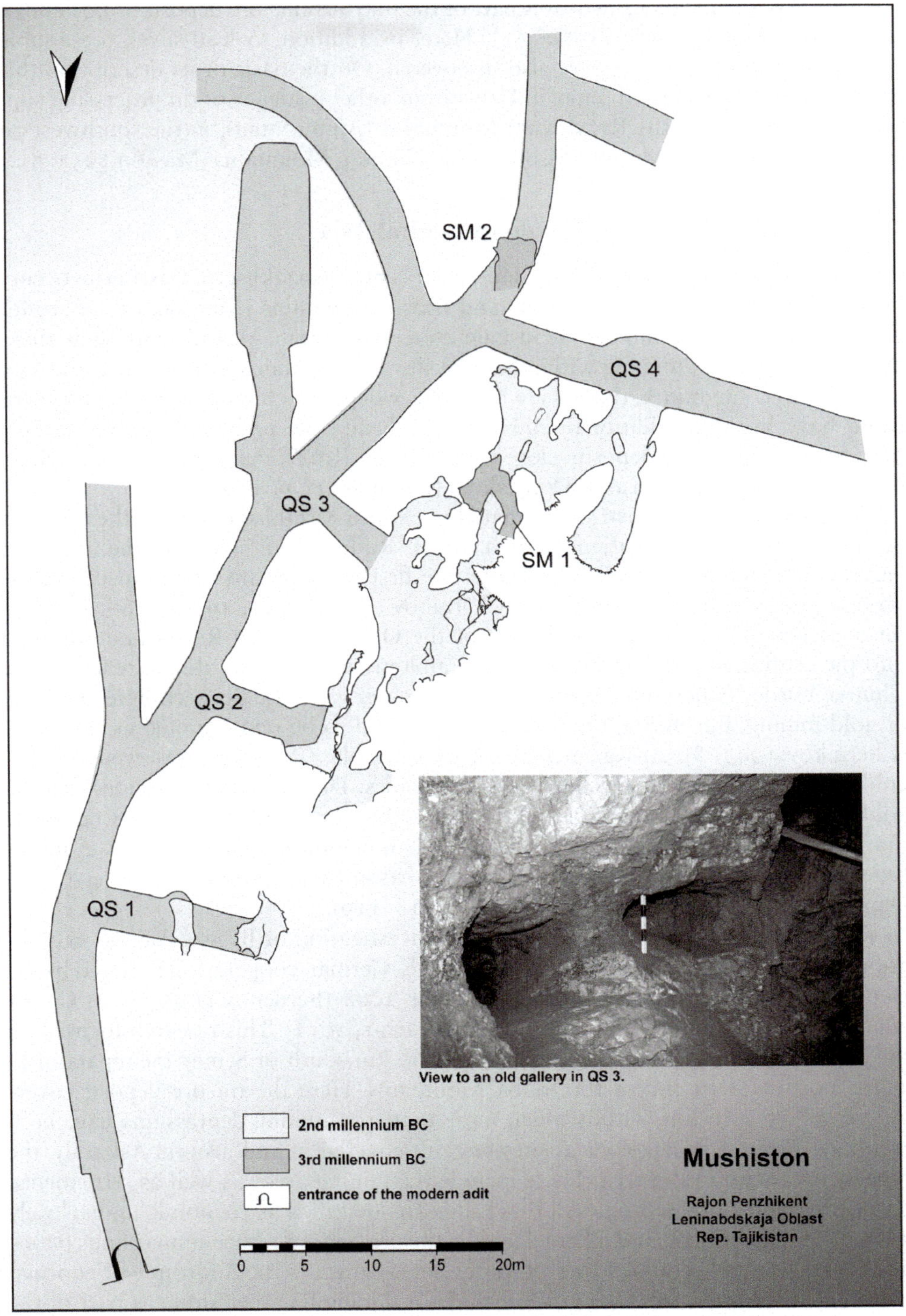

Figure 28.6 Plan view of the Mushiston (Tajikistan) labyrinth-like underground tin mine (plan by DBM).

ore field "Vozrozhdennoe," which is part of the polymetallic ore deposit of Bostandyk (Ruzanov 1979: 101–102; 1980: 62). Here, in addition to lead-silver ores, some copper ores and cassiterite were also discovered. On the basis of its description this was most likely a medieval mine. In Uzbekistan, other evidence of tin processing supposedly exists within the Kyzyl Kum Mountains (Auminzatau), in the southwestern part of the Kuraminsk Mountains, and in the Nuratau Mountains (Masson 1953: 8).[15]

Northern Central Asia

Relatively little is known about tin sources of central Kazakhstan. Cassiterite occurs in the Ulutau (Berdenov 1998: 185) and Atasu Mountains (Margulan *et al.* 1966: 269–270; Kuznetsova and Teplovodskaja 1994: 64–65, tab. 3), and supposedly there are also traces of old mining within the Kokshetau Mountains (Margulan 1972: 25). In addition to the mines, there were also placer deposits containing gold, and these could have been an additional source of tin during the prehistoric period.[16] This includes the cassiterite deposits close to Issyk-Kul' (Kuz'mina 2004: 60), as well as those in northern Kazakhstan (Donetskoe) (Laumulin *et al.* 1996).

The most famous tin district is situated in eastern Kazakhstan within the foothills of the Kalba-Narym Massif, in the area called "Rudnyj Altai" (Altai of the ores) (see Figure 28.1, mining center XVI). There, the first intensive investigation of prehistoric mining was carried out by S.S. Chernikov (1949; 1951; 1960: 118–121). The tin occurrences start at the confluence of the Oba and Irtysh Rivers and continue into the central and eastern areas of the Kalba mountain ridge, almost reaching the Chinese border (Chernikov 1960: 119). Old mines were usually seen here as relicts of gold mining, but during the 1930s these started to be reinterpreted as tin mines (Chernikov 1949: 8): At Kazanchunkur, for example, Chernikov observed that the gold mines were of the same shape as the tin mines. This led him to consider that the gold mines may date to the same period (Chernikov 1949: 36–37). It is also possible that alluvial tin extraction was linked to gold panning in the placer deposits. A multitude of tin placers are known from along the Irtysh River, although these have been exploited in later periods (Chernikov 1960: 118–119).

There has been a renewed interest in the investigation of Bronze Age tin exploitation in eastern Kazakhstan following a Kazakh–German cooperation initiated by the Deutsches Bergbau-Museum Bochum in 2004–2008 (Berdenov *et al.* 2004; Cierny *et al.* 2005; Garner *et al.* 2007; Stöllner *et al.* 2011, 2013). This research focused on the area of the Delgebetej granite massif ca. 60 km south of Semey (Semipalatinsk) at the northwestern end of the Kalba Mountains. Here the tin ore deposit covers an area of 15 x 10 km, within which more than 100 mining depressions have been identified. Several archaeological sites were discovered in and around Askaraly, the highest mountain in the area. These include the mining areas as well as settlements, cemeteries, and stone settings. All these sites are identified as economic units closely associated with mining, and all are dated by the presence of rather similar finds (stone and bone tools and ceramics) and ^{14}C dates linking them to the Andronovo-Fedorovo culture (see Chapter 26). One grave includes a mining hammer stone as part of the grave inventory (Stöllner *et al.* 2010).

Given the current state of research it is still quite difficult to judge if the tin ore deposits of Kazakhstan played an important role for the BMAC. There is almost no doubt, however, that the currently investigated mining areas and metallurgical centers of central and eastern Kazakhstan can be attributed to the Middle Bronze Age Andronovo culture (Alekseev and Fedorovo phases) and the Late Bronze Age Begazy-Dandybaj culture. Nevertheless, an initial phase of tin exploitation starting sometime during the 3rd millennium BC can also be postulated (Kuznetsova and Teplovodskaja 1994).

COPPER DEPOSITS AND COPPER MINING IN CENTRAL ASIA: AN OVERVIEW

Copper, as has been mentioned above, is ubiquitous within this broad geographical area. During the Bronze Age, the Mesopotamian texts frequently mention the Magan copper sources (i.e., Oman). Our presentation here will concentrate upon the Central Asian deposits, closer to the BMAC, but much less known.

In the case of copper extraction, archaeological investigations are similarly rudimentary. Most of what is known is likewise based upon field surveys and surface finds.[17] Of particular interest here are reports of the discovery of stone hammers. Although it is not always possible to precisely date these finds, they were the typical tools of the Bronze Age miners. Many have a groove running around the middle of the stone to which a leather band or wooden handle could be attached. Due to the large number of copper ore deposits (mostly polymetallic) known, we will mention here only the mines with archaeologically relevant finds (i.e., evidence of prehistoric mining in the form of stone hammers), or that have been archaeologically investigated, or suggested as possible ore sources for the BMAC following geochemical analysis.[18]

Pamir-Hissar Mountains

In the western Pamir Mountains of Tajikistan, there is a multitude of copper mines, most of which date from the medieval to the modern period. However, there are reports of stone hammers, such as at Rosht-Kalinski in the Dzhaoshango valley. The tools were found together with charred wood and other stone tools, which make a prehistoric exploitation of these ore deposits probable (Bubnova 1996: 20; 2007: 225).

The southern Hissar Mountains in Tajikistan (see Figure 28.1, mining center III) are mostly known for their iron ore deposits. However, there are also nonferrous deposits with copper, lead, and tin ores, as well as many well-preserved open-cast mines and slag heaps. For example, between the Shirkent and Karatag Rivers, at the headwaters of the Dzhaupai, and upon the left bank of the River Sandladarja, a number of copper-tin mines have been discovered and investigated. Here pyrite and chalcopyrite can be found together with their oxides. Along an area of greisenized granite there is a 1.5 km long strip of 30 old mines that mostly consist of small chambers around 4 m x 0.5–3m in cross-section. Four of the underground mines seem to be prospection tunnels between 20 and 40 m long. Analysis of the ore has yielded, in addition to copper, some 0.21% of tin, and occasionally some lead, silver, and gold. The presence of metallurgical sites and the results of slag and ore analyses prove that an on-site smelting of this copper-tin ore was undertaken (Novikov *et al.* 1981: 342–343; Novikov and Radililovskij 1990: 86–87).

It had generally been considered that the ores from Almalyk, Naukat (Uzbekistan), or Dzhezkazgan (Kazakhstan) were only locally exploited during the medieval period. However, it is now accepted that the ores from these deposits could have been exploited earlier since they possess native copper that could be the source for the artifacts found within archaeological settlements and graves (Litvinskij *et al.* 1962: 170–172; Margulan 1973; Burjakov 1974: 53–55; Islamov 1976: 35; Kuz'mina 1991: 33; 2008: 79). The analysis of artifacts from the Bronze Age cemetery at Tandyrjul, close to the mines, points at a possible exploitation of local ore sources during the Bronze Age (Novikov and Radililovskij 1990: 63–64). Within the same area of chalcopyrite deposits, upon the floodplains of the Kinyk (a tributary of the Sandaldarja), V.P. Novikov and V.V. Radililovskii discovered five mines in 1981. Three of them are shallow niches, while the two others were *adits* up to 4 m long. Upon the working faces and the backs of these *adits*, parallel grooves and oval depressions left from the use of tools are still visible (Novikov *et al.* 1981: 342). The first of these marks may be from the tools of medieval mining (wedges, hammers, and gad), while the others (i.e., stone hammers) provide evidence of a prehistoric exploitation.

Kyzyl Kum Desert

Copper deposits are known from the Kyzyl Kum Desert in Uzbekistan within the area of the Bukantau, Sangruntau, Sultanuizdag, Auminzatau, and Tamdytau Mountains where Bronze Age copper production is documented on the basis of numerous smelting sites and slag heaps (Baskakov *et al.* 1974; Islamov 1976: 37; Itina 1977: 136; Sverchkov 2009: 155–156) (see Figure 28.1, mining center I), and likewise attributable to the Andronovo-Tazabag'jab culture (Garner 2013: 127–132). It is worth noting here that turquoise was largely exploited there in the 3rd millennium (Vinogradov 1972; see also Chapter 24).

Naukat Mountains

Yet other copper mines have been found in Ferghana at Naukat (Figure 28.1, mining center IV), the latter being considered as a Bronze Age ore source. Geochemical analyses show that the composition of bronze artifacts found on the smelting sites at Kajrak Kum can be correlated with the above deposit. Moreover, large quantities of slag, as well as pieces of ore, copper droplets, and a stone casting mold were found in the Kajrak Kum settlements (Litvinskij *et al.* 1962: 170–175; Kuz'mina 1991: 33; 2008: 79). However, no mining archaeological investigations have yet been carried out within these areas.

Southern Karamazar Mountains

One of the few areas that have been investigated with archaeological soundings is the lead-silver mining district of Kani-Mansur within the southwestern area of the Karamazar Mountains on a Tajik site, between Altyn-Topkan and Adrasman. Here a total of six mining districts are known (Kani-Mansur, Taboshar, Kan-Dzhol, Kalkanata, Altyn-Topkan, and Kansaj). This area was first described by M.E. Masson and B.A. Litvinskij (Masson 1953; Litvinskij 1954), but the earliest archaeological research was carried out by Ju.F. Burjakov at the beginning of the 1960s in the area of Kani-Mansur (Burjakov 1974: 70–75) and Lashkerek (Burjakov 1965a). He was able to document a complex network of mines represented by surface depressions, as well as underground works, that, at the time, he dated to the medieval period on the basis of iron tool finds. To the west and northwest of the archaeological trenches, possible ore-processing dumps were discovered along with a large number of stone tools (grooved hammers and grinding stones). The hammers were of a rough form and weighed between 1.25 and 4 kg each, which points at their use as a tool for ore processing since at the time iron tools were being used in the mine itself. It is explicitly mentioned that it would not be correct to assume that the tools date to the Bronze Age, since they were also found in Kajnar and in the main chamber, both of which date to the medieval period (Burjakov 1974: 73).[19] From the coin finds, Burjakov dates the mine to the 9th century AD, while the ceramics span a larger period (7th to 12th centuries AD). However, he does not exclude the possibility that mining could have started earlier, and, in fact, he later attributed the grooved stone hammers to the 1st millennium BC (Sverchkov 2009: 144). I.O. Islamov, on the other hand, attributes the trenches and crater-shaped mining depressions to a "prehistorical and slave-based society" (Bronze Age), while he interprets the underground mines as medieval (Islamov 1977: 22–24). Today grooved stone hammers are seen as a clear indicator for prehistoric mining, and therefore it is assumed that Kani-Mansur and Kajnar were most probably Bronze Age copper mines. Likewise, in the Ferghana valley, a large numbers of stone tools were found in the mines of Chauyai and Simob (Surgaj 1951: 15; Sverchkov 2009: 147).

Northern Karamazar Mountains

Another important area with traces of potential prehistoric copper production is the Almalyk (Olmaliq) district within the northern foothills of the Kuraminsk Mountains in Uzbekistan (see Figure 28.1, mining center VII). Within the various ore fields of Almalyk there are traces of old mine workings many of which have already been destroyed by modern mining, or entirely covered up by later mining dumps. In the area of Nakpeisaj, not far from Almalyk, some of the mines are still preserved and are attributed to gold, copper, and turquoise extraction. The remains consist of mining depressions or underground works that follow the ore veins and include shafts up to 50–70m deep. Large grooved and notched stone hammers have been found within the dumps close to the turquoise mines. Additionally, a copper knife and a crouched burial were found in the vicinity of the copper mines. This, together with the presence of native copper, makes an exploitation of these mines during the Bronze Age more probable (Burjakov 1974: 53–56; Islamov 1976: 35).

Most of the copper deposits in Central Asia are polymetallic (lead, silver, copper, arsenic, zinc, etc.). This is particularly the case with the sources that would have been of interest to the BMAC, such as those in the Shash-Ilak region towards the upper and middle course of the Angren River in northeast Uzbekistan (see Figure 28.1, mining center VII). There is some evidence to suggest that mining districts such as Takelo, Kan-Dzhol and Kuru-Saj, Kurgashin-Kan, and Kajnar have been exploited since at least the antique period (Burjakov 1965b; 1974: 43–46; Islamov 1977: 41–43). In fact, almost all of these deposits were exploited more or less continually from the antique period right up to the 20th century AD, and, consequently, any earlier prehistoric mining traces are unlikely to have survived. According to the ore analyses, oxidic lead-silver ore was mainly mined in antiquity, although copper ore was also a primary product (Islamov 1977: 51). This is certainly true for the Kajnar mining district located 8 km to the north of Angren. This was the site of a possible silver mine, the ore of which also contained copper (Burjakov 1974: 52–53; Islamov 1977: 43). These mines have been dated to the medieval period on the basis of the ceramics and mining tools, yet grooved stone hammers have also been found upon the mining dumps (Burjakov 1974: 52–53). It is possible therefore that a semicontinuous exploitation began here during the prehistoric period. A rather similar situation is found within the mining district of Kan-Dzhol where, in addition to an iron chisel from the medieval mining phase, a number of grooved stone hammers from a prehistoric phase of working have been identified (Turlychkin 1961: 170, fig. 2).

There is some additional evidence of copper production in the form of old mines and smelting sites within the neighboring Chatkal Mountains, which are located in northeastern Uzbekistan and southern Kyrgyzstan to the north of Chust, most of which are medieval in date. However, the "earlier" population of this area is said to have certainly known the occurrences of these minerals as well (Masson 1953: 25). Grooved stone hammers as well as other stone tools such as a disk-shaped object (for grinding?) are known from Aktashkan (Burjakov 1974: 62–64). Here there are mining depressions as well as underground mines, some of which have been described as "primitive." Combined with other archaeological finds, this has led to the dating of the mines to the Bronze Age. Since this is a sulphidic copper ore deposit (chalcopyrite), and since it is generally accepted that the technology to smelt such ores first appeared in the region during the Late Bronze Age, Aktashkan would thus probably date to this slightly later period (Burjakov 1974: 64, 66, 98).[20] On the basis of metallurgical analyses it is generally thought that the Chatkal Mountains were the main source of raw materials during the period of the Burguljuk and Chust cultures (i.e., the Early Iron Age from ca. 1400 BC) (Ruzanov and Lushpenko 2000).

Talas-Alatau Mountains

At the west end of the Tian-Shan Mountains (Talas-Alatau Mountains) close to the border with Tajikistan/Kyrgyzstan (see Figure 28.1, mining center IX) are the important medieval lead-silver mines of "Sheldzhi" in the Talas valley, investigated by M.A. Bubnova (1963a, 1963b, 1963c). Traces of both open-cast mining depressions and underground workings, galleries, and shafts are still visible here. The lead-silver extraction dates from the 9th century AD, although a Bronze Age exploitation of this mine cannot been dismissed completely due to the finds of a pick made from the antler of a maral deer and of a silver earring attributed to the Andronovo culture in a grave at Tash-Tobe II nearby (Bubnova 1963a: 239; 1963c: 91). Gold has also been found here, and Bubnova mentions that gold mining most likely started sometime during the Bronze Age and then continued right up to the 19th century AD, with most of the older mining traces disappearing under this later work (Bubnova 1991: 4). The same holds true for the lead-silver exploitation. Stone hammers have been found around some of the mines, but these were considered as medieval.[21]

Central Kazakhstan

The question of Bronze Age copper production was investigated in Kazakhstan by A. Ch. Margulan during the 1940s. His research rested upon the evidence from the copper mines of Dzhezkazgan and the nearby metallurgical settlements of Milykuduk, Sorkuduk, and Ajnakul'. He also investigated the copper mines of Sorang, Shankpaktas, Meizek, Sajak, and additional metallurgical centers and settlements at Atasu und Myrzhik (among others) (Margulan 1973, 1979, 2001).[22]

As is the case for tin, it is difficult to say if the copper deposits of Kazakhstan played an important role for the BMAC, yet it is clear that they were already being exploited by Andronovo groups (Alekseev and Fedorovo phases) during the Middle Bronze Age, and possibly even earlier during the 3rd millennium BC (Kuznetsova and Teplovodskaja 1994).

METALLURGICAL STUDIES

Studies concerning the relationship between the ore deposits and the processing sites around or within prehistoric settlements are still rare in Central Asia.

One exception concerns the copper ores of the southern Urals region (see Figure 28.1, mining center XVIII) thanks to the metallurgical studies done in the 1960s by E.N. Chernykh (Chernykh 1970; Kuz'mina 1991: 31–32; 2007: 85, 87, 95). Chernykh has compared a wide spectrum of analyses of metallic items from the western Andronovo-Fedorovo groups with ores from this region. Among the different chemical groups, some correlate well with specific mining districts, thereby helping in the identification of the metallurgical centers and the verification of copper production. From this starting point in the southern Urals, Chernykh expanded his research to all of Eurasia, determining several Bronze Age "metallurgical provinces," two of which are relevant for the BMAC: those of Central Asia and Iran/Afghanistan (Chernykh 1992: 191–195, 264–269; Chernykh *et al.* 2004, 21–25). On the basis of ca. 800 metallurgical analyses he assumed that two production centers existed in the BMAC area, one in Bactria, the other in Margiana/Turkmenistan (Chernykh 1992: 179). He made a further differentiation between Bactria and Turkmenistan, the first area being a metallurgical center with evidence for both mining and smelting on account of finds such as crucibles and slag, the second one being solely a metalworking center. In Bactria, Chernykh (1992: 179) looked at finds from Sapalli and Dzharkutan, where tin bronzes make up 40–45% of the find inventory, and from Dashly 3 where they represent only 9%. In Margiana and Turkmenistan, bronze artifacts are almost completely absent (at Togolok, for example) or extremely rare (at Altyn Depe). Arsenical bronze or copper-lead alloys (rarely more than 8–10% of the metal inventory) present a rather similar picture to the above, pointing to the exploitation of natural polymetallic ores to produce alloys.[23] Pure copper dominates in 50–90% of the investigated objects from the sites of Margiana and South Turkmenistan. This is the reason why Chernykh puts Turkmenistan into the "Iran–Afghanistan metallurgical province" that stretches as far as the Indus valley (Chernykh 1992: 179–180, 182). Recently, however, K. Kaniuth has proposed a revision of the BMAC chronology where northern Bactria (Sapalli and Dzharkutan phases) stands in contrast to the surrounding regions of southern Bactria and Margiana (which used copper-arsenic or copper-lead based material) and is dated to the Late Bronze Age I (SP I[24]). Kaniuth (2006: 166) also revised the metal inventory and reduced it to 25% tin bronze because, according to him, Chernykh had included in his analysis material from a later phase, Bustan, now attributed to the Late Bronze Age II (SP II[25]) (Kaniuth 2006: 166 fn894) (see also the more recent results of metallurgical analyses presented in Chapter 27).

Another exception concerns Uzbekistan and Tajikistan. Ruzanov made a similar study in this region using the geochemical characterization of the different ore deposits in an attempt to match them with metal objects from the various cultures discovered there (Ruzanov 1982). He identified three different ore deposits that he considered particularly important during the Bronze Age as mining and metallurgical centers: the Bukan-Tamdy-Auminzatau center in the Kyzyl Kum (see Figure 28.1, mining center I), the Chatkal-Kuraminsk center in the Ferghana-Ilak region (see Figure 28.1, mining centers VI, VII, and VIII), and the Zirabulak-Karatjube-Zeravshan center (see Figure 28.1, mining center II) – the last being the only one with proof of prehistoric mining. He then divided the metal objects into three main cultural groups: one from the cultures of the oasis (Namazga (NMG) V/VI or BMAC), one from the steppe cultures (Andronovo), and one from the Early Iron Age cultures associated with painted pottery (Chust and Burguljuk). Finally, he carried out 1,535 spectral analyses on them. The results led him to determine, on the one hand, metallic groups (pure copper, tin bronze, lead bronze, arsenic bronze, lead-tin bronze, lead-arsenic bronze, tin-arsenic bronze, tin-lead-arsenic bronze), and, on the other hand, chemical groups that appear to be linked with the various different cultural types (ten groups within the oases cultures, and nine for both the steppe cultures and the cultures with painted pottery). He concluded that a correlation exists between metal objects and the above-mentioned three ore deposits and he proposed seven different metallurgical centers in southern Central Asia (Ruzanov 1982: 12–15). He also determined three metallurgical phases: the first being strongly influenced by the southern Turkmenistan–Iranian province, the second showing an increase in metal production and the formation of the above-mentioned metallurgical centers (although the influences from the south are still present), and the third one pointing at the development of a new center (the Burguljuk/Chust) in northeastern Uzbekistan, where tin bronzes from the more northerly ore sources dominate (Ruzanov 1982: 21).

These studies, however, were carried out almost half a century ago and further research is now needed to confirm these results. Work should also be done on the different trade networks. In fact, the current state of knowledge is only in its infancy. We still know almost nothing about the operational sequence in this area, and although many remains of mining activities have now been recorded, their date, the kinds of ore exploited, and the locations of the smelting sites are still unknown in most of the cases. In order to correctly calculate the potential and also the role of the Central Asian ore deposits as the sources for the metal used in the BMAC, mining archaeological investigations must be intensified in the region and combined with modern analytical methods.

To sum up, at this stage, there are still no known mines that can be attributed to the BMAC with certainty. On the contrary, all the investigated and dated Bronze Age mines, whether for tin or copper, are related to the Andronovo culture, well known for its metallurgical activities and the production of high-quality tin bronzes. Whether or not Andronovo groups were *de facto* the suppliers of tin and copper for the BMAC remains unclear.

NOTES

1 This is also true for the Ferghana valley (Kajrak Kum, a late variant of the Andronovo culture), the occupation of which could be due to the exploitation of its copper deposits (Parzinger 2006: 413).

2 Like the settlements of Kangurttut, Teguzak, or Tashguzor, or the cemetaries of Bustan 6, Tandyrjul', Bishkent, and Kumsaj (Vinogradova 1993; 1997; 1999; 2001; 2004; 2008: 190, 213–217; P'jankova 1999).

3 Near Gonur North (Hiebert 1994: 27, fig. 2, *16*; Hiebert and Moore 2004), at Ojakly (Rouse and Cerasetti 2014), or around Takhirbaj and Auchin (Cerasetti 2012).

4 Except for the research of B.A. Litvinskij (Litvinskij *et al.* 1962) in Uzbekistan, and A.Ch. Margulan (1972, 1973, 2001) and S.S. Chernikov (1949, 1951, 1960) in Kazakhstan.

5 The research about metallurgy from Kajrak Kum by Litvinskij and his colleagues is an exception (Litvinskij *et al.* 1962; Gafurov and Litvinskij 1963).

6 For example, a project about prehistoric tin mining from 1997–1999 in Uzbekistan and Tajikistan under the direction of the Eurasia Department of the German Archaeological Institute Berlin (Boroffka *et al.* 2002; Parzinger and Boroffka 2003; Garner 2013, 2015), or the mining archaeological research of the German Mining Museum Bochum in the Kalba-Narym-Mountains in eastern Kazakhstan 2003–2008 (Stöllner *et al.* 2011, 2013). Since 2013, a research project of the Eurasia Department of the German Archaeological Institute about early mining and metallurgy has been underway in Afghanistan (Thomalsky *et al.* 2013; Thomalsky 2014).

7 For more detailed information about Luristan bronzes, see Overlaet (2016).

8 Old lapis lazuli finds in the 4th millennium are known at Mehrgarh, Pakistan (Tosi and Vidale 1990), Sarazm, Tajikistan (Isakov 1994: 6), and in Egypt (Weisgerber 2004a: 126).

9 A prehistoric mine is characterized by the finds of grooved stone hammers, bone tools, and traces of fire setting (see Craddock 1995: 37–39).

10 This multinational research project was created in conjunction with the German Archaeological Institute Berlin, the University of Freiberg, the German Mining Museum Bochum, the Archaeological Institute of Samarkand of the Science Academy of the Republic of Uzbekistan, and the Institute for History, Archaeology and Ethnology of Tajikistan (Alimov *et al.* 1999; Boroffka *et al.* 2002; Cierny and Weisgerber 2003; Parzinger and Boroffka 2003; Garner 2013, 2015).

11 ^{14}C dates of Karnab (all cal. 2σ), settlement of Karnab-Sichkonchi: fireplace 37: 1890–1690 BC; house 4: 1650–1450 BC; house 2: 1525–1375 BC; house 3: 1520–1260 BC (Parzinger and Boroffka 2003: 73). Karnab mining area: mine 6-1: 1350–1310 BC; mine 5-5: 1620–940 BC, 1420–1020 BC, 1200–825 BC; mine 5-3: 1315–915 BC; mine 3-2: 905–775 BC (Garner 2013: 232; see also the Appendix at the end of this volume).

12 These experiments were done by J. Lutz, H. Herdits, and E. Pernicka.

13 ^{14}C dates of Mushiston (all cal. 2σ), the little galleries, near the surface: mine 2: 2450–1935 BC; mine 1: 2190–1860 BC; mine 10: 2140–1625 BC; mine 6: 1935–1635 BC. The labyrinth-like underground mines: QS 4: 1735–1440 BC; 1690–1405 BC; 1645–1250 BC; QS 3: 1635–1445 BC; 1625–1395 BC; 1120–800 BC; QS 1: 1515–1265 BC (Garner 2013: 234–235; see also the Appendix at the end of this volume).

14 As for Takfon, no evidence of protohistoric tin mining was found during the investigations made in 1998 (Garner 2013: 127).

15 The Nuratau range forms the northwest continuation of the Turkestan Mountains. Geological investigations that started in the 1930s have identified ancient mines of gold, galena, arsenopyrite, and different kinds of copper ores. Tin was found only mineralogically at 11 places, without any traces of mining activities (Sverchkov 2009: 154).

16 That is, the ore placer in the valley of Atasu with the mine of Kalai-Kazgan to the north of Betpak-dala (Margulan *et al.* 1966: 269; Margulan 1973: 5), plus the ore placers at the upper and middle course of the Ishim River, which has barely been investigated (Margulan 1972: 25; Berdenov 1998: 181).

17 Because of these surveys, several references about possible protohistoric copper mining are mentioned: in the Nuratau range, north of the Zeravshan River at Nurata and Ljangar; in the valley of the Angren River, north of Kokand (Ferghana); in the southern Karamazar range, on the site of Tajikistan, not far from Uzhkatly-Miskan, Kansaj, and Adrasman (Burjakov 1974: 5, 53; Islamov 1976: 35; Kuz'mina 1991: 33); in Semirech'e, along the Chu River in the valley of the Talas (Aktashkan); at the Balkash Lake, in Chatkal, not far away of Ketmentjube; and at the Issyk-Kul Lake, in the area of Almaty and Arta (Kuz'mina 1966: 96–102; Ruzanov 1980: 62–63; Kadyrbaev and Kurmankulov 1992).

18 For a detailed compilation of these deposits, see Garner (2013: 246–270).

19 Burjakov assumed that the use of stone hammers was to hit iron tools like chisels when they are actually mostly used for ore processing. His dating of the stone hammers to the medieval period is based on the excavations made by M.E. Masson near the main chamber, which produced pottery dated to the 9th–10th century (Burjakov 1974: 74).

20 E.E. Kuz'mina also considers that the Andronovo culture was not able to smelt sulfide ores (Kuz'mina 1991: 29; 2007: 85). However, in the Mitterberg region (Austria), huge underground mining (Arthurstollen) and open-cast mining (Mitterberger Hauptgang) for chalcopyrite are dated to the Middle Bronze Age. Furthermore the smelting sites are so specialized for smelting sulfide ores, that "Mitterberger Prozess" has become an expression (Pernicka and Lutz 2015; Stöllner 2015a; 2015b: 99–101). Other examples are attested in Kitzbühel, Austria (Koch Waldner 2013; Koch Waldner and Klaunzer 2015). In the Mauken valley, Radfeld/Brixlegg, North Tyrol, the people exploited and smelted even fahlore during the Late Bronze Age, which is more complicated to smelt than sulfides (Goldenberg 2015).

21 Personal communication to G. Weisgerber in 1998.

22 See also S. Berdenov with a summarized overview on the state of research on old mines in Kazakhstan (Berdenov 1998).

23 According to Ruzanov, arsenic rich copper (group KD) was imported from Iran or Afghanistan. Arsenic poor copper may have come from the Bukan-Tamdy-Auminzatau deposit in the Kyzyl Kum Desert (Ruzanov 1982: 10).

24 SP I means "Spätbronzezeit I" = Late Bronze Age I, 2000–1700 BC; and SP II "Spätbronzezeit II" = Late Bronze Age II, 1700–1500 BC (Kaniuth 2006: 3–6, 158).

25 Kaniuth only used artifacts from SP I (Sapalli, Dzharkutan). In order to make the analyses comparable, a major problem is the proper definition of tin bronze, tin copper, arsenic bronze, arsenic copper, etc. Kaniuth, for instance, considers tin bronze, arsenic bronze, and lead bronze as having a content of 3% or more of the alloyed element; with 1–2.99% it becomes a tin copper, arsenic copper, and lead copper, while with a content of these elements under 1% it is then referred to as pure copper (Kaniuth 2006: 161). In contrast, Ruzanov considered copper objects with a content of 0.8% Sn as tin bronzes, but copper objects with arsenic or lead only become bronze when there is a 3% intentional alloy (Ruzanov 1987).

REFERENCES

Alimov, K., N. Boroffka, Ju. Burjakov, J. Cierny, J. Lutz, H. Parzinger, E. Pernicka, V. Ruzanov, T. Shirinov, and G. Weisgerber 1999. "Research at Karnab, Uzbekistan: Preliminary Notes from the 1997 Campaign." *Istorija Material'noj Kultury Uzbekistana* 30: 80–87.

Babu, T.M. 2003. "Advent of the Bronze Age in the Indian Subcontinent." In *Mining and Metal Production Through the Ages*, ed. P. Craddock and J. Lang. London: British Museum, pp. 174–180.

Baskakov, Ju.F., Ju.T. Abramov, and E.B. Pruger 1974. "Drevnie medeplavil'ni v Kyzylkumakh" [Ancient copper smelting in the Kyzylkum]. *Uzbekskij Geologicheskij Zhurnal* 4: 76–77.

Begemann, F., E. Haerinck, B. Overlaet, S. Schmitt-Strecker, and F. Tallon 2008. "An Archaeo-Metallurgical Study of the Early and Middle Bronze Age in Luristan, Iran." *Iranica Antiqua* 43: 1–66.

Begemann, F. and S. Schmitt-Strecker 2009. "Über das frühe Kupfer Mesopotamiens." *Iranica Antiqua* XLIV: 1–45.

Berdenov, S.A. 1998. "Kazakhstanskaja gorno-metallurgicheskaja oblast'" [The smith-metallurgical region of Kazakhstan]. *Voprosy Arkheologii Kazakhstana* 2: 180–190.

Berdenov, S., Z. Samashev, T. Shtol'ner, Ja. Cherny, A. Ermolaeva, and G. Kushch 2004. "Drevnee gornoe delo i metallurgija Vostochnogo Kazakhstana" [Old mining and metallurgy in East Kazakhstan]. *Voprosy Istorii i Arkheologii Zapadnogo Kazakhstana* 3: 154–170.

Berthoud, T., R. Besenval, J.-P. Carbonnel, F. Cesbron, and J. Liszak-Hours 1978. *Les anciennes mines d'Afghanistan – Rapport Préliminaire. Prospection 1977*. Paris: Unité de Recherche Archéologique 7, Commissariat à l'Energie Atomique, Laboratoire de Recherche des Musées de France.

Bobomulloev, S. 1997. "Ein bronzezeitliches Grab aus Zardča Chalifa bei Penžikent (Zervšantal)." *Archäologische Mitteilungen aus Iran und Turan* 29: 121–134.

Boroffka, N., J. Cierny, J. Lutz, H. Parzinger, E. Pernicka, and G. Weisgerber 2002. "Bronze Age Tin from Central Asia: Preliminary Notes." In *Ancient Interactions: East and West in Eurasia*, ed. K. Boyle, C. Renfrew, and M. Levine. McDonald Institute Monographs. Oxford: McDonald Institute for Archaeological Research, pp. 135–159.

Bubnova, M.A. 1963a. *Dobycha serebrjano-svintsovykh rud v Shel'zhi v IX–XII vv* [Exploitation of silver and lead ore in Shel'zhi during the 9th–12th century]. Arkheologicheskie pamjatniki Talasskoj doliny, Frunze. Adademija Nauk Kirgizskoj SSR: Institut Istorii, pp. 225–262.

Bubnova, M.A. 1963b. "Gorno-metallurgicheskaja oblast' Shel'zhi v IX-XII vv.n.e." [The mining-metallurgical district of Shel'zhi during 9th–12th century AD]. Dissertation abstract, Akademija Nauk SSSR, Institut Arkheologii, Moscow.

Bubnova, M.A. 1963c. "K istorii srednevekovoj gornoj promyshlennosti v Srednej Azii" [On the history of medieval mining industries]. *Sovetskaja Arkheologija* 2: 85–94.

Bubnova, M.A. 1991. *Issledovanija drevnikh rudnikov Gorno-Badakhshanskoj Avtonomnoj oblasti Tadzhikskoj SSR* [Investigation about ancient mining in Gorno-Badakhshansk autonomic district in Tajikistan]. Dushanbe: Akademija Nauk Tadjikskoj SSR, Institut Istorii.

Bubnova, M.A. 1996. *Okhrana prirodno-kul'turnykh zapovednikov doliny Shakhdary (Zapadnyj Pamir)* [Preservation of nature and culture in the protected area of the Shadhdary valley (West Pamir)]. Dushanbe: Donish.

Bubnova, M.A. 2007. *Arkheologicheskaja karta Gorno-Badakhshanskoj avtonomnoj oblasti zapadnyj Pamir (pamjatniki kamennogo veka – XX v.)* [Archaeological map of Gorno-Badakhshansk autonomic district in West Pamir (monuments from the Stone Age up to the 20th century)]. Dushanbe: Donish.

Burjakov, Ju.F. 1965a. "Drevnij serebrjanyj rudnik Lashkerek" [Ancient silver mining of Lashkerek]. *Sovetskaja Arkheologija* 1: 282–289.

Burjakov, Ju.F. 1965b. "O mestonakhozhdenii 'serebrjanogo rudnika Shasha'" [The deposits of the "silver mines of Shasha"]. *Obshchestvennye Nauki v Uzbekistane* 12: 28–30.

Burjakov, Ju.F. 1974. *Gornoe delo i metallurgija srednevekovogo Ilaka* [Mining and metallurgy during the Middle Age in Ilak]. Moscow: Nauka.

Cerasetti, B. 2012. "Remote Sensing and Survey of the Murghab Alluvial Fan, Southern Turkmenistan: the Coexistence of Nomadic Herders and Sedentary Farmers in the Late Bronze Age and Early Iron Age." In *Proceedings of the 7th ICAANE 2010*, vol. 1, ed. R. Matthews and J. Curtis. Wiesbaden: Harrassowitz, pp. 539–558.

Chernikov, S.S. 1949. *Drevnjaja metallurgija i gornoe delo zapadnogo Altaja* [Ancient metallurgy and mining in West Altai]. Alma-Ata: Akademija Nauk Kazakzskoj SSR, Institut Istorii Material'noj Kult'ury.

Chernikov, S.S. 1951. "K voprosu o sostave drevnikh bronz Kazakhzstana" [Question on the composition of ancient bronzes of Kazakhstan]. *Sovetskaja Arkheologija* 15: 140–161.

Chernikov, S.S. 1960. *Vostochnyj Kazakhstan v epokhu bronzy* [East Kazakhstan during the Bronze Age]. Materialy i Issledovanija po Arkheologii SSSR 88. Moscow: Akademii Nauk SSSR.

Chernykh, E.N. 1970. *Drevnejshaja metallurgija Urala i Povolzh'ja* [Ancient metallurgy in Ural and Povolzhye]. Materialy i Issledovanija po Arkheologii SSSR 172. Moscow: Nauka.

Chernykh, E.N. 1992. *Ancient Metallurgy in the USSR: The Early Metal Age*. Cambridge: Cambridge University Press.

Chernykh, E.N., S.V. Kuzminykh, and L.B. Orlovskaja 2004. "Ancient Metallurgy in Northeast Asia: From the Urals to the Saiano-Altai." In *Metallurgy in Ancient Eastern Eurasia from the Urals to the Yellow River*, ed. K.M. Linduff. Lewiston, NY: Edwin Mellen Press, pp. 15–36.

Cierny, J., T. Stöllner, and G. Weisgerber 2005. "Zinn in und aus Mittelasien." In *Das Schiff von Uluburun. Welthandel vor 3000 Jahren, Ausstellungskatalog*, ed. Ü. Yalçın, C. Pulak, and R. Slotta. Bochum: Deutsches Bergbau-Museum Bochum, pp. 431–448.

Cierny, J. and G. Weisgerber 2003. "Bronze Age Tin Mines in Central Asia." In *Le problème de l'étain à l'origine de la métallurgie. The Problem of Early Tin*, ed. A. Giumlia-Mair and F. Lo Schiavo. BAR International Series 1199. Oxford: Archaeopress, pp. 23–31.

Craddock, P.T. 1995. *Early Metal Mining and Production*. Washington, DC: University Press.

Crawford, H.E.W. 1974. "The Problem of Tin in Mesopotamian Bronzes." *World Archaeology* 6: 242–247.

Dales, G.F. 1971. "Early Human Contacts from the Persian Gulf Through Baluchistan and Southern Afghanistan." In *Food, Fiber and the Arid Lands*, ed. W.G. McGinnies, B.J. Goldman, and P. Paylore. Tuscon, AZ: Springer Science and Business Media, pp. 146–170.

Dales, G.F. 1972. "Prehistoric Research in Southern Afghan Seistan." *Afghanistan* 25(4): 14–40.

Dales, G.F. 1977. "Shifting Trade Patterns Between the Iranian Plateau and the Indus Valley in the Third Millennium BC." In *Le plateau iranien et l'Asie Centrale des origines à la conquête islamique*, ed. J. Deshayes. Paris: Centre National de la Recherche Scientifique, pp. 67–78.

Dales, G.F. 1992. "A Line in the Sand: Explorations in Afghan Seistan." In *South Asian Archaeology Studies*, ed. G. Possehl. New Delhi: Society for South Asian Studies, pp. 19–32.

Dayton, J.E. 2003. "The Problem of Tin in the Ancient World." In *Le problème de l'étain à l'origine de la métallurgie. The Problem of Early Tin*, ed. A. Giumlia-Mair and F. Lo Schiavo. BAR International Series 1199. Oxford: Archaeopress, pp. 49–104.

Francfort, H.-P. 1989. *Fouilles de Shortugaï. Recherches sur l'Asie centrale protohistorique*. Paris: de Boccard.

Gafurov, B.G. and B.A. Litvinskij 1963. *Istorija Tadzhikskogo Naroda: S drevnejshikh vremen do V v. n.e.* [The history of the Tajik people from the most ancient times to the 5th century AD], vol. 1. Moscow: Akademija Nauk Tadjikskoj SSR, Institut Istorii.

Garner, J. 2013. *Das Zinn der Bronzezeit in Mittelasien*. Vol. II, *Die Montanarchäologischen Forschungen der Zinnlagerstätten*. Archäologie in Iran und Turan 12, Veröffentlichungen aus dem Deutschen Bergbau-Museum 194. Bochum: von Zabern.

Garner, J. 2015. "Bronze Age Tin Mines in Central Asia." In *Archaeometallurgy in Europe*, vol. III, ed. A. Hauptmann and D. Modarressi-Tehrani. Der Anschnitt, Beiheft 26. Bochum: Deutsches Bergbau-Museum Bochum, pp. 135–143.

Garner, J., A. Gorelik, G. Kush, V. Merts, Z. Samashev, and T. Shtel'ner 2007. "Issledovanie shakht po dobyche olovjannoj rudy, poselenija i nekropolja epokhi bronzy v gornom massive Del'gebetej (Askaraly II) v Vostochnom Kazakhstane" [Investigations of a bronze age tin mine, settlement and cemetery at the mountain of Del'gebetej (Askaraly 2) in East Kazakhstan]. Arkheologijalyk zertteyler zhajly esep – po gosudarstvennoj programme "Kul'turnoe nasledie" 2006: 123–124.

Goldenberg, G. 2015. "Prähistorische Kupfergewinnung aus Fahlerzen der Lagerstätte Schwaz-Brixlegg im Unterinntal, Nordtirol." In *Bergauf – Bergab. 10000 Jahre Bergbau in den Ostalpen*, ed. T. Stöllner and K. Oeggl. Bochum: Marie Leidorf, pp. 151–163.

Gülçur, S. 2005. "Handelsbeziehungen im Vorderen Orient." In *Anatolian Metal*, vol. III, ed. Ü. Yalçin. Der Anschnitt, Beiheft 18. Bochum: Deutsches Bergbau-Museum Bochum, pp. 11–16.

Hedge, K.T.M. 1978. "Sources of Ancient Tin in India." In *The Search for Ancient Tin*, ed. A.D. Franklin, J.S. Olin, and T.A. Wertime. Washington DC: Smithsonian Institution Press, pp. 39–42.

Helwing, B. 2009. "Rethinking the Tin Mountains: Patterns of Usage and Circulation of Tin in Greater Iran from the 4th to the 1st Millennium BC." *TÜBA-AR* 12: 188–207.

Hiebert, F.T. 1994. *Origins of the Bronze Age Oasis Civilization in Central Asia*. American School of Prehistoric Research 42. Harvard, MA: Peabody Museum Press.

Hiebert, F.T. and K.M. Moore 2004. "A Small Steppe Site Near Gonur." In *Near the Sources of Civilization*, ed. M.F. Kosarev, P.M. Kozhin, and N.A. Dubova. Moscow: Staryj Sad, pp. 294–302.

Isakov, A.I. 1994. "Sarazm: An Agricultural Center of Ancient Sogdiana." *Bulletin of the Asia Institute* 8: 1–12.

Isakov, A.I. 1998. *Verkhov'ja Zarafshana vo II tysjacheletii do n.e.* [The upper course of the Zarafshan River during the 2nd millennium BC]. Dushanbe: Institut Istorii, Arkheologii i Etnofgrafii Im. A. Donisda AN, Republikii Tadzhikistana.

Islamov, O.I. 1976. *Iz istorii geologicheskikh znanij v Srednej Azii* [On the knowledge of geological history in Central Asia]. Tashkent: Fan.

Islamov, O.I. 1977. *Iz istorii geologicheskikh znanij v Srednej Azii. Istorija geologicheskikh znanij v feodal'nyj period (V v.n.e.–nachalo XIX v)* [On the knowledge of geological history in Central Asia: the feudal period (5th century–early 19th century)]. Tashkent: Fan.

Itina, M.A. 1977. *Istorija stepnykh plemen Juzhnogo Priaral'ja (II–nachalo I tysjacheletija do n. e.)* [The history of the steppe people in the southern Aral-area (2nd–beginning of the 1st millennium BC)]. Trudy Khorezmskoj arkheologo – etnograficheskoj ekspeditsii X. Moscow: Nauka.

Kadyrbaev, M.K. and Zh. Kurmankulov 1992. *Kul'tura drevnikh skotovodov i metallurgov Sary-Arki* [The culture of the old stockbreeders and metallurgists in Sary-Arka]. Alma-Ata: Gylym.

Kaniuth, K. 2006. *Metallobjekte der Bronzezeit aus Nordbaktrien*. Archäologie in Iran und Turan 6. Mainz: Philipp von Zabern.

Kochhar, N., R. Kochhar, and D.K. Chakrabarti 1999. "A New Source of Primary Tin Ore in the Indus Civilization." *South Asian Studies* 15: 115–118.

Koch Waldner, T. 2013. "Prehistoric Copper Production in the Region of Kitzbühel, North Tyrol: Mining and Smelting." In *Mining in European History and Its Impact on Environment and Human Societies*, ed. P. Anreiter, G. Goldenberg, K. Hanke, R. Krause, W. Leitner, F. Mathis, K. Nicolussi *et al.* Innsbruck: Innsbruck University Press, pp. 73–78.

Koch Waldner, T. and M. Klaunzer 2015. "Das prähistorische Bergbaugebiet in der Region Kitzbühel." In *Bergauf – Bergab. 10000 Jahre Bergbau in den Ostalpen*, ed. T. Stöllner and K. Oeggl. Bochum: Marie Leidorf, pp. 165–173.

Kohl, P.L. 2005. "Afghanistan and the Bronze Age Limning an Image Dimly Perceived." In *Afghanistan ancien carrefour entre l'est et l'ouest.* Indicopleustoi Archaeologies of the Indian Ocean 3, ed. O. Bopearachchi and M.-F. Boussac. Turnhout: Brepols, pp. 65–79.

Kohl, P.L. 2007. *The Making of Bronze Age Eurasia*, Cambridge: Cambridge University Press.

Kuz'mina, E.E. 1966. "Metallicheskie izdelija eneolita i bronzovogo veka v Srednej Azii" [Metallurgical artifacts of the Chalcolithic and Bronze Age in Central Asia]. In *Arkheologija SSSR svod arkheologicheskikh istochnikov*, issue B4-9, ed. V.I. Moshinskaja. Moscow: Nauka.

Kuz'mina, E.E. 1991. "Die urgeschichtliche Metallurgie der Andronovo-Kultur. Bergbau, Metallurgie und Metallbearbeitung." *Zeitschrift für Archäologie* 25: 29–48.

Kuz'mina, E.E. 1994. *Otkuda prishli Indoarii? Material'naja kul'tura plemen andronovskoj obshchnosti i proiskhozhdenie indoirantsev* [Where from came the Indo-Ariens? Material culture of the Andronovians and origin of the Indo-Aryens]. Moscow: Kalina.

Kuz'mina, E.E. 2004. "Historical Perspectives on the Andronovo and Metal Use in Eastern Asia." In *Metallurgy in Ancient Eastern Eurasia from the Urals to the Yellow River*, ed. K.M. Linduff. Lewiston, NY: Edwinn Mellen Press, pp. 37–84.

Kuz'mina, E.E. 2007. *The Origin of the Indo-Iranians.* Leiden Indo-European Etymological Dictionary Series 3. Leiden: Brill.

Kuz'mina, E.E. 2008. *The Prehistory of the Silk Road.* Philadelphia: University of Pennsylvania Press.

Kuznetsova, E.F. and T.M. Teplovodskaja 1994. *Drevnjaja metallurgija i goncharstvo Tsentral'nogo Kazakhstana* [Ancient metallurgy and pottery in Central Kazakhstan]. Almaty: Gylym.

Laing, L.R. 1968. "A Greek Tin Trade with Cornwall?" *Cornish Archaeology* 7: 15–23.

Lamberg-Karlovsky, C.C. 1967. "Archeology and Metallurgical Technology in Prehistoric Afghanistan, India, and Pakistan." *American Anthropologist* 69(1): 145–162.

Laumulin, T.M., F.G. Gubaidulin, and A.P. Gulyaev 1996. "Tin Deposits of North Kazakhstan." In *Granite-Related Ore Deposits of Central Kazakhstan and Adjacent Areas*, ed. V. Shatov, R. Seltmann, A. Kremenetsky, V. Popov, and P. Ermolov. St. Petersburg: Glagol Publishing House, pp. 187–196.

Litvinskij, B.A. 1950. "K istorii dobychi olova v Uzbekistane" [On the history of ore exploitation in Uzbekistan]. *Trudy Sredneaziatskogo Gosudarstvennogo Universiteta* XI: 51–68.

Litvinskij, B.A. 1954. "Drevnejshie stranitsy istorii gornogo dela Tadzhikistana i drugikh respublik Srednej Azii" [History of ancient mining in Tadzhikistan and other republics of middle Asia]. In *Nauchno-populjarnaja biblioteka*, vol. 19. Stalinabad: Adademija Nauk Tadzhikskoj SSR, Institut Istorii, Archeologii i Etnografii, pp. 5–46.

Litvinskij, B.A., A.P. Okladnikov, and V.A. Ranov 1962. *Drevnosti Kajrak-Kumov (drevnejshaja istorija Severnogo Tadzhikistana)* [Kajrak-Kum in antiquity (ancient history of north Tadzhikistan)]. Trudy, t. XXXIII. Dushanbe: Adademija nauk Tadzhikskoj SSR, Institut Istorii, Archeologii i Etnografii.

Lyonnet, B. 2005. "Another Possible Interpretation of the Bactro-Margiana Culture (BMAC) of Central Asia: The Tin Trade." In *South Asian Archaeology 2001*. Vol. I, *Prehistory*, ed. C. Jarrige and V. Lefèvre. Paris: ERC, pp. 191–200.

Maddin, R., T. Stech Wheeler, and J.D. Muhly 1977. "Tin in the Ancient Near East: Old Questions and New Finds." *Expedition* 19(2): 35–47.

Margulan, A.Kh. 1972. "Gornoe delo v Tsentral'nom Kazakhstane v drevnie i srednie veka" [Mining in Central Kazakhstan in ancient time and Middle Age]. *Poiski i raskopki v Kazakhstane*. Alma-Ata: Nauka, pp. 3–30.

Margulan, A.Kh. 1973. "Dzhezkazgan – drevnij metallurgicheskij tsentr (Gorodishche Milykuduk)" [Dzhezkazgan: ancient center of metallurgy (settlement of Milykuduk)]. *Arkheologicheskie Issledovanija v Kazakhstane*. Alma-Ata: Nauka Kazakhskoj SSR, pp. 3–42.

Margulan, A.Kh. 1979. *Begazy-Dandybaevskaja kul'tura Tsentral'nogo Kazakhstana* [The Begazy-Dandybaj culture in Central Kazakhstan]. Alma-Ata: Nauka.

Margulan, A.Kh. 2001. *Sarjarka. Gornoe delo i metallurgija v epokhy bronzy. Dzhezkazgan – drevnij i srednevekovij metallurgicheskij tsentr (Gorodishche Milyduduk)* [Saryarka: mining and metallurgy during the Bronze Age. Dzhezkazgan – ancient and Middle Age center of metallurgy (settlement of Milykuduk)]. Sochinenija 2. Almaty: Dajk-Press.

Margulan, A.Kh., M.K. Kadyrbaev, and A.M. Orazbaev 1966. *Drevnjaja kul'tura Tsentral'nogo Kazakhstana* [The ancient culture of Central Kazakhstan]. Alma-Ata: Nauka.

Masson, M.E. 1930. "Arkheologicheskie materialy k istorii gornogo dela v Srednej Azii" [Archaeological materials for the history of mining in Middle Asia]. *Bjulleten'* 2: 38–49.

Masson, M.E. 1935. "K istorii gornoj promyshlennosti Karamazara" [History of the mining industry of Karamazar]. *Trudy Tadzhikskoj Bazy Akademii Nauk SSSR* 4: 215–232.

Masson, M.E. 1953. *K istorii gornogo dela na territorii Uzbekistana* [History of mining in Uzbekistan]. Tashkent: Akademija Nauk UZSSR.

Maxwell-Hyslop, K.R. 1982. "The Khosh Tapa-Fullol Hoard." *Afghan Studies* 3-4: 25–37.

Moorey, P.R.S. 1994. *Ancient Mesopotamian Materials and Industries: The Archaeological Evidence*. Oxford: Oxford University Press.

Muhly, J.D. 1973. "Tin Trade Routes of the Bronze Age." *American Scientist* 61: 404–413.

Muhly, J.D. 1985. "Sources of Tin and the Beginnings of Bronze Metallurgy." *American Journal of Archaeology* 89: 275–291.

Nezafati, N., M. Momenzadeh, and E. Pernicka 2008. "New Insights into the Ancient Mining and Metallurgical Researches in Iran." In *Ancient Mining in Turkey and the Eastern Mediterranean*, ed. Ü. Yalçin, H. Özbal, and A.G. Paşamehmetoğlu. Ankara: Atilim University, pp. 307–328.

Nezafati, N., E. Pernicka, and M. Momenzadeh 2009. "Introduction of the Den Hosein Ancient Tin-Copper Mine, Western Iran: Evidence from Geology, Archaeology, Geochemistry and Lead Isotope Data." *TÜBA-AR* 12: 223–236.

Novikov, V.P. and V.V. Radililovskij 1990. "K istorii gornorudniykh promyslov i metallurgii drevnego Gissara" [On the history of mining industry and metallurgy in ancient Hissar]. *Ocherki po istorii geologii Tadzhikistana*, vol. I. Dushanbe: Donish, pp. 54–103.

Novikov, V.P., V.V. Radililovskj, and V.V. Suprychev 1981. "Geologo-arkheologicheskie issledovanija v verkhov'jakh r. Shirkent (1981 g.)" [Geological and archaeological investigations at the upper course of the River Shirkent]. *Arkheologicheskie Raboty v Tadzhikistane* XXI, pp. 337–347.

Overlaet, B. 2016. "Luristan Bronzes I: The Field Research." Encyclopædia Iranica, online edition. www.iranicaonline.org/articles/luristan-bronzes-i-the-field-research.

Parzinger, H. 1997. "Sejma-Turbino und die Anfänge des sibirischen Tierstils." *Eurasia Antiqua* 3: 223–247.

Parzinger, H. 2006. *Die frühen Völker Eurasiens – Vom Neolithikum bis zum Mittelalter.* München: C.H. Beck.

Parzinger, H. and N. Boroffka 2003. *Das Zinn der Bronzezeit in Mittelasien*. Vol. I, Die siedlungsarchäologischen Forschungen im Umfeld der Zinnlagerstätten. Archäologie in Iran und Turan 5. Mainz: von Zabern.

Penhallurick, R.D. 1986. *Tin in Antiquity: Its Mining and Trade Throughout the Ancient World with Particular Reference to Cornwall.* London: Institute of Metals.

Pernicka, E. and J. Lutz 2015. "Fahlerz- und Kupferkiesnutzung in der Bronze- und Eisenzeit." In *Bergauf – Bergab. 10000 Jahre Bergbau in den Ostalpen*, ed. T. Stöllner and K. Oeggl. Bochum: Marie Leidorf, pp. 107–111.

Piggott, S. 1977. "A Glance at Cornish Tin." In *Ancient Europe and the Mediterranean*, ed. V. Markotic. Warminster: Aris & Phillips, pp. 141–145.

Pigott, V.C. 1999. "A Heartland of Metallurgy: Neolithic/Chalcolithic Metallurgical Origins on the Iranian Plateau." In *The Beginning of Metallurgy*, ed. A. Hauptmann, E. Pernicka, T. Rehren, and Ü. Yalçin. Der Anschnitt, Beiheft 9. Bochum: Deutsches Bergbau-Museum Bochum, pp. 107–120.

Pigott, V.C. 2018. "The Bactria-Margiana Archaeological Complex (BMAC), the Seima-Turbino Horizon and a Possible Eastward Transmission of Tin-Bronze Technology in Later Third and Early Second Millennium BCE Inner Asia." In *How Objects Tell Stories: Essays in Honor of E.C. Bunker*, ed. K.M. Linduff and K.S. Rubinson, Inner and Central Asian Art and Archaeology I. Turnhout: Brepols, pp. 191–221.

P'jankova, L.T. 1999. "Keramik aus der Siedlung Tašguzor. Zur Frage der Enddatierung der Vachš-Kultur." *Archäologische Mitteilungen aus Iran und Turan* 31: 49–67.

Roden, C. 1985. "Montanarchäologische Quellen des ur- und frühgeschichtlichen Zinnerzbergbaus in Europa." *Der Anschnitt* 37(2–3): 50–80.

Ruzanov, V. 1979. "O nekotorykh drevnikh olovorudnykh istochnikakh na territorii Uzbekistana" [About some ancient tin mines as tin-source in Uzbekistan]. *Istorija Material'noj Kul'ury Uzbekistana* 15: 98–104.

Ruzanov, V. 1980. "K voprosu o metalloobrabotke u plemen chustskoj kul'ury" [On the question of the Chust culture metalworking]. *Sovetskaja Arkheologija* 4: 55–64.

Ruzanov, V. 1982. "Istorija drevnej metallurgii i gornogo dela Uzbekistana v epokhu bronzy i rannego zheleza" [The history of ancient metallurgy and mining in Uzbekistan during the Bronze Age and Early Iron Age]. Dissertation abstract, Istoricheskie Nauki, Moscow.

Ruzanov, V. 1987. "Drevnie splavy Srednej Azii" [Ancient alloys in Middle Asia]. *Istorija Material'noj Kul'ury Uzbekistana* 21: 192–203.

Ruzanov, V. 1999a. "K voprosu o istokakh olovjanistykh bronz v metalloobrabatyvajushchikh proizvodstakh plemen juga Srednej Azii v epokhi bronzy" [On the question of the tin bronzes sources for the metal production of southern Middle Asia populations during the Bronze Age]. *Istorija Material'noj Kul'ury Uzbekistana* 30: 37–40.

Ruzanov, V. 1999b. "Zum frühen Auftreten der Zinnbronze in Mittelasien." In *The Beginnings of Metallurgy*, ed. A. Hauptmann, E. Pernicka, T. Rehren, and Ü. Yalçin. Der Anschnitt, Beiheft 9. Bochum: Deutsches Bergbau-Museum Bochum, pp. 103–105.

Ruzanov, V.D. and Ju.F. Burjakov 1997. "Drevnie rudniki i pamjatniki metallurgii v gorakh Kutitangtau" [Ancient mines and metallurgical sites in the mountains of Kutitangtau]. *Uzbekiston Moddij Madanijati Tarikhi* 28: 175–178.

Ruzanov, V.D. and O.H. Lushpenko 2000. "Nekotorye rezul'taty izuchenija khimicheskogo sostava metalla burguljukskoj kul'tury" [Results from studies on the chemical composition of the Burguljuk culture metals]. *Istorija Material'noj Kul'tury Uzbekistana* 31: 77–85.

Sarianidi, V.I. 1971. "The Lapis Lazuli Route in the Ancient East." *Archaeology* 24(1): 12–15.

Stech, T. and V.C. Pigott 1986. "The Metals Trade in Southwest Asia in the Third Millennium B.C." *Iraq* 48: 39–64.

Stöllner, T. 2015a. "Der Mitterberg als Großproduzent für Kupfer in der Bronzeezeit." In *Bergauf – Bergab. 10000 Jahre Bergbau in den Ostalpen*, ed. T. Stöllner and K. Oeggl. Bochum: Marie Leidorf, pp. 175–185.

Stöllner, T. 2015b. "Die alpinen Kupfererzreviere: Aspekte ihrer zeitlichen, technologischen und wirtschaftlichen Entwicklung im zweiten Jahrtausen vor Christuns." In *Bergauf – Bergab.*

10000 Jahre Bergbau in den Ostalpen, ed. T. Stöllner and K. Oeggl. Bochum: Marie Leidorf, pp. 99–105.

Stöllner, T., Z. Samaschev, S. Berdenov, J. Cierny, M. Doll, J. Garner, A. Gontscharov *et al.* 2011. "Tin from Kazakhstan: Steppe Tin for the West." In *Anatolian Metal*, ed. Ü. Yalçın. Der Anschnitt, Beiheft 24. Bochum: Deutsches Bergbau-Museum Bochum, pp. 231–251.

Stöllner, T., Z. Samaschev, S. Berdenov, J. Cierny, M. Doll, J. Garner, A. Gontscharov *et al.* 2013. "Zinn und Kupfer aus dem Osten Kasachstans. Ergebnisse eines deutsch-kasachischen Projektes 2003–2008." In *Unbekanntes Kasachstan – Archäologie im Herzen Asiens.* Band I, ed. T. Stöllner and Z. Samašev. Bochum: Katalog der Ausstellung des Deutschen Bergbau-Museums Bochum, pp. 357–382.

Stöllner, T., Z. Samaschev, S. Berdenov, J. Cierny, J. Garner, A. Gorelik, and G.A. Kusch 2010. "Bergmannsgräber im bronzezeitlichen Zinnrevier von Askaraly, Ostkasachstan?" *Der Anschnitt* 62(3): 86–98.

Surgaj, V.T. 1951. *K istorii gornogo promysla v Kirgizii* [On the history of mining industry in Kyrgyztan]. Frunze: Kirgizkogo filizla Akademii Nauk SSSR, Geologicheskij Institut.

Sverchkov, L.M. 2009. "A History of Research on Ancient Mining in Uzbekistan." *Archäologische Mitteilungen aus Iran und Turan* 41: 141–164.

Teufer, M. 2003. "Kulturkontakte zwischen Sogdien und Baktrien am Beginn der Spätbronzezeit. Dargestellt am Grab von Zardča Chalifa." *Mitteilungen der Berliner Gesellschaft für Anthropologie, Ethnologie und Urgeschichte* 24: 121–140.

Thomalsky, J. 2014. "Afghanistan: Ancient Mining and Metallurgy: Initial Project Stage." In *Proceedings of the 9th ICAANE*, vol. 3, ed. R.A. Stucky, O. Kaelin, and H.P. Mathys. Basel: Harrassowitz, pp. 647–661.

Thomalsky, J., B. Bräutigam, M. Karaucak, and S. Kraus 2013. "Early Mining and Metal Production in Afghanistan: The First Year of Investigations." *Archäologische Mitteilungen aus Iran und Turan* 45: 199–213.

Tosi, M. and M. Vidale 1990. "4th Millennium BC Lapis Lazuli Working at Mehrgarh, Pakistan." *Paléorient* 16(2): 89–99.

Turlychkin, V.M. 1961. "Geologo-arkheologicheskoe obsledovanie drevnikh vyrabotok Kandzhola" [Geological and archaeological investigations about ancient mining in Kandzhol]. *Arkheologicheskie Raboty v Tadzhikistane* 7: 167–171.

Vinogradov, A.V. 1972. "Birjuza, pervobytnaja moda, etnogenez" [Turquoise, a primitive fashion, ethnogenesis]. *Sovetskaja Etnografija* 5: 120–130.

Vinogradova, N. 1993. "Interrelation Between Farming and 'Steppe' Tribes in the Bronze Age South Tajikistan." In *South Asian Archaeology 1991*, ed. A.J. Gail and G.J.R. Mevissen. Stuttgart: Franz Steiner Verlag, pp. 289–301.

Vinogradova, N. 1997. "Spätbronzezeitliche Kulturkontakte in der baktrischen Flußoase nach den Befunden der Nekropole Bustān 6." *Archäologischen Mitteilungen aus Iran und Turan* 29: 145–178.

Vinogradova, N.M. 1999. "Die spätbronzezeitliche Siedlung Tašguzor in Südtadžikistan." *Archäologische Mitteilungen aus Iran und Turan* 31: 33–48.

Vinogradova, N. 2001. "Südtadžikistan, in der Spätbronze- und Früheisenzeit." In *Migration und Kulturtransfer. Der Wandel vorder- und zentralasiatischer Kulturen im Umbruch vom 2. zum 1. vorchristlichen Jahrtausend*, ed. R. Eichmann and H. Parzinger. Bonn: Habelt, pp. 199–219.

Vinogradova, N.M. 2004. *Jugo-Zapadnij Tadzhikistan v epokhu pozdnej Bronzy* [Southwestern Tajikistan during the Late Bronze Age]. Moscow: Rossijskaja Akademija Nauk Institut Vostokovedenija.

Vinogradova, N.M. 2008. "Zemledel'cheskoe poselenie i mogil'nik Kangurtut – pamjatnik epokhi pozdenej Bronzy" [The settlement and cemetery of Kangurtut, sites of farmers of the Late Bronze Age]. In *Pamjatniki Kangurtuta v jugo-zapadnom Tadzhikistane (epokha neolita i bronzovyj vek)* [Neolithic and Bronze Age sites at Kangurtut in southwest Tajikistan], ed. N.M. Vinogradova, V.A. Ranov, and T.G. Filimonova. Moscow: Rossijskaja Akademija Nauk Institut Vostokovedenija, pp. 161–454.

Weisgerber, G. 2004a. "Prähistorischer und historischer Bergbau in Afghanistan: Part 1." *Der Anschnitt* 56(4): 126–151.

Weisgerber, G. 2004b. "Prähistorischer und historischer Bergbau in Afghanistan: Part 2." *Der Anschnitt* 56(5–6): 190–211.

CHAPTER TWENTY-NINE

THE ACQUISITION OF TIN IN BRONZE AGE SOUTHWEST ASIA

*Vincent C. Pigott**

Dedicated to the memory of my good friend and colleague, Tamara Stech.

INTRODUCTION

As of late, the much discussed issue of the transmission of tin bronze technology in Southwest and Central Asia during the Bronze Age (ca. 3rd and 2nd millennia BC) has been revisited by a number of scholars who have adopted broad regional (e.g., Rahmstorf 2011, 2017; Wilkinson 2014), if not a trans-Eurasian perspective (e.g., Sherratt 2006; Beaujard 2011; Pigott 2012 a, b, 2018; Potts 2012; Nessel *et al.* 2018) (Figure 29.1).[1] Emerging evidence in support of this technology's transmission, acceptance, and impact across Eurasia currently suggests a west to east chronological sloping horizon (e.g., Roberts *et al.* 2009: 1014, fig. 1a, b; Roberts 2011). This transmission is thought to have proceeded in a nonuniform, nonlinear fashion by the early 3rd millennium in Southwest Asia eastward to China's central plain by the early 2nd millennium BC. This ensued within a relatively short span of human cultural interaction, approximately the duration of a single millennium. When, after spreading across Eurasia to the east, this technology ultimately arrived in its continental extreme (i.e., Southeast Asia), remains under debate with dates ranging from the early to mid-2nd millennium BC (White and Hamilton 2009) to the late 2nd (e.g., Ciarla 2007, 2013; Pigott and Ciarla 2007; Higham *et al.* 2011).

In the Old World, an earlier (5th millennium) production of tin bronze has been documented in the Balkans and it is linked to the innovation of smelting in that region, the earliest evidence for which is currently found in eastern Serbia (Radivojević *et al.* 2010, 2013, 2018; Radivojević 2015; Radivojević and Rehren 2016). However, this Balkan tin bronze production, based on the smelting of mixed copper-tin ores that yielded "natural bronzes," did not result in the spread of tin bronze technology beyond the immediate region. Just how intentional these tin bronzes were is subject to some debate (cf. Radivojević *et al.* 2018).

In the continuing effort to understand the early Southwest Asian development of tin bronze technology, emphasis will be placed herein on current evidence for exploitation of regional tin sources, trade in tin, and transmission of the technology. It is important to note that one of the best means to an enhanced understanding of such

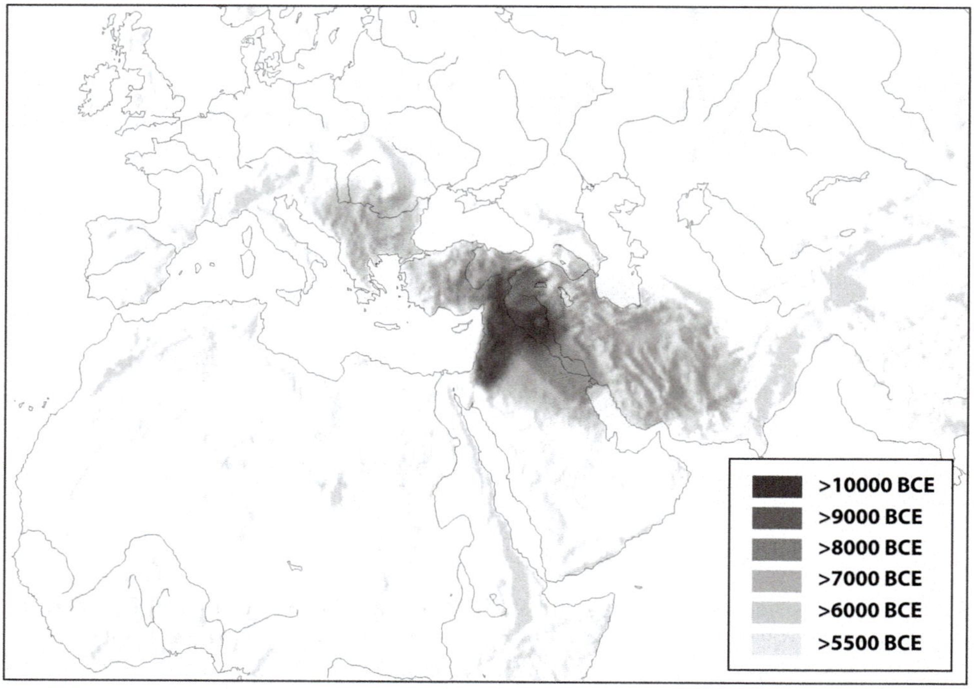

Figure 29.1 Map showing the "southwest Eurasian metallurgical heartland" wherein the earliest exploitation of copper ores and native copper began. (Map provided courtesy of Dr. Ben Roberts.)

issues is to scrutinize the substantive evidence from the 2nd millennium BC with the intent to ascertain how it may relate to sparser evidence from the 3rd millennium.

THE SOUTHWEST EURASIAN METALLURGICAL HEARTLAND

Smelting was a cross-craft innovation, built on millennia of prior, often aesthetically driven, pyrotechnological experimentation with materials (e.g., clay, plaster, colorful ores and related minerals, flint, obsidian). It first occurred, based on current evidence, during the 5th millennium and, over the course of a little more than 1,000 years, often among tell-centered societies that were spread across southwest Eurasia, from the Balkans to the Indo-Iranian borderlands (Figure 29.2). Once innovated, smelting and the casting of more complex shapes in copper-base metals, spread across a similar geographical expanse, including the Balkans, Greece, the Aegean, the Levant, Anatolia, northern Mesopotamia, the Caucasus, the Iranian Plateau, Turkmenistan, and Pakistan (Thornton 2002: 22–26, tabs. 1, 2; also Roberts *et al.* 2009). Broadly speaking, this region can be seen as the "southwest Eurasian metallurgical heartland." Across this region "the application of heat in a controlled reducing atmosphere led to the smelting of metallic ores to produce lead, copper, copper alloys and eventually silver" (Roberts *et al.* 2009: 1013) (see Figure 29.2). We argued that the innovation of metallurgy was transmitted by individual metalworkers or groups of metalworkers

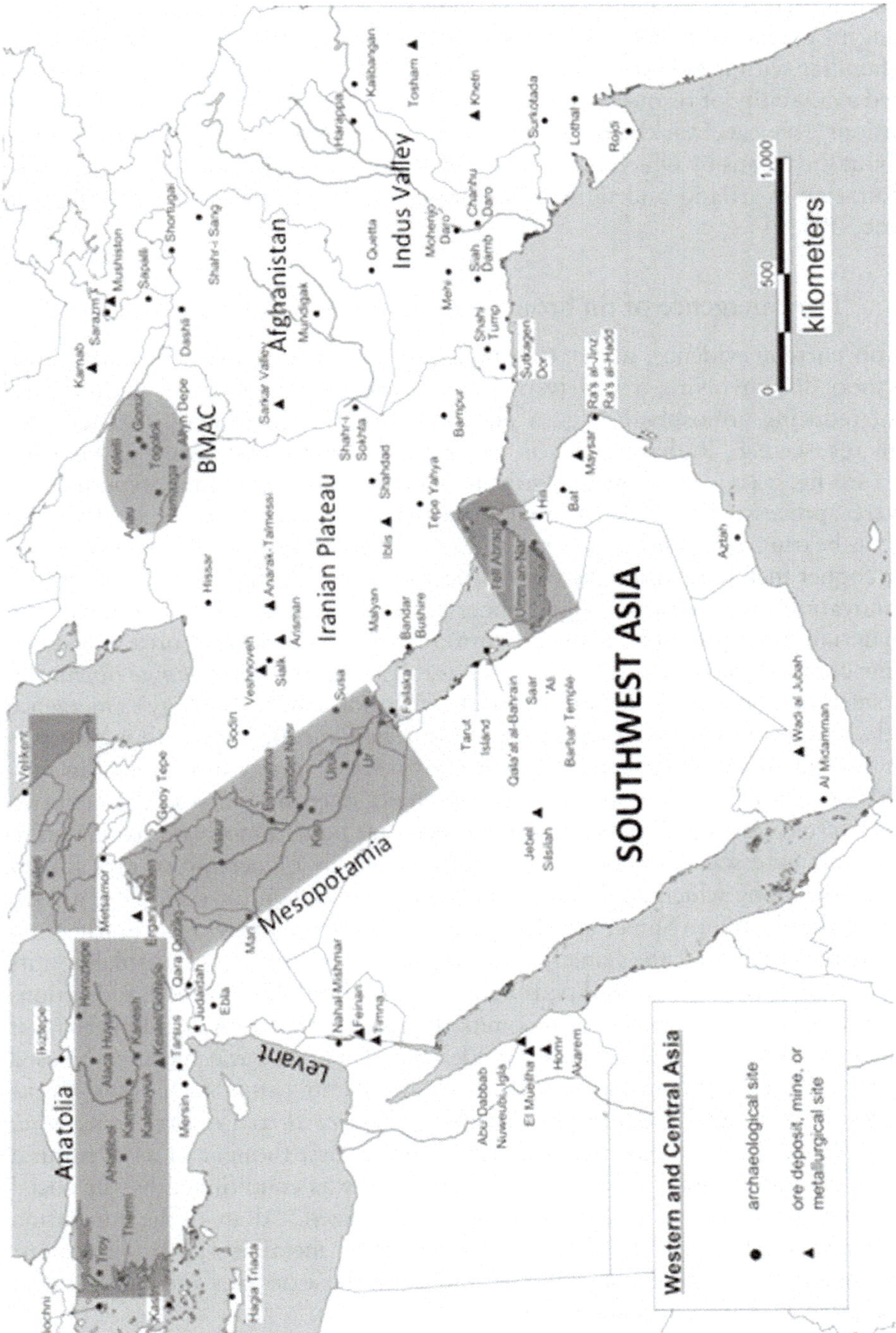

Figure 29.2 Map of Southwest and Central Asian regions where the earliest tin bronze artifacts occur, beginning in the first half of the 3rd millennium BC and continuing later. (Map provided courtesy of Dr. Lloyd Weeks.)

on a face-to-face basis between and among often contiguous and receptive cultures over time (Roberts *et al.* 2009). Technological know-how and expertise would have been transmitted between cultures by the movement of these individuals who were engaging in extended contact. Adoption of metallurgical technology by groups as yet unfamiliar with it would not have been wholesale or inevitable, but would have reflected availability of resources and cultural receptivity (i.e., a willingness to accept innovations) (see e.g., Stockhammer and Maran 2017). Interestingly, the same can be said about the means by which tin bronze metallurgy was to spread across the southwest Eurasian heartland and ultimately Eurasia itself over the next two millennia post-3000 BC.

The emergence of tin bronze metallurgy in Southwest Asia

Based on current evidence, about two millennia of smelting experimentation (ca. 5000–3000 BC) involving a variety of metal-bearing ores contained within some form of reducing atmosphere (e.g., a portable crucible, a stationary furnace) (for the Balkans, see e.g., Radivojević *et al.* 2010; for Southwest Asia, see e.g., Helwing 2017: 166, fig. 14.3). Fundamental to this haphazard but persistent experimentation were "processual precedents" that ultimately led to innovation. At the point where tin-bearing ores and/or tin metal were, in the presence of copper-bearing ores or copper metal, intentionally placed together in a reducing context, tin bronze the "innovation" was born (see e.g., Lehner *et al.* 2015: 204–207). It resulted as a new material, most probably, from the introduction of tin-bearing ores to a smelt involving copper-base ores as part of a well-established production *chaîne opératoire* for the smelting of copper in various forms. Furthermore, the technology's emergence marked a significant technological transition to a new metal, bronze, though it was not a replacement for copper and its other alloys. For a variety of reasons this new material was gradually to become the metal of choice over the course of the next two millennia across Eurasia. Significant among them are human motivations related to the desire for elite social status and prestige brought by ownership of rare luxury commodities among which gold, lapis, and bronze are known for their close associations in 3rd-millennium Sumerian Mesopotamia.[2] There is little evidence, where tin bronze is concerned, that the Sumerians chose to use tin bronze based solely on its mechanical properties of castability, toughness, and durability. Instead, indications are that it was chosen as a high-status commodity, as a simulacrum of gold perhaps if tin contents were sufficient to produce a golden hue. Tin bronze was meant to be used and displayed in elite circumstances (e.g., Stech and Pigott 1986; Stech 1999; Weeks 2003: 188–191; see also Thornton *et al.* 2005; Lehner *et al.* 2015: 195; Yener *et al.* 2015). As Weeks (2003: 200) suggests, "tin-bronze is best thought of as a 'cultural alloy' (see Hamilton 1991) whose adoption and use was conditioned by the social contexts in which it was produced, exchanged and utilized." All in all the innovation of tin bronze resulted in a significant and long-lasting metallurgical tradition, one that was, over time, to exercise a profound impact on those peoples and cultures who were receptive to it and chose to accept it.

Scholars continue to explore just what the particular impetus or impetuses were, not only for the innovation of tin bronze, but also for such a rapid expansion of the technology over such a broad and remarkably complex mosaic of peoples and cultures. By the early to mid-3rd millennium tin bronze artifacts are found at sites across Southwest Asia (e.g., Rahmstorf 2011: 105, 106, fig. 9.1; 2017: 193, figs. 16.5, 16.6; Lehner *et al.* 2015: 205 citing Pernicka 1998) (Figure 29.2).

Two contrasting models have been proposed that, it is argued, can explain the adoption of tin bronze at this time (Lehner *et al.* 2015: 206–207).

The first or "standard model" as espoused by James D. Muhly (e.g., Muhly 1973a, 1973b, 1976, 1985) argues that "the irregular distribution and size of copper and tin resources across Eurasia and the nature of bronze technologies determined the geographic scope of the metals trade" (Lehner *et al.* 2015: 206). The Anatolian tin sources, to be discussed later in the chapter, are discounted as not economically viable or apparent to local mining communities. The result is a model built on a long-distance trade in tin that fed into pre-existing "regional copper exchange systems" ultimately responsible for the innovation and production of tin bronze and overtime its spread across Eurasia (Lehner *et al.* 2015: 206). Evidence in support of the long-distance trade model is discussed at the close of this chapter.

A second model is proposed as an alternative to, and offering a better explanation for, the origins of tin bronze technology than the "standard model." This "highland production model" (e.g., Yener 2000; Yener *et al.* 2015) is built on the hypothesis that it was the utilization of local Near Eastern tin sources (e.g., those now known from central Anatolia; possibly even the recently documented deposits at Deh Hosein in western central Iran?) that resulted in the innovation of the earliest tin bronzes in concert with the development of tin metallurgy. In this model, these innovations are associated with the proximity of excavated early tin bronzes, many (but not all) from greater northern Mesopotamia, and the Anatolian tin-mining and tin-processing sites at Kestel/Göltepe, and those on the Erciyes Dağ and near Kayseri.

These two models do not necessarily need to be seen as mutually exclusive, but further discussion will have to be pursued elsewhere. Lehner *et al.* (2015: 207) suggest that, in all probability, "multiple sources of tin were probably used over time" and Brügmann *et al.* (2017: 112) argue that tin isotopy of tin bronzes from several Mesopotamian sites (e.g., Ur, Tell Asmar, Tell es-Suleimeh, Khafajah) now suggests that tin came from several sources. Given the identification of several tin sources in central Anatolia it is now hard to ignore this region as a core-area producer of tin. Just how far and wide tin bronzes made with this tin actually were distributed in the 3rd millennium (and perhaps later) is a subject for future investigation. By the early 2nd millennium BC textual evidence for the Old Assyrian long-distance trade bringing tin into central Anatolia arguably from southern Mesopotamia and points further to the east can be seen as supporting the validity of the "standard model" proposed above.

THE OCCURRENCE OF EARLY TIN BRONZE ARTIFACTS IN THE ANCIENT NEAR EAST

Compiled lists of the early tin bronze artifacts demonstrate just how widespread they were (see Figure 29.1). Tin bronzes come from the early to around the mid-3rd millennium sites known in Iran (e.g., Susa (IVb), Tepe Giyan, Tepe Sialk, Tepe Yahya);[3] in Mesopotamia (e.g., Tepe Gawra, the "Y" cemetery at Kish, Abu Salabikh, Tell Razuk, Tell Agrab, Tell Qara Quzaq, Tell Kharab Sayyar, Tell Bderi, Tell Judaidah,

Tell Sweyhat); in the Caucasus (e.g., Talin in Armenia); in Anatolia (e.g., Tarsus, Ahlatlibel, Mahmatlar, Alişar Höyük, Çadir Höyük, Gedikli, Horoztepe, Kalinkaya, Resuloğlu, Ikiztepe, Troy, Beşik-Yassiktepe, Norşuntepe); in the Aegean (e.g., Poliochni on Lemnos, Thermi on Lesbos, Kastri on Syros); and in the southern Gulf (e.g., Al Sufouh, Unar 1, Unar 2) (see, e.g., Stech and Pigott 1986: 42–48; Pernicka 1998; Weeks 2003: 173–178; Thornton 2007: 129–130; Helwing 2009: 212, 215 n. 9, 10, 12, 13; Nezafati *et al.* 2011: 212–213; Oudbashi *et al.* 2012; Courcier 2014; Cuénod *et al.* 2015; Lehner *et al.* 2015: 207–209).[4] Of course artifacts will continue to be added to the burgeoning list of the earliest Southwest Asian tin bronzes (e.g., see Rahmstorf 2017; Franke 2018 for updates on the earliest tin bronze occurrences).

Finally, in the southwest Central Asian region of Turkmenistan, though sparsely distributed, tin bronze artifacts do occur in its Namazga sequence and date to the early 3rd millennium (Kohl 1984: 113; Ruzanov 1999; Thornton *et al.* 2005; Pigott 2018; see also Chapter 27). For 3rd-millennium tin bronze at sites east of Afghanistan reaching to the Indus valley, see, for example, Kenoyer and Miller (1999), Weeks (2003: 176–177), Hoffman and Miller (2014), and Hoffman 2018.

On a broader scale and later, from the mid- to later 3rd millennium, Lorenz Rahmstorf (2011: 105, 106, fig. 9.1; 2017: 193, figs. 16.5, 16.6) maintains that there are now about 100 sites that can be identified across the region ranging from the Aegean to Bulgaria and to Pakistan/northwest India that have yielded tin bronze artifacts. He also itemizes only a few tin bronze-yielding sites from central and western Europe dating to prior to ca. 2200 BC (Rahmstorf 2011: 105), suggesting thus that European deposits could not have served as sources of tin for the Early Bronze Age of the Aegean or Southwest Asia (Rahmstorf 2011: 105).

Rahmstorf (2011: 105) also states that some 30 Mesopotamian sites dating to the Early Dynastic III and Akkadian periods of the mid- to later 3rd millennium have yielded tin bronze artifacts. This suggests a high concentration of this metal in this region and reflects their proximity to the early core area of production. He summarizes the distribution of early tin bronzes by saying that beginning about 2600 BC there was a "sudden increase in the number of sites with tin bronzes" (Rahmstorf 2011: 105; see also Hauptmann and Pernicka 2004). This date may well mark the "tipping point" of the florescence of tin bronze technology across greater Southwest Asia, but it is those tin bronzes that can be dated prior to this date that help to focus on what Rahmstorf sees as the heartland of tin bronze production; namely, the region spanning the Aegean, Anatolia, and Mesopotamia (Rahmstorf 2011: 105–106, fig. 9.1). Thus, the paucity of early 3rd-millennium tin bronze in Europe in concert with this millennium's concentration of early tin bronze-bearing sites, from the Aegean across Anatolia to the Indus valley and spread south along the length of Mesopotamia to the Gulf, lends credence to a consensus that sees this region of southwest Eurasia as the heartland in which tin bronze technology was innovated and from whence it spread.[5]

The quantities of tin bronze artifacts that have been excavated across Eurasia suggest that tin ore, though sporadically distributed, was at the same time relatively accessible (Rahmstorf 2011). Considerable recent research has given us information on sources of tin ore exploited in Bronze Age Southwest Asia (e.g., Kestel-Göltepe and the recently documented mined sources on the volcanic slopes of Erciyes Dağ in central Anatolia); Deh Hosein in western Iran; innumerable deposits across Afghanistan; and in Central Asia (e.g., Karnab in Uzbekistan, Mushiston in Tajikistan, and at a multitude of deposits with ancient workings in East Kazakhstan) (see Chapter 28) (Figure 29.3). In recent decades a dramatic uptick in excavations in concert with related analytical research programs including lead isotope analysis (LIA), which is helping to demonstrate isotopic signature compatibility (or lack thereof) between ore fields, deposits, and final products, are providing the basis for hypotheses coupling technological and cultural interactions across Bronze Age Eurasia.

Figure 29.3 Map showing locations (1–7) of documented prehistoric tin mines or mining regions mentioned in the chapter. Map also shows the location of the Uluburun shipwreck, as well as the layered ecological zonation of northern Eurasia. (Map provided courtesy of Dr. Thomas Stöllner.)

SOUTHWEST ASIAN TIN SOURCES MINED DURING THE BRONZE AGE

Anatolia

Kestel-Göltepe (3rd millennium, south-central region)

It has been argued that the 3rd-millennium mine at Kestel and its associated ore-processing settlement of Göltepe in the Bolkardağ mining district of south-central Anatolia offer the earliest evidence for large-scale tin (cassiterite – SnO_2) mining and processing in Eurasia (e.g., Yener 2000, 2008). Evidence suggests that gold was not being mined here (Hanilçi *et al.* 2010). These two associated sites were investigated during the late 1980s and the 1990s by K. A. Yener and colleagues. Kestel and Göltepe form an industrial complex that spreads over some 60 ha. The mine itself has at least 1.5 miles of underground workings that would have yielded some 4,600 cubic meters of ore and hostrock (Figure 29.4).

It has been estimated that, based on a low-grade ore (1% tin), this mine could have produced some 115 tons of tin metal. The tin ore cassiterite was bound up in a hematite-rich hostrock and was beneficiated at Göltepe by being crushed to a fine powder with hand-held hammer stones. A total of some 55,000 ore crushing tools were collected from Göltepe's surface and 5,000 such tools from its workshops (Figure 29.5). This tool count and the ton of tin-rich vitrified material excavated from the workshops attest to the magnitude of the production being undertaken.

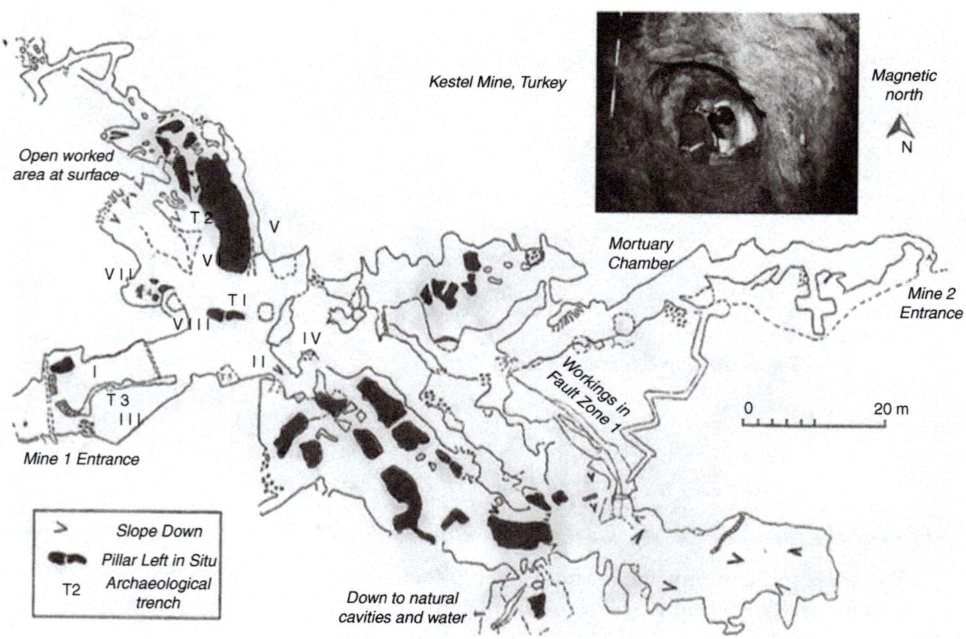

Figure 29.4 The Kestel tin mine, Bolkardağ region, south-central Anatolia, 3rd millennium BC. 1.5 miles of tunnels yielding ca. 4,600 m³ of tin ore that could be smelted to 115 tons of tin metal. (Plan provided courtesy of Dr. K.A. Yener.)

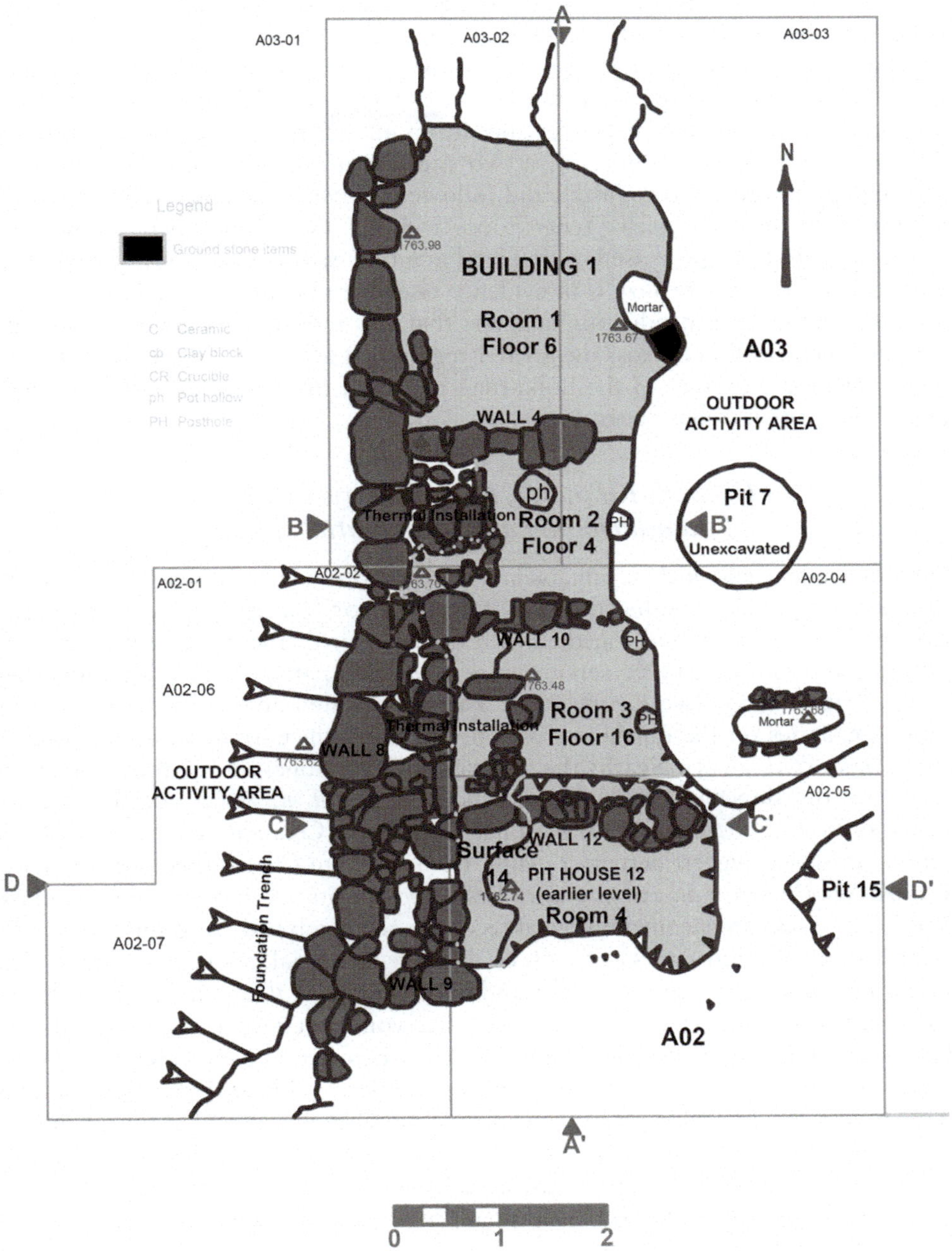

Figure 29.5 Plan for Göltepe building 1 (Area A) depicting ore processing, workshop stalls, thermal installations, ore grinders, and crucibles. This building dates to EBIII, Period 2; but note Pit House 12 underneath from Period 3 EBII at the lower end of the plan. Some 5,000 ore crushing tools were excavated from the workshops along with crucibles, powdered tin ore and ca. 1 ton of vitrified tin-rich material. (Plan provided courtesy of Dr. K.A. Yener. Yener in press.)

The tin-rich powders crushed from the mined ore, estimated as high as 2% tin, were smelted to tin metal in ceramic crucibles (Hanilçi *et al.* 2010).

Thus, indications are that at Kestel-Göltepe the earliest evidence for mining and smelting tin exist on a substantial scale. This industrial complex is certainly among the most significant yet revealed from anywhere in prehistoric Southwest Asia and, on this basis alone, these two sites merit our attention. The miners at 3rd-millennium Kestel recognized and followed ore veins, as suggested by the configuration of the mine itself (Yener 2000: 198, fig. 15). Mining here would have yielded tin ore to be processed to tin metal at a time when conscious alloying of tin with copper to make bronze is in evidence elsewhere in Anatolia and neighboring regions. The scale of production suggests that this industrial complex could have supplied metalworkers across the greater region and LIA links sites in Anatolia and Syria with the signature of the Bolkardağ region of the Taurus Mountains (Sayre *et al.* 1992, 2001; Yener 2000).[6]

Hisarcık-Kıranardı, Kayseri Erciyes Dağ area (3rd millennium and later, central region)

Subsequent to geological reconnaissance and excavations at Kestel-Göltepe in the Bolkardağ region of south-central Anatolia, in 2005 Turkish geologists identified cassiterite deposits (bound together with oxides of iron and arsenic) near modern Hisarcik and Zincidere in Kayseri province in central Anatolia (e.g., Yalçin and Özbal 2009; Yazgan 2015) (see Figure 29.3). They are located among numerous ancient mining galleries on the northeastern slopes of the extinct stratovolcano of Erciyes Dağ (Yener *et al.* 2015). Nearby the galleries grinding stones for ore processing were found on the slopes at the Senir Sırtu locus (Yener *et al.* 2015: 600, fig. 1), evidence suggesting the presence of seasonal workshops. Ceramics associated with the mining evidence suggest activity ranging from the Late Chalcolithic into the Early Bronze Age (Yener *et al.* 2015: 604). These ore deposits are currently unique in the Southwest Asian archaeometallurgical record as the combined "iron-tin-arsenic ores were deposited in fumaroles as a soft, easily carved material, allowing the mineral to be readily extracted, even with antler and horn" (Yener *et al.* 2015: 599; 601, fig. 2). These are mostly nonvein-type ore deposits that would not have necessitated arduous, hard rock, shaft, and gallery mining known elsewhere across the greater region (e.g., Kestel, Veshnoveh). Moreover, these deposits were accessible at the surface and may have been easily locatable given the probable presence of colorful arsenical sulfides including realgar (α-As_4S_4) and orpiment (As_2S_3). It is also the case that elemental analyses from innumerable tin bronze artifacts from contexts across the greater ancient Near East have yielded the combined presence of tin and arsenic, two elements that occur only rarely with one another geologically.

The reducibility of the Erciyes Dağ complex ores is problematic, as discussed by Yener *et al.* (2015: 603–604), but the authors do suggest that the tin-rich ores could have been co-smelted with a copper ore to produce a tin bronze of uncontrolled tin content (Johnson *et al.* 2017). Such a possibility is also suggested by Rovira and colleagues (Rovira and Montero 2003; Rovira *et al.* 2009; see also Smith 1996; Chirikure *et al.* 2010; Berger *et al.* 2016; Figueiredo *et al.* 2016). Thus, in antiquity, if these deposits were viable for the production of tin bronze then Bronze Age metalworkers would have had to have perfected the means to deal with such difficult to smelt complex ores.

Some 25 km from the Old Assyrian *kārum* of Kültepe-Kanesh and 2 km distant from the Senir Sirtu mines, a fortified, second-tier site focused on ore processing was located at Teknekayası Höyük. A surface survey yielded ceramics ranging from the Early Bronze Age to the Iron Age along with mortars, hammer stones, multi-hollow anvils, and grinding stones for ore crushing.

The highland production model described above and operating in the Ericyes volcano's hinterland, which includes the not too distant Kültepe-Kanesh habitations, is seen to have had a three-tiered settlement hierarchy linked to a complex settlement pattern (Yener *et al.* 2015: 608). This involves the interaction of mines, ore processing, and urban habitation that indications suggest were established during the 3rd millennium/Early Bronze Age period, and involved the early production of tin and presumably bronze. The analysis of an Early Bronze Age Kültepe crucible suggests that it was used for the melting of copper containing tin and arsenic, consistent with metal possibly produced from local ores (Yener *et al.* 2015). Moreover, this three-tiered system appears to have persisted into the 2nd millennium BC at least, and helped to create "the backdrop of the subsequent Assyrian merchant trade" as typified at *kārum*-Kanesh well known as a destination entrepôt for a long-distance trade in tin (presumably as smelted metal) coming, as has been argued, from points to the south of Anatolia during the Old Assyrian period of the early 2nd millennium BC (Yener *et al.* 2015). Metal workshops documented by excavation existed in Kanesh's lower town, but it is not clear just how early they may have begun (Özguç 1955; Lehner 2014). Thus, while at this juncture, just what the significance may have been for the tin trade with regard to the apparent availability of tin (rich in arsenic and iron) from the Kültepe vicinity remains a matter for further discussion. Tin bronze manufacture based on these proximate ores appears to have been practiced in the region prior to the establishment of renowned tin trade. Moreover, strong indications point to Early Bronze Age Kültepe as the market hub for regional ores and metal products made from these ores by "first-tier processing centers in the hinterland of the Erciyes volcano" (Yener *et al.* 2015: 609).

At the same time, as it applies to 3rd millennium tin bronze in Anatolia, in the view of Yener and colleagues (2015: 606–607, emphasis mine):

> The highland production model is one that involves metals, precious not only in terms of economic value but also critical beyond calculation for their role in the negotiation of status and power, and their embodiment of a value-added commodity for trade. These new ways of making things and organising production reveal much about the nature of society ... Once metal became locked into a cultural system as an indicator of wealth, disparities in access to labour and resources influenced how individuals operated within and between communities of producers and consumers. These relations often linked distant groups together into cooperative agreements, *hence the Assyrian commercial network*. Therefore, metal technologies are strategically placed in complex networks and institutions of production, exchange and consumption that effectively unite disparate highland resource areas and agricultural lowlands (Lehner and Yener 2014).

What we take away from the emerging Anatolian evidence for tin bronze production discussed above is not only an affirmation of the highland production model, but also that these south and central regions had accessible, mined tin ores and were bronze producing from the emergence of the technology over the course of the 3rd millennium in the ancient Near East. It is perhaps worth noting that, as the crow flies, Kültepe-Kanesh is not that far from the Bolkardağ region and the Kestel-Göltepe complex. Those seeking tin and/or tin bronze would have gravitated to these Anatolian production zones and/or would have negotiated trade and exchange in order to facilitate access. Subsequently, during the early 2nd millennium BC, tin apparently was being traded in remarkably large quantities from southern Mesopotamia into central Anatolia (e.g., Lehner 2014; Barjamovic 2018) to supply a burgeoning tin bronze industry satisfying an increasing demand by elites across what must have been a considerable region. We turn to a short discussion of this trade in tin now.

Kültepe and Kārum Kanesh (3rd and 2nd millennium BC, central region)

Kültepe was an important city in the ancient Near East with occupations in the 3rd and 2nd millennium BC and this is reflected in its remarkable size (ca. 150 ha) as an archaeological site to this day. The site itself has been the subject of at least 70 excavation seasons and digging continues. Its lower town ringing the main mound (*tepe*) is some 2.5 km in diameter and was home to a population estimated at ca. 25,000 people.

A massive archive of some 23,500 merchant records written in cuneiform (only partially studied) has given Assyriologists an intimate look into the commercial life of the early 2nd millennium BC city, *kārum* Kanesh and its hinterland, and they have characterized business interactions involving the Old Assyrian period trade in metals including gold, silver, copper, and, most importantly for this discussion, tin, which was arguably coming from the south (e.g., Muhly 1973b; Helwing 2009; Pigott 2011; Atici *et al.* 2014; Lehner 2014; Larsen 2015; Lehner *et al.* 2015; Barjamovic 2018). Steinkeller (2016: 132–133) makes it clear that lapis lazuli was also moving from Assur into Anatolia. This luxury mineral seems to have joined tin among commodities being traded out of Elam and into the southern route headed north to Assur and beyond. At the same time there is no indication that tin and lapis lazuli were traveling together from their Afghan source areas across the Iranian Plateau to Elam, whereas, from Elam north, texts indicate they seem to be associated in trade.

The trade is mainly attested in the period ca. 1895–1865 BC (Yoffee and Barjamovic 2018: 816) (Figures 29.6 and 29.7). Suffice it to say that it has been theorized that "the development of long-distance exchange happened because of the increased interest among competing resident elite social groups to acquire exotic materials, including most importantly metals and textiles" (Bachhuber 2011 in Lehner *et al.* 2015: 148).

The continuing enigma concerning just where the tin entering the long-distance trading system in Mesopotamia may have been coming from is the focus of the following discussion. Not only do the sources of this tin (most probably from the East) remain elusive, but so also does the evidence in the East for the mining, smelting, and transporting of such massive amounts of this metal.[7] According to Gojko Barjamovic (2018: 141): "One may carefully suggest that some 1,500 donkeys made the journey

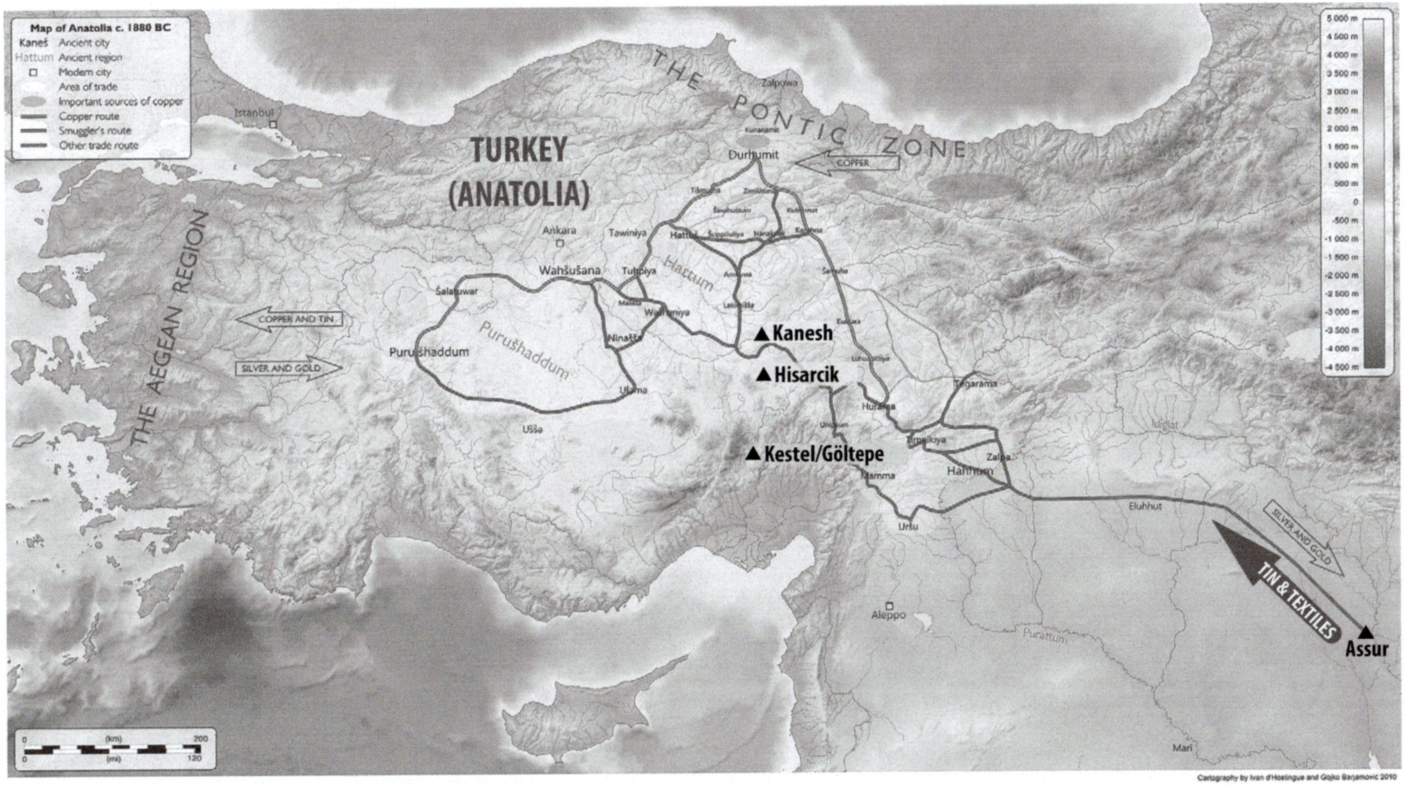

Figure 29.6 The Assyrian network of trade in Anatolia ca. 1880 BC. Note tin coming from southern Mesopotamia. (Cartography by Ivan d'Hostingue and Gojko Barjamovic 2010. After map provided courtesy of Dr. Gojko Barjamovic. Adaptation by Carolin Frank and Ardeth Anderson.)

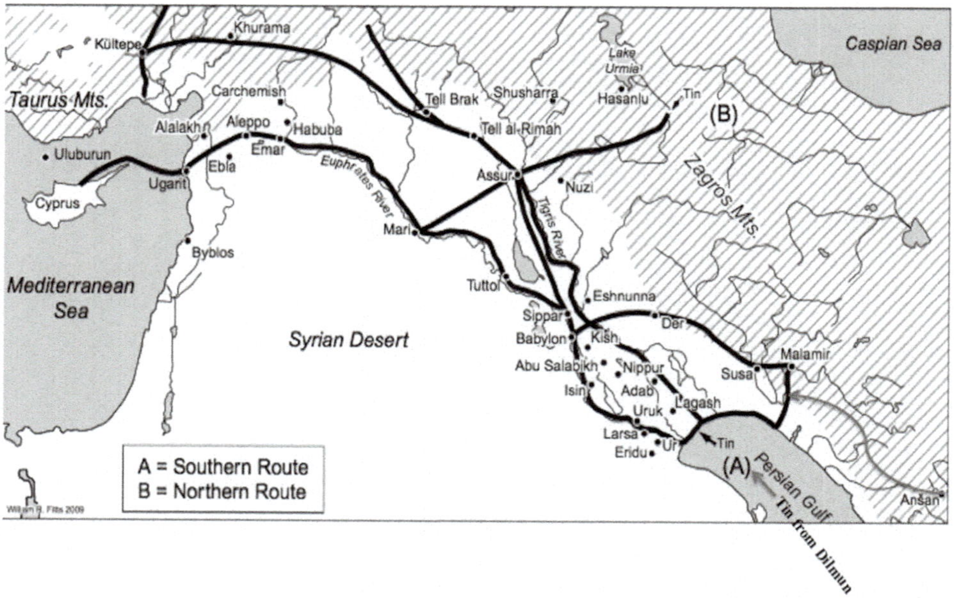

Figure 29.7 Map showing (A) the southern route of the Assyrian trade network and (B) the overland northern potential route used to transport tin to the ancient Near East from sources east of Mesopotamia. (After map in Muhly 1973b.)

to Anatolia each year during the best documented period carrying with them 15 tons of tin and 32,000 textiles" (Figure 29.8). In terms of the routes south of Anatolia by which tin could have been moving north, a ruler of the period, Ilušuma, discussing trade in copper (not tin) describes

> what appears to be three main routes leading to Aššur from the south one from the "edge of the marshes" via Ur and Nippur, one from Awal and Kismar in the Hamrin region, and one from Der on the road to Susa.
>
> (Barjamovic 2018: 143)

Clearly multiple routes to the south were in play at the time and one would assume that multiple areas to the east were annually producing (e.g., mining, smelting) and supplying commodities including tin and lapis lazuli to flow along them (Pigott 2011: 277–282; Barjamovic 2018; see map of southern routes in Muhly 1973b: 407, fig. 1; 1976: 97–104; 1985; Pigott 2011: 278, fig. 27.2; see map of northern end of routes in Barajamovic 2018: 135, fig. 1) (see Figures 29.6 and 29.7).

Research and fieldwork in recent decades has documented tin deposits to the east of Mesopotamia with dated ancient mining evidence in at least four Southwest and Central Asian locales: Deh Hosein on the eastern border of Luristan in western central Iran (e.g., Nezafati *et al.* 2009, 2011); the tin (and gold) placers of Afghanistan dotting its riverbeds in some profusion (e.g., Cleuziou and Berthoud 1982; Stech and Pigott 1986: 40; Thomalsky *et al.* 2013; Jansen *et al.* 2016; Steinkeller 2016; Thomalsky 2016); the hard rock tin mines (e.g., Karnab, Mushiston) of the Zeravshan

Figure 29.8 A modern donkey caravan transporting slabs of salt from the salt-mining area of Hamed Ela and Lake Asale in the Afar region of Ethiopia. It is a two-day trip to market and travel is mostly at night due to extreme heat. (Image courtesy of Bruno Oudmayer © 2018 All rights reserved. www.theonearmedcrab.com.)

valley region in Uzbekistan and Tajikistan (most recently, e.g., Garner 2013, 2015); and a cluster of mined tin deposits in East Kazakhstan (e.g., Stöllner *et al.* 2011; Stöllner and Samašev 2013, see also Chapter 28) (see Figure 29.3). All of these major Central Asian mining districts also have contemporaneous settlements conducting tin smelting and, along with Göltepe in Anatolia, they represent the only solid evidence we have for the *in situ* production of tin in western Eurasia.

Uluburun shipwreck (13th century BC, southeast coastal region)

Though not a mining site, the remarkable shipwreck finds from Uluburun are mentioned here only in passing. They included about nine tons of copper and one ton of tin ingots (e.g., Bass 1986; Pulak 1988, 1997, 1998; Bass *et al.* 1989; Yalçin *et al.* 2005) (Figure 29.9). This underwater site off the southeastern coast of Turkey, the wreck, its contents, and laboratory analyses have been discussed widely in scholarly circles (e.g., Stos-Gale *et al.* 1998; Pulak 2000; Hauptmann *et al.* 2002; Pigott 2011) (see Figure 29.3) At the same time such massive amounts of metal (especially tin) are hard to ignore only a few centuries following the winding down of the Old Assyrian trade.

Interestingly, the preliminary LIA that have been conducted on the tin ingots from the wreck seem to indicate that they are comprised of a minimum of two distinct clusters suggesting in turn at least two separate sources for the tin (Stos-Gale *et al.*

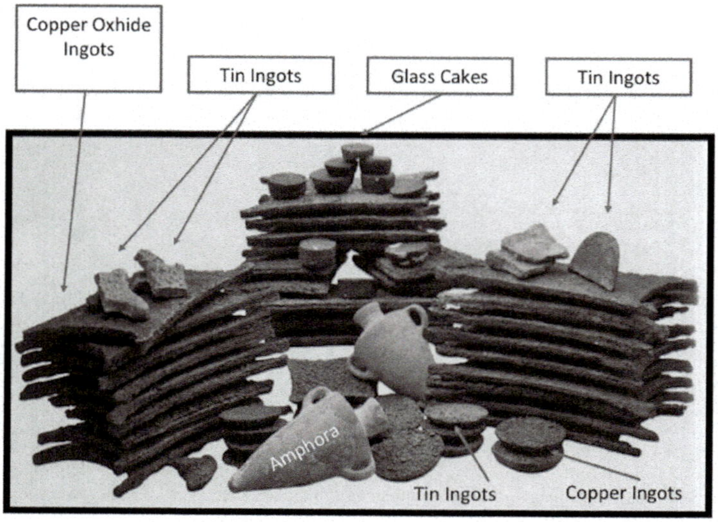

Figure 29.9 Cargo excavated from the shipwreck at Uluburun, off the southern coast near Kaş, Turkey, late 14th century BC. It included, as seen here, ca. 1 ton of tin ingots, ca. 9 tons of copper ingots, glass cakes, amphorae, and many important, often luxury, commodities. (After Yalcin 2005: 20, Abb. 2.)

1998: 119, 123; see also Pulak 2000: 153–155, fig. 23). Current indications are that one of the high-lead clusters shows overlap with the Taurus IA isotopic field that encompasses the Bolkardağ mining district of the Taurus Mountains in south-central Turkey (see Yener *et al.* 1991, 546–548; Sayre *et al.* 1992: 74–75; Pulak 2000: 155). This is the location of the 3rd-millennium Kestel-Göltepe first-tier, tin-mining industrial complex, but one of the multiple possible tin sources to be found in this mining district alone (Yener 2000: 72; Yener *et al.* 2015). The low-lead LIA cluster is not only not compatible with any Anatolian field, but also not with any other currently known field. This opens the door to speculation that these tin ingots came from afar, traded possibly from points to the east (Pigott 2011: 274–277).

Iran

Deh Hosein (3rd?–2nd millennium BC, western-central region)

The site of Deh Hosein is located in western Central Iran on the eastern border of Luristan province (Figure 29.10). The mining area is comprised of a massive concentration of open-pit mines dotting the landscape and it was first revealed to the archaeological community in 2002 (Momenzadeh *et al.* 2002). The cluster of 75-plus ancient workings, some with enormous spoil heaps, extends over an area 4.5 x 6 km^2 (see Nezafati *et al.* 2011: figs. 4 and 5) (Figure 29.11). In addition to these open-pit mines, Iranian geologists have recently identified extensive underground workings where ancient miners would have been following ore veins (Nezafati, pers. comm.).

The Deh Hosein deposits are polymetallic with mixed copper-tin ores including stannite, and also the tin ore, cassiterite (see Nezafati *et al.* 2011: 217, tab. 2; 208,

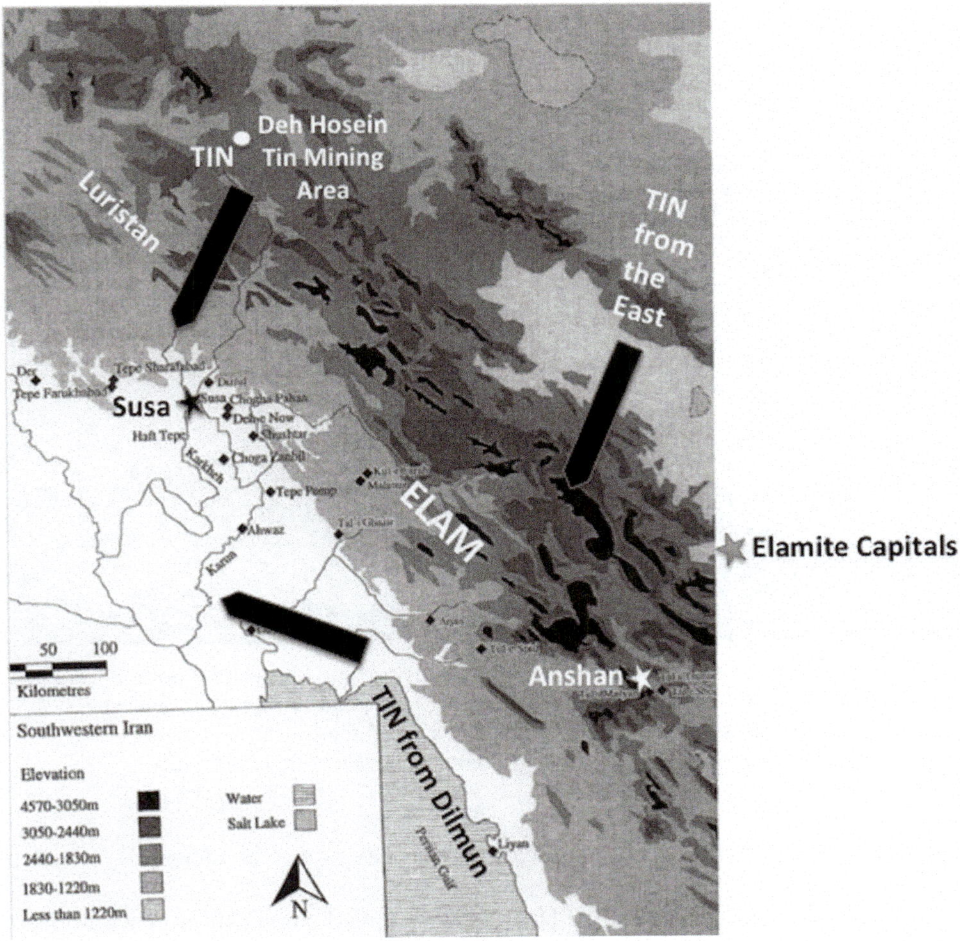

Figure 29.10 Map of Elam and its two capitals at Susa and Anshan (Tal-i Malyan), the neighboring Deh Hosein mining area, and the potential directions from which tin was entering Elam and southern Mesopotamia. (After Potts 1999: 190, fig. 7.1.)

tab. 3; 224, tab. 5 for examples of ores and minerals present). Geological studies show that the cassiterite appears as a "rather abundant ore mineral" occurring as disseminated "grains up to 250μ ... in association with oxidized copper minerals and gossan" (Nezafati *et al.* 2011: 220). Such fine-grained dissemination of cassiterite in association with copper ores may have meant that this mix of ores could have smelted unintentionally to a natural bronze. Furthermore, such ore occurrences would lie at or near the deposits' surface and would have lent themselves to the open-pit mining seen in multiple locations at Deh Hosein today.

Nezafati and colleagues (2011: 12) have suggested that the Deh Hosein deposits were probably first exploited for their copper and that the tin-containing copper ores were encountered in that process. Thus far no smelting evidence has been found in the Deh Hosein area, but the presence of cassiterite placers in the vicinity meant that at some stage this ore was available and that quite possibly tin metal

Figure 29.11 Open-pit-mining depressions and spoil heaps that dot the landscape of the Deh Hosein tin-mining area. The cluster of 75-plus ancient workings, some with enormous spoil heaps, extends over an expanse of 4.5 x 6 km². (Photo provided courtesy of Dr. Nima Nezafati.)

was isolable by smelting this ore at locations that have yet to be identified (Nezafati *et al.* 2011: 220).

Metalworkers choosing to smelt Deh Hosein's mixed copper-tin ores would have made use of a mix of oxidic (e.g., cassiterite, malachite, iron oxides) and sulfidic ores (e.g., stannite, stannite bound to chalcopyrite). This combination of ores would have allowed for the single-step production of metal by the process known as "co-smelting" (e.g., Rostoker *et al.* 1989; Rovira *et al.* 2009), which could yield a natural tin bronze marked by an uncontrolled tin content and high sulfur content (see also Thornton 2014; Pigott 2018). Elemental analyses of 25 excavated and unprovenanced "Luristan bronzes" (stylistically dated to 3rd–1st millennium BC) from Iranian museums conducted by Nezafati and colleagues, as well as analyses conducted by the Penn Museum (MASCA) team on the Belgian Archaeological Mission in Iran (BAMI), indicated that their tin contents were uncontrolled (Fleming *et al.* 2005, 2006; Nezafati *et al.* 2008: 223; see also Begemann *et al.* 2008). These studies have raised the possibility that Deh Hosein ores may have been used to produce Luristan bronzes from sites in relative proximity to its mined deposits.

Currently, the only calibrated ^{14}C date from Deh Hosein (from an intermediate layer in an open-pit mine layer) spans ca. 1775–1522 BC (2-sigma confidence level with 95% probability) (Nezafati *et al.* 2008: 220). The suggestion of earlier exploitation of the deposit is based on LIA results. LIA was conducted on 18 Deh Hosein ore samples and on the above-mentioned 25 samples taken from "Luristan bronzes."

LIA of the ores "shows a rather small variation of lead isotope ratios" (Nezafati *et al.* 2008: 225), making for a more accurate signature of this deposit. The ores were said to have LIA signatures compatible with those from the "Luristan bronzes" dated stylistically to the 3rd millennium. Of the "Luristan bronzes" dated to the 2nd and 1st millennium BC, half also had signatures compatible with those of the Deh Hosein ore samples. Samples of artifacts submitted to LIA from various other Near Eastern sites, ranging from western Anatolia to the southern Gulf and dating to the 3rd millennium and later, also signaled compatibility with Deh Hosein ores. These LIA results led the researchers to argue that Deh Hosein was a major regional source of tin and/or tin bronze from the 3rd millennium onward (Nezafati *et al.* 2008: 225–227).

Nezafati and colleagues (2008: 213, 225–226) also argue that Deh Hosein could easily be the source of what cuneiform texts refer to as "tin from the east." Other than the Anatolian sources discussed above, this is the only other known tin-bearing deposit that is relatively close to the tin bronze consumers of Mesopotamia, Luristan, and elsewhere in the Near East that was possibly utilized during the 3rd millennium BC and later. We still need greater certainty as to just how accessible the tin ores at Deh Hosein actually were and that there is *in situ* evidence to suggest it was mined as early as the 3rd millennium. LIA does appear to link the tin bronze of that period from these regions to Deh Hosein ores. Interestingly, excavations at Bronze Age settlements of consumers of tin bronze in and across the ancient Near East have revealed only sparse physical evidence for the production of tin bronze artifacts, nor has Deh Hosein or its hinterland offered up any evidence for the smelting of its ores.

As current arguments would have it, Deh Hosein and its vicinity have ore deposits sufficient to have served as a major ancient tin source for the greater region (see also Pigott 2011, 2012a). Modern heavy-mineral prospecting by Iranian geologists in streams in and around Deh Hosein has yielded nodules of cassiterite (Nezafati *et al.* 2011: 220), and the Geological Survey of Iran has recently located additional placer deposits in the greater Deh Hosein region as well as *in situ* cassiterite occurrences in nearby Luristan near Azna (Nezafati, pers. comm.). In addition, given that Deh Hosein went undetected for so long, there is every reason to believe, given the presence of appropriate geology in western Iran, that other tin sources may well exist along the mineral-rich Sanandaj-Sirjan fold belt that extends the length of the Zagros Mountains (Nezafati *et al.* 2011: 227) (Figure 29.12). This belt is a region with millennia of metallurgical activity and abundant mineral resources that by all indications could have supplied tin and tin bronze on a significant scale. Deh Hosein itself could well have been among the regional suppliers in provisioning tin/tin ore either directly to the Old Assyrian period traders in southern Mesopotamia or to those polities and centers in the immediate region (e.g., Elam, Susa, Anshan) (see Figure 29.10) that may have been processing ores to tin to be fed into the trade, as is discussed later in the chapter. At a minimum these entities clearly had a role in facilitating the movement of tin from local sources or that coming from points further east to caravans headed to Assur and north.

Afghanistan

As discussed above, and argued by its researchers, from the 3rd millennium onward the mines at Deh Hosein in western central Iran have the potential to have been a

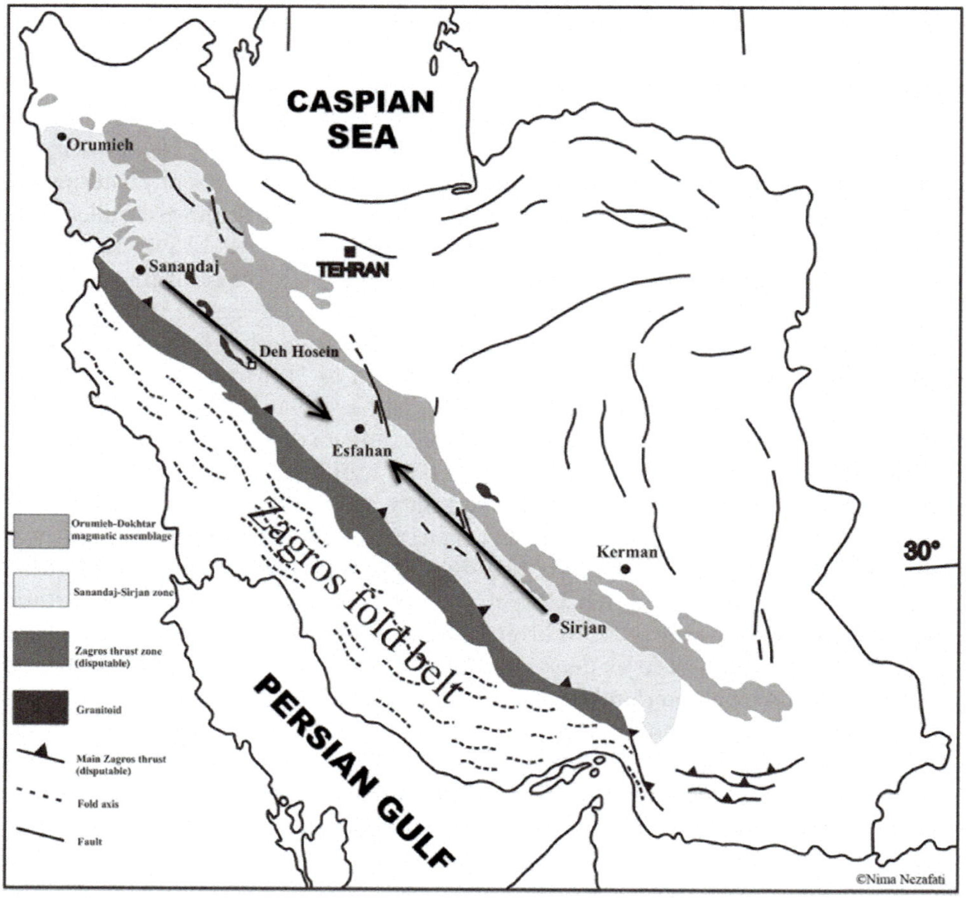

Figure 29.12 Map of major Zagros orogenic belts in western Iran. Important to the discussion is the Sanandaj-Sirjan fold belt (see arrows) in which Deh Hosein (box) is situated. This belt holds the prospect of yielding even more such ore deposits that may have been exploited in prehistory. (Map provided courtesy of Dr. Nima Nezafati.)

supplier of tin to consumers of tin bronze across the Near East proper. To the east of Deh Hosein on the Iranian Plateau, however, one encounters one of the region's metallurgical enigmas, namely, that tin bronze does not appear there until the very end of the 3rd/early 2nd millennium BC and, even then, mostly in southern (e.g., at Tepe Yahya), southwestern (e.g., at Susa and Anshan), and western Iran (Luristan) (e.g., Stech and Pigott 1986; Tallon 1987; Pigott *et al.* 2003; Thornton *et al.* 2005; Helwing 2009; Pigott 2009; Thornton 2014; Cuénod *et al.* 2015). A detailed discussion of just how tin bronze metallurgy may have arrived on the Iranian Plateau, involving peoples from the Bactria-Margiana Archaeological Complex (BMAC) from the north, as has been argued is beyond the scope of this chapter, but relevant discussions of the issue and its context are available (e.g., Hiebert 1994, 1998; Anthony 2007; Lamberg-Karlovsky 2012; Thornton and Giardino 2012; Thornton 2014; Pigott 2018; Pigott in press). Regional cultural interactions that would have facilitated the adoption of tin bronze technology on the Iranian Plateau as well as regional sources of tin, its key raw material, are beginning to be better understood. Scholars have long looked at Afghanistan for explanations as this neighboring region is rich in tin reserves, but archaeological documentation of their exploitation in antiquity has remained elusive (Cleuziou and Berthoud 1982; Penhallurick 1986/2008; Stech and Pigott 1986; Weeks 2003; Kaniuth 2007; Pigott 2018).

However, research by the German Archaeological Institute beginning in 2013 has made significant advances towards documenting the prehistoric and later exploitation of Afghanistan's remarkable mineral reserves in relation to its archaeological sites (e.g., Thomalsky *et al.* 2013; Thomalsky 2016). Detailed maps in both of these publications locate large numbers of prehistoric sites in proximity to economically viable ore deposits (silver, gold, tin, copper, lead) that have been documented through the use of site gazetteers, recent modern geological surveys, and the use of CORONA and Landsat remote-sensing data. Notably we see concentrations of tin deposits at a number of locales spread across the country (Thomalsky *et al.* 2013: 212, fig. 12).

There are several points that merit discussion here. First, Afghanistan has ample tin, copper, and gold reserves (see Chapter 28). However, only a small number of excavated Bronze Age sites in the country have yielded modest numbers of tin bronzes and most of these date to the end of the 3rd millennium and later. While Afghanistan has often been raised as a source area for tin for the metal industries of the ancient Near East, Thornton and Giardino (2012: 257) have argued that evidence in support of this hypothesis is sparse and that it was not until the end of the 3rd millennium or more likely the beginning of the 2nd millennium BC that Afghan tin began to be exploited on any scale. Furthermore, it may have been BMAC peoples intruding from the north who, recognizing the wealth of mineral sources that lay to the south, began serious exploitation. Can it be posited that BMAC peoples in turn were involved in a regional tin trade based on areal sources from the Zeravshan valley region to the northeast as well as those from Afghanistan to the south? (e.g., Lyonnet 2005: 191–93, 197; Wilkinson 2014: 214, fig. 5.47; 223, fig. 5.52; Steinkeller 2016: 136–137; Pigott 2018; Pigott in press; *contra* Kaniuth 2007: 34).

Second, Thomalsky *et al.* (2013: 200) make mention of distinctive deposits of complex – that is, combined copper and tin ores – (e.g., stannite, Cu_2FeSnS_4) in the country. Such ores, it has been noted with some frequency as of late, can be smelted directly in a one-step operation to produce a natural alloy of uncontrolled tin content (e.g., Rovira and Montero 2003). Thomalsky (2016: 651) sees the presence of such ores as a "precondition" to the coming of tin bronze. These ores and their significance are discussed further in the "concluding observations" section at the end of the chapter.

Third, the issue of the early distribution of Afghan lapis lazuli across Southwest Asia also has been raised. Thomalsky (2016: 651, fig. 3) maps its rather widespread distribution in the 4th and 3rd millennium across greater Southwest Asia. For a semi-precious mineral to have been traded or trans-shipped as far afield as pre-dynastic Egypt as early as the 4th millennium is quite remarkable (e.g., most recently, Casanova 2013). However, there are no indications that lapis lazuli and tin (in some form) initially moved together at least east of Mesopotamia, but in Elam and Mesopotamia by all indications they traveled in tandem. The absence until relatively late of tin bronze on the Iranian Plateau at Bronze Age sites (e.g., Tepe Hissar, Shahr-i Sokhta) where lapis was being processed is but one piece of supporting data in this regard (e.g., Thornton 2014: 673–674, tab. 23.1; 686). If Afghan tin was reaching foundries in the ancient Near East, how it was being transported before reaching Mesopotamian is subject to debate (but see Chapter 1).

However, circumstantial evidence continues to build in support of Afghanistan's role in supplying Near Eastern elites with highly prestigious luxury commodities like lapis, tin(?), and gold. Muhly (1977: 76) indicated that

> tin and gold appear together in the archaeological record. There is very little gold before 3000 B.C., just as there is little or no tin bronze. After 3000 B.C. both gold and tin bronze appear together at a number of sites over a wide geographical area, from the Aegean to southern Mesopotamia.

Nowhere is this truer than among the regal contents of the Early Dynastic IIIA royal tombs at Ur from the mid-3rd millennium wherein lapis and gold were paired in magnificent ornaments in tombs that contained ample numbers of artifacts in tin bronze (e.g., Stech and Pigott 1986: 45–48; Zettler and Horne 1998; Stech 1999; Hauptman and Pernicka 2004; Müller-Karpe 2004; Helwing 2009: 215 n4). This so-called "Sumerian trio" of luxury materials has renewed significance based on recent research on gold artifacts from Ur by Moritz Jansen and colleagues (2016). This is discussed further later in the chapter.

As is well established geologically, both tin ore and gold have high specific gravities and tend to concentrate alluvially in placer deposits often marked by another heavy ore mineral, the black sands of iron (mostly magnetite). Hence, in Afghanistan, an abundance of placer deposits of tin and gold suggests they may have been encountered together by local peoples (see map in Stech and Pigott 1986: 40). In this regard, Berger and colleagues (2014: 79) have suggested that after collecting "gold dust" from such tin-rich placers (e.g., by panning or greased sheep fleece in the stream bed) the two minerals may have been heated together in a crucible, the aim being to concentrate the gold, the tin being an accidental accompaniment into the crucible. If reducing conditions prevailed in the depths of the crucible then a gold-bearing alloy would be produced. They suggest this may have led to additional experimentation such as mixing this alloy with copper that in turn could yield a tin bronze. As we saw earlier in the chapter, in the view of the researchers, the processing of mined ores rich in tin-gold from 3rd-millennium Kestel at neighboring Göltepe is what may be reflected in the crucibles excavated and analyzed from the site.

One of the markers of placer gold is the occurrence of platinum group elements (PGEs, osmium, iridium, ruthenium) that pass into cast gold artifacts unmodified by the high temperature of melting (ca. 1100°C). Jansen and team (2016: 12–13) used osmium isotope ratios in concert with LIA, trace element analysis, and archaeological data to "narrow down the potential ore sources for gold." It has been known for sometime that gold artifacts from Ur have macroscopic PGE inclusions (Young 1972; Jansen *et al.* 2016: 14, fig. 1). As Jansen *et al.* (2016: 13, 21) point out, lapis known from Badakhshan province in northeastern Afghanistan occurs close to the Amu Darya River, which itself is known for rich placer deposits of gold. Placers also occur in western parts of the province. In the estimation of the research team, this region, based on its Paleozoic ophiolite formations and abundant gold (near

Samti) comprises one of the potential source areas for the gold used in the manufacture of the Ur artifacts under study (Jansen *et al.* 2016: 20–22, fig. 6). A second potential source area is around Takab in northwestern Iran located in Anguran Chay, Zanjan province. It is future research focused on the geochemical fingerprints of the locales mentioned here that should refine our understanding of sources in relation to excavated Sumerian gold even further. This research will at the same time offer new insights into just how high-prestige luxury commodities may have moved from their source areas to production loci and on to elites in the urban centers of the ancient Near East.

CONCLUDING OBSERVATIONS

In closing, a few comments are offered here to amplify some of the points raised in the above discussions. First, mixed copper-tin ores, as mentioned previously occurring in Afghanistan, have been documented at Bronze Age mining sites elsewhere in Southwest Asia (e.g., the Bolkardağ mining district, Deh Hosein) and in Central Asia (e.g., Mushiston) (Nezafati *et al.* 2008: 77, 89; Roberts *et al.* 2009: 1017–1018; Pigott 2012b: 91; Garner 2013; Yener *et al.* 2015: 597, 608; Pigott 2018). The initial smelting of such ores that may have yielded Southwest Asia's earliest tin bronzes and their smelting may well have occurred as prehistoric peoples chanced upon such unique deposits and experimented with smelting the mixed ores. Tin bronze, the alloy, cast easily, exhibited hardness, durability, and, depending on tin content, a golden hue; all highly desirable properties (cf. Radivojević *et al.* 2018). In this regard, Lehner *et al.* (2015: 206) raise the question with regard to the earliest tin bronzes as to why there is little to no evidence of "local parallel innovations in early tin metallurgy." Is it possible that for some period the initial smelting of mixed copper-tin ores that yielded a quite desirable natural alloy obviated the parallel innovation of tin smelting? Until recently evidence for tin smelting has been very rare in western Eurasian Asian Bronze Age contexts. Some of the best evidence for actual ancient tin-smelting sites is now coming from research by the German and local teams who have conducted fieldwork near mines in tin-rich mineral districts dated to the 2nd millennium BC in the greater Zeravshan valley region and in East Kazakhstan (e.g., Stöllner *et al.* 2011; Garner 2013; Stöllner and Samašev 2013). It should not escape our attention that this tin smelting is occurring at settlements in relative proximity to abundant tin sources and one might expect to find similar evidence near the other Southwest Asian tin mines, unless, of course, this Central Asian tin-smelting evidence is among the earliest to have occurred in western Eurasia. However, the newly emerging tin-mining evidence from the Erciyes Dağ volcano area of central Anatolia, some of which dates to the 3rd millennium, has the potential to be earlier than that from Central Asia.

Second, Central Asia and its purported role as among the suppliers of tin/tin bronze to elite urban consumers in the ancient Near East have been discussed by various scholars (e.g., Parzinger and Boroffka 2003; Barjamovic 2018: 142, 156). Hauptmann and Stöllner (Stöllner *et al.* 2011: 247–248, fig. 18), in the process of research related to the documentation of a number of 2nd-millennium BC tin mines and tin-smelting loci in East Kazakhstan (Stöllner and Samašev 2013), conducted LIA on 15 excavated tin bronzes and one tin bead. The indications are that these artifacts derive from older geological formations characteristic of their research area, including the Askaraly tin ore-mining fields. These artifacts exhibit an isotopic signature compatible with those established for artifacts from Troy (IIg) and the Troas in northwestern Anatolia (see also Pernicka *et al.* 2003: 165–167; *contra* Kaniuth 2007; Thornton 2007: 129–130; Helwing 2009). It must be said that such LIA results are not definitive indicators of evidence linking the two regions. However, they do offer tantalizing clues to the possibility that Central Asian tin may have been reaching markets far to the west and, as such, may define new levels of granularity in emerging insights into the nature of Trans-Eurasian exchange in the Bronze Age and later periods (e.g., Sherratt 2006).

Third, analytical evidence from the South Caucasus, it has been argued, supports the supposition that tin used in regional bronze making may well have had its origin in Afghanistan. Tin bronze artifacts were excavated from Velikent in Daghestan dating, it appears, to the second half of the 3rd millennium BC (Courcier 2014: 585, fig. 22.2; 635, 641; *contra* Peterson 2003: 1, 25; Kohl 2007: 108–109). Based on LIA conducted by Lloyd Weeks (2002: 178–183; see also Kohl 2002) on five of these tin bronzes, he concludes from this quite small sample that the tin in these bronzes may well have been of eastern (e.g., Afghanistan) origin (cf. Peterson 2007: 199–200; see also Courcier 2014: 640–641). Weeks (2002: 178–183) also observes that the linear isotopic array of these samples is "matched by tin-bronzes from other sites (e.g., Poliochni, Thermi, Kastri), and is unlikely to have occurred accidentally or fortuitously. It strongly suggests some link between the tin-bronze being used at these widely separated third millennium sites." These sites are situated on islands in the Aegean Sea more than 2,000 miles/over 3,000 km west of Afghanistan.

Fourth, concerning the Elamites, the southern route, and tin, Barbara Helwing (2009: 210–211) wrote about a purported source area in the Iranian highlands being comprised of the lands of Zabshali/Shimashki in the southern Zagros, the location of the latter having been discussed by Dan Potts (1999: 141–143) as lying somewhere between the more southerly Elamite capital of Anshan located at the modern site of Tal-i Malyan in Fars province and the Caspian Sea (but see Chapter 2 *contra*). Helwing (2009) states that these lands were known for the rich booty of tin taken by the Ur III king Shu-Sin, and also from the lands of Anshan as well as from its hinterland (see Muhly 1973a: 289). Anshan, and the other Elamite capital to the northwest at Susa, are documented textually as road stations for donkey caravans bound for Mari well to the northeast on the middle Euphrates via the city and kingdom of Eshnunna and routes along the Diyala valley (Joannès 1991). Helwing (2009: 211) reports that texts in the Mari archive record that such caravans were not only operated by agents with Elamite names, but also were transporting tin ingots of a standard weight (see Moorey 1994: 298; Potts 2016: 159, tab. 6.2).

Furthermore, textual evidence suggests that peoples occupying the mountainous Zagros region of western Iran during the 3rd (e.g., Gutians and Lullubeans) and 2nd millennia BC (e.g., Turukkaeans) were intercepting tin supplies passing through their regions (Muhly 1973a: 320). This, in turn, implies the possible proximity of a tin source, and Deh Hosein on Luristan's eastern border is one such candidate (for both tin and gold). Moreover, if tin was being trans-shipped across Iran and through the Zagros from points much further east (e.g., Afghanistan) would it not have been possible to redirect the trade around any potential disruption (Pigott 2011: 286–287)?

In addition, in discussing Mari in the early 2nd millennium BC and the tin trade, Muhly (1973a: 293) proposed that

> together with the reference in the "itinerary" of tin sent to Mari from Šeplarpak (to be identified with Siwe-palar-huppak, king of Anšan) it suggests an eastern source for tin, coming from Susa to Ešnunna via Der and then from Ešnunna to Mari via Sippar. Presumably the tin came to Susa from somewhere farther east or to the north.

Thus, in sum, much of the evidence for the movement of tin lacks geographical precision. The implication from what is discussed above and in related publications, is that Elam with its two capitals of Anshan and Susa, named tin suppliers, or sources of tin as booty along with Elamite hegemony, appear to have played an influential role in the control of tin moving through the empire into Mesopotamia (Barjamovic 2018: 148, see also fig. 5). The Elamites, by all indications were in charge of provisioning tin to caravans moving north along the southern route of the Old Assyrian tin trade.

Finally, Barjamovic (2018: 134, see also 148) offers a closing thought on the Old Assyrian tin trade in a Eurasian context:

> However, a growing understanding of the chronology and geography of the period precludes the interpretation of the commercial phenomenon at Kültepe as a local anomaly. Instead, a rapid influx of new data, and a collaborative and cross-disciplinary effort to deal with this evidence, now allows us to link the Assyrian trade to a chain of contemporary interlocking commercial networks that ultimately connected the Chinese frontier with the Balkans around 1900 BCE.

In this regard, it is of interest to note that post ca. 3000 BC the technology of tin bronze appears to undergo transmission moving from Southwest Asia eastward across the remainder of Eurasia (e.g., Sherratt 2006; Roberts *et al.* 2009; Nessel *et al.* 2018; Pigott 2018 in press). In contrast, by all current indications, some amount of tin ore/tin metal used in the ancient Near East was moving westward along the above-mentioned extensive networks (Wilkinson 2014; Barjamovic 2018: 148, fig. 5) from what we now know to be multiple greater Southwest/Central Asian ore deposits that were mined in the Bronze Age (see Figure 29.3).[8] These ores, in turn, had to be processed to metal, at as yet mostly unidentified loci, as it is metallic tin that appears to have been the highly desired trade commodity. Just how far, and in how many directions, such a metal flow, some in tin(?), may have proceeded is currently the source of persistent speculation and remains a focal point for research by an ever-increasing scholarly community. With ongoing fieldwork and ever-improving research and technique development, including those in the realms of isotopy (e.g., lead, tin, copper), we can expect to have our current assumptions overturned with great regularity.

NOTES

* I would first like to thank the editors for their assistance in the process of bringing this manuscript to publication. The following individuals have given graciously of their time, knowledge, and pdfs, all of which added significantly to the discussions over-time of various

topics incorporated into this chapter: David Anthony, Gojko Barjamovic, Daniel Berger, Antoine Courcier, Kristina Franke, Barbara Helwing, Moritz Jansen, Joseph Lehner, Bertille Lyonnet, Nima Nezafati, Ernst Pernicka, Dan Potts, Miljana Radivojević, Judith Thomalsky, Christopher Thornton, Aslihan Yener, and Norm Yoffee. I wish also to thank Nima Nezafati, Thomas Stöllner, Bruno Oudmayer, Ben Roberts, Lloyd Weeks, and Aslihan Yener for permission to publish select images that they made available to me. Any errors, omissions, or misinterpretations remain the sole responsibility of the author.

1 In the discussion of bronze in this chapter, the term will refer only to the alloy of copper and tin. In this and my previous publications I do not use the term "bronze" to refer to any complex alloy of copper (e.g., arsenical copper, antimonial copper).

2 James D. Muhly (1973b: 406) states: "From the early third millennium on, Sumerian and Akkadian texts describe the mixture of 'copper (Sumerian URUDU, Akkadian *erû* and tin (Sumerian AN.NA, Akkadian *annaku*)' in order to produce bronze (Sumerian UD.KA.BAR, Addadian *siparru*)." See also Moorey (1994: 252), Weeks (2003: 178), and Franke (2018).

3 In the oft-cited Luristan graveyard of Kalleh Nisar, due to the reuse of tombs in the AI area of the site, the bronze artifacts found in the tombs cannot be securely dated to the early to mid-3rd millennium (Haerinck and Overlaet 2008: 26).

4 This is a list of sites located mostly in the ancient Near East and that date presumably to the early 3rd millennium. The point here is to demonstrate the relative abundance of such sites in the region at an early date. It is also the case that excavations now decades old may well have chronologies that are no longer necessarily accurate by the standards of today's understanding of context.

5 Cf. Rahmstorf (2011: 106) who proposes that the distribution of tin bronze as seen on his map suggests that they were the product of multiple independent innovations across Europe and Southwest Asia, often the product of mixed-copper tin ores, hence they were unintentional bronzes.

6 For more detailed arguments in support of the Kestel-Göltepe industrial complex as focused on the mining and processing of tin-rich ores, see e.g., Yener (2008).

7 Helwing (2009: 215 n7) states that an "overview of the sources referring to eastern tin" can be found in Potts (1994: 156, tab. 4.2) and Parzinger and Boroffka (2003: 1–4).

8 In addition, there is insufficient space remaining in this chapter to deal with two additional issues regarding the movement of tin to Mesopotamia; namely, the role of ancient Dilmun (region of Bahrain) as a trading entrepôt for tin, as well as the potential role played by Harappan middlemen in trading tin (from Afghanistan?) from the Indus valley (Meluhha) to the Gulf region. As Ben Foster (1997: 59) has stated, "Sumerian and Babylonian merchants went to Dilmun to buy copper and tin, while traders from Dilmun came to Mesopotamia, for example, to Sargonic Umma, Lagash, and Agade as well as to Susa when it was under Sargonic rule." Thus, tantalizing glimpses exist into these somewhat enigmatic players in Bronze Age economic interactions in the Gulf and east. Most recently Steffen Laursen and Piotr Steinkeller (2017) have provided an up-to-date overview of these issues, but see also, e.g., Muhly (1973a), Crawford (1974), Penhallurick ([1986] 2008: 21–32), D.T. Potts (1993), T.F. Potts (1993), Possehl (1996, 1997), and Pigott (2011: 277–281).

REFERENCES

Anthony, D.W. 2007. *The Horse, the Wheel, and Language*. Princeton, NJ: Princeton University Press.

Atici, L., F. Kulakoğlu, G. Barjamovic, and A. Fairbairn (eds.) 2014. *Current Research at Kültepe-Kanesh: An Interdisciplinary and Integrative Approach to Trade Networks, Internationalism, and Identity*. Journal of Cuneiform Studies Supplemental Series No. 4. Atlanta, GA: Lockwood.

Bachhuber, C.S. 2011. "Negotiating Metal and Metal Form in the Royal Tombs of Alacahöyük in North-Central Anatolia." In *Interweaving Worlds: Systemic Interactions in Eurasia, 7th to 1st Millennium BC*, ed. T.C. Wilkinson, S. Sherratt, and J. Bennet. Oxford: Oxbow Books, pp. 158–174.

Barjamovic, G. 2018. "Interlocking Commercial Networks and the Infrastructure of Trade in West Asia During the Bronze Age." In *Trade and Civilisation in the Pre-Modern World*, ed. K. Kristiansen, T. Lindkvist, and J. Myrdal. Cambridge: Cambridge University Press, pp. 133–167.

Bass, G.F. 1986. "A Bronze Age Shipwreck at Ulu Burun (Kaş): 1984 Campaign." *American Journal of Archaeology* 90: 269–296.

Bass, G.F., C. Pulak, D. Collon, and J. Weinstein 1989. "The Bronze Age Shipwreck at Ulu Burun: 1986 Campaign." *American Journal of Archaeology* 93: 1–29.

Beaujard, P. 2011. "Evolutions and Temporal Delimitations of Bronze Age World-Systems in Western Asia and the Mediterranean." In *Interweaving Worlds: Systemic Interactions in Eurasia, 7th to 1st Millennia BC*, ed. T.C. Wilkinson, S. Sherratt, and J. Bennet. Oxford: Oxbow Books, pp. 7–26.

Begemann, F., E. Haerinck, B. Overlaet, S. Schmitt-Strecker, and F. Tallon 2008. "An Archaeo-Metallurgical Study of the Early and Middle Bronze Age in Luristan, Iran." *Iranica Antiqua* 43: 1–66.

Berger, D., G. Brugmann, E. Figueiredo, and E. Pernicka 2016. "Zinnisotopenverhältnisse von Verhüttungsprodukten von Kassiterit und Ihre Bedeutung für die Herkunftbestimmung von Zinn." *Metalla* 8: 194–197.

Berger, D., E. Pernicka, B. Nessel, G. Brugmann, C. Frank, and N. Lockhoff 2014. "Neue Wege zur Herkunftsbestimmung des bronzezeitlichen Zinns." *Blickpunkt Archäologie* 4: 76–82.

Brügmann, G., D. Berger, C. Frank, J. Marahrens, B. Nessel, and E. Pernicka 2017. "Tin Isotope Fingerprints of Ore Deposits and Ancient Bronze." In *The Tinworking Landscape of Dartmoor in a European Context: Prehistory to 20th Century*, ed. Ph. Newman. Tavistock, Devon: Dartmoor Tinworking Research Group, pp. 103–114.

Casanova, M. 2013. *Le Lapis-Lazuli dans l'Orient Ancien: Production et Circulation du Néolithique au IIe Millénaire av. J.-C.* Documents Préhistoriques 27. Paris: Editions du Comité des travaux historiques et scientifiques.

Chirikure, S., R.B. Heimann, and D. Killick 2010. "The Technology of Tin Smelting in the Rooiberg Valley, Limpopo Province, South Africa, ca. 1650–1850 CE." *Journal of Archaeological Science* 37: 1656–1669.

Ciarla, R. 2007. "Rethinking Yuanlongpo: The Case for Technological Links Between the Lingnan (PRC) and Central Thailand in the Bronze Age." *East and West* 57(1–4): 305–328.

Ciarla, R. 2013. "Interazioni culturali e tecnologiche tra Cina meridionale e Sudest asiatico continentale tra la fine del II millennio a. C. e l'inizio del I millennio a. C.: la dispersione meridionale della tecnologia del rame/bronzo." Unpublished PhD dissertation, Universitá Ca' Foscari, Venice.

Cleuziou, S. and T. Berthoud 1982. "Early Tin in the Near East." *Expedition* 25(1): 14–19.

Courcier, A. 2014. "Ancient Metallurgy in the Caucasus from the Sixth to the Third Millennium BCE." In *Archaeometallurgy in Global Perspective: Methods and Syntheses*, ed. B.W. Roberts and C.P. Thornton. New York: Springer, pp. 579–664.

Crawford, H.E.W. 1974. "The Problem of Tin in Mesopotamian Bronzes." *World Archaeology* 6: 242–247.

Cuénod, A., P. Bray, and M.A. Pollard 2015. "The "Tin Problem" in the Prehistoric Near East. Further Insights from a Study of Chemical Datasets on Copper Alloys from Iran and Mesopotamia." *Iran* 53: 29–48.

Figueiredo, E., A. Lackinger, B. Comendador Rey, R.J.C. Silva, J.P. Veiga, and J. Mirão 2016. "An Experimental Approach for Smelting Tin Ores from Northwestern Iberia." *Materials and Manufacturing Processes* 32(7–8): 765–774.

Fleming, S.J., V.C. Pigott, C.P. Swann, and S.K. Nash 2005. "Bronze in Luristan: Preliminary Analytical Evidence from Copper/Bronze Artifacts Excavated by the Belgian Mission in Iran." *Iranica Antiqua* 40: 35–64.

Fleming, S.J., V.C. Pigott, C.P. Swann, S.K. Nash, E. Haerinck, and B. Overlaet 2006. "The Archaeo-Metallurgy of War Kabud, Western Iran." *Iranica Antiqua* 41: 31–57.

Foster, B. 1997. "A Sumerian Merchant's Account of the Dilmun Trade." *Acta Sumerologica Japonica* 19: 53–62.

Franke, K. 2018. "Metals Without Ores: The Metallurgy of the 3rd Millennium Upper Mesopotamia with Special Focus on the Jezirah." Unpublished PhD thesis, Institute of Archaeology, University College, London.

Garner, J. 2013. *Das Zinn der Bronzezeit in Mittelasien II. Die Montanarchaeologischen Forschungen an den Zinn Lagerstätten.* Archäologie in Iran and Turan 12. Berlin: Deutsches Archäologisches Institut, Deutsches Bergbau-Museum Bochum.

Garner, J. 2015. "Bronze Age Tin Mines in Central Asia." In *Archaeometallurgy in Europe*, vol. III, ed. A. Hauptmann and D. Modarressi-Tehrani. Bochum: Deutsches Bergbau-Museum, pp. 135–143.

Hamilton, E.G. 1991. "Metallurgical Analyses and the Bronze Age of Bohemia: Or, Are Cultural Alloys Real?" *Archeomaterials* 5: 75–89.

Hanilçi, N., C. Altaylı, S. Altuncu, and H. Öztürk 2010. "Was the Göl Tepe (Niğde, Central Anatolia, Turkey), a Tin Processing Site During the Early Bronze Age? Preliminary Finds from Soil Geochemistry." *Geologie* (IASI, Special Issue): 57–59.

Hauptmann, A., R. Maddin, and M. Prange 2002. "On the Structure and Composition of Copper and Tin Ingots Excavated from the Shipwreck of Uluburun." *Bulletin of the American Schools of Oriental Research* 328: 1–30.

Hauptmann, H. and E. Pernicka (eds.) 2004 *Die Metallindustrie Mesopotamiens von den Anfängen bis zum 2. Jahrtausend v. Chr.* Orient-Archäologie Band 3. Rahden: Marie Leidorf.

Haerinck, E. and B. Overlaet 2008. *The Kalleh Nisar Bronze Age Graveyard in the Chavar Area, Pusht-I Kuh, Luristan.* Luristan Excavation Documents VII, Acta Iranica 46. Leuven: Peeters.

Helwing, B. 2009. "Rethinking the Tin Mountains: Patterns of Usage and Circulation of Tin in Greater Iran from the 4th to the 1st Millennium BCE." *TÜBA-AR* 12: 209–221.

Helwing, B. 2017. "A Comparative View on Metallurgical Innovations in South-Western Asia: What Came First?" In *Appropriating Innovations: Entangled Knowledge in Eurasia, 5000–1500 BCE*, ed. P.W. Stockhammer and J. Maran. Oxford: Oxbow Books, pp. 161–170.

Hiebert, F.T. 1994. *Origins of the Bronze Age Oasis Civilization in Central Asia.* American School of Prehistoric Research Bulletin 42. Cambridge, MA: Harvard University Press.

Hiebert, F.T. 1998. "Central Asians on the Iranian Plateau: A Model for Indo-Iranian Expansionism." In *The Bronze Age and Early Iron Age Peoples of Eastern Central Asia*, vol. 1, ed. V.H. Mair. Philadelphia: Institute for the Study of Man and University of Pennsylvania Museum, pp. 148–161.

Higham, C., T. Higham, R. Ciarla, K. Douka, A. Kijngam, and F. Rispoli 2011. "The Origins of the Bronze Age in Southeast Asia." *Journal of World Prehistory* 24: 227–274.

Hoffman, B.C. 2018. "Indus Copper and Bronze: Traditional Perspectives and New Interpretations." In *Walking with the Unicorn: Social Organization and Material Culture in Ancient South Asia*, ed. D. Frenez, G.M. Jamison, R.W. Law, M. Vidale, and R.H. Meadow. Oxford: Archaeopress, pp. 251–264.

Hoffman, B.C. and H.M.-L. Miller 2014. "Production and Consumption of Copper-Base Metals in the Indus Civilization." In *Archaeometallurgy in Global Perspective: Methods and Syntheses*, ed. B.W. Roberts and C.P. Thornton. New York: Springer, pp. 697–727.

Jansen, M., S. Aulbach, A. Hauptmann, H.E. Höfer, S. Klein, M. Krüger, and R.L. Zettler 2016. "Platinum Group Placer Minerals in Ancient Gold Artifacts: Geochemistry and Osmium Isotopes of Inclusions in Early Bronze Age Gold from Ur/Mesopotamia." *Journal of Archaeological Science* 68: 12–23.

Joannès, F. 1991. "L'étain, de Elam à Mari." In *Mésopotamie et Elam*, ed. L.D. Meyer and H. Gasche. Mesopotamian History and Environment Occasional Publications 1. Ghent: REP/Cultura, pp. 67–76.

Johnson, M.A., F. Kulakoğlu, K.A. Yener, G. Dardeniz, and E. Yazgan 2017. "Experimental Smelting of Tin from Senir Sırtu and Hısarcık near Kayseri: Mostly Heartbreak, Some Elation." In *Movements, Resources, Interaction*, ed. F. Kulakoğlu and G. Barjamovic, KIM 2. Subartu 34. Turnhout: Brepols, pp. 117–129.

Kaniuth, K. 2007. "The Metallurgy of the Late Bronze Age Sapalli Culture (Southern Uzbekistan) and Its Implications for the 'Tin Question.'" *Iranica Antiqua* 47: 23–39.

Kenoyer, J.M. and H.M.-L. Miller 1999. "Metal Technologies of the Indus Valley Tradition in Pakistan and Western India." In *The Archaeometallurgy of the Asian Old World*, ed. V.C. Pigott. University Museum Monograph 89. MASCA Research Papers in Science and Archaeology, vol. 16. Philadelphia: University Museum, University of Pennsylvania, pp. 107–151.

Kohl, P.L. 1984. *Central Asia: Palaeolithic Beginnings to the Iron Age*. Paris: Editions Recherche sur les Civilisations.

Kohl, P.L. 2002. "Bronze Production and Utilization in Southeastern, Daghestan, Russia: c. 3600–1900 BC (with an Appendix by Lloyd Weeks, Summary of Velikent Compositional and Lead Isotope Analysis)." In *Die Anfänge der Metallurgie in der Alten Welt*, ed. M. Barthelheim, E. Pernicka, and R. Krause. Rahden/Westf: Marie Leidorf, pp. 161–183.

Kohl, P.L. 2007. *The Making of Bronze Age Eurasia*. Cambridge: Cambridge University Press.

Lamberg-Karlovsky, C.C. 2012. "Commentary: The Oxus Civilization (aka the Bactria-Margiana Archaeological Complex)." *Review of Archaeology* 30: 59–75.

Larsen, M.T. 2015. *Ancient Kanesh: A Merchant Colony in Bronze Age Anatolia*. New York: Cambridge University Press.

Laursen, S. and P. Steinkeller 2017. *Babylonia, the Gulf Region and the Indus: Archaeological and Textual Evidence for Contact in the Third and Early Second Millennia BC*. Mesopotamian Civilizations 21. Winona Lake, IN: Eisenbrauns.

Lehner, J.W. 2014. "Metal Technology, Organization, and the Evolution of Long-Distance Trade at Kültepe." In *Current Research at Kültepe/Kanesh: An Interdisciplinary and Integrative Approach to Trade Networks, Internationalism, and Identity*, ed. L. Atıcı, F. Kulakoğlu, G. Barjamovic, and A. Fairbairn. Boston, MA: American School of Oriental Research, pp. 135–155.

Lehner, J.W., E. Yazgan, E. Pernicka, and F. Kulakoğlu 2015. "Continuity of Tin Bronze Consumption During the Late 3rd Millennium BC at Kültepe." In *Proceedings of the 1st Kultepe International Meeting, Kültepe, 2013: Studies Dedicated to Kutlu Emre*, ed. F. Kulakkoglu and C. Michel. Subartu 35, Kültepe International Meetings 1. Turnhout: Brepols, pp. 195–217.

Lehner, J.W. and K.A. Yener 2014. "Organization and Specialization of Early Mining and Metal Technologies in Anatolia." In *Archaeometallurgy in Global Perspective: Methods and Syntheses*, ed. B.W. Roberts and C.P. Thornton. New York: Springer, pp. 529–557.

Lyonnet, B. 2005. "Another Possible Interpretation of the Bactro-Margiana Culture (BMAC) of Central Asia: The Tin Trade." In *South Asian Archaeology 2001*. Vol. 1, *Prehistory*, ed. C. Jarrige and V. Lefèvre, Paris: Éditions Recherche sur les Civilisations, pp. 190–200.

Momenzadeh, M., N. Nezafati, and E. Pernicka 2002. "First Indication of Tin at the Ancient Mining Site near Deh Hosein." In *Proceedings of the 33rd International Symposium on*

Archaeometry, ed. H. Kars and E. Burke. Geoarchaeological and Bioarchaeological Studies 3. Amsterdam: Institute for Geo- and Bioarchaeology, Vrije Universiteit, pp. 116–117.

Moorey, P.R.S. 1994. *Ancient Mesopotamian Materials and Industries*. Oxford: Clarendon.

Muhly, J.D. 1973a. *Copper and Tin: The Distribution of Mineral Resources and the Nature of the Metals Trade in the Bronze Age*. Hamden, CT: Archon Books.

Muhly, J.D. 1973b. "Tin Trade Routes of the Bronze Age." *American Scientist* 61(4): 404–413.

Muhly, J.D. 1976. *Supplement to Copper and Tin: The Distribution of Mineral Resources and the Nature of the Metals Trade in the Bronze Age*. Transactions of the Connecticut Academy of Arts 46. Hamden, CT: Archon Books.

Muhly, J.D. 1977. "The Copper Ox-Hide Ingots and the Bronze Age Metals Trade." *Iraq* 39: 73–82.

Muhly, J.D. 1985. "Sources of Tin and the Beginnings of the Bronze Metallurgy." *American Journal of Archaeology* 89: 275–291.

Müller-Karpe, M. 2004. "Katalog I: Untersuchte Metallobjekte aus Mesopotamien." In *Die Metallindustrie Mesopotamiens von den Anfängen bis zum 2. Jahrtausend v. Chr*, ed. H. Hauptmann and E. Pernicka. Orient-Archäologie 3. Rahden/Westf.: Verlag Marie Leidorf GmbH, pp. 1–92.

Nessel, B., G. Brügmann, D. Berger, C. Frank, J. Marahrens, and E. Pernicka 2018. "Bronze Production and Tin Provenance: New Thoughts about the Spread of Metallurgical Knowledge." In *Metals, Minds and Mobility. Integrating Scientific Data with Archaeological Theory*, ed. X.-L. Armada, M. Murillo-Barroso, and M. Charlton. Oxford: Oxbow Books, pp. 67–84.

Nezafati, N., E. Pernicka, and M. Momenzadeh 2008. "Iranian Ore Deposits and Their Role in the Development of the Ancient Cultures." In *Anatolian Metal* vol. IV, ed. Ü Yalçin. Der Anschnitt, Beiheft 21. Bochum: Deutsches Bergbau-Museum, pp. 77–90.

Nezafati, N., E. Pernicka, and M. Momenzadeh 2009. "Introduction of the Deh Hosein Ancient Tin-Copper Mine, Western Iran: Evidence from Geology, Archaeology, Geochemistry and Lead Isotope Data. *TÜBA-AR* 12: 223–236.

Nezafati, N., E. Pernicka, and M. Momenzadeh 2011. "Early Tin-Copper Ore from Iran, a Possible Clue for the Enigma of Bronze Age Tin." In *Anatolian Metal*, vol. V, ed. Ü Yalçin. Der Anschnitt, Beiheft 24. Bochum: Deutsches Bergbau-Museum, pp. 211–230.

Oudbashi, O., M. Emami, and P. Davami 2012. "Bronze in Archaeology: A Review of the Archaeometallurgy of Bronze in Ancient Iran." In *Copper Alloys: Early Applications and Current Performance – Enhancing Processes*, ed. L. Collini. IntechOpen. doi: 10.5772/32687, pp. 153–178. www.intechopen.com/books/copper-alloys-early-applications-and-current-performance-enhancing-processes/bronze-in-archaeology-a-review-on-archaeometallurgy-of-bronze-in-ancient-iran.

Özguç, T. 1955. "Report on a Workshop Belonging to the Late Phase of the Colony Period IB." *Belleten* 73: 77–80.

Parzinger, H. and N. Boroffka 2003. *Das Zinn der Bronzezeit in Mittelasien I*. Archäologie in Iran und Turan 5. Mainz am Rhein: Philipp von Zabern.

Penhallurick, R.D. 1986/2008. *Tin in Antiquity: Its Mining and Trade Throughout the Ancient World with Particular Reference to Cornwall*. London: Maney Main.

Pernicka, E. 1998. "Die Ausbreitung der Zinnbronze im 3. Jt." In *Mensch und Umwelt in der Bronzezeit Europas*, ed. B. Hänsel, Kiel: Oetler-Voges, pp. 135–147.

Pernicka, E., C. Eibner, O. Oztunali, and G.A. Wagner 2003. "Early Bronze Age Metallurgy in the Northeast Aegean." In *Troia and Troad*, ed. G.A. Wagner, E. Pernicka, and H.P. Uerpmann. New York: Springer, pp. 143–172.

Peterson, D.L. 2003. "Ancient Metallurgy in the Mountain Kingdom: The Technology and Value of Early Bronze Age Metalwork from Velikent, Daghestan." In *Archaeology in the Borderlands. Investigations in Caucasia and Beyond*, ed. A.T. Smith and K.S. Rubinson. Monograph 47. Los Angeles, CA: Cotsen Institute of Archaeology, pp. 22–37.

Peterson, D.L. 2007. "Changing Technologies and Transformations of the Value in the Middle Volga and the Northeastern Caucasus, circa 3000–1500 B.C." PhD dissertation, Department of Anthropology, University of Chicago. University Microfilms International/ProQuest.

Pigott, V.C. 2009. "'Luristan Bronzes' and the Development of Metallurgy in the West-Central Zagros, Iran." In *Metals and Societies: Studies in Honour of Barbara S. Ottaway*, ed. T.L. Kienlin and B.W. Roberts. Universitätsforschungen zur Prähistorischen Archäologie Band 169. Bonn: Dr. Rudolf Habelt, pp. 369–382.

Pigott, V.C. 2011. "Sources of Tin and the Tin Trade in Southwest Asia: Recent Research and Its Relevance to Current Understanding." In *Metallurgy: Understanding How, Learning Why: Studies in Honor of James D. Muhly*, ed. P.P. Betancourt and S.C. Ferrence. Prehistory Monograph 29. Philadelphia, PA: INSTAP Academic Press, pp. 273–291.

Pigott, V.C. 2012a. "On Ancient Tin and Tin-Bronze in the Asian Old World: Further Comments." In *Eastern Mediterranean Metallurgy and Metalwork in the Second Millennium BC*, ed. V. Kassianidou and G. Papasavvas. Oxford: Oxbow Books, pp. 222–236.

Pigott, V.C. 2012b. "Looking East: The Coming of Tin-Bronze in the Context of Bronze Age Trans-Eurasian Exchange – An Overview." In *Scientific Research on Ancient Asian Metallurgy*, ed. P. Jett, B. McCarthy, and J. Douglas. London: Archetype Publications, pp. 85–100.

Pigott, V.C. 2018. "The Bactria-Margiana Archaeological Complex (BMAC), the Seima-Turbino Horizon and a Possible Eastward Transmission of Tin-Bronze Technology Through Later Third Millennium BCE Central Asia to the Altai Mountains/Eastern Steppe." *How Objects Tell Stories: Essays in Honor of Emma C. Bunker*, ed. K.M. Linduff and K.S. Rubinson. Inner and Central Asian Art and Archaeology, vol. I. Turnhout: Brepols.

Pigott, V.C. in press. "Tin-Bronze in Bronze Age Southwest Asia and the Possible Transmission of Its Technology into Central Asia," In *Bronze Age Tin: Geological Sources, Production, and Distribution of Tin in Bronze Age Eurasia*, ed. E. Pernicka, D. Berger, and C. Frank. International Workshop sponsored by the Curt-Englehorn-Zentrum Archaeometrie gGmbH, Mannheim, Germany, March 14–16, 2018.

Pigott, V.C. and R. Ciarla 2007. "On the Origins of Metallurgy in Prehistoric Southeast Asia: The View from Thailand." In *Metals and Mines: Studies in Archaeometallurgy*, ed. S. La Niece, D. Hook, and P. Craddock. London: Archetype Publications in association with the British Museum, pp. 76–88.

Pigott, V.C., H.C. Rogers, and S.K. Nash 2003. "Archaeometallurgical Investigations at Tal-e Malyan: The Evidence for Tin-Bronze in the Kaftari Phase." In *Yeki Bud, Yeki Nabud: Essays on the Archaeology of Iran in Honor of William M. Sumner*, ed. N.F. Miller and K. Abdi. Philadelphia: University of Pennsylvania Museum, pp. 161–175.

Possehl, G.L. 1996. "Meluhha." In *The Indian Ocean in Antiquity*, ed. J. Reade. London: Kegan Paul International in association with the British Museum, pp. 133–208.

Possehl, G.L. 1997. "Seafaring Merchants of Meluhha." In *South Asian Archaeology 1995*, ed. F.R. Allchin and B. Allchin. New Delhi: Oxford & IBH Co., pp. 87–100.

Potts, D.T. 1993. "Rethinking Some Aspects of Trade in the Arabian Gulf." *World Archaeology* 24: 423–439.

Potts, D.T. 1999. *The Archaeology of Elam: Formation and Transformation of an Ancient Iranian State*. Cambridge: Cambridge University Press.

Potts, D.T. 2012. "Technology Transfer and Innovation in Ancient Eurasia." In *The Globalization of Knowledge in History*, ed. J. Renn. Berlin: Edition Open Access, pp. 105–123.

Potts, D.T. 2016. *The Archaeology of Elam. Formation and Transformation of an Ancient Iranian State*, 2nd ed. New York: Cambridge University Press.

Potts, T.F. 1993. "Patterns of Trade in Third Millennium B.C. Mesopotamia and Iran." *World Archaeology* 24(3): 379–402.

Potts, T.F. 1994. *Mesopotamia and the East: An Archaeological and Historical Study of Foreign Relations ca. 3400/2000 BC*. Oxford University Committee for Archaeology, Monograph 37. Oxford: Oxbow Books.

Pulak, C. 1988. "The Bronze Age Shipwreck at Ulu Burun, Turkey: 1985 Campaign." *American Journal of Archaeology* 92: 1–37.

Pulak, C. 1997. "The Uluburun Shipwreck." In *Res Maritimae: Cyprus and the Eastern Mediterranean from Prehistory to Late Antiquity*, ed. S. Swiny, R.L. Hohlfelder, and H.W. Swiny. Cyprus American Archaeological Research Institute Monograph Series, vol. 1; American Schools of Oriental Research Archaeological Reports, no. 4. Atlanta, GA: Scholars Press, pp. 233–262.

Pulak, C. 1998. "The Uluburun Shipwreck: An Overview." *International Journal of Nautical Archaeology* 27: 188–224.

Pulak, C. 2000. "The Copper and Tin Ingots from the Late Bronze Age Shipwreck at Uluburun." In *Anatolian Metal*, vol. I, ed. Ü. Yalçin. Der Anschnitt, Beiheft 13. Bochum: Deutsches Bergbau-Museum, pp. 137–157.

Radivojević, M. 2015. "Inventing Metallurgy in Western Eurasia: A Look Through the Microscope Lens." *Cambridge Archaeological Journal* 25: 321–338.

Radivojević, M., J. Pendić, A. Srejić, M. Korać, C. Davey, A. Benzonelli, M. Martinon-Torres, N. Jovanović, and Z. Kamberović 2018. "Experimental Design of the Cu-As-Sn Ternary Colour Diagram." *Journal of Archaeological Science* 90: 106–119.

Radivojević, M. and T. Rehren 2016. "Paint It Black: The Rise of Metallurgy in the Balkans." *Journal of Archaeological Method and Theory* 23: 200–237.

Radivojević, M., T. Rehren, J. Kuzmanović-Cvetković, M. Jovanović, and P. Northover 2013. "Tainted Ores and the Rise of Tin Bronzes in Eurasia c. 6500 Years Ago." *Antiquity* 87: 1030–1045.

Radivojević, M., T. Rehren, E. Pernicka, D. Sljivar, M. Brauns, and D. Borić 2010. "On the Origins of Extractive Metallurgy: New Evidence from Europe." *Journal of Archaeological Science* 37: 2775–2787.

Rahmstorf, L. 2011. "Re-Integrating 'Diffusion': The Spread of Innovations Among the Neolithic and Bronze Age Societies of Europe and the Near East." In *Interweaving Worlds: Systematic Interactions in Eurasia, 7th to the 1st Millennia BCE*, ed. T.C. Wilkinson, S. Sherratt, and J. Bennet. Oxford: Oxbow Books, pp. 100–119.

Rahmstorf, L. 2017. "The Use of Bronze Objects in the 3rd Millennium BC: A Survey between Atlantic and Indus." In *Appropriating Innovations. Entangled Knowledge in Eurasia, 5000–1500 BCE*, ed. P.W. Stockhammer and J. Maran. Oxford: Oxbow Books, pp. 184–210.

Roberts, B.W. 2011. "Ancient Technology and Archaeological Cultures: Understanding The Earliest Metallurgy in Eurasia." In *Investigating Archaeological Cultures*, ed. B.W. Roberts and M. Vander Linden. New York: Springer, pp. 137–150.

Roberts, B.W., C.P. Thornton, and V.C. Pigott 2009. "Development of Metallurgy in Eurasia." *Antiquity* 83(322): 1012–1022.

Rostoker, W., V.C. Pigott, and J.R. Dvorak 1989. "Direct Reduction to Copper Metal by Oxide-Sulfide Interaction." *Archaeomaterials* 3: 69–87.

Rovira, S.Y. and I. Montero 2003. "Natural Tin-Bronze Alloy in Iberian Peninsula Metallurgy: Potentiality and Reality." In *Le problème de l'étain à l'origine de la métallurgie/The Problem of Early Tin*, ed. A. Giumlia-Mair and F. Lo Schiavo. BAR-IS 1199. Oxford: Archaeopress, pp. 15–22.

Rovira, S., I. Montero-Ruiz, and M. Renzi 2009. "Experimental Co-Smelting to Copper-Tin Alloys." In *Metals and Societies: Studies in Honour of Barbara S. Ottaway*, ed. T.L. Kienlin and B.W. Roberts. Universitätsforschungen zur Prähistorischen Archäologie Band 169. Bonn: Dr. Rudolf Habelt, pp. 407–414.

Ruzanov, V.D. 1999. "Zum frühen Auftreten der Zinnbronze in Mittelasien." In *The Beginnings of Metallurgy*, ed. A. Hauptmann, E. Pernicka, T. Rehren, and Ü. Yalçin. Der Anschnitt, Beiheft 9. Bochum: Deutsches Bergbau-Museum, pp. 103–106.
Sayre, E.V., E.C. Joel, M.J. Blackman, K.A. Yener, and H. Özbal 2001. "Stable Lead Isotope Studies of Black Sea Anatolian Ore Sources and Related Bronze Age and Phrygian Artefacts from Nearby Archaeological Sites. Appendix: New Central Taurus Ore Data." *Archaeometry* 43(1): 77–115.
Sayre, E.V., K.A. Yener, E.C. Joel, and L.I. Barnes 1992. "Statistical Evaluation of the Presently Accumulated Lead Isotope Data from Anatolia and Surrounding Regions." *Archaeometry* 34: 73–105.
Sherratt, A. 2006. "The Trans-Eurasian Exchange: the Prehistory of Chinese Relations with the West." In *Contact and Exchange in the Ancient World*, ed. V.H. Mair. Honolulu: University of Hawaii Press, pp. 30–61.
Smith, R. 1996. "An Analysis of the Processes for Tin Smelting." *Bulletin of the Peak District Mines Historical Society* 13(2): 91–99.
Stech, T. 1999. "Aspects of Early Metallurgy in Mesopotamia and Anatolia." In *The Archaeometallurgy of the Asian Old World*, ed. V.C. Pigott. University Museum Monograph 89. MASCA Research Papers in Science and Archaeology, vol. 16. Philadelphia: University Museum, University of Pennsylvania, pp. 59–71.
Stech, T. and V.C. Pigott 1986. "The Metals Trade in Southwest Asia in the Third Millennium BCE." *Iraq* 48: 39–64.
Steinkeller, P. 2016. "The Role of Iran in the Inter-Regional Exchange of Metals: Tin, Copper, Silver and Gold in the Second Half of the Third Millennium BC." In *Ancient Iran: New Perspectives from Archaeology and Cuneiform Studies*, ed. K. Maekawa. Ancient Text Studies in the National Museum, vol. 2. Iran–Japan Project of Ancient Texts. Kyoto: Kyoto University, Center for Eurasian Cultural Studies.
Stockhammer, P.W. and J. Maran (eds.) 2017. *Appropriating Innovations. Entangled Knowledge in Eurasia, 5000–1500 BCE.* Oxford: Oxbow Books, pp. 161–170.
Stöllner, T. and Z. Samašev 2013. *Unbekanntes Kasachstan.* Band I–II. Katalog der Ausstellung des Deutschen Bergbau-Museums vom 26. Bochum: Deutsches Bergbau-Museums.
Stöllner, T., Z. Samašev, S. Berdenov, J. Cierny, M. Doll, J. Garner, A. Gontsharov *et al.* 2011. "Tin from Kazakhstan: Steppe Tin for the West?" In *Anatolian Metal*, vol. V, ed. Ü Yalçin. Der Anschnitt, Beiheft 24. Bochum: Deutsches Bergbau-Museum, pp. 231–251.
Stos-Gale, Z.A., N.H. Gale, G. Bass, C. Pulak, E. Galili, and J. Sharvit 1998. "The Copper and Tin Ingots of the Late Bronze Age Mediterranean: New Scientific Evidence." In *Proceedings of the Fourth International Conference on the Beginning of the Use of Metals and Alloys (*BUMA-IV*)*. Matsue Shimane: Japan Institute of Metals, pp. 115–126.
Tallon, F. 1987. *Métallurgie susienne I: De la fondation de Suse au XVIIIe siècle avant J.C.* Paris: Notes et documents des Musées de France 15.
Thomalsky, J. 2016. "Afghanistan: Ancient Mining and Metallurgy: Initial Project Stage." In *Proceedings of the 9th International Congress on the Archaeology of the Ancient Near East*, vol. 3, ed. R.A. Stucky, O. Kaelin, and H.-P. Mathys. Weisbaden: Harrassowitz, pp. 647–661.
Thomalsky, J., J.B. Braütigam, M. Karaucak, and S. Kraus 2013. "Early Mining and Metal Production in Afghanistan: The First Year of Investigations." *Archäologische Mitteilungen aus Iran und Turan* 45: 199–230.
Thornton, C.P. 2002. "The Domestication of Metal: a Reassessment of the Early Use of Copper Minerals and Metal in Anatolia and Southeastern Europe." MPhil thesis, Cambridge University, Cambridge, UK.

Thornton, C.P. 2007. "Of Brass and Bronze in Prehistoric Southwest Asia." In *Metals and Mines: Studies in Archaeo-Metallurgy*, ed. S. La Niece, D. Hook, and P. Craddock. London: Archetype Press in association with the British Museum, pp. 123–135.

Thornton, C.P. 2014. "The Emergence of Complex Metallurgy on the Iranian Plateau: Escaping the Levantine Paradigm." In *Archaeometallurgy in Global Perspective: Methods and Syntheses*, ed. B.W. Roberts and C.P. Thornton. New York: Springer, pp. 529–557.

Thornton, C.P. and C. Giardino 2012. "Serge Cleuziou and the 'Tin Problem.'" In *Aux Marges de l'archéologie: Hommage à Serge Cleuziou*, ed. J. Giraud and G. Gernez. Paris: De Boccard, Travaux de la Maison René Ginouvès, pp. 253–260.

Thornton, C.P., C.C. Lamberg-Karlovsky, M. Liezers, and S.M.M. Young 2005. "Stech and Pigott Revisited: New Evidence Concerning the Origins of Tin Bronze in the Light of Chemical and Metallographic Analyses of the Metal Artefacts from Tepe Yahya, Iran." In *Proceedings of the 33rd International Symposium on Archaeometry, Amsterdam*, ed. H. Kars and E. Burke. Geoarchaeological and Bioarchaeological Studies 3. Amsterdam: VU University of Amsterdam, pp. 395–398.

Weeks, L. 2002. "Appendix: Summary of Velikent Compositional and Lead Isotope Analysis." In *Die Anfänge der Metallurgie in der Alten Welt*, ed. M. Barthelheim, E. Pernicka, and R. Krause. Rahden/Westf: Marie Leidorf, pp. 178–183.

Weeks, L. 2003. *Early Metallurgy of the Persian Gulf*. Leiden: Brill Academic.

White, J.C. and E.G. Hamilton 2009. "The Transmission of Early Bronze Technology to Thailand: New Perspectives." *Journal of World Prehistory* 22(4): 357–397.

Wilkinson, T.C. 2014. *Tying the Threads of Eurasia: Trans-Regional Routes and Material Flows in Transcaucasia, Eastern Anatolia and Western Central Asia, c.3000–1500 BC*. Leiden: Sidestone Press.

Yalçin, Ü. 2005. "Ein Schiff Macht Geschichte." In *Das Schiff von Uluburun: Welthandel vor 3000 Jahren*, ed. Ü. Yalçin, C. Pulak, and R. Slotta. Bochum: Deutsches Bergbau-Museums, pp. 19–24.

Yalçin, Ü. and H. Özbal 2009. "Ein Neues Zinnvorkommen in Kayseri-Hisarcık, Zentralanatolien: Ein Vorbericht." *TUBA-AR* 12: 117–122.

Yalçin, Ü., C. Pulak, and R. Slotta (eds.) 2005. *Das Schiff von Uluburun: Welthandel vor 3000 Jahren*. Bochum: Deutsches Bergbau-Museums.

Yazgan, E. 2015. "Cassiterite (Tin) Mineralization Related with Erciyes Volcanic Activities and the Mode of Formation of the Hematite-Cassiterite-Yazganite-Tridymite Paragenesis and its Implication for Bronze Alloys." In *Proceedings of the 1st Kültepe International Meeting. Studies Dedicated to Kutlu Emre*, ed. F. Kulakoğlu and C. Michel. Subartu 35. Turnhout: Brepols, pp. 183–193.

Yener, K.A. 2000. *The Domestication of Metals: The Rise of Complex Metal Industries in Anatolia*. Leiden: Brill.

Yener, K.A. 2008. "Revisiting Kestel Mine and Göltepe: The Dynamics of Local Provisioning of Tin during the Early Bronze Age." In *Ancient Mining in Turkey and the Eastern Mediterranean*, ed. Ü. Yalçın, H. Özbal, and A.G. Paşamehmetoğlu. Atılım University Turkey Historical Research Applications and Research Center Publications 2. Ankara: Atılım University, pp. 57–64.

Yener, K.A. in press. *Tin Production at Göltepe: Excavations at an Early Bronze Age Mining Town in the Central Taurus Mountains, Turkey*, vol. 1, parts 1 and 2. Philadelphia: Institute for Aegean Prehistory Academic Press.

Yener, K.A., F. Kulakoğlu, E. Yazgan, R. Kontani, Y.S. Hayakawa, J.W. Lehner, G. Dardeniz *et al.* 2015. "New Tin Mines and Production Sites near Kültepe in Turkey: A Third-Millennium BC Highland Production Model." *Antiquity* 89(345): 596–612.

Yener, K.A., E.V. Sayre, E. Joel, H. Özbal, I.L. Barnes, and R.H. Brill 1991. "Stable Lead Isotope Studies of Central Taurus Ore Sources and Related Artifacts from the Eastern Mediterranean Chalcolithic and Bronze Age Sites." *Journal of Archaeological Science* 18: 541–577.
Yoffee, N. and G. Barjamovic 2018. "Old Assyrian Trade and Economic History." In *Hans Neumann Festschrift: Grenzüberschreitungen. Studien zur Kulturgeschichte des Alten Orients*, ed. K. Kleber, G. Neumann, and S. Paulus. Münster: Zaphon, pp. 815–824.
Young, W.I. 1972. "The Fabulous Gold of the Pactolos Valley." *Bulletin of the Museum of Fine Arts* 70(359): 5–13.
Zettler, R.L. and L. Horne (eds.) 1998. *Treasures from the Royal Tombs of Ur.* Seattle, WA: Marquand Books.

APPENDIX

RADIOCARBON DATES RELATED TO THE BMAC/OXUS CIVILIZATION

Michel Fontugne, Ganna I. Zajtseva, Bertille Lyonnet, Nadezhda A. Dubova, and Natalja D. Burova

Conventional ^{14}C ages are expressed in years before present (BP), with 1σ error. These ages were calibrated using OxCal v4.3n (Bronk Ramsey and Lee 2013): r5 IntCal 13 atmospheric curve (Reimer *et al.* 2013). The calibrated dates are given with a level of confidence of 95.4% (2σ range). Many more radiocarbon dates have been provided for some of the sites mentioned here (except for Gonur Depe, where we give all of them), or for other sites that we mention in the book. However, not all could be (re) calibrated because of a lack of part of the information. Moreover, for sites excavated long ago, it is sometimes difficult to be certain of the provenance, stratigraphy, and/ or the analyses, and we preferred not to take those into consideration.

Lab. reference	*Site name*	*Provenience*	*Date BP*	*Material*	*Publication*	*Median*	*Cal. BC 1σ (68.2%)*	*Cal. BC 2σ (95.4%)*
Margiana								
Hel-4619	Gonur, Large Necropolis	Necropolis, grave 2283	3460 ± 100	Charcoal	Jungner 2004	1783	1897–1645	2031–1527
Hel-672	Gonur, Large Necropolis	Necropolis, grave 2304	3585 ± 45	Charcoal	Jungner 2004	1940	2017–1886	2119–1774
PSUAMS-2066	Gonur, Large Necropolis	Necropolis, grave 1340	3630 ± 30	Human tooth	Narasimhan *et al.* 2019	1992	2030–1948	2127–1905
Hel-673	Gonur, Large Necropolis	Necropolis, grave 2315	3690 ± 45	Charcoal	Jungner 2004	2081	2141–1984	2201–1951
PSUAMS-2804	Gonur, Large Necropolis	Necropolis, grave 1899	3705 ± 20	Human tooth	Narasimhan *et al.* 2019	2085	2137–2041	2193–2031
PSUAMS-2310	Gonur, Large Necropolis	Necropolis, grave 1315	3680 ± 20	Human tooth	Narasimhan *et al.* 2019	2086	2131–2029	2139–1981
PSUAMS-2335	Gonur, Large Necropolis	Necropolis, grave 1506	3715 ± 20	Human tooth	Narasimhan *et al.* 2019	2087	2187–2042	2196–2034
PSUAMS-2316	Gonur, Large Necropolis	Necropolis, grave 1311	3720 ± 20	Human tooth	Narasimhan *et al.* 2019	2092	2192–2045	2198–2036
Poz-83485	Gonur, Large Necropolis	Necropolis, grave 2380	3720 ± 30	Human bone	Narasimhan *et al.* 2019	2106	2194–2041	2201–2031
PSUAMS-2806	Gonur, Large Necropolis	Necropolis, grave 1300	3770 ± 20	Human tooth	Narasimhan *et al.* 2019	2179	2269–2142	2285–2135
Hel-675	Gonur, Large Necropolis	Necropolis, grave 2041	3810 ± 40	Charcoal	Jungner 2004	2253	2334–2149	2456–2137
Hel-4634	Gonur, Large Necropolis	Necropolis, grave 95	3810 ± 80	Charcoal	Jungner 2004	2260	2436–2139	2471–2032
GX-20645	Gonur, Large Necropolis	Necropolis, grave 4	3880 ± 95	?	Kaniuth 2006: 50	2351	2473–2206	2619–2039

Le-9120	Gonur North	"Royal Necropolis," grave 3915, from the pit above the grave	3160 ± 320	Charcoal	Zajtseva *et al.* forthcoming	1440	1871–1006	2397–595
Le-7998	Gonur North	"Royal Necropolis," grave 3210	3470 ± 110	Charcoal	Zajtseva *et al.* 2008	1797	1926–1645	2121–1512
Le-9970	Gonur North	"Royal Necropolis," grave 3915	3630 ± 55	Charcoal	Zajtseva *et al.* forthcoming	1999	2121–1919	2195–1786
UBA-7888	Gonur North	"Royal Necropolis," grave 3245	3691 ± 42	Human bone	Zajtseva *et al.* 2008	2082	2141–2023	2200–1957
Poz-83487	Gonur North	"Royal Necropolis," grave 3240(9)	3840 ± 35	Human tooth	Narasimhan *et al.* 2019	2302	2397–2207	2458–2202
Le-8782	Gonur North	"Royal Necropolis," grave 3900	3940 ± 240	Animal (dog) bones	Zajtseva *et al.* forthcoming	2443	2863–2136	3091–1768
Le-2408	Gonur North	Kremlin, room 5 (= period 1 according to Hiebert)	3510 ± 40	Charcoal	Hiebert 1994: 78–79; Zajtseva *et al.* 2008	1830	1890–1771	1941–1700
Le-9470	Gonur North	Kremlin, southeastern tower, fire layer	3570 ± 70	Charcoal	Zajtseva *et al.* forthcoming	1919	2024–1779	2134–1700

(*continued*)

Lab. reference	*Site name*	*Provenience*	*Date BP*	*Material*	*Publication*	*Median*	*Cal. BC 1σ (68.2%)*	*Cal. BC 2σ (95.4%)*
Le-9468	Gonur North	Kremlin, southern wall, fire layer	3570 ± 100	Charcoal	Zajtseva *et al.* forthcoming	1922	2035–1760	2200–1664
Le-1207	Gonur North	Kremlin, jarus 1 (= period 1 according to Hiebert)	3580 ± 70	Charcoal	Hiebert 1994: 78–79; Zajtseva *et al.* 2008	1933	2031–1781	2136–1746
Le-9465	Gonur North	Kremlin, northern wall, fire layer, 4th pilaster from the east	3600 ± 65	Charcoal	Zajtseva *et al.* forthcoming	1961	2117–1881	2139–1770
Le-9471	Gonur North	Kremlin, eastern wall, fire layer. Upper part of the surface under the 12th pilaster from the north	3650 ± 45	Charcoal	Zajtseva *et al.* forthcoming	2023	2126–1951	2141–1900
Le-9977	Gonur North	Kremlin, 13th tower, oven	3710 ± 45	Charcoal	Zajtseva *et al.* forthcoming	2099	2194–2033	2275–1961
Le-9978	Gonur North	Kremlin, 2nd tower, oven	3765 ± 45	Charcoal	Zajtseva *et al.* forthcoming	2182	2281–2062	2336–2033
Le-9973	Gonur North	Kremlin, 16th tower, oven	3780 ± 60	Charcoal	Zajtseva *et al.* forthcoming	2212	2298–2056	2456–2032

Le-9980	Gonur North	Kremlin, under the 2nd buttress to the left of the main entrance	3780 ± 60	Charcoal	Zajtseva *et al.* forthcoming	2212	2298–2056	2456–2032
Le-9974	Gonur North	Kremlin, 10th tower, oven	3800 ± 110	Charcoal	Zajtseva *et al.* forthcoming	2245	2455–2056	2566–1941
Le-9467	Gonur North	Kremlin, eastern wall, fire layer lower part of the surface under the 12th pilaster from the north	3860 ± 100	Charcoal	Zajtseva *et al.* forthcoming	2325	2469–2201	2581–2027
Le-9466	Gonur North	Kremlin, northern wall, fire layer, 4th pilaster (to the south from the entrance)	3870 ± 70	Charcoal	Zajtseva *et al.* forthcoming	2343	2463–2233	2561–2141
Beta-35125	Gonur North	Northeast ext. kremlin, room 2, locus 45 under floor = period 1	3630 ± 90	Charcoal	Hiebert 1994: 78–79; Zajtseva *et al.* 2008	2003	2135–1890	2280–1747
Le-8639	Gonur North	Palace, room 54, ritual pit 2	3360 ± 50	Charcoal	Zajtseva *et al.* forthcoming	1652	1738–1565	1762–1511
Beta-35356	Gonur North	Palace, fireplace in wall	3470 ± 100	Charcoal	Hiebert 1994: 78–79; Zajtseva *et al.* 2008	1796	1914–1664	2109–1528

(continued)

Lab. reference	*Site name*	*Provenience*	*Date BP*	*Material*	*Publication*	*Median*	*Cal. BC 1σ (68.2%)*	*Cal. BC 2σ (95.4%)*
Hel-4628	Gonur North	Palace, room 189	3670 ± 60	Charcoal	Jungner 2004	2055	2137–1971	2206–1890
Hel-4620	Gonur North	Palace, int. hall, sup. part of north wall	3670 ± 80	Charcoal	Jungner 2004	2057	2194–1942	2293–1781
Beta-33562	Gonur North	Palace, room 48, pit under the wall (before the construction)	3700 ± 60	Charcoal	Hiebert 1994: 78–79; Zajtseva *et al.* 2008	2091	2197–1983	2284–1928
Le-8641	Gonur North	Palace, room 54, ritual pit 1	3710 ± 110	Charcoal	Zajtseva *et al.* forthcoming	2117	2282–1949	2465–1782
Le-9114	Gonur North	Palace, room 189 (niche with "apples")	3740 ± 90	Charcoal	Zajtseva *et al.* forthcoming	2156	2290–2024	2458–1936
Le-2411	Gonur North	Palace, room 1	4200 ± 40	Tree	Zajtseva *et al.* 2008	2781	2889–2701	2900–2639
Le-2409	Gonur North	Palace, room 1	4290 ± 40	Tree	Zajtseva *et al.* 2008	2907	2928–2879	3022–2779
Beta-33563	Gonur North	Palace, room 2, hearth on floor	4330 ± 70	Charcoal	Hiebert 1994: 78–79; Zajtseva *et al.* 2008	2976	3082–2888	3327–2709
PSUAMS-2065	Gonur North	Area 5, grave 3012	3550 ± 30	Human tooth	Narasimhan *et al.* 2019	1897	1945–1784	2009–1772

Hel-4629	Gonur North	Area 5, grave 2900	3350 ± 70	Charcoal	Jungner 2004	1641	1737–1533	1876–1457
UBA-7993	Gonur North	Area 5, grave 2900	3565 ± 42	Human bone, ind. 2	Zajtseva *et al.* 2008	1914	2009–1829	2028–1772
Hel-4627	Gonur North	Area 5, room 70, under animal burial of grave 2900	3580 ± 70	Charcoal	Jungner 2004	1933	2031–1781	2136–1746
UBA-7994	Gonur North	Area 5, grave 2900	3609 ± 48	Human bone, ind. 1	Zajtseva *et al.* 2008	1971	2029–1907	2136–1783
PSUAMS-2315	Gonur North	Area 5, grave 2871	3670 ± 20	Human tooth	Narasimhan *et al.* 2019	2064	2129–1984	2136–1977
Hel-4671	Gonur North	Area 5, grave 2953	3695 ± 45	Charcoal	Jungner 2004	2085	2190–2026	2203–1954
Poz-83484	Gonur North	Area 5, grave 3454	3725 ± 35	Human tooth	Narasimhan *et al.* 2019	2120	2197–2042	2275–2024
PSUAMS-2314	Gonur North	Area 5, grave 3007	3735 ± 20	Human tooth	Narasimhan *et al.* 2019	2143	2198–2058	2203–2041
Hel-4674	Gonur North	Area 5, grave 2922	3810 ± 40	Charcoal	Jungner 2004	2253	2334–2149	2456–2137
Hel-4625	Gonur North	Area 5, room 87, oven	3460 ± 80	Charcoal	Jungner 2004	1782	1886–1686	2014–1561
Hel-4626	Gonur North	Area 5, room 84 (Trudy 2, 2008) (or room 77 (Jungner 2004)	3720 ± 70	Charcoal	Jungner 2004	2121	2271–1984	2340–1921

(*continued*)

Lab. reference	*Site name*	*Provenience*	*Date BP*	*Material*	*Publication*	*Median*	*Cal. BC 1σ (68.2%)*	*Cal. BC 2σ (95.4%)*
Hel-4616	Gonur North	Area 5, room 78	3810 ± 70	Charcoal	Jungner 2004	2260	2400–2140	2467–2041
Hel-4621	Gonur North	Area 5, room 84, oven	3870 ± 80	Charcoal	Jungner 2004	2340	2465–2211	2570–2061
Hel-4617	Gonur North	Area 5, room 99, near grave 2938, in wall of hearth	3950 ± 80	Charcoal	Jungner 2004	2447	2571–2309	2838–2200
PSUAMS-2317	Gonur North	Area 6, grave 3042	3645 ± 20	Human tooth	Narasimhan *et al.* 2019	2006	2034–1966	2125–1945
PSUAMS-2309	Gonur North	Area 6, grave 3049	3785 ± 20	Human tooth	Narasimhan *et al.* 2019	2213	2278–2147	2288–2142
Le-9119	Gonur North	Area 6, room 20, first period	3550 ± 100	Charcoal	Zajtseva *et al.* forthcoming	1894	2023–1752	2194–1637
Le-9117	Gonur North	Area 6, room 11/19	3590 ± 120	Charcoal	Zajtseva *et al.* forthcoming	1953	2132–1771	2288–1638
Le-9118	Gonur North	Area 6, room 3969	3710 ± 90	Charcoal	Zajtseva *et al.* forthcoming	2112	2273–1965	2454–1884
PSUAMS-2313	Gonur North	Area 8, grave 3201	3580 ± 20	Human bone	Narasimhan *et al.* 2019	1930	1950–1896	2011–1886
Le-9121	Gonur North	Area 8, in the center, grave 3912	3720 ± 100	Charcoal	Zajtseva *et al.* forthcoming	2129	2284–1974	2458–1889
UBA-8016	Gonur North	Area 8, grave 3280	3768 ± 45	Charcoal	Zajtseva *et al.* 2008	2187	2284–2065	2338–2034

Le-9983	Gonur North	Area 8, accumulation of pot shards and stones, destroyed (washed) oven in the northern wall, center	2370 ± 50	Charcoal	Zajtseva *et al.* forthcoming	470	536–392	751–363
Le-8008	Gonur North	Area 8, east part, room 48	3320 ± 30	Charcoal	Zajtseva *et al.* 2008	1596	1637–1534	1683–1521
Le-9982	Gonur North	Area 8, accumulation of pot shards and stones, destroyed (washed) oven in the northern wall	3470 ± 60	Charcoal	Zajtseva *et al.* forthcoming	1795	1881–1698	1936–1641
UBA-8011	Gonur North	Area 8, room 48	3555 ± 43	Charcoal	Zajtseva *et al.* 2008	1900	1957–1780	2023–1766
UBA-8017	Gonur North	Area 9, east part, grave 3271	3770 ± 40	Charcoal	Zajtseva *et al.* 2008	2190	2282–2137	2335–2037
Le-8011	Gonur North	Area 9, room 112	3290 ± 60	Charcoal	Zajtseva *et al.* 2008	1571	1632–1501	1730–1438
Le-8010	Gonur North	Area 9, room 108	3500 ± 50	Charcoal	Zajtseva *et al.* 2008	1823	1886–1756	1947–1691
Le-8009	Gonur North	Area 9, room 109	3510 ± 120	Charcoal	Zajtseva *et al.* 2008	1845	2016–1687	2194–1531

(*continued*)

Lab. reference	*Site name*	*Provenience*	*Date BP*	*Material*	*Publication*	*Median*	*Cal. BC 1σ (68.2%)*	*Cal. BC 2σ (95.4%)*
PSUAMS-2313	Gonur North	Area 10, grave 3466	3650 ± 20	Human tooth	Narasimhan *et al.* 2019	2015	2113–1976	2130–1948
Poz-83491	Gonur North	Area 10, grave 3465	3815 ± 35	Human tooth	Narasimhan *et al.* 2019	2257	2333–2155	2452–2140
Hel-4623	Gonur North	Area 10, encircling wall, near grave 2951	3720 ± 70	Charcoal	Jungner 2004	2121	2271–1984	2340–1921
UBA-8014	Gonur North	Area 10, room 319	3774 ± 46	Charcoal	Zajtseva *et al.* 2008	2198	2287–2137	2344–2035
Hel-4622	Gonur North	Area 10, encircling wall, oven	3820 ± 80	Charcoal	Jungner 2004	2274	2452–2144	2474–2035
Hel-4618	Gonur North	Area 10, encircling wall, oven	3930 ± 70	Charcoal	Jungner 2004	2413	2559–2300	2619–2202
Hel-4624	Gonur North	Area 10, encircling wall, oven	4000 ± 90	Charcoal	Jungner 2004	2534	2834–2348	2871–2235
PSUAMS-2311	Gonur North	Area 12, grave 4258	3660 ± 20	Human tooth	Narasimhan *et al.* 2019	2033	2121–1979	2134–1957
PSUAMS-2312	Gonur North	Area 12, grave 4384	3700 ± 20	Human tooth	Narasimhan *et al.* 2019	2086	2135–2039	2190–2029
Poz-83486	Gonur North	Area 12, grave 4329	3735 ± 35	Human tooth	Narasimhan *et al.* 2019	2139	2200–2046	2277–2030

UBA-8015	Gonur North	Area 12, grave 3361	3787 ± 37	Charcoal	Zajtseva *et al.* 2008	2220	2286–2144	2343–2049
Le-9986	Gonur North	Area 12, in the layer with grave 4180	3940 ± 60	Charcoal	Zajtseva *et al.* forthcoming	2429	2563–2343	2581–2210
PSUAMS-2152	Gonur North	Area 13, grave 3453	3680 ± 20	Human tooth	Narasimhan *et al.* 2019	2086	2131–2029	2139–1981
UBA-8013	Gonur North	Area 14, grave 3551	3548 ± 49	Charcoal	Zajtseva *et al.* 2008	1889	1953–1776	2022–1750
Le-8196	Gonur North	Area 16, grave 3732	3250 ± 20	Charcoal	Zajtseva *et al.* 2008	1520	1600–1497	1610–1454
UBA-8012	Gonur North	Area 16, grave 3718	3517 ± 41	Charcoal	Zajtseva *et al.* 2008	1836	1897–1771	1950–1700
Le-8640	Gonur North	Area 16, grave 3828	3560 ± 70	Charcoal	Zajtseva *et al.* forthcoming	1905	2016–1775	2132–1696
Le-9125	Gonur North	Area 16, grave 3740	3590 ± 120	Charcoal	Zajtseva *et al.* forthcoming	1953	2132–1771	2288–1638
UBA-8019	Gonur North	Area 16, grave 3652	3601 ± 37	Charcoal	Zajtseva *et al.* 2008	1960	2019–1912	2121–1880
Le-8778	Gonur North	Area 16, grave 3854	3700 ± 200	Charcoal	Zajtseva *et al.* forthcoming	2118	2455–1831	2836–1564
Le-8003	Gonur North	Area 16, room 8, upper layer, in pit in the floor (50 cm deep)	3360 ± 60	Charcoal	Zajtseva *et al.* forthcoming	1652	1741–1546	1871–1502
Le-8195	Gonur North	Area 16, preliminary sounding to the east of the temple	3420 ± 60	Charcoal	Zajtseva *et al.* 2008	1728	1869–1637	1892–1560

(*continued*)

Lab. reference	*Site name*	*Provenience*	*Date BP*	*Material*	*Publication*	*Median*	*Cal. BC 1σ (68.2%)*	*Cal. BC 2σ (95.4%)*
Le-8194	Gonur North	Area 16, room 101, inf. level	3530 ± 30	Tree	Zajtseva *et al.* 2008	1847	1917–1777	1943–1763
UBA-8018	Gonur North	Area 16, room 85	3586 ± 46	Charcoal	Zajtseva *et al.* 2008	1942	2018–1887	2120–1774
Le-8002	Gonur North	Area 16, room 2, 30 cm above grave 3640	3960 ± 200	Charcoal	Zajtseva *et al.* forthcoming	2469	2857–2201	3011–1901
Le-8004	Gonur North	Area 16, room 48	3960 ± 160	Charcoal	Zajtseva *et al.* forthcoming	2470	2836–2206	2890–2036
Le-9123	Gonur North	Area 17, grave 3935	3510 ± 120	Charcoal	Zajtseva *et al.* forthcoming	1845	2016–1687	2194–1531
Le-8779	Gonur North	Area 17, from the black ash layer	3620 ± 100	Charcoal	Zajtseva *et al.* forthcoming	1991	2140–1786	2286–1696
Le-8780	Gonur North	Area 17, from the gray ash layer	3720 ± 80	Charcoal	Zajtseva *et al.* forthcoming	2124	2275–1981	2430–1896
Le-9691	Gonur North	Area 18, grave 4015	3220 ± 280	Charcoal	Zajtseva *et al.* forthcoming	1508	1878–1128	2268–821
Le-9686	Gonur North	Area 18, grave 4140	3650 ± 45	Charcoal	Zajtseva *et al.* forthcoming	2023	2126–1951	2141–1900
Le-9115	Gonur North	Area 18, room 54, from the oven	3370 ± 70	Charcoal	Zajtseva *et al.* forthcoming	1665	1748–1546	1878–1503

Le-9124	Gonur North	Area 18, room 67, from the wall near the oven	3480 ± 100	Charcoal	Zajtseva *et al.* forthcoming	1807	1932–1683	2118–1532
GN-2012/1[a]	Gonur North	Area 18, northern part	3695 ± 30	Charcoal	Sarianidi *et al.* 2014: 137	2086	2134–2036	2196–1980
Poz-83490	Gonur North	Area 19, grave 4290	3605 ± 35	Charcoal	Sarianidi *et al.* 2014: 137	1964	2021–1918	2118–1883
GN-2012/3	Gonur North	Area 19, grave 4148	3670 ± 40	Charcoal	Sarianidi *et al.* 2014: 137	2054	2134–1979	2195–1939
GN-2012/7	Gonur North	Area 19, grave 4149	3670 ± 35	Charcoal	Sarianidi *et al.* 2014: 137	2055	2132–1981	2190–1946
GN-2012/6	Gonur North	Area 19, grave 4155	3675 ± 35	Charcoal	Sarianidi *et al.* 2014: 137	2064	2134–1982	2192–1952
GN-2012/2	Gonur North	Area 19, under grave 4164	3685 ± 30	Charcoal	Sarianidi *et al.* 2014: 137	2080	2134–2029	2194–1972
GN-2012/5	Gonur North	Area 19, room 26	3670 ± 35	Charcoal	Sarianidi *et al.* 2014: 137	2055	2132–1981	2190–1946
GN-2012/4	Gonur North	Area 19, room 19	3685 ± 35	Charcoal	Sarianidi *et al.* 2014: 137	2078	2135–2028	2196–1960
Le-9988	Gonur North	Area 22, grave 4185, in the dromos, +50 cm above the level of the burial floor	3320 ± 75	Charcoal	Zajtseva *et al.* forthcoming	1605	1686–1511	1862–1433
Le-9987	Gonur North	Area 22, oven near burial 4185	3670 ± 80	Charcoal	Zajtseva *et al.* forthcoming	2057	2194–1942	2293–1781

(continued)

Lab. reference	Site name	Provenience	Date BP	Material	Publication	Median	Cal. BC 1σ (68.2%)	Cal. BC 2σ (95.4%)
Beta-33559	Gonur North, outside Kremlin	Deep sounding, hearth upon virgin soil	3310 ± 80	Too little carbon	Hiebert 1994: 80	1595	1685–1504	1685–1421
Beta-33561	Gonur North, outside Kremlin	Deep sounding, layer 6 (1989)	3520 ± 60	Charcoal	Hiebert 1994: 78–79	1843	1921–1762	2020–1692
Beta-33560	Gonur North, outside Kremlin	Deep sounding, layer 3 = period 1 (1988)	3580 ± 60	Charcoal	Hiebert 1994: 78–79	1933	2028–1786	2131–1751
Hel-2970	Gonur South	Temenos, room 266, upper level deposit with Takhirbaj ceram	3380 ± 90	Charcoal	Sarianidi 1993	1681	1860–1534	1905–1456
Hel-2967	Gonur South	Temenos, room 178, upper level floor with Takhirbaj ceram	3380 ± 110	Charcoal	Sarianidi 1993	1685	1871–1530	1948–1433
Hel-2966	Gonur South	Temenos, room 226B, phase 1	3410 ± 80	Charcoal	Sarianidi 1993	1719	1875–1618	1914–1517
Hel-2969	Gonur South	Temenos, room 208, chimney, phase 1	3480 ± 90	Charcoal	Sarianidi 1993	1806	1914–1689	2035–1546
Hel-2963	Gonur South	Temenos, room 200, fireplace, phase 1	3540 ± 80	Charcoal	Sarianidi 1993	1878	1971–1754	2132–1667

Hel-2968	Gonur South	Temenos, room 134, small pit beneath arch	3600 ± 80	Charcoal	Sarianidi 1993	1961	2127–1784	2197–1745
GX-19834	Gonur South	At the base of room 629, in alluvial deposit	3608 ± 75		Cremaschi 1998	1972	2129–1881	2196–1754
Hel-2964	Gonur South	Temenos, room 65, small pit beneath the architecture	3750 ± 90	Charcoal	Sarianidi 1993	2171	2296–2026	2459–1946
Hel-2965	Gonur South	Temenos, room 226A, hearth, phase 1	3750 ± 90	Charcoal	Sarianidi 1993	2171	2296–2026	2459–1946
Beta-33558	Gonur South	Hearth on floor	4360 ± 60	Contaminated with coal or bitumen?	Hiebert 1994: 80	2995	3083–2905	3324–2883
Le-9475	Gonur 20	Grave 58	3390 ± 80	Charcoal	Zajtseva *et al.* forthcoming	1692	1869–1561	1891–1503
Le-9968	Gonur 20	Area 2, grave 40	3460 ± 90	Charcoal	Zajtseva *et al.* forthcoming	1782	1890–1665	2021–1534
Le-9122	Gonur 20	Area 2, grave 45	3540 ± 120	Charcoal	Zajtseva *et al.* forthcoming	1885	2032–1696	2267–1547
GN-65	Gonur 20	Area 2, grave 65	3610 ± 25	Charcoal	Zajtseva *et al.* forthcoming	1969	2020–1935	2031–1897
Le-9966	Gonur 20	Area 2, room 22 (pit in the center of the room)	3560 ± 65	Charcoal	Zajtseva *et al.* forthcoming	1905	2015–1776	2126–1699

(continued)

Lab. reference	*Site name*	*Provenience*	*Date BP*	*Material*	*Publication*	*Median*	*Cal. BC 1σ (68.2%)*	*Cal. BC 2σ (95.4%)*
Le-9474	Gonur 21	Grave 8	3450 ± 140	Charcoal	Zajtseva *et al.* forthcoming	1777	1949–1566	2137–1447
Le-9473	Gonur 21	Grave 7	3500 ± 110	Charcoal	Zajtseva *et al.* forthcoming	1831	1965–1686	2136–1535
GX-20858	Site M9 between Gonur and Kelyadji	Under Late Bronze Age buried soil, in alluvial sand and silt from carbonized swamp plants	4017 ± 51	Charcoal	Cremaschi 1998	2543	2617–2472	2853–2351
SC-532	Adji Kui 1	Middle levels, hearth SU 12	3135 ± 110	Charcoal	Salvatori 2002	1391	1521–1235	1663–1092
SC-531	Adji Kui 1	Middle levels, hearth SU 10a	3680 ± 85	Charcoal	Salvatori 2002	2072	2198–1946	2339–1780
SC-534	Adji Kui 9	Phase 3b, SU 9	3710 ± 60	Charcoal	Salvatori 2002	2103	2200–2026	2287–1944
SC-533	Adji Kui 9	Phase 3a, SU 7	3810 ± 60	Charcoal	Salvatori 2002	2259	2390–2142	2463–2051
SC-535c	Adji Kui 9	Phase 2, SU 17a	4420 ± 110	Charcoal	Salvatori 2002	3107	3326–2920	3496–2872
?	Adji Kui 9	Level 1	3785 ± 35		Rossi Osmida 2011: 294	2215	2284–2144	2339–2051
?	Adji Kui 9	Level 2	3795 ± 35		Rossi Osmida 2011: 294	2231	2287–2151	2397–2061
?	Adji Kui 9	Level 3	3800 ± 35		Rossi Osmida 2011: 294	2239	2290–2151	2429–2065

?	Adji Kui (1?)	?	3760 ± 10	Salvatori 2010	2169	2200–2143	2205–2138
Le-2678	Togolok 21	Room 60	3270 ± 40	Sarianidi 1990: 73	1552	1610–1504	1632–1449
Beta-33564	Togolok 21	Large altar, under 2nd floor	3470 ± 100	Hiebert 1994: 79	1783	1897–1645	2031–1527
Le-2683	Togolok 21	Room 64	3540 ± 40	Sarianidi 1990: 73	1877	1941–1777	2009–1751
Le-2685	Togolok 21	Room 47	3540 ± 40	Sarianidi 1990: 73	1877	1941–1777	2009–1751
Le-2684	Togolok 21	Room 57	4270 ± 40	Sarianidi 1990: 73	2895	2919–2876	3011–2705
Le-2681	Togolok 21	Room 44	4610 ± 50	Sarianidi 1990: 73	3413	3509–3197	3622–3112
Le-2679	Togolok 21	Hearth	4620 ± 40	Sarianidi 1990: 73	3449	3498–3356	3520–3138
Le-2682	Togolok 21	Room 61	5060 ± 40	Sarianidi 1990: 73	3867	3942–3800	3963–3766
Bo-0000	Takhirbaj 1	SU 125	3311 ± 75	Cattani 2008: 147	1595	1682–1506	1768–1428
GX-20648	Takhirbaj 1	A5, SU 125	3375 ± 80	Cattani 2008: 147	1673	1756–1534	1886–1498
GX-30713	Murghab site 1219	SU 4	2960 ± 140	Cattani 2008: 147	1175	1385–1006	1496– 841
GX-30714	Murghab site 1219	SU 15	3300 ± 40	Cattani 2008: 147	1576	1623–1528	1684– 1498

(*continued*)

Lab. reference	Site name	Provenience	Date BP	Material	Publication	Median	Cal. BC 1σ (68.2%)	Cal. BC 2σ (95.4%)
GX-30715	Murghab site 1211	SU 71	3140 ± 50		Cattani 2008: 147	1411	1495–1309	1508– 1276
GX-19833	Murghab site 55	In bioturbated sand with a lens of charcoal and burnt bones under the alluvial deposits with Bronze Age pot sherds	3880 ± 125		Cremaschi 1998	2352	2562–2146	2851– 1979
Beta-280705	Ojakly	Kiln fragments	3260 ± 40		Rouse and Cerasetti 2014	1540	1610–1499	1624–1445
OS-92543	Ojakly	Kiln ashy fill	3270 ± 25		Rouse and Cerasetti 2014	1552	1608–1506	1617–1498
OS-92546	Ojakly	Refuse fill	3280 ± 30		Rouse and Cerasetti 2014	1561	1610–1519	1630–1497
OS-92536	Ojakly	Refuse fill	3290 ± 25		Rouse and Cerasetti 2014	1568	1611–1531	1623–1507
OS-92542	Ojakly	Kiln fill	3310 ± 25		Rouse and Cerasetti 2014	1579	1623–1534	1658–1516
OS-93540	Ojakly	Kiln fill	3310 ± 25		Rouse and Cerasetti 2014	1579	1623–1534	1658–1516
OS-92539	Ojakly	Kiln fill	3310 ± 25		Rouse and Cerasetti 2014	1579	1623–1534	1658–1516

AA-90389	Ojakly	Kiln fill	3329 ± 38		Rouse and Cerasetti 2014	1613	1662–1535	1729–1511
OS-87205	Ojakly	Refuse fill	3340 ± 30		Rouse and Cerasetti 2014	1630	1683–1565	1729–1531
AA-90370	Ojakly	Ashy fill	3352 ± 37		Rouse and Cerasetti 2014	1644	1691–1611	1741–1531
OS-92537	Ojakly	Fill above floor	3370 ± 25		Rouse and Cerasetti 2014	1663	1688–1631	1740–1614
OS-92541	Ojakly	Pit 23	3370 ± 25	Millet grain	Rouse and Cerasetti 2014	1663	1688–1631	1740–1614
OS-92538	Ojakly	Fill above floor	3490 ± 25		Rouse and Cerasetti 2014	1816	1878–1770	1887–1745
Turkmenistan								
R-1305	Namazga Depe	NMG V	3380 ± 140		Hiebert 1994: 82–83	1691	1878–1517	2117–1322
R-1309	Namazga Depe	NMG V	3550 ± 40		Hiebert 1994: 82–83	1893	1949–1780	2016–1756
R-1306	Namazga Depe	NMG V	3760 ± 50		Hiebert 1994: 82–83	2176	2280–2051	2343–2026
R-1308	Namazga Depe	NMG V	3770 ± 40		Hiebert 1994: 82–83	2190	2282–2137	2335–2037
R-1307	Namazga Depe	NMG V	3860 ± 40		Hiebert 1994: 82–83	2339	2455–2235	2464–2206

(*continued*)

Lab. reference	*Site name*	*Provenience*	*Date BP*	*Material*	*Publication*	*Median*	*Cal. BC 1σ (68.2%)*	*Cal. BC 2σ (95.4%)*
R-1306a	Namazga Depe	NMG V	4080 ± 50		Hiebert 1994: 82–83	2643	2850–2499	2866–2485
R-1303	Namazga Depe	NMG VI lower architecture	2680 ± 50		Hiebert 1994: 82–83; Kircho and Popov 1999	846	895–803	929–790
R-1297	Namazga Depe	NMG VI upper architecture	2870 ± 50		Hiebert 1994: 82–83; Kircho and Popov 1999	1047	1118–946	1208–917
R-1300a	Namazga Depe	NMG VI upper architecture	2880 ± 60		Hiebert 1994: 82–83; Kircho and Popov 1999	1063	1187–945	1229–906
R-1301	Namazga Depe	NMG VI upper architecture	2960 ± 50		Hiebert 1994: 82–83; Kircho and Popov 1999	1171	1261–1093	1374–1015
Le-665	Namazga Depe	Tower, topmost level, oven 3, late NMG VI	2980 ± 60		Kohl 1984: 234–5; Kircho and Popov 1999	1202	1369–1112	1392–1024
R-1300	Namazga Depe	NMG VI upper architecture	3050 ± 50		Hiebert 1994: 82–83; Kircho and Popov 1999	1307	1394–1233	1427–1131

R-1302	Namazga Depe	NMG VI lower architecture	3220 ± 9		Hiebert 1994: 82–83; Kircho and Popov 1999	1474	1503–1459	1508–1448
R-1298	Namazga Depe	NMG VI upper architecture	3220 ± 40		Hiebert 1994: 82–83; Kircho and Popov 1999	1490	1526–1441	1610–1418
R-1299	Namazga Depe	NMG VI upper architecture	3240 ± 50		Hiebert 1994: 82–83; Kircho and Popov 1999	1517	1607–1449	1625–1421
R-1304	Namazga Depe	NMG VI lower architecture	3350 ± 50		Hiebert 1994: 82–83; Kircho and Popov 1999	1639	1732–1549	1749–1509
Le-767	Altyn Depe	NMG V, exc. 7 ("Cult Center"), east of room 7 and 10, near ext. wall	3140 ± 50		Kohl 1984: 234; Kircho and Popov 1999	1411	1495–1309	1508–1276
Le-769	Altyn Depe	NMG V, exc. 7, wall	3260 ± 60		Kohl 1984: 234; Kircho and Popov 1999	1542	1614–1461	1682–1421
Le-2442	Altyn Depe	NMG V, exc. 9 ("Elite Quarter"), niv. 2, inf. floor room 314	3410 ± 40		Kircho and Popov 1999	1711	1751–1644	1876–1617

(*continued*)

Lab. reference	Site name	Provenience	Date BP	Material	Publication	Median	Cal. BC 1σ (68.2%)	Cal. BC 2σ (95.4%)
Le-1097	Altyn Depe	NMG V, exc. 9 ("Elite Quarter"), niv. 1, room 26, hearth	3540 ± 80		Kohl 1984: 234; Kircho and Popov 1999	1878	1971–1754	2132–1667
BLN-717	Altyn Depe	NMG V early, exc. 7 ("Cult Center")	4025 ± 100		Kohl 1984: 234; Kircho and Popov 1999	2572	2857–2458	2876–2298
Le-4119	Altyn Depe	NMG V, exc. 9 ("Elite Quarter"), niv. 2, room 368, over a niche	4060 ± 210		Kircho and Popov 1999	2608	2892–2310	3326–2025
BLN-716	Altyn Depe	NMG V early, exc. 7 ("Cult Center")	4120 ± 100		Kohl 1984: 234; Kircho and Popov 1999	2696	2871–2578	2916–2461
Le-4118	Altyn Depe	NMG V, exc. 9 ("Elite Quarter"), niv. 2, hearth on floor 2, room 360	4150 ± 90		Kircho and Popov 1999	2726	2875–2630	2908–2488
Le-1527	Altyn Depe	NMG V, exc. 9 ("Elite Quarter"), niv. 1, room 52	5320 ± 40		Kircho and Popov 1999	4148	4233–4061	4316–4042

Le-257	Geoksjur 1	?	3750 ± 90	Charcoal	Kohl 1984: 233; Possehl 1993 for material	2171	2296–2026	2459–1946
Le-647	Geoksjur 1	?	4440 ± 180	Charcoal	Kohl 1984: 233; Possehl 1993 for material	3137	3359–2906	3633–2636
BLN-720	Geoksjur 1	?	4710 ± 100	Charcoal	Kohl 1984: 233; Possehl 1993 for material	3490	3632–3374	3704–3117
Afghanistan								
Le-976	Dashly 1	Room 18	3200 ± 45	Charcoal	Sarianidi 1977: 169	1474	1509–1426	1611–1396
Le-975	Dashly 1	Room 24	3520 ± 45	Charcoal	Sarianidi 1977: 169	1839	1908–1772	1965–1698
Le-977	Dashly 2	Pit	3340 ± 45	Charcoal	Sarianidi 1977: 169	1627	1687–1546	1741–1510
Le-1175	Dashly 3	Oval temple, room 2	3066 ± 70	?	Sarianidi 1977: 169	1317	1414–1231	1496–1121
Le-1251	Dashly 3	Palace, room 50	3250 ± 40	?	Sarianidi 1977: 169	1527	1608–1460	1617–1440
Le-978	Dashly 3	Burial	3440 ± 50	Charcoal	Sarianidi 1977: 169	1753	1875–1685	1885–1632
Le-1252	Dashly 3	Palace	3670 ± 50	?	Sarianidi 1977: 169	2054	2135–1977	2199–1921
Le-1253	Dashly 3	Palace, room 50	4060 ± 70	?	Sarianidi 1977: 169	2621	2840–2482	2872–2467

(*continued*)

Lab. reference	*Site name*	*Provenience*	*Date BP*	*Material*	*Publication*	*Median*	*Cal. BC 1σ (68.2%)*	*Cal. BC 2σ (95.4%)*
Le-1254	Dashly 3	Palace	4230 ± 70		Sarianidi 1977: 169	2794	2912–2680	3011–2586
GX-0910	Dara-i Kur	Goat cult Neolithic	3780 ± 130	Bone	Possehl 1993	2218	2451–2034	2578–1832
MC-1727	Shortughaï	Mound A, level 6, period I	3570 ± 95	?	Francfort 1989: 241	1921	2032–1771	2198–1683
Ny-427	Shortughaï	Mound A, level 2, period II	3710 ± 100	Charcoal	Francfort 1989: 241; Possehl 1993 for the material	2115	2278–1954	2460–1882
MC-2447	Shortughaï	Mound A, level 6, period I	3725 ± 80	?	Francfort 1989: 241	2131	2278–1983	2435–1905
MC-1726	Shortughaï	Mound A, level 6, period I	3875 ± 95	?	Francfort 1989: 241	2345	2471–2206	2617–2037
MC-2446	Shortughaï	Mound A, level 6, period I	3890 ± 80	?	Francfort 1989: 241	2364	2473–2213	2575–2140
MC-2445	Shortughaï	Mound A, level 3, period II	3890 ± 80	?	Francfort 1989: 241	2364	2473–2213	2575–2140
Ny-425	Shortughaï	Mound A, level 6, period I	4040 ± 100	Charcoal	Francfort 1989: 241; Possehl 1993 for the material	2596	2856–2467	2881–2306
Ny-430	Shortughaï	Mound A, level 5, period I	4075 ± 95	Charcoal	Francfort 1989: 241; Possehl 1993 for the material	2646	2858–2490	2893–2350

Ny-428	Shortughaï	Mound A, level 2, period II	4190 ± 125	Charcoal	Francfort 1989: 241; Possehl 1993 for the material	2762	2906–2582	3311–2459
Ny-429	Shortughaï	Mound A, level 4, period II	4375 ± 160	Charcoal	Francfort 1989: 241; Possehl 1993 for the material	3056	3339–2882	3508–2580
Ny-422	Shortughaï	Mound B, level 2, period III	3050 ± 250	Charcoal	Francfort 1989: 241; Possehl 1993 for the material	1292	1605–944	1959–768
Ny-424	Shortughaï	Mound B, level 3, period III	3180 ± 335	Charcoal	Francfort 1989: 241; Possehl 1993 for the material	1469	1878–1019	2463–757
Ny-421	Shortughaï	Mound B, level 4, period IV	3535 ± 165	Charcoal	Francfort 1989: 241; Possehl 1993 for the material	1889	2131–1661	2346–1460
MC-1729	Shortughaï	Mound B, level 2, period III	3620 ± 105	?	Francfort 1989: 241	1992	2187–1782	2290–1693
MC-1730	Shortughai	Mound B, level 1, period IV	3640 ± 95	?	Francfort 1989: 241	2017	2187–1889	2287–1751
MC-1728	Shortughaï	Mound B, level 4, period III	3975 ± 90	?	Francfort 1989: 241	2492	2620–2310	2860–2206
GSY-53	Mundigak	III.5	4185 ± 150		Salvatori and Tosi 2005	2762	2924–2496	3329–2346

(*continued*)

Lab. reference	*Site name*	*Provenience*	*Date BP*	*Material*	*Publication*	*Median*	*Cal. BC 1σ (68.2%)*	*Cal. BC 2σ (95.4%)*
Uzbekistan								
Le-1078	Sapalli tepa	Exc. 3	3450 ± 50		Askarov 1977	1767	1876–1691	1891–1638
Le-916	Sapalli tepa	Exc. 3	3640 ± 90	Charcoal	Askarov 1977	2017	2137–1896	2283–1754
Bln-5017	Dzharkutan IV	Settled area, niv. inf.	3474 ± 29	Charcoal	Görsdorf and Huff 2001	1805	1877–1747	1886–1696
Bln-5008	Dzharkutan IV	Settled area, niv. inf.	3505 ± 37	Charcoal	Görsdorf and Huff 2001	1825	1886–1771	1929–1700
Bln-5302	Dzharkutan IV	Settled area, niv. inf.	3524 ± 33	Charcoal	Görsdorf and Huff 2001	1839	1903–1775	1940–1752
Bln-5303	Dzharkutan IV	Settled area, niv. inf.	3590 ± 29	Charcoal	Görsdorf and Huff 2001	1944	2007–1900	2025–1885
Bln-5009	Dzharkutan Citadel	Exc. 4 pit Dzharkutan period	3465 ± 37	Charcoal	Görsdorf and Huff 2001	1791	1877–1700	1886–1690
Lyon-8211	Dzharkutan Citadel	Exc. 7 silo Dzharkutan period	3640 ± 35	Charcoal	Bendezu-Sarmiento and Mustafakulov 2013	2005	2113–1949	2135–1912
Beta-33557	Dzharkutan VI Temple (A)	Large altar	3540 ± 70	Charcoal	Hiebert 1994: 79	1876	1957–1766	2118–1690
Bln-5012	Dzharkutan VI	Pre-temple period	3274 ± 33	Charcoal	Görsdorf and Huff 2001	1556	1609–1509	1626–1460
Bln-5301	Dzharkutan VI	Pre-temple period	3323 ± 47	Charcoal	Görsdorf and Huff 2001	1604	1660–1532	1737–1502
Bln-5304	Dzharkutan VI	Pre-temple period	3459 ± 29	Charcoal	Görsdorf and Huff 2001	1777	1873–1698	1880–1692

Bln-5300	Dzharkutan VI	Pre-temple period	3494 ± 29	Charcoal	Görsdorf and Huff 2001	1818	1879–1771	1896–1701
Bln-5140	Dzharkutan VI	Pre-temple period	3544 ± 31	Charcoal	Görsdorf and Huff 2001	1887	1939–1782	1971–1688
Bln-5299	Dzharkutan VI Temple (B)	Pre-temple period	3639 ± 32	Charcoal	Görsdorf and Huff 2001	2002	2111–1948	2134–1916
Bln-5144	Dzharkutan VI	Kuzali material from the temple period	3256 ± 34	Charcoal	Görsdorf and Huff 2001	1533	1608–1498	1616–1449
Bln-5014	Dzharkutan VI (C)	Kuzali material from the temple period	3285 ± 39	Charcoal	Görsdorf and Huff 2001	1565	1613–1521	1659–1458
Bln-5015	Dzharkutan VI	Temple period	3322 ± 32	Charcoal	Görsdorf and Huff 2001	1600	1641–1534	1686–1520
Bln-5010	Dzharkutan VI	Temple period	3336 ± 31	Charcoal	Görsdorf and Huff 2001	1624	1682–1560	1727–1529
Bln-5143	Dzharkutan VI	Kuzali material from the temple period	3337 ± 29	Charcoal	Görsdorf and Huff 2001	1626	1683–1562	1691–1530
Bln-5016	Dzharkutan VI	Temple period	3375 ± 33	Charcoal	Görsdorf and Huff 2001	1669	1731–1628	1751–1562
Bln-5145	Dzharkutan VI	Temple period	3388 ± 32	Charcoal	Görsdorf and Huff 2001	1682	1736–1641	1755–1614
Bln-5013	Dzharkutan VI	Temple period	3414 ± 29	Charcoal	Görsdorf and Huff 2001	1713	1748–1666	1867–1630
Bln-5011	Dzharkutan VI	Temple period	3418 ± 37	Charcoal	Görsdorf and Huff 2001	1719	1763–1661	1876–1625

(continued)

Lab. reference	*Site name*	*Provenience*	*Date BP*	*Material*	*Publication*	*Median*	*Cal. BC 1σ (68.2%)*	*Cal. BC 2σ (95.4%)*
Bln-5142	Dzharkutan VI	Temple period	3433 ± 29	Charcoal	Görsdorf and Huff 2001	1735	1771–1688	1877–1659
Bln-5146	Dzharkutan VI	Temple period	3505 ± 29	Charcoal	Görsdorf and Huff 2001	1824	1884–1774	1911–1746
Bln-5141	Dzharkutan VI	Temple period	3510 ± 35	Charcoal	Görsdorf and Huff 2001	1829	1889–1772	1931–1744
Lyon-9843/SacA 31620	Dzharkutan	Grave 1045	3600 ± 30	Human bone	Bendezu-Sarmiento and Mustafakulov 2013	1957	2016–1913	2031–1888
Beta-259705	Dzharkutan	Grave DZH07-3-1001B	3600 ± 40	Human bone	Bendezu-Sarmiento and Mustafakulov 2013	1959	2019–1906	2126–1784
MAMS-13262	Tilla-Bulak	Phase 1 destruction level	3489 ± 28	Charred seeds	Kaniuth 2016	1815	1878–1768	1891–1701
KIA-36744	Tilla-Bulak	Phase 2?	3510 ± 25	Charred grain	Kaniuth 2016	1827	1886–1775	1906–1751
Hd-26247	Tilla-Bulak	TB07-157-HK01 (hearth in chimney from wall 128. More recent than TB07-006)	3527 ± 32	Charcoal	Kaniuth 2007, 2016	1842	1912–1776	1940–1755
MAMS-13263	Tilla-Bulak	Phase 1 destruction level	3530 ± 28	Charred seeds	Kaniuth 2016	1846	1916–1778	1941–1767

MAMS-13272	Tilla-Bulak	Phase 2?	3540 ± 33	Charred grain	Kaniuth 2016	1880	1933–1781	1960–1756
MAMS-13268	Tilla-Bulak	Phase 2?	3541 ± 33	Charred grain	Kaniuth 2016	1882	1935–1781	1962–1760
MAMS-13270	Tilla-Bulak	Phase 2?	3559 ± 34	Charred grain	Kaniuth 2016	1908	1956–1785	2018–1773
MAMS-13266	Tilla-Bulak	Phase 2?	3578 ± 28	Charred grain	Kaniuth 2016	1930	1958–1889	2024–1831
MAMS-13264	Tilla-Bulak	Phase 1 destruction level	3590 ± 27	Charred seeds	Kaniuth 2016	1944	1973–1901	2024–1886
Hd-28467	Tilla-Bulak	TB08-322, phase 2?	3594 ± 27	Charcoal	Kaniuth 2010, 2016	1949	2010–1911	2024–1889
MAMS-13269	Tilla-Bulak	Phase 2?	3602 ± 32	Charred grain	Kaniuth 2016	1960	2018–1917	2036–1884
MAMS-13265	Tilla-Bulak	Phase 2?	3614 ± 26	Charred grain	Kaniuth 2016	1973	2021–1940	2034–1897
Hd-26067	Tilla-Bulak	TB07-006-HK02 (phase B)	3615 ± 23	Charcoal	Kaniuth 2007, 2016	1974	2021–1941	2033–1907
MAMS-13267	Tilla-Bulak	Phase 1 destruction level	3617 ± 27	Charred seeds	Kaniuth 2016	1977	2022–1942	2111–1896
MAMS-13262	Tilla-Bulak	Phase 1 destruction level	3626 ± 28	Charcoal from beam	Kaniuth 2016	1987	2026–1951	2121–1903
MAMS-13271	Tilla-Bulak	Phase 1 destruction level	3645 ± 33	Charred seeds	Kaniuth 2016	2012	2115–1954	2135–1923
Hd-26250	Tilla-Bulak	TB07-160-HK01 (Phase A, constr. phase)	3714 ± 20	Charcoal	Kaniuth 2007, 2016	2087	2141–2041	2196–2034

(*continued*)

Lab. reference	Site name	Provenience	Date BP	Material	Publication	Median	Cal. BC 1σ (68.2%)	Cal. BC 2σ (95.4%)
Hd-28466	Tilla-Bulak	TB08-211	3719 ± 19	Charcoal	Kaniuth 2010, 2016	2089	2191–2045	2198–2036
MAMS-13274	Tilla-Bulak	Phase 1 destruction level	3722 ± 22	Charcoal from beam	Kaniuth 2016	2100	2194–2044	2199–2036
MAMS-13273	Tilla-Bulak	Not in situ	3883 ± 62	Ivory	Kaniuth 2016	2361	2465–2290	2562–2151
Le-4991	Bustan VI	Crematorium 3	3540 ± 70		Vinogradova 2004	1876	1957–1766	2118–1690
Gif-12097	Bustan VI	Crematorium V	3575 ± 40	Charcoal	Avanesova 2006, unpublished	1928	2010–1884	2032–1776
Le-4994	Bustan VI	Crematorium I	3580 ± 70		Vinogradova 2004	1933	2031–1781	2136–1746
Gif-1201	Bustan VI	Grave 315	3605 ± 45	Charcoal	Avanesova 2006, unpublished	1966	2025–1909	2134–1782
Le-4995	Bustan VI	Crematorium II	3620 ± 40		Vinogradova 2004	1983	2031–1927	2132–1886
Gif-12098	Bustan VI	Crematorium VIII	4070 ± 65	Charcoal	Avanesova 2006, unpublished	2634	2851–2491	2871–2472
Gif-12099	Bustan VI	Autel 316	4701 ± 55	Charcoal	Avanesova 2006, unpublished	3483	3626–3375	3633–3369
Beta- (1999)	Karnab	03, Grube 2	2650 ± 60		Cierny and Weisgerber (thanks to J. Garner)	826	895–790	974–590

Beta- (1999)	Karnab	5-5-IV (E1)	2840 ± 70		Cierny and Weisgerber (thanks to J. Garner)	1011	1110–916	1208–841
Beta- (1999)	Karnab	5-3	2920 ± 70		Cierny and Weisgerber (thanks to J. Garner)	1119	1215–1015	1373–920
Beta- (1999)	Karnab	5-5-I (Feuerstelle)	3010 ± 70		Cierny and Weisgerber (thanks to J. Garner)	1246	1385–1128	1423–1047
Beta- (1999)	Karnab	5-5-III	3070 ± 130		Cierny and Weisgerber (thanks to J. Garner)	1307	1494–1128	1621–976
Beta- (1999)	Karnab	6-1 (9.5m)	3190 ± 60		Cierny and Weisgerber (thanks to J. Garner)	1467	1530–1405	1616–1302
Zeravshan								
Gif-11780	Tugai	Metallurgical furnace	3555 ± 40	Charcoal?	Avanesova 2003, unpublished	1901	1955–1781	2020–1768
Gif-11781	Tugai	Metallurgical furnace	3665 ± 40	Charcoal?	Avanesova 2003, unpublished	2045	2133–1977	2194–1930

(continued)

Lab. reference	*Site name*	*Provenience*	*Date BP*	*Material*	*Publication*	*Median*	*Cal. BC 1σ (68.2%)*	*Cal. BC 2σ (95.4%)*
Gif-11779	Tugai	Settlement	3710 ± 40	Ashes	Avanesova 2003, unpublished	2098	2191–2036	2270–1977
Gif-11778	Tugai	Hearth	3775 ± 55	Charcoal	Avanesova 2003, unpublished	2202	2291–2061	2451–2030
Le-1420	Sarazm	Exc. 3, level 2	3790 ± 80		Isakov *et al.* 1987	2231	2346–2050	2468–1985
Le-1808	Sarazm	Exc. 3, level 2	4230 ± 40		Isakov *et al.* 1987	2803	2902–2760	2913–2678
Le-1806	Sarazm	Exc. 3, level 1	4460 ± 50		Isakov *et al.* 1987	3170	3330–3027	3348–2938
Gif-7398	Sarazm	Exc. 7, level IV.1, room 404	3800 ± 70		Besenval and Isakov 1989	2246	2399–2136	2464–2037
Gif-7397	Sarazm	Exc. 7, level IV.1, room 402	3850 ± 90		Besenval and Isakov 1989	2314	2458–2206	2568–2037
Gif-7201	Sarazm	Exc. 7, level III.3	3990 ± 70		Besenval and Isakov 1989	2518	2621–2350	2855–2291
Gif-7202	Sarazm	Exc. 7, level II, 1	4130 ± 70		Besenval and Isakov 1989	2715	2866–2620	2889–2496
Gif-7560	Sarazm	Exc. 7, level I.1 (late)	4380 ± 70		Besenval and Isakov 1989	3028	3096–2906	3333–2889
Gif-7561	Sarazm	Exc. 7, level I.1 (early)	4450 ± 60		Besenval and Isakov 1989	3144	3330–3020	3341–2929
Le-2173	Sarazm	Exc. 4, level 1	4880 ± 30		Isakov *et al.* 1987	3667	3695–3642	3708–3637

Le-2172	Sarazm	Exc. 4, level 1	5050 ± 60	Isakov *et al.* 1987	3853	3944–3791	3965–3709
Le-1807	Sarazm	Exc. 2, level 3	3840 ± 40	Isakov *et al.* 1987	2304	2401–2206	2461–2154
Le-2174	Sarazm	Exc. 2, level 1	4940 ± 30	Isakov *et al.* 1987	3709	3761–3661	3777–3654
Beta- (1999)	Mushiston	QS-3, Holz oben	2770 ± 80	Garner 2013	935	1004–833	1127–798
Beta- (1999)	Mushiston	SM1, Stempel	3140 ± 60	Garner 2013	1408	1496–1306	1531–1233
Beta- (1999)	Mushiston	QS-4, Stempel, K1	3180 ± 90	Garner 2013	1453	1603–1306	1660–1225
Beta- (1999)	Mushiston	QS-3, Abbauort	3270 ± 40	Garner 2013	1552	1610–1504	1632–1449
Beta- (1999)	Mushiston	QS-4, Stempel, K2	3270 ± 70	Garner 2013	1554	1625–1456	1734–1415
Beta- (1999)	Mushiston	SM1, 2. Sohle, vorne	3290 ± 90	Garner 2013	1577	1682–1457	1871–1327
Beta- (1999)	Mushiston	QS-4, Geleucht	3320 ± 60	Garner 2013	1602	1681–1527	1746–1454
Beta- (1999)	Mushiston	SM1, 2. Sohle, links	3640 ± 70	Garner 2013	2014	2133–1920	2205–1776
Beta- (1999)	Mushiston	SM2, hinten, vor Ort	3760 ± 80	Garner 2013	2184	2292–2036	2458–1975
Southwest Tadjikistan							
Le-717	Tigrovaja Balka	?	3350 ± 60	Vinogradova 2004	1639	1734–1535	1869–1498

(continued)

Lab. reference	*Site name*	*Provenience*	*Date BP*	*Material*	*Publication*	*Median*	*Cal. BC 1σ (68.2%)*	*Cal. BC 2σ (95.4%)*
Le-715	Tigrovaja Balka	Kurgan 7 (A)	3350 ± 60		Kaniuth 2006: 51 (from P'jankova 1986: 58)	1639	1734–1535	1869–1498
Le-1407	Tigrovaja Balka	Kurgan 8	2810 ± 40		Kaniuth 2006: 51 (from P'jankova 1986: 58)	963	1009–913	1073–843
GIN-2524	Tigrovaja Balka	Kurgan 8	3920 ± 100		Kaniuth 2006: 51 (from P'jankova 1986: 58)	2402	2568–2215	2850–2059
GIN-2526	Tigrovaja Balka	Kurgan 8	4020 ± 80		Kaniuth 2006: 51 (from P'jankova 1986: 58)	2561	2840–2460	2871–2307
GIN-2525	Tigrovaja Balka	Kurgan 8	4510 ± 340		Kaniuth 2006: 51 (from P'jankova 1986: 58)	3201	3650–2709	4037–2299
MAMS-15064	Gelot	Necropole 6, burial 2	3667 ± 24		Teufer *et al.* 2014	2050	2129–1981	2136–1965
MAMS-15065	Gelot	Necropole 4, burial 3	3667 ± 21		Teufer *et al.* 2014	2050	2128–1982	2136–1966
MAMS-15067	Gelot	Necropole 6, burial 13	3732 ± 26		Teufer *et al.* 2014	2135	2197–2050	2204–2036
MAMS-15072	Darnajchi	Necropole 2, burial 2	3814 ± 36		Teufer *et al.* 2014	2256	2335–2153	2453–2140

Gif-8414	Kangurttut	Exc. VI	3150 ± 60	Charcoal	Vinogradova *et al.* 2008	1420	1499–1311	1601–1261
Gin-9479	Kangurttut	Exc. XI	3300 ± 110		Vinogradova *et al.* 2008	1591	1732–1451	1885–1309
Gin-9480	Kangurttut	Exc. XI	3800 ± 160		Vinogradova *et al.* 2008	2248	2468–2030	2847–1772
Gin-9482	Kangurttut	Exc. XI	4770 ± 500		Vinogradova *et al.* 2008	3512	4222–2896	4718–2231
Steppes								
Le-1141	Sintashta I (Ural)	Burial mound V, grave 2	3440 ± 60	Wood	Kuz'mina 2007: 467	1755	1877–1669	1908–1616
KI-862	Sintashta	Grave 5	3360 ± 70	Wood	Kuz'mina 2007: 467	1653	1742–1535	1877–1499
KI-863	Sintashta	Grave 7	3420 ± 70	Wood	Kuz'mina 2007: 467	1730	1871–1632	1903–1533
KI-864	Sintashta	Grave 19	3560 ± 180	Wood	Kuz'mina 2007: 467	1926	2191–1683	2456–1503
KI-657	Sintashta	Grave 28	3760 ± 120	Wood	Kuz'mina 2007: 467	2189	2398–1984	2565–1880
KI-652	Sintashta	Grave 7	3870 ± 215	Wood	Kuz'mina 2007: 467	2347	2625–2026	2901–1769
KI-658	Sintashta	Grave 39	4100 ± 170		Kuz'mina 2007: 467	2662	2896–2466	3264–2142
KI-706	Sintashta	Grave 22	4090 ± 480	Wood	Kuz'mina 2007: 467	2667	3338–2037	3941–1497

(*continued*)

Lab. reference	*Site name*	*Provenience*	*Date BP*	*Material*	*Publication*	*Median*	*Cal. BC 1σ (68.2%)*	*Cal. BC 2σ (95.4%)*
KI-650	Sintashta	Grave 2	4160 ± 105		Kuz'mina 2007: 467	2732	2883–2622	3011–2469
KI-653	Sintashta	Grave 11	4200 ± 100		Kuz'mina 2007: 467	2768	2903–2632	3023–2491
RUL-276	Sintashta, Tsarev Kurgan, burial mound	Petrovka, central grave	3520 ± 65		Kuz'mina 2007: 467	1845	1926–1754	2024–1690
RUL-275	Sintashta, Tsarev Kurgan, burial mound	Petrovka, central grave	3570 ± 60		Kuz'mina 2007: 467	1919	2021–1782	2127–1746
UBA-30783	Adunqiaolu (northwest China)	Petrovka, house F1	3090 ± 28	Sheep dung	Jia *et al.* 2017	1346	1410–1303	1424–1280
UBA-30781	Adunqiaolu	Petrovka, house F1	3189 ± 37	Sheep dung	Jia *et al.* 2017	1464	1498–1433	1595–1396
UBA-30786	Adunqiaolu	Petrovka, house F1	3251 ± 33	Sheep dung	Jia *et al.* 2017	1526	1608–1461	1614–1449
UBA-19168	Adunqiaolu	East Fedorovo, burial SM1	3253 ± 27	Human bone	Jia *et al.* 2017	1527	1607–1497	1613–1453
UBA-30789	Adunqiaolu	East Fedorovo, house F1	3265 ± 32	Charcoal	Jia *et al.* 2017	1546	1608–1503	1621–1455
UBA-21986	Adunqiaolu	East Fedorovo, burial SM50	3266 ± 34	Charcoal	Jia *et al.* 2017	1548	1608–1503	1624–1453
UBA-19164	Adunqiaolu	East Fedorovo, house F1	3270 ± 27	Charcoal	Jia *et al.* 2017	1552	1608–1506	1621–1465

UBA-19163	Adunqiaolu	East Fedorovo, house F1	3331 ± 38	Charcoal	Jia *et al.* 2017	1616	1664–1535	1731–1516
UBA-21985	Adunqiaolu	East Fedorovo, burial SM4	3337 ± 32	Wood	Jia *et al.* 2017	1625	1683–1561	1729–1529
UBA-19165	Adunqiaolu	East Fedorovo, house F1	3403 ± 28	Charcoal	Jia *et al.* 2017	1700	1742–1663	1766–1626
UBA-19167	Adunqiaolu	East Fedorovo, burial SM9	3434 ± 28	Wood	Jia *et al.* 2017	1736	1769–1690	1876–1661
UBA-19166	Adunqiaolu	East Fedorovo, burial SM9	3447 ± 31	Wood	Jia *et al.* 2017	1758	1870–1692	1880–1686
Pakistan and India								
Beta-2691	Sibri (Pakistan)	Mehrgarh VIIC or later	3220 ± 60	?	Possehl 1993	1497	1602–1428	1637–1322
Beta-163713	Harappa (Pakistan)	5	3440 ± 40		Meadow and Kenoyer 2005	1750	1871–1688	1881–1658
WIS-2074	Harappa	Urban Harappa, hearth	3700 ± 60	Charcoal	Possehl 1993	2091	2197–1983	2284–1928
WIS-2144	Harappa	Urban Harappa	3720 ± 100	Charcoal	Possehl 1993	2129	2284–1974	2458–1889
QL-4484	Harappa	Urban phase	3730 ± 30	Charcoal	Possehl 1993	2131	2198–2046	2266–2032
QL-4488	Harappa	Urban phase	3750 ± 40	Charcoal	Possehl 1993	2162	2269–2050	2286–2035
Beta-163720	Harappa	3B/3C	3760 ± 40		Meadow and Kenoyer 2005	2174	2277–2060	2293–2036
WIS-2043	Harappa	Urban Harappa, upper levels	3770 ± 70	Wood, charcoal	Possehl 1993	2197	2295–2042	2458–1984

(continued)

Lab. reference	*Site name*	*Provenience*	*Date BP*	*Material*	*Publication*	*Median*	*Cal. BC 1σ (68.2%)*	*Cal. BC 2σ (95.4%)*
QL-4377	Harappa	Transitional Early/Urban Harappa, foundation trench	3770 ± 100	Charcoal	Possehl 1993	2202	2340–2035	2473–1936
Beta-163712	Harappa	2	3780 ± 40		Meadow and Kenoyer 2005	2206	2283–2141	2342–2042
QL-4483	Harappa	Urban phase	3784 ± 30	Charcoal	Possehl 1993	2212	2281–2145	2299–2061
Beta-163721	Harappa	3C	3800 ± 40		Meadow and Kenoyer 2005	2240	2293–2146	2452–2059
QL-4374	Harappa	Urban Harappa, in front of kiln	3800 ± 50	Charcoal	Possehl 1993	2243	2334–2141	2457–2050
QL-4487	Harappa	Urban phase	3816 ± 25	Charcoal	Possehl 1993	2251	2290–2206	2346–2146
QL-4376	Harappa	Urban Harappa	3810 ± 50	Charcoal	Possehl 1993	2257	2339–2146	2460–2064
WIS-2220	Harappa	Urban phase	3815 ± 60		Hiebert 1994: 86	2267	2397–2145	2465–2054
WIS-2139	Harappa	Urban Harappa, hearth	3820 ± 60	Carbonized wood	Possehl 1993	2275	2401–2146	2466–2060
WIS-2143	Harappa	Urban Harappa, hearth	3825 ± 60	Charcoal	Possehl 1993	2282	2433–2151	2468–2063

WIS-2075	Harappa	Urban Harappa, floor levels	3830 ± 60	Charcoal	Possehl 1993	2290	2436–2154	2470–2064
QL-4378	Harappa	Urban Harappa, near kiln	3850 ± 50	Charcoal	Possehl 1993	2322	2453–2209	2468–2151
WIS-2219	Harappa	Urban phase	3910 ± 65		Hiebert 1994: 86	2388	2474–2295	2571–2204
Beta-163714	Harappa	3A	3910 ± 40		Meadow and Kenoyer 2005	2394	2468–2346	2550–2234
WIS-2141	Harappa	Urban Harappa	3920 ± 70	Carbonized wood	Possehl 1993	2400	2488–2294	2580–2155
WIS-2053	Harappa	Urban Harappa, upper levels	3920 ± 210	Unknown	Possehl 1993	2413	2847–2061	3006–1778
Beta-163711	Harappa	2	3930 ± 40		Meadow and Kenoyer 2005	2415	2477–2346	2565–2294
WIS-2221	Harappa	Urban phase	3940 ± 120		Hiebert 1994: 86	2436	2617–2209	2867–2061
QL-4380	Harappa	Transitional Early/Urban Harappa	3950 ± 80	Charcoal	Possehl 1993	2447	2571–2309	2838–2200
Beta-163718	Harappa	2	4000 ± 40		Meadow and Kenoyer 2005	2528	2569–2474	2831–2356
WIS-2145	Harappa	Urban Harappa	4020 ± 60	Charcoal	Possehl 1993	2553	2623–2469	2860–2348
Beta-163710	Harappa	2	4040 ± 40		Meadow and Kenoyer 2005	2563	2620–2488	2840–2469
WIS-2142	Harappa	Transitional Early/Urban Harappa	4135 ± 65	Carbonized wood	Possehl 1993	2722	2867–2625	2888–2500

(continued)

Lab. reference	*Site name*	*Provenience*	*Date BP*	*Material*	*Publication*	*Median*	*Cal. BC 1σ (68.2%)*	*Cal. BC 2σ (95.4%)*
WIS-2140	Harappa	Urban Harappa	4290 ± 70	Charcoal	Possehl 1993	2916	3022–2779	3263–2639
Beta-163717	Harappa	1A	4390 ± 40		Meadow and Kenoyer 2005	3005	3084–2925	3310–2904
Beta-163715	Harappa	1A	4430 ± 40		Meadow and Kenoyer 2005	3077	3310–2936	3331–2922
Beta-163719	Harappa	1A	4480 ± 50		Meadow and Kenoyer 2005	3195	3335–3093	3359–3013
TF-75	Mohenjo-daro (Pakistan)	Late levels from Marshall's exc. (Late Harappan or Jhukar period)	3600 ± 110	Charred grain	Possehl 1993; Hiebert 1994: 86	1965	2134–1776	2287–1683
P-1182A	Mohenjo-Daro	Harappan Urban phase	3702 ± 63	Charred straw	Possehl 1993	2094	2198–1984	2287–1928
P-1176	Mohenjo-Daro	Harappan Urban phase	3801 ± 59	Ash	Possehl 1993	2246	2339–2140	2461–2044
P-1178A	Mohenjo-Daro	Harappan Urban phase	3802 ± 59	Charred straw	Possehl 1993	2248	2339–2141	2461–2045
P-1180	Mohenjo-Daro	Harappan Urban phase	3828 ± 61	Charred wood	Possehl 1993	2287	2436–2152	2469–2062
P-1179	Mohenjo-Daro	Harappan Urban phase	3913 ± 64	Charred wood	Possehl 1993	2392	2475–2296	2570–2206
P-1177	Mohenjo-Daro	Harappan Urban phase	3985 ± 64	Charcoal	Possehl 1993	2510	2617–2351	2848–2290

TF-135	Lothal	IIA, Lothal A, Urban Harappa	3405 ± 125	Charcoal	Possehl 1993	1719	1882–1546	2028–1432
TF-19	Lothal	VA, Lothal B, post-Urban Harappa	3650 ± 135	Charcoal	Possehl 1993	2038	2272–1782	2457–1693
TF-23	Lothal	VA, Lothal B, post-Urban Harappa	3705 ± 105	Charcoal	Possehl 1993	2109	2278–1947	2461–1783
TF-29	Lothal	IVA, Lothal A, Urban Harappa	3740 ± 110	Charcoal	Possehl 1993	2160	2333–1977	2473–1882
TF-133	Lothal	IIA, Lothal A, Urban Harappa	3740 ± 110	Charcoal	Possehl 1993	2160	2333–1977	2473–1882
TF-26	Lothal	IIIB, Lothal A, Urban Harappa	3830 ± 120	Charcoal	Possehl 1993	2285	2468–2136	2619–1938
TF-27	Lothal	IIIB, Lothal A, Urban Harappa	3840 ± 110	Charcoal	Possehl 1993	2298	2466–2145	2580–1966
TF-22	Lothal	IIIB, Lothal A, Urban Harappa	3845 ± 110	Charcoal	Possehl 1993	2305	2467–2146	2617–1974
TF-136	Lothal	IA, Lothal A, Urban Harappa	3915 ± 130	Charcoal	Possehl 1993	2400	2576–2202	2864–2034

(*continued*)

Lab. reference	*Site name*	*Provenience*	*Date BP*	*Material*	*Publication*	*Median*	*Cal. BC 1σ (68.2%)*	*Cal. BC 2σ (95.4%)*
Iran								
TUNC-61	Shahr-i Sokhta	I, pre-phase 10	4480 ± 100		Salvatori and Tosi 2005	3176	3346–3028	3496–2905
Beta-25899	Shahr-i Sokhta	I, phases 10–9	4150 ± 80		Salvatori and Tosi 2005	2729	2875–2632	2901–2496
P-25543 = R-629	Shahr-i Sokhta	I, phases 10–9	4200 ± 60		Salvatori and Tosi 2005	2773	2893–2680	2910–2601
P-2086	Shahr-i Sokhta	II, phase 7	4080 ± 60		Salvatori and Tosi 2005	2649	2853–2498	2871–2476
P-2541	Shahr-i Sokhta	II, phase 7	4080 ± 70		Salvatori and Tosi 2005	2652	2856–2496	2872–2476
P-2081B	Shahr-i Sokhta	II, phase 7	4150 ± 70		Salvatori and Tosi 2005	2732	2872–2636	2896–2500
R-638	Shahr-i Sokhta	II, phase 7	4150 ± 50		Salvatori and Tosi 2005	2740	2871–2640	2881–2581
P-2076	Shahr-i Sokhta	II, phase 7	4160 ± 60		Salvatori and Tosi 2005	2744	2876–2668	2889–2581
P-2546	Shahr-i Sokhta	II, phase 7	4170 ± 70		Salvatori and Tosi 2005	2748	2880–2666	2904–2573
R-633A	Shahr-i Sokhta	I, phase 7	4170 ± 50		Salvatori and Tosi 2005	2756	2877–2679	2891–2601
R-623	Shahr-i Sokhta	II, phase 6	4050 ± 50		Salvatori and Tosi 2005	2590	2832–2488	2859–2469
P-2079	Shahr-i Sokhta	II, phase 6	4060 ± 70		Salvatori and Tosi 2005	2621	2840–2482	2872–2467
P-2070	Shahr-i Sokhta	II, phase 6	4070 ± 60		Salvatori and Tosi 2005	2631	2850–2492	2866–2473

P-2542	Shahr-i Sokhta	II, phase 5	3990 ± 60		Salvatori and Tosi 2005	2518	2619–2409	2837–2299
R-628 = R-641 = R-637A	Shahr-i Sokhta	II, phase 5	4000 ± 50		Salvatori and Tosi 2005	2529	2574–2470	2835–2346
R-627	Shahr-i Sokhta	II, phase 5	4020 ± 50		Salvatori and Tosi 2005	2546	2617–2473	2855–2409
P-2544	Shahr-i Sokhta	II, phase 5	4060 ± 70		Salvatori and Tosi 2005	2621	2840–2482	2872–2467
R-626	Shahr-i Sokhta	III, phase 4	3890 ± 50		Salvatori and Tosi 2005	2372	2463–2306	2481–2205
TUNC-24	Shahr-i Sokhta	III, phase 4	3943 ± 70		Salvatori and Tosi 2005	2434	2566–2340	2623–2206
TUNC-27	Shahr-i Sokhta	III, phase 3 (Rud-i Biyaban 2)	3890 ± 90		Salvatori and Tosi 2005	2363	2475–2209	2619–2049
Beta-18844	Shahr-i Sokhta	III, phase 3	4010 ± 80		Salvatori and Tosi 2005	2547	2836–2369	2865–2296
Beta-18842	Shahr-i Sokhta	III, phase 3	4030 ± 70		Salvatori and Tosi 2005	2572	2834–2469	2868–2348
Beta-18843	Shahr-i Sokhta	III, phase 3	4070 ± 70		Salvatori and Tosi 2005	2636	2851–2491	2872–2471
TUNC-22	Shahr-i Sokhta	III, phase 2 (Rud-i Biyaban 2)	3829 ± 61		Salvatori and Tosi 2005	2288	2436–2153	2469–2063
R-898	Shahr-i Sokhta	IV, phase 1, burnt building	3680 ± 50	Charcoal	Salvatori and Tosi 2005	2069	2138–1980	2202–1931
R-900	Shahr-i Sokhta	IV, phase 1, burnt building	3730 ± 50	Charcoal	Salvatori and Tosi 2005	2131	2201–2039	2289–1979
TUNC-63	Shahr-i Sokhta	IV, phase 0 later use of burnt building	3430 ± 70		Salvatori and Tosi 2005	1743	1876–1642	1922–1545

(continued)

Lab. reference	*Site name*	*Provenience*	*Date BP*	*Material*	*Publication*	*Median*	*Cal. BC 1σ (68.2%)*	*Cal. BC 2σ (95.4%)*
R-901A	Shahr-i Sokhta	IV, phase 0 later use of burnt building	3540 ± 50		Salvatori and Tosi 2005	1875	1944–1775	2022–1744
TUNC-20	Hissar	IIIC, Treasure Hill, below top two building levels	3483 ± 63	Charcoal	Possehl 1993; Hiebert 1994: 87	1808	1887–1700	1956–1639
P-2620	Hissar	IIIC, main mound, upper strata	3610 ± 70	Charcoal	Possehl 1993; Hiebert 1994: 87	1974	2122–1885	2195–1767
P-2701	Hissar	III	3860 ± 60	Charcoal	Possehl 1993	2333	2457–2233	2476–2142
SI-6015	Hissar	III	4110 ± 60	Charcoal	Possehl 1993	2693	2859–2579	2878–2496
Beta-6473	Tepe Yahya IVA		3300 ± 70		Prickett 1986: 421	1582	1661–1502	1744–1433
Beta-6474	Tepe Yahya IVB1		3240 ± 120		Prickett 1986: 421	1524	1683–1401	1875–1221
TF-1140	Tepe Yahya IVB1		3560 ± 110		Prickett 1986: 421	1910	2110–1746	2203–1627
Beta-6470	Tepe Yahya IVB1		3610 ± 70		Prickett 1986: 421	1974	2122–1885	2195–1767
P-2063	Tal-i Malyan/ Anshan	Kaftari period	3430 ± 60	Charcoal	Possehl 1993	1741	1873–1659	1903–1566
TUNC-28	Tal-i Malyan/ Anshan	Kaftari period	3510 ± 63	Charcoal	Possehl 1993	1834	1912–1750	2020–1685

TUNC-29	Tal-i Malyan/ Anshan	Kaftari period	3526 ± 61	Charcoal	Possehl 1993	1851	1934–1768	2025–1693
TUNC-30	Tal-i Malyan/ Anshan	Kaftari period	3531 ± 63	Charcoal	Possehl 1993	1860	1941–1770	2030–1692
P-3065	Tal-i Malyan/ Anshan	Early Kaftari period	3550 ± 70		Petrie *et al.* 2005	1891	2009–1772	2126–1693
P-2062	Tal-i Malyan/ Anshan	Kaftari period	3560 ± 60	Charcoal	Possehl 1993	1905	2014–1777	2120–1701
P-3066	Tal-i Malyan/ Anshan	Early Kaftari period	3560 ± 60		Petrie *et al.* 2005	1905	2014–1777	2120–1701
P-3071	Tal-i Malyan/ Anshan	Middle Kaftari period	3560 ± 60		Petrie *et al.* 2005	1905	2014–1777	2120–1701
P-3070	Tal-i Malyan/ Anshan	Early Kaftari period	3590 ± 60		Petrie *et al.* 2005	1947	2034–1834	2134–1767
Bahrain and Failaka								
K-1576	Barbar Temple IIb	Contemp. Q. al-B. IIb	3600 ± 100		Højlund 2007	1963	2132–1779	2275–1690
K-5965	Qala'at al-Bahrain	Period IIb	3610 ± 105		Højlund 2007	1978	2137–1782	2285–1691
K-1575	Barbar Temple IIb	Contemp. Q. al-B. IIb	3630 ± 100		Højlund 2007	2005	2191–1882	2290–1700
K-5966	Qala'at al-Bahrain	Period IIc	3650 ± 60		Højlund 2007	2026	2133–1942	2201–1884
K-5152	Failaka period 1?	Contemp. Q. al-B. IIb	3570 ± 60		Højlund 2007	2164	2278–2041	2400–1972

(*continued*)

Lab. reference	*Site name*	*Provenience*	*Date BP*	*Material*	*Publication*	*Median*	*Cal. BC 1σ (68.2%)*	*Cal. BC 2σ (95.4%)*
K-5962	Qala'at al-Bahrain	Period IIa	3780 ± 85		Højlund 2007	2216	2341–2042	2468–1978
K-5963	Qala'at al-Bahrain	Period IIa	3780 ± 85		Højlund 2007	2216	2341–2042	2468–1978
K-5964	Qala'at al-Bahrain	Period IIa	3810 ± 85		Højlund 2007	2260	2451–2139	2479–1984
Mesopotamia								
BM-64	Ur	Royal graves, skeleton of Mes-Kalam-Sar	3920 ± 150	Bone	Possehl 1993	2409	2620–2147	2872–2032
P-724	Ur	Royal graves, handles, and sockets	3959 ± 59	Wood	Possehl 1993	2468	2570–2348	2625–2236
BM-76	Ur	Royal graves, skeleton of Pu-Abi	3990 ± 150	Bone	Possehl 1993	2516	2851–2293	2901–2054
BM-70	Ur	Royal graves, coffin	4030 ± 150	Bone	Possehl 1993	2572	2866–2350	2923–2136
GIF-723	Mari	Burnt palace or palace room 219	3720 ± 150	Carbonized wood and grain	Possehl 1993	2137	2342–1920	2567–1700
GIF-722	Mari	Burnt palace or palace room 219	3820 ± 120	Carbonized wood	Possehl 1993	2272	2465–2135	2581–1921

[a] GN-2012/1 Gonur North Area 18 is without precise information as to the origin of the sample since it was not given to V. Sarianidi and N. Dubova by N. Boroffka.

REFERENCES

Askarov, A.A. 1977. *Drevnezemledel'cheskaja kultura juga Uzbekistana* [Bronze Age ancient farming culture of southern Uzbekistan]. Tashkent: Fan.

Bendezu-Sarmiento, J. and S. Mustafakulov 2013. "Le site proto-urbain de Dzharkutan durant les âges du bronze et du fer. Recherches de la Mission archéologique franco-ouzbèque-Protohistoire." *Cahiers d'Asie centrale* 21–22: 207–236.

Besenval, R. and A.I. Isakov 1989. "Sarazm et les débuts du peuplement agricole dans la région de Samarkand." *Arts Asiatiques* 44: 5–20.

Bronk Ramsey C. and S. Lee 2013. "Recent and Planned Developments of the Program Oxcal." *Radiocarbon* 55(2–3): 720–730.

Cattani, M. 2008. "The Final Phase of the Bronze Age and the 'Andronovo Question' in Margiana." In *The Bronze Age and Early Iron Age in the Margiana Lowlands: Facts and Methodological Proposals for a Redefinition of the Research Strategies*. Vol. II, *Archaeological Map of the Murghab Delta Studies and Reports*, ed. S. Salvatori and M. Tosi. BAR International Series 1806. Oxford: Archaeopress, pp. 133–151.

Cremaschi, M. 1998. "Palaeohydrography and Middle Holocene Desertification in the Northern Fringe of the Murghab Delta." In *The Archaeological Map of the Murghab Delta: Preliminary Reports 1990–95*, ed. A Gubaev, G. Koshelenko, and M. Tosi. Rome: IsIAO, pp. 15–25.

Francfort, H.-P. 1989. *Fouilles de Shortughaï. Recherches sur l'Asie centrale protohistorique*. Paris: de Boccard.

Garner, J. 2013. *Das Zinn der Bronzezeit in Mittelasien*. Vol. II, *Die Montanarchäologischen Forschungen der Zinnlagerstätten*. Archäologie in Iran und Turan 12, Veröffentlichungen aus dem Deutschen Bergbau-Museum 194. Bochum: von Zabern.

Görsdorf, J. and D. Huff 2001. "^{14}C-Datierung von Materialien aus der Grabung Džarkutan, Uzbekistan." *Archäologische Mitteilungen aus Iran und Turan* 33: 75–87.

Hiebert, F.T. 1994. *Origins of the Bronze Age Oasis Civilization in Central Asia*. American School of Prehistoric Research, Bulletin 42. Cambridge, MA: Peabody Museum of Archaeology and Ethnology, Harvard University.

Højlund, F. 2007. *The Burial Mounds of Bahrain. Social Complexity in Early Dilmun*. Jutland Archaeological Society Publications, vol. 58. Højbjerg: Jutland Archaeological Society.

Isakov, A., P.L. Kohl, C.C. Lamberg-Karlovsky, and R. Maddin 1987. "Metallurgical Analysis from Sarazm, Tadjikistan SSR." *Archaeometry* 29(1): 90–102.

Jia P.W., A. Betts, D. Cong, X. Jia, and P.D. Dupuy 2017. "Adunqiaolu: New Evidence for the Andronovo in Xinjiang, China." *Antiquity* 91(357): 621–639.

Jungner, H. 2004. "A Collaboration for Developing Archaeological Research in Turkmenistan." In *U Istokov Tsivilizatsii* [Near the sources of civilizations], ed. M.F. Kosarev, P.M. Kozhin, and N.A. Dubova. Moscow: Staryj Sad, pp. 290–291.

Kaniuth, K. 2006. *Metallobjekte der Bronzezeit aus Nordbaktrien*. Archäologie in Iran und Turan, 6. Mainz: Ph. von Zabern.

Kaniuth, K. 2007. "The Metallurgy of the Late Bronze Age Sapalli Culture (Southern Uzbekistan) and Its Implications for the 'Tin Question.'" *Iranica Antiqua* XLII: 23–40.

Kaniuth, K. 2010. "Tilla Bulak 2009 – Vorbericht zur dritten Kampagne." *Archäologische Mitteilungen aus Iran und Turan* 42: 129–163.

Kaniuth, K. 2016. "The Late Bronze Age Settlement of Tilla Bulak (Uzbekistan): A Summary of Four Years' Work." In *South Asian Archaeology and Art 2012*, vol. 1, ed. V. Lefèvre, A. Didier, and B. Mutin. Turnhout: Brepols, pp. 117–128.

Kircho, L.B. and S.G. Popov 1999. "K voprosu o radiouglerodnoj khronologii drevnejshikh tsivilizatsij srednej Azii" [Questions on the radiocarbon chronology of the ancient civilizations of Central Asia]. *Stratum Plus* 2: 350–361.

Kohl, P.L. 1984. *Central Asia, Palaeolithic Beginnings to the Iron Age*. Paris: Recherche sur les Civilisations.

Kuz'mina, E.E. 2007. *The Origin of the Indo-Iranians*. Leiden: Brill.

Meadow, R.H. and J.M. Kenoyer 2005. "Excavations at Harappa 2000–2001: New Insights on Chronology and City Organization." *South Asian Archaeology 2005*, vol. 1, ed. C. Jarrige and V. Lefèvre. Paris: Erc, pp. 207–225.

Narasimhan V.M., N.J. Patterson, P. Moorjani, N. Rohland, R. Bernardos, S. Mallick, I. Lazaridis *et al.* 2019. "The Formation of Human Populations in South and Central Asia." *Science* 365(6457), eaat7487.

Petrie, C.A., A.A. Chaverdi, and M. Seyedin 2005. "From Anshan to Dilmun and Magan: The Spatial and Temporal Distribution of Kaftari and Kaftari-Related Ceramic Vessels." *Iran* 43: 49–86.

Possehl, G.L. 1993. *Radiometric Dates for South Asian Archaeology*. Philadelphia: University Museum, University of Pennsylvania.

Prickett, M.E. 1986. "Man, Land and Ware: Settlement Distribution and the Development of Irrigation Agriculture in the Upper Rud-i Gushk Drainage, Southeastern Iran." PhD dissertation, Harvard University, Cambridge, MA.

P'jankova, L.T. 1986. *Jungbronzezeitliche Gräberfelder im Vachš-Tal, Süd-Tadžikistan*. München: C.H. Beck.

Reimer, P., E. Bard, A. Bayliss, J. Warren Beck, P.G. Blackwell, C. Bronk Ramsey, C.E. Buck *et al.* 2013. "IntCal 13 and Marine 13 Radiocarbon Age Calibration Curves 0–50,000 years Cal BP." *Radiocarbon* 55(4), 1869–1887.

Rossi Osmida, G. 2011. *Adji Kui Oasis. La cittadella delle statuette*, vol. II. Padova: Il Punto.

Rouse, L.M. and B. Cerasetti 2014. "Ojakly: A Late Bronze Age Mobile Pastoralist Site in the Murghab Region, Turkmenistan." *Journal of Field Archaeology* 39(1): 1–19.

Salvatori, S. 2002. "The Project for the Archaeological Map of the Murghab Delta (Turkmenistan): Stratigraphical Trial Trenches at Adzhi Kui 1 and 9." *Ancient Civilizations from Scythia to Siberia* 8(1–2): 107–178.

Salvatori, S. 2010. "Thinking around Grave 3245 in the 'Royal Graveyard' of Gonur (Murghab Delta, Turkmenistan). In *Na puti otkrytija tsivilizatsii.* [On the track of uncovering a civilization]. *Trudy Margianskoj Arkheologicheskoj Ekspeditsii*, vol. 3, ed. P.M. Kozhin, M.F. Kosarev, and N.A. Dubova. St. Petersburg: Aletejja, pp. 244–257.

Salvatori, S. and M. Tosi 2005. "Shahr-i Sokhta Revised Sequence." *South Asian Archaeology 2001*, vol. I, ed. C. Jarrige and V. Lefèvre. Paris: Erc, pp. 281–292.

Sarianidi, V.I. 1977. *Drevnie zemledel'tsy Afganistana* [Ancient farmers of Afghanistan]. Materialy Sovetsko-Afganskoj ekspeditsii 1969–1974 gg. Moscow: Nauka.

Sarianidi, V.I. 1990. *Drevnosti strany Margush* [Ancient Margush]. Ashkhabad: Ylym.

Sarianidi, V.I. 1993. "Margiana on the Ancient Near East." *Information Bulletin of the International Association for the Study of Cultures of Central Asia* 19: 5–28.

Sarianidi, V.I., N. Boroffka, and N.A. Dubova 2014. "Kul'turnye kontakty Margiany (Turkmenistan) v III tys. do n.e. Novye dannye po Gonur-Depe (pogrebenie NO. 4150)" [Margiana (Turkmenistan) cultural contacts in the 3rd millennium BC: new data from Gonur Depe (tomb no. 4150)]. In *Trudy Margianskoj Arkheologicheskoj Ekspeditsii*, vol. 4, ed. V.I. Sarianidi. Moscow: Staryj Sad, pp. 127–137.

Teufer, M., N.V. Vinogradova, and Ju.G. Kutimov 2014. "K voprosu ob absoljutnoj datirovke pogrebenii bronzovogo veka s nekropolej Gelot i Darnajchi" [Question on the absolute chronology of the Gelot and Darnaichi necropolises]. *Arkheologicheskie Raboty v Tadzhikistane* 37: 110–119.

Vinogradova, N.M. 2004. *Jugo-Zapadnij Tadzhikistan v epokhu pozdnej Bronzy* [Southwestern Tadjikistan in the Bronze Age]. Moscow: Institut Vostokovedenija.

Vinogradova, N.M., V.A. Ranov, and T.G. Filimonova 2008. *Pamjatniki Kangurttuta v jugo-zapadnom Tadzhikistane (epokha neolita i bronzovogo veka)* [The site of Kangurttut in southwestern Tadjikistan (Neolithic and Bronze Age)]. Moscow: Institut Vostokovedenija.

Zajtzeva, G.I., N.A. Dubova, and N. Burova forthcoming. "New Radiocarbon Dates from Gonur Depe." *Trudy Margianskoj Arkheologicheskoj Ekspeditsii*, vol. 9. Moscow: Staryi Sad.

Zajtseva, G.I., N.A. Dubova, A.A. Sementsov, P. Rejmar, Dzh. Mellori, and Kh. Jungner 2008. "Radiouglerodnaja khronologija pamjatnika Gonur Depe" [The radiocarbon-based chronology of Gonur Depe]. In *Trudy Margianskaja Arkheologicheskaja Ekspeditsija*, vol. 2, ed. V.I. Sarianidi. Moscow: Staryj Sad, pp. 166–179.

INDEX

Note: Entries that are underlined indicate archaeological sites that contain numbers